ELECTRICAL PHENOMENA AT INTERFACES AND BIOINTERFACES

ELECTRICAL PHENOMENA AT INTERFACES AND BIOINTERFACES

Fundamentals and Applications in Nano-, Bio-, and Environmental Sciences

EDITED BY

HIROYUKI OHSHIMA

WILEY

A JOHN WILEY & SONS, INC., PUBLICATION

Published by John Wiley & Sons, Inc., Hoboken, New Jersey
Published simultaneously in Canada

Library of Congress Cataloging-in-Publication Data:
Electrical phenomena at interfaces and biointerfaces : fundamentals and applications in nano-, bio-, and environmental sciences / edited by Hiroyuki Ohshima.
p. cm.
Includes index.
ISBN 978-0-470-58255-8 (cloth)
1. Biological interfaces. 2. Surface chemistry. 3. Electric double layer. 4. Biotechnology. 5. Nanotechnology. 6. Environmental sciences. I. Ohshima, Hiroyuki, 1944-
QP517.S87E44 2012
612'.01583–dc23

2011028225

Printed in the United States of America

10 9 8 7 6 5 4 3 2 1

CONTENTS

PREFACE

This book is based on a joint project of the Center for Colloid and Interface Science, Research Institute for Science and Technology, Tokyo University of Science, and the Electrokinetic Society of Japan. Kunio Furusawa and I edited *Electrical Phenomena at Interfaces* (1990; 2nd Edition, 1998); although this book has a similar title, it is on completely different concepts. This book is written for scientists, engineers, and graduate students who want to study theoretical and experimental aspects of electrical phenomena at interfaces and biointerfaces. The principal purpose of this book is to bridge three different fields: nano-, bio-, and environmental sciences. As a basis of these three different fields, the understanding of electrical phenomena at interfaces and biointerfaces is becoming more and more important.

This book is divided into three parts. Part I contains the fundamentals of electrical phenomena at interfaces and biointerfaces. Parts II and III treat many topics in this field, including applications in nano- and environmental sciences (Part II) and applications in biosciences (Part III).

I would like to gratefully acknowledge the assistance provided by Ms. Anita Lekhwani, Senior Acquisitions Editor, and Ms. Rebekah Amos, Editorial Program Coordinator, at John Wiley & Sons.

Hiroyuki Ohshima

Faculty of Pharmaceutical Sciences and Center for Colloid and Interface Science
Research Institute for Science and Technology
Tokyo University of Science, Japan

CONTRIBUTORS

Masahiko Abe, Department of Pure and Applied Chemistry, Faculty of Science and Technology, Tokyo University of Science, 2641 Yamazaki, Noda, Chiba 278-8510, Japan

Yasuhisa Adachi, Graduate School of Life and Environmental Science, University of Tsukuba, 1-1-1, Tennnoudai, Tsukuba-shi, Ibaraki 305-8572 Japan

Silvia Ahualli, Department of Applied Physics, School of Sciences, Campus Fuentenueva, University of Granada, 18071 Granada, Spain

Takanori Akagi, Department of Bioengineering, School of Engineering, The University of Tokyo, 2-11-16 Yayoi, Bunkyo-ku, Tokyo 113-8656, Japan

Klaus Arnold, Institute for Medical Physics and Biophysics, Medical Faculty, University of Leipzig, Leipzig 04103, Germany

Ángel V. Delgado, Department of Applied Physics, School of Sciences, Campus Fuentenueva, University of Granada, 18071 Granada, Spain

Gjergi Dodbiba, Department of Systems Innovation, Graduate School of Engineering, The University of Tokyo, 7-3-1 Hongo, Bunkyo-Ku, Tokyo 113-8656, Japan

Stanislav S. Dukhin, New Jersey Institute of Technology, Newark, NJ 07102-1982, USA

Kazunaka Endo, Center for Colloid and Interface Science, Research Institute for Science and Technology, Tokyo University of Science, 12-1 Ichigaya-funagawara, Shinjuku, Tokyo 162-0826, Japan

Toyohisa Fujita, Department of Systems Innovation, Graduate School of Engineering, The University of Tokyo, 7-3-1 Hongo, Bunkyo-Ku, Tokyo 113-8656, Japan

Fumio Fukai, Faculty of Pharmaceutical Sciences and Center for Colloid and Interface Science, Research Institute for Science and Technology, Tokyo University of Science, 2641 Yamazaki, Noda, Chiba 278-8510, Japan

Fernando González-Caballero, Department of Applied Physics, School of Sciences, Campus Fuentenueva, University of Granada, 18071 Granada, Spain

Hiroshi Hayashi, Department of Resources and Environmental Engineering, School of Creative Science and Engineering, Faculty of Science and Engineering, Waseda University, 3-4-1 Okubo, Shinjuku-ku, Tokyo 169-8555, Japan

Jyh-Ping Hsu, Department of Chemical Engineering, National Taiwan University Taipei, Taiwan 10617

Takanori Ichiki, Department of Bioengineering, School of Engineering, The University of Tokyo, 2-11-16 Yayoi, Bunkyo-ku, Tokyo 113-8656, Japan

María Luisa Jiménez, Department of Applied Physics, School of Sciences, Campus Fuentenueva, University of Granada, 18071 Granada, Spain

Kang Kim, Institute for Molecular Science, Okazaki 444-8585, Japan

Takeshi Kawai, Department of Industrial Chemistry, Faculty of Engineering and Center for Colloid and Interface Science, Research Institute for Science and Technology, Tokyo University of Science, 12-1 Ichigaya-funagawara, Shinjuku, Tokyo 162-0826, Japan

Motoyoshi Kobayashi, Graduate School of Life and Environmental Sciences, UniversityofTsukuba,1-1-1,Tennoudai,Tsukuba-shi,Ibaraki305-8572,Japan

Takeshi Kondo, Department of Industrial Chemistry, Faculty of Engineering and Center for Colloid and Interface Science, Research Institute for Science and Technology, Tokyo University of Science, 12-1 Ichigaya-funagawara, Shinjuku, Tokyo 162-0826, Japan

Yukishige Kondo, Department of Industrial Chemistry, Faculty of Engineering and Center for Colloid and Interface Science, Research Institute for Science and Technology, Tokyo University of Science, 12-1 Ichigaya-funagawara, Shinjuku, Tokyo 162-0826, Japan

Shinichi Komaba, Department of Applied Chemistry, Faculty of Science and Center for Colloid and Interface Science, Research Institute for Science and Technology, Tokyo University of Science, Kagurazaka 1-3, Shinjuku, Tokyo 162-8601, Japan

Friedrich Kremer, Institute of Experimental Physics I, University of Leipzig, Linnéstr. 5, 04103, Leipzig, Germany

Jun Kuwano, Department of Industrial Chemistry, Faculty of Engineering and Center for Colloid and Interface Science, Research Institute for Science and Technology, Tokyo University of Science, 12-1 Ichigaya-funagawara, Shinjuku, Tokyo 162-0826, Japan

Kimiko Makino, Faculty of Pharmaceutical Sciences, Center for Colloid and Interface Science, Center for Physical Pharmaceutics, Research Institute for Science and Technology, and Center for Drug Delivery Research, Tokyo University of Science, 2641 Yamazaki, Noda, Chiba 278-8510, Japan

Hisao Morisaki, Department of Biotechnology, College of Life Sciences, Ritsumeikan University, 1-1-1 Noji-higashi, Kusatsu, Shiga 525-8577, Japan

Hidetoshi Matsumoto, Department of Organic and Polymeric Materials, Tokyo Institute of Technology, 2-12-1-S8-27 Ookayama, Meguro-Ku, Tokyo 152- 8552, Japan

Kaoru Nakamura, Science Shop, Graduate School of Human Development and Environment, Kobe University, 3-11Tsurukabuto, Nada, Kobe 657-8501 Japan

Kazuho Nakamura, Department of Chemical Engineering, Yokohama National University, 79-5 Tokiwadai, Hodogaya-ku, Yokohama 240-8501, Japan

Yasuya Nakayama, Department of Chemical Engineering, Kyushu University, Fukuoka 819-0395, Japan

Naoto Nishida, Department of Applied Chemistry, Tokyo University of Science Yamaguchi, SanyoOnoda-shi, Yamaguchi 756-0884, Japan

Satoshi Nishimura, Nanotechnology Research Institute, National Institute of Advanced Industrial Science and Technology (AIST), Tsukuba, Higashil-1-1, Tsukuba, Ibaraki, 305-8565, Japan

Masashi Nojima, Center for Colloid and Interface Science, Research Institute for Science and Technology, Tokyo University of Science, 2641 Yamazaki, Noda, Chiba 278-8510, Japan

Shinpei Ohki, Department of Physiology and Biophysics, School of Medicine and Biomedical Sciences, State University of New York at Buffalo, Buffalo, NY 14214, USA

Hiroyuki Ohshima, Faculty of Pharmaceutical Sciences and Center for Colloid and Interface Science, Research Institute for Science and Technology, Tokyo University of Science, 2641 Yamazaki, Noda, Chiba 278-8510, Japan

Kunihiko Okano, Department of Pure and Applied Chemistry, Faculty of Science and Technology and Center for Colloid and Interface Science, Research Institute of Science and Technology, Tokyo University of Science, 2641 Yamazaki, Noda, Chiba 278-8510, Japan

Hidenori Otsuka, Department of Applied Chemistry, Faculty of Science and Center for Colloid and Interface Science, Research Institute for Science and Technology, Tokyo University of Science, 1-3 Kagurazaka, Shinjuku-ku, Tokyo 162-8601, Japan

Akira Otsuki, Chemical and Biomolecular Engineering, University of Melbourne, Parkville, VIC, 3010, Australia

Toshiyuki Owaki, Faculty of Pharmaceutical Sciences, Tokyo University of Science, 2641 Yamazaki, Noda, Chiba 278-8510, Japan

Raúl A. Rica, Department of Applied Physics, School of Sciences, Campus Fuentenueva, University of Granada, 18071 Granada, Spain

Walter Richtering, Lehrstuhl für Physikalische Chemie II, RWTH Aachen University, Landoltweg 2, D-52056 Aachen, Germany

Morihiro Saito, Department of Molecular Chemistry and Biochemistry, Faculty of Science and Engineering, Doshisha University, Kyotanabe, Kyoto 610-0321, Japan

Yohei Saito, Faculty of Pharmaceutical Sciences, Tokyo University of Science, 2641 Yamazaki, Noda, Chiba 278-8510, Japan

Hideki Sakai, Department of Pure and Applied Chemistry, Faculty of Science and Technology and Center for Colloid and Interface Science, Research Institute for Science and Technology, Tokyo University of Science, 2641 Yamazaki, Noda, Chiba 278-8510, Japan

Joshua R. Sangoro, Institute of Experimental Physics I, University of Leipzig, Linnéstr. 5, 04103, Leipzig, Germany

Hiroshi Sasaki, Department of Resources and Environmental Engineering, School of Creative Science and Engineering, Faculty of Science and Engineering, Waseda University, 3-4-1 Okubo, Shinjuku-ku, Tokyo 169-8555, Japan

Anatoli Serghei, Université Lyon 1, CNRS, UMR 5223, Ingénierie des Matériaux Polymères, F-69622 Villeurbanne, France

Yukihide Shiraishi, Department of Applied Chemistry, Tokyo University of Science Yamaguchi, SanyoOnoda-shi, Yamaguchi 756-0884, Japan, and

Center for Colloid and Interface Science, Research Institute for Science and Technology, Tokyo University of Science, 1-3 Kagurazaka, Shinjuku-ku, Tokyo 162-8601, Japan

Makoto Tadokoro, Department of Chemistry, Faculty of Science and Center for Colloid and Interface Science, Research Institute for Science and Technology, Tokyo University of Science, 1-3 Kagurazaka, Shinjuku-ku, Tokyo 162-8601, Japan

Tharwat Tadros, 89 Nash Grove Lane, Wokingham, Berkshire RG40 4HE, UK

Youichi Takata, Department of Chemical and Biological Engineering, Ube National College of Technology, Tokiwadai 2-14-1, Ube, Yamaguchi 755-8555, Japan

Shin-ichi Takeda, Takeda Colloid Techno-Consulting Co., Ltd., Senriyamanishi 1-41-14, Suita, Osaka 565-0851, Japan

Tetsuo Takemura, Department of Chemistry, Faculty of Science and Center for Colloid and Interface Science, Research Institute for Science and Technology, Tokyo University of Science, 1-3 Kagurazaka, Shinjuku-ku,Tokyo 162-8601, Japan

Akihiko Tanioka, Department of Organic and Polymeric Materials, Tokyo Institute of Technology, 2-12-1-S8-27 Ookayama, Meguro-Ku, Tokyo 152-8552, Japan

Chiharu Tokoro, Department of Resources and Environmental Engineering, School of Creative Science and Engineering, Faculty of Science and Engineering, Waseda University, 3-4-1 Okubo, Shinjuku-ku, Tokyo 169-8555, Japan

Naoki Toshima, Department of Applied Chemistry, Tokyo University of Science Yamaguchi, SanyoOnoda-shi, Yamaguchi 756-0884, Japan

Tomoya Tsuchikawa, Department of Applied Chemistry, Faculty of Science and Center for Colloid and Interface Science, Research Institute for Science and Technology, Tokyo University of Science, Kagurazaka 1-3, Shinjuku, Tokyo 162-8601, Japan

Koji Tsuchiya, Department of Applied Chemistry, Faculty of Science and Center for Colloid and Interface Science, Research Institute for Science and Technology, Tokyo University of Science, Kagurazaka 1-3, Shinjuku, Tokyo 162-8601, Japan

Katsumi Uchida, Department of Applied Chemistry, Faculty of Science and Center for Colloid and Interface Science, Research Institute for Science and Technology, Tokyo University of Science, Kagurazaka 1-3, Shinjuku, Tokyo 162-8601, Japan

Julián López-Viota, Department of Physics, Polytechnic School, University of Jaén, Campus Linares, 23700 Linares, Jaén, Spain

Carsten Werner, Leibniz Institute of Polymer Research Dresden & The Max Bergmann Center of Biomaterials Dresden, Hohe Strasse 6, 01069 Dresden, Germany and Technische Universität Dresden, Center of Regenerative Therapies Dresden, Tatzberg 47, 01307 Dresden, Germany

John Erik Wong, RWTH Aachen University, Chemical Process Engineering, Turmstrasse 46, 52064 Aachen, Germany

Naoaki Yabuuchi, Department of Applied Chemistry, Faculty of Science and Center for Colloid and Interface Science, Research Institute for Science and Technology, Tokyo University of Science, 1-3 Kagurazaka, Shinjuku-ku, Tokyo 162-8601, Japan

Hirofumi Yajima, Department of Applied Chemistry, Faculty of Science and Center for Colloid and Interface Science, Research Institute for Science and Technology, Tokyo University of Science, 1-3 Kagurazaka, Shinjuku, Tokyo 162-8601, Japan

Kiyofumi Yamagiwa, Department of Industrial Chemistry, Faculty of Engineering and Center for Colloid and Interface Science, Research Institute for Science and Technology, Tokyo University of Science, 12-1 Ichigaya-funagawara, Shinjuku, Tokyo 162-0826, Japan

Ryoichi Yamamoto, Department of Chemical Engineering, Kyoto University, Kyoto 615-8510, Japan

Takashi Yamashita, Department of Pure and Applied Chemistry, Faculty of Science and Technology and Center for Colloid and Interface Science, Research Institute for Science and Technology, Tokyo University of Science, 2641 Yamazaki, Noda, Chiba 278-8510, Japan

Li-Hsien Yeh, Department of Chemical Engineering, National Taiwan University Taipei, Taiwan 10617

Hiroharu Yui, Department of Chemistry, Faculty of Science and Center for Colloid and Interface Science, Research Institute for Science and Technology, Tokyo University of Science, 1-3 Kagurazaka, Shinjuku-ku, Tokyo 162-8601, Japan

Ralf Zimmermann, Leibniz Institute of Polymer Research Dresden & The Max Bergmann Center of Biomaterials Dresden, Hohe Strasse 6, 01069 Dresden, Germany

PART I

FUNDAMENTALS

1

POTENTIAL AND CHARGE OF A HARD PARTICLE AND A SOFT PARTICLE

HIROYUKI OHSHIMA

1.1 INTRODUCTION

When a charged colloidal particle is immersed in an electrolyte solution, mobile electrolyte ions form an ionic cloud around the particle. As a result of electrostatic interaction between electrolyte ions and particle surface charges, in the ionic cloud the concentration of counterions (electrolyte ions with charges of the sign opposite to that of the particle surface charges) becomes very high, while that of coions (electrolyte ions with charges of the same sign as the particle surface charges) is very low. Figure 1.1 schematically shows the distribution of ions around a charged spherical particle of radius a. The ionic cloud together with the particle surface charge forms an electrical double layer. Such an electrical double layer is often called an electrical diffuse double layer since the distribution of electrolyte ions in the ionic cloud takes a diffusive structure due to the thermal motion of ions. The electrostatic interaction between colloidal particles and the motion of colloidal particles in an external field (e.g., electric field and gravitational field) depend strongly on the distributions of electrolyte ions and of the electric potential across the electrical double layer around the particle surface [1–5].

Electrical Phenomena at Interfaces and Biointerfaces: Fundamentals and Applications in Nano-, Bio-, and Environmental Sciences, First Edition. Edited by Hiroyuki Ohshima.

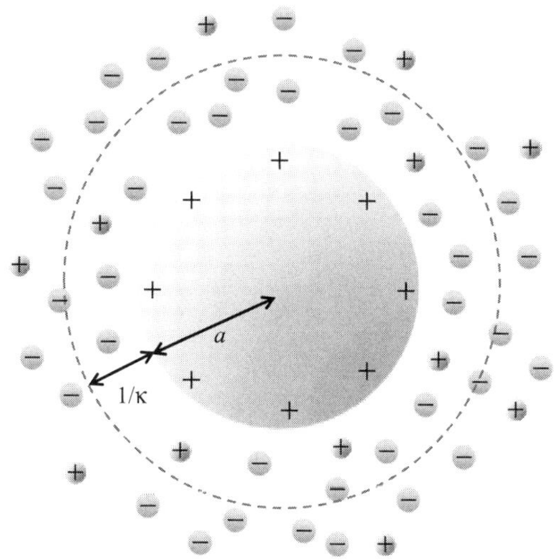

Figure 1.1. Electrical double layer of thickness $1/\kappa$ around a spherical charged particle of radius a.

1.2 THE POISSON–BOLTZMANN EQUATION

Consider a uniformly charged particle immersed in a liquid containing N ionic species with valence z_i and bulk concentration (number density) n_i^∞ ($i = 1, 2 \dots N$) (in units of cubic meter). From the electroneutrality condition, we have

$$\sum_{i=1}^{N} z_i n_i^\infty = 0. \tag{1.1}$$

The electric potential $\psi(\boldsymbol{r})$ at position $\boldsymbol{r}$ outside the particle, measured relative to the bulk solution phase, where ψ is set equal to zero, is related to the charge density $\rho_{\rm el}(\boldsymbol{r})$ at the same point by the Poisson equation, viz.,

$$\Delta\psi(\boldsymbol{r}) = -\frac{\rho_{\rm el}(\boldsymbol{r})}{\varepsilon_{\rm r}\varepsilon_{\rm o}}, \tag{1.2}$$

where Δ is the Laplacian, $\varepsilon_{\rm r}$ is the relative permittivity of the electrolyte solution, and $\varepsilon_{\rm o}$ is the permittivity of a vacuum. We assume that the distribution of the electrolyte ions $n_i(\boldsymbol{r})$ obeys Boltzmann's law, viz.,

$$n_i(\boldsymbol{r}) = n_i^\infty \exp\left(-\frac{z_i e\psi(\boldsymbol{r})}{kT}\right), \tag{1.3}$$

where $n_i(\boldsymbol{r})$ is the concentration (number density) of the ith ionic species at position $\boldsymbol{r}$, e is the elementary electric charge, k is Boltzmann's constant, and T is the absolute temperature. The charge density $\rho_{\mathrm{el}}(\boldsymbol{r})$ at position $\boldsymbol{r}$ is thus given by

$$\rho_{\mathrm{el}}(\boldsymbol{r}) = \sum_{i=1}^{N} z_i e n_i(\boldsymbol{r}) = \sum_{i=1}^{N} z_i e n_i^{\infty} \exp\left(-\frac{z_i e \psi(\boldsymbol{r})}{kT}\right). \tag{1.4}$$

Combining Equations 1.2 and 1.4 gives

$$\Delta\psi(\boldsymbol{r}) = -\frac{1}{\varepsilon_{\mathrm{r}}\varepsilon_{\mathrm{o}}} \sum_{i=1}^{N} z_i e n_i^{\infty} \exp\left(-\frac{z_i e \psi(\boldsymbol{r})}{kT}\right). \tag{1.5}$$

This is the Poisson–Boltzmann equation for the potential distribution $\psi(\boldsymbol{r})$, which is subject to the following boundary conditions:

$$\psi(\boldsymbol{r}) = \psi_{\mathrm{o}} \text{ at the particle surface} \tag{1.6}$$

and

$$\psi(\boldsymbol{r}) \to 0 \text{ as } r = |\boldsymbol{r}| \to \infty. \tag{1.7}$$

If the internal electric fields inside the particle can be neglected, then the surface charge density σ of the particle is related to the potential derivative normal to the particle surface as

$$\frac{\partial\psi}{\partial n} = -\frac{\sigma}{\varepsilon_{\mathrm{r}}\varepsilon_{\mathrm{o}}}, \tag{1.8}$$

where n is the outward normal at the particle surface.

1.3 LOW POTENTIAL CASE

If the potential ψ is low, viz.,

$$\left|\frac{z_i e \psi}{kT}\right| \ll 1, \tag{1.9}$$

then Equation 1.5 reduces to the following linearized Poisson–Boltzmann equation (Debye–Hückel equation):

$$\Delta\psi = \kappa^2\psi \tag{1.10}$$

with

$$\kappa = \left(\frac{1}{\varepsilon_{\mathrm{r}}\varepsilon_{\mathrm{o}} kT} \sum_{i=1}^{N} z_i^2 e^2 n_i^{\infty}\right)^{1/2}, \tag{1.11}$$

TABLE 1.1. Debye–Hückel Parameter for Various Electrolytes

Symmetrical electrolyte of valence z and bulk concentration n $$\kappa=\left(\frac{2z^2e^2n}{\varepsilon_r\varepsilon_o kT}\right)^{1/2}$$	1-1 symmetrical electrolyte of bulk concentration n $$\kappa=\left(\frac{2e^2n}{\varepsilon_r\varepsilon_o kT}\right)^{1/2}$$
2-1 electrolyte of bulk concentration n $$\kappa=\left(\frac{6e^2n}{\varepsilon_r\varepsilon_o kT}\right)^{1/2}$$	3-1 electrolyte of bulk concentration n $$\kappa=\left(\frac{12e^2n}{\varepsilon_r\varepsilon_o kT}\right)^{1/2}$$
Mixed solution of 1-1 electrolyte of bulk concentration n_1 and 2-1 electrolyte of bulk concentration n_2 $$\kappa=\left(\frac{2(n_1+3n_2)e^2}{\varepsilon_r\varepsilon_o kT}\right)^{1/2}$$	Mixed solution of 1-1 electrolyte of concentration n_1 and 3-1 electrolyte of concentration n_2 $$\kappa=\left(\frac{2(n_1+6n_2)e^2}{\varepsilon_r\varepsilon_o kT}\right)^{1/2}$$

where κ is called the Debye–Hückel parameter. The reciprocal of κ (i.e., $1/\kappa$), which is called the Debye length, corresponds to the thickness of the double layer. Note that n_i^∞ in Equations 1.5 and 1.10 is given in units of cubic meter. If one uses the units of M (mole per liter), then n_i^∞ must be replaced by 1000 $N_A n_i^\infty$, N_A being Avogadro's number. Expressions for κ for various types of electrolytes are explicitly given in Table 1.1.

Linearized Equation 1.10 can be solved for particles of various shapes. Table 1.2 gives the potential distribution for a planar surface, a sphere of radius a, and a cylinder of radius a, which can be obtained by solving Equation 1.10 (with $\Delta = d^2/dx^2$ for a planar surface, $\Delta = d^2/dr^2 + 2/r{\cdot}d/dr$ for a sphere, and $\Delta = d^2/dr^2 + 1/r{\cdot}d/dr$ for a cylinder) subject to Equations 1.6 and 1.7, where x is the distance from the planar surface located at $x = 0$ and r is the distance from the sphere center or the cylinder axis. Table 1.2 also shows the surface potential ψ_o/surface charge density σ relationship, which can be obtained by substituting ψ into Equation 1.8.

1.4 ARBITRARY POTENTIAL CASE

The nonlinear Poisson–Boltzmann Equation 1.5 (with $\Delta = d^2/dx^2$) for a planar surface can be solved analytically. For a planar surface in contact with a z-z symmetrical electrolyte solution, a 2-1 electrolyte solution, or a mixed solution of 1-1 electrolyte of bulk concentration n_1 and 2-1 electrolyte of bulk concentration n_2, the potential distribution $\psi(x)$ and the surface potential ψ_o/surface charge density σ relationship are given in Table 1.3.

Consider the asymptotic behavior of the potential distribution at large distances, which will also be used for calculating the electrostatic inter-

TABLE 1.2. Solution to the Linearized Poisson–Boltzmann Equation

	Potential Distribution	Surface Potential ψ_o/Surface Charge Density σ Relationship
Planar surface 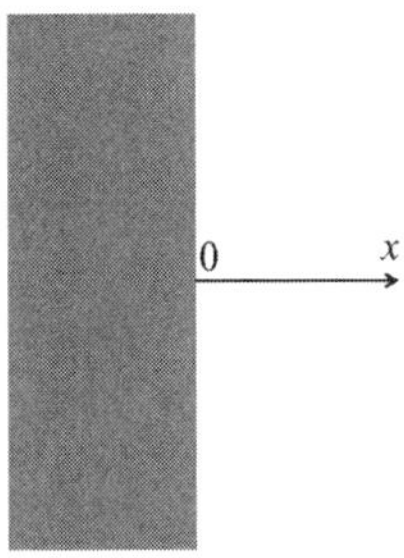	$\psi(x)=\psi_o e^{-\kappa x}$	$\psi_o=\dfrac{\sigma}{\varepsilon_r\varepsilon_o\kappa}$
Sphere of radius a 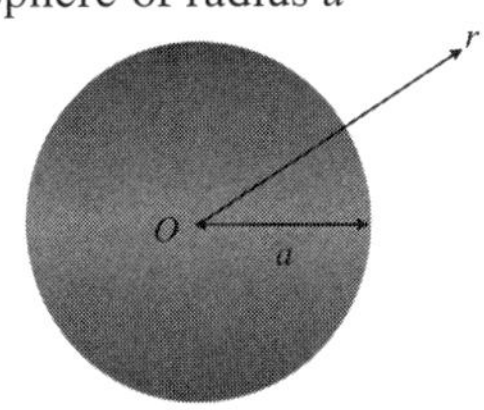	$\psi(r)=\psi_o\dfrac{a}{r}e^{-\kappa(r-a)}$	$\psi_o=\dfrac{\sigma}{\varepsilon_r\varepsilon_o\kappa(1+1/\kappa a)}$
Cylinder of radius a 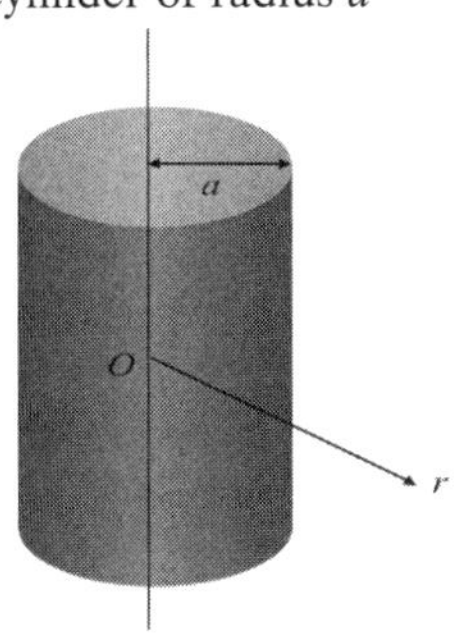	$\psi(r)=\psi_o\dfrac{K_0(\kappa r)}{K_0(\kappa a)}$	$\psi_o=\dfrac{\sigma}{\varepsilon_r\varepsilon_o\kappa}\dfrac{K_0(\kappa a)}{K_1(\kappa a)}$

Note: x (>0) is the distance from the planar surface and r (>a) is the distance from the center O of the sphere or that from the axis of the cylinder. $K_n(z)$ is the modified Bessel function of the second kind of order n.

action between two particles. When a planar surface is in contact with a z-z symmetrical electrolyte, the potential distribution $\psi(x)$ around the surface (see Table 1.3) in the region far from the surface, that is, at large κx, takes the form

$$\psi(x)=\frac{4kT}{ze}\gamma e^{-\kappa x}=\frac{4kT}{ze}\tanh\left(\frac{ze\psi_o}{4kT}\right)e^{-\kappa x}. \tag{1.12}$$

TABLE 1.3. Potential Distribution $\psi(x)$, Surface Potential ψ_o/Surface Charge Density σ Relationship, and Effective Surface Potential ψ_{eff} for a Planar Surface with Arbitral Surface Potential in an Electrolyte Solution

Electrolytes	$\psi(x)$	ψ_o/σ Relationship	ψ_{eff}
z-z	$\psi(x)=\dfrac{2kT}{ze}\ln\left(\dfrac{1+\gamma e^{-\kappa x}}{1-\gamma e^{-\kappa x}}\right)$	$\sigma=\dfrac{2\varepsilon_r\varepsilon_o\kappa kT}{ze}\sinh\left(\dfrac{ze\psi_o}{2kT}\right)$	$\psi_{eff}=\dfrac{4kT}{ze}\gamma$
2-1	$\psi(x)=\dfrac{kT}{e}\ln\left[\dfrac{3}{2}\left(\dfrac{1+\frac{2}{3}\gamma' e^{-\kappa x}}{1-\frac{2}{3}\gamma' e^{-\kappa x}}\right)^2-\dfrac{1}{2}\right]$	$\sigma=\dfrac{\varepsilon_r\varepsilon_o\kappa kT}{e}\left\{1-\exp\left(-\dfrac{e\psi_o}{kT}\right)\right\}$ $\times\left\{\dfrac{2}{3}\exp\left(\dfrac{e\psi_o}{kT}\right)+\dfrac{2}{3}\right\}^{1/2}$	$\psi_{eff}=\dfrac{4kT}{e}\gamma'$
1-1 plus 2-1	$\psi(x)=\dfrac{kT}{e}\ln\left[\left(\dfrac{1}{1-\eta/3}\right)\left\{\dfrac{1+\left(1-\frac{\eta}{3}\right)\gamma'' e^{-\kappa x}}{1-\left(1-\frac{\eta}{3}\right)\gamma'' e^{-\kappa x}}\right\}-\left(\dfrac{\eta/3}{1-\eta/3}\right)\right]$	$\sigma=\dfrac{\varepsilon_r\varepsilon_o\kappa kT}{e}(1-e^{-y_o})$ $\times\left[\left(1-\dfrac{\eta}{3}\right)e^{y_o}+\dfrac{\eta}{3}\right]^{1/2}$	$\psi_{eff}=\dfrac{4kT}{e}\gamma''$

Note: $\gamma=\tanh\left(\dfrac{ze\psi_o}{4kT}\right)=\dfrac{\exp(ze\psi_o/2kT)-1}{\exp(ze\psi_o/2kT)+1}$

$$\gamma'=\frac{3}{2}\left\{\frac{\left(\frac{2}{3}e^{y_o}+\frac{1}{3}\right)^{1/2}-1}{\left(\frac{2}{3}e^{y_o}+\frac{1}{3}\right)^{1/2}+1}\right\},\ \gamma''=\left(\frac{1}{1-\eta/3}\right)\left[\frac{\left\{\left(1-\frac{\eta}{3}\right)e^{y_o}+\frac{\eta}{3}\right\}^{1/2}-1}{\left\{\left(1-\frac{\eta}{3}\right)e^{y_o}+\frac{\eta}{3}\right\}^{1/2}+1}\right],\ \eta=\frac{3n_2}{n_1+3n_2},\ y_o=\frac{e\psi_o}{kT}.$$

Comparing Equation 1.12 with the linearized form $\psi(x) = \psi_o \exp(-\kappa x)$ (see Table 1.2), we find that the effective surface potential ψ_{eff} of the plate is given by

$$\psi_{\text{eff}} = \frac{4kT}{ze}\gamma = \frac{kT}{ze}\cdot 4\tanh\left(\frac{ze\psi_o}{4kT}\right). \tag{1.13}$$

This result, together with those for other types of electrolytes, is given in Table 1.3.

For a sphere, the nonlinear Poisson–Boltzmann equation has not been solved analytically. Loeb et al. [6] tabulated numerical computer solutions to the nonlinear spherical Poisson–Boltzmann equation and approximate analytic solutions are given in References 7–9 (Table 1.4). For the case of an infinitely long cylindrical particle of radius a, approximate solutions are derived in References 7 and 10 (Table 1.5).

TABLE 1.4. Potential Distribution $\psi(r)$ and Surface Potential ψ_o/Surface Charge Density σ Relationship for a Sphere of Radius a with Arbitrary Surface Potential

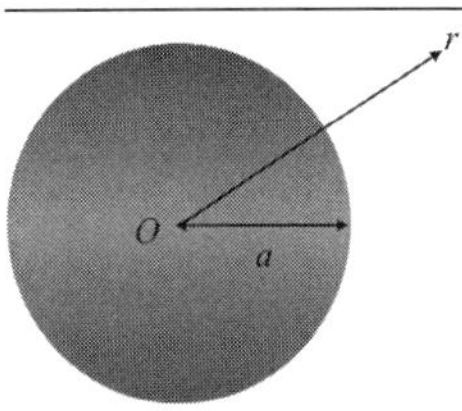

Potential distribution $\psi(r) = \dfrac{2kT}{ze}\ln\left[\dfrac{(1+Bs)\left(1+\dfrac{Bs}{2\kappa a+1}\right)}{(1-Bs)\left(1-\dfrac{Bs}{2\kappa a+1}\right)}\right]$

Surface potential ψ_o/surface charge density σ relationship

$$\sigma = \frac{2\varepsilon_r\varepsilon_o\kappa kT}{ze}\sinh\left(\frac{ze\psi_o}{2\kappa T}\right)$$
$$\times\left[1+\frac{1}{\kappa a}\frac{2}{\cosh^2(ze\psi_o/4kT)}+\frac{1}{(\kappa a)^2}\frac{8\ln[\cosh(ze\psi_o/4kT)]}{\sinh^2(ze\psi_o/2kT)}\right]^{1/2} \quad (z\text{-}z)$$

$$\sigma = \frac{\varepsilon_r\varepsilon_o\kappa kT}{e}pq\left[1+\frac{4}{\kappa a}\frac{(3-p)q-3}{(pq)^2}+\frac{4}{(\kappa a)^2(pq)^2}\left\{6\ln\left(\frac{q+1}{2}\right)+\ln(1-p)\right\}\right]^{1/2} \quad (2\text{-}1)$$

$$\sigma = \frac{\varepsilon_r\varepsilon_o\kappa kT}{e}\left[pt+\frac{2}{\kappa apt}\left\{(3-p)t-3-\frac{3^{1/2}(1-\eta)}{2\eta^{1/2}}\ln\left(\frac{\{1+(\eta/3)^{1/2}\}\{t-(\eta/3)^{1/2}\}}{\{1-(\eta/3)^{1/2}\}\{t+(\eta/3)^{1/2}\}}\right)\right\}\right]$$
(1-1 plus 2-1)

Note: $s = \dfrac{a}{r}\exp(-\kappa(r-a))$, $B = \dfrac{\left(\dfrac{2\kappa a+1}{\kappa a+1}\right)\tanh\left(\dfrac{ze\psi_o}{4kT}\right)}{1+\left\{1-\dfrac{2\kappa a+1}{(\kappa a+1)^2}\tanh^2\left(\dfrac{ze\psi_o}{4kT}\right)\right\}^{1/2}}$, $p = 1-\exp(-e\psi_o/kT)$,

$q = \left[\dfrac{2}{3}\exp\left(\dfrac{e\psi_o}{kT}\right)+\dfrac{1}{3}\right]^{1/2}$, $t = \left[\left(1-\dfrac{\eta}{3}\right)\exp\left(\dfrac{e\psi_o}{kT}\right)+\dfrac{\eta}{3}\right]^{1/2}$, $\eta = \dfrac{3n_2}{n_1+3n_2}$.

TABLE 1.5. Potential Distribution $\psi(r)$ and Surface Potential ψ_o/Surface Charge Density σ Relationship for a Cylinder of Radius a with Arbitrary Surface Potential

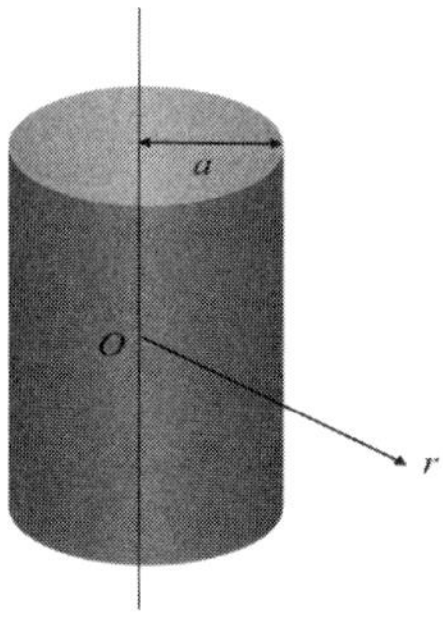

Potential distribution

$$\psi(r) = \frac{2kT}{ze}\ln\left[\frac{(1+Dc)\left\{1+\left(\dfrac{1-\beta}{1+\beta}\right)Dc\right\}}{(1-Dc)\left\{1-\left(\dfrac{1-\beta}{1+\beta}\right)Dc\right\}}\right]$$

Surface potential ψ_o/surface charge density σ relationship

$$\sigma = \frac{2\varepsilon_r\varepsilon_o\kappa kT}{ze}\sinh\left(\frac{y_o}{2}\right)\left[1+\left(\frac{1}{\beta^2}-1\right)\frac{1}{\cosh^2(ze\psi_o/4kT)}\right]^{1/2} \quad (z\text{-}z)$$

$$\sigma = \frac{\varepsilon_r\varepsilon_o\kappa kT}{e}pq\left[1+2\left(\frac{1}{\beta^2}-1\right)\frac{(3-p)q-3}{(pq)^2}\right]^{1/2} \quad (2\text{-}1)$$

$$\sigma = \frac{\varepsilon_r\varepsilon_o\kappa kT}{e}pt\left[1+\frac{2(\beta^{-2}-1)}{(pt)^2}\left\{(3-p)t-3-\frac{\sqrt{3}(1-\eta)}{2\sqrt{\eta}}\ln\left(\frac{\left(t-\sqrt{\frac{\eta}{3}}\right)\left(1+\sqrt{\frac{\eta}{3}}\right)}{\left(t+\sqrt{\frac{\eta}{3}}\right)\left(1-\sqrt{\frac{\eta}{3}}\right)}\right)\right\}\right]^{1/2}$$

(1-1 plus 2-1)

Note: $c = \dfrac{K_0(\kappa r)}{K_0(\kappa a)}$, $\beta = \dfrac{K_0(\kappa a)}{K_1(\kappa a)}$, $D = \dfrac{(1+\beta)\tanh\left(\frac{y_o}{4}\right)}{1+\left\{1-(1-\beta^2)\tanh^2\left(\frac{y_o}{4}\right)\right\}^{1/2}}$, $t = \sqrt{\left(1-\dfrac{\eta}{3}\right)e^{y_o}+\dfrac{\eta}{3}}$,

$\eta = \dfrac{3n_2}{n_1+3n_2}$, $y_o = \dfrac{e\psi_o}{kT}$.

1.5 SOFT PARTICLES

We consider the case where the particle core is covered by an ion-penetrable surface layer of polyelectrolytes, which we term a surface charge layer (or, simply, a surface layer). Polyelectrolyte-coated particles are often called soft particles (Fig. 1.2) [3–5]. Soft particles serve as a model for biocolloids such as cells. Figure 1.3 gives a schematic representation of ion and potential distributions around a hard surface (Fig. 1.3a) and a soft surface (Fig. 1.3b), which shows that the potential deep inside the surface layer is practicably equal to the Donnan potential ψ_{DON}, if the surface layer is much thicker than the Debye length $1/\kappa$. Also we term $\psi_o \equiv \psi(0)$ (which is the potential at the boundary

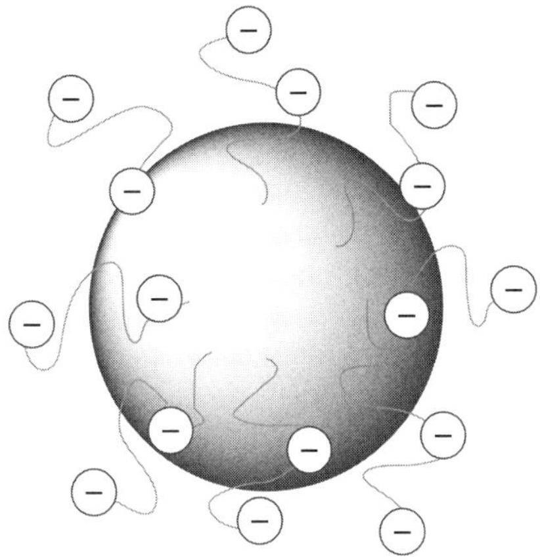

Figure 1.2. Soft particle (polyelectrolyte-coated particle).

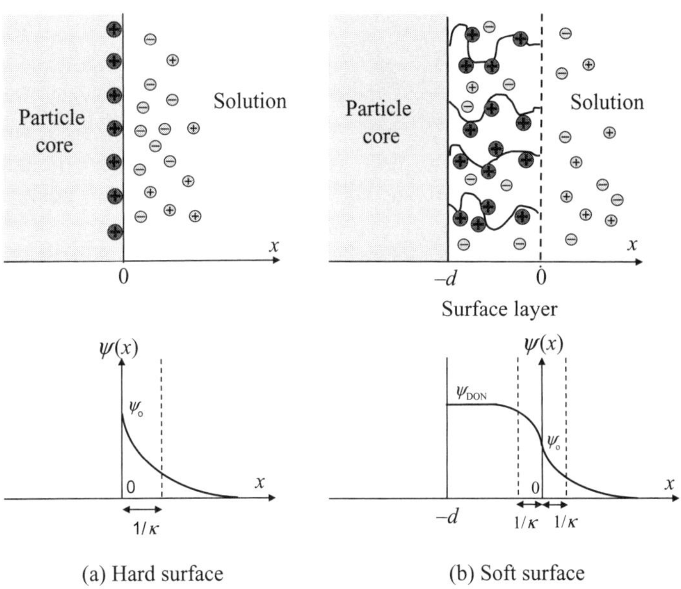

Figure 1.3. Ion and potential distribution around a hard surface (a) and a soft surface (b). When the surface layer is thick, the potential deep inside the surface layer becomes the Donnan potential.

between the surface layer and the surrounding electrolyte solution) the surface potential of the polyelectrolyte layer.

Consider a surface charge layer of thickness d coating a planar hard surface in a general electrolyte solution containing M ionic species with valence z_i and bulk concentration (number density) n_i^∞ ($i = 1, 2, \ldots, M$). We treat the case where fully ionized groups of valence Z are distributed at a uniform density of N in the surface charge layer and the particle core is uncharged. We take an x-axis perpendicular to the surface charge layer with its origin $x = 0$ at the boundary between the surface charge layer and the surrounding electrolyte solution so that the surface charge layer corresponds to the region $-d < x < 0$ and the electrolyte solution to $x > 0$ (Fig. 1.3b). The Poisson–Boltzmann equations for the regions inside and outside the surface charge layer are given by

$$\frac{d^2\psi}{dx^2} = -\frac{1}{\varepsilon_r\varepsilon_o}\sum_{i=1}^{M} z_i e n_i^\infty \exp\left(-\frac{z_i e\psi(x)}{kT}\right), 0 < x < +\infty \tag{1.14}$$

and

$$\frac{d^2\psi}{dx^2} = -\frac{1}{\varepsilon_r\varepsilon_o}\sum_{i=1}^{M} z_i e n_i^\infty \exp\left(-\frac{z_i e\psi(x)}{kT}\right) - \frac{ZeN}{\varepsilon_r\varepsilon_o}, d < x < 0. \tag{1.15}$$

We have here assumed that the relative permittivity ε_r takes the same value in the regions inside and outside the surface charge layer. Note that the right-hand side of Equation 1.15 contains the contribution of the fixed charges of density $\rho_{fix} = ZeN$ in the polyelectrolyte layer. The boundary conditions are

$$\left.\frac{d\psi}{dx}\right|_{x=-d^+} = 0, \tag{1.16}$$

$$\psi(-0^-) = \psi(-0^+) \text{ and } \left.\frac{d\psi}{dx}\right|_{x=-0^-} = \left.\frac{d\psi}{dx}\right|_{x=-0^+}, \tag{1.17}$$

and

$$\psi(x) \to 0 \text{ and } \frac{d\psi}{dx} \to 0 \text{ as } x \to \infty. \tag{1.18}$$

Equation 1.16 corresponds to the situation in which the particle core is uncharged.

Table 1.6 gives the potential distribution and the surface potential ψ_o/charge density N for the low potential case. Table 1.6 also gives the results for a soft sphere or a soft cylinder (i.e., a hard sphere or cylinder of radius a covered by a surface layer of thickness $d = b - a$). Table 1.7 gives the results for the case

TABLE 1.6. Solution to the Linearized Poisson–Boltzmann Equation for Soft Particles with Low N

	Potential Distribution	Surface Potential ψ_o/Charge Density N Relationship
Soft planar surface 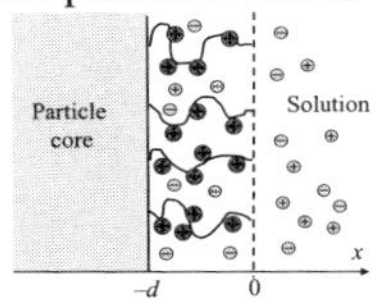	$\psi(x)=\frac{ZNkT}{4z^2ne}(1-e^{-2\kappa d})e^{-\kappa x},\ x>0$ $\psi(x)=\frac{ZNkT}{2z^2ne}\left\{1-\frac{e^{\kappa x}+e^{-\kappa(x+2d)}}{2}\right\},\ -d<x<0$	$\psi_o=\frac{ZeN}{2\varepsilon_r\varepsilon_o\kappa^2}(1-e^{-2\kappa d})$
Soft sphere 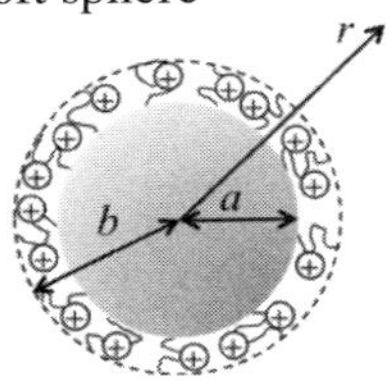	$\psi(r)=\frac{ZeN}{\varepsilon_r\varepsilon_o\kappa^2}\left[1-\left(\frac{1+\kappa b}{1+\kappa a}\right)e^{-\kappa(b-a)}\times\left\{\frac{\sinh[\kappa(r-a)]}{\kappa r}+\frac{a\cosh[\kappa(r-a)]}{r}\right\}\right],\ a<r<b$ $\psi(r)=\psi_o\frac{b}{r}\exp[-\kappa(r-b)],\ r>b$	$\psi_o=\frac{ZeN}{2\varepsilon_r\varepsilon_o\kappa^2}\left\{1-\frac{1}{\kappa b}+\frac{(1-\kappa a)(1+\kappa b)}{(1+\kappa a)\kappa b}e^{-2\kappa(b-a)}\right\}$
Soft cylinder 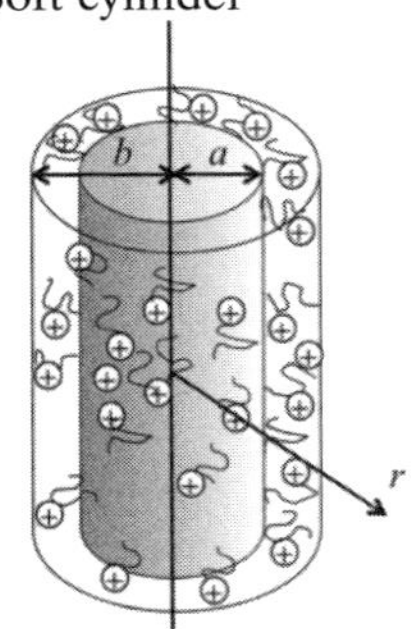	$\psi(r)=\frac{ZeN}{\varepsilon_r\varepsilon_o\kappa^2}\left[1-\kappa bK_1(\kappa b)\times\left\{I_0(\kappa r)+\frac{I_1(\kappa a)}{K_1(\kappa a)}K_0(\kappa r)\right\}\right],\ a<r<b$ $\psi(r)=\psi_o\frac{K_0(\kappa r)}{K_0(\kappa b)},\ r>b$	$\psi_o=\frac{ZeN}{\varepsilon_r\varepsilon_o\kappa^2}\kappa bK_0(\kappa b)\times\left\{I_1(\kappa b)-\frac{I_1(\kappa a)}{K_1(\kappa a)}K_1(\kappa b)\right\}$

TABLE 1.7. Potential Distribution $\psi(r)$, Surface Potential ψ_o/Charge Density N Relationship, and the Effective Surface Potential ψ_{eff} for a Planar Soft Surface with a Thick Surface Charge Layer Carrying Arbitrary N

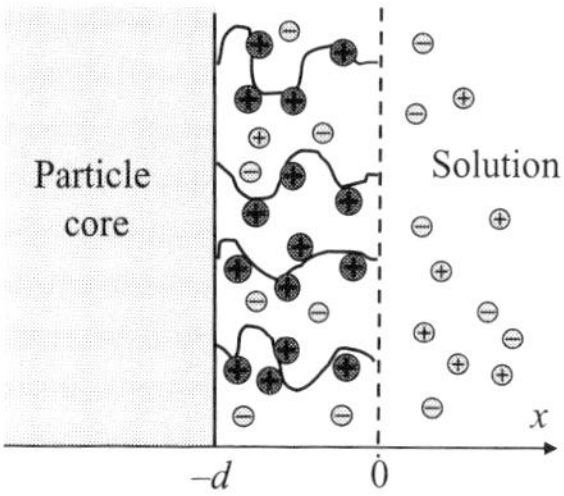

$$\psi(x) = \psi_{DON} + (\psi_o - \psi_{DON})e^{-\kappa_m|x|}, \quad x \le 0$$

$$\psi(x) = \frac{2kT}{ze}\ln\left(\frac{1+\gamma e^{-\kappa x}}{1-\gamma e^{-\kappa x}}\right), \quad x \ge 0$$

with

$$\kappa_m = \kappa\left[1+\left(\frac{ZN}{2zn}\right)^2\right]^{1/4}$$

$$\psi_{DON} = \left(\frac{kT}{ze}\right)\mathrm{arcsinh}\left(\frac{ZN}{2zn}\right) = \left(\frac{kT}{ze}\right)\ln\left[\frac{ZN}{2zn}+\left\{\left(\frac{ZN}{2zn}\right)^2+1\right\}^{1/2}\right]$$

Surface potential ψ_o/charge density N relationship

$$\psi_o = \psi_{DON} - \left(\frac{kT}{ze}\right)\tanh\left(\frac{ze\psi_{DON}}{2kT}\right) = \left(\frac{kT}{ze}\right)\left(\ln\left[\frac{ZN}{2zn}+\left\{\left(\frac{ZN}{2zn}\right)^2+1\right\}^{1/2}\right]+\frac{2zn}{ZN}\left[1-\left\{\left(\frac{ZN}{2zn}\right)^2+1\right\}^{1/2}\right]\right)$$

Effective surface potential

$$\psi_{eff} = \frac{4kT}{e}\gamma$$

Note: $\gamma = \tanh\left(\frac{ze\psi_o}{4kT}\right) = \frac{\exp(ze\psi_o/2kT)-1}{\exp(ze\psi_o/2kT)+1}$

and κ_m is the Debye–Hückel parameter in the surface charge layer.

where a planar soft surface is in contact with a z-z symmetrical electrolyte solution and the thickness of the surface layer d is much greater than the Debye length $1/\kappa$.

The potential distribution outside the surface charge layer of a soft particle with surface potential ψ_o is the same as the potential distribution around a hard particle with a surface potential ψ_o. The asymptotic behavior of the potential distribution around a soft particle and that around a hard particle are the same, and thus their effective surface potentials are also the same provided they have the same surface potential ψ_o (Table 1.7). It must be stressed here that for a hard plate, ψ_o is related to the surface charge density, σ, while for a soft plate, ψ_o is related to the volume charge density $\rho_{fix} = ZeN$.

REFERENCES

1 B. V. Derjaguin, L. Landau. *Acta Physicochim.* 14 (1941) 633.

2 E. J. W. Verwey, J. Th. G. Overbeek. *Theory of the stability of lyophobic colloids.* Elsevier, Amsterdam, 1948.

3 H. Ohshima, K. Furusawa (eds.), *Electrical phenomena at interfaces, fundamentals, measurements, and applications*, 2nd ed., revised and expanded. Dekker, New York, 1998.

4 H. Ohshima. *Theory of colloid and interfacial electric phenomena.* Elsevier/Academic Press, Amsterdam, 2006.

5 H. Ohshima. *Biophysical chemistry of biointerfaces.* John Wiley & Sons, Hoboken, NJ, 2010.

6 A. L. Loeb, J. Th. G. Overbeek, P. H. Wiersema. *The electrical double layer around a spherical colloid particle.* MIT Press, Cambridge, MA, 1961.

7 H. Ohshima, T. W. Healy, L. R. White. *J. Colloid Interface Sci.* 90 (1982) 17.

8 H. Ohshima. *J. Colloid Interface Sci.* 171 (1995) 525.

9 H. Ohshima. *J. Colloid Interface Sci.* 174 (1995) 45. Effective surface potential.

10 H. Ohshima. *J. Colloid Interface Sci.* 200 (1998) 291.

2

ELECTROSTATIC INTERACTION BETWEEN TWO COLLOIDAL PARTICLES

HIROYUKI OHSHIMA

2.1 INTRODUCTION

The Derjaguin–Landau–Verwey–Overbeek (DLVO) theory reveals that when two charged particles approach each other on an electrolyte solution, the electrical double layers around the particles overlap, resulting in an electrostatic interaction between them [1–9]. In this chapter, we derive expressions for the potential energy of the electrostatic interaction between charged particles for various cases.

2.2 INTERACTION BETWEEN TWO COLLOIDAL PARTICLES: LOW POTENTIAL CASE

Consider first two parallel dissimilar plates, plates 1 and 2, separated by a distance h immersed in a liquid containing N ionic species with valence z_i and bulk concentration (number density) n_i^∞ ($i = 1, 2 \ldots N$). We take an x-axis perpendicular to the plates with its origin at the right surface of plate 1, as in

Electrical Phenomena at Interfaces and Biointerfaces: Fundamentals and Applications in Nano-, Bio-, and Environmental Sciences, First Edition. Edited by Hiroyuki Ohshima.

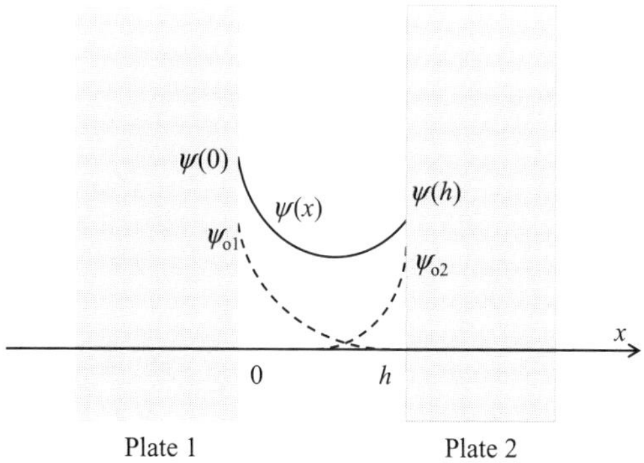

Figure 2.1. Interaction between two parallel dissimilar plates, plates 1 and 2.

Figure 2.1. We assume that the electric potential $\psi(x)$ outside the plates $(0 < x \leq h)$ obeys the following one-dimensional planar Poisson–Boltzmann equation:

$$\frac{d^2\psi(x)}{dx^2} = -\frac{1}{\varepsilon_r\varepsilon_o}\sum_{i=1}^{N} z_i e n_i^\infty \exp\left(-\frac{z_i e\psi(x)}{kT}\right) \quad \text{for } 0 < x < h. \tag{2.1}$$

The boundary conditions at the plate surface depends on the type of the double-layer interaction between plates 1 and 2. If the surface potentials of plates 1 and 2 remain constant at their unperturbed surface potentials ψ_{o1} and ψ_{o2} during interaction, respectively (interaction at constant surface potential), then

$$\psi(0) = \psi_{o1} \tag{2.2}$$

and

$$\psi(h) = \psi_{o2}. \tag{2.3}$$

On the other hand, if the surface charge densities of plates 1 and 2 remain constant at σ_1 and σ_2 (electrostatic interaction at constant surface charge density), then

$$\left.\frac{d\psi}{dx}\right|_{x=0^+} = -\frac{\sigma_1}{\varepsilon_r \varepsilon_o} \tag{2.4}$$

and

$$\left.\frac{d\psi}{dx}\right|_{x=h^-} = \frac{\sigma_2}{\varepsilon_r \varepsilon_o}, \tag{2.5}$$

where we have assumed that the influence of the external fields within the plates can be neglected.

The interaction force P can be calculated by integrating the excess osmotic pressure and the Maxwell stress over an arbitrary closed surface, Σ, enclosing either one of the two interacting plates. As Σ, we choose two planes located at $x = -\infty$ (in the bulk solution far from the plates) and $x = x'$ $(0 < x' < h)$ enclosing plate 1. Here x' is an arbitrary point near the midpoint in the region $0 < x < h$ between plates 1 and 2. Thus, the force $P(h)$ of the double-layer interaction per unit is given by [1–5]

$$P_{\mathrm{pl}}(h) = \sum_{i=1}^{N} n_i^{\infty} kT \left[\exp\left(-\frac{z_i e \psi(x')}{kT} \right) - 1 \right] - \frac{1}{2} \varepsilon_r \varepsilon_o \left(\left.\frac{d\psi}{dx}\right|_{x=x'} \right)^2, \tag{2.6}$$

which, for the low potential case, can be linearized to

$$P_{\mathrm{pl}}(h) = \frac{1}{2} \varepsilon_r \varepsilon_o \left\{ \kappa^2 \psi^2(x') - \left(\left.\frac{d\psi}{dx}\right|_{x=x'} \right)^2 \right\}. \tag{2.7}$$

Here $P(h) > 0$ corresponds to repulsion and $P(h) < 0$ to attraction. The corresponding interaction energy $V_{\mathrm{pl}}(h)$ between two parallel plates per unit area can be obtained by integrating $P_{\mathrm{pl}}(h)$ with the result that

$$V_{\mathrm{pl}}(h) = \int_h^{\infty} P_{\mathrm{pl}}(h) dh. \tag{2.8}$$

For the low potential case, Equation 2.1 can be linearized so that we can easily obtain expressions for the interaction energy between two parallel plates at separation h.

Once the interaction energy $V_{\mathrm{pl}}(h)$ between two parallel plates at separation h is obtained, one can derive the expression for the interaction energy $V_{\mathrm{sp}}(H)$ between two spheres of radii a_1 and a_2 at separation H between their

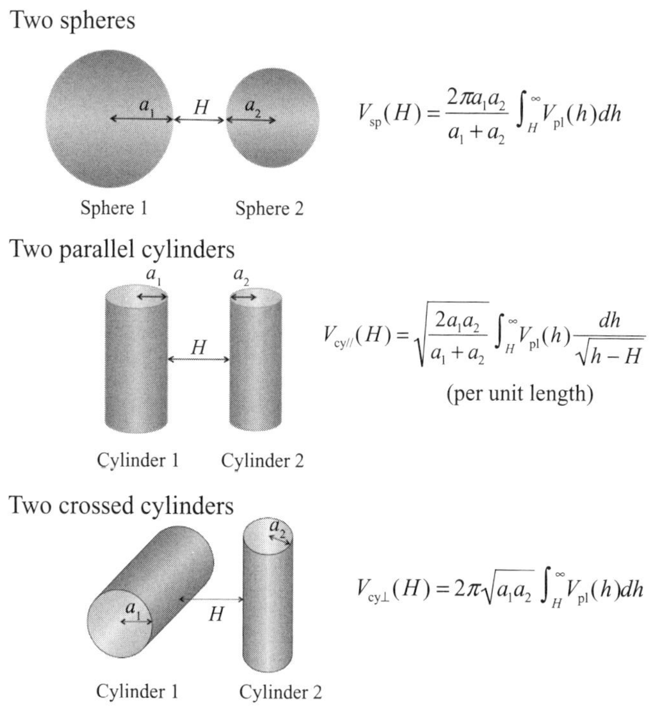

Figure 2.2. Derjaguin's approximation for sphere–sphere interaction [10] and cylinder–cylinder interaction [11, 12].

surfaces and the interaction energy $V_{cy//}(H)$ or $V_{cy\perp}(H)$ between two parallel or crossed cylinders of radii a_1 and a_2 at separation H with the help of Derjaguin's approximation [10–12], provided that $\kappa a_1 \gg 1$, $\kappa a_2 \gg 1$, and $H \ll a_1, a_2$. Figure 2.2 shows Derjaguin's approximation to calculate the sphere–sphere interaction energy [10] and the cylinder–cylinder interaction energy [11, 12]. Figures 2.3 and 2.4 give expressions for $V_{pl}(h)$, $V_{sp}(H)$, $V_{cy//}(H)$, and $V_{cy\perp}(H)$ for the electrostatic interaction energy between two various hard particles. Figures 2.5 and 2.6 give expressions for $V_{pl}(h)$, $V_{sp}(H)$, $V_{cy//}(H)$, and $V_{cy\perp}(H)$ for the electrostatic interaction energy between two various soft particles carrying fixed charge densities ρ_{fix1} and ρ_{fix2}.

2.3 LINEAR SUPERPOSITION APPROXIMATION (LSA)

Although it is generally difficult to derive the interaction energy between two particles with an arbitrary surface potential, a simple approximation method (the LSA) is available to derive the interaction energy between particles at

Two parallel hard plates

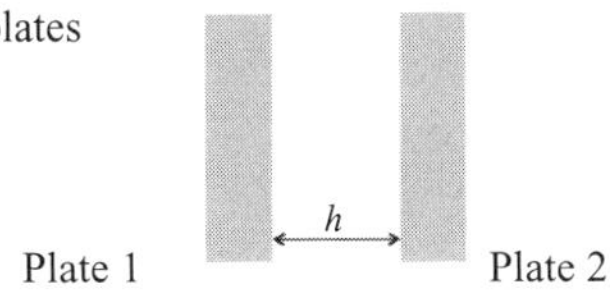

$$V_{\rm pl}^{\psi}(h)=\varepsilon_r\varepsilon_o\kappa\left[\left(\frac{\psi_{o1}+\psi_{o2}}{2}\right)^2\left\{1-\tanh\left(\frac{\kappa h}{2}\right)\right\}-\left(\frac{\psi_{o1}-\psi_{o2}}{2}\right)^2\left\{\coth\left(\frac{\kappa h}{2}\right)-1\right\}\right] \quad \text{(per unit area)}$$

$$V_{\rm pl}^{\sigma}(h)=\varepsilon_r\varepsilon_o\kappa\left[\left(\frac{\psi_{o1}+\psi_{o2}}{2}\right)^2\left\{\coth\left(\frac{\kappa h}{2}\right)-1\right\}-\left(\frac{\psi_{o1}-\psi_{o2}}{2}\right)^2\left\{1-\tanh\left(\frac{\kappa h}{2}\right)\right\}\right] \quad \text{(per unit area)}$$

$$V_{\rm pl}^{\psi-\sigma}(h)=\varepsilon_r\varepsilon_o\kappa\left[\frac{\psi_{o1}\psi_{o2}}{\cosh(\kappa h)}+\frac{1}{2}(\psi_{o1}^2-\psi_{o2}^2)\{1-\tanh(\kappa h)\}\right] \quad \text{(per unit area)}$$

Two hard spheres

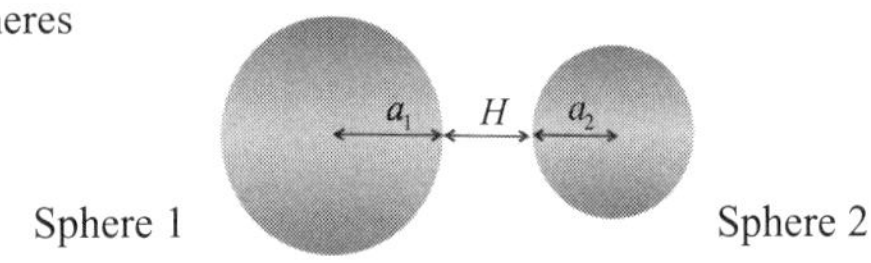

$$V_{\rm sp}^{\psi}(H)=\pi\varepsilon_r\varepsilon_o\frac{a_1a_2}{a_1+a_2}\left\{(\psi_{o1}+\psi_{o2})^2\ln(1+e^{-\kappa H})+(\psi_{o1}-\psi_{o2})^2\ln(1-e^{-\kappa H})\right\}$$

$$V_{\rm sp}^{\sigma}(H)=\pi\varepsilon_r\varepsilon_o\frac{a_1a_2}{a_1+a_2}\left\{-(\psi_{o1}+\psi_{o2})^2\ln(1-e^{-\kappa H})-(\psi_{o1}-\psi_{o2})^2\ln(1+e^{-\kappa H})\right\}$$

$$V_{\rm sp}^{\psi-\sigma}(H)=4\pi\varepsilon_r\varepsilon_o\frac{a_1a_2}{a_1+a_2}\left\{\psi_{o1}\psi_{o2}\arctan(e^{-\kappa H})+\frac{1}{4}(\psi_{o1}^2-\psi_{o2}^2)\ln(1+e^{-2\kappa H})\right\}$$

Figure 2.3. Interaction energies between two parallel plates and between two hard spheres. The superscripts ψ, σ, and ψ–σ, respectively, corresponds to interactions at constant surface potential [13], constant surface charge density [14], and the mixed case in which particle 1 is at constant surface potential and particle 2 is at constant surface charge density [15]. $\psi_{o1} = \sigma_1/\varepsilon_r\varepsilon_o\kappa$ and $\psi_{o2} = \sigma_2/\varepsilon_r\varepsilon_o\kappa$, respectively, are the unperturbed surface potentials of plates 1 and 2 or spheres 1 and 2 with $\kappa a_1 \gg 1$ and $\kappa a_2 \gg 1$.

large particle separations, $\kappa h \gg 1$ or $\kappa H \gg 1$. Expressions for the interaction energy obtained with the help of the LSA, which do not depend on the type of the double-layer interaction, can be applied both to the constant surface potential and constant surface charge density cases as well as their mixed case. In addition, the results obtained on the basis of the LSA can be applied not only to hard particles but also to soft particles.

We now derive the interaction energy between two parallel plates at separation h with arbitrary (unperturbed) surface potentials (Fig. 2.7). We take an x-axis perpendicular to the plates with its origin at the surface of plate 1. The asymptotic forms of the unperturbed potentials $\psi_1(x)$ and $\psi_2(x)$ produced at large distances by plates 1 and 2, respectively, in the absence of interaction are given by Equation 1.12, viz.,

Two parallel hard cylinders

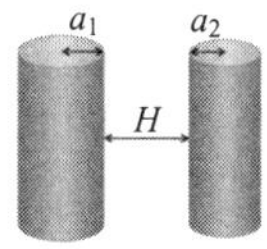

$$V_{\mathrm{cy}//}^{\psi}(H) = -2\varepsilon_{\mathrm{r}}\varepsilon_{\mathrm{o}}\sqrt{\kappa}\sqrt{\frac{2\pi a_1 a_2}{a_1+a_2}}\left[\left(\frac{\psi_{\mathrm{o}1}+\psi_{\mathrm{o}2}}{2}\right)^2 \mathrm{Li}_{1/2}(-e^{-\kappa H}) + \left(\frac{\psi_{\mathrm{o}1}-\psi_{\mathrm{o}2}}{2}\right)^2 \mathrm{Li}_{1/2}(e^{-\kappa H})\right] \quad \text{(per unit length)}$$

$$V_{\mathrm{cy}//}^{\sigma}(H) = 2\varepsilon_{\mathrm{r}}\varepsilon_{\mathrm{o}}\sqrt{\kappa}\sqrt{\frac{2\pi a_1 a_2}{a_1+a_2}}\left[\left(\frac{\psi_{\mathrm{o}1}+\psi_{\mathrm{o}2}}{2}\right)^2 \mathrm{Li}_{1/2}(e^{-\kappa H}) - \left(\frac{\psi_{\mathrm{o}1}-\psi_{\mathrm{o}2}}{2}\right)^2 \mathrm{Li}_{1/2}(-e^{-\kappa H})\right] \quad \text{(per unit length)}$$

$$V_{\mathrm{cy}\perp}^{\psi-\sigma}(H) = 4\pi\varepsilon_{\mathrm{r}}\varepsilon_{\mathrm{o}}\sqrt{a_1 a_2}\left\{\psi_{\mathrm{o}1}\psi_{\mathrm{o}2}\arctan(e^{-\kappa H}) + \frac{1}{4}(\psi_{\mathrm{o}1}^2 - \psi_{\mathrm{o}2}^2)\ln(1+e^{-2\kappa H})\right\} \quad \text{(per unit length)}$$

Two crossed hard cylinders

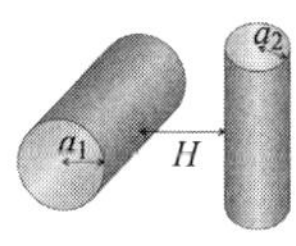

$$V_{\mathrm{cy}\perp}^{\psi}(H) = 4\pi\varepsilon_{\mathrm{r}}\varepsilon_{\mathrm{o}}\sqrt{a_1 a_2}\left[\left(\frac{\psi_{\mathrm{o}1}+\psi_{\mathrm{o}2}}{2}\right)^2 \ln(1+e^{-\kappa H}) + \left(\frac{\psi_{\mathrm{o}1}-\psi_{\mathrm{o}2}}{2}\right)^2 \ln\left(\frac{1}{1-e^{-\kappa H}}\right)\right]$$

$$V_{\mathrm{cy}\perp}^{\sigma}(H) = 4\pi\varepsilon_{\mathrm{r}}\varepsilon_{\mathrm{o}}\sqrt{a_1 a_2}\left[\left(\frac{\psi_{\mathrm{o}1}+\psi_{\mathrm{o}2}}{2}\right)^2 \ln\left(\frac{1}{1-e^{-\kappa H}}\right) - \left(\frac{\psi_{\mathrm{o}1}-\psi_{\mathrm{o}2}}{2}\right)^2 \ln(1+e^{-\kappa H})\right]$$

$$V_{\mathrm{cy}//}^{\psi-\sigma}(H) = \sqrt{2}\varepsilon_{\mathrm{r}}\varepsilon_{\mathrm{o}}\sqrt{\kappa}\sqrt{\frac{2\pi a_1 a_2}{a_1+a_2}}\left[\psi_{\mathrm{o}1}\psi_{\mathrm{o}2}e^{-\kappa H}\Phi\left(-e^{-\kappa H},\frac{1}{2},\frac{1}{2}\right) - \frac{1}{2}(\psi_{\mathrm{o}1}^2 - \psi_{\mathrm{o}2}^2)\mathrm{Li}_{1/2}(-e^{-2\kappa H})\right]$$

Figure 2.4. Interaction energies between two parallel or crossed hard cylinders [11, 12]. $\Phi(z,s,a) = \sum_{k=0}^{\infty} z^k/(k+a)^s$ is the Lerch transcendent.

$$\psi_1(x) = \psi_{\mathrm{eff}1}e^{-\kappa x} \tag{2.9}$$

and

$$\psi_2(x) = \psi_{\mathrm{eff}2}e^{-\kappa(h-x)}, \tag{2.10}$$

where ψ_{eff1} and ψ_{eff2}, respectively, are the effective surface potentials of plates 1 and 2 (given in Chapter 1, Table 1.3). The potential distribution $\psi(x)$ around the midpoint between plates 1 and 2 can thus be approximated by

$$\psi(x) = \psi_1(x) + \psi_2(x) = \psi_{\mathrm{eff}1}e^{-\kappa x} + \psi_{\mathrm{eff}2}e^{-\kappa(h-x)}. \tag{2.11}$$

By substituting Equation 2.11 into Equation 2.7, we obtain

$$P(h) = 2\varepsilon_{\mathrm{r}}\varepsilon_{\mathrm{o}}\kappa^2\psi_{\mathrm{eff}1}\psi_{\mathrm{eff}2}e^{-\kappa h}. \tag{2.12}$$

Two parallel soft plates

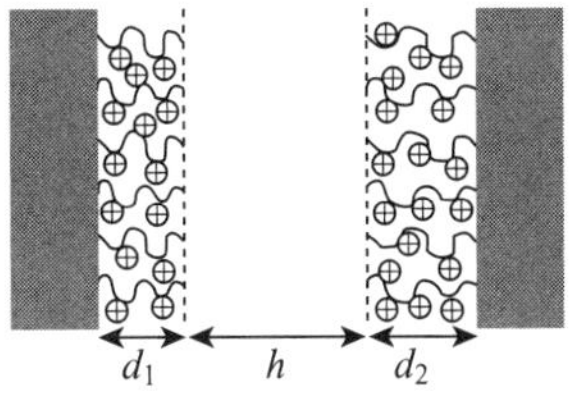

$$V_{\mathrm{pl}}(h)=\frac{1}{4\varepsilon_r\varepsilon_o\kappa^3}\left[\{\rho_{\mathrm{fix1}}\sinh(\kappa d_1)+\rho_{\mathrm{fix2}}\sinh(\kappa d_2)\}^2\left\{\coth\left(\frac{\kappa(h+d_1+d_2)}{2}\right)-1\right\}\right.$$
$$\left.+\{\rho_{\mathrm{fix1}}\sinh(\kappa d_1)-\rho_{\mathrm{fix2}}\sinh(\kappa d_2)\}^2\left\{1-\tanh\left(\frac{\kappa(h+d_1+d_2)}{2}\right)\right\}\right] \quad \text{(per unit area)}$$

Two soft spheres

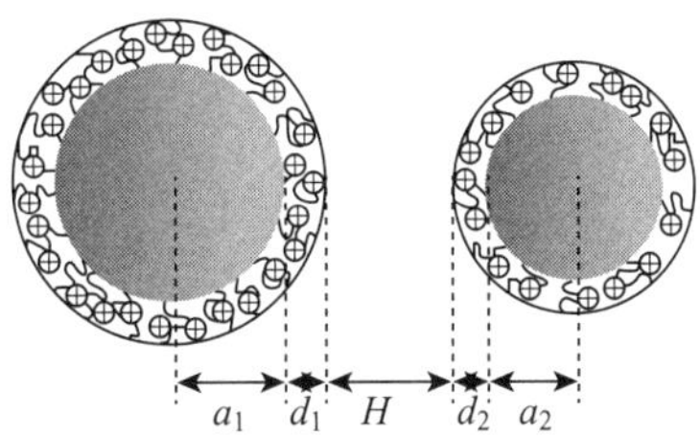

$$V_{\mathrm{sp}}(H)=\frac{1}{\varepsilon_r\varepsilon_o\kappa^4}\left(\frac{\pi b_1 b_2}{b_1+b_2}\right)\left[\{\rho_{\mathrm{fix1}}\sinh(\kappa d_1)+\rho_{\mathrm{fix2}}\sinh(\kappa d_2)\}^2\ln\left(\frac{1}{1-e^{-\kappa(H+d_1+d_2)}}\right)\right.$$
$$\left.+\{\rho_{\mathrm{fix1}}\sinh(\kappa d_1)-\rho_{\mathrm{fix2}}\sinh(\kappa d_2)\}^2\ln(1+e^{-\kappa(H+d_1+d_2)})\right]$$

Figure 2.5. Interaction energies between two parallel soft plates [16] and between two soft spheres [17]. ρ_{fix1} and ρ_{fix2}, respectively, are the fixed charge densities in the surface layers on particles 1 and 2. $b_1 = a_1 + d_1$ and $b_2 = a_2 + d_2$.

The interaction energy $V(h)$ per unit area of plates 1 and 2 is thus

$$V_{\mathrm{pl}}(h)=\int_h^\infty P_{\mathrm{pl}}(h)dh=2\varepsilon_r\varepsilon_o\kappa\psi_{\mathrm{eff}1}\psi_{\mathrm{eff}2}e^{-\kappa h}. \tag{2.13}$$

From Equation 2.13 as combined with Derjaguin's approximation, the interaction energy between two spheres or cylinders of radii a_1 and a_2 at large separations can be derived. The results are given in Figure 2.8. These results are also applicable to the interaction between soft spheres or cylinders consisting of hard spheres or cylinders of radii a_1 and a_2 covered with surface charge layers of thicknesses d_1 and d_2, in which case a_1 and a_2 must be replaced by $b_1 = a_1 + d_1$ and $a_2 + d_2$.

Two parallel soft cylinders

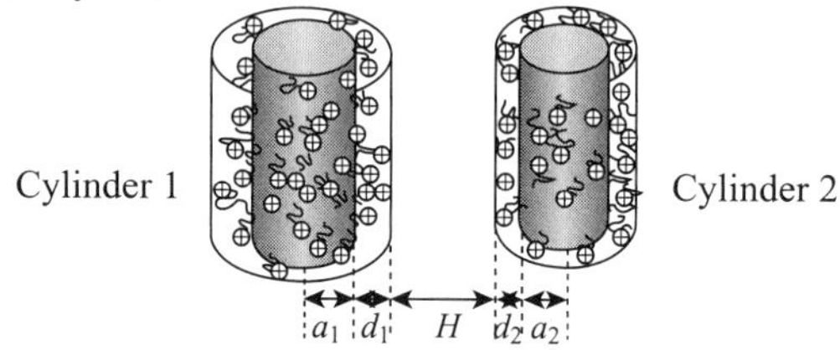

$$V_{\text{cy//}}(H) = \frac{1}{2\varepsilon_r\varepsilon_o\kappa^{7/2}}\sqrt{\frac{2\pi b_1 b_2}{b_1+b_2}}\Big[\{\rho_{\text{fix1}}\sinh(\kappa d_1)+\rho_{\text{fix2}}\sinh(\kappa d_2)\}^2 \text{Li}_{1/2}(e^{-\kappa(H+d_1+d_2)})$$

$$+\{\rho_{\text{fix1}}\sinh(\kappa d_1)-\rho_{\text{fix2}}\sinh(\kappa d_2)\}^2 \text{Li}_{1/2}(-e^{-\kappa(H+d_1+d_2)})\Big] \quad \text{(per unit length)}$$

Two crossed soft plates

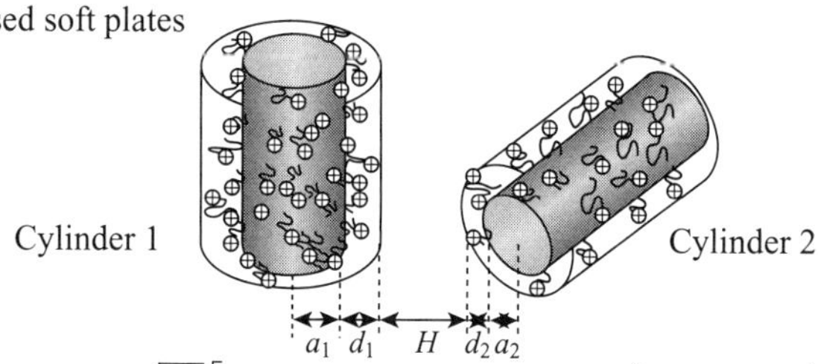

$$V_{\text{cy}\perp}(H) = \frac{\pi\sqrt{b_1 b_2}}{\varepsilon_r\varepsilon_o\kappa^4}\left[\{\rho_{\text{fix1}}\sinh(\kappa d_1)+\rho_{\text{fix2}}\sinh(\kappa d_2)\}^2 \ln\left(\frac{1}{1-e^{-\kappa(H+d_1+d_2)}}\right)\right.$$

$$\left.+\{\rho_{\text{fix1}}\sinh(\kappa d_1)-\rho_{\text{fix2}}\sinh(\kappa d_2)\}^2 \ln(1+e^{-\kappa(H+d_1+d_2)})\right]$$

Figure 2.6. Interaction energies between two parallel soft cylinders 1 and 2 and between two crossed soft cylinders 1 and 2 [12]. ρ_{fix1} and ρ_{fix2}, respectively, are the fixed charge densities in the surface layers on particles 1 and 2. $b_1 = a_1 + d_1$ and $b_2 = a_2 + d_2$. $\text{Li}_s(z)$ is the polylogarithm function, defined by $\text{Li}_s(z) = \sum_{k=1}^{\infty} z^k / k^s$.

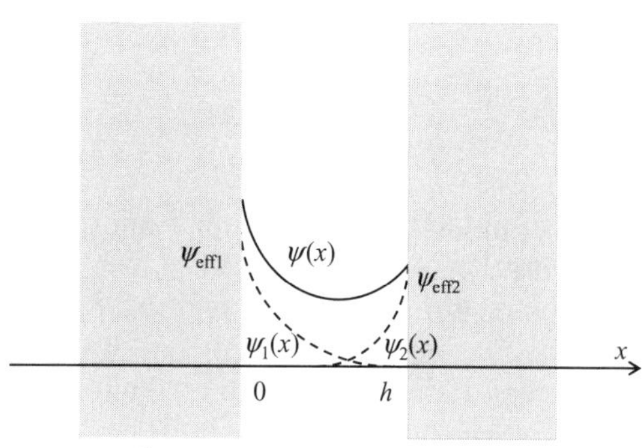

Figure 2.7. Linear superposition approximation.

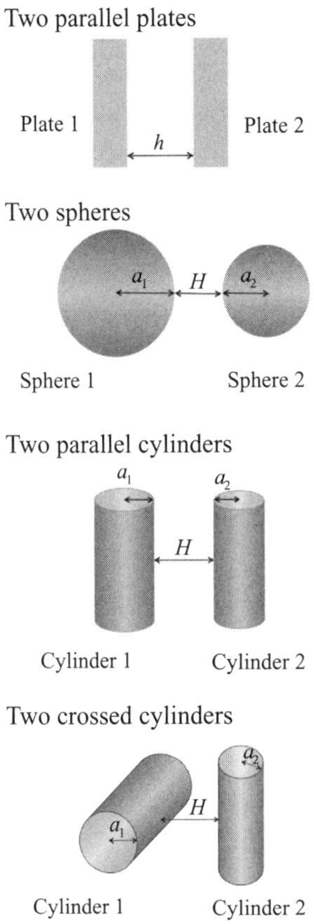

$$V_{pl}(h) = 2\varepsilon_r\varepsilon_o\kappa\psi_{eff1}\psi_{eff2}e^{-\kappa h}$$

(per unit area)

$$V_{sp}(H) = \frac{4\pi a_1 a_2}{a_1 + a_2}\varepsilon_r\varepsilon_o\psi_{eff1}\psi_{eff2}e^{-\kappa H}$$

$$V_{cy//}(H) = 2\sqrt{\frac{2\pi a_1 a_2}{a_1 + a_2}}\sqrt{\kappa}\varepsilon_r\varepsilon_o\psi_{eff1}\psi_{eff2}e^{-\kappa h}$$

(per unit length)

$$V_{cy\perp}(H) = 4\pi\sqrt{a_1 a_2}\varepsilon_r\varepsilon_o\psi_{eff1}\psi_{eff2}e^{-\kappa H}$$

Figure 2.8. Interaction energy $V_{pl}(h)$ between two parallel plates, plates 1 and 2, with effective surface potentials ψ_{eff1} and ψ_{eff2} at separation h. $V_{sp}(H)$ and $V_{cy}(H)$ are obtained with the help of Derjaguin's approximation.

REFERENCES

1 B. V. Derjaguin, L. Landau. *Acta Physicochim.* 14 (1941) 633.

2 E. J. W. Verwey, J. Th. G. Overbeek. *Theory of the stability of lyophobic colloids.* Elsevier, Amsterdam, 1948.

3 B. V. Derjaguin. *Theory of stability of colloids and thin films.* Consultants Bureau, New York, London, 1989.

4 J. N. Israelachvili. *Intermolecular and surface forces*, 2nd ed., Academic Press, New York, 1992.

5 J. Lyklema. *Fundamentals of interface and colloid science, solid-liquid interfaces*, vol. 2. Academic Press, New York, 1995.

6 T. F. Tadros (ed.), *Colloid stability, the role of surface forces—Part 1.* Wiley-VCH, Weinheim, 2007.

7 H. Ohshima, K. Furusawa (eds.), *Electrical phenomena at interfaces, fundamentals, measurements, and applications*, 2nd ed., revised and expanded. Dekker, New York, 1998.
8 H. Ohshima. *Theory of colloid and interfacial electric phenomena*. Elsevier/Academic Press, Amsterdam, 2006.
9 H. Ohshima. *Biophysical chemistry of biointerfaces*. John Wiley & Sons, Hoboken, 2010.
10 B. V. Derjaguin. *Kolloid Z.* 69 (1934) 155.
11 M. J. Sparnaay. *Recueil* 712 (1959) 6120.
12 H. Ohshima, A. Hyono. *J. Colloid Interface Sci.* 333 (2009) 202.
13 R. Hogg, T. W. Healy, D. W. Fuerstenau. *Trans. Faraday Soc.* 62 (1966) 1638.
14 G. R. Wiese, T. W. Healy. *Trans. Faraday Soc.* 66 (1970) 490.
15 G. Kar, S. Chander, T. S. Mika. *J. Colloid Interface Sci.* 44 (1973) 347.
16 H. Ohshima, K. Makino, T. Kondo. *J. Colloid Interface Sci.* 116 (1987) 196.
17 H. Ohshima. *J. Colloid Interface Sci.* 328 (2008) 3.

3

THE DERJAGUIN–LANDAU–VERWEY–OVERBEEK (DLVO) THEORY OF COLLOID STABILITY

Hiroyuki Ohshima

3.1 INTRODUCTION

The stability of colloidal systems consisting of charged particles can be essentially explained by the Derjaguin–Landau–Verwey–Overbeek (DLVO) theory [1–8]. According to this theory, the stability of a suspension of colloidal particles is determined by the balance between the electrostatic interaction (Chapter 2) and the van der Waals interaction between particles. In this chapter, we first consider the van der Waals interaction between particles.

3.2 THE VAN DER WAALS INTERACTION BETWEEN MOLECULES

A neutral molecule can have a fluctuating instantaneous electric dipole originating from quantum mechanics. When two molecules are approaching each other, attractive intermolecular forces, which are called the van der Waals forces, are acting due to interactions between the fluctuating dipoles. The interaction energy $u(r)$ between molecules 1 and 2 at separation r (Fig. 3.1) is given by

Electrical Phenomena at Interfaces and Biointerfaces: Fundamentals and Applications in Nano-, Bio-, and Environmental Sciences, First Edition. Edited by Hiroyuki Ohshima.

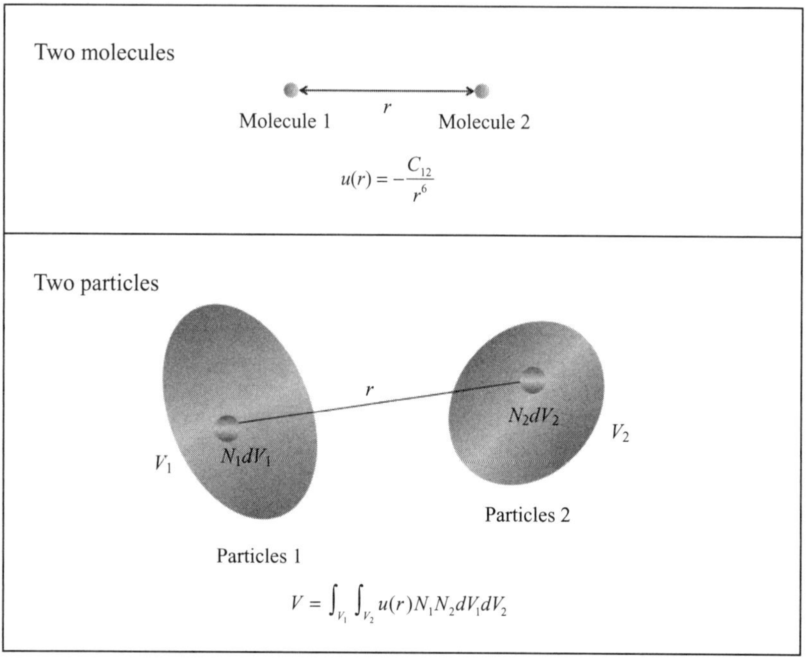

Figure 3.1. The van der Waals interaction energy u between molecules 1 and 2 at separation r and the interaction energy V between particles 1 and 2 of volumes V_1 and V_2 and molecular densities N_1 and N_2, respectively. C_{12} is the London–van der Waals constant.

$$u(r) = -\frac{C_{12}}{r^6} \tag{3.1}$$

with

$$C_{12} = \frac{3\alpha_1\alpha_2 h}{2(4\pi\varepsilon_o)^2}\left(\frac{\nu_1\nu_2}{\nu_1+\nu_2}\right), \tag{3.2}$$

where C_{12} is called the London–van der Waals constant; α_i and ν_i, respectively, are the electronic polarizability and the characteristic frequency of molecule i ($i = 1, 2$); and h is the Planck constant. Thus, $h\nu_i$ corresponds to the ionization energy of molecule i.

3.3 THE VAN DER WAALS INTERACTION BETWEEN PARTICLES

A most remarkable characteristic of the van der Waals interaction is that the additivity of interactions approximately holds. The interaction energy between two particles in a vacuum may thus be calculated approximately by a summation (or by an integration) of the interaction energies for all molecular pairs formed by two molecules belonging to different particles [9] (Fig. 3.1); that is, the interaction energy V between particles 1 and 2, containing N_1 and N_2 molecules per unit volume, respectively, can be obtained by the integration of the interaction energy u between two volume elements, dV_1 and dV_2, at separation r, containing N_1dV_1 and N_2dV_2 molecules, respectively, over volumes V_1 and V_2 of particles 1 and 2, viz.,

$$V = \int_{V_1}\int_{V_2} u(r)N_1N_2dV_1dV_2. \tag{3.3}$$

Expressions for the van der Waals interaction energy V between a molecule and a particle and between various particles have been derived [1–13] and are given in Figures 3.2–3.4.

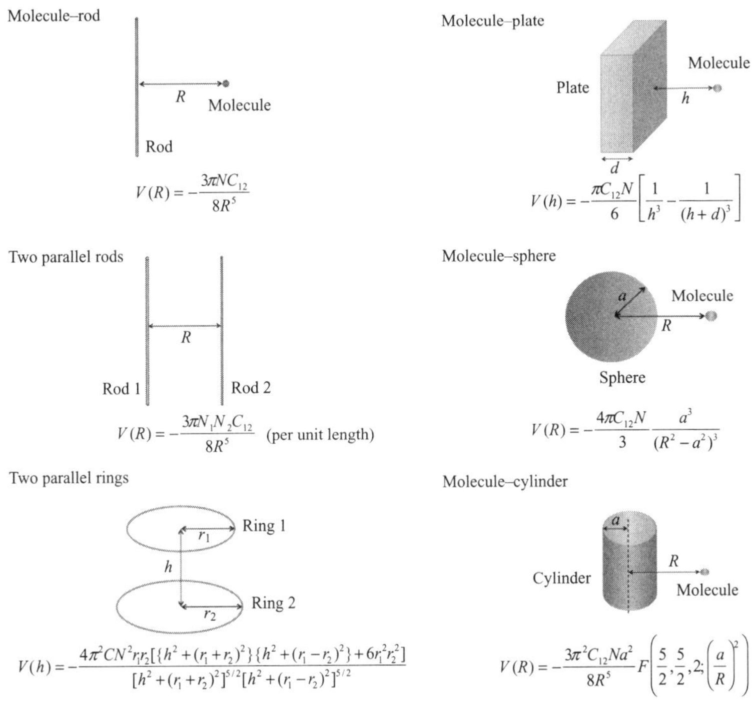

Figure 3.2. The van der Waals interaction energy V between various particles and a molecule and between various particles. N, N_1, and N_2 are the line densities of molecules along rods and rings or the volume densities of molecules in a plate, a sphere, or a cylinder. C_{12} is the London–van der Waals constant. $F(\alpha, \beta, \gamma, x)$ is the hypergeometrical function defined by $F(\alpha,\beta,\gamma,z) = \{\Gamma(\gamma)/\Gamma(\alpha)\Gamma(\beta)\}\sum_{n=0}^{\infty}\Gamma(\alpha+n)\Gamma(\beta+n)z^n/\Gamma(\gamma+n)n!$, where $\Gamma(z)$ is the gamma function.

Two parallel plates

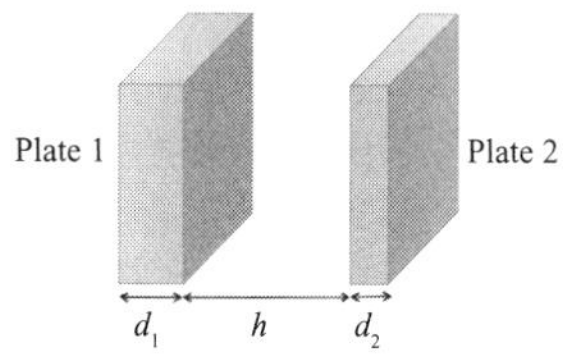

For arbitrary h

$$V(h) = -\frac{A}{12\pi}\left[\frac{1}{h^2} - \frac{1}{(h+d_1)^2} - \frac{1}{(h+d_2)^2} + \frac{1}{(h+d_1+d_2)^2}\right]$$

For small h ($h \ll d_1, d_2$)

$$V(h) = -\frac{A}{12\pi h^2}$$

Two spheres

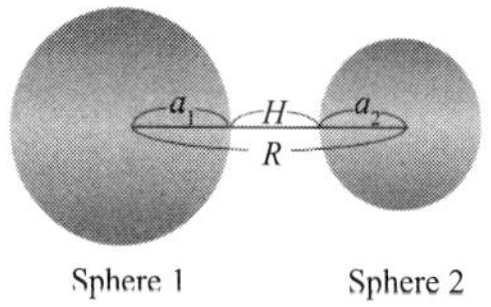

For arbitrary R

$$V(R) = -\frac{A}{6}\left\{\frac{2a_1a_2}{R^2-(a_1+a_2)^2} + \frac{2a_1a_2}{R^2-(a_1-a_2)^2} + \ln\left[\frac{R^2-(a_1+a_2)^2}{R^2-(a_1-a_2)^2}\right]\right\}$$

For small H ($H \ll a_1, a_2$)

$$V(H) = -\frac{Aa_1a_2}{6(a_1+a_2)H}$$

Figure 3.3. The van der Waals interaction energy V between two parallel plates, plates 1 and 2, of thicknesses d_1 and d_2, respectively, separated by distance h, and between spheres 1 and 2 of radii a_1 and a_2, respectively, separated by a distance R between their centers (or H between their closest surfaces). A is the Hamaker constant for the interaction between particles 1 and 2.

Two parallel cylinders

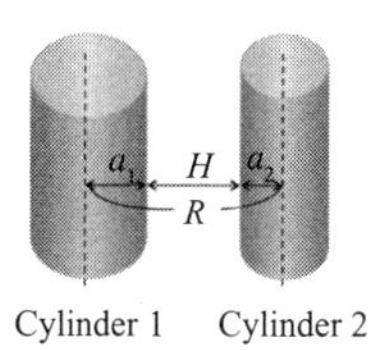

For arbitrary R

$$V(R) = -\frac{2A}{3R}\sum_{m=0}^{\infty}\sum_{n=0}^{\infty}\frac{\Gamma^2(m+n+1/2)}{m!n!(m-1)!(n-1)!}\left(\frac{a_1}{R}\right)^{2m}\left(\frac{a_2}{R}\right)^{2n} \quad \text{(per unit length)}$$

For small H ($H \ll a_1, a_2$)

$$V(H) = -\frac{A}{12\sqrt{2}H^{3/2}}\left(\frac{a_1a_2}{a_1+a_2}\right)^{1/2} \quad \text{(per unit length)}$$

Two crossed cylinders

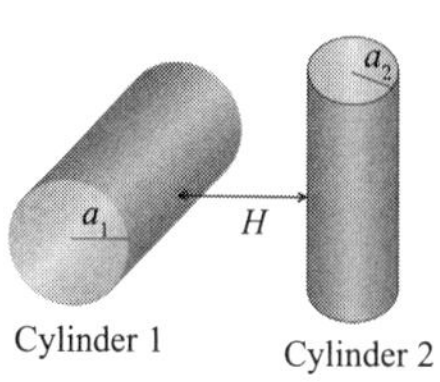

For small H ($H \ll a_1, a_2$)

$$V(H) = -\frac{A\sqrt{a_1a_2}}{6H}$$

Figure 3.4. The van der Waals interaction energy V between two parallel or crossed cylinders, cylinders 1 and 2, of radii a_1 and a_2, respectively, separated by distance R between their axes (or H between their closest surfaces). A is the Hamaker constant.

Equation 3.2 is based on the assumption that the interacting particles are in a vacuum. If the medium between the particles is no longer a vacuum but a second substance (e.g., water), we must account for the fact that each particle replaces an equal volume of water. It follows from the assumption of the additivity of the van der Waals interaction energies that the Hamaker constant A_{132} for the interaction of two different particles of substances 1 and 2 immersed in a medium of substance 3 is

$$A_{132} = \left(\sqrt{A_1} - \sqrt{A_3}\right)\left(\sqrt{A_2} - \sqrt{A_3}\right), \tag{3.4}$$

where A_1 and A_2, respectively, are referred to the interaction between two similar particles of substances 1 and 2 immersed in a vacuum. Thus, if two particles of substances 1 and 2 are interacting in medium 3, then A_{12} appearing in Figures 3.3 and 3.4 must be replaced by A_{132}.

3.4 DLVO THEORY OF COLLOID STABILITY

A number of studies on colloid stability are based on the DLVO theory. As an example, we consider the interaction between two identical spherical particles of radius a carrying unperturbed surface potential ψ_o at separation H between their surfaces in a z-z symmetrical electrolyte solution of bulk concentration (number density) n (Fig. 3.5). The electrostatic interaction energy $V_{el}(H)$ between the spheres is given by (from Fig. 2.8 in Chapter 2 and Table 1.3 in Chapter 1)

$$V_{el}(H) = 2\pi a \varepsilon_r \varepsilon_o \psi_{eff}^2 e^{-\kappa H} = \frac{64\pi ankT\gamma^2}{\kappa^2} e^{-\kappa H}, \tag{3.5}$$

where $\gamma = \tanh(ze\psi_o/kT)$ and κ is the Debye–Hückel parameter. The van der Waals interaction energy $V_{vdW}(H)$ between the spheres is given by (from Fig. 3.3)

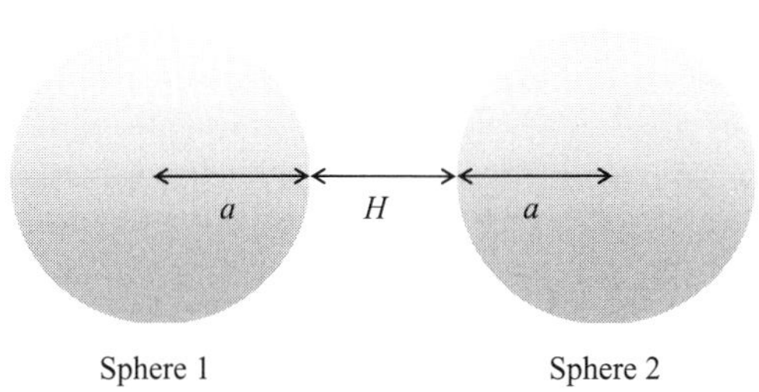

Figure 3.5. Interaction between two identical hard spheres of radius a in an electrolyte solution.

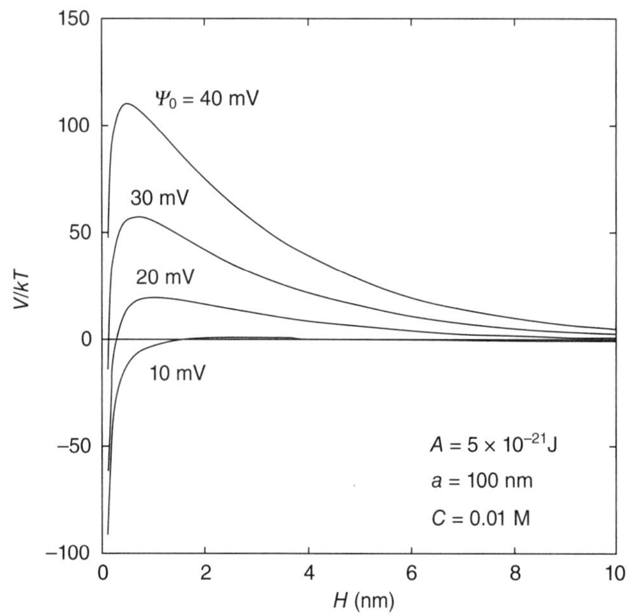

Figure 3.6. Interaction energy V between two identical hard spheres of radius a = 100 nm at C = 0.01 M for various values of the particle surface potential ψ_o. A = 5×10^{-21} J.

$$V_{vdW}(H) = -\frac{Aa}{12H}. \tag{3.6}$$

The total interaction energy $V_t(H) = V_{el}(H) + V_{vdW}(H)$ between the spheres is thus given by

$$V_t(H) = \frac{64\pi ankT\gamma^2}{\kappa^2} e^{-\kappa H} - \frac{Aa}{12H}. \tag{3.7}$$

Figures 3.6 and 3.7 show examples of the calculation of $V(H)$ for the interaction between two spheres of radius a = 100 nm and $A = 5 \times 10^{-21}$ J, where the electrolyte concentration n (in units of cubic meter) is converted to C (in units of molar, M) by $n = 1000\, N_A C$, where N_A = Avogadro's constant. Figure 3.6 shows the interaction energy V at C = 0.01 M for various values of the particle surface potential ψ_o. Figure 3.7 plots the interaction energy V at ψ_o = 20 mV for various values of electrolyte concentration C. As seen from Figures 3.6 and 3.7, there can exist a maximum V_{max} in the potential curves for particles with high surface potentials in low electrolyte concentrations. The DLVO theory predicts that a suspension of colloidal particles should be stable if the potential

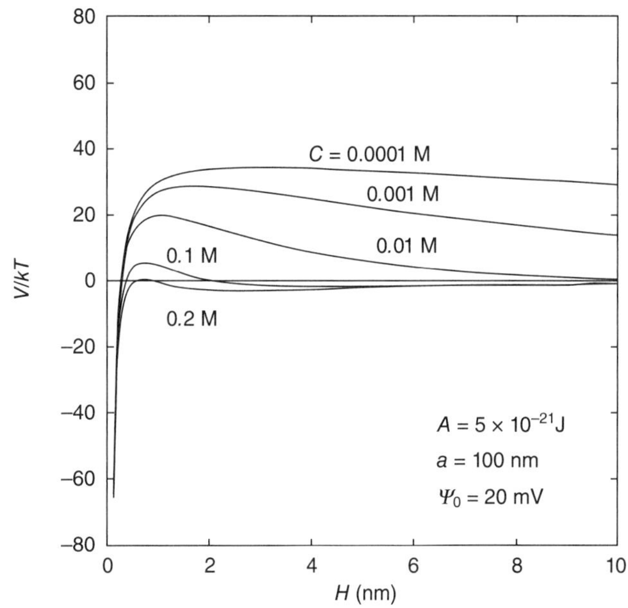

Figure 3.7. Interaction energy V between two identical hard spheres of radius a = 100 nm at ψ_o = 20 mV for various values of electrolyte concentration C. $A = 5 \times 10^{-21}$ J.

maximum V_{max} is sufficiently higher than the thermal energy kT (which amounts to ca. 4×10^{-21} J at room temperature).

As shown in Figure 3.7, at a certain electrolyte concentration C_{cr}, the potential maximum becomes zero. Thus, at this concentration, which is called the critical aggregation concentration, the suspension becomes unstable, resulting in aggregation of the suspension. The critical coagulation concentration C_{cr} is given by the solution to the following coupled equations: $V(H) = 0$ and $dV/dH = 0$, with the solution

$$C_{cr} = \frac{(384)^2 \pi^2 \gamma^4 (kT)^5 (\varepsilon_r \varepsilon_o)^3}{2000 A^2 e^6 \exp(2) z^6 N_A}. \tag{3.8}$$

For particles with high surface potentials ψ_o, γ becomes unity. Thus, from Equation 3.8, we find that

$$C_{cr} \propto \frac{1}{z^6}, \tag{3.9}$$

which is in consistent with Schulze–Hardy's empirical formula.

REFERENCES

1 B. V. Derjaguin, L. D. Landau. *Acta Physicochim.* 14 (1941) 633.

2 E. J. W. Verwey, J. Th. G. Overbeek. *Theory of the stability of lyophobic colloids.* Elsevier/Academic Press, Amsterdam, 1948.

3 B. V. Derjaguin. *Theory of stability of colloids and thin films.* Consultants Bureau, New York, London, 1989.

4 J. N. Israelachvili. *Intermolecular and surface forces*, 2nd ed. Academic Press, New York, 1992.

5 J. Lyklema. *Fundamentals of interface and colloid science, solid-liquid interfaces*, vol. 2. Academic Press, New York, 1995.

6 H. Ohshima, in H. Ohshima, K. Furusawa (eds.), *Electrical phenomena at interfaces, fundamentals, measurements, and applications*, 2nd ed., revised and expanded. Dekker, New York, 1998, Chapter 3.

7 T. F. Tadros (ed.), *Colloid stabolity, the role of surface forces—Part 1.* Wiley-VCH, Weinheim, 2007.

8 H. Ohshima. *Biophysical chemistry of biointerfaces.* John Wiley & Sons, Hoboken, 2010.

9 H. C. Hamaker. *Physica* 4 (1937) 1058.

10 M. J. Sparnaay. *Recueil* 78 (1959) 680.

11 D. Langbein. *Theory of van der Waals attraction.* Springer, Berlin, 1974.

12 V. A. Kirsch. *Adv. Colloid Interface Sci.* 104 (2003) 311.

13 H. Ohshima, A. Hyono. *J. Colloid Interface Sci.* 332 (2009) 251.

4

ELECTROPHORETIC MOBILITY OF CHARGED PARTICLES

HIROYUKI OHSHIMA

4.1 INTRODUCTION

When an external electric field is applied to a suspension of charged colloidal particles in a liquid, the particles in the stationary state move with a constant velocity as a result of the balance between the applied field and a viscous resistance exerted by the liquid on the particles. This phenomenon is called electrophoresis. The particle velocity, which is called electrophoretic velocity, depends strongly on the electrical diffuse double layer around the particles and the particle zeta potential ζ [1–27]. The zeta potential ζ is defined as the potential at the plane where the liquid velocity relative to the particle is zero. This plane is called the slipping plane or the shear plane. The slipping plane does not necessarily coincide with the particle surface. Only if the slipping plane is located at the particle surface, the zeta potential ζ becomes equal to the surface potential ψ_o. In this chapter, we treat the case where $\zeta = \psi_o$ and where particles are symmetrical (spherical or cylindrical). Usually, we treat the case where the magnitude of the applied electric field E is not very large. In such cases, the electrophoretic velocity $\boldsymbol{U}$ of the particles is proportional to $\boldsymbol{E}$ in magnitude. The ratio of the magnitude of the velocity U to that of the applied field E is called electrophoretic mobility, μ, which is defined by $\mu = U/E$ (where $U = |\boldsymbol{U}|$ and $E = |\boldsymbol{E}|$). In this chapter, we derive equations for the electrophoretic mobilities of hard and soft colloidal particles.

Electrical Phenomena at Interfaces and Biointerfaces: Fundamentals and Applications in Nano-, Bio-, and Environmental Sciences, First Edition. Edited by Hiroyuki Ohshima.

4.2 GENERAL THEORY OF ELECTROPHORETIC MOBILITY OF HARD PARTICLES

Full electrokinetic equations determining electrophoretic mobility of spherical particles with arbitrary values of κa and ζ were derived independently by Overbeek [4] and Booth [5]. Wiersema, Loeb, and Overbeek [6] solved the equations numerically using an electronic computer. The computer calculation of the electrophoretic mobility of a spherical particle was considerably improved by O'Brien and White [9]. Approximate analytic mobility expressions, on the other hand, have been proposed by various authors [1–5, 7, 8, 10, 12, 17, 20–27].

Consider a spherical hard particle of radius a moving with a velocity $\boldsymbol{U}$ in a liquid containing a general electrolyte composed of N ionic species with valence z_i and bulk concentration (number density) n_i^∞, and drag coefficient λ_i ($i = 1, 2 \ldots N$). The origin of the spherical polar coordinate system (r, θ, ϕ) is held fixed at the center of the particle and the polar axis ($\theta = 0$) is set parallel to $\boldsymbol{E}$ (Fig. 4.1). For a spherical particle, $\boldsymbol{U}$ takes the same direction as $\boldsymbol{E}$.

The main assumptions are as follows: (i) The Reynolds number of the liquid flow is small enough to ignore inertial terms in the Navier–Stokes equation and the liquid can be regarded as incompressible; (ii) the applied field E is weak so that the particle velocity $\boldsymbol{U}$ is proportional to $\boldsymbol{E}$ and terms of higher order in $\boldsymbol{E}$ may be neglected; (iii) the slipping plane (at which the liquid velocity relative to the particle becomes zero) is located on the particle surface (at $r = a$); (iv) no electrolyte ions can penetrate the particle surface.

Electrokinetic equations that govern the motion of charged colloidal particles in a liquid containing an electrolyte caused by an externally applied electric field consist of the Navier–Stokes equation for the liquid flow around the particle, continuity equations for the liquid flow and the flow of the ionic species, and the Poisson equation connecting the electric potential and the

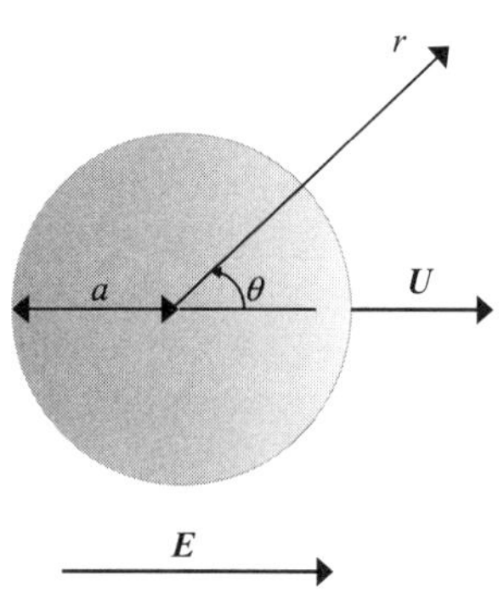

Figure 4.1. Electrophoresis of a spherical particle of radius a with a velocity $\boldsymbol{U}$ in an applied electric field $\boldsymbol{E}$.

concentrations of the ionic species. The Navier–Stokes equation for the liquid flow $\boldsymbol{u}(\boldsymbol{r})$ around the particle is given by

$$\eta\nabla\times\nabla\times\boldsymbol{u}(\boldsymbol{r})+\nabla p(\boldsymbol{r})+\rho_{\mathrm{el}}(\boldsymbol{r})\nabla\psi(\boldsymbol{r})=\boldsymbol{0},\, r>a, \tag{4.1}$$

where η is the viscosity of the electrolyte solution, $p(\boldsymbol{r})$ is the pressure at position $\boldsymbol{r}$, $\rho_{\mathrm{el}}(\boldsymbol{r})$ is the charge density resulting from the mobile charged ionic species, $\psi(\boldsymbol{r})$ is the electric potential, and $r = |\boldsymbol{r}|$.

Ohshima et al. [12] derived the general expression for the electrophoretic mobility of a sphere:

$$\mu=\frac{a^2}{9}\int_a^\infty\left(1-\frac{3r^2}{a^2}+\frac{2r^3}{a^3}\right)G(r)dr \tag{4.2}$$

with

$$\begin{aligned} y&=\frac{e\psi^{(0)}}{kT}\\ G(r)&=-\frac{e}{\eta r}\frac{dy}{dr}\sum_{i=1}^{N}z_i^2n_i^\infty e^{-z_iy}\phi_i, \end{aligned} \tag{4.3}$$

where $\psi^{(0)}$ is the equilibrium potential, y is the corresponding scaled quantity, and ϕ_i is related to the deviation $\delta\mu_i$ of the electrochemical potential of ith ionic species by

$$\delta\mu_i(\boldsymbol{r})=-z_ie\phi_i(r)E\cos\theta \tag{4.4}$$

and

$$\phi_i(r)=r+\frac{a^3}{2r^3}+f_i(r), \tag{4.5}$$

where the first term r on the right-hand side corresponds to the potential of the applied external field $\boldsymbol{E}\cdot\boldsymbol{r}$, the second term $a^3/2r^2$ to the deviation of the external field due to the presence of the particle core, and the third term $f_i(r)$ to the contribution of the relaxation effect of electrolyte ions.

4.3 SMOLUCHOWSKI'S, HÜCKEL'S, AND HENRY'S EQUATIONS

It can be shown that for the limiting case of $\kappa a \to \infty$, ϕ_i becomes

$$\phi_i(r)=\frac{3a}{2}, \tag{4.6}$$

from which we obtain

$$\mu = \frac{\varepsilon_r \varepsilon_o}{\eta} \zeta. \tag{4.7}$$

This is Smoluchowski's equation [1]. In the opposite limit of $\kappa a \to \infty$, ϕ_i becomes

$$\phi_i(r) = r, \tag{4.8}$$

so that we have

$$\mu = \frac{2\varepsilon_r \varepsilon_o}{3\eta} \zeta, \tag{4.9}$$

which is with Hückel's equation [2]. For spherical particles with low ζ and arbitrary κa, ϕ_i becomes

$$\phi_i(r) = r + \frac{a^3}{2r^2}, \tag{4.10}$$

which gives the following Henry equation [3]:

$$\mu = \frac{\varepsilon_r \varepsilon_o}{\eta} \zeta f(\kappa a), \tag{4.11}$$

where $f(\kappa a)$ is called Henry's function. Accurate approximate expressions for the mobilities of a sphere and a cylinder are derived in References 22 and 23. Figure 4.2 summarizes the aforementioned results. Similarly, we can derive the corresponding equations for the electrophoretic mobility of an infinitely long cylinders moving in a tangential or transverse electric field. The results are shown in Figure 4.3.

4.4 MOBILITY EQUATIONS TAKING INTO ACCOUNT THE RELAXATION EFFECT

Henry's mobility equation, Equation 4.10, assumes that the double layer potential distribution around a spherical particle remains unchanged during electrophoresis (Fig. 4.4). For high zeta potentials, however, the double layer is no longer spherically symmetrical. This effect is called the relaxation effect (the double layer polarization or the surface conduction) (Fig. 3.8). Henry's equation, Equation 4.10, does not take into account

$\kappa a \gg 1$

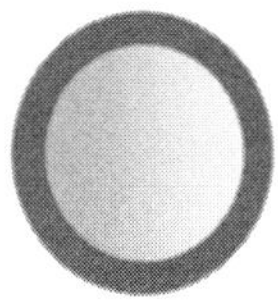

Smoluchowski's equation

$$\mu = \frac{\varepsilon_r \varepsilon_o}{\eta} \zeta$$

$\kappa a \ll 1$

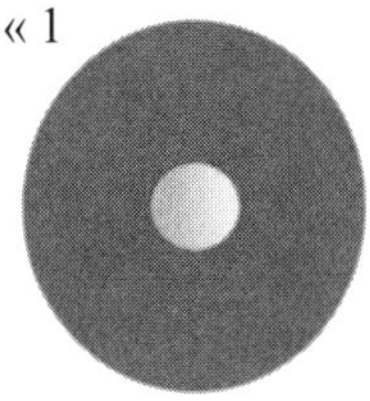

Hückel's equation

$$\mu = \frac{2\varepsilon_r \varepsilon_o \zeta}{3\eta}$$

Arbitrary κa

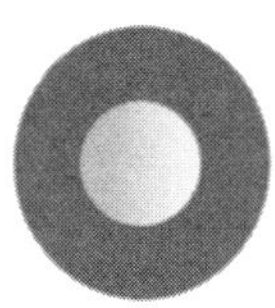

Henry's equation

$$\mu = \frac{\varepsilon_r \varepsilon_o}{\eta} \zeta f(\kappa a)$$

with $$f(\kappa a) = \frac{2}{3}\left[1 + \frac{1}{2\left(1 + \frac{2.5}{\kappa a\{1 + 2\exp(-\kappa a)\}}\right)^3}\right]$$

Figure 4.2. Mobility equations for a sphere of radius a. Smoluchowski's equation for $\kappa a \gg 1$, Hückel's equation for $\kappa a \ll 1$, and Henry's equation for arbitrary κa and low ζ.

the relaxation effect, and thus, this equation is correct to the first order of zeta potential ζ (Fig. 4.5). If the relaxation effect is taken into account, then the electrophoretic mobility μ becomes dependent on the ionic drag coefficient λ_i. We introduce the scaled ionic drag coefficient m_i, defined by

$$m_i = \frac{2\varepsilon_r \varepsilon_o kT}{3\eta z_i^2 e^2} \lambda_i. \tag{4.12}$$

For an aqueous KCl solution at 25°C, $m = 0.176$ for K^+ ions and $m = 0.169$ for Cl^- ions. In Figure 3.9, we plot the electrophoretic mobility μ of a positively charged sphere in an aqueous KCl solution at 25°C ($\varepsilon_r = 78.5$ and $\eta = 0.891$ mPa/s) as a function of scaled zeta potential $e\zeta/kT$ for several values of κa calculated by the method of O'Brien and White [9]. Here E_m is the scaled electrophoretic mobility defined by (Fig. 4.6)

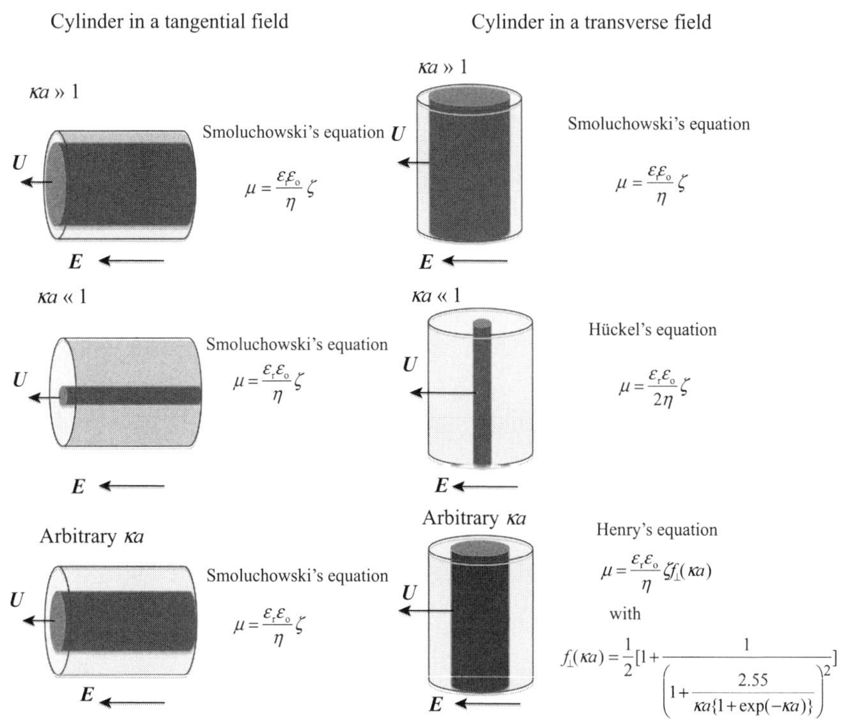

Figure 4.3. Mobility equations for a cylinder of radius a. Smoluchowski's equation for κa » 1, Hückel's equation for κa « 1, and Henry's equation for arbitrary κa and low ζ.

$$E_m = \frac{3\eta e}{2\varepsilon_r\varepsilon_o kT}\mu. \tag{4.13}$$

The cases of $\kappa a = \infty$ and $\kappa a = 0$, which are both straight lines, correspond to Smoluchowski's equation, Equation 4.7, and Hückel's equation, Equation 4.9, respectively. It is seen that there is a mobility maximum due to the relaxation effect.

Accurate approximate mobility expressions of a sphere of radius a correct to order $1/\kappa a$ applicable for $\kappa a \geq 10$ and arbitrary ζ [12] and that correct to order ζ^3 [24] applicable for $ze\zeta/kT \leq 3$ and arbitrary κa are given in Table 4.1. Approximate mobility expressions for a sphere with $\kappa a \geq 30$ and arbitrary ζ [25] is given in Table 4.2 (see also Fig. 4.7). In Figure 3.15, we show the range of validity of various mobility expressions: Smoluchowski's equation (Eq. 4.7) [1], Hückel's equation (Eq. 4.9) [2], Henry's equation (Eq. 4.11) [3], Ohshima–Healy–White's equation (Eq. (1) in Table 4.1) [12], and Ohshima's equation (Eq. (2) in Table 4.1) [24].

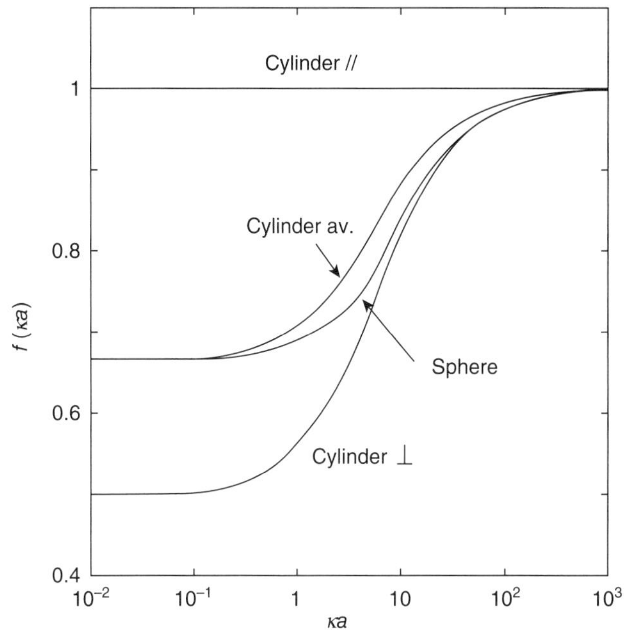

Figure 4.4. Henry's function $f(\kappa a)$ for a sphere [22] and a cylinder [23] of radius a as a function of scaled radius κa. The cylinder is oriented parallel ($\mu_{//}$) or perpendicular ($\mu_{\perp}$) to the applied electric field. For a cylindrical particle oriented at an arbitrary angle between its axis and the applied electric field, its electrophoretic mobility averaged over a random distribution of orientation is given by $\mu_{av} = (\mu_{//} + 2\mu_{\perp})/3$ [26, 27]. From Reference 20.

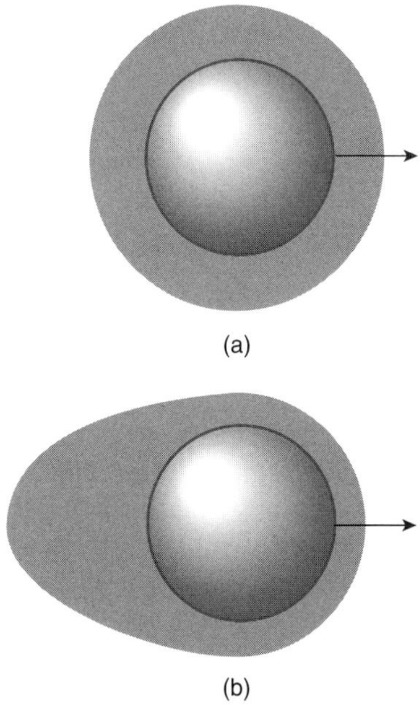

Figure 4.5. Relaxation effect. The electrical double layer around a sphere with low ζ is spherically symmetrical (a) but loses its spherical symmetry for a sphere with high ζ (b).

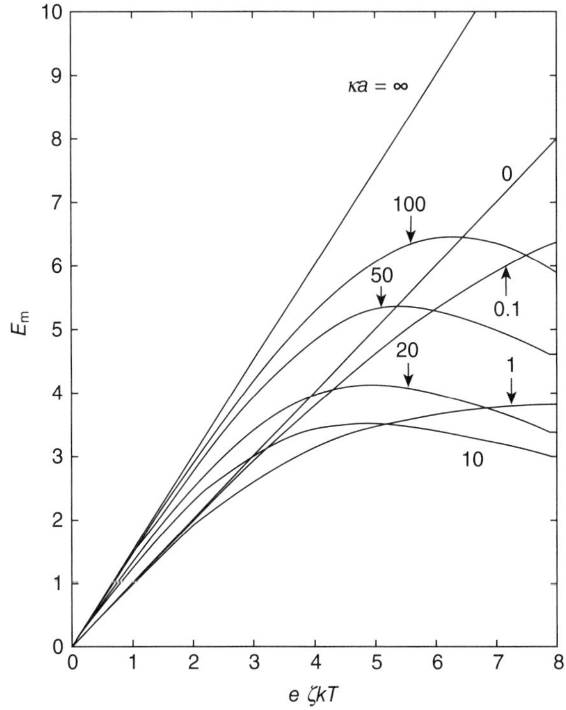

Figure 4.6. The scaled electrophoretic mobility E_m of a positively charged spherical colloidal particle of radius a in an aqueous KCl solution at 25°C as a function of scaled zeta-potential $e\zeta/kT$ for various values of κa. Calculated by the method of O'Brien and White [9]. E_m is defined by Equation 4.13. From Reference 20.

TABLE 4.1. Accurate Mobility Expressions for a Hard Sphere of Radius *a*

Hard sphere of radius a with arbitrary ζ $\kappa a \geq 10$ in a symmetrical electrolyte:

$$\mu = \mathrm{sgn}(\zeta)\frac{\varepsilon_r\varepsilon_o}{\eta}\left\{|\zeta| - \frac{2F}{1+F}\left(\frac{kT}{ze}\right)H\right\} + \mathrm{sgn}(\zeta)\frac{2\varepsilon_r\varepsilon_o kT}{3\eta ze}\left[\frac{1}{\kappa a}\left\{-18\left(t+\frac{t^3}{9}\right)K + \frac{15F}{1+F}\left(t+\frac{7t^2}{20}+\frac{t^3}{9}\right) - 6(1+3\bar{m})(1-e^{-\bar{\zeta}/2})G + \frac{12F}{(1+F)^2}H + \frac{9\bar{\zeta}}{1+F}(\bar{m}G+mH) - \frac{36F}{1+F}\left(\bar{m}G^2 + \frac{m}{1+F}H^2\right)\right\}\right] \quad (1)$$

with

$$\bar{\zeta} = \frac{ze|\zeta|}{kT},\ F = \frac{2}{\kappa a}(1+3m)(e^{\bar{\zeta}/2}-1),\ G = \ln\left(\frac{1+e^{-\bar{\zeta}/2}}{2}\right),\ H = \ln\left(\frac{1+e^{\bar{\zeta}/2}}{2}\right),$$

$$m = \frac{2\varepsilon_r\varepsilon_o kT}{3\eta z^2e^2}\lambda,\ \bar{m} = \frac{2\varepsilon_r\varepsilon_o kT}{3\eta z^2e^2}\bar{\lambda},\ K = 1 - \frac{25}{3(\kappa a+10)}\exp\left[-\frac{\kappa a}{6(\kappa a-6)}\bar{\zeta}\right],\ t = \tanh(\bar{\zeta}/4)$$

Hard sphere of radius a with arbitrary κa and $ze\zeta/kT \leq 3$ in a symmetrical electrolyte:

$$\mu = \frac{2\varepsilon_r\varepsilon_o\zeta}{3\eta}\left(1 + \frac{1}{2[1+2.5/\{\kappa a(1+2e^{-\kappa a})\}]^3}\right) - \frac{2\varepsilon_r\varepsilon_o\zeta}{3\eta}\left(\frac{ze\zeta}{kT}\right)^2\left[\frac{\kappa a\{\kappa a + 1.3\exp(-0.18\kappa a)+2.5\}}{2\{\kappa a+1.2\exp(-7.4\kappa a)+4.8\}^3} + \left(\frac{m+\bar{m}}{2}\right)\frac{9\kappa a\{\kappa a+5.2\exp(-3.9\kappa a)+5.6\}}{8\{\kappa a-1.55\exp(-0.32\kappa a)+6.02\}^3}\right] \quad (2)$$

Note: Equation (1) is correct to order $1/\kappa a$ [12] and Equation (2) is correct to order ζ^3 [24].

TABLE 4.2. Approximate Mobility Expressions for a Hard Sphere of Radius a with $\kappa a \geq 30$ [24]

	Electrophoretic Mobility μ	Expressions for F	Drag Coefficient of Counterions
Charged particles in a z-z symmetrical electrolyte	$\mu = \mathrm{sgn}(\zeta)\dfrac{\varepsilon_r\varepsilon_o}{\eta}\left\{\lvert\zeta\rvert - \dfrac{2F}{1+F}\left(\dfrac{kT}{ze}\ln\left[\dfrac{1+\exp(ze\lvert\zeta\rvert/2kT)}{2}\right]\right)\right\}$	$F = \dfrac{2}{\kappa a}(1+3m)\left\{\exp\left(\dfrac{ze\lvert\zeta\rvert}{2kT}\right)-1\right\}$	$m = \dfrac{2\varepsilon_r\varepsilon_o kT}{3\eta z^2 e^2}\lambda$
Positively charged particles ($z > 0$) in a 2-1 electrolyte	$\mu = \dfrac{\varepsilon_r\varepsilon_o}{\eta}\left\{\zeta - \dfrac{2F}{1+F}\left(\dfrac{kT}{e}\right)\ln\left[\dfrac{1}{2}+\dfrac{1}{2}\sqrt{\dfrac{2}{3}+\exp\left(\dfrac{e\zeta}{kT}\right)}+\dfrac{1}{3}\right]\right\}$	$F = \dfrac{\sqrt{6}}{\kappa a}(1+3m_-)\left\{\exp\left(\dfrac{e\zeta}{2kT}\right)-1\right\}$	$m_- = \dfrac{2\varepsilon_r\varepsilon_o kT}{3\eta e^2}\lambda_-$
Negatively charged particles ($\zeta < 0$) in a 2-1 electrolyte	$\mu = \dfrac{\varepsilon_r\varepsilon_o}{\eta}\left\{\zeta + \dfrac{2F}{1+F}\left(\dfrac{kT}{e}\right)\right.$ $\left.\times\ln\left[\dfrac{1}{2}\exp\left(-\dfrac{e\zeta}{2kT}\right)+\dfrac{1}{2}\sqrt{\dfrac{1}{3}\exp\left(-\dfrac{e\zeta}{kT}\right)+\dfrac{2}{3}}\right]\right\}$	$F = \dfrac{\sqrt{3}}{\kappa a}(1+3m_+)\left\{\exp\left(\dfrac{e\lvert\zeta\rvert}{kT}\right)-1\right\}$	$m_+ = \dfrac{\varepsilon_r\varepsilon_o kT}{6\eta e^2}\lambda_+$
Positively charged particles ($\zeta > 0$) in a 1-2 electrolyte	$\mu = \dfrac{\varepsilon_r\varepsilon_o}{\eta}\left\{\zeta - \dfrac{2F}{1+F}\left(\dfrac{kT}{e}\right)\right.$ $\left.\times\ln\left[\dfrac{1}{2}\exp\left(\dfrac{e\zeta}{2kT}\right)+\dfrac{1}{2}\sqrt{\dfrac{1}{3}\exp\left(\dfrac{e\zeta}{kT}\right)+\dfrac{2}{3}}\right]\right\}$	$F = \dfrac{\sqrt{3}}{\kappa a}(1+3m_-)\left\{\exp\left(\dfrac{e\zeta}{kT}\right)-1\right\}$	$m_- = \dfrac{\varepsilon_r\varepsilon_o kT}{6\eta e^2}\lambda_-$
Negatively charged particles ($\zeta < 0$) in a 1-2 electrolyte	$\mu = \dfrac{\varepsilon_r\varepsilon_o}{\eta}\left\{\zeta + \dfrac{2F}{1+F}\left(\dfrac{kT}{e}\right)\ln\left[\dfrac{1}{2}+\dfrac{1}{2}\sqrt{\dfrac{2}{3}\exp\left(\dfrac{e\lvert\zeta\rvert}{kT}\right)}+\dfrac{1}{3}\right]\right\}$	$F = \dfrac{\sqrt{6}}{\kappa a}(1+3m_+)\left\{\exp\left(\dfrac{e\lvert\zeta\rvert}{2kT}\right)-1\right\}$	$m_+ = \dfrac{2\varepsilon_r\varepsilon_o kT}{3\eta e^2}\lambda_+$

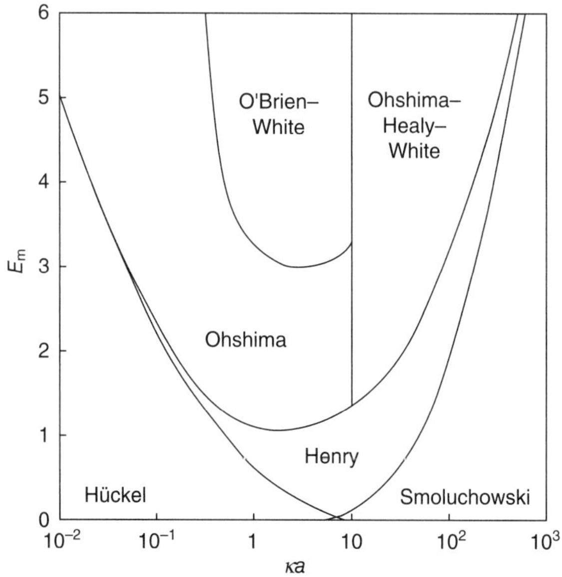

Figure 4.7. Range of validity of several mobility expressions. Smoluchowski's equation (Eq. 4.7) [1], Hückel's equation (Eq. 4.9) [10], Henry's equation (Eq. 4.11) [12], Ohshima–Healy–White's equation (Eq. (2) in Table 4.1) [19], and Ohshima's equation (Eq. (2) in Table 4.1) [22]. Each area corresponds to the region with relative errors less than 1%. In the region labeled O'Brien–White, only a computer solution is available [16]. From Reference 20.

4.5 ELECTROPHORETIC MOBILITY OF SOFT PARTICLES

Consider now a spherical soft particle, that is, a charged spherical particle covered with an ion-penetrable layer of polyelectrolytes moving with a velocity $\boldsymbol{U}$ in a symmetrical electrolyte solution of valence z and bulk concentration (number density) n in an applied electric field $\boldsymbol{E}$ [28–44]. We suppose that the uncharged particle core of radius a is coated with an ion-penetrable layer of polyelectrolytes of thickness d and that ionized groups of valence Z are distributed within the polyelectrolyte layer at a uniform density N so that the polyelectrolyte is later uniformly charged at a constant density, $\rho_{\text{fix}} = ZeN$. The polymer-coated particle has thus an inner radius a and an outer radius $b = a + d$ (Fig. 4.8). Let the drag coefficient of coions be λ_+ and that of counterions be λ_-. We adopt the model of Debye–Bueche [45] wherein the polymer segments are regarded as resistance centers distributed uniformly in the polyelectrolyte layer, exerting frictional forces on the liquid flowing in the polyelectrolyte layer.

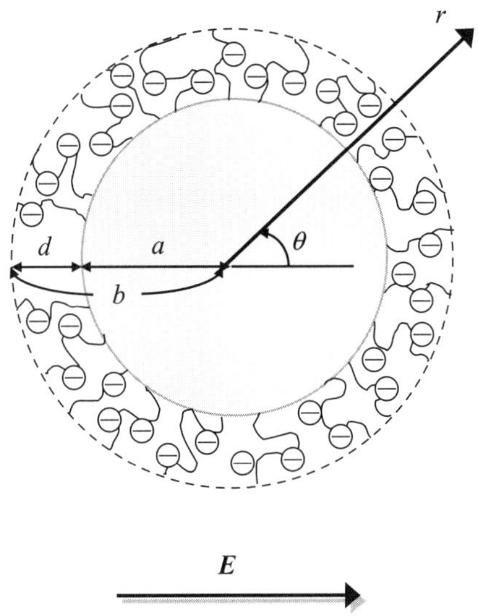

Figure 4.8. A spherical soft particle in an external applied electric field $\boldsymbol{E}$. a = radius of the particle core and d = thickness of the polyelectrolyte layer coating the particle core. $b = a + d$.

The fundamental electrokinetic equations for the liquid velocity $u(\boldsymbol{r})$ at position r are [20, 21, 28, 32]

$$\eta\nabla\times\nabla\times\boldsymbol{u}+\nabla p+\rho_{\mathrm{el}}\nabla\psi=\boldsymbol{0}, r>b \tag{4.14}$$

and

$$\eta\nabla\times\nabla\times\boldsymbol{u}+\gamma\boldsymbol{u}+\nabla p+\rho_{\mathrm{el}}\nabla\psi=\boldsymbol{0}, a<r<b. \tag{4.15}$$

Equation 4.14 is the usual Navier–Stokes equation. The term $\gamma\boldsymbol{u}$ on the left-hand side of Equation 4.15 represents the frictional forces exerted on the liquid flow by the polymer segments in the polyelectrolyte layer, and γ is the frictional coefficient. If it is assumed that each resistance center corresponds to a polymer segment, which in turn is regarded as a sphere of radius a_{p} and the polymer segments are distributed at a uniform volume density of N_{p} in the polyelectrolyte layer, then each polymer segment exerts the Stokes resistance $6\pi\eta a_{\mathrm{p}}\boldsymbol{u}$ on the liquid flow in the polyelectrolyte layer so that

$$\gamma=6\pi\eta a_{\mathrm{p}}N_{\mathrm{p}}. \tag{4.16}$$

The general mobility expression for a soft sphere is given by [32]

$$\mu = \frac{b^2}{9}\int_b^\infty \left[3\left(1-\frac{r^2}{b^2}\right) - \frac{2L_2}{L_1}\left(1-\frac{r^3}{b^3}\right)\right] G(r)dr + \frac{2L_3}{3\lambda^3 L_1}\int_a^\infty \left(1+\frac{r^3}{2b^3}\right) G(r)dr$$
$$-\frac{2}{3\lambda^2}\int_a^b \left[1-\frac{3a}{2\lambda^2 b^3 L_1}\{(L_3+L_4\lambda r)\cosh[\lambda(r-a)]\}\right. \tag{4.17}$$
$$\left. -\{-(L_4+L_3\lambda r)\sinh[\lambda(r-a)]\}\right] G(r)dr,$$

where

$$\lambda = (\gamma/\eta)^{1/2} \tag{4.18}$$

$$L_1 = \left(1+\frac{a^3}{2b^3}+\frac{3a}{2\lambda^2 b^3}-\frac{3a^2}{2\lambda^2 b^4}\right)\cosh[\lambda(b-a)]$$
$$-\left(1-\frac{3a^2}{2b^2}+\frac{a^3}{2b^3}+\frac{3a}{2\lambda^2 b^3}\right)\frac{\sinh[\lambda(b-a)]}{\lambda b} \tag{4.19}$$

$$L_2 = \left(1+\frac{a^3}{2b^3}+\frac{3a}{2\lambda^2 b^3}\right)\cosh[\lambda(b-a)] + \frac{3a^2}{2b^2}\frac{\sinh[\lambda(b-a)]}{\lambda b} - \frac{3a}{2\lambda^2 b^3} \tag{4.20}$$

$$L_3 = \cosh[\lambda(b-a)] - \frac{\sinh[\lambda(b-a)]}{\lambda b} - \frac{a}{b} \tag{4.21}$$

and

$$L_4 = \sinh[\lambda(b-a)] - \frac{\cosh[\lambda(b-a)]}{\lambda b} + \frac{\lambda a^2}{3b} + \frac{2\lambda b^2}{3a} + \frac{1}{\lambda b}. \tag{4.22}$$

If we neglect the relaxation effect, we approximate $\phi_{\pm}(r)$, where $\phi_{+}(r)$ and $\phi_{-}(r)$, respectively, are the ϕ_i function for coions and counterions in Equation 4.5 by

$$\phi_{\pm}(r) = \phi_{\pm}(b) = b + \frac{a^3}{2b^3}. \tag{4.23}$$

In order to take into account the relaxation effect outside the surface charge layer, we approximate $\phi_{\pm}(r)$ by [44]

$$\phi_{+}(r) = \phi_{+}(b) = b + \frac{a^3}{2b^2} \tag{4.24}$$

TABLE 4.3. Approximate Mobility Expression for a Soft Sphere Consisting of the Particle Core of Radius *a* Coated with a Surface Layer of Thickness of *d*

Soft particles in a *z*-*z* symmetrical electrolyte solution with $\kappa a \gg 1$ and $\kappa d \gg 1$, and moderate Donnan potential for $N \leq 0.1$ M and $n \geq 10^{-3}$ M:

$$\mu = \frac{2\varepsilon_r\varepsilon_o}{3\eta}\left(1+\frac{a^3}{2b^3}\right)\frac{\psi_o/\kappa_m + \psi_{DON}/\lambda}{1/\kappa_m + 1/\lambda} + \frac{ZeN}{\eta\lambda^2}$$

Soft particles in a *z*-*z* symmetrical electrolyte solution with $\kappa a \gg 1$ and $\kappa d \gg 1$, and arbitrary Donnan potential (for arbitrary N and n):

$$\mu = \frac{2\varepsilon_r\varepsilon_o}{3\eta}\left(1+\frac{a^3}{2b^3}\right)\frac{\psi_o/\kappa_m + \psi_{DON}/\lambda}{1/\kappa_m + 1/\lambda} + \frac{ZeN}{\eta\lambda^2}\left(1-\frac{F}{1+F}\cdot\frac{1}{1+e^{-|y_{DON}|}}\right)$$
$$-\frac{F}{1+F}\cdot\frac{2\varepsilon_r\varepsilon_o}{3\eta}\left(1+\frac{a^3}{2b^3}\right)\left(\frac{kT}{ze}\right)\left[2\ln\left(\frac{1+e^{|y_o|/2}}{2}\right)+\frac{\kappa}{\lambda}\left(e^{|y_o|/2}-1\right)-\frac{\kappa^2}{2\lambda(\lambda+\kappa_m)}\cdot\frac{e^{|y_{DON}|}-1}{1+e^{-|y_{DON}|}}\right]$$

Note: ψ_o and ψ_{DON}, respectively, are the surface potential and the Donnan potential of a soft particle, and κm is the Debye–Hückel parameter in the surface charge layer (see Table 1.7 in Chapter 1). $y_{DON} = ze\psi_{DON}/kT$, $y_o = ze\psi_o/kT$.

and

$$\phi_-(r) = \phi_-(b) = \left(b + \frac{a^3}{2b^2}\right)\cdot\frac{1}{1+F} \tag{4.25}$$

with

$$F = \frac{2}{\kappa b}(1+3m_-)(e^{|y_o|/2}-1). \tag{4.26}$$

Here F corresponds to Dukhin's number, which expresses the relaxation effect with respect to counterions and m_- is the scaled drag coefficient of counterions (anions). By using Equations 4.23 or 4.24 and 4.25 for $\phi_\pm(r)$ and equations in Table 1.7 (Chapter 1) for the potential distribution across the surface charge layer under the conditions $\kappa a \gg 1$, $\kappa d \gg 1$, $\lambda a \gg 1$, and $\lambda d \gg 1$, we finally obtain the electrophoretic mobility expressions given in Table 4.3.

REFERENCES

1 M. von Smoluchowski, in L. Greatz (ed.), *Handbuch der Elektrizität und des Magnetismus*, vol. 2. Barth, Leipzig, 1921, p. 366.

2 E. Hückel. *Phys. Z.* 25 (1924) 204.

3 D. C. Henry. *Proc. R. Soc. Lond A* 133 (1931) 106.

4 J. Th. G. Overbeek. *Kolloid-Beihefte* 54 (1943) 287.

5 F. Booth. *Proc. R. Soc. Lond A* 203 (1950) 514.

6 P. H. Wiersema, A. L. Loeb, J. Th. G. Overbeek. *J. Colloid Interface Sci.* 22 (1966) 78.

7 S. S. Dukhin, N. M. Semenikhin. *Kolloid Zh.* 32 (1970) 360.

8 S. S. Dukhin, B. V. Derjaguin, in E. Matievic (ed.), *In surface and colloid science*, vol. 7. Wiley, New York, 1974.

9 R. W. O'Brien, L. R. White. *J. Chem. Soc. Faraday Trans. 2* 74 (1978) 1607.

10 R. W. O'Brien, R. J. Hunter. *Can. J. Chem.* 59 (1981) 1878.

11 R. J. Hunter. *Zeta potential in colloid science*. Academic Press, New York, 1981.

12 H. Ohshima, T. W. Healy, L. R. White. *J. Chem. Soc. Faraday Trans. 2* 79 (1983) 1613.

13 T. G. M. van de Ven. *Colloid hydrodynamics*. Academic Press, New York, 1989.

14 R. J. Hunter. *Foundations of colloid science*, vol. 2. Clarendon Press University Press, Oxford, 1989, Chapter 13.

15 S. S. Dukhin. *Adv. Colloid Interface Sci.* 44 (1993) 1.

16 J. Lyklema. *Fundamentals of interface and colloid science, solid-liquid interfaces*, vol. 2. Academic Press, New York, 1995.

17 H. Ohshima, K. Furusawa. *Electrical phenomena at interfaces, fundamentals, measurements, and applications*, 2nd ed., revised and expanded. Dekker, New York, 1998.

18 A. V. Delgado (ed.), *Electrokinetics and electrophoresis*. Dekker, New York, 2000.

19 A. Spasic, J.-P. Hsu (eds.), *Finely dispersed particles. Micro-. nano-, atto-engineering*. CRC Press, Boca Raton, FL, 2005.

20 H. Ohshima. *Theory of colloid and interfacial electric phenomena*. Elsevier, Amsterdam, 2006.

21 H. Ohshima. *Biophysical chemistry of biointerfaces*. John Wiley & Sons, Hoboken, 2010.

22 H. Ohshima. *J. Colloid Interface Sci.* 168 (1994) 269.

23 H. Ohshima. *J. Colloid Interface Sci.* 180 (1996) 299.

24 H. Ohshima. *J. Colloid Interface Sci.* 239 (2001) 587.

25 H. Ohshima. *Colloids Surf. A Physicochem. Eng. Aspects* 267 (2005) 50.

26 W. P. J. T. van der Drift, A. de Keizer, J. Th. G. Overbeek. *J. Colloid Interface Sci.* 71 (1979) 67.

27 A. de Keizer, W. P. J. T. van der Drift, J. Th. G. Overbeek. *Biophys. Chem.* 3 (1975) 107.

28 H. Ohshima. *J. Colloid Interface Sci.* 163 (1994) 474.

29 H. Ohshima. *Colloids Surf. A Physicochem. Eng. Asp.* 103 (1995) 249.

30 H. Ohshima. *Adv. Colloid Interface Sci.* 62 (1995) 189.

31 H. Ohshima. *Electrophoresis* 16 (1995) 136.

32 H. Ohshima. *J. Colloid Interface Sci.* 228 (2000) 190.

33 J. J. López-García, C. Grosse, J. Horno. *J. Colloid Interface Sci.* 265 (2003) 327.

34 R. J. Hill, D. A. Saville. *Colloids Surf. A Physicochem. Eng. Asp.* 267 (2005) 31.

35 S. S. Dukhin, R. Zimmermann, C. Werner. *Adv. Colloid Interface Sci.* 122 (2006) 93.

36 J. F. L. Duval, H. Ohshima. *Langmuir* 22 (2006) 3533.

37 H. Ohshima. *Electrophoresis* 27 (2006) 526.
38 H. Ohshima. *Colloid Polym. Sci.* 285 (2006) 1411.
39 H. Ohshima. *Sci. Technol. Adv. Mater.* 10 (2009) 063001.
40 S. Ahualli, M. L. Jiménez, F. Carrique, A. V. Delgado. *Langmuir* 25 (2009) 1986.
41 C. Grosse, A. V. Dergado. *Curr. Opin. Colloid Interface Sci.* 15 (2010) 145.
42 J. F. L. Duval, F. Gaboriaud. *Curr. Opin. Colloid Interface Sci.* 15 (2010) 184.
43 H. Ohshima. *J. Colloid Interface Sci.* 349 (2010) 641.
44 H. Ohshima. *Colloids Surf. A Physicochem. Eng. Asp.* 376 (2011) 72.
45 P. Debye, A. Bueche. *J. Chem. Phys.* 16 (1948) 573.

5

ELECTROPHORETIC MOBILITY OF GOLD NANOPARTICLES

Kimiko Makino and Hiroyuki Ohshima

5.1 INTRODUCTION

When the electrophoretic mobility μ of a particle in an electrolyte solution is measured, the obtained electrophoretic mobility values are usually converted to the particle zeta potential ζ (which is practically equal to the particle surface potential) with the help of a proper relationship between μ and ζ, depending on the ratio of the particle size to the Debye–Hückel parameter κ and on the magnitude of the particle zeta potential [1–17]. For a particle with constant surface charge density, however, the surface charge density σ should be more of characteristic quantity than the zeta potential ζ, since for such particles, the zeta potential is not a constant quantity but depends on the electrolyte concentration. In this chapter, a systematic method is proposed to determine the surface charge density of a spherical colloidal particle on the basis of the particle electrophoretic mobility data [18]. This method is based on two analytic equations, that is, the relationship between the electrophoretic mobility and the zeta potential of the particle and the relationship between the zeta potential and the surface charge density of the particle. The measured mobility values are analyzed with these two equations. As an example, the present method is applied to electrophoretic mobility data on gold nanoparticles [19].

Electrical Phenomena at Interfaces and Biointerfaces: Fundamentals and Applications in Nano-, Bio-, and Environmental Sciences, First Edition. Edited by Hiroyuki Ohshima.

5.2 ELECTROPHORETIC MOBILITY–ZETA POTENTIAL RELATIONSHIP

Consider a charged spherical colloidal particle of radius a moving with a constant velocity in a symmetrical electrolyte solution of valence z and bulk (number) concentration n^∞ in an applied electric field. The drag coefficient of cations λ_+ and that of anions λ_- may be different. The drag coefficients $\lambda_\pm$ of cations and anions are related to their limiting conductances $\Lambda_\pm^0$ by

$$\lambda_\pm = \frac{N_A e^2 z}{\Lambda_\pm^0}, \tag{5.1}$$

where N_A is Avogadro's number and e is the elementary electric charge.

As mentioned in the "Introduction" in Section 5.1, usually, the measured value of the electrophoretic mobility μ of a particle is converted to the particle zeta potential ζ with the help of an appropriate equation relating μ to ζ. This equation can be written as $\mu = \mu\ (\zeta,\ \kappa a)$ for a spherical particle of radius a, given by

$$\kappa = \left(\frac{2n^\infty z^2 e^2}{\varepsilon_r \varepsilon_o kT} \right)^{1/2}, \tag{5.2}$$

where k is Boltzmann's constant and T is the absolute temperature. The most widely used relationship between the electrophoretic mobility μ of a spherical colloidal particle of radius a and its zeta potential ζ is Smoluchowski's mobility formula [1],

$$\mu = \frac{\varepsilon_r \varepsilon_o}{\eta} \zeta, \tag{5.3}$$

where ε_r and η, respectively, are the relative permittivity and viscosity of the electrolyte solution, and ε_o is the permittivity of a vacuum. This formula is applicable when κa is sufficiently large, that is, $\kappa a \gg 1$. For arbitrary values of κa, the mobility becomes a function of κa. For low ζ, the mobility is given by Henry's formula [3]:

$$\mu = \frac{\varepsilon_r \varepsilon_o \zeta}{\eta} \left[1 - e^{\kappa a} \{ 5E_7(\kappa a) - 2E_5(\kappa a) \} \right], \tag{5.4}$$

where $E_n(\kappa a)$ is the exponential integral of order n and is defined by

$$E_n(\kappa a) = (\kappa a)^{n-1} \int_{\kappa a}^{\infty} \frac{e^{-t}}{t^n} dt = a^{n-1} \int_a^\infty \frac{e^{-\kappa r}}{r^n} dr. \tag{5.5}$$

Ohshima [11] showed that Equation 5.4 can be approximated with negligible errors by

$$\mu = \frac{2\varepsilon_r\varepsilon_o\zeta}{3\eta}\left(1 + \frac{1}{2[1+2.5/\{\kappa a(1+2e^{-\kappa a})\}]^3}\right). \quad (5.6)$$

In the limit of large κa, Henry's formula (Eq. 5.4) tends to Smoluchowski's formula (Eq. 5.3), while in the opposite limit ($\kappa a \to 0$), it becomes Hückel's formula [2], given by

$$\mu = \frac{2\varepsilon_r\varepsilon_o\zeta}{3\eta}. \quad (5.7)$$

Equations 5.2, 5.4 (or Eq. 5.6), and 5.7 are correct to the order of ζ so that the relaxation effect is not taken into account. Overbeek [4] derived a mobility expression correct to the order of ζ^3, which takes into account the relaxation effect. Following Overbeek [4], we write a mobility expression correct to the order of ζ^3 in the following form:

$$\mu = \frac{2\varepsilon_r\varepsilon_o\zeta}{3\eta}\left[f_1(ka) - \left(\frac{ze\zeta}{kT}\right)^2\left\{f_2(ka) + \left(\frac{m_+ + m_-}{2}\right)f_3(ka)\right\}\right], \quad (5.8)$$

where $f_1(\kappa a) - f_3(\kappa a)$ are some functions of κa, and m_+ and m_- are dimensionless ionic drag coefficients, defined by

$$m_\pm = \frac{2\varepsilon_r\varepsilon_o kT}{3\eta z^2 e^2}\lambda_\pm. \quad (5.9)$$

Ohshima [12] derived the following approximate mobility expression for Equation 5.8:

$$\begin{aligned}\mu = {} & \frac{2\varepsilon_r\varepsilon_o\zeta}{3\eta}\left(1 + \frac{1}{2[1+2.5/\{\kappa a(1+2e^{-\kappa a})\}]^3}\right) \\ & - \frac{2\varepsilon_r\varepsilon_o\zeta}{3\eta}\left(\frac{ze\zeta}{kT}\right)^2\left[\frac{\kappa a\{\kappa a + 1.3\exp(-0.18\kappa a) + 2.5\}}{2\{\kappa a + 1.2\exp(-7.4\kappa a) + 4.8\}^3}\right. \\ & \left. + \left(\frac{m_+ + m_-}{2}\right)\frac{9\kappa a\{\kappa a + 5.2\exp(-3.9\kappa a) + 5.6\}}{8\{\kappa a - 1.55\exp(-0.32\kappa a) + 6.02\}^3}\right],\end{aligned} \quad (5.10)$$

which is a good approximation for $ze|\zeta|/kT| \leq 4$. An accurate expression for the electrophoretic mobility applicable for arbitrary ζ and $\kappa a \geq 10$ is derived in Reference 9. In the case of a particle in a general electrolyte solution, the readers should be referred to Reference 13.

Equations 5.3, 5.4 (or Eq. 5.6), 5.7, and 5.10 express the electrophoretic mobility μ as a function of ζ and κa, that is, $\mu = \mu\ (\zeta, \kappa a)$. Figure 5.1 shows the scaled electrophoretic mobility $E_m = 3\eta e\mu/2\varepsilon_r\varepsilon_o kT$ of a spherical particle of

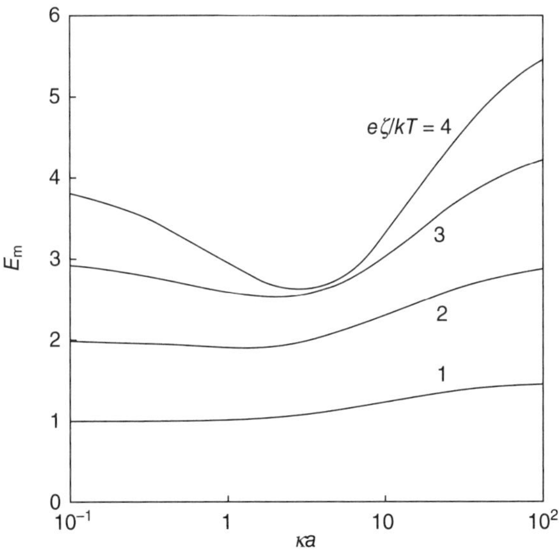

Figure 5.1. Scaled electrophoretic mobility $E_m = 3\eta e\mu/2\varepsilon_r\varepsilon_o kT$ of a spherical colloidal particle of radius a in an aqueous KCl solution at 25°C ($m_+ = 0.176$ and $m_- = 0.169$) as a function of κa for several values of the scaled particle zeta potential $e\zeta/kT$. From Reference 18.

radius a in a KCl solution at 25°C (m_+ for K^+ ions = 0.176 and m_- for Cl^- ions = 0.169) calculated via Equation 5.10 as a function of κa for four values of the scaled zeta potential $e\zeta/kT$. The κa dependence of μ represents the electrolyte concentration dependence of μ if a is regarded as constant.

5.3 ZETA POTENTIAL–SURFACE CHARGE DENSITY RELATIONSHIP

Most colloidal particles, however, carry not a constant surface potential (which depends on the electrolyte concentration) but a constant surface charge density, σ. For such a particle, the surface charge density σ should be more characteristic quantity than the zeta potential ζ, which thus becomes a function of σ and κa. In order to obtain the zeta potential ζ–surface charge density σ relationship, one must solve the spherical Poisson–Boltzmann equation for the electric potential around a spherical particle in an electrolyte solution. Numerical tables for the solution to the spherical Poisson–Boltzmann equation are given by Loeb et al. [16] An acculturate approximate expression for the ζ–σ relationship was derived by Ohshima et al. [17] with the result that

$$\sigma = \frac{2\varepsilon_r\varepsilon_o\kappa kT}{ze}\sinh\left(\frac{ze\zeta}{2kT}\right)\left[1+\frac{1}{\kappa a}\frac{2}{\cosh^2(ze\zeta/4kT)} + \frac{1}{(\kappa a)^2}\frac{8\ln[\cosh(ze\zeta/4kT)]}{\sinh^2(ze\zeta/2kT)}\right]^{1/2}. \tag{5.11}$$

The maximum relative error of Equation 5.11 is 4% at $\kappa a = 0.1$ and is less than 1% for $\kappa a \geq 1$. Figure 5.2 gives the scaled zeta potential $e\zeta/kT$ of a sphere of radius a as a function of κa, calculated from Equation 5.11 with $z = 1$. Figure 5.2 shows the significant dependence of ζ upon κa (i.e., upon electrolyte concentration n). Thus, for a particle with a constant surface charge density, σ, independent of electrolyte concentration, the particle surface potential becomes a function of the electrolyte concentration. For such a situation, in order to estimate the electrolyte concentration dependence of μ, Equations 5.2, 5.4 (or Eq. 5.6), 5.7, and 5.10 should be rewritten as functions of σ and κa, that is, $\mu = \mu$ (σ, κa), instead of as functions of ζ and κa, that is, $\mu = \mu$ (ζ, κa).

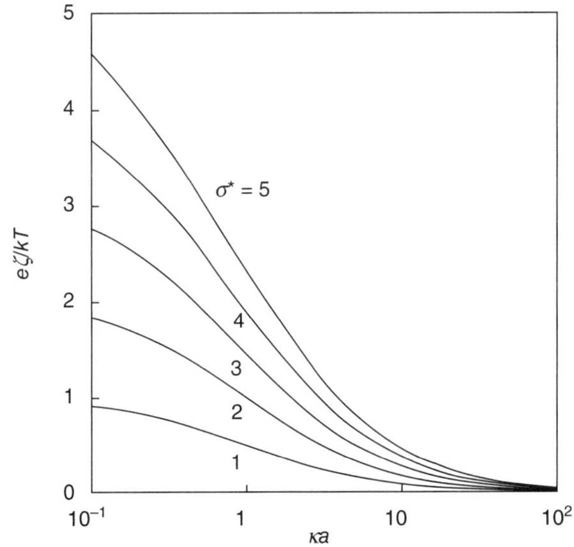

Figure 5.2. Scaled zeta potential $e\zeta/kT$ of a particle of radius a as a function of κa for several values of scaled surface charge density $\sigma^* = ea\sigma/\varepsilon_r\varepsilon_o kT$. From Reference 18.

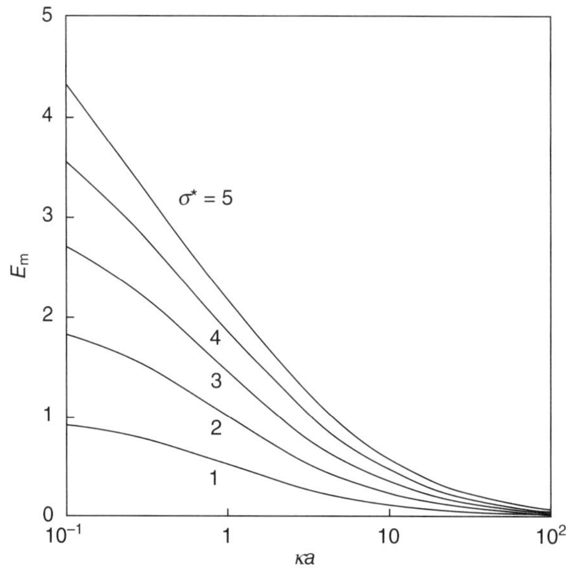

Figure 5.3. Scaled electrophoretic mobility $E_m = 3\eta e\mu/2\varepsilon_r\varepsilon_o kT$ of a spherical colloidal particle of radius a in an aqueous KCl solution at 25°C ($m_+ = 0.176$ and $m_- = 0.169$) as a function of κa at several values of the scaled surface charge density $\sigma^* = ea\sigma/\varepsilon_r\varepsilon_o kT$ for several values of the particle zeta potential $e\zeta/kT$. From Reference 18.

5.4 ELECTROPHORETIC MOBILITY–SURFACE CHARGE DENSITY RELATIONSHIP

Therefore, Figure 5.1 does not properly represent the electrolyte concentration dependence of the electrophoretic mobility μ of a spherical particle with a constant surface charge density, σ. We plot the electrophoretic mobility μ as a function of κa ($\mu = \mu$ (σ, κa)) for several values of the scaled surface charge density, $\sigma^* = ea\sigma/\varepsilon_r\varepsilon_o kT$, in Figure 5.3. In view of the fact that most colloidal particles carry a constant surface charge density rather than a constant surface potential, one should analyze the measured values of electrophoretic mobility on the basis of Equations 5.10 and 5.11, that is, $\mu = \mu(\kappa a, \sigma)$ instead of $\mu = \mu(\kappa a, \zeta)$.

5.5 ANALYSIS OF ELECTROPHORETIC MOBILITY OF GOLD NANOPARTICLES

In this section, we analyze the electrolyte concentration dependence of the electrophoretic mobility of gold nanoparticles and determine the surface charge densities of gold nanoparticles of different sizes [18]. Figures 5.4–5.6

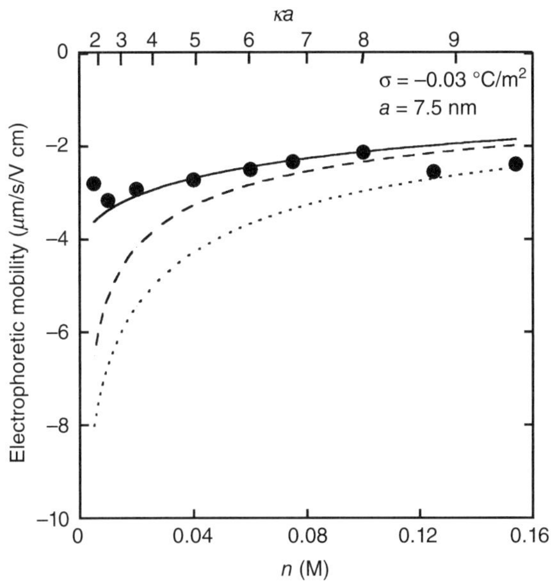

Figure 5.4. Electrophoretic mobility μ of gold nanoparticles of radius 7.5 nm in an aqueous KCl solution as a function of electrolyte concentration n (M) (or κa) in comparison with three theoretical predictions. Solid lines are results for $\sigma = -0.03/\text{m}^2$ calculated via Equation 5.9 as combined with Equation 5.10; a dotted line corresponds to Smoluchowski's formula (Eq. 5.3) and a dashed line to Henry's formula (Eq. 5.4 or 5.6). From Reference 18.

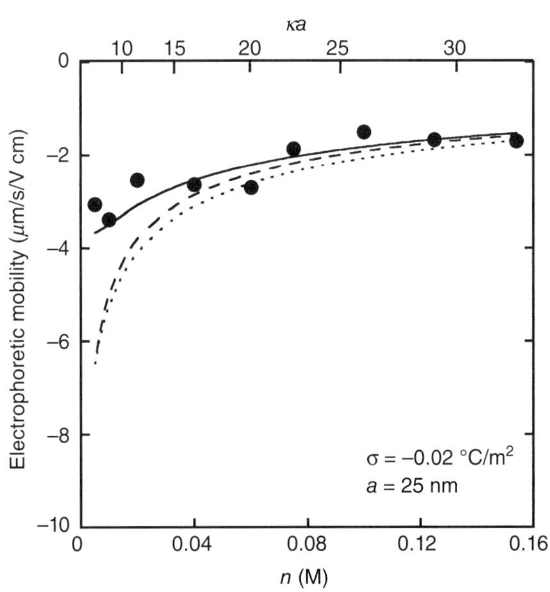

Figure 5.5. Same as Figure 5.4, but $a = 25$ nm and $\sigma = -0.02/\text{m}^2$. From Reference 18.

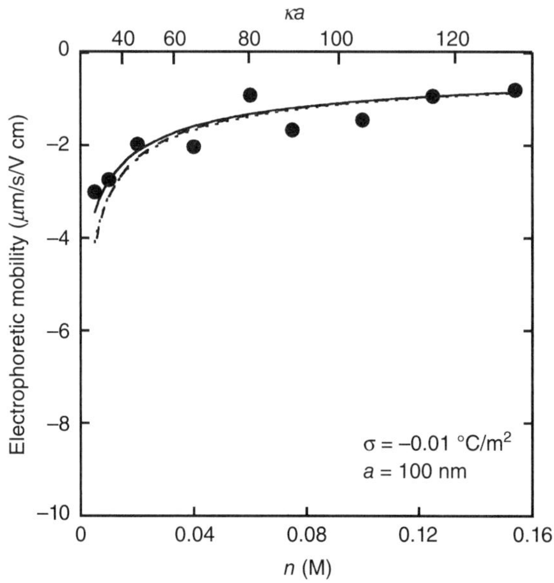

Figure 5.6. Same as Figure 5.4, but a = 100 nm and $\sigma = -0.01/\mathrm{m}^2$. From Reference 18.

exhibit the electrophoretic mobility data on gold nanoparticles of different sizes: a = 7.5 nm (Fig. 5.4), 25 nm (Fig. 5.5), and 100 nm (Fig. 5.6), as a function of electrolyte concentration n (molar, M) in an aqueous KCl solution at 25°C (experimental data are taken from Reference 19) in comparison with the results calculated with Equation 5.10 as combined with Equation 5.11 for $\sigma = -0.03$ C/m² (Fig. 5.4), $\sigma = -0.02$ C/m² (Fig. 5.5), and $\sigma = -0.01$ C/m² (Fig. 5.6) [18]. The values of κa are plotted on the upper abscissa of each of Figures 5.4–5.6 and are shown as functions of electrolyte concentration n in Figure 5.7. It is seen that Equations 5.10 and 5.11 with appropriate values of σ are in good agreement with the experimental results. In these figures, the results calculated with Smoluchowski's formula (Eq. 5.3) and Henry's formula (Eq. 5.6), both combined with Equation 5.10, are also given, showing only poor agreement with the experimental results. It is also seen that deviation among these three formulas (Eqs. 5.3, 5.6, and 5.10) becomes small for a large particle (Fig. 5.5) but is large for a small particle (Fig. 5.4). In Figure 5.8, the zeta potential ζ predicted for three different gold nanoparticles is given as a function of electrolyte concentration n. The significant dependence of ζ on the electrolyte concentration n is again seen. The observed poor agreement of Smoluchowski's formula (Eq. 5.3) and Henry's formula (Eq. 5.6) with the experimental data is due to the fact that the applicability of these formulas with respect to κa and ζ lies outside the region in which these formulas are applicable.

Thus, on the basis of the plot of the measured electrophoretic mobility values as a function of the electrolyte concentration in comparison with cal-

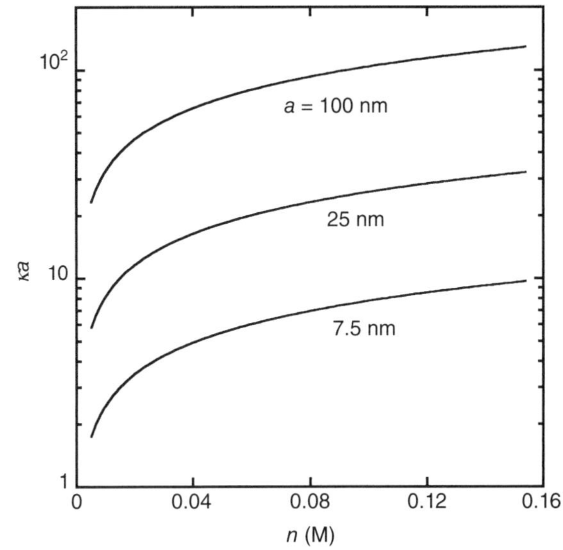

Figure 5.7. κa for a spherical particle of radius a in a monovalent electrolyte solution carrying the Debye–Hückel parameter κ as a function of electrolyte concentration n (M) for three values of a.

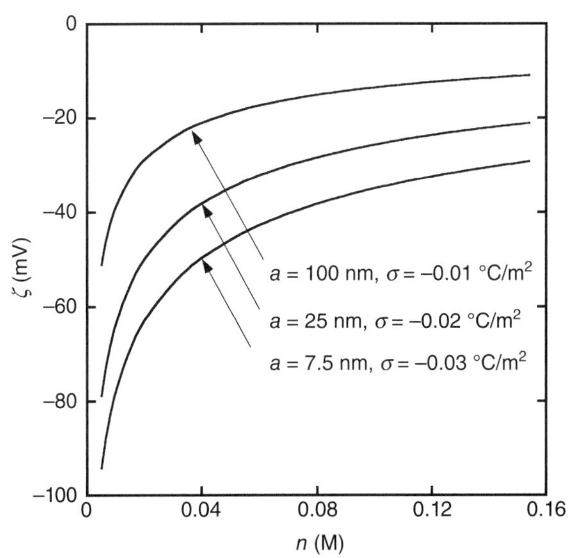

Figure 5.8. Scaled zeta potential $e\zeta/kT$ of a spherical particle of radius a in a KCl solution of concentration n as a function of n (M) for three sets of a and σ, corresponding to Figures 5.4–5.6, calculated from Equation 5.11.

culated results, by using Equations 5.10 and 5.11, one can determine the surface charge density σ of a spherical particle by a curve-fitting procedure.

REFERENCES

1 M. Von Smoluchowski, in L. Greatz (ed.). *Handbuch der Elektrizitt und des Magnetismus*, vol. 2. Barth, Leipzig, 1921, p. 366.

2 E. Hückel. *Phys. Z.* 25 (1924) 204.

3 D. C. Henry. *Proc. R. Soc. Lond A* 133 (1931) 106.

4 J. Th. G. Overbeek. *Kolloid-Beihefte* 54 (1943) 287.

5 F. Booth. *Proc. R. Soc. Lond A* 203 (1950) 514.

6 P. H. Wiersema, A. L. Loeb, J. Th. G. Overbeek. *J. Colloid Interface Sci.* 22 (1966) 78.

7 S. S. Dukhin, N. M Semenikhin. *Kolloid Zh.* 32 (1970) 360.

8 R. W. O'Brien, L. R. White. *J. Chem. Soc. Faraday Trans. 2* 74 (1978) 1607.

9 H. Ohshima, T. W. Healy, L. R. White. *J. Chem. Soc. Faraday Trans. 2* 79 (1983) 1613.

10 S. S. Dukhin. *Adv. Colloid Interface Sci.* 44 (1993) 1.

11 H. Ohshima. *J. Colloid Interface Sci.* 168 (1994) 269.

12 H. Ohshima. *J. Colloid Interface Sci.* 239 (2001) 587.

13 H. Ohshima. *Colloids Surf. A Physicochem. Eng. Aspects* 267 (2005) 50.

14 H. Ohshima. *Theory of colloid and interfacial electric phenomena*. Elsevier, Amsterdam, 2006.

15 H. Ohshima. *Biophysical chemistry of biointerfaces*. John Wiley & Sons, Hoboken, NJ, 2010.

16 A. L. Loeb, J. Th. G. Overbeek, P. H. Wiersema. *The electrical double layer around a spherical colloid particle*. MIT Press, Cambridge, MA, 1961.

17 H. Ohshima, T. W. Healy, L. R. White. *J. Colloid Interface Sci.* 90 (1982) 17.

18 K. Makino, H. Ohshima. *Langmuir* 26 (2010) 18016.

19 S. M. Agnihotri, H. Ohshima, H. Terada, K. Tomoda, K. Makino. *Langmuir* 25 (2009) 4804.

6

ELECTROPHORESIS OF SOFT PARTICLES IN A CONFINED SPACE

LI-HSIEN YEH AND JYH-PING HSU

6.1 INTRODUCTION

Electrophoresis, one of the basic electrokinetic phenomena, has been analyzed and applied to numerous fields of both fundamental and practical significance. Although its principle is straightforward, namely, the movement of a charged entity driven by an externally applied electrical field, several problems must be considered in practice. These include, for example, the presence of neighboring particles (concentration effect), the influence of the container wall (boundary effect), and the possible presence of the electroosmotic flow arising from a charged boundary. In addition to being used as an effective technique for the separation and purification of both organic and inorganic entities, electrophoresis is also widely applied as a convenient and efficient tool to characterize the surface properties of particles of colloidal size. Modern techniques such as electrophoresis display, protein separation, and DNA analysis, to name a few, also involve the application of electrophoresis operation. A systematic approach for the description of an electrophoresis problem under general conditions and for solving the corresponding governing equations is highly desirable.

Smoluchowski [1] conducted a comprehensive theoretical analysis of electrophoresis through considering the motion of an isolated, rigid, spherical, nonconducting particle with constant surface potential in an infinite electrolyte solution driven by an applied uniform electrical field. Assuming symmetric

Electrical Phenomena at Interfaces and Biointerfaces: Fundamentals and Applications in Nano-, Bio-, and Environmental Sciences, First Edition. Edited by Hiroyuki Ohshima.

electrolytes, infinitely thin double layer, low surface (zeta) potential ζ_p (i.e., $ez\zeta_p/k_BT < 1$, with e, z, k_B, and T being the elementary charge, the valence of electrolytes, the Boltzmann constant, and the absolute temperature, respectively) he was able to derive the following expression:

$$\mu = \frac{U_p}{E} = \frac{\varepsilon\zeta_p}{\eta}, \tag{6.1}$$

where U_p, μ, E, ε, and η are the electrophoretic velocity and the electrophoretic mobility of the particle, the strength of the applied electrical field, the dielectric constant, and the viscosity of the electrolyte solution, respectively. For the case where the double layer is infinitely thick, Hückel [2] showed that the mobility is two-thirds of that expressed in Equation 6.1. The problem of an arbitrarily thick double layer was considered by Henry [3] to arrive at

$$\mu = \frac{\varepsilon\zeta_p}{\eta} f(\kappa a), \tag{6.2}$$

where the dependence of the electrophoretic mobility of a particle on the thickness of the double layer is incorporated in the function $f(\kappa a)$ with κ^{-1} and a being the reciprocal Debye length and the linear size of the particle, respectively. To apply Equation 6.2, the surface potential of the particle must be low. This restriction implies that as the particle moves, its double layer is not distorted by the neighboring flow field.

Although Equation 6.2 is very concise and convenient for predicting the behavior of the electrophoretic mobility of a particle and/or interpreting the corresponding experimentally data, the limitations of rigid particles and low surface potentials are undesirable in practice. This is because while inorganic entities are often of rigid nature, it is not the case for many organic and biological entities. For example, polymer gels and polyelectrolytes [3–8], where the gel-like structure is penetrable to fluid, ions, and even small colloids, should not be considered as rigid. Soft or fuzzy particles [9–17], which consist of a rigid core and an ion-penetrable membrane or polymer layer, are another example where assuming rigid entity is inappropriate. The electrophoretic behaviors of soft particles are usually more complicated than those of the corresponding rigid particles because the presence of the membrane layer in the former affects appreciably both the neighboring electrical and hydrodynamic fields [18–22]. The assumption of low surface potential, also known as the Debye–Hückel condition, is satisfied if the surface potential is lower than ca. 25.7 mV, which can easily be violated in many cases of practical significance. In these cases, the double-layer polarization, which arises from the convective motion of the ionic species, becomes significant. The effect of double-layer polarization was first analyzed by Overbeek [23] and Booth [24] through considering the electrophoresis of a nonconducting spherical particle under the conditions of low surface potential. Taking account of the effect of

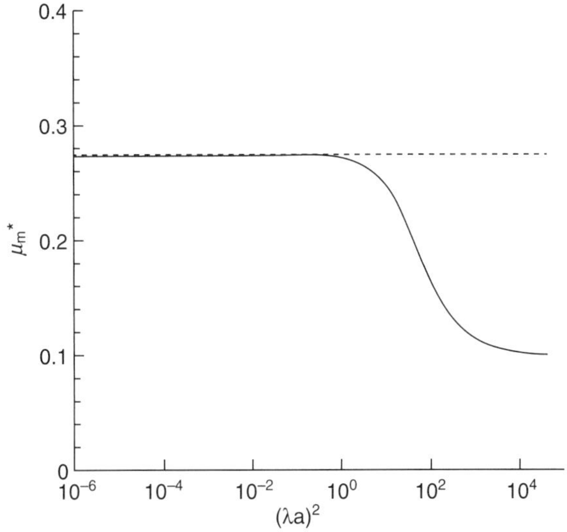

Figure 6.1. Variation of the dimensionless electrophoretic mobility μ_m^* as a function of $(\lambda a)^2$ for a dispersion of spherical charged particles covered with an uncharged ion-penetrable membrane layer [22], where the dashed line represents the scaled mobility of the corresponding rigid particles [27].

ionic convection, their analyses were extended by Wiersema et al. [25] to the case of high surface potentials through solving numerically the governing equations for the case of binary electrolytes. The effect of double-layer polarization was also modeled by O'Brien and White [26] under the conditions of arbitrary level of surface potential and thickness of the double layer. Figure 6.1 shows the typical variation of the dimensionless electrophoretic mobility μ_m^* as a function of $(\lambda a)^2$, where λa is the scaled friction coefficient of the membrane layer, for a spherical dispersion of particles covered with an uncharged ion-penetrable membrane layer in the case where the effect of double-layer polarization can be significant. This figure indicates that as $(\lambda a)^2 \to 0$, that is, the membrane layer becomes frictionless to the fluid flow, then the present case should reduce to the corresponding case of a rigid spherical dispersion [27].

One of the key effects that often need to be considered in conducting electrophoresis is the presence of a boundary. This effect is present and should not be ignored, for instance, in capillary electrophoresis and electrophoresis of particles through a porous medium, where the operation is conducted in a narrow space. In addition to making the mathematical analysis more difficult, the presence of the boundary effect may also lead to profound electrophoretic behavior where the mobility of a particle is influenced both quantitatively and qualitatively. This suggests that if the boundary effect can be significant, then care must be taken in the elaboration of the experimental data gathered. In particular, if the boundary is charged, an electroosmotic flow is present, which is also capable of affecting the electrophoretic behavior of a particle appreciably. A considerable amount of effort has been made in the last few decades to model the influence of a boundary on the electrophoretic behavior of rigid [28–57] and soft [58–61] particles through considering various types of geometry. These include, for example, a sphere normal to a plane [28–34, 58], a sphere in a spherical cavity [35–42, 59], and a sphere [28–30, 43–46, 60], a cylinder [47–53], an ellipsoid [54–56], and a toroid [57, 61] in a cylindrical pore. Note that the presence of a boundary can influence the electrical, concentration, and flow fields surrounding a particle, thereby affecting both the electrical force and the hydrodynamic forces acting on the particle.

This chapter aims to introduce a systematic mathematical approach for the evaluation of the electrophoretic mobility of a soft particle for the case where the presence of a boundary can be important through a perturbation approach [26, 62], which is sufficiently efficient and accurate in many cases. For illustration, we assume that the particle is spherical. The solution procedure presented is similar to that for rigid particles, except that the key properties of the membrane layer of a particle, such as its fixed charge density and frictional coefficient, are incorporated. Because both the electrical and flow fields near the particle are influenced, the presence of the membrane layer plays an important role, and the corresponding electrophoretic behavior is profound and interesting. This chapter is structured as follows. The general electrokinetic equations and the associated boundary conditions for the case of an isolated soft particle in an infinite electrolyte solution are introduced first. Then, the corresponding problem for the case where the boundary effect is significant is introduced, followed by an introduction of various types of boundary conditions. The special case of low surface potential is then discussed.

6.2 ELECTROPHORESIS OF A SOFT PARTICLE

Referring to Figure 6.2, let us consider first the electrophoresis of a soft spherical particle of radius c comprising a rigid core of radius a and an ion-penetrable membrane layer of thickness d, $c = a + d$, in an infinite electrolyte solution medium containing N kinds of ionic species of valence z_j and bulk concentration n_{j0}. A uniform electrical field $\mathbf{E}$ of strength E is applied in the z-direction, resulting in the electrophoretic velocity U_p in the same direction. We assume that the rigid core of the particle is nonslip. For convenience, we let the particle be fixed and the liquid far away from the particle moves with a relative velocity $\mathbf{u}_\infty = -U_p\mathbf{e}_z$ with $\mathbf{e}_z$ being the unit vector in the z-direction. Note that, if the rigid core of a soft particle shrinks to a point, it becomes a completely porous particle. On the other hand, if its ion-penetrable membrane layer vanishes, then it reduces to a rigid particle.

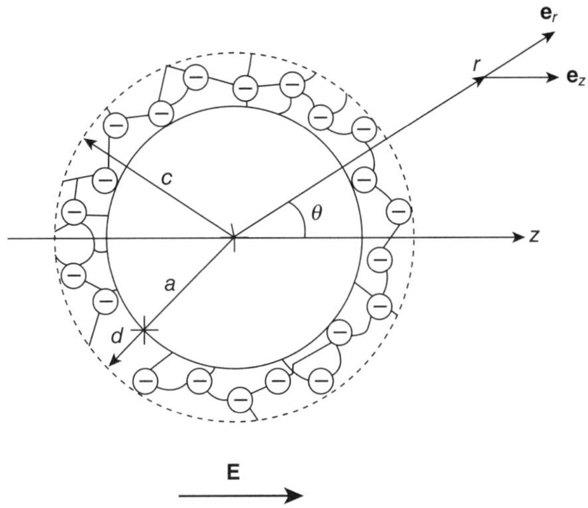

Figure 6.2. Electrophoresis of a soft spherical particle of radius c comprising a rigid core of radius a and an ion-penetrable membrane layer of thickness d, $c = a + d$, in an applied uniform electrical field $\mathbf{E}$; $\mathbf{e}_r$ and $\mathbf{e}_z$ are the unit vectors in the r- and z-directions, respectively.

6.2.1 General Electrokinetic Equations

The equations governing the problem illustrated in Figure 6.1 comprise those for the electrical field, the concentration field, and the flow field, usually called the electrokinetic equations. These equations are coupled differential equations and are highly nonlinear, implying that solving them simultaneously under general conditions is nontrivial, if not impossible. Applying Gauss's law, it can be shown that the electrical potential ϕ can be described by the Poisson equation:

$$\nabla^2\phi = -\frac{\rho_e + h(\bar{r})\rho_{fix}}{\varepsilon}, \quad h(\bar{r}) = 0, 1, \tag{6.3}$$

where ∇^2 is the Laplace operator; $h(\bar{r})$is a unit step region function ($h(\bar{r}) = 0$ for the liquid phase outside the particle, $h(\bar{r}) = 1$ inside the membrane layer); ε is the permittivity of the liquid phase; $\rho_e = \sum_{j=1}^{N} z_j e n_j$ is the space charge density of mobile ions; ρ_{fix} is the fixed charged density of the membrane layer; e is the elementary charge; n_j and z_j are the number concentration and the valence of ionic species j, respectively.

In the absence of any chemical reactions in the liquid phase, the conservation of each ionic species yields

$$\nabla \cdot \mathbf{f}_j = 0, j = 1, 2, \ldots, N, \tag{6.4}$$

where ∇ is the gradient operator and $\mathbf{f}_j$ is the flux of ionic species j, which can be described by the Nernst–Planck equation:

$$\mathbf{f}_j = n_j \mathbf{v}_j = n_j \mathbf{u} - D_j \left(\nabla n_j + \frac{z_j e n_j}{k_B T} \nabla \phi \right), \; j = 1, 2, \ldots, N. \tag{6.5}$$

Here, n_j and D_j are the number concentration and the diffusion coefficient of ionic species j, respectively; $\mathbf{u}$ and $\mathbf{v}_j$ are the flow velocity of the fluid and that of the jth mobile ionic species, respectively; k_B is the Boltzmann constant; and T is the absolute temperature. Equation 6.5 indicates that the mobile ion species are driven by the liquid flow (convective process), the concentration gradient (diffusion process), and the electrical field (electromigration process). This implies that even in the case where the friction coefficient of the membrane layer is very large ($\gamma \to \infty$) such that the liquid velocity inside vanishes ($\mathbf{u} = \mathbf{0}$), the ionic species can still be driven by the diffusion and the electromigration processes.

Since the Reynolds number in electrophoresis is on the order of 10^{-4}, typically, the fluid flow is in the creeping flow regime, and therefore, the flow field at steady state can be described by the following continuity equation and the modified Navier–Stokes equation:

$$\nabla \cdot \mathbf{u} = 0 \tag{6.6}$$

and

$$-\nabla P + \eta \nabla^2 \mathbf{u} - h(\bar{r}) \gamma \mathbf{u} - \rho_e \nabla \phi = 0, \quad h(\bar{r}) = 0, 1. \tag{6.7}$$

In these expressions, η and P are the viscosity and the pressure of the liquid phase, respectively; γ is the hydrodynamic frictional coefficient inside the membrane layer per unit volume of the fluid; $-\rho_e \nabla \phi$ is the electrical body force acting on the liquid; and $-\gamma \mathbf{u}$ represents the frictional force acting on the polymer segments in the membrane layer [62–64]. Equation 6.7 is also known as the Brinkman equation. Note that γ can be expressed as $6\pi\eta a_S N_S$ in the free-draining limit, where a_S and N_S are the Stokes radius and the number density of the hydrodynamic frictional polymer segments of the porous particle [65, 66].

6.2.2 Boundary Conditions

The boundary conditions associated with Equations 6.3–6.7 can be written on the basis of the following assumptions. If the charge density on the surface of the rigid core of the particle is constant, then

$$\mathbf{n} \cdot \nabla \phi \big|_{r=a} = -\frac{\sigma_p}{\varepsilon}, \tag{6.8}$$

where $\mathbf{n}$ is the unit outer normal vector on the surface of the rigid core and σ_p is the surface charge density. Here, we must assume that the permittivity of the rigid core is much smaller than that of the liquid phase. On the other hand, if the surface potential of the rigid core surface remains constant, then the corresponding boundary condition becomes

$$\phi\big|_{r=a} = \zeta_p. \tag{6.9}$$

Since the rigid core is impenetrable to ionic species, we must have

$$\mathbf{n}\cdot\mathbf{f}_j\big|_{r=a} = \mathbf{n}\cdot\mathbf{v}_j\big|_{r=a} = 0. \tag{6.10}$$

The nonslip nature of the surface of the rigid core yields

$$\mathbf{u}\big|_{r=a} = 0. \tag{6.11}$$

This equation implies that the slipping (or shear) plane on which $\mathbf{u} = \mathbf{0}$ locates on the surface of the rigid core.

Because the particle is in an infinite liquid medium, the other boundary is at a point far away from the particle ($r \to \infty$), where the following conditions apply:

$$\nabla\phi\big|_{r\to\infty} = -\mathbf{E}, \tag{6.12}$$

$$n_j\big|_{r\to\infty} = n_{j0}, \tag{6.13}$$

and

$$\mathbf{u}\big|_{r\to\infty} = -U_p\mathbf{e}_z. \tag{6.14}$$

These expressions state that the electrical, concentration, and flow fields at a point far away from the particle are not influenced by the particle. Note that $\sum_{j=1}^{N} z_j e n_{j0} = 0$; that is, the electroneutrality must be satisfied in the bulk phase.

For simplicity, we assume that the values of ε and η for the liquid phase inside the membrane layer are the same as those outside the particle, and ρ_{fix} is position independent. Under these conditions, the electrical potential, the electrical field, the ionic concentration, the ionic flux, and the flow field are continuous on the particle–liquid interface ($r = b$); that is,

$$\phi\big|_{r=b^-} = \phi\big|_{r=b^+}, \tag{6.15}$$

$$\mathbf{n}\cdot\nabla\phi\big|_{r=b^-} = \mathbf{n}\cdot\nabla\phi\big|_{r=b^+}, \tag{6.16}$$

$$n_j\big|_{r=b^-} = n_j\big|_{r=b^+}, \tag{6.17}$$

$$\mathbf{f}_j\big|_{r=b^-} = \mathbf{f}_j\big|_{r=b^+}, \tag{6.18}$$

$$\mathbf{u}\big|_{r=b^-} = \mathbf{u}\big|_{r=b^+}, \tag{6.19}$$

$$\mathbf{n}\cdot\mathbf{u}\big|_{r=b^-} = \mathbf{n}\cdot\mathbf{u}\big|_{r=b^+}, \tag{6.20}$$

$$\mathbf{n}\times\mathbf{u}\big|_{r=b^-} = \mathbf{n}\times\mathbf{u}\big|_{r=b^+}, \tag{6.21}$$

$$\mathbf{n}\cdot(\boldsymbol{\sigma}^{\mathbf{H}}\cdot\mathbf{n})\big|_{r=b^-} = \mathbf{n}\cdot(\boldsymbol{\sigma}^{\mathbf{H}}\cdot\mathbf{n})\big|_{r=b^+}, \tag{6.22}$$

and

$$\mathbf{n}\times(\boldsymbol{\sigma}^{\mathbf{H}}\cdot\mathbf{n})\big|_{r=b^-} = \mathbf{n}\times(\boldsymbol{\sigma}^{\mathbf{H}}\cdot\mathbf{n})\big|_{r=b^+}. \tag{6.23}$$

In theses expressions, $\boldsymbol{\sigma}^{\mathbf{H}} = -P\mathbf{I} + 2\eta\boldsymbol{\Delta}$ is the scaled hydrodynamic stress tensor with $\mathbf{I}$, $\boldsymbol{\Delta} = (\nabla\mathbf{u} + [\nabla\mathbf{u}]^{\mathrm{T}})/2$, and the superscript T being the unit tensor, the rate of deformation tensor, and matrix transpose, respectively.

To evaluate the electrophoretic velocity of the particle, Equations 6.3–6.7 need be solved simultaneously subject to the conditions expressed in Equations 6.8–6.23. As pointed out previously, solving this problem analytically under general conditions is almost impossible. In practice, this difficulty can be circumvented by resorting to the perturbation approach below [26, 62, 67].

6.2.3 Perturbation Approach

Suppose that the applied electrical field is relatively weak compared with that established by the soft particle. Typically, the surface potential of a particle is lower than ca. 250 mV and the Debye length ranges from 10 nm to 1 μm. Therefore, the strength of the electrical field established by a particle ranges from 250 to 2.5×10^4 kV/m. Because the strength of the applied electrical field is much lower than this level, the assumption of weak applied electrical field is realistic. Under this condition, all the relevant dependent variables, $\mathbf{u}$, P, ϕ, n_j, and ρ_e are first partitioned into an equilibrium component and a perturbed component, namely,

$$\mathbf{u} = \mathbf{u}^{(\mathrm{eq})} + \delta\mathbf{u}, \tag{6.24}$$

$$P = p^{(\mathrm{eq})} + \delta p, \tag{6.25}$$

$$\phi = \phi^{(\mathrm{eq})} + \delta\phi, \tag{6.26}$$

$$n_j = n_j^{(\mathrm{eq})} + \delta n_j, \tag{6.27}$$

and

$$\rho_{\mathrm{e}} = \rho_{\mathrm{e}}^{(\mathrm{eq})} + \delta\rho_{\mathrm{e}}. \tag{6.28}$$

In these expressions, a term with a superscript (eq) denotes the equilibrium property arising from the presence of the particle in the absence of the applied electrical field $\mathbf{E}$, and a term with a prefix δ represents the perturbed property coming from the application of $\mathbf{E}$. The perturbed terms are proportional to the strength of the applied electrical field. Since the particle is assumed to be stagnant when $\mathbf{E}$ is not applied, $\mathbf{u}^{(\mathrm{eq})} = \mathbf{0}$, and therefore, $\mathbf{u} = \delta\mathbf{u}$. Then, it can be shown that the governing equations, Equations 6.3–6.7, and the associated boundary conditions, Equations 6.8–6.23, can be approximated by two sets of linear expressions, one describes the equilibrium dependent variables and the other describes the perturbed dependent variables [26, 62, 67].

6.2.3.1 Equilibrium Problem. Collecting terms involving the equilibrium variables, we obtain the following set of equations for the equilibrium electrical, concentration, and flow fields:

$$\nabla^2\phi^{(\mathrm{eq})} = -\frac{\rho_{\mathrm{e}}^{(\mathrm{eq})} + h(\bar{r})\rho_{\mathrm{fix}}}{\varepsilon} = -\sum_{j=1}^{N}\frac{z_j e n_j^{(\mathrm{eq})}}{\varepsilon} - \frac{h(\bar{r})\rho_{\mathrm{fix}}}{\varepsilon}, \quad h(\bar{r}) = 0, 1; \tag{6.29}$$

$$\nabla n_j^{(\mathrm{eq})} + \frac{z_j e n_j^{(\mathrm{eq})}}{k_{\mathrm{B}}T}\nabla\phi^{(\mathrm{eq})} = 0; \tag{6.30}$$

and

$$-\nabla p^{(\mathrm{eq})} - \rho_{\mathrm{e}}^{(\mathrm{eq})}\nabla\phi^{(\mathrm{eq})} = -\nabla p^{(\mathrm{eq})} + [\varepsilon\nabla^2\phi^{(\mathrm{eq})} + h(r)\rho_{\mathrm{fix}}]\nabla\phi^{(\mathrm{eq})} = 0. \tag{6.31}$$

Integrating Equation 6.30 from a point far away from the particle, where $\phi^{(\mathrm{eq})} = 0$ and $n_j^{(\mathrm{eq})} = n_{j0}$, to a point near the particle, where $\phi^{(\mathrm{eq})} = \phi^{(\mathrm{eq})}$ and $n_j^{(\mathrm{eq})} = n_j^{(\mathrm{eq})}$, yields the Boltzmann distribution:

$$n_j^{(\mathrm{eq})} = n_{j0}\exp\left(-\frac{z_j e\phi^{(\mathrm{eq})}}{k_{\mathrm{B}}T}\right), \; j = 1, 2, \ldots, N. \tag{6.32}$$

Substituting this expression into Equation 6.29 gives

$$\nabla^2\phi^{(\mathrm{eq})} = -\sum_{j=1}^{N}\frac{z_j e n_{j0}}{\varepsilon}\exp\left(-\frac{z_j e\phi^{(\mathrm{eq})}}{k_{\mathrm{B}}T}\right) - \frac{h(\bar{r})\rho_{\mathrm{fix}}}{\varepsilon}, \quad h(\bar{r}) = 0, 1. \tag{6.33}$$

Combining Equations 6.29 and 6.31 yields

$$-\nabla p^{(\mathrm{eq})} - \sum_{j=1}^{N} z_j e n_j^{(\mathrm{eq})}\nabla\phi^{(\mathrm{eq})} = 0. \tag{6.34}$$

Integrating this expression from $p^{(\mathrm{eq})} = p_\infty$ and $n_j^{(\mathrm{eq})} = n_{j0}$ to $p^{(\mathrm{eq})} = p^{(\mathrm{eq})}$ and $n_j^{(\mathrm{eq})} = n_j^{(\mathrm{eq})}$, we obtain

$$p^{(\mathrm{eq})} - p_\infty = \sum_{j=1}^{N} k_{\mathrm{B}}T\left(n_j^{(\mathrm{eq})} - n_{j0}\right), \tag{6.35}$$

where p_∞ is the equilibrium pressure of the fluid at a point far from the particle.

6.2.3.2 Perturbed Problem. If the effect of double-layer polarization is important, then the electrical double layer surrounding the particle is no longer spherically symmetric. This effect is found to have a significance influence on the electrophoretic behavior of a particle [25, 26, 37]. Because the polarized double layer induces an internal electrical field, the direction of which is opposite to that of the applied electrical field, the electrophoretic velocity of the particle is slower than that for the case where the double layer is not polarized. A convenient way to take this effect into account is to assume the following distribution for each mobile ionic species [26, 37]:

$$n_j = n_{j0}\exp\left(-\frac{z_j e\left(\phi^{(\mathrm{eq})}+\delta\phi+g_j\right)}{k_{\mathrm{B}}T}\right),\ j=1,2,\ldots,N, \tag{6.36}$$

where g_j is a hypothetical perturbed potential function. Note that although this expression is of Boltzmann type, depending upon the value of g_j, the actual distribution of n_j can deviate appreciably from that distribution.

Applying Equations 6.33 and 6.36 and the relation $\phi = \phi^{(\mathrm{eq})} + \delta\phi$, we obtain the following governing equation for the perturbed potential $\delta\phi$:

$$\begin{aligned}\nabla^2\delta\phi &= \nabla^2\phi - \nabla^2\phi^{(\mathrm{eq})}\\ &= -\sum_{j=1}^{N}\frac{z_j e n_{j0}}{\varepsilon}\left[\exp\left(-\frac{z_j e\left(\phi^{(\mathrm{eq})}+\delta\phi+g_j\right)}{k_{\mathrm{B}}T}\right)-\exp\left(-\frac{z_j e\phi^{(\mathrm{eq})}}{k_{\mathrm{B}}T}\right)\right].\end{aligned} \tag{6.37}$$

Equations 6.27, 6.32, and 6.36 lead to

$$\delta n_j = n_{j0}\exp\left(-\frac{z_j e\left(\phi^{(\mathrm{eq})}+\delta\phi+g_j\right)}{k_{\mathrm{B}}T}\right)-n_{j0}\exp\left(-\frac{z_j e\phi^{(\mathrm{eq})}}{k_{\mathrm{B}}T}\right),\ j=1,2,\ldots,N. \tag{6.38}$$

Substituting Equations 6.24, 6.26, and 6.38 into Equations 6.4 and 6.5, we obtain

$$\begin{aligned}&\nabla^2 g_j - \frac{z_j e}{k_{\mathrm{B}}T}\nabla\phi^{(\mathrm{eq})}\cdot\nabla g_j\\ &= \frac{1}{D_j}\delta\mathbf{u}\cdot\nabla\phi + \frac{1}{D_j}\delta\mathbf{u}\cdot\nabla g_j + \frac{z_j e}{k_{\mathrm{B}}T}\nabla\delta\phi\cdot\nabla g_j + \frac{z_j e}{k_{\mathrm{B}}T}\nabla g_j\cdot\nabla g_j,\ j=1,2,\ldots,N.\end{aligned} \tag{6.39}$$

If the applied electrical field is weak relative to that established by the particle and/or the boundary, then, because $\phi^{(\mathrm{eq})} >> \delta\phi$ and $\phi^{(\mathrm{eq})} >> \delta\phi + g_j$ [26], Equation 6.36 can be approximated by

$$n_j \cong n_{j0}\left[\exp\left(-\frac{z_j e\phi^{(\mathrm{eq})}}{k_{\mathrm{B}}T}\right)\right]\left[1-\frac{z_j e}{k_{\mathrm{B}}T}(\delta\phi+g_j)\right],\ j=1,2,\ldots,N. \tag{6.40}$$

Since $\delta\mathbf{u}$, δp, $\delta\phi$, and δn_j are all on the order of E, substituting Equation 6.40 into Equations 6.37 and 6.39 and neglecting terms involving the product of two or more perturbed terms, we obtain the following approximated equations for the perturbed electrical field:

$$\nabla^2\delta\phi = -\sum_{j=1}^{N}\frac{z_j e n_{j0}}{\varepsilon}\left[-\exp\left(-\frac{z_j e\phi^{(\mathrm{eq})}}{k_\mathrm{B}T}\right)\right]\left[-\frac{z_j e}{k_\mathrm{B}T}(\delta\phi + g_j)\right] \tag{6.41}$$

$$\nabla^2 g_j - \frac{z_j e}{k_\mathrm{B}T}\nabla\phi^{(\mathrm{eq})}\cdot\nabla g_j = \frac{1}{D_j}\delta\mathbf{u}\cdot\nabla\phi^{(\mathrm{eq})},\ j = 1, 2, \ldots, N. \tag{6.42}$$

Substituting Equations 6.24–6.26 and 6.28 into Equations 6.6 and 6.7 and collecting terms of the order of E, we obtain the following perturbed expressions for the flow field:

$$\nabla\cdot\delta\mathbf{u} = 0 \tag{6.43}$$

and

$$-\nabla\delta p + \eta\nabla^2\delta\mathbf{u} - h(\bar{r})\gamma\delta\mathbf{u} + [\varepsilon\nabla^2\phi^{(\mathrm{eq})} + h(\bar{r})\rho_{\mathrm{fix}}]\nabla\delta\phi + \varepsilon\nabla^2\delta\phi\nabla\phi^{(\mathrm{eq})} = 0,\ h(\bar{r}) = 0, 1. \tag{6.44}$$

The linearized expressions, Equations 6.33 and 6.41–6.44, are applicable to arbitrary surface potential, double-layer thickness, and geometry. Note that after applying the perturbation approach, the sum of the last two terms on the left-hand side of Equation 6.44, $[\varepsilon\nabla^2\phi^{(\mathrm{eq})} + h(\bar{r})\rho_{\mathrm{fix}}]\nabla\delta\phi + \varepsilon\nabla^2\delta\phi\nabla\phi^{(\mathrm{eq})}$, which represents the electrical body force acting on the liquid, is different from that in the case of rigid particles, for example, $-\rho_\mathrm{e}\nabla\phi = \varepsilon\nabla^2\phi\nabla\phi$ [28, 35–38, 59, 67] or $(\varepsilon\nabla^2\phi^{(\mathrm{eq})}\nabla\delta\phi + \varepsilon\nabla^2\delta\phi\nabla\phi^{(\mathrm{eq})})$ [29, 30, 34, 39, 41, 45, 47]. Therefore, care must be taken in choosing an appropriate expression because it is possible that the electrophoretic behavior of the particle can be wrongly interpreted both quantitatively and qualitatively.

6.3 ELECTROPHORESIS OF A SOFT PARTICLE IN A CONFINED SPACE

Let us consider next the case where the presence of a boundary can be important. For illustration, we choose to work on the electrophoresis of a soft sphere at the center of a spherical cavity; the corresponding problem of a rigid particle was analyzed by Zydney [35]. Referring to Figure 6.3, the soft particle comprises a rigid core of radius a and an ion-penetrable membrane layer of thickness d $(c = a + d)$. The cavity is filled with an aqueous, incompressible Newtonian fluid containing z_1:z_2 electrolytes with z_1 and z_2 being the valences of cations and anions, respectively. A uniform electrical field $\mathbf{E}$ of strength E

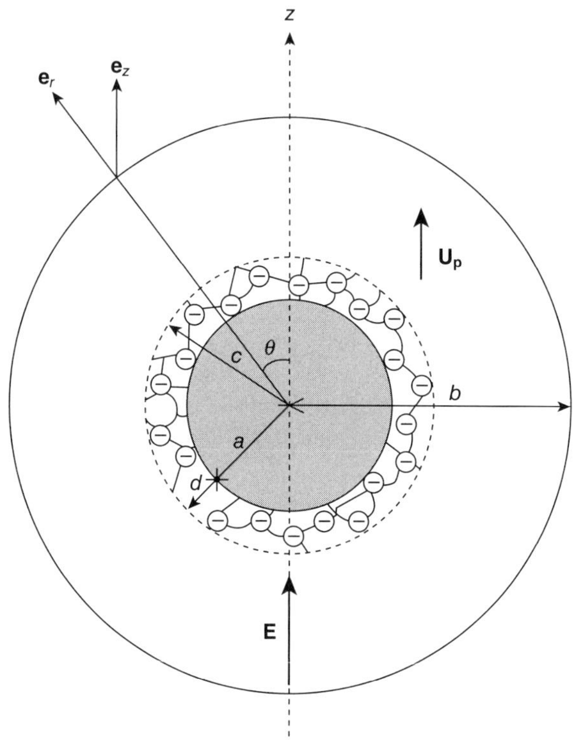

Figure 6.3. Electrophoresis of a soft spherical particle at the center of a spherical cavity, where $\mathbf{E}$ is an applied uniform electrical field parallel to the z-axis; $\mathbf{e}_r$ and $\mathbf{e}_z$ are the unit vectors in the r- and z-directions, respectively; $\mathbf{U_p}$ is the electrophoretic velocity of the particle.

is applied in the z-direction, and $\mathbf{U_p}$ is the electrophoretic velocity of the particle; $\mathbf{e}_r$ and $\mathbf{e}_z$ are the unit vectors in the r- and z-directions, respectively. For convenience, we let the slipping plane, where the liquid velocity relative to the particle vanishes, be on the surface of the rigid core of the particle; that is, the cavity moves with a relative velocity of $-U_p\mathbf{e}_z$. Suppose that the strength of the applied electrical field is relatively weak compared with that established by the particle and/or the boundary. The membrane layer is uniformly structured and carries a fixed charge, which may arise, for example, from the dissociation of the functional groups.

6.3.1 Governing Equations

For a more concise presentation, the subsequent discussions are based on scaled symbols; that is, the governing equations and the corresponding boundary conditions are written in a dimensionless form. To this end, the following characteristic quantities are chosen: the radius of the particle rigid core, a; the bulk number concentration of the ionic species j, n_{j0}; the thermal electrical potential per valence of species 1, $\phi_{ref} = k_B T/ez_1$; and the magnitude of the velocity of the particle predicted by Smoluchowski's theory when an electrical field of strength (ϕ_{ref}/a) is applied, $U_{ref} = \varepsilon(\phi_{ref})^2/\eta a$. Based on these, we define the following scaled variables: $r^* = r/a$; $z^* = z/a$; $\phi^{(eq)*} = \phi^{(eq)}/\phi_{ref}$; $\delta\phi^* = \delta\phi/\phi_{ref}$; $g_j^* = g_j/\phi_{ref}$; $E^* = E/(\phi_{ref}/a)$; $n_j^* = n_j/n_{j0}$; $Q_{fix} = \rho_{fix}a^2/\varepsilon\phi_{ref}$; $p^* = \delta p/[\varepsilon(\phi_{ref})^2/a^2]$; and $\mathbf{u}^* = \delta\mathbf{u}/U_{ref}$.

Using the scaled symbols, the electrokinetic governing equations, Equations 6.33 and 6.40–6.44, can be rewritten as

$$\nabla^{*2}\phi^{(eq)*} = -\frac{(\kappa a)^2}{1+\alpha}[\exp(-\phi^{(eq)*}) - \exp(\alpha\phi^{(eq)*})] - h(\bar{r})Q_{fix}, \quad h(\bar{r}) = 0, 1; \quad (6.45)$$

$$\nabla^{*2}\delta\phi^* = \frac{(\kappa a)^2}{1+\alpha}[(\delta\phi^* + g_1^*)\exp(-\phi^{(eq)*}) + \alpha(\delta\phi^* + g_2^*)\exp(\alpha\phi^{(eq)*})]; \quad (6.46)$$

$$\nabla^{*2}g_1^* - \nabla^*\phi^{(eq)*}\cdot\nabla^* g_1^* = Pe_1\mathbf{u}^*\cdot\nabla^*\phi^{(eq)*}; \quad (6.47)$$

$$\nabla^{*2}g_2^* + \alpha\nabla^*\phi^{(eq)*}\cdot\nabla^* g_2^* = Pe_2\mathbf{u}^*\cdot\nabla^*\phi^{(eq)*}; \quad (6.48)$$

$$\nabla^*\cdot\mathbf{u}^* = 0; \quad (6.49)$$

$$-\nabla^* p^* + \nabla^{*2}\mathbf{u}^* + [(\nabla^{*2}\phi^{(eq)*} + h(\bar{r})Q_{fix})\nabla^*\delta\phi^* + \nabla^{*2}\delta\phi^*\nabla^*\phi^{(eq)*}] - h(\bar{r})(\lambda a)^2\mathbf{u}^* = 0, \quad h(\bar{r}) = 0, 1; \quad (6.50)$$

$$n_1^* = \exp(-\phi^{(eq)*})[1 - (\delta\phi^* + g_1^*)]; \quad (6.51)$$

and

$$n_2^* = \exp(\alpha\phi^{(eq)*})[1 + \alpha(\delta\phi^* + g_2^*)]. \quad (6.52)$$

In these expressions, $\alpha = -z_2/z_1$, and $\nabla^* = a\nabla$ and $\nabla^{*2} = a^2\nabla^2$ are the scaled gradient operator and the scaled Laplace operator, respectively. $(\lambda a) = (\gamma a^2/\eta)^{1/2}$ is the scaled friction coefficient of the membrane layer. Here, λ^{-1} represents the shielding length characterizing the extent of flow penetration inside the membrane layer [8, 68–70]. $(\lambda a)^2\mathbf{u}^*$ is the scaled hydrodynamic drag acting on the membrane layer by the interstitial fluid. Note that in the present case, the slipping plane is on the surface of the rigid core of the particle, where the fluid velocity relative to the particle vanishes. $\kappa = \left[\sum_{j=1}^{2} n_{j0}(ez_j)^2/\varepsilon k_B T\right]^{1/2}$ is the reciprocal Debye length and $Pe_j^* = \varepsilon(\phi_{ref})^2/\eta D_j$ is the electric Peclet number of ionic species j. As an example, for an aqueous KCl solution of at $T = 298$ K, $D_j \cong 2 \times 10^{-9}$ m^2/s [71]; $\varepsilon = 8.854 \times 10^{-12} \times 80$ C/(Vm); $k_B T/ez_1 = 0.02568$ V; and $\eta = 0.993 \times 10^{-3}$ kg/(ms), yielding $Pe_j = 0.235$.

An important example for the charged conditions of the membrane layer of a soft particle is that it is of a charge-regulated nature. This occurs often in biocolloids such as cells and microorganisms, where the distribution of fixed charge depends largely on the degree of dissociation/association reactions of the functional groups contained in its membrane layer. For illustration, let us consider the case where the membrane layer contains both acidic functional groups AH and basic functional groups BH^+, and the dissociation/association reactions of these functional groups can be expressed by [72]

$$\mathrm{AH} \leftrightarrow \mathrm{A}^- + \mathrm{H}^+ \tag{6.53}$$

and

$$\mathrm{BH}^+ \leftrightarrow \mathrm{B} + \mathrm{H}^+. \tag{6.54}$$

Let K_A and K_B be the dissociation constants of these reactions; that is,

$$K_\mathrm{A} = \frac{[\mathrm{A}^-][\mathrm{H}^+]}{[\mathrm{AH}]} \tag{6.55}$$

and

$$K_\mathrm{B} = \frac{[\mathrm{B}][\mathrm{H}^+]}{[\mathrm{BH}^+]}. \tag{6.56}$$

Here, a symbol with square brackets represents the number concentration. If we let N_A and N_B be the total number of acidic and basic functional groups per unit volume of the membrane layer, respectively, then

$$N_\mathrm{A} = [\mathrm{A}^-] + [\mathrm{AH}] \tag{6.57}$$

and

$$N_\mathrm{B} = [\mathrm{BH}^+] + [\mathrm{B}]. \tag{6.58}$$

Suppose that the spatial concentration of H^+ follows the Boltzmann distribution, that is,

$$[\mathrm{H}^+] = [\mathrm{H}^+]_\mathrm{b} \exp\left(\frac{-ez_1\phi^{(\mathrm{eq})}}{k_\mathrm{B}T}\right), \tag{6.59}$$

where $[H^+]_b$ is the bulk concentration of H^+. It can be shown that Equations 6.55, 6.57, and 6.59 lead to

$$[\mathrm{A}^-] = \frac{N_\mathrm{A}}{1 + \left(\dfrac{[\mathrm{H}^+]_\mathrm{b}}{K_\mathrm{A}}\right)\exp\left(-\dfrac{ez_1\phi^{(\mathrm{eq})}}{k_\mathrm{B}T}\right)}. \tag{6.60}$$

Therefore, the fixed charge density in the membrane layer, $\rho_{\text{fix, A}} = -e[\text{A}^-]$, is

$$\rho_{\text{fix,A}} = -\frac{eN_{\text{A}}}{1+\left(\frac{[\text{H}^+]_{\text{b}}}{K_{\text{A}}}\right)\exp\left(-\frac{ez_1\phi^{(\text{eq})}}{k_{\text{B}}T}\right)}. \tag{6.61}$$

Similarly, combining Equations 6.56, 6.58, and 6.59 with $\rho_{\text{fix, B}} = e[\text{BH}^+]$, the charge density in the membrane layer arising from BH^+, $\rho_{\text{fix, B}}$, can be expressed as

$$\rho_{\text{fix,B}} = \frac{eN_{\text{B}}\left(\frac{[\text{H}^+]_{\text{b}}}{K_{\text{B}}}\right)\exp\left(-\frac{ez_1\phi^{(\text{eq})}}{k_{\text{B}}T}\right)}{1+\left(\frac{[\text{H}^+]_{\text{b}}}{K_{\text{B}}}\right)\exp\left(-\frac{ez_1\phi^{(\text{eq})}}{k_{\text{B}}T}\right)}. \tag{6.62}$$

Therefore, the net fixed charge density in the membrane layer is

$$\rho_{\text{fix}} = -\frac{eN_{\text{A}}}{1+\left(\frac{[H^+]_{\text{b}}}{K_{\text{A}}}\right)\exp\left(-\frac{ez_1\phi^{(\text{eq})}}{k_{\text{B}}T}\right)} + \frac{eN_{\text{B}}\left(\frac{[H^+]_{\text{b}}}{K_{\text{B}}}\right)\exp\left(-\frac{ez_1\phi^{(\text{eq})}}{k_{\text{B}}T}\right)}{1+\left(\frac{[H^+]_{\text{b}}}{K_{\text{B}}}\right)\exp\left(-\frac{ez_1\phi^{(\text{eq})}}{k_{\text{B}}T}\right)}, \tag{6.63}$$

where $\rho_{\text{fix}} = \rho_{\text{fix, A}} + \rho_{\text{fix, B}}$. In terms of scaled symbols, this expression can be written as

$$Q_{\text{fix}} = -\frac{A}{1+B\exp(-\phi^{(\text{eq})*})} + \frac{\Phi AB\exp(-\phi^{(\text{eq})*})}{\Omega[1+\Phi B\exp(-\phi^{(\text{eq})*})]}. \tag{6.64}$$

Here, $A = e^2a^2z_1N_{\text{A}}/\varepsilon k_{\text{B}}T$, $B = [\text{H}^+]_{\text{b}}/K_{\text{A}}$, $\Omega = N_{\text{A}}/N_{\text{B}}$, and $\Phi = K_{\text{A}}/K_{\text{B}}$ [73, 74]. Note that the constant fixed charge density model can be recovered from the charge-regulated model, Equation 6.64, by letting $A \to 0$ and choosing either a sufficiently small (<<0, 01) or a sufficiently large (>>100) value of B [73, 74].

6.3.2 Boundary Conditions

Let us introduce next the boundary conditions associated with the governing equations for the electrical and flow fields for a soft particle in the presence of a boundary. In practice, polymer adsorbents are often introduced into a dispersion containing rigid colloidal particles to prevent aggregation between particles, thereby increasing the stability of the dispersion. This yields a dispersion of soft particles where a particle comprises a rigid core and a porous polymer layer, which is penetrable to liquid, ionic species, and other small entities. Therefore, two types of soft particle are usually considered in the literature, namely, the particle that comprises a charged rigid core and an ion-penetrable polymer layer with either $Q_{\text{fix}} = 0$ or $Q_{\text{fix}} \neq 0$ [14–22, 58–61, 65,

75–77] and the particle that comprises an uncharged rigid core and a charged polymer layer ($Q_{\text{fix}} = 0$) [15–17, 65, 74].

6.3.2.1 Charged Colloidal Particle Covered with an Ion-Penetrable Polymer Layer. In this case, it is usually assumed that the surface of the rigid core of a soft particle is maintained either at a constant potential or at a constant charge density. We assume the following: (i) Both the surface of the rigid core of the particle and that of the cavity boundary are nonslip, nonconductive, and impermeable to ionic species; (ii) the concentration of ionic species j reaches the corresponding equilibrium value on the cavity surface; (iii) both the electrical potential and the corresponding electrical field are continuous over the membrane layer–liquid interface; (iv) both the concentration and the flux of ionic species j are continuous over the membrane layer–liquid interface; and (v) the fluid velocity and both the normal and the tangential components of the hydrodynamic stress tensor are continuous over the membrane layer–liquid interface. Under these conditions, the boundary conditions expressed in Equations 6.65–6.72 and in 6.75–6.81 on the surface of the rigid core of the soft particle, the cavity surface, and the membrane layer–liquid interface must be satisfied.

Let Ω_p and Ω_b be the surface of the rigid core of the particle and that of the cavity, respectively. Suppose that Ω_p and Ω_b are maintained at potentials ζ_p and ζ_b, respectively. Then the boundary conditions associated with the electrical field and the flow field on Ω_p and Ω_b can be expressed as

$$\phi^{(\text{eq})*} = \zeta_\text{p}/\phi_\text{ref} \text{ on } \Omega_\text{p}; \tag{6.65}$$

$$\mathbf{n}\cdot\nabla^*\delta\phi^* = 0 \text{ on } \Omega_\text{p}; \tag{6.66}$$

$$\mathbf{n}\cdot\nabla^* g_j^* = 0,\ j = 1, 2 \text{ on } \Omega_\text{p}; \tag{6.67}$$

$$\mathbf{u}^* = \mathbf{0} \text{ on } \Omega_\text{p}; \tag{6.68}$$

$$\phi^{(\text{eq})*} = \zeta_\text{b}/\phi_\text{ref} \text{ on } \Omega_\text{b}; \tag{6.69}$$

$$\mathbf{n}\cdot\nabla\delta\phi^* = -E^*\cos\theta \text{ on } \Omega_\text{b}; \tag{6.70}$$

$$g_j^* = -\delta\phi^*,\ j = 1, 2 \text{ on } \Omega_\text{b}; \tag{6.71}$$

and

$$\mathbf{u}^* = -U_p^*\mathbf{e}_z \text{ on } \Omega_\text{b}. \tag{6.72}$$

Here, $E^* = E/(\phi_\text{ref}/a)$ and $U_p^* = U_p/U_\text{ref}$, where U_p is the electrophoretic velocity of the particle in the z-direction. Equation 6.70 states that the local electrical field on Ω_b comes from the applied electrical field [35, 78]. Equation 6.71 implies that the concentration of ionic species j on Ω_b reaches the Boltzmann distribution at equilibrium state; that is,

$$n_j = n_{je} = n_{j0} \exp[\phi^{(\mathrm{eq})*}], \tag{6.73}$$

where n_{je} is the bulk number concentration of ionic species j at equilibrium.

If the surface of the rigid core of a soft particle is maintained at a constant charge density σ_p, then applying Gauss's law around Ω_p yields

$$[\mathbf{n}\cdot\nabla^*\phi^{(\mathrm{eq})*}]_{\Omega_\mathrm{p}^+} - \frac{\varepsilon_\mathrm{p}}{\varepsilon}[\mathbf{n}\cdot\nabla^*\phi^{(\mathrm{eq})*}]_{\Omega_\mathrm{p}^-} = -\sigma_\mathrm{p}^*, \tag{6.74}$$

where $\sigma_\mathrm{p}^* = \sigma_\mathrm{p}/(\varepsilon\phi_\mathrm{ref}/a)$ is the scaled σ_p, $<\varepsilon_\mathrm{p}>$ is the permittivity of the rigid core, and Ω_p^- and Ω_p^+ denote the boundary immediately inside and outside Ω_p, respectively. Because the permittivity of the rigid core is much smaller than that of the liquid phase, Equation 6.74 can be approximated by

$$\mathbf{n}\cdot\nabla^*\phi^{(\mathrm{eq})*} = -\sigma_\mathrm{p}^* \text{ on } \Omega_\mathrm{p}. \tag{6.75}$$

Similarly, if the boundary is a rigid surface, which is maintained at a constant charge density σ_b, then

$$\mathbf{n}\cdot\nabla^*\phi^{(\mathrm{eq})*} = -\sigma_\mathrm{b}^* \text{ on } \Omega_\mathrm{b}, \tag{6.76}$$

where $\sigma_\mathrm{b}^* = \sigma_\mathrm{b}/(\varepsilon\phi_\mathrm{ref}/a)$ is the scaled form of σ_b.

Let Ω_m be the membrane layer–liquid interface. For simplicity, we assume that the permittivity ε and the viscosity η of the liquid phase inside the membrane layer are identical to those outside the membrane layer [15–17]. This leads to the following continuous boundary conditions on Ω_m:

$$\phi^{(\mathrm{eq})*}\big|_{\Omega_\mathrm{m}^-} = \phi^{(\mathrm{eq})*}\big|_{\Omega_\mathrm{m}^+} \text{ and } \mathbf{n}\cdot\nabla^*\phi^{(\mathrm{eq})*}\big|_{\Omega_\mathrm{m}^-} = \mathbf{n}\cdot\nabla^*\phi^{(\mathrm{eq})*}\big|_{\Omega_\mathrm{m}^+}; \tag{6.77}$$

$$\delta\phi^*\big|_{\Omega_\mathrm{m}^-} = \delta\phi^*\big|_{\Omega_\mathrm{m}^+} \text{ and } \mathbf{n}\cdot\nabla^*\delta\phi^*\big|_{\Omega_\mathrm{m}^-} = \mathbf{n}\cdot\nabla^*\delta\phi^*\big|_{\Omega_\mathrm{m}^+}; \tag{6.78}$$

$$g_j^*\big|_{\Omega_\mathrm{m}^-} = g_j^*\big|_{\Omega_\mathrm{m}^+} \text{ and } \mathbf{n}\cdot\nabla^* g_j^*\big|_{\Omega_\mathrm{m}^-} = \mathbf{n}\cdot\nabla^* g_j^*\big|_{\Omega_\mathrm{m}^+}; \tag{6.79}$$

$$\mathbf{u}^*\big|_{\Omega_\mathrm{m}^-} = \mathbf{u}^*\big|_{\Omega_\mathrm{m}^+}, \mathbf{n}\cdot\mathbf{u}^*\big|_{\Omega_\mathrm{m}^-} = \mathbf{n}\cdot\mathbf{u}^*\big|_{\Omega_\mathrm{m}^+}, \text{ and } \mathbf{n}\times\mathbf{u}^*\big|_{\Omega_\mathrm{m}^-} = \mathbf{n}\times\mathbf{u}^*\big|_{\Omega_\mathrm{m}^+}; \tag{6.80}$$

and

$$\mathbf{n}\cdot(\boldsymbol{\sigma}^{\mathbf{H}*}\cdot\mathbf{n})\big|_{\Omega_\mathrm{m}^-} = \mathbf{n}\cdot(\boldsymbol{\sigma}^{\mathbf{H}*}\cdot\mathbf{n})\big|_{\Omega_\mathrm{m}^+} \text{ and } \mathbf{n}\times(\boldsymbol{\sigma}^{\mathbf{H}*}\cdot\mathbf{n})\big|_{\Omega_\mathrm{m}^-} = \mathbf{n}\times(\boldsymbol{\sigma}^{\mathbf{H}*}\cdot\mathbf{n})\big|_{\Omega_\mathrm{m}^+}, \tag{6.81}$$

where $\boldsymbol{\sigma}^{\mathbf{H}*} = \boldsymbol{\sigma}^{\mathbf{H}}/[\varepsilon(\phi_\mathrm{ref})^2/a^2]$ is the scaled hydrodynamic stress tensor.

6.3.2.2 Uncharged Colloidal Particle Covered with a Charged Polymer Layer. Biocolloids such as cells and microorganisms can be classified as this type of particle. Since the surface of the rigid core of the particle is free of charge, the boundary conditions associated with the equilibrium potential on Ω_p, Equations 6.65 and 6.75, should be replaced by [15–17, 65, 74]

$$\mathbf{n} \cdot \nabla^* \phi^{(\mathrm{eq})*} = 0 \text{ on } \Omega_\mathrm{p}. \tag{6.82}$$

Note that the potential on the surface of the rigid core of the particle at equilibrium, known as the Donnan potential, needs be determined [65].

6.3.3 Forces Acting on a Particle

Solving the electrokinetic governing equations expressed in Equations 6.45–6.52 subject to appropriately assumed boundary conditions yields the spatial distributions of the electrical potential, the fluid velocity, and the concentration of ionic species. These provide necessary information for the evaluation of the forces acting on the particle. Then, the electrophoretic mobility of the particle, defined as its velocity per unit strength of the applied electrical field, can be evaluated based on a force balance on the membrane layer. This, however, usually involves a tedious trial-and-error procedure [79]. The forces acting on the particle include the electrical force $\mathbf{F_E}$ and the hydrodynamic force $\mathbf{F_H}$. Suppose that only the motion of the particle in the z-direction is important. Then, only the z-components of $\mathbf{F_E}$ and $\mathbf{F_H}$, F_E and F_H, respectively, need be considered. Under the conditions that the applied electrical field is weak relative to that established by the particle and the boundary and the equilibrium electrical field has no contribution to the movement of the particle, F_E [62, 80] and F_H [62, 81] can be obtained by integrating the Maxwell stress tensor and the hydrodynamics stress tensor over the particle surface Ω_m; that is,

$$F_\mathrm{E} = \iint_{\Omega_\mathrm{m}} (\boldsymbol{\sigma}^\mathrm{E} \cdot \mathbf{n}) \cdot \mathbf{e}_z d\Omega_\mathrm{m} \tag{6.83}$$

and

$$F_\mathrm{H} = \iint_{\Omega_\mathrm{m}} (\boldsymbol{\sigma}^\mathrm{H} \cdot \mathbf{n}) \cdot \mathbf{e}_z d\Omega_\mathrm{m}. \tag{6.84}$$

Here, $\boldsymbol{\sigma}^\mathrm{E} = \varepsilon\mathbf{EE} - (1/2)\varepsilon E^2\mathbf{I}$ is the Maxwell stress tensor with $\mathbf{E} = -\nabla\phi = \mathbf{n}(\partial\phi/\partial n) + \mathbf{t}(\partial\phi/\partial t)$ and $E^2 = \mathbf{E} \cdot \mathbf{E}$; $\mathbf{t}$ is the unit tangential vector on Ω_m; n and t are the magnitude of $\mathbf{n}$ and that of $\mathbf{t}$, respectively. Substituting $\boldsymbol{\sigma}^\mathrm{E}$ into Equation 6.83, it can be shown that

$$F_\mathrm{E} = \iint_{\Omega_\mathrm{m}} \left(\varepsilon \frac{\partial\phi}{\partial n}\frac{\partial\phi}{\partial z} - \frac{1}{2}\varepsilon\left[\left(\frac{\partial\phi}{\partial n}\right)^2 + \left(\frac{\partial\phi}{\partial t}\right)^2\right] n_z \right) d\Omega_\mathrm{m}, \tag{6.85}$$

where n_z is z-component of the unit normal vector $\mathbf{n}$. Substituting $\phi = \phi^{(\mathrm{eq})} + \delta\phi$ into this expression, neglecting the terms associated with $[\partial\phi^{(\mathrm{eq})}/\partial n]^2$, $[\partial\phi^{(\mathrm{eq})}/\partial t]^2$, and $[\partial\phi^{(\mathrm{eq})}/\partial n][\partial\phi^{(\mathrm{eq})}/\partial z]$ because the particle cannot be driven by its equilibrium electrical field, and the terms involving the product of two perturbed quantities, $(\partial\delta\phi/\partial n)^2$, $(\partial\delta\phi/\partial t)^2$, and $(\partial\delta\phi/\partial n)(\partial\delta\phi/\partial z)$ because they are relatively small, and collecting the rest of the terms that are on the order of E, we obtain [62]

$$F_{\mathrm{E}} = \iint_{\Omega_{\mathrm{m}}} \left\{ \varepsilon \left[\frac{\partial \phi^{(\mathrm{eq})}}{\partial n} \frac{\partial \delta\phi}{\partial z} + \frac{\partial \delta\phi}{\partial n} \frac{\partial \phi^{(\mathrm{eq})}}{\partial z} \right] - \varepsilon \left[\frac{\partial \phi^{(\mathrm{eq})}}{\partial n} \frac{\partial \delta\phi}{\partial n} + \frac{\partial \phi^{(\mathrm{eq})}}{\partial t} \frac{\partial \delta\phi}{\partial t} \right] n_z \right\} d\Omega_{\mathrm{m}}. \tag{6.86}$$

This expression is applicable to arbitrary geometry, arbitrary level of surface potential, and all possible types of surface conditions [62, 80].

Equation 6.84 is suitable for various kinds of particle surface. If we let $\mathbf{u} = \mathbf{n}u_{\mathrm{n}} + \mathbf{t}u_{\mathrm{t}} = \mathbf{n}(\mathbf{u}\cdot\mathbf{n}) + \mathbf{t}(\mathbf{u}\cdot\mathbf{t})$, then it can be shown that it reduces to

$$F_{\mathrm{H}} = \iint_{\Omega_{\mathrm{m}}} \left[t_z \eta \left(\frac{\partial u_{\mathrm{n}}}{\partial t} + \frac{\partial u_{\mathrm{t}}}{\partial n} \right) + n_z \left(-\delta p + 2\eta \frac{\partial u_{\mathrm{n}}}{\partial n} \right) \right] d\Omega_{\mathrm{m}}, \tag{6.87}$$

where t_z is the z-component of the unit tangent vector $\mathbf{t}$. Equation 6.87 can be partitioned into a viscous component $F_{\mathrm{H(v)}}$ and a pressure component $F_{\mathrm{H(p)}}$ as

$$F_{\mathrm{H}} = F_{\mathrm{H(v)}} + F_{\mathrm{H(p)}} \tag{6.88}$$

with

$$F_{\mathrm{H(v)}} = \iint_{\Omega_{\mathrm{m}}} \left[t_z \eta \left(\frac{\partial u_{\mathrm{n}}}{\partial t} + \frac{\partial u_{\mathrm{t}}}{\partial n} \right) + n_z \left(2\eta \frac{\partial u_{\mathrm{n}}}{\partial n} \right) \right] d\Omega_{\mathrm{m}} \tag{6.89}$$

and

$$F_{\mathrm{H(p)}} = \iint_{\Omega_{\mathrm{m}}} (-\delta p n_z) d\Omega_{\mathrm{m}}. \tag{6.90}$$

In terms of the scaled symbols, $F_{\mathrm{E}}^* = F_{\mathrm{E}} / \varepsilon(\phi_{\mathrm{ref}})^2$ and $F_{\mathrm{H}}^* = F_{\mathrm{H}} / \varepsilon(\phi_{\mathrm{ref}})^2$, we have

$$\begin{aligned} F_{\mathrm{E}}^* &= \iint_{\Omega_{\mathrm{m}}^*} (\boldsymbol{\sigma}^{\mathbf{E}*} \cdot \mathbf{n}) \cdot \mathbf{e}_z d\Omega_{\mathrm{m}}^* \\ &= \iint_{\Omega_{\mathrm{m}}^*} \left\{ \left[\frac{\partial \phi^{(\mathrm{eq})*}}{\partial n} \frac{\partial \delta\phi^*}{\partial z^*} + \frac{\partial \delta\phi^*}{\partial n} \frac{\partial \phi^{(\mathrm{eq})*}}{\partial z^*} \right] - \left[\frac{\partial \phi^{(\mathrm{eq})*}}{\partial n} \frac{\partial \delta\phi^*}{\partial n} + \frac{\partial \phi^{(\mathrm{eq})*}}{\partial t} \frac{\partial \delta\phi^*}{\partial t} \right] n_z \right\} d\Omega_{\mathrm{m}}^* \end{aligned} \tag{6.91}$$

and

$$F_{\mathrm{H}}^* = \iint_{\Omega_{\mathrm{m}}^*} (\boldsymbol{\sigma}^{\mathbf{H}*} \cdot \mathbf{n}) \cdot \mathbf{e}_z d\Omega_{\mathrm{m}}^*. \tag{6.92}$$

Here, $\boldsymbol{\sigma}^{\mathbf{E}*} = \boldsymbol{\sigma}^{\mathbf{E}}/[\varepsilon(\phi_{\text{ref}})^2/a^2]$ is the scaled Maxwell stress tensor and $\Omega_{\text{m}}^* = \Omega_{\text{m}} / a^2$ is the scaled surface area of the soft particle. The electrophoretic mobility of the particle can be determined from the fact that the net force acting on it in the z-direction vanishes at steady state; that is,

$$F_{\text{E}}^* + F_{\text{H}}^* = 0. \tag{6.93}$$

6.3.4 Solution Procedure

Under general conditions, the procedure of solving Equations 6.45–6.52 subject to the prespecified boundary conditions, such as those expressed in Equations 6.65–6.72 and 6.75–6.82, involves a tedious trial-and-error numerical procedure. However, because the present problem is of a linear nature, that procedure can be avoided by adopting a supposition method [37, 40]. In this section, a systematic numerical approach for the resolution of the present problem is introduced briefly.

Based on the methodology proposed by O'Brien and White [26], the present electrophoresis problem is decomposed into two hypothesized subproblems. In the first subproblem, a particle moves with a constant velocity U_{p} in the absence of the applied electrical field $\mathbf{E}$, and in the second subproblem, $\mathbf{E}$ is applied but the particle is fixed in the space. Let $F_{\text{E},\,i}$ and $F_{\text{H},\,i}$ be the magnitudes of the z-components of $\mathbf{F_E}$ and $\mathbf{F_H}$ in subproblem i, respectively, and $F_{\text{E},i}^* = F_{\text{E},i} / \varepsilon(\phi_{\text{ref}})^2$ and $F_{\text{H},i}^* = F_{\text{H},i} / \varepsilon(\phi_{\text{ref}})^2$ are the corresponding scaled quantities. Then, the total force acting on the particle in the z-direction in subproblem i, F_i, is

$$F_i^* = F_{\text{E},i}^* + F_{\text{H},i}^*, i = 1, 2, \tag{6.94}$$

where $F_i^* = F_i / \varepsilon(\phi_{\text{ref}})^2$ is the scaled F_i. If we let $F_i^* = F_i / \varepsilon(\phi_{\text{ref}})^2$ be the scaled F_i, then $F_1^* = \chi_1 U_{\text{p}}^*$ and $F_2^* = \chi_2 E^*$, where the proportional constants χ_1 and χ_2 are independent of U_{p}^* and E^*, respectively. Because the total force acting on the particle in the z-direction must vanish at steady state, $F_1^* + F_2^* = 0$, yielding

$$\mu^* = \frac{U_{\text{p}}^*}{E^*} = -\frac{\chi_2}{\chi_1}, \tag{6.95}$$

where μ^* is the scaled electrophoretic mobility of the particle. Therefore, the solution procedure includes the following steps [37, 40]: (i) Assume an arbitrary value for U_{p}^* in Equation 6.72 and solve the governing equations for the electrical, concentration, and flow fields, Equations 6.45–6.52 numerically subject to the associated boundary conditions, Equations 6.65–6.72 and 6.75–6.82; (ii) calculate $F_{\text{E},i}^*$, $F_{\text{H},i}^*$, and F_i^*, $i = 1, 2$; (iii) apply $F_1^* = \chi_1 U_{\text{p}}^*$ and $F_2^* = \chi_2 E^*$ to evaluate χ_1 and χ_2; and (iv) use Equation 6.95 to calculate μ^*.

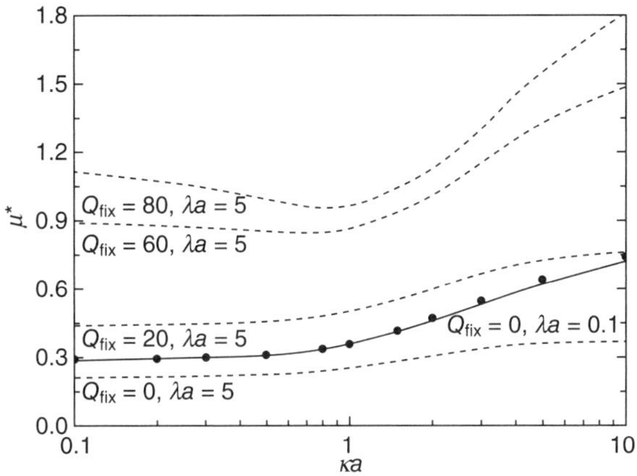

Figure 6.4. Variation of the scaled electrophoretic mobility μ^* as a function of κa at various combinations of Q_{fix} and λa for the case of a positively charged soft sphere at the center of an uncharged spherical cavity at $a/b = 0.4$ and $Pe_1 = Pe_2 = 0.235$. Solid curves are results based on the present approach; discrete symbols are the analytical result of Zydney [35].

As an example, the numerical procedure introduced earlier is applied to the case of a soft sphere at the center of an uncharged spherical cavity. Figure 6.4 shows the calculated scaled electrophoretic mobility of the particle μ^* as a function of the thickness of double-layer κa at various combinations of Q_{fix} and λa. For comparison, the corresponding analytical results of Zydney [35] for the case of a rigid sphere are also presented; these results are based on the Debye–Hückel condition (i.e., low surface potential). As seen in Figure 6.4, at $Q_{fix} = 0$ and $\lambda a = 0.1$, that is, the uncharged membrane layer is nearly frictionless to the fluid flow, the present results are consistent with the corresponding results of Zydney [35], indicating that the numerical procedure introduced in this section is satisfactory. A slight difference between the present results and those of Zydney is observed at larger values of κa. This is expected because the effect of double-layer polarization, which is neglected by Zydney, is considered in the present study [37]. Figure 6.4 also suggests that the qualitative behavior of the mobility of a soft particle at a large Q_{fix} is different from that at a small Q_{fix}. If Q_{fix} is small, μ^* increases with increasing κa. On the other hand, if Q_{fix} is large, then μ^* has a local minimum as κa varies. A similar phenomenon was also observed by O'Brien and White [26] and by Lee et al. [37] in the cases of a rigid sphere in an infinite fluid medium and a rigid sphere at the center of a spherical cavity. The difference in the electrophoretic behavior of a soft particle at a large Q_{fix} and that at a small Q_{fix} arises mainly from the effect of double-layer polarization (relaxation) [22, 26, 37]. The larger the Q_{fix}, the higher is the electrical potential and the more significant the effect of double-layer polarization, which has the effect of reducing the mobility of a particle. The effect of double-layer polarization is most important when the thickness of the double layer surrounding a particle is comparable to its linear size, that is, $\kappa a \cong 1$. Figure 6.4 also reveals that μ^* decreases with an increasing λa. This is because the larger the λa, the more compact the structure of the membrane layer, and therefore, the greater the hydrodynamic drag acting on the particle.

6.4 SPECIAL CASE OF LOW SURFACE POTENTIAL

If the surface potential is low, because the effect of double-layer polarization becomes negligible, the solution procedure described earlier can be simplified considerably. The assumption of low surface potential, that is, ζ_p is lower than ca. 25.7 mV, is often made in the literature (e.g., see References 14–21, 60, 65, and 73–77) to arrive at an approximate solution. Although this assumption can easily be violated in practice, the resulting approximate result still provides some useful information.

If the surface potential is low, it can be shown that the original electrokinetic equations, Equations 6.45–6.52, can be simplified by letting $\delta n_j = 0$ [3] as [60, 73, 74]

$$\nabla^{*2}\phi^{(eq)*} = (\kappa a)^2 \phi^{(eq)*} - h(\bar{r})Q_{fix}, \quad h(\bar{r}) = 0, 1, \tag{6.96}$$

$$\nabla^{*2}\delta\phi^* = 0, \tag{6.97}$$

$$\nabla^* \cdot \mathbf{u}^* = 0, \tag{6.98}$$

and

$$-\nabla^* p^* + \nabla^{*2}\mathbf{u}^* + [(\nabla^{*2}\phi^{(eq)*} + h(\bar{r})Q_{fix})\nabla^*\delta\phi^*] - h(\bar{r})(\lambda a)^2 \mathbf{u}^* = 0, \quad h(\bar{r}) = 0, 1. \tag{6.99}$$

For the case of a charge-regulated membrane containing both acidic function groups AH and basic functional groups BH^+, the condition of low surface potential leads to the following approximate expression for the scaled fixed charged density [33]:

$$\begin{aligned} Q_{fix} &= Q_{fix}\big|_{\phi^{(eq)*}=0} + \left.\frac{dQ_{fix}}{d\phi^{(eq)*}}\right|_{\varphi^{(eq)*}=0} [\phi^{(eq)*} - 0] \\ &= -\frac{A}{1+B} - \frac{AB}{(1+B)^2}\phi^{(eq)*} + \frac{\Phi AB}{\Omega(1+\Phi B)} - \frac{\Phi AB}{\Omega(1+\Phi B)^2}\phi^{(eq)*}. \end{aligned} \tag{6.100}$$

The procedure of evaluating the electrophoretic velocity (or mobility) of a soft particle for both the case where a rigid, charged core is covered by an ion-penetrable polymer layer and that for the case where a rigid, uncharged core is covered by a charged polymer layer can be summarized as the following. In the former, Equations 6.96–6.99 need be solved subject to the boundary conditions expressed in Equations 6.65, 6.66, 6.68–6.70, 6.72, 6.77, 6.78, 6.80, and 6.81. The governing equations and the associated boundary conditions of the latter are the same as those of the former except that Equation 6.65 needs be replaced by Equation 6.82. As in the case of high surface potentials, we focused only on the z-components of the electrical force F_{E}^{*} and the hydrodynamic force F_{H}^{*} acting on a particle. At steady state, the sum of these two forces must vanish; that is, $F_{\mathrm{E}}^{*}+F_{\mathrm{H}}^{*}=0$, suggesting that a trial-and-error procedure [38, 47] can be applied to estimate the electrophoretic velocity of the particle. Instead of using that approach, a more efficient method is available where the original problem is decomposed mathematically into two subproblems [39, 60, 74]. In the first subproblem, the particle moves with a constant velocity U_{p} in the absence of **E** and, therefore, it experiences a scaled conventional hydrodynamic force, $F_{\mathrm{H,1}}^{*}=-[U_{\mathrm{p}}/\varepsilon(\phi_{\mathrm{ref}})^{2}]D$. Note that the drag coefficient D is positive and depends only on the geometry of the system under consideration and the frictional coefficient of the membrane layer, but is independent of ka and Q_{fix}. In the second subproblem, **E** is applied, but the particle remains fixed. In this case, it experiences a scaled electrical force, F_{E}^{*}, and a scaled hydrodynamic force, $F_{\mathrm{H,2}}^{*}$; the latter comes from the motion of the mobile ions in the electrical double layer when **E** is applied. At steady state, substituting the relations $F_{\mathrm{H}}^{*}=F_{\mathrm{H,1}}^{*}+F_{\mathrm{H,2}}^{*}$ and $F_{\mathrm{H,1}}^{*}=-[U_{\mathrm{p}}/\varepsilon(\phi_{\mathrm{ref}})^{2}]D$ into $F_{\mathrm{E}}^{*}+F_{\mathrm{H}}^{*}=0$ yields

$$U_{\mathrm{p}}^{*}=\frac{F_{\mathrm{E}}^{*}+F_{\mathrm{H,2}}^{*}}{D^{*}}, \tag{6.101}$$

where $D^{*}=D/[\varepsilon(\phi_{\mathrm{ref}})^{2}/U_{\mathrm{ref}}]$ is the scaled drag coefficient, and subscripts 1 and 2 denote quantities in subproblems 1 and 2, respectively. Equation 6.101 suggests that the scaled electrophoretic mobility of a particle can be determined through the following steps. (i) Because $\phi^{(\mathrm{eq})*}=0$ and $\delta\phi^{*}=0$ in the first subproblem, the scaled electrical body force $[\nabla^{*2}\phi^{(\mathrm{eq})*}+h(\bar{r})Q_{\mathrm{fix}}]\nabla^{*}\delta\phi^{*}$ can be eliminated from Equation 6.99. In this case, assume an arbitrary value for U_{p}^{*} in Equation 6.68 and solve Equations 6.98 and 6.99 simultaneously for the flow field. (ii) Evaluate $F_{\mathrm{H,1}}^{*}$ (or $D^{*}=-F_{\mathrm{H,1}}^{*}/U_{\mathrm{p}}^{*}$) by Equation 6.92. (iii) In the second subproblem, $\phi^{(\mathrm{eq})*}$ and $\delta\phi^{*}$ are evaluated first by solving Equations 6.96 and 6.97 subject to assumed boundary conditions, for example, Equations 6.65, 6.66, 6.69, and 6.70, for a soft particle in a charged spherical cavity, followed by evaluating F_{E}^{*} through Equation 6.91. (iv) Substitute the resultant $\phi^{(\mathrm{eq})*}$ and $\delta\phi^{*}$ into Equation 6.99; replace the boundary condition expressed in Equation 6.68 by letting $U_{\mathrm{p}}^{*}=0$; and solve Equations 6.98 and 6.99 simultaneously for the flow field. (v) Calculate $F_{\mathrm{H,2}}^{*}$ by Equation 6.92. (vi) Substitute D^{*}, F_{E}^{*}, and $F_{\mathrm{H,2}}^{*}$ into Equation 6.101 to evaluate U_{p}^{*}.

Note that the solution procedure for the case of high surface potential and that of low surface potential are different. In the latter, because the effect of double-layer polarization arising from the convective motion of ionic species can be neglected, that is, $F_{\mathrm{E,1}}^{*}=0$, the governing equations for the electrical field and those used to evaluate $F^*_{H,1}$ (or D^*) are decoupled. However, this is not the case in the former. In addition, if the surface potential is low, then F^*_E, $F^*_{H,1}$ (or D^*), and $F^*_{H,2}$ can be evaluated separately, which is not the case when the surface potential is high. As an example, let us consider the electrophoresis of a membrane-coated sphere at the center of an uncharged spherical cavity. Figure 6.5 shows the variations of the scaled electrophoretic mobility U^*_p and the corresponding scaled forces, D^*, F^*_E, and $F^*_{H,2}$ as a function of the relative size of the cavity (a/b) at various values of the relative thickness of the membrane layer (c/a). Here, because a is fixed, the larger the (a/b), the more significant the boundary effect is. As expected, the more significant the presence of the boundary (cavity), the larger the D^* and, therefore, the smaller the U^*_p. As seen in Figure 6.5a, U^*_p increases with increasing (c/a). This is because although both F^*_E and D^* increase with increasing (c/a), since the thicker the member layer, the greater the total amount of fixed charge and the friction force for the fluid flow inside, the former is more significant than the latter; that is, the behavior of U^*_p as (c/a) varies is dominated by the electrical driving force. A similar result is also observed by Hsu et al. [60] in a study of the electrophoresis of a membrane-coated sphere along the axis of a cylindrical pore. Note that, whereas F^*_E is always positive, $F^*_{H,2}$, which comprises a viscous component and a pressure component, can be negative when (a/b) is sufficiently small. This is because the viscous component is negative and the pressure component is positive. If (a/b) is sufficiently small, that is, the boundary effect is unimportant, then the former dominated, leading to a negative of $F^*_{H,2}$ [60].

6.5 CONCLUSIONS

We have introduced briefly a systematic approach for the analysis of the electrophoresis of soft particles in a confined space under the conditions where the effect of double-layer polarization (relaxation) can be significant. The special case of low surface potential is also discussed. The soft particle, which comprises a rigid core and an ion-penetrable membrane layer, is capable of simulating a wide class of colloidal particle. In particular, by varying the relative magnitudes of its inner core and the outer membrane layer, the conventional rigid particles and entirely porous particles can be recovered as the limiting cases of the present soft particle model. In addition to the friction coefficient of the membrane layer, which affects the mobility of the particle hydrodynamically, its electrophoretic behavior is also influenced electrically by the fixed charge carried by the membrane layer. In general, the higher the density of the fixed charge in the membrane layer, the more significant is the effect of double-layer polarization, which reduces the mobility of the particle. The mobility is seen to have a local minimum as the thickness of the double layer varies, which is consistent with the experimental observations in the literature. The present model can be applied directly to interpret the observed

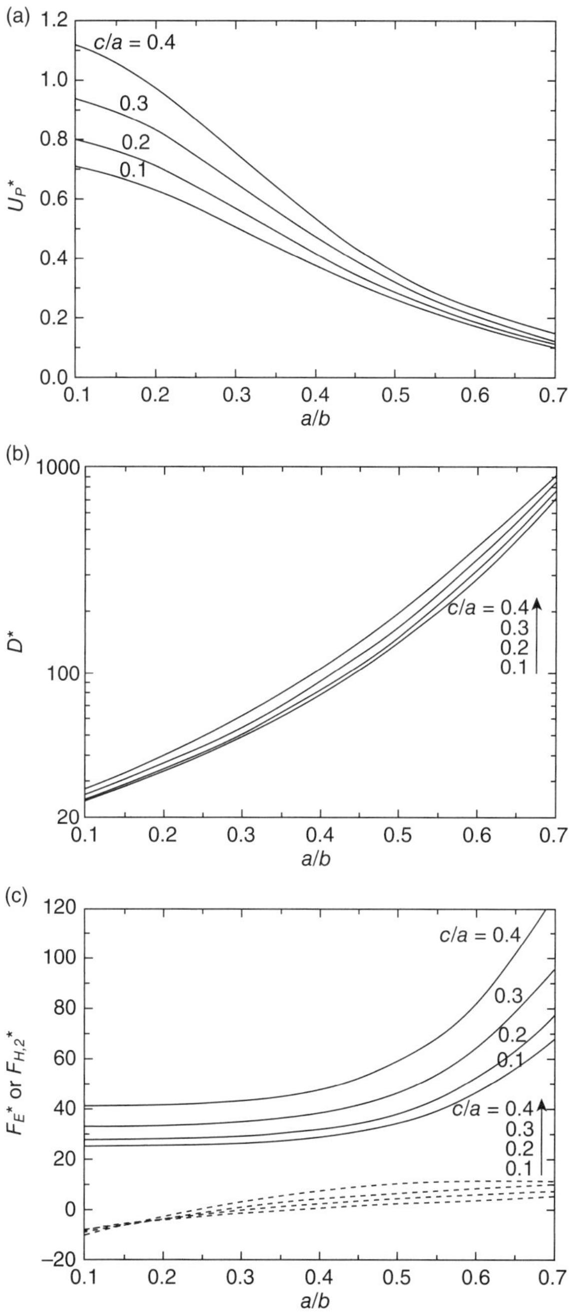

Figure 6.5. Variation of the scaled electrophoretic mobility U_{p}^* (a), scaled drag coefficient D^* (b), and scaled electrical force F_{E}^* (solid lines) and scaled hydrodynamic force $F_{\mathrm{H},2}^*$ (dashed lines) (c) as a function of (a/b) at various values of (c/a) for the case of a positively charged soft sphere at the center of an uncharged spherical cavity at $Q_{\mathrm{fix}} = 10$, $\lambda a = 3$, and $\kappa a = 1$.

experimental data, to estimate the key adjustable parameters in the model, and to design an electrophoresis apparatus.

6.6 NOMENCLATURE

a	Particle radius or the radius of the rigid core of a soft particle (m)
a_S	Stokes radius of the hydrodynamic frictional polymer segments in the membrane layer (m)
$A = e^2a^2z_1N_A/\varepsilon k_BT$ (–)	
A^-	Dissociated form of the functional group HA (–)
$[A^-]$	Number of A^- per unit volume of membrane layer ($1/m^3$)
AH	Functional group (–)
[AH]	Number of AH per unit volume of mem brane layer ($1/m^3$)
b	Radius of cavity (m)
B	functional group (–)
[B]	Number of B per unit volume of membrane layer ($1/m^3$)
BH^+	Associated form of B (–)
$[BH^+]$	Number of BH^+ per unit volume of membrane layer ($1/m^3$)
$B = [H^+]_b/K_A$ (–)	
c	Radius of a soft particle (m)
d	Thickness of membrane layer (m)
D	Drag coefficient (kg/m)
$D^* = D/[\varepsilon(\phi_{ref})^2/U_{ref}]$	Scaled drag coefficient (–)
D_j	Diffusivity of ionic species j (m^2/s)
e	Elementary charge (C)
$\mathbf{e}_z$	Unit vector in the z-direction (–)
$\mathbf{e}_r$	Unit vector in the radial direction (–)
E	Strength of the applied electrical field (V/m)
$E^* = E/(\phi_{ref}/a)$	Scaled E (–)
$\mathbf{E}$	Applied uniform electrical field (V/m)
F_E	Electrical force acting on a particle in the z-direction (N)
$\mathbf{F_E}$	electrical force acting on a particle (N)
F_E^*	Scaled form of F_E defined by Equation 6.91 (N)
$F_{E,i}$	z-Component of electrical force in subproblem i (N)

$F_{E,i}^{*}$	Scaled form of $F_{E,\ i}$ (–)
F_H	Hydrodynamic force acting on a particle in the z-direction (N)
$\mathbf{F_H}$	Hydrodynamic force acting on a particle (N)
$F_{H(v)}$	Viscous component of F_H (N)
$F_{H(p)}$	Pressure component of F_H (N)
F_H^{*}	Scaled form of F_H defined by Equation 6.92 (–)
$F_{H,\ i}$	z-Component of F_H in subproblem i (N)
$F_{H,i}^{*}$	Scaled form of $F_{H,i}$ (–)
F_i	Total force acting on a particle in the z-direction in subproblem i (N)
F_i^{*}	Scaled form of F_i defined by Equation 6.94 (–)
$\mathbf{f}_j$	Flux of ionic species j (1/s/m^2)
$f(ka)$	Henry's function (–)
g_j	Perturbed potential associated with ionic species j (V)
$g_j^{*} = g_j / \phi_{\mathrm{ref}}$	Scaled form of g_j (–)
$h(\bar{r})$	Unit step region function ($h(\bar{r}) = 0$ for the phase outside the particle, $h(\bar{r}) = 1$ inside the membrane layer) (–)
H^+	Hydrogen ion (–)
$[H^+]$	Number concentration of H^+ (mol/m^3)
$[H^+]_b$	Bulk number concentration of H^+ (mol/m^3)
$\mathbf{I}$	Unit tensor (–)
j	Index for ionic species (–)
k_B	Boltzmann constant (J/K)
K_A	Equilibrium dissociation constant defined by Equation 6.55 (mol/m^3)
K_B	Equilibrium dissociation constant defined by Equation 6.56 (mol/m^3)
n	Magnitude of unit normal vector **n** (–)
n_z	z-Component of unit normal vector **n** (–)
n_j	Number concentration of ionic species j (1/m^3)
n_{je}	Bulk number concentration of ionic species j at equilibrium (1/m^3)
n_{j0}	Bulk number concentration of ionic species j (1/m^3)
$n_j^{*} = n_j / n_{j0}$	Scaled number concentration of ionic species j (–)
$\mathbf{n}$	Unit outer normal vector on the particle surface (–)
N	Total number of ionic species j (–)

N_A	Number of acidic functional group per unit volume of membrane layer ($1/m^3$)
N_B	Number of basic functional group per unit volume of membrane layer ($1/m^3$)
N_S	Number density of the hydrodynamic frictional polymer segments in the membrane layer ($1/m^3$)
p_∞	Equilibrium pressure at a point far from the particle (Pa)
$p^* = \delta p/[\varepsilon(\phi_{ref})^2/a^2]$	Scaled perturbed hydrodynamic pressure (–)
P	Hydrodynamic pressure (Pa)
$P_{ej} = \varepsilon(\phi_{ref})^2/\eta D_j$	Electric Peclet number of ionic species j (–)
$Q_{fix} = \rho_{fix}a^2/\varepsilon\phi_{ref}$	Scaled fixed charged density (–)
r	Radial coordinate (m)
t	Magnitude of unit tangential vector **t** (–)
t_z	z-Component of unit tangential vector **t** (–)
$\mathbf{t}$	Unit tangential vector on particle surface (–)
T	Absolute temperature (K)
T	Matrix transpose (–)
u_n	Normal component of liquid velocity on the particle surface (m/s)
u_t	Tangential component of liquid velocity on the particle surface (m/s)
$\mathbf{u}$	Velocity of liquid phase (m/s)
$\mathbf{u}_\infty$	Liquid velocity at a point far from the particle (m/s)
$\mathbf{u}^* = \delta\mathbf{u}/U_{ref}$	Scaled velocity of liquid phase (–)
U_p	z-Component of particle velocity (m/s)
$U_p^* = U_p/U_{ref}$	Scaled electrophoretic mobility defined in Equation 6.101 (–)
$\mathbf{U_p}$	Particle velocity (m/s)
$U_{ref} = \varepsilon(\phi_{ref})^2/\eta a$ (–)	
$\mathbf{v}_j$	Velocity of ionic species$j$ (m/s)
z	Axial coordinate (m)
z^*	Scaled z-component of unit normal vector (–)
z_j	Valence of ionic species j (–)
z_1	Valence of cations (–)
z_2	Valence of anions (–)

Greek Letters

$\alpha = -z_2/z_1$ (–)	
χ_1, χ_2	Proportional constant (–)

$\delta\phi$	Perturbed potential (V)
$\delta\phi^* = \delta\phi/\phi_{\text{ref}}$	Scaled perturbed potential (–)
∇	Gradient operator (1/m)
$\nabla^* = a\nabla$	Scaled gradient operator (–)
∇^2	Laplace operator (1/m^2)
$\nabla^{*2} = a^2\nabla^2$	Scaled Laplace operator (–)
ε	Permittivity of liquid phase (C^2/N/m^2)
ε_p	Permittivity of the rigid core of a soft particle (C^2/N/m^2)
ϕ	Total electrical potential (V)
$\phi^{(\text{eq})}$	Equilibrium potential (V)
$\phi^{(\text{eq})*} = \phi^{(\text{eq})}/\phi_{\text{ref}}$	Scaled equilibrium potential (–)
$\phi_{\text{ref}} = k_\text{B}T/ez_1$	Reference potential (V)
$\Phi = K_\text{A}/K_\text{B}$ (–)	
$\gamma = 6\pi\eta a_\text{S}N_\text{S}$	Hydrodynamic frictional coefficient inside membrane layer per unit volume of the fluid (kg/m^3/s)
η	Viscosity of liquid phase (kg/m/s)
$\kappa = \left[\sum_{j=1}^{2} n_{j0}(ez_j)^2/\varepsilon k_\text{B}T\right]^{1/2}$	Reciprocal Debye length (m^{-1})
$\lambda a = (\gamma a^2/\eta)^{1/2}$	Scaled friction coefficient of the membrane layer (–)
μ	Electrophoretic mobility (m^2/V/s)
μ^*	Scaled electrophoretic mobility defined in Equation 6.95 (–)
μ_m^*	Dimensionless electrophoretic mobility used in Reference 22 (–)
θ	Angular coordinate (–)
ρ_e	Space charge density (C/m^3)
ρ_fix	Fixed charge density in the membrane layer of a soft particle (C/m^3)
$\rho_{\text{fix, A}}$	Fixed charge density of A^- (C/m^3)
$\rho_{\text{fix, B}}$	Fixed charge density of BH^+ (C/m^3)
σ_b	Charge density on the boundary surface (C/m)
$\sigma_\text{b}^* = \sigma_\text{b}/(\varepsilon\phi_{\text{ref}}/a)$	Scaled charge density on the boundary surface (–)
σ_p	Charge density on the surface of the rigid core of a soft particle (C/m)
$\sigma_\text{p}^* = \sigma_\text{p}/(\varepsilon\phi_{\text{ref}}/a)$	Scaled charge density on the surface of the rigid core of a soft particle (–)
$\boldsymbol{\sigma}^\mathbf{E}$	Maxwell stress tensor (N/m^2)
$\boldsymbol{\sigma}^{\mathbf{E}*} = \boldsymbol{\sigma}^\mathbf{E}/[\varepsilon(\phi_{\text{ref}})^2/a^2]$	Scaled Maxwell stress tensor (–)
$\boldsymbol{\sigma}^\mathbf{H}$	Hydrodynamic stress tensor (N/m^2)

$\boldsymbol{\sigma}^{\mathbf{H}*} = \boldsymbol{\sigma}^{\mathbf{H}}/[\varepsilon(\phi_{\text{ref}})^2/a^2]$	Scaled hydrodynamic stress tensor (–)
$\Omega = N_A/N_B$ (–)	
Ω_p	Surface (area) of the rigid core of a soft particle (m^2)
Ω_b	Surface (area) of boundary (m^2)
Ω_m	Surface (area) of a soft particle (m^2)
$\Omega_m^* = \Omega_m / a^2$	Scaled surface area of a soft particle (–)
ζ_p	Surface potential of the rigid core of a soft particle (V)
ζ_w	Surface potential of the boundary (V)

Superscripts

(eq)	Equilibrium property
*	Scaled property

Subscripts

i	Subproblem i

Prefix

δ	Perturbed property

REFERENCES

1 M. Smoluchowski. Von Versuch einer mathematischen theorie der koagulationskinetik kolloider lösungen. *Z. Phys. Chem.* 92 (1918) 129–168.

2 E. Hückel. Die kataphorese der kugel. *Physik. Z.* 25 (1924) 204–210.

3 D. C. Henry. The cataphoresis of suspended particles, Part 1. The equation of cataphoresis. *Proc. Roy. Soc. Lond A* 133 (1931) 106–129.

4 M. Ciszkowska, M. D. Guillaume. Transport of ions and molecules in biopolymeric gels: electroaalytical Studies. *J. Phys. Chem. A* 103 (1999) 607–613.

5 A. Y. Grosberg, T. T. Nguyen, B. I. Shklovskii. Colloquium: the physics of charge inversion in chemical and biological systems. *Rev. Mod. Phys.* 74 (2002) 329–345.

6 M. Zembala. Electrokinetics of heterogeneous interfaces. *Adv. Colloid Interface Sci.* 112 (2004) 59.

7 J. F. L. Duval, K. J. Wilkinson, H. P. van Leeuwen, J. Buffle. Humic substances are soft and permeable: evidence from their electrophoretic mobilities. *Environ. Sci. Technol.* 39 (2005) 6435.

8 J. J. Lietor-Santos, A. Fernandez-Nieves. Motion of microgels in electric fields. *Adv. Colloid Interface Sci.* 147–148 (2009) 178–185.

9 E. Donath, V. Pastuschenko. Electrophoretical study of cell-surface properties—Influence of the surface-coat on the electric-potential distribution and on general

electrokinetic properties of animal cells. *Bioelectrochem. Bioenerg.* 6 (1979) 543–554.

10 I. S. Jones. Theory of electrophoresis of layer colloid particles with adsorbed polyelectrolyte. *J. Colloid Interface Sci.* 68 (1979) 451–461.

11 R. W. Wunderlich. The effects of surface-structure on the electrophoretic mobilities of large particles. *J. Colloid Interface Sci.* 88 (1982) 385–397.

12 S. Levine, M. Levine, K. A. Sharp, D. E. Brooks. Theory of the electrokinetic behavior of human erythrocytes. *Biophys. J.* 42 (1983) 127–135.

13 K. A. Sharp, D. E. Brooks. Calculation of the electrophoretic mobility of a particle bearing bound poly-electrolyte using the nonlinear Poisson–Boltzmann equation. *Biophys. J.* 47 (1985) 563–566.

14 H. Ohshima, T. Kondo. On the electrophoretic mobility of biological cells. *Biophys. Chem.* 39 (1991) 191–198.

15 H. Ohshima. Electrophoretic mobility of soft particles. *J. Colloid Interface Sci.* 163 (1994) 474–483.

16 H. Ohshima. On the general expression for the electrophoretic mobility of a soft particle. *J. Colloid Interface Sci.* 228 (2000) 190–193.

17 H. Ohshima. Electrophoresis of soft particles: analytic approximations. *Electrophoresis* 27 (2006) 526–533.

18 D. A. Saville. Electrokinetic properties of fuzzy colloidal particles. *J. Colloid Interface Sci.* 222 (2000) 137–145.

19 R. J. Hill, D. A. Saville, W. B. Russel. Electrophoresis of spherical polymer-coated colloidal particles. *J. Colloid Interface Sci.* 258 (2003) 56–74.

20 J. J. López-García, C. Grosse, J. Horno. Numerical study of colloidal suspensions of soft spherical particles using the network method—1. DC electrophoretic mobility. *J. Colloid Interface Sci.* 265 (2003) 327–340.

21 J. J. López-García, C. Grosse, J. Horno. Numerical study of colloidal suspensions of soft spherical particles using the network method—2. AC electrokinetic and dielectric properties. *J. Colloid Interface Sci.* 265 (2003) 341–350.

22 E. Lee, K. T. Chou, J. P. Hsu. Electrophoresis of a concentrated dispersion of spherical particles covered by an ion-penetrable membrane layer. *J. Colloid Interface Sci* 280 (2004) 518–526.

23 J. Th. G. Overbeek. Theory of the relaxation effect in electrophoresis. *Kolloide Beihefte* 54 (1943) 287–364.

24 F. Booth. The cataphoresis of spherical, solid non-conducting particles in a symtetrical electrolyte. *Proc. Roy. Soc. London Ser. A* 203 (1950) 514–533.

25 P. H. Wiersema, A. L. Loeb, J. Th. G. Overbeek. Calculation of the electrophoretic mobility of a spherical colloid particle. *J. Colloid Interface Sci.* 22 (1966) 78–99.

26 R. W. O'Brien, L. R. White. Electrophoretic mobility of a spherical colloidal particle. *J. Chem. Soc. Faraday Trans. 2* 74 (1978) 1607–1626.

27 E. Lee, J. W. Chu, J. P. Hsu. Electrophoretic mobility of a concentrated suspension of spherical particles. *J. Colloid Interface Sci.* 209 (1999) 240–246.

28 H. J. Keh, J. L. Anderson. Boundary effects on electrophoretic motion of colloidal spheres. *J. Fluid Mech.* 153 (1985) 417–439.

29 J. Ennis, J. L. Anderson. Boundary effects on electrophoretic motion of spherical particles for thick double layers and low zeta potential. *J. Colloid Interface Sci.* 185 (1997) 497–514.
30 A. A. Shugai, S. L. Carnie. Electrophoretic motion of a spherical particle with a thick double layer in bounded flows. *J. Colloid Interface Sci.* 213 (1999) 298–315.
31 Y. P. Tang, M. H. Chih, E. Lee, J. P. Hsu. Electrophoretic motion of a charge-regulated sphere normal to a plane. *J. Colloid Interface Sci.* 242 (2001) 121–126.
32 M. H. Chin, E. Lee, J. P. Hsu. Electrophoresis of a sphere normal to a plane at arbitrary electrical potential and double layer thickness. *J. Colloid Interface Sci.* 248 (2002) 383–388.
33 J. P. Hsu, M. H. Ku, C. C. Kuo. Electrophoresis of a charge-regulated sphere normal to a large disk. *Langmuir* 21 (2005) 7588–27597.
34 J. P. Hsu, L. H. Yeh, S. J. Yeh. Electrophoresis of a rigid sphere in a Carreau fluid normal to a large charged disk. *J. Phys. Chem. B* 111 (2007) 12351–12361.
35 A. L. Zydney. Boundary effects on the electrophoretic motion of a charged particle in a spherical cavity. *J. Colloid Interface Sci.* 169 (1995) 476–485.
36 E. Lee, J. W. Chu, J. P. Hsu. Electrophoretic mobility of a spherical particle in a spherical cavity. *J. Colloid Interface Sci.* 196 (1997) 316–320.
37 E. Lee, J. W. Chu, J. P. Hsu. Electrophoretic mobility of a sphere in a spherical cavity. *J. Colloid Interface Sci.* 205 (1998) 65–76.
38 J. P. Hsu, S. H. Hung, C. Y. Kao. Electrophoresis of a sphere at an arbitrary position in a spherical cavity. *Langmuir* 18 (2002) 8897–8901.
39 J. P. Hsu, L. H. Yeh, Z. S. Chen. Effect of a charged boundary on electrophoresis: a sphere at arbitrary position in a spherical cavity. *J. Colloid Interface Sci.* 310 (2007) 281–291.
40 J. P. Hsu, Z. S. Chen, M. H. Ku, L. H. Yeh. Effect of a charged boundary on electrophoresis: sphere in spherical cavity at arbitrary potential and double-layer thickness. *J. Colloid Interface Sci.* 314 (2007) 256–263.
41 J. P. Hsu, L. H. Yeh. Electrophoresis of a charged boundary on electrophoresis in a Carreau fluid: a sphere at an arbitrary position in a spherical cavity. *Langmuir* 23 (2007) 8637–8646.
42 J. P. Hsu, C. Y. Chen, D. J. Lee, S. Tseng, A. Su. Electrophoresis of a charge-regulated sphere at an arbitrary position in a charged spherical cavity. *J. Colloid Interface Sci.* 325 (2008) 516–525.
43 J. P. Hsu, M. H. Ku, C. Y. Kao. Electrophoresis of a spherical particle along the axis of a cylindrical pore: effect of osmotic flow. *J. Colloid Interface Sci.* 276 (2004) 248–254.
44 C. Ye, X. Xuan, D. Li. Eccentric electrophoretic motion of a sphere in circular cylindrical microchannels. *Microfluid Nanofluid* 1 (2005) 234–241.
45 J. P. Hsu, L. H. Yeh, M. H. Ku. Electrophoresis of a spherical particle along the axis of a cylindrical pore filled with a Carreau fluid. *Colloid Polym. Sci.* 284 (2006) 886–892.
46 J. P. Hsu, Z. S. Chen. Electrophoresis of a sphere along the axis of a cylindrical pore: effect of double-layer polarization and electroosmotic flow. *Langmuir* 23 (2007) 6198–6204.
47 J. P. Hsu, C. Y. Kao. Electrophoresis of a finite cylinder along the axis of a cylindrical pore. *J. Phys. Chem. B* 106 (2002) 10605–10609.
48 H. Liu, H. Bau, H. Hu. Electrophoresis of concentrically and eccentrically positioned cylindrical particles in a long tube. *Langmuir* 20 (2004) 2628–2639.

49 J. P. Hsu, M. H. Ku. Boundary effect on electrophoresis: finite cylinder in a cylindrical pore. *J. Colloid Interface Sci.* 283 (2005) 592–600.

50 J. P. Hsu, C. C. Kuo. Electrophoresis of a finite cylinder positioned eccentrically along the axis of a cylindrical pore. *J. Phys. Chem. B* 110 (2006) 17607–17615.

51 S. M. Davison, K. V. Sharp. Boundary effects on the electrophoretic motion of cylindrical particles: concentrically-positioned particles in a capillary. *J. Colloid Interface Sci.* 303 (2006) 288–297.

52 J. P. Hsu, Z. S. Chen, D. J. Lee, S. Tseng, A. Su. Effect of double-layer polarization and electroosmotic flow on the electrophoresis of finite cylinder along the axis of a cylindrical pore. *Chem. Eng. Sci.* 63 (2008) 4561–4569.

53 L. H. Yeh, J. P. Hsu. Electrophoresis of a finite rod along the axis of a long cylindrical microchannel filled with Carreau fluids. *Microfluid Nanofluid* 7 (2009) 383–392.

54 J. P. Hsu, S. H. Hung. Electrophoresis of a charge-regulated spheroid along the axis of a cylindrical pore. *J. Colloid Interface Sci.* 264 (2003) 121–127.

55 J. P. Hsu, S. H. Hung, C. Y. Kao, S. Tseng. Electrophoresis of a spheroid along the axis of a cylindrical pore. *Chem. Eng. Sci.* 58 (2003) 5339–5347.

56 S. Tseng, C. H. Cho, Z. S. Chen, J. P. Hsu. Electrophoresis of an ellipsoid along the axis of a cylindrical pore: effect of a charged boundary. *Langmuir* 24 (2008) 2929–2937.

57 J. P. Hsu, C. C. Kuo, M. H. Ku. Electrophoresis of a toroid along the axis of a cylindrical pore. *Electrophoresis* 27 (2006) 3155–3165.

58 W. L. Cheng, Y. Y. He, E. Lee. Electrophoresis of a soft particle normal to a plane. *J. Colloid Interface Sci.* 335 (2009) 130–139.

59 E. Lee, Y. P. Tang, J. P. Hsu. Electrophoresis of a membrane-coated sphere in a spherical cavity. *Langmuir* 20 (2004) 9415–9421.

60 J. P. Hsu, Z. S. Chen, S. Tseng. Effect of electroosmotic flow on the electrophoresis of a membrane-coated sphere along the axis of a cylindrical pore. *J. Phys. Chem. B* 113 (2009) 7701–7708.

61 J. P. Hsu, C. C. Kuo. Electrophoresis of a soft toroid coaxially along the axis of a cylindrical pore. *Chem. Eng. Sci.* 64 (2009) 5247–5254.

62 J. P. Hsu, L. H. Yeh, M. H. Ku. Evaluation of the electric force in electrophoresis. *J. Colloid Interface Sci.* 305 (2007) 324–329.

63 H. C. Brinkman. A calculation of the viscous force exerted by a flowing fluid on a dense swarm of particles. *Appl. Sci. Res. A* 1 (1947) 27–34.

64 P. Debye, M. Bueche. Intrinsic viscosity, diffusion, and sedimentation rate of polymers in solution. *J. Chem. Phys.* 16 (1948) 573–579.

65 H. Ohshima. Electrophoresis of soft particles. *Adv. Colloid Interface Sci.* 62 (1995) 189–325.

66 Y. K. Wei, H. J. Keh. Diffusiophoretic mobility of charged porous spheres in electrolyte gradients. *J. Colloid Interface Sci.* 269 (2004) 240–250.

67 J. H. Masliyah, S. Bhattacharjee. *Electrokinetic transport phenomena*, 4th ed. Wiley, New York, 2006.

68 S. Kawahata, H. Ohshima, N. Muramatsu, T. Kondo. Charge-distribution in the surface region of human erythrocytes as estimated from electrophoretic mobility data. *J. Colloid Interface Sci.* 138 (1990) 182–186.

69 K. Morita, N. Muramatsu, H. Ohshima, T. Kondo. Electrophoretic behavior of rat lymphocyte subpopulations. *J. Colloid Interface Sci.* 147 (1991) 457–461.

70 O. Aoyanagi, N. Muramatsu, H. Ohshima, T. Kondo. Electrophoretic behavior of polyA-graft-polyB-type microcapsules. *J. Colloid Interface Sci.* 162 (1994) 222–226.

71 B. Chun, A. J. C. Ladd. The electroviscous force between charged particles: beyond the thin-double-layer approximation. *J. Colloid Interface Sci.* 274 (2004) 687–694.

72 B. W. Ninham, V. A. Parsegian. Electrostatic potential between surfaces bearing ionizable groups in ionic equilibrium with physiologic saline solution. *J. Theor. Biol.* 31 (1971) 405–428.

73 J. P. Hsu, C. C. Kuo, M. H. Ku. Electrophoresis of a charge-regulated toroid normal to a large disk. *Electrophoresis* 29 (2008) 348–357.

74 X. Zhang, J. P. Hsu, Z. S. Chen, L. H. Yeh, M. H. Ku. Electrophoresis of a charge-regulated soft sphere in a charged cylindrical pore. *J. Phys. Chem. B* 114 (2010) 1621–1631.

75 H. Ohshima. Electrophoretic mobility of a charged spherical colloidal particle covered with an uncharged polymer layer. *Electrophoresis* 23 (2002) 1995–2000.

76 H. Ohshima. Modified Henry function for the electrophoretic mobility of a charged spherical colloidal particle covered with an ion-penetrable uncharged polymer layer. *J. Colloid Interface Sci.* 252 (2002) 119–125.

77 H. Ohshima. Electrophoretic mobility of a charged cylindrical colloidal particle covered with an ion-penetrable uncharged polymer layer. *J. Colloid Interface Sci.* 258 (2003) 252–258.

78 S. Levine, G. H. Neale. The prediction of electrokinetic phoenmena with multiparticle systems. I. Electrophoresis and electroosmosis. *J. Colloid Interface Sci.* 47 (1974) 520–529.

79 R. J. Hunter. *Foundations of colloid science*, 2nd ed. Oxford University Press, Oxford, 2001.

80 J. P. Hsu, L. H. Yeh. Comparison of three methods for the evaluation of the electric force on a particle in electrophoresis. *J. Chin. Inst. Chem. Eng.* 37 (2006) 601–607.

81 J. Happel, H. Brenner. *Low Reynolds number hydrodynamics*. Academic Press, New York, 1983.

7

SURFACE CONDUCTIVITY

STANISLAV S. DUKHIN, RALF ZIMMERMANN, AND CARSTEN WERNER

7.1 INTRODUCTION

The concept of specific surface conductivity was developed by considering charge transfer along almost flat double layers of cylindrical capillaries. It retains its meaning for dispersed particles, although the theory becomes more complicated.

If an electric field is applied to a particle dispersion or a liquid-filled capillary system, two transport phenomena can be observed: the transport of matter (electroosmosis and electrophoresis) and the transport of charge (the excessive electric current due to presence of excess ions in the interfacial region, i.e., surface conduction). Both phenomena are closely related to the formation of electrical double layers and can be used to characterize the interfacial charge by two independent parameters: zeta potential, ζ, and surface conductivity, K^{σ}. The transport of matter and the transport of charge are coupled in a simple manner in the case of straight capillaries, while the coupling is rather complicated in the case of curved particle surfaces. In the latter case, double-layer deformation (polarization) has to be taken into account. For this reason, double-layer characterization by both ζ and K^{σ} was first established for locally almost flat double layers of cylindrical capillaries. Afterward, research was performed mainly in two directions:

1. *Development of Theories for Double-Layer Polarization and Nonequilibrium Electrosurface Phenomena.* The theoretical efforts were combined

---

Electrical Phenomena at Interfaces and Biointerfaces: Fundamentals and Applications in Nano-, Bio-, and Environmental Sciences, First Edition. Edited by Hiroyuki Ohshima.

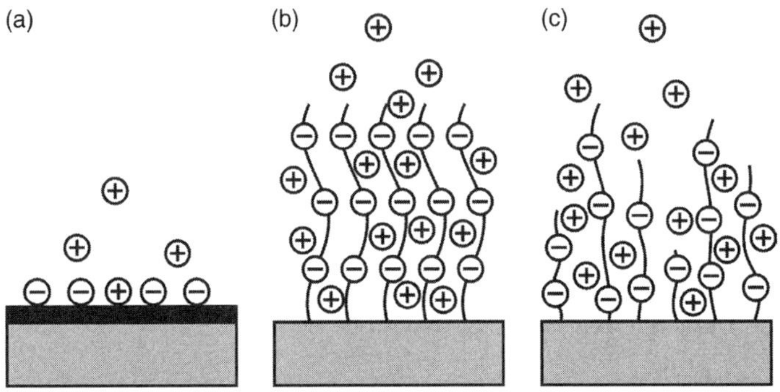

Figure 7.1. Schematic representation of different types of surfaces. (a) Hard surface: the volume fraction of the bulk material or surface coating is equal to 1. In the case of soft surfaces, polymer chains or a polymeric meshwork is attached to a hard surface. In contrast to hard surfaces, the volume fraction of soft surfaces is smaller than 1. As a consequence, water molecules and electrolyte ions can penetrate into the polymer coating. One can distinguish between soft surfaces with homogeneous (b) and inhomogeneous (c) polymer/charge distributions.

with experimental strategies to obtain information about both ζ and K^σ of dispersed particles.

2. *Soft Surface Electrokinetics.* This direction includes the extension of theories for surfaces that cannot be characterized by the standard electrokinetic model (SEM), that is, surfaces that are not molecularly smooth, homogeneous, and impermeable for ions. Polyelectrolyte layers (PLs) immobilized on a hard surface, for example, glass or silicon, are a typical example of a soft surface permeable for counterions.

At the transition from a hard surface to a soft surface, K^σ may increase by orders of magnitude, which predetermines the larger role of surface conductivity in the field of soft surface electrokinetics. Taking this into account, the surface conductivity of hard and soft surfaces (Fig. 7.1) is considered separately in Sections 7.2 and 7.3 of this chapter. In the recently published IUPAC report [1], large attention was paid to the measurement of K^σ for the consistent electrokinetic characterization of surfaces. In the next section, we follow the argumentation within this report.

7.2 SURFACE CONDUCTIVITY OF HARD SURFACES

7.2.1 Surface Conductivity and Electrokinetic Potential

Surface conduction is an excess electric conduction that takes place in capillaries and dispersed systems by the presence of electric double layers. Excess ions may move under the influence of electric fields applied tangentially to the surface. The phenomenon is quantified in terms of the surface conductivity, K^σ, which is the surface equivalent to the bulk conductivity, K_L. K^σ is a surface excess quantity like the surface concentration Γ_i of a certain species, i. Whatever the charge distribution, K^σ can always be defined through the two-dimensional analog of Ohm's law:

$$j^\sigma = K^\sigma E, \tag{7.1}$$

where j^σ is the (excess) surface current density (ampere per meter).

A measure of the importance of surface conductivity for the total conductivity of a capillary system or a particle dispersion is given by the dimensionless Dukhin number [1], Du, relating the surface conductivity K^σ and the bulk conductivity K_L:

$$Du = \frac{K^\sigma}{K_L a}, \tag{7.2}$$

where a is a characteristic length of the system, for example, the radius of a particle or capillary. Dukhin himself called the ratio expressed by Equation 7.2 Rel [2]. For any system, the total conductivity K can be expressed as the sum of a solution contribution (K_L) and a surface contribution (K^σ). For instance, for a cylindrical capillary, the following expression results [1, 2]:

$$K = \left(K_L + 2K^\sigma / a\right) = K_L(1 + 2Du). \tag{7.3}$$

The factor 2 in Equation 7.3 applies for cylindrical geometry. For other geometries, the value may be different.

Surface conduction is not considered within the Helmholtz–Smoluchowski (HS) theory [3]. In this theory, only the solution conductivity K_L is taken into account to derive the tangential electric field within the double layer. Thus, in addition to the condition of large electrokinetic radius κa, that is, the Debye length κ^{-1} is small as compared with the capillary or particle radius a,

$$\kappa a >> 1, \tag{7.4}$$

the applicability of the HS theory requires

$$Du << 1. \tag{7.5}$$

The surface conductivity can have contributions owing to the diffuse layer charge outside the plane of shear, $K^{\sigma d}$, and to the stagnant layer, $K^{\sigma i}$ [4]:

$$K^\sigma = K^{\sigma d} + K^{\sigma i}. \tag{7.6}$$

Accordingly, Du can be written as

$$Du = \frac{K^{\sigma d}}{K_L a} + \frac{K^{\sigma i}}{K_L a} = Du^d + Du^i. \qquad (7.7)$$

The contribution of the diffuse layer $K^{\sigma d}$ is called the *Bikerman surface conductivity* [2, 4–6] after Bikerman, who derived an equation for the conductivity within this part of the double layer (see further discussion). The stagnant layer conductivity (SLC) [2, 7, 8] may include a contribution due to the specifically adsorbed ions and another one due to the part of the diffuse layer charge that is caused by ions that reside behind the plane of shear. The intrinsic charge on the solid surface, that is, the charge caused by ionized groups, is generally assumed to be immobile and does not contribute to K^σ.

The conductivity in the diffuse part of the double layer outside the plane of shear, $K^{\sigma d}$, consists of two parts: a migration contribution, caused by the movement of ions with respect to the liquid, and a convective contribution, due to the electroosmotic liquid flow beyond the shear plane, which gives rise to an additional mobility of the ions and hence leads to an extra contribution to $K^{\sigma d}$. For the calculation of $K^{\sigma d}$, the Bikerman equation (see Eq. 7.8) can be used. This equation expresses $K^{\sigma d}$ as a function of the electrolyte concentration and double-layer parameters. For a symmetrical z-z electrolyte, the equation can be written as

$$K^{\sigma,d} = \frac{2e^2 N_A z^2 c}{kT\kappa}\left[D_+\left(e^{-\frac{ze\zeta}{2kT}} - 1\right)\left(1 + \frac{3m_+}{z^2}\right) + D_-\left(e^{\frac{ze\zeta}{2kT}} - 1\right)\left(1 + \frac{3m_-}{z^2}\right)\right], \qquad (7.8)$$

where e is the elementary charge, N_A is the Avogadro constant, z is the valency, c is the electrolyte concentration, k is the Boltzmann constant, and T is the absolute temperature. The parameters m_+ and m_- characterize the relative contribution of electroosmosis to the surface conductivity for cations and anions, respectively:

$$m_\pm = \frac{2}{3}\left(\frac{kT}{e}\right)^2 \frac{\varepsilon_0 \varepsilon_r}{\eta D_\pm}, \qquad (7.9)$$

where $D_\pm$ is the ionic diffusion coefficient.

The extent to which K^σ influences the electrokinetic behavior of the systems depends on the value of Du. For a symmetrical z-z electrolyte and identical cation and anion diffusion coefficients ($m_+ = m_- = m$), the conductivity in the diffuse part of the double layer can be expressed as [1, 2, 4]

$$Du^d = \frac{K^{\sigma d}}{K_L a} = \frac{2}{\kappa a}\left(1 + \frac{3m}{z^2}\right)\left[\cosh\left(\frac{ze\zeta}{2kT}\right) - 1\right]. \qquad (7.10)$$

From this equation, it follows that Du^d is small under conditions of large κa ($\kappa a >> 1$) and at low zeta potentials. Substitution of this expression for Du^d in Equation 7.7 yields [1, 2, 4]

$$Du = \frac{2}{\kappa a}\left(1+\frac{3m}{z^2}\right)\left[\cosh\left(\frac{ze\zeta}{2kT}\right)-1\right]\left(1+\frac{K^{\sigma i}}{K^{\sigma d}}\right). \tag{7.11}$$

This equation shows that Du depends in general on the zeta potential, on the ion mobility in the bulk electrolyte, and on the ratio $K^{\sigma i}/K^{\sigma d}$. Now, the condition $Du << 1$, required for application of the HS theory, is achieved for $\kappa a >> 1$, rather low values of ζ, and $K^{\sigma i}/K^{\sigma d} < 1$.

7.2.2 Discovery of Surface Conduction and its Implications for Electrokinetic Phenomena and Double-Layer Polarization

Smoluchowski pointed out that an excess of ions near a charged surface, forming the outer side of the double layer, makes a definite contribution to the conductivity; this is called the surface conductance. Although the first experimental indications of this effect were obtained at the turn of the 19th century, systematic investigations were not performed before the end of the 1920s. Smoluchowski [9] assumed that the ions are mobile only in the hydrodynamic mobile part of the double layer. Consequently, he expressed the specific surface conductivity in his formula by the electrokinetic potential. In deriving this formula, Smoluchowski used Helmholtz's formal theory of the double layer. Later, Abramson [10] had to modify it, expressing the double-layer thickness, which is an indeterminate magnitude in Helmholtz's theory, by means of the equation for the Debye length.

Using the experimental data obtained by Briggs [11] as well as by Bull and Gortner [12] for the concentration dependence of the zeta potential and the surface conductivity, Ambramson gave a qualitative corroboration of Smoluchowski's formula. Cole [13] interpreted the data of these studies on the basis of ideas regarding the proportionality of the surface conductivity to the charge of the diffuse part of the double layer. These first experimental studies and their comparison with the theory confirmed the common cause of surface conductivity and electrokinetic phenomena and pointed out the necessity of developing a more rigorous theory, which was soon proposed by Bikerman [14]. A great number of experiments were subsequently conducted on various systems by diverse methods, some of them in good agreement with Bikerman's theory [15].

In 1952, Overbeek [16] showed, on the basis of experiments by Wijga [17] and calculations by Overbeek and Wijga [18], that Smoluchowski's equation for electroosomosis cannot be applied in case of considerable surface conductivity of the diaphragm, and moreover, that it is impossible to introduce corrections as in the case of a single capillary. The difficulty of taking into account the effect of surface conductivity on electrokinetic phenomena in strongly charged diaphragms also involves the limitations of measuring surface conductivity in diaphragms exactly. Actually, the surface conductivity can be measured with sufficient accuracy only if it is commensurate with the bulk conductivity. But in that case, K^{σ} strongly affects the potential distribution in the pores, so that it is impossible to correlate the experimentally measured parameter with the surface conductivity exactly. Because of these difficulties, the research on surface conductivity and electrokinetic phenomena in strongly charged diaphragms, very intensely conducted in the 1930s, was not continued with the same intensity in the 1940s and 1950s. An exception was the systematic research of a group of Soviet scientists. They concluded that the measurement of surface conductivity and the determination of its effect on electrokinetic phenomena in diaphragms are possible without excessive error [19].

As the surface current (caused by the surface conductivity) influences the potential distribution in the vicinity of dispersed particles, surface conductivity also affects their migration in an external electric field. This is not taken into account in the classical electrophoresis theory. In 1948, Booth [20] and Henry [21] independently proposed extended equations for the electrophoresis, enabling for correction of surface conductivity effects. According to this theory, the magnitude of *Du* determines the degree of the effect of the surface conductivity in electrophoresis as well.

It should be noted that the theories for surface conductivity and its incorporation in electrokinetic phenomena are usually based on the simplifying assumption, originally made by Helmholtz and later retained by Smoluchowski, of a strictly equilibrium and undeformed state of the double layer under the imposition of an external field. In the mid-1930s, some researcher attempted to describe the deformation of the double layer by the external field and its effect on electrophoretic movement theoretically. However, only Overbeek [22] and Booth [23] succeeded in formulating a correct theory for the double-layer polarization during electrophoresis of spherical particles.

A comparison of the Overbeek–Booth and Henry–Booth theories leads to the conclusion that whenever the surface conductivity is relevant for the electrophoresis, double-layer polarization is also an essential factor. Therefore, theories taking into account surface conductivity and neglecting double-layer polarization provide only a first and very rough approximation to the electrophoresis of strongly charged particles.

It may be inferred that the magnitude of *Du* is also a characteristic of the degree of double-layer polarization. Since electroosmosis in a diaphragm with isometric particles may be approximately considered as an inverse phenomenon of electrophoresis in a concentrated system, it follows that *Du* also characterizes the degree of double-layer polarization inside the diaphragm and, accordingly, the degree of the effect of polarization on electroosmosis. Dukhin [2] suggested to distinguish two states of electrokinetic phenomena by double-layer polarization for a deeper analysis of the problem: uncomplicated (polarization can be ignored) [24] and complicated (polarization significantly determines the properties of the system) [25].

7.2.3 Electrokinetic Models and Surface Conductivity

7.2.3.1 SEM and Generalized Standard Electrokinetic Models (GSEMs). The Smoluchowski theory can be applied for electrosurface characterization if a surface satisfies the condition of so-called SEMs [3, 26], that is, if it is homogeneous, molecularly smooth, nonporous, and impermeable for ions. Systems that meet these requirements and that have been used to check the standard model are, for example, oil droplets in water [27], silver iodide crystals [28, 29], hydrophobic polymer surfaces [30], partially fused molecularly smooth quartz [31], and mica [32].

Lyklema and coworkers [33–36] proved the existence of stagnant liquid layers at solid surfaces and that "at least in a number of systems, conduction behind the shear plane does take place." This defines a new standard [37] for electrokinetic transport modeling that can be called the GSEM [38]. Its range of applicability is wider since, unlike the standard model (SEM), the existence of SLC is considered, and there are no requirements concerning surface smoothness on the molecular level. Thus, the double layer is described by at least two parameters (ζ and $K^{\sigma i}$) in the GSEM, while only ζ is used in the SEM.

7.2.3.2 Non-SEM and Surface Conductivity. Because the majority of real surfaces are rough, porous, or permeable for ions, the GSEM is not universal. The Bikerman equation describes only a part of the total surface conductivity in the majority of systems. However, the additional surface conductivity [2], caused either by surface roughness or surface permeability for ions, also has to be characterized [2, 38]. Its measurement allows a more complete characterization [2, 38] of real surfaces than that with the use of the electrokinetic potential only. Moreover, the consistent calculation of the zeta potential without the simultaneous measurement of K^{σ} is impossible [2, 38] when the SEM is not valid. This is the reason why integrated electrosurface investigations were proposed in References 2, 37, and 38.

It is not sufficient to state that K^{σ} is essential for a more complete surface characterization. The measurement of K^{σ} is necessary to extend consistent electrosurface characterization over the majority of real surfaces and colloid particles. For example, Tadros and Lyklema [39] recommended measuring the surface conductance rather than the potential for determining the surface charge of a porous glass electrode. A similar recommendation can be given in case of large surface roughness [38].

Because the shape of the boundaries between the Stern layer and the diffuse layer, as well as between the stagnant and mobile liquids, is extremely complicated in case of surface roughness, the notions of slipping plane and outer Helmholtz plane (OHP) are oversimplification with respect to the electrokinetic characterization of these surfaces and can be used only as a very crude approximation. Correspondingly, the interpretation of the effective values of the Stern and electrokinetic potentials is not clear at large roughness. In contradistinction, the mobile charge can provide a more precise evaluation of the surface charge in the case of a rough surface because the electroosmotic component of the surface conductivity is small [38].

It was shown [40] that small protrusions on the particle surface with dimensions much smaller than the particle dimension, but in the order of κ^{-1}, may have a dramatic influence on the Derjaguin–Landau–Verwey–Overbeek (DLVO) interaction energy. As a consequence, the impact of surface heterogeneities as roughness on colloidal forces attracted large attention [41]. However, so far, the generalization of electrokinetic theories for rough surfaces remains an important task.

7.2.3.3 Mobile Charge and Surface Charge. In many studies, the convective transport of the surface charge, that is, the term for electroosmosis in the Bikerman equation, is disregarded. This simplification is justified due to the small value of the electrokinetic charge and allows for the determination of the charge in the diffuse layer [2]:

$$\sigma_{md} = \frac{K^{\sigma d}}{u_i}. \tag{7.12}$$

For the application of Equation 7.12, the use of bulk mobility for the mobility u_i of the counterions in the diffuse layer is justified. Because there is little difference in the mobility of adsorbed and free monovalent counterions, Equation 7.20 can be transformed into an equation for the entire mobile charge:

$$\sigma_m = \frac{K^{\sigma d} + K^{\sigma i}}{u_i} = \frac{K^{\sigma}}{u_i}. \tag{7.13}$$

As K^{σ} is a measurable quantity, Equation 7.13 can be applied to estimate the surface charge:

$$\sigma_o = -\sigma_m. \tag{7.14}$$

The traditional approach to determine σ_o is titration:

$$\sigma_o = -\sigma_t, \tag{7.15}$$

where σ_t is the charge determined by the titration experiment. The justification for the application of Equation 7.14 is the small difference between σ_m and σ_t found for many systems.

7.2.4 Experimental Determination of the Surface Conductivity

Surface conductivities are excess quantities that cannot be measured directly. In the past, various methods have been developed to estimate K^{σ} for different systems:

1. For capillaries, K^σ can be deduced from the radius dependence of the streaming potential and the definition of *Du* (Eq. 7.2). This method is rather direct but requires a range of capillaries with different radii but identical surface properties [16, 17].
2. In the case of plugs, the determination of K^σ is relatively straightforward. The plug conductivity K_{plug} has to be measured as a function of K_L for neutral electrolytes (no ion adsorption). The latter can be changed by adjusting the electrolyte concentration. The plot of K_{plug} versus K_L has a large linear range that can be extrapolated to $K_L = 0$. The intercept of this curve represents K^σ.
3. In the case of suspensions, the electric conductance has to be measured as a function of the volume fraction. *Du* and K^σ can be derived based on the theory published by Dukhin and Derjaguin [24].
4. An alternative method for the determination of the K^σ of suspensions is to measure the dielectric dispersion as a function of the volume fraction. In this case, *Du* and K^σ can be calculated based on the equations derived in References 42 and 43.
5. In the case of flat surfaces, streaming current I_S and streaming potential U_S can be measured across rectangular streaming channels formed by two parallel sample surfaces at various separation distances h [44, 45]. As U_S depends on the channel conductivity ($Kh + 2K^\sigma$), while I_S is independent, K^σ can be derived from the plot of the channel conductance (I_S/U_S) versus h by a linear regression based on the following equation:

$$\frac{I_S}{U_S}\frac{L}{2b} = h\frac{K_L}{2} + K^\sigma, \qquad (7.16)$$

where L and b are the length and the width of the channel wall, respectively.

7.2.5 Integrated Electrosurface Investigations

7.2.5.1 Straight Capillaries. Integrated electrosurface investigations were accomplished for the first time at straight glass capillaries of different radii [16, 17]. Conductivity measurements and the application of Equation 7.3 allowed the direct determination of K^σ. The zeta potential was calculated from the streaming potentials measured across capillaries of different radii to predict K^σ according to the Bikerman equation. Thus, Bikerman's theory was convincingly confirmed for KCl concentrations below 0.1 mM [16, 17, 46]. The measured surface conductivity systematically exceeded that calculated according to the Bikerman equation at higher electrolyte concentrations, suggesting additional surface conductivity. It was assumed that the mobile counterions within the gel layer of glass contribute to the surface conductivity, while the hydrodynamic flow almost does not penetrate into this region.

7.2.5.2 Plugs. Fridrikhsberg and Barkovsky [47] measured three different parameters to characterize the electrical properties of diaphragms prepared from $BaSO_4$ microcrystals: the surface conductivity, the streaming potential, and the ionic adsorption. Comparing the results, these authors concluded that the tangential ion stream runs through the mobile part of the double layer as well as the stagnant layer at the wall, and that the mobility of ions in the stagnant layer may be close to the bulk values. Evidently, the trivial explanation attributing the substantial excess of the mobile charge (compared to the electrokinetic charge) to the surface roughness was excluded by the fact that the thermal dependence of the ionic mobility in the double layer differs from that in the bulk.

Since the electrokinetic theory was developed for spherical particles, monodisperse lattices have been extensively studied. Van der Put and Bijsterbosch [48, 49] performed conductivity measurements at dilute and concentrated latexes and concluded that the SEM cannot describe their electrokinetic behavior. They attributed the observed differences to the presence of a flexible "hairy" layer whose thickness increases with decreasing ionic strength.

7.2.5.3 Diluted Suspensions. Measurements of the electric conductance of dilute lattices as a function of the volume fraction were conducted in References 50–52. Data interpretation was based on the Dukhin theory [24], which made it possible to calculate Du, K^{σ}, and σ_m. The latter was found to be close to σ_t. The low values of the electrokinetic charge, σ_{ζ}, as compared to σ_m found in this work revealed that K^{σ} is the most important characteristic of these surfaces. This conclusion was confirmed by the fact that values of σ_m close to σ_t have been obtained in many studies [50–54]. At the low-frequency end of the dispersion curve, the double-layer relaxation effect occurs [53] over the frequency range 1–20 MHz, giving information about Du and σ_m, respectively. Values of σ_m obtained in this way for three samples of polystyrene lattices were found to be very close to σ_t. The agreement with the theory was demonstrated even more convincingly by interpreting the frequency relationship in Reference [54]. Measurements of the low-frequency dielectric dispersion on lattices [55–59] allowed to conclude on only small differences between of σ_t and σ_m.

Combined measurements of σ_t, K^{σ}, and the electrophoretic mobility have been conducted by Baran et al. [50] and Sidorova et al. [51, 52] The plot of the zeta potential calculated according to the Smoluchowski equation versus the concentration showed an extreme caused by the polarization of the double layer. The calculation of the zeta potential applying the Dukhin–Semenikhin equation [60], which accounts for additional surface conductivity, revealed monotonously decreasing zeta potentials with an increasing electrolyte concentration. It is important to mention that differences of the Stern potential calculated from K^{σ} and σ_t have been rather small. The efficiency of such an approach was confirmed in References 61–63, which have also found an extreme in the concentration dependence of the zeta potential. The authors have calculated Ψ_d from titration data and substituted it into the Dukhin–Semenikhin equation. This approach resulted also in a monotonous decrease of ζ and with increasing electrolyte concentration.

7.2.5.4 Smooth Hard Surfaces. For smooth surfaces without adsorbed macromolecules, σ_ζ characterizes the diffuse layer charge and the SLC the charge of the Stern layer [33–36]. Combined measurements comprising two electrical surface phenomena allow the determination of the charge distribution between the diffuse layer and the Stern layer. This new level in surface characterization provides accurate information on the charge distribution if the mobilities of adsorbed and bulk counterions are similar. The ion mobility at interfaces was extensively studied by Lyklema and coworkers [33–36] and was generalized in Reference 64. Their method for the determination of the adsorbed counterion mobility u_i is based on the measurement of σ_ζ, σ_t, and σ_m. An alternative method was proposed by Matsumura et al. [65, 66], who derived u_i from the surface conductivity at the isoelectric point.

The information about u_i is important both for the further development of electrokinetic theories and for data interpretation. As u_i also affects the double-layer relaxation during particle interaction [67], it is important for the further development of the DLVO theory [68] as well.

The progress in the determination of u_i is achieved due to disregarding the possible difference between Ψ_d and ζ and the identification of the SLC as that caused by the mobility of adsorbed counterions. The initial interpretation of the SLC assumed either conductivity due to diffuse layer ions between the OHP and the slipping plane [2] or due to the mobility of adsorbed counterions [7, 8] or both. Zukoski and Saville [69] suggested to disregard the role of the immobile portion of the diffuse layer without any analysis. To our opinion, the question whether the ion transport along the Stern layer exceeds very much that between the OHP and the slipping plane remains open.

7.3 SURFACE CONDUCTIVITY OF SOFT SURFACES

7.3.1 Charge Density Distribution at Interfaces between PLs and an Aqueous Solution

Charge formation within surface-confined PLs, including biopolymer films, is of highest importance for the application of such materials in demanding products. However, due to the lack of adequate analytical tools, the impact of electrical charging on the intra- and intermolecular structures of surface-confined PLs so far remained poorly understood.

The traditional characterization of hard surfaces by electrokinetic (zeta potential) measurements cannot be applied for the characterization of the internal structure of thick PLs, although the traditional electrokinetics remains important for characterizing PL/electrolyte interfaces. Systematic investigations revealed that surface conductivity measurements provide a unique opportunity for the characterization of the PL, including the determination of Donnan and surface potentials, maximal PL charge at complete dissociation, fractional PL charge, counterion condensation, and even PL thickness. This was achieved through advanced electrokinetic measurements in microchannels and through an extension of the related theoretical modeling.

Charging is usually the result of the separation of positive and negative charges in space. Accordingly, the pattern of charge formation and compensation at solid/liquid interfaces depends on structural characteristics of the interface. In the case of polymers, one can distinguish between hard and soft surfaces (Fig. 7.1). If a polymer surface is nonporous and not penetrable for both ions and water, the polymer/electrolyte interface is designated as "hard" (Fig. 7.1a). On the contrary, polymer surfaces are considered "soft" if water and electrolyte ions are also located between polymer chains grafted to a (hard) substrate (Fig. 7.1b,c). PLs [70] are often used as models to reveal the structural characteristics of surface-confined biopolymers, for example, to understand protein adsorption [71].

7.3.1.1 Local and Nonlocal Screening. The thickness d of PLs often exceeds the screening length κ_m^{-1} of the PL [72]:

$$d > \kappa_m^{-1}, \tag{7.17}$$

where

$$\kappa_m^{-1} = \kappa^{-1}\left(\frac{2FC_o}{|\rho|}\right)^{1/2}, \tag{7.18}$$

and κ^{-1} is the Debye length. If the condition (Eq. 7.17) is fulfilled, d does not affect the charge formation in the layer [72], which essentially simplifies the theory.

The second simplification usually used for the development of electrokinetic theories for PLs [72–75] is the assumption of complete dissociation of functional groups. In this case, the charge density per unit volume within PLs can be expressed as follows:

$$\rho = nF, \tag{7.19}$$

where n is the concentration of charged segments per unit volume. If the condition (Eq. 7.17) is fulfilled, the charge of the polymer chains is mainly screened by mobile counterions within the inner part of the PL (Fig. 7.2), characterized by the Donnan potential Ψ_D. At the PL/electrolyte solution boundary, the charge of the polymer is screened only partially; that is, there is a locally unscreened charge within a thin layer of the thickness κ_m^{-1}. This charge

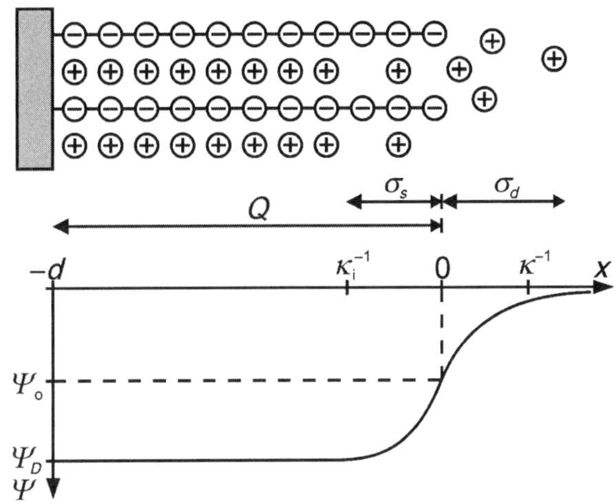

Figure 7.2. Illustration of the charge distribution in a PL. In the inner part of the PL, the charge of the polymer chains is completely locally screened by counterions, while at the PL/electrolyte solution interface, the charge is screened nonlocally. According to Equation 2.3, the complete local screening determines the value of Donnan potential Ψ_D. The surface charge σ_s is formed within the zone of partial local screening near the PL/electrolyte solution interface. The charge σ_s is screened nonlocally due to the diffuse countercharge σ_d. Because of the double-layer neutrality, $\sigma_s + \sigma_d = 0$. The absolute value of the potential decreases within the zone of partial local screening from $|\Psi_D|$ to $|\Psi_o|$. The surface potential Ψ_o is an analog of the surface potential of hard surfaces.

is called the surface charge (σ_s) in distinction from the intrinsic charge $Q = \rho d$. Specific properties caused by the counterion exchange between the PL and the electrolyte will be discussed next.

There are numerous similarities between the charge of PLs and the surface charge of hard surfaces. Similar to hard surfaces, the surface charge σ_s is screened nonlocally by the diffuse countercharge σ_d within an electrical double layer (Fig. 7.2). Accordingly, electrokinetic measurements provide information about the charging of PLs.

Note that the majority of efforts related to the characterization of soft surfaces are focused on the characterization of the double layer by traditional electrokinetic methods (electrophoresis, streaming potential, etc.). Because the structure of PLs is mainly determined by the intrinsic charge, the question arises whether traditional electrokinetics can be applied to obtain information about the intrinsic charge of PLs. At first glance, an attempt to characterize PLs using electrokinetic (surface charge) measurements may be justified when the PL is uniform; that is,

$$n(x) = n_o, \tag{7.20}$$

where x is the coordinate perpendicular to the substrate of the PL (Fig. 7.2). The surface charge depends on the segment density at the PL/electrolyte interface ($x = 0$). Consequently, surface charge measurements provide information about $n|_{x=0}$. If the thickness of uniform PL is measured in addition, for example, by ellipsometry, the intrinsic charge

$$Q = \rho d_o = n_o d_o F \tag{7.21}$$

may be estimated on the basis of traditional electrokinetic measurements. Unfortunately, this approach is inaccurate without information on the PL structure. As traditional electrokinetics cannot be applied for this purpose, an approach based on the electrokinetic determination of $n|_{x=0}$ may either provide Q, if the PL is uniform, or yield a crude estimation for Q, if the PL is nonuniform.

The theoretical background for the characterization of uniform PL on the basis of electrokinetic measurements was elaborated by Ohshima [72–75]. He derived an analytical solution of the Poisson–Boltzmann (PB) equation for the case of complete dissociation. Based on Ohshima's theory, it is possible to calculate the Donnan potential Ψ_D and the surface potential Ψ_o (see Fig. 7.2). He succeeded to solve the hydrodynamic problem as well and obtained an equation for the calculation of Ψ_D and Ψ_o based on the experimentally accessible electroosmotic velocity [72–75]. In addition, the intrinsic charge can be determined using the well-known relation between the Donnan potential Ψ_D and the charge density ρ. Therefore, the approach derived by Ohshima permits a comprehensive characterization of uniform PL by ρ, Ψ_o, and Ψ_D.

The equilibrium in the system surface charge–countercharge is reached, when their absolute values are equal; that is, an electrical double layer is formed. The electrical field is only localized inside the double layer. As a consequence, there is no ion transport between the double layer and the electrolyte, which is a necessary equilibrium condition for the double layer. The second condition is the absence of ionic fluxes inside the double layer. This is achieved due to the nonuniform ion distributions inside the double layer, which satisfy the condition of mutual compensation of ionic fluxes caused by the electric field and by ion diffusion. These conditions correspond to Boltzmann distributions for cations C^+ and anions C^- inside the double layer:

$$C^{\pm}(x) = C_o^{\pm} e^{\pm z^{\pm}\tilde{\Psi}(x)}, \tag{7.22}$$

where x is the distance to the surface, $z^{\pm}$ are the ion valences, $C_o^{\pm}$ are the cation and anion electrolyte concentrations in the bulk, $\tilde{\Psi} = F\Psi / RT$ is the dimensionless potential, and R is the universal gas constant. The substitution of Equation 7.22 into the Poisson equation leads to the well-known PB equation [2–4, 75]:

$$\frac{d^2\tilde{\Psi}}{dx^2} = \frac{1}{\varepsilon\varepsilon_o}\left(z^+ e^{-z^+\tilde{\Psi}(x)} - z^- e^{z^-\tilde{\Psi}(x)}\right). \tag{7.23}$$

The linearized form of this equation permits to derive analytical expressions for the potential distribution inside the double layer and for the characteristic thickness of the diffuse layer (Debye length) [2–4, 75]. For the sake of simplicity, subsequently the equations are specified for the case of a 1-1 electrolyte; that is, $z^+ = z^- = 1$.

The polyelectrolyte chains within the PL bear many identical functional groups. Their dissociation is related to a three-dimensional charge distribution within the PL (Fig. 7.2). If the PL is thick [72–74, 76], the local screening causes an almost complete screening of the intrinsic charge, except in a small zone at the PL/electrolyte interface. The potential difference between the electroneutral zone of the PL and the bulk electrolyte is characterized by the Donnan potential Ψ_D (Fig. 7.2). The magnitude of Ψ_D is determined by the condition of local electroneutrality:

$$\rho + FC_o\left(e^{-\tilde{\Psi}_D} - e^{\tilde{\Psi}_D}\right) = 0 \tag{7.24}$$

As ρ usually exceeds FC_o, the counterion concentration within the PL exceeds that in the electrolyte. The counterion concentration decreases within a thin layer near the PL/electrolyte interface (located at $x = 0$) by the diffusion of ions into the electrolyte. As a consequence, the local screening within this layer is not complete; that is, a charged surface layer with the charge density σ_s is formed (Fig. 7.2). The potential of the PL decreases within the charged surface layer. Accordingly, the absolute value of the surface potential $|\Psi_o| = |\Psi(x = 0)|$ is smaller than $|\Psi_D|$ (Fig. 7.2). The deficit of the local screening in this zone is compensated by nonlocal screening [77], that is, by the formation of the diffuse countercharge (σ_d) outside the PL:

$$\sigma_d = -\sigma_s. \tag{7.25}$$

The structure of this diffuse layer is described by Equation 7.23 and is identical to that in the case of hard surfaces. Equation 7.23 can be generalized for the calculation of the potential distribution between Ψ_D and Ψ_o as well as for the calculation of the surface charge σ_s.

7.3.1.2 *Intrinsic Charge and Surface Charge.*

The calculation of σ_s and Ψ_o is based on the joint solution of the PB equation (Eq. 7.23) for the diffuse layer and the PB equation for the zone of partial local screening (Eq. 7.26) [75]. This new type of PB equation accounts for ρ and the local screening of charge:

$$\frac{d^2\tilde{\Psi}}{dx^2} = -\frac{1}{\varepsilon_0\varepsilon}\left[\rho + F\left(C^+(x) - C^-(x)\right)\right]. \tag{7.26}$$

With respect to PLs with uniform density, Ohshima and Kondo [78] derived an analytical expression for the potential distribution and for the surface potential:

$$\tilde{\Psi}_o = \tilde{\Psi}_D - \tanh\left(\frac{\tilde{\Psi}_D}{2}\right). \tag{7.27}$$

As the diffuse layer can be described by the same PB equation (Eq. 7.23), the equation for double-layer charge σ_s is identical to that obtained by the Gouy–Chapman theory for hard surfaces [2–4]:

$$\sigma_s = -\sigma_d = \frac{4FC_o}{\kappa^{-1}}\sinh\left(\frac{\tilde{\Psi}_o}{2}\right). \tag{7.28}$$

As $d >> \kappa^{-1}$ and $\exp(-\tilde{\Psi}_D) >> \exp(-\tilde{\Psi}_o/2)$, the comparison of Equation 7.28, specified with Equation 7.27, and the equation for the intrinsic charge $Q = \rho d$, specified with Equation 7.24, leads to the conclusion that the surface charge density is at least one order of magnitude smaller than the intrinsic charge. This important conclusion is supported by measurements of both σ_s and Q for various types of soft surfaces [79, 80]. This comparison also reveals that the experimental determination of Ψ_o is not sufficient for the characterization of the internal structure of thick PLs [81–83]. Despite this limitation, Ψ_o provides qualitative information on the ionization of PLs. Altogether, the comprehensive characterization of electrosurface phenomena within grafted PLs requires both the determination of the intrinsic charge and the surface charge.

7.3.2 Characterization of PLs with Uniform Segment Distribution by Surface Conductivity Measurements

The surface conductivity is caused by the accumulation of counterions $C^+(x) - C_o$ and by the deficit of coions $C^-(x) - C_o$. For a binary electrolyte, the equation for the surface conductivity caused by the migration of ions can be written in the form [2–4]

$$K_m^\sigma = F\left[u^+\int_{-d}^{\infty}(C^+(x)-C_o)dx + u^-\int_{-d}^{\infty}(C^-(x)-C_o)dx\right], \tag{7.29}$$

where u^+ and u^- are the cation and anion mobilities. Both components of the double layer contribute to K_m^σ and their contributions are proportional to σ_s and σ_d, respectively [81]. These contributions cancel each other because of the double-layer electroneutrality. As σ_s does not contribute to K_m^σ, the entire PL can be considered as completely locally screened; that is, $C^+(x)$ and $C^-(x)$ may be replaced by $C^+(-\Psi_D)$ and $C^-(-\Psi_D)$, respectively, within the range $-d < x < 0$

[81]. As a consequence, the measurement of the surface conductivity K_m^σ provides information about the Donnan potential of PLs [81]:

$$K_m^\sigma = \frac{K_L d}{2} e^{-\tilde{\Psi}_D}, \tag{7.30}$$

where K is the bulk conductivity of the electrolyte solution.

Because of the electroneutrality within the PL and the weak influence of coions, K_m^σ can be expressed through the total intrinsic charge Q:

$$K_m^\sigma = u^+ Q. \tag{7.31}$$

Q is maximal at complete dissociation:

$$Q_{\max} = \rho d_o = F n_o d_o. \tag{7.32}$$

As a consequence, the Donnan potential and the surface conductivity are maximal in the case of complete dissociation.

Dissociation of functional groups is the main process that controls the intrinsic charge of PLs. Therefore, detailed knowledge about the dependence of Ψ_D and Ψ_o on the pH and electrolyte concentration C_o is of highest interest. The well-known equation for the fractional polyelectrolyte charge [84, 85],

$$\alpha = \frac{1}{1 + 10^{\mathrm{p}K - \mathrm{pH}} e^{-\tilde{\Psi}(x)}}, \tag{7.33}$$

allows to determine the fixed charge density at noncomplete dissociation:

$$\rho(\mathrm{pH}, C_o) = \rho_{\max}\alpha. \tag{7.34}$$

Equation 7.34 corresponds to a chemical reaction involving a proton, H^+, and an anionic site, A^-:

$$\mathrm{HA} \Leftrightarrow \mathrm{H}^+ + \mathrm{A}^-. \tag{7.35}$$

The pK value of this reaction corresponds to the pH of the solution at which the dissociation of the groups is just 50%. Since the electric potential difference between the bulk and the inner PL interferes with the proton distribution, the equilibrium of the reaction described by Equation 7.35 depends also on the potential distribution.

An equation for the pH dependence of the Donnan potential can be derived based on Equation 7.24 if the density of ionized groups is expressed by Equation 7.33 and 7.34 [81]:

$$e^{-\tilde{\Psi}_D} = \frac{10^{\mathrm{pH} - \mathrm{p}K}}{2} \left(\sqrt{1 + 4 \frac{|\rho_{\max}|}{C_o F} 10^{\mathrm{p}K - \mathrm{pH}}} - 1 \right). \tag{7.36}$$

The substitution of ρ in Equation 7.26 by Equations 7.33 and 7.34 yields the PB equation for the zone of partial local screening at any degree of dissociation. The solution of this equation and of the PB equation for the diffuse layer yields [81]

$$\tilde{\Psi}_o - \tilde{\Psi}_D = \ln \frac{e^{\alpha} + \alpha - 1}{\alpha}, \tag{7.37}$$

where

$$\alpha = 2\left(1 + \sqrt{1 + \beta n}\right)^{-1} \tag{7.38}$$

and

$$\beta = \frac{4 \times 10^{\mathrm{p}K - \mathrm{pH}}}{C_o}. \tag{7.39}$$

Consequently, the fractional charge of the PL Q and the reduced fractional charge can be written as

$$Q = Q_{\max} \alpha \tag{7.40}$$

and

$$\frac{Q}{Q_{\max}} = \alpha \tag{7.41}$$

As β is known, the measurement of α allows n_o to be determined. Subsequently, the PL thickness d_o can be obtained using Equation 7.32. The theory described earlier was recently successfully applied for the characterization of Na acrylate gels [86].

7.3.3 Generalized Donnan Potential Concept for PLs with Nonuniform Segment Distribution

Although the measurement of K_m^{σ} offers a unique opportunity for the characterization of the internal structure of PLs and the developed theory is well confirmed by experiments, the broad application of this achievement will become possible only when a more general and realistic model for the segment distribution is incorporated into the theory.

The idea of a gradual decrease of the segment concentration was explored by some authors with respect to the electrokinetic theory. Ohshima [87] assumed an exponential distribution of the segment density in his theory for the electrophoresis of colloidal particles coated with charged polymers. Lopez-Garcia et al. [88, 89] solved the PB equation numerically for the case of linear and exponential distributions of functional groups.

The generalized Donnan potential concept [77] allows to obtain the potential distribution analytically for any distribution of fixed charges if the segment density does change gradually with the distance from the surface. If this prerequisite is fulfilled, the notion of the local Donnan potential (which is a function of the distance x) can be introduced,

$$\tilde{\Psi}_D(x) = -\ln\frac{n(x)}{C_o}, \tag{7.42}$$

because the deviation from the local electroneutrality $\delta\rho(x)$ may be small. In this case, Equation 7.24 may be generalized as

$$\rho(x) + FC_o\left(e^{-\tilde{\Psi}_D(x)} - e^{\tilde{\Psi}_D(x)}\right) = \delta\rho(x) \tag{7.43}$$

with the assumption that

$$\delta\rho(x) << \rho(x). \tag{7.44}$$

If this condition (Eq. 7.44) is fulfilled and the (dimensionless) Donnan potential exceeds approximately 2, Equation 7.42 follows from Equation 7.43. For potentials smaller than approximately 2, the expansion of the exponents in a Taylor series yields, in the case of a linear approximation,

$$\tilde{\Psi}_D(x) = -\frac{n(x)}{C_o}. \tag{7.45}$$

The exact solution of the PB equation for PL with nonuniform charge density $\rho(x)$ differs from the dependence of $\Psi_D(x)$. However, this difference can be considered small very often. The condition for the applicability of this approach is derived in the "Appendix" of Reference 77 on the basis of the PB equation.

When the segment concentration decreases gradually with the distance x from the solid surface, the concept of the local Donnan potential is valid, except for the thin charged layer near the PL/electrolyte interface with thickness κ_m^{-1}. The condition is

$$h >> \kappa_m^{-1}, \tag{7.46}$$

where h is the characteristic length of the decay of the segment concentration. According to Equation 7.18, κ_m^{-1} does not depend on the solution concentration C_o and remains rather small even at low electrolyte concentration. If n_o is very large as compared with C_o, κ_m^{-1} is very small in comparison with κ^{-1}, that is, in the most interesting case of high Donnan potentials.

The application of the concept of the local Donnan potential was proposed in Reference 90 as an assumption without an estimation of the boundary conditions. It was not revealed that a charged zone with thickness κ_m^{-1} preserves at the PL/electrolyte interface and a double layer is formed. Accordingly, this concept was not considered in electrokinetic theories.

7.3.4 Determination of Nonuniform Segment Distribution by Surface Conductivity Measurements

7.3.4.1 Generalization of the Equation for the Surface Conductivity of PLs. The equation for the surface conductivity of PL becomes very simple when the Donnan potential is high and the approximation of the excluded coions can be applied (Eq. 7.30). An analog expression can be derived for the case of nonuniform segment distribution:

$$K_m^\sigma = \frac{K}{2}\int_{-d}^{0} e^{-\tilde{\Psi}_D(x)}dx. \qquad (7.47)$$

When the coion and counterion mobilities are equal ($u^+ = u^- = u$), the conductivity of the electrolyte can be written as

$$K_L = 2FC_0u. \qquad (7.48)$$

Usually, the solution of the PB equation is necessary to obtain the potential distribution within PLs and to calculate the counterion concentration. However, the introduced concept of the local Donnan potential does not require the solution of the PB equation. The concept of local potentials can be applied when the deviation from the local electroneutrality,

$$Fn(x) - FC^+(x) = 0, \qquad (7.49)$$

is very small. The substitution of $C^+(x)$ according to Equation 7.49 into Equation 7.48 yields an even more valuable equation for the surface conductivity in the approximation of excluded coions:

$$K_m^\sigma = u^+Q_{\max}, \qquad (7.50)$$

where

$$Q_{\max} = \int_{-d}^{0} \rho(x)dx = F\int_{-d}^{0} n(x)dx \qquad (7.51)$$

is the intrinsic charge in the case of complete dissociation. In Reference 81, it was shown that the intrinsic charge of uniform PL can be determined based on surface conductivity measurements. Now this restriction is eliminated due to the introduction of the concept of local electroneutrality and the local Donnan potential [77]. The boundary for the applicability of this concept is simultaneously the boundary for the exact measurement of the intrinsic charge.

This means that the intrinsic charge can be determined by surface conductivity measurements when the segment distribution decreases gradually with the distance from the surface, that is, when the condition (Eq. 7.46) is satisfied.

7.3.4.2 Intrinsic Charge at Noncomplete Dissociation and Nonuniform Segment Distribution. When the segment distribution is not uniform, both the segment concentration n and the fractional PL charge α are functions of the distance x. The dependence of α from x is given by Equation 7.38 and n_o can be replaced by $n(x)$. Using these expressions, Equation 7.51 can be rewritten as

$$Q = \int_{-d}^{0} n(x)\alpha(x)dx = F\int_{-d}^{0} \frac{2n(x)dx}{1+\sqrt{1+\beta n(x)}}. \tag{7.52}$$

At exponential segment distribution (see eq. A.17 in Reference 77), the integral in Equation 7.52 can be solved analytically:

$$Q/F = h\int_{n_1}^{n_0} \frac{dn}{1+\sqrt{1+\beta n}} = \frac{4h}{\beta}\left(\sqrt{1+\beta n_o} - \sqrt{1+\beta n_1} - \ln\frac{1+\sqrt{1+\beta n_o}}{1+\sqrt{1+\beta n_1}}\right), \tag{7.53}$$

where the relation between h, n_o, n_1, and the PL thickness d is given by equation A.19 in [77]. The substitution of the exponential segment distribution into Equation 7.51 yields

$$\frac{Q_{\max}}{F} = h(n_o - n_1). \tag{7.54}$$

7.3.4.3 Determination of Structural PL Characteristics. In Section 7.3.2, it was shown that the thickness of uniform PL can be determined based on surface conductivity measurements. For the practical application of this option, an experimental criterion is necessary to distinguish between systems with uniform and nonuniform segment distributions. The equation derived for the fractional charge at nonuniform segment distribution (Eq. 7.53) is more general than Equation 7.38. For practical purposes, it is convenient to introduce the reduced fractional charge:

$$\frac{Q_{\beta h}}{Q_{\max h}} = \frac{4}{\beta(n_o - n_1)}\left(\sqrt{1+\beta n_o} - \sqrt{1+\beta n_1} - \ln\frac{1+\sqrt{1+\beta n_o}}{1+\sqrt{1+\beta n_1}}\right). \tag{7.55}$$

In distinction from $Q_{\max h}$ and $Q_{\beta h}$, the reduced charge does not depend on h. A simple examination confirms that the right-hand side (RHS) of Equation 7.55 transforms into the RHS of Equation 7.41 when $n_1 \to n_o$. The comparison of Equations 7.55 and 7.41 shows an essential difference in the dependence of

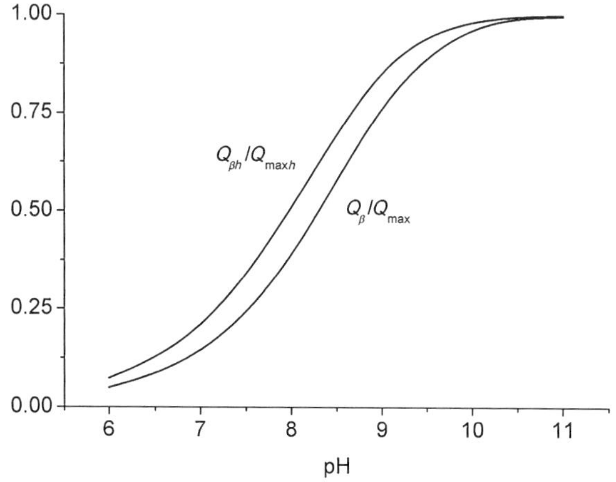

Figure 7.3. pH dependence of the reduced fractional charge at uniform ($Q_\beta/Q_{\max}$) and nonuniform ($Q_{\beta h}/Q_{\max h}$) segment distributions ($n_o/C_o = 10^3$, pK = 5.6).

the reduced charge on β for uniform ($n_1 = n_o$) and nonuniform ($n_1 << n_o$) segment distributions. This difference is illustrated in Figure 7.3 for the conditions $n_1 = 0.1n_o$, $n_o/C_o = 10^3$, and pK = 5.6. This difference is about 30% at $\beta n_1 >> 1$. According to Reference 86, the difference between predicted and measured values of Ψ_D and consequently of the surface conductivity may be only a few percent.

At first glance, the deduction of information about the segment concentration n_1 at the PL/electrolyte solution interface from surface conductivity measurements seems to be impossible because the contributions of different sublayers with the thickness Δx to the overall surface conductivity are additive. This additivity is inherent in the definitions of the maximal surface conductivity (Eq. 7.50) and of the maximal charge (Eq. 7.51). To overcome this problem, it was proposed to perform surface conductivity measurements at a small degree of dissociation ($\beta n_o >> 1$). The charge at noncomplete dissociation (Eq. 7.52) depends on the fractional charge $\alpha(x)$, which increases with increasing distance to the surface. The maximum of the fractional charge coincides with the PL/electrolyte solution interface. As a consequence, the dependence of the fractional charge on the segment density n_1 is much stronger at noncomplete dissociation. This fact offers the opportunity to estimate the segment distribution based on surface conductivity measurements. The regularities of charging under conditions of low and complete dissociation manifest themselves in asymptotic simplifications of Equation 7.55:

$$\frac{Q_{\beta h}}{Q_{\max h}} = 1 - \frac{\beta}{8}(n_o + n_1)(\beta n_o << 1) \tag{7.56}$$

and

$$\frac{Q_{\beta h}}{Q_{\max h}} = \frac{4}{\sqrt{\beta n_o}\left(1+\sqrt{\frac{n_1}{n_o}}\right)} (\beta n_o >> 1). \tag{7.57}$$

Equation 7.56 shows that n_o and n_1 are additive at almost complete dissociation. Accordingly, $Q_{\max h}$ provides mainly information about n_o. On the contrary, the term $(n_1/n_o)^{0.5}$ in the equation for the fractional charge (Eq. 7.57) is favorable for the estimation of n_1/n_o by means of surface conductivity measurements, because the smaller the ratio n_1/n_o, the larger the ratio $(n_1/n_o)^{0.5}$ in comparison with n_1/n_o.

The options arising for the characterization of PLs based on surface conductivity measurements become more obvious if the limited access of traditional electrokinetics to the structural features of PLs is taken into account. With respect to the latter statement, a recent publication about the electrokinetic mobilities of fibrillated and nonfibrillated oral streptococcal strains provides useful arguments [91]. The polydisperse fibril layer of the bacteria *Streptococcus salivarius* HB determines its electrophoretic velocity [91]. Because of the large similarity in the mechanism of electroosmotic flow within a fibrillar layer and within a PL, the electrokinetic model for fibrillar layers published in Reference 91 and that for PL are mainly identical. It is remarkable that the authors do not apply the electrophoresis theory for "diffuse soft particles" [92] and how they justify the application of Ohshima's theory for uniform soft surfaces despite the nonuniform structure of the fibrillar layer: "However, the electrophoretic motion of individual *S. salivarius* HB bacteria is determined by its outer corona, where the electroosmotic velocity is the most significant. In other words, the mobility of *S. salivarius* HB is solely determined by the outer tails of its longest fibrils, electrokinetic detection of the other fibril types being impossible because of the above argument." This completely agrees with our statement that electrokinetics is useful for the characterization of the PL/electrolyte solution interface, while the application of the theory for the characterization of the inner part of nonuniform thick charged layers is impossible. However, it could be shown that surface conductivity measurements allow to distinguish between uniform and nonuniform PLs.

Unfortunately, the determination of the segment distribution on the basis of intrinsic charge measurements has many limitations. They are mainly related to the determination of the maximal charge. Their influence on the pH dependence of the fractional charge is weak and can be disregarded in some cases. Accordingly, the consideration of these limitations below will lead to the conclusion that the discrimination between uniform and nonuniform structures of PLs on the basis of surface conductivity measurement is possible for some systems.

7.3.5 Subprocesses That Complicate the Quantification of Surface Conductivity

7.3.5.1 Counterion Condensation. Counterion condensation is basically determined by the average distance between the fixed charge b [93]. In Reference 94, b was estimated for poly(acrylic acid) chains grafted to a solid substrate. Afterward, the linear charge density parameter $\xi = l_B/b$ (l_B is Bjerrum length [93]) and the fraction of uncondensed counterions $f = 0.886\xi^{-1}$ [93] were determined. Surprisingly, a value of about 0.2 was obtained for f, which is in good agreement with the measured decrease of K_m^σ. The interpretation of these data agrees well with data obtained by the measurement of the osmotic pressure for polyacrylic acid (PAA) in solution [95], which can be attributed to the mobile counterions of the polyelectrolyte remaining under conditions of the counterion condensation. Counterion condensation is observed for PLs with high segment density. The effect does not occur at low densities of ionizable groups, for example, in loose gels [86], and at sufficient low pH, when only a portion of acid groups is dissociated and accordingly, the distance between them (similar to b) increases. Hence, the maximal charge can be underestimated when the distance between fixed charges is small. On the contrary, the counterion condensation may not affect the measurement at a small fractional charge. This is important because the pH dependence of the fractional charge offers an opportunity to estimate the ratio n_1/n_o. On other hand, the knowledge of $Q_{\max h}$ is necessary for the calculation of the ratio $Q_h/Q_{\max h}$. If counterion condensation prevents the determination of $Q_{\max h}$, the pH dependence of the fractional charge can be used for the characterization of the PL structure instead of the pH dependence of the reduced fractional charge. In this case, the constant multiplier in the equation for the fractional charge (Eq. 7.52) does not cause an essential problem in the determination of n_o and n_1/n_o by comparing the measured pH dependence of Q_h with the pH dependence given by Equation 7.53.

7.3.5.2 Counterion Binding. The influence of counterion binding on the conductivity of polyelectrolyte solutions was analyzed by Lyklema et al. [70]. Counterion binding changes the charge and mobility of mobile polyelectrolyte macroions. In turn, this decreases the contribution of the macroion to the conductivity of the polyelectrolyte solution.

The number of mobile couterions within PLs can be also decreased by counterion binding. According to Lyklema et al. [70], the counterion mobility within PL can be considered (as a first approximation) identical with the bulk mobilities. In order to estimate the influence of counterion binding on the determination of $n(x)$, a multiplier for the fraction of mobile counterions (according to Lyklema et al. [70], we use f^{cond}) has to be introduced in Equations

7.29–7.31, 7.47, and 7.50. The multiplier cancels in Equations 7.41 and 7.55. Hence, counterion binding does not affect the determination of $n(x)$ on the basis of the reduced fractional charge. However, the determination of the reduced fractional charge may be inconsistent because counterion binding may influence the determination of $Q_{\max}$ and $Q_{\max h}$. As Equation 7.53 has to be applied for the determination of $n(x)$ in this case, it is necessary to replace h in Equation 7.53 by the product hf^{cond}. Both h and f^{cond} are unknown parameters. Therefore, the introduction of f^{cond} in Equation 7.53 does not deteriorate the possibilities to discriminate between uniform and nonuniform segment distributions. However, the unknown value of f^{cond} causes an uncertainty in the determination of h.

7.3.5.3 Influence of the Segment Volume Fraction on the Counterion Mobility. The polymer chains of a PL manifest themselves as obstacles for the counterion migration [96, 97]. For small volume fractions, the drop of the conductivity of a macromolecular gel, K_G, in comparison with the conductivity of the electrolyte, K_L, linearly depends on the volume fraction ϕ of the gel [96, 97]:

$$K_G = K_L(1 - B\phi), \tag{7.58}$$

where B is the shape factor of the polymer chains ($1.5 < B < 2.5$). Conductivity measurements at loose gels showed [86] that at volume fractions smaller than 0.07, the influence of the polymer chains on ion mobility is weak and in good agreement with Equation 7.58. At a higher volume fraction, this influence becomes nonlinear [98]. However, because of the onset of additional mechanisms of charging at a higher volume fraction, we restrict our discussion here to PLs of low density. For dense PLs, a difference in the chemical potentials of entrapped electrolyte ions occurs as compared to the bulk solution [99]. These values are different for cations and for anions, which causes the differences in ion concentrations between the polymer layer and the bulk even in the absence of any fixed charge [99, 100]. The phenomenon contains a component caused by the difference in dielectric permittivity between the polymer chains [101] and water, a component caused by the change in short-range ion–solvent interaction, and a component caused by the steric factor; that is, the dense PL may be only partially penetrable for inorganic ions.

7.3.5.4 Influence of the Fixed Charge Distribution on the Ion Mobility. In concentrated disperse systems, for example, in ionites, polyelectrolyte solutions, or grafted PL, the diffusion coefficients of small inorganic ions may be smaller than in an electrolyte solution. Electrostatic interactions between the mobile ions and the fixed charges of the disperse system or macroions are one of the major factors causing the decrease of the ion mobility. The nonuniform distribution of the fixed charge leads to the nonuniform distribution of the electric potential that hinders the ion motion. The influence of the fixed charge heterogeneity on the diffusion coefficient has been studied experimentally and theoretically for ionites [102, 103] and biopolymers [104].

A general theory for the influence of the fixed charge distribution on the counterion mobility is described in References 105 and 106. In Reference 106, it is shown that this influence increases with the increasing concentration of the fixed charges when the density of the charged groups becomes higher than 0.1 mol/L. Therefore, a general theory for surface conductivity has to account also for electrostatic interactions between ions and fixed charges.

7.3.6 Experiment

Surface conductivity measurements were demonstrated to provide valuable information on the charging of soft surface coatings [86, 94, 107]. The properties of these films (e.g., swelling) are often determined by the three-dimensional distribution of ionizable groups within the polymer meshwork, which are not accessible by traditional electrokinetic experiments. In this case, surface conductivity measurements allow, for example, the study of the ionization characteristics within the soft surface coating, the quantification of the density of ionizable groups, and conclusion on effects like ion condensation. Next, we discuss some recent results obtained for soft surfaces of different complexity.

Zimmermann et al. [107] studied the charging of grafted poly(L-glutamic acid) brushlike films by streaming current and surface conductivity measurements in KCl solutions of varied ionic strengths. Isoelectric points of 2.7 ± 0.1 derived from pH-dependent streaming current measurements indicated the ionization of carboxylic acid groups to be the major charge formation process at the outermost region of the films. In order to characterize the charging within the polypeptide layers, the surface conductivity and Dukhin number were determined in 0.1, 1.0, and 10.0 mM KCl solutions (Fig. 7.4). In general, the grafted polypeptide layers exhibited considerably higher values of the surface conductivity K^σ when compared with the hard surfaces [107]. This can be attributed to the accumulation of mobile counterions within the three-dimensional structure of the brush. Furthermore, the surface conductivity and Dukhin number were found to be strongly dependent on the pH and electrolyte concentration of the solution (Fig. 7.4). In 0.1 mM KCl solution, the ratio of $K^\sigma(\mathrm{pH} = 9.0)/K^\sigma(\mathrm{pH} = 6.0)$ of 35.1 was determined. For higher salt concentrations, the corresponding values were 16.6 (1 mM KCl) and 5.2 (10 mM KCl), respectively. In line with the decreasing ratio of the surface conductivity values, the increase of *Du* was observed at lower pH values at higher salt concentrations (Fig. 7.4). The plot of *Du* versus pH and the decreasing ratio of the surface conductivity revealed that the ionization of the carboxylic acid groups within the brush occurs in the neutral and weak alkaline pH range (depending on the ionic strength). Furthermore, in combination with results from swelling experiments, which showed an increase of the film thickness in

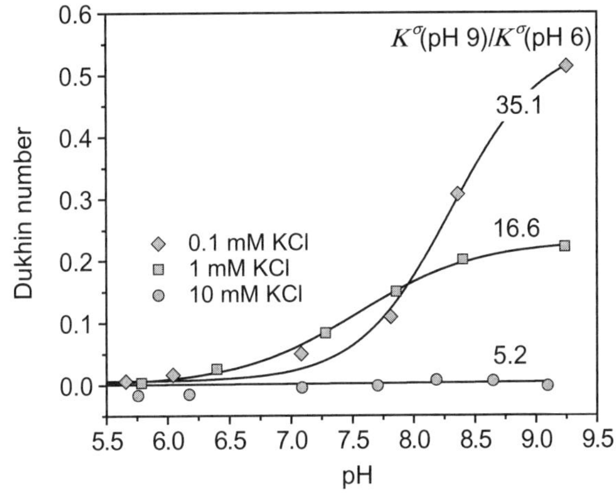

Figure 7.4. Dukhin number versus solution pH for grafted poly(L-glutamic acid) layers in 0.1, 1.0, and 10.0 mM KCl solutions at a channel height of 50 μm.

the same pH range, the data [107] allowed to conclude on a charge-induced helix–coil transition within the polypeptide brush.

As discussed previously, the surface conductivity of the poly(L-glutamic acid) films increased at lower pH values in solutions of higher KCl concentration [107]. This phenomenon can be explained by the increasing ionization of the carboxylic acid groups within the layer with the increasing ionic strength of the electrolyte solution; that is, the grafted poly(L-glutamic acid) chains behave more acidic at lower Debye screening radii. This finding demonstrates that surface conductivity measurements also provide information on electrostatic interactions between ionizable groups within PLs.

Yezek [86] carried out conductivity measurements on a family of polyacrylamide-*co*-sodium acrylate gels at various ionic strengths. The conductivity data allowed him to estimate the Donnan potential between the gel and the electrolyte. The experimental results showed excellent agreement with the theory for the surface conductivity of weak PLs [81, 86].

Dukhin et al. [94] demonstrated that surface conductivity measurements also provide information about counterion condensation with highly charged PLs. In this study [94], the surface conductivity data measured at grafted PAA brushes were used to estimate the density of COOH groups. The experimental determined surface conductivity at alkaline pH was found to be only 14% of the theoretical values expected from the brush structure for the case of complete ionization. In view of this large deviation and limited sterical constraints within the brushes, the authors concluded that the number of freely moving counterions is decreased due to counterion condensation. This interpretation agreed very well with the measurement of the osmotic pressure for PAA solution [95], which can be exclusively attributed to mobile counterions.

7.4 SUMMARY

Although excess conduction at interfaces was discovered about a century ago [9], the surface conductivity (K^{σ}) was considered for a very long time as just a simple correction term for the evaluation of electrokinetic experiments. Meanwhile, surface conductivity is an established parameter in the field of interface and colloid science. Sophisticated theories were developed for surface conduction at hard and soft surfaces.

In the GSEM [38], the electrical double layer at hard surfaces is described by at least two parameters (ζ and $K^{\sigma i}$), while only ζ is used in the SEM [3, 26]. Surface conductivity measurements allow experimental access to the intrinsic charge located at the solid surface and to derive information on the pattern of charge compensation, as well as in combination with titration experiments, to the ion mobility at interfaces.

While the surface conductivity is considered within this chapter mainly with respect to surface characterization, it also plays a large role in the formation of induced dipole moments of colloid particles [24, 25, 83] and double-layer polarization [25]. In turn, these phenomena are the basis for a large family of "nonequilibrium electrosurface phenomena" (NESP) [37, 38, 109]. As NESP phenomena are beyond the scope of this chapter, we mention only that surface conductivity attracted large attention during the past several decades due to its role in NESP. Surface conduction also achieved large attention in the field of micro- and nanofluidics, where the characteristic dimensions of the systems are small. Accordingly, K^{σ} and Du are not negligible and significantly determine electrosurface phenomena in these systems.

During the past several years, large progress was achieved in the electrokinetic characterization of soft surfaces. In this case, surface conductivity measurements and traditional electrokinetics provide information about charging in different zones at the interface. While surface conductivity mainly provides information on the thick internal electroneutral zone of PL, electrokinetics is sensitive to the structure of the PL/electrolyte solution interface. Based on the pioneering work of Ohshima [72], the charging within completely ionized PE layers can be characterized based on Donnan and surface potentials. Meanwhile, advanced theories [81] also exist for PE layers with an arbitrary degree of ionization. Systematic investigations revealed that surface conductivity measurements provide a unique opportunity for the characterization of PL, including the determination of Donnan and surface potentials, maximal PL charge at complete dissociation, fractional PL charge, counterion condensation, and even PL thickness. This was achieved through advanced electrokinetic measurements in microchannels and an extension of the related theoretical modeling.

The concept of local Donnan potential [77] has been demonstrated to be useful for the characterization of PE layers with nonuniform segment distribution. It permits to approximate the real potential distribution by the local Donnan potential, to determine the local fractional charge, and to estimate the maximal intrinsic charge at complete dissociation. Furthermore, at favorable conditions, surface conductivity measurements permit to discriminate between uniform and nonuniform segment distributions.

Further progress in the characterization of correlations between the charge and structure of soft surfaces may be achieved by the combination of electrokinetic and surface conductivity measurements and the rigorous application of numerical methods for data evaluation [108]. This advanced methodology provides a versatile tool for unraveling electrohydrodynamic phenomena and structural characteristics at soft polymer coatings with applications in cutting-edge technologies as, for example, in nanofluidics and biomaterial research.

REFERENCES

1 A. V. Delgado, F. Gonzales-Caballero, R. J. Hunter, L. K. Koopal, J. Lyklema. *J. Colloid Interface Sci.* 309 (2007) 194.

2 S. S. Dukhin, in E. Matijevich (ed.), *Surface and colloid science*, vol. 7. Wiley, New York, 1974, Chapter 1.

3 M. Smoluchowski. In W. Graetz (ed.), *Handbuch der Elektrizität und des Magnetismus*, vol. 11. Barth, Leipzig, 1914, p. 336.

4 J. Lyklema. *Fundamentals of colloid and interface science*, vol. 2. Academic Press, London, 1995, Chapter 4.

5 J. J. Bikerman. *J. Phys. Chem.* 46 (1942) 725.

6 J. J. Bikerman. *Physical surfaces*. Academic Press, New York, 1970, Chapter 8.

7 S. S. Dukhin, in B. V. Derjaguin (ed.), *Research in surface forces*, vol. 3. Plenum Press, New York, 1971.

8 S. S. Dukhin, V. N. Shilov. *Adv. Colloido Interface Sci.* 13 (1980) 153.

9 M. Smoluchowski. *Z. Phys. Chem.* 92 (1918) 129.

10 H. A. Abramson. *J. Physiol. Chem.* 36 (1932) 2141.

11 D. R. Briggs. *J. Phys. Chem.* 32 (1928) 641.

12 H. B. Bull, R. A. Gortner. *Colloid symposium monograph*, vol. 8. Chemical Catalog Company, New York, 1932, p. 309.

13 O. K. S. Cole. *Physics* 3 (1932) 114.

14 J. J. Bikerman. *Z. Phys. Chem. A* 163 (1933) 378.

15 N. Street. *Aust. J. Chem.* 9 (1956) 333.

16 H. R. Kruyt, in H. R. Kruyt (ed.), *Colloid science*, vol. 1. Elsevier, Amsterdam, 1952.

17 P. W. Wijga. Thesis, Utrecht, 1946.

18 J. Th. G. Overbeek, P. W. Wijga. *Rec. Trac. Chim. Pay-Bas* 65 (1946) 556.

19 O. N. Grigorov, L. P. Kuzmina, A. V. Markovic, D. A. Fridrikhsberg. *Electrokinetic properties of capillary systems*, Monograph Collection. Academy Press, Moscow, 1956.

20 F. Booth. *Trans. Faraday Soc.* 44 (1948) 955.

21 D. C. Henry. *Trans. Faraday Soc.* 44 (1948) 1021.

22 J. Th. G. Overbeek. *Kolloidchem Beih.* 54 (1943) 287.

23 F. Booth. *Proc. R. Soc. A* 203 (1950) 514.

24 S. S. Dukhin, B. V. Derjaguin, in E. Matijevic (ed.), *Surface and colloid science*, vol. 7. John Wiley, New York, Toronto, 1974, Chapter 2.

25 B. V. Derjaguin, S. S. Dukhin, in E. Matijevic (ed.), *Surface and colloid science*, vol. 7. John Wiley, New York, Toronto, 1974, Chapter 3.

26 J. Th. G. Overbeek. *Pure Appl. Chem.* 10 (1965) 339.

27 D. A. Haydon. *Proc. R. Soc.* 258A (1960) 319.

28 J. Lyklema. *J. Colloid Interface Sci.* 58 (1977) 242.

29 B. H. Bijsterbosch, J. Lyklema. *Adv. Colloid Interface Sci.* 9 (1978) 147.

30 R. Zimmermann, U. Freudenberg, R. Schweiss, D. Küttner, C. Werner. *Curr. Opin. Colloid Interface Sci.* 15 (2010) 196–202.

31 B. V. Derjaguin, N. V. Churaev. *Progress in surface and membrane science*, vol. 4. Academic Press, New York, London, 1981, p. 69.

32 G. B. Westermann-Clark, J. L. Anderson. *J. Electrochem. Soc.* 130 (1983) 839.

33 J. Lyklema, M. Minor. *Colloids Surf. A* 149 (1998) 33.

34 M. Löbbus, H. P. Leeuwen, J. Lyklema. *Colloids Surf. A* 161 (2000) 103.

35 M. Minor, A. J. van der Linde, J. Lyklema. *J. Colloid Interface Sci.* 203 (1998) 177.

36 J. Lyklema. *Oil Gas Sci. Technol.* 56 (2001) 41.

37 S. S. Dukhin. *Adv. Colloid Interface Sci.* 44 (1993) 1.

38 S. S. Dukhin. *Adv. Colloid Interface Sci.* 61 (1995) 17.

39 Th. F. Tadros, J. Lyklema. *J. Electroanal. Chem.* 17 (1968) 267.

40 S. Bhattacharjee, C.-H. Ko, M. Elimelech. *Langmuir* 14 (1998) 3365.

41 J. Y. Walz. *Adv. Colloid Interface Sci.* 74 (1998) 119.

42 V. N. Shilov, A. V. Delgado, F. Gonzalez-Caballero, C. Grosse. *Colloids Surf.* 192 (2001) 253.

43 S. S. Dukhin, V. N. Shilov. *Dielectric phenomena and the double layer in disperse system and polyelectrolytes*. John Wiley, New York, Toronto, 1974.

44 C. Werner, R. Zimmermann, T. Kratzmüller. *Colloids Surf. A* 192 (2001) 205.

45 C. Werner, H. Körber, R. Zimmermann, S. S. Dukhin, H.-J. Jacobasch. *J. Colloid Interface Sci.* 208 (1998) 329.

46 A. Wattillon. *J. Chim. Phys.* 54 (1957) 130.

47 D. A. Fridrikhsberg, V. Y. Barkovsky. *Kolloidn. Zh.* 26 (1964) 722.

48 A. G. Van der Put, B. H. Bijsterbosch. *J. Colloid Interface Sci.* 92 (1983) 499.

49 A. G. Van der Put, B. H. Bijsterbosch. *J. Colloid Interface Sci.* 75 (1980) 512.

50 A. A. Baran, L. M. Dudkina, N. M. Soboleva. *Kolloidn. Zh.* 43 (1981) 211.

51 D. A. Fridrichsberg, M. P. Sidorova, V. E. Shubin, L. E. Ermakova. *Kolloidn. Zh.* 48 (1986) 967.

52 M. P. Sidorova, I. B. Dmitrieva, G. P. Golub. *Colloidn. Zh.* 41 (1979) 488.
53 B. R. Midmore, R. J. Hunter. *J. Colloid Interface Sci.* 122 (1988) 521.
54 B. R. Midmore, D. Diggins, R. J. Hunter. *J. Colloid Interface Sci.* 129 (1989) 153.
55 J. Lyklema, S. S. Dukhin, N. V. Shilov. *J. Electroanal. Chem. Interfacial Electrochem.* 143 (1983) 1.
56 M. M. Springer, A. Korteweg, J. Lyklema. *J. Electroanal. Chem. Interfacial Electrochem.* 153 (1983) 55.
57 M. Minor. Electrodynamics of Colloids, Thesis, Wageningen, 1998.
58 L. A. Rosen, D. A. Saville. *Langmuir* 7 (1991) 36.
59 J. Kijlstra. Double Layer relaxation in Colloids, Thesis, Wageningen, 1992.
60 S. S. Dukhin, N. M. Semenikhin. *Kolloidn. Zh.* 32 (1970) 366.
61 R. Hidalgo Alvarez, F. J. de las Nieves, A. J. van der Linde, B. H. Bijsterbosch. *Colloids Surf.* 21 (1986) 259.
62 R. Hidalgo Alvarez, J. A. Molon, F. J. de las Nieves, B. H. Bijsterbosch. *J. Colloid Interface Sci.* 149 (1992) 23.
63 V. E. Shubin, M. P. Sidorova, O. S. Chechik, N. A. Sacharova. *Kolloidn. Zh.* 53 (1991) 187.
64 H. Lyklema, in A. Delgado (ed.), *Interfacial electrokinetics and electrophoresis*. Marcel Decker, New York, Basel, 2002.
65 S. V. Verbich, S. S. Dukhin, H. Matsumura. *J. Disp. Sci. Technol.* 20 (1999) 83.
66 H. Matsumura, S. V. Verbich, S. S. Dukhin. *Colloids Surf.* 159 (1999) 271.
67 S. S. Dukhin, J. Lyklema. *Langmuir* 3 (1987) 94.
68 S. S. Dukhin, J. Lyklema. *Faraday Discuss. Chem. Soc.* 90 (1990) 261.
69 C. F. Zukoski, D. A. Saville. *J. Colloid Interface Sci.* 107 (1985) 322.
70 M. Cohen Stuart, R. De Vries, J. Lyklema, in J. Lyklema (ed.), *Fundamentals of colloid and interface science*, vol. 5. Elsevier, Amsterdam, 2005.
71 W. Norde, J. Buijs, H. Lyklema, in J. Lyklema (ed.), *Fundamental of colloid and interface science*, vol. 5. Elsevier, Amsterdam, 2005.
72 H. Ohshima. *Colloids Surf. A* 103 (1995) 249.
73 H. Ohshima. *Adv. Colloid Interface Sci.* 62 (1995) 189.
74 H. Ohshima. *J. Colloid Interface Sci.* 163 (1994) 474.
75 H. Ohshima. *Theory of colloid and interface electric phenomena*. Elsevier, Amsterdam, 2006.
76 E. S. Romm, A. A. Rubashkin. *Kolloidn. Zh.* 47 (1985) 545.
77 S. S. Dukhin, R. Zimmermann, C. Werner. *J. Colloid Interface Sci.* 328 (2008) 217.
78 H. Ohshima, T. Kondo. *Biophys. Chem.* 38 (1990) 117.
79 R. Schweiss, P. B. Welzel, C. Werner, W. Knoll. *Colloids Surf. A* 195 (2001) 97.
80 R. Zimmermann, T. Kratzmüller, D. Erickson, D. Li, H.-G. Braun, C. Werner. *Langmuir* 20 (2005) 2369.
81 S. S. Dukhin, R. Zimmermann, C. Werner. *J. Colloid Interface Sci.* 274 (2004) 309.
82 S. S. Dukhin, R. Zimmermann, C. Werner. *J. Colloid Interface Sci.* 286 (2005) 761.
83 S. S. Dukhin, R. Zimmermann, C. Werner. *Adv. Colloid Interface Sci.* 122 (2006) 93.

84 A. Voigt, E. Donath, G. Kretzschmar. *Colloids Surf. A* 47 (1990) 23.
85 J. Lyklema. *Fundamentals of colloid and interface science*, vol. 2. Academic Press, London, 1995, Chapter 3.
86 L. P. Yezek. *Langmuir* 21 (2005) 10054.
87 H. J. Ohshima. *J. Colloid Interface Sci.* 185 (1997) 269.
88 J. J. Lopez-Garcia, C. Grosse, J. Horno. *J. Colloid Interface Sci.* 265 (2003) 327.
89 J. J. Lopez-Garcia, C. Grosse, J. Horno. *J. Colloid Interface Sci.* 265 (2003) 341.
90 Y. U. V. Lyatskaya, F. A. M. Leermakers, G. J. Fleer, E. B. Zhulina, T. M. Birshtein. *Macromolecules* 28 (1995) 3562.
91 J. F. L. Duval, H. J. Busscher, B. van de Belt-Gritter, H. van der Mei, W. Norde. *Langmuir* 21 (2005) 11268.
92 J. F. L. Duval, H. Ohshima. *Langmuir* 22 (2006) 3533.
93 G. Manning. *J. Phys. Chem.* 79 (1975) 262.
94 S. S. Dukhin, R. Zimmermann, C. Werner. *J. Phys. Chem.* 111 (2007) 979.
95 J.-Ph. Boisvert, A. Malgat, I. Pochard, C. Daneault. *Polymer* 43 (2002) 141.
96 A. C. Lagdon, H. C. Thomas. *J. Phys. Chem.* 75 (1971) 1821.
97 P. H. Elworthy, A. T. Florence, A. Rahman. *J. Phys. Chem.* 76 (1972) 1763.
98 S. S. Dukhin, V. N. Shilov. *Dielectric phenomena and the double layer in disperse systems and polyelectrolytes*. Wiley, New York, 1974, Chapters 6 and 7.
99 S. S. Dukhin, N. V. Churaev, V. M. Starov, V. N. Shilov. *Russ. Chem. Rev.* 57 (1988) 1010 (Engl. Transl).
100 S. S. Dukhin, M. P. Sidorova, A. E. Yaroshchuk. *Membrane electrochemistry and reverse osmosis*. Khimiya Publishers, Leningrad, 1991 (in Russian).
101 A. E. Yaroshchuk. *Adv. Colloid Interface Sci.* 85 (2000) 193.
102 H. L. Yeager, A. J. Steek. *Electrochem. Soc.* 128 (1981) 1880.
103 V. I. Zabolotsky, V. V. Nikonenko. *Ion transfer in the membranes*. Nauka, Moscow, 1996 (in Russian).
104 H. Magdelenat, P. Turq, P. Tivant, M. Chemla, R. Menezand, M. Drifford. *Biopolymer* 18 (1979) 187.
105 V. N. Shilov, N. P. Bondarenko. *Kolloidn. Zh.* 50 (1988) 948.
106 N. P. Bondarenko, V. N. Shilov. *J. Colloid Interface Sci.* 220 (1999) 443.
107 R. Zimmermann, T. Osaki, R. Schweiß, C. Werner. *Microfluidics Nanofluidics* 2 (2006) 367.
108 J. F. L. Duval, R. Zimmermann, A. L. Cordeiro, N. Rein, C. Werner. *Langmuir* 25 (2009) 10691.
109 S. S. Dukhin, V. N. Shilov, in A. Delgado (ed.), *Interfacial electrokinetics and electrophoresis*. Marcel Decker, New York, 2002.

8

COMPUTER SIMULATIONS OF CHARGED COLLOIDS

1. MESOSCOPIC MODELING

YASUYA NAKAYAMA, KANG KIM, AND RYOICHI YAMAMOTO

8.1 INTRODUCTION

Understanding the electrokinetics of solutions of colloids and/or polymers has been a challenging problem in colloid and interface science [1, 2]. Theory and numerical simulations both play major roles in unveiling the basic aspects of electrokinetic phenomena including electrophoresis, electroosmotic flow, electroacoustics, and more. However, theoretical consideration is often limited to rather simplified systems. To understand more complex systems, modeling and numerical simulations of electrokinetic phenomena are required.

The interactions between colloidal particles are critical for modeling electrokinetic phenomena. Interparticle interactions in colloidal dispersions consist of thermodynamic potential interactions [1–3] and hydrodynamic interactions (HIs) [1, 4, 5]. In electrokinetics, the thermodynamic potential interactions are specifically called electric double-layer (EDL) interactions and occur in both static and dynamic situations. On the other hand, the HIs occur solely in dynamic situations, in which most relevant processes are conducted, for example, separation of macromolecules, electrophoretic deposition, electrophoretic display, and rheological processing of slurry. Both

Electrical Phenomena at Interfaces and Biointerfaces: Fundamentals and Applications in Nano-, Bio-, and Environmental Sciences, First Edition. Edited by Hiroyuki Ohshima.

HIs and the EDL interaction are solvent mediated, meaning that they are dependent on solvent dynamics. In electrokinetic simulations, it is essential to include the solvent hydrodynamics and the coupling between HIs and the EDL interaction. The effects of HI–EDL coupling should be more pronounced in nanoparticle suspensions and/or solvents of polymer solutions in which the size of the EDL is close to the sizes of the particles and polymers.

Because the HI in dispersions is essentially a long-range, many-body effect, it is extremely difficult to study its role using analytic methods alone. Various numerical simulations for HI-only problems have been developed. There are specialized schemes to model the HI between spherical particles in Newtonian solvents, such as Stokesian dynamics (SD) [6] and Brownian dynamics (BD) with an Oseen tensor or a Rotne–Prager–Yamakawa mobility tensor. In these schemes, the solvent hydrodynamics are not explicitly solved and the HI between particles is included through certain interaction tensors.

For more complex boundaries between particles and solvents, the solvent hydrodynamics should be directly computed. One straightforward approach to solving the HI based on the Navier–Stokes equation is the arbitrary Lagrangian–Eulerian (ALE) method [7]. However, the ALE method requires too much computational cost to simulate many-colloid systems. Other techniques use an Eulerian grid to reduce computational costs. In the distributed Lagrangian multiplier (DLM) method [8], fluid particle dynamics (FPD) [9], the force coupling method (FCM) [10], and the method in Reference 11, the external forces on particles are distributed within the particle domain, and the hydrodynamic forces on particles are implicitly taken into account. In the immersed boundary method (IBM) [12] and the smoothed profile method (SPM) [13], the hydrodynamic forces on colloids are explicitly computed, and the body force is introduced to impose the rigidity of the particles in the continuum equations. Lattice Boltzmann (LB) simulation is another continuum discretization scheme on an Eulerian grid based on the distribution function [14]. In LB for suspensions, the rigidity of the particles is embedded in the LB equations and the hydrodynamic force is explicitly computed [15].

Mesoscopic modeling of HIs has been developed based not on the Navier–Stokes equation but solely on momentum conservation for virtual particles in the fluid. Such schemes include dissipative particle dynamics (DPD) [16] and stochastic rotation dynamics (SRD) [17], which describe compressible flow and recover long-time hydrodynamics. Mesoscopic fluid modeling presents an inconvenience in that macroscopic fluid parameters such as viscosity are indirectly controlled by the interactions between virtual particles.

The numerical schemes for solving HI presented earlier are supposed to be able to deal with HI–EDL coupled problems, but some of them have been applied to electrokinetic phenomena. LB simulation was applied to the electrokinetic equations in References 18–20. In References 21 and 22, the particles are represented by a point particle network and are frictionally coupled to an LB treatment of the Navier–Stokes equation. The counterions are treated by molecular dynamics. Many-colloid systems coupled with the electrokinetic equations are solved using FPD [23, 24] and the SPM [29].

In this chapter, we explain the SPM for electrokinetics. The SPM is a scheme for coupling a solid particle to continuum hydrodynamic equations. The numerical efficiency and accuracy of the SPM have been proven in References 26, 27, and 29. Section 8.5 describes a free software package named KAPSEL (Kyoto Advanced Particle Simulator for Electrohydrodynamics), which implements the SPM for the electrokinetics of colloidal dispersions.

8.2 DYNAMICS OF AN ELECTROLYTE SOLVENT AND COLLOIDS

8.2.1 Hydrodynamics of Multicomponent Fluids

We start with a brief description of the hydrodynamic equations for an electrolyte solution. Consider N ionic solutes that each satisfy a law of conservation for the concentration C_α, of the αth species:

$$\partial_t C_\alpha + \nabla \cdot C_\alpha \boldsymbol{v}_\alpha + \nabla \cdot \boldsymbol{g}_\alpha = 0, \tag{8.1}$$

where $\boldsymbol{v}_\alpha$ is the velocity of the αth solute and $\boldsymbol{g}_\alpha$ is a random current that represents the effect of thermal agitation on C_α. Because the inertial time scales of the solute molecules are extremely small, the velocity of the αth solute can be decomposed into the velocity of the solvent $\boldsymbol{v}$ and the diffusive current arising from the chemical potential gradient $\nabla\mu_\alpha$ as follows:

$$\boldsymbol{v}_\alpha = \boldsymbol{v} - \Gamma_\alpha \nabla \mu_\alpha, \tag{8.2}$$

where $\Gamma_\alpha k_B T = D_\alpha$ is the diffusivity of the αth ion, k_B is the Boltzmann constant, and T is the temperature. The random current should satisfy the following fluctuation–dissipation relation [28]:

$$< g_{\alpha,i}(\boldsymbol{x}, t) g_{\beta,j}(\boldsymbol{x}', t') > = 2k_B T D_\alpha \delta_{\alpha\beta} \delta_{ij} \delta(\boldsymbol{x} - \boldsymbol{x}') \delta(t - t'). \tag{8.3}$$

The conservation of momentum implies that the velocity of the solvent follows the Navier–Stokes equation for incompressible flow with Maxwell stress:

$$\nabla \cdot \boldsymbol{v} = 0 \tag{8.4}$$

and

$$\rho(\partial_t + \boldsymbol{v} \cdot \nabla)\boldsymbol{v} = -\nabla p + \eta \nabla^2 \boldsymbol{v} - \sum_\alpha C_\alpha \nabla \mu_\alpha + \nabla \cdot \boldsymbol{s}, \tag{8.5}$$

where ρ is the total mass density of the fluid, p is the pressure, η is the shear viscosity of the fluid, and $\boldsymbol{s}$ is a random stress satisfying the fluctuation–dissipation relation [28]:

$$< s_{ik}(\boldsymbol{x},t)s_{jl}(\boldsymbol{x}',t') > = 2k_{\mathrm{B}}T\eta(\delta_{ij}\delta_{kl} + \delta_{il}\delta_{kj})\delta(\boldsymbol{x}-\boldsymbol{x}')\delta(t-t'). \tag{8.6}$$

For a chemical potential, μ_α, we consider the Poisson–Nernst–Planck equation:

$$\mu_\alpha(\{C_1,\ldots,C_N\}) = k_{\mathrm{B}}T\log C_\alpha + Z_\alpha e(\Phi - \boldsymbol{E}^{\mathrm{ext}}\cdot\boldsymbol{x}) \tag{8.7}$$

and

$$\varepsilon_r\varepsilon_0\nabla^2\Phi = -\rho_e, \tag{8.8}$$

where Z_α is the valence of the αth ion, e is the elementary charge, Φ is the electrostatic potential, E^{ext} is the external field, ε_0 is the permittivity of the vacuum, ε_r is the relative dielectric permittivity (assumed to be uniform), and ρ_e is the charge density field. This equation describes the Poisson–Boltzmann distribution for ion concentrations at equilibrium as

$$C_\alpha = \bar{C}_\alpha \exp\left[-\frac{Z_\alpha e(\Phi - \boldsymbol{E}^{\mathrm{ext}}\cdot\boldsymbol{x})}{k_{\mathrm{B}}T}\right]. \tag{8.9}$$

The bulk concentration of ions, $\bar{C}_\alpha$, is related to the Debye length, $\kappa^{-1} = \left\{4\pi\lambda_{\mathrm{B}}\sum_\alpha Z_\alpha^2\bar{C}_\alpha\right\}^{-1/2}$, where the Bjerrum length is $\lambda_{\mathrm{B}} = e^2/4\pi\varepsilon_r\varepsilon_0 k_{\mathrm{B}}T$, which is about 0.72 nm for water at 25°C. This set of equations describes the electrokinetic phenomena of a simple electrolyte solution.

8.2.2 Colloids in Electrolyte Solutions

The colloidal dynamics are dependent on the force exerted by the solvent. Consider monodisperse spherical colloids with radius a, mass M_p, and moment of inertia $\boldsymbol{I}_p$. Momentum conservation between the fluid and the ith colloid implies the following hydrodynamic force and torque:

$$\boldsymbol{F}_i^H = \int(d\boldsymbol{S}_i\cdot\boldsymbol{\sigma}),\ \mathbf{N}_i^H = \int(\boldsymbol{x}-\boldsymbol{R}_i)\times(d\boldsymbol{S}_i\cdot\boldsymbol{\sigma}), \tag{8.10}$$

where $\boldsymbol{R}_i(t)$ is the center of mass, $\int d\boldsymbol{S}_i(\ldots)$ indicates the surface integral over the ith colloid, and $\boldsymbol{\sigma}$ is the stress tensor of the fluid. In terms of the electrokinetic equations, the stress reads

$$\boldsymbol{\sigma} = -p\boldsymbol{I} + \boldsymbol{\sigma}' + \boldsymbol{\sigma}^{st} + \boldsymbol{s}, \tag{8.11}$$

in which $\sigma' = \eta(\nabla\boldsymbol{v} + (\nabla\boldsymbol{v})^{\mathrm{T}})$ is the dissipative stress and $\sigma^{st} = \varepsilon\{\boldsymbol{EE} - (|\boldsymbol{E}|^2/2)\boldsymbol{I}\}$ is the Maxwell stress of the electric field $E = -\nabla\Phi + E^{\mathrm{ext}}$, where $\boldsymbol{I}$ is the unit tensor. All of the solvent-mediated effects, including many-body HIs, EDL interactions, electroviscous drag, electrokinetic lift, and the Brownian effect are accounted for in $\boldsymbol{F}_i^H$ and $\boldsymbol{N}_i^H$. The evolution of the colloids follows Newton's equations:

$$\dot{\boldsymbol{R}}_i = \boldsymbol{V}_i, \tag{8.12}$$

$$M_p \dot{\boldsymbol{V}}_i = \boldsymbol{F}_i^H + \boldsymbol{F}_i^c + \boldsymbol{F}_i^{\text{ext}}, \tag{8.13}$$

and

$$\boldsymbol{I}_p \cdot \dot{\boldsymbol{\Omega}}_i = \boldsymbol{N}_i^H + \boldsymbol{N}_i^{\text{ext}}, \tag{8.14}$$

where $\boldsymbol{V}_i(t)$ and $\Omega_i(t)$ are the translational and angular velocities, respectively, of the ith colloid; $\boldsymbol{F}_i^{\text{ext}}$ and $\boldsymbol{N}_i^{\text{ext}}$ are the external forces and torques, respectively; $\boldsymbol{F}_i^c$ is the force that arises from the core potential of the particles, which prevents the colloids from overlapping. The soft-core truncated Lennard–Jones potential, which is often called the Weeks–Chandler–Andersen (WCA) potential, is adopted for $\boldsymbol{F}_i^c$.

8.2.3 Particle–Solvent Boundary Conditions

The hydrodynamic equations require relevant boundary conditions at the particle–solvent interface. For the solvent velocity, the nonslip condition is assigned, so $\boldsymbol{v} = \boldsymbol{V}_i + \Omega_i \times \boldsymbol{r}_i$ with $\boldsymbol{r}_i = \boldsymbol{x} - \boldsymbol{R}_i$ for the ith colloid. For the concentration fields, the no-penetration condition is enforced as $\boldsymbol{n} \cdot \nabla \mu_\alpha = 0$, where $\boldsymbol{n}$ represents the unit normal to the surface of the colloids. Coupling of the hydrodynamics of the solvent to the dynamics of the colloids defines the moving boundary condition problem.

The usual numerical techniques of partial differential equations require a body-fitted grid at the particle–solvent interface. The body-fitted grid is necessary not only to enforce explicitly the boundary conditions but also to compute the surface integral of the hydrodynamic force and torque (Eq. 8.10). This approach cannot deal with the dynamic evolution of a many-colloid system, as the movement of a body-fitted grid requires remeshing at every discrete time step as the colloids move. The shifting boundary conditions lead to a huge computational cost, which is the main difficulty in simulation of many-colloid systems.

In contrast, the SPM provides an efficient scheme for this kind of moving boundary condition problem and incurs the same level of computational cost that is required for solving a fluid-only problem [26].

8.3 COMPUTATIONAL METHOD: SPM

8.3.1 Smoothed Profile for Particle–Solvent Boundary

In the SPM, each particle is represented by a smoothed profile that equals unity in the particle domain, zero in the fluid domain, and varies smoothly between zero and unity around the particle–solvent boundary, which is called the interfacial domain. Any function with the above properties can be used as a smoothed profile. Several analytic forms are presented in Reference 13. One convenient choice for a spherical particle is

$$\phi_i(\boldsymbol{x},t)=\frac{1}{2}\left[\tanh\left(\frac{a-|\boldsymbol{x}-\boldsymbol{R}_i(t)|}{\xi}\right)+1\right], \tag{8.15}$$

where ξ is the thickness of the interfacial domain. An extension of the smoothed profile to general nonspherical particles can be found in Reference 27, In the sharp interface limit of $\xi \to 0$, $\phi_i \to H(a - |\boldsymbol{x} - \boldsymbol{R}_i(t)|)$, where $H(\ldots)$ is the Heaviside step function.

Field variables are defined over the entire domain, which consists of the fluid domain and the particle domain. To designate the particle domain, we use a concentration field for the colloids given by $\phi(\boldsymbol{x},t)=\sum_{i=1}^{N_p}\phi_i(\boldsymbol{x},t)$. With the field ϕ, the total velocity field and concentration fields of the ionic solutes are defined as

$$\boldsymbol{v}=(1-\phi)\boldsymbol{v}_f+\phi\boldsymbol{v}_p, \tag{8.16}$$

and

$$C_\alpha=(1-\phi)C_\alpha^*, \tag{8.17}$$

where $(1 - \phi)\boldsymbol{v}_f$ represents the velocity field of the fluid, and

$$\phi\boldsymbol{v}_p(\boldsymbol{x},t)=\sum_{i=1}^{N_p}\phi_i(\boldsymbol{x},t)\left[\boldsymbol{V}_i(t)+\boldsymbol{\Omega}_i(t)\times\boldsymbol{r}_i(t)\right] \tag{8.18}$$

is the velocity field of the colloids. The auxiliary concentration field C_α^*, which can have a finite value in the particle domain, is introduced; C_α, the physical concentration field, is forced to be zero in the particle domain through multiplication by $(1 - \phi)$.

Another smoothed profile is introduced for the surface charge density of the colloids, which becomes $\delta(a - |\boldsymbol{x} - \boldsymbol{R}_i(t)|)$ in the sharp interface limit. One convenient choice is $|\nabla\phi|$. The charge density field is represented as

$$\rho_e=(1-\phi)\sum_\alpha Z_\alpha eC_\alpha^*+|\nabla\phi|e\sigma_e, \tag{8.19}$$

where σ_e is the surface charge density of the colloids.

8.3.2 Basic Equations for Colloidal Dispersions with Smoothed Profile

The hydrodynamic equations for Equations 8.16 and 8.17 defined on the entire domain are

$$\nabla \cdot \boldsymbol{v} = 0, \tag{8.20}$$

$$\rho(\partial_t + \boldsymbol{v} \cdot \nabla)\boldsymbol{v} = -\nabla p + \eta \nabla^2 \boldsymbol{v} - \sum_{\alpha} C_\alpha \nabla \mu_\alpha, \\ + \nabla \cdot (1-\phi)\boldsymbol{s} + \phi \boldsymbol{f}_p, \tag{8.21}$$

$$\partial_t C_\alpha^* = -\nabla \cdot C_\alpha^* \boldsymbol{v}_\alpha^* - \nabla \cdot \boldsymbol{g}_\alpha, \tag{8.22}$$

and

$$\boldsymbol{v}_\alpha^* = \boldsymbol{v} + (\boldsymbol{I} - \boldsymbol{nn}) \cdot (-\Gamma_\alpha \nabla \mu_\alpha), \tag{8.23}$$

with μ_α from Equation 8.7 and the Poisson Equation 8.8 for Φ, where $\boldsymbol{n}(\boldsymbol{x}, t)$ is the unit surface normal defined on the interfacial domain. In these equations, the variables to be solved for are $\boldsymbol{v}$ and C_α^*, but not $\boldsymbol{v}_f$ and C_α. The boundary conditions at the particle–solvent interface are implicitly taken into account in the SPM equations. The body force, $\phi \boldsymbol{f}_p$, is introduced to impose the particle velocity field onto the total velocity field. The nonslip condition for $\boldsymbol{v}$ is accounted for by the viscous force, which acts in both the fluid and interfacial domains. The no-penetration condition for the concentration field is directly accounted for by the projection operator $(\boldsymbol{I} - \boldsymbol{nn})$ in the interfacial domain. The expressions for $\phi \boldsymbol{f}_p$, $\boldsymbol{F}_i^H$, and $\boldsymbol{N}_i^H$ should be defined by the momentum conservation law between a particle and the solvent, which is consistently achieved in the time-discretized scheme described in the following subsection.

8.3.3 Temporal Discretization

The time-discretized evolution equations are outlined as follows. To simplify the presentation, we neglect the random currents and focus on an explicit marching algorithm. It is possible to formulate the implicit marching algorithm to enhance the stability, as presented in Reference 27. For the initial conditions at the nth discretized time step, the position, velocity, and angular velocity of the colloids, $\{\boldsymbol{R}_i^n, \boldsymbol{V}_i^n, \boldsymbol{\Omega}_i^n\}(i = 1, \ldots, N_p)$, are mapped onto ϕ^n and $\phi^n \boldsymbol{v}_p^n$, and the following conditions are imposed: $\boldsymbol{v}^n = (1-\phi^n)\boldsymbol{v}_f^n + \phi^n \boldsymbol{v}_p^n$ must satisfy the incompressibility condition on the entire domain, $\nabla \cdot \boldsymbol{v}^n = 0$, and $C_\alpha^{*,n}$ must satisfy the charge neutrality condition, $\int dx \rho_e^n = 0$. The auxiliary concentrations are advected by Equation 8.23 as

$$C_\alpha^{*,n+1} = C_\alpha^{*,n} - \int_{t_n}^{t_n+h} ds \nabla \cdot C_\alpha^* \boldsymbol{v}_\alpha^*, \tag{8.24}$$

where h is the time increment and $t_n = nh$ is the nth discretized time. The total velocity field is updated using a fractional step approach. First, the advection and the viscous force are solved:

$$\boldsymbol{v}^\dagger = \boldsymbol{v}^n + \int_{t_n}^{t_n+h} ds \nabla \cdot \left[\frac{1}{\rho} \{-p\boldsymbol{I} + \eta(\nabla \boldsymbol{v} + (\nabla \boldsymbol{v})^{\mathrm{T}})\} - \boldsymbol{vv} \right] \tag{8.25}$$

and

$$\boldsymbol{R}_i^{n+1} = \boldsymbol{R}_i^n + \int_{t_n}^{t_n+h} ds\, \boldsymbol{V}_i, \tag{8.26}$$

with the incompressibility condition $\nabla \cdot \boldsymbol{v}^{\dagger} = 0$. Along with the advection of the total velocity, the particle position is updated using the particle velocity. The electrostatic potential for the updated particle configuration is determined by solving the following Poisson equation,

$$\varepsilon_r \varepsilon_0 \nabla^2 \Phi^{n+1} = -\rho_e^{n+1}, \tag{8.27}$$

with the charge density field at the $(n + 1)$th time. The momentum change as a result of the electrostatic field is solved with

$$\boldsymbol{v}^{\ddagger} = \boldsymbol{v}^{\dagger} - h\rho_e^{n+1} \nabla \Phi^{n+1}. \tag{8.28}$$

At this point, the momentum conservation equations are solved for the total velocity field. The rest of the updating procedure assigns the particle velocity field.

The hydrodynamic force and torque exerted by the solvent on the colloids are derived from momentum conservation. Momentum conservation between the colloids and the fluid dictates that the hydrodynamic force and torque integrated over a time period, h, are equal to the momentum change over the particle domain during the time interval h:

$$\left[\int_{t_n}^{t_n+h} ds \boldsymbol{F}_i^H(s) \right] = \int dx \rho \phi_i^{n+1} \left(\boldsymbol{v}^{\ddagger} - \boldsymbol{v}_p^n \right) \tag{8.29}$$

and

$$\left[\int_{t_n}^{t_n+h} ds \boldsymbol{N}_i^H(s) \right] = \int dx \left[\boldsymbol{r}_i^{n+1} \times \rho \phi_i^{n+1} \left(\boldsymbol{v}^{\ddagger} - \boldsymbol{v}_p^n \right) \right]. \tag{8.30}$$

With this and the other forces on the colloids, the particle velocity and angular velocity are updated as follows:

$$\boldsymbol{V}_i^{n+1} = \boldsymbol{V}_i^n + M_p^{-1} \left[\int_{t_n}^{t_n+h} ds \boldsymbol{F}_i^H \right] + M_p^{-1} \int_{t_n}^{t_n+h} ds \left(\boldsymbol{F}_i^c + \boldsymbol{F}_i^{\text{ext}} \right) \tag{8.31}$$

and

$$\boldsymbol{\Omega}_i^{n+1} = \boldsymbol{\Omega}_i^n + \boldsymbol{I}_p^{-1} \cdot \left[\int_{t_n}^{t_n+h} ds \boldsymbol{N}_i^H \right] + \boldsymbol{I}_p^{-1} \cdot \int_{t_n}^{t_n+h} ds \boldsymbol{N}_i^{\text{ext}}. \tag{8.32}$$

The resultant particle velocity field $\phi^{n+1}\boldsymbol{v}_p^{n+1}$ is directly imposed on the total velocity field as

$$\boldsymbol{v}^{n+1} = \boldsymbol{v}^{\ddagger} + \left[\int_{t_n}^{t_n+h} ds\phi\boldsymbol{f}_p\right] \tag{8.33}$$

and

$$\left[\int_{t_n}^{t_n+h} ds\phi\boldsymbol{f}_p\right] = \phi^{n+1}\left(\boldsymbol{v}_p^{n+1} - \boldsymbol{v}^{\ddagger}\right) - \frac{h}{\rho}\nabla p_p. \tag{8.34}$$

The pressure due to the rigidity of the particle is determined by the incompressibility condition $\nabla\cdot\boldsymbol{v}^{n+1} = 0$, which leads to the following Poisson equation for p_p, viz.,

$$\nabla^2 p_p = \frac{\rho}{h}\nabla\cdot\left[\phi^{n+1}\left(\boldsymbol{v}_p^{n+1} - \boldsymbol{v}^{**}\right)\right].$$

We note that on the left-hand side of Equations 8.29, 8.30, and 8.34, the integrands $\boldsymbol{F}_i^H$, $\boldsymbol{N}_i^H$, and $\phi\boldsymbol{f}_p$ are not explicitly calculated, but their time integrals, namely, the momentum impulses, are solved. The body force $\phi\boldsymbol{f}_p$ formally introduced in the total momentum Equation 8.21 is realized in this time-discretized scheme. Furthermore, the hydrodynamic force and torque are defined as volume integrals rather than as surface integrals, which are suitable for spatial discretization on a regular grid. The SPM does not require a body-fitted irregular grid, which is the main source of difficulty and inefficiency in modeling many-colloid systems.

8.3.4 How to Choose an Interface Thickness, ξ

The SPM explicitly defines the interfacial domain with a thickness of ξ so that the moving boundary between the particle and the solvent can be handled on a particle-independent fixed grid. The interfacial domain should be resolved on the grid with a spacing of Δ, giving a lower bound to the quantity ξ/Δ. For the smoothed profile of Equation 8.15, at least $\xi/\Delta \geq 1$ is required. In electrokinetics problems, for approximate delta functions like $|\nabla\phi|$ to be resolved, $\xi/\Delta \geq 2$ is required. For an appropriately chosen small ξ/Δ, the SPM produces accurate hydrodynamic forces [13, 27, 29] in various problems.

8.3.5 Restrictions on the Time Increment

To enable spatial discretization of the hydrodynamic equations, any standard scheme can be used, such as the finite difference method, the finite volume method, the finite element method, the spectral method, LB discretization, and so forth. The SP method basically defines a coupling scheme between the hydrodynamic equations for the solvent and the equations for the discrete colloids with finite volume without requiring a body-fitted grid. Under spatial discretization with a typical spacing of Δ, some general restrictions on the time increment of the hydrodynamic equations should be satisfied to guarantee stable computation of the solvent hydrodynamics:

$$h < \frac{\Delta}{v},\ h < \frac{\Delta^2 \rho}{\eta},\ h < \frac{\Delta^2}{\max_\alpha D_\alpha}, \tag{8.35}$$

where v represents a characteristic velocity. The first inequality in Equation 8.35 is from the so-called Courant condition, which arises from the advection terms in Equations 8.21 and 8.22. The second and third inequalities are from velocity diffusion and solute diffusion, respectively. Typically, the diffusion conditions control the time increment in colloidal dispersion flow because the Reynolds number, $\rho v\Delta/\eta$, is much less than unity.

Another important restriction arises from the particle kinetics:

$$h < \frac{\Delta}{V}, \tag{8.36}$$

where V is the velocity of the colloidal particle and Δ/V is the Stokes time based on grid spacing Δ, which represents the time required for the particle to move a distance of Δ. The above-mentioned restrictions on h are common in direct numerical simulations of particulate flow and simply indicate that smaller values of h give better accuracy, assuming that there is enough precision in the finite arithmetic.

The SPM body force $\phi \boldsymbol{f}_p$ in Equation 8.21 imposes another restriction on h. In every time step, $\phi \boldsymbol{f}_p$ impulsively assigns a particle velocity in the particle domain. During the time step h, the effect of the momentum impulse diffuses around the particle domain in the range of $\sqrt{h\eta/\rho}$, whose domain is called the Stokes layer. To resolve the Stokes layer by $\phi \boldsymbol{f}_p$ in the interfacial domain, the relation

$$\sqrt{\frac{h\eta}{\rho}} > \text{cst.} \times \xi \tag{8.37}$$

is required, which places a lower bound on h. Detailed error analysis of the SPM in Reference 27 revealed that there is an optimal h around $\sqrt{h\eta/\rho}/\xi = O(1)$.

8.4 RESULTS AND DISCUSSION

8.4.1 One-Body HI in a Finite Volume Fraction

We apply the SPM to the sedimentation of spheres of radius a arranged in a periodic cubic lattice of length L. For a colloidal dispersion, the Reynolds number $Re = aV/\nu$ based on the velocity of a colloid is much less than unity, and thus the settling velocity V is proportional to the external force F:

$$F = fV. \tag{8.38}$$

For neutral colloids in a Newtonian solvent, the friction coefficient f depends on the volume fraction, $\varphi = 4\pi a^3/3L^3$, as follows:

$$f = \frac{6\pi\eta a}{Q(\varphi)}. \tag{8.39}$$

For a low volume fraction, $Q(\varphi)$ was derived by Hasimoto [30] as

$$Q(\varphi) = \frac{1}{1 - 1.7601\varphi^{1/3} + \varphi - 1.5593\varphi^2 + O(\varphi^{8/3})}, \tag{8.40}$$

and the analytic solution for arbitrary φ was given by Zick and Homsy [31]. The result from the SPM shown in Figure 8.1 is in agreement with the analytic solutions [26]. By applying the Stokes–Einstein relation, the contribution of

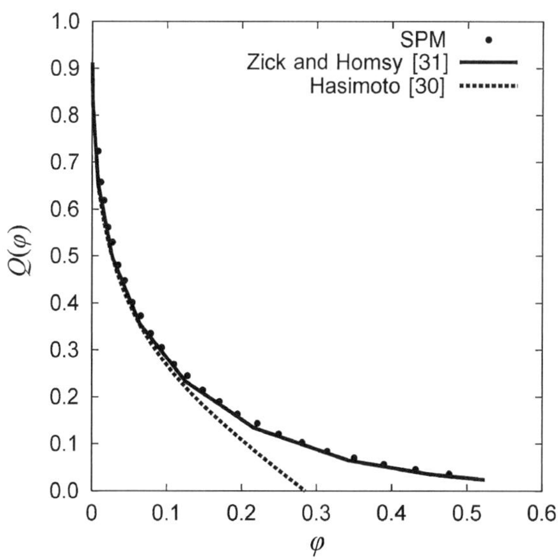

Figure 8.1. The inverse drag coefficient Q, defined in Equation 8.39, for a periodic array of spheres as a function of volume fraction φ. The result from SPM with $\xi = 1$ (lattice unit) is compared with the analytic result [31] (solid line) and low-φ asymptotics [30] (dashed line). Reproduced from Reference 26, with kind permission from Springer Science and Business Media.

the one-body HI to the diffusion coefficient of the colloid in a finite volume fraction becomes

$$D(\varphi) = \frac{k_B T}{f} = \frac{k_B T}{6\pi\eta a} Q(\varphi), \tag{8.41}$$

where $k_B T/6\pi\eta a$ is the diffusion coefficient at $\varphi \to 0$ and $Q(\varphi)$ is identified as the reduction factor of the diffusion coefficient compared to the infinite dilution. Figure 8.1 shows that $Q(\varphi)$ sharply decreases for low volume fractions. The friction coefficient is highly affected by the HI, as is the diffusion coefficient. Even at $\varphi = 0.1$, the diffusion coefficient is estimated to be about one-fourth of that at infinite dilution.

8.5 A SOFTWARE FOR ELECTROKINETICS OF COLLOIDAL DISPERSIONS: KAPSEL

For readers interested in electrokinetic simulations by SPM, we introduce a software called KAPSEL, which implements SPM by the authors and their coworkers. KAPSEL is available for free at http://www-tph.cheme.kyoto-u.ac.jp/kapsel/, in which some examples and movies are also available.

8.6. SUMMARY

In this chapter, we described the basic equations for colloidal particles in an electrolyte solution and the difficulties of direct simulation of many-colloid systems in traditional numerical schemes. To address this problem, an efficient and accurate numerical scheme, the SPM, was introduced. The SPM was developed by the authors and their coworkers and has been applied to a number of problems [13, 26, 29, 32–39].

Basically, the SPM is a framework for coupling rigid objects and continuum hydrodynamics. SPM is designed to accurately resolve the solid–fluid interaction and is suited for numerical simulations in which space and time can be discretized in several manners. Application of the SPM is not limited to electrolyte solutions and spherical colloids; it can be applied to any constitutive equation and to different colloidal particle shapes.

REFERENCES

1 W. B. Russel, D. A. Saville, W. R. Schowalter. *Colloidal dispersions*. Cambridge University Press, Cambridge, England, 1989.

2 R. F. Probstein. *Physicochemical hydrodynamics: an introduction*, 2nd ed. John Wiley & Sons, New York, 2003.

3 C. N. Likos. Effective interactions in soft condensed matter physics. *Phys. Rep.* 348 (2001) 267–439.

4 J. Happel, H. Brenner. *Low Reynolds number hydrodynamics: with special applications to particulate media*, 2nd ed. Martinus Nijhoff, Dordrecht, 1983.

5 S. Kim, S. J. Karrila. *Microhydrodynamics: principles and selected applications*. Butterworth-Heinemann, London, 1991.

6 J. F. Brady, G. Bossis. Stokesian dynamics. *Annu. Rev. Fluid Mech.* 20 (1988) 111–157.

7 H. H. Hu, N. A. Patankar, M. Y. Zhu. Direct numerical simulations of fluid–solid systems using the arbitrary Lagrangian–Eulerian technique. *J. Comput. Phys.* 169 (2001) 427–462.

8 R. Glowinski, T. W. Pan, T. I. Hesla, D. D. Joseph, J. Périaux. A fictitious domain approach to the direct numerical simulation of incompressible viscous flow past moving rigid bodies: application to particulate flow. *J. Comput. Phys.* 192 (2001) 363–426.

9 H. Tanaka, T. Araki. Simulation method of colloidal suspensions with hydrodynamic interactions: fluid particle dynamics. *Phys. Rev. Lett.* 85 (2000) 1338–1341.

10 S. Lomholt, M. R. Maxey. Force-coupling method for particulate two-phase flow: stokes flow. *J. Comput. Phys.* 184 (2003) 381–405.

11 N. Sharma, N. A. Patankar. A fast computation technique for the direct numerical simulation of rigid particulate flows. *J. Comput. Phys.* 205 (2005) 439–457.

12 C. Peskin. Flow patterns around heart valves: a numerical method. *J. Comput. Phys.* 10 (1972) 252–271.

13 Y. Nakayama, R. Yamamoto. Simulation method to resolve hydrodynamic interactions in colloidal dispersions. *Phys. Rev. E* 71 (2005) 036707.

14 S. Succi. *The lattice Boltzmann equation for fluid dynamics and beyond*. Clarendon Press, Oxford, 2001.

15 A. J. C. Ladd, R. Verberg. Lattice-Boltzmann simulations of particle fluid suspensions. *J. Stat. Phys.* 104 (2001) 1191–1251.

16 P. J. Hoogerbrugge, J. M. V. A. Koelman. Simulating microscopic hydrodynamic phenomena with dissipative particle dynamics. *Europhys. Lett.* 19 (1992) 155–160.

17 A. Malevanets, R. Kapral. Mesoscopic model for solvent dynamics. *J. Chem. Phys.* 110 (1999) 8605–8613.

18 J. Horbach, D. Frenkel. Lattice-Boltzmann method for the simulation of transport phenomena in charged colloids. *Phys. Rev. E* 64 (2001) 061507.

19 F. Capuani, I. Pagonabarraga, D. Frenkel. Discrete solution of the electrokinetic equations. *J. Chem. Phys.* 121 (2004) 973–986.

20 F. Capuani, I. Pagonabarraga, D. Frenkel. Lattice-Boltzmann simulation of the sedimentation of charged disks. *J. Chem. Phys.* 124 (2006) 124903.

21 V. Lobaskin, B. Dünweg, C. Holm. Electrophoretic mobility of a charged colloidal particle: a computer simulation study. *J. Phys.: Condens. Matter* 16 (2004) S4063–S4073.

22 A. Chatterji, J. Horbach. Combining molecular dynamics with lattice Boltzmann: a hybrid method for the simulation of (charged) colloidal systems. *J. Chem. Phys.* 122 (2005) 184903.

23 H. Kodama, K. Takeshita, T. Araki, H. Tanaka. Fluid particle dynamics simulation of charged colloidal suspensions. *J. Phys. Condens. Matter* 16 (2004) L115–L123.

24 T. Araki, H. Tanaka. Physical principle for optimizing electrophoretic separation of charged particles. *EPL* 82 (2008) 18004.

25 Y. W. Kim, R. R. Netz. Electroosmosis at inhomogeneous charged surfaces. *Europhys. Lett.* 72 (2005) 837–843.

26 Y. Nakayama, K. Kim, R. Yamamoto. Simulating (electro)hydrodynamic effects in colloidal dispersions: smoothed profile method. *Euro. Phys. J. E* 26(4) (2008) 361–368.

27 X. Luo, M. Maxey, G. Karniadakis. Smoothed profile method for particulate flows: error analysis and simulations. *J. Comput. Phys.* 228(5) (2009) 1750–1769.

28 L. D. Landau, E. M. Lifshitz. *Fluid mechanics*. Pergamon Press, London, 1959.

29 K. Kim, Y. Nakayama, R. Yamamoto. Direct numerical simulations of electrophoresis. *Phys. Rev. Lett.* 96 (2006) 208302.

30 H. Hasimoto. On the periodic fundamental solutions of the Stokes equations and their application to viscous flow past a cubic array of spheres. *J. Fluid Mech.* 5 (1959) 317–328.

31 A. A. Zick, G. M. Homsy. Stokes flow through periodic arrays of spheres. *J. Fluid Mech.* 115 (1982) 13–26.

32 R. Yamamoto. Simulating particle dispersions in nematic liquid-crystal solvents. *Phys. Rev. Lett.* 87 (2001) 075502.

33 R. Yamamoto, Y. Nakayama, K. Kim. A smooth interface method for simulating liquid crystal colloidal dispersions. *J. Phys. Condens. Matter* 16 (2004) S1945–S1955.

34 K. Kim, R. Yamamoto. Efficient simulations of charged colloidal suspensions: a density functional approach. *Macromol. Theory Simul.* 14 (2005) 278–284.

35 R. Yamamoto, K. Kim, Y. Nakayama, K. Miyazaki, D. R. Reichman. On the role of hydrodynamic interactions in colloidal gelation. *J. Phys. Soc. Jpn.* 77 (2008) 084804.

36 T. Iwashita, Y. Nakayama, R. Yamamoto. A numerical model for Brownian particles fluctuating in incompressible fluids. *J. Phys. Soc. Jpn.* 77 (2008) 074007.

37 T. Iwashita, Y. Nakayama, R. Yamamoto. Velocity autocorrelation function of fluctuating particles in incompressible fluids—Toward direct numerical simulation of particle dispersions. *Prog. Theor. Phys.* Suppl. 178 (2009) 86–91.

38 T. Iwashita, R. Yamamoto. Short-time motion of Brownian particles in a shear flow. *Phys. Rev. E* 79 (2009) 031401.

39 T. Iwashita, R. Yamamoto. Direct numerical simulations for non-Newtonian rheology of concentrated particle dispersions. *Phys. Rev. E* 80 (2009) 061402.

9

COMPUTER SIMULATIONS OF CHARGED COLLOIDS

2. ELECTROPHORESIS AND SEDIMENTATION

KANG KIM, YASUYA NAKAYAMA, AND RYOICHI YAMAMOTO

9.1 INTRODUCTION

Electrohydrodynamic phenomena are of great importance in the physical, chemical, and biological sciences, and also in several engineering fields [1]. For example, in the case of electrophoresis of charged particles, the particles start to move upon the application of external electric fields. The electric double layer (EDL), that is, the cloud of counterions around charged particles, tends to be deformed and its distribution tends to be an isotropic because of the applied external field and also because of the friction between the ions and the fluid. The electrophoretic mobility of a single colloidal particle is then determined by balancing the electrostatic driving force and the hydrodynamic frictional force acting on the particle. In this situation, the time evolutions of the states of the colloidal particles, the ions, and the host fluids are described by coupled hydrodynamic (Navier–Stokes) and electrostatic (Poisson) equations, with proper boundary conditions imposed at the surfaces of the colloidal particles. However, the usual numerical techniques for solving partial differential equations cannot deal with the dynamic evolution of many-particle

Electrical Phenomena at Interfaces and Biointerfaces: Fundamentals and Applications in Nano-, Bio-, and Environmental Sciences, First Edition. Edited by Hiroyuki Ohshima.

systems as the shifting particle–fluid boundary conditions must be considered at each discrete time step.

In this chapter, to demonstrate the reliability of the smoothed profile method (SPM) [2–4], we present results of numerical simulations of electrophoretic motions of charged colloids under an external electric field. Our results are also compared with those of classical approximation theories [5–8] with respect to the relationship between the zeta potential and the electrophoretic mobility. Next, comparisons are made for the electrophoretic mobilities of dense dispersions, and the simulation results show notable deviations from a mean field-type theory such as the cell model [9, 10]. In addition, electrohydrodynamic effects are investigated in the absence of external electric fields. The electroviscous drag on a charged colloid and the lubrication interaction between two approaching charged colloids in an electrolyte solution are computed. Unlike the drag on a neutral colloid, the valence of the colloid and the thickness of the EDL affect the electroviscous drag and the lubrication interaction.

9.2 NUMERICAL CALCULATIONS

Simulations were performed in a three-dimensional cubic box with periodic boundary conditions. The linear dimension is $L/\Delta = 64$, where the lattice spacing Δ is chosen as the unit of length and is defined relative to the Bjerrum length as $\Delta = 4\pi\lambda_B$. We use a particle radius of $a = 5$ and a thickness parameter of $\xi = 2$ throughout these simulations. The host fluid contains a 1:1 electrolyte composed of monovalent counterions ($\alpha = +$) and coions ($\alpha = -$). The units of energy and electrostatic potential are k_BT and k_BT/e, respectively. The latter corresponds to 25.7 mV at 25°C. The nondimensional parameter $m_\alpha = 2\varepsilon k_B T f_\alpha / 3\eta e^2$ is set to $m_+ = m_- = 0.184$, which corresponds to the value of KCl solution at 25°C. Our unit of time $\tau = \Delta^2 f_+ / k_B T$ corresponds to 0.44 μs.

9.2.1 Electrophoretic Mobility and ζ Potential

We first consider a single charged particle moving with the drift velocity $V = (-V, 0, 0)$ in a constant electric field $E = (E, 0, 0)$. The electrophoretic mobility V/E is related to the zeta potential ζ, which is defined as the electrostatic potential on a slipping plane, as

$$V/E = f\varepsilon\zeta/\eta, \tag{9.1}$$

when ζ is small [1]. The Smolchowski equation $f = 1$ is valid in the limit $\kappa a \rightarrow \infty$ [5], while the Hückel equation $f = 2/3$ is valid in the opposite limit $\kappa a \rightarrow 0$ [6]. Henry equation derived an expression, $f = f_H(\kappa a)$, that is valid for a general value of κa [7]. These equations indicate that the mobility is proportional to ζ; however, this relation tends to fail for larger ζ, when the relaxation effect due to deformations of the EDL becomes notable. O'Brien and White proposed an approximation theory that is also valid for larger ζ [8].

We performed simulations of electrophoresis of a single particle in linear response regimes, $E \leq 0.15$, and compared them with results obtained based on the O'Brien–White theory. A constant, uniform electric field $E = 0.1$ corresponding to 2.85×10^3 V/cm was applied. The terminal V was calculated for $50 \leq -Z \leq 750$ with $\kappa^{-1} = 10$ and $100 \leq -Z \leq 1000$ with $\kappa^{-1} = 5$. The corresponding salt concentrations were 44 μmol/L for $\kappa^{-1} = 5$ and 11 μmol/L for $\kappa^{-1} = 10$ at 25°C in water. We chose $\nu = \eta/\rho = 5$, so the Reynolds number $\mathrm{Re} = aV/\nu$ remains small. In both the O'Brien–White theory and our simulations, the zeta potential is assumed to be the electrostatic potential at the particle surface, $\zeta = \Psi|_{\mathrm{surface}}$. Here, the dimensionless zeta potential $y \equiv e\zeta/k_B T$ is introduced. The relationship between the surface charge $|Z|e$ and the dimensionless zeta potential y is shown in Figure 9.1a, where our numerical results are plotted along with the analytic solution to the nonlinear Poisson–Boltzmann (PB) equation [11]. We see that our numerical results are consistent with the nonlinear PB theory for $\kappa a = 0.5$ and 1.0. In Figure 9.1b, the dimensionless mobility $E_m \equiv e\eta V/\varepsilon k_B TE$ is plotted as a function of the dimensionless zeta potential for $\kappa a = 0.5$ and 1.0. It is clearly demonstrated that our method reproduced the predictions of the O'Brien–White theory almost perfectly, including the nonlinear regime $y \geq 2$, with only a few percent error. We emphasize that such precise agreement with theory has never been obtained by any proposed simulation method. The distributions of the charge density due to the counterions and the coions are shown in Figure 9.1c for $Z = -100$ and in Figure 9.1d for $Z = -500$ with $\kappa a = 0.5$. The corresponding zeta potential is $y = 1.044$ for $Z = -100$ and $y = 3.692$ for $Z = -500$. It can be seen that the EDL is considerably deformed in the nonlinear regime (Fig. 9.1d), while it is almost isotropic in the linear regime (Fig. 9.1c). As mentioned before, the relaxation effect due to the deformed EDL causes the nonlinearity in E_m.

9.2.2 Electrophoretic Mobility in Concentrated Dispersions

Our simulation method is easily applicable to dense dispersions consisting of many particles, so we examined the effect of the particle concentration on the electrophoretic mobility. The linearized theory for a single particle Equation 9.1 is still valid for dense dispersions when E is small; however, f depends on both κa and φ in this regime. Simulations were carried out with $Z = -100$ and $E = 0.1$ for various particle volume fractions $\varphi \equiv 4\pi a^3 N/3L^3$ to calculate $f(\kappa a, \varphi) = \eta V/\varepsilon\zeta E$. We used Debye lengths κ^{-1} of 5 and 10, which correspond to $\kappa a = 1.0$ and 0.5, respectively. Figure 9.2 shows typical snapshots of the systems for (i) face-centered cubic (FCC), (ii) body-centered cubic (BCC), and (iii) random configurations. The horizontal color map represents the charge density for $\kappa a = 1$ on a cross-section perpendicular to the z-axis. In the FCC and BCC cases, E was applied perpendicular to both the (1, 0, 0) and (1, 1, 1) faces, and we obtained very small differences, within 1%.

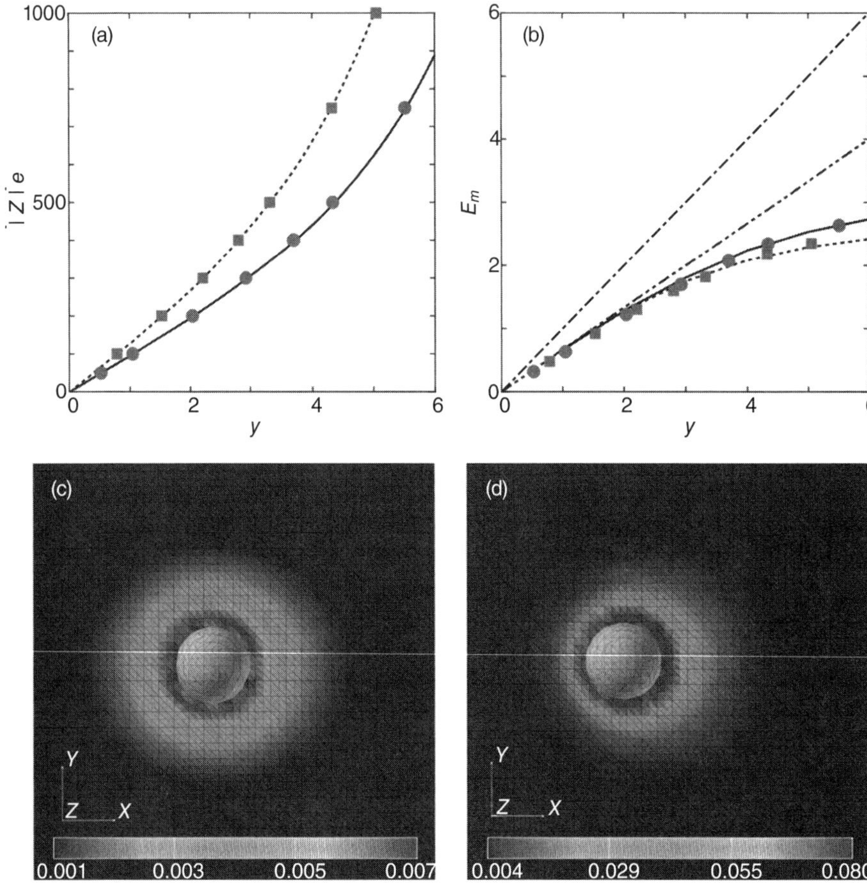

Figure 9.1. (a) Relationship between the surface charge $|Z|e$ and the dimensionless zeta potential y for $\kappa a = 0.5$ (circles) and $\kappa a = 1.0$ (squares). Our numerical data are in good agreement with the analytic solution of the nonlinear PB equation for $\kappa a = 0.5$ (solid curve) and $\kappa a = 1.0$ (dashed curve) [11]. (b) The dimensionless mobility E_m of a single particle is plotted as a function of the dimensionless zeta potential y for $\kappa a = 0.5$ (circles) and $\kappa a = 1.0$ (squares). For comparison, the results of Smoluchowski (dot-dashed line), Henry (two dot-dashed lines), and O'Brien–White for $\kappa a = 0.5$ (solid curve) as well as those of O'Brien–White for $\kappa a = 1.0$ (dashed curve) are shown. The color contours in (c) and (d) represent the total ionic charge density $\Sigma_\alpha eZ_\alpha C_\alpha$ around a single particle for (c) $Z = -100$ and (d) $Z = -500$ with $\kappa a = 0.5$. The electric field is applied in the horizontal ($+x$) direction.

The mobility coefficient $f(\kappa a, \varphi)$ for $\kappa a = 1.0$ and 0.5 is plotted as a function of φ in Figure 9.3a,b. We found that f decreases rapidly with increasing φ. Furthermore, the overall behavior looks almost independent of the particle configuration. A theoretical model for the electrophoretic mobility of dense dispersions has been proposed by Levine and Neale using the cell model [9].

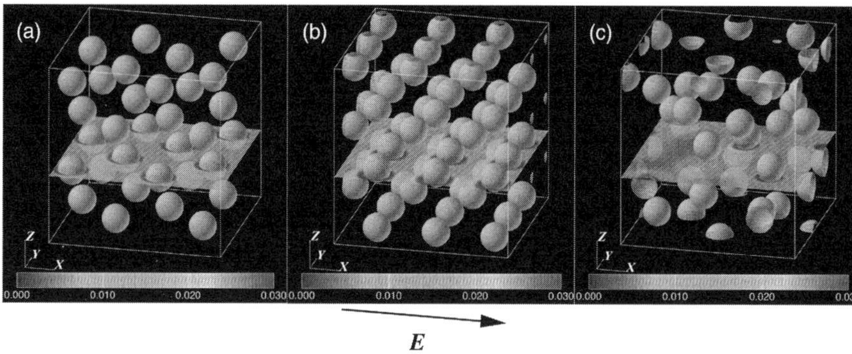

Figure 9.2. Snapshots of the electrophoresis of dense dispersions with (a) face-centered cubic (FCC), (b) body-centered cubic (BCC), and (c) random particle configurations. The color map represents the total ionic charge density $\Sigma_\alpha eZ_\alpha C_\alpha$ in a plane perpendicular to the z-axis. The electric field is applied in the $+x$ direction, normal to the (1, 0, 0) face for FCC and BCC.

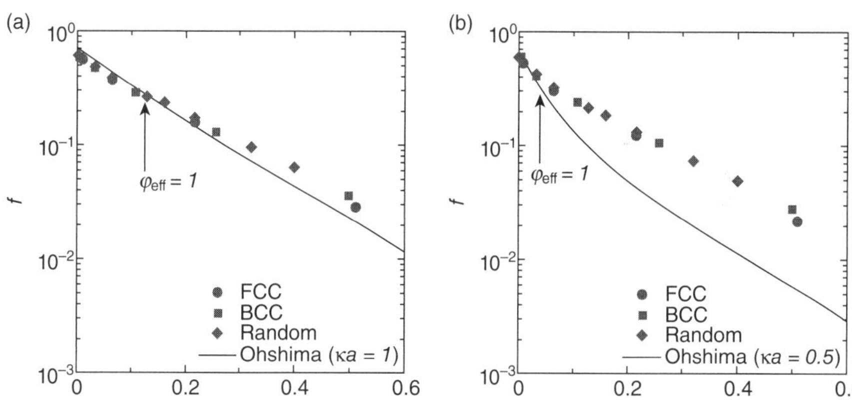

Figure 9.3. The mobility coefficient $f(\kappa a, \varphi)$ as a function of the volume fraction φ for (a) $\kappa a = 1.0$ and (b) $\kappa a = 0.5$. The solid lines represent the approximation theory proposed by Ohshima [10]. The theory is confirmed to be accurate for $\varphi_{\text{eff}} \leq 1$; however, it tends to deviate from our numerical results for $\varphi_{\text{eff}} > 1$, where overlapping of the EDLs becomes notable.

They assumed a situation in which a spherical particle with radius a is located at the center of a spherical container (cell) with radius b and calculated V as a function of κa and $\varphi = (a/b)^3$. Ohshima proposed a simpler expression for the mobility coefficient f according to the cell model [10]. Ohshima's prediction is shown in Figure 9.3a,b along with our numerical results. The overall agreement between our simulation and Ohshima's theory is better in Figure 9.3a, with a smaller Debye length $\kappa^{-1} = 5 = a$, than in Figure 9.3b, with a larger one $\kappa^{-1} = 10$. In Figure 9.3a,b, the simulation results tend to become larger than the theoretical values as φ increases. We expect that the deviation arises from the occurrence of overlapping EDLs for larger φ because this effect is totally neglected in the theory. To this end, we estimated the effective radius $a + \kappa^{-1}$ of the ionically dressed particles and defined the effective volume fraction $\varphi_{\text{eff}} \equiv 4\pi(a + \kappa^{-1})^3 N/3L^3 = (1 + (\kappa a)^{-1})^3\varphi$. As clearly shown in Figure 9.3a,b, our results agree well with Ohshima's theory for $\varphi_{\text{eff}} \leq 1$, where the effect of overlapping is small. However, for $\varphi_{\text{eff}} > 1$, where the overlapping of the EDLs becomes large, deviations between our simulations and the theory become notable. We emphasize that the present study produced the first successful simulations to provide the quantitative data necessary to examine the reliability of Ohshima's cell model calculations, including their boundary conditions, for electrophoresis in dense colloidal dispersions. Our results are consistent with recent studies that also accounted for the effects of overlapping EDLs [12–14].

9.2.3 One-Body Electrohydrodynamic Interaction: Diffusion Coefficient of A Charged Colloid

The SPM is applied to the sedimentation of charged spheres of radius a arranged in a periodic cubic lattice of length L [3]. In colloidal dispersions, the terminal velocity V of a colloid is proportional to the drag force, $F = fV$. For a charged colloid in electrolyte solution, the friction coefficient f is supposed to be larger than that of a neutral colloid due to the friction between the ions and the solvent. Therefore, the friction coefficient depends not only on the volume fraction, $\varphi = 4\pi a^3/3L^3$ as for a neutral colloid, but also on the Debye length and the valence of the colloid (or the zeta potential) as

$$f = \frac{6\pi\eta a}{Q(\varphi, \kappa a, Z)}, \tag{9.2}$$

where a 1:1 electrolyte is considered to be the solvent. The friction coefficient is related to the diffusion coefficient via the Stokes–Einstein relation as

$$D(\varphi, \kappa a, Z) = \frac{k_B T}{6\pi\eta a} Q(\varphi, \kappa a, Z). \tag{9.3}$$

Figure 9.4 shows the diffusion coefficient of a charged sphere. In this case, the nondimensional ion friction coefficients of the cation and the anion are arbitrarily set to $k_B T/(6\pi\eta D_\pm l_B) = 2\varepsilon k_B T f_\pm/[3\eta e^2] = 1.333$. For the limits of the Debye length $\kappa a \to 0$ and $\kappa a \to \infty$, the electroviscous drag approaches that of a neutral colloid, irrespective of Z. For a finite κa, the electroviscous drag is larger than that of a neutral colloid. In other words, the diffusion constant of

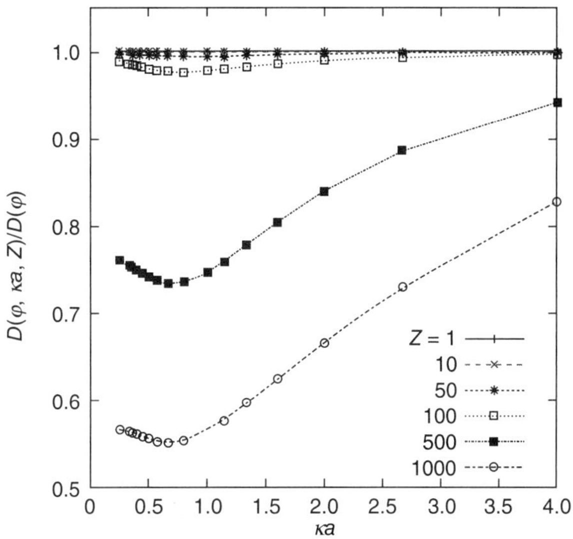

Figure 9.4. Diffusion coefficients of charged colloids of different valences, Z, at a volume fraction of $\varphi = 0.008$ computed from the friction coefficient of a sphere arranged in a periodic cubic lattice via the Stokes–Einstein relation. The ordinate is normalized to the diffusion coefficient of a neutral colloid at the same volume fraction. The lines are guides for the eye.

a charged colloid is smaller than that of a neutral colloid. This electroviscous effect on the drag is maximized at a certain value of the Debye length, $\kappa a = O(1)$. These qualitative aspects have also been indicated for the analytic solution at infinite dilution [15, 16] and by experimental results [17].

These findings are explained by considering a small deformation of the EDL. Ion friction mainly occurs in the domain with a gradient of ion concentration $|\nabla C_\alpha| > 0$ and not near the surface of the solid, as the slip velocity of the ions develops at a distance from the solid surface. This domain is the region within the EDL. In the thin double-layer limit ($\kappa a \to \infty$), the size of the EDL becomes zero and the electroviscous effect is suppressed. In the other limit of $\kappa a \to 0$, the EDL is extended to the whole solvent domain, and the gradient of the ion concentration vanishes. Thus, the source of the electroviscous effect vanishes. For a finite EDL, flow perturbation by the solid surface deforms the EDL and causes the electroviscous effect on the drag.

9.2.4 Two-Body Electrohydrodynamic Interaction: Pair Lubrication Interaction

The lubrication interaction is a hydrodynamic frictional interaction between nearby particles undergoing relative motion. In a Newtonian solvent, the friction coefficient diverges at a rate inversely proportional to the gap between surfaces, h, as

$$F = -\left(\Gamma_1 + \frac{\Gamma_2}{h}\right)\Delta V, \tag{9.4}$$

$$\Gamma_1 = 6\pi\eta a \gamma_1(\varphi), \tag{9.5}$$

and

$$\Gamma_2 = \frac{3\pi\eta a^2}{2}\gamma_2(\varphi), \tag{9.6}$$

where ΔV is the relative velocity, F is the drag force, and γ_1, γ_2 represent the volume fraction effect and satisfy $\gamma_1(0) = \gamma_2(0) = 1$. The SPM was applied to this problem and the volume fraction dependence of the friction coefficients was determined [3].

For charged colloids, the coupling between the EDL and hydrodynamic interactions affects the lubrication interaction, though it has often been taken into account independently in prior mesoscopic models. Figure 9.5 shows the effect of the Debye length on the pair lubrication interaction between two colloids of $Z = -500$, which, as shown in Figure 9.1, gives a rather higher zeta potential for which the linear ζ–$|Z|$ relationship does not hold. For a thin EDL

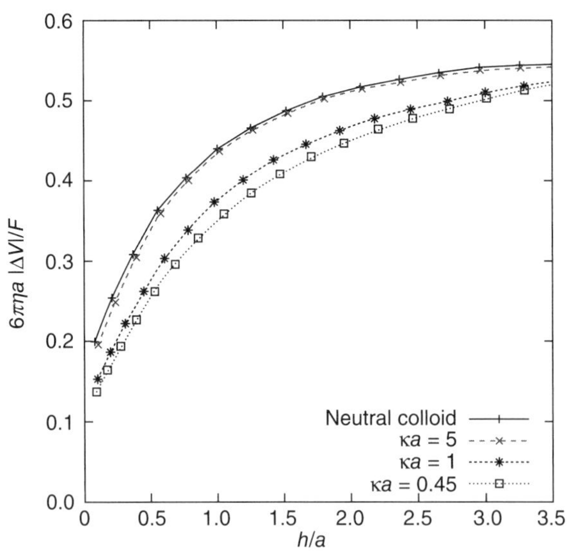

Figure 9.5. Mobility between two approaching spheres versus the gap between the sphere surfaces at a volume fraction of 0.016 for charged colloids with a valence of $Z = -500$.

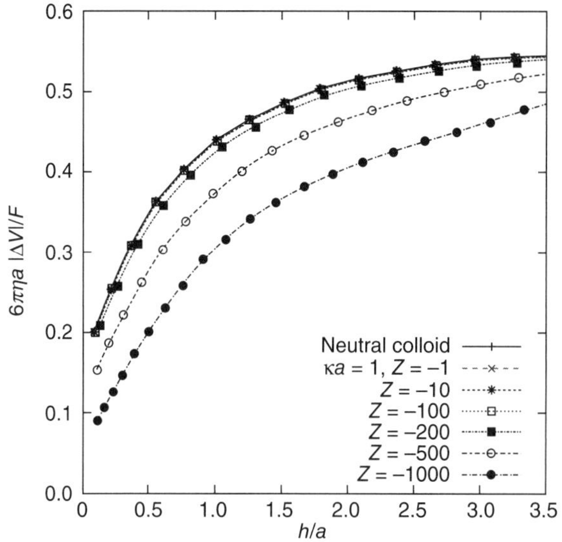

Figure 9.6. The same plot as Figure 9.6 but for different charged colloid valences at $\kappa a = 1$.

of $\kappa a = 5$, the mobility is close to that of neutral colloids. However, as the Debye length becomes larger, the mobility becomes smaller.

Figure 9.6 shows the effect of the valence of colloids on the pair lubrication interaction at a Debye length of $\kappa a = 1$. For low values of the zeta potential ($|Z| \leq 200$), the reduction of the mobility compared to a neutral colloid is small. In contrast, for higher values of the zeta potential ($|Z| \geq 500$), a larger reduction of the mobility is observed. The pair lubrication interaction is highly dependent on both the zeta potential and the Debye length. These results suggest that the electrohydrodynamic interaction has large effects on the dynamic behaviors of charged colloidal dispersions.

9.3 SUMMARY

In summary, we have developed a unique numerical method for simulating electrohydrodynamic phenomena in colloidal dispersions. The method was first applied to simulate the electrophoresis of a single spherical particle, and we found that our method can quantitatively reproduce the reliable analytic theory proposed by O'Brien and White. Simulations were then performed for the electrophoresis of colloids in dense dispersions, and we compared our results with a theoretical analysis based on the cell model. We found that the cell model is reliable when overlapping of EDLs is small, but becomes less reliable as overlapping becomes important. Furthermore, we applied our method to the diffusion coefficients of sedimenting charged colloids and the lubrication interaction between two charged colloids. We found the electrohydrodynamic coupling of the momentum transfer leads to differences between the solvent flow pattern around the colloid and that around a neutral colloid.

For electrokinetic phenomena, solvent-mediated electrohydrodynamic interactions are important for understanding the dynamic processes of colloids. Direct simulation allows us to access the basic physical processes in colloidal dynamics and will be utilized to obtain constitutive relations for both inhomogeneous particle dispersions and particle–particle interactions to allow mesoscopic modeling.

REFERENCES

1 W. B. Russel, D. A. Saville, W. R. Schowalter. *Colloidal dispersions*. Cambridge University Press, Cambridge, 1989.

2 Y. Nakayama, R. Yamamoto. Simulation method to resolve hydrodynamic interactions in colloidal dispersions. *Phys. Rev. E* 71 (2005) 036707.

3 Y. Nakayama, K. Kim, R. Yamamoto. Simulating (electro)hydrodynamic effects in colloidal dispersions: smoothed profile method. *Eur. Phys. J. E* 26 (2008) 361.

4 K. Kim, Y. Nakayama, R. Yamamoto. Direct numerical simulations of electrophoresis of charged colloids. *Phys. Rev. Lett.* 96 (2006) 208302.

5 M. von Smoluchowski. Mathematical theory of the kinetics of the coagulation of colloidal solutions. *Z Phys. Chem.* 92 (1918) 129.

6 E. Hückel. The cataphoresis of the sphere. *Phys. Z* 25 (1924) 204.

7 D. C. Henry. The cataphoresis of suspended particles Part I—The equation of cataphoresis. *Proc. R. Soc. London Ser. A* 133 (1931) 106.

8 R. W. O'Brien, L. R. White. Electrophoretic mobility of a spherical colloidal particle. *J. Chem. Soc. Faraday Trans. 2* 74 (1978) 1607.

9 S. Levine, G. H. Neale. The prediction of electrokinetic phenomena within multiparticle systems. I. Electrophoresis and electroosmosis. *J. Colloid Interface Sci.* 47 (1974) 520.

10 H. Ohshima. Electrophoretic mobility of spherical colloidal particles in concentrated suspensions. *J. Colloid Interface Sci.* 188 (1997) 481.

11 H. Ohshima, T. W. Healy, L. R. White. Accurate analytic expressions for the surface charge density/surface potential relationship and double-layer potential distribution for a spherical colloidal particle. *J. Colloid Interface Sci.* 90 (1982) 17.

12 F. Carrique, F. J. Arroyo, M. L. Jiménez, Á. V. Delgado. Influence of double-layer overlap on the electrophoretic mobility and DC conductivity of a concentrated suspension of spherical particles. *J. Phys. Chem. B* 107 (2003) 3199.

13 T. Palberg, M. Medebach, N. Garbow, M. Evers, A. B. Fontecha, H. Reiber, E. Bartsch. Electrophoresis of model colloidal spheres in low salt aqueous suspension. *J. Phys. Condens. Matter* 16 (2004) S4039.

14 V. Lobaskin, B. Dünweg, M. Medebach, T. Palberg, C. Holm. Electrophoresis of colloidal dispersions in the low-salt regime. *Phys. Rev. Lett.* 98 (2006) 176105.

15 F. Booth. Sedimentation potential and velocity of solid spherical particles. *J. Chem. Phys.* 22 (1954) 1956–1968.

16 H. Ohshima, T. W. Healy, L. R. White, R. W. O'Brien. Sedimentation velocity and potential in a dilute suspension of charged spherical colloidal particles. *J. Chem. Soc. Faraday Trans.* 2(80) (1984) 1299–1317.

17 G. A. Schumacher, T. G. M. vande Ven. Brownian motion of charged colloidal particles surrounded by electric double layers. *Faraday Discuss. Chem. Soc.* 83 (1987) 75–85.

10

ELECTROSTATIC AND STERIC STABILIZATION OF COLLOIDAL DISPERSIONS

THARWAT TADROS

10.1 INTRODUCTION

Colloidal dispersions occur in many industrially manufactured products, for example, paints, printing inks, paper coatings, cosmetics, agrochemicals, and pharmaceuticals. Stabilizing these dispersions against aggregation during their preparation by nucleation and growth, or by subdivision of preformed particles by dispersion and milling in aqueous or nonaqueous media is crucial. It is also important to maintain colloid stability on storage for up to 2 years under different conditions such as temperature changes and during transportation. To achieve the stability of these dispersions, it is essential to create an effective repulsion mechanism that is sufficient to overcome the ever present van der Waals attraction between the particles (see below). There are generally two types of repulsion mechanisms: (i) Electrostatic repulsion is produced by creating an electrical double layer around the particles (see below) as a result of charge separation. When two particles with extended double layers (that are produced at low electrolyte concentrations) approach each other to a separation distance whereby the double layers begin to overlap, strong repulsion occurs and this can overcome the van der Waals attraction (see below). (ii) Steric repulsion is produced by adsorbed nonionic surfactant or polymer layers. The latter can extend from the particle surfaces giving an adsorbed

Electrical Phenomena at Interfaces and Biointerfaces: Fundamentals and Applications in Nano-, Bio-, and Environmental Sciences, First Edition. Edited by Hiroyuki Ohshima.

layer thickness, δ. When two particles approach to a distance of separation h that is smaller than 2δ, the adsorbed layers begin to overlap, resulting in very strong repulsion as a result of the unfavorable mixing of the adsorbed chains (when these are in good solvent conditions) and loss of configurational entropy of the chains on significant overlap. This steric repulsion is very effective in stabilizing colloidal dispersions (see below) against any aggregation. Sometimes, a combination of electrostatic and steric repulsion may occur, for example, when using mixtures of ionic surfactants and nonionic polymers or when using polyelectrolytes. This stabilization mechanism is referred to as electrosteric.

In this chapter, I will discuss the various interaction forces existing between colloidal particles that are the main forces controlling dispersion stability. This is followed by a discussion of electrostatic and steric stabilization and other factors that control the stability of colloidal dispersions. A brief description will be given of electrosteric stabilization. The last part of the chapter will deal with the flocculation of colloidal dispersions and the factors that determine this process.

10.2 INTERACTION FORCES BETWEEN PARTICLES IN COLLOIDAL DISPERSIONS

To control the stability of colloidal dispersions, one has to control the interparticle interactions [1]. Four different types of interactions can be distinguished as illustrated in Figure 10.1 and these are summarized next.

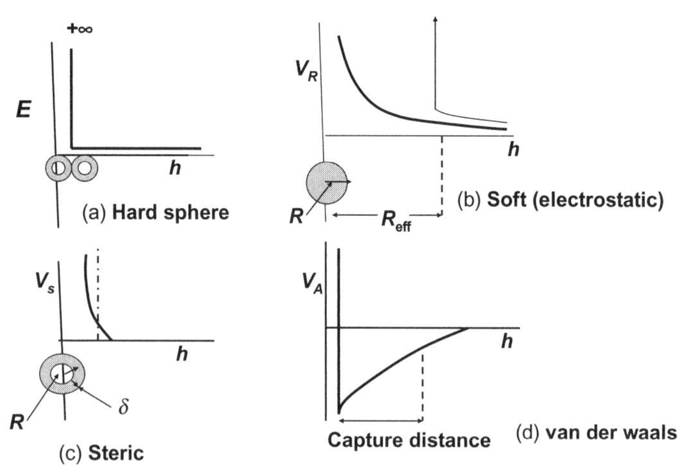

Figure 10.1. Types of interaction forces.

10.2.1 Hard-Sphere Interaction (Figure 10.1a)

The particles are considered to behave as "hard spheres" with a radius R_{HS} that is slightly larger than the core radius R. When the particles reach a center-to-center distance that is smaller than 2 R_{HS}, the interaction increases very sharply, approaching ∞. These hard-sphere dispersions are seldom encountered in practice and they are sometimes prepared by screening the electrostatic interaction (using a less polar medium than water, e.g., benzyl alcohol) for fundamental studies.

10.2.2 "Soft" or Electrostatic Interaction (Figure 10.1b)

The particles in this case have a surface charge, either by ionization of surface groups as in the case of oxides or in the presence of adsorbed ionic surfactants. The surface charge σ_o is compensated by an unequal distribution of counterions (opposite in charge to the surface) and coions (same sign as the surface) that extend to some distance from the surface [2]. This is schematically represented in Figure 10.2.

The potential decays exponentially with distance x. At low potentials,

$$\psi = \psi_o \exp -(\kappa x). \tag{10.1}$$

Note that when $x = 1/\kappa$, $\psi_x = \psi_o/e$; $1/\kappa$ is referred to as the "thickness" of the double layer.

The double-layer extension depends on the electrolyte concentration and valency of the counterions:

$$\left(\frac{1}{\kappa}\right) = \left(\frac{\varepsilon_r \varepsilon_o kT}{2n_o Z_i^2 e^2}\right)^{1/2}, \tag{10.2}$$

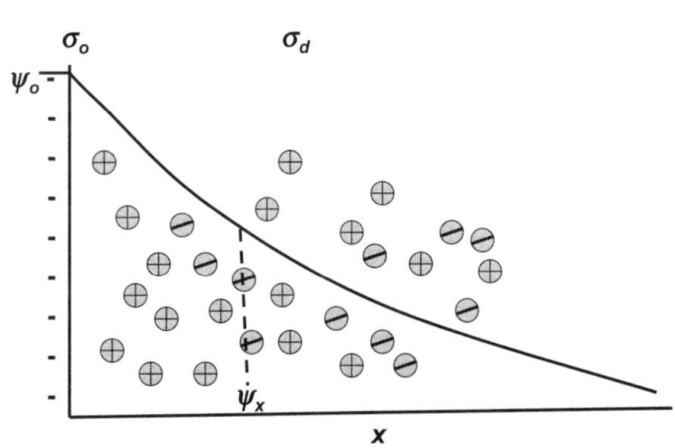

Figure 10.2. Schematic representation of the diffuse double layer according to Gouy and Chapman [2].

TABLE 10.1. Values of (1/κ) for 1:1 Electrolyte at 25°C

$C/mol/dm^3$	10^{-5}	10^{-4}	10^{-3}	10^{-2}	10^{-1}
$(1/\kappa)/nm$	100	33	10	3.3	1

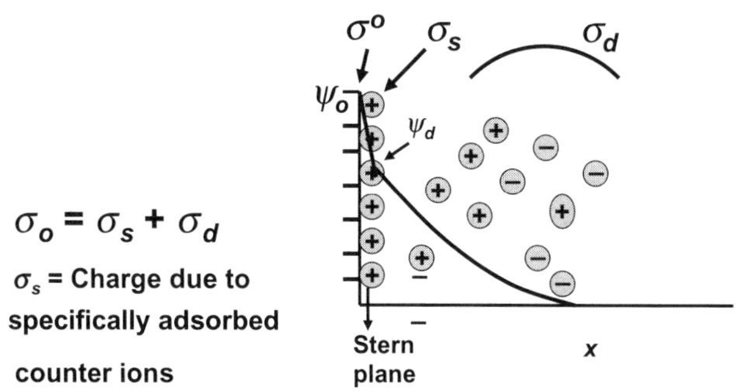

Figure 10.3. Schematic representation of the double layer according to Stern [3] and Grahame (cited in [4]).

where ε_r is the permittivity (dielectric constant), 78.6 for water at 25°C; ε_o is the permittivity of free space; k is the Boltzmann constant; T is the absolute temperature; n_o is the number of ions per unit volume of each type present in the bulk solution; Z_i is the valency of the ions; and e is the electronic charge.

Values of $(1/\kappa)$ for a 1:1 electrolyte (e.g., KCl) are given in Table 10.1.

The double-layer extension increases with a decrease in electrolyte concentration.

Stern [3] introduced the concept of the nondiffuse part of the double layer for specifically adsorbed ions, the rest being diffuse in nature. This is schematically illustrated in Figure 10.3

The potential drops linearly in the Stern region and then exponentially. Grahame distinguished two types of ions in the Stern plane: physically adsorbed counterions (outer Helmholtz plane) and chemically adsorbed ions (that lose part of their hydration shell) (inner Helmholtz plane).

When charged colloidal particles in a dispersion approach each other such that the double layers begin to overlap (particle separation becomes less than twice the double-layer extension), repulsion occurs. The individual double layers can no longer develop unrestrictedly since the limited space does not allow complete potential decay [3].

This is illustrated in Figure 10.4 for two flat plates.

The potential $\psi_{H/2}$ halfway between the plates is no longer zero (as would be the case for isolated particles at $x \to \infty$).

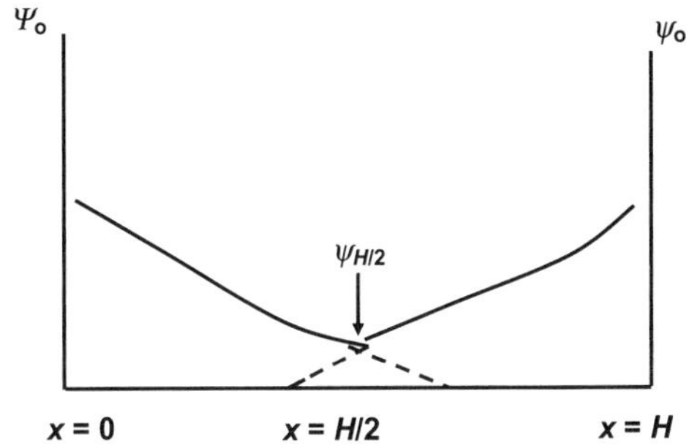

Figure 10.4. Schematic representation of double-layer interaction for two flat plates.

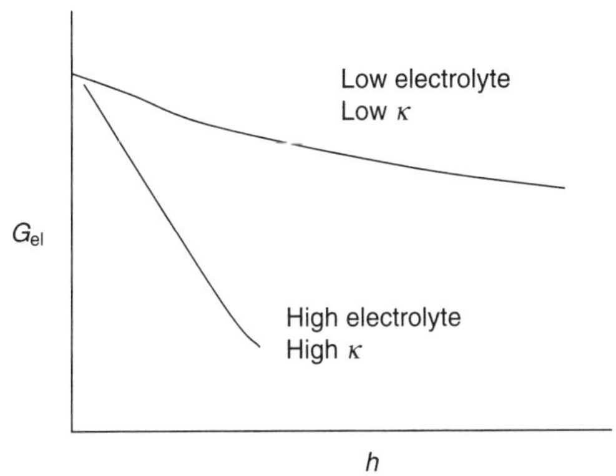

Figure 10.5. Variation of G_{el} with h at different electrolyte concentrations.

For two spherical particles of radius R, surface potential ψ_o, and condition $\kappa R < 3$, the expression for the electrical double-layer repulsive interaction is given by [4]

$$G_{el} = \frac{4\pi\varepsilon_r\varepsilon_o R^2 \psi_o^2 \exp -(\kappa h)}{2R+h}, \tag{10.3}$$

where h is the closest distance of separation between the surfaces.

The above-mentioned expression shows the exponential decay of G_{el} with h. The higher the value of κ (i.e., the higher the electrolyte concentration), the steeper the decay, as schematically shown in Figure 10.5. This means that at any given distance h, the double-layer repulsion decreases with an increase in electrolyte concentration.

10.2.3 Steric Interaction (Figure 10.1c)

This occurs when the particles contain adsorbed nonioinic surfactant or polymer layers of the A-B, A-B-A block, or BA_n graft types, where B is the "anchor" chain that has a high affinity to the surface (strongly adsorbed) and A is the "stabilizing" chain that is highly soluble in the medium and is strongly solvated by its molecules. One can define an adsorbed layer thickness δ_h for the surfactant or polymer and hence an effective radius $R_{\text{eff}} = R + \delta_h$.

When two particles each with a radius R and containing an adsorbed polymer layer with a hydrodynamic thickness δ_h approach each other to a surface–surface separation distance h that is smaller than 2 δ_h, the polymer layers interact with each other, with the result that (i) the polymer chains may overlap with each other or (ii) the polymer layer may undergo some compression [5]. In both cases, there will be an increase in the local segment density of the polymer chains in the interaction region. This is schematically illustrated in Figure 10.6. In real situations, the polymer chains may undergo some interpenetration and some compression.

Provided the dangling chains (the A chains in A-B, A-B-A block, or BAn graft copolymers) are in a good solvent, this local increase in segment density in the interaction zone will result in strong repulsion as a result of two main effects: (i) One effect is an increase in the osmotic pressure in the overlap region as a result of the unfavorable mixing of the polymer chains when these are in good solvent conditions. This is referred to as osmotic repulsion or mixing interaction and is described by a free energy of interaction, G_{mix}. (ii) The other effect is the reduction of the configurational entropy of the chains in the interaction zone; this entropy reduction results from an decrease in the volume available for the chains when these are either overlapped or compressed. This is referred to as a volume restriction, entropic or elastic interaction, and it is described by a free energy of interaction, G_{el}.

The combination of G_{mix} and G_{el} is usually referred to as the steric interaction free energy G_s; that is,

$$G_s = G_{\text{mix}} + G_{\text{el}}. \tag{10.4}$$

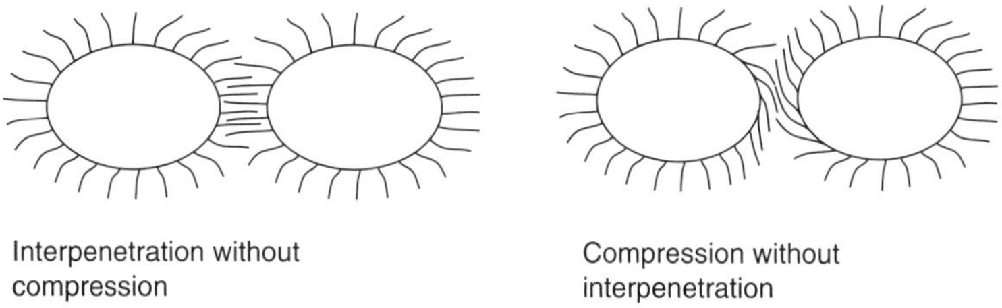

Figure 10.6. Schematic representation of the interaction between particles containing adsorbed polymer layers.

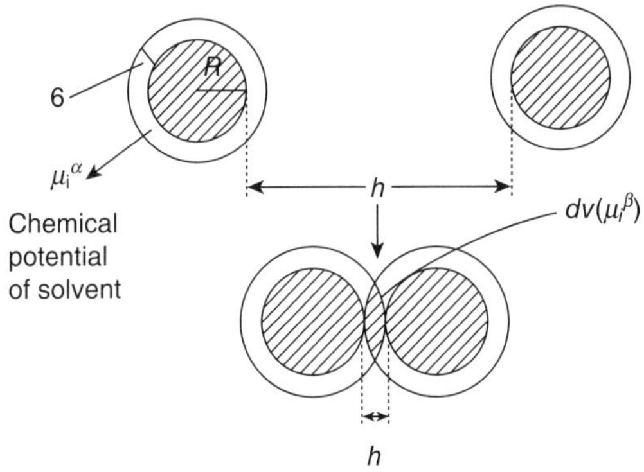

Figure 10.7. Schematic representation of a polymer layer overlap.

The sign of G_{mix} depends on the solvency of the medium for the chains. If in a good solvent, that is, where the Flory–Huggins interaction parameter χ is less than 0.5, then G_{mix} is positive and the mixing interaction leads to repulsion (see below). In contrast, if $\chi > 0.5$ (i.e., the chains are in a poor solvent condition), G_{mix} is negative and the mixing interaction becomes attractive. G_{el} is always positive, and hence in some cases, one can produce stable dispersions in a relatively poor solvent (enhanced steric stabilization).

As mentioned earlier, the mixing interaction results from the unfavorable mixing of the polymer chains when these are in good solvent conditions. This is schematically shown in Figure 10.7.

Consider two spherical particles with the same radius and each containing an adsorbed polymer layer with thickness δ. Before overlap, one can define in each polymer layer a chemical potential for the solvent μ_i^α and a volume fraction for the polymer in the layer ϕ_2.

In the overlap region (volume element dV), the chemical potential of the solvent is reduced to μ_i^β. This results from the increase in polymer segment concentration in this overlap region.

In the overlap region, a strong repulsive energy arises because the chemical potential of the polymer chains is now higher than in the rest of the layer (with no overlap). This amounts to an increase in the osmotic pressure in the overlap region. As a result, the solvent will diffuse from the bulk to the overlap region, thus separating the particles. The repulsive energy can be calculated by considering the free energy of mixing of two polymer solutions, as, for example, treated by Flory and Krigbaum [6]. The free energy of mixing is given by two terms: (i) an entropy term that depends on the volume fraction of polymer and solvent and (ii) an energy term that is determined by the Flory–Huggins interaction parameter χ.

Using the above-mentioned theory and assuming a uniform segment density distribution in each layer, one can derive an expression for the free energy of mixing of two polymer layers surrounding two spherical particles as a function of the separation distance h between the particles [4]. The expression for G_{mix} is

$$\frac{G_{mix}}{kT} = \left(\frac{2V_2^2}{V_1}\right) v_2^2 \left(\frac{1}{2} - \chi\right)\left(\delta - \frac{h}{2}\right)^2 \left(3R + 2\delta + \frac{h}{2}\right), \tag{10.5}$$

where k is the Boltzmann constant, T is the absolute temperature, V_2 is the molar volume of polymer, V_1 is the molar volume of solvent, and v_2 is the number of polymer chains per unit area.

The sign of G_{mix} depends on the value of the Flory–Huggins interaction parameter χ: If $\chi < 0.5$, G_{mix} is positive and the interaction is repulsive; if $\chi > 0.5$, G_{mix} is negative and the interaction is attractive; if $\chi = 0.5$, $G_{mix} = 0$ and this defines the θ-condition.

The elastic interaction arises from the loss in configurational entropy of the chains upon the approach of a second particle. As a result of this approach, the volume available for the chains becomes restricted, resulting in a loss in the number of configurations. This can be illustrated by considering a simple molecule, represented by a rod that rotates freely in a hemisphere across a surface (Fig. 10.8). When the two surfaces are separated by an infinite distance ∞, the number of configurations of the rod is $\Omega(\infty)$, which is proportional to the volume of the hemisphere. When a second particle approaches to a distance h such that it cuts the hemisphere (losing some volume), the volume available to the chains is reduced and the number of configurations becomes $\Omega(h)$, which is less than $\Omega(\infty)$.

For two flat plates, G_{el} is given by the following expression [5]:

$$\frac{G_{el}}{kT} = 2v_2 \ln\left[\frac{\Omega(h)}{\Omega(\infty)}\right] = 2v_2 R_{el}(h), \tag{10.6}$$

where $R_{el}(h)$ is a geometric function whose form depends on the segment density distribution. It should be stressed that G_{el} is always positive and could

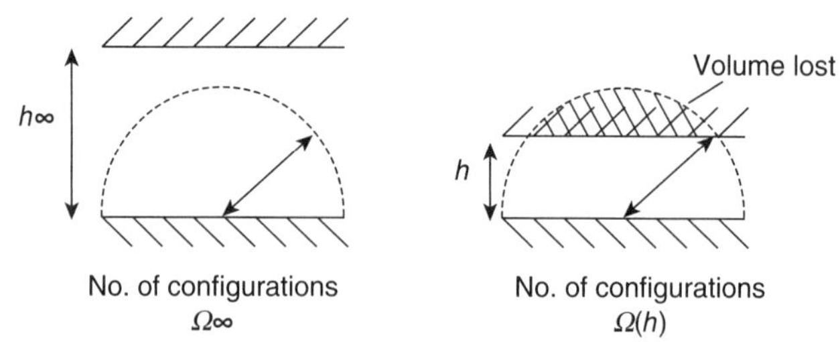

Figure 10.8. Schematic representation of configurational entropy loss on the approach of a second particle.

play a major role in steric stabilization. It becomes very strong when the separation distance between the particles becomes comparable to the adsorbed layer thickness δ.

10.2.4 van der Waals Attraction (Figure 10.1d)

As is well-known, atoms or molecules always attract each other at short distances of separation. The attractive forces are of three different types: dipole–dipole interaction (Keesom), dipole-induced dipole interaction (Debye), and London dispersion force. The London dispersion force is the most important since it occurs for polar and nonpolar molecules. It arises from fluctuations in the electron density distribution.

At small distances of separation r in vacuum, the attractive energy between two atoms or molecules is given by

$$G_{\mathrm{aa}} = -\frac{\beta_{11}}{r^6}, \tag{10.7}$$

where β_{11} is the London dispersion constant.

For colloidal particles that are made of atom or molecular assemblies, the attractive energies may be added, and this results in the following expression for two spheres (at small h values) [7]:

$$G_{\mathrm{A}} = -\frac{A_{11(2)}R}{12h}, \tag{10.8}$$

where $A_{11(2)}$ is the effective Hamaker constant of two identical particles with Hamaker constant A_{11} in a medium with Hamaker constant A_{22}. When the particles are dispersed in a liquid medium, the van der Waals attraction has to be modified to take into account the effect of the medium. When two particles are brought from an infinite distance to h in a medium, an equivalent amount of medium has to be transported the other way round. Hamaker forces in a medium are excess forces.

The effective Hamaker constant for two identical particles 1 and 1 in medium 2 is given by

$$A_{11(2)} = A_{11} + A_{22} - 2A_{12} = (A_{11}^{1/2} - A_{22}^{1/2})^2. \tag{10.9}$$

Equation 10.8 shows that two particles of the same material attract each other unless their Hamaker constants exactly match each other. The Hamaker constant of any material is given by

$$A = \pi q^2 \beta_{ii}, \tag{10.10}$$

where q is number of atoms or molecules per unit volume.

As shown in Figure 10.1d, V_A increases very sharply with h at small h values. A capture distance can be defined at which all the particles become strongly attracted to each other (coagulation).

10.3 ELECTROSTATIC STABILIZATION

The combination of G_{el} and G_A follows the well-known Derjaguin–Landau–Verwey–Overbeek (DLVO) theory of colloid stability [8, 9]:

$$G_T = G_{el} + G_A. \tag{10.11}$$

A plot of G_T versus h is shown in Figure 10.9, which represents the case at low electrolyte concentrations, where there exists strong electrostatic repulsion between the particles. G_{el} decays exponentially with h; that is, $G_{el} \rightarrow 0$ as h becomes large. G_A is ∞ $1/h$, meaning that G_A does not decay to 0 at large h values.

At long distances of separation, $G_A > G_{el}$, resulting in a shallow minimum (secondary minimum). At very short distances, $G_A >> G_{el}$, resulting in a deep primary minimum.

At intermediate distances, $G_{el} > G_A$, resulting in energy maximum G_{max}, whose height depends on ψ_o (or ψ_d) and the electrolyte concentration and valency.

At low electrolyte concentrations ($<10^{-2}$ mol/dm^3 for a 1:1 electrolyte), G_{max} is high (>25 kT) and this prevents particle aggregation into the primary minimum. The higher the electrolyte concentration (and the higher the valency of the ions), the lower the energy maximum.

Under some conditions (depending on electrolyte concentration and particle size), flocculation into the secondary minimum may occur. This flocculation is weak and reversible. By increasing the electrolyte concentration, G_{max} decreases until at a given concentration it vanishes and particle coagulation

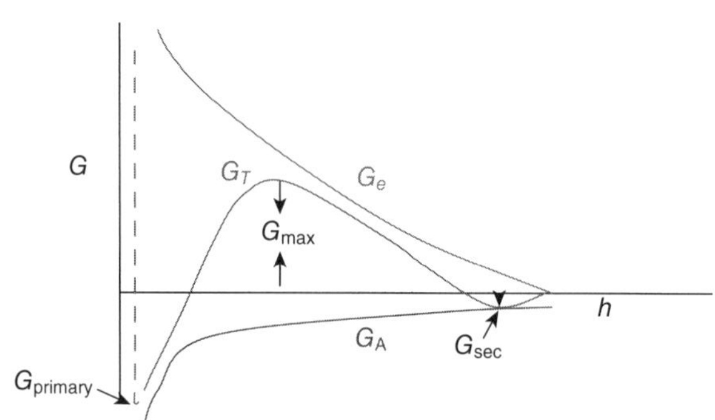

Figure 10.9. Schematic representation of the variation of G_T with h according to the DLVO theory.

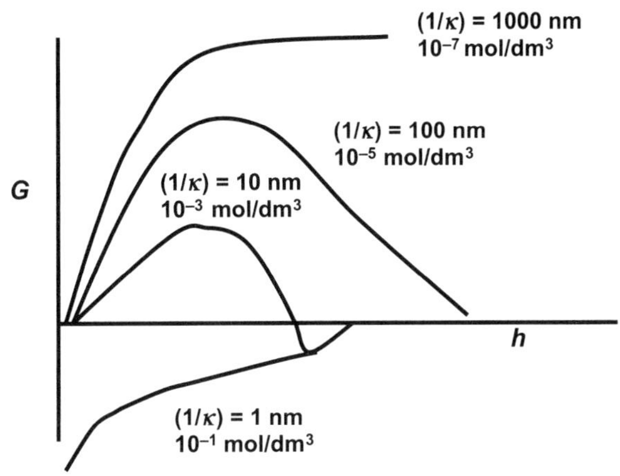

Figure 10.10. Variation of G with h at various electrolyte concentrations.

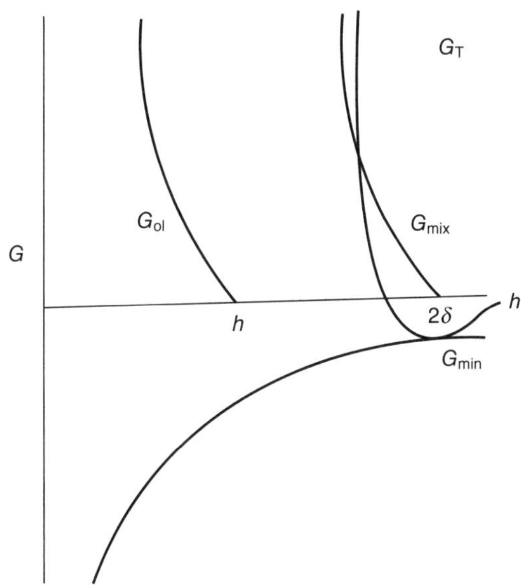

Figure 10.11. Energy–distance curves for sterically stabilized systems.

occurs. This is illustrated in Figure 10.10, which shows the variation of G_T with h at various electrolyte concentrations.

10.4 STERIC STABILIZATION

The combination of G_{mix} and G_{el} with G_A (the van der Waals attractive energy) gives the total free energy of interaction G_T (assuming there is no contribution from any residual electrostatic interaction) (Fig. 10.11); that is,

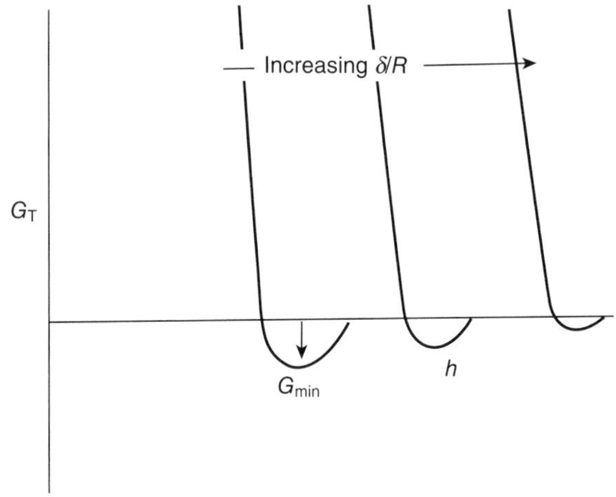

Figure 10.12. Variation of G_T with h at various δ/R values.

$$G_T = G_{mix} + G_{el} + G_A. \tag{10.12}$$

A schematic representation of the variation of G_{mix}, G_{el}, G_A, and G_T with surface–surface separation distance h is shown in Figure 10.12.

G_{mix} increases very sharply with a decrease in h when $h < 2\delta$. G_{el} increases very sharply with a decrease in h when $h < \delta$. G_T versus h shows a minimum, G_{min}, at separation distances comparable to 2δ. When $h < 2\delta$, G_T shows a rapid increase with a decrease in h.

The depth of the minimum depends on the Hamaker constant A, the particle radius R, and the adsorbed layer thickness δ. G_{min} increases with an increase in A and R. At a given A and R, G_{min} increases with a decrease in δ (i.e., with a decrease of the molecular weight, M_w, of the stabilizer). This is illustrated in Figure 10.12, which shows the energy–distance curves as a function of δ/R. The larger the value of δ/R, the smaller the value of G_{min}. In this case, the system may approach thermodynamic stability as is the case with nanodispersions.

10.5 ELECTROSTERIC STABILIZATION

The combination of electrostatic repulsion, steric repulsion, and van der Waals attraction is referred to as electrosteric stabilization. This is the case when using a mixture of ionic and nonionic stabilizers or when using polyelectrolytes. In this case, the energy–distance curve has two minima, one shallow maximum (corresponding to the DLVO type) and a rapid increase at small h corresponding to steric repulsion. This is illustrated in Figure 10.13.

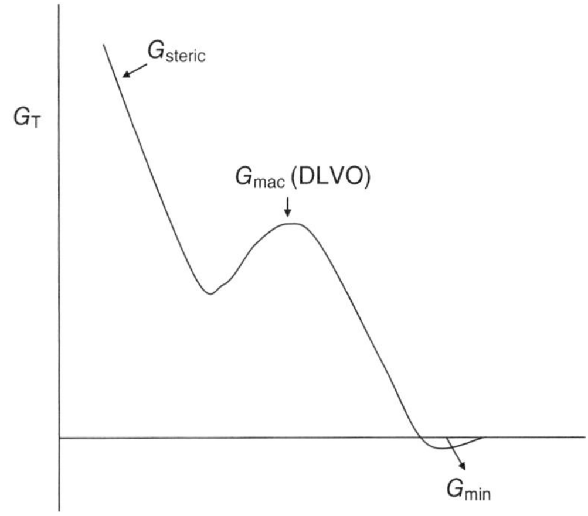

Figure 10.13. Energy–distance curve for electrosteric stabilization.

10.6 FLOCCULATION OF DISPERSIONS AND ITS PREVENTION

Flocculation is the result of van der Waals attraction, which is universal for all disperse systems. The van der Waals attraction G_A was described before. G_A is inversely proportional to the particle–particle distance of separation h and depends on the effective Hamaker constant A of the suspension. One way to overcome the van der Waals attraction is by electrostatic stabilization using ionic surfactants, which results in the formation of electrical double layers that introduce a repulsive energy that overcomes the attractive energy. Suspensions stabilized by electrostatic repulsion become flocculated at intermediate electrolyte concentrations (see below).

The second and most effective method of overcoming flocculation is by "steric stabilization" using nonionic surfactants or polymers. Stability may be maintained in electrolyte solutions (as high as 1 mol/dm^3, depending on the nature of the electrolyte) and up to high temperatures (in excess of 50°C) provided the stabilizing chains (e.g., polyethylene oxide [PEO]) are still in better than θ-conditions ($\chi < 0.5$).

10.7 MECHANISM OF FLOCCULATION

This can occur if the energy barrier is small or absent (for electrostatically stabilized dispersions) or when the stabilizing chains reach poor solvency (for sterically stabilized dispersions, i.e., $\chi > 0.5$). For convenience, I will discuss flocculation of electrostatically and sterically stabilized suspensions separately.

10.7.1 Flocculation of Electrostatically Stabilized Suspensions

As discussed before, the condition for kinetic stability is $G_{max} > 25\ kT$. When $G_{max} < 5\ kT$, flocculation occurs.

Two types of flocculation kinetics may be distinguished: fast flocculation, when no energy barrier exists, and slow flocculation, when an energy barrier exists. Fast flocculation kinetics was treated by Smoluchowski [10], who considered the process to be represented by second-order kinetics, and the process is simply diffusion controlled.

The number of particles n at any time t may be related to the initial number (at $t = 0$) n_o:

$$n = \frac{n_o}{1 + kn_o t}, \tag{10.13}$$

where k is the rate constant for fast flocculation that is related to the diffusion coefficient of the particles, D:

$$k = 8\pi DR, \tag{10.14}$$

and D is given by the Stokes–Einstein equation:

$$D = \frac{kT}{6\pi\eta R}. \tag{10.15}$$

Combining Equations 10.14 and 10.15,

$$k = \frac{4}{3}\frac{kT}{\eta} = 5.5 \times 10^{-18}\ \text{m}^3/\text{s for water at 25°C}. \tag{10.16}$$

The half-life $t_{1/2}$ ($n = (1/2)\,n_o$) can be calculated at various n_o or volume fraction ϕ as given in Table 10.2.

Slow flocculation kinetics was treated by Fuchs [11], who related the rate constant k to the Smoluchowski rate by the stability ratio W:

$$W = \frac{k_o}{k}. \tag{10.17}$$

TABLE 10.2. Half-Life of Dispersion Flocculation

R/μm	$\phi = 10^{-5}$	$\phi = 10^{-2}$	$\phi = 10^{-1}$	$\phi = 5 \times 10^{-1}$
0.1	765 seconds	76 ms	7.6 ms	1.5 ms
1.0	21 hours	76 seconds	7.6 seconds	1.5 seconds
10.0	4 months	21 hours	2 hours	25 m

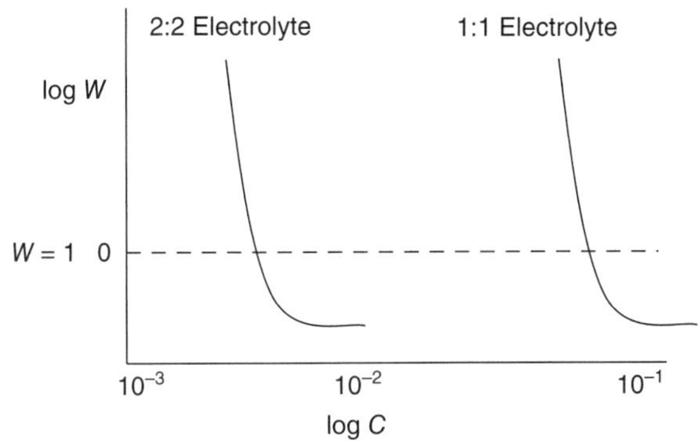

Figure 10.14. log W–log C curves.

W is related to G_{max} by the following expression [12]:

$$W = \frac{1}{2}\exp\left(\frac{G_{max}}{kT}\right). \tag{10.18}$$

Since G_{max} is determined by the salt concentration C and valency, one can derive an expression relating W to C and Z:

$$\log W = -2.06\times 10^{9}\left(\frac{R\gamma^2}{Z^2}\right)\log C, \tag{10.19}$$

where γ is a function that is determined by the surface potential ψ_o:

$$\gamma = \left[\frac{\exp(Ze\psi_o/kT)-1}{\exp(ZE\psi_o/kT)+1}\right]. \tag{10.20}$$

Plots of log W versus log C are shown in Figure 10.14 for 1:1 and 2:2 electrolytes.

The condition log $W = 0$ ($W = 1$) is the onset of fast flocculation. The electrolyte concentration at this point defines the critical flocculation concentration (CFC). Above the CFC, $W < 1$ due to the contribution of van der Waals attraction, which accelerates the rate above the Smoluchowski value. Below the CFC, $W > 1$, and it increases with a decrease in electrolyte concentration. Figure 10.14 also shows that the CFC decreases with an increase of valency in accordance to the Schultze–Hardy rule.

10.7.1.1 Weak (Reversible) Flocculation. Another mechanism of flocculation is that involving the secondary minimum (G_{min}), which represents only a few kT units. In this case, flocculation is weak and reversible, and hence one must consider both the rate of flocculation (forward rate k_f) and deflocculation (backward rate k_b). The rate of decrease in particle number with time is given by the expression

$$-\frac{dn}{dt} = -k_f n^2 + k_b n. \tag{10.21}$$

The backward reaction (breakup of weak flocs) reduces the overall rate of flocculation.

10.7.1.2 Orthokinetic (Shear-Induced) Flocculation. Another process of flocculation occurs under shearing conditions and is referred to as orthokinetic (to distinguish it from the diffusion-controlled perikinetic process). In this case, the rate of flocculation is related to the shear rate γ by the expression

$$-\frac{dn}{dt} = \frac{16}{3}\alpha^2 \dot{\gamma} R^3, \tag{10.22}$$

where α is the frequency of collisions that result in permanent aggregates.

10.7.2 Flocculation of Sterically Stabilized Dispersions

Flocculation can be weak and reversible or strong and irreversible depending on the magnitude of the attractive energy. Weak flocculation can occur when the thickness of the adsorbed layer becomes small such that the attractive minimum (G_{min}) in the energy–distance curve reaches several kT units. Strong flocculation occurs when the solvency of the medium for the chains reaches the θ-point (i.e., the Flory–Huggins interaction parameter $\chi > 0.5$). This situation results in catastrophic flocculation, referred to as incipient flocculation.

10.8 WEAK FLOCCULATION

The $G_T - h$ for systems that are sterically stabilized shows only one minimum, G_{min}, followed by sharp increase in G_T with a decrease in h (when $h < 2\delta$). The depth of the minimum depends on the Hamaker constant A, the particle radius R, and the adsorbed layer thickness δ; G_{min} increases with an increase in A and R. At a given A and R, G_{min} increases with a decrease in δ (i.e., with decrease of the molecular weight M_w of the stabilizer). This was illustrated in Figure 10.12, which shows the energy–distance curves as a function of δ/R. The smaller the value of δ/R, the larger the value of G_{min}. When G_{min} reaches a critical value that depends on the volume fraction of the suspension, weak flocculation occurs. The above flocculation is reversible, and when the suspension is gently stirred, the particles become dispersed. On removal of the shear, flocculation occurs.

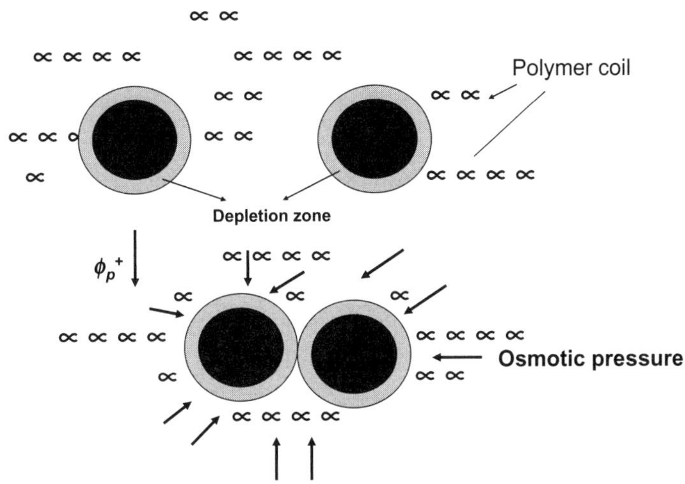

Figure 10.15. Schematic representation of depletion flocculation.

10.9 DEPLETION FLOCCULATION

This results when using "free" (nonadsorbing) polymer [13, 14]. When the concentration of the polymer exceeds a certain critical concentration (that depends on the molecular weight of the polymer), the chains become "squeezed out" from between the particles. As a result, the osmotic pressure outside the particle surfaces is higher than in between the particles, and this results in weak attraction. A schematic representation of depletion flocculation is given in Figure 10.15.

10.10 INCIPIENT FLOCCULATION

This occurs when the solvency of the medium for the chain becomes worse than a θ-solvent ($\chi > 0.5$). Under these conditions, G_{mix} becomes negative, that is, attractive, and a deep minimum is produced, resulting in catastrophic flocculation (referred to as incipient flocculation). This is schematically represented in Figure 10.16.

With many systems, good correlation between the flocculation point and the θ-point is obtained. For example, the suspension will flocculate at a temperature referred to as the critical flocculation temperature (CFT), which is equal to the θ-temperature of the stabilizing chain. The suspension may flocculate at a critical volume fraction of a nonsolvent (CFV), which is equal to the volume of nonsolvent that changes it to a θ-solvent.

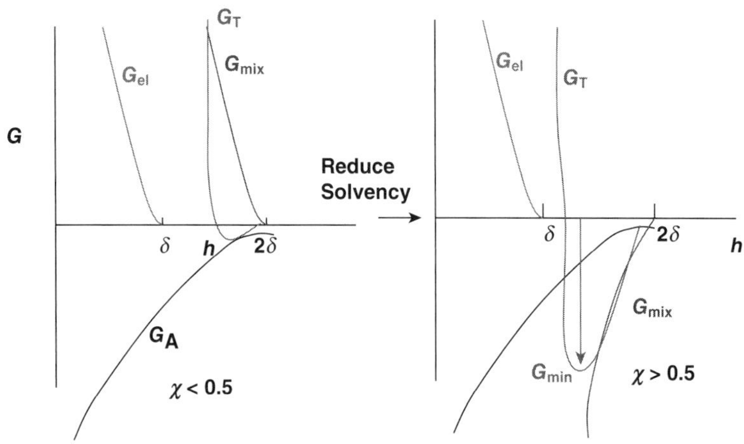

Figure 10.16. Schematic representation of flocculation of sterically stabilized emulsions.

10.11 BRIDGING OR CHARGE NEUTRALIZATION BY POLYMERS

Bridging flocculation occurs under conditions of incomplete coverage of the particles by the polymer chains [15]. Segments of a polymer chain adsorb on different particles, thus linking them together. Adsorption is an essential step and this requires some favorable interaction between polymer segments and particle surface. Typical polymeric flocculants have molecular dimensions comparable to the size of colloidal particles (0.1–1.0 μm). The flocs produced by polymer bridging are considerably stronger than aggregates formed by the coagulation of dispersions by electrolytes.

With polyelectrolytes, the picture is more complicated since the dimensions of the polyelectrolyte chains depend on the ionic strength. At low ionic strength, screening of the charges on the chains is limited and the polyelectrolyte adopts an extended configuration, which makes bridging more likely. At high ionic strength, screening of the charge produces a more compact configuration, thus reducing the chance of bridging. However, increasing the electrolyte concentration also results in reduction in the range of interparticle repulsion, and hence compact chains could still bridge the particles.

At high electrolyte concentration, adsorption of the polyelectrolyte with the same charge as that of the particle surface is enhanced, and this also increases the likelihood of bridging.

Polyelectrolytes with a charge opposite to that of the particles are very effective in the flocculation of dispersions. With oppositely charged polyelectrolytes, it is likely that adsorption gives a rather flat adsorbed chain configuration as a result of the strong electrostatic interaction between the ionic groups on the polyelectrolyte chain and charge sites on the surface. This probably reduces the probability of bridging. The electrostatic patch model can explain the flocculation by polyelectrolytes. Patches of excess positive charge corresponding to the adsorbed polyelectrolyte chains (possibly in a flat configuration) are surrounded by areas of negative charge, representing the original particle surface. The particles having this patchy or mosaic-type surface may interact in such a way that positive and negative patches come in contact, resulting in a quite strong attachment (although not a strong as that yielded by bridging flocculation).

10.12 GENERAL RULES FOR REDUCING (ELIMINATING) FLOCCULATION

10.12.1 Charge Stabilized Suspensions (e.g., Using Ionic Surfactants)

The most important criterion is to make G_{max} as high as possible. The condition for high kinetic stability is $G_{max} > 25\ kT$. This is achieved under three main conditions: high surface or zeta potential, low electrolyte concentration, and low valency of ions.

10.12.2 Sterically Stabilized Suspensions

Four main criteria are necessary:

1. Complete coverage of the particles by the stabilizing chains
2. Firm attachment (strong anchoring) of the chains to the particles or droplets. Firm attachment requires the chains to be insoluble in the medium and to have strong affinity to the surface (Fig. 10.16). However, this is incompatible with stabilization, which requires a chain that is soluble in the medium and strongly solvated by its molecules. These conflicting requirements are resolved by the use of A-B, A-B-A block, or BA_n graft copolymers (B is the anchor chain and A is the stabilizing chain). Examples of B chains for hydrophobic particles are polystyrene, polymethylmethacrylate, polypropylene oxide, and alkyl polypropylene oxide. For the A chains, PEO or polyvinyl alcohol are good examples.
3. Thick adsorbed layers; that is, adsorbed layer thickness should be in the region of 5–10 nm. This means that the molecular weight of the stabilizing chains could be in the region of 1000–5000.
4. The stabilizing chain should be maintained in good solvent conditions ($\chi < 0.5$) under all conditions of temperature change upon storage.

REFERENCES

1 Th. F. Tadros. *Adv. Colloids Interface Sci.* 68 (1996) 97.

2 (a) G. Gouy. *J. Phys.* 9(4) (1910) 457; (b) G. Gouy. *Ann. Phys.* 7(9) (1917) 129; (c) D. L. Chapman. *Phil. Mag.* 25(6) (1913) 475.

3 O. Stern. *Z. Elektroche* 30 (1924) 508.

4 B. H. Bijsterbosch, in Th. F. Tadros(ed.), *Solid/Liquid dispersions*. Academic Press, London, 1987.

5 D. H. Napper. *Polymeric stabilisation of colloidal dispersions*. Academic Press, London, 1983.

6 P. J. Flory, W. R. Krigbaum. *J. Chem. Phys.* 18 (1950) 1086.

7 H. C. Hamaker. *Physica (Utrecht)* 4 (1937) 1058.

8 B. V. Deryaguin, L. Landau. *Acta Physicochem. USSR* 14 (1941) 633.

9 E. J. W. Verwey, J. Th. G. Overbeek. *Theory of stability of lyophobic colloids*. Elsevier, Amsterdam, 1948.

10 M. V. Smoluchowski. *Z. Phys. Chem.* 92 (1927) 129.

11 N. Fuchs. *Z. Physik* 89 (1936) 736.

12 H. Reerink, J. Th. G. Overbeek. *Discuss. Faraday Soc.* 18 (1954) 74.

13 S. Asakura, F. Osawa. *J. Phys. Chem.* 22 (1954) 1255.

14 S. Asakura, F. Osawa. *J. Polym. Sci.* 33 (1958) 183.

15 G. Gregory, Th. F. Tadros (ed.), *Solid/Liquid dispersions*. Academic Press, London, 1987.

11

AGGREGATION KINETICS OF COLLOIDAL PARTICLES

Motoyoshi Kobayashi

11.1 INTRODUCTION

Colloidal particles such as clay minerals and organic matter are abundant in natural water environments [1, 2]. The particles are able to adsorb many substances owing to their high specific surface area. In order to control and predict the results of adsorption for various purposes, it is essential to control and predict the transport of colloidal particles.

One of the important properties of colloidal particles is their aggregation, also called coagulation or flocculation. Aggregation of colloidal particles increases the hydrodynamic size of the transport unit as well as the apparent volume fraction in colloidal dispersions. Increased particle size enhances the sedimentation rate of particles and causes a drastic change in the transport properties of colloidal dispersions. Understanding aggregation is of vital importance in the field of environmental science and technology. For the purpose of water purification, for example, aggregation induced by the addition of coagulants and/or flocculants is an important part of the process of solid–liquid separation.

The kinetics of aggregation is the key to understanding the whole aggregation process. Therefore, this chapter describes fundamental theories of aggregation kinetics of colloidal particles and discusses the applicability of the theories to the model colloidal dispersion. The discussion in this chapter is

Electrical Phenomena at Interfaces and Biointerfaces: Fundamentals and Applications in Nano-, Bio-, and Environmental Sciences, First Edition. Edited by Hiroyuki Ohshima.

limited to charge-stabilized particles in an aqueous solution of a simple electrolyte, where the Derjaguin–Landau–Verwey–Overbeek (DLVO) force (Chapter 3) is dominant in physicochemical particle–particle interactions. The effect of polymers on flocculation has been the focus of intensive research from scientific and technological points of view. This topic is discussed in Chapters 10 and 17.

11.2 POPULATION BALANCE EQUATION

Colloidal particles suspended in water are transported by Brownian motion (diffusion), the flow of water surrounding the particles, gravitational sedimentation, and so on. The relative motion of the particles induces collision and contact among the particles (Fig. 11.1). When the attractive force between particles at contact is strong enough, the particles stick to each other and form flocs (aggregates of particles). Sequential aggregation processes change the size distribution of flocs (Fig. 11.2).

Temporal variations in the size and number of particles induced by aggregation have been detected by means of several techniques: direct microscopic

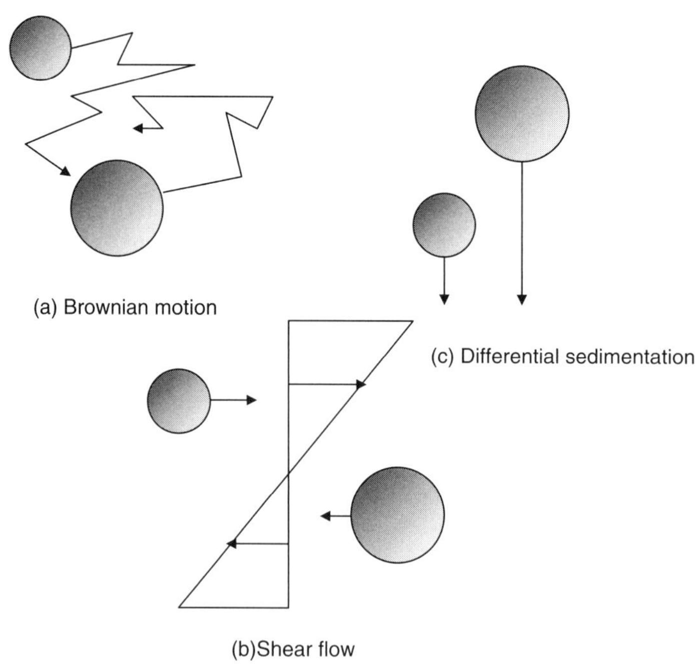

Figure 11.1. Schematic illustration of collision mechanisms.

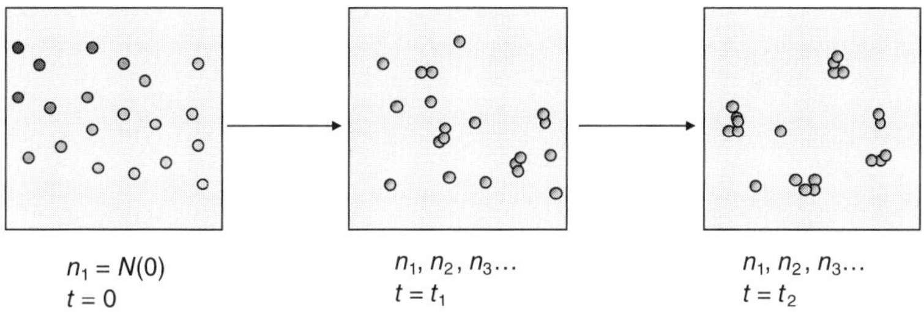

Figure 11.2. Schematic illustration of change in the number and size of particles (flocs).

particle counting and sizing [3, 4], turbidity change [5, 6], electronic particle sizing and counting [7, 8], single particle optical sizing [9], and static and dynamic light scattering [10–12].

A mathematical treatment of the aggregation process was undertaken by Smoluchowski [13], who derived the following equation:

$$\frac{dn_k}{dt} = \frac{1}{2}\sum_{\substack{i=1 \\ i+j=k}}^{i=k-1} \beta_{ij} n_i n_j - n_k \sum_{j=1}^{\infty} \beta_{kj} n_j, \tag{11.1}$$

which assumes an aggregation achieved through the binary collision between flocs composed of monodisperse primary particles [14–16]. The equation indicates that the temporal change in the number concentration of k flocs (aggregates composed of k primary particles) is determined by the formation of k flocs as a result of collision between ith and jth ($i + j = k$) flocs and the loss of k flocs due to the collision of kth flocs with other flocs. Equation 11.1 does not take into account the disaggregation of flocs and is thus only applicable in the early stage of aggregation. In Equation 11.1, β_{ij} is called an aggregation rate constant between the ith and jth flocs and reflects the collision mechanism and interactions between colliding particles and flocs. In the following sections, theoretical expressions of β_{ij} between spherical particles are given.

11.3 AGGREGATION DUE TO BROWNIAN MOTION

11.3.1 Brownian Aggregation Rate Constant

Consider a colloidal suspension in stationary condition. The colloidal particles in the suspension are moved by Brownian motion, which gives rise to collision between the particles. Also, the interaction potential between colliding particles induces their motion in response to forces of attraction or repulsion. The aggregation rate constant between the ith and jth flocs, $\beta_{B,ij}$, is obtained by calculating the total flux: the diffusion flux and the flux induced by the field

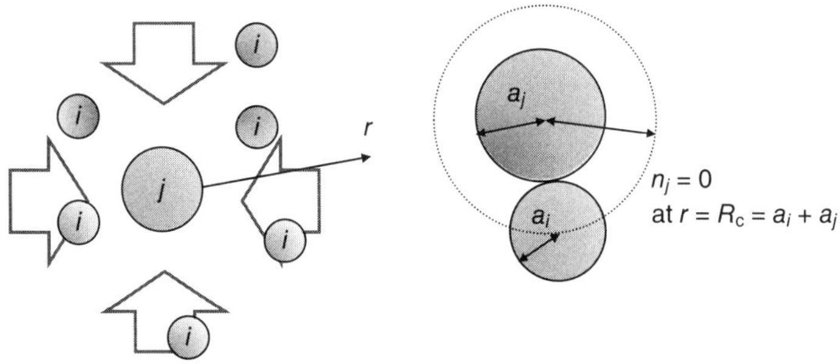

Figure 11.3. Schematic illustration of the flux of *i*th spheres (flocs) toward a reference *j*th sphere (floc) and the assumed coordinate system.

force of *i*th flocs with a radius of a_i toward a reference *j*th floc with a radius of a_j. The flux can be given by [16, 17]

$$J(r) = D_{ij}\left(\frac{dn_i}{dr} + \frac{1}{k_B T}\frac{d\Phi}{dr}n_i\right), \tag{11.2}$$

taking the *j*th floc as the center of the reference and an origin of spherical coordinates, and the *i*th flocs are distributed in spherical symmetry (Fig. 11.3). In the above expression, D_{ij} is a relative diffusion constant, n_i is the number of *i*th flocs per unit volume, k_B is the Boltzmann constant, T is the absolute temperature, and Φ is the physicochemical interaction potential energy between the *i*th and *j*th flocs. The first and second terms on the right-hand side of Equation 11.2 correspond to the diffusion flux and the flux induced by the interaction force (potential), respectively. The number of *i*th flocs colliding toward the reference floc per unit time Z at a steady state is given by

$$Z = 4\pi r^2 D_{ij}\left(\frac{dn_i}{dr} + \frac{1}{k_B T}\frac{d\Phi}{dr}n_i\right) = \text{constant}. \tag{11.3}$$

The application of boundary conditions $n_i = 0$ at $r = a_i + a_j$, $n_i = n_{i,b}$ as $r \to \infty$ (Fig. 11.3) leads to

$$Z = 4\pi n_{i,b}\left[\int_{a_i+a_j}^{\infty} \frac{1}{D_{ij}r^2}\exp\left(\frac{\Phi(r)}{k_B T}\right)dr\right]^{-1}, \tag{11.4}$$

where $n_{i,b}$ is the concentration of *i*th flocs in the bulk of suspension. The number of collisions between *i*th and *j*th flocs per unit time and volume is given by multiplying Equation 11.4 by n_j:

$$4\pi n_i n_j\left[\int_{a_i+a_j}^{\infty} \frac{1}{D_{ij}r^2}\exp\left(\frac{\Phi(r)}{k_B T}\right)dr\right]^{-1}, \tag{11.5}$$

where $n_{i,b}$ is replaced by n_i for simplicity. Finally, the Brownian aggregation rate constant, $\beta_{\mathrm{B},ij}$, is written as

$$\beta_{\mathrm{B},ij} = 4\pi\left[\int_{a_i+a_j}^{\infty}\frac{1}{D_{ij}r^2}\exp\left(\frac{\Phi(r)}{k_{\mathrm{B}}T}\right)dr\right]^{-1}. \tag{11.6}$$

Furthermore, the relative diffusion constant D_{ij} with hydrodynamic interaction is given by

$$D_{ij} = \left(\frac{k_{\mathrm{B}}T}{6\pi a_i\mu} + \frac{k_{\mathrm{B}}T}{6\pi a_j\mu}\right)\frac{1}{B_{ij}(h)} \tag{11.7}$$

and

$$B_{ij}(h) = \frac{6(h/a_{ij})^2 + 13(h/a_{ij}) + 2}{6(h/a_{ij})^2 + 4(h/a_{ij})} \text{ with } a_{ij} = \frac{2a_ia_j}{a_i + a_j}, \tag{11.8}$$

where μ is the viscosity of medium and $h = r - a_i - a_j$ is the surface–surface distance between flocs approximated as spheres [16–18]. The rate constant $\beta_{\mathrm{B},ij}$ is then rewritten as

$$\beta_{\mathrm{B},ij} = \frac{2k_{\mathrm{B}}T}{3\mu}\left(\frac{1}{a_i} + \frac{1}{a_j}\right)(a_i + a_j)\alpha_{\mathrm{B},ij} \tag{11.9}$$

and

$$\alpha_{\mathrm{B},ij} = \left[(a_i + a_j)\int_0^{\infty}\frac{B_{ij}(h)}{(a_i + a_j + h)^2}\exp\left(\frac{\Phi(h)}{k_{\mathrm{B}}T}\right)dh\right]^{-1}. \tag{11.10}$$

With Equation 11.9, one can calculate aggregation rate constants with interaction potential energy Φ. To do this calculation, we often use the DLVO potential given by the sum of the van der Waals attractive potential Φ_{vdW} and the electric double-layer potential, Φ_{edl}: $\Phi = \Phi_{\mathrm{vdW}} + \Phi_{\mathrm{edl}}$. Properties and applicability of the rate constant calculated with the DLVO theory are discussed later. Here, one can derive the original Smoluchowski rate constant as

$$\beta_{\mathrm{B},ij}^{\mathrm{SM}} = \frac{2k_{\mathrm{B}}T}{3\mu}\left(\frac{1}{a_i} + \frac{1}{a_j}\right)(a_i + a_j), \tag{11.11}$$

neglecting any interactions; $\Phi = 0$, $B_{ij} = 1$, and thus $\alpha_{\mathrm{B},ij} = 1$.

11.3.2 The Smoluchowski Solution

In the early stage of the aggregation of monodisperse spheres with a radius of a, we assume $a_i \cong a_j \cong a$. The Brownian aggregation rate constant is thus approximated as

$$\beta_{B,11} = \alpha_{B,11}\frac{8k_BT}{3\mu} \tag{11.12}$$

and

$$\alpha_{B,11} = \left[2a\int_0^\infty \frac{B_{11}(h)}{(2a+h)^2}\exp\left(\frac{\Phi(h)}{k_BT}\right)dh\right]^{-1} \tag{11.13}$$

for every collision. In this case, Equation 11.1 can be simplified as follows [13, 15]:

$$\begin{aligned}
\frac{dn_1}{dt} &= \beta_{B,11}\left(-n_1\sum_{i=1}^{\infty}n_i\right)\\
\frac{dn_2}{dt} &= \beta_{B,11}\left(\frac{1}{2}n_1n_1 - n_2\sum_{i=1}^{\infty}n_i\right)\\
\frac{dn_3}{dt} &= \beta_{B,11}\left(\frac{1}{2}(n_1n_2+n_2n_1) - n_3\sum_{i=1}^{\infty}n_i\right).\\
&\vdots
\end{aligned} \tag{11.14}$$

Taking the summation of Equation 11.14, the temporal variation of the total number of flocs, $N(t) = \sum_{i=1}^{\infty} n_i$, is given by

$$\frac{dN(t)}{dt} = -\frac{1}{2}\beta_{B,11}N(t)^2. \tag{11.15}$$

Solving Equation 11.15 with an initial condition $N(t) = N(0)$ at $t = 0$ leads to

$$\frac{N(t)}{N(0)} = \frac{1}{1+\frac{\beta_{B,11}}{2}N(0)t} \quad \text{or} \quad \frac{1}{N(t)} - \frac{1}{N(0)} = \frac{\beta_{B,11}}{2}t, \tag{11.16}$$

which shows the reduction of $N(t)$ with time due to aggregation. Using this expression, one can obtain the characteristic aggregation time or half-life in aggregation, τ_B:

$$\tau_B = \frac{2}{\beta_{B,11}N(0)}; \tag{11.17}$$

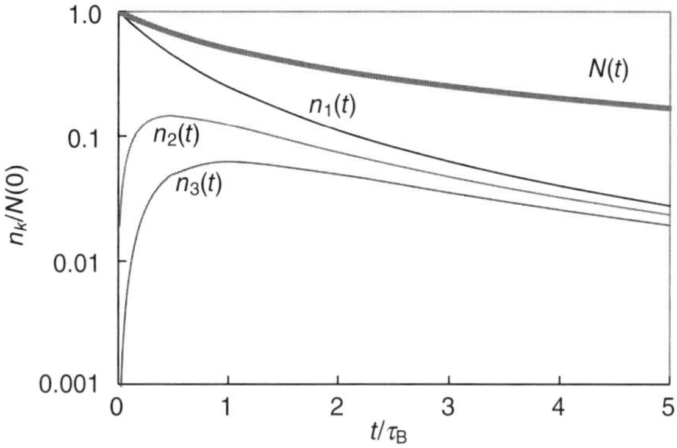

Figure 11.4. Temporal variation in the total number concentration of flocs $N(t)$ and the concentration of kth flocs n_k calculated by the Smoluchowski solution. $N(t)$ and n_1 decrease monotonically, and n_2 and n_3 show the maximum.

that is, $N(\tau_B) = N(0)/2$. Furthermore, substituting Equation 11.16 into Equation 11.14 and then solving the obtained first-order ordinary differential equations from $i = 1$ to $2, 3 \ldots$ yields the temporal change of the size distribution of flocs:

$$\frac{n_i(t)}{N(0)} = \frac{\left(\dfrac{t}{\tau_B}\right)^{i-1}}{\left(1+\dfrac{t}{\tau_B}\right)^{i+1}}. \tag{11.18}$$

Figure 11.4 demonstrates the variations of $n_i(t)$ and $N(t)$ with dimensionless time t/τ_B drawn by Equations 11.16 and 11.18. Experimental studies [19, 20] verified the linear relationship of $1/N(t)$ versus t as shown by Equation 11.16, indicating the usefulness of the Smoluchowski solution. However, the Smoluchowski rate constant $\alpha_{B,11} = 1$ overestimates the rate constant even in a rapid aggregation regime, where the repulsive electric double-layer force is negligible (Fig. 11.5). In order to obtain the quantitative agreement between the theory and experiments, we need to assume that the capture efficiency in the fast regime, $\alpha^f_{B,11}$, is in the range between 0.2 and 1.0 [18, 20]. Some researchers suggested that a reasonable value of $\alpha^f_{B,11}$ can be calculated using

$$\alpha^f_{B,11} = \left[2a\int_0^\infty \frac{B_{11}(h)}{(2a+h)^2}\exp\left(\frac{\Phi_{vdW}(h)}{k_B T}\right)dh\right]^{-1} \tag{11.19}$$

by choosing a proper value of the Hamaker constant, which reflects the magnitude of the van der Waals attraction [14]. However, other studies [4, 20]

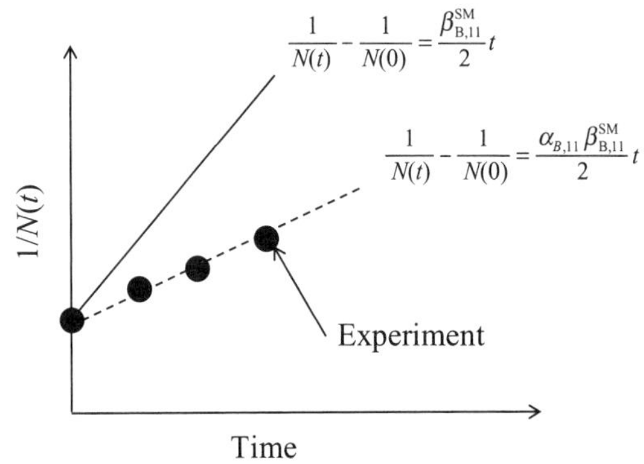

Figure 11.5. Schematic representation of the temporal variation of the total number concentration of flocs $N(t)$ due to Brownian aggregation. The Smoluchowski rate constant solution overestimates the fast aggregation rate.

claimed that $\alpha^f_{B,11}$ depends on $N(0)$, which cannot be explained by Equation 11.19. Further studies are still required to clarify this point. As for the size distribution, experimental data show the usefulness of Equation 11.18 for choosing proper values of $\alpha^f_{B,11}$ for the description of the early-stage aggregation process, when the electric double-layer force is negligible [9, 19]. In the later stage of aggregation, however, the effect of the development of floc structure should be taken into account [3, 4].

11.3.3 The Effect of the DLVO Potential on Brownian Aggregation

The DLVO theory is the most fundamental theory describing the interaction potential and force between colloidal particles dispersed in aqueous solution. With the DLVO potential and numerical integration of Equation 11.13, one can calculate the aggregation rate constant $\beta_{B,11}$. Figure 11.6 shows a schematic graph of the aggregation rate constant as a function of salt concentration. As shown in the figure, the rate increases with increasing salt concentration C and reaches a constant rate. The minimum salt concentration, above which the rate becomes constant, is called the critical coagulation concentration (CCC). The CCC separates two types of aggregation regimes: the fast aggregation regime above CCC and the slow aggregation regime below CCC [15, 21]. In the former regime, the rate is controlled by the collision frequency and the van der Waals attractive potential. The aggregation rate constant in the rapid regime is called the rapid aggregation rate constant, $\beta^f_{B,11}$. In the latter regime, the repulsive electric double-layer force reduces the rate of aggregation. According to the DLVO theory, the magnitude of the electric repulsive force decreases with increasing salt concentration. Thus, the rate increases with increasing salt concentration.

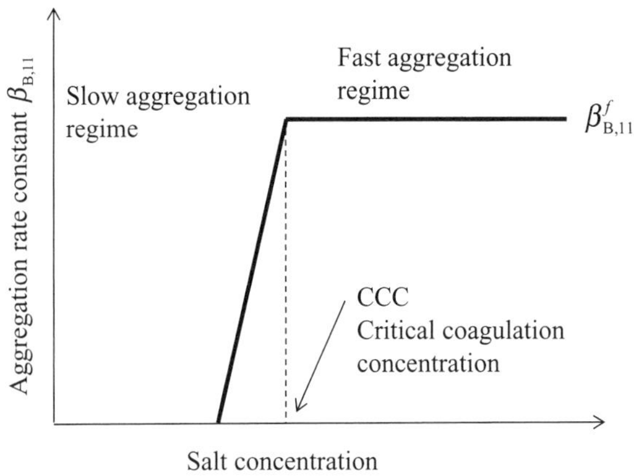

Figure 11.6. Schematic representation of the Brownian aggregation rate constant as a function of salt concentration.

It is practically important to judge whether a colloidal dispersion is stable or not. Stable dispersions are attained unless the dispersed particles aggregate. On the other hand, in unstable colloidal dispersions, suspended particles rapidly aggregate. The stability ratio defined by

$$W = \beta^f_{B,11} / \beta_{B,11} \tag{11.20}$$

has been used to discuss the stability of colloidal dispersion; $W = 1$ means the dispersion is unstable, and $W >> 1$ indicates the dispersion is stable [21].

Many experimental data on the relationship between the stability ratio W and the salt concentration C of charge-stabilized colloidal dispersions confirm the qualitative validity of the DLVO theory; a slow aggregation regime, a CCC, and a fast aggregation regime do exist [14]. However, it is well-known that there has been a discrepancy between theory and experimental data in describing the slope of the stability curve $d \log W/d \log C$ in the slow aggregation regime [21, 22] (Fig. 11.7a). To reduce this discrepancy, several additional mechanisms that are not included in the classical DLVO theory were proposed, for example, surface heterogeneity, and roughness and shift of the onset of the van der Waals and electric forces [5, 15, 21, 23]. Recently, Behrens et al. [22, 24, 25] found that the classical DLVO theory successfully describes a whole stability curve of weakly charged latex spheres without using any fit parameters (Fig. 11.7b). They suggested that the DLVO theory quantitatively works when the surface–surface distance at the maximum of interaction potential energy is beyond 1–2 nm. This condition is fulfilled for weakly charged particles. On the other hand, the discrepancy between theory and experiment is recovered for highly charged particles. In addition, monodisperse spherical silica particles show anomalous stability behaviors that are

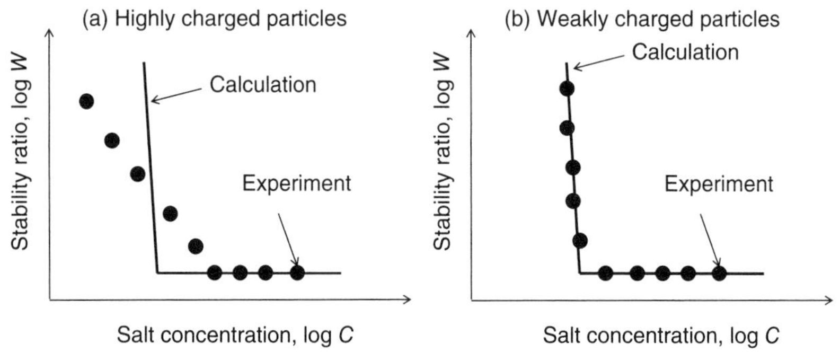

Figure 11.7. Schematic representation of stability ratio W as a function of salt concentration C by Behrens et al. [25]: (a) highly charged particles and (b) weakly charged particles.

completely different from the DLVO prediction [11, 12, 26]. Decisive explanations for these discrepancies have not yet been given.

11.4 AGGREGATION IN FLOW FIELDS

11.4.1 Collision Frequency in a Shear Flow

In many natural and engineered systems, colloidal particles are subjected to fluid motion. A laminar shear flow is the simplest flow field, where the linear distribution of flow velocity is characterized by the shear rate of the flow. Smoluchowski [13] considered colliding two spheres placed in the shear flow with a shear rate of G by neglecting both physicochemical and hydrodynamic interactions between colliding particles until collision and possible disturbances of the flow field by the presence of the particles (Fig. 11.8a). The particles in this case move along the stream line of the flow and aggregate by the interception. The number of ith flocs (spheres) with a radius of a_i attaching to a jth floc (sphere) with a radius of a_i, Z can be obtained by multiplying the volumetric flow rate passing through the projected circle area with a collision radius $R_c = a_i + a_j$ (Fig. 11.9a) by the number of ith flocs per unit volume:

$$\begin{aligned} Z &= 4Gn_i \int_0^{R_c} \int_0^{x_c(z)} x\,dx\,dz \\ &= 4Gn_i \int_0^{R_c} \frac{x_c^2}{2} dz = 2n_i G \int_0^{R_c} \left(R_c^2 - z^2\right) dz \\ &= \frac{4}{3} G R_c^3 n_i = \frac{4}{3} G\left(a_i + a_j\right)^3 n_i. \end{aligned} \tag{11.21}$$

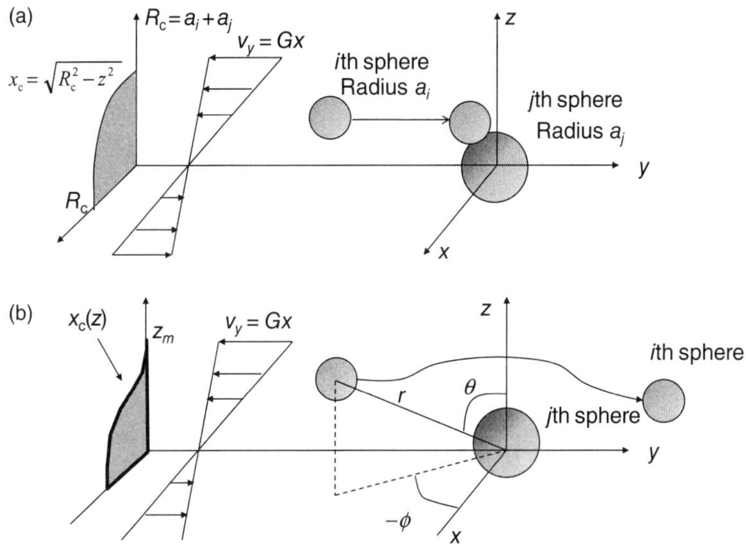

Figure 11.8. Schematic illustration of two colliding spheres in a shear flow: (a) without any interactions and (b) with hydrodynamic and physicochemical interactions.

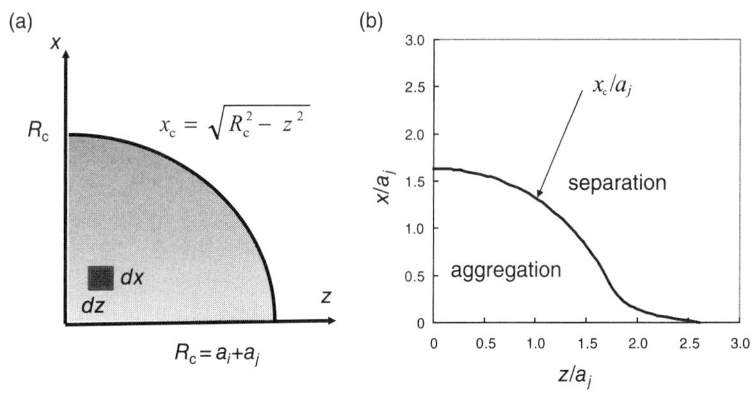

Figure 11.9. Geometry for the calculation of the rate of shear aggregation: (a) without interactions and (b) with interactions.

The Smoluchowski collision frequency in a shear flow, $\beta_{S,ij}^{SM}$,, is thus derived:

$$\beta_{S,ij}^{SM} = \frac{4}{3} G (a_i + a_j)^3. \tag{11.22}$$

The expression demonstrates the stronger dependence of the shear aggregation on the size of particles. The relative importance of shear aggregation compared to a Brownian aggregation can be evaluated by the Peclet number, *Pe*, defined as

$$Pe = \frac{\beta_{S,11}^{SM}}{\beta_{B,11}^{SM}} = \frac{\frac{32}{3}Ga^3}{\frac{8k_BT}{3\mu}} = \frac{4\mu Ga^3}{k_BT} \tag{11.23}$$

for the early stage of aggregation of monodisperse particles; $a_i = a_j = a$ [15, 20].

From Equations 11.15 and 11.22, one can obtain the temporal change of the number of flocs:

$$\frac{dN(t)}{dt} = -\frac{16}{3}Ga^3N(t)^2. \tag{11.24}$$

Solving Equation 11.24 with an initial condition $N(t) = N(0)$ at $t = 0$ and assuming the constant volume fraction $\phi = 4\pi a^3N(0)/3$ lead to [3, 15]

$$\ln\frac{N(t)}{N(0)} = -\frac{4\phi Gt}{\pi}. \tag{11.25}$$

This equation shows that the degree of aggregation $N(0)/N(t)$ in a shear flow is determined by the product of particle concentration, shear rate, and time $G\phi t$. Using this nondimensional parameter, $G\phi t$ is practically useful to design flocculation units in industrial plants [15].

11.4.2 The Effect of Interactions

The original Smoluchowski model does not consider any interaction between colliding particles. In a real situation, however, the hydrodynamic, electric double layer and the van der Waals interactions between colliding particles do come into play. Such interactions induce a curved colliding trajectory and thus affect the aggregation rate constant (Fig. 11.8b). The modification due to the interactions can be expressed by introducing the capture efficiency $\alpha_{S,ij}$ in the shear aggregation [15, 20, 27]:

$$\beta_{S,ij} = \alpha_{S,ij}\beta_{S,ij}^{SM}. \tag{11.26}$$

Theoretical evaluation of capture efficiency in a simple shear flow was initiated by Van de Ven and Mason [27] and Zeichner and Schowalter [28] by numerically solving trajectory equations given by [29, 30]

$$\frac{dr}{dt} = Gr(1 - A')\sin^2\theta\sin\phi\cos\phi + \frac{C'}{6\pi\mu a_j}(F_{edl} + F_{vdW}), \tag{11.27}$$

$$\frac{d\theta}{dt} = G(1 - B')\sin\theta\cos\theta\sin\phi\cos\phi, \tag{11.28}$$

and

$$\frac{d\phi}{dt} = G\left(\cos^2\phi - \frac{B'}{2}\cos 2\phi\right) \tag{11.29}$$

in a spherical coordinate and setting the origin as the center of the jth sphere (Fig. 11.8b). In these equations, A', B', and C' are hydrodynamic interaction functions, and F_{edl} and F_{vdW} are the electric double layer and the van der Waals forces, respectively. The trajectory equations describe the relative motion of colliding spheres. In the calculation of colliding trajectories, an ith sphere is released from an initial position and moves following the trajectory equations. Then, the released sphere attaches to the reference jth sphere (aggregation) or goes downstream (nonaggregation). Repeating the calculations from many trial starting points, one can obtain the shape of the boundary of capture cross section $x_{\mathrm{c}}(z)$ at the released y-position (Fig. 11.9b). All spheres started from the initial positions within the boundary result in aggregation. The shape of the capture cross section enables us to evaluate the capture efficiency $\alpha_{\mathrm{S},ij}$ from the following equation [29, 30]:

$$\alpha_{\mathrm{S},ij} = \frac{3}{2(a_i + a_j)^3}\int_0^{z_m} [x_c(z)]^2\, dz. \tag{11.30}$$

Calculated values of capture efficiency for the aggregation without the electric double-layer force, $\alpha^f_{\mathrm{S},11}$, can be correlated using dimensionless parameters C_A and H_A, which mean the ratio of the van der Waals force to the hydrodynamic force. For equal-sized spheres, one can use the following equation:

$$\alpha^f_{\mathrm{S},11} = f(a)C_A^{0.18},\ C_A = \frac{A}{36\pi\mu a^3 G}, \tag{11.31}$$

where A is the Hamaker constant reflecting the magnitude of the van der Waals force, and a size-dependent coefficient $f(a) = 0.79$, 0.87, and 0.95 for $a = 2.0$, 1.0, and 0.5 μm [27]. For the heteroaggregation between unequal-sized spheres, the capture efficiency $\alpha^f_{\mathrm{S},ij}$ is summarized as

$$\alpha^f_{\mathrm{S},ij} = \frac{8}{(1+\lambda)^3}10^{\left(a'+b'\lambda+c'\lambda^2+d'\lambda^4\right)},\ H_A = \frac{A}{18\pi\mu G(2a_j)^3}, \tag{11.32}$$

where λ $(= a_i/a_j,\ a_i < a_j)$ is the size ratio, and a', b', c', and d' are constants depending on H_{A} [31]. Adachi [20] reviewed the published data on the rapid aggregation rate of monodisperse latex spheres in a laminar shear flow and concluded that Equations 11.26 and 11.31 quantitatively describe the rate constant by assuming $A \cong 10^{-21}$ J. Similar conclusions were reported by Sato et al. [3] and Kobayashi [32] (Fig. 11.10). The value of A is smaller than the theoretical one ($A \cong 10^{-20}$ J) [24, 33]. The reduction of A is probably due to an

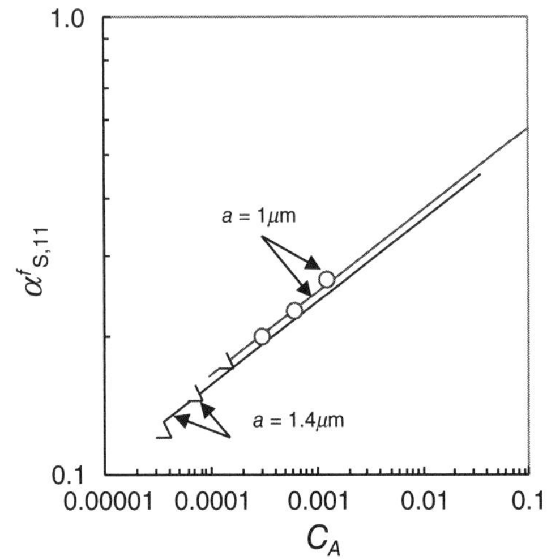

Figure 11.10. Capture efficiency of shear aggregation without EDL force. Experiments (symbols) are compared with calculations (lines). To obtain better fits, the Hamaker constant $A(10^{-21}\ \mathrm{J}) = 3$ and 1 are assumed for the radius of particles $a = 1$ and 1.4 μm, respectively. Taken from Kobayashi [32].

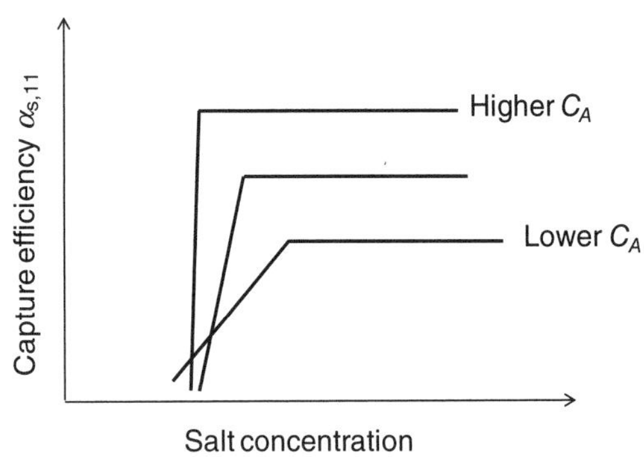

Figure 11.11. Schematic representation of the capture efficiency of shear aggregation in the presence of an electric double-layer force calculated by Van de Ven and Mason [27].

unknown structure of the interface and/or liquid structure adjacent to the interface [3, 33, 34].

The capture efficiency $\alpha_{S,ij}$ is schematically shown in Figure 11.11 in the presence of repulsive electric double-layer force and van der Waals force [27]. The figure shows the typical features of the calculated capture efficiency of shear aggregation: CCCs increase by increasing the shear rate of flow, and the slope in the slow aggregation regime becomes gradual for higher shear rates. Experimental data by Sato et al. [8] qualitatively agree with the theoretical features. However, quantitative comparison of theory with experiment in a slow shear aggregation has not yet been performed. Further studies will be required for this aspect.

11.4.3 Aggregation in a Turbulent Flow

In many natural and industrial processes, aggregation proceeds in turbulent flow fields, where flow velocity fluctuates temporally and spatially. It has been recognized that a fluctuated turbulent flow is composed of many eddies of different length scales [35]. Larger eddies produce smaller eddies, which produce smaller and smaller eddies (Fig. 11.12). The process is called a cascade process, and through it kinetic energy is transferred from larger eddies to smaller ones, and directional bias accompanied with larger eddies is gradually lost at a smaller scale of eddies. Kolmogorov proposed a hypothesis of the local isotropy of turbulence. Quantities related to turbulence in the local area become statistically isotropic, and finally, the transferred energy dissipates into heat due to viscosity at the smallest scale of turbulence. In the range of local isotropy, statistical physical parameters can be given by two quantities: kinematic viscosity ν and the rate of energy dissipation per unit mass ε. The smallest scale of turbulence, called the Kolmorov microscale η_K, is given by

$$\eta_K = \left(\frac{\nu^3}{\varepsilon}\right)^{\frac{1}{4}} \tag{11.33}$$

from dimensional analysis [35]. Aggregation of colloidal particles takes place in the range smaller than the Kolmogorov microscale, where the hydrodynamic viscous force becomes predominant [9, 20, 36].

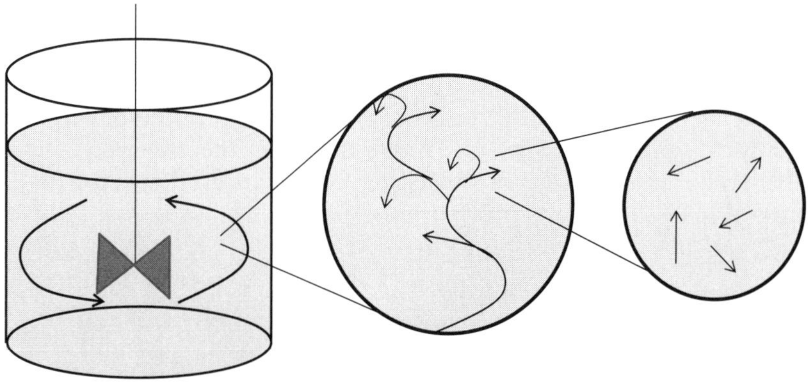

Figure 11.12. Schematic illustration of a turbulent flow.

Saffman and Turner [37] carried out the rigorous derivation of the collision frequency in an isotropic turbulence. They obtained the average number of ith flocs (spheres) with a radius of a_i colliding to a reference jth floc (sphere) with a radius of a_i, Z, calculating the average inward flux across the surface of the collision sphere with a radius of $R_c(= a_i + a_j)$:

$$Z = 2\pi R_c^2 \cdot \overline{\left|\frac{\partial u_r}{\partial r}\right|} R_c n_i. \tag{11.34}$$

The mean absolute value of the longitudinal derivative of the radial component of the relative velocity $\overline{|\partial u_r / \partial r|}$ is given by using the result of Taylor's statistical theory of isotropic turbulence and the assumption that the velocity gradient is approximately normally distributed; that is,

$$\overline{\left|\frac{\partial u_r}{\partial r}\right|} = \sqrt{\frac{2\varepsilon}{15\pi\nu}}. \tag{11.35}$$

Therefore, the aggregation rate constant in a turbulent flow by Saffman and Turner, $\beta_{T,ij}^{\mathrm{ST}}$, is written as

$$\beta_{T,ij}^{\mathrm{ST}} = \sqrt{\frac{8\pi\varepsilon}{15\nu}} R_c^3. \tag{11.36}$$

The rate constant of Saffman and Turner does not take into account any interaction between colliding particles. However, as considered in the shear aggregation, hydrodynamic as well as physicochemical interactions modify the rate constant. The modification due to the effects of the interparticle interactions is expressed by introducing the capture efficiency $\alpha_{T,ij}$ in the turbulent flow [9, 36]:

$$\beta_{T,ij} = \alpha_{T,ij} \beta_{T,ij}^{\mathrm{ST}}. \tag{11.37}$$

At the moment, it is still difficult to rigorously evaluate the capture efficiency in a turbulent flow. Therefore, as a first approximation, the capture efficiency obtained for the shear aggregation has been used for the turbulent aggregation by replacing the shear rate G with the mean absolute value of the lateral derivative of relative velocity G_T [9, 36]; that is, in the absence of electric double-layer force, the capture efficiency for equal-sized spheres, $\alpha_{T,11}^f$, is expressed as

$$\alpha_{T,11}^f = f(a) C_A^{0.18}, C_A = \frac{A}{36\pi\mu a^3 G_T}, G_T = \sqrt{\frac{4\varepsilon}{15\pi\nu}}. \tag{11.38}$$

The practical usefulness of Equation 11.38 was experimentally verified by several researchers [9, 36, 38] who measured the turbulent aggregation rate of

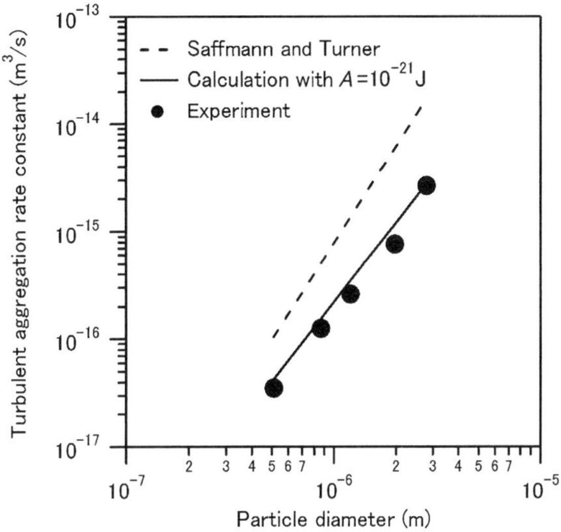

Figure 11.13. Fast aggregation rate constant in a turbulent flow. Experiments (circles) are compared with theory (line) calculated with the Hamaker constant $A = 10^{-21}$ J. Taken from Kobayashi and Ishibashi [38].

latex spheres in the fast aggregation regime. Figure 11.13 shows the fast aggregation rate constant in a turbulent flow as a function of particle size [38]. This figure demonstrates that Equations 11.37 and 11.38 reasonably describe experimental data.

As for the effect of the electric double-layer force on turbulent aggregation, Kobayashi et al. [7] measured the rate of turbulent aggregation of monodisperse latex spheres as a function of turbulent intensity and salt concentration. The obtained capture efficiency in the turbulent flow, $\alpha_{T,ij}$, shows a trend similar to that predicted for a laminar shear flow (Fig. 11.11). These findings suggest that the flow field in the Kolmogolov microscale can be treated as a sort of laminar shear flow. Again, however, quantitative comparison between the theory and experiments has been lacking. We still need further studies on the initial kinetics of aggregation in flow fields.

REFERENCES

1 K. J. Wilkinson, J. R. Lead (eds.), *Environmental colloids and particles*. John Wiley & Sons, West Sussex, UK, 2007.

2 I. G. Droppo, G. G. Leppard, S. N. Liss, T. G. Milligan (eds.), *Flocculation in natural and engineered environmental systems*. CRC Press, Boca Raton, FL, 2004.

3 D. Sato, M. Kobayashi, Y. Adachi. Effect of floc structure on the rate of shear coagulation. *J. Colloid Interface Sci.* 272 (2004) 345–351.

4 T. Fukasawa, Y. Adachi. Effect of floc structure on the rate of Brownian coagulation. *J. Colloid Interface Sci.* 304 (2006) 115–118.

5 N. H. G. Penners, L. K. Koopal. The effect of particle size on the stability of haematite (α-Fe_2O_3) hydrosols. *Colloids Surf.* 28 (1987) 67–83.

6 Z. Sun, J. Liu, S. Xu. Study on improving the turbidity measurement of the absolute coagulation constant. *Langmuir* 22 (2006) 4946–4951.

7 M. Kobayashi, T. Maekita, Y. Adachi, H. Sasaki. Colloid stability and coagulation rate of polystyrene latex particles in a turbulent flow. *Int. J. Miner. Process.* 73 (2004) 177–181.

8 D. Sato, M. Kobayashi, Y. Adachi. Capture efficiency and coagulation rate of polystyrene latex particles in a laminar shear flow: effects of ionic strength and shear rate. *Colloids Surf. A* 266 (2005) 150–154.

9 Y. Adachi, M., A. Cohen Stuart, R. Fokkink. Kinetics of coagulation studied by means of end-over-end rotation. *J. Colloid Interface Sci.* 165 (1994) 310–317.

10 H. Holthoff, S. U. Egelhaaf, M. Borkovec, P. Schurtenberger, H. Sticher. Coagulation rate measurements of colloidal particles by simultaneous static and dynamic light scattering. *Langmuir* 12 (1996) 5541–5549.

11 M. Kobayashi, F. Juillerat, P. Galletto, P. Bowen, M. Borkovec. Aggregation and charging of colloidal silica particles: effect of particle size. *Langmuir* 21 (2005) 5761–5769.

12 M. Kobayashi, M. Skarba, P. Galletto, D. Cakara, M. Borkovec. Effects of heat treatment on the aggregation and charging of Stober-type silica. *J. Colloid Interface Sci.* 292 (2005) 139–147.

13 M. von Smoluchowski. Versuch einer mathematischen thorie der koagulationkinetic kolloider loesungen. *Z. Phy. Chem.* 92 (1917) 129–168.

14 W. B. Russel, D. A. Saville, W. R. Schowalter. *Colloidal dispersions*. Cambridge University Press, Cambridge, UK, 1989.

15 M. Elimelech, J. Gregory, X. Jia, R. A. Williams. *Particle deposition and aggregation*. Butterworth-Heinemann, Woburn, MA, 1998.

16 J. H. Masliyah, S. Bhattacharjee. *Electrokinetic and colloid transport phenomena*. John Wiley & Sons, Hoboken, NJ, 2006.

17 P. C. Hiemenz, R. Rajagopalan. *Principles of colloid and surface chemistry*, 3rd ed. Marcel Dekker, New York, 1997.

18 W. Lin, M. Kobayashi, M. Skarba, C. Mu, P. Galletto, M. Borkovec. Heteroaggregation in binary mixtures of oppositely charged particles. *Langmuir* 22 (2006) 1038–1047.

19 K. Higashitani, Y. Matsuno. Rapid Brownian coagulation of colloidal dispersion. *J. Chem. Eng. Japan* 12 (1979) 460–465.

20 Y. Adachi. Dynamic aspects of coagulation and flocculation. *Adv. Colloid Interface Sci.* 56 (1995) 1–31.

21 M. Borkovec, S. H. Behrens. Stabilization of aqueous colloidal dispersions: electrostatic and steric forces, in P. Somasundaran, A. Hubbard (eds.), *Encyclopedia of surface and colloid science*, 2nd ed. Marcel Dekker, New York, 2006, pp. 4795–4805.

22 S. H. Behrens, M. Semmler, M. Borkovec. Aggregation in sulfate latex suspensions: the role of charge for stability predictions. *Prog. Colloid Polymer Sci.* 110 (1998) 66–69.

23 H. Kihira, N. Ryde, E. Matijević. Kinetics of heterocoagulation. Part 2—The effect of the discreteness of surface charge. *J. Chem. Soc., Faraday Trans.* 88 (1992) 2379–2386.

24 S. H. Behrens, D. I. Christl, R. Emmerzael, P. Schurtenberger, M. Borkovec. Charging and aggregation properties of carboxyl latex particles: experiment versus DLVO theory. *Langmuir* 16 (2000) 2566–2575.

25 S. H. Behrens, M. Borkovec, P. Schurtenberger. Aggregation in charge-stabilized colloidal suspensions revisited. *Langmuir* 14 (1998) 1951–1954.

26 R. K. Iler. *The chemistry of silica*. John Wiley & Sons, New York, 1979.

27 T. G. M. Van de Ven, S. G. Mason. The microrheology of colloidal dispersions VII. Orthokinetic doublet formation of spheres. *Colloid Polymer Sci.* 255 (1977) 468–479.

28 G. R. Zeichner, W. R. Schowalter. Use of trajectory analysis to study stability of colloidal dispersions in flow fields. *AIChE J.* 23 (1977) 243–254.

29 Q. Wang. A study on shear coagulation and heterocoagulation. *J. Colloid Interface Sci.* 150 (1992) 418–427.

30 M. Vanni, J. B. Baldi. Coagulation efficiency of colloidal particles in shear flow. *Avd. Colloid Interface Sci.* 97 (2002) 151–177.

31 M. Han, D. F Lawler. The (relative) insignificance of G in flocculation. *J. Am. Water Works Assoc.* 84(10) (1992) 79–91.

32 M. Kobayashi. Aggregation of unequal-sized and oppositely charged colloidal particles in a shear flow. *J. Appl. Mech. JSCE* 11 (2008) 517–523.

33 J. N. Israelachvili. *Intermolecular and surface forces*, 2nd ed. Academic Press, London, 1992.

34 I. U. Vakarelski, K. Ishimura, K. Higashitani. Adhesion between silica particle and mica surfaces in water and electrolyte solution. *J. Colloid Interface Sci.* 227 (2000) 111–118.

35 S. B. Pope. *Turbulent flows*. Cambridge University Press, Cambridge, UK, 2000.

36 K. Higashitani, K. Yamauchi, Y. Matsuno, G. Hosokawa. Turbulent coagulation of particles dispersed in a viscous fluid. *J. Chem. Eng. Japan* 16 (1983) 299–304.

37 P. G. Saffman, J. S. Turner. On the collision of drops in turbulent clouds. *J. Fluid Mech.* 1 (1956) 16–30.

38 M. Kobayashi, D. Ishibashi. Absolute rate of turbulent coagulation from turbidity measurement. *Colloid Polym. Sci.* 289 (2011) 831–836.

12

ELECTROACOUSTIC THEORIES AND MEASUREMENT TECHNIQUES

SHIN-ICHI TAKEDA

12.1 INTRODUCTION

Electrokinetic measurements have long been used to probe interfacial phenomena and the stability of colloidal systems. In particular, electroacoustics offers distinct advantages over conventional electrokinetic techniques such as microelectrophoresis:

1. Colloids containing particles with sizes ranging from a few nanometers to a few hundred micrometers can be analyzed.
2. Samples having a wide range of particle concentrations can be measured, from around a 0.5% range to essentially volume-filling systems, that is, up to 90% by weight particles.
3. Measurements can be made in flowing streams relevant to online applications.
4. Opaque or photosensitive materials can be analyzed because no optical imaging is required.
5. Dispersed systems consisting of nonaqueous media with low permittivity can be analyzed even though microelectrophoresis cannot apply.
6. Mobile living organisms can be analyzed.

Electrical Phenomena at Interfaces and Biointerfaces: Fundamentals and Applications in Nano-, Bio-, and Environmental Sciences, First Edition. Edited by Hiroyuki Ohshima.

In this chapter, the historical background and theory of the colloid vibration current (CVI) method are reviewed, and some of the data obtained by this method are shown in order to demonstrate the advantages of electroacoustics over conventional techniques.

12.2 HISTORICAL BACKGROUND

In 1933, Debye [1] proposed a method using ultrasonic waves for the determination of the relative masses of electrolyte ions in solution. If an electrolytic solution is irradiated with ultrasonic waves, periodic differences in potential should appear within the solution, primarily as a result of the dynamic reaction of the ions to the material wave. The existence of the effect, however, was not established experimentally for 15 years following its prediction because of the small magnitude of the alternating potentials (10^{-6}V for unit velocity amplitude) and the problem of preventing electromagnetic coupling between the ultrasonic generator and the detection apparatus. In 1949, Yeager et al. [2] succeed in detecting the effect with standing wave techniques but were limited to semiquantitative measurements because of the problem of determining the acoustical parameters in a complex field. However, this phenomena itself can yield useful information about the properties of ions and is usually referred to as an *ion vibration potential* (IVP). A thorough theoretical treatment of the IVP phenomenon was given by in 1947 by Bugosh et al. [3]. There was much interest in this effect in the 1950s and 1960s because it was considered to be a very promising tool for characterizing ion solvation [4–12]. Zana and Yeager [12] summarized the results of this two-decade effort in a review. Sadly, this phenomenon has been largely forgotten, and virtually all of the interest in electroacoustic phenomena has shifted from pure electrolyte solutions to colloids.

Hermans [13] and Rutgers and Rigole [14] in 1938 were the first scientists to report a colloid vibration potential (CVP). Since that time, there have been several hundred experimental and theoretical works published. For brevity, I mention here only a few key papers. Enderby and Booth [15, 16] developed the first theory for CBP in the early 1950s. The first quantitative experiments were made in 1960 by Yeager's group [12]. In the early 1980s, Cannon and coauthors [17] discovered an inverse electroacoustic effect, which they termed *electrosonic amplitude* (ESA). The first commercially available electroacoustic instrument was developed by Pen Kem, Inc. [18]. There are now several commercially available instruments based on both electroacoustic effects. After the basic works by Enderby and Booth [15, 16], the electroacoustic theory for colloids has been developing in two quite different directions. The original Enderby–Booth theory was very complex as a result of considering both surface conductivity and low-frequency effects. In addition, it did not take particle–particle interactions into account and consequently was valid only for dilute colloids. Hence, the Enderby–Booth theory required modifications and simplifications. The first extension of the Enderby–Booth theory was performed by Marlow et al. [19]. They attempted to generalize it for concentrated systems using a Levine cell model [20]. This approach leads to somehow complicated mathematical formulas, and perhaps this was the reason why it was abandoned. An alternative approach to the electroacoustic theory was later proposed by O'Brien [21]. He introduced the concept of a *dynamic electrophoretic mobility*, μ_d, and suggested the simple relationship between this parameter and measured electroacoustic parameters, such as CVI or ESA. Later, O'Brien stated that his relationship is valid for concentrated as well as diluted systems [22].

According to the O'Brien relationship, the average dynamic electrophoretic mobility, μ_d, is defined as

$$\mu_d = \frac{\mathrm{ESA}(\omega)\rho_m}{A(\omega)F(Z_T, Z_s)\varphi(\rho_p - \rho_m)}, \tag{12.1}$$

where the ESA is normalized by the applied external electric field, $A(\omega)$ is an instrument constant determined by calibration procedure, and $F(Z_T, Z_s)$ is a function of the acoustic impedances of the transducer and the dispersion under investigation.

A similar expression can be used for the CVI mode:

$$\mu_d = \frac{\mathrm{CVI}(\omega)\rho_m}{A(\omega)F(Z_T, Z_s)\varphi(\rho_p - \rho_m)}. \tag{12.2}$$

Here, the CVI is assumed to be normalized by the gradient of pressure (grad P), which, in the case of CVI, is the external driving force.

According to the O'Brien theory, a complete functional dependence of ESA (CVI) on key parameters, such as the ζ-potential, particle size, and frequency, is incorporated into the dynamic electrophoretic mobility. O'Brien mentioned that for all considered cases, the coefficient of proportionality between ESA (CVI) and μ_d is frequency independent and, in addition, is independent of particle size and ζ-potential. This feature made the dynamic electrophoretic mobility an important and central parameter of the electroacoustic theory.

The first theory that relates this dynamic electrophoretic mobility to the properties of the dispersion medium and dispersed particles was initially formulated by O'Brien, but it neglected particle–particle interactions and was therefore limited to the diluted case (*dilute O'Brien theory*). Later, Rider and O'Brien [23], Ohshima [24], and Ennis et al. [25, 26] proposed modifications to extend this theory to concentrated systems.

Recently, an international group of scientists, Shilov from Ukraine, Ohshima from Japan, and Dukhin and Goetz from the United States [27–29]. This group has generalized the Enderby–Booth approach to the point that it should now be equivalent to the latest version of O'Brien's theory. What is of most importance is that the group has also developed an electroacoustic analog of Smoluchowski's equation. This equation is known to be valid for any concentration and any particle shape, provided that the double layer is thin ($ka \gg 1$) and the surface conductivity is negligible ($Du \ll 1$). By making these same two assumptions, this group derived a general theory for electroacoustics that is valid for any concentration or particle shape. The theory requires no other assumptions, which is called the *Smoluchowski dynamic electroacoustic limit* (SDEL). The SDEL is a low-frequency asymptotic solution and is a natural test for all possible electroacoustic theories. Hence, any other proposed theory should reduce to the SDEL under the conditions of $ka \gg 1$ and $Du \ll 1$, at sufficiently low frequencies. The SDEL serves as a verification criterion for any proposed electroacoustic theory in the same manner that Smoluchowski's equation serves for electrophoretic theories.

12.3 THEORY OF THE CVI IN CONCENTRATED SYSTEMS

The modern theory for the CVI has been formulated by the collaboration between A. Dukhin and V. Shilov. Now, we retain the O'Brien expression for introducing dynamic mobility as follows:

$$\mathrm{CVI} = A(\omega)F(Z_{\mathrm{T}}, Z_{\mathrm{s}})\varphi\frac{\rho_{\mathrm{p}} - \rho_{\mathrm{m}}}{\rho_{\mathrm{m}}}\mu_{\mathrm{d}}\nabla P. \tag{12.3}$$

We also retain the same structure for the dynamic electrophoretic mobility expression, presenting as separate multipliers both the inertial effects (function G) and the electrodynamic effects (function $1+F$). However, in contrast to the diluted case expressed by Equations 12.1 and 12.2, function G and F for concentrated systems depend upon the particle concentration. There is also an additional density-dependent multiplier,

$$\frac{(\rho_{\mathrm{p}} - \rho_{\mathrm{s}})\rho_{\mathrm{m}}}{(\rho_{\mathrm{p}} - \rho_{\mathrm{m}})\rho_{\mathrm{s}}},$$

which is equal to the ratio of the particle velocity relative to the liquid and to the particle velocity relative to the center of mass of the dispersion. The convenience of the introduction of such a multiplier, which differs from unity only for concentrated systems, follows from the exact structure of Smoluchowski's asymptotic solution for μ_{d}. The corresponding equation, which in a convenient way reflects simultaneously limiting transformations both to Smoluchowski's asymptotic solution and O'Brien's asymptotic solution, is given as follows:

$$\mu_{\mathrm{d}} = \frac{2\varepsilon_0\varepsilon_{\mathrm{m}}\zeta(\rho_{\mathrm{p}} - \rho_{\mathrm{s}})\rho_{\mathrm{m}}}{3\eta(\rho_{\mathrm{p}} - \rho_{\mathrm{m}})\rho_{\mathrm{s}}}G(s, \varphi)(1 + F(Du, \omega', \varphi)). \tag{12.4}$$

The generalization for the case of polydisperse systems is given by

$$\mu_d = \frac{2\varepsilon_0\varepsilon_m\zeta(\rho_p - \rho_s)\rho_m}{3\eta(\rho_p - \rho_s)\rho_m}\sum_{i=1}^{N} G_i(s_i, \varphi)(1 + F_i(Du_i, \omega', \varphi)). \tag{12.5}$$

The new values of the functions G and F are given by the following equations:

$$G_i(s, \varphi) = \frac{\dfrac{9\varphi_i h(s_i)\rho_s}{4j\varphi(1-\varphi)s_i I(s_i)\left(\rho_p - \rho_m\left(\dfrac{3H_i}{2I_i}+1\right)\right)}}{1 - \dfrac{\rho_p}{1-\varphi}\displaystyle\sum_{i=1}^{N}\dfrac{\varphi_i\left(\dfrac{3H_i}{2I_i}+1\right)}{\rho_p - \rho_m\left(\dfrac{3H_i}{2I_i}+1\right)}} \tag{12.6}$$

and

$$F_i(Du_i, \omega', \varphi) = \frac{(1-2Du_i)(1-\varphi) + j\omega'\left(1 - \dfrac{\varepsilon_p}{\varepsilon_m}\right)(1-\varphi)}{2(1 + Du_i + \varphi(0.5 - Du_i)) + j\omega'\left(2 + \dfrac{\varepsilon_p}{\varepsilon_m} + \varphi\left(1 - \dfrac{\varepsilon_p}{\varepsilon_m}\right)\right)}, \tag{12.7}$$

where

$$s_i^2 = \frac{a_i^2\omega\rho_m}{2\eta_m};\ \omega' = \frac{\omega}{\omega_{MW}},\ \omega_{MW} = \frac{K_m}{\varepsilon_0\varepsilon_m},\ \varphi_i \text{ and } Du_i = k^{\sigma}/K_m\, a_i$$

are the volume fraction and the Dukhin number for the ith fraction of the polydisperse colloid, correspondingly, and ϕ is the total volume fraction of the dispersed particle.

These equations are restricted to the case of the Maxwell–Wagner relaxation range. They take into account both hydrodynamic and electrodynamic particle interaction and are valid for polydisperse systems without making any superposition assumption.

12.4 INSTRUMENT FOR MEASURING CVI

The Dispersion Technology DT-300 is an example of a modern instrument for determining the zeta potential of colloidal dispersions by the measurement of electroacoustic properties and the subsequent analysis of these data. This instrument, which consists of a CVI probe and an electronic unit, measures the CVI. A photo of a typical system is shown in Figure 12.1.

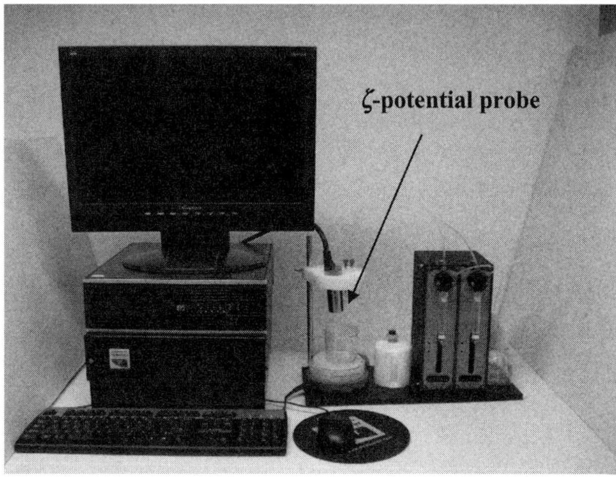

Figure 12.1. The probe for measuring colloid vibration current with titration setup.

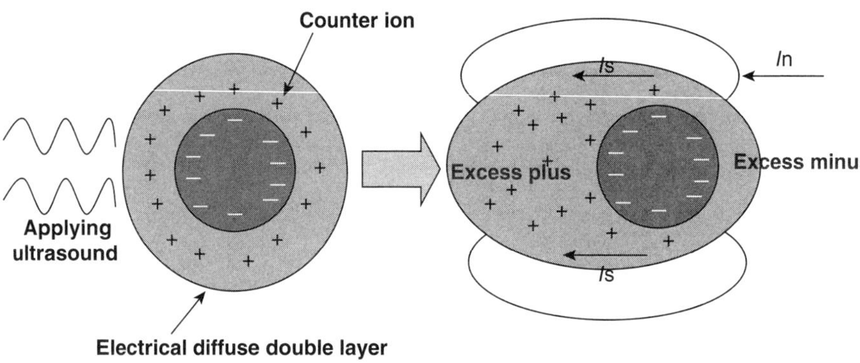

Figure 12.2. Polarization of the particle double layer with the relative liquid flow.

The CVI probe consists of two parts: a transmitting transducer and a receiving transducer. The transmitting transducer consists of a piezoelectric transducer to convert a radio frequency (RF) pulse into acoustic energy and a delay rod that launches this acoustic pulse into the slurry after a suitable delay. This acoustic excitation of the slurry causes the particles to gain induced dipole moments. The dipole moments from the particles add up to create an electric field, which can be sensed by a receiver.

Figure 12.2 illustrates how a particle creates an electric field by forming induced dipole moments. This figure shows a particle having a double layer and moving relative to the liquid. This motion includes the counterions of the double layer. In this example of a negatively charged particle, we need to consider only the positive counterions opposing the negatively charged particle surface. The hydrodynamic surface current, I_s, reduces the number of positive ions near the right-side particle pole and enriches the double layer with extra ions near the left-side pole. As a result, the double layer shifts from the original equilibrium. A negative surface charge dominates at the right pole, whereas an extra positive diffuse charge dominates at the left pole. The net result is that the particle motion has an induced dipole moment. This induced dipole moment generates an electric field usually referred to as the CVP. This CVP is external to the particle double layer. It affects ions in the bulk of the electroneutral solution beyond the double layer and generates an electric current, I_n. This electric current serves a very important purpose. It compensates for the surface current, I_s, and makes the whole system self-consistent.

Now, we go back to explain the role of a receiver. The receiver, in essence, consists of a two-element antenna immersed in the sample slurry. The electric field changes the potential of one element with respect to the other. If the impedance of the measuring circuit associated with the antenna is relatively high with respect to the colloid, then the antenna senses an open circuit voltage, which is typically referred to as the CVP. Alternatively, in the preferred configuration, the impedance of the circuitry associated with the antenna circuit is low compared to that of the colloid, and therefore the electric field causes a current to flow in the antenna. This short circuit current is referred to as the CVI. The probe can be dipped into a beaker for laboratory use, immersed in a storage tank, or inserted into a pipe for the online monitoring of a flowing process stream. The structure of the probe in detail can be seen in the literature [30].

12.5 MEASUREMENT TECHNIQUES

12.5.1 Characterization of Aggregation and Flocculation: pH Titration

Determining the stability of dispersions is one of the central problems in colloid science. Accurate and nondestructive stability characterization is the key to the successful formulation and use of many kinds of colloids. Ultrasound-based techniques offer several advantages over traditional approaches for stability characterization. Eliminating the dilution process is one obvious advantage. Another important advantage is the ability to distinguish between aggregation and flocculation.

Aggregation is a process that results in the buildup of rigid agglomerates of the initial primary particles. These agglomerates are essentially new, inherently large particles, which move as rigid entities in the liquid under stress. On the other hand, flocculation is a process in which the original particles retain, to some degree, their independent motion. The particle size distribution remains the same as in the original stable system, but the particles are now weakly linked together by specific colloidal–chemical forces. The nature of these forces might arise from various sources, including perhaps the formation of polymer chains that form bridges between the particles, or electrodynamic forces corresponding to particles being held in the secondary minima.

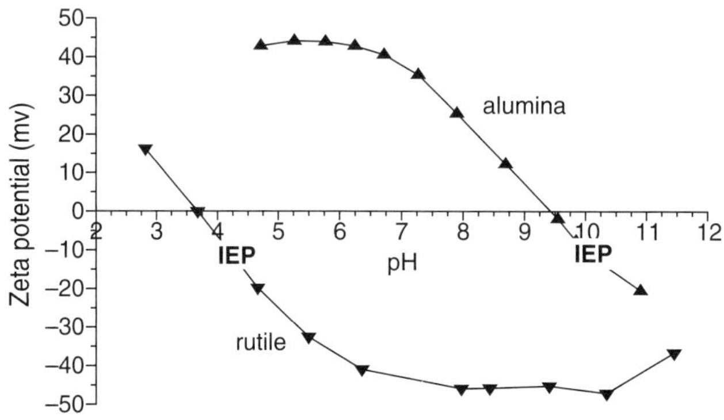

Figure 12.3. Electroacoustic pH titrations of rutile (7 vol %) and alumina (4 vol %).

A typical example of the aggregation/flocculation behavior is shown here using systems of alumina and rutile slurries: The rutile sample was prepared from a commercially available aqueous rutile slurry, R746 from DuPont. This slurry is supplied at 76.5 wt % (44.5 vol %) but was diluted to 7 vol % using distilled water. The alumina slurry was prepared at a concentration of 11.6 wt % (4.0 vol %) from dry powder. The initial pH was 4.0.

Titration of these rutile and alumina slurries and the corresponding measurements of the ζ-potential confirmed the expected *isoelectric points* (IEPs) for each material, as shown in Figure 12.3. According to the general colloid chemical principles, a dispersed system typically loses stability when the magnitude of the ζ-potential decreases to less than approximately 25 mV. As a result, there will be some region surrounding a given IEP for which the system is not particularly stable. Within this unstable region, the particles may agglomerate, thereby increasing the particle size, which can be evaluated by ultrasound attenuation spectroscopy as shown in Figure 12.4.

For the rutile slurry, unimodal distribution of the particle size was seen at pH = 8.9 and bimodal distribution of the particle size was seen at pH = 3.9, in which ζ-potential decreases less than 25 mV. Similarly, for the alumina slurry, unimodal distribution of the particle size was seen at pH = 4.0 and bimodal distribution of the particle size was seen at pH = 9.0. As was seen for rutile slurry, alumina particles agglomerate near the IEP. We can summarize this experiment on rutile and alumina slurries with the conclusion that the combination of acoustics and electroacoustics allows us to characterize the aggregation of concentrated colloids. Electroacoustics yields information about electric surface properties such as ζ-potential.

12.5.2 Surface Characterization by Monitoring CVI: Surfactant Titration

Automatic "surfactant titration" is a simple way for determining the optimum dose of surfactant required to stabilize the system. In this case, a surfactant

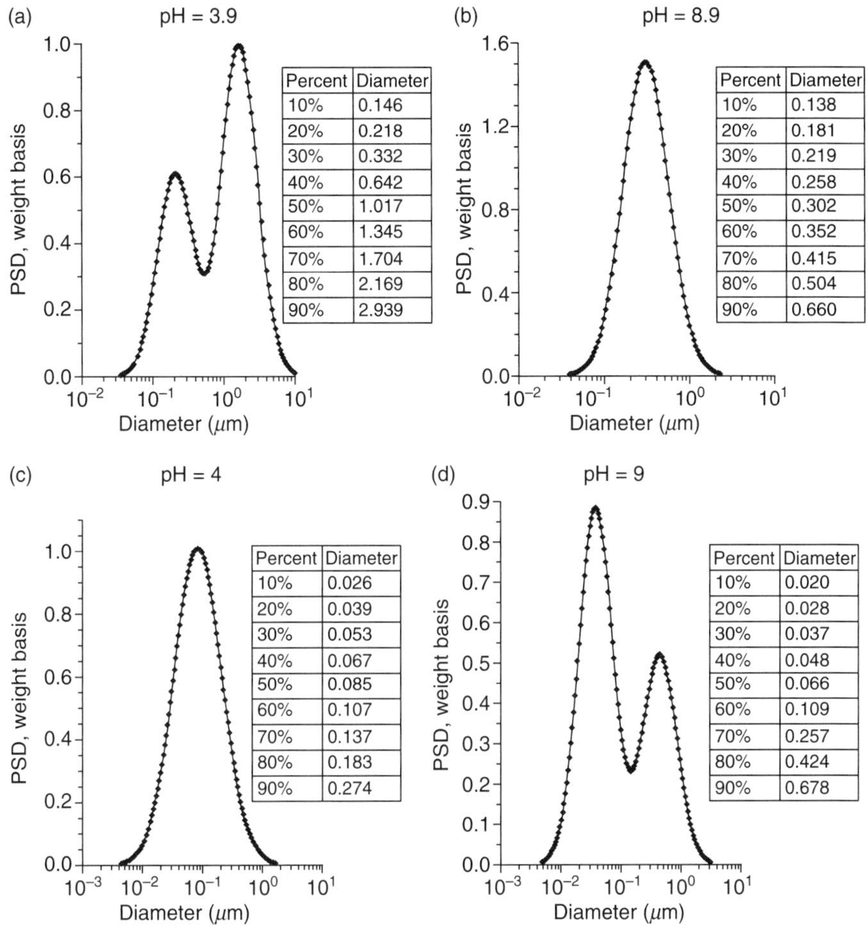

Figure 12.4. Particle size distribution of stable and unstable rutile (7 vol %) and alumina (4 vol %), respectively. (a) Rutile at pH=3.9, (b) rutile at pH=8.9, (c) alumina at pH=4.0, and (d) alumina at pH=9.0. PSD, particle size distribution.

solution of specific concentration is automatically dispersed by a burette. Figure 12.5 demonstrates an example of titration for kaolin concentrated slurry. This method allows us to compare the efficiency of the surfactant at different pH values by directly monitoring the surface electrical property, that is, the ζ-potential.

It is clearly seen that the ζ-potential reaches the maximum value at a dose of 0.6 wt % for starting pH=10.5, where hexametaphosphate (HMPH) efficiency improves with increasing initial pH. At the same time, the addition of HMPH changes the pH value. This leads us to the conclusion that surfactant titration and "pH titration" are in many cases related. In order to represent

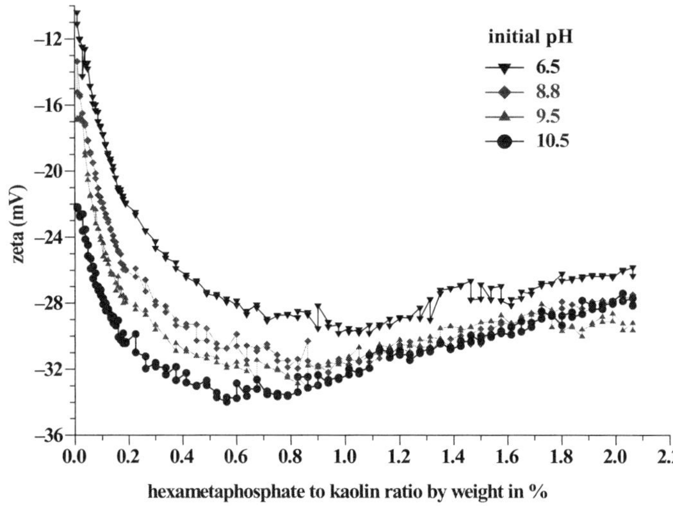

Figure 12.5. Titration of a 40 wt % kaolin slurry using hexametaphosphate (HMPH) with different starting pH values.

the surface modification properly or to characterize the surface property, we should use three dimensions, plotting the ζ-potential as a function of pH and the dose of surfactant.

12.5.3 Time Titration: Kinetics of the Surface–Bulk Equilibration

There is one important requirement for pH/surfactant titration: It must go through points where the surface chemistry is in equilibrium with the liquid bulk chemistry. This is the only case where pH/surfactant titration is meaningful and reproducible. It is in this situation that the IEP demarcates the instability range of the system. This implies that a true titration is actually an "equilibrium pH titration." This means that in order to run pH/surfactant titration properly, we should have some idea about chemical kinetics and about the time required for reaching surface–bulk equilibrium after the addition of chemicals. We have come across situations in which investigators neglected this kinetic aspect of the pH/surfactant titration and therefore measured completely meaningless, nonreproducible titration curves.

Figure 12.6 demonstrates variations in the ζ-potential and pH of an yttria-stabilized zirconia (YSZ) slurry with time following a step change of pH. We call this a "time titration." It is seen that equilibration takes more than half an hour. In contrast, the equilibration of 10 wt % alumina slurries in various pH values takes less than a couple of minutes, as shown in Figure 12.7. This difference between YSZ and alumina requires a large difference in the respective titration protocols used to obtain the data.

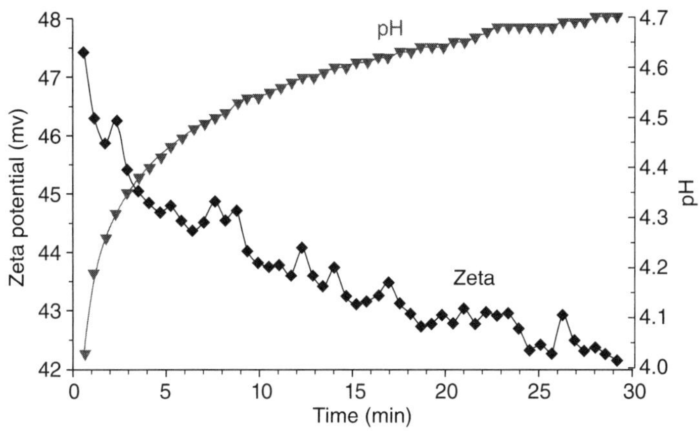

Figure 12.6. Equilibration of 3 vol % YSZ slurry prepared in 0.01 mol/L KCl.

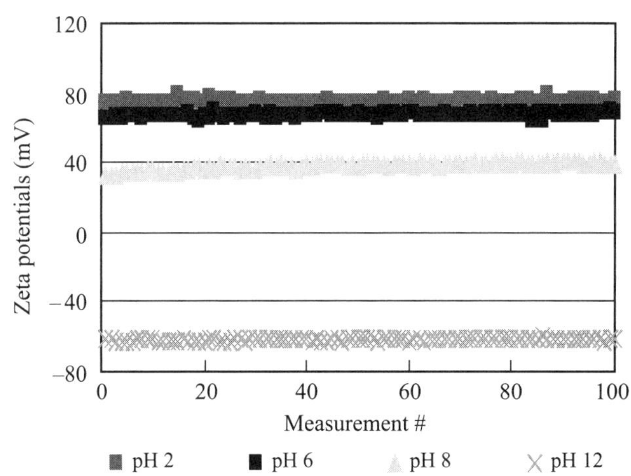

Figure 12.7. Zeta potential for 10 mass % alumina as a function of time. Equilibration of 10 mass % alumina slurry prepared in 0.01 mol/L KCl: One measurement requires 30 seconds.

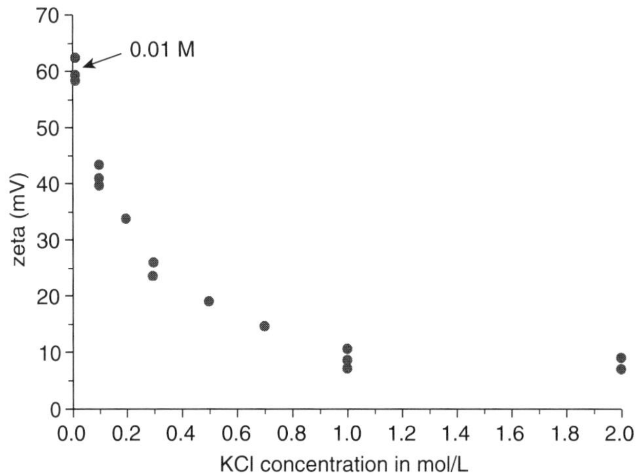

Figure 12.8. ζ-Potential of alumina AKP-30 at 3 vol % versus various KCl concentrations.

12.5.4 ζ-Potential Measurement under High Ionic Strength

Ionic strength is important not just as a factor affecting ζ-potential. Figure 12.8 provides a typical example showing the capability of acoustics to measure the CVI over an extremely wide range of ionic strength. The above-mentioned result shows an example that emphasizes one important advantage of the electroacoustic method. It is capable of ζ-potential measurements over an extremely wide range of conductivities, from nonpolar liquids like toluene up to several moles per liter of aqueous solutions. This is an important feature of acoustics because the capability of measuring the ζ-potential in aqueous/nonaqueous solutions with nonpolar and high ionic strength can open the way to apply the many kinds of practical systems.

12.5.5 Comparison of ζ-Potential with Other Techniques

Figure 12.9 shows representative samplings of a base titration curve for 5 vol % concentration for each of the three techniques involved in the study. Each corresponding *y*-axis has been scaled to provide a better indication of the degree of correlation that can be expected between the techniques over the tested pH range. For the purpose of comparison, the ζ-potential–pH curve determined at dilute concentrations by laser Doppler electrophoresis (LDE) is included. The LDE curve has been plotted on the same scale used for CVI data.

The results of this investigation lead us to conclude that there is a quite good agreement between the data obtained from ESA and CVI.

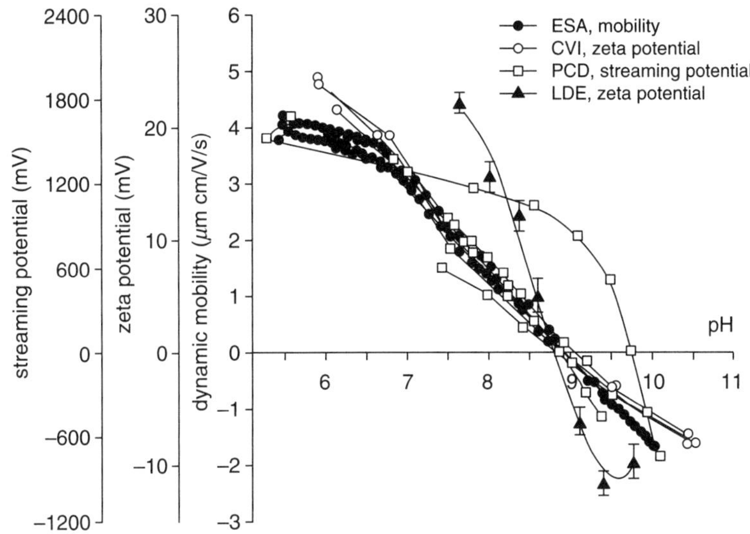

Figure 12.9. Electrokinetic curves for 5% AKP-30 alumina measured by CVI, ESA, and particle charge detection (PCD) (streaming potential). Zeta potential measured by laser Doppler electrophoresis (LDE) at a dilute concentration is shown for comparison. Three curves for each technique (except LDE) were more or less randomly selected to represent the typical spread in measured elektrokinetic titrations.

REFERENCES

1 P. Debye. A method for the determination of the mass of electrolyte ions. *J. Chem. Phys.* 1 (1933) 13–16.

2 E. Yeager, J. Bugosh, F. Hovorka, J. McCarthy. The application of ultrasound to the study of electrolyte solutions. II. The detection of Debye effect. *J. Chem. Phys.* 17 (1949) 411–415.

3 J. Bugosh, E. Yeager, F. Hovorka, J. McCarthy. The application of ultrasonic wave to the study of electrolyte. I, a modification of Debye's equation for the determination of the masses of electrolyte ions by means of ultrasonic waves. *J. Chem. Phys.* 15 (1947) 592–597.

4 B. Derouet, F. Denizot. The measurement of ultrasonic trolytic solutions. IV. Vibration potentials in electrolytes. *C. R. Acad. Sci. Paris* 233 (1951) 368.

5 E. Yeager, F. Hovorka. The application of ultrasonics to study of electrolyte solutions. III. The effect of acoustical waves on the hydrogen electrode. *J. Chem. Phys.* 17 (1949) 416–417.

6 E. Yeager, F. Hovorka. Ultrasonic waves and electrochemistry. I. A survey of the electrochemical applications of ultrasonic waves. *J. Acoust. Am.* 25 (1953) 443–455. *J. Chem. Phys.* 17 (1949) 416–417.

7 E. Yeager, H. Dietrich, F. Hovorka. Ultrasonic waves and electrochemistry. II. Colloidal and ionic vibration potentials. *J. Acoust. Soc. Am.* 25 (1953) 456.

8 R. Zana, E. Yeager. Determination of ionic partial molar volumes from ionic vibration potentials. *J. Phys. Chem.* 70 (1966) 954–956.

9 R. Zana, E. Yeager. Quantitaive studies of ultrasonic vibration potentials in polyelectrolyte solutions. *J. Phys. Chem.* 71 (1967) 3502–3520.

10 R. Zana, E. Yeager. Ultrasonic vibration potentials in tetraalkylammonium halide solutions. *J. Phys. Chem.* 71 (1967) 4241–4244.

11 R. Zana, E. Yeager. Ultrasonic vibration potentials and their use in the determination of ionic partial molal volumes. *J. Phys. Chem.* 71 (1967) 521–535.

12 R. Zana, E. Yeager. Ultrasonic vibration potentials. *Mod. Aspects Electrochem.* 14 (1982) 3–60.

13 J. Hermans. Charged colloid particles in an ultrasonic field. *Phils. Mag.* 25 (1938) 426.

14 A. J. Rutgers, W. Rigole. Ultrasonic vibration potentials in colloid solutions, in solutions of electrolytes and pure liquids. *Trans. Faraday Soc.* 54 (1958) 139–143.

15 J. A. Enderby. On electrical effects due to sound waves in colloidal suspensions. *Proc. Roy. Soc. Lond.* A207 (1951) 329–342.

16 F. Booth, J. Enderby. On electrical effects due to sound waves in colloidal suspensions. *Proc. Am. Phys. Soc.* 208A (1952) 32.

17 T. Oja, G. Petersen, D. Cannon. Measurement of electric-kinetic properties of a solution. US Patent 4,497,208, 1985.

18 B. J. Marlow, T. Oja, P. J. Goetz. Colloid analyzer. US Patent 4,907,453, 1990.

19 B. J. Marlow, D. Fairhurst, H. P. Pendse. Colloid vibration potential and the electrokinetic characterization of comcemtrated colloids. *Langmuir* 4 (1983) 611–626.

20 S. Levine, G. H. Neale. The prediction of electrokinetic phenomena within multiparticle systems. 1. Electrophoresis and electroosmosis. *J. Colloid Interface Sci.* 47 (1974) 520–532.

21 R. W. O'Brien. Electro-acoustic effects in a dilute suspension of spherical particles. *J. Fluid Mech.* 190 (1988) 71–86.

22 R. W. O'Brien, B. R. Midmore, A. Lamb, R. J. Hunter. Electroacoustic studies of moderately concentrated colloidal suspensions. *Faraday Discuss. Chem. Soc.* 90 (1990) 1–11.

23 P. F. Rider, R. W. O'Brien. The dynamic mobility of particles in a non-dilute suspension. *J. Fluid Mech.* 257 (1993) 607–636.

24 H. Ohshima. Dynamic electrophoretic mobility of spherical particles in concentrated suspensions. *J. Colloid Interface Sc.* 195 (1997) 137–148.

25 J. P. Ennis, A. A. Shugai, S. L. Carnie. Dynamic mobility of two spherical particles with thick double layers. *J. Colloid Interface Sci.* 223 (2000) 21–36.

26 J. P. Ennis, A. A. Shugai, S. L. Carnie. Dynamic mobility of two spherical particles with thick double layers in a non-dilute suspension. *J. Colloid Interface Sci.* 223 (2000) 37–53.

27 A. S. Dukhin, V. N. Shilov, Y. U. Borkovskaya. Dynamic electropholetic mobility in concentrated dispersed systems. Cell model. *Langmuir* 15 (1999) 3452–3457.

28 A. S. Dukhin, V. N. Shilov, H. Ohshima, P. J. Goetz. Electroacoustics phenomana in concentrated dispersions. New theory and CVI experiment. *Langmuir* 15 (1999) 6692–6706.

29 A. S. Dukihin, V. N. Shilov, H. Ohshima, P. J. Goetz. Electroacoustic phenomena in concentrated dispersions. Effcct of the surface conductivity. *Langmuir* 16 (2000) 2615–2620.

30 A. S. Dukihin, P. J. Goetz. *Characterization of liquids, nano- and microparticles, and porous bodies using ultrasound*, 2nd ed. Elsevier, Oxford, UK., 2010, p. 293.

13

COLLOID VIBRATION POTENTIAL AND ION VIBRATION POTENTIAL IN SURFACTANT SOLUTIONS

YOUICHI TAKATA AND HIROYUKI OHSHIMA

13.1 INTRODUCTION

The mutual interaction acting between colloidal particles in a medium plays an important role for determining the stability of colloidal dispersions such as suspensions, emulsions, and so on. The Derjaguin–Landau–Verwey–Overbeek (DLVO) theory states that the attractive van der Waals interaction and the repulsive electrical double-layer interaction are the principal determinants for the stability of a colloidal dispersion [1]. The electrical double-layer interaction depends on the surface charge density or zeta potential of colloidal particles. In general, such electrostatic properties have been evaluated experimentally by taking advantage of the electrokinetic phenomena in a *dilute* suspension of colloidal particles, for example, electrophoresis, sedimentation potential, streaming potential, and electroosmosis [2]. Recently, however, an ultrasonic vibration method based on an electroacoustic phenomenon has been developed for zeta-potential measurement in a *concentrated* suspension of colloidal particles [3].

In this chapter, the ultrasonic vibration method is applied to an aqueous solution of ionic surfactants, and we discuss the electrokinetic phenomenon of ultrasound with respect to surfactant aggregates.

Electrical Phenomena at Interfaces and Biointerfaces: Fundamentals and Applications in Nano-, Bio-, and Environmental Sciences, First Edition. Edited by Hiroyuki Ohshima.

13.2 THEORETICAL BACKGROUND OF ULTRASONIC VIBRATION POTENTIAL

The basic concept of ultrasonic vibration potential resembles sedimentation potential. Instead of measuring the sedimentation potential obtained by moving the charged colloidal particles under a gravitational field or centrifugal force, one measures the potential difference induced by the irradiation of ultrasonic waves. The propagation of ultrasonic waves in an electrolyte solution causes the vibration of both cations and anions. Because of the difference in their masses, the motion of cations and that of anions may differ from each other. This brings about the periodic change in the density of either cations or anions at a given point in the solution (Fig. 13.1). The resulting potential difference between two points in the solution can be measured by placing electrodes into position at half the wavelength. The potential difference and corresponding current are called the ion vibration potential (IVP) and the ion vibration current (IVI), respectively. Debye first predicted such an electroacoustic phenomenon and derived an approximate expression for IVP [4]. Then, Debye's theory for IVP was extended by some researchers [5, 6].

A similar effect occurs in a suspension of colloidal particles. In the case of colloidal suspension, the mass of the particles is obviously much heavier than that of electrolyte ions in the double layer. Therefore, the distortion of an electrical double layer formed around the colloidal particles, in contrast to the relative motion of cations and anions in the electrolyte solution, is caused by the propagation of ultrasonic waves in the colloidal dispersion and consequently brings about a difference in the potential (Fig. 13.2). This is the origin of colloid vibration potential (CVP) and colloid vibration current (CVI). The theory for CVP has been developed by Enderby et al. [7, 8], Marlow et al. [9],

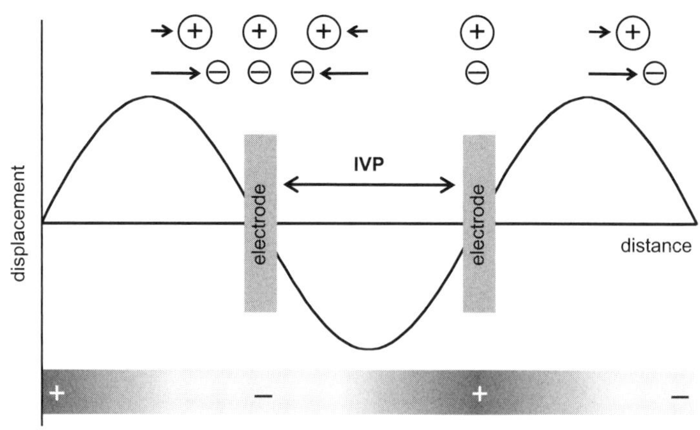

Figure 13.1. Ion vibration potential caused by the relative motion of cations and anions.

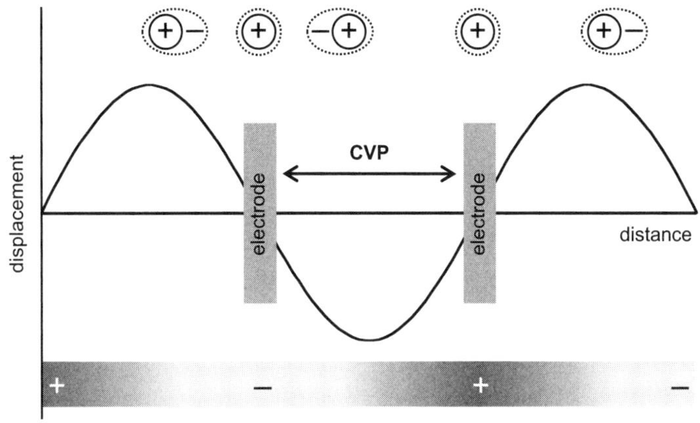

Figure 13.2. Colloid vibration potential caused by the asymmetry of an electrical double layer around the colloidal particles.

O'Brien [10], and Hunter [11]. From a practical point of view, furthermore, Dukhin et al. have derived the theoretical equation of CVP in a concentrated colloidal dispersion [12–16].

In the above-mentioned theoretical studies, IVP and CVP have been treated separately. Taking into account that IVP and CVP are generated simultaneously in real colloidal dispersions in electrolyte solution, however, a more general unified theory is required for application to the experimental results. In order to shed light on the relationship among the theories for IVP and CVP, Ohshima has derived a general electroacoustic theory for the potential difference caused by ultrasonic waves in a dilute suspension of spherical colloidal particles in an electrolyte solution by considering explicitly the presence of both particles and electrolyte ions, and has proved that the ultrasonic vibration potential is the sum of IVP and CVP [17, 18]. The derivation of this theory enables us to analyze the ultrasonic vibration potential in systems including both particles and electrolyte ions.

13.3 ULTRASONIC VIBRATION CURRENT IN SURFACTANT SOLUTIONS

Surfactant molecules in aqueous solution form a micelle above a defined concentration, that is, a critical micelle concentration (CMC). In this regard, it is of much interest to investigate electroacoustic effects in aqueous solutions of ionic surfactants because the solution behaves like an electrolyte solution below the CMC but may be regarded as a kind of dilute dispersion of colloidal particles above the CMC. Once the ultrasonic vibration potential is measured as a function of the concentration of an aqueous solution of ionic surfactants across the CMC, therefore, the results associated with the micelle formation are expected to provide information on the relationship between IVP and CVP.

13.3.1 Theoretical Equation of Ultrasonic Vibration Current

For a dilute suspension of spherical colloidal particles in an electrolyte solution, Ohshima derived a theoretical equation for the total vibration current (TVI) as the sum of IVI and CVI given by [17, 18]

$$\mathrm{TVI} = \mathrm{IVI} + \mathrm{CVI} = \sum_j \left(\frac{z_j e n_j^\infty}{\lambda_j} \right) \left(\frac{m_j - \rho_0 V_j}{\rho_0} \right) \Delta P + \frac{\phi(\rho_p - \rho_0)}{\rho_0} \mu(\omega) \Delta P, \quad (13.1)$$

where e, ρ_0, ρ_p, ϕ, $\mu(\omega)$, and ΔP represent the elementary electric charge, the liquid mass density, the particle mass density, the particle volume fraction, the dynamic electrophoretic mobility depending on frequency ω, and the pressure difference between two points in the solution, respectively. Then, z_j, n_j^∞, m_j, and V_j, respectively, are the ionic valence, the bulk concentration, the mass, and the partial molar volume of ionic species j. Also, the drag coefficient λ_j of ionic species j is defined by the following equation:

$$\lambda_j = \frac{N_A e^2 |z_j|}{\Lambda_j^0}, \quad (13.2)$$

where N_A and Λ_j^0 stand for the Avogadro's number and the limiting molar conductivity of ionic species j, respectively. Equation 13.1 tells us that the value of IVI depends on the ionic species existing in the solution, that is, the values of z_j, m_j, V_j, and Λ_j^0, whereas that of CVI depends on the size and/or number of colloidal particles, ϕ; particle mass density, ρ_p; and the electrokinetic property of particle, $\mu(\omega)$. In the following sections, we discuss the effect of these parameters on the ultrasonic vibration current by comparing the experimental results for various ionic surfactants.

13.3.2 Effect of Micelle Formation on Ultrasonic Vibration Current

Figure 13.3 shows the result of ultrasonic vibration current measurement in an aqueous solution of cationic surfactant, dodecyltrimethylammonium bromide (DTAB) [19]. It must be kept in mind that the value of the vertical axis in Figure 13.3 obtained from the measurement is the amplified signal magnitude, not the current itself. The values of TVI increase linearly as the concentration of the aqueous DTAB solution is increased, and then a break point appears at around 15 mM. Considering that the CMC of DTAB in an aqueous solution estimated by other techniques such as surface tensiometry and conductometry is about 15 mM, the occurrence of a break point on the TVI versus concentration curve suggests that the ultrasonic vibration current

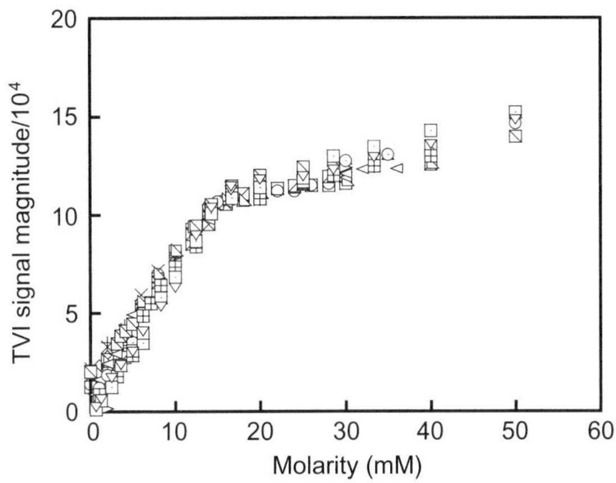

Figure 13.3. Plots of TVI signal magnitude against molarity of an aqueous DTAB solution. The concentration at a break point is close to the CMC of DTAB obtained from the conductivity measurement.

of an aqueous surfactant solution is clearly affected by micelle formation. The TVI value continues to increase across a break point, but its slope becomes gradual.

For the aqueous surfactant solution, we can distinguish formally between IVI and CVI on the basis of the following: (i) In a concentration regime lower than the CMC, only cations and anions originated from surfactant molecules are present in the solution; namely, TVI is equivalent to IVI; (ii) in a concentration regime higher than the CMC, the micelle is formed in the solution, and therefore we need to consider the contribution of both IVI and CVI to the ultrasonic vibration current. Taking into account that the concentration of monomer is approximately constant at concentrations higher than the CMC when the aggregation number of micelles is sufficiently large [20–22], however, the contribution of IVI to TVI may be taken as constant above the CMC. Summarizing these points, the concentration dependence of ultrasonic vibration current can be presumed as in Figure 13.4. Therefore, it is found that the increase in TVI values above the CMC is attributable to the micelle particles existing in the aqueous solution. Since the CVI value in Equation 13.1 includes the dynamic electrophoretic mobility of particles, it is expected that the concentration dependence of TVI above the CMC yields the micelle zeta potential.

As another experimental output other than the ultrasonic vibration current, we can obtain the phase of electroacoustic signals. By calibrating using negatively charged silica particles before the measurement [19], the phase for a negatively charged particle is in general assigned to 180°, and that for a positively charged one is 0° or 360° [23]. It is expected that the latter should be the case for a micelle of cationic surfactants because its surface has a positive

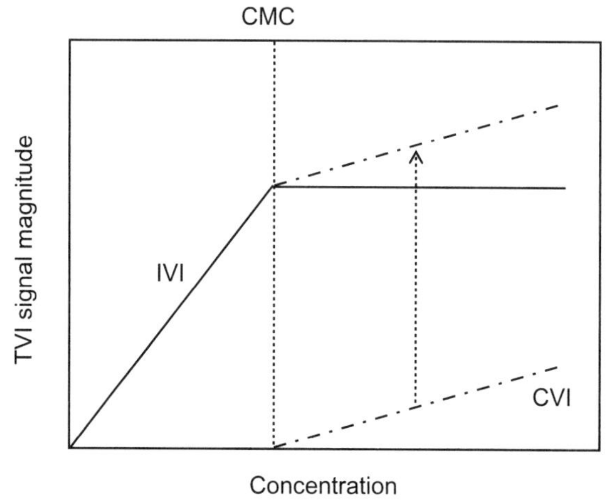

Figure 13.4. Scheme of concentration dependence of ultrasonic vibration current in the system associated with micelle formation.

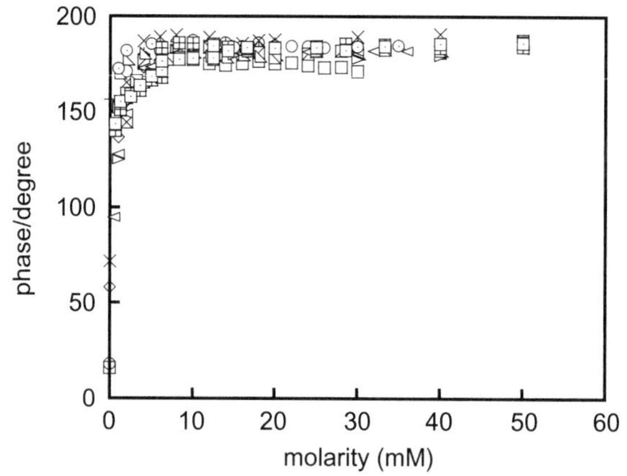

Figure 13.5. Plots of phase against molarity of an aqueous DTAB solution. The values of phase approach 180° by increasing the molarity.

charge. Figure 13.5 represents the concentration dependence of phase in an aqueous DTAB solution. The values for phase steeply increase with increasing concentration, and then are held constant at around 180° above the CMC. Although this result seems to be in conflict with the above expectation that the phase for a cationic surfactant micelle becomes 0 or 360°, this is due to the density contrast between particles and the surrounding medium. The silica particles for calibration are heavier than the surrounding medium. Assuming that the density of the micelle is regarded as that of a hydrocarbon having the same carbon number as the hydrophobic group of the surfactant (dodecane in the case of DTAB), the micelle particles are lighter than the surrounding medium. In principle, the reversal of density contrast causes the phase to shift by 180°. Therefore, the results shown in Figure 13.5 also suggest that the DTAB micelle possesses a positive charge.

13.3.3 Counterion Effect on the Ultrasonic Vibration Current

In order to investigate the effect of counterions on the electrokinetics of surfactant micelles, we measured the ultrasonic vibration current in an aqueous solution of DTAB and dodecyltrimethylammonium chloride (DTAC). Since these surfactant molecules possess the common surfactant ion (DTA^+) and different counterions (Br^- and Cl^-), it is possible to examine not only the effect of different constituent ions on IVI but also the direct influence of counterions on the zeta potential of micelles because the dissociation degree of bromide ions from the positively charged surface is different from that of chloride ions [24–26].

Figure 13.6 shows the results of ultrasonic vibration current measurement for the DTAB and DTAC systems [19]. In analogy with the DTAB system, the TVI values in the DTAC system increase linearly with increasing molarity, and then the concentration dependence (we call this "slope" hereafter) of the TVI abruptly changes upon passing through a break point, although this point is obscure as compared to the DTAB system. As illustrated in Figure 13.6, it is found that the different constituent ions, even though they are only counterions, cause a remarkable change in the ultrasonic vibration current of the aqueous solution of ionic surfactants. The concentration C_{break} at the break point on the TVI versus molarity curve, the slope below and above C_{break}, and the CMC determined from the conductivity measurement are tabulated in

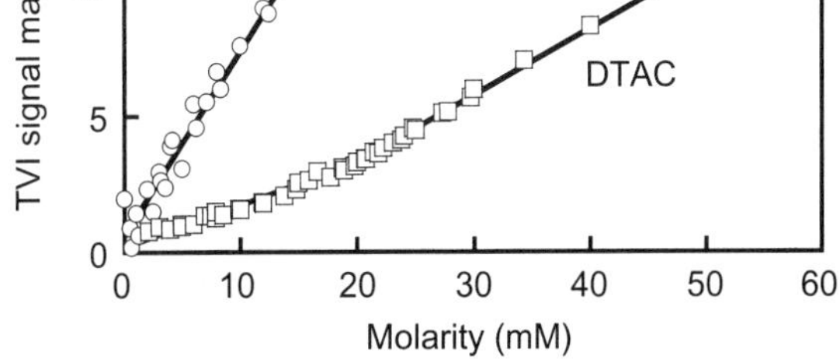

Figure 13.6. Plots of TVI signal magnitude against molarity of aqueous DTAB and DTAC solutions.

TABLE 13.1. Parameters Determined from Ultrasonic Vibration Current Measurement in Each Surfactant System

	CMC (mM)	C_{break} (mM)	Slope (<CMC)	Slope (>CMC)
DTAB	15.46	15.27	6486	936
DTAC	23.00	19.09	1604	2447
DeTAB	64.80	61.14	4693	505
DTAB	15.46	15.27	6486	936
TTAB	3.77	3.65	7657	953
LiDS	8.58	9.06	−1294	2,742
LiFOS	7.63	7.42	13,897	45,353

Table 13.1. The deviation of C_{break} from the CMC in the DTAC system is probably an artifact introduced by a similar slope below and above C_{break}.

Let us focus on the ultrasonic vibration current below the CMC. As mentioned earlier, TVI is equivalent to IVI in this concentration regime because only cations and anions originated from the surfactant molecules are present in the aqueous solution. This means that the theoretical values of IVI are calculated using the first term of Equation 13.1, where the partial molar volumes of cations and anions are the unknown parameter. The detailed calculation has been reported elsewhere [19], and the results are shown in Figure 13.7. The partial molar volume determined in the course of theoretical calculation is 233.0 cm^3/mol for the dodecyltrimethylammonium ion. Therefore, the partial molar volumes of surfactant molecules, which are defined as the sum of the partial molar volume of cations and anions, are 263.2 and 256.7 cm^3/mol for DTAB and DTAC, respectively. Thus, the IVI enables us to determine the partial molar volume of ions in the solution.

Next, we consider the ultrasonic vibration current in the concentration regime higher than the CMC. Under this condition, the micelle particles are formed in the aqueous solution and necessarily coexist with the monomer of the surfactant. Here, we assumed that the monomer concentration is approximately constant, at least near the CMC, on the basis of the mass action model [20–22], and therefore the value of IVI after micelle formation is kept constant at the value of TVI at C_{break}. According to this assumption, it is suggested that the increment in TVI, alternatively the slope of the TVI, above the CMC is caused by the micelle particles and assigned to the CVI. As shown in Figure 13.6, the TVI values in both the DTAB and DTAC systems still increase linearly above the CMC, but their slopes differ from each other. This difference can be qualitatively explained according to Equation 13.1. Supposing that the micelle mass density, ρ_p, is the same in the DTAB and DTAC systems because the surfactant molecules contain the common surfactant ion (dodecyltrimethylammonium ion), which mainly constitutes the core of micelle, and furthermore, the volume fraction of micelle particles, φ, is proportional to the surfactant concentration, it is noted from Equation 13.1 that the slope of CVI

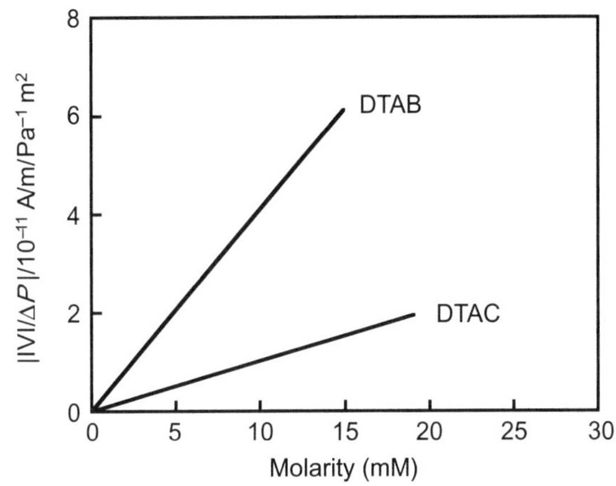

Figure 13.7. Magnitude of IVI calculated by using Equations 13.1 and 13.2 as a function of the molarity of aqueous DTAB and DTAC solutions. Note that the vertical axis is the absolute value of IVI, and its order is extremely smaller than the TVI signal magnitude in Figure 13.6.

solely depends on the dynamic electrophoretic mobility, $\mu(\omega)$, of micelle particles. Therefore, the concentration dependence of CVI is correlated to the zeta potential of micelles. From the results presented in Figure 13.6, we can easily speculate that the zeta potential of DTAC micelles becomes higher than that of DTAB micelles. This is predictable from the well-known fact that the degree of dissociation of bromide ions from micelle surfaces is less than that of chloride ions [24–26], and a qualitatively reasonable conclusion.

The qualitative understanding of the electrokinetics of micelle particles above is further proved by the following sections.

13.3.4 Alkyl Chain Length Effect on the Ultrasonic Vibration Current

Counterions directly participate in the electrokinetic phenomena of charged particles, and therefore, it is natural that their effect on the ultrasonic vibration current is prominent, as shown in Figure 13.6. Next, it is interesting to investigate how the hydrophobic group of surfactant ions affects the ultrasonic vibration current. To this end, we measured the ultrasonic vibration current of an aqueous solution of cationic surfactants having different alkyl chain lengths, where the hydrophilic group and counterion are common. The alkyl chain length brings about the differences in the size and aggregation number of micelle particles [27] in addition to differences in the partial molar volume of surfactant ions. Consequently, the values of V_j and ϕ in Equation 13.1 should affect IVI and CVI, respectively.

Figure 13.8 shows the results of ultrasonic vibration current measurements for decyltrimethylammonium bromide (DeTAB), DTAB, and

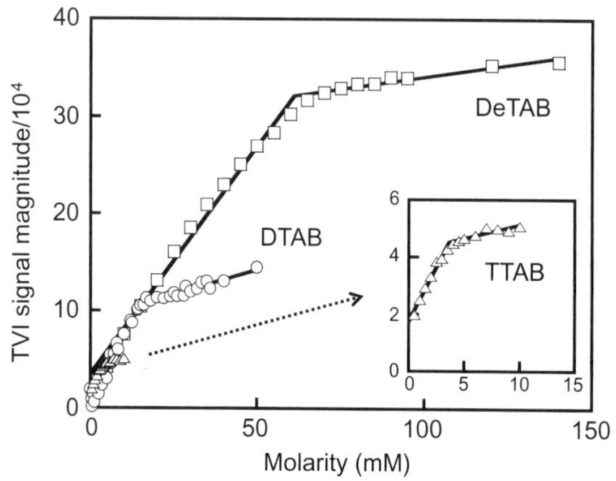

Figure 13.8. Plots of TVI signal magnitude against molarity of aqueous DeTAB, DTAB, and TTAB solutions. The concentration dependence of TVI exhibits similar behavior in these systems.

tetradecyltrimethylammonium bromide (TTAB) systems. In all three systems, the TVI values increase linearly by increasing the molarity, and moreover, there exists a break point on the TVI versus molarity curves. The parameters obtained in this figure are summarized in Table 13.1. The agreement between the values of C_{break} and CMC leads to the conclusion that the break point on the TVI versus molarity curve corresponds to the CMC. Focusing on the slope below and above the CMC in each system, it seems that the TVI is not influenced by the differences in the alkyl chain length as compared to the counterion effect in the DTAB and DTAC systems.

Here let us estimate the partial molar volume of surfactant ions using the slope of IVI given in Figure 13.8. The theoretical calculation yielded 164.4 and 294.8 cm^3/mol for the partial molar volumes of decyltrimethylammonium and tetradecyltrimethylammonium ions, respectively. Then, the partial molar volumes of DeTAB and TTAB becomes 194.6 and 325.0 cm^3/mol. These values are far from the partial molar volumes determined by the density measurement, for example, 256.9 cm^3/mol for DeTAB [28]. When the slope of IVI is close to each other, as in Figure 13.8, the results of theoretical calculation become fairly sensitive to the slight variation in the slope caused by an experimental error of ultrasonic vibration current. Although there is still room for improvement regarding the theoretical calculation of IVI, it is certain that there is a difference between the partial molar volumes of surfactant ions having different alkyl chain lengths. Therefore, the results shown in Figure 13.8 suggest that the potential difference induced by the vibration of surfactant ions and counterions in an aqueous solution greatly depends on the kind of counterions involved but is hardly affected by the hydrocarbon chain in spite of the change in the partial molar volume associated with the extension of the alkyl chain.

In the concentration regime higher than the CMC, the different counterions cause a remarkable change in the slope of CVI (Fig. 13.6), but different alkyl chain lengths do not (Fig. 13.8). In the DTAB and DTAC systems, the effect of counterions on CVI is dominated by the dynamic electrophoretic mobility, $\mu(\omega)$, in Equation 13.1. Furthermore, the change in the size and/or aggregation number of micelle particles, which relates to the volume fraction, ϕ, in Equation 13.1, with an increase in the methylene group is not a crucial factor contributing to CVI. As described later, the similarity in the slope of CVI also implies that the effect of ρ_p in Equation 13.1 on CVI is negligibly small in this case.

13.3.5 Hydrophobic Group Effect on the Ultrasonic Vibration Current

CVI occurs as a result of the vibrating of ionic species (the charged colloidal particles and counterions) in an aqueous solution. There are two main factors contributing to this electroacoustic phenomenon. One is the electric charge of the species and the resulting interaction among them and the other is the density of the species, more accurately the density contrast between medium and colloidal particles. The former is supported by the results of the counterion effect on CVI as shown in Figure 13.6. For the latter, it is anticipated that the heavier the colloidal particle is, the larger the CVI becomes. Assuming that the density of a micelle particle can be regarded as that of a hydrocarbon that has the same carbon number as the hydrophobic chain of the surfactant molecules, the results in Figure 13.8 indicate that the difference in the density between decane (~0.73 g/cm^3) and tetradecane (~0.76 g/cm^3) is not enough to affect CVI. This is the reason why we chose a fluorocarbon surfactant for the following step. In general, the density of fluorocarbons is markedly different from that of hydrocarbons. Furthermore, fluorocarbons are heavier than water, whereas hydrocarbons are lighter. Therefore, the novel information on the electrokinetic phenomena of surfactant micelles can be obtained from the standpoint of density contrast.

The concentration dependence of TVI in an aqueous solution of fluorocarbon surfactant, lithium heptadecafluorooctane sulfonate (LiFOS), and that of a hydrocarbon surfactant, lithium dodecyl sulfate (LiDS), is shown in Figure 13.9. For comparison, the results obtained for the DTAB system is inserted in this figure. First of all, it should be noted that there is a break point on the TVI versus molarity curve in each system. Again, the concentration at the break point is close to the CMC estimated from the conductivity measurement (Table 13.1), and this is consistent with the above results. With respect to the unique behavior of the negative slope of IVI found in the LiDS system, this may be due to the balance of the contribution of cations and anions to IVI caused by the negative value of the partial molar volume of lithium ions [6].

The most striking result in Figure 13.9 is the fairly larger values of TVI and slope in the LiFOS (fluorocarbon surfactant) system, compared to the LiDS and DTAB (hydrocarbon surfactant) systems. In particular, the slope of TVI above the CMC strongly reflects the density contrast between the medium

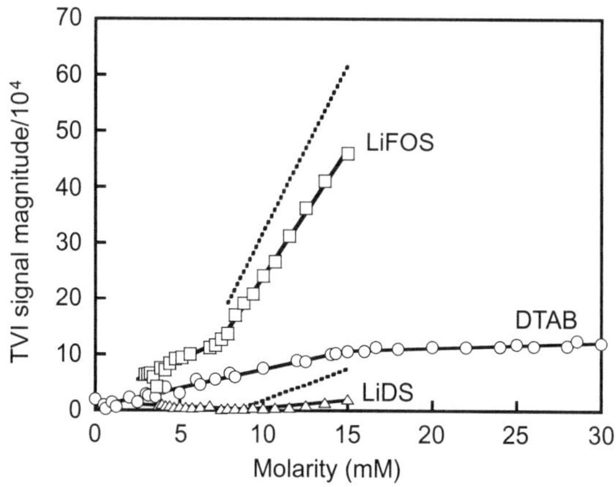

Figure 13.9. Plots of TVI signal magnitude against molarity of aqueous LiDS and LiFOS solutions. For reference, the result of the DTAB system is inserted. The dotted lines show the CVI versus molarity curves reduced by the term $(\rho_p - \rho_0)/\rho_0$ in Equation 13.1.

(water) and the micelle particles (fluorocarbon). Here let us eliminate the contribution of the density contrast by considering the term $(\rho_p - \rho_0)/\rho_0$ in Equation 13.1. The dotted lines in Figure 13.9 stand for the TVI values reduced by the density contrast. The ratio of the slope of the LiFOS system to that of the LiDS system significantly decreases from 16.540 (= 45,353/2742) to 5.402 (= 60,049/11,116) before and after the correction by the density contrast. Actually, the aggregation number and shape of micelle particles formed in the solution depend on the kind of surfactant molecules, and therefore it is probable that these factors affect the slope of CVI through the volume fraction ϕ in Equation 13.1. Nevertheless, it should be emphasized that the CVI values are highly influenced by the density if there is enough difference in the density between medium and colloidal particles.

13.4 CONCLUSION

We applied the ultrasonic vibration technique to aqueous solutions of various ionic surfactants and investigated the effect of counterions, aggregation number and size, and core density of micelle on the ultrasonic vibration current. A break point on the TVI versus molarity curves around the CMC suggested the close relationship between ultrasonic vibration current and micelle formation. The slope of IVI, depending on the ionic species in the solution, allowed us to estimate the partial molar volume of surfactant ions. The slope of CVI was sensitive to the change in the micelle zeta potential associated with the different counterion binding and, furthermore, amplified remarkably when the density contrast between the micelle particles and the surrounding medium was prominent. Thus, the ultrasonic vibration technique provides a useful tool for studying the electrokinetic properties of surfactant aggregates.

REFERENCES

1 J. N. Israelachvili. *Intermolecular and surface forces*, 2nd ed. Academic Press, London, 1992, Chapter 12.
2 D. F. Evans, H. Wennerström. *The colloidal doamin*, 2nd ed. Wiley-VCH, New York, 1999, Chapter 8.
3 A. S. Dukhin, P. J. Goetz. *Ultrasound for characterizing colloids*. Elsevier, Amsterdam, 2006.
4 P. Debye. *J. Chem. Phys.* 1 (1933) 13.
5 J. Bugosh, E. Yeager, F. Hovorka. *J. Chem. Phys.* 15 (1947) 592.
6 R. Zana, E. Yeager. *J. Phys. Chem.* 71 (1967) 521.
7 J. A. Enderby. *Proc. R. Soc. Lond. A* 207 (1951) 329.
8 F. Booth, J. A. Enderby. *Proc. Phys. Soc. A* 65 (1952) 321.
9 B. J. Marlow, D. Fairhurst, H. P. Pendse. *Langmuir* 4 (1988) 611.
10 R. W. O'Brien. *J. Fluid Mech.* 190 (1988) 71.
11 R. J. Hunter. *Colloids Surf. A* 141 (1998) 37.
12 H. Ohshima, A. S. Dukhin. *J. Colloid Interface Sci.* 212 (1999) 449.
13 A. S. Dukhin, H. Ohshima, V. N. Shilov, P. J. Goetz. *Langmuir* 15 (1999) 3445.
14 A. S. Dukhin, V. N. Shilov, H. Ohshima, P. J. Goetz. *Langmuir* 15 (1999) 6692.
15 A. S. Dukhin, V. N. Shilov, H. Ohshima, P. J. Goetz. *Langmuir* 16 (2000) 2615.
16 V. N. Shilov, Y. B. Borkovskaja, A. S. Dukhin. *J. Colloid Interface Sci.* 277 (2004) 347.
17 H. Ohshima. *Langmuir* 21 (2005) 12100.
18 H. Ohshima. *Theory of colloid and interfacial electric phenomena*. Elsevier, Amsterdam, 2006, Chapter 10.
19 Y. Takata, T. Miyayama, T. Nagahashi, A. Hyono, H. Ohshima. *J. Oleo Sci.* 58 (2009) 557.
20 K. J. Mysels. *J. Colloid Sci.* 10 (1955) 507.
21 J. N. Phillips. *Trans. Faraday Soc.* 51 (1955) 561.
22 A. W. Adamson, A. P. Gast. *Physical chemistry of surfaces*, 6th ed. John Wiley & Sons, New York, 1997, Chapter 13.
23 A. S. Dukhin. *J. Colloid Interface Sci.* 310 (2007) 270.
24 L. Sepúlveda, J. Cortés. *J. Phys. Chem.* 89 (1985) 5322.
25 C. Gamboa, H. Ríos, L. Sepúlveda. *J. Phys. Chem.* 93 (1989) 5540.
26 M. Aratono, M. Villeneuve, T. Takiue, N. Ikeda, H. Iyota. *J. Colloid Interface Sci.* 200 (1998) 161.
27 C. Tanford. *J. Phys. Chem.* 76 (1972) 3020.
28 M. Yamanaka, S. Kaneshina. *J. Solution Chem.* 19 (1990) 729.

14

INTERFACIAL TENSION OF AQUEOUS ELECTROLYTE SOLUTIONS: ION-FREE LAYER

Youichi Takata and Hiroyuki Ohshima

14.1 INTRODUCTION

Surface tension reflects the properties of gas/liquid and liquid/liquid interfaces. For example, the surface tension of pure water drastically decreases with the addition of surfactants. This is because the surfactant molecules with hydrophilic and hydrophobic groups gather around the interface between hydrophilic (water) and hydrophobic (air) phases. This phenomenon is called *positive* adsorption. When electrolyte concentration increases, the surface tension of aqueous solutions of electrolytes, which refer to inorganic salts here, gradually increases. Thermodynamically, the positive slope on the surface tension versus solute concentration curve suggests the expulsion of solutes from the interfacial region (see the following section). Therefore, this phenomenon is called *negative* adsorption.

Although many fundamentals have been established in the field of surface chemistry, there still remain many interesting and unresolved issues concerning the surface tension of aqueous electrolyte solutions and the distribution of electrolyte ions in (thermodynamic) negative adsorption. For a long time, the decrease in surface tension at extremely low electrolyte concentrations has been known as the Jones–Ray effect [1, 2]. Positive adsorption of electrolyte ions was only suspected initially because of indirect information given by

Electrical Phenomena at Interfaces and Biointerfaces: Fundamentals and Applications in Nano-, Bio-, and Environmental Sciences, First Edition. Edited by Hiroyuki Ohshima.

surface tension measurement; then Petersen et al. directly confirmed the Jones–Ray effect through an experiment that applied second harmonic generation (SHG) to the aqueous electrolyte solution surface [3, 4]. According to some researchers [5, 6], the layer of hydroxide anions formed by their positive adsorption to the air/water surface plays a key role for the Jones–Ray effect. With respect to the positive adsorption of hydroxide anions, the problem of whether the water surface is acidic or basic is under discussion. The SHG measurement [7, 8] and molecular dynamics simulation [9] suggest the existence of hydronium (H_3O^+) cations near the air/aqueous electrolyte solution surface, whereas the negative zeta potentials of the air bubble [10] and the oil droplet [11, 12] support the adsorption of hydroxide ions.

Understanding of the nature of the air/aqueous electrolyte solution surface is important from a practical point of view as well as for the purpose of fundamental study. For example, this understanding may be related to the stability of a bubble [13] and an emulsion [12, 14] in the absence of a surfactant. This chapter presents a theoretical discussion on the interfacial tension of aqueous electrolyte solution, and by applying theory to experimental results, we consider the adsorption behavior of various electrolytes at various interfaces.

14.2 THEORETICAL CONSIDERATION ON INTERFACIAL TENSION OF AQUEOUS ELECTROLYTE SOLUTIONS

Theoretical studies on the surface tension of aqueous electrolyte solutions have been performed by some researchers. Wagner [15] and Onsager and Samaras [16] have shown that negative adsorption is caused by the electrostatic repulsive interaction acting between electrolyte ions in water phase and their images in air phase. The theoretical equation based on this concept described the experimental surface tension at very low electrolyte concentrations (~0.2 M). For further improvement on the Onsager–Samaras theory, Levin and Flores-Mena [17, 18] adopted the canonical approach (temperature T, volume V, number of molecules N ensemble) instead of the grand canonical approach (T, V, chemical potential μ ensemble) by Onsager and Samaras, and introduced an ion-free layer of thickness, δ, which is regarded as the radius of hydrated electrolyte ions, into the theoretical calculation. The agreement between theory and experiment was extended to a relatively high electrolyte concentration of 1 M. On the basis of the theory of Levin and Flores-Mena, Ohshima and Matsubara [19] have recently derived a simple analytic expression for the surface tension of aqueous electrolyte solutions that is applicable not only to the air/water surface but also to the interface between two arbitrary media.

In this section, we review the thermodynamic significance of negative adsorption. A schematic comprehension of the Gibbs dividing surface is useful for making the adsorption behavior of electrolytes clearer. Next, we address the concept of an ion-free layer defined on the basis of the Gibbs dividing surface. Finally, using Oshima's method [20, 21], we show the analytic equations for the interfacial tension acting between arbitrary fluid phases and aqueous solutions of general electrolytes including asymmetric electrolytes as well as symmetrical ones.

14.2.1 Gibbs Dividing Surface and Negative Adsorption

To understand the thermodynamics of an interface, we begin with considering the mechanical energy associated with the interface in the first law of thermodynamics. Assuming that the system is composed of three components (air, water, and nonionic solute) and two phases (air and aqueous solutions), the total differential of internal energy dU is given by the following equation:

$$dU = TdS - PdV + \gamma d\sigma + \mu_a dn_a + \mu_w dn_w + \mu_1 dn_1, \tag{14.1}$$

where T, S, P, V, γ, and σ stand for the temperature, entropy, pressure, volume, interfacial tension, and interfacial area, respectively. μ_i and n_i are the chemical potential and number of moles of component i. The subscripts a, w, and 1 mean the air, the water, and the solute, respectively. From Equation 14.1, the total differential of Gibbs free energy dG is derived as

$$dG = -SdT + VdP - \sigma d\gamma + \mu_a dn_a + \mu_w dn_w + \mu_1 dn_1. \tag{14.2}$$

Using Euler's theorem, we can obtain the Gibbs–Duhem equation for the system with the interface:

$$\sigma d\gamma = -SdT + VdP - n_a d\mu_a - n_w d\mu_w - n_1 d\mu_1. \tag{14.3}$$

For the homogeneous regions in the air and water phases at a great distance from the interface, on the other hand, we have the corresponding equations:

$$s^A dT - dP + C_a^A d\mu_a + C_w^A d\mu_w + C_1^A d\mu_1 = 0 \tag{14.4}$$

and

$$s^W dT - dP + C_a^W d\mu_a + C_w^W d\mu_w + C_1^W d\mu_1 = 0, \tag{14.5}$$

where s and C_i stand for the entropy and number of moles of component i per unit volume, respectively. The superscripts A and W mean respectively the air and water phases. Multiplying Equations 14.4 and 14.5 by the volume of the air and water phases, V^A and V^W, respectively, and then combining their equations with Equation 14.3, the following equation can be obtained:

$$d\gamma = -s^\sigma dT + v^\sigma dP - \Gamma_a^\sigma d\mu_a - \Gamma_w^\sigma d\mu_w - \Gamma_1^\sigma d\mu_1, \tag{14.6}$$

where y^{σ} ($y = s, v$) is the surface excess thermodynamic quantity defined by

$$y^{\sigma} = \frac{Y - V^{A}y^{A} - V^{W}y^{W}}{\sigma}. \tag{14.7}$$

Similarly, the surface excess number of moles of component i, Γ_i^{σ}, is defined by

$$\Gamma_i^{\sigma} = \frac{n_i - V^{A}C_i^{A} - V^{W}C_i^{W}}{\sigma}. \tag{14.8}$$

As the degree of freedom in this system is three, we need to eliminate two variables from Equation 14.6. According to Gibbs' method, the following conditions are imposed on Equation 14.6:

$$v^{\sigma} = 0 \tag{14.9}$$

and

$$\Gamma_{w}^{\sigma} = 0. \tag{14.10}$$

These conditions suggest that the thickness of the fluid/liquid interface can be regarded as zero, and the position of such a two-dimensional plane interface is chosen so that the surface excess number of moles of water becomes zero. By combining Equations 14.9 and 10 with Equation 14.6, we have

$$d\gamma = -s^{G}dT - \Gamma_{a}^{G}d\mu_{a} - \Gamma_{1}^{G}d\mu_{1}. \tag{14.11}$$

The superscript G represents the Gibbs' method. Assuming that the aqueous solution is ideally dilute, Equation 14.11 yields the Gibbs adsorption equation:

$$\Gamma_1^{G} = -\left(\frac{\partial \gamma}{\partial \mu_1}\right)_T = -\frac{C_1^{W}}{RT}\left(\frac{\partial \gamma}{\partial C_1^{W}}\right)_T. \tag{14.12}$$

Thus, the surface density of the nonionic solute, Γ_1^{G}, can be obtained from the concentration dependence of surface tension. Here, it should be noted that the variable p apparently vanishes in Equation 14.12. This is not favorable for an exact thermodynamic treatment because the surface tension measurement is performed under constant atmospheric pressure. This point is logically resolved by introducing two dividing planes that are chosen so that the surface excess of moles of air and water becomes zero into Equation 14.6. The details of this convention are summarized elsewhere [22–24].

Let us consider the Gibbs dividing surface schematically. The Gibbs dividing surface is defined by the condition of Equation 14.10:

$$\Gamma_{w}^{\sigma} = \frac{n_{w} - V^{A}C_{w}^{A} - V^{W}C_{w}^{W}}{\sigma} = \frac{n_{w}}{\sigma} - l^{A}C_{w}^{A} - l^{W}C_{w}^{W} = 0, \tag{14.13}$$

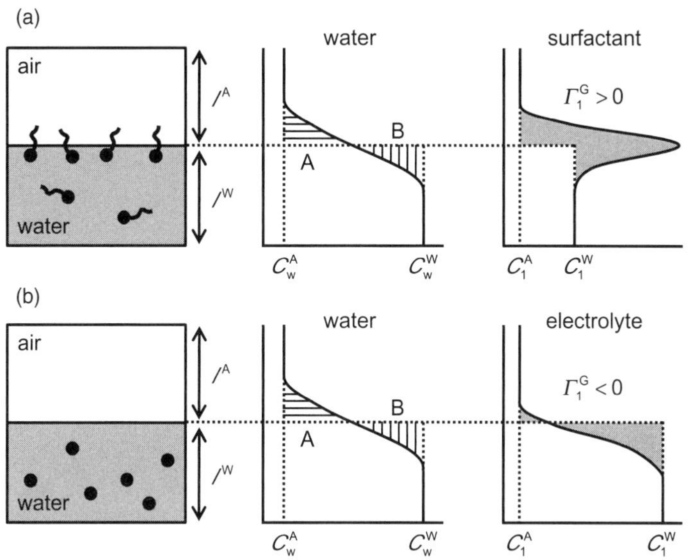

Figure 14.1. Schematic illustration of air/aqueous solution of (a) surfactant and (b) electrolyte surface. Concentration of water in air and water phases across the surface (center). The position of the Gibbs dividing surface is determined so that the areas of regions A and B become equal to each other. Concentration of solute in air and water phases across the surface (right).

where l^{A} and l^{W}, respectively, are the lengths of the air and water phases as shown in Figure 14.1, which shows the concentration change of water (w) and nonionic solute (1) in the system. From Figure 14.1, we can say that the condition of Equation 14.13 is to satisfy the equality of the area of regions A and B. Thus, the position of the Gibbs dividing surface is determined.

The surface density of nonionic solute Γ_{1}^{G} is defined by the following equation:

$$\Gamma_{1}^{G} = \frac{n_1 - V^{A}C_{1}^{A} - V^{W}C_{1}^{W}}{\sigma} = \frac{n_1}{\sigma} - l^{A}C_{1}^{A} - l^{W}C_{1}^{W} \tag{14.14}$$

and is schematically obtained on the basis of the Gibbs dividing surface. In the case of an aqueous surfactant solution, the adsorption of surfactant molecules brings about an increase in the solute concentration near the air/water surface, compared to that in the bulk phase as shown in Figure 14.1a. The surface density of the surfactant on the air/water surface is given by the shaded area in Figure 14.1a, and the Γ_{1}^{G} value is positive. If Γ_{1}^{G} takes negative values, on the other hand, the solute molecules in the aqueous solution should be distributed, for example, as in Figure 14.1b, namely, the solute molecules seem to be excluded from the surface region. The negative adsorption is thus interpreted from a schematic point of view.

There are some reports on the positive adsorption of electrolyte ions on the air/water surface [25–27]. Although this seems to be in conflict with the above-mentioned consideration, we can say that this is the difference between the macroscopic thermodynamic viewpoint and microscopic behavior of each ion.

14.2.2 Ion-Free Layer

Levin and Flores-Mena [17, 18] suggested that the Onsager–Samaras theory [16], which did not explain the experimental surface tension of aqueous electrolyte solutions at especially higher concentrations, did not take into account a space where the center of electrolyte ions cannot exist because of the finite radius and hydration of electrolyte ions on the air/aqueous solution surface. Such a space is an ion-free layer and is defined as shown in Figure 14.2. Here, we take a z-axis perpendicular to the air/water surface with its origin $z = 0$ at the Gibbs dividing surface so that the region $z < 0$ corresponds to the air phase and $z > 0$ to the aqueous solution phase. Following the theory of Levin and Flores-Mena [17, 18], an ion-free layer is formed just below the Gibbs dividing surface at the air/aqueous solution surface. We designate the thickness of the ion-free layer as δ, which corresponds to the average radius of hydrated electrolyte ions. Since the center of electrolyte ions cannot enter the ion-free layer region ($0 < z < \delta$), the following relation holds with respect to the potential energy $W_i(z)$ of an ionic species i at position z:

$$W_i(z) = \infty \quad (0 < z < \delta). \tag{14.15}$$

14.2.3 Analytic Equations of Interfacial Tension

Consider the surface tension γ of an aqueous electrolyte solution. Let the electrolytes be composed of N ionic species of valence ν_i and bulk concentra-

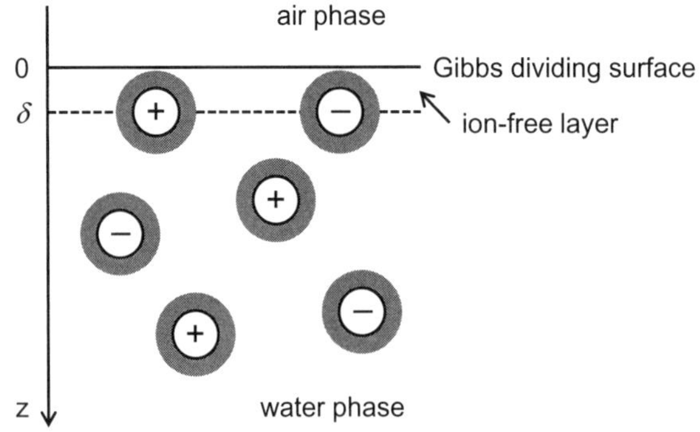

Figure 14.2. Schematic illustration of ion-free layer based on the Gibbs dividing surface. The thickness of ion-free layer δ corresponds to the radius of hydrated ions.

tion (number density) n_i ($i = 1, 2, \ldots, N$). We start with the Gibbs adsorption isotherm

$$d\gamma = -kT\sum_{i=1}^{N} \Gamma_i^{\mathrm{G}} d\ln n_i \tag{14.16}$$

with

$$\Gamma_i^{\mathrm{G}} = n_i \int_0^\infty \left[\exp\left(-\frac{v_i e\phi(z) + W_i(z)}{kT}\right) - 1\right] dz, \tag{14.17}$$

where Γ_1^{G} is the number of ions of species i adsorbed (negatively) at the interface per unit area between the air and the aqueous electrolyte solution; e is the elementary electric charge; k is the Boltzmann constant; and $\phi(z)$ is the mean potential. By integrating Equation 14.16 after substituting Equation 14.17, we obtain

$$\gamma = \gamma_0 - kT\sum_{i=1}^{N} \int_0^{n_i}\int_0^\infty \left[\exp\left(-\frac{v_i e\phi(z) + W_i(z)}{kT}\right) - 1\right] dz dn_i, \tag{14.18}$$

where γ_0 is the surface tension of pure water. Considering the ion-free layer shown in Figure 14.2 and the condition of Equation 14.15, the following equation is obtained:

$$\gamma = \gamma_0 + kT\delta\sum_{i=1}^{N} n_i - kT\sum_{i=1}^{N} \int_0^{n_i}\int_\delta^\infty \left[\exp\left(-\frac{v_i e\phi(z) + W_i(z)}{kT}\right) - 1\right] dz dn_i. \tag{14.19}$$

Further, we linearize the integrand in the last term on the right-hand side of Equation 14.19 with respect to $\phi(z)$ and $W_i(z)$:

$$\exp\left(-\frac{v_i e\phi(z) + W_i(z)}{kT}\right) - 1 = -\frac{v_i e\phi(z) + W_i(z)}{kT}. \tag{14.20}$$

This equation is valid if

$$\frac{|v_i e\phi(z) + W_i(z)|}{kT} << 1. \tag{14.21}$$

Then, Equation 14.19 becomes

$$\begin{aligned} \gamma &= \gamma_0 + kT\delta\sum_{i=1}^{N} n_i + \sum_{i=1}^{N} \int_0^{n_i}\int_\delta^\infty [v_i e\phi(z) + W_i(z)] dz dn_i \\ &= \gamma_0 + kT\delta\sum_{i=1}^{N} n_i + \sum_{i=1}^{N} \int_0^{n_i}\int_\delta^\infty W_i(z) dz dn_i, \end{aligned} \tag{14.22}$$

where we utilize the electroneutrality condition $\Sigma \nu_i n_i = 0$. We therefore see that the contribution from the mean potential $\phi(z)$ can be neglected as long as the linearization of Equation 14.20 is allowed.

The expression of $W_i(z)$ for the region outside the ion-free layer ($z > \delta$) was obtained by Ohshima and Matsubara [19] using the method of Levin and Flores-Mena [18]:

$$W_i(z) = \frac{\nu_i^2 e^2}{8\pi\varepsilon_r\varepsilon_0} \times \int_0^\infty e^{-2p(z-\delta)} \frac{q\{(\varepsilon_r+\varepsilon_r')(p-q)\exp(q\delta)+(\varepsilon_r-\varepsilon_r')(p+q)\exp(-q\delta)\}}{p\{(\varepsilon_r+\varepsilon_r')(p+q)\exp(q\delta)+(\varepsilon_r-\varepsilon_r')(p-q)\exp(-q\delta)\}} dq, \tag{14.23}$$

where ε_0, ε_r, and ε_r' represent the permittivity of vacuum and the relative permittivity of water and air (or oil), respectively. In addition, the following relation is defined:

$$p = \sqrt{q^2 + \kappa^2}, \tag{14.24}$$

where κ is the Debye–Hückel parameter defined by

$$\kappa = \left(\frac{\sum_{i=1}^{N} n_i \nu_i^2 e^2}{\varepsilon_r \varepsilon_0 kT} \right)^{\frac{1}{2}}. \tag{14.25}$$

Equation 14.22, when combined with Equation 14.23, can be further simplified into

$$\gamma = \gamma_0 + kT\left[\delta \sum_{i=1}^{N} n_i + \left(\sum_{i=1}^{N} \nu_i^2 n_i\right)\lambda_B \left\{\frac{2\ln 2 - 1}{8} + f\left(\kappa\delta, \frac{\varepsilon_r'}{\varepsilon_r}\right)\right\}\right] \tag{14.26}$$

with

$$f\left(\kappa\delta, \frac{\varepsilon_r'}{\varepsilon_r}\right) = \int_0^\infty \ln\left[1 + \left(\frac{1-\varepsilon_r'/\varepsilon_r}{1+\varepsilon_r'/\varepsilon_r}\right)\left(\sqrt{1+t^2} - t\right)^2 \exp(-2\kappa\delta t)\right] t\,dt \tag{14.27}$$

and the Bjerrum length

$$\lambda_B = \frac{e^2}{4\pi\varepsilon_r\varepsilon_0 kT}. \tag{14.28}$$

Equation 14.26 is the required expression for the interfacial tension of aqueous solution of general electrolytes.

The expressions of γ for 1-1 and 2-1 electrolytes are explicitly given next:

1. *1-1 Electrolyte of Concentration n:*

$$\gamma = \gamma_0 + nkT\left[2\delta + 2\lambda_{\mathrm{B}}\left\{\frac{2\ln 2 - 1}{8} + f\left(\kappa\delta, \frac{\varepsilon_r'}{\varepsilon_r}\right)\right\}\right] \tag{14.29}$$

with

$$\kappa = \left(\frac{2ne^2}{\varepsilon_r\varepsilon_0 kT}\right)^{\frac{1}{2}}; \tag{14.30}$$

2. *2-1 Electrolyte of Concentration n*:

$$\gamma = \gamma_0 + nkT\left[3\delta + 6\lambda_{\mathrm{B}}\left\{\frac{2\ln 2 - 1}{8} + f\left(\kappa\delta, \frac{\varepsilon_r'}{\varepsilon_r}\right)\right\}\right] \tag{14.31}$$

with

$$\kappa = \left(\frac{6ne^2}{\varepsilon_r\varepsilon_0 kT}\right)^{\frac{1}{2}}. \tag{14.32}$$

In Equations 14.29 and 14.31, the concentration n is given in units of cubic meter. The electrolyte concentration C, which is expressed in units of molar, is related to n by $n = 1000N_{\mathrm{A}}C$, N_{A} being Avogadro's number.

14.3 EXPERIMENTAL RESULTS OF INTERFACIAL TENSION OF AQUEOUS ELECTROLYTE SOLUTIONS

We can estimate the thickness of an ion-free layer by applying Equation 14.26 to the experimental results of interfacial tension of aqueous electrolyte solutions. In Equation 14.26, there are two important points: One is to contain the valence of ionic species i, and the other is the dielectric constant of the phase against the aqueous electrolyte solution. The former means that it is possible to discuss the interfacial tension of aqueous solutions of various electrolytes irrespective of their valence, as shown in Equations 14.29 and 14.31. The latter indicates the effect of the dielectric constant of the solvent on the interfacial tension of aqueous electrolyte solutions by the use of Equation 14.27. In the following sections, we discuss the negative adsorption behavior of electrolytes from these two points of view.

14.3.1 Air/Water Interface

Figure 14.3 shows the excess surface tension (= $\gamma - \gamma_0$) of aqueous solutions of 1-1 (NaCl, KCl) and of 2-1 ($CaCl_2$, $MgCl_2$) electrolytes against molarity. The excess surface tension linearly increases with increasing molarity. From the Gibbs adsorption equation, the surface density of electrolytes, Γ_1^G, is found to be negative, which suggests the negative adsorption of electrolytes. In addition, the surface tension in the 2-1 electrolyte system largely increases compared to that in the 1-1 electrolyte system. This is related to the difference in the repulsive interaction due to the image charge and the thickness of the ion-free layer associated with the increase in the valence and radius of hydrated ion i.

In order to discuss the above-mentioned expectation quantitatively, we need to estimate the theoretical surface tension. In a previous work by Ohshima [20, 21], the simple formula for $f(\kappa\delta, \varepsilon_r'/\varepsilon_r)$ was given by

$$f\left(\kappa\delta, \frac{1}{78.54}\right) = 0.0996 \times \ln\left[0.9288 + \frac{0.5365}{\kappa\delta} + \frac{0.3741}{(\kappa\delta)^2}\right] \tag{14.33}$$

for the case of an air/water surface ($\varepsilon_r' = 1$, $\varepsilon_r = 78.54$) at 298.15 K. By using this formula with Equations 14.29 and 14.31, the theoretical surface tensions are obtained as the solid lines in Figure 14.3, where we employed the thickness of the ion-free layer as a parameter. The δ values determined here are summarized in Table 14.1. They are ~0.27 nm for 1-1 electrolytes, and ~0.40 nm for 2-1 electrolytes, respectively. In Table 14.1, the radius of bare (r_b) and hydrated (r_h) ions is also summarized [28]. As compared to these values, it is found that

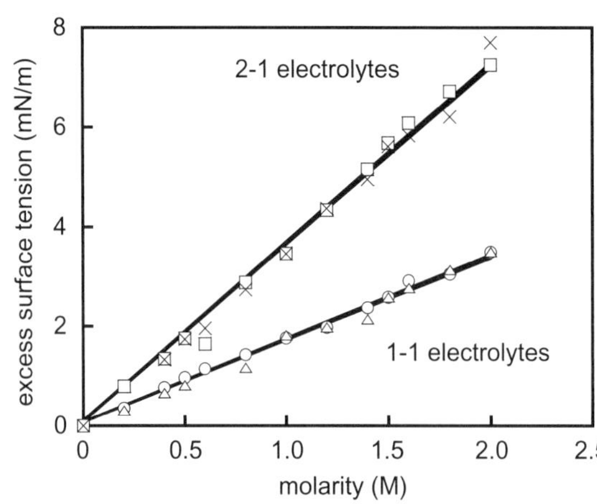

Figure 14.3. Plots of excess surface tension of aqueous electrolyte solution against molarity: sodium chloride (open circle), potassium chloride (open triangle), calcium chloride (open square), and magnesium chloride (cross). The solid lines are obtained by theoretical calculation using Equations 14.29 and 14.31.

TABLE 14.1. Values of Thickness of Ion-Free Layer and Radius of Bare (r_b) and Hydrated (r_h) Ions at the Air/Water Surface

		r_b (nm)	δ (nm)	r_h (nm)
NaCl	Na^+	0.095	0.278	0.36
	Cl^-	0.181		0.33
KCl	K^+	0.133	0.269	0.33
	Cl^-	0.181		0.33
$MgCl_2$	Mg^{2+}	0.065	0.395	0.43
	Cl^-	0.181		0.33
$CaCl_2$	Ca^{2+}	0.099	0.401	0.41
	Cl^-	0.181		0.33

the ion-free layer is at least thicker than the radius of bare ions. Taking into account that the thickness of the ion-free layer corresponds by definition to the average radius ($\bar{r}_h$) of hydrated ions, the values of δ are smaller than those of $\bar{r}_h$ in the 1-1 electrolyte systems and larger in the 2-1 electrolyte systems. Even in the latter case, however, the thickness of the ion-free layer is below the radius of the larger hydrated ions (Mg^{2+} and Ca^{2+} in this case). The result that the ion-free layer as a whole has a smaller thickness compared to the radius of hydrated ions may be attributable to the fact that the fluid/liquid interface is not considered as a perfectly sharp plane.

Equation 14.26 for the calculation of theoretical surface tension is rewritten as follows:

$$\begin{aligned}\gamma-\gamma_0 &= kT\delta\sum_{i=1}^{N} n_i + kT\left(\sum_{i=1}^{N} v_i^2 n_i\right)\lambda_B\frac{2\ln 2-1}{8} + kT\left(\sum_{i=1}^{N} v_i^2 n_i\right)\lambda_B f\left(\kappa\delta,\frac{\varepsilon_r'}{\varepsilon_r}\right)\\ &= \gamma_a+\gamma_b+\gamma_c. \end{aligned} \tag{14.34}$$

The terms γ_a, γ_b, and γ_c respectively indicate the contribution from the ion-free layer, the image interaction between the electrolyte ions and ion-free layer (as if the thickness of the latter were infinitely thick), and the image interaction between the electrolyte ions and air. Figure 14.4 shows the contribution of each term to the excess surface tension. It was found that the contribution from the ion-free layer (curve 1) becomes significant for the surface tension with increasing molarity. As suggested by Levin and Flores-Mena [17, 18], consideration of the ion-free layer is indispensable for determining the surface tension of aqueous electrolyte solutions. In the 2-1 electrolyte system, furthermore, the values of γ_a and γ_b increase definitely, as shown in Figure 14.4b. This proves quantitatively the increase in both the repulsive interaction due to the image charge and the thickness of the ion-free layer by replacing monovalent ions with divalent ones.

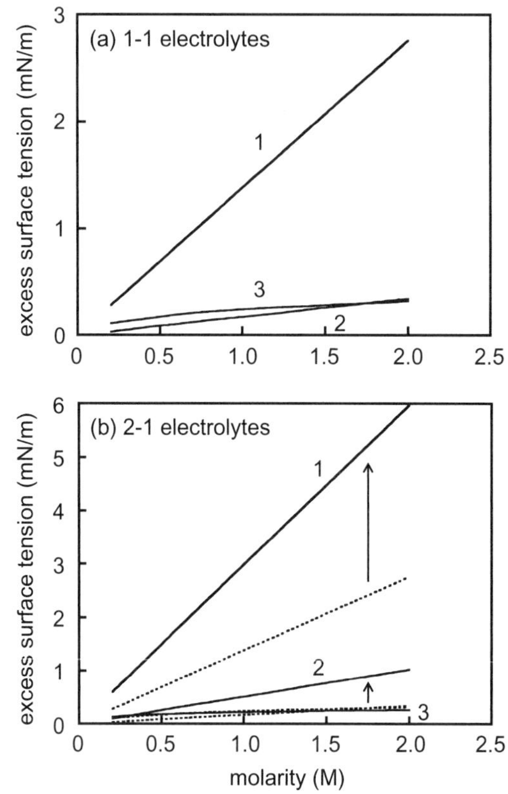

Figure 14.4. Dependence of theoretical excess surface tension on concentration of aqueous solution of (a) 1-1 and (b) 2-1 electrolytes: contribution from (1) ion-free layer, (2) image interaction between electrolyte ions and the ion-free layer (as if the thickness of the latter were infinitely thick), and (3) image interaction between electrolyte ions and air.

14.3.2 Oil/Water Interface

The interfacial tensions of hexane and benzene against water are 50.20 and 34.10 mN/m, respectively. On the basis of these values, the results of excess interfacial tension measurement are shown in Figure 14.5. It should be noted that the excess interfacial tensions are independent of the kinds of electrolytes. This result is clearly different from that presented in Figure 14.3, which showed the remarkabe difference between 1-1 and 2-1 electrolytes. Furthermore, it seems that the solvent effect on the interfacial tension does not appear on Figure 14.5a,b.

Here, the thickness of the ion-free layer determined from the theoretical curves (solid lines in Fig. 14.5) is summarized in Table 14.2. The δ values in 1-1 electrolyte systems increase compared to those at the air/water surface and

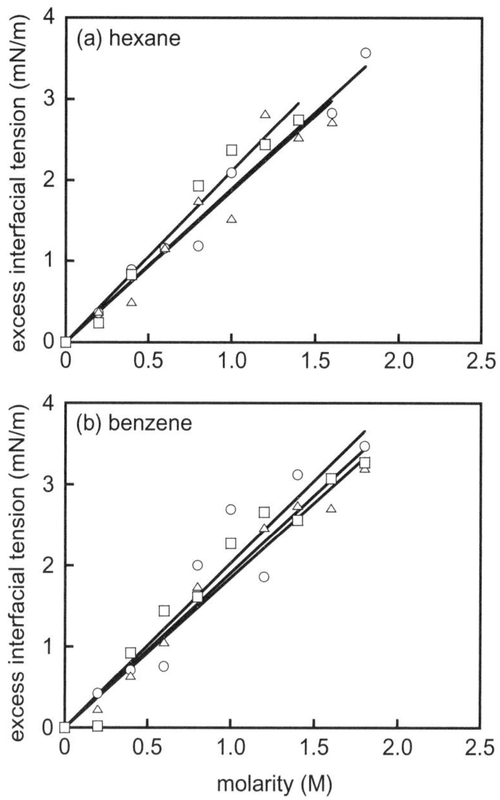

Figure 14.5. Plots of excess interfacial tension of aqueous electrolyte solution against molarity at the (a) hexane/water and (b) benzene/water interfaces: sodium chloride (open circle), potassium chloride (open triangle), and magnesium chloride (open square). The solid lines are obtained by theoretical calculation using Equations 14.29 and 14.31.

TABLE 14.2. Values of Thickness of Ion-Free Layers at Hexane/Water and Benzene/Water Interfaces

	δ_{hexane} (nm)	$\delta_{benzene}$ (nm)
NaCl	0.329	0.359
KCl	0.320	0.319
$MgCl_2$	0.158	0.100

approach the radius of hydrated ions. In 2-1 electrolyte systems, on the one hand, the δ values become extremely small. These values are close to the radius of bare ions. Although not conclusively proven, this may be attributable to the increase in relative permittivity of oils and mutual solubility between oil and water because the former contributes to the decrease in the repulsive

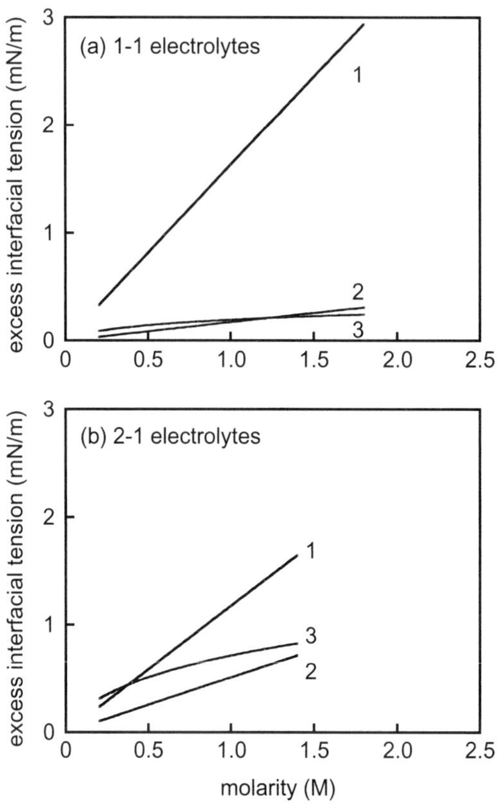

Figure 14.6. Dependence of theoretical excess interfacial tension on the concentration of aqueous solution of (a) 1-1 and (b) 2-1 electrolytes at the hexane/water interface: contribution from the (1) ion-free layer, (2) image interaction between electrolyte ions and ion-free layer (as if the thickness of the latter were infinitely thick), and (3) image interaction between electrolyte ions and hexane.

interaction between electrolyte ions and image charges, and the latter contributes to the fluctuation of the oil/water interface.

Finally, the contribution of γ_a, γ_b, and γ_c to the excess interfacial tension is shown in Figures 14.6 and 14.7. Taking Figure 14.4 as well as these figures into consideration, we can state the following facts in regard to the 2-1 electrolyte systems: (i) The contribution of γ_b is approximately the same irrespective of the systems. This suggests that the image interaction between electrolyte ions and the ion-free layer is not influenced by the fluid phase against the aqueous electrolyte solution; (ii) the contribution from the ion-free layer decreases by decreasing the δ values in Table 14.2; and (iii) as described earlier, it was expected that the contribution from image interaction between electrolyte ions and the oil phase would decrease by increasing the relative permittivity

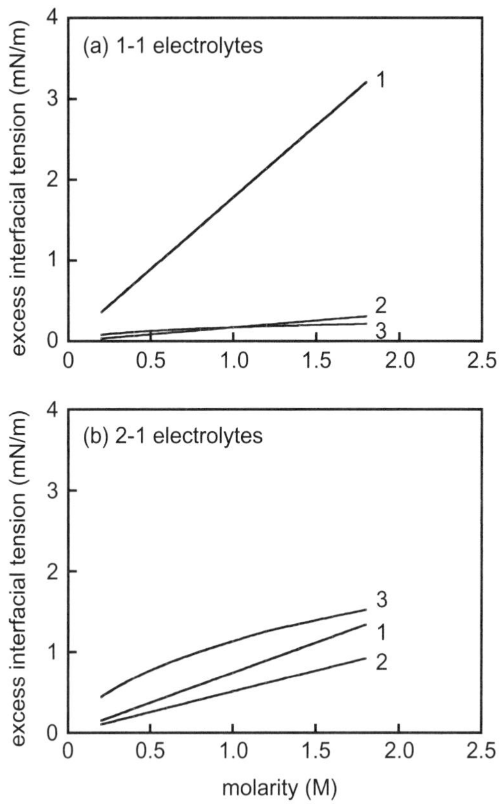

Figure 14.7. Dependence of theoretical excess interfacial tension on the concentration of aqueous solution of (a) 1-1 and (b) 2-1 electrolytes at the benzene/water interface: contribution from the (1) ion-free layer, (2) image interaction between electrolyte ions and ion-free layer (as if the thickness of the latter were infinitely thick), and (3) image interaction between electrolyte ions and benzene.

of oils. Actually, the γ_c values increase as opposed to such expectation. This implies that the thickness of the ion-free layer is possibly underestimated because of the fluctuation of the oil/water interface.

14.4 CONCLUSION

In this chapter, a theoretical equation for the interfacial tension of aqueous electrolyte solutions against any fluid phase was derived by considering an ion-free layer, the thickness of which corresponds to the radius of hydrated ions, according to the method of Levin and Flores-Mena. The application of the theoretical equation to the experimental interfacial tension provided the thickness of an ion-free layer. The increase in interfacial tension associated with the electrolyte concentration was composed of the contribution from (i) the ion-free layer, (ii) the image interaction between the electrolyte ions and the ion-free layer (as if the thickness of the latter were infinitely thick), and (iii) the image interaction between the electrolyte ions and fluid phases. The importance of the ion-free layer was discussed. The air/water surface tension clearly reflected the influence of the valence and radius of the electrolyte ions. On the other hand, the oil/water interfacial tension has no effect on the kinds of electrolytes used in our experiment because of the thermal fluctuation of the oil/water interface and some mutual solubility between oil and water. Therefore, the values of the ion-free layer at the oil/water interface are still to be defined.

REFERENCES

1 G. Jones, W. A. Ray. *J. Am. Chem. Soc.* 59 (1937) 187.

2 M. Dole, J. A. Swartout. *J. Am. Chem. Soc.* 62 (1940) 3039.

3 P. B. Petersen, J. C. Johnson, K. P. Knutsen, R. J. Saykally. *Chem. Phys. Lett.* 397 (2004) 46.

4 P. B. Petersen, R. J. Saykally. *J. Am. Chem. Soc.* 127 (2005) 15446.

5 K. A. Karraker, C. J. Radke. *Adv. Colloid Interface Sci.* 96 (2002) 231.

6 M. Manciu, E. Ruckenstein. *Adv. Colloid Interface Sci.* 105 (2003) 63.

7 P. B. Petersen, R. J. Saykally. *J. Phys. Chem. B* 109 (2005) 7976.

8 P. B. Petersen, R. J. Saykally. *J. Phys. Chem. B* 110 (2006) 14060.

9 M. K. Petersen, S. S. Iyengar, T. J. F. Day, G. A. Voth. *J. Phys. Chem. B* 108 (2004) 14804.

10 M. Takahashi. *J. Phys. Chem. B* 109 (2005) 21858.

11 V. Knecht, H. J. Risselada, A. E. Mark, S. J. Marrink. *J. Colloid Interface Sci.* 318 (2008) 477.

12 J. K. Beattie, A. M. Djerdjev, G. G. Warr. *Faraday Discuss.* 141 (2009) 31.

13 V. S. J. Craig, B. W. Ninham, R. M. Pashley. *Nature* 364 (1993) 317.

14 T. Sakai. *Curr. Opin. Colloid Interface Sci.* 13 (2008) 228.

15 C. Wagner. *Phys. Z.* 25 (1924) 474.

16 L. Onsager, N. N. T. Samaras. *J. Chem. Phys.* 2 (1934) 528.

17 Y. Levin. *J. Chem. Phys.* 113 (2000) 9722.

18 Y. Levin, J. E. Flores-Mena. *Europhys. Lett.* 56 (2001) 187.

19 H. Ohshima, H. Matsubara. *Colloid Polym. Sci.* 282 (2004) 1044.

20 H. Ohshima. *Colloid Polym. Sci.* 283 (2004) 119.

21 H. Ohshima. *Theory of colloid and interfacial electric phenomena.* Academic Press, London, 2006, Chapter 19.

22 R. S. Hansen. *J. Phys. Chem.* 66 (1962) 410.

23 K. Motomura. *J. Colloid Interface Sci.* 64 (1978) 348.

24 K. Motomura, M. Aratono. *Langmuir* 3 (1987) 304.

25 P. Jungwirth, D. J. Tobias. *J. Phys. Chem. B* 105 (2001) 10468.

26 B. Minofar, R. Vácha, A. Wahab, S. Mahiuddin, W. Kunz, P. Jungwirth. *J. Phys. Chem. B* 110 (2006) 15939.

27 T. Ishiyama, A. Morita. *J. Phys. Chem. C* 111 (2007) 721.

28 J. N. Israelachvili. *Intermolecular and surface forces*, 2nd ed. Academic Press, London, 1992, Chapter 4.

PART II

APPLICATIONS IN NANO- AND ENVIRONMENTAL SCIENCES

15

BROADBAND DIELECTRIC SPECTROSCOPY ON ELECTRODE POLARIZATION AND ITS SCALING

Anatoli Serghei, Joshua R. Sangoro, and Friedrich Kremer

15.1 INTRODUCTION

Charge transport at the interfaces between electronic and ionic conductors is a topic of special technological and fundamental importance in contemporary science. The quest for the quantitative understanding of the extent and nature of these interfaces is necessitated by the need for more efficient and optimal power sources, among others. For practical purposes, the dynamics of the charge carriers around these boundaries needs to be understood over the broadest range of frequencies and temperatures possible. Due to its ability to measure the complex dielectric function (and hence the complex conductivity) over many orders of magnitude in frequency and in a wide temperature interval, broadband dielectric spectroscopy (BDS) turns out to be an ideal experimental tool in this field [1–11].

The typical conductivity spectra of ion-conducting materials consist essentially of two regimes, apparently related to two different phenomena. At high frequencies (and low temperatures), the conductive properties are governed by the transport of the charge carriers in the bulk. The underlying mechanism of this contribution is theoretically well understood [1, 12–16]. This part of the spectra can be used to extract important molecular parameters characterizing the hopping mechanism of the charge carriers [5, 17, 18]. At lower frequencies

Electrical Phenomena at Interfaces and Biointerfaces: Fundamentals and Applications in Nano-, Bio-, and Environmental Sciences, First Edition. Edited by Hiroyuki Ohshima.

(and higher temperatures), pronounced changes in the complex conductivity function are detected due to the presence of electrodes. This part, exhibiting a characteristic dependence on the geometry of the measurement cell (e.g., the cell length) *and* on the material used for the electrodes, was considered for a long time to be a parasitic effect ("electrode polarization") and usually was cut off from the bulk contribution of the spectra [5, 10].

Remarkable effort aimed at providing complete and accurate quantitative description of the electrode polarization phenomenon (including its spatial, frequency, and temperature dependence) in the last few decades has been made by a number of researchers. The most notable models have been proposed by MacDonald [19], Chang and Jaffé [20], Schwan and coworkers [6–8, 21], Uemura [22], DeLacey and White [23], Liu [24], Klein et al. [4], Sawada [25, 26], Alexe-Ionescu et al. [27], and Borukhov et al. [28]. However, none of these approaches can explain all the salient experimental features of electrode polarization.

In this chapter, BDS is employed to analyze electrode polarization in detail by examining its scaling with respect to the frequency, temperature, concentration, and length of the sample cell. This leads to a quantitative description of electrode polarization. Furthermore, a formula is derived enabling one to deduce the bulk DC conductivity of the material under study from the spectral characteristics of electrode polarization.

15.2 EXPERIMENTAL

Several ionic liquids (purchased from Solvent Innovation GmbH and Iolitec GmbH) are investigated in the current work: 1-hexyl-3-methylimidazolium chloride (HMIM-Cl), 1-hexyl-3-methylimidazolium hexafluorophosphate [HMIM] [PF_6], 1,3-dimethylimidazolium dimethylphosphate [HMIM] [Me_2PO_4], and trioctylmethylammonium bis(trifluoromethylsulfonyl)imide [OMA] [FMSI]. Sodium chloride solutions of different concentrations (in Millipore water) are examined as well. The dielectric measurements are carried out using a high-resolution Alpha Analyzer (Novocontrol GmbH, Hundsangen, Germany) and an HP Impedance Analyzer, both assisted by Quatro temperature controllers. The samples are measured in parallel-plate geometry, using circular metallic plates of 20-mm diameter as electrodes and small, flat Teflon pieces as spacers (having a well-controlled thickness on the submillimeter length scale). The applied voltages (always chosen in the linear regime) are in the order of 0.1–0.2 V.

15.3 CHARGE TRANSPORT PROPERTIES IN THE BULK

The dielectric properties of a material are characterized by the complex permittivity function $\varepsilon^* = \varepsilon' - i\varepsilon''$ or by the complex conductivity function $\sigma^* = \sigma' + i\sigma''$. Given the linear dependence between permittivity and conduc-

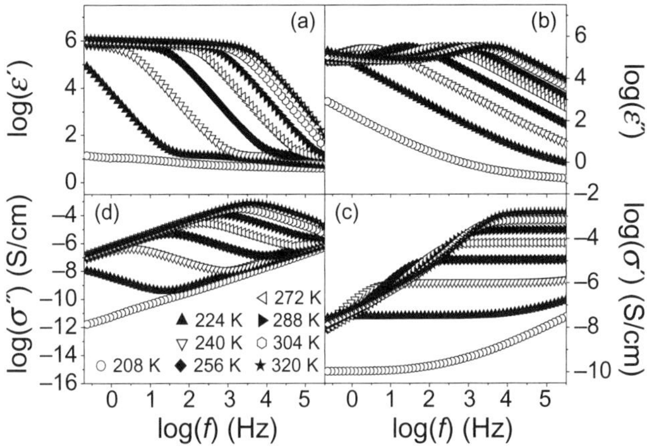

Figure 15.1. Complex dielectric permitivitty $\varepsilon^* = \varepsilon' + i\varepsilon''$ and conductivity $\sigma^* = \sigma' + i\sigma''$ of the ionic liquid [HMIM] [PF_6] versus frequency at different temperatures, as indicated.

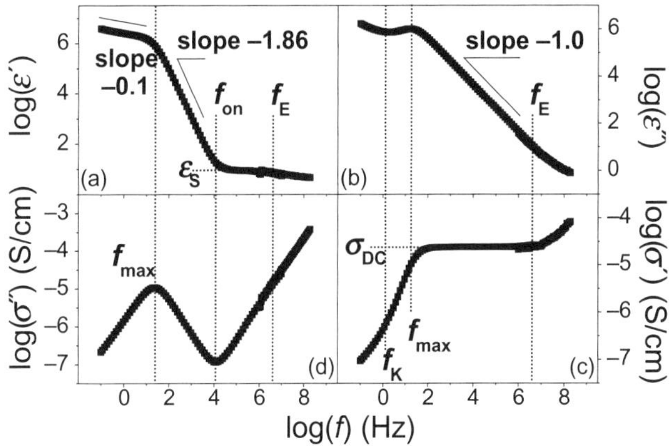

Figure 15.2. Complex dielectric permittivity $\varepsilon^* = \varepsilon' + i\varepsilon''$ and conductivity $\sigma^* = \sigma' + i\sigma''$ of the ionic liquid [HMIM] [PF_6] versus frequency at a temperature of 264 K. The dotted lines indicate the main experimental features characterizing the dielectric spectra of conductive materials as discussed in the text.

tivity, $\sigma^* = i\varepsilon_0\omega\varepsilon^*$ (with $\omega = 2\pi f$, f being the frequency of the applied electric field), these two representations are fully equivalent. Typical dielectric spectra of ionic liquids are shown in Figure 15.1. As a first observation, the intensity of the measured dielectric responses varies over many orders of magnitude upon varying the frequency of the applied electric field and the temperature. Depending on these two parameters—frequency and temperature—distinct characteristic features are observed in the dielectric spectra of conductive materials. These features are discussed in detail in Figure 15.2, showing the complex permittivity and conductivity function of [HMIM] [PF_6] measured in a broad frequency range and at a constant temperature of 264 K. At high frequencies (and low temperatures), the net dielectric response is dominated by contributions from charge transport processes in the bulk. These contributions reflect well-defined molecular physical quantities described by recent models of charge transport in disordered systems [1, 14, 16]:

1. The plateau observed in σ' (corresponding to a linear dependence with a slope of –1 in the spectra of ε'') represents the contribution of the DC conductivity σ_{DC} (Fig. 15.2c). This contribution is related to the effective number density of the charge carriers, n, the ionic mobility μ, and the electrical charge q as follows:

$$\sigma_{\mathrm{DC}} = qn\mu. \tag{15.1}$$

2. The characteristic hopping rate of the charge carrier f_e is given by the frequency position where the plateau of the DC conductivity turns into a steep increase in the high-frequency limit of σ' (Fig. 15.2c). The hopping rate f_e, marking the transition between the subdiffusive and the diffusive regimes, is also observed as a step function in the spectra of ε' (Fig. 15.2a) or as a deviation from a linear dependence in the spectra of ε'' in the high-frequency limit (Fig. 15.2b).

A linear relation between the DC conductivity and the hopping rate of the charge carriers $\sigma_{DC} \sim f_e$ (known as the Barton–Nakajima–Namikawa relation) has been experimentally proven since decades and has recently been explained by considering the Einstein and Einstein–Smoluchowski equations [29]. It has been also shown that dielectric measurements on the conductivity of ionic liquids can be used to determine the diffusion coefficient of the charge carriers—in excellent agreement with data obtained by pulsed field gradient nuclear magnetic resonance [5, 29, 30].

15.4 ELECTRODE POLARIZATION EFFECTS IN DIELECTRIC SPECTRA: EXPERIMENTAL FEATURES

As the frequency of the applied electrical field decreases, deviations from the bulk characteristics of ε^* and σ^*—marking the onset and the development of electrode polarization effects—show up. These effects lead to changes by many orders of magnitude in the intensity of the measured responses. It is important to notice that these values (e.g., $\varepsilon' \cong 10^6$) do not have a direct molecular interpretation any longer but reflect the net impedance of the measured cell Z^*_{net}, where the contribution of the interfaces Z^*_{net} must be added to that of the bulk Z^*_{bulk} (thus, $Z^*_{\mathrm{net}} = Z^*_{\mathrm{bulk}} + Z^*_{\mathrm{int}}$). The electrode polarization phenomenon is mani-

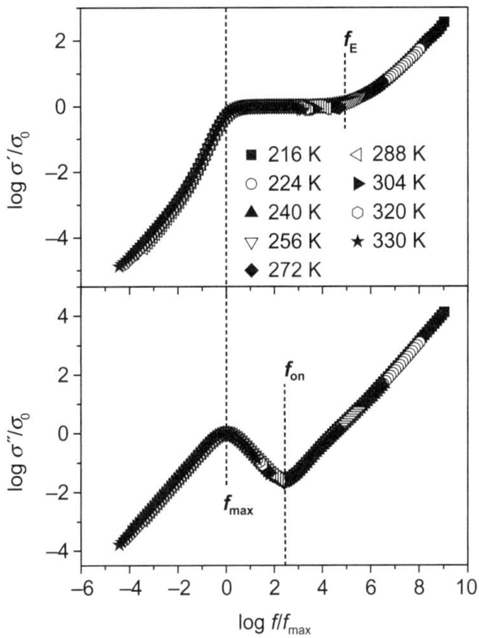

Figure 15.3. Normalized conductivity function of the ionic liquid [HMIM] [PF_6] measured at different temperatures. The experimental curves collapse onto a single plot. The normalization was done in respect to the position and the intensity of the maximum observed in σ''.

fested first as a steep increase in ε' observable at a certain frequency, f_{on} ("onset of electrode polarization effects"). At the same frequency f_{on}, a minimum in σ'' is detected. Note that no changes in ε'' or σ' are observed at f_{on}. Decreasing the frequency further, a plateau in ε' starts to develop, corresponding to a maximum in σ''. At the same frequency position f_{max} ("full development of electrode polarization effects"), the value of σ' starts to decrease and a peak in ε'' is observed. At even lower frequencies ($f = f_k$), the dielectric loss ε'' increases again (but with a slope different from −1) and a change in slope is detected in σ'. Despite this complicated frequency dependence, a perfect scaling is observed in both σ' and σ'' (or, correspondingly, in ε' and ε'') upon varying the temperature: All measured curves (Fig. 15.1) collapse onto a single plot when normalized (Fig. 15.3). This finding indicates a similar temperature dependence of the parameters characterizing the dielectric response of conductive materials (i.e., σ_{DC}, f_e, f_{on}, f_{max}, and f_k) and implies the existence of scaling laws relating electrode polarization to the conductivity contribution of the material in the bulk. This can be expressed more comprehensibly by plotting together f_{max}, f_{on}, and σ_{DC} as a function of inverse temperature (Fig. 15.4). Parallel Vogel–Fulcher–Tammann-type temperature dependences are obtained

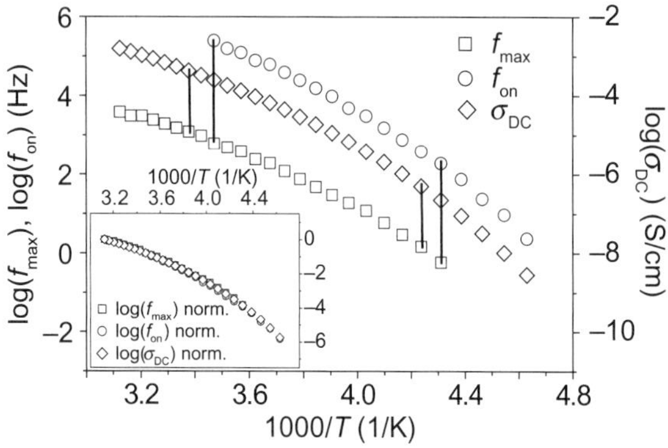

Figure 15.4. f_{max}, f_{on}, and σ_{DC} as a function of inverse temperature for the ionic liquid [HMIM] [PF_6]. The same number of decades is plotted on the left and right *y*-axes of the graph to indicate parallel Vogel–Fulcher–Tammann dependences. Inset: normalized Vogel–Fulcher–Tammann-type dependences collapse onto a single plot. The normalization was done in respect to the values of f_{max}, f_{on}, and σ_{DC} measured at 250 K.

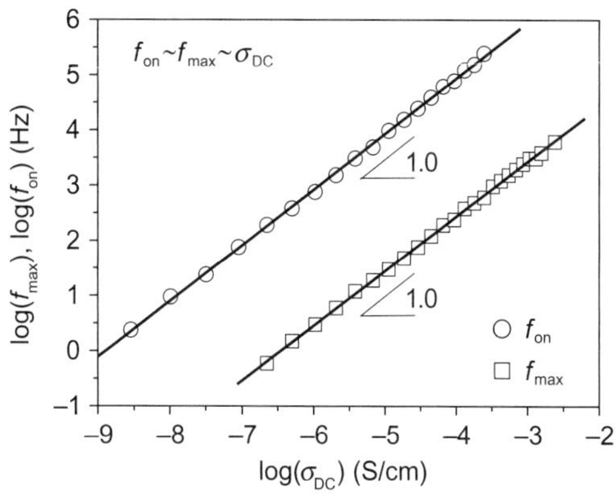

Figure 15.5. f_{max} and f_{on}, presented in Figure 15.4, are replotted as a function of σ_{DC}. Linear dependences with a slope of 1 are obtained, indicating the simple scaling law $f_{on} \sim f_{max} \sim \sigma_{DC}$.

(inset, Fig. 15.4). By replotting these data as functions of σ_{DC} (Fig. 15.5), a simple linear relation—holding in a frequency range covering many orders of magnitude—is revealed:

$$f_{on} \sim f_{max} \sim \sigma_{DC}. \tag{15.2}$$

Other scaling laws characterizing the phenomenon of electrode polarization are revealed by analyzing the spectra of ε'. The steep increase detected after

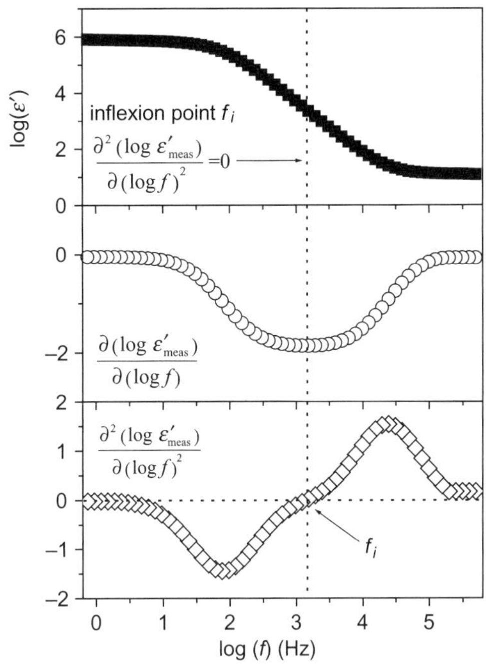

Figure 15.6. Increase in ε′ observed after the onset of the electrode polarization effects at $f < f_{on}$. An inflexion point is obtained at f_i, where

$$\left|\frac{\partial^2(\log \varepsilon'_{meas})}{\partial(\log f)^2}\right|_{f=fi} = 0.$$

the onset of electrode polarization effects at $f < f_{on}$ shows an inflexion point at $f = f_i$, where the second derivative

$$\left|\frac{\partial^2(\log \varepsilon'_{net})}{\partial(\log f)^2}\right|_{f=fi} = 0$$

(Fig. 15.6).

Plotting the square of f_i in dependence on the product between f_{on} and f_{max} in a double logarithmic graph (Fig. 15.7), a linear dependence with a slope of 1 is found over many decades in frequency. A quantitative analysis of these data gives $f_i^2 = f_{on} f_{max}$. Furthermore, the slope characterizing the increase in ε′ does not show, in wide ranges, a significant temperature dependence (Fig. 15.8). This slope was calculated as the first derivative

$$\frac{\partial(\log \varepsilon'_{net})}{\partial(\log f)}$$

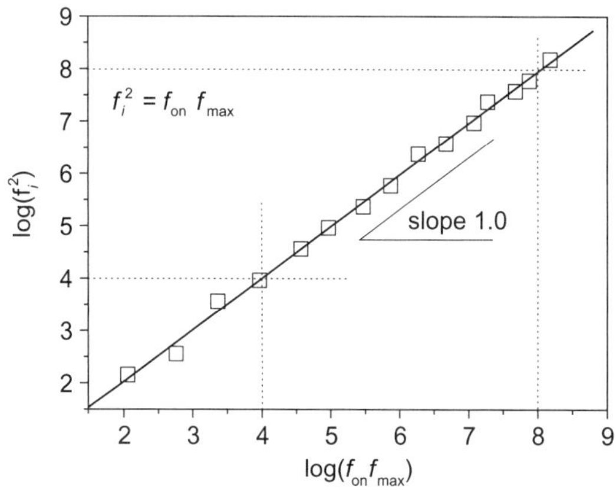

Figure 15.7. f_i^2 as a function of $f_{on}f_{max}$ for the ionic liquid [HMIM] [PF_6] measured at different temperatures. A linear dependence with a slope of 1 is obtained. A quantitative analysis renders $f_i^2 = f_{on}f_{max}$.

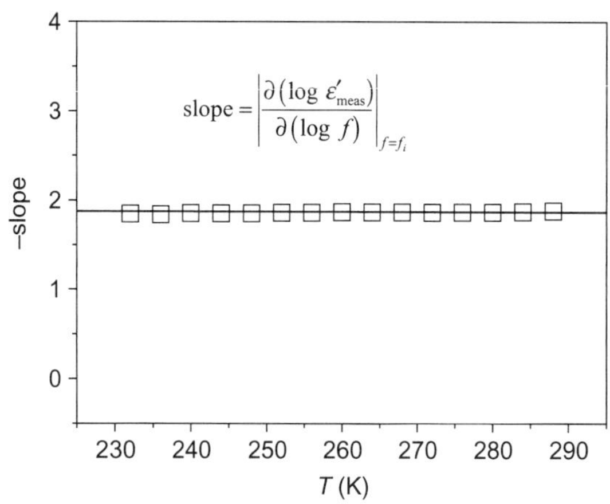

Figure 15.8. The slope

$$\left|\frac{\partial(\log \varepsilon'_{meas})}{\partial(\log f)}\right|_{f=fi}$$

observed in the increase of ε' of the ionic liquid [HMIM] [PF_6] after the onset of the electrode polarization effects. A negligible temperature dependence is found.

at the inflexion frequency f_i, giving the middle point of the plateau shown in Figure 15.6. Another particular feature characterizing the phenomenon of electrode polarization is revealed by plotting together the frequency positions of the maxima observed in ε'' and σ'' (Fig. 15.9). In a broad frequency range, coinciding values for the position of these two maxima are obtained.

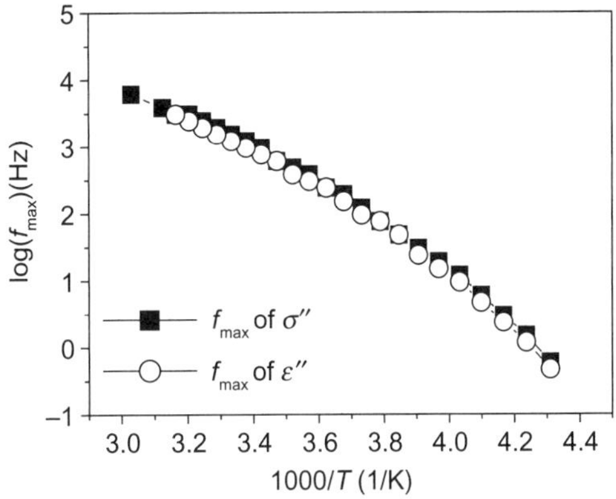

Figure 15.9. The frequency position of the maxima observed in ε'' and σ'' in the dielectric spectra of the ionic liquid [HMIM] [PF_6] as functions of inverse temperature. In a broad temperature range, coinciding values are obtained.

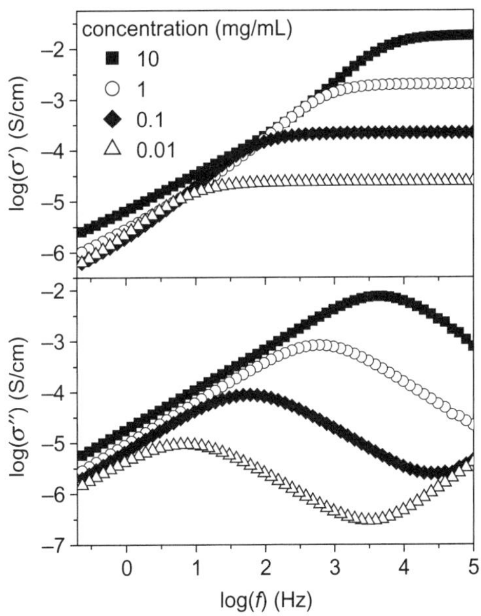

Figure 15.10. Complex conductivity spectra of aqueous solutions of sodium chloride at different salt concentrations. The measurements were carried out at 300 K.

The influence of the concentration of charge carriers on the electrode polarization effects can be analyzed in detail by investigating aqueous solutions of sodium chloride. With increasing salt concentration, the DC conductivity increases and the characteristic frequencies f_{on} and f_{max} are shifted to higher values (Fig. 15.10). When normalized, the raw experimental data shown in

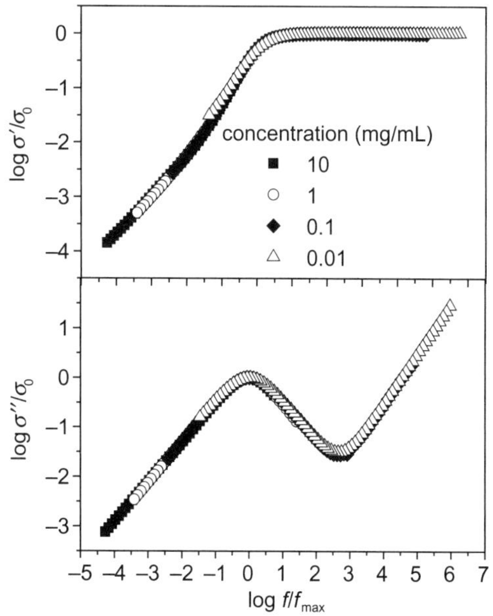

Figure 15.11. The data in Figure 15.10 are presented normalized (in respect to the position and the intensity of the maximum observed in σ''). The experimental curves collapse onto a single plot.

Figure 15.10 collapse onto a single plot (Fig. 15.11). By replotting the data as a function of σ_{DC} (Fig. 15.12), a similar scaling law as that obtained for the temperature dependence is revealed:

$$f_{on} \sim f_{max} \sim \sigma_{DC}. \qquad (15.2a)$$

Varying length L of the sample cell leads to pronounced changes in the low-frequency region of the dielectric spectra, where electrode polarization effects play the dominant role. This is analyzed in Figure 15.13, showing the frequency dependence of σ' and σ'' measured with sample cells having different thicknesses (while all the other parameters are kept constant). The thickness range was chosen so that $L \ll \sqrt{D}$, where D is the diameter of the electrodes. Fulfilling this condition ensures that the distortions of the electric field at the edges of the sample cell, known to affect predominantly the low-frequency range of the dielectric measurements, are negligible. Changing the length of the sample cell does not affect, as expected, the dielectric response in the high-frequency limit (where the bulk contribution is measured), but with decreasing frequency, pronounced shifts in f_{on} and f_{max} are observed. As opposed to the temperature or concentration dependence, normalizing the experimental curves upon varying the sample length L does not lead to coinciding plots (Fig. 15.13, inset). This implies different scaling laws for f_{on} and f_{max} as a function of L. Indeed,

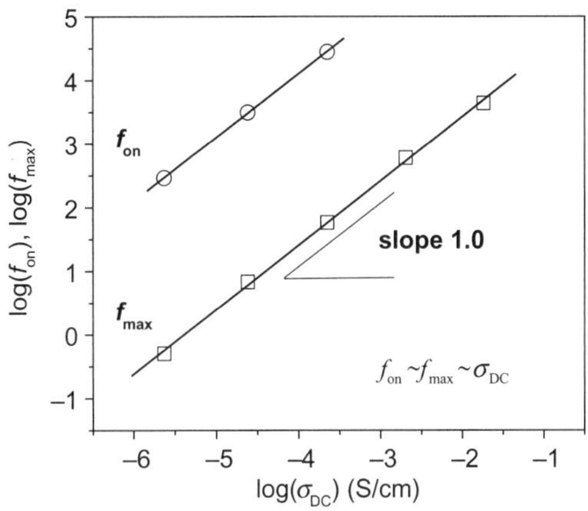

Figure 15.12. f_{max}, f_{on} corresponding to different salt concentrations plotted as a function of σ_{DC}. Linear dependences in log–log representation with slopes of 1 are obtained, indicating the scaling relation $f_{on} \sim f_{max} \sim \sigma_{DC}$.

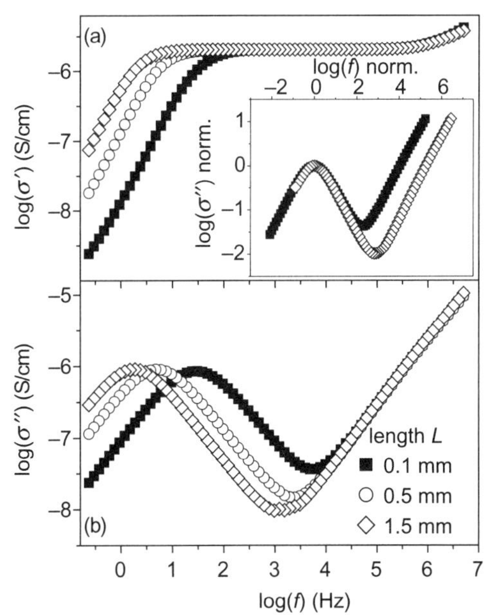

Figure 15.13. Complex conductivity spectra of the ionic liquid [OMA] [FMSI] measured at 260 K using different cell thicknesses L, as indicated. Inset: normalized data (with respect to the position and the intensity of the maximum observed in σ'').

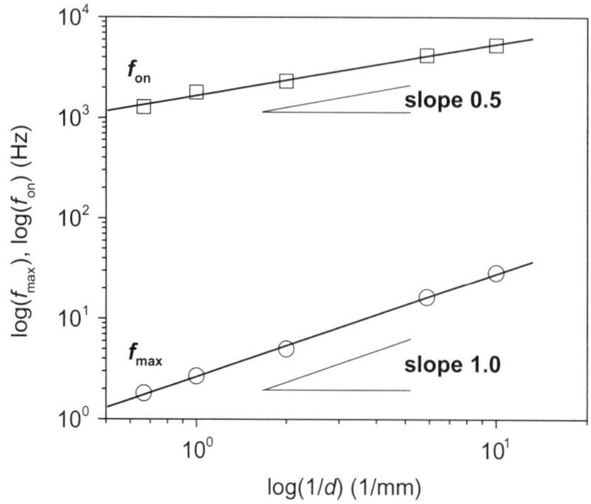

Figure 15.14. The values of f_{on} and f_{max} for the ionic liquid [OMA] [FMSI] measured at 260 K are represented as functions of inverse sample thickness $1/L$.

$$f_{max} \sim 1/L^{1.0\pm0.1}, \tag{15.3a}$$

while

$$f_{on} \sim 1/L^{0.5\pm0.1} \tag{15.3b}$$

is determined from the experimental data (Fig. 15.14).

Changing the material of the electrodes used for the sample cell is known to significantly affect the electrode polarization effects. This is exemplified in Figure 15.15, showing the frequency dependence of σ' and σ'' measured using electrodes of different materials (brass, stainless steel, aluminum). Pronounced changes are detected in the low-frequency region (i.e., shifts in f_{on} and f_{max}), where the electrode polarization effects dominate. Similar to the length dependence, normalizing the experimental curves upon varying the material of the electrodes does not lead to coinciding plots (Fig. 15.15, inset).

The surface roughness is another important physical parameter characterizing the electrodes believed to affect the electrode polarization effects. This is analyzed in detail in Figure 15.16, showing the frequency dependence of σ' and σ'' measured using electrodes with different surface roughnesses (with all other parameters kept constant). The root mean square roughness R_{RMS} was varied by polishing the surface of the electrodes with sandpaper having different grain sizes. The absolute values for R_{RMS} were then determined by means of a mechanical profilometer (Alpha Stepper) upon performing a line scan of 1-mm length. Changing the surface roughness within a range wider than one decade (between 0.1 and 2.5 μm) did not lead to significant changes

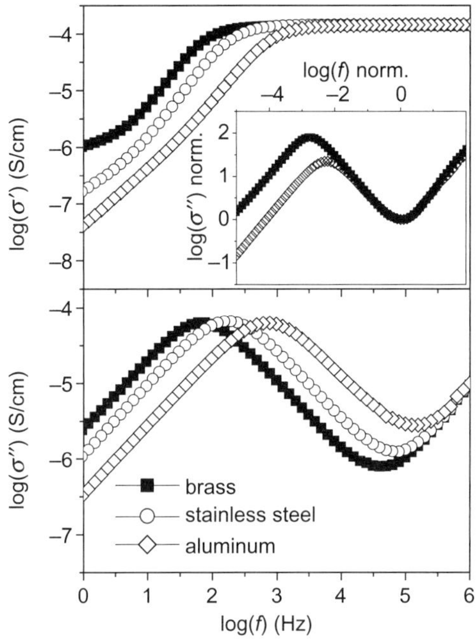

Figure 15.15. Complex conductivity spectra of the ionic liquid [MMIM] [Me_2PO_4] measured at 270 K using different materials for the electrodes. Inset: normalized data (with respect to the position and the intensity of the minimum observed in σ'').

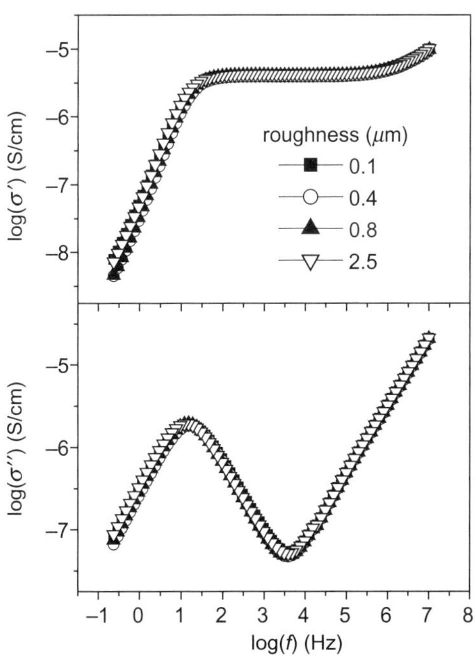

Figure 15.16. Complex conductivity spectra of the ionic liquid [HMIM] [PF_6] measured at 260 K using electrodes with different surface roughnesses, as indicated.

in the dielectric spectra (Fig. 15.16). The length of the sample cell L was 0.5 mm in these experiments. In contrast to reports by Liu [24], it is concluded that the surface roughness, R_{RMS}, does not affect the dielectric spectra as long as it is much smaller than the distance between the electrodes ($R_{\text{RMS}} << L$).

15.5 SUMMARY OF THE EXPERIMENTAL RESULTS

Electrode polarization effects are manifested by pronounced changes in the dielectric response of conductive materials, particularly in the low-frequency region of the dielectric spectra. These changes depend on the frequency of the applied electric field, temperature, length of the sample cell, and material of the electrodes. The following scaling relations are found:

1. $f_{\text{on}} \sim f_{\text{max}} \sim \sigma_{\text{DC}}$, upon varying the temperature or the concentration of the charge carriers

2. $f_i^2 = f_{\text{on}} f_{\text{max}}$

3. The slope characteristic for the increase in ε' measured at frequencies $f < f_{\text{on}}$ is temperature independent.
4. The maxima detected in ε'' and in σ'' have the same frequency position.
5. $f_{\text{max}} \sim 1/L^{1.0\pm0.1}$, while $f_{\text{on}} \sim 1/L^{0.5\pm0.1}$. This finding explains why the experimental curves do not scale upon normalization when the length of the sample cell is varied.
6. No scaling upon normalization is obtained by changing the material of the electrodes.
7. The surface roughness R_{RMS} shows a negligible influence as long as it is much smaller than the length of the sample cell ($R_{\text{RMS}} << L$).

15.6 ELECTRODE POLARIZATION AND CHARGE TRANSPORT AT SOLID INTERFACES

The scaling laws of the electrode polarization effects can be understood by analyzing the hopping process of the charge carriers in the nanometric vicinity of the metal/ionic conductor interfaces (Fig. 15.17). Basic considerations of electrochemistry evidence the fact that due to the strong coulombic forces between the ions and the charged metallic electrodes upon applying a voltage across the sample cell, interaction energies ΔE_c larger than the thermal energy $k_{\text{B}}T$ are reached ($\Delta E_c >> k_{\text{B}}T$). As it is well-known [12, 31] that the hopping time of the charge carrier τ_e depends exponentially on the activation energy E,

$$\tau_e = \tau_0 \exp\left(\frac{E}{k_{\text{B}}T}\right), \tag{15.4}$$

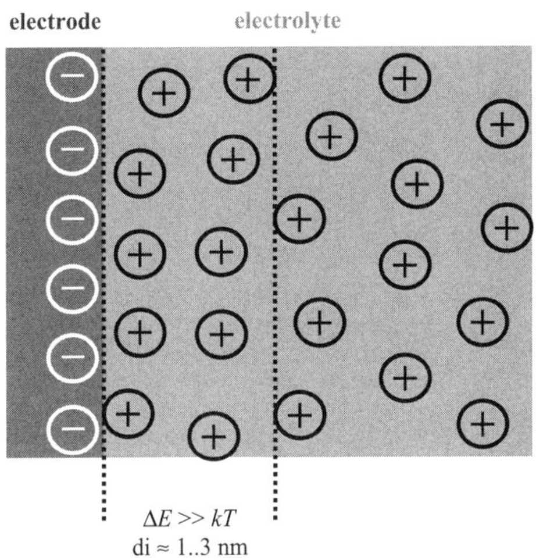

Figure 15.17. Interface metal/ionic liquid (only the positive ions are drawn for clarity). Due to the coulombic forces, interaction energies larger than $k_B T$ are achieved in the nanometric vicinity of metal interfaces.

where τ_0 is the hopping time in the high-temperature limit, the coulombic interactions manifested at the metallic interfaces will lead to a tremendous increase in the hopping time:

$$\frac{\tau_e\,(\text{interface})}{\tau_e\,(\text{bulk})} \cong \exp\left(\frac{\Delta E_C}{k_B T}\right) \gg 1. \tag{15.5}$$

This will result in a correspondingly large decrease in the values of the DC conductivity σ_{DC} at the interface:

$$\frac{\sigma_{DC}\,(\text{bulk})}{\sigma_{DC}\,(\text{interface})} \gg 1, \tag{15.6}$$

since, according to the Barton–Nakajima–Namikawa relation, the DC conductivity is directly proportional to the hopping rate of the charge carriers

$$\sigma_{DC} \sim f_e = \frac{1}{2\pi\tau_e}.$$

The measured complex permittivity $\varepsilon^*_{\text{net}}$ can be calculated by taking into account the dielectric function of the ionic liquid in the bulk $\varepsilon^*_{\text{bulk}} = \varepsilon'_{\text{bulk}} + i\varepsilon''_{\text{bulk}}$ and that of the ionic liquid at the interface $\varepsilon^*_{\text{int}} = \varepsilon'_{\text{int}} + i\varepsilon''_{\text{int}}$:

$$\frac{L}{\varepsilon^*_{\text{net}}} = \frac{2d_i}{\varepsilon^*_{\text{int}}} + \frac{L-2d_i}{\varepsilon^*_{\text{bulk}}}, \tag{15.7}$$

where L is the length of the sample cell and d_i is the thickness of the interfacial layers. Separating the real and the imaginary part in Equation 15.7 leads to

$$\varepsilon'_{\text{net}} = (x+1)\left((\varepsilon'_{\text{bulk}})^2 + (\varepsilon''_{\text{bulk}})^2\right)\frac{x\varepsilon'_{\text{bulk}} + y\varepsilon'_{\text{int}}}{(x\varepsilon'_{\text{bulk}} + y\varepsilon'_{\text{int}})^2 + (x\varepsilon''_{\text{bulk}} + y\varepsilon''_{\text{int}})^2} \tag{15.8a}$$

and

$$\varepsilon''_{\text{net}} = (x+1)\left((\varepsilon'_{\text{bulk}})^2 + (\varepsilon''_{\text{bulk}})^2\right)\frac{x\varepsilon''_{\text{bulk}} + y\varepsilon''_{\text{int}}}{(x\varepsilon'_{\text{bulk}} + y\varepsilon'_{\text{int}})^2 + (x\varepsilon''_{\text{bulk}} + y\varepsilon''_{\text{int}})^2}, \tag{15.8b}$$

with

$$x = \frac{L - 2d_i}{2d_i}.$$

and

$$y = \frac{(\varepsilon'_{\text{bulk}})^2 + (\varepsilon''_{\text{bulk}})^2}{(\varepsilon'_{\text{int}})^2 + (\varepsilon''_{\text{int}})^2}.$$

For the bulk, analytical functions derived in a recently developed microscopic model of charge transport in disordered systems are employed [12, 15]. This model considers hopping of the charge carriers in a randomly varying energy landscape. It leads to analytical formulae for the complex conductivity (or correspondingly permittivity) functions based on two physical parameters: the DC conductivity σ_{DC} and the characteristic time τ_e corresponding to the attempt rate to overcome the energy barriers determining the DC conductivity. Within this framework, the complex conductivity function is given by

$$\sigma^*(\omega) = \sigma_{\text{DC}}\left[\frac{i\omega\tau_e}{\ln(1+i\omega\tau_e)}\right]. \tag{15.9}$$

Separating Equation 15.9 into real and imaginary parts, one obtains

$$\sigma'(\omega) = \frac{\sigma_{\text{DC}}\omega\tau_e \arctan(\omega\tau_e)}{0.25\ln^2(1+\omega^2\tau_e^2) + (\arctan(\omega\tau_e))^2} \tag{15.10a}$$

and

$$\sigma''(\omega) = \frac{\sigma_{\text{DC}}\omega\tau_e \ln(1+\omega^2\tau_e^2)}{0.5\ln^2(1+\omega^2\tau_e^2) + 2(\arctan(\omega\tau_e))^2}. \tag{15.10b}$$

For the interfacial regions (having thicknesses on the length scale defined by the screening length of the coulombic interactions, i.e., ~1 nm), a formula similar to Equation 15.9 is used. However, in accordance to Equations 15.5 and 15.6, the following condition is imposed:

$$\frac{\tau_e\,(\text{interface})}{\tau_e\,(\text{bulk})} \cong \frac{\sigma_{\text{DC}}\,(\text{bulk})}{\sigma_{\text{DC}}\,(\text{interface})} \gg 1. \tag{15.11}$$

It is important to mention at this point that Equation 15.9 was developed to express the conductivity of ion conductive materials in the *bulk* and therefore might not be applicable to describe the hopping mechanism at interfaces, especially in the case when the interaction energies become larger than the thermal energy k_BT. However, *no* explicit theoretical developments aiming to deliver conductivity (or permittivity) functions at the interfaces exist at present in the literature. Additionally, it can be easily shown that other approaches used to describe the conductivity in the bulk (as, for instance, the universal model of Jonscher [2, 3, 32]) can be alternatively employed with similar conclusions on the electrode polarization effects, as long as the condition (Eq. 15.11) is imposed.

Given the fact that the current density $\vec{j} = \sigma^* \vec{E}$ is constant across the length of the sample cell, changes in the conductivity function at the interface will automatically imply redistributions of the local field strengths. In our particular case, the conductivity decreases, leading to an increase in the strength of the local electric field E in the interfacial regions.

The calculations of the complex permittivity and conductivity according to Equation 15.7 are shown in Figure 15.18. Interfacial regions of 1-nm thickness and—in accordance to Equation 15.11—a decrease by about eight orders of magnitude in the hopping rate (and correspondingly in the DC conductivity) are assumed. In a wide frequency range covering ~10 decades and with dielectric responses varying over six orders of magnitudes in intensity, the spectral dependence of the electrode polarization effects is *quantitatively* reproduced.

Approximations of the current model can be made to obtain simpler functions suitable for analytical studies of two important aspects of electrode polarization effects: (i) the scaling relations resulting from the variation of temperature, concentration, length of the sample cell, and material of the electrodes and (ii) the influence of intermediate "transition" regions where the local dielectric properties gradually vary from the interface to the bulk region. Since

$$\frac{\tau_e\,(\text{interface})}{\tau_e\,(\text{bulk})} \cong \frac{\sigma_{\text{DC}}\,(\text{bulk})}{\sigma_{\text{DC}}\,(\text{interface})} \gg 1,$$

it immediately follows that

$$\frac{\varepsilon''\,(\text{interface})}{\varepsilon''\,(\text{bulk})} \ll 1.$$

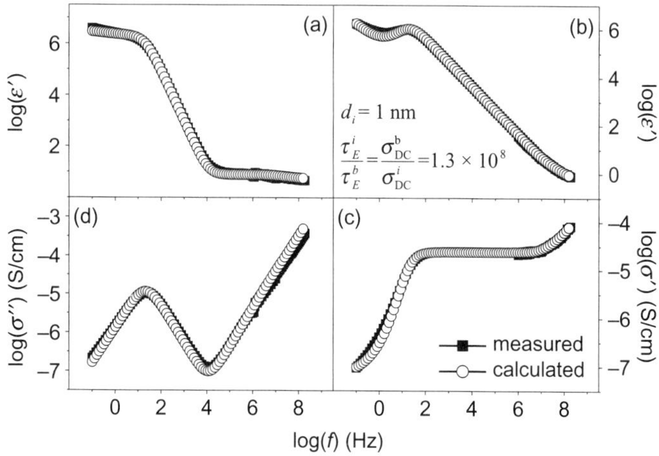

Figure 15.18. Measured and calculated (Eqs. 15.7, 15.9, and 15.11) complex dielectric permittivity $\varepsilon^* = \varepsilon' + i\varepsilon''$ and conductivity $\sigma^* = \sigma' + i\sigma''$ of the ionic liquid [HMIM] [PF_6] versus frequency at 264 K. In addition to the bulk contribution, interfacial regions of d_i = 1-nm thickness were assumed in the calculations, with

$$\frac{\tau_e(\text{interface})}{\tau_e(\text{bulk})} = \frac{\sigma_{\text{DC}}(\text{bulk})}{\sigma_{\text{DC}}(\text{interface})} = 1.3 \times 10^8.$$

This can be approximated by setting $\varepsilon''(\text{interface}) \cong 0$. The results are shown in Figure 15.19. Even under these oversimplified conditions, the onset, as well as the full development of the electrode polarization effects, is correctly reproduced.

In addition to the interfacial regions (where $\varepsilon''(\text{interface}) \to 0$ is used for mathematical simplicity), transition regions where the local physical quantities gradually vary from the interface to the bulk can be taken into account. Instead of Equation 15.7, one has to use

$$\frac{L/2}{\varepsilon^*_{\text{net}}} = \frac{d_i}{\varepsilon^*_{\text{int}}} + \int_{di}^{L/2} \frac{dx}{\varepsilon^*(x)}, \tag{15.12}$$

where $\varepsilon^*(x)$ represents the local dielectric function characterizing the gradient regions between the bulk and the interfaces. For the local dielectric loss $\varepsilon''(x)$, an exponential decrease with a decay factor of $\kappa = 1/d_D$ (d_D being the effective thickness of the gradient regions) between the bulk value $\varepsilon''_{\text{bulk}}$ and the interfacial value ($\varepsilon''(\text{interface}) = 0$) is assumed:

$$\varepsilon''(x) = \varepsilon''_{\text{bulk}}\left(1 - e^{-kx}\right). \tag{15.13}$$

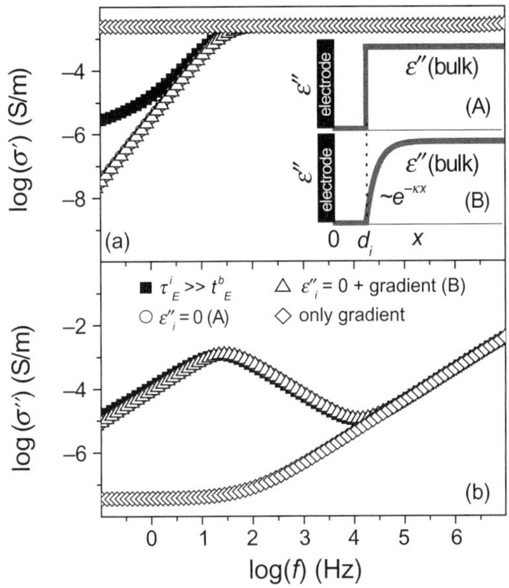

Figure 15.19. Calculated complex conductivity $\sigma^* = \sigma' + i\sigma''$ (according to Equation 15.7) assuming (○) 1-nm interfacial regions with ε''(interface) = 0; (■) 1-nm thick interfacial regions with

$$\frac{\tau_e(\text{interface})}{\tau_e(\text{bulk})} = \frac{\sigma_{\text{DC}}(\text{bulk})}{\sigma_{\text{DC}}(\text{interface})} = 10^8$$

and σ^*(interface) given by Equation 15.9; (Δ) 1-nm interfacial regions with ε''(interface) = 0 and transition regions where ε' and ε'' are exponentially varying between the interfacial and bulk values with a decay coefficient of $\kappa = 1$ nm^{-1}; (◊) only transition regions where ε' and ε'' are exponentially varying between the interfacial and bulk values with a decay coefficient of $\kappa = 1$ nm^{-1}. Inset (4a): Gradients in the local dielectric properties are schematically illustrated.

Imposing the Kramers–Kronig relation for the local dielectric function $\varepsilon^*(x)$,

$$\varepsilon'(x) = \varepsilon_\infty + \frac{2}{\pi}\int_0^\infty \frac{\omega\varepsilon''(x)}{\omega^2 - \omega_0^2}\,d\omega, \tag{15.14}$$

as well as for the bulk dielectric function $\varepsilon^*_{\text{bulk}}$,

$$\varepsilon'_{\text{bulk}} = \varepsilon_\infty + \frac{2}{\pi}\int_0^\infty \frac{\omega\varepsilon''_{\text{bulk}}}{\omega^2 - \omega_0^2}\,d\omega, \tag{15.15}$$

leads to

$$\varepsilon'(x) = \varepsilon'_{\text{bulk}} - e^{-kx}\left(\varepsilon'_{\text{bulk}} - \varepsilon_\infty\right). \tag{15.16}$$

Thus, within this framework, the real and imaginary parts of the local dielectric function $\varepsilon^*(x)$ decrease exponentially with the same decay factor $\kappa = 1/d_D$. The measured (net) dielectric response is expressed as

$$\begin{aligned}\frac{L/2}{\varepsilon^*_{\text{net}}} &= \frac{d_{\text{i}}}{\varepsilon^*_{\text{int}}} + \int_{di}^{L/2} \frac{dx}{\varepsilon^*(x)} \\ &= \frac{d_i}{\varepsilon^*_{\text{int}}} + \int_{di}^{L/2} \frac{dx}{\varepsilon'(x) + i\varepsilon''(x)} \\ &= \frac{d_i}{\varepsilon^*_{\text{int}}} + \int_{di}^{L/2} \frac{dx}{\varepsilon'_{\text{bulk}} - e^{-kx}(\varepsilon'_{\text{bulk}} - \varepsilon_\infty) + i\varepsilon''_{\text{bulk}}(1 - e^{-kx})}.\end{aligned} \tag{15.17}$$

This yields analytical formulae for $\varepsilon'_{\text{net}}$ and $\varepsilon''_{\text{net}}$.

The influence of the gradient regions between the bulk and the interfaces can be understood by considering four different cases:

1. No gradient regions are taken into account between the bulk and the interfaces with ε''(interface) = 0 and d_i = 1 nm. This situation is depicted in Figure 15.19a, inset A. The net response is characterized by two dielectric functions, $\varepsilon^*_{\text{bulk}}$ and $\varepsilon^*_{\text{int}}$. The calculations show that even under these oversimplified conditions, the main experimental features of the electrode polarization effects (the onset and the full development) are accurately described (Fig. 15.19, open circles).
2. No gradient regions are taken into account between the bulk and the interfaces, but ε''(interface) = 0 is no longer imposed (black squares in Fig. 15.19). The permittivity function at the interface $\varepsilon^*_{\text{int}}$ is calculated using Equation 15.7, with

$$\frac{\tau_e\,(\text{interface})}{\tau_e\,(\text{bulk})} \cong \frac{\sigma_{\text{DC}}\,(\text{bulk})}{\sigma_{\text{DC}}\,(\text{interface})} = 1.3 \times 10^8$$

and d_i = 1 nm. In addition to the main features of the electrode polarization effects, further subtle dependencies on the low-frequency side of the spectra (at frequencies $f << f_{\text{max}}$) show up as well (e.g., the change in slope detected in σ' at $f = f_k$, Fig. 15.2).
3. Gradient regions are taken into account between the bulk and the interfaces with ε''(interface) = 0, d_i = 1 nm and d_D = 1 nm (open triangles in Fig. 15.19). This is depicted in inset B of Figure 15.19a. Surprisingly, no significant changes are observed as compared to the case when only the interfacial regions are taken into account without the gradient regions (open circles in Fig. 15.19). Varying the effective thickness of the gradient regions d_D did not lead to changes as well.

4. Only gradient regions with exponential decays are taken into account (with $d_D = 1$ nm) without any interfacial layers (open rhomboids). The main experimental features of the electrode polarization phenomenon are not reproduced in this case.

This analysis clearly shows that the low-frequency side of the dielectric spectra is dominated by the conduction process in the interfacial regions. The condition $\varepsilon''(\text{interface}) \to 0$, which describes the main features of electrode polarization as demonstrated in Figure 15.19, significantly simplifies the analytical calculations necessary to examine the scaling relations with respect to variations in temperature, concentration, length of the sample cell, and material of the electrodes. These calculations can be performed by finding the analytical solutions of the equation

$$\frac{d}{d\omega}(\sigma''_{\text{net}}) = 0$$

and by analyzing the resulting formula for f_{on} and $f_{\max}$ in dependence on σ_0, L, and d_i. Thus, under the conditions $\varepsilon''(\text{interface}) = 0$ and

$$\varepsilon''_{\text{bulk}} = \frac{\sigma_{\text{DC}}}{\varepsilon_0 \omega},$$

the following analytical solutions are found:

$$f_{\text{on}} = \frac{\sigma_{\text{DC}}}{2\pi\varepsilon_0\varepsilon'_{\text{int}}} \times \frac{1}{\sqrt{\dfrac{\dfrac{L}{2d_i}+\dfrac{\varepsilon'_{\text{bulk}}}{\varepsilon'_{\text{int}}}-1}{2\dfrac{\varepsilon'_{\text{int}}}{\varepsilon'_{\text{bulk}}}}\left[\dfrac{\varepsilon'_{\text{int}}}{\varepsilon'_{\text{bulk}}}\left(\dfrac{L}{2d_i}+\dfrac{\varepsilon'_{\text{bulk}}}{\varepsilon'_{\text{int}}}-1\right)-3-\sqrt{\left(\dfrac{\varepsilon'_{\text{int}}}{\varepsilon'_{\text{bulk}}}\left(\dfrac{L}{2d_i}+\dfrac{\varepsilon'_{\text{bulk}}}{\varepsilon'_{\text{int}}}-1\right)\right)^2-10\dfrac{\varepsilon'_{\text{int}}}{\varepsilon'_{\text{bulk}}}\left(\dfrac{L}{2d_i}+\dfrac{\varepsilon'_{\text{bulk}}}{\varepsilon'_{\text{int}}}-1\right)+9}\right]}} \tag{15.18a}$$

$$f_{\max} = \frac{\sigma_{\text{DC}}}{2\pi\varepsilon_0\varepsilon'_{\text{int}}} \times \frac{1}{\sqrt{\dfrac{\dfrac{L}{2d_i}+\dfrac{\varepsilon'_{\text{bulk}}}{\varepsilon'_{\text{int}}}-1}{2\dfrac{\varepsilon'_{\text{int}}}{\varepsilon'_{\text{bulk}}}}\left[\dfrac{\varepsilon'_{\text{int}}}{\varepsilon'_{\text{bulk}}}\left(\dfrac{L}{2d_i}+\dfrac{\varepsilon'_{\text{bulk}}}{\varepsilon'_{\text{int}}}-1\right)-3+\sqrt{\left(\dfrac{\varepsilon'_{\text{int}}}{\varepsilon'_{\text{bulk}}}\left(\dfrac{L}{2d_i}+\dfrac{\varepsilon'_{\text{bulk}}}{\varepsilon'_{\text{int}}}-1\right)\right)^2-10\dfrac{\varepsilon'_{\text{int}}}{\varepsilon'_{\text{bulk}}}\left(\dfrac{L}{2d_i}+\dfrac{\varepsilon'_{\text{bulk}}}{\varepsilon'_{\text{int}}}-1\right)+9}\right]}}. \tag{15.18b}$$

Under identical conditions

$$\left(\varepsilon''(\text{interface}) = 0, \quad \varepsilon''_{\text{bulk}} = \frac{\sigma_{\text{DC}}}{\varepsilon_0 \omega}\right),$$

the inflexion point f_i characterizing the increase in $\varepsilon'_{\text{net}}$ observed after the onset of electrode polarization effects (Fig. 15.6) can be calculated by solving the equation

$$\frac{\partial^2 (\log \varepsilon'_{\text{net}})}{\partial (\log f)^2} = 0.$$

It turns out that

$$f_i = \frac{\sigma_{\text{DC}}}{2\pi\varepsilon_0} \frac{1}{\sqrt[4]{(\varepsilon'_{\text{int}})^3 \varepsilon'_{\text{bulk}} \left(\dfrac{L}{2d_i} - 1 + \dfrac{\varepsilon'_{\text{bulk}}}{\varepsilon'_{\text{int}}}\right)^3}}. \tag{15.19}$$

Without any further simplifications, combining Equations 15.18a,b and 15.19 leads directly to $f_i^2 = f_{\text{on}} f_{\text{max}}$, which is identical to the scaling relation proven experimentally in Figure 15.7.

Given that

$$\frac{L}{2d_i} \gg 1$$

and $1 < \varepsilon'_{\text{int}} < \varepsilon''_{\text{bulk}}$, Equation 15.18a,b simplifies to

$$f_{\text{on}} \cong \frac{\sigma_{\text{DC}}}{2\pi\varepsilon_0} \frac{1}{\sqrt{\varepsilon'_{\text{bulk}} \varepsilon'_{\text{int}}}} \sqrt{\frac{2d_i}{L}} \tag{15.20a}$$

$$f_{\text{max}} \cong \frac{\sigma_{\text{DC}}}{2\pi\varepsilon_0} \frac{1}{\varepsilon'_{\text{int}}} \frac{2d_i}{L}. \tag{15.20b}$$

It is immediately obvious that

1. $f_{\text{on}} \sim f_{\text{max}} \sim \sigma_{\text{DC}}$, which is identical to the scaling laws experimentally proven upon varying the temperature (Fig. 15.5) or the concentration of the charge carriers (Fig. 15.12).
2. $f_{\text{on}} \sim \dfrac{1}{\sqrt{L}}$,

   while

$$f_{\text{max}} \sim \frac{1}{L},$$

which is identical to the scaling law experimentally proven upon varying the length of the sample cell. This confirms the experimental finding that the normalized measured curves do not coincide.

3. $f_{on} \sim \sqrt{d_i}$, while $f_{max} \sim d_i$, which delivers a qualitative explanation of the dependence on the material of the electrodes, since varying the material of the electrodes is expected to change the thickness of the interfacial layers. In addition, possible effects arising due to differently thick oxide layers should also be considered. Furthermore, the experimental observation that the dielectric response does not scale with respect with the variation of the electrode material is obtained as well.
4. $f \cong f_{max}$ is a solution for both

$$\frac{d}{d\omega}(\sigma''_{net}) = 0$$

and

$$\frac{d}{d\omega}(\varepsilon''_{net}) = 0,$$

which explains the experimental finding that the maxima detected in σ''_{net} and ε''_{net} appear at the same frequency position.

5.

$$\sigma_{DC} = 2\pi\varepsilon_0\varepsilon'_{bulk}\frac{(f_{on})^2}{f_{max}},$$

which allows one to determine the DC conductivity of a material using parameters extracted from the electrode polarization effects. This formula is tested in Figure 15.20 for different ionic liquids as well as for different material of the electrodes. An excellent agreement between experiment and calculations is found, even in the case when f_{on} and f_{max} strongly depend on the material of the electrodes (Fig. 15.15). This indicates that the current theoretical approach renders the correct explanation for the dependence on the material of the electrodes.

15.7 THE PHYSICAL SIGNIFICANCE OF f_{on} AND f_{max}

The physical significance of f_{on} and f_{max} is understood by analyzing the net impedance of the sample cell:

$$Z^*_{net} = Z^*_{bulk} + Z^*_{int}. \tag{15.21}$$

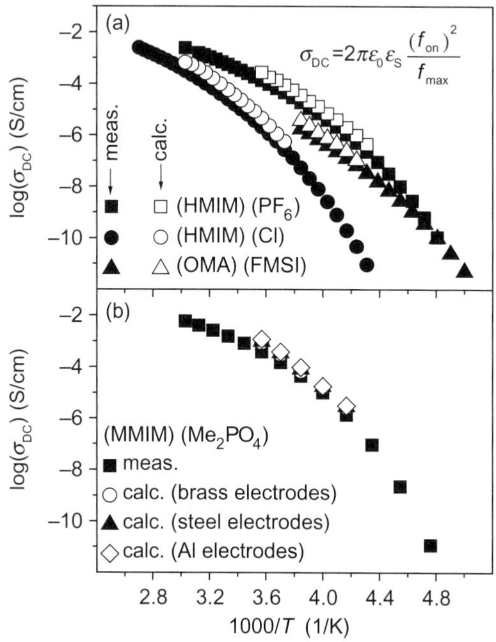

Figure 15.20. (a) Measured and calculated DC conductivities

$$\left(\sigma_{DC} = 2\pi\varepsilon_0\varepsilon'_{bulk}\frac{(f_{on})^2}{f_{max}}\right).$$

for a variety of ionic liquids; (b) DC conductivities calculated from the measurements shown in Figure 15.15 carried out using different materials for the electrodes. The calculated DC conductivities are in full agreement with the direct measurements even when the characteristic frequencies f_{on} and f_{max} strongly depend on the material of the electrodes.

Generally, the impedance is related to the geometry of the measurement cell as follows:

$$Z^* = \frac{1}{\sigma^*}\frac{d}{A} = \frac{\sigma' - i\sigma''}{(\sigma')^2 + (\sigma'')^2}\frac{L}{A} = Z' + iZ'', \qquad (15.22)$$

where L is the thickness and A is the area of the electrodes. With $\sigma^* = i\varepsilon_0\omega\varepsilon^*$, this gives

$$Z' = \frac{\varepsilon''}{\varepsilon_0\omega\left((\varepsilon')^2 + (\varepsilon'')^2\right)}\frac{L}{A} \qquad (15.23a)$$

$$Z'' = \frac{\varepsilon'}{\varepsilon_0\omega\left((\varepsilon')^2 + (\varepsilon'')^2\right)}\frac{L}{A}. \qquad (15.23b)$$

The approximation $\varepsilon''_{int} = 0$, proven to correctly recover the main features of electrode polarization effects (the onset and full development), implies $Z'_{int} = 0$. Thus,

$$Z'_{bulk} = \frac{\varepsilon''_{bulk}}{\varepsilon_0 \omega \left((\varepsilon'_{bulk})^2 + (\varepsilon''_{bulk})^2 \right)} \frac{L}{A} \tag{15.24a}$$

$$Z''_{bulk} = \frac{\varepsilon'_{bulk}}{\varepsilon_0 \omega \left((\varepsilon'_{bulk})^2 + (\varepsilon''_{bulk})^2 \right)} \frac{L}{A} \tag{15.24b}$$

$$Z'_{int} \cong 0 \tag{15.24c}$$

$$Z''_{int} = \frac{1}{\varepsilon_0 \omega \varepsilon'_{int}} \frac{2d_i}{A}. \tag{15.24d}$$

For simplicity, two identical interfaces having a total thickness of $2d_i$ are assumed. Under these conditions, the measured phase angle between the applied voltage and the resulting current approaches the value of –90° in the low-frequency limit:

$$\sin(\varphi_{net}) = \frac{-Z''_{net}}{|Z^*_{net}|} = \frac{-Z''_{net}}{\sqrt{(Z'_{net})^2 + (Z''_{net})^2}} = \frac{-1}{\sqrt{1 + \left(\frac{Z'_{net}}{Z''_{net}} \right)^2}}$$

with

$$\frac{Z'_{net}}{Z''_{net}} = \frac{Z'_{bulk}}{Z''_{bulk} + Z''_{int}} = \frac{\dfrac{\varepsilon''_{bulk}}{\varepsilon_0 \omega \left((\varepsilon'_{bulk})^2 + (\varepsilon''_{bulk})^2 \right)} \dfrac{L}{A}}{\dfrac{\varepsilon'_{bulk}}{\varepsilon_0 \omega \left((\varepsilon'_{bulk})^2 + (\varepsilon''_{bulk})^2 \right)} \dfrac{L}{A} + \dfrac{\varepsilon'_{int}}{\varepsilon_0 \omega (\varepsilon'_{int})^2} \dfrac{2d_i}{A}}$$
$$= \frac{\varepsilon''_{bulk} (\varepsilon'_{int})^2 L}{\varepsilon'_{bulk} (\varepsilon'_{int})^2 L + 2\varepsilon'_{int} d_i \left((\varepsilon'_{bulk})^2 + (\varepsilon''_{bulk})^2 \right)}.$$

In the low-frequency limit,

$$\frac{Z'_{net}}{Z''_{net}} \xrightarrow[\omega \to 0]{} 0 \Rightarrow \sin(\varphi_{net}) \xrightarrow[\omega \to 0]{} -1 \Rightarrow \varphi_{net} \xrightarrow[\omega \to 0]{} -90°.$$

This is indeed close to the experimental observation (Fig. 15.21).

Let us now consider the ratio

$$\frac{Z''_{int}}{Z''_{bulk}} = \frac{2\varepsilon'_{int} d_i}{\varepsilon'_{bulk} L} \frac{(\varepsilon'_{bulk})^2 + (\varepsilon''_{bulk})^2}{(\varepsilon'_{int})^2} = \frac{2\varepsilon'_{int} d_i (\varepsilon'_{bulk})^2}{\varepsilon'_{bulk} L (\varepsilon'_{int})^2} \left(1 + \frac{(\varepsilon''_{bulk})^2}{(\varepsilon'_{bulk})^2} \right).$$

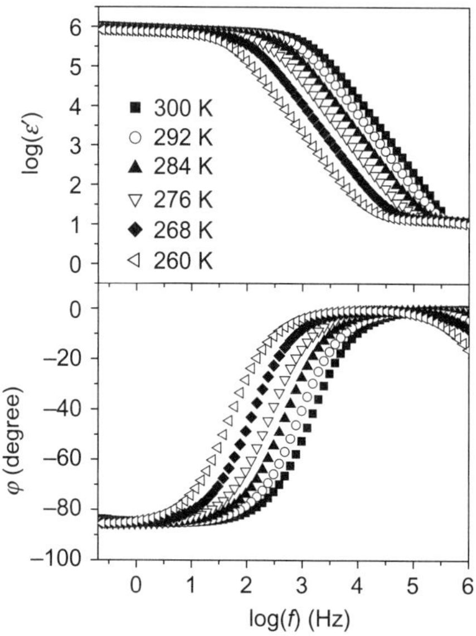

Figure 15.21. Measured permittivity ε' (a) and phase angle between the applied voltage and the resulting current (b) for the ionic liquid [HMIM] [PF_6] as a function of frequency at different temperatures, as indicated. The phase angle approaches a value of –90° in the low-frequency limit.

Since $\frac{(\varepsilon''_{\text{bulk}})^2}{(\varepsilon'_{\text{bulk}})^2} \ll 1$ when $\omega \to \infty$ and

$$\frac{d_i}{L} \ll 1,$$

it immediately follows that $Z''_{\text{int}} \ll Z''_{\text{bulk}}$ in the high-frequency limit. Considering also that $Z'_{\text{int}} = 0$ and $Z'_{\text{bulk}} > 0$, $Z^*_{\text{net}} \cong Z^*_{\text{bulk}}$ results when $\omega \to \infty$. As expected, interfacial effects do not play a significant role in the high-frequency limit where the bulk contribution is dominant. But what happens when $Z''_{\text{int}} = Z''_{\text{bulk}}$? If

$$\frac{Z''_{\text{int}}}{Z''_{\text{bulk}}} = \frac{2\varepsilon'_{\text{int}} d_i}{\varepsilon'_{\text{bulk}} L} \frac{(\varepsilon'_{\text{bulk}})^2 + (\varepsilon''_{\text{bulk}})^2}{(\varepsilon'_{\text{int}})^2} = 1$$

with

$$\varepsilon''_{\text{bulk}} = \frac{\sigma_{\text{DC}}}{\varepsilon_0 \omega},$$

then

$$f = \frac{\sigma_{DC}}{2\pi\varepsilon_0}\frac{1}{\sqrt{\varepsilon'_{int}\varepsilon'_{bulk}}}\sqrt{\frac{2d_i}{L}}.$$

The latter expression is identical to Equation 15.20a, which gives f_{on}. The onset of the electrode polarization effects f_{on} is therefore the frequency, where $Z''_{int} = Z''_{bulk}$. Given that $Z'_{int} = 0$, $|Z^*_{int}| < |Z^*_{bulk}|$ at the onset of electrode polarization affects f_{on}. Again, one has to interpret the case when $|Z^*_{int}| = |Z^*_{bulk}|$. The latter relation implies that

$$\frac{L^2}{(\varepsilon'_{bulk})^2 + (\varepsilon''_{bulk})^2} = \frac{2d_i}{(\varepsilon'_{int})^2},$$

which leads to

$$f = \frac{\sigma_{DC}}{2\pi\varepsilon_0\varepsilon'_{int}}\frac{2d_i}{L}.$$

This is identical to Equation 15.20b, which gives f_{max}. The full development of the electrode polarization effect f_{max} is therefore the frequency where $|Z^*_{int}| = |Z^*_{bulk}|$.

This discussion is summarized in Figure 15.22:

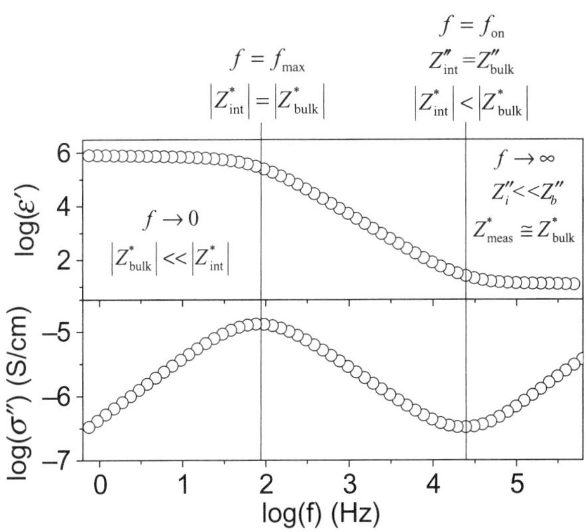

Figure 15.22. Measured permittivity ε' and conductivity σ' of the ionic liquid [HMIM] [PF_6] at 262 K as a function of frequency. Three main regions in the dielectric spectra, marked by the characteristic frequencies f_{max} and f_{on}, are highlighted.

1. In the high-frequency limit $Z''_{\text{int}} \ll Z''_{\text{bulk}}$ and $Z^*_{\text{net}} \cong Z^*_{\text{bulk}}$, the measurements reflect predominantly the bulk contribution.
2. At the onset of the electrode polarization, $Z''_{\text{int}} = Z''_{\text{bulk}}$, but $|Z^*_{\text{int}}| < |Z^*_{\text{bulk}}|$ still. Both Z^*_{bulk} and Z^*_{int} contribute to the measured response.
3. At the full development of the electrode polarization, $|Z^*_{\text{int}}| = |Z^*_{\text{bulk}}|$. Both Z^*_{bulk} and Z^*_{int} contribute to the measured response.
4. In the low-frequency limit $|Z^*_{\text{bulk}}| \ll |Z^*_{\text{int}}|$, the measurements reflect predominantly the interfacial contributions.

15.8 THE DIELECTRIC FUNCTION OF THE INTERFACIAL LAYERS

In the low-frequency limit, $Z''_{\text{net}} \cong Z''_{\text{int}}$ and $Z'_{\text{net}} \ll Z''_{\text{net}}$ hold, implying $C'_{\text{net}} \cong C'_{\text{int}}$. This immediately leads to

$$d_i \cong \frac{\varepsilon'_{\text{int}}}{\varepsilon'_{\text{net}}} L.$$

Since $1 < \varepsilon'_{\text{int}} < \varepsilon'_{\text{bulk}}$ (due to Kramers–Kronig relation and $\varepsilon''_{\text{int}} < \varepsilon''_{\text{bulk}}$), it turns out that

$$\frac{L}{\varepsilon''_{\text{net}}} \le d_i \le \frac{\varepsilon''_{\text{bulk}}}{\varepsilon''_{\text{net}}} L, \tag{15.25}$$

which gives estimative limits for the thickness d_i of the interfacial layers. Using the experimental data presented in Fig. 15.2 to estimate d_i according to Equation 15.25, reasonable values are obtained: 0.29 nm $\le d_i \le$ 3.4 nm.

Equation 15.7 gives the net permittivity of the sample cell and can be inverted to calculate the dielectric function of the interfacial layer $\varepsilon^*_{\text{int}}$ based on the measured permittivity $\varepsilon^*_{\text{net}}$ and the permittivity of the bulk $\varepsilon^*_{\text{bulk}}$. These calculations are performed here in the framework of a layered model. Taking into account interfacial gradients requires elaborate analytical or numerical calculations especially when the Kramers–Kronig relations are imposed. This is not the goal of the current contribution. The complex permittivity $\varepsilon^*_{\text{int}}$ should therefore be regarded as representing the "effective" dielectric properties of the interfaces.

Thus,

$$\varepsilon^*_{\text{int}} = \frac{\varepsilon^*_{\text{net}} \varepsilon^*_{\text{bulk}}}{\varepsilon^*_{\text{bulk}} - \left(\frac{L}{2d_i} - 1\right) \varepsilon^*_{\text{net}}}. \tag{15.26}$$

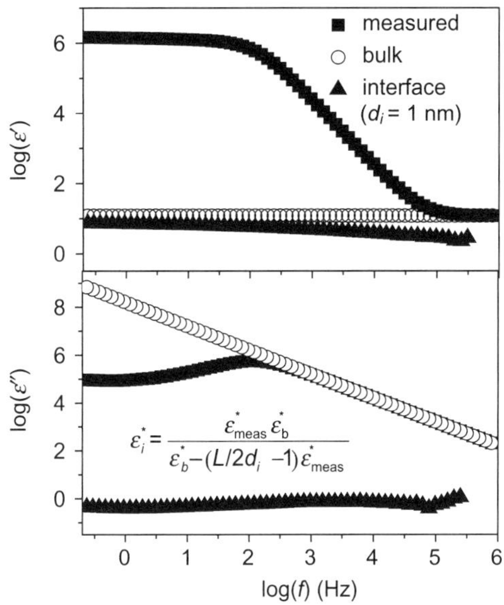

Figure 15.23. Complex permittivity of the interfacial layer ε_{int}^* for the ionic liquid [HMIM] [PF_6] at 276 K calculated—according to Equation 15.26—based on the measured permittivity ε_{meas}^* and the bulk permittivity ε_{bulk}^*. Interfacial layers of 1-nm thickness are assumed.

In accordance to Equation 15.25, nanometric thicknesses are assumed for the interfacial layers. The results of the calculations are shown in Figure 15.23. It is immediately evident that

$$\frac{\varepsilon''(\text{interface})}{\varepsilon''(\text{bulk})} \ll 1,$$

in agreement with Equation 15.11, which represented the main "assumption" of the proposed theoretical approach. It is also obvious that

$$\frac{\varepsilon'(\text{interface})}{\varepsilon'(\text{bulk})} < 1,$$

which was also derived in the current model by considering the Kramers–Kronig relation [33, 34].

15.9 FINAL CONCLUSIONS

Electrode polarization significantly affects the dielectric spectra of conductive materials. Its contribution depends on the frequency of the applied electric field, temperature, concentration of the charge carriers, length of the sample cell, and material of the electrodes. In the present contribution, it is shown that this phenomenon can be traced back to the mechanism of charge transport in the immediate vicinity of solid interfaces. Based on the fact that, due to coulombic interactions, the mobility of charge carriers is drastically slowed down at the metal/ionic conductor interface, this approach quantitatively reproduces the observed scaling laws and enables one to deduce—by the use of a novel formula—the *bulk* conductivity of the ionic charge carriers under study.

REFERENCES

1 F. Kremer, A. Schoenhals. *Broadband dielectric spectroscopy*. Springer, Berlin, 2003.
2 A. K. Jonscher. *Dielectric relaxations in solids*. Chelsea Dielectric Press, London, 1983.
3 A. K. Jonscher. *Universal relaxation law*. Chelsea Dielectrics Press, London, 1996.
4 R. J. Klein, S. Zhang, S. Dou, B. H. Jones, R. H. Colby, J. Runt. *J. Chem. Phys.* 124 (2006) 144903.
5 J. R. Sangoro, A. Serghei, S. Naumov, P. Galvosas, J. Karger, C. Wespe, F. Bordusa, F. Kremer. *Phys. Rev. E* 77 (2008) 051202.
6 H. P. Schwan. *Biophysik* 3 (1966) 181.
7 H. P. Schwan. *Ann. N. Y. Acad. Sci.* 148 (1968) 191.
8 H. P. Schwan. *Ann. Biomed. Eng.* 20 (1992) 269.
9 M. Scott, R. Paul, K. V. I. S. Kaler. *J. Colloid Interface Sci.* 230 (2000) 377.
10 A. Serghei, M. Tress, J. R. Sangoro, F. Kremer. *Phys. Rev. B* 80 (2009) 184301.
11 E. Riande, R. Diaz-Calleja. *Electrical properties of polymers*. Marcel Dekker, New York, 2004.
12 J. C. Dyre. *J. Appl. Phys.* 64 (1988) 2456.
13 J. C. Dyre, P. Maass, B. Roling, D. L. Sidebottom. *Rep. Prog. Phys.* 72 (2009) 046501.
14 J. C. Dyre, T. B. Schroeder. *Rev. Mod. Phys.* 72 (2000) 873.
15 T. B. Schroder, J. C. Dyre. *Phys. Rev. Lett.* 101 (2008) 025901.
16 D. L. Sidebottom. *Rev. Mod. Phys.* 81 (2009) 999.
17 J. Sangoro, et al. *J. Chem. Phys.* 128 (2008) 214509.
18 J. R. Sangoro, G. Turky, M. A. Rehim, C. Iacob, S. Naumov, A. Ghoneim, J. Karger, F. Kremer. *Macromolecules* 42 (2009) 1648.
19 J. R. Macdonald. *Phys. Rev.* 92 (1953) 4.
20 H.-C. Chang, G. Jaffé. *J. Chem. Phys.* 20 (1952) 1071.
21 H. P. Schwan. *Blut* 46 (1983) 185.
22 S. Uemura. *J. Polym. Sci. Polym. Phys. Ed.* 10 (1972) 2155.
23 E. H. B. DeLacey, L. R. White. *J. Chem. Soc. Faraday Trans.* 78 (1981) 457.
24 S. H. Liu. *Phys. Rev. Lett.* 55 (1985) 529.

25 A. Sawada. *J. Chem. Phys.* 126 (2007) 224515.

26 A. Sawada. *J. Chem. Phys.* 129 (2008) 064701.

27 A. L. Alexe-Ionescu, G. Barbero, F. Ciuchi, A. Mazzulla, N. Scaramuzza. *J. Appl. Phys.* 106 (2009) 044508.

28 I. Borukhov, D. Andelman, H. Orland. *Phys. Rev. Lett.* 79 (1997) 435.

29 J. R. Sangoro, C. Iacob, A. Serghei, C. Friedrich, F. Kremer. *Phys. Chem. Chem. Phys.* 11 (2009) 913.

30 J. Sangoro, et al. *J. Chem. Phys.* 128 (2008) 214509.

31 P. Bruce. *Chemistry of solid state materials : solid state electrochemistry*. Cambrigde University Press, Cambridge, 1995.

32 A. K. Jonscher. *Nature* 267 (1977) 673.

33 H. A. Kramers. *Atti Cong. Intern. Fisica* (Transactions of Volta Centenary Congress) Como 2 (1927) 545.

34 R. d. L. Kronig. *J. Opt. Soc. Am.* 12 (1926) 547.

16

LAYER-BY-LAYER ASSEMBLY ON STIMULI-RESPONSIVE MICROGELS

John Erik Wong and Walter Richtering

16.1 INTRODUCTION

In the early 1990s, Decher [1, 2] developed a simple and yet extremely powerful approach to the formation of controlled architecture of multilayer polymer films based upon the layer-by-layer (LbL) assembly of oppositely charged polyelectrolytes on a charged surface, irrespective of the shape and size of the substrate. Research was then focused principally on parameters that influence the formation and growth of polyelectrolyte multilayers, such as type of polyelectrolytes, pH, salts, ionic strength, and degree of charge [3–14]. Once thought to be exclusively electrostatically driven through charge reversal, polyelectrolyte multilayer formation has also been achieved through secondary cooperative interactions such as hydrogen bonding and hydrophobic–hydrophobic interaction [15–17]. Linear and exponential growths [18, 19] were identified as the two major modes of film growth, with both modes yielding extremely interdigitated layers [20].

The rapid success of the LbL assembly of polyelectrolytes on flat and rigid substrates was extended to 3-D hard colloidal particles, which became the precursor route for the fabrication of hollow capsules [21–23]. Soon it became evident that these hollow capsules could be potentially used as drug vehicles to store, carry, target, and deliver drugs. Research was focused on understanding the relationship between the internal structure and the diffusive properties

Electrical Phenomena at Interfaces and Biointerfaces: Fundamentals and Applications in Nano-, Bio-, and Environmental Sciences, First Edition. Edited by Hiroyuki Ohshima.

of polyelectrolyte multilayer of thin films. However, the drawback of using hollow capsules to incorporate macromolecules from the surrounding medium by switching the permeability of the polyelectrolyte shell is the low incorporation efficiency, and for encapsulation of biomacromolecules, the harsh conditions usually used for core dissolution can be detrimental to the bioactivity of the materials.

The use of porous colloidal particles with interfaces characterized by a large surface-to-volume ratio [24–27] wherein encapsulated biomacromolecules can retain their activity suddenly became attractive as new types of microsupports for LbL. $CaCO_3$ proved to be an excellent candidate: biocompatible, inexpensive, easy to produce, highly porous with a large surface area for adsorption of substances of interest, and also easy to decompose at neutral pH under mild conditions. LbL adsorption of polyelectrolytes into mesoporous $CaCO_3$ [24, 25] followed by dissolution of the core leads to the formation of interconnecting networks (matrix-like structure) made of polyelectrolyte complexes. The structure provides one of the most versatile techniques for encapsulating therapeutics with a high loading of the latter due to the high surface area of the porous particle and is applicable to a wide range of materials of different sizes, from proteins to low-molecular-weight drugs.

From planar and hard substrates, spherical and rigid particles to hard and porous particles, the complexity of the shape and size of templates have taken LbL into new dimensions with the advent of a new class of soft nanomaterials obtained from more conventional polymeric structures. The ultimate challenge is to perform LbL on soft and porous templates [28–34], such as microgels. LbL engineering of the surface could impart new thermoresponsive properties to microgels and improve their stability. With the proper choice of parameters of polyelectrolyte pairs, the trafficking of macromolecules across the polyelectrolyte multilayer barrier, selective permeability, as well as sustained release can be achieved. Nanoparticles have also been LbL deposited onto microgels to make hybrid core–shell structures [35–39]. LbL-coated microgels exhibit potential features for loading, encapsulation (storage), and delivery of small (dye) molecules or drugs. Although no dissolution of the core is needed for encapsulation purposes, one could also use biodegradable microgels [40–43] when desired.

In the last couple of years, the LbL technique has been the subject of numerous reviews or communications [44–53]. Although there have been several works reported on the use of microgels in LbL [54–63], there has not been relatively much work done on LbL onto the microgel itself [28–38, 40–43]. This chapter reviews recent advances on the LbL assembly on soft and porous stimuli-responsive microgels with particular emphasis on the challenges microgels present as building blocks and not as depositing species; for example, the electrokinetics of such systems, composed of a water-swollen soft core with a soft polyelectrolyte shell, will need to be formulated [64, 65].

16.2 MICROGELS

Smart microgels are a stimuli-responsive 3-D soft and porous polymeric network [66] that can be either neutral or charged and that swell and contract in response to external stimuli such as temperature, pH, ionic strength, light, and electromagnetic field. Charges can be extrinsic or intrinsic to the microgel [67] depending on whether they are accounted for due to the presence of charged species such as initiators introduced during the preparation of the microgel or due to ionizable (or functional) groups attached chemically to the polymer backbone as main or side chain (copolymerization with charged comonomers such as methacrylic acid [MAA]). Poly(N-isopropylacrylamide) (PNiPAM) is a well-studied smart microgel [68–74], while in a copolymer such as poly(N-isopropylacrylamide-*co*-methacrylic acid) (P(NiPAM-*co*-MAA)), the principal monomer, N-isopropylacrylamide (NiPAM), lends thermosensitivity to these systems, while the comonomer, MAA, allows for pH tunability. Thermoresponsive microgels are characterized by the so-called volume phase transition temperature (VPTT)—at which temperature the microgel undergoes a transition from a swollen to a collapsed state during heating. Below the VPTT, the solubility of PNiPAM in water is ensured by the formation of numerous H-bonds between amide groups of PNiPAM and water molecules. These interactions prevent the precipitation of this relatively hydrophobic polymer. Above the VPTT, PNiPAM collapses and precipitates in water. This phase transition is commonly related to the breaking of the stabilizing H-bonds between water and PNiPAM. Amide groups can then dissociate by forming inter- and intramolecular hydrogen bonds, allowing attractive hydrophobic interactions to dominate. However, in the collapsed state, microgels still contain water [34], which makes them more hydrophilic than polystyrene latex particles, rendering them quite a complex system. Such complexity is also observed during the LbL assembly of polyelectrolyte multilayers on microgels.

16.2.1 Effect of Internal Architecture of Microgels: Core and Core–Shell

With soft and porous templates such as microgels, the layer thickness, the mechanism of multilayer growth, and the surface conformation of polyelectrolyte multilayers are no longer independent of the underlying substrate size and topology [29]. Microgels can be either of the core or core–shell type [40, 75–78]. The core can be either neutral [29, 42] or negatively [29, 43, 79, 80] or positively charged [41, 81–83]. Core–shell [75–77] can benefit from various combinations and permutations of the three situations to confine charges either in the core or in the shell. The notion of core–shell has been extensively used by several authors to define various systems; here, a core–shell microgel is a system with a defined core made up of one type of microgel and a shell made up of a second type of microgel. Such core–shell microgels are usually obtained by a two-step synthesis where the core is first polymerized and then used as seeds in a second polymerization reaction to yield core–shell microgels. Polyelectrolytes can penetrate microgels and induce new surface properties. The degree of interpenetration is dictated by a large extent to the architecture of the microgel, the location of charges, the mesh size of the polymeric network of the microgel (related to the degree of cross-linking), and also by the molecular weight (MW) and degree of branching of the polyelectrolyte [29, 84]. The larger the pore or mesh size, the easier it is for the polyelectrolyte to ease its way within the microgel. Similarly, a high-MW or a highly branched polyelectrolyte may have a lower degree of interpenetration with the microgel. It is important to remember that the mesh size within the microgel core is not monodisperse due to the inhomogeneity of the segment density inside the microgel (some authors also refer to such core system as a dense core with a less dense shell).

Wong and Richtering [29] have demonstrated that the LbL assembly on uncharged PNiPAM core microgels is possible, while De Geest et al. [43] have shown that it is possible to adsorb polycation as well as polyanion as the first layer on uncharged dextran-hydroxyethyl methacrylate (dex-HEMA) microgel. De Geest et al. reported the deposition of poly(allylamine hydrochloride) (PAH) and poly(styrene sulfonate) (PSS) on three different microgels based on dex-HEMA [42]: neutral dex-HEMA, positively [41] and negatively [43] charged microgels obtained by copolymerizing dex-HEMA with dimethylaminoethyl methacrylate (DMAEMA, pK_a 8.3), and MAA (pK_a 4.5), respectively. Figure 16.1 shows the ζ-potential as a function of the number of layers. While Figure 16.1a clearly shows charge reversal upon successive deposition, as expected, Figure 16.1b demonstrates that it is feasible to coat neutral dex-HEMA with either positively or negatively charged polyelectrolyte as the first layer. In the case of charged core (*cc*) microgels, Figure 16.1a is indicative of electrostatic interactions being the main driving force for polyelectrolyte adsorption on charged microgels, while in the case of neutral core (*nc*) microgels, hydrophobic interactions and physical entanglements between the polyelectrolytes and the microgel appear to dominate [29].

16.2.1.1 Effect of the Location of Charge. The interaction of a polyelectrolyte with two sets of core–shell microgels composed of temperature-sensitive PNiPAM with different spatial distribution of pH-sensitive MAA groups has been investigated [84]. The cores consist either of an *nc* of PNiPAM or a *cc* of P(NiPAM-*co*-MAA). A charged shell of P(NiPAM-*co*-MAA) was added to the neutral PNiPAM core (yielding neutral core–charged shell [*nccs*]), while a neutral shell of PNiPAM was added to a *cc* of P(NiPAM-*co*-MAA) (charged core–neutral shell [*ccns*]). The charged groups within the microgels are located either in the inside of the microgels (*ccns*) or exposed (*cc*, *nccs*); therefore, their interaction with oppositely charged polyelectrolytes is expected to be different and might also depend on the polyelectrolyte's chain length. Complexes of these microgels with positively charged

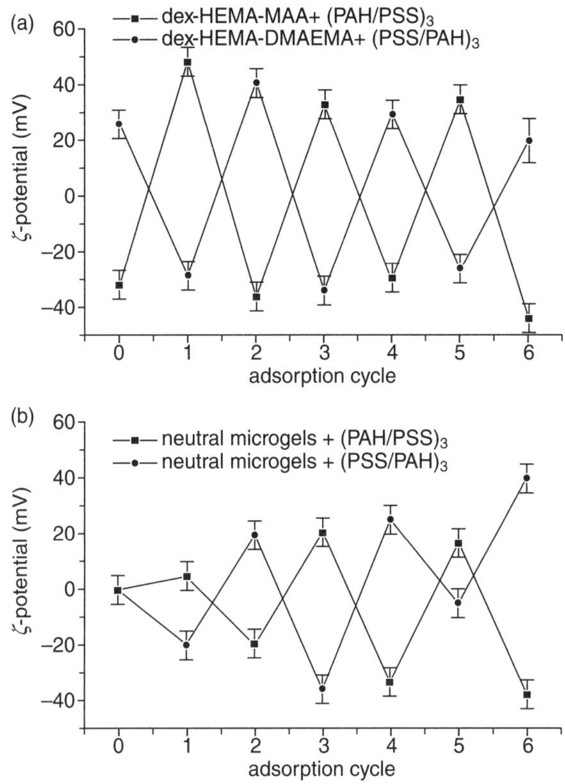

Figure 16.1. Variation of the ζ-potential, as a function of layer number, during the LbL coating of (a) positively (dex-HEMA-DMAEMA) and negatively (dex-HEMA-MAA) charged microgels and (b) neutral dex-HEMA microgels. Reprinted from Reference 43 with permission, © 2006 Elsevier.

poly(diallyldimethylammonium chloride) (PDADMAC) of different MWs were prepared: $PDADMAC_{10-20}$, a low-MW PDADMAC of 10–20 KDa, and $PDADMAC_{400-500}$, a high-MW PDADMAC of 400–500 KDa. The amount of bound polyelectrolyte was quantified and the microgel–polyelectrolyte complexes were characterized with respect to electrophoretic mobility and hydrodynamic radius, R_h [84]. The amount of negative charges of the microgels at a certain concentration can be known from titrations while polyelectrolyte and argentiometric titrations allow the determination of the concentration of PDADMAC. The ratio between positive charges from the PDADMAC and the negative charges of the microgel is named *initial charge ratio* (*icr*). Adding PDADMAC to the microgel does not necessarily lead to complete binding of PDADMAC, so the true composition of the microgel–polyelectrolyte complex can vary from the *icr*. By knowing the initial amount of PDADMAC given to the microgel and the amount of PDADMAC remaining in the supernatant

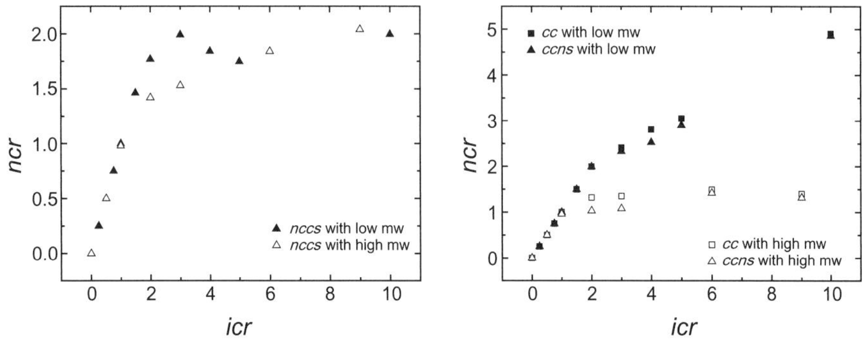

Figure 16.2. Composition of the complexes obtained from low- and high-MW PDADMAC and different microgels. The microgels were fully deprotonated (pH 9). Filled and open symbols show the low-MW and high-MW PDADMAC, respectively. Complexes obtained from *nccs* microgels are shown on the left, while those obtained from *cc* and *ccns* microgels are shown on the right. Redrawn from Reference 84 with permission, © 2010 American Chemical Society.

after centrifugation, the amount of PDADMAC actually bound to the microgel can be calculated. The composition of such complexes is defined by the *nominal charge ratio* (*ncr*). The use of *icr* and *ncr* allows comparing the binding of different microgels with polyelectrolytes. When $ncr = icr$, all PDADMAC is bound to the microgel.

As shown in Figure 16.2, the *nccs* microgel is able to bind similar quantities of low- and high-MW PDADMAC. All offered PDADMAC is bound to the microgel if $icr < 1$, but the binding is no longer quantitative if more PDADMAC is offered to the microgel. The amount of bound PDADMAC reaches a plateau at $ncr \approx 2$; thus, the complexes contain twice as much positive charges (PDADMAC) as negative charges (MAA in the microgel). In contrast, the *cc* and the *ccns* microgels show different binding for the two PDADMAC samples. $PDADMAC_{400-500}$ binds quantitatively at $icr \leq 1.4$, then a plateau is reached. However, for $PDADMAC_{10-20}$, the *ncr* equals the *icr* up to a value of 2. If more $PDADMAC_{10-20}$ is offered to both the *cc* and the *ccns* microgels, the binding is not quantitative anymore and a plateau is never reached. The latter microgels (*cc* and *ccns*) bind less of $PDADMAC_{400-500}$ than of $PDADMAC_{10-20}$, suggesting that $PDADMAC_{400-500}$ is less able to penetrate into the microgel as compared with the shorter $PDADMAC_{10-20}$. Indeed, electrophoretic mobilities of the complexes of *cc* and *ccns* with $PDADMAC_{400-500}$ turn from negative to positive at $ncr = 1$, while those of complexes with $PDADMAC_{10-20}$, however, hardly turn positive (*cc*) or even remain negative (*ccns*). These results are clear indications that both polyelectrolytes are able to penetrate through the noncharged shell albeit to different extents; $PDADMAC_{10-20}$ penetrates fully the *ns* to reach the *cc* while the $PDADMAC_{400-500}$ does not penetrate fully the *ns*. Consequently, the low-MW polyelectrolyte is *ab*sorbed in the microgel without

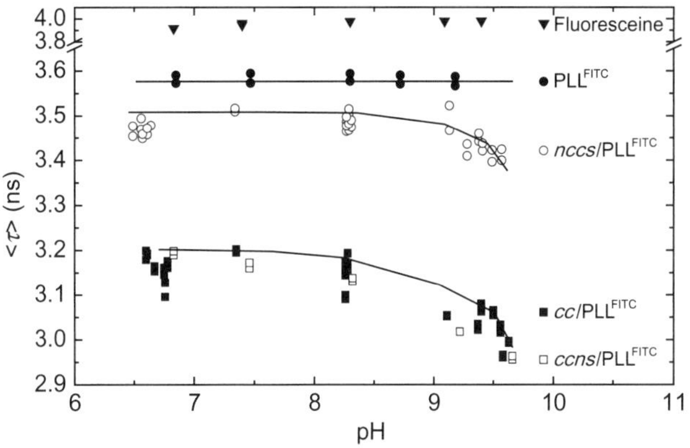

Figure 16.3. The pH dependence of the mean lifetime for complexes between PLL^{FITC} and microgels with different architectures. The lines are only guide for the eyes. Redrawn from Reference 84 with permission, © 2010 American Chemical Society.

causing charge reversal, while the high-MW one is *ad*sorbed on the microgel and leads to charge reversal.

The penetration of polyelectrolytes into microgels was also confirmed by using a fluorescently labeled polyelectrolyte, fluorescein isothiocyanate-labeled poly(L-lysine) (PLL^{FITC}), and by monitoring its lifetime. The lifetime of fluorescent dyes can depend on the dye's local environment and thus lifetime data can provide interesting information on the interpenetration of polyelectrolytes in microgels. The pH dependence of the lifetime of free fluoresceine and of PLL^{FITC} is displayed in Figure 16.3. Free fluoresceine shows a lifetime of around 4 ns comparable to values in the literature and is almost pH independent for the investigated pH range. PLL^{FITC} shows a slightly lower lifetime of around 3.5–3.6 ns, but the lifetime is still independent of pH. The lifetime is also independent of the poly(L-lysine) (PLL) concentration. Figure 16.3 also shows the lifetime as a function of the pH for complexes formed between PLL^{FITC} and microgels of different architectures. The experimental observation that the PLL^{FITC} shows the same lifetime for the *cc* and *ccns* microgels (but a different lifetime as compared to the complexes with the *nccs* microgel; see Fig. 16.3) indicates that the polyelectrolyte is in the same environment for both *cc* and *ccns* microgels. This strongly indicates that the PLL is located in the *cc* for these two microgels. For the *ccns* microgel, this requires a penetration of the PLL through the uncharged shell to the negative charges located in the core.

16.2.2 Effect of Polyelectrolyte

The choice of the polyelectrolyte pair system has a strong influence on the overall thermosensitivity of the coated microgels. Polyelectrolytes can be either synthetic or natural and can be further classified as being strong or weak depending whether they are fully or partially dissociated. Weak polyelectrolytes possess the advantage of having their degree of charge tuned by pH [85]. Natural polypeptides or polysaccharide chains with different amino acid sequences are capable of accessing a very diverse range of configurations. The folding and unfolding of chains through an optimal balance via hydrogen bonds and molecular, electrostatic, and other interactions provide fascinating properties of natural polyelectrolytes to interact with microgels via LbL assembly [34].

An interesting comparative study between multilayers of polysaccharides and polypeptides revealed important differences between their interactions with P(NiPAM-*co*-MAA) microgels [34]. Polypeptide multilayers of PLL and poly-L-glutamic acid (PGA) layers were deposited from freshly prepared buffer solutions (25 mM TRIS, 10 mM 2-(*N*-morpholino)ethanesulfonic acid (MES) and 0.15 M NaCl, pH 7.5). At neutral pH, PLL ($pK_a \sim 10.5$) and PGA ($pK_a \sim 4.3$) are positively and negatively charged, respectively. Chitosan (85% deacetylated) (CHIT) and dextran sulfate (DEX) were employed to build up polysaccharide multilayers. CHIT ($pK_a \sim 6.0$) layers were deposited from an aqueous solution with 25% acetic acid (pH 1.3) and DEX ($pK_a \sim 4.0$) layers were assembled from an aqueous solution (pH 6.2). Figure 16.4 shows the monitoring of the LbL assembly of both PLL/PGA and CHIT/DEX multilayers using dynamic light scattering (DLS) as a function of the number of layers. Both systems demonstrate the "odd–even" effect in the R_h depending on the type of polyelectrolyte in the outermost layer, as already observed with synthetic polyelectrolytes [29–32]. While the swelling and deswelling of microgels coated with polysaccharides are almost reversible, DLS curves for those coated

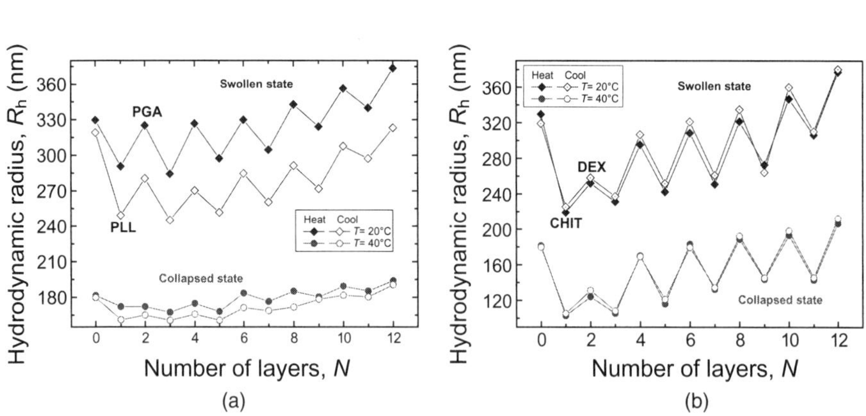

Figure 16.4. The size of the microgels as a function of the number of (a) PLL/PGA and (b) CHIT/DEX layers for the swollen (T = 20°C) and collapsed (T = 40°C) states. N = 0 correspond to the uncoated microgel, and the odd and even numbers indicate a polycation- and a polyanion-terminated microgel, respectively. Redrawn from Reference 34 with permission, © 2010 Elsevier.

with polypeptides reveal strong hysteresis between the heating and the cooling cycles (see Fig. 16.4). Interestingly, in a PLL/PSS system on the same P(NiPAM-*co*-MAA) microgels, the hysteresis is observed only when the last layer is PLL, while PSS-terminated microgels show no hysteresis [33]. P(NiPAM-*co*-MAA) microgels coated with CHIT/DEX multilayers demonstrate a unique and interesting feature: The odd–even effect of the R_h is pronounced in both the swollen and collapsed states; such behavior has never been observed in any other polyelectrolyte-coated microgel systems. Also, upon deposition of the first layer of CHIT (R_h ~ 320 nm), the size of the coated microgel (R_h ~ 220 nm) is reduced to a size barely larger than the size of the uncoated microgel in the collapsed state (R_h ~ 190 nm). Even more amazing is the further collapse of the one-layer CHIT-coated microgel on heating to R_h ~ 110 nm. This is the first report whereby the collapse of a microgel is taken beyond its original collapsed size, proving that in the collapsed state, the microgel still contains water [34].

16.2.2.1 Effect of MW. De Geest et al. [42] substituted synthetic polyelectrolytes by biopolyelectrolytes in an attempt to achieve self-rupture of the membranes at physiological conditions. With poly-L-arginine (pARG) as polycation, the authors successfully achieved LbL on dex-HEMA microgel using several polyanions, such as DEX, chondroitin sulfate (CHON), poly-L-aspartic acid (pASP), but also with low-molecular-weight poly-L-glutamic acid ($pGLU_{low}$) and high-molecular-weight poly-L-glutamic acid ($pGLU_{high}$). Figure 16.5a shows the confocal microscopy images of dex-HEMA microgels coated with different combinations of biopolyelectrolytes. A clear ring of red fluorescence originating from the rhodamine B isothiocyanate-labeled ($pARG^{RITC}$) surrounding the microgels is indicative of a successful LbL coating of the dex-HEMA microgels. Exposure of these coated microgels to a 0.1 M phosphate buffer (pH 7.4) at 37°C for 5 days (the time it takes to fully degrade the dex-HEMA microgel core at this pH and temperature) reveals the different mechanical strength and permeability of the core–shell system relative to the composition of the shell. Figure 16.5b (A) shows that all microcapsules were ruptured, while Figure 16.5b (B) shows that some microcapsules were ruptured while others were still intact and filled with a fluorescein isothiocyanate-labeled dextran (DEX^{FITC}). Figure 16.5b (C) shows that hollow capsules were obtained; that is, the microcapsules did not rupture but released the 150 KDa DEX^{FITC}, as no significant fluorescence was detected within the microcapsules. De Geest et al. [42] also reported the influence of the MW of the polyelectrolytes on the behavior of the core–shell ensemble. Core–shell obtained with $pGLU_{low}$/pARG all ruptured, while those obtained with $pGLU_{high}$/pARG did not all explode. This indicates that the increase in MW of the pGLU changes the interplay between the mechanical properties and the permeability of the biopolyelectrolyte coating [29].

Wong et al. studied the buildup of PDADMAC/PSS multilayers on P(NiPAM-*co*-MAA) microgel using PDADMAC of different MW:

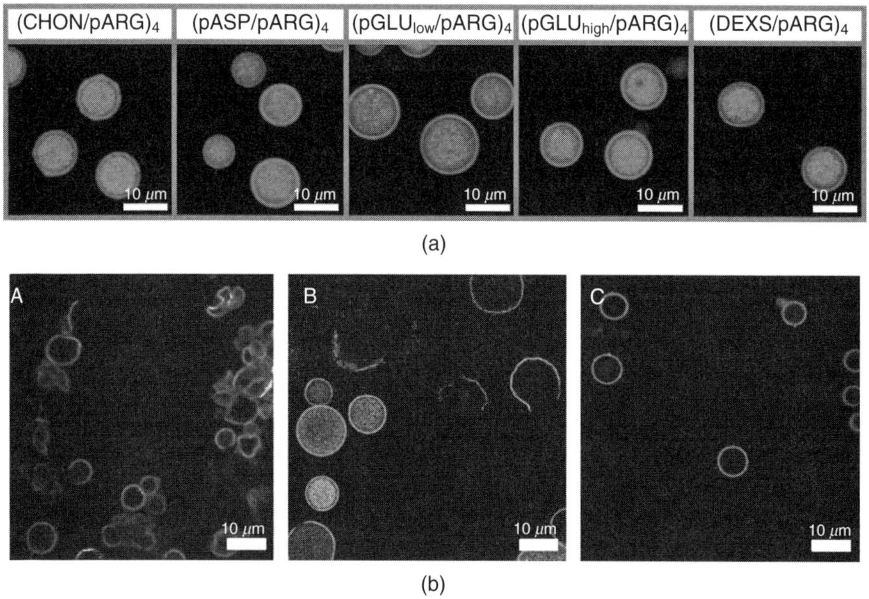

Figure 16.5. (a) Confocal microscopy images of microcapsules obtained after LbL coating of the 150 KDa FITC-dextran fluorescently labeled dex-HEMA microgels (green)with four biopolyelectrolyte bilayers, wherein $pARG^{RITC}$ (RITC, red) was used. (b) Confocal images of dex-HEMA microgels fluorescently labeled with 150 KDa DEX^{FITC} coated with (A) $(CHON/pARG)_4$, (B) $(pGLU_{high}/pARG)_4$, and (C) $(DEX/pARG)_4$ after degradation of the microgel core. In (A), all microcapsules were broken and released their contents. In (B), both broken as well as intact (still filled with 150 KDa DEX^{FITC}) microcapsules were observed. The capsules in (C) remained intact but released their contents by diffusion through the biopolyelectrolyte coating. In all cases, the pARG was fluorescently labeled with RITC. Reprinted from 42 with permission, ©2007 Wiley-VCH. See color figure on companion ftp site.

$PDADMAC_1$ (1 KDa) and $PDADMAC_{100-200}$ (100–200 KDa). For both MW and at $T <$ VPTT, upon multilayer deposition, DLS study reveals a significant decrease and increase of the R_h each time the outermost layer is PDADMAC and PSS, respectively. In spite of the extreme difference in MW, the R_h obtained for LbL-coated microgel with both MW polyelectrolytes are quasi-identical. A particularly noteworthy point to remember is that the DLS study gives no indication of the amount of PDADMAC adsorbed onto the microgel nor the resulting charge on the coated microgel. Figure 16.6 shows the electrophoretic measurements as a function of the number of layers deposited. Despite the fact that the coated microgels have similar sizes as determined by DLS, there are discrepancies in the magnitude of the charge reversal if $PDADMAC_{100-200}$ ($+3.3 \times 10^{-8}$ $m^2/V/s$) or $PDADMAC_1$ ($+2.2 \times 10^{-8}$ $m^2/V/s$) is deposited. Such observation is a clear difference compared to the LbL deposition on hard and

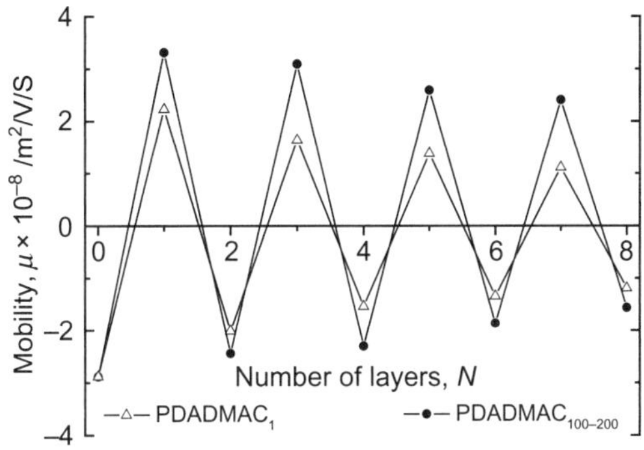

Figure 16.6. Electrophoretic mobility measurements (at T = 20°C) as a function of the number of layers deposited on P(NiPAM-*co*-MAA) (N = 0): comparison between high- and low-MW PDADMAC. An odd and even number of layers correspond to PDADMAC and PSS, respectively. The polyelectrolytes were deposited from 0.25 M NaCl, then washed and redispersed in water. Reprinted from [32] with permission, © 2009 American Chemical Society.

rigid particles, whereby the charge reversal is almost constant, and so is the systematic decrease in mobility (or net surface charge) with increasing number of layers deposited on soft and porous microgels, irrespective of the MW of the PDADMAC used (see Fig. 16.6). $PDADMAC_1$, being of smaller chain length than $PDADMAC_{100-200}$, could interpenetrate more the microgel, neutralizing more charges, thereby conferring a lower surface charge density as revealed by the mobility. Indeed, electrophoretic measurements as a function of temperature [32] for P(NiPAM-*co*-MAA) microgels coated with various numbers of layers reveal that, with $PDADMAC_1$, hysteresis between the heating and cooling curves is more pronounced than in the case with $PDADMAC_{100-200}$. $PDADMAC_1$ has shorter chains and possesses greater mobility and consequently a higher degree of freedom for rearrangement and, in doing so, has a higher possibility of neutralizing charges in the microgel, which were previously not accessible, and hence causing the effective charge at the surface to drop (in magnitude). With $PDADMAC_{100-200}$, only slight rearrangement is allowed, hence the almost complete reversibility of the mobility curves versus temperature.

Wong and Richtering [29] observed a drastic change in the thermoresponsive behavior of the coated microgel when PNiPAM was coated with a priming layer of polyethylenimine (PEI), a highly branched high-MW (MW = 750 KDa) polyelectrolyte known for its high surface charge density and excellent anchoring properties. Due to the high branching and high MW, PEI is assumed quasi impenetrable in PNiPAM, which consequently simply sits on the microgel, and this is translated as a noticeable increase in the particle size (from R_h = 140–150 nm) in the swollen state. As the temperature is increased, the collapse of the PNiPAM microgel is significantly restricted by the electrostatic interaction of the PEI layer with the core, and the thermoresponsive behavior is almost suppressed; the PEI-coated PNiPAM microgels collapse to only 120 nm, which is about twice the size of the uncoated PNiPAM in the collapsed state (around 60 nm). The flow of water out of PEI-coated PNiPAM microgels is drastically impeded by the branched network around the core and also the elastic properties conferred by the interactions between PNiPAM core and PEI chains. Interestingly, deposition of a PSS layer on the PEI-coated PNiPAM microgel amazingly recovers partially the native thermoresponsivity of PNiPAM [29].

16.2.2.2 Effect of the Ionic Strength. Figure 16.7 shows the R_h of the LbL-coated microgels at two distinct temperatures (20 and 40°C, below and above the VPTT, respectively) as a function of the number of layers deposited, for $PDADMAC_{100-200}$/PSS systems whereby the polyelectrolytes were deposited from various salt concentrations [32]. It has to be stressed that, while the coating is done in polyelectrolyte solutions with various salts, the coated microgel is washed and redispersed in water for DLS measurements. Comparing identical numbers of layers deposited on the microgel from 0 (water), 0.1, 0.25 to 0.5 M NaCl concentrations, an increase in R_h of the coated microgel with increasing salt concentrations is observed. For example, when one layer of $PDADMAC_{100-200}$ is deposited on the microgel from 0 (water), 0.1, 0.25, and 0.5 M NaCl, the R_h of the coated microgel is 250, 270, 275, and 280 nm, respectively. Under the assumption that the interaction between the microgel and the different deposited conformations, or different amount of polyelectrolytes

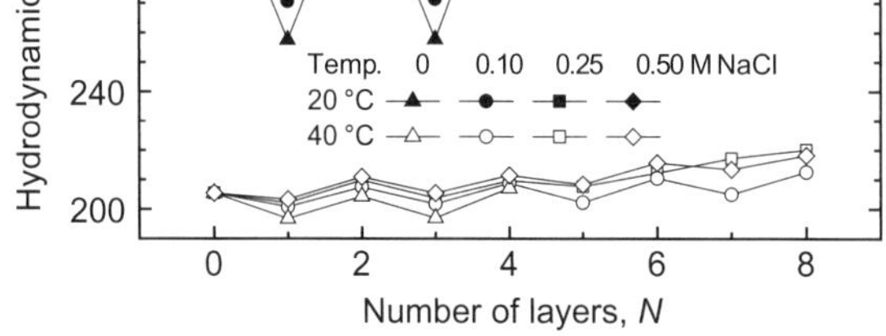

Figure 16.7. Hydrodynamic radius of LbL-coated P(NiPAM-*co*-MAA) microgels, deposited from polyelectrolyte solutions of different salt concentration, as a function of the number of layers measured at two temperatures: below (20°C) and above (40°C) the VPTT of the microgel (~32°C). $N = 0$ is the microgel, and odd number of layers are PDADMAC-terminated microgels and even number layers are PSS-terminated microgels. Reprinted from Reference 32 with permission, © 2009 American Chemical Society.

adsorbed, all being equal in all cases, the increase in the hydrodynamic size can be translated as an increase in the film thickness. In other words, the higher the salt concentration of the polyelectrolyte solutions, the thicker is the film deposited. This is in agreement with what is commonly observed for deposition on hard and rigid substrates. In aqueous or salt solutions of low ionic strength, the polyelectrolyte adopts a stretched conformation due to the repulsion of charges along the polyelectrolyte backbone, while in salt solutions of high ionic strength, charges of the polyelectrolyte are partially screened by counterions, thereby reducing both inter- and intramolecular repulsions, making the polyelectrolyte adopt a more coiled and compact conformation. Consequently, multilayers deposited at low ionic strength are thinner than those deposited at high ionic strength.

16.2.2.3 Effect of the pH. The effect of pH *during* multilayer formation on P(NiPAM-*co*-MAA) microgel has been studied with the following polyelectrolytes: PAH, poly(acrylic acid) (PAA) and PSS. Figure 16.8 shows the size of the coated microgel as a function of the number of (i) PAH/PSS and (ii) PAH/PAA layers for various pH deposition conditions [85]. In both systems, the coated microgels retain their thermoresponsive property; their swelling and deswelling processes are reversible; and there is an odd–even effect of the R_h of the coated microgels depending on the type of polyelectrolytes in the outermost layer: PAA- or PSS-terminated microgels are more swollen and more temperature sensitive than PAH-terminated microgels. Electrophoretic mobility studies reveal charge reversal after each layer deposition, even in the case where the polyelectrolyte bears few charges during buildup, and the

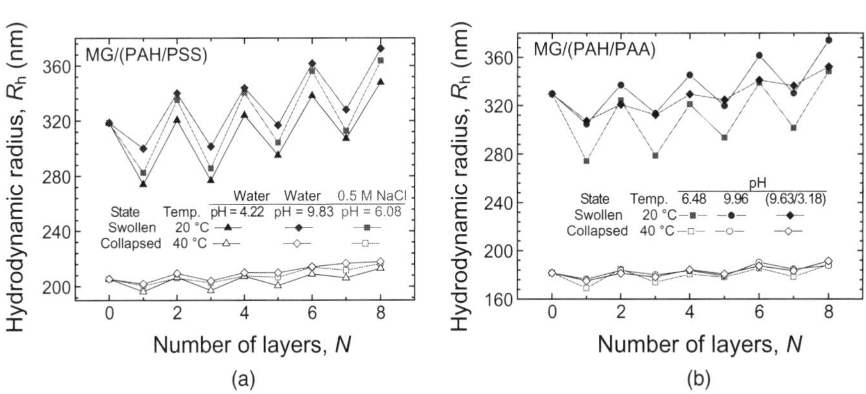

Figure 16.8. Hydrodynamic radius of (a) PAH/PSS and (b) PAH/PAA, LbL-coated P(NiPAM-*co*-MAA) microgels (MG) deposited from polyelectrolyte solutions of different pH, as a function of the number of layers measured at two temperatures: below (20°C) and above (40°C) the VPTT of the microgel (~32°C). $N = 0$ is the microgel, and odd number of layers are PAH-terminated microgels and even number layers are either PSS- or PAA-terminated microgels.

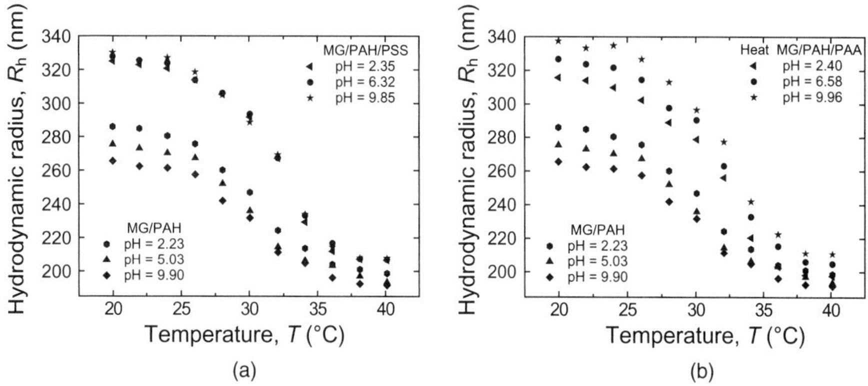

Figure 16.9. Influence of the pH on the thermoresponsive behavior of coated microgels as a function of temperature. A comparative study between (a) weak/strong and (b) weak/weak polyelectrolyte pairs.

magnitude of which decreases with an increasing number of layers. Confining PAH and PAA on the microgel, the effective pK_a and pK_b of the bound polyelectrolyte in the multilayer is shifted to higher and lower values, respectively [85].

The effect of pH on LbL-coated microgels has been studied for the following systems: PDADMAC/PSS [32], PAH/PSS [85], and PAH/PAA [85]. It has been demonstrated that the nature of the last or outermost layer determines the pH sensitivity of the ensemble (microgel + polyelectrolytes). Depositing just one layer of a strong polyelectrolyte onto a pH-sensitive microgel (e.g., PDADMAC onto P(NiPAM-*co*-MAA)) is sufficient to render the coated microgel insensitive to pH [32]. Figure 16.9 shows the influence of pH on coated microgels with the last layer being either a strong or a weak polyelectrolyte. It can be seen that coated microgels terminating with a weak polyelectrolyte such as PAH or PAA retain the pH sensitivity of the ensemble, but with a different thermoresponsivity and pH sensitivity to that of the uncoated microgel, while those terminating with a strong polyelectrolyte (such as PSS) are no longer pH sensitive. With proper choice of the polyelectrolyte, the thermoresponsivity of the coated microgel can be accordingly tuned.

16.3 STABILITY OF LBL-COATED MICROGELS

The stability of LbL-coated microgels was monitored by following the change in their size as a function of time [29, 32]. Figure 16.10 shows the thermoresponsive behavior of neutral PNiPAM coated with PEI and PSS over different periods of time. In between experiments, the samples were stored at room temperature. Greinert and Richtering [28] previously reported hysteresis on

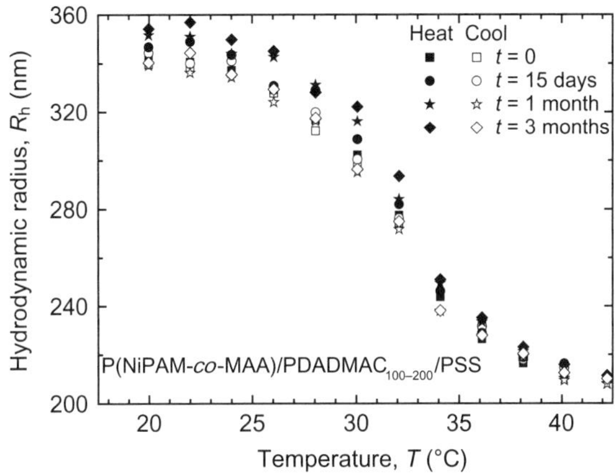

Figure 16.10. Temporal stability of MG/PDADMAC$_{100-200}$/PSS microgels as monitored by DLS as a function of temperature over a period of 3 months. Reprinted from Reference 32 with permission, © 2009 American Chemical Society.

a two-layer coated core–shell microgels (neutral PNiPAM core and a charged P(NiPAM-*co*-AA) shell), where they found that the first heating cycle results in major alteration of the microgel and that extended time periods under harsher conditions lead only to small changes. In the case of PNiPAM/PEI/PSS, the highly branched positively charged PEI causes fast and permanent alteration upon its deposition, but the deposition of the PSS layer recovers the thermoresponsive properties of the microgels [29]. The only difference after 1 month was that, in the swollen state (below the VPTT), the particles size was bigger than that originally prepared. However, subjecting these microgels after 1 month to a heating and cooling cycle recovers the original behavior of the coated microgels. Similar observations were recorded on LbL-coated charged microgels. For example, the R_h of P(NiPAM-*co*-MAA) microgels coated with PDADMAC and PSS after $t = 0$, 15 days, 1 month, and 3 months were 342, 347, 352, and 354 nm at 20°C, respectively. These values can be considered as almost constant and, after time t, when the coated microgel is heated again (independently of the size it reached or, in other words, irrespective of its history), it always collapses to the same size as when it was first coated ($t = 0$) and exhibits the same VPTT. On cooling, they all swell to the original R_h of P(NiPAM-*co*-MAA)/PDADMAC$_{100-200}$/PSS (~340 nm) as on the day it was prepared.

De Geest et al. [43] subjected the LbL-coated microgels to change in pH to investigate whether self-rupturing occurs or not. Dex-HEMA microgels coated with $(PSS/PAH)_3$ shows that rupturing did not occur at physiological pH value (pH 7.4) but only when incubated at pH 9. The authors attributed this to the pH-dependent permeability of the $(PSS/PAH)_3$ membrane; at pH 7, the $(PSS/PAH)_3$ membrane is permeable to the degradation products of the dex-HEMA microgels (mainly dextran chains of 19 KDa), whereas the membrane is impermeable at pH 9. So at pH 9, the products remained inside the microgel, and the increased osmotic/swelling pressure of the core eventually causes the $(PSS/PAH)_3$ membrane to burst. They also investigated the release of 158 KDa tetramethyl rhodamine B isothiocyanate-dextrans (DEX^{TRITC}) from uncoated and $(PSS/PAH)_3$-coated dex-HEMA-DMAEMA microgels [42]. Uncoated microgels show a substantial burst release followed by a continuous release until the microgels are completely degraded after 20 hours. In the case of $(PSS/PAH)_3$-coated dex-HEMA-DMAEMA microgels, the release is much more pulsatile: The release of DEX^{TRITC} remains low even during the first 10 hours of degradation, and after this initial phase, the majority of the encapsulated DEX^{TRITC} molecules are released in a couple of hours.

16.4 PROOF OF LBL COATING ON MICROGELS

For microgels under the micrometer size, it is extremely difficult to actually visualize the exact location of the LbL coating using fluorescently labeled species. Single focus fluorescence correlation spectroscopy (FCS) [86, 87] has been used to demonstrate succesful LbL assembly of polyelectrolyte multilayers on PNiPAM microgels [30]. Two differently fluorescently labeled polyelectrolytes, rhodamine B-labeled PDADMAC ($PDADMAC^{Rh}$) and PLL^{FITC}, were incorporated in a PNiPAM/$PDADMAC^{Rh}$/PSS/PLL^{FITC} multilayer configuration. Single or autocorrelated experiments performed on each species at different layering stages yield similar curves, revealing diffusion times of the same order of magnitude, and strongly suggesting that both labeled polyelectrolytes are anchored on the same species and are moving together. To definitely remove any doubt if the polyelectrolytes are anchored to the microgel or not, one needs (i) dual-focus fluorescence correlation spectroscopy (2*f*-FCS) [88–91] for a more precise and quantitative analysis and (ii) a microgel, which is chemically tagged with a fluorescent species and achieve LbL buildup using a different fluorescently labeled polyelectrolyte [33].

Figure 16.11 shows the normalized 2*f*-FCS auto- and cross-correlation curves of NG^{Rh}/PLL^{FITC}, NG^{Rh}/$(PLL/PSS)_{1.0}$/PLL^{FITC}, and NG^{Rh}/$(PLL/PSS)_{2.0}$/PLL^{FITC} nanogels for two different temperatures: $T = 25.0$ and 40.7°C; NG^{Rh} being poly(N-isopropylacrylamide-*co*-acrylic acid-*co*-rhodamine). Some correlation curves reveal a broad decay at long delay times, indicating the presence of some large aggregates in addition to the individual coated nanogels. The size of the coated nanogels was obtained from fitting (line) the experimental data (points) with a bimodal particle size effect (PSE) model [88] for the auto-correlation functions (ACFs) and cross-correlation functions (CCFs) obtained from the rhodamine moieties contained in the nanogel core and the FITC moieties contained in the multilayer shell. At 25°C, the ACF and CCF curves for NG^{Rh}/PLL^{FITC}, NG^{Rh}/$(PLL/PSS)_{1.0}$/PLL^{FITC}, and NG^{Rh}/$(PLL/PSS)_{2.0}$/PLL^{FITC} nanogels are nicely fitted and the R_h obtained are 245, 254, and

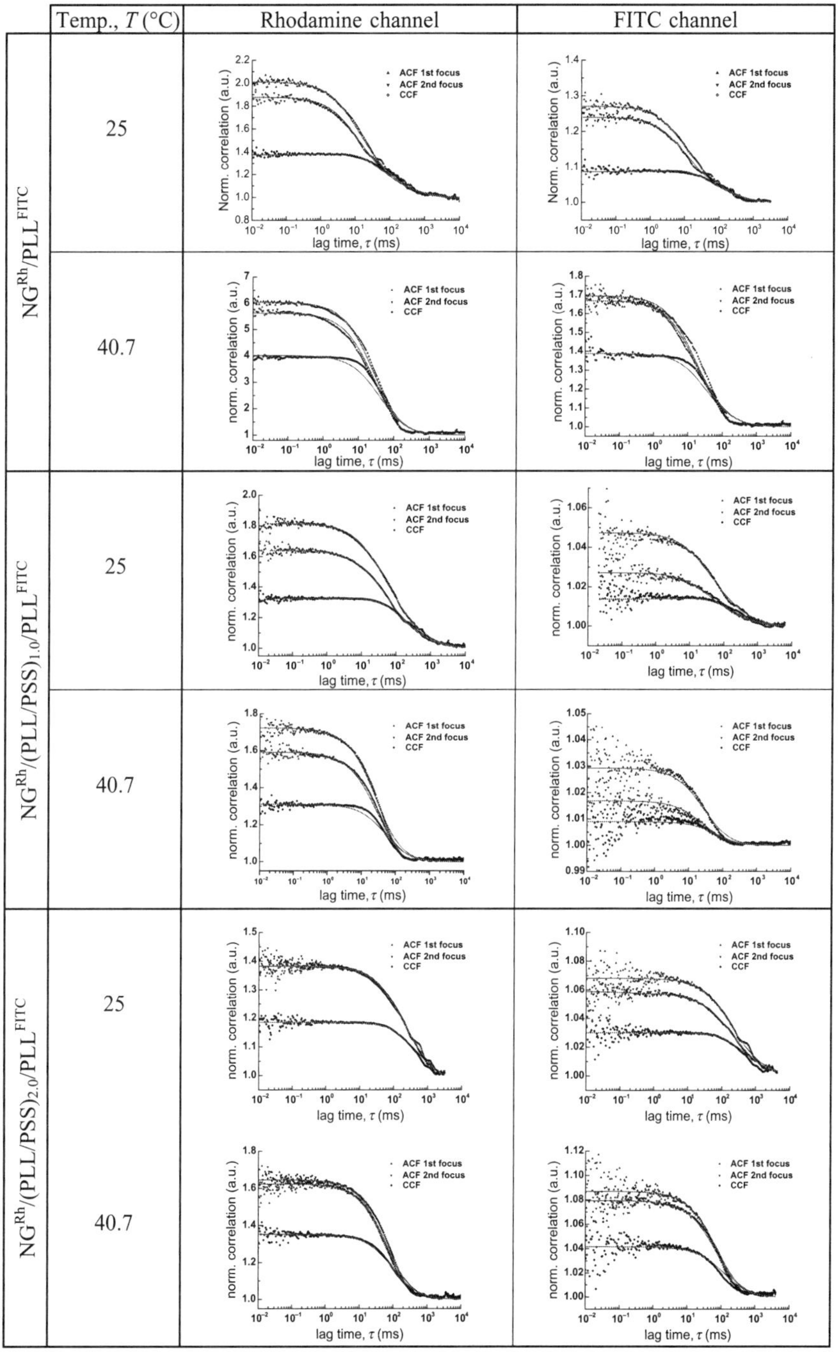

Figure 16.11. The 2*f*-FCS auto- and cross-correlation curves of the NG^{Rh}/PLL^{FITC}, NG^{Rh}/(PLL/PSS)1.0/PLL^{FITC}, and NG^{Rh}/(PLL/PSS)2.0/PLL^{FITC} as detected in the rhodamine and FITC channel at $T = 25°C$ and $T = 40.7°C$. Experimental data points are fitted (line) using the dual particle size effect (PSE) model, and the R_h obtained is in good agreement with that obtained from DLS. Reprinted from Reference 33 with permission, © 2009 American Chemical Society.

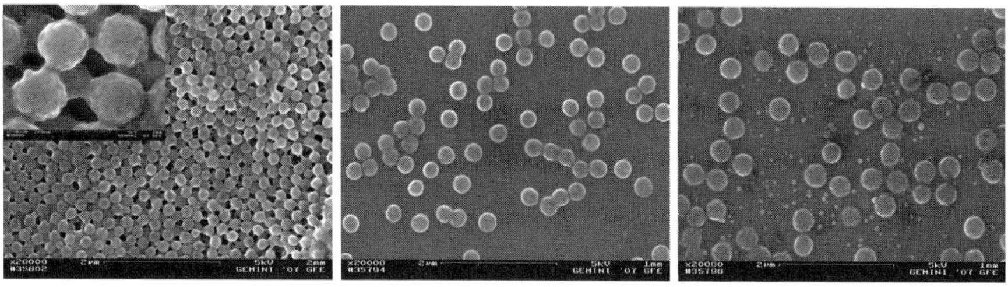

(a) (b) (c)

Figure 16.12. SEM micrographs of (a) MG, (b) MG/PDADMAC$_{100-200}$, and (c) MG/PDADMAC$_{100-200}$/PSS in the dry state. Inset in (a) is a magnification to show bridging between MG. Reprinted from Reference 32 with permission, © 2009 American Chemical Society.

258 nm, respectively. These results are in excellent agreement with those obtained using DLS, indicating clearly that the polyelectrolyte shell is bound to the nanogel core and the whole ensemble is moving as one entity. At $T = 40.7°C$, the R_h obtained for NG^{Rh}/PLL^{FITC}, $NG^{Rh}/(PLL/PSS)_{1.0}/PLL^{FITC}$, and $NG^{Rh}/(PLL/PSS)_{2.0}/PLL^{FITC}$ nanogels are 170, 168, and 166 nm, respectively. Although there are some discrepancies between the experimental and the fit curves concerning the steepness of the decay (observed only at such high temperatures), the fits yield R_h, which are in good agreement with those obtained by DLS. These results point to the fact that in the collapsed state, the polyelectrolyte multilayer is still bound to the collapsed nanogel core. Furthermore, during the heating and cooling cycles, no polyelectrolyte release is observed because ACFs and CCFs show no extra decay, which could have been correlated to the free labeled polyelectrolyte. This is the first report of the fate of the polyelectrolyte multilayers on thermoresponsive LbL-coated microgels during the VPTT [33].

Figure 16.12 shows scanning electron microscope (SEM) images from the dried microgel (MG), MG/PDADMAC$_{100-200}$, and MG/PDADMAC$_{100-200}$/PSS [32], MG being P(NiPAM-*co*-MAA). Unlike hollow capsules (templated from ~5 μm silica) that usually shrink to flat structures with creases and folds, microgels tend to keep their spherical shape upon drying. The dried uncoated and coated microgels shrunk during drying to different sizes. It has to be stressed that all samples are in the dry state, which is very different from being in aqueous medium in the collapsed state. Although SEM micrographs do not allow for direct size comparison with that obtained with DLS, they provide a visual representation of the morphology of the samples. Figure 16.12a shows a typical SEM image of the uncoated microgel where an individual microgel is "interconnected" by sticking bridge, clearly visible in the inset. Figure 16.12b,c shows individual LbL-coated microgels, with clear surface modification of the microgel: MG/PDADMAC$_{100-200}$ still "stick" to one or two particles, but MG/PDADMAC$_{100-200}$/PSS are more individual particles.

16.5 NANOPARTICLE–MICROGEL HYBRID

Nanoparticles have received a great deal of interest for the construction of novel organic–inorganic hybrid structures, and their incorporation into microgels to exploit and combine their optical, electronic, or magnetic properties with the thermoresponsive behavior of smart microgels has recently attracted much attention [35–38, 92–96] in areas ranging from sensors to drug delivery systems. Several approaches to the preparation of hybrid structures made of inorganic nanoparticles and microgels have been reported [35–38, 92–96], but we shall focus only on work wherein LbL assembly was employed, in one way or another, to achieve such hybrid structures.

Karg et al. [35, 36] reported on the deposition of gold nanorods, coated with a polyelectrolyte bilayer (PSS/PAH) to achieve a positively charged surface, onto a negatively charged PNiPAM microgels simply by mixing. By varying the concentration of the coated gold nanorods, various surface coverage was achieved (no complete surface coverage was achieved even with the highest loading), but distinct differences were observed between low and high surface coverage in terms of optical properties. For low surface coverage, UV-vis spectra recorded at 15 and 50°C (fully swollen and fully collapsed state) show that an increase in temperature causes (i) an increase in the absorbance at low wavelengths (up to 650 nm) and (ii) a red shift and increase in the intensity of the longitudinal plasmon band. Both effects were attributed to an increase in the refractive index of the microgel (during collapse) and a local increase near the nanorods, but because of the low surface coverage, any electronic interactions among nanorods were excluded. For high surface coverage, the longitudinal plasmon band is quite different, with a pronounced red shift, damping, and broadening of the band. These effects were attributed to a decrease in the interparticle distances upon reduction of the particle volume when the microgel collapses. The nanorods are close enough to each other for plasmon coupling to occur. Independently of the surface coverage, temperature and optical response of the hybrid microgels were found to be fully reversible, indicating that after heating and cooling, the original rod distribution is recovered.

Wong et al. [37, 38] reported on the LbL assembly of magnetic nanoparticles (MNPs) onto PNiPAM microgels whose surfaces have been modified with polyelectrolyte multilayers (PDADMAC and PSS). PDADMAC- and PSS-terminated microgels were used for the LbL assembly of negatively and positively charged MNPs, respectively. Positively charged MNPs were obtained by the coprecipitation technique, while negatively charged MNP were obtained through surface modification of the MNP either via *in situ* precipitation of the MNP in the presence of a polyanion or postmodification by the LbL technique [97]. DLS measurements reveal that repeated swelling and deswelling of the hybrid microgel is completely reversible and reproducible, and is indicative that the MNP do not get detached from the microgel during the volume phase transition of the microgel. The VPTT of the hybrid microgel was similar to that of the pure microgel. Electrophoretic studies reveal charge reversal at each deposition step, polyelectrolytes as well as MNPs, confirming the successful buildup of the hybrid microgel.

16.6 CLOSING REMARKS AND OUTLOOK

Since the advent of the LbL technique, it is only recently that its use on soft and porous microgels has been achieved. Microgels as building templates constitute a challenge in themselves, and their different architectural ensemble needs to be further investigated for a better understanding of the nature of the surface they present for interaction with other species such as polyelectrolytes. More in-depth and systematic studies regarding the formation and growth of polyelectrolyte multilayers on soft and porous microgels should be carefully undertaken with respect to new parameters such as temperature [98], density of cross-linking in the template or location of charges within the template [99], as well as with respect to conventional parameters such as type of polyelectrolyte pairs, pH, ionic strength, or type of salts. We predict that in the near future, much work will be devoted to the use of fluorescently labeled templates (microgels) as well as adsorbing species, such as polyelectrolytes, labeled markers, or quantum dots in an attempt to understand the internal structure and restructuring of the ensemble (to external stimuli) for better commercial exploitation of the properties of these species. This chapter has focused essentially on the very recent development of the LbL assembly of polyelectrolytes (and nanoparticles) on microgels as templates. Upcoming challenges include potential use of biodegradable microgels and shells, efficient loading of the microgels with drugs for storage and transport, efficient release of the drug, or incorporation of functionality to site specifically target delivery. It is hoped that this chapter will stimulate research toward the enormous potential of microgels as templates in the LbL technique for the manipulation and creation of new hybrid structures with endless possibilities.

ACKNOWLEDGMENT

The authors thank the *Deutsche Forschungsgemeinschaft* (DFG) for financial support of their research.

REFERENCES

1 G. Decher, J. Schmitt. *Prog. Colloid Polym. Sci.* 89 (1992) 160.
2 G. Decher. *Science* 277 (1997) 1232.
3 D. Yoo, S. S. Shiratori, M. F. Rubner. *Macromolecules* 31 (1998) 4309.
4 S. S. Shiratori, M. F. Rubner. *Macromolecules* 33 (2000) 4213.

5 S. T. Dubas, J. B. Schlenoff. *Macromolecules* 32 (1999) 8153.
6 R. v. Klitzing, J. E. Wong, W. Jaeger, R. Steitz. *Curr. Opin. Colloid Interface Sci.* 9 (2004) 158.
7 M. Salomäki, P. Tervasmäki, S. Areva, J. Kankare. *Langmuir* 20 (2004) 3679.
8 M. Salomäki, I. A. Vinokurov, J. Kankare. *Langmuir* 21 (2005) 11232.
9 B. Schoeler, E. Poptoshev, F. Caruso. *Macromolecules* 36 (2003) 5258.
10 R. Steitz, W. Jaeger, R. von Klitzing. *Langmuir* 17 (2001) 4471.
11 K. Glinel, A. Moussa, A. M. Jonas, A. Laschewsky. *Langmuir* 18 (2002) 1408.
12 U. Voigt, W. Jaeger, G. H. Findenegg, R. v. Klitzing. *J. Phys. Chem. B* 107 (2003) 5273.
13 J. E. Wong, F. Rehfeldt, P. Haenni, M. Tanaka, R. v. Klitzing. *Macromolecules* 37 (2004) 7285.
14 J. E. Wong, H. Zastrow, W. Jaeger, R. von Klitzing. *Langmuir* 25 (2009) 14061.
15 W. B. Stockton, M. F. Rubner. *Macromolecules* 30 (1997) 2717.
16 N. A. Kotov. *Nanostruct. Mater.* 12 (1999) 789.
17 S. A. Sukhishvili, S. Granick. *Macromolecules* 35 (2002) 301.
18 C. Picart, J. Mutterer, L. Richert, Y. Luo, G. D. Prestwich, P. Schaaf, J.-C. Voegel, P. Lavalle. *Proc. Natl. Acad. Sci. U.S.A.* 99 (2002) 12531.
19 P. Lavalle, C. Picart, J. Mutterer, C. Gergeley, H. Reiss, J. C. Voegel, B. Senger, P. Schaaf. *J. Phys. Chem. B* 108 (2004) 635.
20 M. Lösche, J. Schmitt, G. Decher, W. G. Bouwman, K. Kjaer. *Macromolecules* 31 (1998) 8893.
21 E. Donath, G. B. Sukhorukov, F. Caruso, S. A. Davis, H. Möhwald. *Angew. Chem. Int. Ed.* 37 (1998) 2202.
22 F. Caruso, R. A. Caruso, H. Möhwald. *Science* 282 (1998) 1111.
23 F. Caruso, H. Lichtenfeld, E. Donath, H. Möhwald. *Macromolecules* 32 (1999) 2317.
24 G. B. Sukhorukov, D. V. Volodkin, A. M. Günther, A. I. Petrov, D. B. Shenoy, H. Möhwald. *J. Mater. Chem.* 14 (2004) 2073.
25 A. I. Petrov, D. V. Volodkin, G. B. Sukhorukov. *Biotechnol. Prog.* 21 (2005) 918.
26 Y. Wang, A. Yu, F. Caruso. *Angew. Chem. Int. Ed.* 44 (2005) 2888.
27 A. Yu, Y. Wang, E. Barlow, F. Caruso. *Adv. Mater.* 17 (2005) 1737.
28 N. Greinert, W. Richtering. *Colloid Polym. Sci.* 282 (2004) 1146.
29 J. E. Wong, W. Richtering. *Prog. Colloid Polym. Sci.* 133 (2006) 45.
30 J. E. Wong, C. B. Müller, A. Laschewsky, W. Richtering. *J. Phys. Chem. B* 111 (2007) 8527.
31 J. E. Wong, W. Richtering. *Curr. Opin. Colloid Interface Sci.* 13 (2008) 403.
32 J. E. Wong, A. M. Díez-Pascual, W. Richtering. *Macromolecules* 42 (2009) 1229.
33 J. E. Wong, C. B. Müller, A. M. Díez-Pascual, W. Richtering. *J. Phys. Chem. B* 113 (2009) 15907.
34 A. M. Díez-Pascual, J. E. Wong. *J. Colloid Interface Sci.* 347 (2010) 79.
35 M. Karg, I. Pastoriza-Santos, L. M. Liz-Marzán, T. Hellweg. *ChemPhysChem* 7 (2006) 2298.
36 M. Karg, I. Pastoriza-Santos, J. Pérez-Juste, T. Hellweg, L. M. Liz-Marzán. *Small* 3 (2007) 1222.

37 J. E. Wong, A. K. Gaharwar, D. Müller-Schulte, D. Bahadur, W. Richtering. *J. Magn. Magn. Mater.* 311 (2007) 219.

38 J. E. Wong, A. K. Gaharwar, D. Müller-Schulte, D. Bahadur, W. Richtering. *J. Colloid Interf. Sci.* 324 (2008) 47.

39 M. Karg, T. Hellweg. *Curr. Opin. Colloid Interface Sci.* 14 (2009) 438.

40 S. Nayak, D. Gan, M. J. Serpe, L. A. Lyon. *Small* 1 (2005) 416.

41 B. G. De Geest, C. Déjugnat, G. B. Sukhorukov, K. Braeckmans, S. C. De Smedt, J. Demeester. *Adv. Mater.* 17 (2005) 2357.

42 B. G. De Geest, C. Déjugnat, M. Prevot, G. B. Sukhorukov, J. Demeester, S. C. De Smedt. *Adv. Funct. Mater.* 17 (2007) 531.

43 B. G. De Geest, C. Déjugnat, E. Verhoeven, G. B. Sukhorukov, A. M. Jonas, J. Demeester, S. C. De Smedt. *J. Control. Release* 116 (2006) 159.

44 G. B. Sukhorukov, A. L. Rogach, B. Zebli, T. Liedl, A. G. Skirtach, K. Köhler, A. A. Antipov, N. Gaponik, A. S. Susha, M. Winterhalter, W. J. Parak. *Small* 1 (2005) 194.

45 A. P. R. Johnston, C. Cortez, A. S. Angelatos, F. Caruso. *Curr. Opin. Colloid Interface Sci.* 11 (2006) 203.

46 J. A. Jaber, J. B. Schlenoff. *Curr. Opin. Colloid Interface Sci.* 11 (2006) 324.

47 R. v. Klitzing. *Phys. Chem. Chem. Phys.* 8 (2006) 5012.

48 A. S. Angelatos, K. Katagiri, F. Caruso. *Soft Matter* 2 (2006) 18.

49 S. E. Moya, J. L. Toca-Herrera. *J. Nanosci. Nanotechnol.* 6 (2006) 2329.

50 K. Ariga, J. P. Hill, Q. M. Ji. *Phys. Chem. Chem. Phys.* 9 (2007) 2319.

51 G. B. Sukhorukov, H. Möhwald. *Trends Biotechnol.* 25 (2007) 93.

52 B. G. De Geest, N. N. Sanders, G. B. Sukhorukov, J. Demeester, S. C. De Smedt. *Chem. Soc. Rev.* 36 (2007) 636.

53 J. F. Quinn, A. P. R. Johnston, G. K. Such, A. N. Zelikin, F. Caruso. *Chem. Soc. Rev.* 36 (2007) 707.

54 R. Steitz, V. Leiner, K. Tauer, V. Khrenov, R. von Klitzing. *Appl. Phys. A: Mater. Sci. Process.* 74 (2002) S519.

55 K. Glinel, G. B. Sukhorukov, H. Moehwald, V. Khrenov, K. Tauer. *Macromol. Chem. Phys.* 204 (2003) 1784.

56 M. J. Serpe, C. D. Jones, L. A. Lyon. *Langmuir* 19 (2003) 8759.

57 S. Nayak, L. A. Lyon. *Chem. Mater.* 16 (2004) 2623.

58 C. M. Nolan, M. J. Serpe, L. A. Lyon. *Macromol. Symp.* 227 (2005) 285.

59 D. B. Shenoy, G. B. Sukhorukov. *Macromol. Biosci.* 5 (2005) 451.

60 M. Prevot, C. Déjugnat, H. Mohwald, G. B. Sukhorukov. *ChemPhysChem* 7 (2006) 2497.

61 M. Rusu, D. Kuckling, H. Möhwald, M. Schönhoff. *J. Colloid Interface Sci.* 298 (2006) 124.

62 H. Alem, F. Blondeau, K. Glinel, S. Demoustier-Champagne, A. M. Jonas. *Macromolecules* 40 (2007) 3366.

63 C. D. Sorrell, L. A. Lyon. *J. Phys. Chem. B* 111 (2007) 4060.

64 H. Ohshima. *Colloid Polym. Sci.* 285 (2007) 1411.

65 H. Ohshima. *J. Colloid Interface Sci.* 328 (2008) 3.

66 S. Nayak, L. A. Lyon. *Angew. Chem. Int. Ed.* 44 (2005) 7686.

67 A. Pich, W. Richtering. *Adv. Polym. Sci.* 234 (2011) 1.
68 R. H. Pelton, P. Chibante. *Colloids Surf.* 20 (1986) 247.
69 H. G. Schild. *Progr. Polym. Sci.* 17 (1992) 163.
70 B. R. Saunders, B. Vincent. *Adv. Colloid Interface Sci.* 80 (1999) 1.
71 R. Pelton. *Adv. Colloid Interface Sci.* 85 (2000) 1.
72 H. Kawaguchi. *Progr. Polym. Sci.* 25 (2000) 1171.
73 T. Hoare, R. Pelton. *Polymer* 46 (2005) 1139.
74 B. H. Tan, K. C. Tam. *Adv. Colloid Interface Sci.* 136 (2008) 25.
75 I. Berndt, J. S. Pedersen, W. Richtering. *J. Am. Chem. Soc.* 127 (2005) 9372.
76 I. Berndt, J. S. Pedersen, W. Richtering. *Angew. Chem. Int. Ed.* 45 (2006) 1737.
77 I. Berndt, J. S. Pedersen, P. Lindner, W. Richtering. *Langmuir* 22 (2006) 459.
78 K. M. Ho, W. Y. Li, C. H. Wong, P. Li. *Colloid Polym. Sci.* 288 (2010) 1503.
79 K. Kratz, T. Hellweg, W. Eimer. *Colloids Surf. A* 170 (2000) 137.
80 H. Senff, W. Richtering. *J. Chem. Phys.* 111 (1999) 1705.
81 F. Meunier, A. Elaissari, C. Pichot. *Polym. Adv. Technol.* 6 (1995) 489.
82 F. Meunier, A. Elaissari, C. Pichot. *Macromol. Symp.* 150 (2000) 283.
83 L. Zha, J. Hu, C. Wang, S. Fu, A. Elaissari, Y. Zhang. *Colloid Polym. Sci.* 280 (2002) 1.
84 J. Kleinen, A. Klee, W. Richtering. *Langmuir* 26 (2010) 11258.
85 A. M. Díez-Pascual, J. E. Wong, W. Richtering. In preparation.
86 D. Madge, W. W. Webb, E. Elson. *Phys. Rev. Lett.* 29 (1972) 705.
87 E. L. Elson, D. Madge. *Biopolymers* 13 (1974) 1.
88 C. B. Müller, A. Loman, W. Richtering, J. Enderlein. *J. Phys. Chem. B* 112 (2008) 8236.
89 C. B. Müller, W. Richtering. *Colloid Polym. Sci.* 286 (2008) 1215.
90 P. Schwille, K. G. Heinze. *ChemPhysChem* 2 (2001) 269.
91 T. Dertinger, V. Pacheco, I. von der Hocht, R. Hartmann, I. Gregor, J. Enderlein. *ChemPhysChem* 8 (2007) 433.
92 D. Suzuki, H. Kawaguchi. *Langmuir* 22 (2006) 3818.
93 Y. Lu, Y. Mei, M. Ballauff, M. Drechsler. *J. Phys. Chem. B* 110 (2006) 3930.
94 Y. Lu, Y. Mei, M. Drechsler, M. Ballauff. *Angew. Chem. Int. Ed.* 45 (2006) 813.
95 S. Bhattacharya, F. Eckert, V. Boyko, A. Pich. *Small* 3 (2007) 650.
96 M. Das, N. Sanson, D. Fava, E. Kumacheva. *Langmuir* 23 (2007) 196.
97 J. E. Wong, A. K. Gaharwar, D. Müller-Schulte, D. Bahadur, W. Richtering. *J. Nanosci. Nanotechnol.* 8 (2008) 4033.
98 J. Kleinen, W. Richtering. Rearrangements in and release from responsive microgel–polyelectrolyte complexes induced by temperature and time. *J. Phys. Chem. B* 115 (2011) 3804–3810.
99 J. Kleinen, W. Richtering. Polyelectrolyte microgels based on poly-N-isopropylacrylamide: influence of charge density on microgel properties, binding of poly-diallyldimethylammonium chloride, and properties of polyelectrolyte complexes. *Colloid Polym. Sci.* 289 (2011) 739–749.

17

DYNAMICS OF POLYMERS AND POLYELECTROLYTES AT COLLOIDAL INTERFACE AND SUBSEQUENT FLOCCULATION

YASUHISA ADACHI

17.1 INTRODUCTION

The use of water-soluble polymers to control the stability of colloids can be dated back many years ago. One of the earliest uses appears in ancient Egypt and China around 2500 BC in the preparation of ink [1, 2] and is essentially the same technique we know as steric stabilization today [3]. It is also known that polymers can destabilize colloidal suspensions. The description of this effect can be found in Sanskrit literature (ca. 2000 BC) [4]. Nowadays, this process is commonly used in conventional water and wastewater treatment processes to enhance the precipitation of suspended matter. Irrespective of the stabilization or destabilization ability of the colloidal suspension, the interaction between polymer molecules and the colloidal surface is the fundamental issue [5, 6]. However, scientific understanding of the dynamics of polymer and polyelectrolyte chains at their interfaces in relation to the behavior of colloidal suspensions is still under discussion at an elementary level. In this section, our strategy of using the flocculation of a monodispersion of colloidal latex [7] is introduced after a basic description of polymer flocculation.

Electrical Phenomena at Interfaces and Biointerfaces: Fundamentals and Applications in Nano-, Bio-, and Environmental Sciences, First Edition. Edited by Hiroyuki Ohshima.

17.2 MECHANISMS OF FLOCCULATION INDUCED WITH WATER-SOLUBLE POLYMERS AND POLYELECTROLYTES

17.2.1 Adsorption versus Depletion

Two types of interactions can be expected for soluble polymers against an interface. When a segment of a polymer chain present in a solution is more preferable to the interface, adsorption takes place and an adsorbing layer of polymer is formed (Fig. 17.1). Vice versa, a depletion layer is formed. As will be described later, flocculation of colloidal particles can be induced for both cases.

A so-called train–loop–tail configuration (Fig. 17.2) is often used to describe the conformation of a polymer chain adsorbing on the surface. A train segment is the part of the polymer chain in proximate contact with the interface mainly by enthalpic effect, while the loop and tail retain the entropic nature of a flexible coil with thermal fluctuation. With a model of the train–loop–tail configuration, the density profile of a segment near the interface can be numerically calculated on the basis of self-consistent mean field theory [5]; however, a more simplified view is available based on a scaling argument using a self-similar grid [6]. The former gives a precise profile of segment density but cannot stand for the system of large fluctuation that is expected for the outside fringe of the adsorbing polymer layer composed of long chains. This picture is possible when bridging flocculation starts to take place. From the measurement of hydrodynamic layer thickness, a relatively thicker layer that scales the size of macromolecules in solution is reported for polyethylene oxide (PEO) adsorbed onto polystyrene latex particles [8, 9].

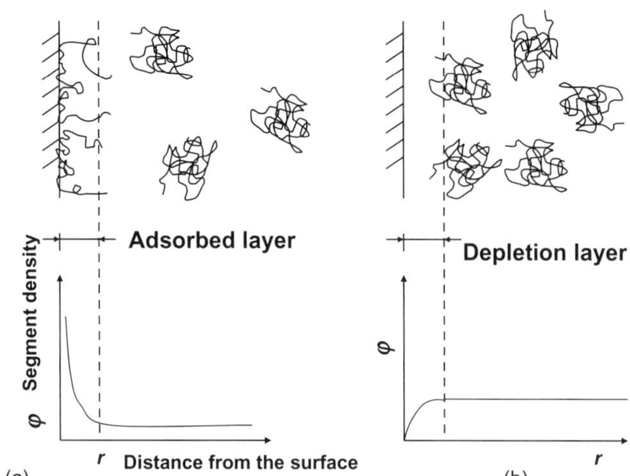

Figure 17.1. Qualitative picture of polymers near a colloidal interface. (a) Adsorption and (b) depletion.

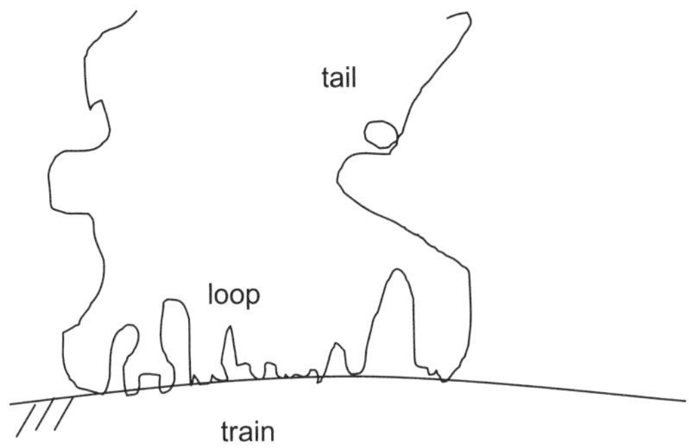

Figure 17.2. Adsorbed polymer chain at the colloidal interface.

In contrast to adsorption, the depletion zone of a polymer segment is formed when the segments prefer the solution rather than the interface. The thickness of the depletion layer in the first approximation is the size of the polymer chain in solution, but it is a decreasing function of polymer concentration [10].

17.2.2 Osmotic Effect of Colloidal Stability

Derjaguin–Landau–Verwey–Overbeek (DLVO) theory predicts the repulsive force of two colloidal particles by the effect of excess osmotic pressure of water located in the gap between two colloidal particles [11]; that is, the repulsion is stemmed out from the increase in osmotic pressure by the incrementation of ion concentration in the gap region due to the overlap of the diffusive part of electrical double layers. This overlap induces incrementation of osmotic pressure, thus preventing the approach of the two surfaces. A similar picture can be applied for the case of interaction between two surfaces that are covered with adsorbing layers of polymers; that is, a polymer segment located in the part of the tail or loop moving with random Brownian motion is an origin of the increment of osmotic pressure when two surfaces with an adsorbing polymer layer approach. The overlaps of tails or loops of adsorbed polymer taking place at the approach of the two surfaces will bring about a similar situation to that of the diffusive part of electrical double layers (Fig. 17.3).

Contrary to the case of the approach of two adsorbing layers of polymer, an attraction is expected when two depletion layers approach each other. In this case, the osmotic pressure of the region between two surfaces with the depletion layer is lower than elsewhere. This will be the driving force of flocculation. It is reported that depletion flocculation, that is, flocculation induced by this mechanism, is relatively weak and reversible [12]. These aspects will be useful to prevent strong flocculation, which is formed in the sediment of industrial products stored for a long period [13].

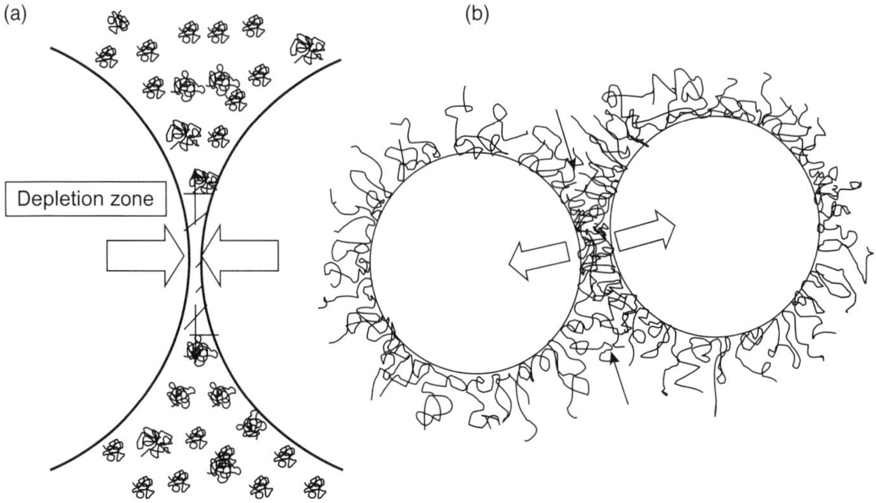

Figure 17.3. Interaction of approaching particles with depletion zone of polymer chains. The solid arrow denotes the migration of solvent and the open arrow denotes the resulting forces exerted on the surfaces. (a) Overlap of two depletion layers will induce an attractive force between two interfaces enclosing the depletion zone. (b) Interference of the protruding part of two adsorbing layers will induce the increment of osmotic pressure of solvent, resulting in repulsion between the two surfaces.

Another issue that should be noted in the mechanism of depletion flocculation is that the depletion flocculation can be induced also by adsorbing polymers; that is, if we continue to inject adsorbing polymers to colloidal dispersion, the surface will be saturated. At this stage, no more adsorption will take place and the surface can be regarded as a sort of depletion layer. Actually, Furusawa et al. [14] confirmed that in a certain range of polyelectrolyte concentration, depletion flocculation can be induced when additional polyelectrolytes are added to the sterically stabilized system.

17.2.3 Bridging versus Charge Neutralization

In the previous section, the entropic aspect of flocculation was described. However, in most practical situations, flocculation with a polymer is induced by the enthalpic force of adhesion. Bridging and charge neutralization are the two main mechanisms of flocculation induced with polymer and polyelectrolyte. One can easily imagine that immediately after the input of polymers that have a strong affinity to the colloidal surface into the colloidal dispersion, the polymers start to adsorb onto the starved surface of colloidal particles. Bridging is formed when one polymer chain adsorbs more than two particles simultaneously. This concept was proposed by Ruehrwein and Ward [15] for the formation of soil aggregate. In their experiment, the optimum dosage for the degree of flocculation was demonstrated. The presence of optimum dosage was explained by the successful collision between one polymer-coated surface and another uncoated surface. With these results, the concept of bridging was established [16].

When the polymer chains are conveying opposite charges to the charge of the colloidal surface, the adsorption will result in the reduction of the total amount of charge being conveyed with the colloidal particle and will eventually reverse the sign of the colloidal particle. Flocculation takes place near the condition where the total amount of charge is neutralized. This mechanism is called charge neutralization. By the combination of the measurement of rate of flocculation and the electrophoretic mobility, Gregory proposed the concept of heterogeneous patches as depicted in Figure 17.4c [17]. In this case, additional adhesive force will arise around the formation of patches [18]. This can be confirmed by the recent atomic force microscopy (AFM) measurement on the patched carpet of cationic dendrimers reported by Borkovec et al. [19].

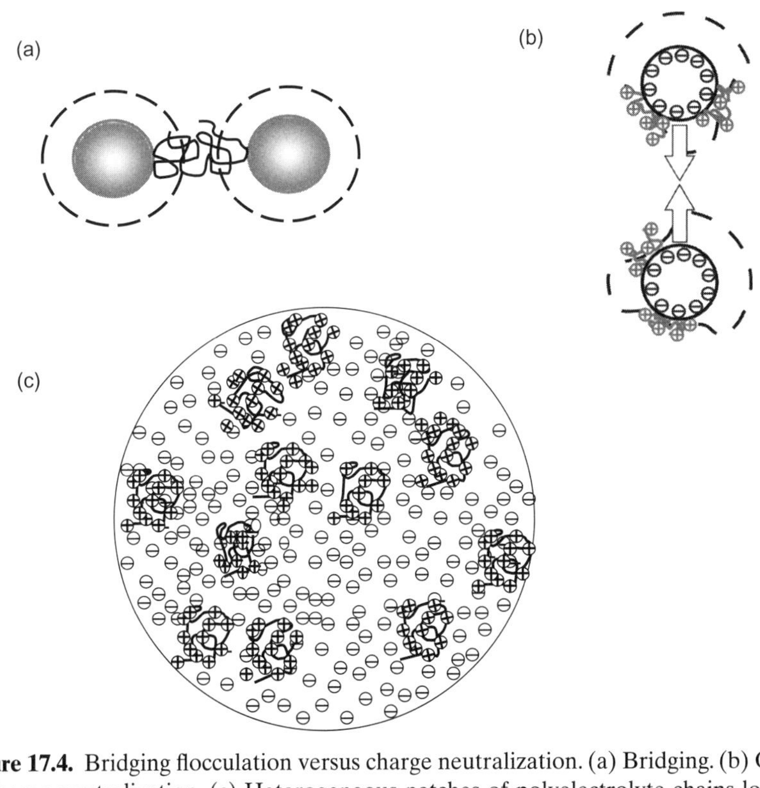

Figure 17.4. Bridging flocculation versus charge neutralization. (a) Bridging. (b) Onset of charge neutralization. (c) Heterogeneous patches of polyelectrolyte chains located on oppositely charged surfaces in order to avoid overlapping each other.

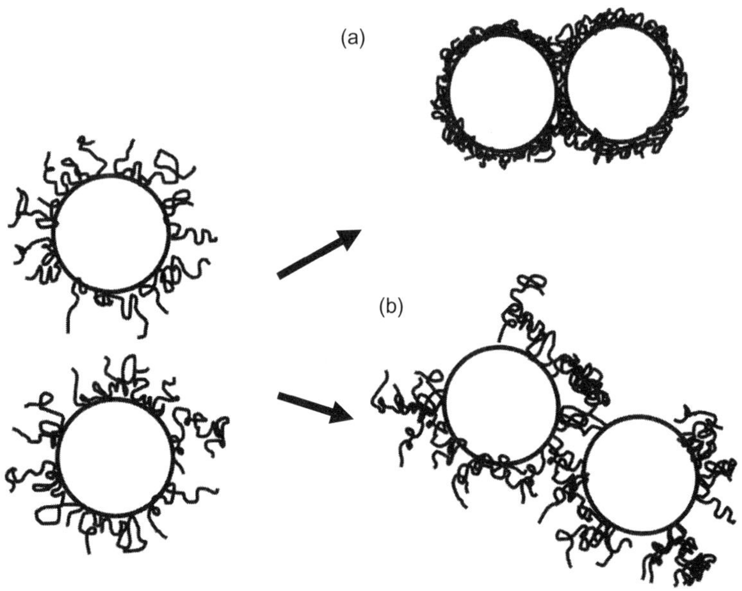

Figure 17.5. Flocculation induced by the hydrophobic interaction of adsorbing polymers. (a) With strong attractive force between chains and surface. (b) In the absence of interaction between chains and surface.

17.2.4 Hydrophobic Interaction of Two Adsorbing Layers

Let us begin with the situation that colloidal dispersion is stabilized by the steric effect of adsorbing polymers with thermosensitive properties [20]. Polymers reach a swollen state when solvent temperature is lower than that of phase transition. When the system is heated up to above the phase transition temperature, the hydrophobic interaction between segments overcomes the entropic repulsive interaction. Loop and tail will start to shrink, resulting in an increase of the hydrophobic domain. If the domain is successfully confined to the colloidal surface, a water-repellent compact structure will be formed as demonstrated in Figure 17.5a [21]. However, in some cases, hydrophobic aggregates remain in the bulky layer of the adsorption layer (Fig. 17.5b) [22].

17.3 ANALYSIS OF FLOCCULATION DYNAMICS BY MEANS OF THE STANDARDIZED COLLISION PROCESS

17.3.1 Schematic Diagram of Flocculation with Polymer Flocculants

A schematic diagram of flocculation induced by an addition of polymer flocculants is depicted in Figure 17.6. Upon the addition of polymer flocculants, the system is brought to a state that is far from equilibrium. It can easily be

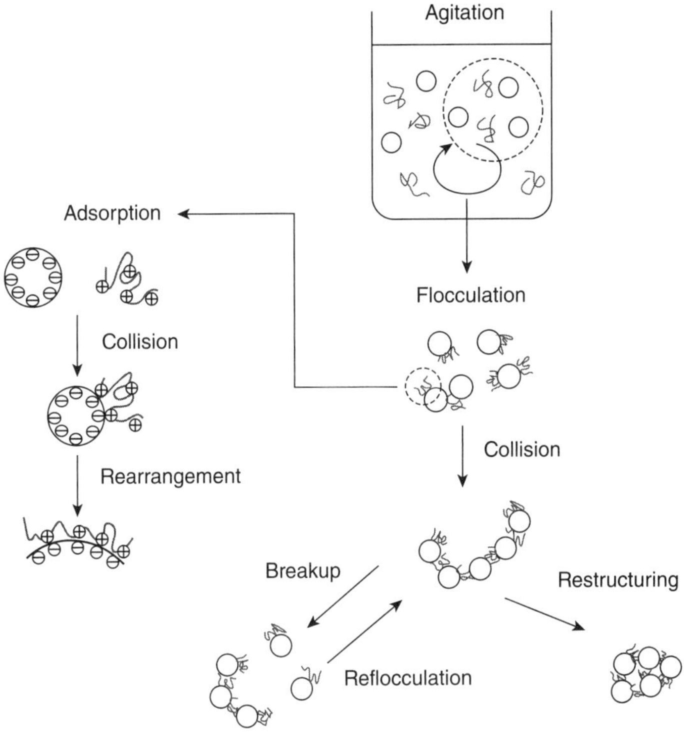

Figure 17.6. Schematic diagram of flocculation with adsorbing polymer.

imagined that the flocculation process consists of several elementary processes. These are the following:

1. Dilution of the flocculants into a homogeneous solution
2. Collision between colloidal particles
3. Transportation of polymer flocculants toward the surface of colloidal particles
4. Reconformation of adsorbed polymers on the surface of colloidal particles
5. Formation of a bond (or bridge) between colloidal particles
6. Rearrangement and breakup of the structure of a floc

All these elementary processes occur simultaneously under the mixing condition, which is usually turbulent flow. Completely different results for the flocculated state were reported for different ways of colloid mixing [23, 24]. This means the most important event take place at the early stage of incubation. An analysis of the dynamic aspects listed earlier are strongly required [25].

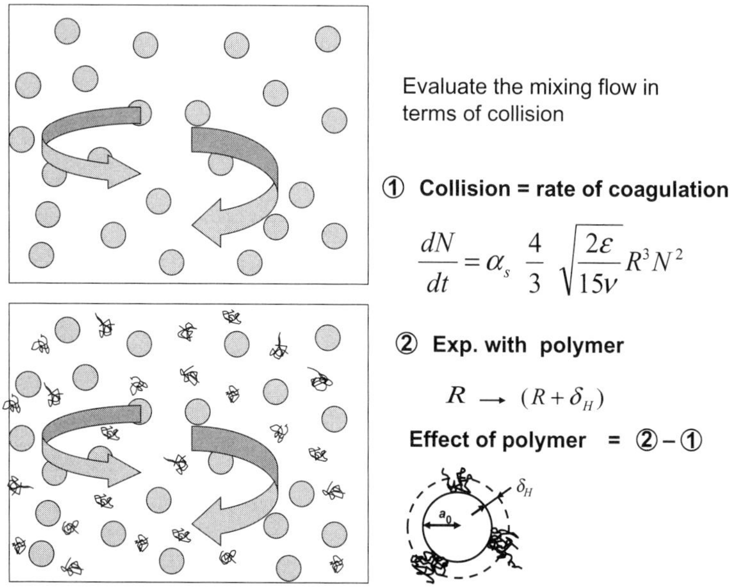

Figure 17.7. The normalization of mixing conditions in terms of collision between colloidal particles applied for the analysis of the influence of polymer addition.

17.3.2 Normalization of Mixing (NM) in Terms of the Collision Process

In order to perform a kinetic analysis of flocculation, we proposed the normalizing method of turbulent flow in terms of the collision process between colloidal particles; that is, if we can assume that all collisions between colloidal particles in the turbulent mixing flow lead to coagulation, the turbulent flow can be evaluated in terms of the collision process by measuring the rate of coagulation. After getting the data for the mixing process, we investigate the effect of the addition of polymer to the same mixing system (Fig. 17.7).

Let us assume the monodispersion of colloidal latex undergoing rapid coagulation is realized by the addition of a sufficient amount of salt. The rate of coagulation in the mixing liquid was worked out by Higashitani et al. [26] and Adachi et al. [27], who confirmed the importance of the correction of collision efficiency due to hydrodynamic interaction in the derived equation by Saffman and Turner [28] invoking the hypothesis of local isotropy of turbulence proposed by Kolmogorov. The progress of coagulation can be expressed by the temporal variation of the total number of particle concentration, $N(t)$; that is,

$$\frac{dN(t)}{dt} = -\frac{1}{2}\alpha_T\sqrt{\frac{8\pi\varepsilon}{15\nu}}(2a_0)^3 N(t)^2, \tag{17.1}$$

where ε, ν, and a_0 are the rate of energy dissipation per unit mass of fluid, kinematic viscosity, and the radius of colloidal particles. The collision efficiency of coagulation is α_T and is determined by reflecting the hydrodynamic interaction. In the case of rapid coagulation, an approximate expression can be derived substituting the effective shear rate of turbulent flow in the numerical result of calculation of Van de Ven and Mason [29], who took into account the van der Waals attractive force in the trajectory analysis of two colloidal particles in the laminar shear field; that is,

$$\alpha_T = \left(\frac{A}{36\pi\mu a_0^3 \sqrt{\frac{4\varepsilon}{15\pi\nu}}} \right)^{0.18}, \tag{17.2}$$

where A is the Hammaker constant and μ is the viscosity. If we restrict the discussion to the early stage of coagulation where we can assume

$$\phi = N(t) \times 4\pi / 3 \times a_0^3, \tag{17.3}$$

Equation 17.1 will yield the following approximate solution:

$$\ln \frac{N(t)}{N(0)} = -\alpha_T \sqrt{\left(\frac{24\varepsilon}{5\pi\nu} \right)} \phi t. \tag{17.4}$$

Evaluation of mixing flow is possible simply by measuring the number concentration of coagulation of colloidal particles as a function of elapsed time using the monodispersion of latex particles under the condition of rapid coagulation.

In the experiment, an end-over-end mixing device is developed (Fig. 17.8) [30]. Coagulation or flocculation is performed by simply turning over the

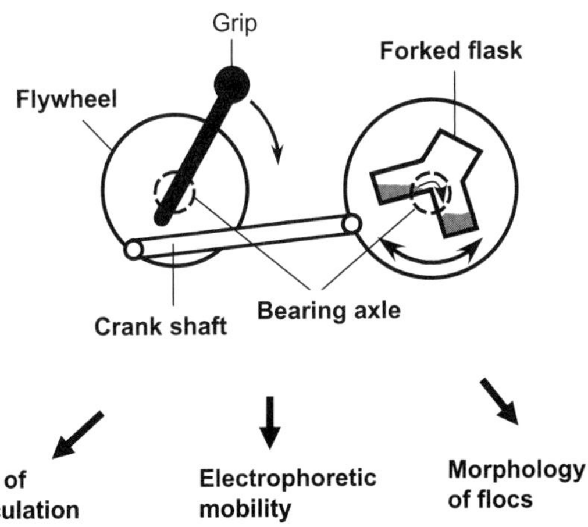

Figure 17.8. The end-over-end rotation apparatus to induce standardized mixing.

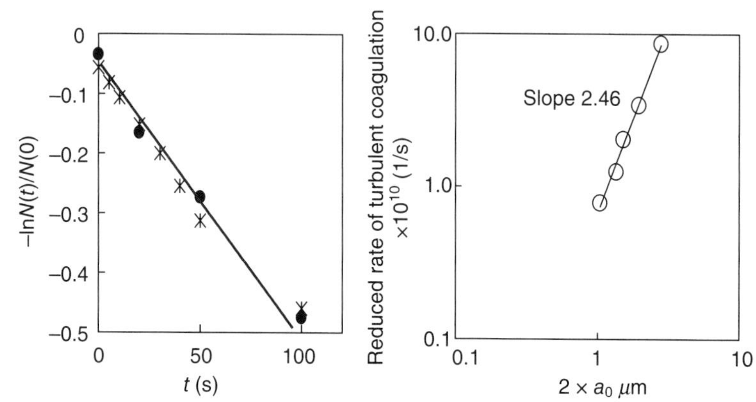

Figure 17.9. The result of the progress of salt-induced rapid coagulation in standardized mixing. (a) Temporal evolution of the number concentration of flocs. (b) Reduced rate of coagulation as a function of particle diameter.

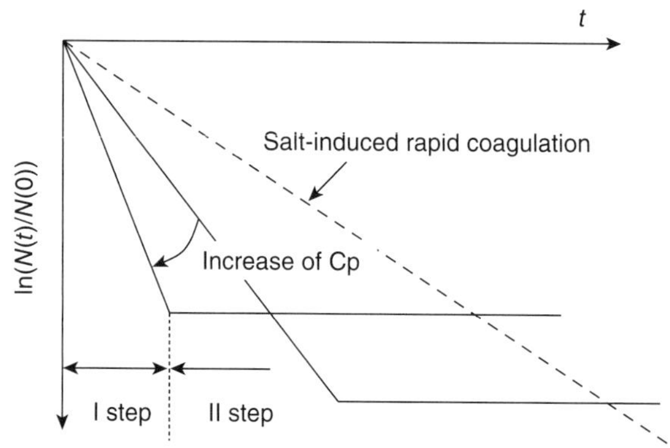

Figure 17.10. Concept of the progress of flocculation with an addition of an excess amount of adsorbing polymers.

forked flask to pour the dispersion from one side to the other. Although this method is primitive, it allows us to monitor the flocculation immediately after the start of mixing, not only the rate of flocculation but the electrophoretic mobility of formed complex and the morphology of formed flocs. A typical result of the progress of coagulation obtained using a homemade end-over-end rotation device is given in Figure 17.9 [31]. As shown in this figure, the prediction on the basis of Equation 17.3 is confirmed.

17.3.3 Evaluation of Adsorbing Polymer Layer by NM

Observing the effect of the application of polymer flocculants can be done simply to compare the monitored progress of flocculation with and without polymer flocculants (Fig. 17.10). Immediately after the incubation of polymer in the dispersion, flocculation is enhanced by the attachment of polymers with protruding conformation. However, this enhancement is disturbed by the saturation of the polymers on the colloidal surface. The period of enhancement is denoted as stage I and the period after saturation as stage II. There are mainly two factors on which one should focus. One is the rate of flocculation, which is reflected in the slope of the line in stage I. The other is the required time that is necessary for the saturation of adsorption.

As written in Equation 17.1, the rate of flocculation is largely dependent on the size of the colloidal particles. The effective thickness of a polymer layer δ_H can be obtained from the ratio of the rate of flocculation with and without polymer by the following equation:

$$\eta = \frac{(a_0 + \delta_H)^3}{\alpha_T a_0^3}. \tag{17.5}$$

The value of δ_H detected will be small if the polymer is relaxed. When polymer concentration is high, the probability of finding particles with freshly adsorbing polymers is also high; that is, the effective collision radius will increase with an increase in polymer concentration. However, the required time for saturation will decrease with an increase in polymer concentration.

The concept of collision process can also be applied to the collision between colloidal particles and polymer flocculants. The flux of polymer to a reference colloidal particle can be expressed by the sum of the diffusion flux by Brownian motion and the collision induced by mixing flow; that is,

$$J_p = 4\pi D_{op} R_{op} N_p + \sqrt{\frac{8\pi\varepsilon}{15\nu}} (a_0 + a_p)^3 N_p, \tag{17.6}$$

where D_{op}, R_{op}, a_p, and N_p are the relative diffusion constant between polymer and colloidal particles, the collision radius between polymer and colloidal particles, the radius of polymer flocculant, and the number concentration of polymer flocculant, respectively.

17.3.4 Application of NM to the Dynamics of the Bridging Flocculation of Polystyrene Latex (PSL) with PEO

Typical data for the progress of the flocculation of PSL particles induced by the excess addition of PEO with nominal Mw of 5.0×10^6 g/mol is demonstrated in Figure 17.11 [32]. As indicated in this figure, the slope of the initial stage flocculation increases with an increase of PEO concentration. This result means the effective value of δ_H for the collision radius will increase with an increase in PEO concentration. If we ascribe the reason for this phenomenon to the detection of chain length before relaxation on the colloidal surface, we can estimate timescale of relaxation as a function of the interval of polymer supply to the surface, assuming a certain area. On the basis of Equation 17.6

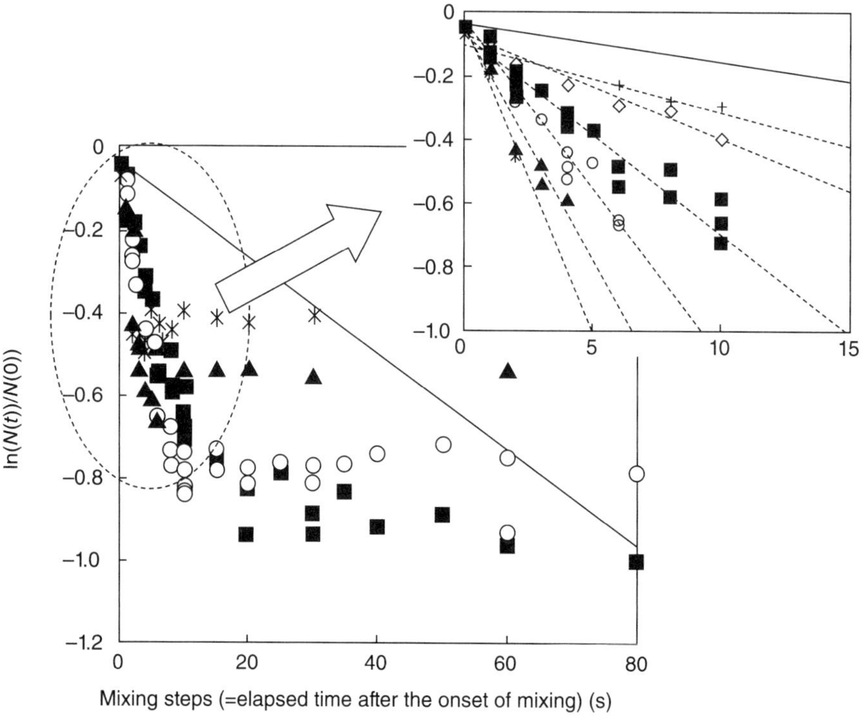

Figure 17.11. $\ln(N(t)/N(0))$ versus t for flocculation with PEO.

and defining the area as a square of the diameter of PEO (210×210 nm^2), δ_H is expressed as a function of the interval of polymer supply in Figure 17.12. It should be noted that a timescale of a few seconds is obtained.

17.3.5 Application of NM to the Dynamics of Flocculation of PSL with Polyelectrolytes

Most polymers being practically applied as flocculants are polyelectrolytes; that is, segments of a polymer chain carrying a charged group of opposite sign to that of the surface of the colloidal particles will be effective in decreasing the repulsion between colloid particles, thus inducing or enhancing flocculation [33]. A polyelectrolyte chain in aqueous solution will swell due to the repulsive forces generated between charged segments. The degree of swelling is a function of ionic strength and decreases with an increase in ionic strength. It is also interesting to examine this effect by means of NM. In Figure 17.13, the result of the progress of flocculation obtained by applying NM is demonstrated as a function of Mw and ionic strength. As shown in the figure, a similar trend in the case of PEO is indicated; that is, polyelectrolytes with larger size (high Mw and swollen conformation under lower ionic strength) enhance the

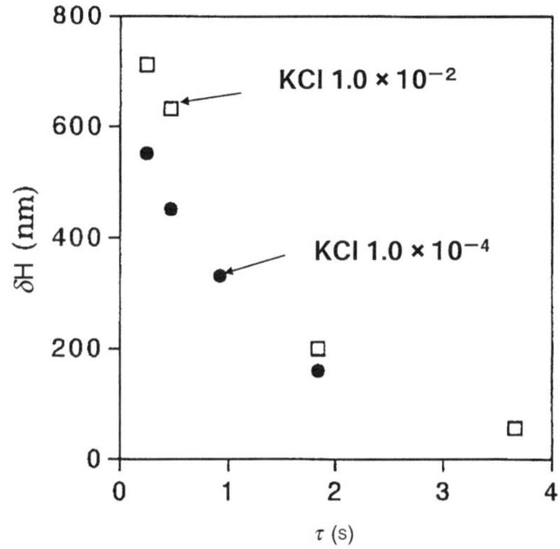

Figure 17.12. $\delta(t)$ as a function of the interval of polymer supply.

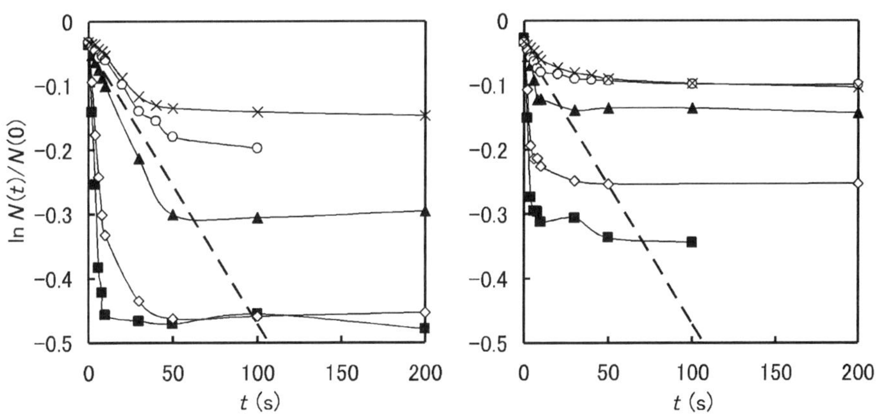

Figure 17.13. $\ln(N(t)/N(0))$ versus t for flocculation with polyelectrolytes. (a) KCl = 1.0×10^{-2}. (b) KCl = 1.0×10^{-4} M.

rate of the initial stage of flocculation more effectively, but progress stops in a shorter period. So, we can conclude that the conformation of polyelectrolytes in solution correlates with increment in the collision radius of colloidal particles. This can be also regarded as evidence of bridging.

In the case of polyelecltrolytes, the adsorption process can be monitored by electrophoretic mobility. Interesting results are demonstrated in Figure 17.14; that is, under the condition of lower ionic strength, a polyelectrolyte of higher Mw does not relax within the period of observation [30, 31]. The

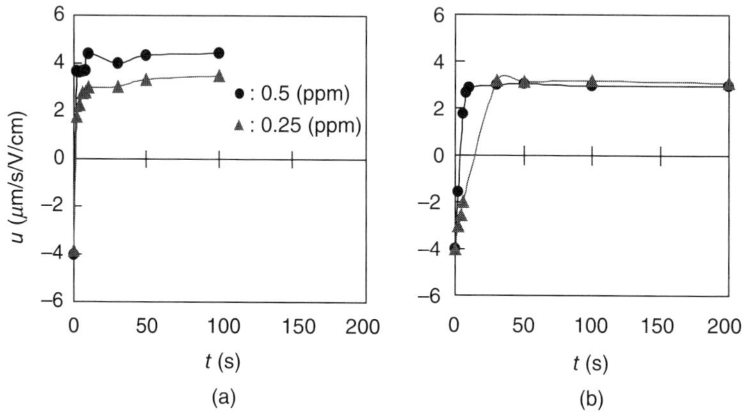

Figure 17.14. The electrophoretic mobility of the primary PSL particle versus the number of mixing steps. The KCl concentration is (a) 1.0×10^{-4}, (b) KCl = 1.0×10^{-4} (M). The molecular weight of a polyelectrolyte is 4,900,000.

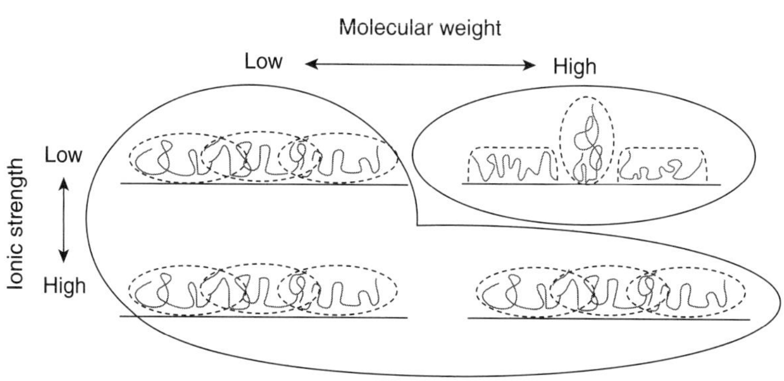

Figure 17.15. Schematic illustration of an adsorbing polyelectrolyte as a function of molecular weight and ionic strength.

ultimate value of electrophoretic mobility is kinetically controlled. However, in the case of lower Mw or higher ionic strength, no clear difference in the ultimate value of the mobility can be found. The expected picture reflecting this result is summarized in Figure 17.15. Such a difference in the dynamic aspect of adsorption conformation will affect the morphology of formed flocs; that is, an application of a polyelectrolyte of higher molecular weight under lower ionic strength will result in the formation of a bulky structure as depicted in Figure 17.16 [34, 35].

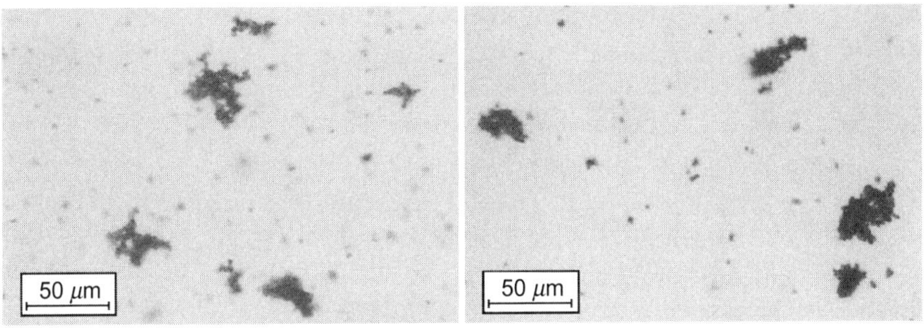

Figure 17.16. Morphology of PSL-flocs formed with polyelectrolytes under the conditions of high and low ionic strengths. Mw = 4.9×10^6 g/mol; Cp = 0.6 mg/L.

17.4 REMARKS FOR FUTURE WORK

Because of the significance of the dynamics of flocculation in the early stage of incubation, we have reviewed the current state of the art. As confessed in the introduction to this chapter, our understanding is still elementary. There are several types of polymer–surface interactions and subsequent flocculation. Our experiments in the last decade simply confirmed the significance of polymer size in solution just before collision and the transient behavior of polymers after collision. The conventional method of electrokinetics can provide some information, but this tool is not sufficient for further analysis. More experimental efforts are necessary to characterize the size of polymers, charge density, distribution within the adsorbed chain, and stiffness of the chain on the basis of underlying principles that are not yet clear. Another important factor is the problem of pure hydrodynamics and the interaction of permeable porous structures. A lot still remains to be done.

ACKNOWLEDGMENTS

The author expresses his thanks to Dr. Kenji Aoki, who provided the data for this discussion in his thesis. This study is financially supported by a Grant-in-Aid for Scientific Research (222408025) in part from the Japanese Society for Promotion of Science..

REFERENCES

1 Anon. *Encyclopaedia Britanica* 17 (1968) 38.
2 Anon. *Encyclopaedia Britanica* 12 (1968) 257.
3 D. H. Napper. *Polymeric stabilization of colloidal dispersion.* Academic Press, New York, 1983.

4 J. M. Cohen, G. A. Rourke, R. L. Woodward. *J. AWWA* 50 (1958) 463.
5 J. M. G. J. Fleer, M. A. Cohen Stuart, J. M. H. M. Scheutjens, T. Cosgrove, B. Vincent. *Polymers at interfaces*. Chapman & Hall, London, 1993.
6 P. G. de Gennes. *Adv. Colloid Interface Sci.* 27 (1987) 189.
7 Y. Adach. *Adv. Colloid Interface Sci.* 56 (1995) 1.
8 T. Kato, K. Nakamura, M. Kawaguchi, A. Takahashi. *Polymer J.* 13 (1981) 1037.
9 M. A. Cohen Stuart, F. H. W. H. Waajen, T. Cosgrove, B. Vincent, T. L. Crowley. *Macromolecules* 17 (1984) 1825.
10 P. G. de Gennes. *Macromolecules* 14 (1981) 1637.
11 C. Hiemenz. *Principles of colloid and surface chemistry*, 2nd ed. Marcel Dekker, Inc., New York, 1986.
12 S. Asakura, F. Oosawa. *J. Chem. Phys.* 22 (1954) 1255.
13 T. F. Tadros, A. Zaednai. *Colloids Surf.* 43 (1990) 105.
14 K. Furusawa, M. Ueda, T. Nashima. *Colloids Surfaces A.* 153 (1999) 575.
15 R. A. Reuhwein, D. W. Ward. *Soil. Sci.* 73 (1952) 485.
16 V. K. La Mer, T. W. Healy. *Rev. Pure Appl. Chem.* 13 (1963) 112.
17 J. Gregory. *J. Colloid Interface Sci.* 42 (1973) 448.
18 Y. Zhou, G. V. Franks. *Langmuir* 22 (2006) 6775.
19 R. Periset-Camara, G. Papastravrou, M. Borkovec. *Macromolecules* 42 (2009) 1749.
20 T. Tanaka. *Sci. Am.* 249 (1981) 124.
21 S. Sakohara, K. Nishikawa. *J. Colloid Interface Sci.* 278 (2004) 304.
22 K. Makino, in N. Kallay (ed.), *Thermosensitive gels*. Interfacial Dynamics, New York, 2000, p. 591.
23 G. J. Fleer, J. Lyklema. *J. Colloid Interface Sci.* 46 (1974) 1.
24 K. Higashitani, T. Kubota. *Powder Technol.* 51 (1987) 61.
25 J. Gregory. *Colloids Surf.* 31 (1988) 231.
26 K. Higashitani, Y. Yamauchi, Y. Matsuno, G. Hosokawa. *J. Chem. Eng. Jpn.* 16 (1983) 299.
27 Y. Adachi, M. A. Cohen Stuart, R. Fokking. *J. Colloid Interface Sci.* 165 (1994) 310.
28 P. G. Saffman, J. S. Turner. *J. Fluid Mech.* 1 (1956) 16.
29 T. G. M. Van de Ven, S. G. Mason. *Colloid and Polymer Sci.* 255 (1977) 468.
30 Y. Adachi, T. Matsumoto. *Colloids Surf. A* 113 (1996) 229.
31 K. Aoki, Y. Adachi. *J. Colloid Interface Sci.* 300 (2006) 69.
32 Y. Adachi, T. Wada. *J. Colloid Interface Sci.* 229 (2000) 148.
33 K. Aoki. Doctoral Thesis, Koubumshi deukaishitsu ni yoru koroid ryushi no gyoushu katei ni kansuru kenkyu (in Japanese). University of Tsukuba, 2007.
34 Y. Adachi, K. Aoki. *Colloids Surf. A* 306 (2009) 137.
35 Y. Adachi. *Bull. Sea Water Sci., Japan* 63 (2009) 76.

18

COLLOIDAL PARTICLE PROCESSING USING HETEROCOAGULATION

HIROSHI SASAKI, CHIHARU TOKORO, AND HIROSHI HAYASHI

18.1 INTRODUCTION

Most mining wastewater, process water, and industrial effluents are treated by precipitation and flocculation followed by sedimentation. However, conventional treatment inevitably generates drainage containing low-concentration ultrafine particles, which are not amenable to clarification by sedimentation due to their negligible settling velocity. Though separation of ultrafine particles is carried out by adding a flocculating reagent, this treatment may be not appropriate considering the effect on the aquatic environment. In other words, rapid solid/liquid separation from dilute colloidal dispersions is very difficult, and an efficient method has yet to be established.

When the particle size range is below 1 μm, methods that rely on the physical property of particle size, that is, gravitational settling, straining, and interception, are not promising approaches to performing particle removal. On the other hand, surface characteristics play greater roles in the stability of ultrafine particles. In the course of the separation process discussed in this work, fine particle electrointerfacial properties are utilized. As a result, flocculation of particles or adding a flocculating reagent is not needed. This technique is based on particulate interactions caused by van der Waals forces and overlapping of the electric double layer. This means that the separation

Electrical Phenomena at Interfaces and Biointerfaces: Fundamentals and Applications in Nano-, Bio-, and Environmental Sciences, First Edition. Edited by Hiroyuki Ohshima.

efficiency greatly depends on the chemical characteristics of suspension such as pH, ionic strength, and the valence of the ions.

This chapter describes an attempt to carry out rapid separation of colloidal particles in accordance with the considerations stated earlier by using a column bed packed with fibrous material (Fe–Ni slag wool). This separation method is very efficient, especially for a large volume of low-concentration suspension containing ultrafine particles, because of low energy costs and lack of contamination. Considering that no chemical bonds between adhered particles and their substrate are formed, particle adhesion to the collector surface is generally interpreted in terms of the heterocoagulation of suspended colloids onto the stationary surface.

18.2 RAPID SEPARATION OF ULTRAFINE PARTICLES FROM DILUTED SUSPENSION

18.2.1 Separation of Ultrafine Particles

Separation tests were performed by the packed column method, using fibrous slag (FS) as a particle collector [1]. Ten grams of FS was soaked in water for 10 minutes and then packed onto a stainless steel disk, yielding a void volume fraction of 0.91. The pH of the suspension was adjusted with 1 mmol/dm^3 HNO_3 and KOH as needed, and the ionic strength of the suspension was kept constant at 1 mmol/dm^3. Suspensions of known particle concentrations were introduced from the top of the column into the bed after agitating in an ultrasonic bath for 10 minutes.

Zeta-potential measurements of ultrafine particles and the FS were carried out using an electrophoretic light scattering spectrophotometer (ELS-800, Otsuka Electronics, Osaka, Japan) in a 1 mmol/dm^3 KNO_3 solution. Figure 18.1

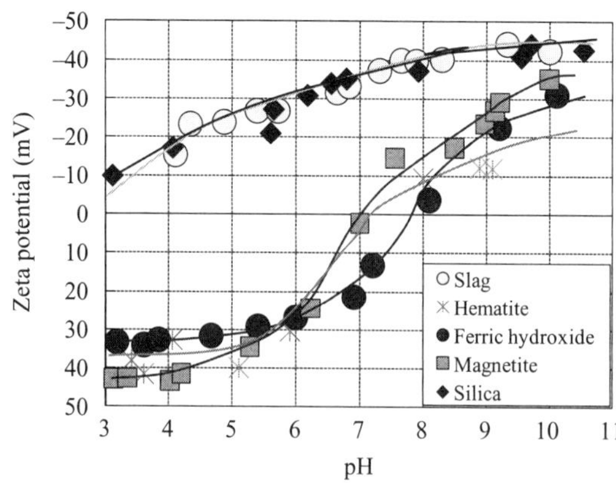

Figure 18.1. Zeta-potential curves of slag, hematite, ferric hydroxide, magnetite, and silica as a function of pH in a 1 mmol/dm^3 KNO_3 solution.

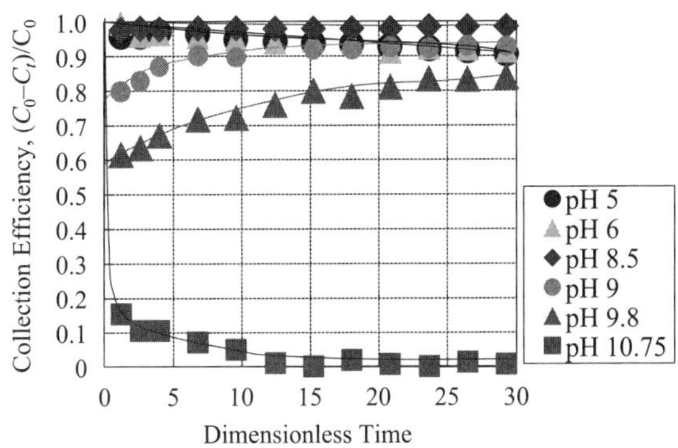

Figure 18.2. Collection efficiency of hematite particles as a function of dimensionless time.

summarizes the zeta-potential curves of FS, hematite, ferric hydroxide, magnetite, and silica as a function of pH. For the measurement of the FS, fine slag particles produced by grinding the original slag were used because it is assumed that the slag particles have the same surface charge density as the fiber. It was found that the isoelectric points (IEPs) were 7.2 for hematite, 7.0 for magnetite, 7.8 for ferric hydroxide, 2.9 for the FS, and 2.7 for silica.

Figure 18.2 shows the separation of hematite particles as a function of dimensionless time at a different pH in a 1 mmol/dm^3 KNO_3 solution. In this separation test, concentrations of hematite (105-nm mean diameter), ferric hydroxide (340 nm), magnetite (300 nm), and silica (200 nm) in the prepared suspensions were 3.4, 3.7, 3.2 (as Fe) and 50 mg/dm^3 (as SiO_2), respectively. The flow rate of the suspension was 250 cm^3/min, and the linear velocity was 2.0 mm/s for all experiments. Samples of effluent suspension were taken at short time intervals. The collection efficiency was defined as the ratio $(C_0 - C_t)/C_0$, where C_0 and C_t represent the initial and outlet concentration of particles in the suspension, respectively. The abscissa expressed as dimensionless time was defined as time divided by the time required for the solution to pass through the bed. Excellent collection of hematite occurred from pH 5 to pH 8.5. In the pH range from 4 to 7, slag and hematite have opposite surface charges (Fig. 18.1), the electrostatic attraction resulting in high collection efficiency. Complete collection was attained at pH 8.5. A noticeable increase in particle collection with time was observed at pH 9.8. Since the IEP of hematite particle existed at pH 7.2, homocoagulation between adhered particles and approaching particles promoted the collection. This phenomenon is referred to as *filter ripening*, indicating that the interaction between approaching particles and already deposited particles is more favorable than the interaction of the approaching particles with the bare collectors. Collection efficiency dramatically decreased at pH 10.75, where the surface charges of both slag and hematite have relatively high negative values.

18.2.2 Coprecipitation Treatment for Wastewater

Coprecipitation treatment of arsenic (As(V)) by not only Fe (III) but also Al salts was investigated [2, 3]. Numerous studies of As(V) adsorption on amorphous oxides, especially those focused on quantitative removal characteristic or surface complexation modeling, have concentrated on hydrous ferric oxide, and little is known about the adsorption mechanism of As(V) on aluminum hydroxide. Since there is a lot of wastewater containing Al, the quantitative confirmation of the effectiveness of Al salts can open up new possibilities for the treatment of As(V) removal using a coprecipitation method.

Generally, amorphous hydroxides such as ferric hydroxide (750-nm median diameter) and aluminum hydroxide (425 nm) are fine particles, and it is known that their solid/liquid separation is very difficult. Both hydroxides had less than 1-μm diameter, which was difficult to separate from the solution by the settling method.

The experimental results of solid/liquid for 50 mg/dm^3 ferrihydrite with 5 mg/dm^3 As(V) are shown in Figure 18.3. In this experiment, ionic strength was 0.01 and superficial velocity was 2.5 mm/s. Dimensionless time in Figure 18.3 is treatment time divided by the transit time of the column bed, and 1 dimensionless time in this experiment is equal to 7.2 seconds. Figure 18.3 shows that residual As(V) concentration was lower than the Japanese effluent standard, and good solid/liquid separation was achieved at pH 5.5 and 7.5.

The experimental results of solid/liquid separation for 10 mg/dm^3 aluminum hydroxide with 5 mg/dm^3 As(V) are shown in Figure 18.4. In this experiment, ionic strength and superficial velocity were the same as in Figure 18.3. Figure 18.4 shows that the residual As(V) concentration was lower than the Japanese effluent standard, and good solid/liquid separation was achieved at pH 7 and 7.7.

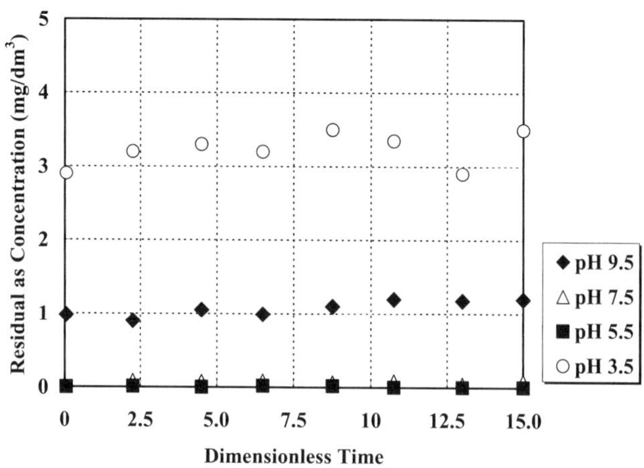

Figure 18.3. Residual As(V) concentration after solid/liquid separation using FS: As(V) = 5 mg/dm^3, Fe(III) = 50 mg/dm^3, I = 0.01, and superficial velocity = 2.5 mm/s.

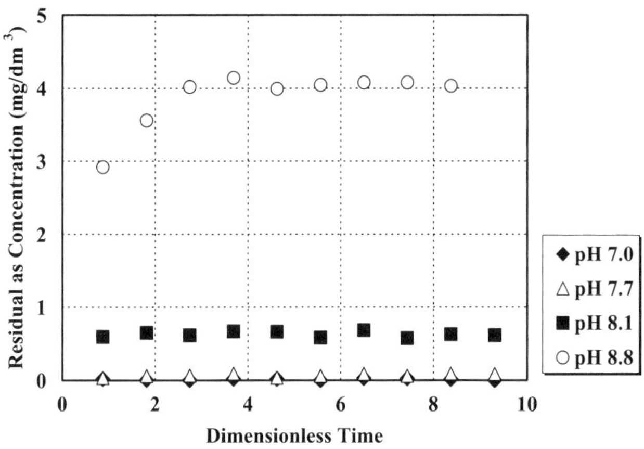

Figure 18.4. Residual As(V) concentration after solid/liquid separation by FS: As(V) = 5 mg/dm^3, Al = 10 mg/dm^3, I = 0.01, and superficial velocity = 2.5 mm/s.

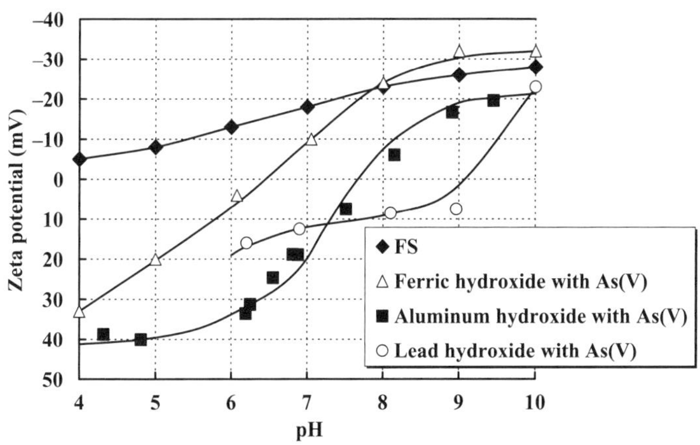

Figure 18.5. Zeta potential of FS, and ferric and aluminum hydroxide with As(V): As(V) = 5 mg/dm^3, Fe(III) = 50 mg/dm^3, Al = 10 mg/dm^3, and I = 0.01.

The zeta potentials of ferric and aluminum hydroxide and the FS are shown in Figure 18.5. In this zeta-potential measurement, ionic strength was 0.01, the As(V) concentration was 5 mg/dm^3, the Fe (III) concentration was 50 mg/dm^3, and the Al concentration was 10 mg/dm^3, which were the same in the solid/liquid separation experiment. Figure 18.5 shows that the zeta potential of the FS was always a negative value; however, the IEP of ferrihydrite with As(V) was at pH 6.5, and aluminum hydroxide with As(V) was at pH 7.7, so the electrostatic attracting force would act between hydroxide, whose surface charge was positive, and the FS, whose surface charge was negative, in the range of a pH of less than the IEP.

18.3 RAPID SEPARATION OF BACTERIAL CELLS FROM A STABLE DISPERSION BY HETEROCOAGULATION TO A FIBROUS COLLECTOR

18.3.1 Rapid Separation of Bacterial Cells

Due to increasing applications requiring microbial cells in many fields of engineering, techniques for the separation of bacteria are of fundamental significance [4, 5]. However, we are confronted with some difficulties when separating bacterial cells, as well as when immobilizing specific cells in order to use their activities, when solid media are used, especially when the bacterial suspension is of low concentration and stably dispersed. Since the physical properties of bacterial cells in aqueous media are insignificant, it is not effective to separate bacteria effectively by gravitational settling, straining, and interception only. In due course, addition of flocculating reagents to facilitate aggregate formation is necessary even though an excessive amount of reagent can alone lead to deterioration of water quality.

Figure 18.6 summarizes the cell collection capacity per unit gram of fibrous slag for the five kinds of strains, *Escherichia coli*, *Pseudomonas putida, Pseudomonas aeruginosa, Alcaligenes faecalis,* and *Alcaligenes* sp., as a function of pH. It is obvious that the cell collection depended significantly on suspension pH, suggesting that surface properties of both cells and FS play important roles in cell attachment to FS. The collection capacity for *E. coli* increased significantly from 0.3 mg/g at pH 7 to 83 mg/g at pH 4. Similarly, the

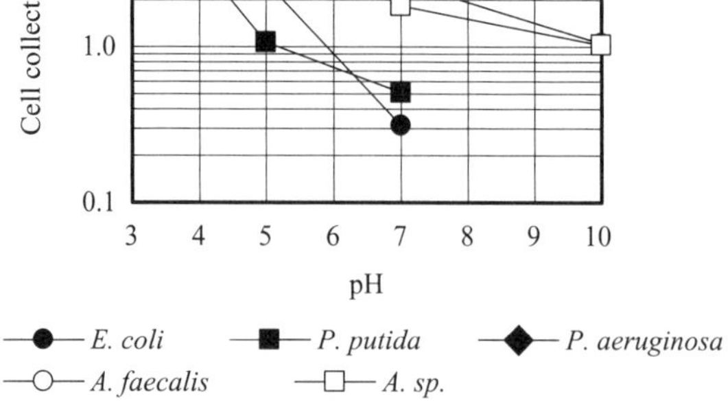

Figure 18.6. Summary of cell collection capacity per unit gram of FS for *E. coli*, *P. putida*, *P. aeruginosa*, *A. faecalis*, and *Alcaligenes sp.*, as a function of pH.

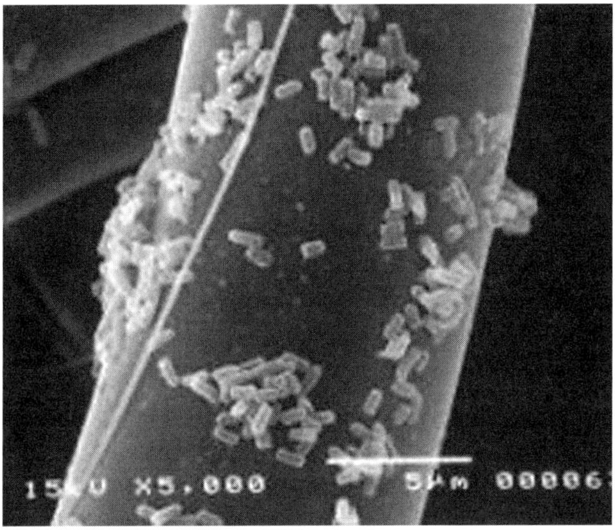

Figure 18.7. Scanning electron microscopic images of *E. coli* deposited on the FS surface.

collection capacity for *A. faecalis* rose from 1.0 mg/g at pH 7 to 74 mg/g at pH 4. In contrast to other strains, good cell recovery was attained for *P. aeruginosa* regardless of suspension pH. These cell collection behaviors were in general agreement with the cell electrophoresis data discussed in the following section.

Figure 18.7 shows the scanning electroscope microscope (SEM) image of *Escherichia coli* attached onto the FS surface. It was found that *E. coli* attached on individual fibers without morphological transformation, suggesting that approaching cells smoothly adhered to the FS surface with their shapes unchanged.

18.3.2 Soft Particle Electrophoresis Analysis

For the analysis of the electrophoretic mobility data of bacterial cells, the soft particle electrophoresis theory was applied. This theory assumes the presence of an ion-penetrable, charged polyelectrolyte layer around a rigid core because biological cells usually have fixed ionogenic groups in their surface region. The approximate mobility formula for soft particles, which was developed by Ohshima and Kondo [6], is expressed as follows:

$$\mu = \frac{\varepsilon_0 \varepsilon_r}{\eta} \frac{\psi_0/\kappa_m + \psi_{DON}/\lambda}{1/\kappa_m + 1/\lambda} + \frac{eZN}{\eta\lambda^2}. \tag{18.1}$$

Here, μ is the electrophoretic mobility, ε_0 is the permittivity of the vacuum, ε_r is the relative permittivity of the medium, η is the viscosity, and e is the elementary charge. ZN and λ are explained later. The Debye–Hückel parameter of the surface region is κ_m; ψ_{DON} is the Donnan potential; and ψ_0 is the potential at the boundary between the surface region and the solution and is recognized as the surface potential of soft particles, which are written in the form

$$\psi_{DON} = \frac{kT}{ze}\ln\left\{\frac{ZN}{2zn} + \left[\left(\frac{ZN}{2zn}\right)^2 + 1\right]^{1/2}\right\}, \tag{18.2}$$

$$\psi_0 = \frac{kT}{ze}\left(\ln\left\{\frac{ZN}{2zn} + \left[\left(\frac{ZN}{2zn}\right)^2 + 1\right]^{1/2}\right\} + \frac{2zn}{ZN}\left\{1 - \left[\left(\frac{ZN}{2zn}\right)^2 + 1\right]^{1/2}\right\}\right), \tag{18.3}$$

and

$$\kappa_m = \kappa\left[1 + \left(\frac{ZN}{2zn}\right)^2\right]^{1/4}. \tag{18.4}$$

Here, k is the Boltzmann constant, T is the absolute temperature, and z and n are the valency and concentration of bulk ions, respectively. From the plot of experimental mobility data against ionic concentration, curve fitting can be carried out to obtain two unknown parameters of ZN and $1/\lambda$. The former represents the spatial charge density in the polyelectrolyte region, and the latter is a parameter characterizing the degree of resistance to liquid flow in the polyelectrolyte region. The reciprocal of λ, that is, $1/\lambda$, has the dimensions of length and can be regarded as the "softness" parameter of the cell surface.

Figure 18.8 shows the results of electrophoretic mobility measurements as a function of pH in 1×10^{-2} mol/dm^3 KNO3. All of the strains showed negative

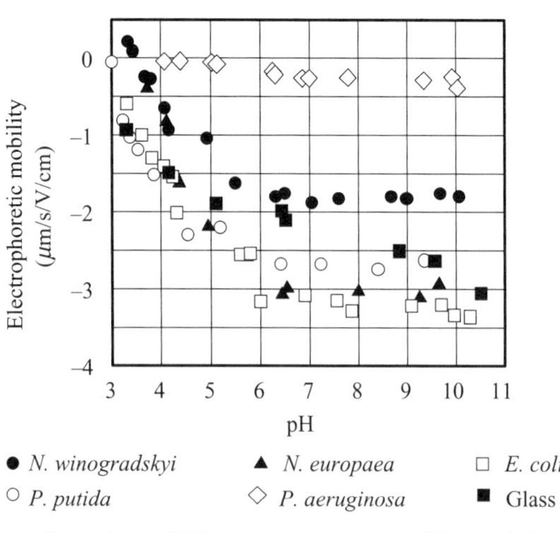

Figure 18.8. Electrophoretic mobility measurements of bacterial cells and glass as a function of pH in 1×10^{-2} mol/dm^3 KNO_3.

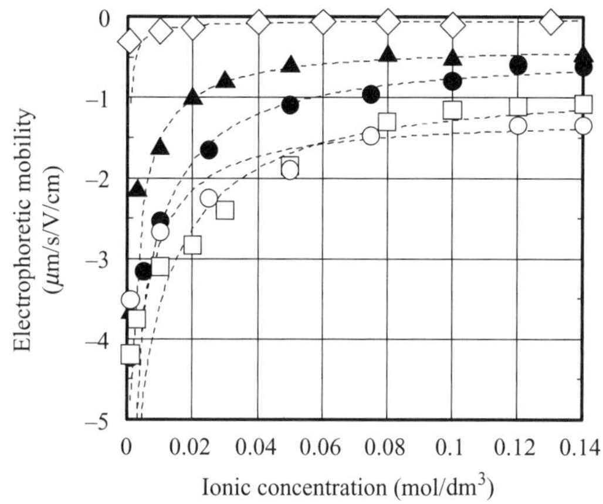

Figure 18.9. Electrophoretic mobility measurements of bacterial cells as a function of ionic concentration at pH 7.0. Dashed lines represent theoretical mobility curves fitted by Equation 18.1.

mobility values, as in the case with most microbes. *E. coli*, *Nitrosomonas europaea*, and *P. putida* showed relatively high negative values, and *Nitrobacter winogradskyi* had a low negative value under neutral and alkaline pH conditions. Most of the cells showed a gradual increment of mobility as pH decreased due to the change in the dissociation density of cell surface groups. In addition, the glass substratum had a negative mobility, and thus glass also had a negative surface charge.

Figure 18.9 shows the electrophoretic mobility of bacterial cells plotted against the ionic concentration. The remarkable point was that all the mobility data converged to nonzero values as the ionic concentration got higher, which could not be explained by the conventional electrophoresis theory. In fact, this was a characteristic distinct to soft particles, suggesting the existence of spatial charge distribution around their outer regions. This result indicates that bacterial cells have soft surfaces and that the application of this new model is reasonable.

Figure 18.10 shows the time course of C_t/C_0 as a function of time (breakthrough curve) for *E. coli* in 1×10^{-2} mol•dm^3 KNO_3. In the pH range from 7.0 to 5.0, little cell attachment occurred, probably because the electrostatic interaction between bacteria and substratum inhibited their mutual adhesion. On the other hand, a gradual increment of cell adhesion was observed as pH decreased, suggesting the reduction of electrorepulsive forces promoted cell adhesion to the substratum.

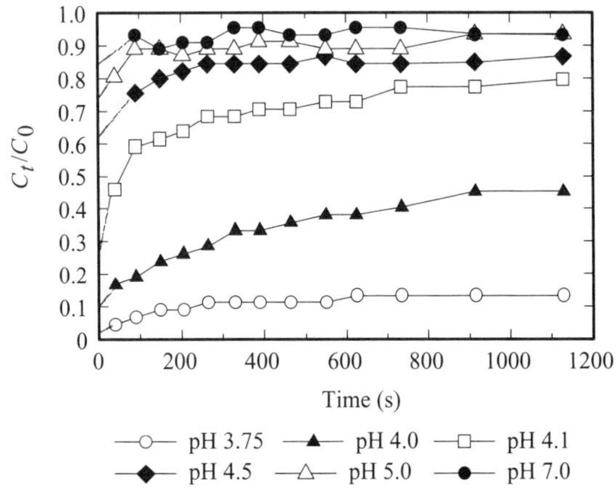

Figure 18.10. Time course of normalized cell concentration in an effluent, C_t/C_0, for *E. coli* in 1×10^{-2} mol/dm^3 KNO_3.

18.3.3 Summary of Rapid Separation Tests of Bacterial Cells

Rapid separation of five bacterial strains was investigated using a column bed packed with FS. The experimental results are summarized below:

1. *E. coli*, *P. putida*, *Alcaligenes faecalis*, and *Alcaligenes* sp. had a negative surface charge at neutral and alkaline conditions, and the magnitude of charge decreased as the pH became lower. However, *P. aeruginosa* exhibited little difference in its mobility, almost zero, suggesting that the net charge of the cell surface was almost neutral.
2. The cell collection ratio was approximated as first-order kinetics, and its decay for each bacterial strain, with the exception of *P. aeruginosa*, was significantly dependent on the chemical properties of the suspension. On the whole, the cell collection ratio increased as the suspension pH became more acidic or the phosphate-buffered saline (PBS) concentration increased. These results indicated that reduction of electrostatic repulsive forces between the cell and fiber led to the promotion of cell recovery by the fibrous medium.
3. Surface treatment of FS with aluminum hydroxide significantly improved the cell collection capacity for *E. coli*, *P. putida*, *A. faecalis*, and *Alcaligenes* sp., probably because deposition of positively charged aluminum hydroxide made the collector surface favorable for cell adhesion. In contrast, there was little difference in the recovery for *P. aeruginosa*, whose cell surface was almost neutral.
4. Cell collection capacity for *E. coli* ranged from 0.3 to 83 mg/g depending on the chemical properties of the suspension. Therefore, in this experiment, it was demonstrated that the surface interaction between the cell and the collector medium is essential for optimum cell recovery by FS.

18.4 RAPID SEPARATION OF OIL PARTICLES FROM LOW-CONCENTRATION OIL-IN-WATER (O/W) EMULSIONS IN THE PRESENCE OF ANIONIC SURFACTANTS

The increasing demand for the development of a sustainable society has encouraged many researchers to design cleaner and more efficient water treatment techniques [7]. In wastewater treatment engineering, some industrial effluents contain low concentrations of and stably dispersed oil droplets. For particle sizes below 1 μm, liquid/liquid separation from stable O/W emulsions cannot be performed by mechanical treatments such as the use of a coalescer, an oil skimmer, or other processes. The addition of flocculating reagents is usually carried out even though an excess amount of them may deteriorate water quality. The rapid treatment of stable O/W emulsions has been very difficult, and therefore it is very important to establish an efficient recovery method for emulsified oil droplets.

As previously described, we proposed a novel technique for separating colloidal particles from a stable dispersion based on surface characteristics. This method is essentially based on the heterocoagulation phenomenon of colloids, that is, spontaneous adhesion of particles onto a macroscopic fibrous collector medium. Particle separation is carried out while passing the solution through a column packed with a fibrous collector at relatively high superficial velocity. Since particle attachment to a solid substrate is influenced by physicochemical properties such as surface potential, ion concentration, solution pH, and adsorption of surfactant, these conditions have substantial effects on particle recovery efficiency.

In Ito et al. [7], the separation of oil droplets in an electrolyte solution without surface-active agents was investigated. We described the liquid/liquid separation of low-concentration decane emulsions (100 v/v ppm) in the presence of an anionic surfactant, sodium dodecyl sulfate (SDS), by the packed bed method, using an FS collector. The effects of SDS concentration on electrointerfacial properties and the stability of n-decane oil droplets were examined, and their relevance in the oil collection ratio was elucidated. The oil collection results could be discussed on the basis of the attachment of oil droplets to the FS according to the DLVO theory.

18.5 RAPID ULTRAFINE PARTICLE PROCESSING (SIZE CLASSIFICATION AND MUTUAL SEPARATION) USING SURFACE CHARACTERISTICS

18.5.1 Size Classification

Size classification of ultrafine particles is one of the most difficult techniques in materials processing [8]. We have developed a novel ultrafine particle classification method based on heterocoagulation phenomena of colloidal

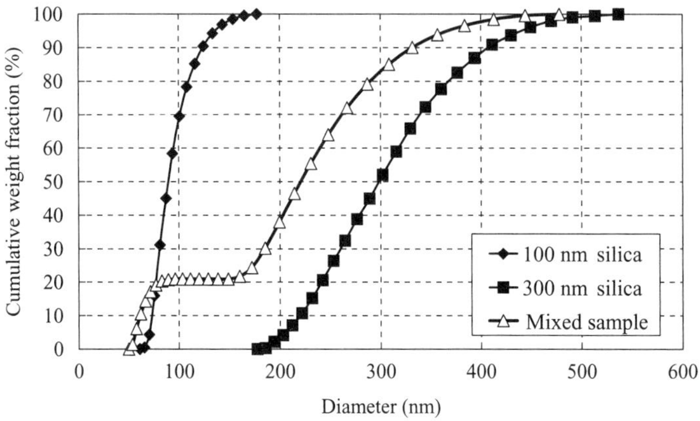

Figure 18.11. Size distribution of silica particles used in the experiment.

particles. In this section, rapid size classification of silica particles with mean diameters of 100 and 300 nm is derived from their mixed suspensions in various solution chemistries using a column bed packed with FS as a fibrous collector medium. In our experiments, many 100-nm silica particles selectively attached to the FS surface as they passed through the FS-packed column, but 300-nm silica remained in an outlet suspension. Newton's efficiency depended on pH and ionic strength; optimal classification was attained at pH 3.5 in 10 mM KNO_3. These results are qualitatively explained by a Derjaguin–Landau–Verwey–Overbeek (DLVO)-type interaction energy curve, indicating that interfacial interaction of both particle and collector determined the classification efficiency. Furthermore, particle concentration and superficial velocity had an influence on classification efficiency. This method proved to be simple, rapid, and cost-effective for classifying ultrafine particles in aqueous media.

Mixed suspensions of 100- and 300-nm silica particles were used in the size classification experiment. Figure 18.11 shows the size distribution of the prepared suspension. Figure 18.12 shows the recovery of total silica as a function of effluent volume at various pH values in 10 mM KNO_3. The vertical line is a normalized silica collection ratio $(C_0\text{-}C_t)/C_0$, where C_0 and C_t represent inlet and outlet silica concentrations, respectively. The silica collection ratio showed a maximum value of 0.62 at the effluent volume of 100 mL when pH was 4.0. The silica collection ratio decreased with increasing pH; the silica collection ratio was 0.22 at pH 5.0. This trend was explained by the fact that the increase of magnitudes of negative potential of silica and FS caused strong repulsive forces, which contributed to the inhibition of particle adhesion to the FS. The pH of the suspension was fixed at 4.0 in the following size classification tests.

Figure 18.13 shows the scanning electron microscopic images of the FS surface after the passing of the effluent through the column bed. Many 100-nm

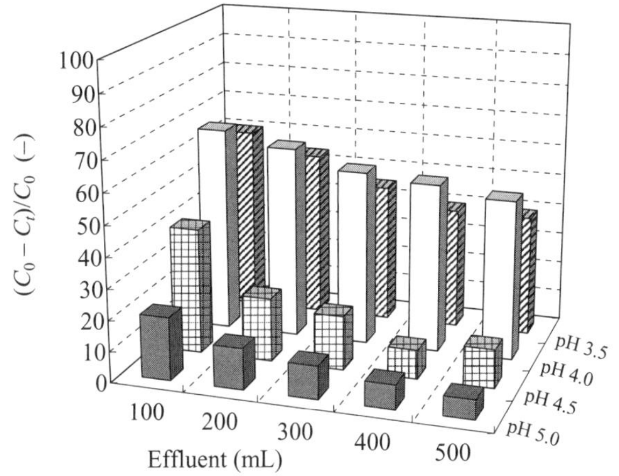

Figure 18.12. Recovery of total silica particles as a function of effluent volume at various pH values. Experiment used a KNO_3 concentration of 10 mM and a silica concentration of 26 mg/L at a superficial velocity of 1.7 mm/s.

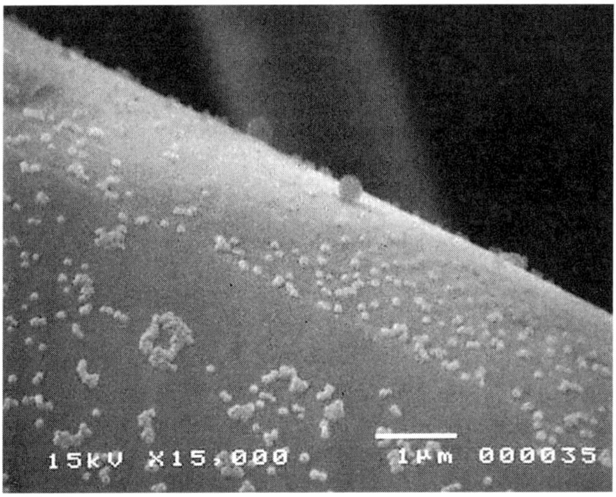

Figure 18.13. Scanning electron microscopic image of silica particles deposited on the FS surface.

silica particles were observed, whereas few 300-nm silica adhered to the FS. This indicated that the 100-nm silica selectively adhered to the FS. Also, Figure 18.14a,b show SEM images of particles in the inlet and outlet suspension. Comparing the proportion of small and large particles, a substantial number of 100-nm silica particles disappeared from the inlet suspension; 100-nm silica particles selectively deposited on the FS surface.

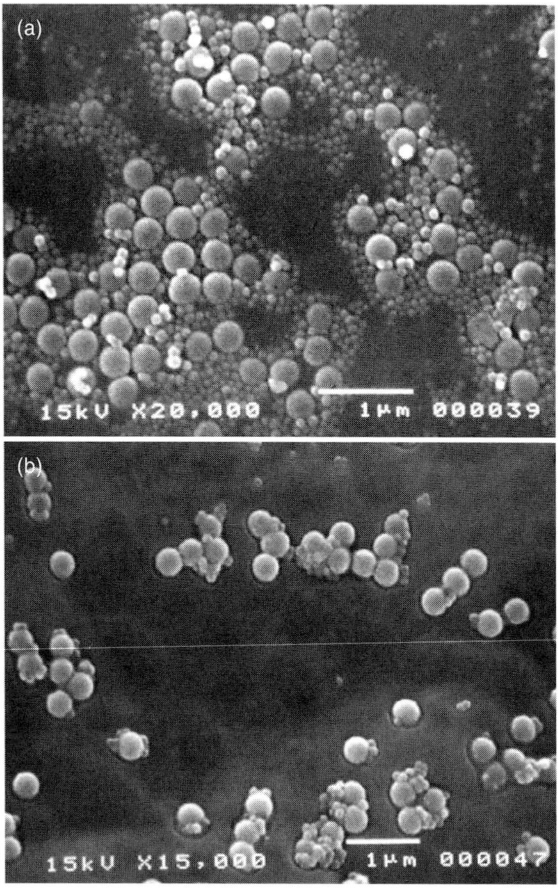

Figure 18.14. Scanning electron microscopic images of silica particles in the (a) influent and (b) effluent samples.

Figure 18.15 shows Newton's efficiency as a function of effluent volume at various pH values. If we are targeting 100-nm silica particles, Newton's efficiency, ξ, is defined in the following form:

$$\begin{aligned}\xi &= [(\text{Collection ratio of 100-nm silica}) \\ &\quad - (\text{Entrainment ratio of 300-nm silica})] \times 100 \\ &= \left[\left(\frac{\text{Number of 100-nm silica attached to FS}}{\text{Number of 100-nm silica in influent}}\right)\right. \\ &\quad \left. - \left(\frac{\text{Number of 300-nm silica attached to FS}}{\text{Number of 300-nm silica in influent}}\right)\right] \times 100. \end{aligned} \tag{18.5}$$

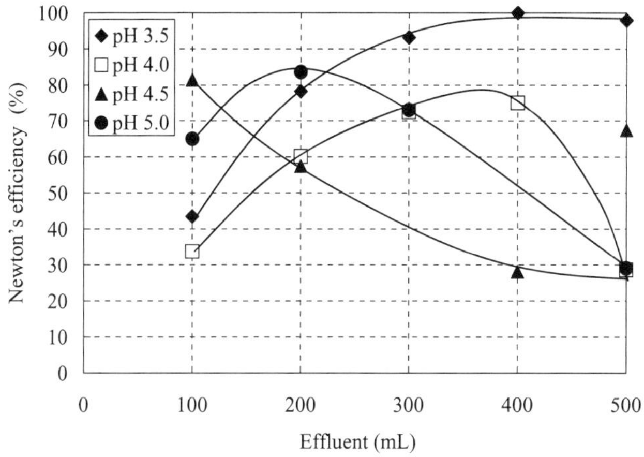

Figure 18.15. Newton's efficiency as a function of effluent volume at various pH values. The experiment used a KNO_3 concentration of 10 mM and a silica concentration of 26 mg/L at a superficial velocity of 1.7 mm/s.

Newton's efficiency reached 100% in the effluent volume of 400 mL when the suspension pH was 3.5. Newton's efficiency decreased at pH 4, where the collection efficiency of total silica particles reached the maximum. This was probably due to the fact that both 100- and 300-nm silica particles were able to deposit on the FS, thereby causing little selectivity in FS attachment. The position of optimal Newton's efficiency shifted to the larger effluent volume as the pH became lower.

18.6 MUTUAL SEPARATION OF ULTRAFINE SILICA AND HEMATITE PARTICLES FROM SUSPENSION USING SURFACE CHARACTERISTICS

Rapid mutual separation of ultrafine particles in suspension is one of the most difficult techniques in materials processing [9]. A column bed method that utilizes the different characteristics of particles was investigated for mutual separation. FS packed in a column was used as a collector. The FS offers a large surface area for selective adhesion of particles, which results in mutual separation of ultrafine particles. In mutual separation of mixed hematite (107-nm median diameter) and silica (143-nm median diameter) suspension, only the hematite attached onto the FS and the silica was separated in the effluent. Mutual separation was achieved at pH 11 by zinc hydroxide modified FS. The limiting particle concentration to achieve mutual separation was 50 mg/dm^3. Mutual separation of large particles (300 nm) was achieved at pH 10. These results demonstrated that the process was simple, quick, and cost-effective.

REFERENCES

1 H. Hayashi, H. Sasaki. Rapid separation of ultrafine particles from low concentrated suspension by surface characteristics. *Proceedings of GME'99 (Global Conference on Environmental Control in Mining and Metallurgy, Beijing)*, 1999.

2 C. Tokoro, Y. Yatsugi, H. Sasaki, S. Owada. As(V) removal by Fe(III) or Al salts and a novel rapid solid-liquid separation in wastewater containing dilute arsenic. European Metallurgical Conference 2007, Vol. I, Dusseldorf, June 11–14, 2007, 445–458.

3 C. Tokoro, Y. Yatsugi, H. Sasaki, S. Owada. A quantitative modeling of co-precipitation phenomena in wastewater containing dilute anions with ferrihydrite using a surface complexation model. *Resour. Recycl.* 55 (2008) 3–8.

4 H. Hayashi, S. Tsuneda, A. Hirata, H. Sasaki. Soft particle analysis of bacterial cells and its interpretation of cell adhesion behaviors in terms of DLVO theory. *Colloids Surf. B Biointerfaces* 22(2) (2001) 149–157.

5 H. Hayashi, T. Nihei, M. Ono, S. Tsuneda, A. Hirata, H. Sasaki. Rapid recovery of bacterial cells from stable dispersion by heterocoagulation to fibrous collector. *J. Colloid Interface Sci.* 242 (2001) 1–7.

6 H. Ohshima, T. Kondo. Electrophoretic mobility and Donnan potential of a large colloidal particle with a surface charge layer. *J. Colloid Interface Sci.* 116 (1987) 305–311.

7 H. Ito, H. Hayashi, H. Sasaki. Rapid separation of oil particles from low-concentrated O/W emulsions in the presence of surfactant using fibrous slag. *J. Colloid Interface Sci.* 252 (2002) 214–221.

8 K. Wakabayashi, S. Sekita, H. Hayashi, H. Sasaki. Rapid size classification of ultra-fine particles using surface characteristics. *KONA* 23 (2005) 173–181.

9 T. Takahashi, W. Todoroki, K. Kazama, H. Hayashi, H. Sasaki. Mutual separation of ultrafine silica and hematite particles from suspension using surface characteristics. *J. Soc. Powder Technol. Japan* 39 (2002) 169–174.

19

ELECTROKINETIC COUPLING IN COLLOIDAL ARRAYS FORMED UNDER AC ELECTRIC FIELDS

Satoshi Nishimura

19.1 INTRODUCTION

When the electric fields are applied to charged particles in aqueous solution, counterions in the electrical double layer (EDL) around the particles accumulate on one side of the EDL and are deficient on the other side. This is referred to as ion concentration polarization of EDL [1]. The induced electrical dipole moments arising from the ion concentration polarization of the EDL cause alignments of colloidal particles and/or dielectrophoresis under AC electric fields [2–4]. These phenomena have been extensively applied to nanoengineering processes such as biosensor, DNA manipulation, cell separation, and laboratory-on-tips where various materials ranging from colloidal materials to polymer molecules are targeted under AC electric field [5–9].

Under AC electric fields, colloidal particles in aqueous media often form complicated and hierarchical arrays of colloidal particles such as band and/or vortex [10–16], and colloidal crystals besides pearl chain [17–19]. In order to explain mechanisms for arrays of colloidal particles, they have tried to reproduce experimental results obtained by microscopic images in terms of theoretical modeling based on electrohydrodynamics. It is true that this approach is effective, but it must be said that experimental evidence other than images

Electrical Phenomena at Interfaces and Biointerfaces: Fundamentals and Applications in Nano-, Bio-, and Environmental Sciences, First Edition. Edited by Hiroyuki Ohshima.

obtained from microscopic observations seem to be insufficient to relate the images to the theoretical modeling.

In this chapter, I describe an experimental approach to compensate gaps between microscopic images and theoretical modeling based on electrohydrodynamic theory. As a result of our trial, I present some experimental evidence for electrokinetic coupling of ion concentration polarization in EDL with electrohydrodynamic effect in the formation of colloidal arrays under an AC electric field.

19.2 ION CONCENTRATION POLARIZATION OF EDL

When the frequency of an applied AC electric field is high enough but below the Maxwell–Wagner relaxation frequency, counterions in EDL around a charged particle accumulate on one side of the EDL and are deficient on the other side, resulting in ion concentration polarization (Fig. 19.1a). When the frequency is low enough, the diffusion effect of counterions becomes dominant, resulting in the relaxation of the ion concentration polarization (Fig. 19.1b). The relaxation frequency for the ion concentration polarization in EDL around a spherical particle (f_0) is given by

$$f_0 = \frac{D}{\pi a^2}, \tag{19.1}$$

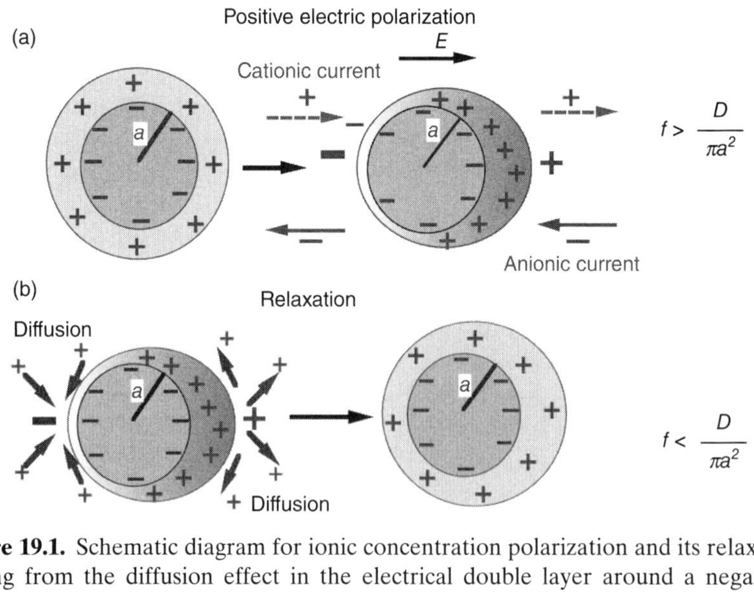

Figure 19.1. Schematic diagram for ionic concentration polarization and its relaxation arising from the diffusion effect in the electrical double layer around a negatively charged particle.

where D is the diffusion constant of the counterion and a is the radius of the particle [20].

In the range of frequencies much higher than the value of f_0, induced dipole moment arising from the ion concentration polarization of EDL around a spherical particle ($d\psi$) is given by the following equations [1]:

$$d\psi = \left[-\frac{1}{2}+\frac{3}{2}\left(\frac{Du}{1+Du}\right)\right]a^3E, \tag{19.2}$$

where Du is the Dukhin number, and

$$Du = \frac{k^\delta}{ak^\mathrm{b}}, \tag{19.3}$$

where k^δ is the surface conductivity and k^b is the bulk conductivity. The conductance against the volume fraction is also described by

$$K^* = 1+3\Phi K\left[-\frac{1}{2}+\frac{3}{2}\left(\frac{Du}{1+Du}\right)\right], \tag{19.4}$$

where K^* is the conductance of suspension, K^* is the conductance of the medium, and Φ is the volume fraction of the particles in suspension. Van der Wal et al. present a frequency dependence of the induced dipole moment as follows [21]:

$$d\psi = -\frac{1}{2}\left[\frac{1-Du}{1+2Du}-3\left(\frac{Du}{1+2Du}\right)^2\left(\frac{\omega\tau\sqrt{\omega\tau}}{\left(1+\sqrt{\omega\tau}\right)(1+\omega\tau)-Du\omega\tau\sqrt{\omega\tau}/(1+2Du)}\right.\right.$$
$$\left.\left.+\frac{j\omega\tau}{(1+\sqrt{\omega\tau})(1+\omega\tau)}\right)\right]a^3E, \tag{19.5}$$

where ω is the circular frequency and τ is the relaxation time of the ion concentration polarization ($\tau = 1/2\pi f_0$). The ratio of the real part of the complex conductance for suspension to its medium is given by

$$\frac{K^*(\omega)}{K(\omega)} = 1+\frac{3}{2}\Phi\left[\frac{Du-1}{1+2Du}+\frac{3}{4}\left(\frac{2Du}{1+2Du}\right)^2\right.$$
$$\left.\times\left(\frac{\omega\tau\sqrt{\omega\tau}}{\left(1+\sqrt{\omega\tau}\right)(1+\omega\tau)-Du\omega\tau\sqrt{\omega\tau}/(1+2Du)}\right)\right]. \tag{19.6}$$

The following equations for two extreme cases correspond to the following equations [22]:

$$\frac{K^*}{K} = 1 + 3\Phi\left[-\frac{1}{2} + \frac{3}{2}\left(\frac{Du}{1+Du}\right)\right], f >> f_0 \tag{19.7}$$

and

$$\frac{K^*}{K} = 1 + 3\Phi\left[-\frac{1}{2} + \frac{3}{2}\left(\frac{Du}{1+2Du}\right)\right], f << f_0. \tag{19.8}$$

Equations 19.7 and 19.8 show a good linear relation in dilute suspension. The slope of K^*/K to Φ corresponds to the polarizability of the induced dipole moment that increases with an increase in the frequency and/or Du.

19.3 HIERARCHICAL ARRAYS OF COLLOIDAL PARTICLES UNDER AN AC ELECTRIC FIELD

Under AC electric fields, complicated and hierarchical arrays of colloidal particles such as band patterns [10–16, 23] and colloidal crystals [17–20, 24] (Fig. 19.2) have been observed. They have not been fully understood because mechanisms for the formation of these colloidal arrays are dominated by many complicated subprocesses including electrohydrodynamics as well as the ionic concentration polarization of EDL.

It has been considered that two-dimensional (2-D) colloidal crystal is formed as a result of the balance between electrohydrodynamic attractive interaction and dipole–dipole repulsion among colloidal particles. Trau et al.

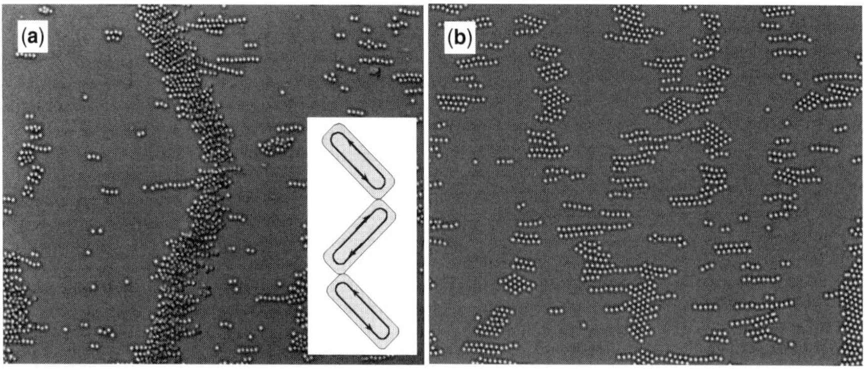

Figure 19.2. Microscopic images of hierarchical arrays of colloidal particles: (a) band and vortices (from Reference 23); (b) 2-D colloidal crystal structure (from Reference 24).

proposed that lateral attraction arises from an electrohydrodynamic effect in which lateral gradients of ionic current density, caused by the presence of colloidal particles near the electrode surface, induce convection flow [17]. Gong et al. suggested a similar mechanism wherein lateral variations in ion concentration polarization, that is, variations in charge density, generate electroosmotic flow near electrode surface. Such an electroosmotic flow causes colloidal particles to draw together [19]. They also noted that the transition from three-dimensional (3-D) to 2-D crystals or from ordered solid to fluid can be reversibly controlled by dipole–dipole repulsion among colloidal particles with changing electric field strength. Besides, a 2-D colloidal crystal structure transforming from bands is also observed when the induced dipole moment is large enough and convection of electroosmotic flow is small [24].

The mechanism of band patterns and vortices flow appears to be more complicated. Experimental factors influencing the formation of bands and vortices are as follows: (i) electric field strength, (ii) frequency of the electric field, (iii) viscosity and/or permittivity of a medium, (iv) salt concentration in suspension, and (v) volume fraction (number density) of particles.

The alignment form of silica particles depends on the strength of the applied electric field (root mean square value, *Erms*): (i) no response of silica particles to the electric field (*Erms* = 28.3 V/cm), (ii) pearl chains (*Erms* = 58.9 V/cm), (iii) bands of pearl chains with vigorous permutation of neighboring particles (*Erms* = 106 V/cm), and (iv) zigzag bands patterns with vortices (*Erms* = 141 V/cm) (Fig. 19.3) [23].

Abe et al. examined the dependence of the band of silica particles on the frequency, viscosity, and permittivity of media and electrolyte concentration [15]. As the frequency of the AC electric field is increased, the space between neighboring zigzag bands and the cycle of zigzag patterns in the bands becomes larger. As for the effect of viscosity and permittivity (dielectric constant) of a medium on the formation of bands, they confirmed that bands disappear as the viscosity increases and/or as the dielectric constant decreases. They also suggested that bands and vortices are transformed via pearl chains into a disordered structure as electrolyte concentration is increased from 0 to 10^{-3} M.

Lele et al. presented an order–disorder phase diagram of polystyrene particle assembly as a function of the frequency and strength of applied AC electric field (Fig. 19.4) [16]. They observed three structures with changing electric field strength and frequency: disordered suspension, pearl chain and 2-D colloidal crystals, and bands and vortices. At frequencies lower than several hundreds of hertz, band patterns are mainly observed. At frequencies higher than ~1 kHz, pearl chains appear under a lower electric field and are transformed into band patterns with an increasing electric field. Phase boundaries between pearl chains and bands shift to higher electric field strength with an increase in the frequency. The formation of colloidal arrays depends on the frequency and strength of the electric field reciprocally. They also present the time dependence of band formation that 2-D crystal structures

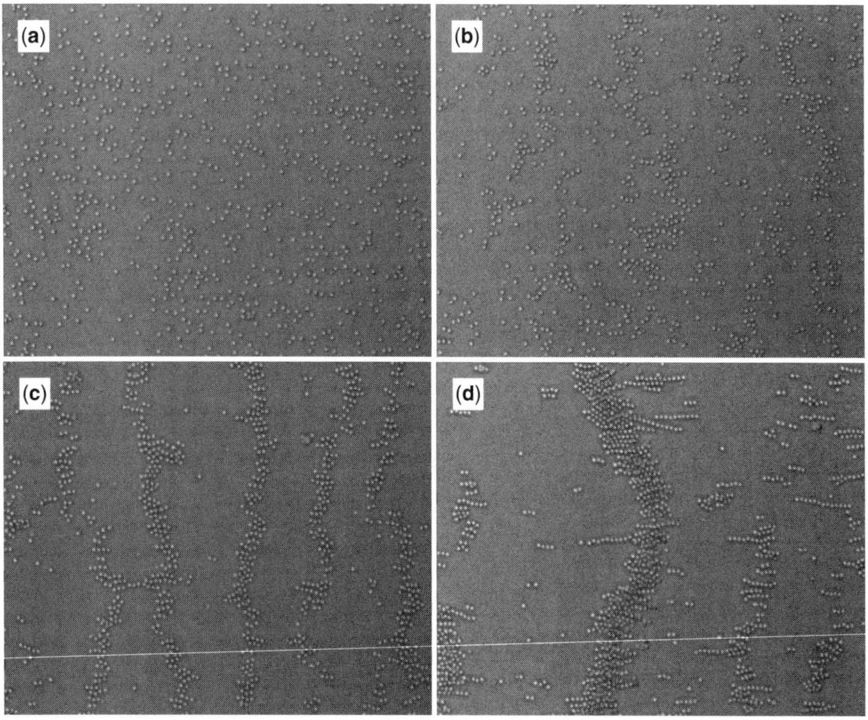

Figure 19.3. Microscopic images of the typical patterns of colloidal arrays of silica particles at various strengths of the applied electric field (root mean square value, *Erms*): (a) *Erms* = 28.3 V/cm; (b) *Erms* = 58.3 V; (c) *Erms* = 70.6–106 V/cm; (d) *Erms* = 141 V/cm (from Reference 23).

are transformed into band patterns after several minuses at higher frequencies near the phase boundary between pearl chains and bands.

The bands and vortices are affected by the effective volume fraction (number density) of silica particles(Φ_e) (Fig. 19.5). The effective volume fraction is defined as the volume fraction of silica particles setting at the bottom in the sedimentation area (see Appendix). The critical electric field, where the bands and vortices are formed, tends to decrease with increasing the effective volume fraction. The vortices in bands are remarkably reduced at $\Phi_e = 0.015$. Neither bands nor vortices are observed, but pearl chains are formed at $\Phi_e = 0.0045$ and *Erms* > 212 V/cm.

There are two explanations for the formation of bands and vortices [16]. One explanation is that they attribute bands and vortices to macroscopic streaming resulting from electrohydrodynamics such as gradient in electrolyte and electroosmotic flow [12, 14]. Another explanation is a result of rotational movement of each particle arising from mutual polarization of particles that is caused by a relaxation of polarization or phase lag between the applied

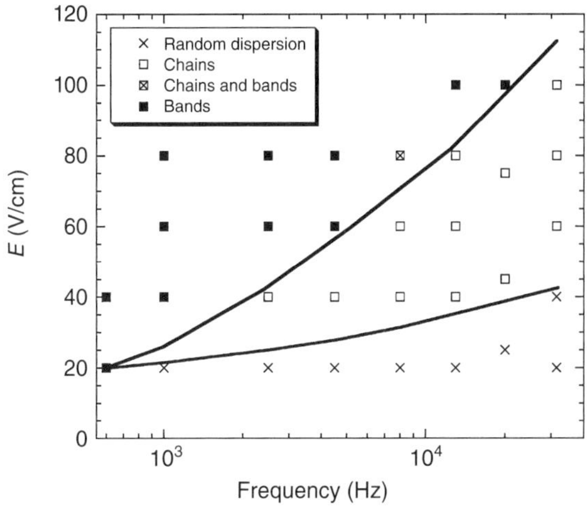

Figure 19.4. Order–disorder phase diagram for polystyrene particles in an AC electric field (redrawn from Reference 16).

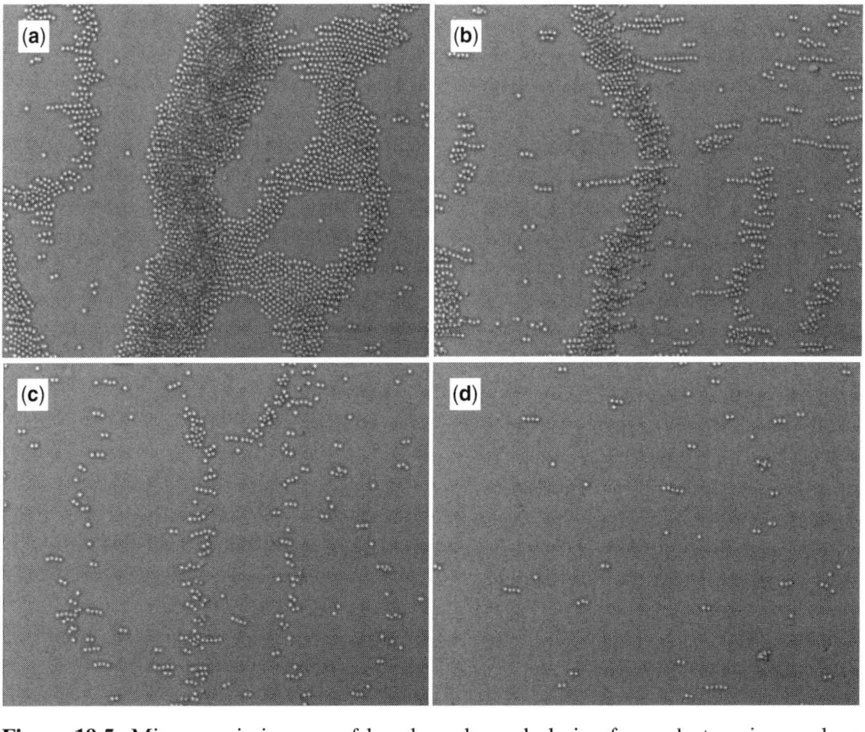

Figure 19.5. Microscopic images of bands and pearl chains formed at various values of the effective volume fraction of silica particles: (a) $\Phi_e = 0.09$ and $Erms = 141$ V/cm; (b) $\Phi_e = 0.045$ and $Erms = 141$ V/cm; $\Phi_e = 0.015$ and $Erms = 212$ V/cm; $\Phi_e = 0.0045$ and $Erms = 212$ V/cm (from Reference 23).

electric field and the induced electric field around dipoles [13, 16]. However, it still remains controversial because there is a lack of experimental evidence other than image data.

19.4 *IN SITU* CONDUCTANCE MEASUREMENTS FOR COLLOIDAL ARRAYS

The ion concentration polarization is dominated by the mobility of counterions in EDL around particles, that is to say, surface conduction of EDL. Electric impedance and dielectric response of suspension provide useful information about the ion concentration polarization of EDL [25–27]. The conductance ratios of the silica suspension (K^*) to its medium (K) are shown as a function of the frequency of applied electric fields (Fig. 19.6). The change in the value of K^*/K can be recognized in the frequency ranging from100 Hz to 1 kHz, reflecting that the diffusion effect of counterions in EDL causes the relaxation of the ion concentration polarization of EDL. The diffusion of counterions in EDL yields ion current flowing against the electric field, and so surface conduction in EDL around particles decreases when the relaxation of the ion concentration polarization of EDL becomes dominant. From Equation 19.1, the relaxation frequency (f_0) for a silica particle ($a = 0.8\ \mu$m) is calculated to be 550~800 Hz using the diffusion constants of $D = 1.1 \sim 1.6 \times 10^{-9}\ m^2/s$ for Na^+ ions (counterions) in EDL. This is within the frequency range where the relaxation of ion concentration polarization is

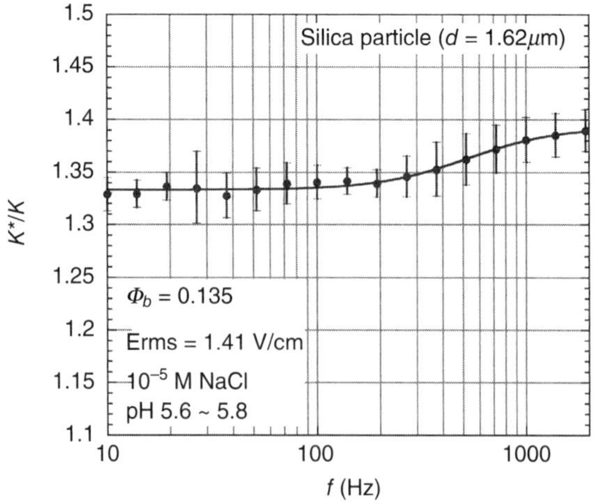

Figure 19.6. Conductance ratios of a silica particle suspension to its medium(K^*/K) as a function of the frequency of applied electric fields (f).

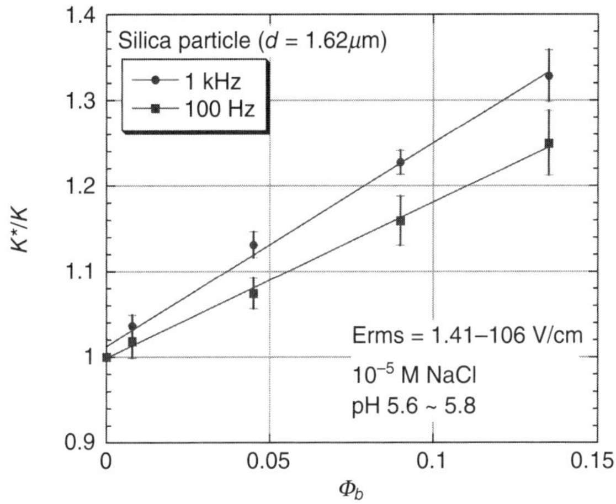

Figure 19.7. Conductance ratios (K^*/K) as a function of the volume fraction (Φ_b) of the silica particles at a concentration of 10^{-5} M NaCl.

recognized (Fig. 19.6). The plot of K^*/K against the volume fraction of particles in suspension (Φ_b) shows a linear relation with an intercept of ~1(Fig. 19.7). Considering that the slope of K^*/K to Φ_b corresponds to the polarizability of the induced dipole moment arising from the ion concentration polarization of EDL, the magnitude of the dipole moment for 1 kHz is larger than that for 100 Hz.

In general, the impedance measurement for colloidal suspension has been performed under a low electric field of ~0.1 V/cm [28]. One is to avoid perturbation, for example, electrolysis and polarization of electrodes. Another is to assess the properties of EDL around an isolated particle in suspension where the interaction among neighboring particles is so small that it can be neglected. However, applied electric field strength required for particle alignments and/or dielectrophosresis must be at least 10 times higher in magnitude than that for the impedance measurement. For the purpose of investigating the properties of EDL under an electric field where the alignments of silica particles are caused, Nishimura et al. have tackled *in situ* electric conductance measurements on aqueous suspension under microscopic observations (Fig. 19.8) [23]. The values of K^*/K are plotted against the strength of the applied electric field (*Erms*) by changing the effective volume fraction of silica particles (Φ_e) at a concentration of 10^{-5} M NaCl (Fig. 19.9). The stepwise increase of K^*/K is recognized. In this region, pearl chains are on the way to being formed from a disordered state. The values of K^*/K are totally increased by increasing the effective volume fraction. The range of the stepwise increase of K^*/K is reduced by decreasing the effective volume fraction. It was also confirmed that the range of the stepwise increase of K^*/K decreases and pearl chains shorten

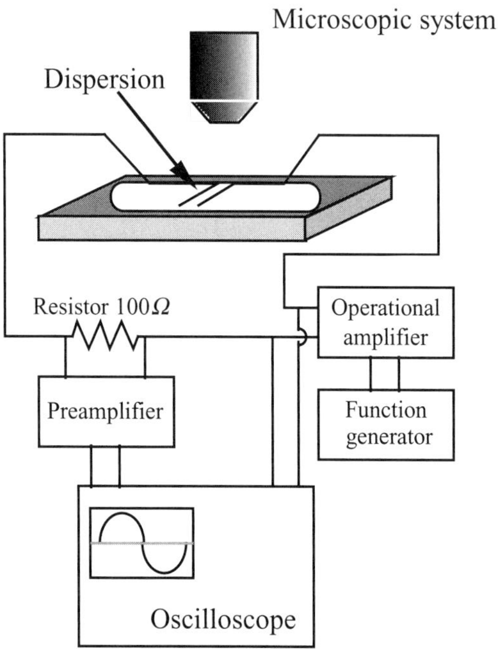

Figure 19.8. Schematic diagram of electronic instruments for *in situ* conductance measurement under a microscopic system (from Reference 23).

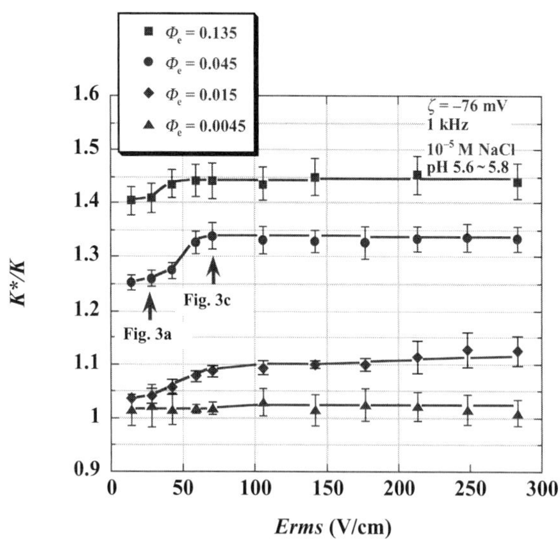

Figure 19.9. Conductance ratios (K^*/K) as a function of the strength of the applied electric field (*Erms*) at various values of the effective volume fraction in 10^{-5} M NaCl aqueous solution (Φ_e) of the silica particles at a concentration of 10^{-5} M NaCl (from Reference 23).

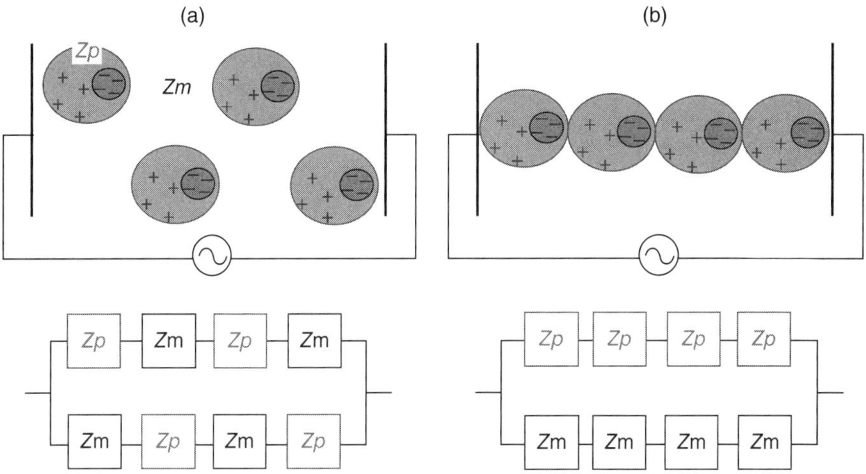

Figure 19.10. Schematic representation of the equivalent circuits to the modes of alignment of silica particles (from Reference 23).

by increasing NaCl concentration [23]. These observations suggested that such a stepwise increase of K^*/K must be closely related to pearl chain formation. The stepwise increase of K^*/K can be quantitatively explained in terms of equivalent circuits (Fig. 19.10). Two patterns of particle alignments and the equivalent circuits are assumed: One is that particles surrounded by EDL are isolated (Fig. 19.10a); the other is that particles form a pearl chain in the direction to the electric field (Fig. 19.10b). The resistive (real) component of the complex impedance of a particle with EDL and a surrounding medium represents Z_p and Z_m, respectively. The imaginary part of the complex impedance measured here can be neglected in comparison to the resistive component at 1 kHz. The *in situ* conductance under microscopic observations, K_a^* and K_b^*, was given by the following equations:

$$\boldsymbol{K_a^*} = \frac{\mathbf{1}}{\mathbf{2(Z_p + Z_m)}} + \frac{\mathbf{1}}{\mathbf{2(Z_p + Z_m)}} = \frac{\mathbf{1}}{\mathbf{Z_p + Z_m}} \tag{19.9}$$

and

$$K_b^* = \frac{1}{4Z_p} + \frac{1}{4Z_m} = \frac{Z_p + Z_m}{4Z_pZ_m}. \tag{19.10}$$

The relation between the geometric mean and the arithmetic mean gives the following relation:

$$\frac{K_b^*}{K_a^*} = \frac{(Z_p + Z_m)^2}{4Z_pZ_m} > 1, Z_p > Z_m. \tag{19.11}$$

This analytic expression justifies that the stepwise increase of K^*/K is a result of pearl chain formation of particles with a surface conduction higher than the conduction of the bulk medium.

At *Erms* > 106 V/cm where the transition from the pearl chain to bands and vortices in zigzag bands is observed, no clear change in K^*/K is detected in the *in situ* conductance measurement. This result implies that the change in surface conduction of EDL accompanying bands and vortices is negligibly small, or bands and vortices would be due to the macroscopic streaming flow arising from electrohydrodynamics or electroosmotic flow. The details remain unclear, but the *in situ* conductance measurement under microscopic observations is expected to be a promising approach to compensate the lack of experimental evidence existing between observed images and theoretical modeling.

19.5 ELECTROKINETIC COUPLING IN PEARL CHAIN FORMATION

As discussed earlier, it has been found that the arrays of colloidal particles observed under an AC electric field cannot be fully understood in terms of the ionic concentration polarization alone.

When pearl chains of negatively charged silica particles are formed in the vicinity of the negatively charged bottom of a glass vessel, the electroosmotic flow at the bottom/water interface enhance the ionic concentration polarization of EDL around particles [29]. The alignments of the silica particles in pearl chains lengthen and stiffen in the direction the electric field when the frequency decreases from 1 kHz to 100 Hz (Fig. 19.11). When the frequency is switched back to1 kHz from100 Hz, the alignments of silica particles in the

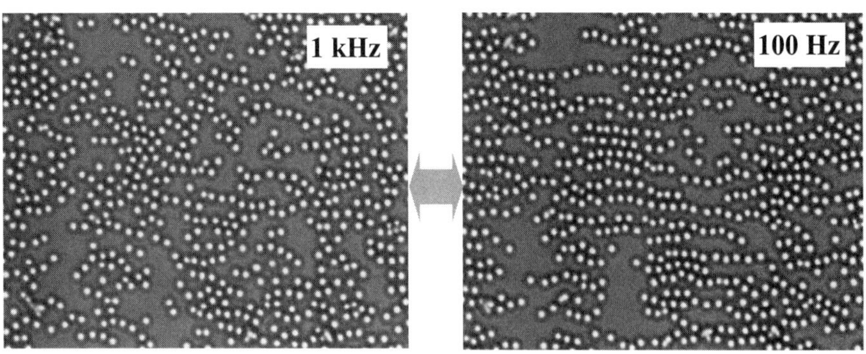

Figure 19.11. Microscopic images of pearl chain formation of silica particles settling at the bottom of a silica vessel at *Erms* = 70.7 v/cm and Φ_e = 0.135 when the frequency of the electric field was switched from 1 kHz to100 Hz (from Reference 29).

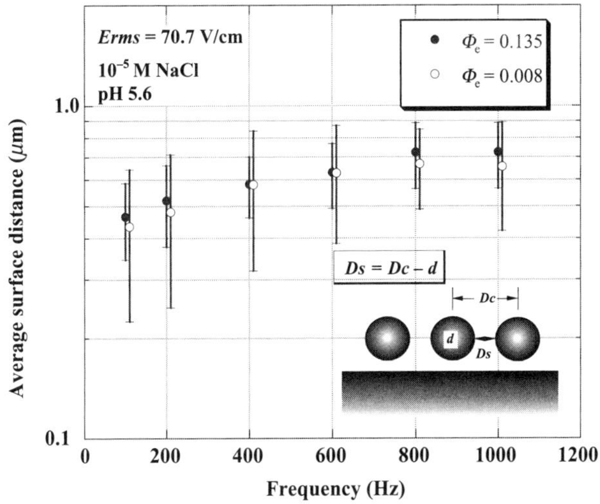

Figure 19.12. Average surface distances between two neighboring silica particles in pearl chains (from Reference 29).

pearl chains are alleviated again. Such a transformation of alignments of silica particles in the pearl chain indicates that attractive dipole–dipole interactions between neighboring silica particles increase with a decrease in the frequency from 1 kHz to 100 Hz. The neighboring particles come up together with decreasing the frequency from 1 kHz to 100 Hz (Fig. 19.12). This also supports that interaction between two neighboring silica particles in pearl chains become more attractive with a decrease in the frequency from 1 kHz to 100 Hz. K^*/K for the *in situ* conductance measurements under microscopic observations and the conductance measurements on uniform suspensions are plotted as a function of effective volume fractions of silica particles at the vessel bottom and the volume fraction of silica particles in the uniform suspension (Φ_b) (Fig. 19.13). The plots of K^*/K against Φ_b for the uniform suspensions show a linear relation at the frequencies of both 1 kHz and 100 Hz as predicted from Equations 19.7 and 19.8. On the other hand, the plots of K^*/K against Φ_e for the *in situ* conductance measurements show a nonlinear relation. The value of K^*/K for the *in situ* conductance measurements is higher than that of K^*/K for the uniform suspensions. More interestingly, the slope of K^*/K to Φ_b for 100 Hz is smaller than that for 1 kHz for the uniform suspensions, whereas the slope of K^*/K to Φ_e for 100 Hz is larger than that for 1 kHz for the *in situ* conductance measurements. The magnitude of induced dipole moment for 100 Hz is larger than that for 1 kHz, although the change in the slope of K^*/K to Φ_e cannot be simply explained.

From Equations 19.2–19.4, the induced dipole moment increases as the difference between surface conductivity of EDL around colloidal particles (k^{δ})

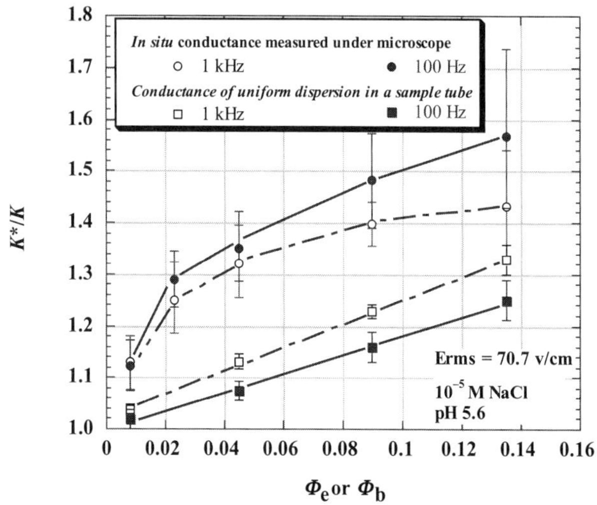

Figure 19.13. Conductance ratios (K^*/K) as a function of the volume fractions (Φ_e or Φ_b) of silica particles (from Reference 29).

and the bulk conductivity of a medium (k^b) enlarges. The above-mentioned results indicate that surface conductivity must increase with lowering the frequency, but the bulk conductivity is constant. The surface conduction is dominated by the concentration and the mobility of counterions in EDL. The concentration of counterions in the system must be constant even if the frequency of the electric field is changed; therefore, the mobility of counterions in EDL would increase with a decrease in the frequency of the electric field.

The mobility of counterions in EDL for surface conduction consists of two contributions: electromigration and electroosmotic flow of medium. The relation between surface conductivity (k^δ) and bulk conductivity (k^b) is expressed for a symmetrical electrolyte as follows [30]:

$$\frac{k^\delta}{k^b} = \frac{2}{\kappa}\left(1 + \frac{3m}{z^2}\right)\left\{\cosh\left(\frac{ze\varsigma}{2kT}\right) - 1\right\}, \tag{19.12}$$

where

$$m = \left(\frac{kT}{e}\right)^2 \left(\frac{2\varepsilon_0\varepsilon_r}{3\eta D}\right). \tag{19.13}$$

In these equations, κ is the reciprocal Debye length, z is the valence of ions, e is the elementary charge, ς is zeta potential, k is the Boltzmann constant, T is the absolute temperature, ε_0 is the dielectric constant in vacuum, ε_r is the relative dielectric constant of water, η is the viscosity of water, D is the diffusion constant for the ions, and m is the contribution of the electroosmotic component to the surface conductivity. As the surface conductivity of a hydrodynamically stagnant layer such as the Stern layer is ignored in this theory, the relative contribution of an electroosmotic component to the surface conductivity in the diffuse part of EDL is given by

$$\alpha = \frac{3m/z^2}{(1+3m/z^2)}. \tag{19.14}$$

α is calculated to be 0.4 using $D = 1.5 \times 10^{-9}$ m^2/s for Na^+ ($z = 1$) from Equations 19.13 and 19.14. The relative contribution of the electroosmotic component to the surface conductivity is as large as that of the electromigration component for Na^+ ions. We need to take account of the electroosmotic component in the mobility of counterions for surface conduction. The total number of counterions in the EDL of the silica particles and the bottom surface should be constant, and so the contribution of electromigration of counterions to the conductance in the vessel is not changed. If the electroosmotic flows in the gap between the silica particles and the bottom are coupled to each other, the electroosmotic flow mobility around the silica particles in the vicinity of the bottom surface would increase.

Minor et al. showed damping oscillation of electroosmotic flow at a planar interface [31]. The velocity of the electroosmotic flow is oscillating with a characteristic penetration length,

$$\sigma = \sqrt{\frac{2\eta}{\omega\rho}}, \tag{19.15}$$

where ρ is the liquid density and η is the liquid viscosity. The characteristic penetration depth decreases with an increase in the frequency of the applied electric field. This suggests that the amplitude of electroosmotic velocity decays in a shorter distance from the vessel bottom as the frequency of the AC electric field is increased. From Equation 19.15, the values of σ for 1 kHz and 100 Hz are calculated to be 18 and 56 μm, respectively (Fig. 19.14a). The damping oscillation of the electroosmotic flow at the bottom extends beyond the region where silica particles are settling at the bottom. It is not surprising that the electroosmotic flow at the bottom can be coupled with the mobility of counterions in the EDL around silica particles. The magnitude of electroosmotic velocity for 100 Hz is larger than that for 1 kHz at a distance of 3 μm from the bottom of the cell, which corresponds to the average diameter of a silica particle plus the surface distance between the particle and the bottom surface. As a result of the coupling of the mobility of counterions in the diffuse part of EDL with the electroosmotic flow at the bottom (Fig. 19.14b), the surface conductivity of the EDL around a silica particle for 100 Hz is higher than that for 1 kHz.

A similar phenomenon can be observed by changing the magnitude of the zeta potential of a platform (vessel bottom) on which silica particles are settling [29]. The mobility of electroosmotic flow (u_{eo}) at the planar interface is directly proportional to the magnitude of the ζ-potential in the Helmholtz–Smoluchowski's equation:

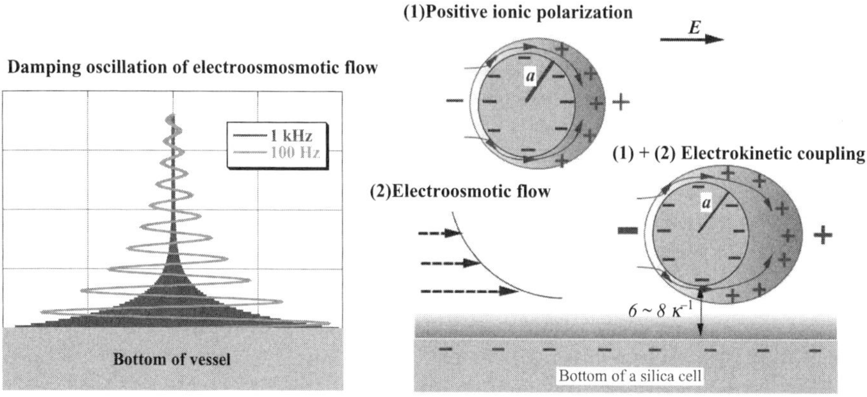

(a) (b)

Figure 19.14. Schematic representation of electrokinetic coupling of ion concentration polarization in EDL with damping oscillation of electroosmotic flow at vessel bottom/water interface.

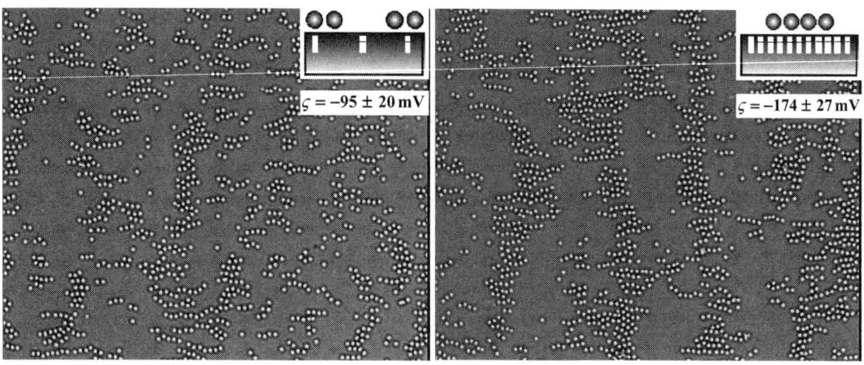

Figure 19.15. Microscopic images of pearl chains of silica particles on glass plates with ζ-potentials of −95 and −174 mV at *Erms* = 57.9 V/cm (1 kHz) and $\Phi_e = 0.045$ (from Reference 24).

$$u_{eo} = -\frac{\varepsilon_0 \varepsilon_r \varsigma}{\eta}, \tag{19.16}$$

where ς is ζ-potentials of the glass plate, ε_0 and ε_r are the permittivity of free space and the relative permittivity of water, and η is the viscosity of water, respectively. The damping oscillation of the electroosmotic flow can be controlled by changing the magnitude of the zeta potential of the surface in the vicinity of the particles. As the magnitude of the ζ-potential of the platform (bottom of the vessel) on which silica particles are settling increases, the alignments of the silica particles in pearl chains lengthen and stiffen in the direction the electric field (Fig. 19.15). Simultaneously, the surface distance between

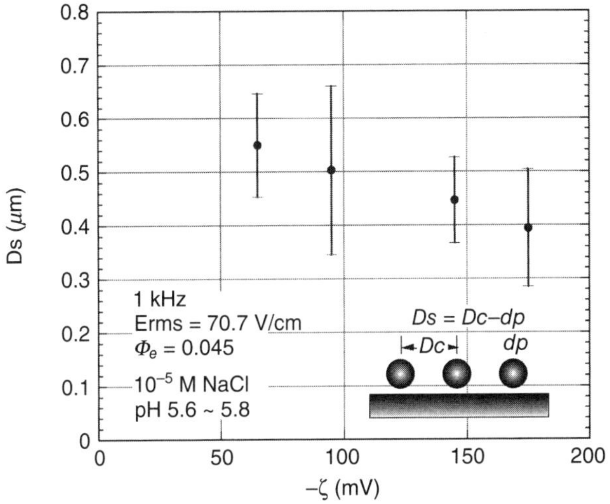

Figure 19.16. Average surface distances between two neighboring silica particles in pearl chains as a function of the ζ-potential at the bottom/water interface (redrawn from Reference 24).

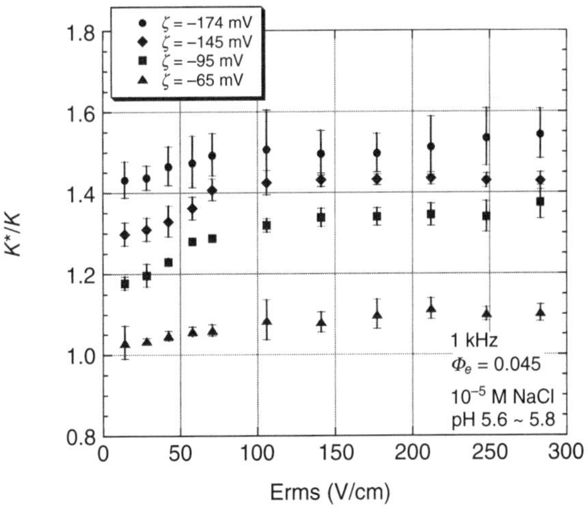

Figure 19.17. Conductance ratios (K^*/K) as a function of the strength of applied electric fields (*Erms*) with changing ζ-potential at the bottom/water interface (from Reference 24).

neighboring particles in pearl chains decreases (Fig. 19.16). K^*/K totally increases with an increase in the magnitude of the zeta potential of the glass platform (Fig. 19.17). Induced dipole moments arising from the ion concentration polarization in the EDL are enhanced as a result of the coupling of the mobility of counterions in the EDL with the electroosmotic flow at the platform/water interface by increasing the magnitude of the zeta potential of the platform.

ACKNOWLEDGMENTS

S.N. thanks Drs. H. Matsumura, K. Kosuge, and T. Yamaguchi at the National Institute of Advanced Industrial Science and Technology (AIST) for their valuable and helpful researching collaborations. S.N. also acknowledges Japan Society for the Promotion of Science (JSPS) Core-to-Core program (Advanced Particle Handling Science), and the Japanese Ministry of Education, Culture, Sports, Science and Technology (MEXT) via Grant-in-Aids for Scientific Research on Innovative Areas "Emergence in Chemistry" (20111007) for organizing the opportunity for valuable discussions.

APPENDIX

The effective volume fraction (Φ_e) was defined as the volume fraction of silica particles settling at the bottom of the vessel (Fig. 19.18) [23]:

$$\Phi_e = \frac{NV_p}{Sd}, \tag{19.17}$$

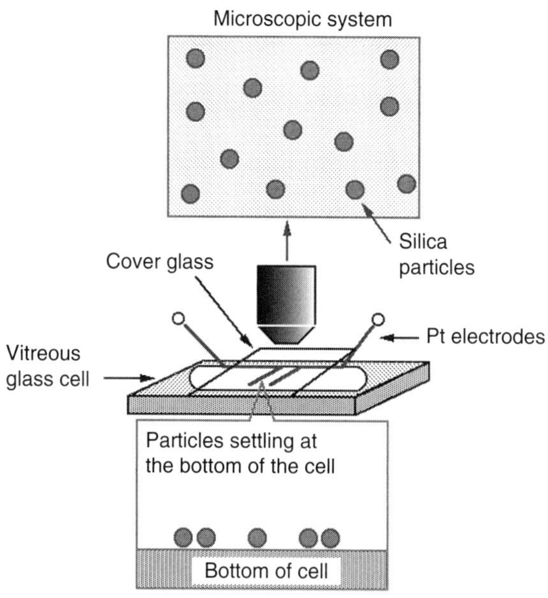

Figure 19.18. The effective volume fraction (Φ_e) of silica particles settling at the bottom of a silica glass cell.

where N is the number of silica particles settling at the bottom in the scope of the microscopic images after an elapsed time of 10 minutes, V_p is the volume of one silica particle, d is the diameter of the silica particle, and S is the area of microscopic images.

REFERENCES

1 S. S. Dukhin, V. N. Shilov. *Adv. Colloid Interface Sci.* 13 (1980) 153.
2 M. P. Hughese. *Nanotechnology* 11 (2000) 124.
3 (a) E. Muth. *Kolloid-Z* 41 (1927) 97; (b) R. Kruyt, J. G. Vogel. *Kolloid-Z* 95 (1941) 2; (c) H. Matsumura *Colloids Surf. A* 104 (1995) 343.
4 A. H. Pohl, Ira. Hawk. *Science* 272 (1996) 706; (b) T. Matsue, N. Matsumoto, A. Uchida. *Electrochim. Acta* 42 (1997) 3251.
5 H. Morgan, N. G. Green. *AC electrokinetics: colloids and nanoparticles*. Research Studies Press, Baldock, Hartfordshire, England, 2003.
6 O. D. Velev, E. W. Laler. *Langmuir* 15 (1999) 3693.
7 (a) M. Washizu, O. Kurosawa, I. Arai, S. Suzuki, N. Shimamoto. *IEEE Trans. Ind. Appl.* 36 (1995) 1010; (b) W. A. Germishuizen, P. Tosch, P. J. Middelberg, C. Wälti, A. G. Dacies, R. Writz, M. J. Pepper. *Appl. Phys.* 97 (2005) 014702; (c) C. Wälti, P. Tosch, W. A. Germishuizen, C. F. Kaminski. *Appl. Phys. Lett.* 88 (2006) 153901.
8 (a) A. H. Pohl, J. K. Pollock, in N. Castimopoolas (ed.), *Method of cell separation*, vol. 1. Plenum Press, New York, 1978, p. 67; (b) A. H. Pohl, J. K. Pollock, in F. Gutmann, H. Keyzer (eds.), *Modern bioelectrochemistry*. Plenum Press, New York, 1986, p. 329, Chapter 12.
9 M. P. Hughes. *Electrophoresis* 23 (2002) 2569.
10 N. I. Gamajunov, V. A Murtsovkin. *Kolloidn Zh.* 45 (1983) 760.
11 R. Isherwood, B. R. Jennings, M. Stankiewicz. *Chem. Eng. Sci.* 42 (1987) 913.
12 (a) B. R. Jennings. In Abstract of 8th Intern. Conf. of Chemistry of Solid/Liquid Interfaces. 1989, 27; (b) B. R. Jennings, M. Stankiewicz. *Proc. R. Soc. Lond A* 427 (1990) 321.
13 Y. Hu, J. L. Glass, A. E. Grifith, S. Fraden. *J. Chem. Phys.* 100 (1994) 4674.
14 H. Isambert, A. Ajdari, J. L. Viovy, J. Prost. *Phys. Rev. E* 56 (1997) 5688.
15 M. Abe, A. Yamamoto, M. Orita, S. Tsukamoto, T. Ohkubo, H. Sakai, N. Momozawa. *Langmuir* 20 (2004) 7021.
16 P. P. Lele, M. Mittal, E. M. Furst. *Langmuir* 24 (2008) 7021.
17 M. Trau, D. A. Saville, I. A. Aksay. *Langmuir* 13 (1997) 6375.
18 M. Abe, M. Orita, H. Yamazaki, S. Tsukamoto, Y. Teshima, T. Sakai, T. Ohkubo, N. Momozawa, H. Sakai. *Langmuir* 20 (2004) 5046.
19 T. Gong, D. T. Wu, W. M. Marr. *Langmuir* 17 (2001) 2301.
20 (a) G. Shwaltz. *J. Phys. Chem.* 66 (1962) 2636; (b) V. N. Shilov, S. S. Dukhin. *Kolloidn Zh* 31 (1969) 706.
21 A. Van der Wal, M. Minor, W. Norde, A. Zehnder, J. Lyklema. *J. Colloid Interface Sci.* 186 (1997) 71.

22 (a) S. S. Dukhin. *Adv. Colloid Interface Sci.* 44 (1993) 1; (b) S. S. Dukhin, A. A. Derjaguin. Equilibrium double layer and electrokinetic phenomena, in E. Matijevic (ed.), *Electrokinetic phenomena, surface and colloid science 7*. John Wiley & Sons, New York, 1974, pp. 217–218, Chapter 2.

23 S. Nishimura, H. Matsumura, K. Kosuge, T. Yamaguchi. *Langmuir* 23 (2007) 6567.

24 S. Nishimura, H. Matsumura, K. Kosuge, T. Yamaguchi. *Langmuir* 24 (2008) 12695.

25 S. S. Dukhin, V. N. Shilov, in A. V. Delgado (ed.), *Interfacial electrokinetics and electrophoresis*. Marcel Dekker, New York, 2002, p. 55.

26 J. Lyklema. *Fundamentals of interface and colloid science*, vol. 2. Academic Press, New York, 1995, Chapter 4.

27 H. Matsumura. *Colloids Surf. A* 104 (1995) 343.

28 (a) J. Kijlstra, H. P. van Leeuwen, J. Lyklema. *J. Chem. Soc. Faraday Trans.* 88 (1992) 3441; (b) J. Kijlstra, H. P. van Leeuwen, J. Lyklema. *Langmuir* 9 (1993) 1625; (c) J. Kijlstra, R. A. J. Wegh, H. P. van Leeuwen, J. J. Lyklema. *Electroanal. Chem.* 366 (1994) 37.

29 S. Nishimura, H. Matsumura, K. Kosuge, T. Yamaguchi. *Langmuir* 23 (2007) 8789.

30 (a) J. J. Bikerman. *Z. Phys. Chem.* A (1933) 163; (b) J. J. Bikerman. *Kolloid Z* 72 (1935) 100; (c) J. J. Bikkerman. *Trans. Fraraday Soc.* 36 (1940) 140.

31 M. Minor, A. J. van der Linde, H. P. van Leeuwen, J. Lyklema. *J. Colloid Interface Sci.* 189 (1997) 370.

20

SIZE DISTRIBUTION MEASUREMENTS OF FINE PARTICLES USING THEIR PEARL CHAIN FORMATIONS UNDER A DC ELECTRIC FIELD

Akira Otsuki, Gjergi Dodbiba, and Toyohisa Fujita

20.1 INTRODUCTION

It is important to manipulate the degree of dispersion and coagulation of fine particles in various media for many industrial processes [1, 2]. Therefore, there is a need for understanding the dispersion and coagulation of fine particles in the media. Many techniques have been developed for evaluating those characteristics of fine particles, that is, size measurement [3], turbidity measurement [4], contact angle measurement [5], zeta-potential measurement [6], force measurement [7, 8], and a combination of these techniques [9, 10]. However, these techniques usually are not suitable for evaluating the dispersion and coagulation of particles in sample solutions of high concentration and/or in solutions with no optical transparency. On the other hand, these kinds of solutions are commonly used in the many industrial procedures, such as separation of fine particles [11, 12] and deposition of fine particles on fibers [13].

In the field of fine particle engineering, many researchers have studied pearl chain formations of dielectric particles under an electric field. Electrorheological

Electrical Phenomena at Interfaces and Biointerfaces: Fundamentals and Applications in Nano-, Bio-, and Environmental Sciences, First Edition. Edited by Hiroyuki Ohshima.

fluid (ERF) is one of the major applications of pearl chain formation. This fluid shows the increase of suspension stress as a result of increasing the applied electric field strength between two electrodes [14]. The mechanism of pearl chain formation has been explained mainly by two models, that is, the dipole–dipole model [15] and the multipole model [16, 17]. Based on these models, calculation of electric field and force on particles under an electric field [16, 17] and dynamic simulation of dielectric particles [18] were carried out. These calculations and simulations use mainly physical properties of particles and medium, experimental conditions, but without experimental data obtained from actual observations of particles. Also, the results were not fully compared with experimental results. Therefore, a methodology development of the *in situ* observation of dielectric particles under an electric field is beneficial to further elucidate the mechanism of pearl chain formation under the electric field.

In this chapter, we focused on particle size measurement for evaluating the degree of dispersion and coagulation of fine particles. Here, our target size range is from single nanometers to 10 micrometers. The techniques can be divided in two categories, that is, measurements in (i) dry condition and (ii) wet condition [2]. In dry condition, microscopic studies based on optical microscopy, scanning electron microscope (SEM), and transmission electron microscope (TEM) are the common techniques for the size measurement of fine particles. In wet condition, techniques using laser source (i.e., dynamic light scattering [DLS] and laser diffraction) are the common techniques. Although these techniques have several advantages, they have some drawbacks. The techniques in dry condition are not applicable for measuring the size of particles in solutions. On the other hand, the techniques in wet condition are not usually suitable for measuring the size of particles in high particle concentrations and/or solutions with no or low optical transparency.

Interactive force apparatus (IFA) was designed for determining the degree of dispersion and coagulation of fine particles suspended in functional fluid under a magnetic or electric field [19, 20]. The apparatus is a direct measurement technique, not depending on both particle concentration and optical transparency of sample solutions. Moreover, measurements can be conducted in various solvents (e.g., aqueous solution, organic solvent). The current experimental setup allows detection of primary particles and/or aggregates of particles in a sample suspension. In the following sections, the methodology will be introduced followed by describing the results of some case studies using the apparatus to measure the size of fine particles in various media under a DC electric field.

20.2 METHODOLOGY

The IFA has three parts, that is, the main part (which consists of an electric balance, a hemisphere, and a flat plate), the control part (i.e., a personal computer, a piezo-stage controller, and a voltage supplier), and the detecting part (i.e., a multimeter and an oscilloscope). The IFA measures the interactive force between two surfaces, that is, a gold-coated glass hemisphere and a brass flat plate, which is fixed at the bottom of the sample cell. The main part measures the weight of the hemisphere immersed in a sample solution by decreasing the distance between the two surfaces at a certain speed. The control part is employed to adjust a supply voltage, regulates the movement of the piezo stage, and collects data from the balance and the piezo-stage controller. The detecting part, on the other hand, measures the contact point where the hemisphere and the flat plate attach.

Figure 20.1 shows the main part of the apparatus. The hemisphere is hung to the electric balance and remains still in the sample solution. The weight of

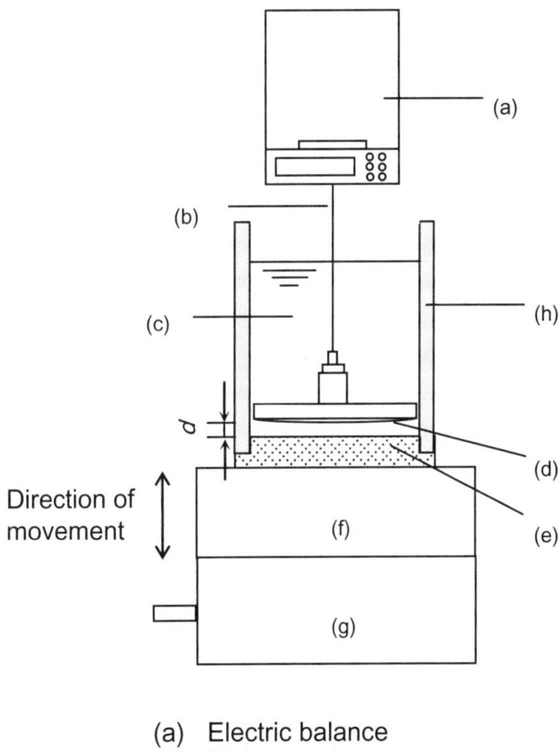

(a) Electric balance
(b) Platinum wire
(c) Sample solution
(d) Hemisphere (gold coated)
(e) Flat plate (brass)
(f) Piezo stage
(g) Z stage
(h) Glass cell

Figure 20.1. Schematic diagram of the main part of the experimental setup for an interactive force apparatus [21].

the hemisphere is measured by using the electric balance and recorded by a personal computer, while the flat plate moves toward the hemisphere, decreasing a distance, d, from a certain distance (e.g., 100 nm) to 0 nm. The hemisphere is selected to make point contact to the flat plate. The movement of the piezo stage by applying voltage is used to regulate the distance between the hemisphere and the flat plate. The rate of movement is determined based on an initial separation distance between the hemisphere and the flat plate when a measurement starts (e.g., 1 nm/s for a 100-nm distance). The piezo stage is located on a z stage. Detecting the contact point of the hemisphere to the flat plate at the bottom of the cell (i.e., the surface distance d is zero) is important for calculating the surface distance because the point determines the certain distance from the contact point for the measurements.

The measured weight is converted to the interactive force by using the Derjaguin equation [22]:

$$\frac{F_{(\mathrm{D})\mathrm{sphere}}}{R} = 2\pi W_{(\mathrm{D})\mathrm{plane}}, \tag{20.1}$$

where $W_{(\mathrm{D})\mathrm{plane}}$ is the interactive free energy, $F_{(\mathrm{D})\mathrm{sphere}}$ is the interactive force between the hemisphere and the flat plate, and R is the curvature radius of the glass hemisphere, respectively. The interactive force was plotted as a function of surface distance between the hemisphere and the flat plate (i.e., force–distance curve) in order to determine the size of fine particles. Figure 20.2

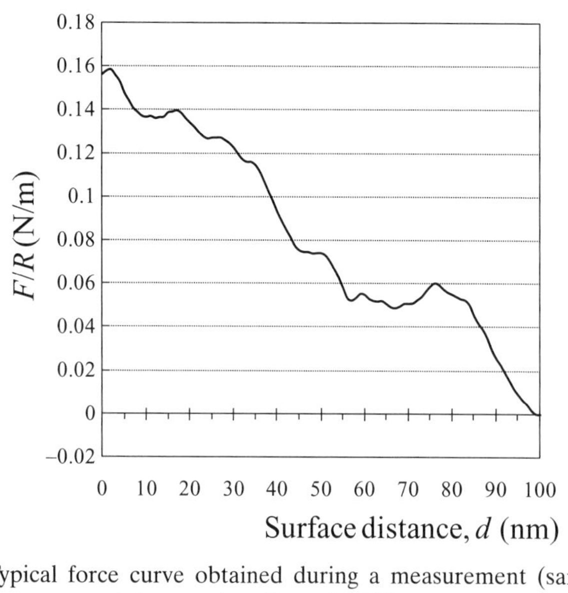

Figure 20.2. Typical force curve obtained during a measurement (sample: aqueous solution of silica nanoparticles; supply voltage: 0.01 V).

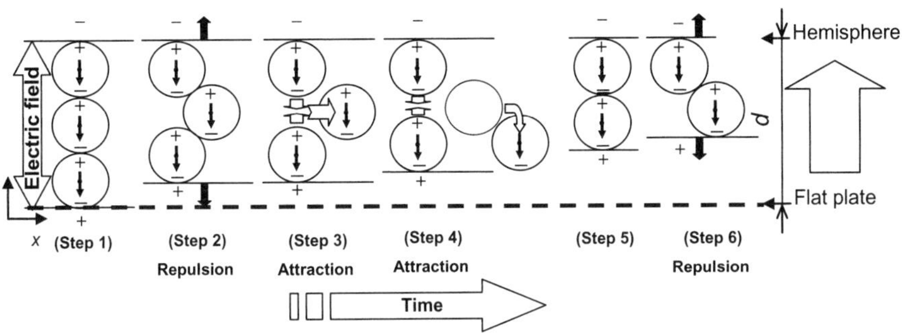

Figure 20.3. Behavior of dielectric particles during measurement under an electric field [21].

shows a typical force curve obtained during a measurement, indicating that repulsive and attractive forces repeatedly appear.

Under an electric field generated between two parallel plates, pearl chains of dielectric particles form toward the direction of the electric field [23]. In terms of the shape of the chains, linear chains form over the triangular chain under the electric field because the former is more stable in terms of potential energy [24]. During the measurement, an electric field is applied between the hemisphere and the flat plate, and thus dielectric particles are arranged toward the direction of the electric field in the area between two plates. When the two plates are close to each other, two different forces (i.e., repulsive and attractive forces) alternately act on the particles due to structural changes of the particles (Fig. 20.3) [21]. The repulsive force occurs when the particle chain structure is stretched by compressive force, whereas the attractive force occurs when the particle chain structure is broken. The cycle of repulsive and attractive forces is a primary size of particle or size of the aggregate, which depends on the degree of agglomeration.

20.3 CASE STUDIES

20.3.1 Evaluation of the Degree of Coagulation of Fine Particles under an Electric or Magnetic Field

Fujita et al. [25] observed an ERF dispersing smectite particles (20–50 nm) in methylphenyl silicone oil. Smectite particles cannot be observed by optical microscope due to their similar refraction index with the one of silicone oil. They applied the IFA to observe particle chain structures under an electric field or no field. Under no field, the degree of repulsive force increased by decreasing the surface distance with only the silicone oil filled in the cell, whereas repulsive and attractive forces repeatedly appeared (150 nm) with the ERF. On the other hand, under an electric field of 1.0 V, the cycle of these forces became longer (200 nm) at a distance of less than 1 μm with the ERF. They explained the difference as aggregate formation of the smectite particles between two electrodes under the electric field by using a multiparticle model. Miyazaki et al. [26] also reported a similar trend with a silicone oil-based magnetic fluid dispersing magnetite particles (diameter D of 10 nm) coated with a surfactant, sodium polyoxyethylene(6) alkyl ether acetate. They indicated that a longer cycle of repulsive and attractive forces was observed under an electric field of 1.0 V compared with the result under no field. A kerosene-based magnetic fluid dispersing the magnetite particles coated with sodium oleate (molecular length δ of 5 nm) was also measured by IFA under a magnetic field (10 mT) generated by an electromagnet or no field (0 T). Under no field, a very slight interactive force appeared between the hemisphere and the flat plate, whereas under the magnetic field, attractive and repulsive forces repeatedly appeared (20–30 nm) at a distance of less than 680 nm. The measured size was equal to the size of the particles coated with sodium oleate, that is, 20 nm (= $D + 2\delta$).

Shibayama et al. [20] measured an ERF dispersing 0.01 wt % melamine particles (median diameter of 1 μm) in SNOWTEX 30 (aqueous solution with 30% SiO_2 provided by Nissan Chemical) using the IFA to determine size of the aggregate under electric fields (0.5, 1.0, and 1.5 V) or no field (0 V). Under no field, the degree of repulsive force increased as the surface distance decreased, whereas attractive and repulsive forces repeatedly appeared (2.5–4.0 μm) at a distance of less than 10 μm. On the other hand, under an electric field of 1.5 V, the cycle of these forces was not shown. They explained the result at 1.5 V as an electric breakdown between two electrodes due to the higher supply voltage than as an electric breakdown voltage of water.

20.3.2 Mechanism Study of Fine Particle Separation

Otsuki et al. [27] evaluated a heterocoagulation of micron-size fluorescent powders and nonpolar oil (n-heptane) droplet immersed in a polar organic solvent (N,N-dimethylformamide (DMF), $(CH_3)_2NCHO$), a high-density organic solvent) in order to elucidate the mechanism of two-liquid flotation when separating a mixture of the fluorescent powders. They measured the zeta potential and aggregate size of the fluorescent powders and applied the heterocoagulation theory by using measured values for evaluating the separation mechanism.

In detail, they measured the size of fluorescent powders (i.e., red, green, and blue powders) with and without a surfactant in a polar solvent DMF using the IFA. The primary particle sizes of the red, green, and blue powders determined by SEM photographs were 2.6, 1.1, and 2.1 μm, respectively. Two different surfactants have been used for the separation: (i) dodecyl amine acetate (DAA, $CH_3(CH_2)_{11}NH_2{\cdot}CH_3COOH$, i.e., a cationic surfactant) to separate green powder from the others in the first step and (ii) sodium

1-octanesulfonate ($CH_3(CH_2)_7SO_3Na$, i.e., an anionic surfactant) to separate blue powder from red powder in the second step. Without surfactant, the size of the fluorescent powders is small since sharp peaks are observed, whereas with the addition of a surfactant, one of the fluorescent powders (green with DAA, blue with sodium 1-octanesulfonate) is aggregated since broad peaks are observed. The size of the aggregate of green particles in the presence of 2×10^{-4} mol/L DAA was approximately 14 μm. In the presence of 20×10^{-4} mol/L sodium 1-octanesulfonate, the size of the aggregate of blue particles was about 3.4 μm.

The total potential energy was calculated according to the heterocoagulation theory applied to a system of micro-sized particles immersed in the polar organic solvent by using the parameters obtained through the series of measurements. The results indicated that the heterocoagulation of one of the fluorescent powders and the n-heptane droplet in the DMF solvent is feasible in the presence of a surfactant. Based on these findings, the mechanism of two-liquid flotation was explained. Fluorescent powders to be separated are first covered by a surfactant and are then aggregated with other powders covered by a surfactant. Next, the aggregates attach on the surface of a nonpolar droplet and then migrate toward the interface of two phases and collect at the interface, whereas the other powders precipitate in the polar solvent.

20.3.3 Measurements of Particle Size Distribution

Otsuki et al. [21] applied the IFA to measure the particle size distribution of fine particles by increasing the number of data acquisition in order to evaluate the availability of IFA for the measurement. In addition to increasing the data acquisition, an improved data handling process was also proposed. In order to determine the cycle precisely, the first derivative of the force is plotted. The distance between points at zero value of the first derivative, where the value turns negative to positive, corresponds to the size of the particles or the aggregates, which is the same as the cycle of repulsive and attractive forces as shown in Figure 20.2. Figure 20.4 shows a typical derivative curve as a function of surface distance analyzed from the data shown in Figure 20.2 when an aqueous solution of silica nanoparticles was measured by IFA under 0.01-V supply voltage. From Figure 20.4, the size of silica nanoparticles was determined in the range from 4 to 23 nm. The wide range of particle size is explained by the aggregate formation of particles due to high concentration of solid in the sample suspension. Five continuous measurements were conducted and then a particle size distribution was drawn based on the measurements.

The results were compared with the ones obtained from TEM photographs and DLS spectroscopy (ELS-8000, Otsuka Electronics Co., Osaka, Japan) to evaluate the availability of the apparatus. An aqueous solution containing silica particles provided by Nissan Chemical (SNOWTEX XS) was used as a sample for the measurement due to the simplicity of controlling the dispersion

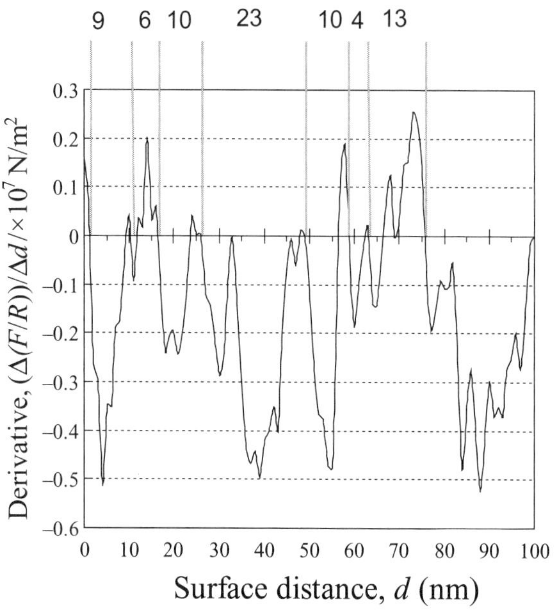

Figure 20.4. Derivative curve of $F_{(D)sphere}$ R^{-1} when an aqueous solution of silica nanoparticles was measured by IFA under 0.01-V supply voltage as a function of surface distance [21].

and coagulation state. The averaged particle size was about 5 nm determined by analyzing a series of photographs taken by TEM. The pH of the solution was 9, and concentration of silica nanoparticles in the solution was 30 wt %.

Figure 20.5 shows that D_{50} measured by IFA was closer to the average particle size determined by TEM. Comparing the results under three different supply voltages, (i) the results at 0.01 and 0.02 V were almost identical, while (ii) these results were different from the one at 0.04 V. The results indicate that breakage of coagulated particles possibly occurs due to electric breakdown. The distribution measured by IFA ($D_{50} = 5$–7 nm) was larger than the one measured by DLS ($D_{50} = 1$ nm). The electric breakdown was explained by curve fitting three different particle size distribution functions (i.e., normal distribution, Rosin–Rammler distribution, and Gaudin–Schuhmann distribution) with the particle size distribution obtained from IFA measurement. Under the supply voltage of 0.01 and 0.02 V, IFA measures the size of particles as it is in aqueous solution, while under the supply voltage of 0.04 V, IFA can measure the size of particles smaller than the size in aqueous solution due to the disintegration of aggregates by electric breakdown between electrodes.

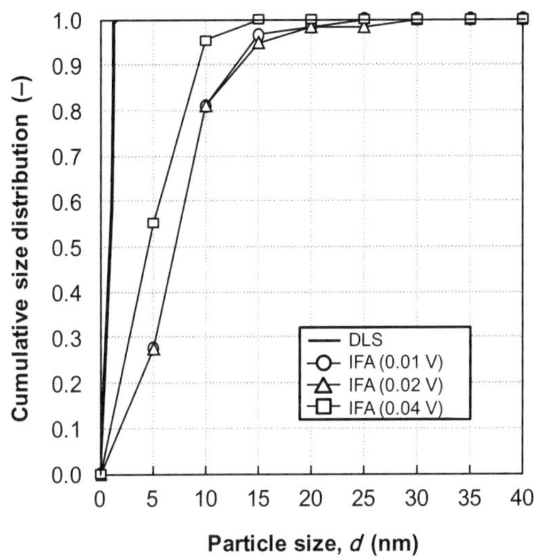

Figure 20.5. Particle size distribution of silica nanoparticles measured by interactive force apparatus under different supply voltages: (a) 0.01 V, (b) 0.02 V and (c) 0.04 V and measured by DLS [21].

REFERENCES

1 M. Elimelech, J. Gregory, X. Jia, R. A. Williams. *Particle deposition and aggregation*. Butterworth-Heinemann, Woburn, MA, 1998, pp. 4–8.

2 T. Allen. Particle size measurement, in T. Allen (ed.), *Powder sampling and particle size measurement*, vol. 1, 5th ed. Chapman and Hall, London, 1997, pp. 44–111.

3 A. N. Nguyen, P. George, G. J. Jameson. Determination of a minimum in the recovery of nanoparticles by flotation: theory and experiment. *Chem. Eng. Sci.* 61 (2006) 2494–2509.

4 S. Ata, P. D. Yates. Stability and flotation behaviour of silica in the presence of a non-polar oil and cationic surfactant. *Colloid. Surf. A* 277 (2006) 1–7.

5 E. Kusaka, H. Tamai, Y. Nakahiro, T. Wakamatsu. Role of surface free energy in a solid surface during collectorless liquid-liquid extraction. *Miner. Eng.* 6 (1993) 455–464.

6 T. K. Mitchell, A. N. Nguyen, G. M. Evans. Heterocoagulation of chalcopyrite and pyrite minerals in flotation separation. *Adv. Colloid Interface Sci.* 114–115 (2005) 227–237.

7 H. J. Butt, B. Cappella, M. Kappl. Force measurements with the atomic force microscope: technique, interpretation and applications. *Surf. Sci. Rep.* 59 (2005) 1–152.

8 J. Drelich, J. Long, Z. Xu, J. Masliyah, C. L. White. Probing colloidal forces between a Si_3N_4 AFM tip and single nanoparticles of silica and alumina. *J. Colloid Interface Sci.* 303 (2006) 627–638.

9 R. H. Yoon, D. H. Flinn, Y. I. Rabinovich. Hydrophobic interactions between dissimilar surfaces. *J. Colloid Interface Sci.* 185 (1997) 363–370.

10 P. D. Yates, G. V. Franks, S. Biggs, G. J. Jameson. Heteroaggregation with nanoparticles: effect of particle size ratio on optimum particle dose. *Colloid. Surf. A* 255 (2005) 85–90.

11 J. Laskowski, J. Iskra. Role of capillary effects in bubble-particle collision in flotation. *Trans. Inst. Min. Metall.* 79 (1970) C6–C10.

12 R. Stratton-Crawly. *Beneficiation of mineral fines—Problems and research needs.* AIME, New York, 1979, pp. 317–330.

13 H. Tamai, T. Hakozaki, T. Suzawa. Deposition of polymethyl methacrylate latex on fibers. *Colloid Polym. Sci.* 258 (1980) 189–200.

14 D. L. Klass, T. W. Martinek. Electroviscous fluids. I. Rheological properties. *J. Appl. Phys.* 38 (1967) 67–74.

15 A. P. Gast, C. F. Zukoski. Electrorheological fluids as colloidal suspensions. *Adv. Colloid Int. Sci.* 30 (1989) 153–202.

16 Y. Nakajima, T. Saito. Calculation of electrostatic force between two charged dielectric spheres by the re-expansion method. *J. Electrostat.* 45 (1999) 213–226.

17 Y. Nakajima, T. Matsuyama. Electrostatic field and force calculation for a chain of identical dielectric spheres aligned parallel to uniformly applied electric field. *J. Electrostat.* 55 (2002) 203–221.

18 A. Limsimarat, B. Techaumna. Dynamic simulation using a multipolar model of particles under dielectrophoretic force. *J. Electrostat.* 65 (2007) 672–679.

19 T. Miyazaki, A. Shibayama, T. Sato, T. Fujita. Measurement of interaction force between small distances sandwiched with magnetic fluid under magnetic field. *J. Magn. Magn. Mater.* 252 (2002) 256.

20 A. Shibayama, T. Otomo, K. Shimada, T. Fujita. Measurement of interactive surface force of suspended particles in ER and MR suspensions under electric and magnetic field. *Int. J. Mod. Phys. B* 19(7–9) (2005) 1177–1183.

21 A. Otsuki, G. Dodbiba, T. Fujita. Measurement of particle size distribution of silica nanoparticles by interactive force apparatus under an electric field. *Adv. Powder Tech.* 21(4) (2010) 419–423.

22 B. V. Derjaguin. Untersuchungen über die Reibung und Adhäsion, IV. *Kolloid Zeits.* 69 (1934) 155–164.

23 T. C. Jordan, M. T. Shaw. Electrorheology. *IEEE Trans. Electr. Insul.* 24 (1989) 849–878.

24 A. Otsuki, G. Dodbiba, T. Fujita. Effect of particle size distribution on formation of linear configuration of dielectric fine particles under the electric field. *J. Phys. Conf. Ser.* 147 (2009) 012003.

25 T. Fujita, T. Miyazaki, T. Tanaka, L. K. Jun, E. Kuzuno, H. Muto, A. Bekkala. Measurements of interaction forces at small distance under electric field in a smectite dispersed silicone oil based ER fluid. *Int. J. Modern Phys. B* 15 (2001) 811–816.

26 T. Miyazaki, A. Shibayama, T. Fujita. Measurement of surface force of fine particles at a short distance in a magnetic fluid under an electric and magnetic field. *J. Magn. Soc. Jpn.* 26 (2002) 153–156. (in Japanese).

27 A. Otsuki, G. Dodbiba, T. Fujita. Two-liquid flotation: heterocoagulation of fine particles in polar organic solvent. *Mater. Trans.* 48 (2007) 1095–1104.

21

ANALYSIS OF FUNCTIONAL GROUPS AT BURIED LIQUID/SOLID INTERFACES UTILIZING POLARIZATION MODULATION INFRARED EXTERNAL REFLECTION SPECTROSCOPY

HIROHARU YUI

Buried liquid/solid interfaces are ubiquitous in various scientific and technological fields and play crucial roles for catalytic reactions, molecular recognitions, and inducing characteristic physical/chemical properties that are not seen in bulk phases. For inducing interface-specific chemical reactions and properties, functional groups at the interfaces play inevitable roles. Thus, analysis on the structure and interactions of these functional groups at the liquid/solid interfaces is essential to understand the mechanisms of the reactions and the origins of the interface-specific properties.

However, at the liquid/solid interfaces, measurements and analysis become quite harder than those for bulk phases. This is because at the liquid/solid interfaces, access of the probing light is generally limited from the liquid phases, and abundant chemical species such as water molecules interfere with the measurement of the slight signal from the minute chemical species at the interfaces. Further, the region of the interface is generally limited to a few nanometer scales, and only surface-selective measurements are allowed to access such limited region. Thus, in spite of the importance of the analysis on

Electrical Phenomena at Interfaces and Biointerfaces: Fundamentals and Applications in Nano-, Bio-, and Environmental Sciences, First Edition. Edited by Hiroyuki Ohshima.

the functional groups at the interfaces, researches have been quite limited to the interfaces that have favorable properties for spectroscopic approaches.

The most successful spectroscopic approach to such interfaces has been vibrational sum frequency generation (VSFG) [1]. The spectroscopic approach in the vibrational energy region provides fruitful information on the structures and interactions of the functional groups. Furthermore, sum frequency generation (SFG) is the surface-specific nonlinear optical phenomena and it inherently meets the requirement of the surface selectivity. The first VSFG spectra of liquid/solid interfaces are on hydrophilic fused silica/water interfaces [2]. The same interface has been studied intensively and other interfaces such as alumina/water and CaF_2/water have been also investigated [3–8]. However, all of these solid materials are transparent and VSFG spectroscopic studies on opaque catalytic and electrode substrates have been quite limited (Cremer; TiO2, 106 Nihonyanagi Gold 29) [9, 10].

Representative catalytic surface that works at a liquid/solid interface is titanium dioxide (TiO_2). TiO_2 is the best-known and most widely applied photocatalyst [11]. UV irradiation onto the TiO_2 surface induces photocatalytic reactions and photoinduced hydrophilicity [11–13]. The photocatalytic reactions based on redox reactions have been widely studied as hydrogen and oxygen production from water molecules adsorbed on the surface with solar energy. Furthermore, the photocatalytic redox reactions have been applied to the photodecomposition of adsorbed organic contaminants [11, 14, 15]. On the other hand, photoinduced hydrophilicity is also a well-known phenomenon in which the hydrophilic nature of the TiO_2 surface increases remarkably after UV irradiation [11, 13]. The improvement of the reaction efficiency toward the application of this phenomenon to many fields has been actively studied [11, 16–21]. To improve the reaction efficiencies and the effective change of the hydrophilicity and to understand the characteristic physical chemistry of the TiO_2 surface, observation of the surface physical properties and the fundamental chemical processes that proceed at the functional groups' level is necessary.

Among the functional groups at the TiO_2 surface, the surface OH group on the TiO_2 surface has been assumed to play important roles in both photocatalytic reactions and photoinduced hydrophilicity [11, 22–28]. In photocatalytic reactions, the surface OH group and adsorbed water molecules react with photogenerated holes and transform into a hydroxyl radical (•OH) with strong oxidation power. This reaction is believed to proceed by the reaction of the •OH and organic species [22, 23]. Regarding the mechanism of photoinduced hydrophilicity, it has been proposed that UV irradiation leads to an increase in the number of the surface OH groups, which in turn increases the hydrophilicity of the TiO_2 surface [11, 13, 24–28]. Hence, spectroscopic measurements of the surface OH groups under the atmospheric or water-immersed conditions in which TiO_2 fulfills its functions are essential in order to understand the mechanism of the characteristic photochemical reactions and property at the TiO_2 surface.

However, selective measurement of the surface OH groups on the TiO_2 surface under atmospheric conditions has been quite difficult due to the interference of the signals by abundant water vapor and adsorbed water molecules. To overcome this difficulty, various measurement techniques have been applied to the surface [24–41]. For example, the adsorbed water layer was observed by infrared reflection absorption spectroscopy (IRRAS) and surface-enhanced infrared adsorption spectroscopy (SEIRAS) [27, 29]. Furthermore, the non-hydrogen-bonded OH groups at the air/adsorbed water layer on the TiO_2 interface were investigated by SFG spectroscopy [30]. However, efficient selectivity to observe the surface OH groups on the TiO_2 surface has not been achieved due to strong interference from abundant water vapor and adsorbed water layer under atmospheric conditions. For these reasons, neither the binding position nor the physical or chemical states of the surface OH groups have been clarified.

In this chapter, we introduce another approach that overcomes strong interference from water vapor and adsorbed water molecules. The approach is polarization modulation infrared external reflection (PM-IER) spectroscopy. In the PM-IER measurement, parallel (p) and perpendicular (s) polarizations to the plane of incident light were rapidly modulated, and their difference was measured in real time. The most important concept adopted here is "real-time cancellation" of the signals from bulk phase and "signal enhancement" based on the external reflection optical configuration. These techniques effectively cancel the signals from the three-dimensional isotropic bulk components, such as water vapor and bulk water adsorbed on the surface, and also enhanced the signal from the surface OH groups that orient to normal directions to the surface. In the following part, we will show the measurements of the rutile TiO_2 surfaces under atmospheric conditions by PM-IER. To clarify the binding position of the surface OH groups, we measured rutile single crystals with different plane directions. We also investigated the local changes in the structure and state of the surface OH groups under UV irradiation to obtain further information on the mechanism of the photoinduced hydrophilicity.

To obtain the difference of the reactions and properties depending on the plane directions, we investigated three different TiO_2 surfaces of rutile (110), (100), and (001) single crystals ($15.0 \times 15.0 \times 0.5$ mm^3, Shinkosha Co., Yokohama, Japan). Infrared external reflection (IER) and PM-IER spectra were acquired using an Fourier transform infrared (FT-IR) spectrometer (Nicolet 8700, Thermo Fisher Scientific, Yokohama, Japan) equipped with a Tabletop Optical Module Box (TOM-BOX). A photoelastic modulator (PEM) module was installed in the TOM-BOX. The infrared (IR) beam from the FT-IR spectrometer was focused onto a TiO_2 surface by a focusing mirror. Only the p-polarized component of the focused IR beam was passed through a wire-grid polarizer. The p-polarized IR beam was then modulated to s-polarizations by the PEM optical head (II/ZS50, 50 kHz, Hinds Instruments, Hillsboro, OR) with ZnSe crystal. Thus, p- and s-polarizations were modulated at 100 kHz. The polarization-modulated IR beam was irradiated onto the TiO_2 surface at an incident angle of 83° from the surfaces. The reflected IR beam was detected using a mercury–cadmium–telluride (MCT) detector (MCT-A*, Thermo Fisher Scientific) cooled with liquid nitrogen. All spectra were recorded with a resolution of 4 cm^{-1}. The IR beam in focus was 3.2 mm in diameter. PM-IER spectra have a convex upward baseline curve represented by the second-order Bessel function originating from the modulation efficiency of the PEM. The modulation efficiency of the PEM depends on the wavelength because the PEM generates p- and s-polarizations most effectively at a set wavelength [42–44]. The PEM maximum efficiency was set for half-wave retardation at 3300 and 1600 cm^{-1} for the analysis of the OH stretching and HOH bending modes, respectively. The PEM maximum efficiency was also set for half-wave retardation at 2900 cm^{-1} for the analysis of the CH stretching modes. All the PM-IER measurements were carried out under atmospheric conditions. IER and PM-IER spectra were acquired with 256 and 128 scans, respectively. The IER spectra were shown with the IER intensity against the p- and s-polarizations under equal intensities of p- and s-polarized irradiated IR beams. The surface wettability was evaluated by water contact angle measurements using a commercial contact angle meter (DM-500, Kyowa Interface Science Co., Ltd.). The droplet size was 2 μL. UV light was radiated onto the above-mentioned rutile TiO_2 single-crystal surfaces using a 100-W super-high-pressure Hg lamp (USH-102D, Ushio, Tokyo, Japan).

Since TiO_2 is nonmetallic, we at first have to confirm the surface selection rules on the TiO_2 single-crystal surfaces. We measured the IER signals for the p- and s-polarized beams and investigated the contribution of the p- and s-polarizations to the signals under a dehydrated condition to avoid interference from the strong vapor signal. The signals from the p- and s-polarizations gave positive absorbance at the (110) surface, whereas no signal was observed at the (001) surface, as shown in Figure 21.1. The IER for the s-polarized beam on the TiO_2 (110) single-crystal surface gave a slight positive signal, indicating

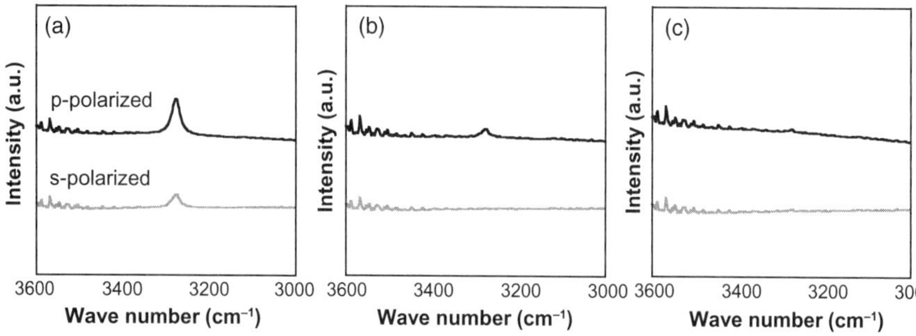

Figure 21.1. IER spectra of p-polarization (black line) and s-polarization (gray line) on the rutile TiO_2 (a) (110), (b) (100), and (c) (001) single-crystal surfaces under a dehydrated condition.

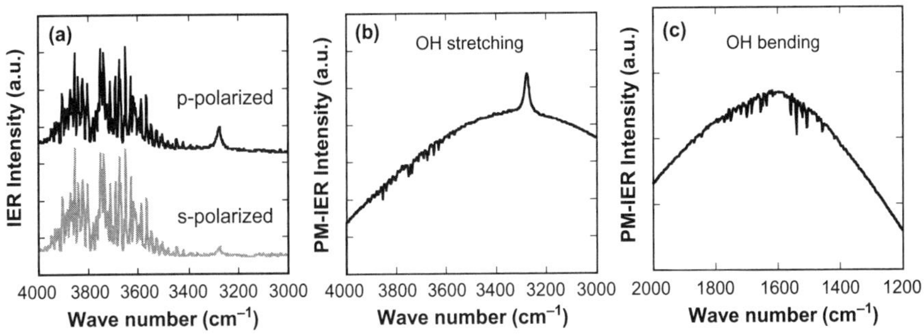

Figure 21.2. (a) IER spectra of p-polarization (black line) and s-polarization (gray line), (b) PM-IER spectrum in the OH stretching vibration region, and (c) PM-IER spectrum in the OH bending mode region of the rutile TiO_2 (110) single-crystal surface under atmospheric conditions.

that the surface selection rule of the TiO_2 (110) surface is similar to metallic surfaces and not nonmetallic ones, which give negative IER signals for an s-polarized beam. In the PM-IER measurements, the subtraction of the IER signals for the p- and s-polarized beams gave the signal intensity. Thus, the slight positive signal for an s-polarized beam would lead to an underestimation of the signals for a p-polarized beam but would not affect the confirmation of the existence of the surface OH groups oriented toward normal directions to the single-crystal surfaces.

Figure 21.2a,b allow a comparison of the IER spectra of p- and s-polarizations and the PM-IER spectrum in the OH stretching vibration region of the TiO_2 (110) single-crystal surface under atmospheric conditions. In the IER measurement, the signal from the water vapor dominated. In contrast, in the PM-IER measurement, a single peak was observed at 3279 cm^{-1}, and the contribution from the isotropic components was effectively eliminated. Candidates for the single peak that appeared above 3000 cm^{-1} are reduced to CH, NH, and OH stretching modes due to the relation between the oscillator frequency and their reduced masses of the composed atoms. However, carbon and nitrogen atoms are absent in the present TiO_2 and water system. Thus, the single peak is assigned to the OH stretching mode. Further, the absence of both symmetrical and asymmetric OH stretching modes and of the bending mode indicates that the OH stretching comes not from H_2O but from a single OH group. Further, no peak corresponding to the HOH bending at 1645 cm^{-1} appeared, as shown in Figure 21.2c. These results lead us to conclude that the single peak appearing at 3279 cm^{-1} is assigned to the OH stretching mode of the surface OH group but not to the water molecule adsorbed on the (110) single-crystal surface. The result of the PM-IER measurement does not preclude the existence of adsorbed water molecules on the surface, but it reveals that there are no aligned water molecules toward the normal direction to the

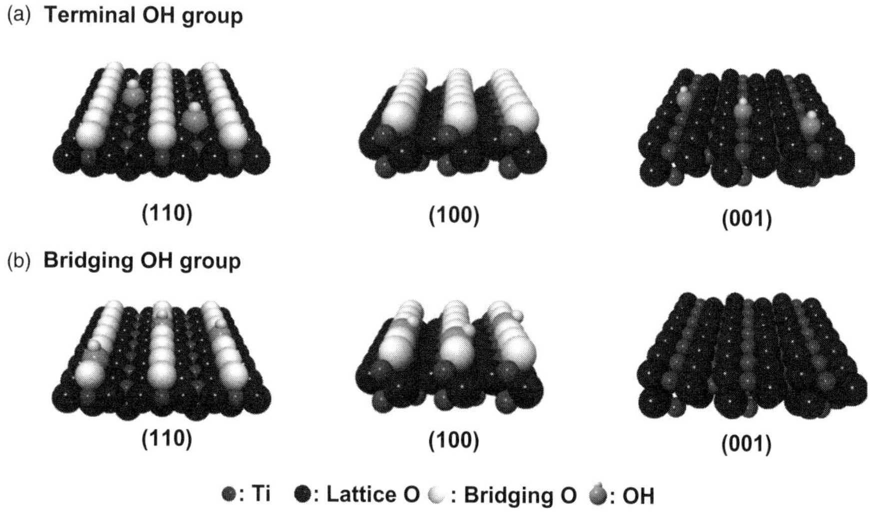

Figure 21.3. Schematic illustrations of the (a) terminal and (b) bridging OH groups on the rutile TiO_2 (110), (100), and (001) single-crystal surfaces.

surfaces. However, only from these results, it is still difficult to discriminate the binding position of the surface OH groups shown schematically in Figure 21.3. The (110) single-crystal surface allows for the binding position of both the terminal OH groups and the bridging OH groups due to its alternate rows of fivefold-coordinated titanium atoms and twofold-coordinated bridging oxygen atoms (Fig. 21.3a) [28, 32, 33, 45].

To elucidate the binding position of the surface OH groups, (100) and (001) single-crystal surfaces were also measured and compared. Figure 21.3 is an illustration of the expected binding position of the surface OH groups on each single crystal. The (100) surface makes the bridging OH groups possible because the bridging oxygen atoms binding the fivefold-coordinated titanium atoms are mainly exposed. The (110) and (100) surfaces have bridging oxygen atoms, while the (001) surface does not. Since the (001) surface has fourfold-coordinated titanium atoms and twofold-coordinated oxygen atoms, the (001) surface makes it possible to have the terminal OH groups. If the surface OH groups locate at the terminal position, the peak of the surface OH groups should appear in the (110) and (001) surfaces. Conversely, if the surface OH groups locate at the bridging position, the peak of the surface OH groups should appear in the PM-IER spectra of the (110) and (100) surfaces. Figure 21.4 shows a comparison of the PM-IER spectra from 3200 to 3350 cm^{-1} of the rutile TiO_2 (110), (100), and (001) single-crystal surfaces. The peak at 3279 cm^{-1} is on the (100) and (110) surfaces, whereas there is no peak on the (001) surface. These results lead us to conclude that the surface OH groups locate at the bridging oxygen sites on the TiO_2 single-crystal surfaces. In addi-

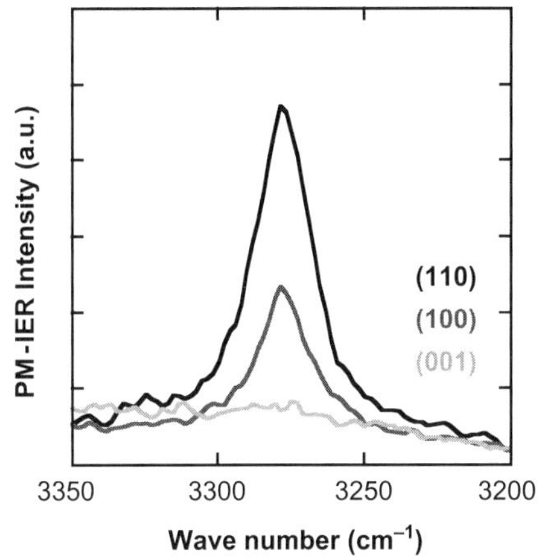

Figure 21.4. PM-IER spectra in the 3200–3350 cm^{-1} region of the rutile TiO_2 (110), (100), and (001) single-crystal surfaces.

tion, the lack of the peak on the (001) surface without the bridging oxygen groups further confirms that the peak at 3279 cm^{-1} is assignable to the surface OH groups. The absence of differences in the frequencies of the peak (3279 cm^{-1}) for the (100) and (110) surfaces, indicating similar physical and chemical environments of the surface OH groups for both surfaces, resulted in no remarkable differences in the interactions with the adsorbed water and/or vapor in the atmosphere.

In high-resolution electron energy-loss spectroscopy (HREELS) and IR studies under ultrahigh vacuum (UHV), OH stretching of the surface OH groups was observed at around 3690 cm^{-1} [37, 39–41]. On the other hand, OH stretching of the bridging OH groups in the atmospheric conditions appeared at 3279 cm^{-1}. It is known that hydrogen bonds between the OH groups decrease the frequency shift. Under UHV, most of the adsorbed water layer on the TiO_2 surface should disappear, and the surface OH groups should have little interaction with other OH groups of the absorbed water molecules. In contrast, under atmospheric conditions, the bridging OH groups should form hydrogen bonds with the neighboring OH groups of the absorbed water molecules. The drastic downshift of 411 cm^{-1} of the OH stretching frequency suggests that a hydrogen-bonding network is developed between the bridging OH group and an adsorbed water molecule on the surface (Fig. 21.5).

Then, the bridging OH groups under UV irradiation were investigated to elucidate the relationship between the photoinduced hydrophilicity and the surface OH groups under UV irradiation. This is because the number of surface OH groups has been believed to increase with UV irradiation and to induce photoinduced hydrophilicity. For example, Hashimoto et al. proposed the following mechanism for highly hydrophilic conversion with

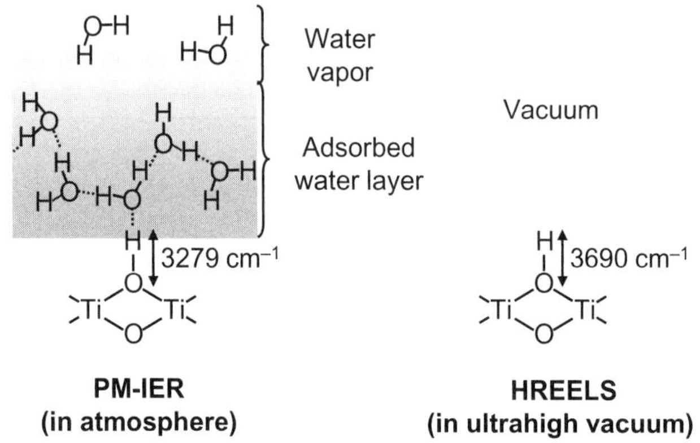

Figure 21.5. Schematic illustrations of the states of the bridging OH groups and corresponding wave numbers of their stretching vibrational modes measured in atmospheric conditions (PM-IER) and in ultrahigh vacuum (HREELS). Hydrogen-bonding network environments are completely different for each condition.

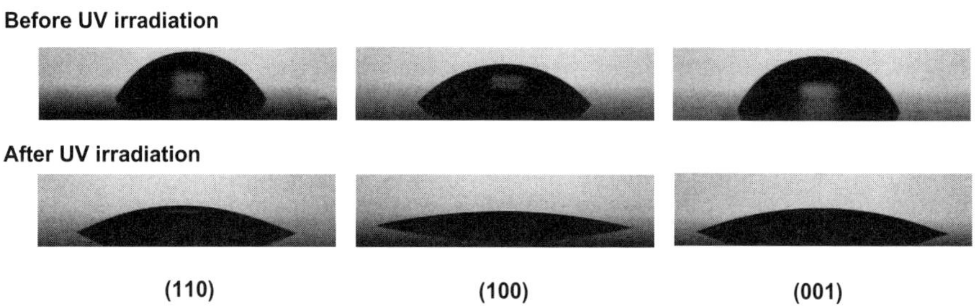

Figure 21.6. Pictures of water droplets by contact angle measurements showing the change in the water wettability of the (110), (100), and (001) rutile TiO_2 single-crystal surfaces. Pictures in the upper line and those in the lower one show the water droplets on the surfaces before and after UV irradiation for 15 minutes, respectively.

UV illumination: The photogenerated holes weaken the bonds between titanium and bridging oxygen atoms, breaking the bonds by the dissociated adsorption of water molecules. As a result, the number of the surface OH groups increases, leading to a highly hydrophilic conversion on a TiO_2 surface [11, 24, 25, 27, 28].

Figure 21.6 shows pictures of the change in wettability on the TiO_2 single-crystal surfaces before and after UV irradiation. The contact angle of water on the TiO_2 single-crystal surfaces decreased approximately 30° after 15 minutes' UV irradiation. Interestingly, the hydrophilicity of the (001) surface without the surface OH groups was also enhanced.

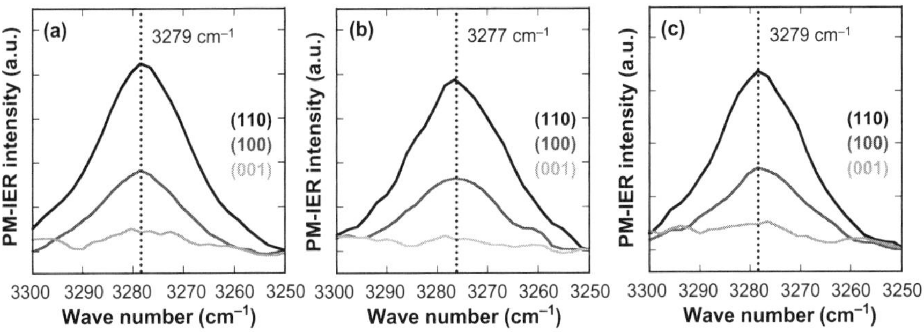

Figure 21.7. PM-IER spectra of the rutile TiO_2 (110), (100), and (001) single-crystal surfaces: (a) before UV irradiation, (b) under 15 minutes' UV irradiation, and (c) 15 minutes after UV irradiation.

To examine the relationship between the highly hydrophilic conversion and the change in the number of the bridging OH groups, the (110), (100), and (001) surfaces before, during, and after UV irradiation were measured by PM-IER (Fig. 21.7). With UV irradiation, the peak frequency of the bridging OH groups shifted from 2 to 3277 cm^{-1}. After UV irradiation was stopped, the shifted peak reverted to the original wave number, 3279 cm^{-1}. These reproducible wave number shifts were also observed on the (100) surface. No remarkable change in the spectrum was observed on the (001) surface, where there was no surface OH group. The wave number shifts were observed only under UV irradiation.

In contrast, the peak intensity and width of the surface OH groups hardly changed. The results of the water contact angle and PM-IER measurements indicated that UV irradiation did not induce a change in the number of surface OH groups, demonstrating that another mechanism, which does not depend on the increase in the number of the surface OH groups, should induce photoinduced hydrophilicity. Recently, Jribi et al. reported that UV irradiation of TiO_2 surface does not produce the surface OH groups by ambient-pressure X-ray photoelectron spectroscopy (APXPS) and second harmonic generation (SHG) [46]. This result is consistent with our PM-IER measurements.

The removal of the adsorbed hydrocarbons by UV irradiation is one possible mechanism of the inducement of hydrophilicity [47, 48]. To examine this possibility, we investigated the relationship between the photocatalytic removal of the adsorbed organic contaminants and the photoinduced highly hydrophilic conversion. However, the variations of the adsorbed hydrocarbon in the atmosphere were almost negligible after 15-minute UV irradiation, as shown in Figure 21.8. The hydrophilic nature of the TiO_2 (110) surface was increased by UV irradiation, while the adsorbed organic contaminants remained on the surface, demonstrating that the removal of the adsorbed organic contaminants was not the main factor in the photoinduced hydrophilicity observed in our

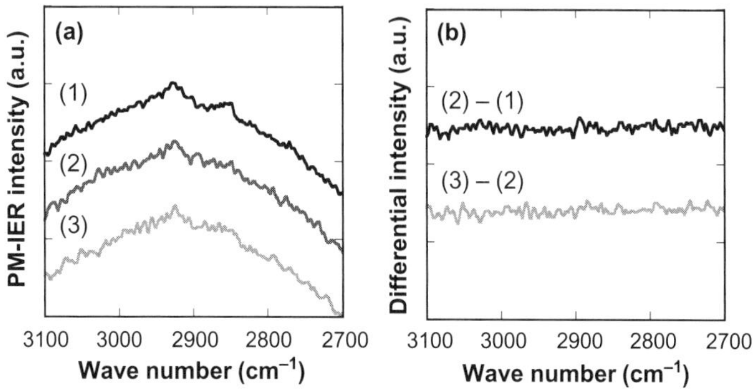

Figure 21.8. (a) PM-IER spectra in the CH stretching vibration region of the rutile TiO_2 (110) single-crystal surface: before UV irradiation (black line), under 15 minutes' UV irradiation (dark gray line), and immediately after UV irradiation (light gray line). (b) Difference spectra between before and under UV irradiation (black line) and before and after UV irradiation (gray line).

cases. Although the precise mechanism of photoinduced hydrophilicity remains unknown, we suggest that photogenerated holes and electrons produce the intermediates, such as hydrogen peroxide (H_2O_2) and/or oxonium ions (H_3O^+) [23]. These intermediates generated at the surfaces may enhance the hydrogen bonds between the adsorbed water molecules, leading to an increase in the amount of adsorbed water and the enhancement of wettability on the TiO_2 single-crystal surfaces [26, 29, 30].

In this chapter, we showed the usefulness of the PM-IER technique to investigate buried liquid/solid interfaces. As such interfaces, we examined TiO_2 surfaces with adsorbed water under atmospheric conditions. The PM-IER technique enabled us to selectively observe the surface OH groups on rutile TiO_2 single crystals covered with adsorbed water and water vapor due to the effective cancellation of their signals [49]. The peak was observed on both (110) and (100) surfaces with the bridging oxygen atoms, whereas no peak was observed on the (001) surface without the bridging oxygen atoms. These results revealed that the surface OH groups are the bridging OH groups located at bridging oxygen sites. In addition, the OH stretching of the bridging OH groups in the atmosphere was observed at a remarkably lower wave number (3279 cm^{-1}) than that of non-hydrogen-bonded OH groups under UHV (3690 cm^{-1}), indicating that the bridging OH groups in the atmosphere formed hydrogen-bonding networks with adsorbed water molecules. The peak of the surface OH groups on the (110) and (100) surfaces shifted to a lower wave number with UV irradiation, but the peak intensity of the bridging OH groups did not change remarkably. Previous explanations for the high hydrophilicity have been based on an increase in the number of the surface OH groups with UV irradiation, but our experimental results show that another mechanism independent of the number of surface OH groups should be taken into account to explain the photoinduced hydrophilicity of TiO_2 surfaces. To clarify the mechanism, we might have to detect transient chemical species generated at the buried interface. Combining the PM technique and time-resolved measurements will be the next challenge to provide further insight on the chemical reactions and curious properties at buried liquid/solid interfaces.

REFERENCES

1 Y. R. Shen, V. Ostroverkhov. *Chem. Rev.* 106 (2006) 1140.
2 Q. Du, E. Freysz, Y. R. Shen. *Phys. Rev. Lett.* 72 (1994) 238.
3 S. Nihonyanagi, S. Ye, K. Uosaki. *Electrochim. Acta* 46 (2001) 3057.
4 J. Kim, G. Kim, P. S. Cremer. *J. Am. Chem. Soc.* 124 (2002) 8751.
5 M. C. Gurau, G. Kim, S. M. Kim, F. Albertorio, H. C. Fleisher, P. S. Cremer. *ChemPhysChem* 4 (2003) 1231.
6 I. Li, J. Bandara, M. J. Shultz. *Langmuir* 20 (2004) 10474.
7 M. S. Yeganeh, S. M. Dougal, H. S. Pink. *Phys. Rev. Lett.* 83 (1999) 1179.
8 K. A. Becraft, F. G. Moore, G. L. Richimond. *Phys. Chem. Chem. Phys.* 6 (2004) 1880.
9 S. Kataoka, M. C. Gurau, F. Arbertorio, M. A. Holden, S. M. Lim, R. D. Yang, P. S. Cremer. *Langmuir* 20 (2004) 1662.
10 S. Nihonyanagi, Y. Shen, K. Uosaki, L. Dreesen, C. Humbert, P. Thiry, A. Peremans. *Surf. Sci.* 573 (2004) 11.
11 K. Hashimoto, H. Irie, A. Fujishima. *Jpn. J. Appl. Phys.* 44 (2005) 8269.
12 A. Fujishima, K. Honda. *Nature* 238 (1972) 37.
13 R. Wang, K. Hashimoto, A. Fujishima, M. Chikuni, E. Kojima, A. Kitamura, M. Shimohigoshi, T. Watanabe. *Nature* 388 (1997) 431.
14 M. A. Fox, M. T. Dulay. *Chem. Rev.* 93 (1993) 341.
15 U. Diebold. *Surf. Sci. Rep.* 48 (2003) 53.
16 R. Blossey. *Nat. Mater.* 2 (2003) 301.
17 D. Luca, D. Mardare, F. Iacomi, C. M. Teodorescu. *Appl. Surf. Sci.* 252 (2006) 6122.
18 A. Hattori, T. Kawahara, T. Uemoto, F. Suzuki, H. Tada, S. Ito. *J. Colloid Interface Sci.* 232 (2000) 410.
19 V. Pore, M. Heikkilae, M. Ritala, M. Leskelae, S. Areva. *J. Photochem. Photobiol. A* 177 (2006) 68.
20 N. Laugel, E. Gonthier, P. Schaaf, J. Hemmerle, N. Ladhari, Y. Arntz, Y. Haikel, J. Voegel, V. Ball. *J. Colloid Interface Sci.* 324 (2008) 127.
21 Z. Liu, X. Zhang, T. Murakami, A. Fujishima. *Sol. Energy Mater. Sol. Cells* 92 (2008) 1434.
22 P. F. Schwarz, N. J. Turro, S. H. Bossmann, A. M. Braun, A.-M. A. A. Wahab, H. Duerr. *J. Phys. Chem. B* 101 (1997) 7127.
23 C. D. Jaeger, A. J. Bard. *J. Phys. Chem.* 83 (1979) 3146.
24 N. Sakai, A. Fujishima, T. Watanabe, K. Hashimoto. *J. Phys. Chem. B* 105 (2001) 3023.

25 N. Sakai, A. Fujishima, T. Watanabe, K. Hashimoto. *J. Phys. Chem. B* 107 (2003) 1028.

26 A. Y. Nosaka, Y. Nosaka, E. Kojima, T. Fujiwara, H. Yagi, H. Akutsu. *J. Phys. Chem. B* 107 (2003) 12042.

27 R. Wang, K. Hashimoto, A. Fujishima, M. Chikuni, E. Kojima, A. Kitamura, M. Shimohigoshi, T. Watanabe. *Adv. Mater.* 10 (1998) 135.

28 R. Wang, N. Sakai, A. Fujishima, T. Watanabe, K. Hashimoto. *J. Phys. Chem. B* 103 (1999) 2188.

29 R. Nakamura, K. Ueda, S. Sato. *Langmuir* 17 (2001) 2298.

30 K. Uosaki, T. Yano, S. Nihonyanagi. *J. Phys. Chem. B* 108 (2004) 19086.

31 S. H. Szczepankiewicz, A. J. Colussi, M. R. Hoffmann. *J. Phys. Chem. B* 104 (2000) 9842.

32 O. Bikondoa, C. A. Muryn, C. L. Pang, H. Onishi, R. Ithnin, G. Thornton. *Nat. Mater.* 5 (2006) 189.

33 C. L. Pang, A. Sasahara, H. Onishi, Q. Chen, G. Thornton. *Phys. Rev. B* 74 (2006) 73411.

34 D. A. Panayotov, J. T. Yates. *Chem. Phys. Lett.* 410 (2005) 11.

35 M. Takeuchi, M. Anpo, G. Martra, S. Coluccia. *J. Phys. Chem. C* 111 (2007) 9811.

36 M. B. Hugenschmidt, L. Gamble, C. T. Campbell. *Surf. Sci.* 302 (1994) 329.

37 M. A. Henderson. *Surf. Sci.* 355 (1996) 151.

38 M. A. Henderson. *Langmuir* 12 (1996) 5093.

39 M. Primet, P. Pichat, M. V. Mathieu. *J. Phys. Chem.* 75 (1971) 1216.

40 D. M. Griffiths, C. H. Rochester. *J. Chem. Soc. Faraday Trans.* 1(73) (1977) 1510.

41 Y. Suda, T. Morimoto. *Langmuir* 3 (1987) 786.

42 J. C. Kemp. *J. Opt. Soc. Am.* 59 (1969) 950.

43 K. W. Hipps, G. A. Crosby. *J. Phys. Chem.* 83 (1979) 555.

44 B. J. Barner, M. J. Green, E. I. Saez, R. M. Corn. *Anal. Chem.* 63 (1991) 55.

45 M. Ramamoorthy, D. Vanderbilt, R. D. King-Smith. *Phys. Rev. B* 49 (1994) 16721.

46 R. Jribi, E. Barthel, E. Sondergard, H. Bluhm, M. Grunze, P. Koelsch, D. Verreault. *J. Phys. Chem. C* 113 (2009) 8273.

47 T. Zubkov, D. Stahl, T. L. Thompson, D. Panayotov, O. Diwald, J. T. Yates Jr. *J. Phys. Chem. B* 109 (2005) 15454.

48 T. L. Thompson, J. T. Yates Jr. *Chem. Rev.* 106 (2006) 4428.

49 K. Takahashi, H. Yui. *J. Phys. Chem.* 113 (2009) 20322.

22

FABRICATION OF LIQUID CRYSTAL DISPLAYS CONTAINING CAPPED NANOPARTICLES AND THEIR ELECTRO-OPTIC PROPERTIES

YUKIHIDE SHIRAISHI, NAOTO NISHIDA, AND NAOKI TOSHIMA

22.1 INTRODUCTION

Metallic nanoparticles have a long history [1–3]. The first scientific report appeared in 1857 by M. Faraday, who prepared gold nanoparticles by the reduction of tetrachloroauric(III) acid by white phosphorus dissolved in carbon disulfide. Gold nanoparticles, prepared conveniently by citrate reduction, can be considered to be dispersed without any stabilizer, but, in practice, citrate ions are adsorbed on the surface of gold colloidal particles and are stabilized by an electrostatic repulsive force. The dispersion is stable, if it is skillfully prepared, but can easily form precipitates without stabilizer by the addition of small amounts of salts. Faraday already found that the addition of gelatin to gold nanoparticles protected them from the action of salts.

The addition of water-soluble natural or synthetic polymers [4, 5], surfactants [6], and organic ligands [7, 8] can stabilize the colloidal dispersion of metals as well. These stabilizers are called "protective colloids." In this case, the protective colloid can be absorbed on the surface of colloidal metal particles and steric repulsion between the protective colloids stabilizes the colloidal dispersions. Recently, much attention has been paid to stabilizer-capped

Electrical Phenomena at Interfaces and Biointerfaces: Fundamentals and Applications in Nano-, Bio-, and Environmental Sciences, First Edition. Edited by Hiroyuki Ohshima.

metallic nanoparticles as nanoparticles based on rapid improvement of analytical technology on fine materials. Such nanoparticles, defined by a diameter of 1–10 nm and a narrow size distribution, are creating a new category of materials, which is different from either conventional bulk materials or atoms, the smallest units of matter. Much attention has been paid to their special properties in physical, biological, and chemical fields. Quantum size effect [9] is one of the greatest problems for physicists. Gold nanoparticles, for example, are interesting as materials of nonlinear optics for third harmonic generation. In the biological field, colloidal gold has been used for labeling the specific position of cells for observation by transmission electron micrograph (TEM). Silver nanoparticles are very popular for the detection of specific proteins in electrophoresis.

The most intensive investigations of applications of metallic nanoparticles are focused on catalyses. Many reports have been published on the application to catalysts [10]. Some applications to industrial processes are known. The catalytic properties of metallic nanoparticles mainly depend on the particle size, the kind of metal, as well as the type of protective colloid. On the other hand, the properties of a practical catalyst can be improved by mixing it with other elements. The additive effect of the second component in metal catalysts is very important not only for industries but also for academia. From this viewpoint, bimetallic nanoparticles have been extensively investigated to improve their preparation procedures as well as their activity and selectivity as catalysts [2, 11–14].

On the other hand, liquid crystal molecules themselves have been investigated as raw materials for an electronic display device for more than three decades due to its electro-optic properties and now construct main parts of information industries. Liquid crystal displays, however, have a disadvantage of slow response compared to electroluminescence displays. Thus, it will be a big impact to design a liquid crystal display with fast response, if liquid crystal sol containing nanoparticles may have novel dynamic properties different from the original liquid crystal medium by giving a perturbation to the self-assemble property of liquid crystals (Fig. 22.1).

The merging of metallic nanoparticles or nanotechnology in a wide sense into self-assembled systems such as liquid crystal devices (LCDs) may attract the attention of researchers who are interested in inaugurating a new kind of combination of different fields [15]. In fact, the number of documents, especially patents on a liquid crystal display concerning nanoparticles, increases rapidly these years. The nanoparticles reported as a dispersed phase in LCDs involve fullerene [16], carbon nanotubes [17, 18], diamond powders [19], MgO or SiO_2 nanoparticles [20], noble metallic nanoparticles [21–25], semiconductor nanoparticles [26], Au nanorods [27, 28], and so on. When used as a dopant for LCDs, they were expected to improve the contrast, decrease the driving voltage, capture ions, and shorten the response time. In this chapter, we would like to describe a new liquid crystal sol composed of self-assembled liquid crystal media and metallic nanoparticles as a dispersion medium and a dis-

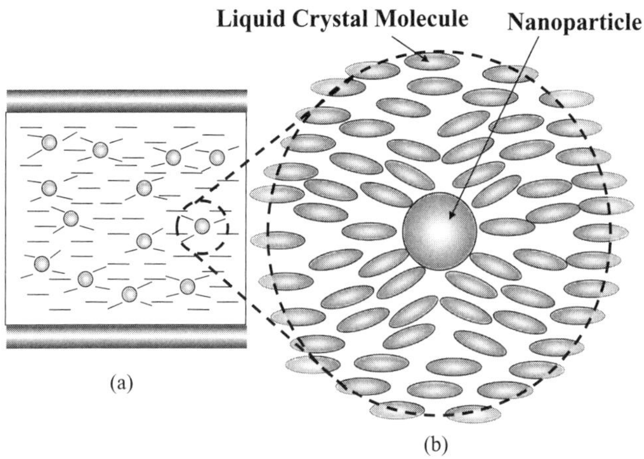

Figure 22.1. Schematic illustration of an LCD cell filling by liquid crystal sol containing nanoparticles. A self-assemble property of liquid crystal is speculatively perturbed by containing nanoparticles in a liquid crystal medium. Liquid crystal molecules and nanoparticles are presented by bars (a) or ellipses (b) and circles, respectively (reprinted from Reference 49, with permission from the American Chemical Society).

persed phase, respectively. The miscibility of two materials in completely different categories is the most important criterion to develop such a liquid crystal sol.

In this chapter, we would like to present a preparation procedure and characteristic aspects of monometallic and bimetallic nanoparticles, and the electro-optic properties of an LCD doped with metallic nanoparticles.

22.2 MONOMETALLIC NANOPARTICLES

22.2.1 Preparation of Monometallic Nanoparticles

Metallic nanoparticles can be prepared by two distinct categories, that is, by degradation of bulk metals (a physical method) and by growth of particles starting from metal atoms, which are obtained from molecular or ionic precursors (a chemical method). The size of the prepared particles by the former method varies widely, primarily from 1 nm to several micrometers, or wider in some cases. In contrast, the latter method is based on the formation of metal atoms by the reduction of metal ions and the subsequent aggregation of the atoms to produce metallic nanoparticles or fine metal particles. Thus, the particle size can be controlled by preparation conditions.

Many kinds of reductants can be used in order to reduce metal ions to produce metallic nanoparticles. Methanol, ethanol, formaldehyde, hydrazine, sodium citrate, hydrogen, white phosphorus, diborane, and sodium tetrahydroborate are examples of such chemical reductants. Radiation energies such as visible light, ultraviolet (UV) light, γ-ray, and ultrasonic waves as well as electrical energy can also be used to cleave metal–ligand bonds to produce colloidal metals via metal atoms.

An alcohol reduction method has the following advantages: (i) The procedure is very simple and reproducible and (ii) the particle size of the obtained metallic nanoparticles is small with a narrow size distribution. The typical process to prepare polymer-capped metallic nanoparticles is as follows: The solution of noble metal salts or complexes in ethanol/water (1/1, v/v) are treated by heat until refluxing for a few hours in the presence of poly(*N*-vinyl-2-pyrrolidone) (PVP) [29]. Ethanol can be replaced with other primary or secondary alcohols, and the alcohol content is changeable. The concentration of noble metals is changeable as well, although the concentration of 0.66 mmol/dm^3 is often used. The degree of polymerization of PVP is also changeable. The K-30 grade, whose degree of polymerization is ca. 360, is often used. The molar ratio of PVP in monomer unit to metal (R) is changeable as well, but is usually 40. If R is less than 1, it is difficult to obtain a stable dispersion.

The same solution of noble metal salts or complexes in ethanol/water as that used for the alcohol reduction can be used for photoreduction [30]. Irradiation of the solution in a Pyrex Schlenck tube with visible light for a few hours usually provides a colloidal dispersion similar to that of alcohol reduction. Any kind of light source is applicable. The photoreduction method has an advantage in preparing metallic nanoparticles even in pure water. When the solution of metal salts in pure water without alcohol is irradiated with visible light, the reduction may take place with more difficulty than in the solution containing alcohol, requiring sometimes continuous irradiation for days. Irradiation with UV light in a quartz Schlenck tube may be required instead of visible light.

22.2.2 Aspect of Monometallic Nanoparticles

The most immediately observable property of metallic nanoparticles is their color. Metallic nanoparticles usually have a transparent dark brown color except gold (red) and silver (yellow), and can keep the colloidal state at room temperature for a long period.

Size and size distribution of metallic nanoparticles can be determined by TEMs. The size obtained by TEM is that of central metal parts. An average particle size of central metal parts is usually 1–8 nm depending on the kind of metal and preparation conditions. For example, stabilizer-capped rhodium nanoparticles have an average diameter of ca. 2.2 and 4.0 nm when prepared by reduction in ethanol/water and methanol/water, respectively.

The whole size involving the envelope of protecting nanoparticles can be measured by scattering tunneling micrographs (STMs) [31] or dispersion rate measurements by the Taylor method [32]. Although the former method gives the whole size of particles not only with central metals but also without metals, the latter method gives the whole size of the particles with central metals by the calculation of the Stokes radii by monitoring the central metals with visible light absorption. Both methods are used for the case of amphiphilic molecules used as the protective colloid. In the case of polymer-capped metallic nanoparticles, small-angle X-ray scattering analysis was carried out and a formation of hierarchy of self-organization is clarified [33]. The hierarchy starts from the smallest "elementary nanoparticles" and is built up to from "fundamental nanoparticles," "superstructure," and "higher-order organization of superstructures."

22.2.3 Electro-optic Properties of LCDDoped with Monometallic Nanoparticles

An investigation was reported for the improvement of a contrast ratio of liquid crystal matrices by the addition of fullerene having a nanoscopic structure [16]. However, this system has a disadvantage of poor miscibility of fullerene molecules in the liquid crystal medium. Metallic nanoparticles stabilized by organic molecules are now creating a new class of materials different from either conventional bulk materials or atoms. 4-Cyano-4′-pentylbiphenyl (5CB)-capped palladium (5CB-Pd) nanoparticles with an average diameter of 1.9 nm were prepared by the photoreduction of a tetrahydrofuran solution of palladium(II) acetate in the presence of 5CB [34]. A guest–host LCD was fabricated by doping liquid crystal host matrices with 5CB-Pd nanoparticles and a dichroic dye. 5CB-Pd nanoparticles have higher solubility than fullerene into the liquid crystal medium (Fig. 22.2). An electro-optic property in the presence of 5CB-Pd nanoparticles is much better than that in its absence (Fig. 22.3).

A twisted nematic (TN)-LCD mode [35] has been widely utilized since 1971 in practical LCDs [36]. Recently, the merging of nanomaterials [37, 38] and nanotechnology into electro-optic device technology such as LCDs may attract the interests of researchers who are interested in inaugurating a new kind of combination of different fields. In this research field, there are two main approaches: One is to modify a surface of alignment layers by nanomaterials for orienting liquid crystal molecules, and the other is to modify a host liquid crystal matrix itself by doping with particles, for example, fullerene or hydrophobic silica particles [39, 40]. We call the nematic liquid crystal (NLC) molecule for the host medium, NLC-1, so as to construct a TN structure, whereas the other NLC molecule used for covering Pd nanoparticles is called NLC-2. As an example, the NLC-1 used in this research was K-15 (Merck, Tokyo, Japan) with positive dielectric anisotropy ($\Delta\varepsilon = 20.1$), while the NLC used as NLC-2 was CCN-47 (Merck) with negative dielectric anisotropy ($\Delta\varepsilon = -8.4$). Both NLC molecules have a cyano group. The colloidal dispersions of Pd nanoparticles were prepared by UV irradiation of an ethanol solution of

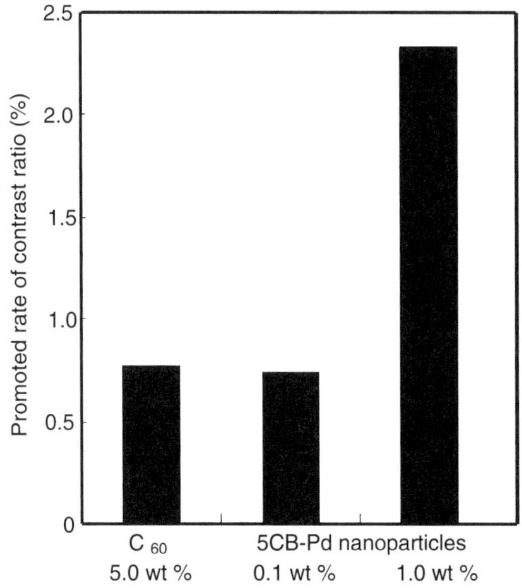

Figure 22.2. Promoted rate of contrast ratio of the guest–host mode liquid crystal cell doped with C_{60} and 5CB-Pd nanoparticles (reprinted from Reference 34, with permission from The Society of Polymer Science, Japan).

(a) C_{60} (b) 5CB-Pd

Figure 22.3. Photographs of the guest–host mode liquid crystal cell doped with (a) C_{60} and (b) 5CB-Pd nanoparticles (reprinted from Reference 34, with permission from The Society of Polymer Science, Japan).

palladium(II) acetate in the presence of NLC-2 with the molar ratio of NLC-2 to Pd of 50. The size of Pd nanoparticles was measured with a transmission electron microscopy to have an average diameter of 2.5 nm and a standard deviation of 1.0 nm. The TN-LCDs were constructed by using NLC-1 and Pd nanoparticles capped by NLC-2. The concentrations of the NLC-2-capped Pd

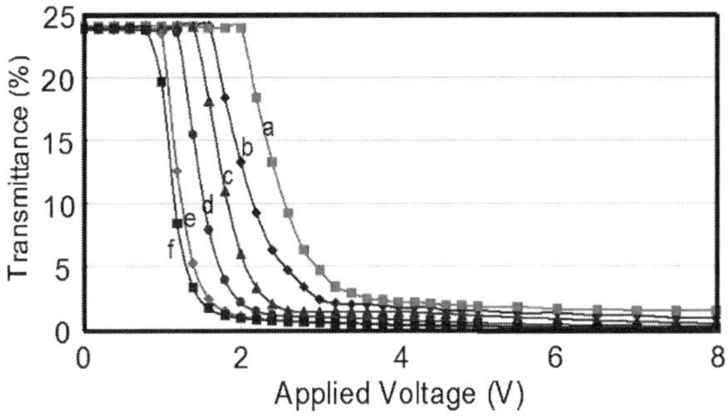

Figure 22.4. Transmittance versus applied voltages, taking their frequencies as a parameter in the FM-TN-LCD (Pd: 1 wt %). (a) 10, (b) 15, (c) 20, (d) 30, (e) 60, and (f) 100 Hz (reprinted from Reference 21, with permission from the American Institute of Physics).

nanoparticles were 1, 5, and 10 wt %. Figure 22.4 shows an example of the optical transmittance versus alternative square voltage waveform applied to a sample cell that contains NLC-2-capped Pd nanoparticles in 1 wt %. In this experiment, the frequencies of the applied alternative square voltage waveforms, whose amplitude is the same as a root mean square value, are taken as a parameter, from 10 to 100 Hz. The normally white mode is adopted to know the effect of the doped Pd nanoparticles on the optical transmission of our TN device at the quiescent condition. For the direct current applied voltage (frequency = zero), the device cell does not show any electro-optic response. As the frequency increases, an electro-optic response starts to occur and the electro-optic characteristics shift toward the lower operation voltage. At last, they tend to be saturated in the range above 100 Hz. The electro-optic characteristics at above 100 Hz are almost exactly the same as that of a pure TN device cell without having doped Pd nanoparticles. In this way, the TN-LCDs fabricated in this research shows a frequency modulation (FM) response. For this reason, this device may be called a frequency-modulated TN-LCD (FM-TN-LCD). The frequency dependence of the threshold voltage shown in Figure 22.4 reflects the frequency dependence of the dielectric anisotropy of the host NLC since the threshold voltage (V_{th}) is given by

$$V_{th} = (\pi/d)\{K_{11} + 1/4(K_{33} - 2K_{22})/\varepsilon_0\Delta\varepsilon\}^{1/2}, \tag{22.1}$$

where K_{ii}, d, and ε_0 are Ossen–Frank elastic constants, a thickness of host NLC, and a dielectric constant in a vacuum state, respectively [41].

The electro-optic response could also be enhanced by addition of metallic nanoparticles, especially Ag nanoparticles (Fig. 22.5). Figure 22.6 demonstrate for the first time examples of FM switching, where the frequencies are switched between 20 and 500 Hz for a sinusoidal voltage waveform and an AC square

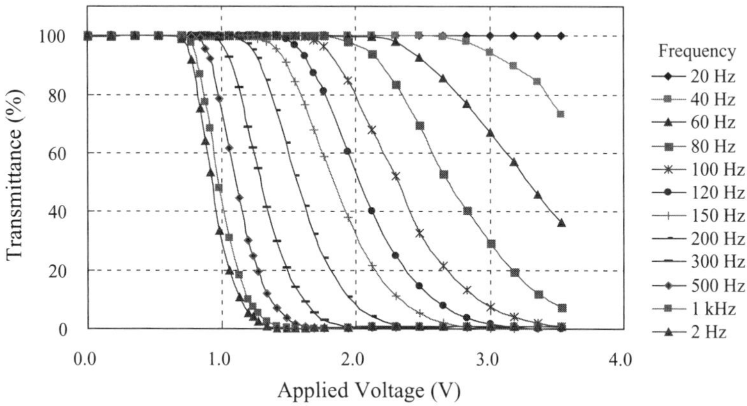

Figure 22.5. Relative optical transmittance versus sinusoidal applied voltage (V–T curves), where the frequencies are taken as a parameter. Sample cell: TN-LCD/Ag/0.12 wt % (reprinted from Reference 22, with permission from The Japan Society of Applied Physics).

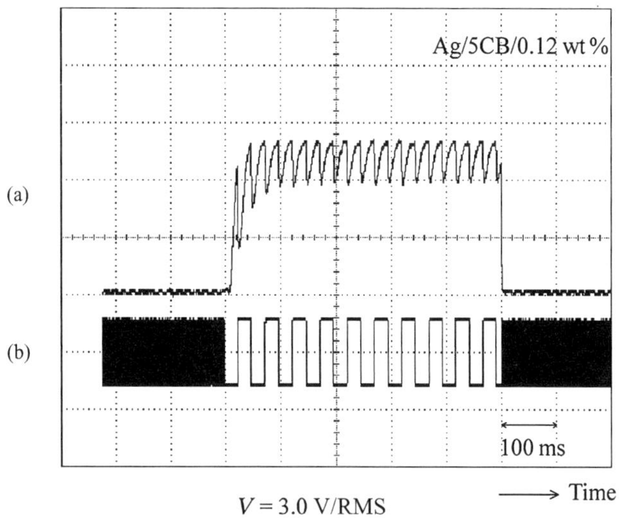

Figure 22.6. An example of FM switching by changing frequencies: low frequency, $f_1 = 20$ Hz, and high frequency, $f_2 = 500$ Hz, where the amplitude of the square wave voltage is 3.0 V; (a) relative optical throughput and (b) operating voltage waveforms (reprinted from Reference 22, with permission from The Japan Society of Applied Physics).

voltage waveform, respectively, with an amplitude of 3.0 V for their root mean square values. In a transition from 500 to 20 Hz, the delay time is 10 ms and the rise time is 11 ms, and in a transition from 20 to 500 Hz, the delay time is almost negligible (below 1 ms) and the fall time is 3 ms. Upon setting $f_1 = 50$ Hz and $f_2 = 2$ kHz, the decay time becomes shorter, being 2 ms. Thus, this device shows a fast response to a burst AC square wave, where the decay time is reduced by several times compared to that of an undoped TN-LCD.

22.2.4 Low Driving Voltage of LCD Doped with Nanoparticles

The energy crisis is a global problem that people all over the world encounter. Recently, large (108 in. at maximum) LCDs with high brightness, high contrast, and high resolution consume huge amounts of energy (500 W to 2 kW), although a decade ago, mainstream CRT TVs only had a power consumption of less than 100 W. Since LCDs are widely used as TV sets, monitors for personal computers, and so on, decreasing the energy consumption of LCDs is an urgent research subject. We have reported that a series of LCDs doped with metallic nanoparticles such as Ag, Pd, Au, Pt, or their alloys may have a much faster response than LCDs without dopants [21–25]. However, the driving voltage (V_d) of LCDs usually increases by doping with metallic nanoparticles. High V_d may provide high electronic power consumption, which should be avoided.

ZnO nanoparticles have attracted increasing attention owing to wide application in anti-UV-radiation cosmetics, highly efficient solar cells, light-emitting device, cancer-detecting biosensors, gas sensors, and degradation of organic toxins. Anyway, the traditional method to prepare ZnO nanoparticles involves precipitation of $Zn(OH)_2$ by the treatment of Zn^{2+} ions with weak alkali and heat treatment of $Zn(OH)_2$ at 400–700°C. This method has some disadvantages: (i) The method requires high temperature and long reaction time, which usually results in large particle sizes, loss of the quantum size effect, and huge power consumption during the annealing process; (ii) ZnO nanoparticles are obtained in the powder state, which is not easy to get good dispersion into a liquid crystal matrix compared with a stable sol state in organic solvents; (iii) since most of the organics attached to the particle surface are removed by heat treatment, ZnO nanoparticles have very poor compatibility with the liquid crystal matrix. Jiang and Toshima reported an improved alco-sol method for the preparation of nanosized ZnO capped oleic acid, which can improve the compatibility of ZnO nanoparticles with 5CB. Electro-optic properties were measured at 25°C in a TN mode with a cell gap of 5 μm using an LCD evaluation system [42]. Voltage–transmittance (V–T) curves of 5CB doped with ZnO nanoparticles having frequency dependence (Fig. 22.7) indicates that the threshold voltage (V_{th}) and the driving voltage (V_d) of doped 5CB decrease with increasing weight percent of ZnO less than 0.5 wt %. The V_d of pure 5CB and 5CB doped with 0.5 wt % of ZnO are 1.7 and 1.2 V, respectively. The doped material has a 30% lower V_d than that of the pure 5CB. The lower V_d may lead to lower power consumption. This is a very interesting phenomenon because the V_d of 5CB doped with metallic nanoparticles (Au, Ag, and Pd) usually increases [25]. Thus, semiconductor nanoparticles may have a unique behavior in the liquid crystal matrix. The elimination of the internal electric field is believed to have contributed

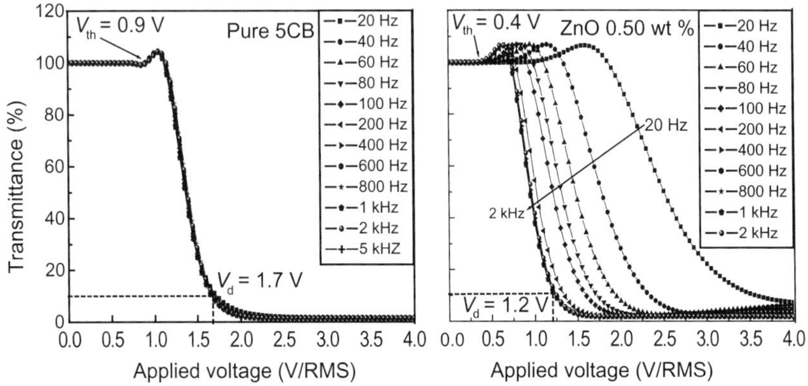

Figure 22.7. *V–T* curve of TN-LCDs of pure 5CB cells and the cells doped with 0.5 wt % of ZnO nanoparticles (reprinted from Reference 42, with permission from The Chemical Society of Japan).

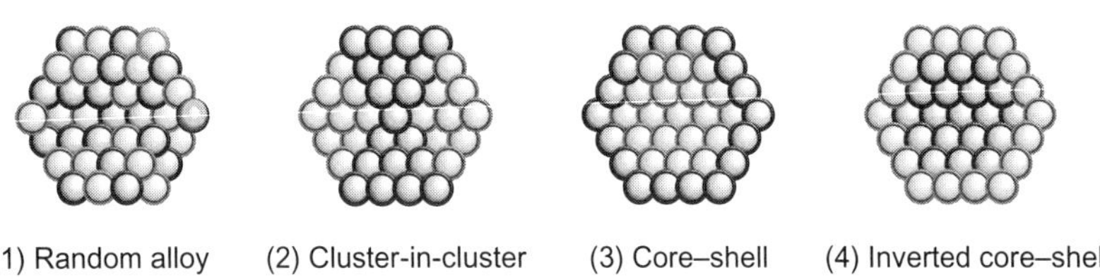

Figure 22.8. Various structural models of bimetallic nanoparticles.

to the decrease of V_d of 5CB cells by doping with the prepared ZnO nanoparticles.

22.3 BIMETALLIC NANOPARTICLES

22.3.1 Preparation of Bimetallic Nanoparticles

The synthesis of bimetallic nanoparticles is mainly divided into two methods, that is, the chemical and physical methods, or the "bottom-up" and "top-down" methods. The chemical method involves simultaneous reduction, successive or two-step reduction of two kinds of metal ions, and "self-organization" of bimetallic nanoparticles by physically mixing two kinds of already-prepared monometallic nanoparticles with or without aftertreatments. In the preparation of bimetallic nanoparticles, one of the most interesting aspects is a core–shell structure, as shown in Figure 22.8. The surface element plays an important role in the functions of metallic nanoparticles like catalytic and optical properties, but these properties can be tuned by the addition of the second element, which may be located on the surface or in the center of the particles adjacent to the surface element.

Polymer-capped bimetallic nanoparticles have been prepared by the simultaneous reduction of two corresponding metal ions with refluxing alcohol in the presence of PVP. For example, PVP-capped Pt/Pd bimetallic nanoparticles can be prepared by refluxing the alcohol/water (1/1, v/v) mixed solution of palladium(II) chloride and hexachloroplatinic(IV) acid in the presence of PVP at about 90–95°C for 1 hour [43]. The special structure has been proposed on the basis of extended X-ray absorption fine structure (EXAFS) analyses [44]. In the PVP-capped Pt-core/Pd-shell model for Pt/Pd (1/4, mole ratio), 13 platinum atoms are preferentially located in the central part (core) of the bimetallic nanoparticles, and 42 palladium atoms surround the core forming a shell with an atomic layer. The formation process of the core–shell structure may be due to orders of reduction potential of metal ions and coordination ability of the protective polymer to reduced atoms. Au/Pd, Au/Pt, and Pt/Rh bimetallic nanoparticles have been prepared in the same method [45]. Torigoe and Esumi prepared Ag–Pd bimetallic nanoparticles by photoreduction of silver(I) bis(oxalato)palladate(II) in water [46]. Ag/Pt bimetallic nanoparticles were obtained by sodium tetrahydroborate reduction of silver(I) bis(oxalato) platinate(II) in ethylene glycol [47]. Colloidal dispersions of Au/Ag bimetallic nanoparticles were prepared by γ-irradiation of the mixed solution for the study of electrochemical reaction [13].

The inverted core–shell, for example, Pd-core/Pt-shell structured bimetallic nanoparticles are difficult to synthesize. The addition of platinum(IV) ions to palladium nanoparticles resulted in a redox reaction to form platinum nanoparticles and palladium(II) ions. In order to avoid this unfavorable reaction, the colloidal dispersions of palladium cores were treated with hydrogen, and then solutions of the second element, platinum(II) ions, were slowly added to the dispersions. This novel method, based on a sacrificial hydrogen strategy, gave inverted core–shell structured bimetallic nanoparticles. The PVP-capped Pd-core/Pt-shell structure was confirmed by Fourier transform infrared (FT-IR) spectrum of adsorbed carbon monoxide [48].

5CB-capped Ag–Pd bimetallic nanoparticles were prepared by the photoirradiation of the tetrahydrofurane solution of silver perchlorate and palladium(II) acetate in the presence of a liquid crystal molecule [49]. The prepared bimetallic nanoparticles had an average diameter of 1.8~3.6 nm. Infrared spectra of carbon monoxide adsorbed on the bimetallic nanoparticles suggested that bimetallic nanoparticles had a random alloy structure.

22.3.2 Electro-Optic Properties of LCD Doped with Bimetallic Nanoparticles

The TN-LCDs doped with metal nanoparticles, fabricated by injecting the 5CB sol containing 5CB-capped Pd nanoparticles into a cell, were found to exhibit

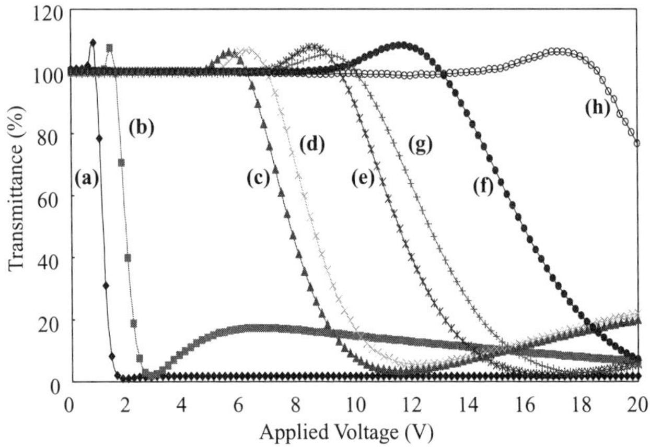

Figure 22.9. *V–T* curve of TN-LCDs at 100 Hz and 25°C. (a) Pure 5CB, (b) 5CB-capped Pd, (c) 5CB-capped Ag–Pd (Ag/Pd = 1/9), (d) 5CB-capped Ag–Pd (1/4), (e) 5CB-capped Ag–Pd (1/1), (f) 5CB-capped Ag–Pd (4/1), (g) 5CB-capped Ag–Pd (9/1), and (h) 5CB-capped Ag (reprinted from Reference 49, with permission from the American Chemical Society).

FM response. Interestingly the electro-optic response could also be enhanced by the addition of metal nanoparticles, especially Ag nanoparticles, resulting in fast response, which could improve one of the disadvantages for the presently available TN-LCDs. During a series of investigations for the effect of metal nanoparticles on the electro-optic properties of TN-LCDs, we have found that the effect of nanoparticles on the degree of the response time drastically depends on the kind of metal. In fact, Ag nanoparticles had the largest effect on the response time among Ag, Au, and Pd nanoparticles. However, the effect of Ag nanoparticles decreased rapidly even during the measurement. Here, we describe the electro-optic properties of TN-LCDs fabricated by liquid crystal sol containing Ag–Pd bimetallic nanoparticles, which are much more stable than Ag nanoparticles.

In Figure 22.9, curves (a)–(h) show the relationships of applied voltage and transmittance of LCDs fabricated by 5CB sol containing 5CB-capped nanoparticles measured at 25°C and 100 Hz. The driving voltage of LCD shifts to a higher voltage by doping with nanoparticles. The shift is small in the sol containing Pd nanoparticles and large in that containing Ag nanoparticles. The extent of the shift depends on the composition of bimetallic nanoparticles, increasing with increasing Ag component in the order of Ag/Pd = 1/9, 1/4, 1/1, 4/1, and 9/1. In Figure 22.10, the Pd content of bimetallic nanoparticles is plotted against the shift of threshold voltage (V_{th}) of TN-LCDs fabricated by 5CB sol containing Ag–Pd bimetallic nanoparticles from that of pure 5CB, which is defined as the shift in the voltage where the transmittance starts to change from 100%. No linear relationship is observed between the shift in V_{th}

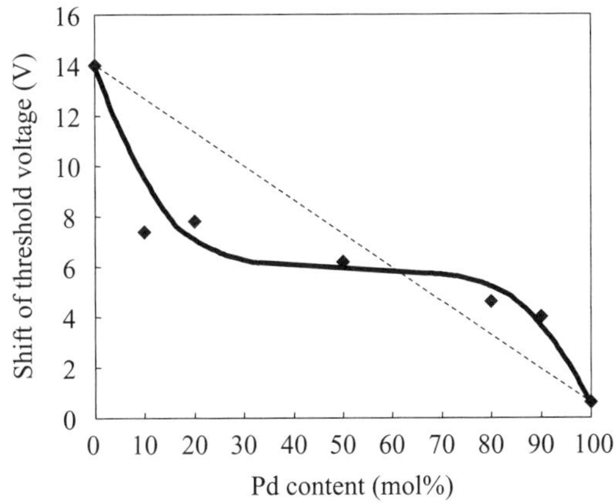

Figure 22.10. Relationship between Pd content and shift of threshold voltage from 5CB in the *V–T* curves of TN-LCDs fabricated by 5CB sol containing Ag–Pd bimetallic nanoparticles. The line is a guide to the eye (reprinted from Reference 49, with permission from the American Chemical Society).

and the Pd content. The shift is larger than that expected from the linear dashed line at the Pd content of 80–90% for bimetallic nanoparticles of Ag/Pd = 1/4 and 1/9, while it is smaller at the Pd content of 10–20% for those of Ag/Pd = 9/1 and 4/1. This nonlinear relationship suggests that the shift in threshold voltage may not depend on the Pd content of a whole particle but on the Pd content of the particle surface. This nonlinear relationship can be explained by the following two ways. The first explanation is based on the difference between the composition of a whole particle and the composition of the particle surface, which can be satisfied by the structure like Pd-core/Ag-shell. The other explanation is based on the variation of the electronic properties of surface Pd depending on the composition of the particle surface. The first explanation based on the core–shell structure is in conflict with the results observed in UV–vis and CO-FT-IR spectra. Thus, an acceptable explanation should be based on the variation of the electronic properties of surface Pd. Here, the bimetallic nanoparticles have a random alloy structure; that is, the Pd content of the particle surface is the same as that of the whole particle. However, atoms of the minor component on the surface of bimetallic nanoparticles change their electronic state owing to charge transfer (electronic) effect by large numbers of adjacent atoms of the main component. This kind of electronic effect was reported for the catalytic property of core–shell structured bimetallic nanoparticles [50, 51].

This charge transfer effect is illustrated as a cartoon in Figure 22.11. In the case of Ag/Pd = 1/4 and 1/9 (Fig. 22.11a), the minor component Ag atom has a rather large positive charge since the Ag atom is surrounded by many Pd atoms. In the case of Ag/Pd = 1/1 (Fig. 22.11b), electric charges of Ag and Pd

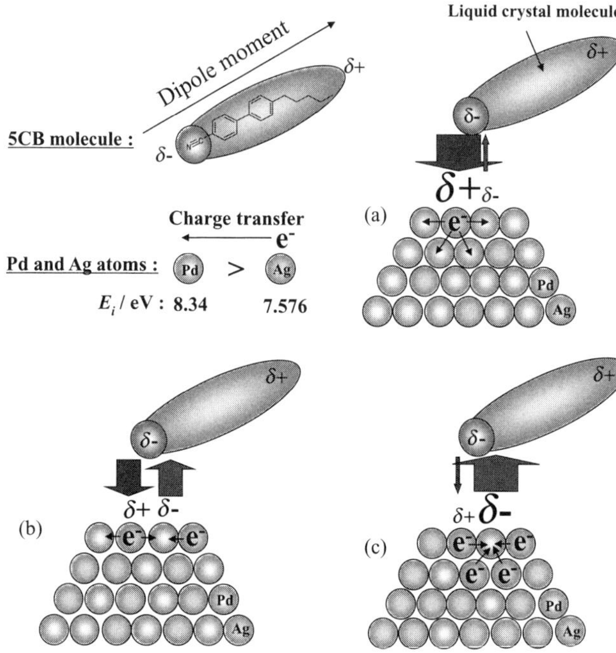

Figure 22.11. Schematic illustration of electronic charge transfer in the 5CB-Ag–Pd (atomic ratio of Ag/Pd = 1/4 and 1/9 (a), 1/1 (b), and 4/1 and 9/1 (c)) bimetallic nanoparticles and Coulomb interaction of a cyano group of liquid crystal molecule 5CB with the charged surface of bimetallic nanoparticles at various Ag/Pd ratios. The green and red arrows mean attractive and repulsive forces, respectively (reprinted from Reference 49, with permission from the American Chemical Society). See color figure on companion ftp site.

atoms are opposite but nearly the same in absolute value because atomic numbers of the surface Ag and Pd are the same. In the case of Ag/Pd = 9/1 and 4/1 (Fig. 22.11c), the minor component Pd atom has a rather large negative charge since the Pd atom is surrounded by several Ag atoms. On the other hand, the liquid crystal molecule 5CB is polarized to result in partial negative charge at the cyano end group, which should interact with the surface of bimetallic nanoparticles. Consequently, the Ag/Pd = 1/4 and 1/9 nanoparticles have a strong Coulomb attraction with the cyano group of 5CB because of a large positive charge at the Ag atom, which can result in a large electro-optic effect by giving a stronger perturbation to the liquid crystal matrix. In contrast, the Ag/Pd = 4/1 and 9/1 nanoparticles have a strong Coulomb repulsion to the cyano group of 5CB because of the large negative charge of the Pd atom, which can result in a small electro-optic effect. Thus, only the small amount of the additive (minor component) can alter the surface properties of nanoparticles.

Although LCDs doped with 5CB-capped Ag nanoparticles provided a big change in the electro-optic properties of LCDs, the change rapidly decreased by deterioration. As mentioned in the previous section, electro-optic properties of nanoparticle-doped LCDs depended on the surface composition of doping nanoparticles. Thus, we examined the accelerated deterioration by heat treatment of LCD cells at 80°C for 28 days. After the heat treatment, the cell fabricated by 5CB sol containing 5CB-capped Ag nanoparticles has black spots and disorder parts in liquid crystal orientation on a photograph taken with a polarization microscope, while that containing 5CB-capped Ag–Pd (1/1) bimetallic nanoparticles did not. Thus, 5CB-capped Ag nanoparticles look like an easy aggregation, which may decrease the practical concentration of dispersed nanoparticles in the liquid crystal medium and the effect of nanoparticles upon the behavior of the liquid crystal as well. In the case of Ag–Pd bimetallic nanoparticles, by contrast, the particles can disperse well in a liquid crystal medium because the surface of bimetallic nanoparticles can be partially covered by Pd, which can suppress the aggregation of nanoparticles. This explanation was confirmed by TEM observation of nanoparticles in the 5CB sol containing Ag nanoparticles and Ag–Pd (Ag/Pd = 1/1) bimetallic nanoparticles after heat treatment at 80°C for 120 hours not in the cells but in separate test tubes. Thus, the TEM photograph of 5CB-capped Ag nanoparticles after heat treatment shows actually the presence of large aggregates with about 30 nm in diameter, while that of 5CB-capped Ag–Pd bimetallic nanoparticles reveals the presence of large single nanoparticles with about 8 nm in diameter along with many small metallic nanoparticles with about 2 nm in diameter. In the latter case, although a kind of so-called Ostwald ripening may have occurred, the resulting dispersions contain a small number of the nanoparticles that are large in size but still are dispersed well. The presence of aggregates in the heat-treated 5CB sol containing Ag nanoparticles will give an explanation for the fact that the cell fabricated by 5CB sol containing Ag nanoparticles has black spots in a polarization microscope. In contrast, heat-treated 5CB sol containing Ag–Pd bimetallic nanoparticles has no aggregates, which may result in a rather clear surface in fabricated cells. Thus, the control of the surface composition and structure of nanoparticles could be of importance to stabilize the liquid crystal sol containing nanoparticles. This kind of stable liquid crystal sol may provide novel functions as hybrid materials containing nanoparticles.

ACKNOWLEDGMENTS

This work was partially supported by Konica Minolta Imaging Science Foundation to Y.S. and a Grant-in-Aid for Scientific Research (C) (No. 20510109, to Y.S.) from the Ministry of Education, Culture, Sports, Science, and Technology (MEXT), Japan.

REFERENCES

1 H. Hirai, N. Toshima, in J. C. Salamone (ed.), *Polymeric materials encyclopedia*, vol. 2/c. CRC Press, Boca Raton, FL, 1996, 1, p. 1310.

2 N. Toshima, T. Yonezawa. *New J. Chem.* 22 (1998) 1179.

3 G. Schmid. *Clusters and colloids: from theory to applications*. VCH, Weinheim, 1994.

4 L. D. Rampino, F. F. Nord. *J. Am. Chem. Soc.* 63 (1941) 2745.

5 H. Hirai, H. Chawanya, N. Toshima. *Reactive Polym.* 3 (1985) 127.

6 C. Petit, A. Taleb, M. P. Pileni. *J. Phys. Chem. B* 103 (1999) 1805.

7 G. Schmid, V. Maihack, F. Lantermann, S. Peschel. *J. Chem. Soc., Dalton Trans.* 27 (1996) 589.

8 N. Toshima, K. Nakata, H. Kitoh. *Inorg. Chim. Acta.* 265 (1997) 149.

9 Y. Volokitin, J. Sinzig, L. J. Jongh, G. Schmid, M. N. Vargaftik, I. I. Moiseev. *Nature* 384 (1996) 621.

10 J. H. Fendler. *Nanoparticles and nanostructured films: preparation, characterization and applications*. Wiley-VCH, Weinheim, 1998.

11 N. Toshima. *Macromol. Symp.* 105 (1996) 111.

12 B. Coq, P. S. Kumbhar, C. Moreau, P. Moreau, F. Figueras. *J. Phys. Chem.* 98 (1994) 10180.

13 P. Mulvaney, M. Giersig, A. Henglein. *J. Phys. Chem.* 97 (1993) 7061.

14 H. Bönnemann, W. Brijoux, R. Brinkmann, R. Fretzen, T. Joussen, R. Köppler, B. Korall, P. Neiteler, J. Richter. *J. Mol. Catal.* 86 (1994) 129.

15 N. Toshima. *Macromol. Symp.* 204 (2003) 219.

16 M. Suzuki, H. Furue, S. Kobayashi. *Mol. Cryst. Liq. Cryst.* 368 (2001) 191.

17 W. Lee, Y.-C. Wang, Y.-C. Shin. *Appl. Phys. Lett.* 85 (2004) 513.

18 Y.-H. Chen, W. Lee. *Appl. Phys. Lett.* 88 (2006) 222105.

19 S.-P. Chen, C.-C. Huang, W.-Y. Liu, Y.-C. Chao. *Appl. Phys. Lett.* 90 (2007) 211111.

20 F. Haraguchi, K. Inoue, N. Toshima, S. Kobayashi, K. Takatoh. *Jpn. J. Appl. Phys.* 46 (2007) 796.

21 Y. Shiraishi, N. Toshima, K. Maeda, H. Yoshikawa, J. Xu, S. Kobayashi. *Appl. Phys. Lett.* 81 (2002) 2845.

22 T. Miyama, J. Thisayukta, H. Shiraki, Y. Sakai, Y. Shiraishi, N. Toshima, S. Kobayashi. *Jpn. J. Appl. Phys.* 43 (2004) 2580.

23 H. Shiraki, S. Kundu, Y. Sakai, Y. Shiraishi, N. Toshima, S. Kobayashi. *Jpn. J. Appl. Phys.* 44 (2004) 5425.

24 Y. Sakai, N. Nishida, H. Shiraki, Y. Shiraishi, T. Miyama, N. Toshima, S. Kobayashi. *Mol. Cryst. Liq. Cryst.* 441 (2005) 143.

25 S. Kobayashi, T. Miyama, N. Nishida, Y. Sakai, H. Shiraki, Y. Shiraishi, N. Toshima. *J. Display Technol.* 2 (2006) 121.

26 Y. Du, N. Toshima. *Bull. Chem. Soc. Jpn.* 80 (2007) 2446.

27 K. C. Chu, C. Y. Chao, Y. F. Chen, Y. C. Wu, C. C. Chen. *Appl. Phys. Lett.* 89 (2006) 103107.

28 P. R. Evans, G. A. Wurtz, W. R. Hendren, R. Atkinson, W. Dickson, A. V. Zayats, R. J. Pollard. *Appl. Phys. Lett.* 91 (2007) 043101.

29 H. Hirai. *J. Macromol. Sci., Chem.* A13 (1979) 633.

30 N. Toshima, T. Takahashi, H. Hirai. *Chem. Lett.* 15 (1985) 1245.

31 M. T. Reetz, W. Helbig, S. A. Quaiser, U. Stimming, N. Breuer, R. Vogel. *Science* 267 (1995) 367.

32 T. Yonezawa, T. Tominaga, N. Toshima. *Langmuir* 11 (1995) 4601.

33 T. Hashimoto, K. Saijo, M. Harada, N. Toshima. *J. Chem. Phys.* 109 (1998) 5627.

34 Y. Shiraishi, S. Sano, A. Baba, S. Kobayashi, N. Toshima. *Kobunshi Ronbunshu* 59 (2002) 753.

35 M. Schadt, W. Helfrich. *Appl. Phys. Lett.* 18 (1971) 127.

36 S. Kobayashi, H. Hori, Y. Tanaka, in P. J. Collings, J. S. Patel (eds.), *Handbook of liquid crystal research*. Oxford University Press, New York, 1997, pp. 220–235. Chapter 10.

37 C. J. Kiely, J. Fink, M. Brust, D. Bethell, D. J. Schiffrin. *Nature* 396 (1998) 444.

38 M. P. Pileni, in J. H. Fendler (ed.), *Nanoparticles and nanostructured films: Preparation, characterization and application*. Wiley-VCH, Weinheim, 1998, p. 71.

39 B. Zhou, G. S. Iannacchione, C. W. Garland, T. Bellini. *Phys. Rev. E* 55 (1997) 2962.

40 S. Lee, C. Park. *Mol. Cryst. Liq. Cryst.* 333 (1999) 123.

41 H. J. Deuling. *Mol. Cryst. Liq. Cryst.* 27 (1974) 81.

42 H. Jiang, N. Toshima. *Chem. Lett.* 38 (2009) 566.

43 N. Toshima, K. Kushihashi, T. Yonezawa, H. Hirai. *Chem. Lett.* 18 (1989) 1769.

44 N. Toshima, M. Harada, T. Yonezawa, K. Kushihashi, K. Asakura. *J. Phys. Chem.* 95 (1991) 7448.

45 N. Toshima. *J. Macromol. Sci., Chem.* A27 (1990) 1225.

46 K. Torigoe, K. Esumi. *Langmuir* 9 (1993) 1664.

47 K. Torigoe, Y. Nakajima, K. Esumi. *J. Phys. Chem.* 97 (1993) 8304.

48 Y. Wang, N. Toshima. *J. Phys. Chem. B* 101 (1997) 5301.

49 N. Nishida, Y. Shiraishi, S. Kobayashi, N. Toshima. *J. Phys. Chem. C* 112 (2008) 20284.

50 N. Toshima, R. Ito, T. Matsushita, Y. Shiraishi. *Catal. Today* 122 (2007) 239.

51 N. Toshima, T. Yonezawa, K. Kushihashi. *J. Chem. Soc., Faraday Trans.* 89 (1993) 2537.

23

FABRICATION OF ORDERED NANOPATTERN STRUCTURES USING TWO-DIMENSIONAL COLLOIDAL MONOLAYERS

Takeshi Kawai

23.1 INTRODUCTION

Fabrication of ordered nanopattern structures has attracted extensive interest because they are promising structures for use as photonic materials, magnetic materials, biosensors, drug delivery systems, and for surface-enhanced Raman scattering [1–3]. A number of techniques for the fabrication of nanopattern structures have been proposed, including electron beam lithography, scanning tunneling microscopic lithography, and atomic force microscopic lithography. These methods can be used to prepare complicated nanostructures with high quality and high resolution, although the apparatus is high cost and the fabrication area is small. To overcome these issues, soft lithography such as microcontact printing has been recently applied to prepare nanopattern structures; however, the templates for soft lithography must be prepared by other lithography techniques, such as electron beam lithography.

Recently, the fabrication of highly ordered colloidal crystals has attracted extensive interest, and a number of techniques for self-assembly fabrication of three-dimensional (3-D) colloidal crystals have been proposed, including gravity sedimentation, slow drying, and vertical deposition [4–6]. The inverse

Electrical Phenomena at Interfaces and Biointerfaces: Fundamentals and Applications in Nano-, Bio-, and Environmental Sciences, First Edition. Edited by Hiroyuki Ohshima.

structure of colloidal crystals has also received much interest as a novel material that can be prepared using polymer and ceramic colloidal crystals as templates. Two-dimensional (2-D) colloidal crystals (colloidal monolayers) are also fascinating functional material for the construction of nanostructures. In this review, we focus on 2-D colloidal monolayers for the fabrication of periodic functional nanopatterned surfaces and thin films. In particular, we review 2-D colloidal monolayers employed as masks and templates for the fabrication of periodic arrays of nanoparticles, nanopillars, and nanorings.

23.2 WETTABILITY CONTROL BY PERIODIC SURFACE ROUGHNESS OF THE COLLOIDAL MONOLAYER

The simplest application of the 2-D colloidal particle monolayer is wettability control by surface roughness. It is well-known that surface roughness can emphasize wettability; a hydrophobic (hydrophilic) surface becomes more hydrophobic (hydrophilic) with an increase in the surface roughness. Thus, superhydrophilic and superhydrophobic surfaces can be obtained by imparting surface roughness. Hydrophobic or hydrophilic 2-D colloidal monolayers can be simply prepared by surface modification with organic materials. For example, a colloidal particle monolayer covered with deposited metal (Fig. 23.1) is functionalized with a self-assembly monolayer (SAM) of thiol compounds. The wettability of the colloidal monolayer can be easily tuned by changing the thiol compounds. Table 23.1 shows the contact angle of water on a polystyrene (PS) particle monolayer covered with a 50-nm Au film functionalized with octadecanethiol (ODT), 1H,1H,2H,2H-perfluorooctanethiol (FOT), and mercaptopropanoic acid (MPA) SAMs. The contact angles for both FOT and ODT, which provide a hydrophobic surface, were ca. 130° and larger than that for the flat surface. In the case of MPA, which yields a hydro-

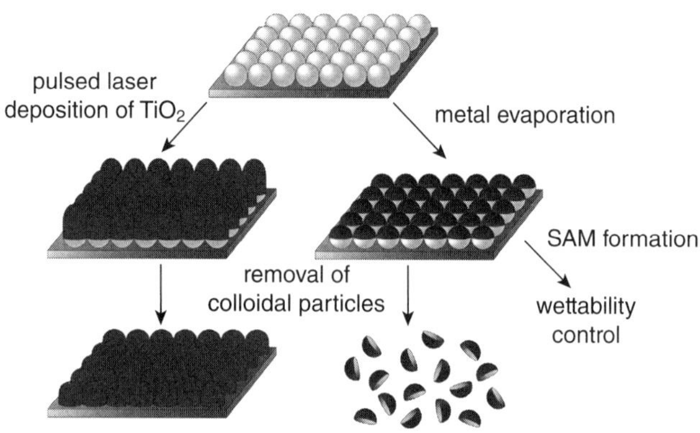

Figure 23.1. Schematic illustration of the functionalization of a colloidal monolayer.

TABLE 23.1. Water Contact Angles on a PS Colloidal Monolayer Covered with a Au Thin Film Functionalized with Various Thiol SAMs

	Contact Angle (Degree)		
Thiol Compound	Flat	PS-200	PS-560
FOT	112	131	133
ODT	105	128	127
MPA	45	65	73

philic surface, the contact angle was ca. 70° and also larger than that of the flat surface. In addition, the contact angles were almost the same between PS-200 and PS-560, which indicates that the hydrophobicity was enhanced by the surface roughness and was independent of the size of the colloidal particles.

The apparent contact angle (θ_R) of rough surfaces can be expressed by the Wenzel equation:

$$\cos\theta_R = r\cos\theta_S, \tag{23.1}$$

where r is the roughness factor (the ratio of the total surface area to the projected area in the horizontal plane) and θ_S is the contact angle measured on the flat surface. The value of r is always greater than 1 because the total surface area is larger than the projected area. Thus, the surface roughness causes enhancement of the hydrophobicity and hydrophilicity of the hydrophobic surface ($\theta_S > 90°$) and the hydrophilic surface ($\theta_S < 90°$), respectively. Therefore, the increase of the contact angle on the colloidal monolayer modified with MPA cannot be explained by the roughness effect from Wenzel's equation.

The increase in the contact angle is explained by the presence of air under the water droplet. Water droplets for contact angle measurements were contacted both with the colloidal particles and air trapped among the particles. Under this condition, the apparent contact angle (θ_R) is estimated by the following Cassie–Baxter equation:

$$\cos\theta_R = \phi(\cos\theta_S + 1) - 1 \tag{23.2}$$

where ϕ is the surface fraction of the particle, and θ_S is the contact angle of water on a flat surface modified with SAMs. The value of ϕ was estimated from plots of cos θ_R against cos θ_S to be 0.52 for FOT and ODT and 0.84 for MPA. This result is consistent with the expectation that the roughness surface modified with hydrophobic SAM tends to have many air gaps.

Although the above-mentioned study clearly demonstrates that the wettability of a colloidal monolayer functionalized with SAMs is enhanced, the enhancement is usually low. This is because the surface roughness of the colloidal monolayer is generally insufficient. As a consequence, it is difficult to prepare superhydrophobic or superhydrophilic surfaces from the original surface of a colloidal monolayer. A possible basis for the formation of superhydrophobic or superhydrophilic surfaces based on a colloidal monolayer is that the surface roughness is increased by the construction of rod or column structures on the colloidal monolayer.

Koshizaki's group has prepared TiO_2 nanocolumns that stand vertically on the colloidal particle tops (Fig. 23.1, left-side protocol) [7]. A hexagonal close-packed, hierarchical amorphous TiO_2 nanocolumn array was fabricated by pulsed laser deposition (PLD) using a PS colloidal monolayer with a diameter of 350 nm as a template. Contact angles on as-prepared TiO_2 nanocolumn arrays were 0° for both water and oil. Generally, TiO_2 films with superamphiphilicity can be obtained by UV irradiation due to hydroxyl groups generated by oxygen defects and dangling bonds on the surface induced by photochemical processes. Interestingly, the as-prepared TiO_2 nanocolumn array film possessed superamphiphilicity without UV irradiation because oxygen vacancies were easily produced during PLD. In addition, periodically rough structures were essential to obtain the superamphiphilic surface because the water and oil contact angles of the TiO_2 film prepared on flat substrates were 15° and 27°, respectively. Furthermore, the amorphous TiO_2 nanocolumn array demonstrated better photocatalytic activity performance than an anatase nanocolumn array due to its large surface area and special microstructures. Significantly, the combination of superamphiphilicity and photocatalytic activity provided the surface with excellent self-cleaning properties.

23.3 TEMPLATE FOR HOLLOW SHELLS

A colloidal monolayer can be used as a template to fabricate a hollow metallic hemisphere film with a well-defined thickness, hemisphere diameter, and composition [8]. Love et al. [8] prepared half-shells of Au, Pt, and Pd with particle diameters ranging from 100 to 500 nm and shell thicknesses of 8–15 nm, as follows. An adhesion layer of titanium or nickel (0.5–0.8 nm) was deposited on a template of silica colloidal monolayer by electron beam evaporation followed by deposition of a thin metal film. The spherical colloid particles were then released from the surface by sonication. The silica core and adhesion layer were dissolved using aqueous hydrofluoric acid (HF), which left the metallic half-shell intact (Fig. 23.1, right-side protocol). They also demonstrated that metallic half-shells functionalized with hexadecanethiolate SAM had superhydrophobic properties. Half-shells of Au, Ag, and Al have potential application in cancer treatment, biomedical microscopy, and infrared (IR) block coatings because their absorption peaks show in the near IR region [9].

It is possible to prepare various shapes of metal shells by changing the deposition conditions of the metal. Pawar and Kretzschmar have applied the glancing angle deposition technique to prepare metal shells with triangular, trapezoidal, and parallelogram shapes. Multistep deposition techniques with a

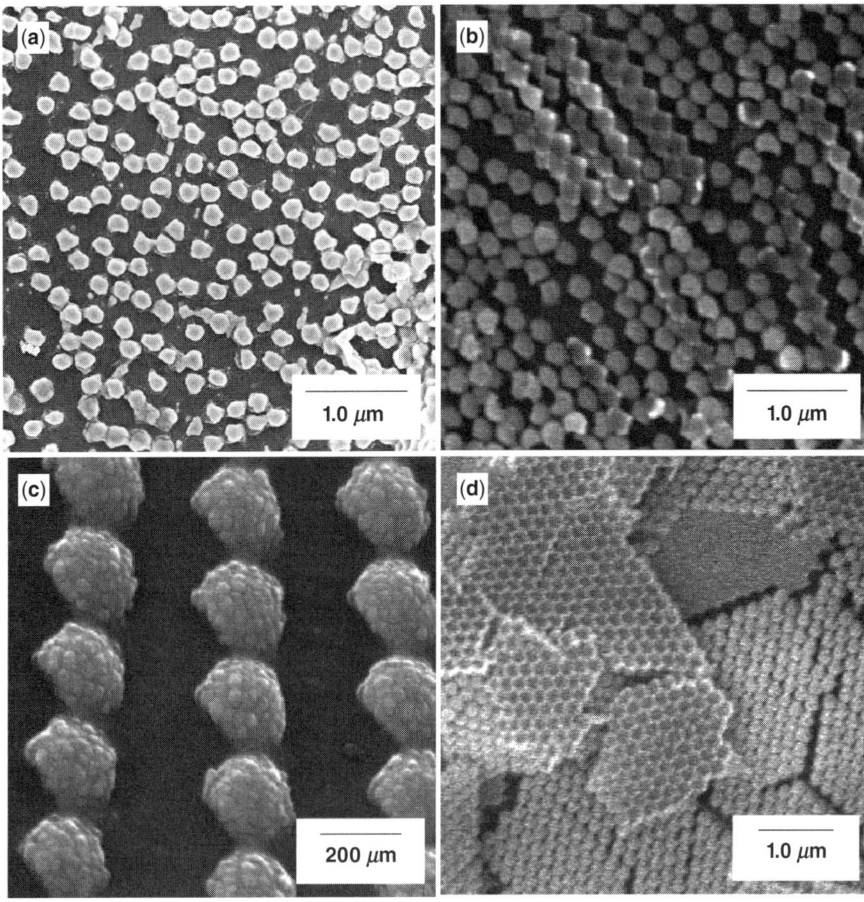

Figure 23.2. Typical SEM images of Au (a) dots, (b,c) 1-D strings, and (d) a 2-D film at a tilt angle of $\theta = 15°$.

combination of varying incident angles and azimuth angles can also provide more complicated metal shell shapes [10, 11].

One-dimensional (1-D) strings of Au half-shells can be also fabricated by the deposition of a Au film onto a particle monolayer, as shown in Figure 23.2b,c [12]. Au is deposited by vacuum evaporation onto a PS particle monolayer, which is tilted at an angle (θ) between the normal of the monolayer plane and the direction of metal deposition. PS particles are then removed from the deposited metal films by ultrasonication of the horizontally positioned glass substrates for a few seconds in toluene. In this approach, the ultrasonication process is very important to obtain metal shells on the glass substrates. Longer ultrasonication times cause the shells to be washed away from the glass substrates, while shorter ultrasonication times result in PS

TABLE 23.2. Morphologies of Remaining Gold Shells Obtained after the Removal of PS Particles for Various Deposition Thicknesses and Tilt Angles

	Thickness of Evaporated Gold (nm)						
Tilt Angle (Degree)	20	30	40	50	80	100	120
15	Dot	1-D	1-D	1-D	2-D	2-D	2-D
45	Dot	1-D	1-D	1-D	1-D	2-D	2-D
75	Dot	Dot	1-D	1-D	1-D	1-D	2-D

The isolated half-shells, 1-D strings of half-shells, and 2-D film of half-shells are denoted as dots, 1-D, and 2-D, respectively.

particles remaining inside the deposited metal. Three types of shells were produced: isolated half-shells (denoted as dots, Fig. 23.2a), a 1-D string of half-shells (Fig. 23.2b,c), and a 2-D film of half-shells (Fig. 23.2d), depending on the tilt angle of the PS particle monolayer and the Au deposition thickness. Table 23.2 summarizes the results of the Au shell types. The boundaries between the 1-D string and dots or the 2-D film were somewhat obscure, and both products coexisted at the boundaries. The main products are summarized in Table 23.2. The increase in the θ value at a constant deposition thickness of 80 nm resulted in the formation of 1-D strings. On the other hand, at a constant θ value, the structure of the Au shells was changed from dots to a 1-D string and a 2-D film with an increase in the deposition thickness. It is reasonable that a 2-D film was obtained above a certain deposition thickness because the increase in deposition thickness tends to cover the entire surface of the PS particle monolayer with Au, even under the condition of a larger tilt angle of the monolayer.

The direction of formation of the 1-D strings, that is, the connecting direction of the half-shells for 1-D strings, is parallel to the tilt axis of the particle monolayer. Therefore, the rotation angle (α) in the plane of hexagonal packing of PS particles (Fig. 23.3a) is also an important factor for the formation of 1-D strings. There are two typical cases: one at $\alpha = 30°$, where one of the contact lines among PS particles is parallel to the tilt axis (y-direction), and the other at $\alpha = 0°$, where the contact line forms an angle of 30° with the y-direction.

At $\alpha = 30°$, the Au shells on each PS particle are expected to be easily connected to each other in the y-direction, while at $\alpha = 0°$, the formation of 1-D strings seems to be difficult because the shortest distance between the particles is not aligned in the y-direction. Figure 23.3b shows representative scanning electron microscope (SEM) images of Au deposited at a thickness of 50 nm on PS particles at $\alpha = 0°$ and $\alpha = 30°$. As expected, the rotation angle α affects the formation of the 1-D strings, as shown in Figure 23.3. The deposited Au shells for $\alpha = 30°$ were connected in the y-direction at tilt angles of $\theta = 15°$, 45°, and 75°, while the shells for $\alpha = 0°$ were not connected to each other. The influence of the rotation angle α is considerably greater at a larger tilt angle

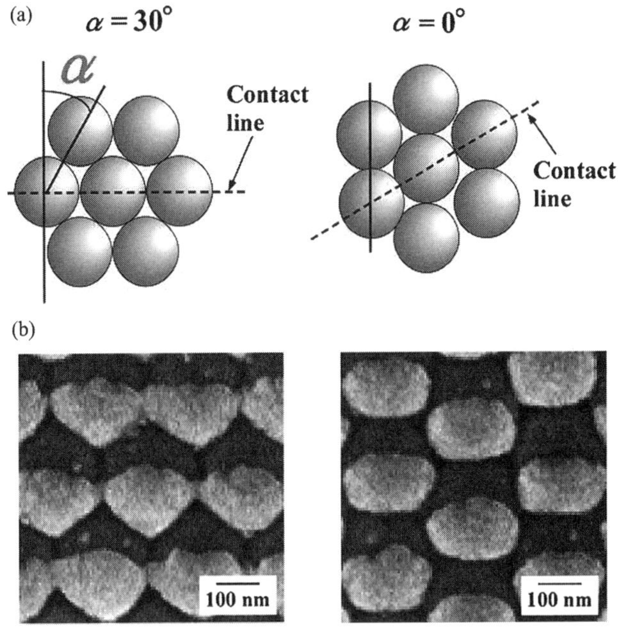

Figure 23.3. Schematic illustration of monolayer orientation with respect to the incident vapor beam. x–y cross-section of a particle monolayer at rotation angles of $\alpha = 0°$ and $\alpha = 30°$. SEM images of the top view of Au shells deposited on particle monolayers for rotation angles of $\alpha = 0°$ and $\alpha = 30°$, under conditions of a deposition thickness of 50 nm and tilt angles of 45°.

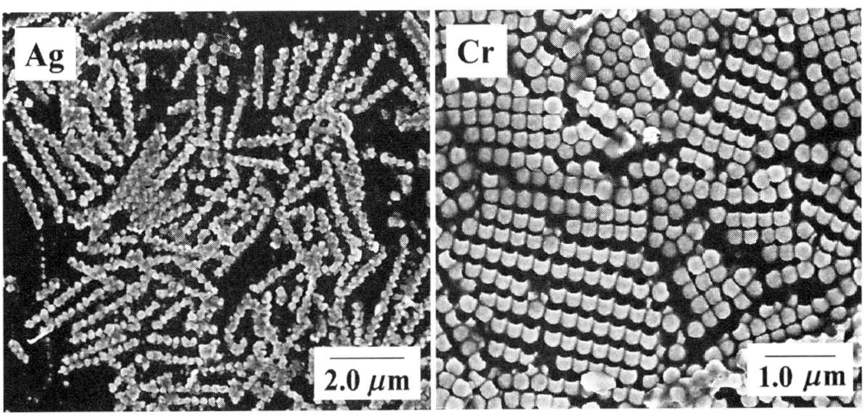

Figure 23.4. SEM images of hybrid 1-D strings of Ag (40 nm) and Cr (50 nm) on predeposited Au (20 nm) at $\theta = 45°$.

of $\theta = 75°$, at which the Au shells on the PS particles were completely isolated from one another. At a smaller tilt angle of $\theta = 15°$, some of the shells were connected between adjacent particles, even at $\alpha = 0°$ (Fig. 23.3b). These results are consistent with the content of dots formed for $\theta = 75°$, which was larger than that for $\theta = 15°$ and 45° in the region of the 1-D string.

It is expected that this method is applicable to other metals; however, in the case of Ag as the deposition metal, only dots were obtained at $\theta = 15°, 45°$, and 75° in the deposition thickness range of 20–120 nm. On the other hand, 2-D films were mainly produced for Cr, except at a larger tilt angle of $\theta = 75°$ and smaller deposition thicknesses of 20–50 nm where dots were obtained. Against expectation, 1-D strings of Ag and Cr shells could not be fabricated due to the morphological differences of the deposited metals. The order of homogeneity of the deposited metals on PS particles is Ag < Au < Cr. Ag tends to form island structures, while Cr completely covers the PS particles, which correlates with the tendency for deposited Cr and Ag to form 2-D films and dots, respectively. The cohesive force and wettability of the metal may also affect the morphology of the deposited films.

The preparation of 1-D strings of Ag and Cr was unsuccessful; therefore, another procedure was attempted to produce Ag and Cr strings by Ag or Cr deposition on a predeposited Au film on PS particles. The thickness of the precoated Au was not sufficient to produce 1-D strings. Figure 23.4 shows representative SEM images of 1-D strings of hybrid metals, Ag on Au and Cr on Au; 1-D strings of Ag and Cr were successfully obtained with the precoated Au film. However, Au deposition on precoated Ag or Cr was not successful.

23.4 NANOSPHERE LITHOGRAPHY

Another application of 2-D colloidal monolayers is as a mask for material deposition to produce nanoparticle arrays, which is referred to as nanosphere lithography. In this approach, both the spaces between close-packed colloidal particles (Fig. 23.5) and the particles in the colloidal monolayer can be used as a template to produce periodic nanostructures. Van Duyne's group demonstrated [13, 14] that periodic arrays of triangular-shaped Ag nanoparticles could be easily fabricated by using spaces in the colloidal monolayer as a nanosphere lithography template. Ag films were deposited over the PS colloidal monolayer mask, and subsequent removal of the mask by dissolution in CH_2Cl_2 under sonication resulted in a single-layer periodic Ag nanoparticle array structure. Figure 23.6 shows SEM images of Au nanoparticles prepared by nanosphere lithography using a colloidal monolayer of 200-nm PS particles. The shape of the Au nanoparticles was changed from triangular to spherical with increasing deposition thickness. It should be noted that in this method, any material that can be deposited from the gas phase can be used to form periodic arrays of nanoparticles. Accordingly, nanoparticles of desired functional materials can be prepared onto various substrates. Furthermore, over-

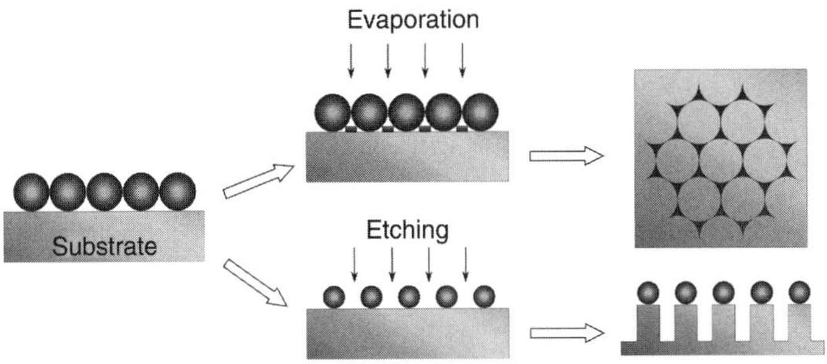

Figure 23.5. Schematic illustration of nanosphere lithography.

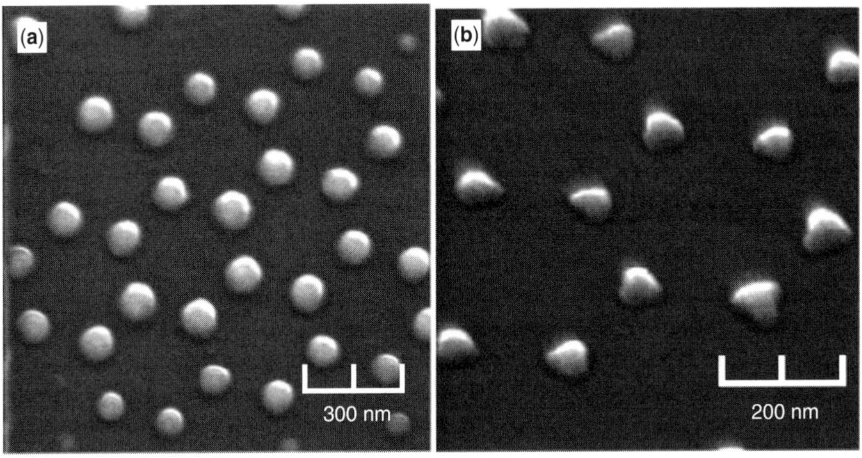

Figure 23.6. Periodic array of Au nanoparticles prepared by nanosphere lithography. Evaporation thickness of (a) 100 nm and (b) 50 nm.

lapped nanostructures, such as two or three overlapped triangular shapes, can be fabricated using a multiangle deposition method. For instance, after a metal is deposited through a nanosphere mask at a deposition angle of 0°, further deposition of the metal or another metal at a different deposition angle produces an overlapped nanostructure of two metal nanoparticles with a triangular shape [15]. As the shape and position of nanoparticles are dependent on the deposition angle, an appropriate combination of first and second deposition angles provides the desired overlapped nanostructure with a regularly ordered array. This technique was developed to prepare periodic arrays of vertically aligned isolated carbon nanofibers. A PS colloidal monolayer was used as a mask to deposit catalytic dots, such as Ni on Cr-coated Si substrates, followed by the complete removal of the PS particles by ultrasonication in toluene. Periodic arrays of carbon nanofibers were fabricated by a combination of Ni dots in a honeycomb lattice pattern on Si and plasma-enhanced chemical vapor deposition [16, 17].

Other studies have exploited the colloidal particles, rather than the spaces between the particles, as a mask to prepare nanostructure arrays. Large-area well-ordered periodic nanopillar arrays have been fabricated using colloidal monolayer masks and etching techniques. As shown in Figure 23.5, when reactive ion etching is applied to a colloidal monolayer on a substrate, the etching rate of that part of the substrate covered with particles becomes slower, which results in the formation of a periodic array of nanopillars [18]. Interestingly, it is possible to control the size, shape, and height of the nanopillar arrays by fine-tuning the etching parameters [19, 20].

23.5 TEMPLATE FOR HONEYCOMB FILM

Thin films with ordered pore arrays, which have a high specific surface area, play an important role in many applications, such as catalysts, gas sensors, and photonic and optoelectronic devices and membranes. The use of a 2-D colloidal monolayer as a template enables the preparation of well-ordered pore arrays of honeycomb structures. There are two typical porous films, as shown in Figure 23.7A,B. Here, types A and B are called honeycomb and porous films.

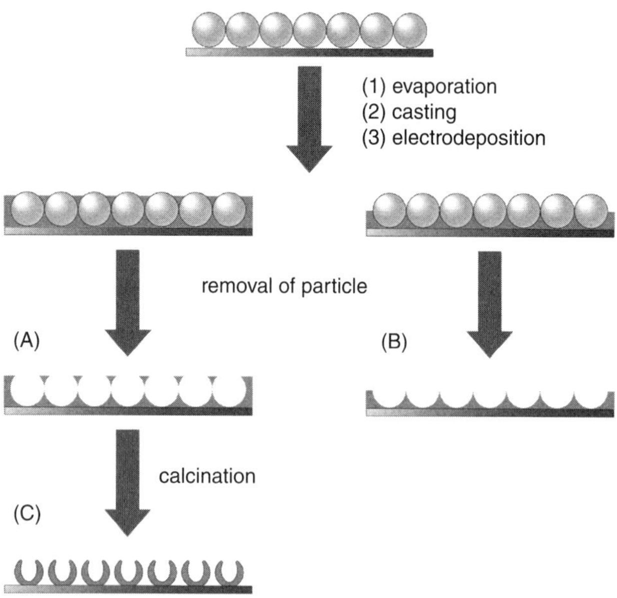

Figure 23.7. Schematic illustration of the fabrication of (A) honeycomb and (B) porous structures using a colloidal monolayer as a template. (C) Ring structure prepared by calcination of porous structure (B).

Filling of pores between particles has been performed using evaporation, casting of nanoparticles [21], spin coating, solution dipping [22], and electrochemical deposition techniques [23, 24].

Li et al. [22] successfully fabricated Ag honeycomb films by the immersion of a PS colloidal monolayer into a silver acetate solution, followed by heating at 360°C. They also demonstrated that the water contact angles of the silver film can be changed from 23° to 169° by chemical modification with hexadecanethiol SAMs. It is also possible to transform the morphology of metal arrays in honeycomb and porous films. For example, Au rings were prepared from a porous polymer film template via filling of the pores with a Au precursor solution followed by calcination (Fig. 23.7C). The diameter, thickness, and height of the rings could be easily controlled by using a porous film with a suitable pore size [25].

Many periodic nanopatterned structures formed with colloidal monolayer templates have been fabricated on rigid substrates. If flexible films with periodic nanopatterns were available, then the application of the wettability control of a surface would be broadened. Ultrathin freestanding porous metal films have been prepared by using colloidal monolayers, where a mixture of a colloidal monolayer and a polymerizable organic liquid is spread on a water surface, followed by photopolymerization of the liquid and removal of the colloidal particles by etching. The resulting porous films can be easily transferred to any desired substrate [26].

Flexible honeycomb films have been also fabricated using a polymer sheet support film [27]. Figure 23.8 shows a schematic of the procedure for the preparation of flexible honeycomb Au films. Important properties for a polymer support sheet are (i) low elasticity for peeling from the substrate and (ii) high solvent resistance during SAM formation. A combination of polyvinyl formal (PVF) and polyvinyl alcohol (PVA) has been experimentally demonstrated as a good supporting polymer. First, Au is deposited onto a PS particle monolayer by vacuum evaporation, and then an aqueous solution of PVA (20 wt %) and a chloroform solution of PVF (20 wt %) are cast onto the Au film and the solvents evaporated. The PVA layer is then peeled from the Au/(PVF + PVA) film after the Au/(PVF + PVA) films were subjected to immersion into various solutions of thiol compounds. As shown in Figure 23.9, the structure of the Au/PVF film was not deformed after peeling off the PVA layer. Although the honeycomb structure was heterogeneous on the microscale, as shown in Figure 23.9, the area of a few square centimeters appears homogeneous to the naked eye. The Au/PVF films were immersed into various solvents, methanol, ethanol, and chloroform, which caused a decrease in the flexibility, curling, and deformation of the films, respectively, while hexane had no effect on the film. Sessile water contact angles of the honeycomb Au film were measured, and the resulting average contact angles are listed in Table 23.3. The contact angles of surfaces covered with FOT and ODT, which yield a hydrophobic surface, were approximately 150° and greater than the values of 112° and 105° for the flat surface. This indicates that surface roughness

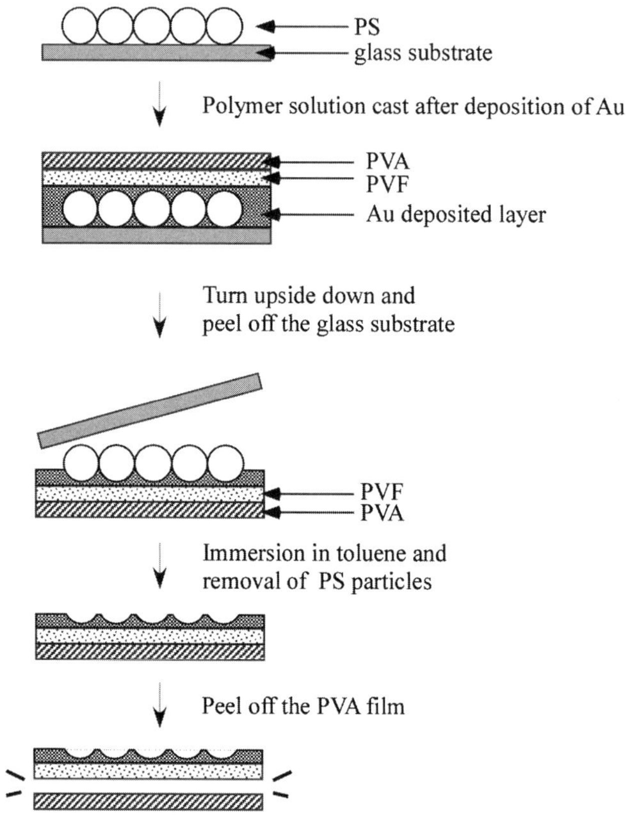

Figure 23.8. Schematic illustration of the experimental procedure used for the fabrication of a flexible honeycomb Au film supported by a polymer sheet.

enhances the hydrophobicity. In contrast, the contact angle for hydrophilic MPA was 8°, much smaller than that for the flat surface. Thus, the hydrophilicity was also emphasized by the surface roughness. In addition, the enhancement of hydrophobicity and hydrophilicity for PS-200 was larger than that for PS-560. The observed values for the FOT and ODT SAMs were consistent with the values predicted from the Cassie–Baxter equation. Thus, for the hydrophobic SAM system, air was present in the honeycomb spaces. For the hydrophilic SAM, the observed angles were close to the calculated angles, while those for the hydrophobic SAM were not close and the contact angles were much larger than the expected values. These results provide evidence that for bare Au, the wettability obeys the Wenzel equation; therefore, no air pockets were present.

Figure 23.9. Photograph and atomic force microscope (AFM) image of a honeycomb Au film.

TABLE 23.3. Contact Angles of Water Droplets for Honeycomb and Flat Surfaces Modified with Various SAMs

	Contact Angle (Degree)				
Samples	FOT	ODT	MHdA	MPA	No Treatment
PS-200 (θ_r)	155	149	68	8	54
PS-560 (θ_r)	150	146	76	13	56
Flat (θ_s)	112	105	75	45	63

MHdA, 16-mercaptohexadecanoic acid.

23.6 COLLOIDAL MONOLAYER ON A LIQUID SURFACE

Although colloidal monolayers on solid substrates have been used to prepare various periodic nanostructures, another study demonstrated the formation of a colloidal monolayer on a liquid surface [28]. For instance, when an aqueous solution of PVA is dropped onto a PS colloidal monolayer on a glass substrate, the monolayer floats on the water surface. The PS particle array in the monolayer is not affected by the floating process and maintains close-hexagonal packing. This approach has a practical advantage for the preparation of polymer honeycomb structures with desired thickness. The thickness can be easily controlled by changing the concentration or amount of PVA solution [28]. Figure 23.10 shows a schematic illustration of the experimental procedure used for the fabrication of polymer films taken from a PVA porous film template. The porous template is prepared by using a PS particle monolayer on a PVA solution. The porous PVA film can be used as a template for fabricating cylinder-like or needlelike nanopatterns of polymers, as shown in Figure 23.11. In addition, when an elastic polymer, such as silicone rubber, is used as a filler, hemisphere arrays of the polymer can be easily obtained. Interestingly, although the diameter of the hemisphere was ca. 520 nm and was larger than that of the honeycomb template pores, the honeycomb film was not damaged.

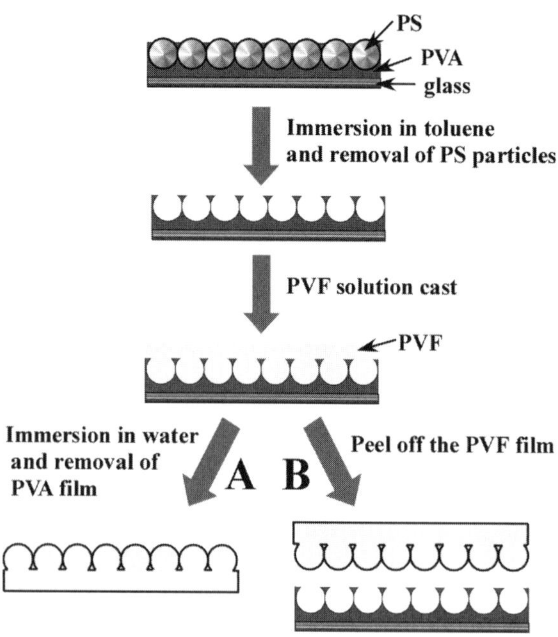

Figure 23.10. Schematic illustration of the experimental procedure used for the fabrication of PVF films taken from a PVA honeycomb film template.

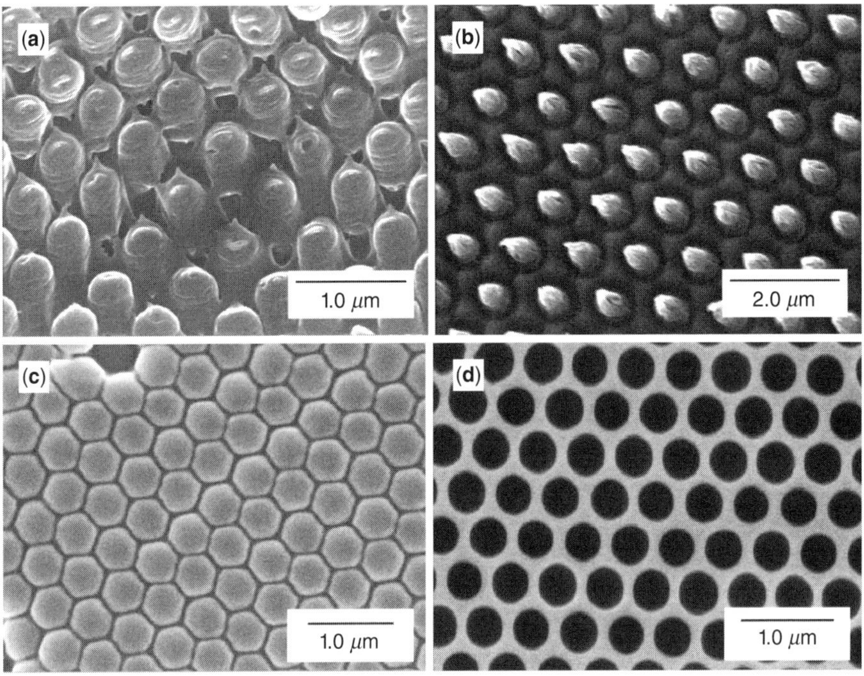

Figure 23.11. SEM images of (a,b) PVF films and (c) silicone rubber film taken from a honeycomb film template, and (d) a PVA film honeycomb template after peeling off a silicone rubber film.

Thus, it would be possible to repeatedly use the PVA porous film as a template.

23.7 CONCLUSION

This review has shown the use of colloidal monolayers as masks and templates as a very simple and inexpensive, yet powerful and versatile process for the fabrication of periodic nanostructures. The application of colloidal monolayers in nanostructure fabrication has been developed over the last 10 years, and the field of colloidal monolayer has witnessed intensive progress. Therefore, the appearance of new nanostructures with desired compositions and novel morphology is naturally expected as the colloidal monolayer technique evolves. To date, the size of colloidal particles used to prepare nanostructures has been greater than ca. 100 nm; however, the use of nanometer-sized particles could introduce the formation of even smaller and more complex nanostructures.

REFERENCES

1 A. Stein, F. Li, N. R. Denny. *Chem. Mater.* 20 (2008) 649.
2 Y. Li, W. Cai, G. Duan. *Chem. Mater.* 20 (2008) 615.
3 F. Caruso (ed.), *Colloids and colloid assemblies*. Wiley-VCH, Weinheim, 2004.
4 K. E. Davis, W. B. Russel, W. J. Glantschnig. *Science* 245 (1989) 507.
5 P. N. Pusey, W. van Megen. *Nature* 320 (1986) 340.
6 P. Jiang, J. F. Bertone, K. S. Hwang, V. L. Colvin. *Chem. Mater.* 11 (1999) 2132.
7 Y. Li, T. Sasaki, Y. Shimizu, N. Koshizaki. *Small* 12 (2008) 2286.
8 J. C. Love, B. D. Gates, D. B. Wolfe, K. E. Paul, G. M. Whitesides. *Nano Lett.* 2 (2002) 891.
9 J. Liu, A. I. Maaroof, L. Wieczorek, M. B. Corite. *Adv. Mater.* 17 (2005) 1276.
10 A. B. Pawar, I. Kretzschmar. *Langmuir* 24 (2008) 355.
11 A. B. Pawar, I. Kretzschmar. *Langmuir* 25 (2009) 9057.
12 Y. Imura, M. Kato, T. Kondo, T. Kawai. *Langmuir* 26 (2010) 11314.
13 A. J. Haes, R. P. Van Duyne. *J. Am. Chem. Soc.* 124 (2002) 10596.
14 T. R. Jensen, M. D. Malinsky, C. L. Haynes, R. P. Van Duyne. *J. Phys. Chem. B* 104 (2000) 10549.
15 C. L. Haynes, R. P. Van Duyne. *J. Phys. Chem. B* 105 (2001) 5599.
16 Y. Wang, J. Rybczynsky, D. Z. Wang, K. Kempa, Z. F. Ren. *Appl. Phys. Lett.* 85 (2004) 4741.
17 K. H. Park, S. Lee, K. H. Koh, R. Lacerda, K. B. K. Teo, W. I. Milne. *J. Appl. Phys.* 97 (2005) 24311.
18 C. L. Cheung, R. J. Nikolic, C. E Reinhardt, T. F. Wang. *Nanotechnology* 17 (2006) 1339.
19 C.-W. Kuo, J.-Y. Shiu, P. Chen. *Chem. Mater.* 15 (2003) 2917.
20 A. Sinitskii, S. Neumeier, J. Nelles, M. Fischler, U. Simon. *Nanotechnology* 18 (2007) 305307.
21 W. Ruan, C. Wang, N. Ji, Z. Lu, T. Zhou, B. Zhao, J. R. Lombardi. *Langmuir* 24 (2008) 8417.
22 Y. Li, C. Li, S. O. Cho, G. Duan, W. Cai. *Langmuir* 23 (2007) 9802.
23 B. Cao, W. Cai, F. Sun, Y. Li, Y. Lei, L. Zhang. *Chem. Commun.* 14 (2004) 1604.
24 F. Sun, W. Cai, Y. Li, B. Cao, F. Lu, G. Duan, L. Zhang. *Adv. Mater.* 16 (2004) 1116.
25 F. Yan, W. A. Goedel. *Nano Lett.* 4 (2004) 1193.
26 H. Xu, W. A. Goedel. *Small* 1 (2005) 808.
27 T. Kawai, M. Suzuki, T. Kondo. *Langmuir* 22 (2006) 9957.
28 Y. Imura, M. Suzuki, C. Morita, T. Kondo, T. Kawai. *Kobunshi Ronbunshu* 64 (2007) 166.

24

LIQUID-PHASE SYNTHESIS OF CARBON NANOTUBES AND OTHER CARBON NANOMATERIALS

Kiyofumi Yamagiwa and Jun Kuwano

24.1 INTRODUCTION

Carbon nanomaterials (CNMs), such as fullerenes, carbon nanotubes (CNTs), carbon nanohorns (CNHs), and so on, have received much attention for their outstanding electrical, mechanical, physical, and chemical properties, which make them potentially useful in a wide range of applications, including microelectronic devices, catalyst supports, electrodes and reinforcement materials, and so on [1–2]. They have been usually synthesized by three main established gas-phase processes (Table 24.1), which can be classified into two groups according to the generation process of carbon atoms: arc discharge and laser ablation methods (sublimation of graphite-condensation) and chemical vapor deposition (CVD) method (decomposition of carbon source compounds) [3].

In this chapter, the authors review the main established gas-phase synthetic methods and introduce a novel synthetic method by the resistance heating of a conductive substrate in a carbon source solvent. This novel method is called "liquid-phase synthesis" from now on. However, we would like to point out in advance that the liquid-phase synthesis does not mean "liquid-phase reaction" or "growth from liquid phase" in the strict sense of the word but "synthesis conducted in a liquid pool" vaguely.

Electrical Phenomena at Interfaces and Biointerfaces: Fundamentals and Applications in Nano-, Bio-, and Environmental Sciences, First Edition. Edited by Hiroyuki Ohshima.

TABLE 24.1. Features of Gas-Phase Synthesis and Liquid-Phase Synthesis

	Gas-Phase Synthesis			
Synthetic Technique	Arc Discharge	Laser Ablation	Chemical Vapor Deposition	Liquid-Phase Synthesis
Synthesis process	Sublimation of graphite-condensation	Sublimation of graphite-condensation	Decomposition of carbon source compounds	Decomposition of carbon source compounds
Temperature/pressure	>3000°C/reduced	>3000°C/reduced	<1200°C/reduced to ambient	<1200°C/ambient
Product	Fullerene, MWCNT, SWCNT (with catalyst)	Primarily SWCNT, fullerene, CNH	MWCNT, SWCNT, HACNTA	MWCNT, SWCNT, HACNTA
Purity/quality of CNT	Low/high (short, few defects)	High/medium	Medium-high/low (long)	Medium-high/low
Production system/cost	Easy design/medium	Elaborate system/high	Medium/low	Easy design/low
Scalability	Medium	Poor	Good	Poor
Advantages/disadvantages	/Costly carbon rod	Good control over diameter/high-power lasers, costly carbon rods	Continuous, large-scale production alignment and patterning/	Alignment/batch process, small scale
Improved method/remarks	Rotating arc, magnetic field assisted/	Ultrafast pulse, continuous wave laser power/	FBCVD, CoMoCAT, HiPCO, DIPS, supergrowth/	One-step process/potentially scalable, continuous process

In the liquid-phase synthesis, CNMs are deposited on a substrate resistance heated in carbon source solvent in most cases. The reaction field is limited to the vicinity of the solvent–substrate boundary in a boiling state (strictly speaking, "pool boiling," according to thermal engineering terminology). When resistance heating starts, the surface temperature of the substrate rises and boiling occurs via three characteristic stages, which are nucleate, transition, and film boiling. In the initial heating stage of the liquid-phase synthesis, the decomposition of a catalyst precursor (CP) and the nucleation of CNMs probably take place in a nucleate or transition boiling state under the coexistence of gas and liquid phases, and almost the whole CNM growth process mainly proceeds in a film boiling stage. In this sense, this process mostly resembles the thermal CVD process of the three main gas-phase processes. The boiling state in each stage is, however, a dynamic, complex physical process involving phase transition, cavitations, liquid convection, vapor flow, and heat transfer. From a technological point of view, the liquid-phase synthesis using resistance heating is thus different from CVD methods and other known gas-phase synthetic processes described next.

24.2 GAS-PHASE SYNTHETIC METHODS OF CNTS

Before moving on to the main subject, we briefly review three conventional gas-phase synthetic methods of CNMs, that is, arc discharge, laser ablation, and CVD methods. Their features are summarized in Table 24.1.

24.2.1 Arc Discharge Method

In this method, CNMs are synthesized through the arc vaporization of two opposing carbon rods separated by ~1 mm. The high-temperature arc plasma (>3000–4000°C) is generated by applying a DC voltage of approximately 30 V (50–120 A) between the rods in an inert gas (He, Ar) with a reduced pressure (between 50 and 700 mbar). The carbon rods vaporize and the vaporized carbon moieties condense very rapidly to form CNMs, including fullerenes (C_{60}, C_{70}, C_{74}, etc.), CNTs, and other carbonaceous by-products. This method has become the most widely used at the first stage because it is the most common and perhaps easiest gas-phase synthetic method. In fact, by using this method, C_{60} was first discovered by Kroto et al. in 1985 [4], multiwalled CNTs (MWCNTs) and single-walled CNTs (SWCNTs) were first discovered by Iijima in 1991 [5] and Iijima and Ichihashi in 1993 [6], respectively. The crude product is a complex mixture, which requires further purification to separate the target CNM from the soot and the residual metallic catalysts. The maximum yield is up to about 30% for MWCNTs. The MWCNTs can be purified by heating the crude product in an oxygen atmosphere. The resultant MWCNTs are high quality and very straight due to the walls with a high crystallinity. The high-power facilities with a 65-mm-diameter rod have been recently constructed. The capacity for MWCNT production is as large as 1 kg/h [7]. For SWCNTs, 6-kW facilities can produce up to 100 g/h [8]. The serious drawbacks are large energy consumption, a relatively low yield, costly graphite rod, and the difficulties of automation and upscaling.

24.2.2 Laser Ablation Method

This method has been used for the production of fullerenes, SWCNTs, and CNHs [9–10]. A high-power yttrium aluminum garnet (YAG) or CO_2 laser is required for the vaporization of a carbon target with a high vaporization temperature. The carbon species are ablated from a graphite composite rod containing preferably Ni and Co, carried downstream by an Ar carrier gas, and deposited on a water-cooled collector. This method is suited to produce high-quality, diameter-controlled SWCNTs. SWCNTs can be usually produced in gram scale with a yield of ~70% by mass.

The production capacity has improved to 45 g/h with ultrafast pulses from a free electron laser (FEL) method with a jet of preheated Ar of 1000°C. It has become possible to supply micron-size graphite powders in the continuous wave laser powder method. However, it has not still been economically advantageous compared with either arc discharge or CVD because the process requires high-purity graphite rods, a vacuum system, and a high-power laser.

24.2.3 CVD Method

The general growth mechanism involves the dissociation of gaseous or volatile carbon sources catalyzed by metallic nanoparticles, and the dissolution and saturation of carbon atoms in the metallic catalyst nanoparticles.

Of the gas-phase methods, CVD techniques show the greatest promise for economically viable, large-scale synthesis. In particular, the fluidized bed chemical vapor deposition (FBCVD) technique [11] is well suited to synthesizing high-quality CNTs inexpensively in large quantities. The selective MWCNT, double-walled CNT and SWCNT synthesis have been well developed using CVD. Another advantage of CVD methods is that they allow more control over the morphology and structure of the produced CNTs.

Different approaches for industrial-scale preparation of CNTs have been shown to be the most promising: FBCVD, high-pressure CO disproportionation (HiPCO) [12], CO CVD [13], and alcohol CVD [14]. Several companies have built/are building over 400 tons/year CNT production plants in 2010 in the world.

The modified direct injection pyrolytic synthesis (DIPS) method with a splay has achieved a high SWCNT purity of 97% and a very high degree of graphitization.

With CVD methods, highly aligned carbon nanotube arrays (HACNTAs) have been prepared on various substrates (Si, quarts, metals, etc.). The "crowding effect" due to the dense growth of CNTs allows them to self-orient and to grow perpendicular to the substrate. It is an outstanding achievement that a Japanese group prepared super-dense single-walled HACNTAs (over 99.98% purity without any purification) with heights up to 2.5 mm by so-called super-growth method [15]. Moreover, highly patterned, organized SWCNT structures were successfully fabricated.

The authors have also reported the synthesis of HACNTAs on stainless steel substrates by a CVD method with the natural plant product camphor [16]. We used stainless steel substrates to make it possible to utilize their anisotropies of thermal conductivity, electrical conductivity, and alignment more effectively in applying to electrodes and catalyst supports, and so on.

24.3 LIQUID-PHASE SYNTHESIS

The liquid-phase synthesis has been conducted for the first time by Ando et al. [17, 18]. The outline is as follows. A Fe metal catalyst is deposited on a small conductive Si substrate by sputtering in a vacuum. The substrate is set between two electric terminals and resistance heated in methanol or ethanol at 500–1000°C. Highly aligned multiwalled carbon nanotube arrays (HAMWCNTAs) are grown on the substrate. The process requires a sputtering system for catalyst preparation and thus involves two steps consisting of catalyst preparation and resistance heating.

24.3.1 One-Step Liquid-Phase (OS-LP) Synthesis

The liquid-phase synthesis is inherently simple and low in cost. The authors have recently developed a novel "one-step" process for liquid-phase synthesis of CNTs and other CNMs [19–22]. In order to take advantage of the inherent feature more, we omitted the preparation of a catalyst by dissolving a CP homogeneously to the carbon source solvent. We made further improvement: We used a metal substrate that is cut preferably from a commercially available metal plate such as stainless steel plates because of their high versatility and application flexibility compared with Si substrates. As-purchased stainless steel plates can be conveniently used without any surface treatments except for degreasing with acetone. One of the important features of our one-step synthetic method of CNMs is that it is the simplest and has the lowest cost among the other CNM synthetic processes. Another advantage is to synthesize a wide variety of CNMs with a minimum modification of the setup and synthetic conditions. Typical products are shown in Figure 24.1a–d. In this section, we mainly focus on the OS-LP synthesis of CNMs.

24.3.1.1 A Typical Example of OS-LP Synthesis. We describe a typical example of OS-LP synthesis using the setup shown in Figure 24.2. Methanol and cobaltocene ($Co(C_5H_5)_2$) were used as carbon and catalyst sources. An as-purchased stainless steel (Japanese Industrial Standards JIS SUS304) plate

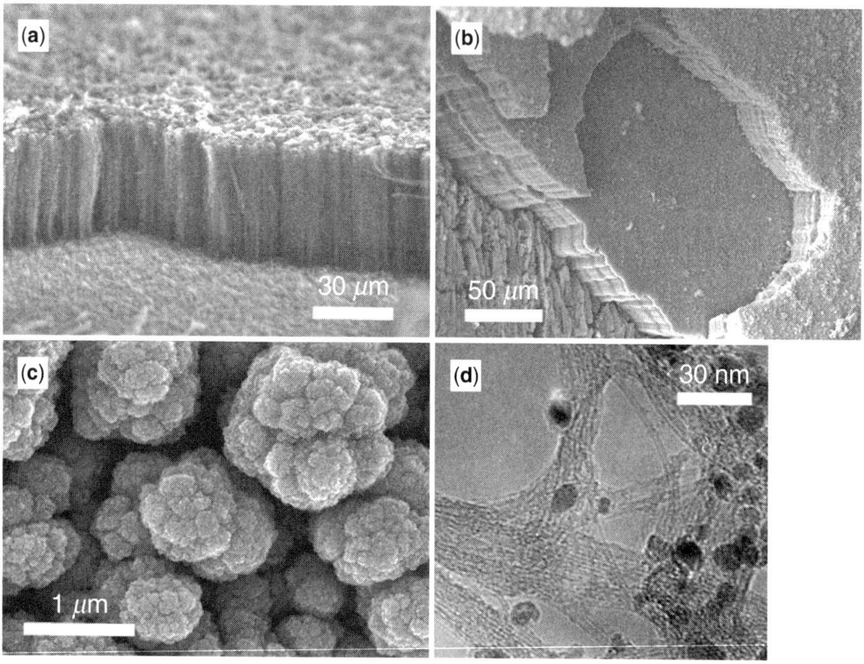

Figure 24.1. Typical CNMs prepared by one-step liquid-phase synthetic method. (a) highly aligned multiwalled CNT arrays (SEM image), (b) multilayer aligned multiwalled CNTs (SEM image), (c) broccoli-like CNTs (SEM image) and (d) single-walled CNT bundles (TEM image).

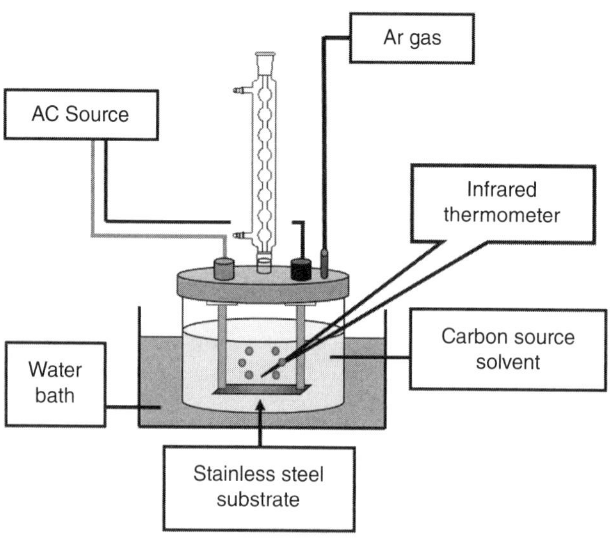

Figure 24.2. Setup for liquid-phase synthesis of CNMs.

was cut into small pieces (5.0 × 25.0 × 0.1 mm), which were used as substrates without any surface treatments except for ultrasonical cleaning in acetone. The substrate was fixed between two poles, and an Ar gas was flown at a rate of 100 mL/min. The substrate was heated at T_{st} = 800°C for t_{rh} = 15 minutes (t_{rh}, resistance-heating time) in pure methanol (200 mL) and $Co(C_5H_5)_2$ methanol solutions (0.1 and 1.0 mM) by applying an alternate current (5–6 V, 30–40 A). The surface temperature (T_{st}) of the substrate was measured by an infrared thermometer.

From pure methanol, small amounts of fibrous products, several dozen nanometers in diameter, were grown randomly on the SUS304 substrate. Nonaligned MWCNTs with an averaged outer diameter (D_o^{av}) of 17 nm were grown from 0.1 mM $Co(C_5H_5)_2$ methanol solution on the substrate. In contrast, HAMWCNTAs with a thickness of 20 μm were grown from 1 mM $Co(C_5H_5)_2$ methanol solution, as shown in Figure 24.3a,b. The tubes were not straight and had somewhat irregular tube walls, as seen from the transmission electron

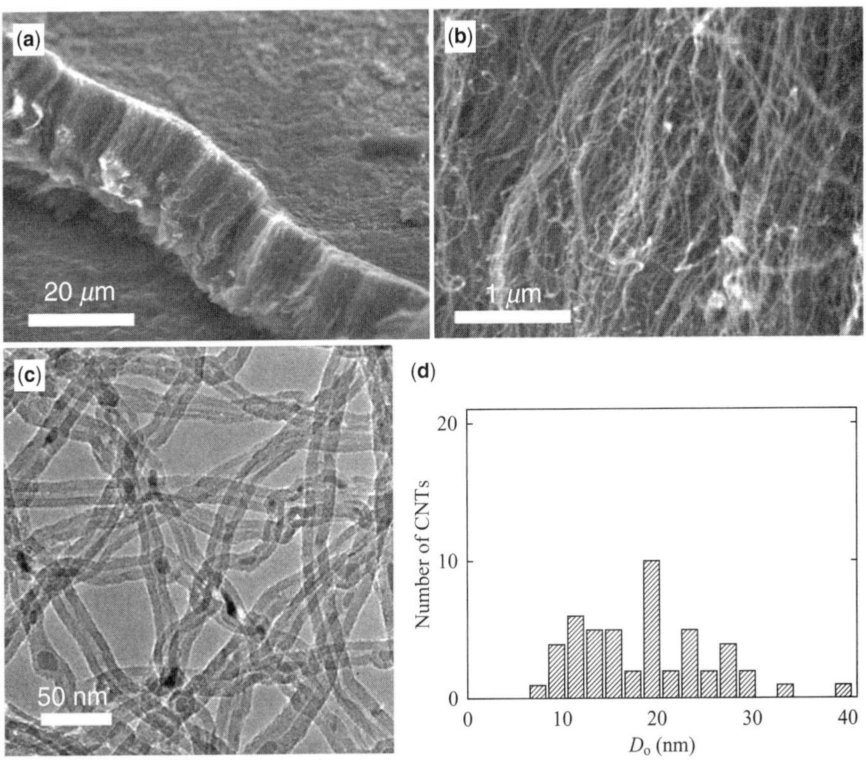

Figure 24.3. Highly aligned multiwalled CNT arrays prepared from 1 mM $Co(C_5H_5)_2$ methanol solution. (a) SEM image, (b) high-resolution SEM image, (c) TEM image, and (d) distribution of outer diameter D_o of multiwalled CNTs.

microscope (TEM) image (Fig. 24.3c). As shown in Figure 24.3d, the outer diameter D_o of MWCNTs had a distribution over a range of 8–39 nm with $D_o^{av} = 19$ nm. Nanoparticles of Co were often observed in the insides and on the outsides of the walls.

The average thicknesses of HAMWCNTAs were 4, 13, 20, and 20 μm for $t_{rh} = 5$, 10, 15, and 30 minutes, respectively. This indicates that the lifetime of the Co catalyst nanoparticles in $Co(C_5H_5)_2$ methanol solution is ~15 minutes.

The CNT growth mechanism presumably involves the following steps:

1. Vaporization of CP and carbon source molecules from the CP–carbon source solution to bubbles or/and the thin-film layer in a boiling state
2. Pyrolysis of CP molecules and deposition of metallic catalyst nanoparticles on the substrate surface
3. Catalytic dissociation of carbon source molecules by the metallic nanoparticles
4. Dissolution of carbon atoms to the metallic nanoparticle and their saturation
5. CNT growth from the saturated metallic particle

24.3.1.2 Catalyst Precursors (CPs). CPs are expected to influence the morphologies and alignment of CNTs considerably, considering the growth mechanism. The SUS304 substrates were heated at $T_{st} = 800°C$ for $t_{rh} = 15$ minutes in methanol solutions of ferrocene $Fe(C_5H_5)_2$, iron pentacarbonyl $Fe(CO)_5$, nickelocene $Ni(C_5H_5)_2$, and tris(acetylacetonato)cobalt $Co(C_5H_7O_2)_3$ ($Co(acac)_3$).

Figure 24.4 shows the scanning electron microscope (SEM) images of CNTs prepared from (a) 1 mM and (b) 10 mM $Fe(C_5H_5)_2$ methanol solutions. In the case of a 1 mM $Fe(C_5H_5)_2$ solution, a small amount of nonaligned MWCNTs was formed, though HACNTAs were grown in 1 mM $Co(C_5H_5)_2$ solution (see Section 24.3.1.1). The $Fe(C_5H_5)_2$ concentration necessary for HACNTA growth was approximately 10 times that of $Co(C_5H_5)_2$. Surprisingly, HACNTAs were prepared even at a very low $Fe(CO)_5$ concentration of 0.01 mM from the $Fe(CO)_5$ methanol solution (Fig. 24.4c), which was much lower than those of $Fe(C_5H_5)_2$ and $Co(C_5H_5)_2$. Such a low optimal $Fe(CO)_5$ concentration is presumably attributed to its decomposition temperature (T_D) as low as below 100°C. It is much lower than those of $Fe(C_5H_5)_2$ (ca. 500°C) and $Co(C_5H_5)_2$ (ca. 250°C). Probably, $Fe(CO)_5$ is able to form numerous Fe catalyst nanoparticles readily enough for HACNTA growth even at a low concentration. In the cases of $Ni(C_5H_5)_2$ and $Co(acac)_3$ methanol solutions, the concentration necessary for HACNTAs growth was as high as that of $Co(C_5H_5)_2$ (~1 mM). The concentration necessary for HACNTA growth has been thus confirmed to decrease with decreasing T_D.

More recently, we have succeeded in controlling D_o^{av} and synthesizing SWCNTs selectively (see Fig. 24.1d) with different CPs.

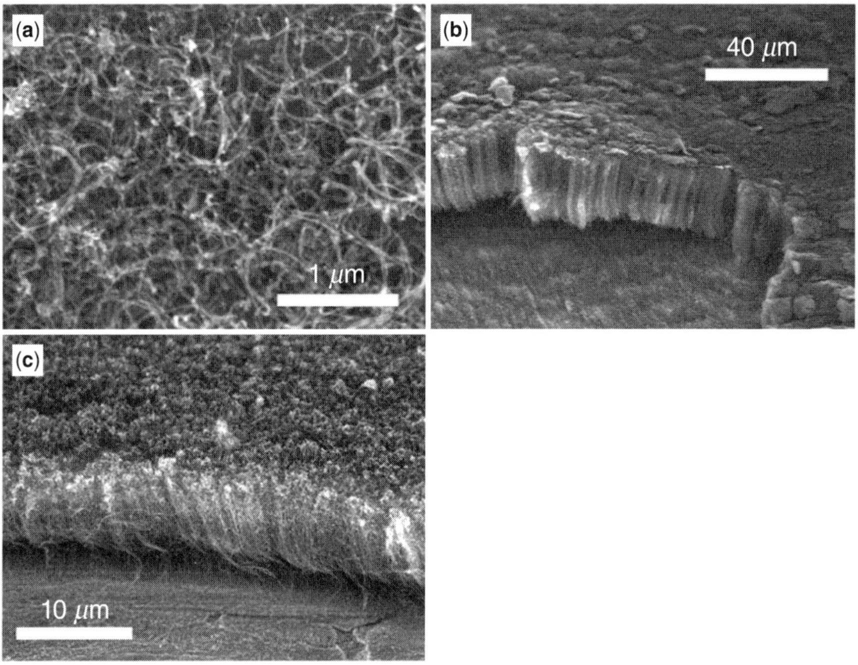

Figure 24.4. SEM images of CNTs prepared from methanol containing (a) 1 mM $Fe(C_5H_5)_2$, (b) 10 mM $Fe(C_5H_5)_2$, and (c) 0.01 mM $Fe(CO)_5$ [21].

The formation of HACNTA and the morphologies of CNTs are thus strongly influenced by CPs.

24.3.1.3 Carbon Sources. In the liquid-phase synthesis, the carbon source is supplied as liquid. In addition, the growth process is a more complicated one involving additional physical and chemical processes associated with coexistence of the liquid and gas phases, compared with those of the gas-phase synthesis. It is expected that the products would be largely influenced by the physical and chemical properties of the carbon sources, compared with the gas-phase synthesis.

Several straight-chain primary alcohols, diols, cyclic alcohols, 2-methyl-2-propanol (*tert*-butanol), and so on, were used as carbon sources to examine the effects of the molecular structures on the morphologies and alignment of CNTs. Homogeneously dissolved $Co(C_5H_5)_2$ in the alcohols was used as a CP. The $R_{Co(C5H5)2}$ ratio ($R_{Co(C5H5)2}$, the molar ratio of $Co(C_5H_5)_2$/alcohol) was set to $R[A] = 5 \times 10^{-5}$ for case A and $R[B] = 5 \times 10^{-5} \times n_c$ for case B (n_c, the number of carbon atoms in an alcohol molecule). The SUS304 substrate was resistance heated at $T_{st} = 800°C$ for $t_{rh} = 15$ minutes.

Aligned MWCNTs were grown from all the straight-chain primary alcohols (*sc-p* ROH: methanol, ethanol, 1-propanol, 1-butanol, 1-hexanol) on the

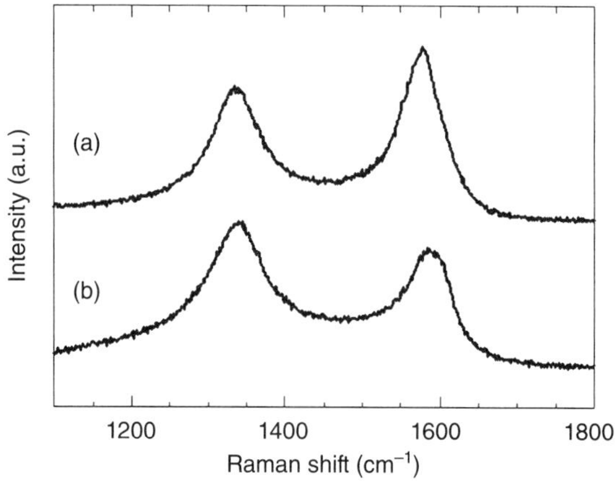

Figure 24.5. Raman spectra of the CNTs prepared from (a) methanol and (b) 1,2-ethanediol (R[A]).

substrates when $R_{\mathrm{Co(C5H5)2}} = R[\mathrm{A}]$. HACNTAs were grown densely and vertically to the substrate from methanol and ethanol (Fig. 24.8a), whereas aligned CNTs were grown somewhat loosely and irregularly from the *sc-p* ROH with $n_c = 3, 4, 6$. The yield of by-products such as amorphous carbons and carbon fibers increased as the chains of carbon sources became long. Their alignment and purity were in the order of methanol > ethanol > *sc-p* ROH with $n_c = 3$, 4 > 1-hexanol. The MWCNTs had a similar D_o^{av} (16–19 nm) and morphology for all the *sc-p* ROH with $n_c = 1$–6. There was a weak tendency that the tubes became straighter and had more regular walls with increasing n_c.

We examined the effects of CP concentration on CNT growth. HACNTAs were grown on the SUS304 substrate at R[B] for all the *sc-p* ROH. The yields of CNTs increased compared with those at R[A]. In particular, thick HACNTAs with a height of about 100 μm were synthesized from *sc-p* ROH with $n_c \geq 3$.

Figure 24.5a shows the Raman spectrum of the HACNTAs prepared from methanol at R[A] (= R[B]). The G-band, assigned to the in-plane vibration of the carbon atoms in graphene sheets, appeared at ~1590 cm^{-1}, and the D-band, due to defects and dangling bonds in the sheets, appeared at ~1350 cm^{-1}. The intensity ratio of both peaks (I_D/I_G) is a measure of the purity of CNTs in the case that CNTs have a similar regularity of CNT walls as described earlier. Figure 24.6 shows the variations of I_D/I_G with n_c for straight-chain primary alcohols ($n_c = 1$–4, 6), 1,2-ethanediol and cyclohexanol. In comparison of the I_D/I_G ratios of the straight-chain primary alcohols, the ratios increased as the chain became long, indicating that the amount of amorphous carbons increased. The methanol carbon source brought the lowest I_D/I_G ratio of all the alcohols.

We also examined the effects of the molecular structures of butanols on CNT growth. Four structural isomers of butanols (1-butanol, 2-butanol, 2-methyl-1-propanol (isobutanol), 2-methyl-2-propanol (*tert*-butanol)) were

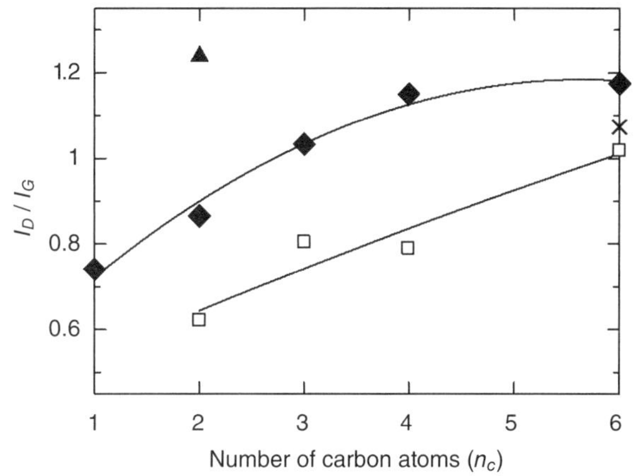

Figure 24.6. Variations of I_D/I_G with n_c. ◆: straight-chain primary alcohols (R[A]), □: straight-chain primary alcohols (R[B]), ▲: 1,2-ethanediol (R[A]), ×: cyclohexanol (R[A]).

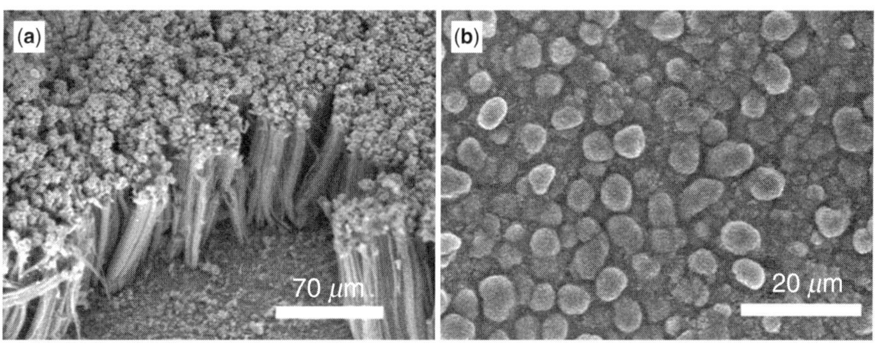

Figure 24.7. SEM images of the deposits prepared from (a) 1-butanol and (b) *tert*-butanol at R[B].

used as a carbon source. Figure 24.7a,b shows the SEM images of the deposits formed from 1-butanol and *tert*-butanol at R[B]. HACNTAs, as typically shown in Figure 24.7a, were grown from all the butanols except *tert*-butanol. In the case of *tert*-butanol, the observed morphology was considerably different from those of the other butanols. A small amount of hemispherical amorphous carbon deposits was produced, as shown in Figure 24.7b. It is well-known that it is relatively hard for tertiary alcohols to be oxidized. This would be associated with this phenomenon.

Figure 24.8b shows the SEM image of the CNTs prepared from 1,2-ethanediol at R[A]. A large amount of the mixture of amorphous carbons and irregularly

Figure 24.8. SEM images of the products prepared from (a) ethanol and (b) 1,2-ethanediol containing $Co(C_5H_5)_2$ (R[A]).

deposited CNTs was produced from 1,2-ethanediol. Their Raman spectra indicate that the products are worse in purity than those from ethanol (see Fig. 24.5b and the data for $n_c = 2$ in Fig. 24.6). This suggests that the high atomic ratio of oxygen or the molecular structure of the diol leads to the formation of a large amount of amorphous carbons.

In the case of cyclohexanol, ACNTAs (i.e., not highly aligned) were grown at R[A]. The arrays were rather densely grown on the whole substrate. The height of the arrays was relatively large, about 70 μm. In comparison with the Raman spectra, the CNTs prepared from cyclohexanol had better purity than those from 1-hexanol at R[A] (see the data for $n_c = 6$ in Fig. 24.6). The results reveal that CNT growth is considerably influenced by the cyclic structure.

24.3.2 Effects of Additives

It is possible to dissolve or disperse various additives in the carbon source solvents in order to enhance the growth rate, to control the morphology, to modify the products, and so on. This is an important advantage of the liquid-phase synthesis and brings a wide variation different from the gas-phase synthesis. We describe some examples of the additives in the OS-LP synthesis.

24.3.2.1 KOH. Potassium hydroxide KOH is a well-known chemical activator for the preparation of activated carbons.

Figure 24.9a shows the SEM image of HACNTAs grown from 10 mM $Fe(C_5H_5)_2$ methanol solution ($T_{st} = 800°C$, $t_{rh} = 15$ minutes) under the presence of the KOH additive. A height of about 100 μm was approximately five times longer than that without KOH (Fig. 24.4b).

Additives of MOH (M = Li, Na, K) and CH_3COOK were added to the solution to elucidate the roles of K and OH ions in HACNTA growth. No such

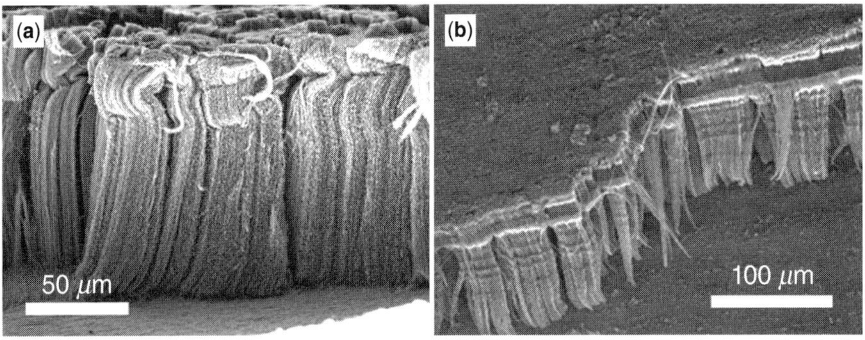

Figure 24.9. SEM images of highly aligned CNT arrays prepared with the addition of (a) KOH and (b) H_2O.

effect was observed when LiOH, NaOH, and CH_3COOK were added. This indicates that neither K nor OH alone accelerates the growth of CNTs. The added KOH presumably acts as a chemical activator: The added KOH would remove amorphous carbons deposited on the surface of the catalyst particles and would probably prevent the inactivation, resulting in the acceleration of HACNTA growth.

24.3.2.2 H_2O. A methanol solution of 5 mM $Co(acac)_3$ was referred to as a "blank solution." A small amount of H_2O (50–1000 μL: 14–278 mM) was added to the blank (200 mL).

Figure 24.9b shows a typical SEM image of HACNTAs grown on the substrate with an optimal amount of 100 μL H_2O ($T_{st} = 700°C$, $t_{rh} = 30$ minutes). The HACNTAs (~100 μm) were twice thicker than those (~50 μm) from the blank solution. Further addition of H_2O gradually lowered the acceleration effect. The growth rate of HACNTAs gradually decreased with t_{rh} and leveled off at about 45 minutes in the blank solution, whereas the growth rate kept constant until 60 minutes in the 100 μL H_2O-containing solution, with formation of over 200 μm HACNTAs.

The added H_2O thus removes the deposited amorphous carbons from the surfaces of catalyst particles and consequently accelerates the growth rate as with the drastic enhancement in the supergrowth CVD method [15].

24.3.2.3 Tetraethoxysilane (TEOS) and Others. TEOS ($Si(OC_2H_5)_4$) was added to a 10 mM $Fe(C_5H_5)_2$ methanol solution.

Figure 24.10 shows TEM images of the products in the cases of (a) no addition and (b) addition of TEOS. HACNTAs were grown by resistance heating at $T_{st} = 800°C$ for $t_{rh} = 15$ minutes in both cases. The averaged inner and outer diameters (D_i^{av} and D_o^{av}) of MWCNTs decreased from (a) $D_i^{av} = 9$ nm, $D_o^{av} = 21$ nm to (b) $D_i^{av} = 6$ nm, $D_o^{av} = 12$ nm.

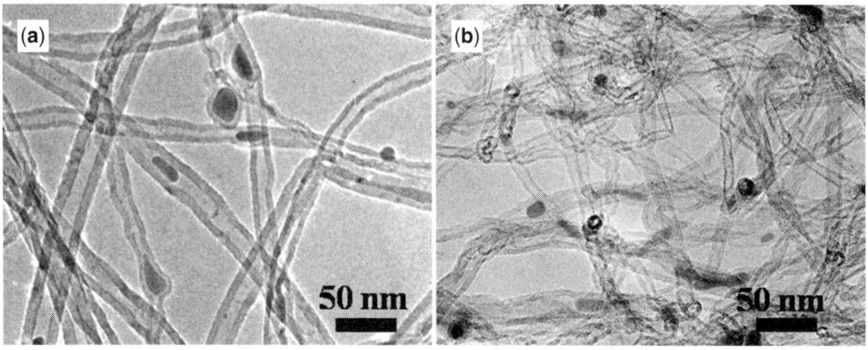

Figure 24.10. TEM images of multiwalled CNTs prepared from a 10 mM $Fe(C_5H_5)_2$ methanol solution. (a) No addition of TEOS and (b) addition of TEOS.

Figure 24.11. SEM images of products prepared from (a) D(+)-glucose and (b) DL-malic acid aqueous solution.

The speculated reason is as follows: In the initial stage of CNT growth, SiO_2 nanoparticles are formed by pyrolysis of TEOS and restrain the agglomerations of the Fe catalyst nanoparticles.

We are intensively examining the effects of other additives. Some of surface active agents have been found to enhance CNT growth and to form metal particle-included carbon shells selectively. The enhanced effect is probably associated with the promotion of bubble generation in the boiling of solvent.

24.3.3 CNT Synthesis in Aqueous Solution

We have developed a novel safe, ecofriendly, and organic solvent-free OS-LP synthesis with a noninflammable aqueous solution containing a powdered organic compound. For example, glucose, harmless and readily soluble in water, is an ideal candidate for the carbon source solid.

Figure 24.11a shows the SEM image of CNTs prepared from 3 M D(+)-glucose aqueous solution containing 10 mM $Co(CH_3COO)_2{\cdot}4H_2O$. A small amount of nonaligned MWCNTs with irregular walls and carbon fibers was grown by resistance heating at $T_{st} = 700°C$ for $t_{rh} = 5$ minutes on the SUS304 substrate.

Figure 24.11b shows the SEM image of CNTs prepared from 5 M DL-malic acid aqueous solution containing 10 mM $Ni(CH_3COO)_2 \cdot 4H_2O$. A small amount of nonaligned MWCNTs and carbon fibers was grown similarly to D(+)-glucose. Optimization of the synthetic conditions has been under progress.

24.4 FUTURE OF LIQUID-PHASE SYNTHESIS

In order to change the batchwise process to a continuous one for scaling up, we plan to design a novel system with a circulation apparatus of carbon source solvent. In this system, a metallic pipe is resistance heated as a carbon source solvent containing a CP is supplied to the heated pipe. The products are collected with a filter in the circulation system.

A broad choice of additives, CPs, carbon sources, and so on, in the OS-LP synthesis makes it possible to enhance the growth rate and purity, to control their morphology, to modify the CNM, to synthesize new CNMs selectively, and so on. In particular, the additives that cannot vaporize will be attractive because the additives cannot be used in the gas-phase synthesis.

Lastly, we emphasize that the setup can be easily modified according to the desired design since the setup is easy to design and the process proceeds under atmospheric pressure.

REFERENCES

1 D. Tomanek, R. J. Enbody (eds.), *Science and application of nanotubes*. Kluwer Academic/Plenum Publishers, New York, 2000.

2 P. J. F. Harris (ed.), *Carbon nanotubes and related structures, new materials for the twenty-first century*. Cambridge University Press, Cambridge, 1999.

3 A. Jorio, M. S. Dresselhaus, G. Dresselhaus (eds.), *Carbon nanotubes, advanced topics in the synthesis, structure, properties and applications*. Springer-Verlag, Berlin, 2008.

4 H. W. Kroto, J. R. Heath, S. C. O'Brien, R. F. Curl, R. E. Smalley. *Nature* 318 (1985) 162.

5 S. Iijima. *Nature* 354 (1991) 56.

6 S. Iijima, T. Ichihashi. *Nature* 363 (1993) 603.

7 M. Meyappan (ed.), *Carbon nanotubes: science and applications*. CRC Press LLC, Boca Raton, FL, 2005.

8 E. G. Rakov. *Nanotechnol. Russ.* 3 (2008) 575.

9 A. Thess, R. Lee, P. Nikolaev, H. Dai, P. Petit, J. Robert, C. Xu, Y. H. Lee, S. G. Kim, A. G. Rinzler, D. T. Colbert, G. E. Scuseria, D. Tomanek, J. E. Fischer, R. E. Smalley. *Science* 273 (1996) 483.

10 S. Iijima, M. Yudasaka, R. Yamada, S. Bandow, K. Suenaga, F. Kokai, K. Takahashi. *Chem. Phys. Lett.* 309 (1999) 165.

11 C. H. See, A. T. Harris. *Ind. Eng. Chem. Res.* 46 (2007) 997.

12 P. Nikolaev, M. J. Bronikowski, R. K. Bradley, F. Rohmund, D. T. Colbert, K. A. Smith, R. E. Smalley. *Chem. Phys. Lett.* 313 (1999) 91.

13 B. Kitiyanan, W. E. Alvarez, J. H. Harwell, D. E. Resasco. *Chem. Phys. Lett.* 317 (2000) 497.

14 S. Maruyama, R. Kojima, Y. Miyauchi, S. Chiashi, M. Kohno. *Chem. Phys. Lett.* 360 (2002) 229.

15 K. Hata, D. N. Futaba, K. Mizuno, T. Namai, M. Yumura, S. Iijima. *Science* 306 (2004) 1362.

16 K. Yamada, K. Abe, M. Mikami, M. Saito, J. Kuwano. *Key Eng. Mater.* 320 (2006) 163.

17 Y. F. Zhang, M. N. Gamo, C. Y. Xiao, T. Ando. *Physica B* 323 (2002) 293.

18 Y. F. Zhang, M. N. Gamo, C. Y. Xiao, T. Ando. *Jpn. J. Appl. Phys.* 41 (2002) L408.

19 K. Yamagiwa, Y. Iwao, M. Mikami, T. Takeuchi, M. Saito, J. Kuwano. *Key Eng. Mater.* 350 (2007) 19.

20 M. Mikami, K. Yamagiwa, T. Takeuchi, M. Saito, J. Kuwano. *Key Eng. Mater.* 350 (2007) 23.

21 K. Yamagiwa, T. Kikitsu, S. Yamashita, J. Kuwano. *Jpn. J. Appl. Phys.* 50 (2011) 01BJ11.

22 S. Yamashita, T. Kikitsu, Y. Yamaguchi, K. Yamagiwa, J. Kuwano. *Key Eng. Mater.* 445 (2010) 201.

25

OXIDE CATHODE ELECTROCATALYSTS FOR FUEL CELLS

Morihiro Saito and Jun Kuwano

25.1 INTRODUCTION

Electrocatalysts with high oxygen reduction reaction (ORR) activities are a key to develop electrochemical devices such as fuel cells, solid oxygen sensors, and reactors [1–3]. The best-known electrocatalyst for practical use is platinum because of its high ORR activity and long-term stability. However, the cost and resource are a critical problem. Therefore, seeking of alternative materials to platinum and improvement of well-known catalysts are of great importance. For this purpose, many researchers in the world have studied new electrocatalysts with high ORR activities. In particular, organic metal complexes supported on carbon [4] and chalcogen compounds such as metal sulfide [5] have been intensively examined as the alternative to carbon-supported platinum (Pt/C) cathode catalysts for proton-exchange membrane fuel cells (PEMFCs). However, the chemical and electrochemical stabilities are not enough in the acidic media and the ORR activities have to be significantly improved for practical use. In recent years, Ota et al. have proposed metal carbonitride-based catalysts such as partially oxidized tantalum and zirconium carbonitride (Ta-CNO and Zr-CNO) and reported that the Ta-CNO and Zr-CNO catalysts exhibited relatively high onset potentials (E_{on}) of ca. 0.9 V versus reversible

Electrical Phenomena at Interfaces and Biointerfaces: Fundamentals and Applications in Nano-, Bio-, and Environmental Sciences, First Edition. Edited by Hiroyuki Ohshima.

hydrogen electrode (RHE) for ORR [6, 7]. Ozaki et al. have also synthesized B- and/or N-doped carbon alloy catalysts with a nanoshell structure of graphitized carbon and found a high E_{on} value of ca. 0.8 V [8, 9]. Both developments have intensively proceeded as the national projects of Japan by the New Energy and Industrial Technology Development Organization (NEDO). However, it is still necessary to improve the ORR activities and their PEMFC performances.

On the other hand, alkaline membrane fuel cells (AMFCs) with anion-exchange membranes (AEMs) as the electrolyte have recently received increasing attention because of the availability of non-Pt-based catalysts and the promotion of ORR at the cathode owing to the alkaline media of electrolyte [10, 11]. Table 25.1 shows known non-Pt-based ORR catalysts in alkaline media. Thus, various types of ORR catalysts such as metals, oxides, metal complexes, and carbons have been so far developed [12], and their potential to applications in energy-related new technologies—brine electrolysis, fuel cells, and metal–air batteries—was examined. In particular, it is well-known that silver dispersed on carbon supports (Ag/C) [10, 11] and Mn oxide-based catalysts [13, 14] have shown potential to use for the cathode catalysts of

TABLE 25.1. Known Non-Pt-Based ORR Catalysts in Alkaline Media [12]

Type		Catalyst
Metal	Noble metal	Ag, Pd, Ru, Au, Rh
	Transition metal	Ni, Co, Cu, Fe
Oxide	Simple	MnO_2, Mn_2O_3, Mn_5O_8, TiO_2, NiO(+Li), RuO_2, CrO_2, Tl_2O_3
	Pyrochlore	$Pb_2Ru_{2-x}Pb_xO_{7-\delta}$, $Bi_2Ru_{2-x}Bi_xO_{7-\delta}$, $Pb_{2-x}Bi_xRuO_{7-\delta}$
	Spinel	Co_3O_4, Co_2NiO_4, $CoAl_2O_4$, $Ni_xCo_{3-x}O_4$, $CoFe_2O_4$, $LiMn_{2-x}M_xO_4$(M = Co, Fe)
	Perovskite	$La_{1-x}Sr_x(Mn, Co, Fe)O_3$, $La_{1-x}Sr_xMn_{1-y}Fe_yO_3$, $LnMnO_3$(Ln = lanthanoid)
Metal complex	T(M)PP[a]	M = Co, Fe, Ni, Cu, Au, Mn
	Pc[b]	M = Co, Ru, Mn, Pb, Fe, Ag
Carbon	–	Graphite, activated carbons, acetylene black, Ketjen black
Others	Hydroxide	$Fe(OH)_2$
	Acetate	$Co(CH_3COO)_2$, $Fe(CH_3COO)_2$
	Chalcogenide	Co_2S, Cu_2S, PbS, VS_2, Co_3S_4, Ni_3S_4, FeS_2, MoS_2, CoS_2, Mo_4Ru_6Se
	Nitride	Mn_4N, Co_3N, Ni_3N, TiN

[a] Tetra(methoxy)phenylporphyrin.
[b] Phtharocianine.

AMFCs and have advantages in resources and costs over conventional Pt-based catalysts. This is another way to solve the intrinsic problems of PEMFCs.

From the viewpoint of these trends on developing new ORR catalysts for the fuel cells, we will mainly introduce two types of metal oxide-based cathode catalysts: One is pyrochlores ($Pb_2Ru_{2-x}M_xO_{7-\delta}$ (PRMx, M = Mn, Fe, Co, Ni) [15], $Ln_2Ru_{2-y}Mn_yO_{7-\delta}$ (LnRMny, Ln = lanthanoid) [16–18], and $Ln_2A_{2-z}Mn_zO_{7-\delta}$ (LnAMnz, Ln = lanthanoid, A = Zr, Sn) [19]) and the other is metal (Ti or Mn) oxide nanosheet-based materials such as nanotubes and randomly restacked nanosheets with various cations (H_3O^+, H^+, and K^+) between the sheets [20–23]. Their ORR activities are shown mainly in alkaline aqueous solution, and strategies on the production of new advanced ORR catalysts are proposed. Moreover, since several oxide-based catalysts exhibit high ORR activities comparable to the conventional Pt/C catalysts, the AMFC single-cell performances are also discussed to demonstrate the possibility to the AMFC application.

25.2 PYROCHLORE-TYPE OXIDES

25.2.1 Exploring Pyrochlores with High ORR Activity

In 1978, Horowitz et al. first reported that the pyrochlores $Pb_2Ru_{2-x}Pb_xO_{7-\delta}$ (PRPbx) synthesized by the precipitation method acted as an ORR active electrocatalyst in alkaline solution as the U.S. patent [24]. Zen et al. revealed that PRPbx ($x = 0.05 - 0.40$) was stable in acid and that the ORR activity depended on the Pb content on the B-site [25]. However, the PRPbx contains a large amount of Pb with chemical toxicity to humans and the environment, and the ORR activity was not still sufficient for the fuel cells. Therefore, the authors have also synthesized two families of pyrochlores, $Pb_2Ru_{2-x}M_xO_{7-\delta}$ (PRMx, M = Mn, Fe, Co, Ni) [15] and $Ln_2Ru_2O_{7-\delta}$ (LnR, Ln = Pr, Nd, Sm, Gd, Dy, Yb) [16–18], by substitution of transition metal ions for Ru on the B-sites and lanthanoid ions for Pb on the A-site, and investigated the ORR activities both in 0.05 M H_2SO_4 and 0.1 M KOH solutions. As a result, the pyrochlores showed high ORR activities particularly in alkaline solution. The ORR activities depended on the kinds of A- and B-site ions. The partial substitution of Mn ion for the Ru-site and the selection of Ln ion with a smaller atomic number were found to be effective in improving the ORR activity in both systems. Figures 25.1 and 25.2 show the semi-steady-state voltammograms (SSVs) and efficiencies (Eff_4) of four-electron reduction of oxygen for LnR and partially Mn-substituted NdR, $Nd_2Ru_{2-y}Mn_yO_{7-\delta}$ (NdRMny), measured with a rotating ring-disk electrode (RRDE) in 0.1 M KOH solution at 70°C [17, 18]. The Eff_4 values were estimated by the following equation [16–18]:

$$Eff_4 = 100 \times \frac{i_D - i_R / N}{i_D + i_R / N}, \tag{25.1}$$

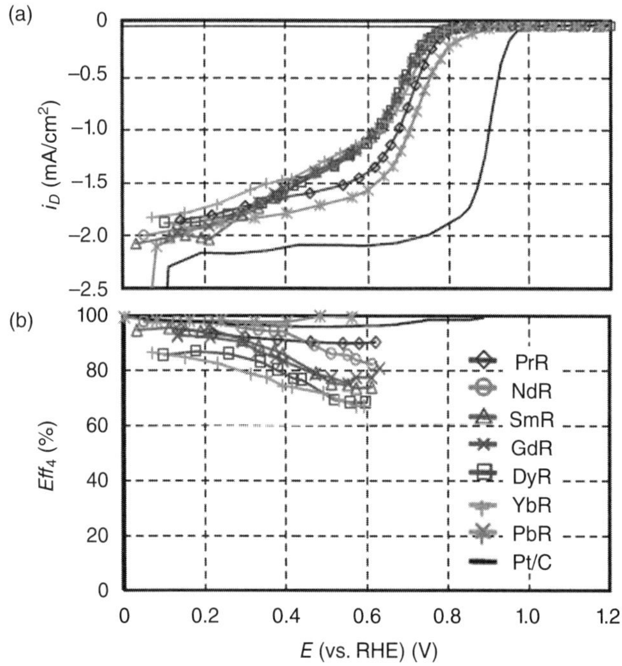

Figure 25.1. (a) Semi-steady-state voltammograms and (b) efficiencies (Eff_4) of four-electron reduction of oxygen for LnR in 0.1 M KOH at 70°C with O_2 bubbling.

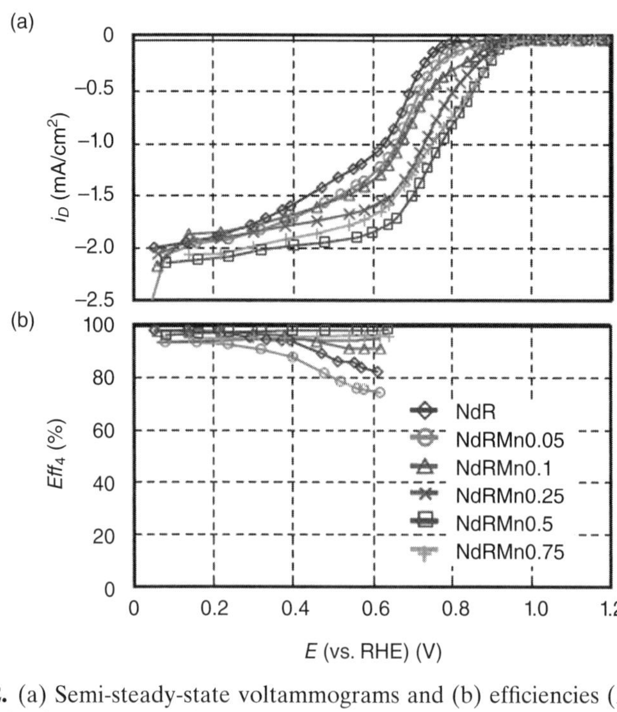

Figure 25.2. (a) Semi-steady-state voltammograms and (b) efficiencies (Eff_4) of four-electron reduction of oxygen for NdR and NdRMnx in 0.1 M KOH at 70°C with O_2 bubbling.

where i_D and i_R are the disk and ring currents, respectively, and N is the collection efficiency. The value of $N = 0.372$ was calculated from the geometric areas of the ring and disk electrodes when the rotating speed of RRDE was set to 300 rpm. As shown in Figures 25.1 and 25.2, the PrR and NdR exhibited higher ORR activities than the other LnR oxides, and their ORR activities increased with an increase in the amount of substituted Mn ion. The solid solution limits of Mn were $x = 0.1$ and 0.25 for PrRMnx and NdRMnx, respectively. In the formation range of the pyrochlore single phase, the Mn-richest NdRMn0.25, therefore, exhibited the highest ORR activity (E_{on} = ca. 0.95 V and $Eff_4 > 90\%$), which was comparable to those of the PRPbx reported by Horowitz et al. [24] and a conventional 20 mass % Pt/C catalyst. A similar trend was confirmed in the PRMx system, where the limited composition PRMn0.5 exhibited the highest ORR activity (E_{on} = ca. 0.95 V and $Eff_4 > 90\%$).

25.2.2 Influences of Alcohol Coexistence on the ORR Activity of NdRMnx

Direct alcohol fuel cells (DAFCs) have been significantly attractive as a power source for portable devices such as cellular phones and laptop computers because the use of liquid fuels makes gas cylinders free and the power source compact compared with the other types using hydrogen gas as the anode fuel. Ogumi et al. have studied an alkaline-type direct alcohol fuel cell (ADAFC) with carbon-supported perovskite oxide $La_{1-x}Sr_xMnO_3$ (LSM/C) as the cathode catalyst and demonstrated that the LSM/C showed high ORR activity and selectivity even in 1 M KOH containing 0.1 M ethylene glycol [26]. They also conducted single-cell tests of the membrane electrode assembly (MEA) fabricated with a commercially available anion-exchange membrane (A-006, Tokuyama Co.), in which a maximum power density of ca. 20 mW/cm^2 was achieved. In our results, the pyrochlore NdRMn0.25 also exhibited a higher ORR selectivity for some alcohols including ethylene glycol (Figs. 25.3 and 25.4) than the Pt/C catalyst, and interestingly, the ORR selectivity is also high for ethanol with much better safety for humans. This indicates that NdRMn0.25 has a potentiality for application to the cathode catalyst of small, low-cost ADAFCs with high safety for mobile devices.

25.2.3 Strategies on Low-Cost Production of New Pyrochlores with High ORR Activities

As mentioned earlier, the Pb-free NdRMn0.25 exhibited the highest ORR activity (E_{on} = ca. 0.95 V and $Eff_4 > 90\%$). However, it still contains the precious metal Ru. It is strongly desired to replace Ru by a cheaper and more resource-abundant metal ion because of its cost and resource. Based on this concept, we synthesized two pyrochlore series of $Ln_2Zr_2O_{7-\delta}$ (LnZ) and $Ln_2Sn_2O_{7-\delta}$ (LnS) (Ln = lanthanoid) and evaluated their ORR activities in 0.1 M KOH [19]. In both the LnZ and LnS series, the E_{on} values were ca. 0.85 V

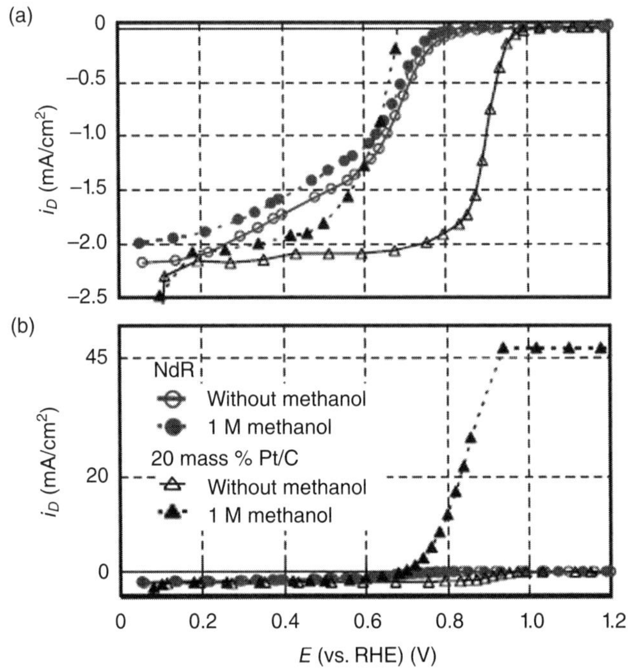

Figure 25.3. Semi-steady-state voltammograms focused on (a) oxygen reduction reaction and (b) methanol oxidation reaction for NdR and 20 mass % Pt/C (E-TEK) in 0.1 M KOH containing and not containing 1 M methanol at 70°C with O_2 bubbling.

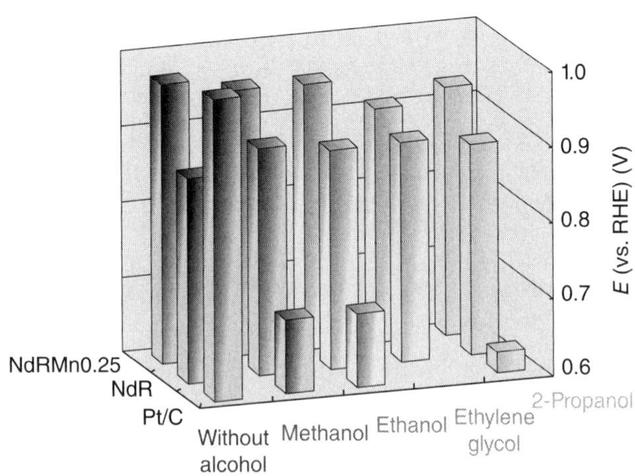

Figure 25.4. E_{on} values evaluated by the SSV measurements for NdR, NdRMn0.25, and 20 mass % Pt/C (E-TEK) in 0.1 M KOH containing various alcohols at 70°C with O_2 bubbling.

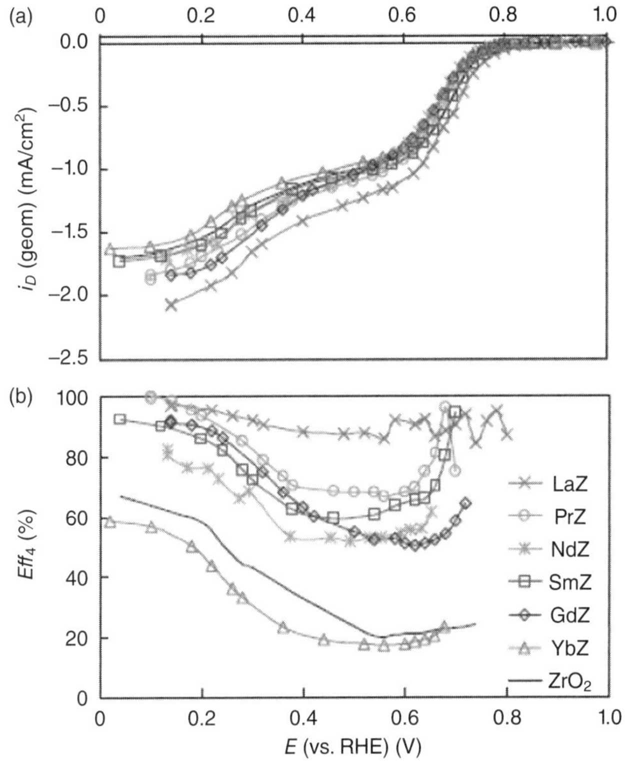

Figure 25.5. (a) Semi-steady-state voltammograms and (b) efficiencies (Eff_4) of four-electron reduction of oxygen for LnZ in 0.1 M KOH at 70°C with O_2 bubbling.

(as shown in Fig. 25.5 for the LnZ system), and the i_D and Eff_4 values increased with decreasing atomic number of the Ln ion in both series. This trend is in good agreement with that of the LnR. As for the effects of the B-site ion, the LnZ series exhibited higher ORR activities than the LnS ones in the case of the same Ln ion: The Eff_4 values of LaZ and LaS were more than 85% and 75%, respectively. On the other hand, the Eff_4 values for both LnZ and LnS series were higher than those of the starting materials ZrO_2 and SnO_2 except for that of YbZ. This indicates that the adoption of the pyrochlore structure leads to the high ORR activity in the LnZ and LnS series. Figure 25.6 shows variations of the ORR activities as a function of the lattice parameter for the LnZ and LnS series along with that for the LnR series. In both LnZ and LnS series, on the whole, the ORR activities (i_D at 0.7 V and Eff_4 values) increase with an increase in the lattice parameter as with that of the LnR series [16–18]. In the case of the LnR series, the electrical conductivity increased with an increase in the lattice parameter [26, 27]. We therefore suggest that the

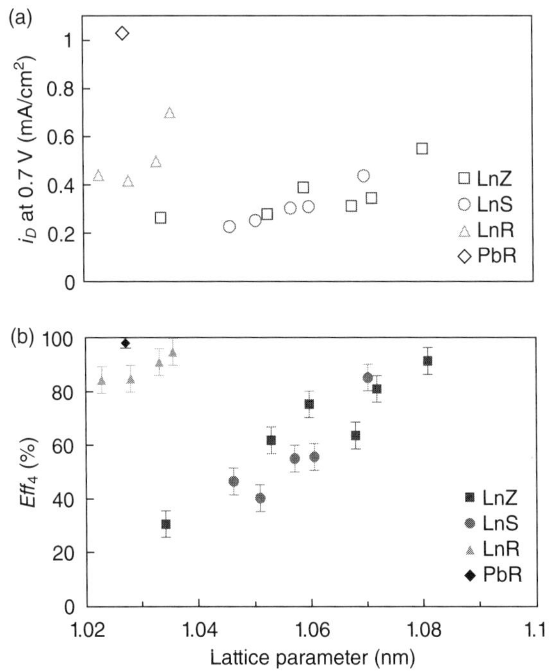

Figure 25.6. Variations in (a) i_D at 0.7 V and (b) Eff_4 for LnZ and LnS against the lattice parameter together with those of PbR and LnR.

electronic conductivity is one of the important factors to determine the ORR activity. However, in the case of the LnZ and LnS series, their electronic conductivities are not so high as compared with that of the LnR series [28]. Therefore, it is very interesting to consider the origin of the high ORR activities. From the X-ray absorption spectroscopy (XAS) analysis for the LnZ, the important determining factors for the ORR activity were suggested to be the electronic state of Zr, the Zr–O bond distance, and the distortion of the local structure around the Zr due to a variation in the ion radius of the Ln ion [19]. It is necessary to investigate the relation between these factors and the ORR activity for the establishment of strategies on the development of new, advanced cathode catalysts based on pyrochlore oxides. Further research is in progress.

On the other hand, cocatalysts consisting of two kinds of oxide are also one of the ways to enhance the ORR activity. Contrary to the LnR series, Zr ion was not completely replaced by Mn ion in the LaZ oxide, and the perovskite-type oxide $LaMnO_3$ gradually appeared as an impurity, with the formation of a mixed oxide containing two phases of the pyrochlore LaZ and the perovskite

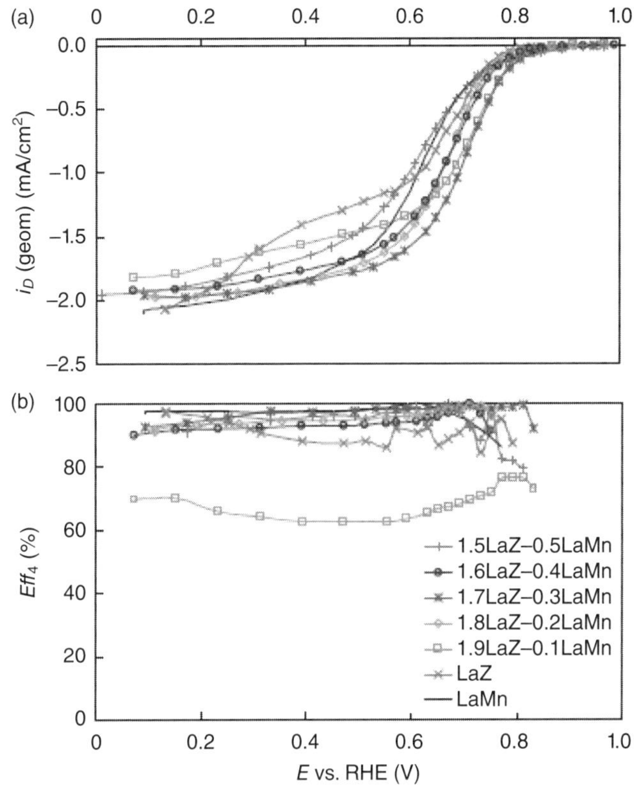

Figure 25.7. (a) Semi-steady-state voltammograms and (b) efficiencies (Eff_4) of four-electron reduction of oxygen for $(2 - x)$LnZ – $x$$LaMnO_3$ in 0.1 M KOH at 70°C with O_2 bubbling.

$LaMnO_3$. However, it is well-known that the $LaMnO_3$ also shows high ORR activity in alkaline solution [11, 12, 26, 29]; the mixing of $LaMnO_3$ is therefore expected to enhance the ORR activity. Figure 25.7 shows the SSV curves of the Mn-incorporated LaZ samples. The E_{on} values apparently became more positive and the magnitudes of I_D increased with an increase in the Mn content up to $x = 0.3$. The further mixing of $LaMnO_3$ caused a decrease in E_{on} against the expectation described earlier while maintaining the Eff_4 value (>95%). The ORR activity for the pure $LaMnO_3$ was rather low as compared with that of 0.7LaZ–0.3LaMn. Therefore, $LaMnO_3$ probably works as a promoter catalyst in the LaZ–LaMn system. Consequently, the ORR activity achieved the highest level, E_{on} = ca. 0.90 V and Eff_4 > 95%. Such a mixed oxide composition should be examined more extensively to elucidate the ORR mechanism for new wave of oxide catalysts.

25.3 METAL OXIDE NANOSHEET-BASED MATERIALS

25.3.1 Finding Unique ORR Properties in Oxide Nanosheet-Based Materials

In recent years, nanomaterials such as oxide nanosheets have received much attention owing to their unique nanostructures and great potential in electronic, optical, mechanical, and bioscience applications [30–32]. They can transform to various forms such as nanotubes and restacked nanosheets by rolling and restacking with cations between the nanosheets. This suggests the possibility of designing unique advanced ORR catalysts that have not yet been developed so far. In our previous studies [20, 21, 23], we synthesized three types of titania nanosheet-based materials, that is, H_3O^+-exchanged rolled nanosheet (titania nanotube, H_3O^+-NT(Ti)), regularly stacked nanosheets (layered titanate, (H_3O^+-RG(Ti))), and randomly restacked nanosheets (H_3O^+-RE(Ti)) by soft chemical methods and examined their ORR activities by SSV measurements with RRDE at 70°C in a 0.05 M H_2SO_4 solution (Fig. 25.8). As a note-

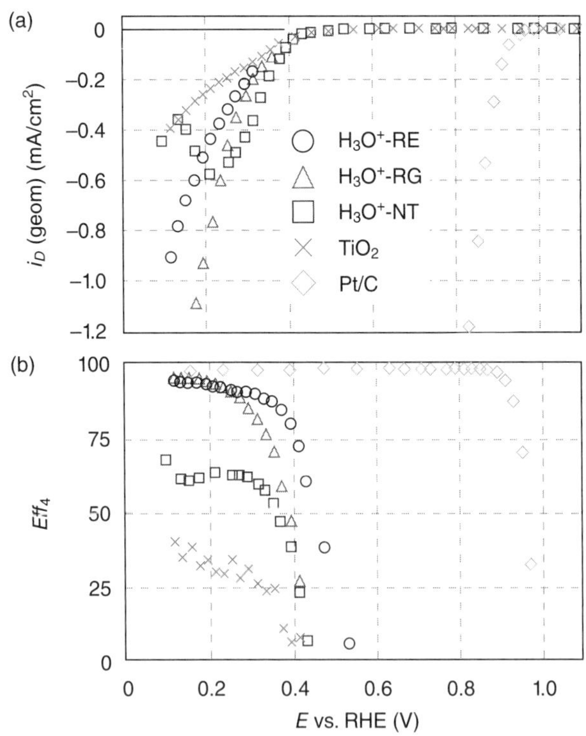

Figure 25.8. (a) Semi-steady-state voltammograms and (b) efficiencies (Eff_4) of four-electron reduction of oxygen for H_3O^+-RG(Ti), H_3O^+-RE(Ti), and H_3O^+-NT(Ti) in 0.05 M H_2SO_4 at 70°C with O_2 bubbling.

TABLE 25.2. Properties of the Titania Nanosheet-Based Materials and the Original Anatase TiO_2

	H_3O^+-RG(Ti)	H_3O^+-RE(Ti)	Cs^+-RG(Ti)	H^+-RE(Ti)	TiO_2
d spacing (nm)	0.94	0.99	0.87	0.73	–
S_{BET} (m^2/g)	18.4	47.0	18.0	17.0	52.4
i_{Disk} at 0.3 V (mA/cm^2)	0.27	0.23	0.04	0.14	0.16
Eff_4 at 0.3 V (%)	85	94	–	45	31
σ_{H+} (S/cm) (25°C)	2.7×10^{-5}	1.0×10^{-4}	–	1.6×10^{-5}	–

worthy result, the ORR activity strongly depended on their proton conductivity. The nanosheet-based material with higher proton conductivity exhibited a higher ORR activity particularly in Eff_4. This trend was systematically confirmed by measuring the proton conductivities and by comparing the ORR activities with those of the starting material anatase TiO_2 and nanosheet-based materials with other cations such as Cs^+ and H^+ (Table 25.2). This suggests that the proton conduction in the H_3O^+-form titania nanosheet-based materials promotes the four-electron reduction of oxygen because of the necessity of H^+ for the ORR in acid solution [20, 21, 23]:

$$O_2 + 4H^+ + 4e^- \rightarrow 2H_2O. \tag{25.2}$$

Otherwise, the two-electron reduction of oxygen occurs as follows:

$$O_2 + 2H^+ + 2e^- \rightarrow H_2O_2 \tag{25.3a}$$

$$H_2O_2 + 2H^+ + 2e^- \rightarrow 2H_2O. \tag{25.3b}$$

Also, the nanosheet-based materials have some advantages proper to the ORR catalysts: a large specific surface area, numerous highly active sites such as the edge, kink, and defect of metal ion in the nanosheets and flexibility of composition with other cations, polymers, and inorganic compounds [31, 32]. For example, all the surface of nanosheets pillared with some compounds would be utilized as the catalytic active site. Oxide nanosheet-based materials have been for the first time demonstrated to show high ORR activity, as described earlier.

25.3.2 The ORR Activity of Manganese Oxide Nanosheet-Based Materials

In alkaline solution, it is well-known that manganese oxides such as MnO_2 and Mn_2O_3 show relatively high ORR activities similarly to Pt catalysts

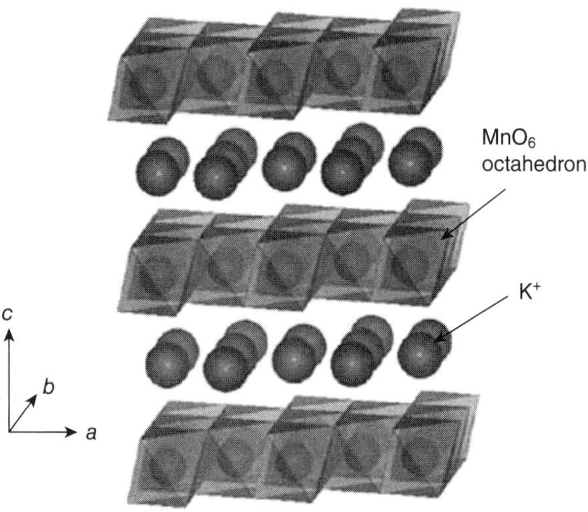

Figure 25.9. Crystal structure of layered manganese oxide $K_{0.5}MnO_2$ (K^+-RG(Mn)).

[12–14]. Therefore, it is of a great interest to focus on the manganese oxide nanosheet-based materials for the development of new ORR catalysts. We synthesized two types of protonated manganese oxide nanosheet-based materials, that is, H_3O^+-form regularly stacked and randomly restacked nanosheets (H_3O^+-RG(Mn) and H_3O^+-RE(Mn)), from the starting layered manganese oxide $K_{0.5}MnO_2$ (K^+-RG(Mn)) (Fig. 25.9), to evaluate their ORR activities in 0.1 M KOH solution at 70°C. Figure 25.10a–c shows the SEM images for the nanosheet-based materials. The K^+-RG(Mn) and H_3O^+-RG(Mn) exhibited a typical morphology of layered oxides. On the other hand, the morphology of H_3O^+-RE(Mn) seemed like a flocculated solid consisting of randomly restacked nanosheets. The morphology of H_3O^+-RE(Mn) is well explainable by the transition electron microscopy (TEM) image of the exfoliated nanosheets (Fig. 25.10d) extracted from the colloidal solution on the way to the synthesis of H_3O^+-RE(Mn). Their morphologies were well supported with the large surface area (S_{BET}) estimated from the Brunauer–Emmett–Teller (BET) method (Table 25.3). Figure 25.11 shows the SSV curves of the manganese oxide nanosheet-based materials. The ORR activities were in the order of H_3O^+-RE(Mn) > H_3O^+-RG(Mn) > K^+-RG(Mn), which agrees well with that in the case of the titania nanosheet-based ones. In addition, both H_3O^+-RE(Mn) and H_3O^+-RG(Mn) showed higher ORR activities than the starting material Mn_2O_3 and the well-known oxide ORR catalyst MnO_2. This indicates that the protonation and restacking of the nanosheets are effective in improving the ORR activity of the transition metal oxides. As a result, the H_3O^+-RE(Mn) showed the highest ORR activity: $E_{on} = 0.97$ and $Eff_4 = 99\%$ at 0.7 V, which was comparable to that of the conventional 20 mass % Pt/C catalyst. In the alkaline

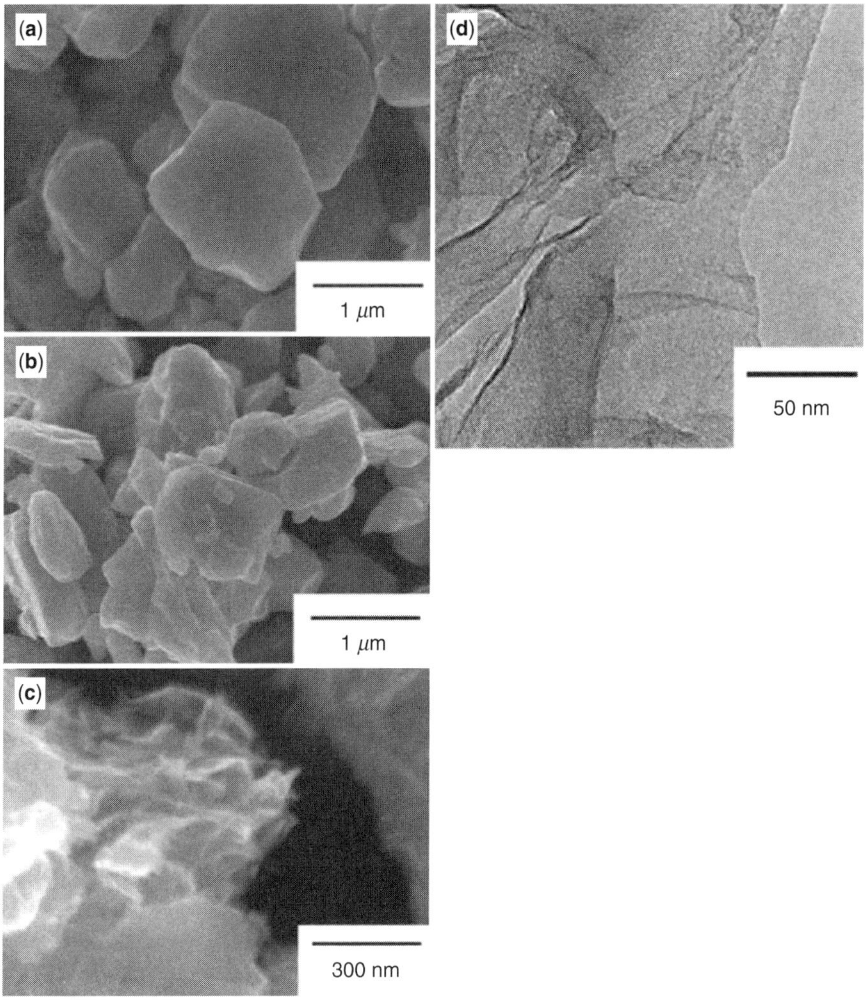

Figure 25.10. SEM images of (a) K^+-RG, (b) H_3O^+-RG and (c) H_3O^+-RE, and TEM image of (d) a manganese oxide nanosheet obtained from the nanosheet colloid solution.

TABLE 25.3. Properties of the Manganese Oxide Nanosheet-Based Materials Together with the Original Mn_2O_3 and MnO_2

	K^+-RG(Mn)	H_3O^+-RG(Mn)	H_3O^+-RE(Mn)	Mn_2O_3	MnO_2
d spacing (nm)	0.706	0.729	0.725	–	–
S_{BET} (m^2/g)	4.1	9.1	141	14.9	46.3
i_D at 0.7 V (mA/cm^2)[a]	0.75/0.50	0.72/0.98	0.43/1.14	0.53/0.57	0.49/0.53
E_{on} at 0.02 mA/cm^2 (V)[a]	0.88/0.93	0.89/0.95	0.86/0.97	0.92/0.92	0.87/0.85
Eff_4 at 0.7 V (%)[a]	86/87	99/91	96/99	94/98	98/93

[a] Potential scan range for the SSV: The top and bottom values mean the values obtained from the SSV curves from 1.0 and 1.2 V, respectively.

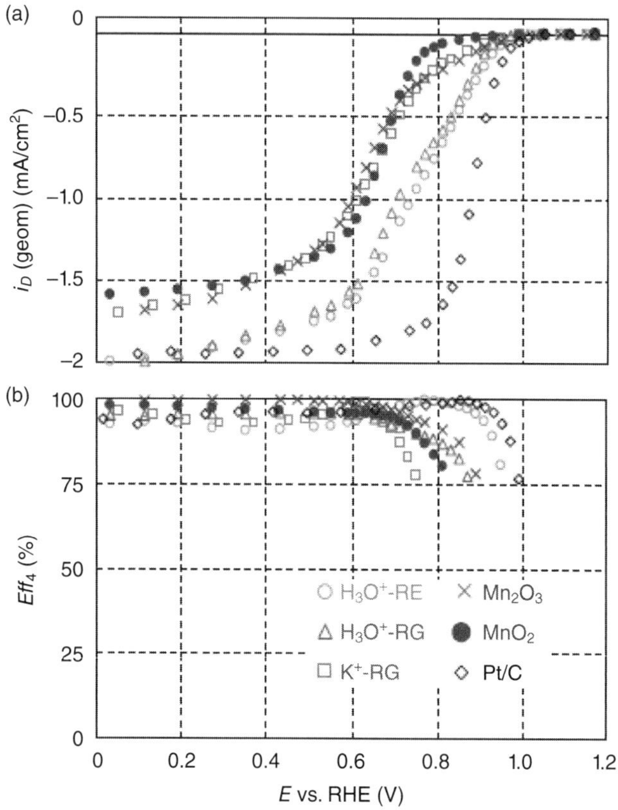

Figure 25.11. (a) Semi-steady-state voltammograms and (b) efficiencies (*Eff*$_4$) of four-electron reduction of oxygen for manganese oxide nanosheet-based materials in 0.1 M KOH at 70°C with O_2 bubbling.

solution, the ORR scheme on a RRDE generally proceeds through the following four-electron (Eq. 25.4) and/or two-electron (Eq. 18.5a,b) pathways [22,23]:

$$O_2 + 2H_2O + 4e^- \rightarrow 4OH^-, \tag{25.4}$$

$$O_2 + H_2O + 2e^- \rightarrow HO_2^- + OH^-, \tag{25.5a}$$

and

$$HO_2^- + H_2O + 2e^- \rightarrow 3OH^-. \tag{25.5b}$$

Both ORR pathways require H_2O in alkaline solution. In particular, the four-electron ORR would be promoted when a large amount of H_2O exists near the ORR active sites. From the X-ray photoelectron spectroscopy (XPS) measurements, H_3O^+-RG(Mn) and H_3O^+-RE(Mn) were found to have larger amounts of H_2O and likely OH^- ion in the interlayers and surfaces of the nanosheets than K^+-RG(Mn) through the protonation of nanosheets. In addition, the amount of H_2O in the nanosheets presumably increases in the aqueous solution rather than that under high vacuum condition for XPS measurements. The H_3O^+-RG(Mn) and H_3O^+-RG(Mn) therefore have an advantage for the four-electron ORR. Moreover, the H_3O^+-RE(Mn) has a very large S_{BET}, which brings about a larger number of ORR active sites. This would enhance the ORR activity together with the large amount of H_2O. Thus, oxide nanosheet-based materials have various advantages as the ORR catalysts. Further research on the effects of the morphological properties (large surface area, catalytically high active sites, and a combination of species in the interlayers) and the electronic structure of the nanosheet-based materials on their ORR activities is expected to open a new field of electrocatalysts.

25.4 APPLICATION OF OXIDE-BASED CATALYSTS TO THE AMFC CATHODE

As mentioned in the previous sections, AMFCs with anion-exchange membranes (AEMs) as the electrolyte have significantly attracted increasing attention owing to the availability of nonplatinum electrocatalysts (Fig. 25.12). We assembled a single cell to test the metal oxide-based ORR catalysts for the cathode. Figure 25.13 shows the *i*–*v* curves of the AMFC single cells with the oxide-based catalysts (PRMn0.5, NdRMn0.25, and H_3O^+-RE(Mn)) and an anion-exchange membrane (A901, Tokuyama Co.). From the results, all the oxide catalysts were found to be available for a cathode catalyst in AMFCs. The order of magnitude for the power densities (P_{cell}) was PRMn0.5 > H_3O^+-RE(Mn) ≥ NdRMn0.25 > MnO_2. In particular, the AMFC with H_3O^+-RE(Mn)

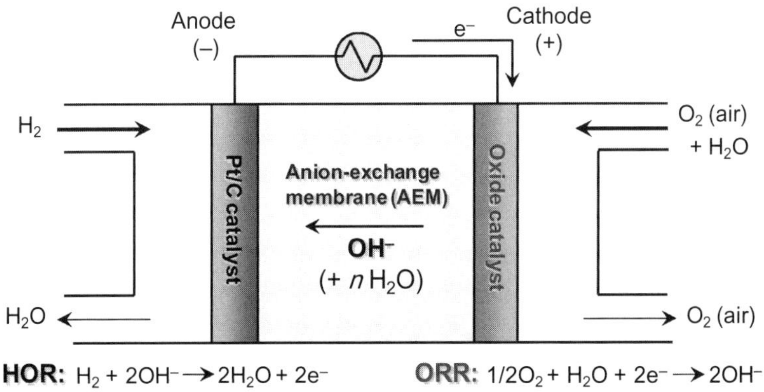

Figure 25.12. Schematic illustration of AMFC with metal oxide-based catalyst for the cathode.

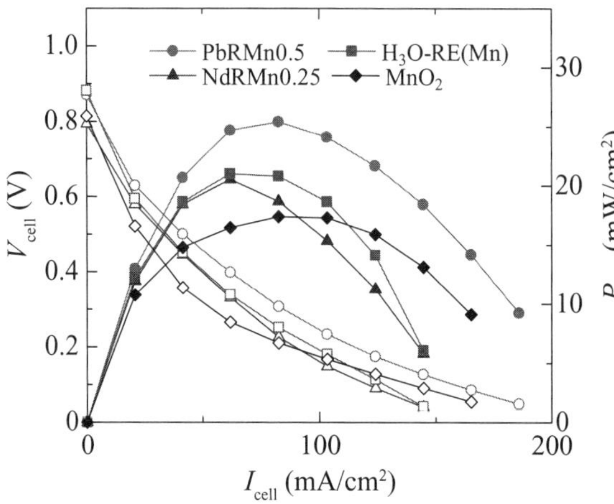

Figure 25.13. Performances of the AMFC single cells with metal oxide-based cathode catalysts at 50°C. Humidified H_2 and O_2 (RH = 100%) were supplied to anode and cathode, respectively.

exhibited a high P_{cell} value of more than 20 mW/cm^2 in spite of being precious metal-free and the first cell assembly. All the synthesized oxides exceeded one of the well-known Pt-alternative cathodes, MnO_2 in the AMFCs. Therefore, PRMn0.5, H_3O^+-RE(Mn) and NdRMn0.25 were found to be good candidates as Pt-free cathode catalysts. Further optimization of the composition and morphologies of the oxide-based electrocatalysts is expected to improve ORR activities and AMFC performances. For practical use in AMFCs, investigations on the durability and influences of CO_2 on cell performances are now in progress.

25.5 SUMMARY AND PERSPECTIVES

In this chapter, two types of oxide-based ORR catalysts (pyrochlores and oxide nanosheet-based materials) have been mainly introduced, and through the results, strategies on the new ORR catalysts for AMFCs have been discussed. In the pyrochlores, a variation in the lattice parameters is expected to influence the local structure around the catalytic active sites, that is, the valence and electronic states of the atoms associated with the catalytic active site, deviation in the M–O bond distance, and distortion of the local structure as well as intrinsic properties, that is, electronic conductivity and band structure. By adjusting the crystal and local structures on the basis of the detailed analysis, the ORR activity would be improved to approach the level required for practical fuel cell applications. Although we have found some pyrochlore ORR catalysts with high activity, there still remains a numerous unexplored combination of the A- and B-site ions even in the pyrochlore structure. Moreover, there are a number of candidates for the oxide-based catalysts such as mixed oxides, perovskites, and spinels. On the other hand, oxide nanosheet-based materials provide a new way to create a new group of oxide-based electrocatalysts. In addition to the unique properties (large surface, number of edge, kink, and defect of metal ion), nanosheets have the potential to create novel composite catalysts with many other materials (cations, polymers, inorganic compounds, etc.). Also, in the battery fields, ORR catalysts have the potential to be utilized as cathode catalysts in metal–air batteries with a larger energy density than that of the conventional lithium ion batteries. In the near future, we expect that a novel, advanced oxide-based electrocatalyst be realized based on the ideas described here.

ACKNOWLEDGMENTS

The authors are deeply grateful to Professor Yoshiharu Uchimoto of Kyoto University and Associate Professor Hidenobu Shiroishi of Tokyo National College of Technology for their technical support and helpful advice on the X-ray absorption fine structure (XAFS) and electrochemical measurements, respectively. The authors wish to thank Tokuyama Co. for providing anionic ionomer (AS-4) and anion-exchange membrane (A901) samples for the fabrication of the AMFCs. Thanks are also due to Mr. Kenji Yoshihara, Mr. Yoshinobu Saito, Mr. Takayuki Konishi, Mr. Yuya Akeboshi, Mr. Naoki Ohno, and Mr. Takeo Hatai of Tokyo University of Science and Dr. Toyuki Okumura of Kyoto University for their cooperation.

REFERENCES

1 W. Vielstich, A. Lamm, H. Gasteiger (eds.), *Handbook of fuel cells: fundamentals technology and applications*, vols. 1–4. John Wiley & Sons, Chichester, 2003.

2 P. J. Gelling, H. J. M. Bouwmeester (eds.), *The handbook of solid state electrochemistry*. CRC Press, Baca Raton, FL, 1997, p. 329, 407.

3 S. M. Haile, D. A. Boysen, C. R. I. Chisholm, R. B. Merle. *Nature* 410 (2001) 910.

4 J. H. Zagal, in W. Vielstich, A. Lamm, H. A. Gasteiger (eds.), *Handbook of fuel cells—Fundamentals, technology and applications*, vol. 2. John Wiley & Sons, Chichester, 2003, Chapter 37.

5 N. Alonso-Vante, in W. Vielstich, A. Lamm, H. A. Gasteiger (eds.), *Handbook of fuel cells—Fundamentals, technology and applications*, vol. 2. John Wiley & Sons, Chichester, 2003, Chapter 36.

6 A. Ishihara, Y. Shibata, S. Mitsushima, K. Ota. *J. Electrochem. Soc.* 155(4) (2008) B400.

7 S. Doi, A. Ishihara, S. Mitsushima, N. Kamiya, K. Ota. *J. Electrochem. Soc.* 154(3) (2007) B362.

8 J. Ozaki, S. Tanifuji, N. Kimura, A. Furuichi, A. Oya. *Carbon* 44 (2006) 1324.

9 Y. Nabae, M. Malon, S. M. Lyth, S. Moriya, K. Matsubayashi, N. Islam, S. Kuroki, M. Kakimoto, J. Ozaki, S. Miyata. 216th ECS Meeting, Abstract #836, 2009.

10 J. R. Varcoe, R. C. T. Slade, G. L. Wright, Y. Chen. *J. Phys. Chem. B* 110 (2006) 21041.

11 K. Matsuoka, Y. Iriyama, T. Abe, M. Matsuoka, Z. Ogumi. *J. Power Sources* 150 (2005) 27.

12 M. Hayashi, T. Shodai. *Electrochemistry* 78(6) (2010) 529.

13 F. H. B. Lima, M. L. Calegaro, E. A. Ticianelli. *Electrochim. Acta* 52 (2007) 3732.

14 L. Mao, D. Zhang, T. Sotomura, K. Nakatsu, N. Koshiba, T. Ohaka. *Electrochim. Acta* 48 (2003) 1015.

15 K. Yoshihara, Y. Saito, M. Saito, J. Kuwano, H. Shiroishi. *Key Eng. Mater.* 350 (2007) 171.

16 Y. Saito, K. Yokota, K. Yoshihara, M. Saito, J. Kuwano, H. Shiroishi. *Key Eng. Mater.* 350 (2007) 167.

17 M. Saito, Y. Saito, T. Konishi, H. Kawai, J. Kuwano, H. Shiroishi, Y. Uchimoto. *ECS Trans.* 16(2) (2008) 891.

18 J. Kuwano, M. Saito. Japan Patent 2008-123460, 2008.

19 T. Konishi, H. Kawai, M. Saito, J. Kuwano, H. Shiroishi, T. Okumura, Y. Uchimoto. *Top. Catal.* 52 (2009) 896.

20 Y. Akeboshi, N. Ohno, H. Tokuno, M. Saito, J. Kuwano, H. Shiroishi. *Key Eng. Mater.* 388 (2009) 73.

21 M. Saito, Y. Akeboshi, N. Ohno, J. Kuwano, H. Shiroishi, Y. Uchimoto. *ECS Trans.* 16(25) (2009) 97.

22 N. Ohno, Y. Akeboshi, M. Saito, J. Kuwano, H. Shiroishi, T. Okumura, Y. Uchimoto. *Top. Catal.* 52 (2009) 903–911.

23 J. Kuwano, M. Saito. Japan Patent 2008-127825, 2008.

24 H. S. Horowitz, J. M. Longo, J. T. Lewandowski. US Patent 4,129,525, 1978.

25 J.-M. Zen, R. Manoharan, J. B. Goodenough. *J. Appl. Electrochem.* 22 (1992) 140.

26 K. Miyazaki, N. Sugimura, K. Matsuoka, Y. Iriyama, T. Abe, M. Matsuoka, Z. Ogumi. *J. Power Sources* 178 (2008) 683.

27 T. Yamamoto, R. Kanno, Y. Takeda, O. Yamamoto, Y. Kawamoto, M. Takano. *J. Solid State Chem.* 109 (1994) 372.

28 H. Yamamura, H. Nishino, K. Kakinuma, K. Nomura. *Solid State Ionics* 158 (2003) 359.

29 S. Imaizumi, K. Shimanoe, Y. Teraoka, N. Yamazoe. *Electrochem. Solid State Lett.* 8 (2005) A270.

30 W. Sugimoto, H. Iwata, Y. Murakami, Y. Takasu. *Angew. Chem. Int. Ed.* 42 (2003) 4092.

31 Y. Ebina, T. Sasaki, M. Harada, M. Watanabe. *Chem. Mater.* 14 (2002) 4390.

32 N. Skai, Y. Ebina, K. Takada, T. Sasaki. *J. Am. Chem. Soc.* 126 (2004) 5851.

26

DYNAMICS AND STRUCTURE OF WATER NANOTUBE CLUSTERS CONFINED TO NANOPOROUS MOLECULAR CRYSTALS

Makoto Tadokoro

Water clusters confined to nanometer-scale surroundings have dynamics and structures distinct from those of bulk water and ice under ambient pressure. For example, theoretically intriguing studies have been carried out for a one-dimensionally ordered chain of water molecules [1] and new ice phases not seen in bulk ice [2] contained inside a carbon nanotube. Experiments have shown that a water decamer, as a fragment model of ice phase I_c, can be stabilized by noncovalent interactions [3]. The depression of the freezing point of water confined to micropores such as zeolite [4] and its unusual viscosity on a thin water membrane bound to hydrophilic slits [5] have revealed the interesting behavior of water related to mutual interactions with the surface of the outer walls. We have programmed to create a nanometer-scale water cluster in a molecule-based nanoporous crystal in order to inspect such dynamics and behaviors on water molecules under a particular condition.

Many artificial hollow compounds with rigid nanometer-scale cavities are synthesized by using metal-organic frameworks [6] and hydrogen-bonded (H-bonded) networks of metal complexes [7] obtained from the self-organization of building blocks. The cavities in such porous crystals can be modified in terms of size, shape, dimensionality, and nature by designing

Electrical Phenomena at Interfaces and Biointerfaces: Fundamentals and Applications in Nano-, Bio-, and Environmental Sciences, First Edition. Edited by Hiroyuki Ohshima.

appropriate molecular building blocks. Therefore, porous crystals, particularly those assembled by crystal engineering [8] using complementarity between building blocks, are important material groups with new functionality.

The field of crystal engineering in supramolecular chemistry attracts considerable attention with regard to the control of a crystal structure by self-organization of artificial molecular building blocks [8]. In order to assemble thermodynamically stable molecular arrays, it is important to control the crystal arrangements in order to utilize the mutual intermolecular interactions such as complementary H-bonds and directional coordination bonds; this requires thermal equilibrium in the solution phase under moderate conditions. Some water clusters occupy the interstitial spaces of such supramolecular crystals. The study of this phenomenon can be developed into a new branch of chemistry that involves the determination of water mobility with regard to spectrochemical characteristics [9], and structural dynamics for their water clusters proceed [10].

By using such crystal engineering, we have recently synthesized new molecular building blocks of H-bonded metal complexes having 2,2′-biimidazolate monoanions ($Hbim^{-1}$) as a ligand [11]. This $Hbim^{-1}$ system controls the multi-dimensionality of molecular aggregations in the crystal by using difunctional ligands that bond to metal with coordinate–covalent interactions and to each other with H-bonding interactions as shown in Figure 26.1a [12]. For example, the $Hbim^{-1}$ works as a difunctional bridging ligand not only to form a stable metal–chelate complex but also new intermolecular complementary H-bonding with two sets of NH donors and N acceptors. The neutral tris-bimidazolate metal complex ($[M^{III}(Hbim)_3]$) forms honeycomb sheets (Fig. 26.2), which stack to form the porous crystal with 1-D channels, a nanometer wide, by H-bonding between $Hbim^{-1}$ ligands [13]. The porous crystal can include solvent molecules in the open channels, similar to zeolite during crystallization, but the accurate location of the confined solvent molecules in the channel cannot be determined by X-ray crystal analysis even at low temperature because of the degree of disorder. To confine water molecules inside a nanometer-wide channel, we propose a new design strategy, which introduces H-bonding donor or acceptor sites onto the surface of the channel framework. Thus, we used

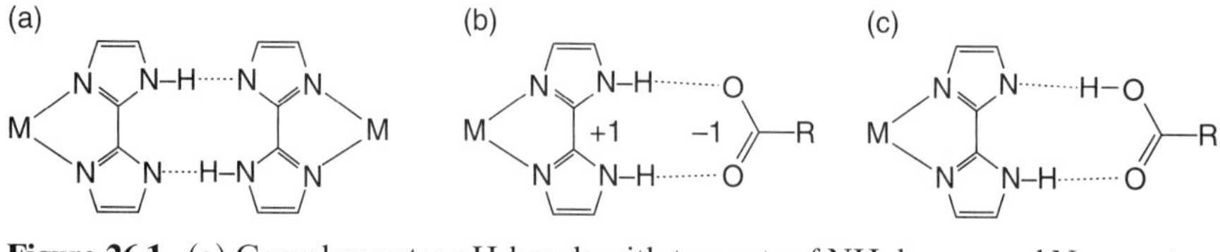

Figure 26.1. (a) Complementary H-bonds with two sets of NH donors and N acceptors between neutral metal complexes with $Hbim^{-1}$ ligands. (b) Complementary neutral H-bonds between a carboxyl acid and a metal complex with $Hbim^{-1}$ ligands. (c) Ionic H-bonds between the anionic acid and the cationic H_2bim metal complex.

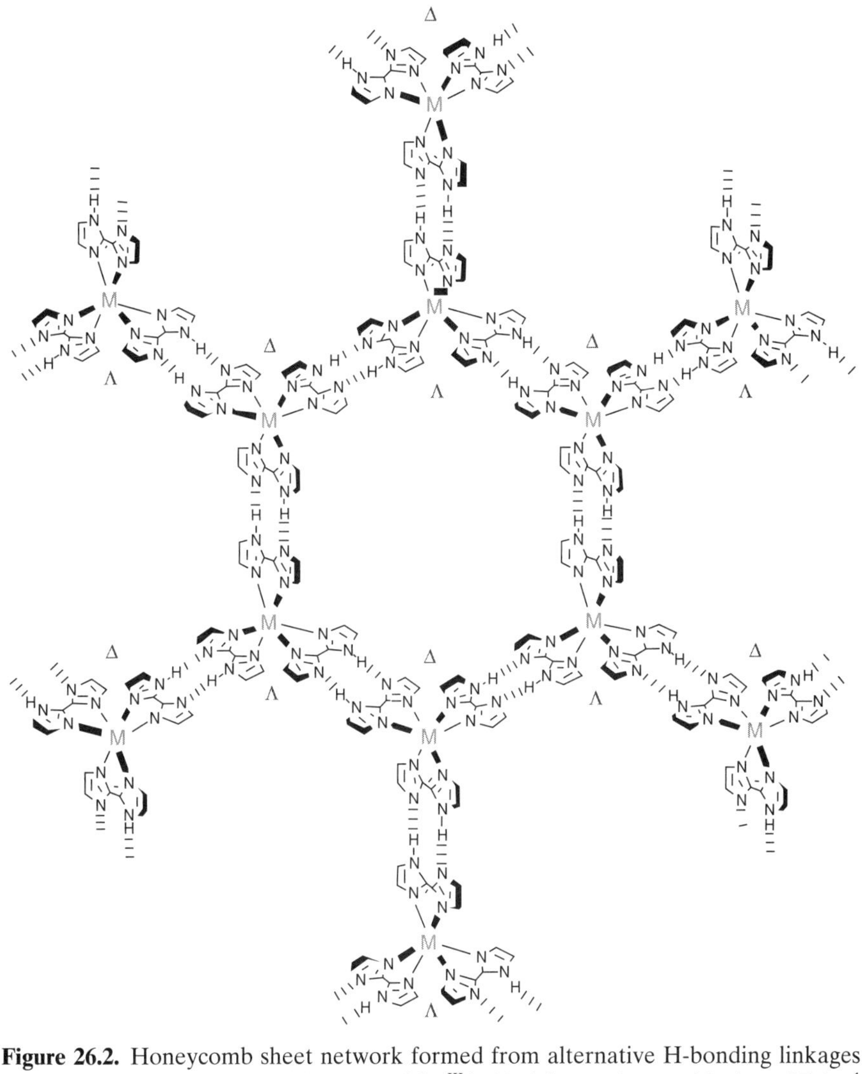

Figure 26.2. Honeycomb sheet network formed from alternative H-bonding linkages between the Δ and Λ optical isomers of $[M^{III}(Hbim)_3]$ complexes with three $Hbim^{-1}$ ligands.

two kinds of building blocks, TMA^{3-} (trimesate) and $[M^{III}(H_2bim)_3]^{3+}$, to construct a framework that has a mixed complementary intermolecular H-bonding between them. By comparing the relative acid–base strength of the corresponding chemical species, we would predict that the complementary H-bonding should be the proton-transferred H-bond (Fig. 26.1b) and not the neutral one (Fig. 26.1c).

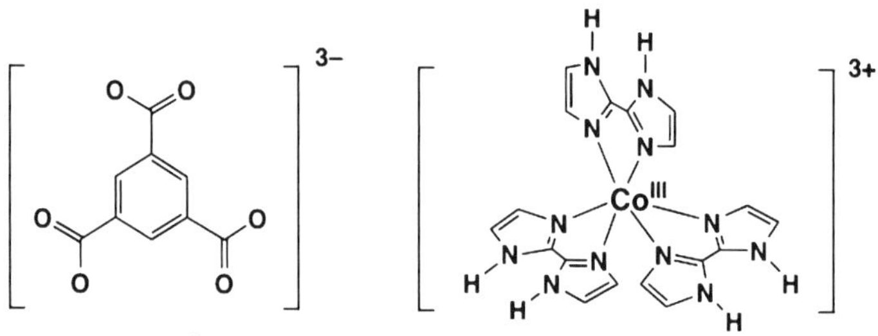

$[TMA]^{3-}$ $[Co^{III}(H_2bim)_3]^{3+}$

Scheme 26.1. Structures of the building blocks of $[TMA]^{3-}$ and $[Co^{III}(H_2bim)_3]^{3+}$.

We have introduced a tris-2,2′-biimidazole cobalt (III) complex $[Co^{III}(H_2bim)_3]^{3+}$ and TMA^{3-} as the building blocks for constructing a nanoporous molecular crystal. The biimidazolate metal complex cannot only form intermolecular H-bonds of the dual N–H···N type by partial deprotonation [14] but can also form complementary H-bonds with a carboxyl group. In addition, the difference between the acidities of the NH group and the carboxylate group results in strong ionic H-bonds based on the two $N^{\delta+}$–H···$O^{\delta-}$ type H-bonds [15]. Therefore, we can prepare the molecular crystal $\{[Co^{III}(H_2bim)_3]\cdot(TMA)\cdot 20H_2O\}_n$ (Scheme 26.1) with one-dimensional (1-D) nanoporous channels by alternative stacking in adjustment with each pore between Δ- and Λ-hexagonal sheets with the (6,6)-net (Fig. 26.3) [16]. The two chiral sheets of the molecular crystal are constructed from alternative H-bonding between the Δ- or Λ-$[Co^{III}(H_2bim)_3]^{3+}$ and TMA^{3-}. A huge tubelike cluster of the water nanotube (WNT) formed from ca. 20 water molecules in the periodical unit is stabilized in each nanochannel pore of crystal 1 [17].

The preparation was carried out by mixing $H_3(TMA)$ and $[Co^{III}(H_2bim)_3](NO_3)_3$ in alkaline water. After standing overnight at 45°C, we obtained orange crystals of 1 containing water molecules as a hexagonal prism with a framework of TMA^{3-} and $[Co^{III}(H_2bim)_3]^{3+}$ in a 1:1 ratio. By a similar procedure, the crystal of 2 containing D_2O was also prepared. The isolated crystals 1 and 2 were susceptible to the escape of confined water (or heavy water) from the channels. Therefore, elemental analyses on the crystals were carried out after dehydrating by vacuum drying at 100°C. The ratio of TMA^{3-} and $[Co^{III}(H_2bim)_3]^{3+}$ is 1:1.

We performed X-ray crystal structure analysis of crystal 1 at room temperature (23°C) by setting crystal 1 into a glass capillary filled with water vapor. Proton-transferred mixed complementary H-bonding generated two types of honeycomb sheet: Δ- and Λ-sheets between the Δ isomer or the Λ isomer of $[Co^{III}(Hbim)_3]$ and trimesic acid, respectively. The Δ- and Λ-sheets stack alternately along the *c*-axis to construct the desired channel frameworks with H-bonding acceptor sites. The channel diameter is estimated to be about 15~16 Å by subtracting the van der Waal radius of the framework. The ratio

(a)

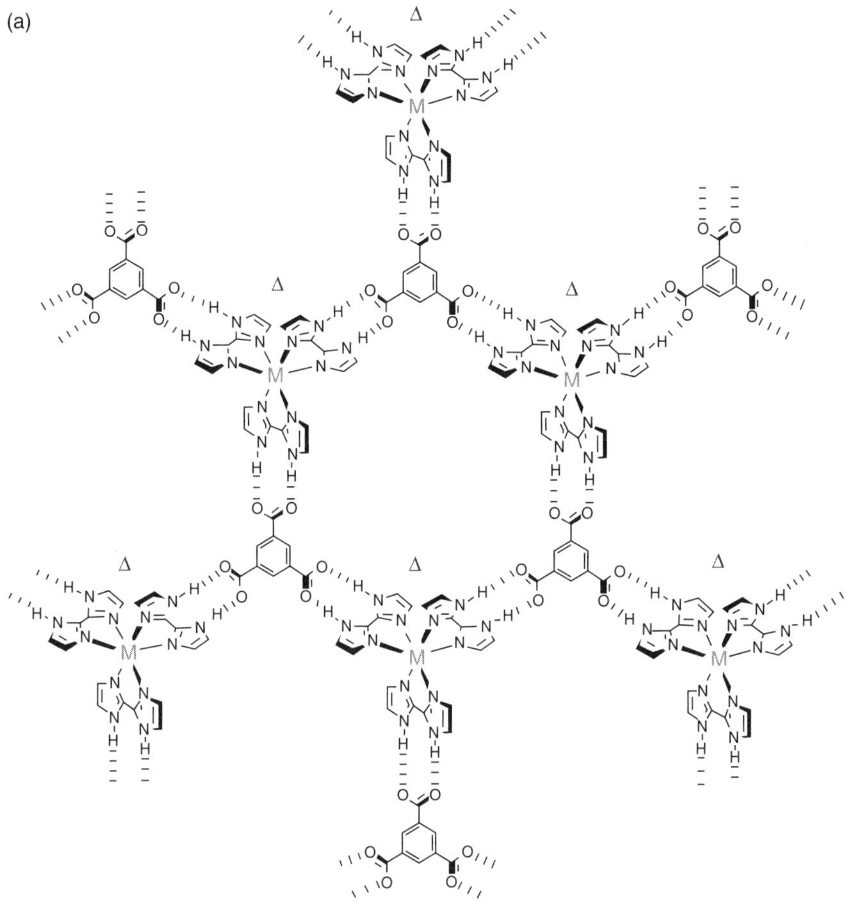

Figure 26.3. Schematic representations of hybrid honeycomb sheet networks formed from new complementary alternative H-bonds between three carboxylic groups of the trimesic acid and three $Hbim^{-1}$ ligands of (a) the Δ optical isomers (Δ-sheet) or (b) the Λ ones (Λ-sheet) of $[Co^{III}(H_2bim)_3]^{3+}$ complexes.

of the channel volume to the unit cell volume is about 63% (Fig. 26.4). The oxygen atoms of the water molecules can be located in the channel despite having high temperature factors. Interestingly, the confined water molecules comprise a novel "three-layered water structure" with nanotube networks of H-bonding composed of the primary water tube (yellow lines), the secondary water layer (red lines), and the tertiary water domain (green spheres) as shown in Figure 26.5a. Such H-bonded WNT networks are held in the channel by additional H-bonds between water molecules of the primary water tube and oxygen atoms (white spheres) of the trimesate on the surface of the framework. In the primary domain, H-bonded networks contain small cyclic

(b)

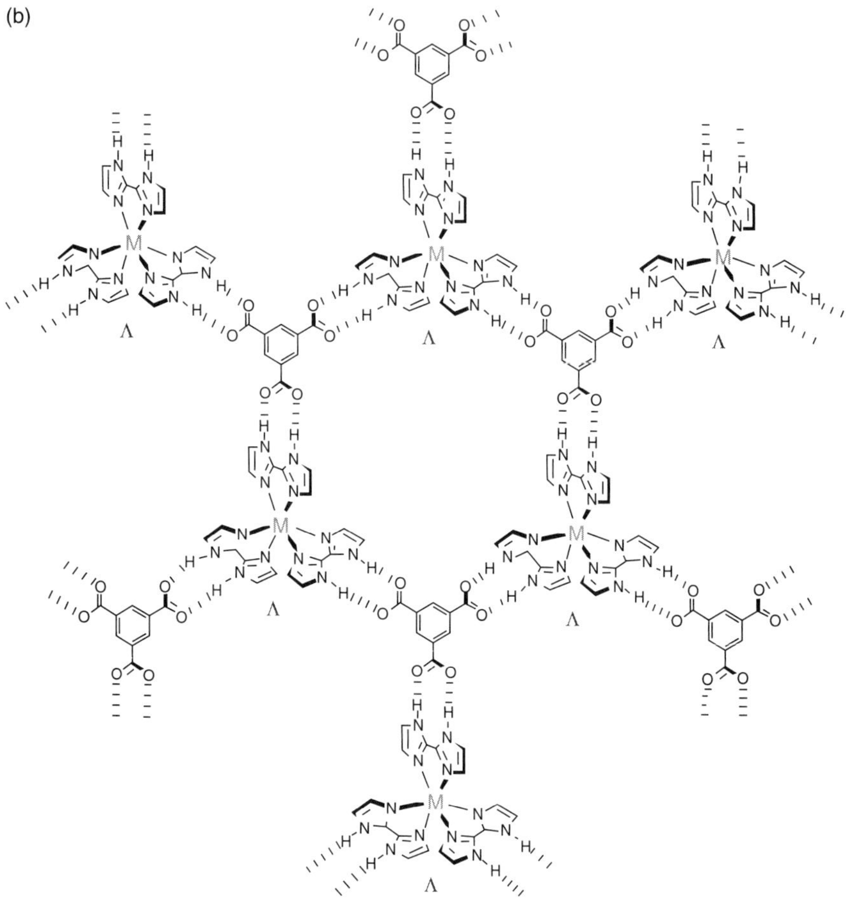

Figure 26.3. *Continued*

structures, pentagons, hexagons, and octagons. The octagons were not found in the clathrate hydrate [18]. In the secondary domain, water molecules form a 1-D column of quasi-planar hexagonal rings with an inclination of ~30° to the *c*-axis (Fig. 26.5c). Each ring is bridged by water molecules in the tertiary domain. The existence of planar hexagonal rings of water molecules is confirmed from both theoretical [19] and experimental [20] studies. The hexagonal rings found in this chapter represent the first example in which stabilization is imparted from water molecules exclusively. Water molecules in the tertiary domain have higher temperature factors than those in other places, and some water molecules in the domain cannot be detected on the Fourier map at room temperature because of the high degree of disorder.

The existence of confined H_2O or D_2O molecules was also confirmed by thermogravimetry (TG) measurements (Fig. 26.6), which were carried out

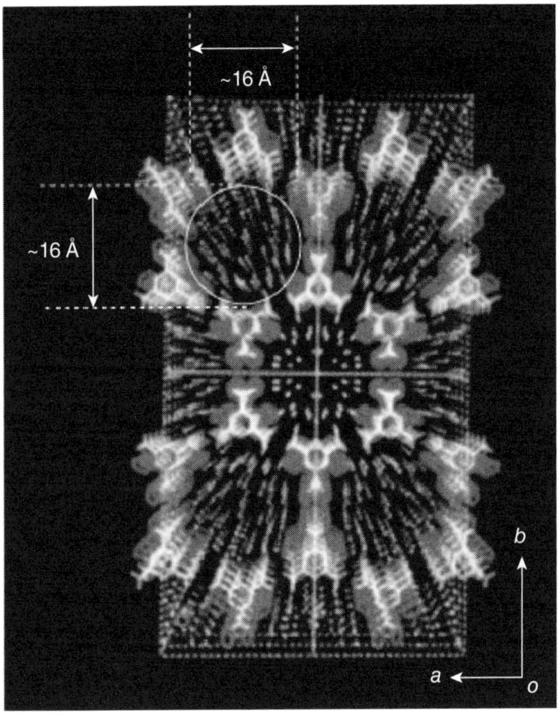

Figure 26.4. Perspective view along the *c*-axis of 1: The red and white lines show the nanochannel framework, and the yellow dots show water oxygen atoms. See color figure on companion ftp site.

after coating the fragile crystals 1 and 2 in air with paraffin liquid. Figure 26.7 shows the differential scanning calorimetry (DSC) thermograms between 263 and 213 K at a scan rate of 10 K/min velocity. We observed a reversible phase change for 1 and 2 showing exothermic peaks at 234 and 240 K during cooling and endothermic peaks at 245 and 250 K during heating, respectively. The difference in the temperatures of these phase transitions for crystals 1 and 2 are almost comparable to the isotope effect (ca. ~5 K under ambient pressure) [21] on the freezing and melting of H_2O and D_2O. This similarity to the isotope effect suggests that the peaks of the phase transitions in DSC are related to condensation and fusion of water molecules inside the channels. The irreversible endothermic peaks of crystals 1 and 2 were also observed at 375 and 378 K, respectively. The freezing enthalpies of crystal 1 (65.89 cal/mol) and crystal 2 (58.22 cal/mol) were quite small compared with that of bulk ice (1436.3 cal/mol) [21], indicating that only a limited number of water molecules in the channels participate in the water–ice transition. Therefore, the strong H-bonded water molecules in the primary water tube are probably less important for this transition since those in the primary domain are strongly H-bonded with the oxygen atoms on the surface of the channels as described earlier.

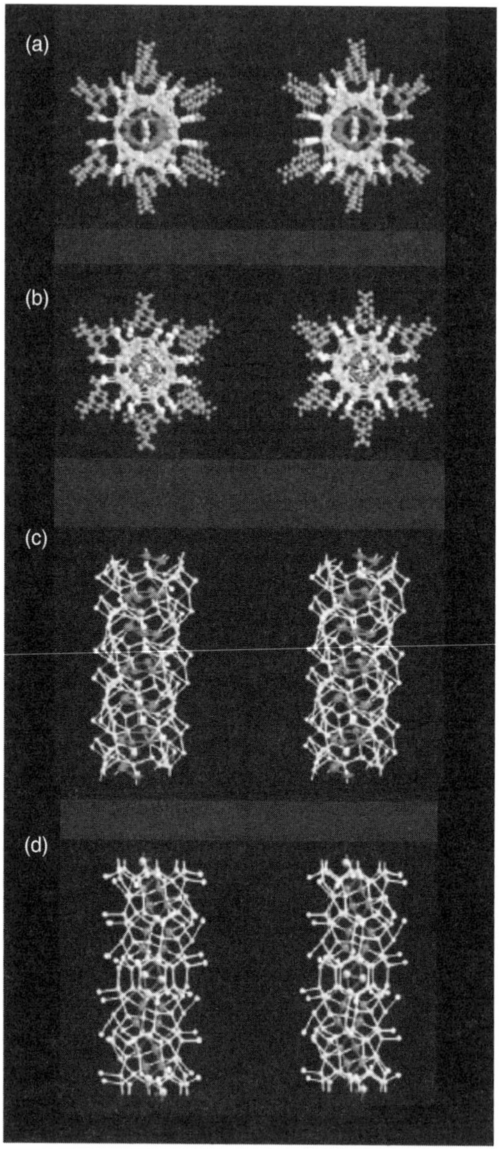
(a)
(b)
(c)
(d)

Figure 26.5. Stereoviews of crystal structures for crystal 1 above the phase transition temperature at 296 K (a,c) and below the phase transition temperature at 198 K (b,d): perspective views of the multilayered structure representing units of the two channels along the *c*-axis confining (a) a one-dimensional hierarchical water tube and (b) a hierarchical icelike tube, respectively. Trimesic acids contained in the outer wall of the channel as building blocks are shown by the blue spheres and solid lines. The yellow spheres and solid lines describe H-bonding between water molecules in the primary water tube stabilized by strong interactions with the oxygen atoms (white spheres) of the trimesate of the outer wall. For clarity, H-bonding greater than 2.7 Å (i.e., the mean value of H-bonding distance) was omitted. The red spheres and solid lines represent H-bonding between water molecules in the secondary water layer, and the green spheres show water molecules in the tertiary water domain, respectively. Stereoviews of the side crystal structures of (c) the hierarchical water tube above the phase transition temperature and (d) the hierarchical icelike tube below the phase transition temperature: Yellow, red, and green spheres indicate water molecules in the primary water tube, secondary water layer, and tertiary water segment, respectively. The solid red lines show H-bonding between water molecules in the secondary and tertiary water domains, and the solid yellow lines also describe it between them in the primary water tube with the oxygen atoms (white spheres) of the trimesic acid (H-bonding greater than 2.7 Å were omitted for clarity). (c) Water clusters of the secondary and tertiary domains with quasi-planar hexagonal rings at an inclination of ~30° bridged by water molecules in the tertiary segment. (d) The icelike tube construct has a one-dimensional poly-spiro chain formed from alternative H-bonds between the C_s water tetramers and the two water hexagons with chairlike structural isomers. See color figure on companion ftp site.

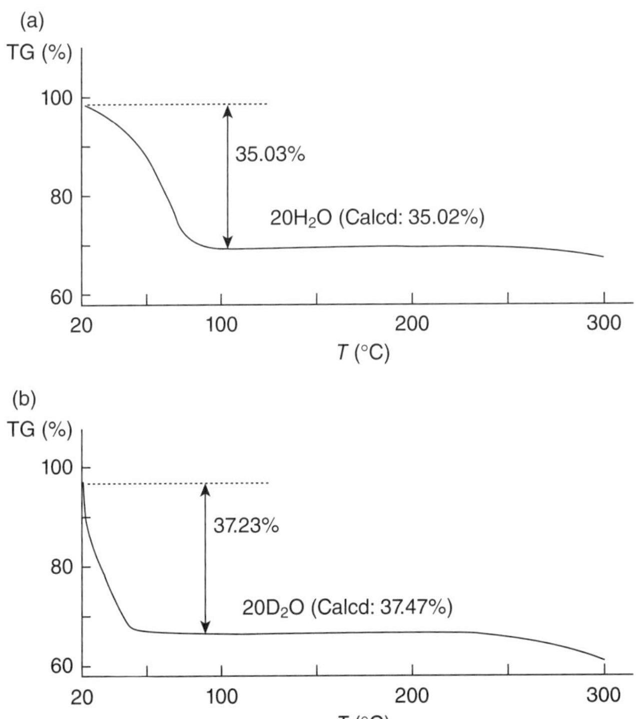

Figure 26.6. The results of thermogravimetry (TG) on crystal 1 containing water molecules (a) and crystal 2 containing heavy water molecules. Both crystals have compositions of approximately 20 H_2O and 20 D_2O for [Co(Hbim)3](TMA), respectively.

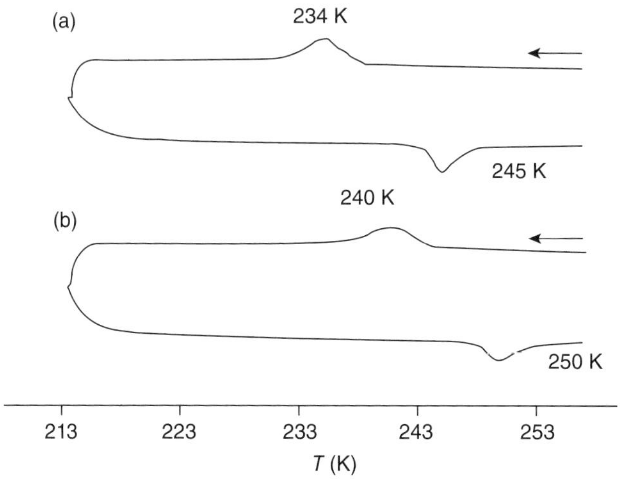

Figure 26.7. DSC traces of (a) crystal 1 (H_2O molecules) and (b) crystal 2 (D_2O molecules) recorded at a heating rate of 10 K/min. The DSC measurements were carried out for crystals 1 and 2 after coating with paraffin liquid. No peak corresponding to paraffin liquid was observed in the range of measurement.

The measurement of ^{2}H solid-state nuclear magnetic resonance (NMR) spectra of crystal 2 reveals that the D_2O molecules around the central axis of the confined water tube network have a highly dynamic disorder above –38°C. This temperature almost coincides with the transition point observed by the DSC measurement and is considered the freezing point of D_2O in a nanometer-scale channel. Figure 26.6 shows the temperature dependence of the ^{2}H-NMR spectra. The NMR data reveal three characteristics of D_2O molecules in the nanometer-scale channel: (i) considerable amount of free heavy water soaking the crystals (Fig. 26.8a), (ii) highly disordered D_2O molecules in the secondary and tertiary water domains but ordered structure for those in the primary water tube (Fig. 26.8a–e), and (iii) appearance of a deutero-icelike state at a very low temperature (–34°C) (Fig. 26.8e–h). The spectra shown in Figure 26.8b–e can be reproduced by simulations using a fixed nuclear quadrupole coupling constant and by varying the asymmetric parameter. These simulations suggest that the fluid molecules move between two sites with a jump angle of about 100° maintaining the rotation around the molecular C_2 axis and implying deuteron diffusion in the channels [22]. The proton conductivity of crystal 1 for the powder pellet at room temperature (1.70×10^{-5} S/cm) [23] is about four orders of magnitude greater than that for bulk ice (1.0×10^{-9} S/cm) and water (5.7×10^{-8} S/cm) (Fig. 26.9) [21]. The conductivity value of crystal 1 is similar to that observed for water molecules confined to some zeolites [24]. The deuteron conductivity of crystal 2 (1.07×10^{-5} S/cm) is lower than the proton conductivity of crystal 1 because of the isotope effect.

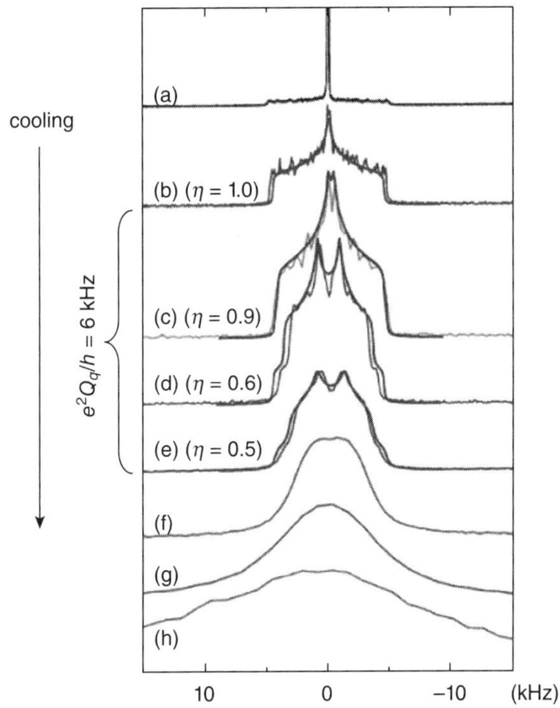

Figure 26.8. The temperature dependent ^{2}H-NMR spectra of crystal 2: The sample crystals soaked into D_2O were measured in a ϕ 5-mm tube. (a) At 285 K, the sharp signal at 0 ppm is due to the fluid D_2O including D_2O used for soaking. The broad signal between +5 and −5 kHz is assigned to the stiff water molecules in the primary hydrate tube. At (b) 265 K, (c) 255 K, (d) 245 K, (e) 235 K, (f) 225 K, the line shape is caused by the anisotropic jumping motion of D_2O molecules in the secondary and tertiary hydrate domains. The signal shows that the deutero-icelike state begins to form from this temperature, (g) −215 K and (h) 205 K. The solid red lines represent a theoretical simulation using a fixed nuclear quadrupole coupling constant (6 kHz) and then varying the asymmetric parameter η from 1.0 at 265 K (b) to 0.5 at 235 K (e). ^{2}H broadline NMR spectra were measured by Bruker MSL-300 NMR system between 155 and 340 K.

We performed X-ray crystal structure analysis below the phase change temperature. Figures 26.5b,d and 26.10 show the crystal structure of the three-layered icelike tube with 1 at −75°C. The unit cell dimensions change in comparison to those at 23°C: The *a*-axis is lengthened by 0.323 Å, while the *b*-axis is shortened by 0.329 Å. The *c*-axis is also shortened by 0.144 Å, which is calculated by normalizing the value of the dimension at low temperature to that at room temperature. The calculated density of crystal 1 decreases from 1.30 to 1.28, indicating the loss of some confined water molecules upon freezing and an increase in the volume of crystal 1 similar to the behavior of bulk water.

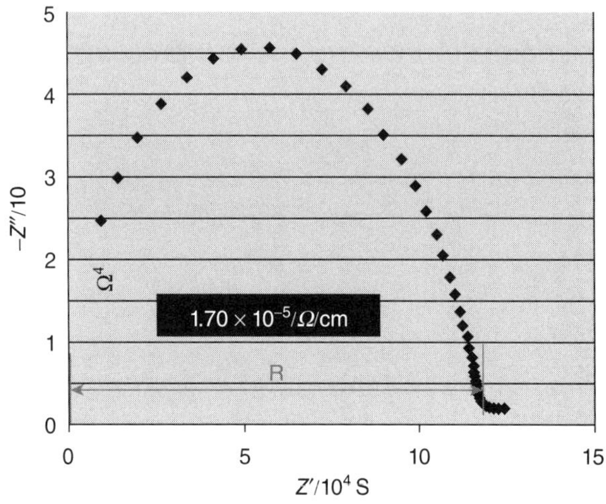

Figure 26.9. Alternating current impedance plots at room temperature in 100% humidity are measured on the pellet disk at the grinding powder of crystal 1 using a Solartron 1260 frequency response analyzer. The proton conductivity (1.70×10^{-5}/S/cm) increases over four and three orders of magnitude more than those of bulk water and ice.

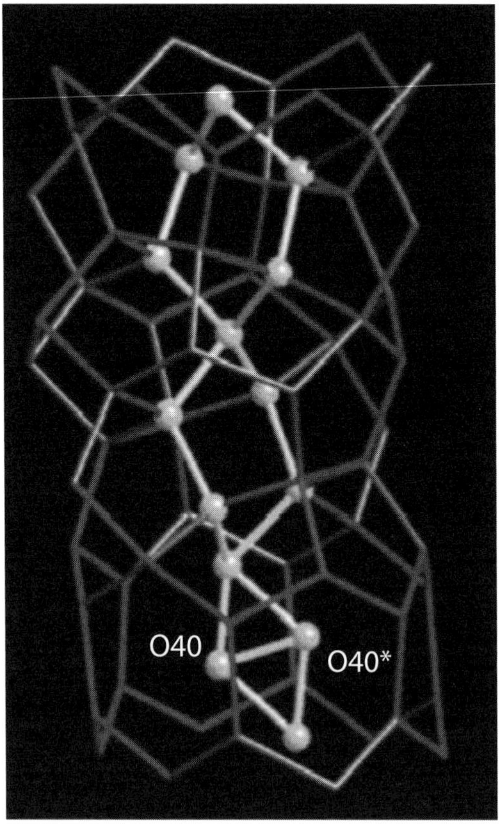

Figure 26.10. A view of the crystal structure for a unit of the three-layered ice nanotube into crystal 1 below the phase transition temperature at 198 K. The red solid lines describe H-bonding between water molecules in the primary water tube. The blue spheres and solid lines represent H-bonding between water molecules in the secondary water layer and the tertiary water domain. See color figure on companion ftp site.

The primary water tube has a structure similar to that at 23°C as described earlier, but the structure displays a more ordered array. The water molecules of the primary water domain are also held by H-bonding to the oxygen atoms on the channel framework and have almost the same structure of water polygons at room temperature (Fig. 26.5b). Interestingly, the secondary and tertiary water domains consist of a poly-spiro chain in which two hexagons and a C_s tetramer are connected to each other by spiro bonds (Fig. 26.10). All hexagons of the spiro chain have a chair–chair conformation, which is found only in the fragment of a cubic ice phase I_c but not in a hexagonal phase I_h, which contains spiro structures with a chair–boat and a boat–boat conformation. The ice phase I_c is known to be stable below 193 K at atmospheric pressure. The spiro-bihexagon unit found in this study will be a novel suitable model for crystal embryo formation of ice I_c during heterogeneous nucleation and may also provide important information concerning ice growth in the I_c phase [25]. Numerous theoretical studies have proposed the geometry of small water clusters [26]. Although the C_s tetramer is predicted to be less stable than the monocyclic tetramer [27], the C_s tetramer fragment found in the present system is the first example of the C_s water tetramer to be characterized by X-ray analysis. The tetramer is stabilized in the unique space surrounded by the cyclic belt-like water clusters consisting of six condensed hexagons in the primary water tube. In summary, at the low temperature phase all the oxygen atoms of the water molecules in the channel are H-bonded to the four nearest-neighbor water molecules except for the C_s water tetramer in which two water molecules H-bond to only three nearest neighbor water molecules. Therefore, structurally the three-layered water tube state becomes the "three-layered icelike tube" state. The water clusters confined to the narrow space have been reported to create structures such as the hexamer [28] octamer [29, 30] and decamer [3] framework. These water clusters do not display the so-called dynamic nature of water because they interact directly and strongly with supramolecules in the crystal.

We obtained some results for the structures of ice nanotubes (INTs), which is frozen WNT in crystal 1, by conducting X-ray crystal analysis below the phase transition temperature of 198 K along the frozen-in direction as follows:

1. The periodic unit of the INT is constructed from 60 water molecules. The length of the *c*-axis of crystal 1 with INT at 198 K varies about three times as long as that with WNT at 296 K because of the expansion of the periodicity for the INT formation, retaining the same space group. Only a limited number of water molecules in the secondary and tertiary regions of the WNT probably participate in the water–ice transition.
2. All the water molecules in the INT form H-bonds with each other, and the INT is separated into three-layered structures with primary and secondary layers and tertiary domains. The INT is also attached to most of the carboxylic O atoms of TMA^{3-} in the channel framework. The H-bonded networks of water molecules in the primary layer are formed

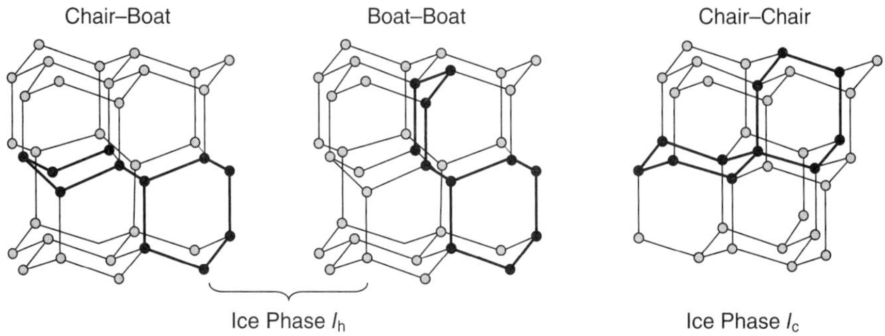

Figure 26.11. The spiro-bihexagon units found in the hydrogen-bonding structure of two ice polymorphs of ice phases I_h and I_c:I_h is constructed from two patterns of the chair–boat and boat–boat conformations, while I_c is constructed from only one pattern of the chair–chair conformation.

from multicyclic water polygons such as pentagons, hexagons, and octagons. By X-ray analysis, it is observed that TMA^{3-} exists as 1.5 molecules in an asymmetry unit; therefore, the carboxylic O atoms exist individually as a total of nine atoms. A carboxylic O(8) atom does not form an H-bond with water oxygen in the primary layer. Thus, no water oxygen exists within less than 4 Å around O(8).

3. In the secondary layer and tertiary regions, a 1-D spirocyclic water–chain cluster is stabilized and constructed from a periodic structure with alternative linkages between a dynamic disordered water molecule (O(40) or O(40)*) and a spiro-bihexagon (a water dodecamer). The dodecamer is recognized as one of the smallest models for a crystal embryo at the cubic ice phase I_c during heterogeneous nucleation (Fig. 26.11). A water molecule in the link between the dodecamers is disordered at two sites of O(40) and O(40)* in the tertiary domain, even at 198 K. The disordered water molecule has an unusual two-handed H-bond and is also located at the center of the cyclic beltlike water hexagons belonging to the primary layer. On the other hand, from the X-ray crystal analysis conducted at 296 K, the electron densities of water oxygens are located around the outer walls of the nanochannel; however, they cannot be accurately ascertained at this temperature due to the electron thermal instability of each water molecule (Fig. 26.12).

In this study, we focus on the structures and properties of a liquidlike WNT by conducting neutron structure analysis and neutron scattering. It has been known that for WNTs confined to single-walled carbon nanotubes with a single-layered structure [31], the nature of WNTs with three-layered structures, which we have observed, would be the essential difference with regard to the WNTs in hydrophobic channels of carbon nanotubes.

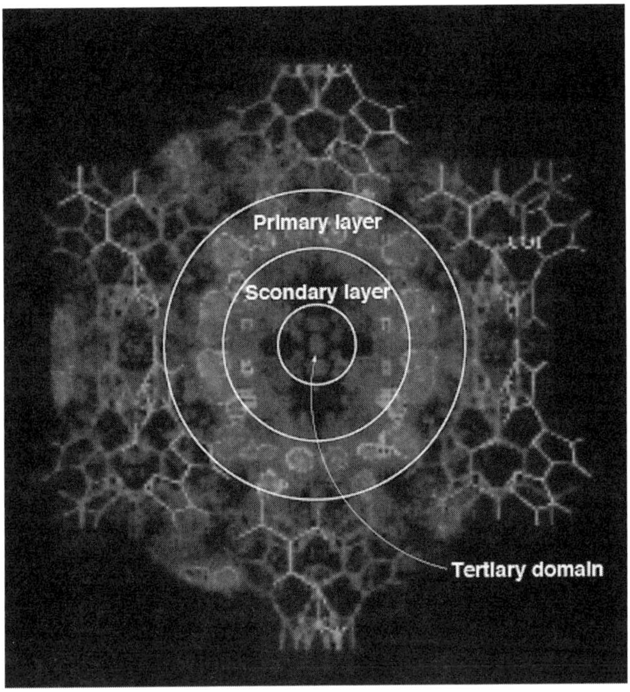

Figure 26.12. The electron density map of the porous channel unit of 1 along the c-axis. The electron densities of structural water in the primary layer are localized near the channel wall but are not observed around the central axis of WNT. Most electron densities of the water molecules belonging to the secondary layer and all those belonging to the tertiary domain in WNT could not be observed by X-ray structure analysis. Nonobservable water molecules around the central axis of WNT show significant movement into the channel maintaining free rotation and causing a liquid state unlike bulk water.

It is also well-known that for hexagonal ice, which is the most typical structure formed as such in the atmosphere, the hydrogen atoms (H atoms) in the network are in a disordered state and that the freezing-in, namely, the glass transition, of the rearrangement of water molecules occurs at $T_g \approx 105$ K [32], at which the characteristic instant of the rearrangement is 1 ks. No phase transition due to the ordering of the H atom positions is observed below T_g because of the immobility of the molecules. The phase transition is realized at 72 K only after introducing hydroxide ions in place of water molecules in the network [33], namely, doping some L-type Bjerrum defects into ice. The phase transition and glass transition behaviors of WNT in 1 are studied by adiabatic calorimetry [34] (Fig. 26.13). The heat capacities of 1 are measured by repeating thermometry at 4–300 K along the melt-in direction; the energy supply

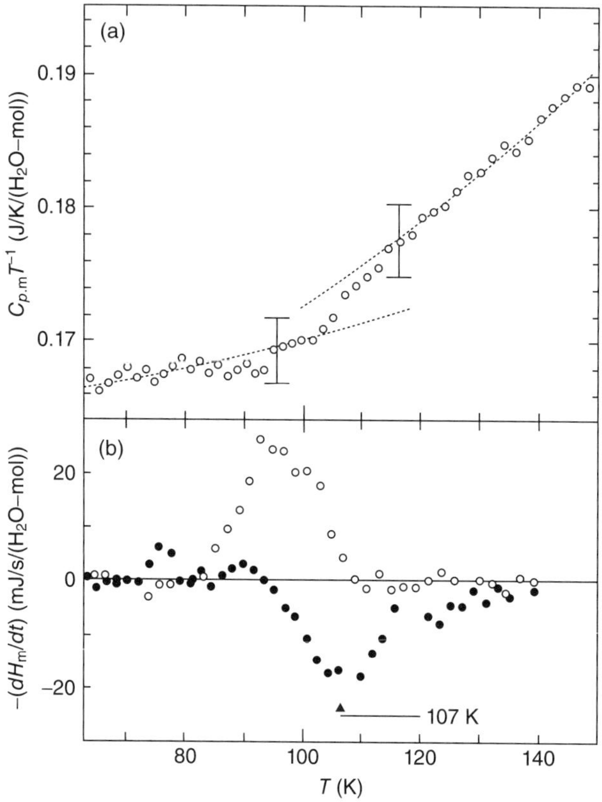

Figure 26.13. Molar encraties (heat capacities divided by the respective temperatures) of the hydrate water (a) and spontaneous enthalpy drift rates (b) on an enlarged scale at 70–150 K. Upper and lower dotted lines in (a) represent the temperature dependences of the heat capacities under equilibrium and frozen-in states, respectively, expected at around a glass transition temperature of 107 K. The error bars stand for the magnitude of the imprecision in the data. Open and closed circles in (b) represent the data for the samples cooled rapidly at 2 K/min and slowly at 20 mK/min, respectively, before the measurements.

yields a temperature increase of 1.0–2.5 K under adiabatic conditions. The glass transition of WNT in 1 is observed at $T_g = 107$ K; this is discussed as a liquid phenomenon of a small number of water molecules around the center of the WNT.

From the neutron crystal structure analysis at 298 K, the dispositions of static structural water in the WNT were determined. Considering the previous results of the X-ray structure analysis mediated by electron densities, which were conducted at 296 K, it is difficult to determine the precise positions of water oxygens in WNT; this is because all the electrons of the water molecules undergo thermally heavy fluctuations by sustaining the disorders and because of large temperatures. Detailed dispositions and H-bonding structures of structural water cannot be precisely determined despite the higher electron densities in structural water. Furthermore, the electron densities of water molecules limited to the secondary layer are hardly observed, while those belonging to the tertiary region around the center of the WNT could not be localized because of the diffusion of molecules particularly during the heavy movement of water molecules. Consequently, it is not possible to draw the entire WNT structure by using X-ray structure analysis based on the diffused electron densities. Hence, we performed the experiment on the basis of structure analysis by using neutron irradiation, which can be used to observe the diffraction intensity for an atomic nucleus as the electron has a thermally more delocalized nature than the nucleus [35]. Figure 26.14 shows a perspective view along the *c*-axis of the WNT confined to a channel unit, and the dispositions of structural water are determined by neutron structure analysis at 293 K. All the water molecules near the channel wall are localized as structural water by H-bonds with the carboxyl O atom of TMA^{3-} forming the channel frameworks. The nuclear densities of the water molecules are localized near the channel walls; however, the densities are not observed around the central axis of WNT (Fig. 26.15). A water molecule comprises covalent bonds between two H atoms and an O atom. The former has negative scattering lengths for a nuclear density when a neutron is irradiated, whereas the latter has positive scattering lengths. Therefore, the heavy mobile water molecule cannot be discriminated by irradiating the neutron because the opposite atomic scattering lengths between H and O atoms nullify each other. We have not observed most of the nuclear densities of water molecules belonging to the secondary layer and all those belonging to the tertiary region in WNT. The nonobservable nuclear densities of water molecules around the center of WNT exhibit significant movement into the WNT channel, thereby maintaining free rotation and resulting in a liquid state, unlike bulk water.

The porous frameworks are confirmed in the structures obtained by X-ray crystal analysis. It is demonstrated that two H atoms participated in the formation of complementary dual H-bonds of $N^{\delta+}$–H···$O^{\delta-}$ types between the carboxylic O atoms of TMA^{3-} and the N atoms of $Hbim^{-}$ (N(4)–H(N4)···O(1)/N(2)–H(N2)···O(2), N(6)–H(N6)···O(3)/N(6)*–H(N6)*···O(3)*) [*: $-x + 1$, y, $-z + 3/2$]. Two H atoms are localized to one side of the N atoms because of the difference in their acidities, and they form strong ionic H-bonds with the −1 charge of two carboxylic O atoms. Three water molecules with O(4), O(9) and O(14), which belong to the primary layer of WNT, form direct intermolecular H-bonds of O–H···O (O(4)–H(4A)···O(3), O(9)–H(9A)···O(1) and O(14)–H(14A)···O(2)) with three carboxylic O atoms of TMA^{3-} of O(4), O(9), and O(14), respectively (Fig. 26.16). Their H-bonds belong to a medium-strength type with a double-well energy potential, which are different from a strong type with a single-well energy potential because each H atom localizes to one side of two water oxygen atoms [36]. Two water molecules of O(5) and O(8), which belong to the secondary layer, bond with two water molecules of

Figure 26.14. Perspective view of the structural water of WNT in a channel unit along the *c*-axis: They are fixed by an anchor effect of hydrogen bonds with carboxylate oxygen atoms, thereby forming the channel framework. The blue closed lines and violet lines show $[Co(Hbim)_3]^{3+}$ and TMA^{3-}, respectively, thereby forming the outer frameworks of the channel. The red spheres and white spheres show the oxygen atoms and hydrogen atoms of the structural water molecules, respectively. The dashed blue lines show the intermolecular hydrogen bonds forming the frameworks. The results of neutron crystal analysis of 1 at 293 K are $C_{27}H_{61}N_{12}O_{26}Co$, $Fw = 542.00$, monoclinic of space group *C*2/c (#15), with cell dimensions as follows: $a = 16.338(2)$ Å, $b = 29.335(2)$ Å, $c = 10.909(2)$ Å, $\beta = 90.336(10)°$, $V = 52{,}228.3(12)$ Å^3, and $Z = 8$; neutron ($\lambda = 1.51000$ Å), $R_1 = 12.33\%$, $wR_2 = 23.20\%$ and GOF = 1.238. See color figure on companion ftp site.

O(4) and O(14) through H(5A) and H(8B) to form two H-bonds of O(4)–H(5A)···O(5) and O(14)–H(8B)···O(8), respectively. H(4A), H(4B), and H(4C) are attached to a water oxygen(4) atom, since H(4B) and H(4C) covalent bonded with O(4) are treated as disorders that contribute toward occupancy factors of 0.33 and 0.67. O(4) is H-bonded with another O(9) through H(4C) to form a sequence of a cyclic H-bonding network of O(3)···H(4A)–O(4)–H(4C)···O(9)–H(9A)···O(1), which inserts two water molecules to either side of O(3) and O(1).

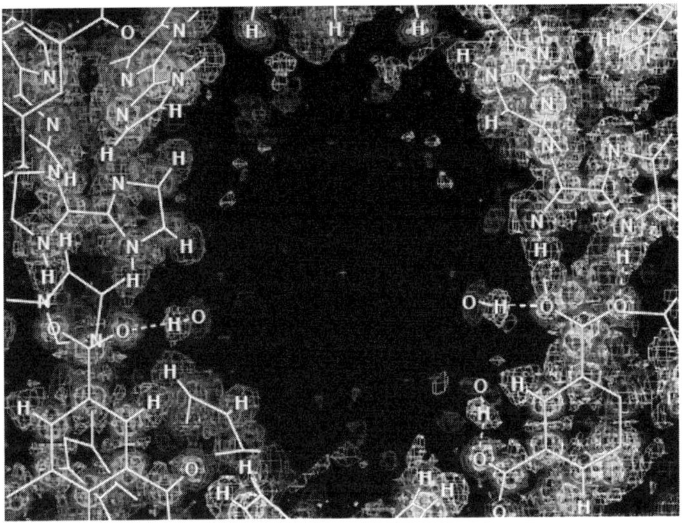

Figure 26.15. Atomic nucleus density map of a channel unit containing WNT along the *c*-axis: The H and O atoms are shown by the white and violet shaded parts, respectively. Molecules are not observed around the center of the channel because of significant free rotation and translational motion of water molecules. The yellow trace lines show the shapes of each molecule and intermolecular hydrogen bonds based on the location of the atomic nucleus density map. See color figure on companion ftp site.

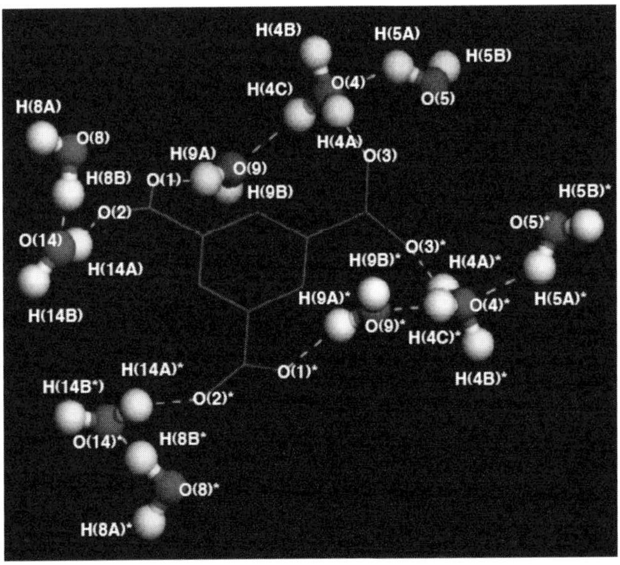

Figure 26.16. Hydrogen-bonding networks with each atom (numbered) of structural water molecules in WNT and some hydrogen-bonded atoms observed by neutron structure analysis: The magenta and deep blue lines show $[TMA]^{3-}$ and $[Co(Hbim)_3]^{3+}$ frameworks, respectively. The H and O atoms of structural water are shown by the white and red circles, respectively. The yellow dotted lines show the hydrogen bonds. O(5) and O(8), which belong to the secondary hydrate layer, connect with two O(4) and O(14) through H(5A) and H(8B). H(4A), H(4B), and H(4C) are attached to the O(4) atom since H(4B) and H(4C) are treated as disorders contributing to each occupancy factors of 0.32746 and 0.67354. O(4) is connected to another O(9) through H(4C) to form a sequence of a cyclic hydrogen-bonding network with O(1) and O(3). See color figure on companion ftp site.

A comparison between the temperature factors of H and O atoms comprising structural water in WNT has indicated that each atom near the channel wall is successively smaller than that near the center of WNT. Thus, the structural water directly connected to the O atoms of TMA^{3-} works as anchors to stabilize the WNT cluster and to restrain the fluctuation of water molecules present in it. For example, in sequences A and B of H-bonded networks of O(2)···H(14A)–O(14)···H(8B)–O(8)–H(8A) and O(3)···H(4A)–O(4)···H(5A)-O(5)–H(5B), respectively, the temperature factors of the former has the following order of isotropic displacement parameters: U_{eq} of 102(3) < 230(14) < 304(15) < 450(40) < 620(50) < 110(200) ($Å^2 \times 10^3$). These are defined as one-third of the trace of the orthogonalized U_{ij} tensor. The latter also has the following order: 54(2) < 140(7) < 224(12) < 231(15) < 335(15) < 550(60) ($Å^2 \times 10^3$), both of which show a gradual increase in their values with increasing distance from the channel wall. In the same manner, the temperature factors of the sequence C of H-bonding atoms of O(1)···H(9A)–O(9)–H(9B) also increase in the order of 77(2) < 213(11) < 347(14) < 430(40), respectively (Fig. 26.17). Interestingly, each TMA^{3-} unit making up the framework simultaneously can stabilize the structural water of the three WNTs confined to the adjacent three different channels by using H-bonds with six carboxylic O atoms. Therefore, all the WNTs are chemically related to each other and must maintain the same periodic structure throughout the entire crystal. As a result, the varying structures of all the WNTs become standardized as a result of the mechanistic anchor effect, and all WNTs simultaneously exhibit macroscopic motions and phase transitions.

The neutron scattering of crystal 1 is measured with an angle focusing cold neutron spectrometer (AGNES). Figure 26.18a shows the elastic scattering intensity of WNT deviated from the nearly straight data for bulk ice above 210 K; this corresponds to the order–disorder transition of water molecules in WNT, which was observed at ca. 210 K in the calorimetric experiments. This temperature dependence of the elastic intensity is similar to that of a biocorrelated material such as protein [37]. The gap in the elastic intensity of WNT at 273 K is due to the melting of a small amount of bulk water that was added to the sample to stabilize WNT crystals. Therefore, the quasi-elastic scattering data were measured at several temperatures below 273 K to account for and to avoid the effect of bulk water.

The quasi-elastic spectra were analyzed using the method established by Teixeira et al. [38], which considers both translational jump diffusion and rotational motions of water molecules. Nine $S(Q, E)$ spectra with different momentum transfers ($Q = 0.31$, 0.42, 0.57, 0.70, 0.88, 1.12, 1.35, 1.57, and 1.80 $Å^{-1}$) and having the same temperature were fitted simultaneously (i.e., global fitting) after taking into consideration the integrated intensity of an elastic peak for each Q value, integrated intensity of quasi-elastic peak at $Q = 0$, mean square displacement, diffusion coefficient, residence time, and rotational relaxation time as flexible parameters. Figure 26.18b shows an example of the fitting for the data at 268 K and 1.57 $Å^{-1}$. As shown in this

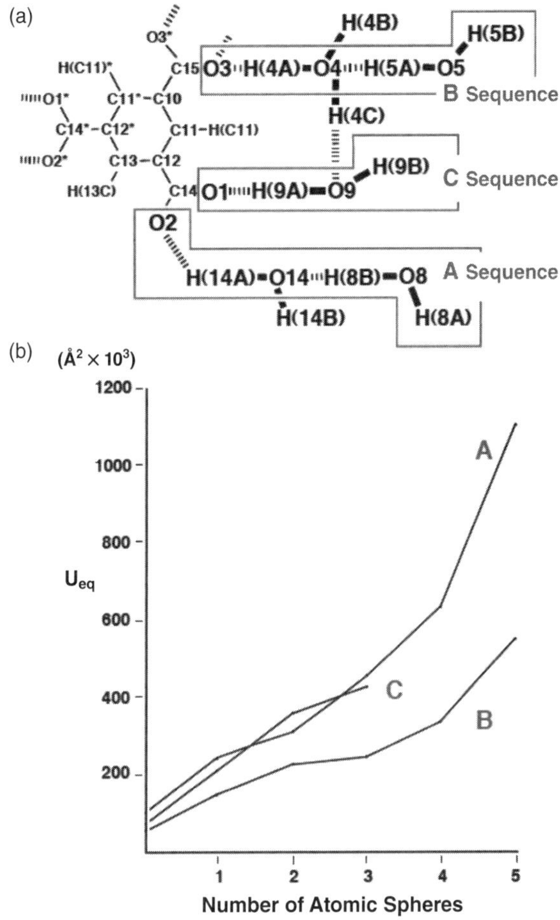

Figure 26.17. (a) Schematic representations with each numbered atom; A, B, and C sequences of hydrogen-bonding networks are shown. (b) Each line graph of A, B, and C sequences describes the relationship between the number of atomic spheres from the carboxylic O atoms in the channels and the values of isotropic displacement parameters U_{eq}, which are defined as one-third of the trace of the orthogonalized U_{ij} tensor. U_{eq} indicates the fluctuations for each atom.

figure, the quasi-elastic component is fitted well by the two Lorentz functions for the zeroth and first terms of a spherical Bessel function [39], each corresponding to the translational diffusion and rotational motions of water molecules, respectively. The second and higher terms were negligible in the fitting. Note that the quasi-elastic component was smaller than the elastic one due to the bulk ice coexisting with WNT and hydrogen atoms in the framework of crystal 1. Because of this undesirable situation, the fitting was successful only

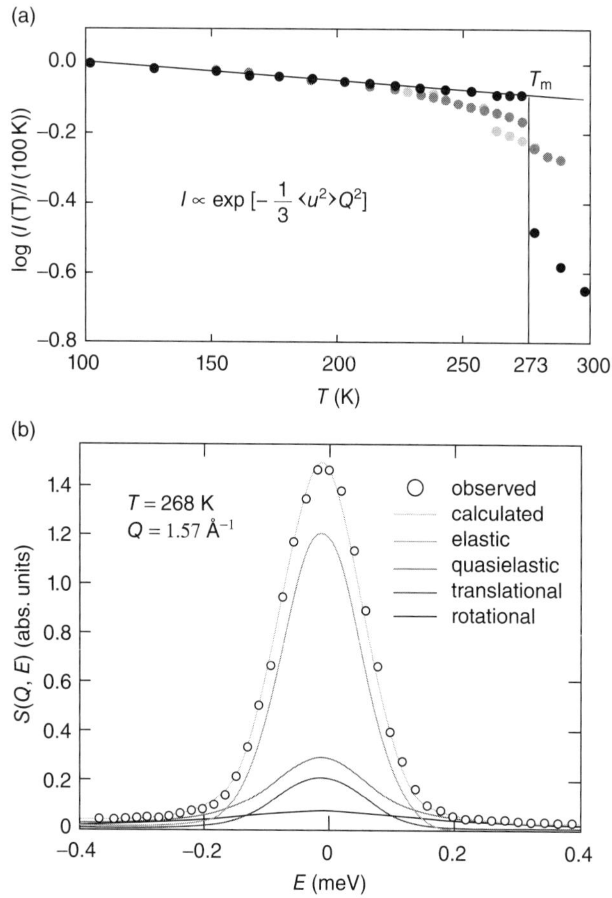

Figure 26.18. (a) Temperature dependence of the elastic intensity normalized by the intensity at 100 K. In bulk water (blue circles), the logarithmic intensity of ice is proportional to temperature as is expected from the harmonic oscillation approximation $I/I_0 = \exp[-1/3\langle u^2\rangle Q^2]$, where Q [$= 4\pi(\sin\theta)/\lambda$] is the momentum transfer (length of scattering vector) and $\langle u^2\rangle$ is the mean square displacement, which is a measure of the average amplitude of motions of water molecules. The elastic intensity changed abruptly at T_m (=273 K), which is the melting temperature of ice; molecules have considerably higher mobility in a liquid phase than in a solid phase. In WNTs (red circles for heating and yellow ones for cooling runs), there are also abrupt jumps around T_m because a small amount of bulk water was added to the sample to stabilize crystal 1. The gradual decrease in the intensity beginning from 210 K in the heating run is due to the order–disorder transition of water molecules in WNT. (b) Neutron quasi-elastic spectrum of crystal 1 measured at $T = 268$ K and $Q = 1.57$ Å^{-1}. The observed data (circles) were fitted to the model function (red line), which is composed of elastic (green line) and quasi-elastic (blue line) components. The quasi-elastic component was further divided into translational (magenta) and rotational (light blue) parts. See text for details of the fitting. See color figure on companion ftp site.

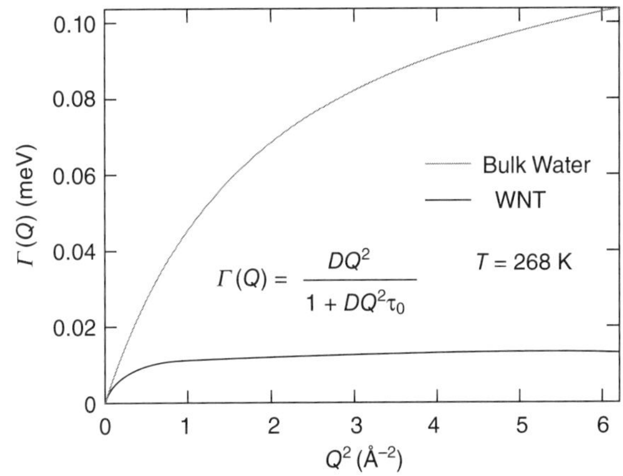

Figure 26.19. Half-width at half-maximum (HWHM) of the Lorentz function for the translational diffusion plotted as a function of squared momentum transfer. The red and blue curves represent the calculated values for bulk water and WNT, respectively. The calculation was performed based on the jump diffusion model with the observed temperature of 268 K and the parameters determined by the fitting described above. $\Gamma(Q)$ is represented as a function of the width of the QENS signal, which is an expression for a random jump diffusion model, where τ_0 represents the residence time. The translational diffusion constant D is expressed in terms of the mean square jump length $[\langle l^2 \rangle_{av}]^{1/2}$ by $D = \langle l^2 \rangle_{av}/6\tau$. See color figure on companion ftp site.

TABLE 26.1. Results of Neutron Scattering Parameters at 268 K

	D / Å^{2}s^{-1}	τ_0 / s	$\langle l \rangle$ / Å
WNT in 1	9.44×10^{10}	4.70×10^{-11}	5.15
Bulk water (this work)	1.01×10^{11}	4.82×10^{-12}	1.73
Bulk water (ref. 38)	0.85×10^{11}	4.66×10^{-12}	1.54
WNT in CNT (ref. 40)*	5.40×10^{10}	12.5×10^{-11}	6.40

* Measured at 260 K.

for the data at 268 K, which has the largest quasi-elastic component. Figure 26.19 shows the half-width at half-maximum (HWHM) curve of the WNT calculated from the parameters determined by the fitting. The curve for the supercooled water, which was measured and analyzed using the same methods as that for WNT, is also shown for comparison. The diffusion coefficient of the translational mode D corresponding to the slope of the curve at $Q = 0$, the residence time τ_0 corresponding to the inverse HWHM at the high-Q limit, and the average jump length $\langle l \rangle$ calculated from the relation $D = \langle l^2 \rangle / 6\tau_0$ are given in Table 26.1.

The quantities of bulk water obtained in this work are essentially the same as those in the previous work [38], demonstrating the reliability of the data and analysis of this work. The characteristic values for τ_0 and $\langle l \rangle$ are also larger than those of bulk water, as shown in Table 26.1, and the value of D is almost identical to that of the bulk one. This result agrees with a possible picture of fused WNT in the channel—that is, the water molecules of WNT are hydrogen bonded to the neighboring molecules for a longer time than those of bulk water; however, once the hydrogen bonds are broken, the water molecules of WNT jump a longer distance than those of bulk water. This picture may be a result of the situation that the density of the water molecules in WNT (0.81 g/cm^3 at 296 K) is less than that of bulk water (0.97 g/cm^3 at 298 K). The determined parameter $\langle l \rangle$ seems to be shorter than the actual jump distance, when the distance between the water molecules is taken into consideration on a 3-D motion as bulk water. This is probably because the actual jump of water molecules in WNT is anisotropic due to the surface effect and geometry of the channel. The relation $D = \langle l^2 \rangle / 2\tau_0$ for 1-D diffusion gives higher values than the relation $D = \langle l^2 \rangle / 6\tau_0$ for 3-D diffusion. From quasi-elastic neutron scattering for water in a single-walled CNT, the diameter is found to be approximately 1.4 Å, which is similar to the WNT of 1 [39–41].

In this study, we have created a novel three-layered water tube network anchored in a 1-D nanometer-scale channel framework. The channels were constructed by designing molecular building blocks for a new supramolecular synthon. The three-layered water molecules within the tube structure show the intriguing hierarchical ice–water phase change. The change is caused by water molecules in the secondary and tertiary domains, where dynamic structural changes take place. In addition, water molecules in the secondary and tertiary domains help generate the spiro-bihexagon and C_s tetramer, which are stabilized by H-bonding below the phase change temperature. Above this temperature, the networks transform to the quasi-planar hexamer and highly disordered species observed by the X-ray analysis and ^{2}H-NMR measurements, which indicate the existence of a restricted 1-D translational motion. The very high proton conductivity of 1 at room temperature is probably due to this translational motion of water molecules in the hierarchical water tube networks with restricted dimensionality. These findings represent the first reported description of this surprising property of water. Studies on confined water is of relevance to a number of biological processes including hydration of a biological surface [42], dynamics of hydrated proteins [43], and the mobility of water within biological pores [44]. Thus, our synthetic efforts using crystal engineering may contribute to a greater understanding of the mechanism of ice formation and the properties of water and ice within a biological context.

In addition, the anomalous water molecules at the center of a WNT are a novel example of the occurrence of a phase transition due to a small number of water molecules; further, a glass transition with a macroscopic behavior similar to that of a liquid phase is also indicated. On the other hand, a characteristic feature is that a freezing INT changes into a melting WNT at 245 K, which is lower than the phase change of bulk ice at 273 K, thereby influencing structural water H-bonding with the channel wall. Furthermore, before the

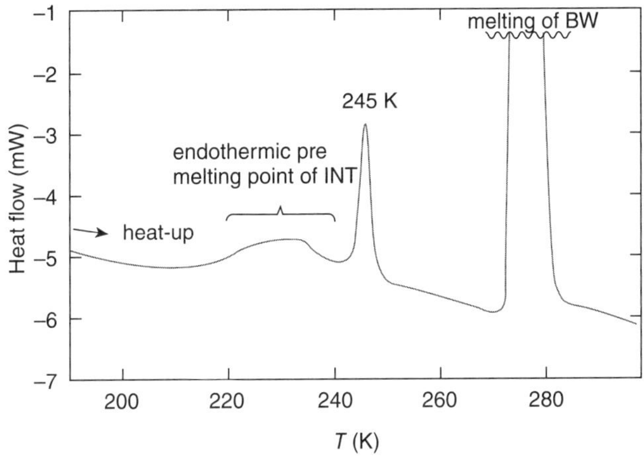

Figure 26.20. DSC spectrum of crystal 1 confined to the WNT through heat-up: The INT melts from 245 K onward. From 210 to 240 K, a broad peak of the premelting phase is also observed. The peak from 273 K is due to melting and consequent transition to bulk water (BW), which is added to stabilize crystal 1 in the aluminum pan (velocity: 5 K/min).

melting point of the INT, a weak endothermic broad peak from 210 to 240 K is observed; this is probably related to the movement of the water molecules in INT (Fig. 26.20). Such premelting reactions appearing prior to the phase transition is a unique characteristic of INTs but not of bulk ice. To our knowledge, employing neutron structure analysis, this study has for the first time confirmed the mechanistic effects of structural water in WNT confined to hydrophilic nanoscale channels with diameters of ca. 1.6 nm. The liquidlike domain in WNT with fast water molecules and the structure of static structural water have been identified simultaneously. The dispositions of structural water in WNT, roughly localized by X-ray analysis, are precisely determined by neutron structure analysis. We have performed neutron scattering on a small amount of anomalous water molecules in WNT. Water molecules restricted by structural water typically move slower than those restricted by bulk water [42, 45, 46] because the H-bonds are highly oriented with the static structural water. However, it is clear that the anomalous water molecules in WNT have at least the same velocity as that of bulk water and move at a faster velocity because they are constructed from incomplete H-bonds and small densities in WNT. For example, in actin filaments interacting with an activated myosin motor domain, it is confirmed that the motion of water molecules in a muscular fiber is impacted by structural water, which moves approximately three times faster than bulk water [47].

Furthermore, for each atom in the WNT, the thermal motion increases with increasing distance from the primary layer near the outer wall toward the secondary and tertiary domains in that order. Hence, this would result in higher flexibility in moving water molecules due to the minimization of the forming of structural water in pseudo-1-D channels. The WNT detected in the experiments may have one of the smallest structural units confined to hydrophilic nanoporous channels that have a liquid behavior and probably reveals a macroscopic phase transition with some water molecules. Further, the 1-D movement of water molecules at the center of the WNT is useful for understanding the model for elucidating the cohesion mechanism involved in transferring water to heights such as 100 m, for example, water transferred through a xylem conduit of a certain tree growing up to a height of approximately 100 m [48, 49]. Moreover, this movement is also useful as a biopore model for the transfer of water molecules through the inner and outer of cells, such as aquaporin-1 [50, 51] of water transporting proteins. Lastly, we hope to find an unprecedented water phase in which water molecules infinitely bond pseudo-one-dimensionally with H-bonds in a future work.

ACKNOWLEDGMENTS

This work was supported by a Grant-in-Aid for Scientific Research (Nos. 18033049 and 20045018) on Priority Areas from the Ministry of Education, Science and Culture, Japan and for REIMEI Research Project for JAERI. The authors thank the Analytical Center in Tokyo University of Science for the use of a CCD single-crystal X-ray diffractometer and a ^{2}H-NMR spectrometer.

REFERENCES

1 G. Hummer, J. C. Rasalah, J. P. Noworyta. *Nature (London)* 414 (2001) 188–190.

2 K. Koga, G. T. Gao, H. Tanaka, X. C. Zeng. *Nature (London)* 412 (2001) 802–805.

3 L. J. Barbour, G. W. Orr, J. L. Atwood. *Nature (London)* 393 (1998) 671–673.

4 U. Raviv, P. Laurat, J. Klein. *Nature (London)* 413 (2001) 51–54.

5 R. G. Horn, D. T. Smith, W. Haller. *Chem. Phys. Lett.* 162 (1989) 404–408.

6 N. L. Rosi, M. Eddaoudi, J. Kim, M. O'Keeffe, O. M. Yaghi. *CrysEngComm* 4 (2002) 401–404.

7 A. M. Beatty. *CrystEngComm* 51 (2001) 1–13.

8 G. R. Desiraju. *Angew. Chem. Int. Ed.* 34 (1995) 2311–2327.

9 M. Henry. *Chem. Phys. Chem.* 3 (2002) 607–616.

10 C. Janiak, T. G. Scharmann. *J. Am. Chem. Soc.* 124 (2002) 14010–14011.

11 M. Tadokoro, K. Nakasuji. *Coord. Chem. Rev.* 198 (2000) 205–218.

12 M. Tadokoro, J. Toyoda, K. Isobe, T. Itoh, A. Miyazaki, T. Enoki, K. Nakasuji. *Chem. Lett.* 24 (1995) 613–614.

13 M. Tadokoro, T. Shiomi, T. Shiromizu, K. Isobe, K. Matsumoto, K. Nakasuji. *Mol. Cryst. Liq. Cryst.* 306 (1997) 235–239.

14 M. Tadokoro, H. Kanno, T. Kitajima, H. Shimada-Umemoto, N. Nakanishi, K. Isobe, K. Nakasuji. *Proc. Natl. Acad. Sci. U.S.A.* 99 (2002) 4950–4955.

15 R. Atencio, M. Chacón, T. González, A. Briceño, G. Agrifoglio, A. Sierraalta. *Dalton Trans.* (2004) 505–513.

16 L. Öhrström, K. Larsson. *Dalton Trans.* (2004) 347–353.

17 M. Tadokoro, S. Fukui, T. Kitajima, Y. Nagao, S. Ishimaru, H. Kitagawa, K. Isobe, K. Nakasuji. *Chem. Commun.* (2006) 1274–1276.

18 C. A. Koh. *Chem. Soc. Rev.* 31 (2002) 157–167.

19 R. Ludwig. *Angew. Chem. Int. Ed.* 40 (2001) 1808–1827.

20 N. J. Moorthy, R. Natarajan, P. Venugopalan. *Angew. Chem. Int. Ed.* 41 (2002) 3417–3420.

21 D. Eisenberg, W. J. Kauzmann. *The structure and properties of water*. Oxford at the Clarendon Press, London, 1969.

22 A. Abragam. *Principles of nuclear magnetism*. Oxford University Press, London, 1961.

23 E. Rodríguez-Castellón, J. Jiménez-Jiménez, A. Jiménez-López, P. Maireles-Torres, J. R. Ramos-Barrado, D. J. Jones, J. Roziére. *Solid State Ionics* 125 (1999) 407–410.

24 E. K. Andersen, I. G. K. Andersen, E. Skou, in P. Colomban (ed.), *Proton conductors*. Cambridge University Press, New York, 1992, pp. 210–223.

25 M. Matsumoto, S. Saito, I. Ohmine. *Nature (London)* 416 (2002) 409–413.

26 J. M. Ugalde, I. Alkorta, J. Elguero. *Angew. Chem. Int. Ed.* 39 (2000) 717–721.

27 J. K. Gregory, D. C. Clary. *J. Phys. Chem.* 100 (1996) 18014–18022.

28 R. Custelcean, C. Afloroaei, M. Vlassa, M. Polverejan. *Angew. Chem. Int. Ed.* 39 (2000) 3094–3096.

29 W. B. Blanton, S. W. Gordon-Wylie, G. R. Clark, K. D. Jordan, J. T. Wood, U. Geiser, T. J. Collins. *J. Am. Chem. Soc.* 121 (1999) 3551–3552.

30 J. L. Atwood, L. J. Barbour, T. J. Ness, C. L. Raston, P. L. Raston. *J. Am. Chem. Soc.* 123 (2001) 7192–7193.

31 Y. Maniwa, K. Matsuda, H. Kyakuno, S. Ogasawara, T. Hibi, H. Kadowaki, S. Suzuki, Y. Achiba, H. Kataura. *Nature Mater.* 6 (2007) 135–141.

32 N. Giovambattista, C. A. Angell, F. Sciortino, H. E. Stanley. *Phys. Rev. Lett.* 93 (2004) 047801–047801.

33 Y. Tajima, T. Matsuo, H. Suga. *Nature* 299 (1982) 810–812.

34 K. Watanabe, M. Oguni, M. Tadokoro, Y. Ohata, R. Nakamura. *J. Phys. Condens. Matter* 18 (2006) 8427–8436.

35 M. Tadokoro, T. Ohhara, Y. Ohhata, T. Suda, Y. Miyazato, T. Yamada, T. Kikuchi, I. Tanaka, K. Kurihara, M. Oguni, K. Nakasuji, O. Yamamuro, R. Kuroki. *J. Phys. Chem. B* 114 (2010) 2091–2099.

36 M. Benoit, D. Marx, M. Parrinello. *Nature* 382 (1998) 258–261.

37 M. Ferrand, A. J. Dianoux, W. Petry, G. Zaccai. *Proc. Natl. Acad. Sci. U.S.A.* 90 (1993) 9668–9672.

38 J. Teixeira, M.-C. Bellissent-Funel, S.-H. Chen, A. J. Dianoux. *Phys. Rev. A* 31 (1985) 1913–1917.

39 M. Bee. *Quasielastic neutron scattering*. Adam Hilger, Bristol, 1988.

40 E. Monontov, C. J. S. Bumham, H. Chen, A. P. Moravsky, C. K. Loong, N. R. de Souza, A. I. Kolesnikov. *J. Chem. Phys.* 124 (2006) 194703–194701.

41 A. I. Kolesnikov, J.-M. Zanotti, C.-K. Loong, P. Thiyagarajan. *Phys. Rev. Lett.* 93 (2004) 0355031–0355034.

42 M. S. P. Sansom, P. C. Biggin. *Nature (London)* 414 (2001) 156–157.

43 M. Settles, W. Doster. *Faraday Discuss.* 103 (1996) 269–279.

44 Y. Zhou, J. H. Morais-Cabral, A. Kaufman, R. MacKinnon. *Nature (London)* 414 (2001) 43–48.

45 S. Takahara, N. Sumiyama, S. Kittaka, T. Yamaguchi, M. C. Bellissent-Funel. *J. Phys. Chem. B* 109 (2005) 11231–11239.

46 E. C. Trantham, H. E. Rorschach, J. S. Clegg, C. F. Hazlewood, R. M. Nicklow, N. Wakabayashi. *Biophys. J.* 45 (1984) 927–938.

47 S. H. Koenig, W. E. Schillinger. *J. Biol. Chem.* 244 (1969) 3283–3289.

48 S. R. Kabir, K. Yokoyama, K. Mihashi, T. Kodama, M. Suzuki. *Biophys. J.* 85 (2003) 3154–3161.

49 M. T. Tyree. *Nature* 423 (2003) 923.

50 W. T. Pockman, J. S. Sperry, J. W. O'Leary. *Nature* 378 (1995) 715–716.

51 P. Agre. *Angew. Chem. Int. Ed.* 43 (2004) 4278–4290.

27

SURFACE ELECTROCHEMISTRY OF ELECTROSPUN NANOFIBERS

HIDETOSHI MATSUMOTO AND AKIHIKO TANIOKA

27.1 INTRODUCTION

Rapid advances in nanotechnology have provided a large variety of nanomaterials with various shapes and superior properties during the past two decades. Among these nanomaterials, one-dimensional (1-D) nanostructures, such as nanofibers, nanotubes, nanowires, nanorods, and nanowhiskers, have attracted much interest because of their unique properties, which can be attributed to their dimensional anisotropy. Nanofibers are nanomaterials known for their cross-sectional direction and a macroscopic material known for its fiber axis. Therefore, nanofibers have both the advantages of nanomaterials due to their nanoscaled diameter and good handling characteristics due to their macroscopic length. Electrospinning is a straightforward and versatile method for forming continuous thin fibers based on an electrohydrodynamic process [1–3]. This method has the following advantages: (i) ability to produce thin fibers with diameters in the micrometer and nanometer ranges and (ii) applicability for a broad spectrum of molecules, such as inorganic molecules, synthetic and biological polymers, proteins, and DNA. Electrospun nanofibers with high surface areas have drawn significant attention for applications, such as high-performance filter media, protective clothes, composites, drug delivery systems, and scaffolds for tissue engineering [4, 5].

Recently, the scope of ion-exchange technology has widened into the fields of biotechnology, pharmaceutical processing, producing ultrapure water for the semiconductor industry, catalytic conversion processing, and battery

Electrical Phenomena at Interfaces and Biointerfaces: Fundamentals and Applications in Nano-, Bio-, and Environmental Sciences, First Edition. Edited by Hiroyuki Ohshima.

technologies [6]. In particular, various ion-exchange fibers have been successfully used due to their large specific surface area, good mechanical properties, good handling characteristics, and flexibility to be processed into diverse forms. The introduction of ionic functional groups to nanofibers is a promising option to provide novel ion exchangers with high adsorption capacity and high catalytic activity. In addition, an analysis of the surface electrochemical properties of the ionic nanofibers is essential for improving the performance of the fibers, such as their separation property, adsorption behavior, and catalytic activity. This chapter describes the fundamental aspects of the preparation and surface electrochemical characterization of ion-exchange nanofibers by electrospinning and their applications, such as adsorbents for bioseparation and catalysts for water splitting in an electrodialysis system.

27.2 ION-EXCHANGE NANOFIBERS BY ELECTROSPINNING

The basic setup for electrospinning is composed of a spinneret (conductive nozzle), a high-voltage power source, and a counter electrode (conductive collector) (Fig. 27.1). The electrospinning process consists of the following steps: (i) A strong electric field is applied between the nozzle containing the spinning solution and the collector, forcing the solution out through the nozzle and deforming the pendant drop at the tip of the nozzle into a conical shape referred to as the "Taylor cone"; (ii) when the voltage reaches a critical value, electrostatic forces overcome the surface tension of the solution, then an electrified jet is ejected out the tip of the nozzle and moves toward the counter electrode; (iii) on the way to the counter electrode, the jet is continuously

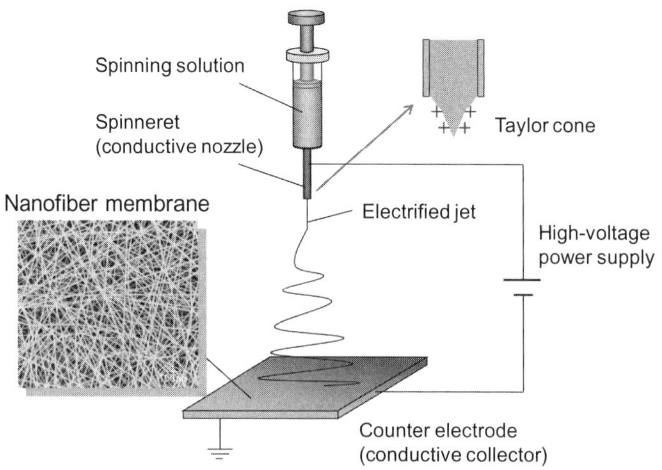

Figure 27.1. Schematic diagram of the basic setup for electrospinning.

stretched due to the electrostatic repulsions between the surface charges and the solvent evaporates; and (iv) a thin fiber is finally collected on the counter electrode [4, 5].

In general, polymer solutions with high electric conductivity (e.g., polyelectrolyte solution) show low electrospinnability because the solution conductivity prevents electric field-induced charging of the solution. Therefore, most of the approaches for forming ion-exchange nanofibers relied on (i) the addition of water-soluble polymers to the spinning solution as the carrier and (ii) electrospinning of a nonionic polymer and successive chemical modification. We prepared ion-exchange nanofibers by electrospinning from two biological polymers—chitosan and chondroitin sulfate—and two synthetic polymers—polystyrene (PS) and poly(4-vinylpyridine) (P4VP). The ionic biopolymer nanofibers were prepared from polysaccharide/polyethylene oxide (PEO) blended solutions. The ionic synthetic polymer nanofibers, on the other hand, were obtained by the sulfonation of the as-spun PS nanofibers and the quaternization of the as-spun P4VP nanofibers. Figure 27.2 shows the surface scanning electron microscopy (SEM) images of the electrospun fibrous membranes from PEO blended solutions [7]. The average diameter of the fibers is about 150 nm. These biological ion-exchange nanofibrous membranes can be applied not only to ion-exchange materials but also to wound healing dressings, scaffolds for tissue engineering, and antimicrobial fabrics. Figure 27.3 shows the surface SEM images of the as-spun and sulfonated PS fibrous membranes and the as-spun and quaternized P4VP fibrous membranes [8]. The average fiber diameters are 1.33 μm and 620 nm for the as-spun PS and P4VP fibers, respectively. These biological and synthetic ion-exchange nanofibrous

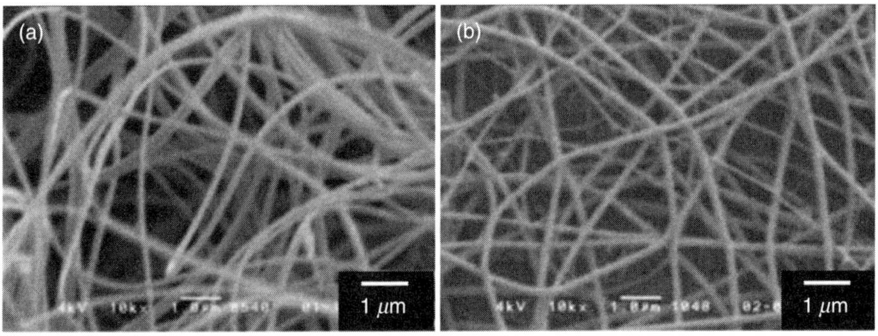

Figure 27.2. SEM images of the electrospun fibrous membranes from polysaccharide/polyethylene oxide (PEO) blended solutions. (a) Cation-exchange nanofibers from the 10 wt % sodium chondroitin sulfate (CHON, M_w = 110,000)/5 wt % PEO (M_w = 500,000) blended solution (the ratio of CHON/PEO = 8/1 w/w) and (b) anion-exchange nanofibers from 10 wt % chitosan (M_w = 36,000, degree of deacetylation = 87%)/5 wt % PEO (M_w = 500,000) blended solution (the ratio of chitosan/PEO = 6/2 w/w) [7].

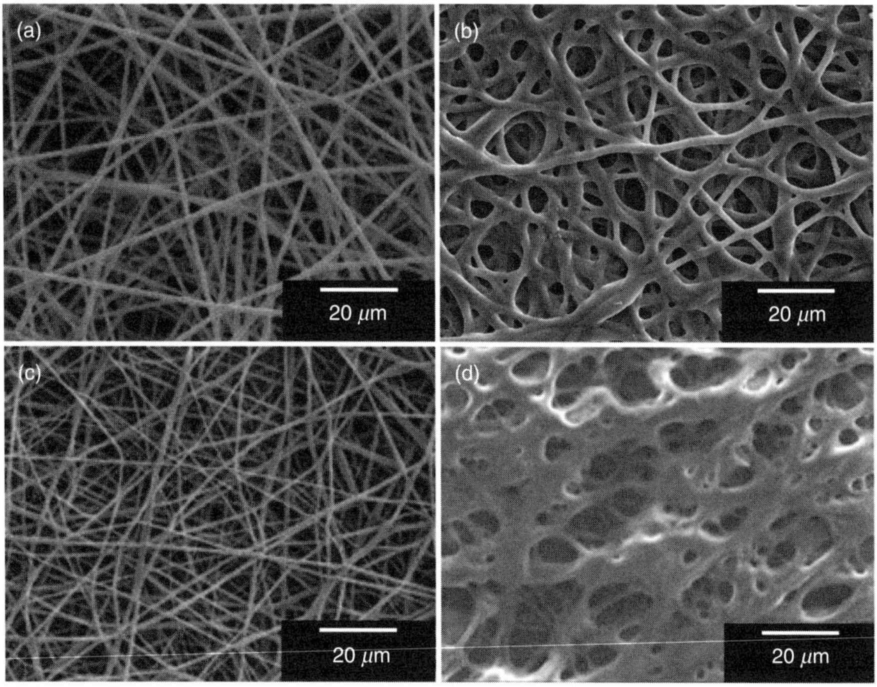

Figure 27.3. SEM images of PS fibrous membranes: (a) the as-spun fibers from 15 wt % PS in the mixture of tetrahydrofuran (THF) and N,N-dimethylformamide (DMF) (50/50 v/v) and (b) cation-exchange fibers after the sulfonation and P4VP fibrous membranes, (c) the as-spun membranes from 15 wt % P4VP in the mixture of EtOH and water (92/8 v/v), and (d) anion-exchange fibers after quaternization [8].

membranes obtained by electrospinning have microscaled interconnected flow-through pores and a high porosity.

27.3 ELECTROKINETIC CHARACTERIZATION OF BIOLOGICAL ION-EXCHANGE NANOFIBERS

Chitosan, which is one of the mucopolysaccharides with a β-1,4-bonded structure of the glucosamine residue having an amino group at the C-2 position and two hydroxyl groups at the C-3 and C-6 positions (Fig. 27.4a), is nontoxic, biocompatible, biodegradable, and bioactive and shows wound healing and antimicrobial effects [9]. The combination of biological polymers and nanofabrication technology, *electrospinning*, is a promising option to provide novel biological ion exchangers. In addition, an analysis of the surface charge of the nanofiber is essential for improving the performance of the ionic fibers or

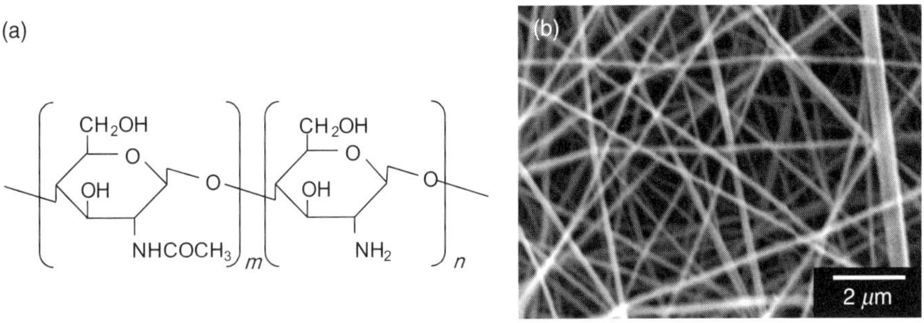

Figure 27.4. (a) Chemical structure of chitosan and (b) SEM image of the electrospun chitosan nanofibrous membrane [10].

TABLE 27.1. Physicochemical Properties of Chitosan Nanofibrous Membranes

Ion-exchange capacity[a] (mmol/g-dry fiber)	5.4
Average pore size of membrane[b] (μm)	0.5
Through-pore specific surface area of membrane[b] (m^2/g)	26
Porosity of membrane[b] (%)	92
Thickness of membrane (μm)	59

[a] Determined by potentiometric titration.
[b] Estimated by bubble-point method.

fibrous membranes. Electrokinetic characterization is a powerful and convenient method to provide information about the *in situ* charge state of material surfaces. It is also applicable to sample surfaces of different geometries including nanofibrous membranes.

Chitosan nanofibers were prepared by electrospinning from a spinning solution containing 12 wt % chitosan (M_w = 36,000, degree of deacetylation = 87%; Kyowa Tecnos, Sanbu, Japan), 1 wt % PEO (M_w = 500,000; Wako, Osaka, Japan), 4 wt % acetic acid (extrapure grade, Wako), and 83 wt % water (deionized water). To improve the water resistance of the as-spun fibers by removing the acetic acid, the fibers were dried in an oven at 105°C for 30 minutes. More detailed information has been reported in Reference 10. Figure 27.4b shows the surface SEM images of the chitosan nanofibrous membrane. The physicochemical properties of the chitosan nanofibers are summarized in Table 27.1. The chitosan nanofibrous membrane has a submicroscaled interconnected flow-through pore structure (average pore size = 0.5 μm) composed of nanoscaled fibers (average diameter = about 100 nm), high porosity (92%), and amino groups on the pore surface (ion-exchange capacity = 5.4 mmol/g), which favorably compares with the reported capacity of 1–4 mmol/g for conventional ion-exchange fibers [8].

The streaming current measurements were carried out using an electrokinetic analyzer (SurPASS, Anton Paar GmbH, Graz, Austria) equipped with a clamping cell [10]. A pair of chitosan nanofibrous membranes having an area of 25 × 55 mm^2 was placed in the measuring cell. The membranes were separated by a spacer that forms a streaming channel. A background electrolyte of a 1 mmol/L KCl solution was used and the pH was adjusted in the range of pH 4–10 with 0.1 mol/L HCl or 0.1 mol/L KOH. The zeta potential was obtained from the streaming current measurements based on the Smoluchowski equation [11]:

$$\zeta = \frac{dI_{\text{str}}}{dp} \frac{\eta}{\varepsilon_0 \varepsilon_r} \frac{L}{A}, \tag{27.1}$$

where I_{str} is the streaming current, p is the pressure drop across the streaming channel, ε_0 is the vacuum permittivity (8.854 10^{-12} J/C^2/m), ε_r is the dielectric constant of the solution (78.3), η is the solution viscosity (0.8902 mPas), and L and A are the length and the cross-section area of the streaming channel, respectively.

Figure 27.5a shows the pH dependence of the zeta potential determined from the streaming current measurement. This pH dependence is commonly observed on basic surfaces and is due to the amino groups on the surface of the chitosan nanofibers. Figure 27.5b shows the pH dependence of the adsorbed amount of DNA on the chitosan nanofibers. Under alkaline conditions, the adsorbed amount of DNA significantly decreased with the pH increase. DNA is a polyanion due to the phosphate groups between each deoxyribose [12]. Therefore, this trend well correlated with the pH dependence of the zeta

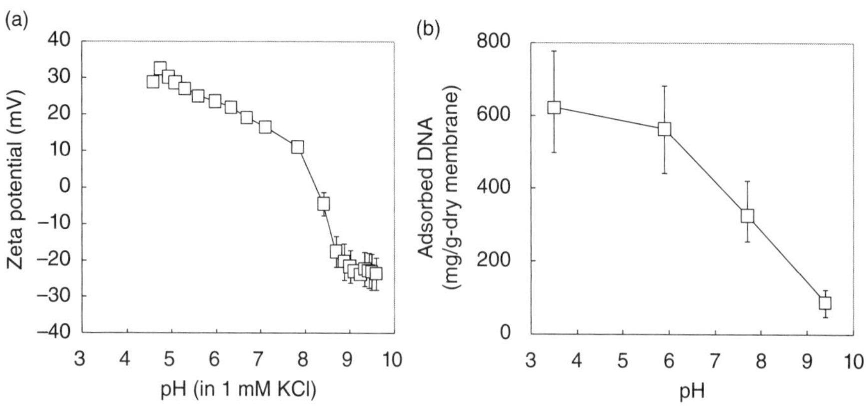

Figure 27.5. (a) pH dependence of zeta potential for chitosan nanofibrous membrane and (b) pH dependence of DNA adsorbed amount on chitosan nanofibrous membrane [10].

TABLE 27.2. Antimicrobial Effect of Chitosan Nanofibrous Membranes

	Bacterial Count[a] (Cells/Vial)	
	Before Incubation	After Incubation
Chitosan nanofiber membrane	5.4×10^5	0
Chitosan microfiber membrane	5.4×10^5	4.2×10^3
PET microfiber membrane	5.4×10^5	2.4×10^5

[a] Determined by Japanese Industrial Standard (JIS L1902) using *Staphylococcus aureus* (NBRC1237). PET, poly (ethylene terephthalate).

potential of the chitosan membrane; the attractive and repulsive electrostatic forces played a more important role under the lower and higher pH conditions than the isoelectric point of the chitosan membrane (pH 8.1). We also confirmed that the DNA adsorbed on the membrane at pH 4.0 could be desorbed at pH 9.2.

The antimicrobial effect of the chitosan nanofibrous membrane is shown in Table 27.2. Two kinds of commercial nonwoven membranes composed of microscaled fibers were tested for comparison. Unlike these controls, no bacteria could be detected on the chitosan nanofiber membrane after incubation. In addition, the chitosan fibrous membrane with a thinner diameter is more effective. This is attributed to the large surface area of the nanofiber membrane.

27.4 CATALYTIC EFFECT OF SYNTHETIC ION-EXCHANGE NANOFIBERS

Bipolar membranes (BPMs), which are composed of cation-exchange layers (CELs) and anion-exchange layers (AELs), joined together in series, show a water-splitting behavior under a reverse-bias condition. The BPM-based electrodialysis is an efficient process for generating acids and bases without by-products and is used for the recovery of acid and base from wastewater, the purification of amino acids, and the recycle system for liquid-crystal display (LCD) manufacturing [13]. The water-splitting phenomenon occurs at the interface between the CEL and the AEL. Therefore, the water-splitting capability of BPMs depends on their interfacial state. It is well accepted that materials immobilized in the intermediate region, such as weak acids (e.g., carboxylic and phosphoric groups), bases (e.g., tertiary and quaternary amino groups), and metallic compounds (e.g., Cr^{3+}, Fe^{3+}), catalyze the water-splitting reaction [14]. Tanioka et al. demonstrated that both the chemical composition of the intermediate region (i.e., a catalytic effect) and the electric field in the intermediate layer influenced the water-splitting capability of the BPMs [15].

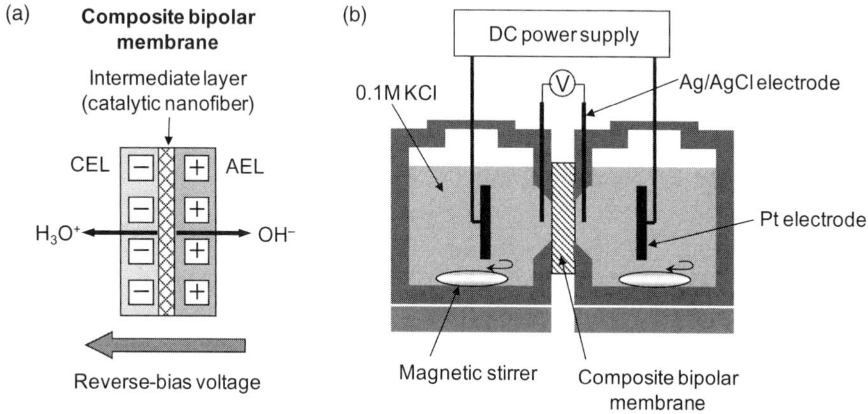

Figure 27.6. (a) Schematic diagram of a catalytic nanofiber layer composite bipolar membrane under the reverse-bias condition. (b) Schematic diagram of apparatus for the current–voltage measurements [13]. CEL, cation-exchange layer; AEL, anion-exchange layer.

TABLE 27.3. Physicochemical Properties of Synthetic Anion-Exchange Nanofibrous Layers

Ion-exchange capacity[a] (mmol/g-dry fiber)	0.8
Average pore size of membrane[b] (μm)	3.5
Through-pore specific surface area of membrane[b] (m^2/g)	14
BET specific surface area of membrane[c] (m^2/g)	600
Porosity of membrane[b] (%)	80
Thickness of membrane (μm)	40

[a] Determined by potentiometric titration.
[b] Estimated by bubble-point method.
[c] Estimated by N_2 adsorption experiments.

In addition, they suggested composite bipolar membranes (CBMs) consisting of a three-layered structure (Fig. 27.6a)—CEL, an intermediate layer with a catalytic activity, and AEL—for the investigation of the effect of the interface component on water splitting [16].

A commercial cation-exchange membrane (Aciplex® K501, Asahi Kasei, Tokyo, Japan) and a commercial anion-exchange membrane (Aciplex A501, Asahi Kasei) were used as the CEL and AEL (Fig. 27.6a), respectively. Anion-exchange nanofibers (AENFs) were used as the intermediate layer of the CBM. The nanofibers were prepared by electrospinning from a 15 wt % P4VP (M_w = 150,000–200,000; Polysciences, Warrington, PA) in a cosolvent mixture of 92/8 v/v ethanol/water (de-ionized water) and the successive simultaneous quaternization and cross-linking of the as-spun P4VP nanofibers in a dibromopropane/*n*-hexane solution (more detailed information has been reported in Reference 8). The physicochemical properties of the prepared AENFs are listed in Table 27.3.

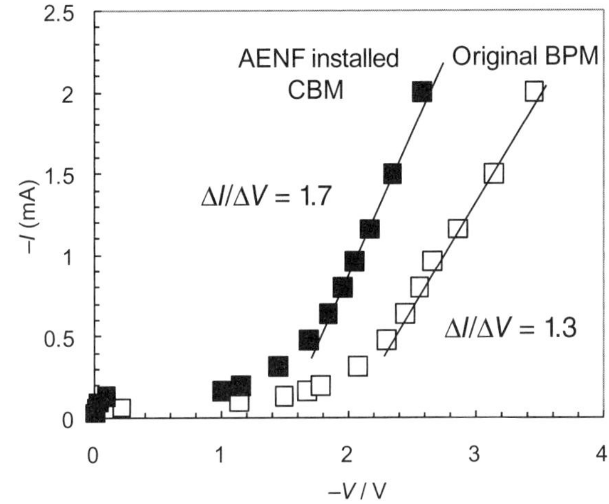

Figure 27.7. Current–voltage characteristics of the CBMs, which consist of CEL/AENF layer/AEL [13].

The current–voltage characteristics of the CBMs were investigated by controlled-current voltammetry using a two-compartment measuring cell (Fig. 27.6b). The CBM was placed between the two compartments, which were both filled with 0.1 mol/L KCl. Both compartments were well stirred using magnetic stirrers. The potential drop across the CBM was measured by Ag/AgCl wire electrodes placed on both sides of the membrane and connected to a voltmeter (Fluke 45 Multimeter, Fluke, Everett, WA) under conditions in which the current was supplied by Pt-black disk electrodes connected to a DC source/monitor (TR6143, Advantest, Tokyo, Japan). All measurements were carried out at 23 ± 0.5°C. The pH change in each compartment was measured by a pH meter (WM-60EG, TOA, Tokyo, Japan) equipped with pH electrodes (GST-5721C, TOA).

Figure 27.7a shows the current–voltage characteristics of the CBMs with the AENF installed (closed squares) and without any (open squares) in the intermediate region under reverse-bias conditions. The electric field-enhanced dissociation can be observed for an applied voltage higher than around 1 V, and the water dissociation is enhanced by the AENF layer, which has microscale flow-through pores (3.5 μm), high porosity (80%), and both tertiary pyridyl groups and quaternary pyridinium groups on the pore surface (total capacity 0.8 mmol/g). The $\Delta I/\Delta V$ value increased from 1.3 to 1.7 (Fig. 27.7b). The pH change in the compartments adjacent to the AENF layer-installed sample was greater than that of the uninstalled one. This is in good agreement with the current–voltage characteristics and suggests that the installed AENF layer enhanced the water dissociation in the intermediate region of the CBM. (Here, the water dissociation occurs at the interface between the AENF layer and the CEL.) This result can be explained by the synergetic effect of the protonation–deprotonation reaction due to the tertiary pyridyl group and the high specific surface area of the AENFs. For the former, Simons suggested that the hydronium and hydroxyl ions are produced from the protonation–deprotonation between some functional groups and water molecules and proposed the following mechanism for the water dissociation reaction [17]:

$$B + H_2O \leftrightarrow BH^+ + OH^- \tag{27.2}$$

and

$$BH^+ + H_2O \leftrightarrow B + H_3O^+, \tag{27.3}$$

where BH^+ represents the catalytic active center for the proton transfer reactions. For the latter, the high specific surface area as a unique property of the AENF layer (i.e., the Brunauer–Emmett–Teller (BET) surface area of 600 m^2/g) would contribute to the increase in the contact surface area between the AENF layer and the CEL. These results indicate that ion-exchange nanofibers with catalytic activity and a high surface area can improve the performance of the BPM-based electrodialysis processes.

27.5 SUMMARY AND FUTURE DIRECTIONS

Electrospun nanofibers have been very successfully applied in the field of air filtration because of their unique nanosize effect (i.e., aerodynamic slip at the nanofiber surface) [5]. The combination of high surface areas of the nanofibers and ionic functional groups will provide improvement in the function of the nanofibers (e.g., extremely rapid kinetics of adsorption or ion exchange, high adsorption or ion-exchange capacity, and high catalytic activity) and will lead to a significant expansion of nanofiber applications. The properties of nanofibers and performances of nanofibrous membranes can be improved by controlling the fiber diameter, surface morphology, and internal structure of the nanofibers. For example, Elabad et al. reported that the proton conductivity of perfluorocarbon polyelectrolyte (Nafion®) nanofibers sharply increased by decreasing the fiber diameter down to a nanometer scale (the maximum conductivity of 1.5 S/cm for the nanofiber with a diameter of 400 nm is higher than that of 0.1 S/cm for the bulk film) [18]. More recently, we have prepared microscaled Nafion hollow fibers by a two-fluid electrospinning process (Fig. 27.8) [19]. The polyelectrolyte (ion-exchange) hollow fiber can be applied to microtubular fuel cells, microfluidic devices [20], and high-capacity membrane modules for water treatment. These electrospun fibers and tubes with ionic functional groups is a promising material for the realization of a safe, sustainable, and healthy society. In addition, electrochemical characterization is a powerful tool for both fundamental research in order to better understand interfacial phenomena on nanofiber surfaces and the development of high-

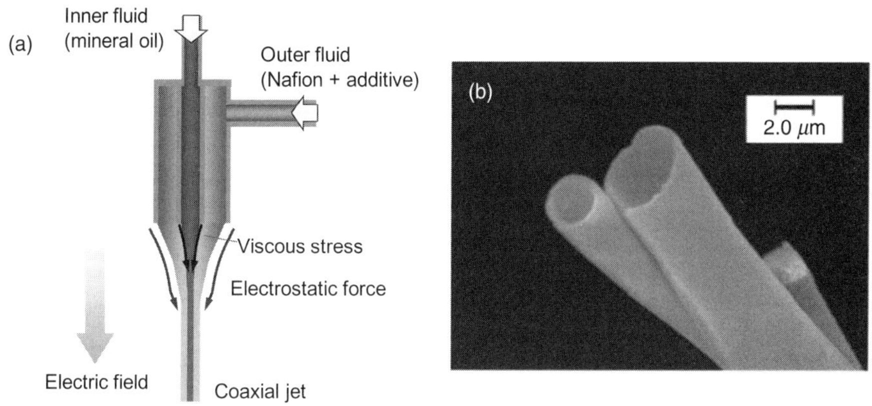

Figure 27.8. (a) Coaxial nozzle for electrospinning and (b) Nafion hollow microfiber prepared by coaxial electrospinning.

functional nanofiber materials. It is expected that the combination of surface electrochemistry and nanofibrous materials including synthetic and biological nanofibers produced by electrospinning will open a new direction in the colloid and interface sciences.

REFERENCES

1 J. Doshi, D. H. Reneker. Electrospinning process and applications of electrospun fibers. *Journal of Electrostatics* 35 (1995) 151–160.

2 D. H. Reneker, H. Fong (eds.), *Polymeric nanofibers*, vol. 918. American Chemical Society, Washington, DC, 2006. ACS Symposium Series.

3 Y. Dzenis. Spinning continuous fibers for nanotechnology. *Science* 304 (2004) 1917–1919.

4 D. Li, Y. Xia. Electrospinning of nanofibers: reinventing the wheel? *Advanced Materials* 16 (2004) 1151–1170.

5 A. Greiner, J. H. Wendorff. Electrospinning: a fascinating method for the preparation of ultrathin fibers. *Angewandte Chemie International Edition* 46 (2007) 5670–5703.

6 M. Streat. Boom time for ion exchange. *Chemistry & Industry (London)* 13 (2004) 20–21.

7 H. Seo, H. Matsumoto, S. Hara, H. Yako, M. Minagawa, A. Tanioka, Y. Yamagata, K. Inoue. Preparation of polysaccharide nanofiber fabrics by electrospray deposition: additive effects of poly(ethylene oxide). *Polymer Journal* 37 (2005) 391–398.

8 H. Matsumoto, Y. Wakamatsu, M. Minagawa, A. Tanioka. Preparation of ion-exchange fiber fabrics by electrospray deposition. *Journal of Colloid and Interface Science* 293 (2006) 143–150.

9 C. J. Brine, P. A. Sandford, J. P. Zikakis (eds.), *Advances in chitin and chitosan*. Elsevier Science, Barking, 1992.

10 H. Matsumoto, H. Yako, M. Minagawa, A. Tanioka. Characterization of chitosan nanofiber fabric by electrospray deposition: electrokinetic and adsorption behavior. *Journal of Colloid and Interface Science* 310 (2007) 678–681.

11 J. Lyklema. Fundamentals of interface and colloid science, in J. Lyklema, *Solid-liquid interface*, vol. II. Academic Press, London, 1995.

12 L. Stryer. *Biochemistry*, 4th ed. W.H. Freeman, New York, 1995.

13 Y. Wakamatsu, H. Matsumoto, M. Minagawa, A. Tanioka. Effect of ion-exchange nanofiber fabric on water splitting in bipolar membrane. *Journal of Colloid and Interface Science* 300 (2006) 442–445.

14 A. J. B. Kemperman (ed.), *Handbook on bipolar membrane technology*. Twente University Press, Enschede, 2000.

15 A. Tanioka, K. Shimizu, T. Hosono, R. Eto, T. Osaki. Effect of interfacial state in bipolar membrane on rectification and water splitting. *Colloids and Surfaces. A* 159 (1999) 395–404.

16 T.-J. Chou, A. Tanioka. Effect of the interface component on current–voltage curves of a composite bipolar membrane for water and methanol solutions. *Journal of Colloid and Interface Sciences* 212 (1999) 576–584.

17 R. Simons. Electric field effects on proton transfer between ionizable groups and water in ion exchange membranes. *Electrochimica Acta* 29 (1984) 151–158.

18 B. Dong, L. Gwee, D. Salas-de la Cruz, K. I. Winey, Y. A. Elabd. Super proton conductive high-purity Nafion nanofibers. *Nano Letters* 10 (2010) 3785–3790.

19 H. Matsumoto, T. Nagata, M. Minagawa, A. Tanioka. Preparation of polyelectrolyte nanofiber by electrospray deposition. *Polymer Preprints, Japan* 55 (2006) 3438.

20 A. L. Yarin, E. Zussman, J. H. Wendorff, A. Greiner. Material encapsulation and transport in core–shell micro/nanofibers, polymer and carbon nanotubes and micro/nanochannels. *Journal of Materials Chemistry* 17 (2007) 2585–2599.

28

SHAVE-OFF PROFILING AS A NANOSCALE 3-D ELEMENT IMAGING TECHNIQUE

Masashi Nojima

28.1 INTRODUCTION

With state-of-the-art of nanotechnology, we can construct and control structures with nanodimensions. The possibilities of nanotechnology are so promising that it has been predicted that the number of transistors on a large-scale integration chip will reach the number of people in the world by 2010 [1].

One possibility for a nanoscale 3-D element imaging technique is to reconstruct all structures of nanodevices completely as a blueprint for the manufacturing process. The other possibility is to search out failures in nanodevices and to find the nature of the failures. Generally, failures in semiconductor packages occur as a result of local errors: migration, dielectric breakdown, or charge/discharge. It is desirable to predict possible failures before the whole device reveals electrical or mechanical errors.

In this study, we discuss the possibility of shave-off profiling as a nanoscale 3-D element imaging technique using shave-off depth profiling [2], multilane shave-off profiling [3], nanoscale element imaging [4], and cross-check analysis using transmission electron microscopy (TEM) [5].

28.2 SHAVE-OFF PROFILING

Shave-off profiling is based on a micromachining process using a focused ion beam (FIB) and secondary ion mass spectrometry (SIMS). Samples are

Electrical Phenomena at Interfaces and Biointerfaces: Fundamentals and Applications in Nano-, Bio-, and Environmental Sciences, First Edition. Edited by Hiroyuki Ohshima.

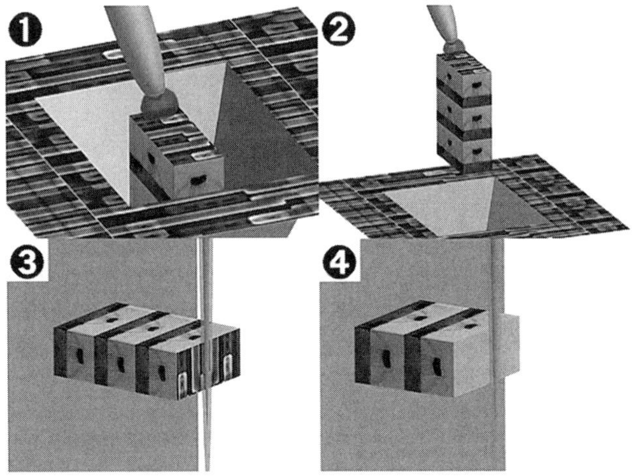

Figure 28.1. Schematic image sequence showing the process of shave-off depth profiling (each number represents a successive stage in the process of shave-off depth profiling).

micromachined and picked up by an FIB microsampling technique. The introduction of a microsampling technique using an FIB extends the applicability and flexibility of shave-off profiling [6]. Nowadays, microsampling is an indispensable technique for a TEM sample preparation. It enables shave-off depth profiling to be a pinpoint depth profiling.

Figure 28.1 represents procedures of shave-off depth profiling. In the first step, a piece (measuring a few cubic microns) is picked up from the surface of a sample using FIB microsampling (1, 2). The picked up piece is planted on a substrate. Then, the substrate with the piece is tilted to be parallel to the stream of the FIB (3). As the FIB reaches the top of the planted piece, the piece is shaved and reduced from the top to the bottom (4). While the piece is shaved, the generated secondary ions are mass separated and monitored as a function of space. Shave-off profiling accepts almost all solid materials even for roughed structure or hetero interface.

The detailed features of shave-off profiling are discussed in Reference 4. The nanobeam SIMS apparatus, our original system, is strongly modified for shave-off profiling [7].

28.2.1 Shave-Off Depth Profiling

Shave-off profiling was first invented for the alternation of conventional depth profiling. Conventional depth profiling with an ion gun sample bombardment came into widespread use by the beginning of the 1970s together with the development of powerful surface analysis tools [8]. The depth is described by a sputtering time or needs to be calibrated, assuming proportionality between depth and time, for example.

Shave-off depth profiling is a process where an FIB is used for micromachining to remove samples in a series of planes normal to the depth axis. This is completely different from depth profiling with a conventional ion bombardment process even if both methods have their origin in the same phenomenon: a sputtering event. Shave-off depth profiling can define the "depth" as a function of magnification projected on a display without a time function. In the shave-off process, the direction of the primary ion beam is parallel to the surface of the sample, which is consequently shaved off while obtaining the depth profile [9].

Here, an example of reconstruction of a nanodevice in a dynamic random access memory (DRAM) chip using shave-off depth profiling is demonstrated [10]. The surface of the chip consists of thousands of via holes, and this perforated structure makes it difficult to maintain a flat surface during a conventional bombardment process. Shave-off depth profiling can eliminate topographical effects and surface roughing. A piece of the chip that was previously micromachined by an FIB was picked up and planted on another substrate in a scanning ion microprobe chamber (SMI3050 SIINT). The piece on the substrate was transferred into the nanobeam SIMS apparatus. The top of the piece revealed sections through the via holes and faced the ion detector. On the bottom of the via holes, via plugs stood in rows. The via plugs consist of aluminum and are covered with a titanium nitride barrier. Tungsten wiring connects the via plugs to the base of the chip and is also covered with the barrier metal. Knowledge of the local coordinates of the titanium is important in estimating the barrier effect and in detecting the diffusion of metals from plugs or electrodes into the dielectric region. The conditions of the shave-off depth profiling are as follows: a 24-keV acceleration energy, a 35-pA beam current of the FIB, a 6-μm^2 sample surface area, and about 60 minutes for total profiling time.

Figure 28.2 shows the result of shave-off depth profiling for a nanodevice. The scanning electron microscope (SEM) image suggests depth information for this profile. When the FIB has reached the top of the sample, the edge of the FIB begins to shave the silicon oxide insulation region, including the via holes. After that, the shaving reveals the face of the barrier metal. The two split peaks of Ti^+ come from the package of via plugs covered with barrier metal. On the top of the Al^+ profile, signals rise gently. This reflects the tapered shape of the via plugs. From the first sharp peak of Ti^+, the thickness of the barrier metal was estimated to be 50 nm in full width at half maximum (FWHM). The value matches with the blueprint of the manufacturing process.

Regarding the shave-off depth profiling, the depth only has a meaning as a virtual structure of the sample. The shave-off direction can be vectored by rotating the scan of the FIB. The shave-off (depth) profiling has free dimensions for profiling.

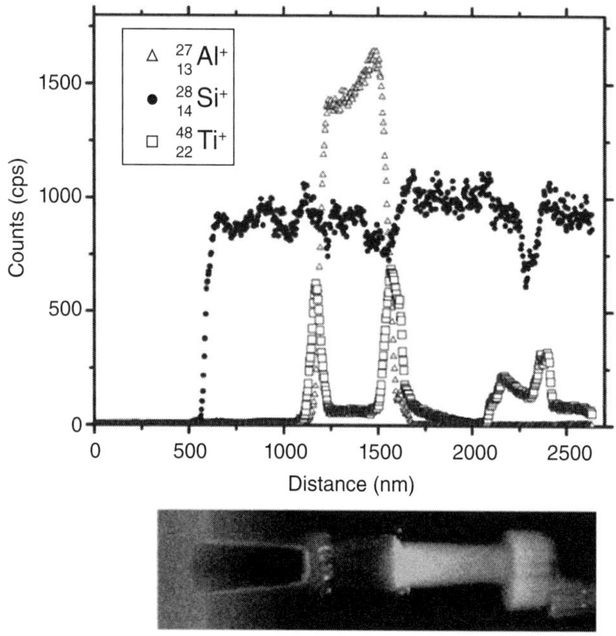

Figure 28.2. Shave-off depth profile of a nanodevice in a DRAM chip (the SEM image below the profile suggests depth information of this profile).

28.2.2 Multilane Shave-Off Profiling

In shave-off depth profiling, the depth is defined as the horizontal position of the FIB. The vertical position of the FIB, however, also has information about elemental distribution. Simple shave-off depth profiling does not take advantage of the vertical information. In order not to waste the vertical information, we have developed the concept of multilane shave-off profiling.

In order to realize multilane shave-off profiling, a new system was developed. The system made it possible to acquire the secondary ion mass signal as a function of the two-dimensional position of the FIB. A shave-off profile can be divided into up to 256 lanes along the vertical direction.

We obtained some multilane shave-off profiles to evaluate the new system. The sample was an Al plate tilted about 15° toward the vertical direction. Figure 28.3 shows the multilane profiles of the sample. The profiles rise according to the approach of the FIB: (4), (3), (2), and (1). This reflects precisely the tilt of the sample. Moreover, each rising edge is parallel and has the same interval. This means that the multilane shave-off profiling method and system can divide a shave-off profile into some profiles according to each lane. Multilane shave-off profiling has made 2-D analysis for shave-off profiling possible.

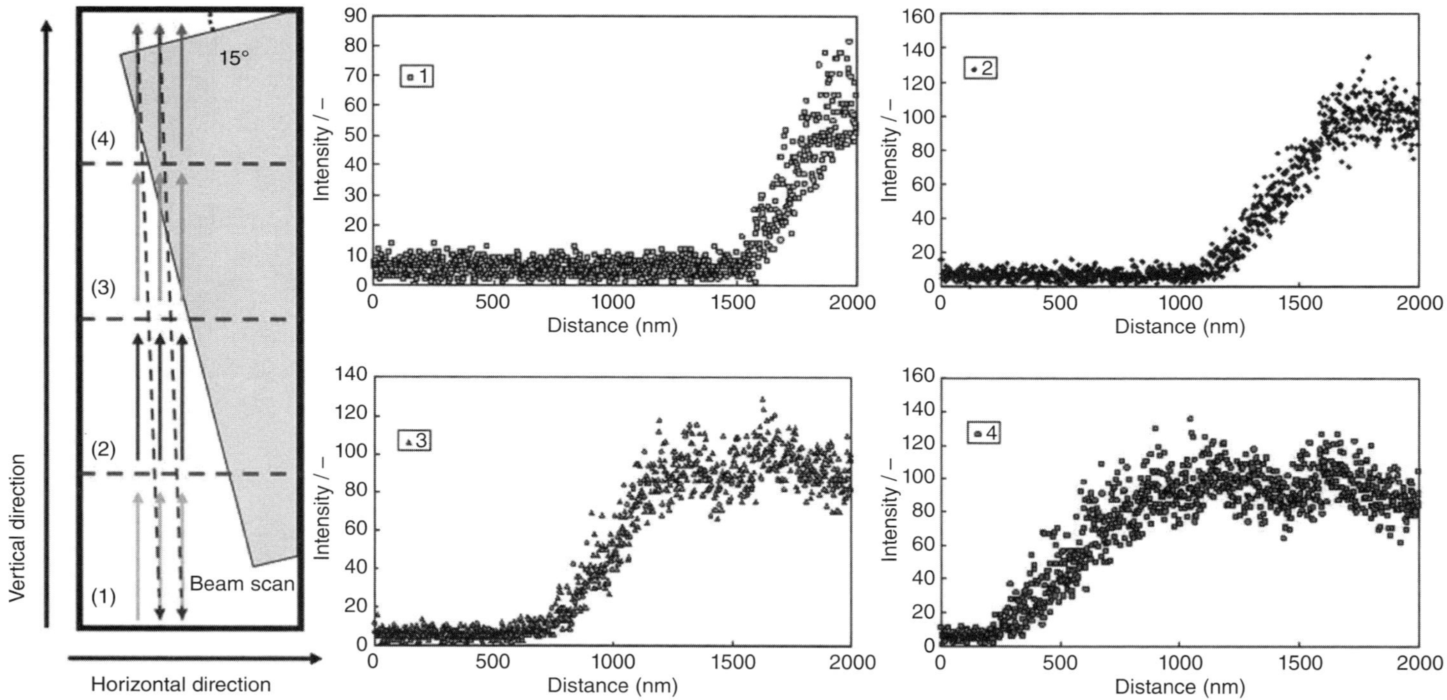

Figure 28.3. The multilane profiles of an Al plate tilted about 15° to the vertical direction.

28.2.3 Nanoscale Element Imaging

By using nanobeam SIMS, we can approach nanodimensions by shave-off profiling and also elemental imaging. We obtained a nanoscale element image of a cross section of a DRAM chip, which is the same sample as the shave-off depth profiling (Chapter 3.1). Previously, we had initially observed a DRAM chip using a scanning transmission electron microscope with an energy dispersive X-ray (STEM-EDX). The STEM-EDX analysis is a powerful method, especially owing to a high spatial resolution of 2 nm. However, the analysis cannot be performed without a long acquisition time owing to its poor count statistics [11]. Nevertheless, ion imaging has the advantages of achieving a high signal-to-background ratio and high throughput.

Figure 28.4 represents elemental images of the cross section of contact holes using a nanobeam SIMS apparatus: (a) sketch, (b) total ions, (c) Si^+, (d) Ti^+, (e) Al^+, and (f) W^+. The primary probe current was 20 pA and the primary beam energy was 25 kV. The acquisition time for imaging was 140 seconds The images were obtained by exposing the film to a cathode ray tube (CRT).

A strong contrast can be observed between the middle region and the lower region of the total ion image. The contrast is caused by chemical conformations: Si oxide and pure Si. A dark band in the middle area could also be caused by the chemical contrast or local electronic charge caused by conformations of the productive processes of the sample. The chemical sensitivity of ion images can provide information about such productive processes.

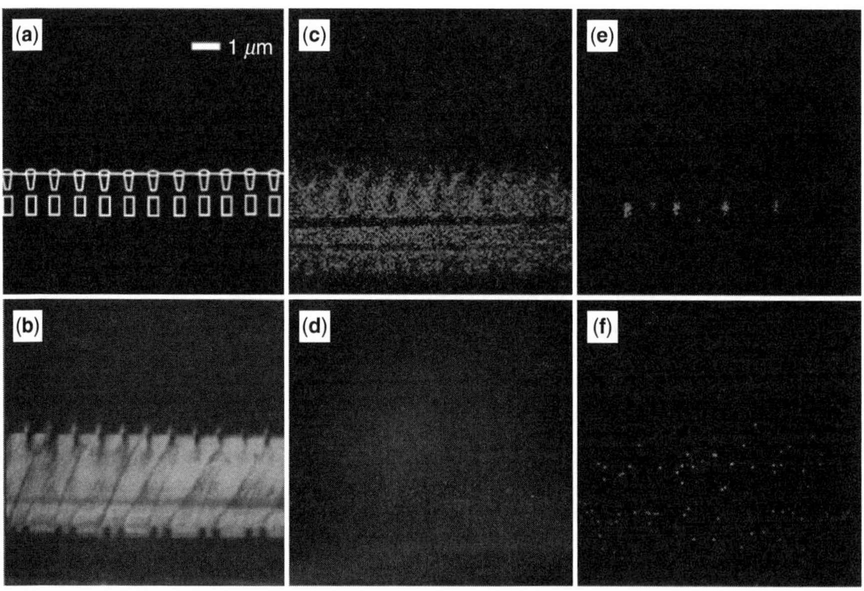

Figure 28.4. Elemental images of the cross section of contact holes using nanobeam SIMS: (a) sketch, (b) total ions, (c) Si^+, (d) Ti^+, (e) Al^+, and (f) W^+.

Clear nanodimensional images of Si^+ and Al^+ were obtained. On the Al^+ image, bright spots were clearly revealed. The appearance of the bright spots may be related to the design of the circuit. It is possible that bright spots were connected to a conductive region, and the other areas were shut off from the other part. The bright spots tended to decrease in size and to disappear from the left side to the right side. This tendency was due to the surface sensitivities of ion images and the inclination of the fracture surface. On the left side, the surfaces of the Al regions were exposed. On the right side, Al regions were buried in the Si oxide layer. The elemental imaging can provide not only elemental distributions but also properties of samples. The faint signals on the Ti^+ and W^+ images gave little information about the sample. On the Ti^+ image, faint signals made it difficult to discriminate signals and background. For evaluating such thin layers, shave-off profiling is powerful.

The elemental imaging can be applied for 3-D element imaging by repeating cross sectioning with an FIB and 2-D imaging. However, the 3-D imaging method has a limitation on volume within nanodimension. And also, the imaging method itself involves many aspects: surface conditions, chemical and electrical properties, geometry between the surface and ion detector, and so on. These aspects may lead to misinterpretation of the 3-D distributions of elements.

28.2.4 Cross-Check Analysis with TEM

We cross-checked the same sample described earlier by TEM imaging and shave-off profiling but at different points. The application of different analytical methods makes clear one specimen from multiple angles. Agreement on cross-check analysis verifies the reliability of each experimental result. Disagreement on cross-check analysis reveals possibilities that can be missed in the other experimental results.

The point of interest is very restricted for failure analysis. It is indispensable that the cross-check analysis can be applied to the same sample and also the same point. We applied two different analytical methods: TEM, which can project nanostructure without destruction, and shave-off profiling, which is destructive and can reconstruct elemental distributions. We cross-checked real TEM samples using shave-off profiling.

A test integrated circuit (IC) package consisted of Cu electrodes, Ti films, silica fillers, and resin. Pairs of electrodes were biased under conditions of high temperature and high moisture (10 V, 400 K, 85% RH). Some packages began to exhibit strange conductivity within 400 hours. Ion migration might occur in their failure packages. A piece including electrodes, fillers, and resin was picked up in the FIB microsampling chamber (FB-2000, Hitachi High-Technologies, Hitachi-maka, Japan). The thickness of the piece was 1 μm, which is a little thick for TEM and enough for shave-off profiling. The piece was manipulated into a scanning transmission electron microscope (STEM) chamber (HD-2000, Hitachi High-Technologies Co. Ltd). Then, the piece with the substrate was

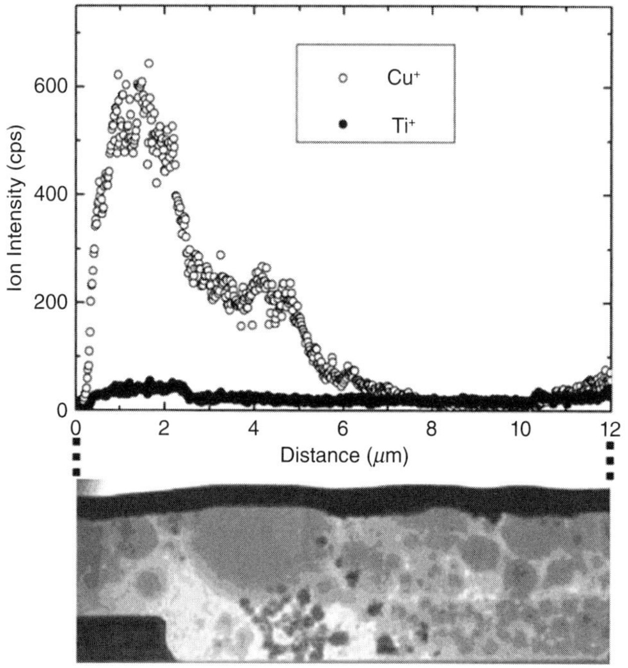

Figure 28.5. Cross-check analysis of IC package, which occurred during electrochemical migration (above: shave-off profile, below: STEM image).

transferred to the nanobeam SIMS chamber. The shave-off profiling conditions were as follows: The acceleration voltage and current of FIB were 24 kV and 35 pA, respectively. Total time for a profiling was 60 minutes for a 12-μm area.

Figure 28.5 shows a shave-off profile and a STEM image at the same point. In both results, the existence of both electrodes and dark dots could be confirmed. A critical point of disagreement on each result is the gradient curve on the shave-off depth profiling from anode to cathode. The gradient curve might have its origin in ion migration. Generally, ionization probability of oxide or complex is much higher than metal on the same elements. It is possible that the high ionization probability of Cu ions amplified the existence of Cu ion migration. In the same piece, shave-off profiling revealed a faint gradient of migrated ions that could not be observed by STEM imaging.

In this case, a piece of 1-μm thickness was introduced for cross-check analysis and made clear the agreements and disagreements of each method. The next step would be how to get along with thinner pieces for shave-off profiling. Our best record thus far is a thickness of 500 nm. The obstacle to achieving thinner pieces for shave-off profiling is not deterioration of signals, but damage from the FIB during tuning of the apparatus. In order to avoid damage of a piece before shave-off profiling, the introduction of a SEM inside the apparatus is anticipated.

It is anticipated that another possibility for nanoscale 3-D element imaging will be the introduction of TEM tomography to the cross-check analysis with shave-off profiling.

28.3 CONCLUDING REMARKS

We discussed the possibility of shave-off profiling as a nanoscale 3-D element imaging technique. The multilane shave-off profiling has made 2-D analysis for shave-off profiling possible. The elemental imaging has many aspects that can be applied to 3-D element imaging. The complementarity of using TEM and shave-off profiling for cross-check analysis presents another possibility for nanoscale 3-D element imaging.

ACKNOWLEDGMENTS

This work was supported in part by a Grant-in-Aid for Creative Scientific Research No.18GS0204 from the Japan Society for the Promotion of Science. We wish to express our gratitude to Ms. Tomoko Arimitsu and Ms. Ikuko Nakatani of Seiko Instruments Nanotechnology for technical support and helpful comments. We also thank Mr. Satoru Gondo of ST-Techno Co., Ltd for technical support and helpful suggestions.

REFERENCES

1 K. Nikawa. *LSI failure analysis*. Kougyo Chousakai Publishing, Inc.
2 M. Nojima, A. Maekawa, T. Yamamoto, B. Tomiyasu, T. Sakamoto, M. Owari, Y. Nihei. *Appl. Surf. Sci.* 7293–7296 (2006) 252.
3 M. Fujii, Y. Ishizaki, M. Nojima, M. Owari, Y. Nihei. *Nucl. Instrum. Methods Phys. Res. B* 267 (2009) 660.
4 M. Nojima, M. Toi, A. Maekawa, B. Tomiyasu, T. Sakamoto, M. Owari, Y. Nihei. *e-J. Surface Sci. Nanotechnol.* 2 (2004) 131.
5 M. Nojima, M. Fujii, Y. Ishizaki, M. Owari, Y. Nihei. *Appl. Surf. Sci.* 255 (2008) 1400.
6 M. Toi, A. Maekawa, T. Yamamoto, B. Tomiyasu, T. Sakamoto, M. Owari, M. Nojima, Y. Nihei. *J. Surf. Anal.* 12(2) (2005) 170.
7 M. Nojima, M. Toi, A. Maekawa, B. Tomiyasu, T. Sakamoto, M. Owari, Y. Nihei. *Appl. Surf. Sci.* 231–232 (2004) 930.
8 S. Hofmann. *Phil. Trans. R. Soc. Lond. A* 362 (2004) 55.
9 H. Satoh, M. Owari, Y. Nihei. *J. Vac. Sci. Technol. B* 6 (1988) 915.
10 M. Nojima, M. Toi, A. Maekawa, T. Yamamoto, T. Sakamoto, M. Owari, Y. Nihei. *Microchim. Acta* 155 (2006) 219.
11 M. Koguchi, H. Kakibayashi, R. Tsuneta, M. Yamaoka, T. Niino, N. Tanaka, K. Kase, M. Iwaki. *J. Electron Microsc.* 50 (2001) 235.

29

INTERFACIAL CHARGE STORAGE OF MANGANESE OXIDE ELECTRODES FOR ELECTROCHEMICAL CAPACITORS

SHINICHI KOMABA, NAOAKI YABUUCHI, AND TOMOYA TSUCHIKAWA

29.1 MANGANESE OXIDES FOR ELECTROCHEMICAL DEVICES

Electrochemical power sources, secondary batteries, and supercapacitors are attractive for the various application in mobile phone, digital camera, laptop computer, electric bicycle, hybrid electric vehicle (HEV), plug-in hybrid electric vehicle (PHEV), pure electric vehicle (EV), smart grid, and so on. In the past century, considerable effort has been devoted to study manganese oxides as electrode materials for power sources. Crystallization of the manganese oxides is highly influenced by many factors, for example, intergrowth of different structural units, cation incorporation to the vacant sites, and structural water, leading to the complexity of the crystal structures. Typically, the manganese oxides constituted from mainly tetravalent manganese ions are simply denoted as "manganese dioxides (MnO_2)." Manganese dioxides are used as a positive electrode with a zinc negative electrode, that is, carbon–zinc cells (dry cell) and alkaline cells. Electrolytic manganese dioxide (γ-type MnO_2) is typically used in the alkaline cell, which can be described as an intergrowth structure between pyrolusite (β-type MnO_2) and ramsdellite. In the alkaline cell, manganese oxides are reduced with proton insertion into the bulk of particles for the charge compensation. This process can be described as follows:

$$Mn^{4+}O_2 + xH^+ + xe^- \leftrightarrow H_xMn^{4+}_{(1-x)}Mn^{3+}_xO_2. \tag{29.1}$$

Electrical Phenomena at Interfaces and Biointerfaces: Fundamentals and Applications in Nano-, Bio-, and Environmental Sciences, First Edition. Edited by Hiroyuki Ohshima.

In theory, one mole of proton can be inserted into MnO_2, which is accompanied by manganese reduction from a tetravalent to a trivalent state. MnO_2 can also accommodate lithium ions, which are much larger than proton, to the vacant sites:

$$Mn^{4+}O_2 + xLi^+ + xe^- \leftrightarrow Li_xMn^{4+}_{(1-x)}Mn^{3+}_xO_2. \quad (29.2)$$

Heat-treated MnO_2 is used as the positive electrode materials for a primary lithium cell. Spinel-type $LiMn_2O_4$ is, furthermore, applied as a positive electrode material in the rechargeable and large-scale lithium ion cells.

Since 1995, amorphous hydrated ruthenium dioxide, $RuO_2{\cdot}nH_2O$, has been extensively studied as the electrode materials for redox capacitors [1–3]. $RuO_2{\cdot}nH_2O$ composite electrode (~100 μm thickness) exhibits an extremely high specific capacitance (typically > 700 F/g) in acidic media (e.g., H_2SO_4 aqueous solution). Although the ruthenium dioxide is one of the attractive electrode materials, the manganese dioxides are also worth studying as the electrode materials for the redox capacitor, that is, material abundance in nature and its environmental benign. The use of manganese dioxides for supercapacitors has been reported by Lee and Goodenough in 1999 [4]. Amorphous hydrated manganese dioxides, $MnO_2{\cdot}nH_2O$, were synthesized by a simple precipitation reaction between $KMnO_4$ and $Mn(CH_3COO)_2$. $MnO_2{\cdot}nH_2O$ has a large surface area (303 m^2/g by Brunauer–Emmett–Teller [BET] measurement), and a capacitor-like electrochemical response in KCl aqueous electrolyte was reported. A specific capacitance of the amorphous hydrated manganese dioxides reached 200 F/g, which corresponds to 66 $\mu F/cm^2$ as normalized capacitance based on the surface area of manganese dioxides. This capacitance has an approximately three times larger value than that of the expected value based on a classical electrical double-layer charging (typically ~20 $\mu F/cm^2$ for a flat electrode). It is noted that the surface area of the activated carbon is ranged from 1000 to 3000 m^2/g [5], from which the area normalized capacitance is calculated to be 10–15 $\mu F/cm^2$. The observed area normalized capacitance for that of $MnO_2{\cdot}nH_2O$ is anomalously large compared to the conventional electrical double-layer charging. From these observations, it is speculated that faradic reaction (or charge transfer reaction) partly contributes to the capacitor-like response of manganese dioxides, that is, "pseudocapacitance." When we assume that one electron transfer per manganese ions with capacitor-like behavior is achieved as shown in Equations 29.1 and 29.2, the specific capacitance would reach more than 1000 F/g. From these results, for the $MnO_2{\cdot}nH_2O$, the faradic reaction occurs only at the surface (Eq. 29.3) or is limited to the near surface with the slow kinetics on the transfer of large cations, such as sodium and potassium ions:

$$(Mn^{4+}O_2)_{surface} + xC^+ + xe^- \leftrightarrow C_x^+(Mn^{4+}_{1-x}Mn^{3+}_xO_2)_{surface}, \quad (29.3)$$

where C^+ denotes alkaline cations, Li^+, Na^+, K^+, and so on, which depend on the electrolyte solution utilized.

Analysis of the manganese dioxide thin films by X-ray photoelectron spectroscopy (XPS) sheds light on the reaction mechanism of the manganese dioxides, especially near the surface [6]. In this case, a manganese 3s and 2p core level shift was observed for the oxidation and reduction states, corresponding to 0.9 and 0.0 V versus Ag/AgCl, respectively. Since the binding energy of the photoelectron for the Mn 3s core level is extremely low (<100 eV), the information of the electronic structure for the manganese ions is relatively limited from the very near-surface of the oxides. It was revealed that the Mn 3s core level shifted, depending on the applied potential. In contrast, less change was observed for the Mn 2p core level, suggesting that manganese ions are predominantly reduced near the surface of the particles. Furthermore, formation of the hydroxide was found from the O 1s core level when the manganese dioxides discharged to 0 V versus Ag/AgCl. Proton partly plays the role of the charge compensation. It is noted that the existence of Na ions was also confirmed by XPS, indicating insertion of Na ions near the surface of manganese dioxides. These observations also prove that the electrochemical performance of manganese oxides is affected not only by its crystal structure but also by the morphology and surface structures of the manganese dioxides.

The concept for the "redox capacitors" in comparison to the conventional electric double-layer capacitor (EDLC) and batteries is summarized in Figure 29.1. The redox capacitor effectively utilizes both the electrical double-layer charging and the reversible faradic reaction of the manganese (metal) oxides as described earlier. There are merits for the redox capacitors; that is, they perform with higher capacity than EDLC and with higher power than batteries.

As the electrode material for the supercapacitor, electrical conductivity is one of the important physical properties to achieve higher power density. Ruthenium oxide shows excellent electrical conductivity, for example, $\sim 2 \times 10^4$ S/cm for the polycrystalline RuO_2 [7], and ~2 S/cm as the hydrous form [8]. The electrical conductivity of the $RuO_2 \cdot nH_2O$ is comparable to the carbon-based electrode materials. In this regard, the major drawback of manganese dioxides as the electrode materials for the redox capacitor seems to be poor electrical conductivity (5×10^{-6} S/cm for cryptomelane [9]). The poor electrical conductivity may suggest that the solid-state diffusion rate for the electrons and/or ions is limited, resulting in reduced power and energy density. In fact, manganese dioxides prepared as the thin-film electrode (<100 nm) demonstrate clear improvement of the specific capacitance of the manganese electrode, which becomes comparable in value (~700 F/g) with that of ruthenium dioxides [10, 11]. This is also consistent with the XPS study of the manganese oxides. The improvement of the energy density has been achieved even as composite electrode by embedding manganese dioxides into the mesoporous carbon [12] and ultraporous carbon [13] frameworks, which allow the three-dimensional path for electron conduction with maximized interface area between electrolyte solution and manganese oxides in the composite structure.

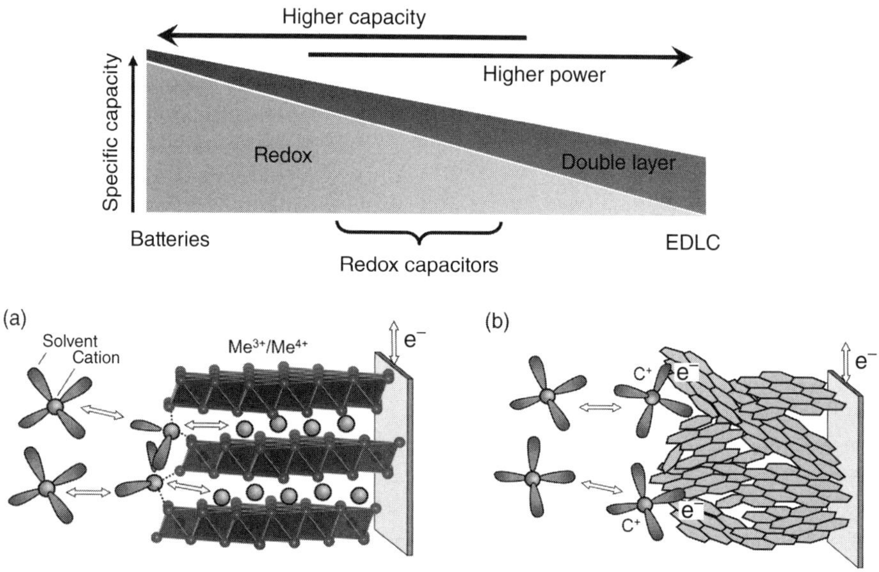

Figure 29.1. A conceptual schematic illustration of the redox capacitor, which effectively utilizes the charge storage mechanisms of both battery and electric double-layer capacitor (EDLC). For the battery (a), an example that transition metals (Me) in $LiMeO_2$ solid are oxidized/reduced accompanied by cation incorporation for the charge compensation. For the EDLC (b), cations/anions are adsorbed at the negatively/positively charged interface during charge/discharge.

The specific capacitance of 600 F/g can be achieved based on the embedded manganese dioxides in the carbon form [12].

29.2 SYNTHESIS OF MANGANESE DIOXIDES AND THEIR APPLICATION IN CAPACITORS

As discussed briefly in the previous section, the selection of electrode materials, their particle size, morphology, surface structures, and crystal structures influence the pseudocapacitance. For the synthesis of the manganese oxides for the supercapacitor, precipitation, sol–gel, and electrodeposition methods are often utilized. For the case of the precipitation method, $KMnO_4$ and divalent manganese salt, which work as oxidation and reduction agents, respectively, are used as starting materials. Amorphous phase or poorly crystallized cryptomelane (α-MnO_2) is typically obtained by the precipitation method. The α-MnO_2 has larger vacant sites (so-called 2×2 tunnel) in comparison to the β (1×1 tunnel) and γ (1×1 and 1×2 tunnels) forms. The cation insertion into the vacant sites, however, is limited near the surface of the manganese in KCl or Na_2SO_4 aqueous solutions. The morphology of the manganese oxides seems to be affected by a source of the divalent manganese salts used, relating to the difference in the pH of solution [4, 14]. Specific capacitance of the manganese dioxides prepared by the precipitation method reach 200 F/g for the composite electrode. Organic chemicals, such as ethanol [15] and ethylene glycol [16], can be used for the reducing agent of permanganate to synthesize nanostructured manganese dioxides, instead of the divalent manganese salts. Morphology of MnO_2 is modified by the selection of the reducing agent [15, 16]. The manganese oxides with high surface area and high porosity can be also prepared by a sol–gel method from $KMnO_4$ [17], and electrochemical deposition method from divalent manganese sources with anodic oxidation [10, 18]. Amorphous and hydrous manganese oxides with nanostructured morphology can be obtained by the electrochemical deposition method [10]. Note that birnessite-type manganese dioxides (δ-MnO_2: layered structure) synthesized at 600°C (BET surface area: 3 m^2/g) showed the specific capacitance of 80 F/g at 5 mV/s in 0.1 mol/dm^3 K_2SO_4 solution, which corresponds to the more than 2000 $\mu F/cm^2$ of area normalized capacitance based on the surface area of manganese dioxide [19], which is much higher than the typical value of a flat substrate, 20 $\mu F/cm^2$. This result indicates that K^+ and/or H^+ are inserted into the interstitial space between MnO_2 slabs. The layered structure with a two-dimensional diffusion path would be able to effectively accommodate large cations, such as potassium ion, in comparison to the materials having a one-dimensional (tunnel type) cation diffusion path.

Recently, our group also reported nanostructured birnessite as the electrode material for the supercapacitor, prepared by electrochemical conversion from spinel-type Mn_3O_4 [20, 21]. Crystalline Mn_3O_4 powder was synthesized by simple heat treatment of electrolytic manganese dioxide in air at 1050°C for 72 hours, and was ground with planetary ball milling for 12 hours. The Mn_3O_4 electrodes were prepared with poly(vinylidene fluoride) binder and carbon (ketjen black) on stainless mesh. The cyclic voltammograms of the Mn_3O_4 electrode containing the carbon additive and polymer binder, in the potential range of –0.1 and 0.9, are shown in Figure 29.2. The Mn_3O_4 electrodes exhibited a pseudocapacitive behavior in the Na_2SO_4 solution in this condition in spite of the general consensus that Mn_3O_4 is not electrochemically active. In addition, the specific capacitance gradually increased during the 300 cycling test, which is similar in phenomena to Mn_3O_4 thin film [22]. Small oxidation and reduction peaks appeared around 0.6 and 0.4 V, respectively. This behavior depended on the following experimental conditions: potential range, scan rate, electrolyte solution, and particle size.

Figure 29.3 shows the variation in the specific capacitance of Mn_3O_4 calculated from the voltammograms in Figure 29.2. The capacitance of the Mn_3O_4 electrodes clearly increased by the successive potential cycling. The Mn_3O_4 demonstrated the maximum value of approximately 190 F/(gram of oxide) after approximately 300 cycles in 1 mol/dm^3 Na_2SO_4 aqueous solution. The electrochemical conversion process depends on the source of the manganese

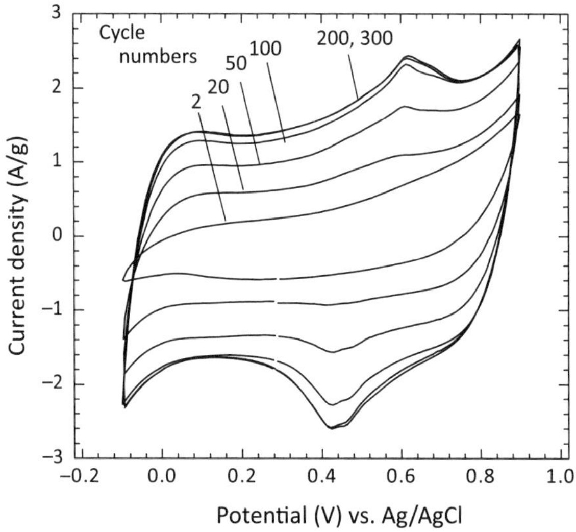

Figure 29.2. Cyclic voltammograms of a Mn_3O_4 electrode in 1 mol/dm^3 Na_2SO_4 aqueous solution at a rate of 10 mV/s.

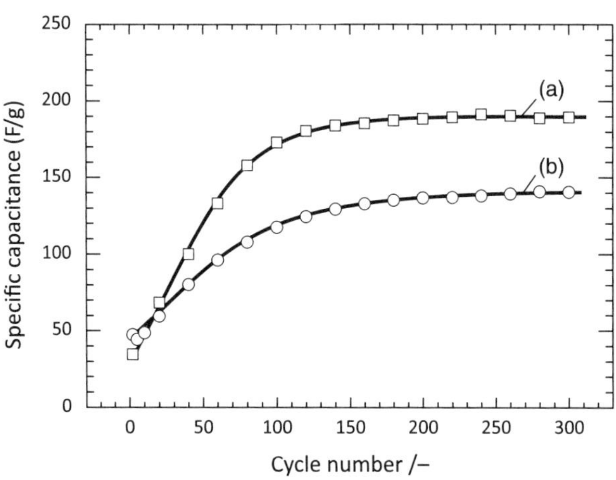

Figure 29.3. The variation of the specific capacitance of Mn_3O_4 electrodes in (a) 1 mol/dm^3 and in (b) 0.1 mol/dm^3 Na_2SO_4 aqueous solutions.

oxides. Mn_2O_3 shows a similar trend for the increase in specific capacitance, which is induced by the electrochemical cycling. To achieve the same capacitance with Mn_3O_4 as the electrode materials, however, much longer electrochemical cycles were needed for Mn_2O_3 (Fig. 29.4). As shown in Figure 29.4, the specific capacitance of spinel Mn_3O_4 reached the maximum value of 140 F/g after approximately 200 cycles in 0.1 mol/dm^3 Na_2SO_4 aqueous solution at

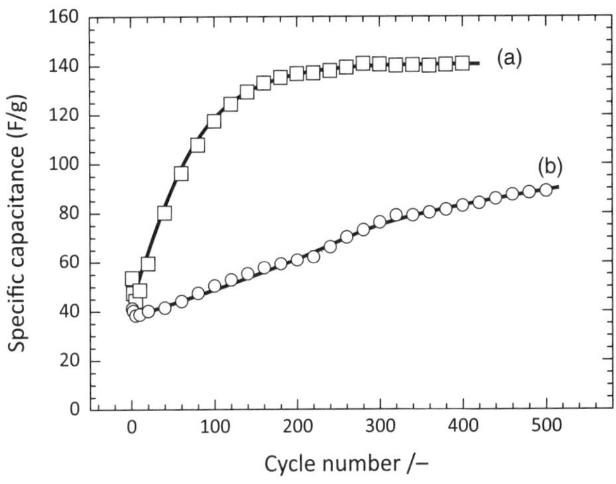

Figure 29.4. Specific capacitance versus cycle number plots of (a) Mn_3O_4 and (b) Mn_2O_3 electrodes treated by ball milling. The capacitances were evaluated from the voltammograms at 10 mV/s in 0.1 mol/dm^3 Na_2SO_4 aqueous solution.

10 mV. The specific capacitance of bixbyite Mn_2O_3 electrodes continuously increased during the 500 cycles under the same experimental condition. The Mn_2O_3 electrode shows 80 F/g after the 500 cycles, which is still a much lower capacitance than that of the Mn_3O_4 electrode.

Broughton et al. described that a Mn thin-film electrode was transformed into MnO, Mn_3O_4, Mn_2O_3, and then finally into a low crystallinity MnO_2 product by electrochemical oxidation, resulting in a high specific capacitance [11, 23], and a Mn_3O_4 layer was transformed into Mn_2O_3 and then MnO_2 when applying an anodic potential [24]. Considering these reports, the Mn_3O_4 powder should also be transformed into hydrated manganese (di)oxide by electro-oxidation during the potential cycling, similar to a thin film of Mn_3O_4 [11]. It is noted that the electrochemical stimulation in 1 mol/dm^3 Na_2SO_4 results in the capacitance of 190 F/g, which is larger than that in 0.1 mol/dm^3 Na_2SO_4 (140 F/g) as seen in Figure 29.3 with the identical electrochemical cycling conditions. The specific capacitance of Mn_3O_4 electrodes is affected not only by the electrolyte solution but also by the thickness of the electrode related to the sample loading condition (weight per unit area) as shown in Figure 29.5. The specific capacitance of ca. 140 F/g for 3- to 5-mg loading reduced to ca. 60 F/g for the 20-mg loading condition in the 0.1 M Na_2SO_4 aqueous solution.

Structural evolution during the electrochemical cycle was examined by the X-ray diffraction (XRD) method in Figure 29.6. It was confirmed that hydrated lamellar manganese dioxide, birnessite, was formed by the electrochemical stimulation of Mn_3O_4 in a Na_2SO_4 solution. As shown in Figure 29.6, diffraction peaks appeared after the electrochemical stimulation and all peaks are indexed

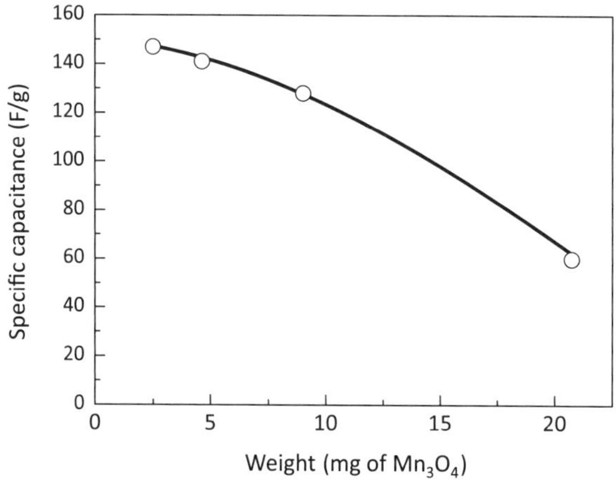

Figure 29.5. Relation between the weight of a Mn_3O_4 electrode and maximum capacitance during 500 cycles at a rate of 10 mV/s in 0.1 mol/dm^3 Na_2SO_4 aqueous solution.

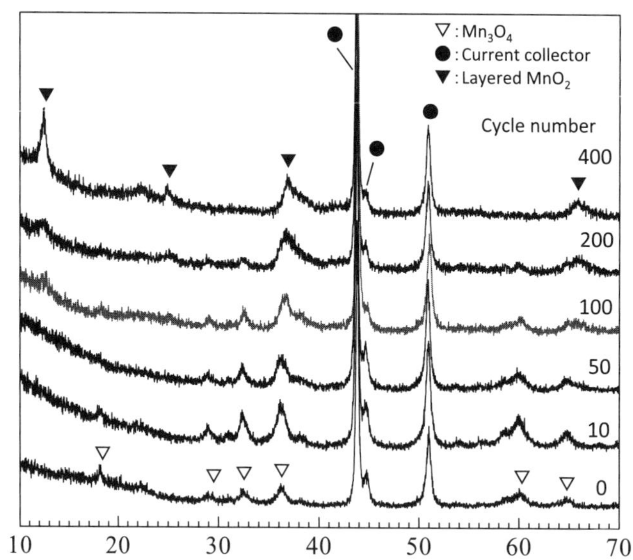

Figure 29.6. X-ray diffraction patterns of the Mn_3O_4 electrodes for different cycling stages obtained in 1 mol/dm^3 Na_2SO_4 aqueous solution.

as birnessite phase [19, 20, 25] except for those of the current collector and conductive agent. Since the broad diffraction peaks were observed, it was difficult to distinguish the stacking sequences of birnessite, such as 2H and 3R [25]. Figure 29.7 compares XRD patterns of birnessite electrodes prepared by different experimental methods: conventional chemical oxidation process of

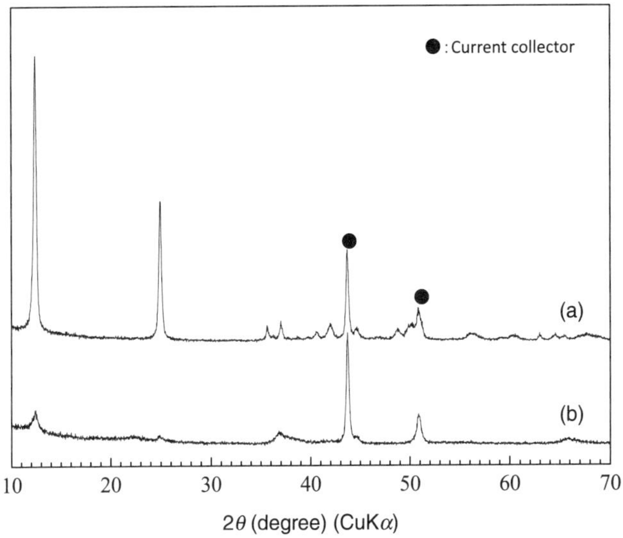

Figure 29.7. X-ray diffraction patterns of birnessite-type MnO_2 electrodes prepared by (a) chemical oxidation of manganese hydroxide and (b) electrochemical stimulation for the Mn_3O_4 electrode.

manganese hydroxide and the electrochemical stimulation for the Mn_3O_4 electrode. The intensity of the 00*l* diffraction lines around $2\theta = 12°$ and 25° is relatively lower than that of the birnessite prepared by the chemical oxidation method, indicating that the *c*-axis orientation perpendicular to the substrate should be less remarkable for the birnessite formed from Mn_3O_4. It is also expected that the crystallite size of the birnessite prepared by the electrochemical stimulation is much smaller than that of pristine Mn_3O_4 from the broader peak profiles.

The morphology of the Mn_3O_4 electrode after cycling was observed by electron microscopy as shown in Figure 29.8. Although the surface of the pristine electrode, which consisted of the Mn_3O_4 particles, was flat (Fig. 29.8a), the successive cycling resulted in a much rougher and porous electrode (Fig. 29.8b–d), which would be due to the electrochemical stimulation as mentioned earlier. This morphology change results in a larger surface area and also causes a remarkable increase in the specific capacitance. The inductively coupled plasma (ICP) analysis of the electrolyte after the electrochemical stimulation revealed that the formation of the birnessite was accompanied with the dissolution of 30% of total manganese in the Mn_3O_4 electrode.

29.3 MANGANESE OXIDE-BASED SUPERCAPACITORS

As the electrolyte solution, concentrated acidic or alkaline media, for example, H_2SO_4, KOH solutions, are often utilized for the carbon and ruthenium dioxide

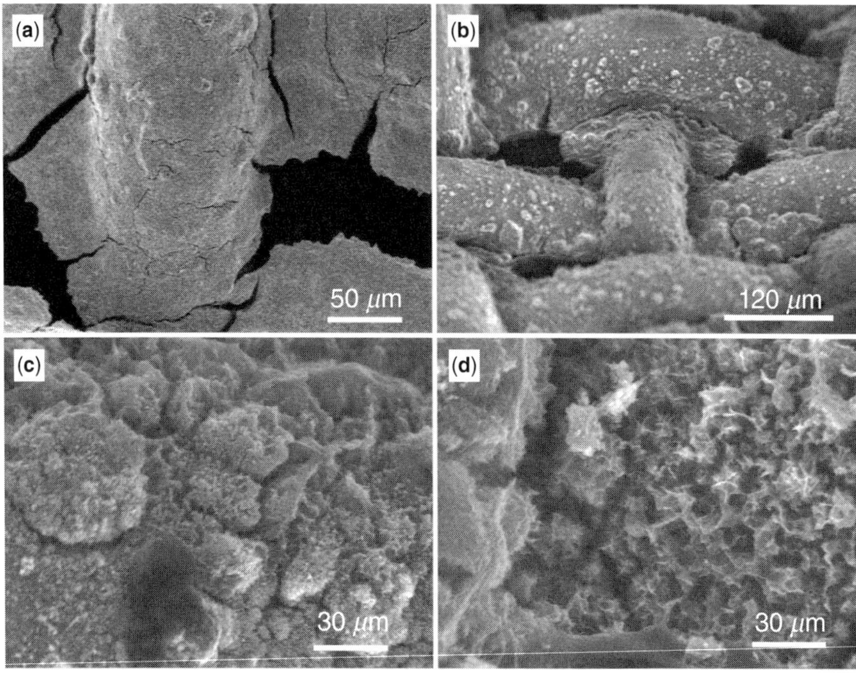

Figure 29.8. Scanning electron microscope (SEM) images of the Mn_3O_4 electrode surface (a) before cycling and after (b) 10, (c) 50, and (d) 200 cycles in 1 mol/dm^3 Na_2SO_4 aqueous solution.

supercapacitor because of the superior ionic conductivity. Manganese dioxides are, however, soluble in the acidic media unlike ruthenium dioxide. Alternatively, 1–2 mol/dm^3 Na_2SO_4 and KCl aqueous solutions are often used as the electrolyte for the manganese dioxide-based supercapacitor. It should be noted that such neutral electrolyte solutions with environmental friendly and safety salts offer great advantage compared to the concentrated acid/alkaline electrolytes.

A potential window of the manganese dioxides in mild aqueous electrolyte solution is typically limited in the range of −0.1 to 0.9 V versus Ag/AgCl because of the decomposition of water. This fact reduces the energy density of the manganese oxides as a conventional symmetrical capacitor with the maximum terminal voltage of 1 V. The energy (E) accumulated in the capacitor is represented by the following formula:

$$E = \frac{1}{2} C V^2, \qquad (29.4)$$

where C and V denote capacitance and voltage of the capacitor, respectively. Since the energy density of the capacitor is proportional to the square of terminal voltage, this results in the lowering energy density for the symmetrical supercapacitor consisting of manganese dioxide electrodes, which is due to the limited potential window. The concept of asymmetric (or hybrid) supercpacitors was therefore introduced to improve the energy/power density as the supercapacitor [26–28]. For the asymmetric capacitor, the manganese dioxide electrode is used as the positive electrode with the different negative electrodes, which can be cycled below 0 V versus Ag/AgCl without hydrogen gas generation. Activated carbon electrodes can be used as the negative electrode material because the electrodes show higher overpotential for hydrogen evolution (~−1.0 V vs. Ag/AgCl in neutral solution).

Cyclic voltammograms of an asymmetric supercapacitor, consisting of the manganese dioxides as positive electrode and the activated carbon as negative electrode, and their half cells are shown in Figure 29.9. The asymmetric supercapacitor can be charged up to approximately 2 V (Fig. 29.9b), and its energy density reaches ca. 21–29 Wh/kg [26–28], which is nearly 10 times larger in energy density than that of symmetrical supercapacitor based on the manganese dioxides. The high oxygen and hydrogen overpotentials for the manganese dioxide and activated carbon electrodes, respectively, allow a 2-V supercapacitor with environmentally benign electrolyte, which contributes to the increase in energy density of the asymmetric supercapacitor as expected from Equation 29.4. Although the specific capacitance of the negative and positive electrodes is lower than a $RuO_2{\cdot}nH_2O$ electrode, the available energy density is comparable with the ruthenium symmetrical supercapacitor (27 Wh/kg) [1] because of its higher terminal voltage among aqueous-based electrolyte systems.

29.4 ELECTROLYTE ADDITIVES FOR THE CAPACITOR

The asymmetric supercapacitor, consisting of manganese dioxides and activated carbon, is one of the most promising electrochemical energy devices as shown in this chapter. However, there is a drawback to the application of manganese oxides for the capacitors, that is, loss of manganese through partial dissolution of the manganese dioxides, which is considered to be due mainly to the disproportionation of trivalent manganese during electrochemical cycling [10, 17]. In general, manganese carbonate and phosphate species are insoluble in water. We expected that, therefore, the carbonate or phosphate additive into an electrolyte can suppress the dissolution of manganese oxides similar to parkerizing (also called phosphating), which is a method of protecting a steel surface from corrosion and increasing its resistance to wear through the application of an electrochemical phosphate conversion coating. Recently, we demonstrated that such electrolyte additives significantly improve the cyclability of the birnessite-type manganese dioxides as active material for supercapacitor, synthesized by the electrochemical stimulation of the Mn_3O_4 powder electrode [21].

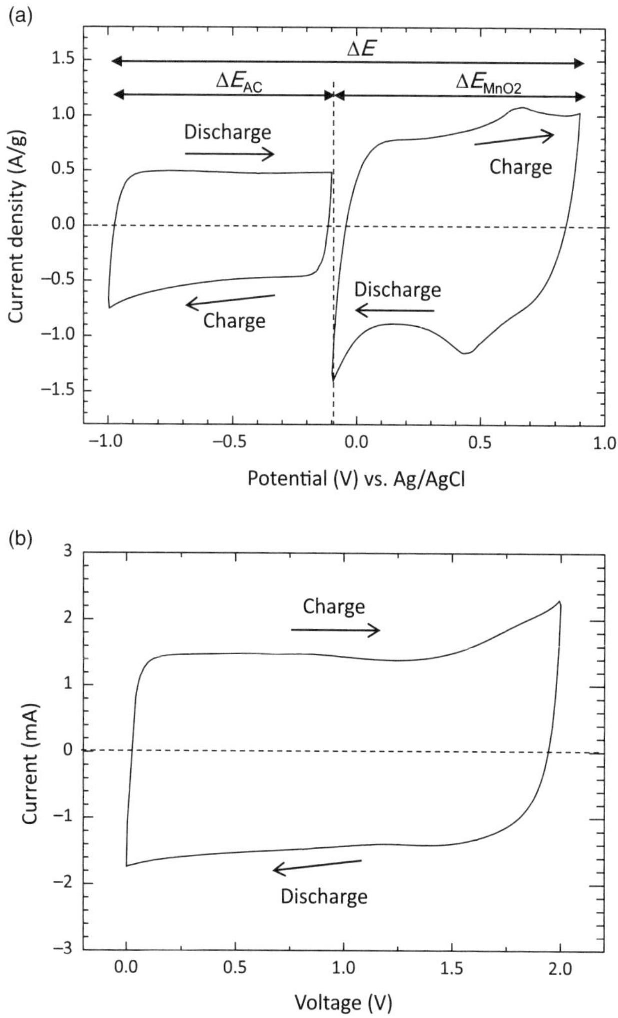

Figure 29.9. Cyclic voltammograms of a (a) birnessite electrode (–0.1~0.9 V) prepared by the electrochemical stimulation method and activated carbon (–1.0~–0.1 V) and (b) the asymmetric supercapacitor of the birnessite/1 mol/dm^3 Na_2SO_4/the activated carbon cell.

Figure 29.10 shows the galvanostatic oxidation/reduction curves of the birnessite electrode in the additive-free and 30 mmol/dm^3 $NaHCO_3$ or Na_2HPO_4 added electrolyte solutions in the potential range of –0.1 to +0.9 V. Among various phosphates and carbonates, $NaHCO_3$ and Na_2HPO_4 were chosen as additives because of the insignificant pH change of the Na_2SO_4 solution and no additional alkali species. When the additive-free (Fig. 29.10a)

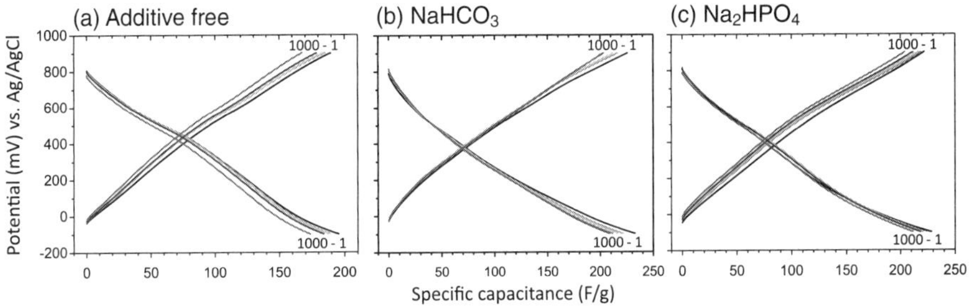

Figure 29.10. Galvanostatic oxidation and reduction curves of the birnessite electrodes synthesized by electrochemical cycling of Mn_3O_4 in (a) additive free, (b) 30 mmol/dm^3 $NaHCO_3$ added, and (c) 30 mmol/dm^3 Na_2HPO_4 added electrolytes at a rate of 1 A/g. The electrolyte used was 1 mol/dm^3 Na_2SO_4 aqueous solution.

electrolyte was used, the potential of the birnessite almost linearly varied at a constant anodic/cathodic current with a capacitance of ca. 190 F/g at the initial cycle. The birnessite electrode exhibited supercapacitive behavior, that is, the faradic capacitance and the electric double-layer capacitance. After 1000 cycling tests, the capacitance decreased and the polarization slightly increased as seen in Figure 29.10a. Manganese oxide precipitation was visually confirmed in the electrolyte after cycling, indicating that the capacity reduction was due mainly to the loss of active material by the partial dissolution of the birnessite. When the birnessite electrode was tested in the $NaHCO_3$ or Na_2HPO_4 added electrolytes, the capacitance was remarkably increased to approximately 230 F/g at the initial cycle as seen in Figure 29.10b,c. There are no additional plateaus in the curves; furthermore, the curves remained unchanged after more than 1000 cycles, indicating no significant degradation of capacity.

The capacitance versus cycle number plots in the electrolytes with and without Na_2HPO_4 are shown in Figure 29.11. The capacitance monotonously decreased in the additive-free electrolyte accompanied with the manganese dissolution. In contrast, the capacitance stayed as a constant value; simultaneously, the Mn dissolution was not confirmed when 30 mmol/dm^3 $NaHCO_3$ and Na_2HPO_4 were dissolved into the electrolyte. The ionic conductivity of the Na_2SO_4 electrolyte, 87 mS/cm, was slightly increased to 92 mS/cm by adding 30 mmol/dm^3 $NaHCO_3$ or Na_2HPO_4. Higher ionic conductivity is important for high power operation as the capacitor. The pH value of the electrolyte slightly changed because of the additives.

Generally, manganese phosphate and carbonate are insoluble in water. Therefore, if Mn^{2+} ions are dissolved from the electrode surface, its phosphate or carbonate should be promptly deposited on the electrode. In other words, the electrode should be covered with the manganese phosphate or carbonate, so that the dissolution would be successfully suppressed by the additive. Figure 29.12 shows the total amount of manganese released from the birnessite

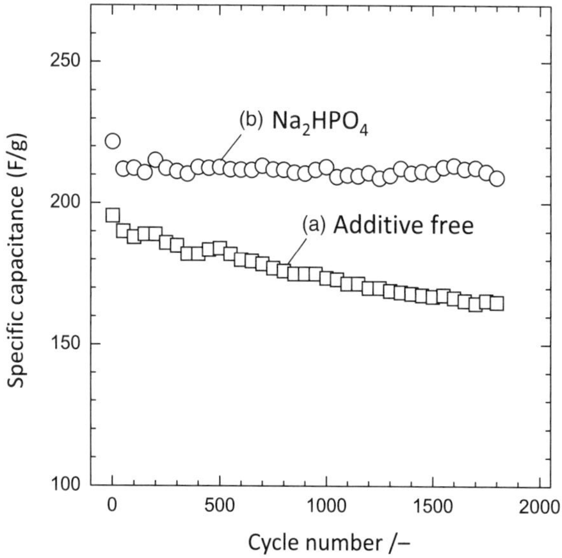

Figure 29.11. Variation in the capacitance of the birnessite electrodes prepared by electrochemical cycling of Mn_3O_4 in (a) additive free or (b) 10 mmol/dm^3 Na_2HPO_4 added electrolytes (1 mol/dm^3 Na_2SO_4).

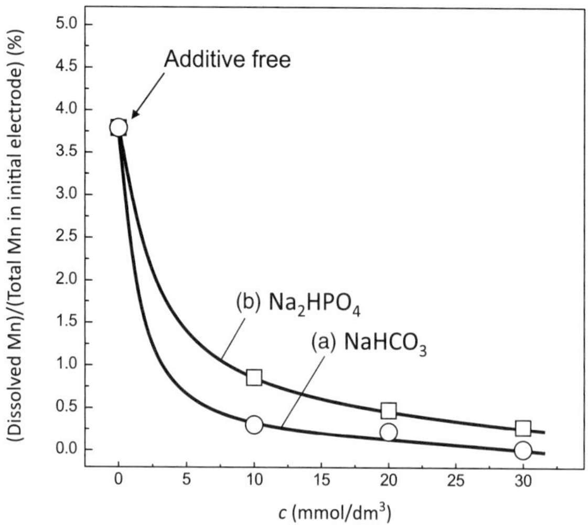

Figure 29.12. Dependence of the amount of dissolved manganese into 1 mol/dm^3 Na_2SO_4 electrolytes after 1000 cycles on the concentration of (a) $NaHCO_3$ and (b) Na_2HPO_4 additives.

electrode into the electrolyte, which was determined by ICP analyses of the electrolyte solution. Prior to the analyses, the HCl and H_2O_2 solution was added to dissolve all precipitation in the electrolyte. The amount of manganese dissolved in the electrolyte was expressed as the ratio of (dissolved Mn)/(total Mn in the initial birnessite electrode). The dissolved manganese amount was dramatically decreased as the $NaHCO_3$ or Na_2HPO_4 concentration increased, and then the cyclability was enhanced. In both additives, the dissolution of manganese is effectively suppressed when the concentration was 10 mmol/dm^3 or higher in the electrolyte. In the case of the additive-free electrolyte, the amount of Mn dissolved was 4%, while the decrease in capacitance was ca. 15% after 1000 cycles (Fig. 29.11a). Furthermore, the morphology of the birnessite is rough and porous [20, 21], suggesting that only the surface part of the electrode is electroactive; that is, all the manganese of the electrode was not utilized in the electrochemical reaction.

Further investigations are needed to understand the influences of the additives and their origin on the enhanced supercapacitive behavior. There is a general consensus that the manganese dissolution is an essential problem to be solved for the application of an electroactive manganese dioxide-based electrode to electrochemical capacitors. We believe that there is a general possibility that the new additives realize the higher capacitance and enhanced cycle life of transition metal oxides as an active material for the electrochemical capacitors.

29.5 CONCLUSIONS AND OUTLOOK

In this chapter, recent progress and development in the manganese oxides have been briefly reviewed as the electrode materials for supercapacitors. Manganese dioxides show a variety of the crystal structures, surface structures, and morphology, depending on the synthesis condition of the materials, which resulted in the distinctive electrochemical properties as the electrode materials. Asymmetric supercapacitors consisting of manganese dioxide and activated carbon electrodes have to be a potentially interesting device, that is, simple, low cost, and environmentally friendly. The major drawbacks as the electrode materials are (i) loss of the manganese during electrochemical cycling and (ii) poor electrical conductivity. It has been demonstrated that the loss of the manganese oxide from the electrode into the electrolyte solution is drastically suppressed by the phosphate and carbonate additives. It is proposed that the insoluble nature of manganese phosphate and carbonate plays an important role in suppressing the dissolution to the electrolyte. Improvement of the intrinsic electrical conductivity of manganese oxides is the great challenge to further enhance energy and power density as the supercapacitor.

REFERENCES

1 J. P. Zheng, P. J. Cygan, T. R. Jow. *Journal of the Electrochemical Society* 142(8) (1995) 2699–2703.

2 J. P. Zheng, T. R. Jow. *Journal of the Electrochemical Society* 142(1) (1995) L6–L8.

3 W. Sugimoto, H. Iwata, Y. Yasunaga, Y. Murakami, Y. Takasu. *Angewandte Chemie (International Edition in English)* 42(34) (2003) 4092–4096.

4 H. Y. Lee, J. B. Goodenough. *Journal of Solid State Chemistry* 144(1) (1999) 220–223.

5 W. Xing, S. Z. Qiao, R. G. Ding, F. Li, G. Q. Lu, Z. F. Yan, H. M. Cheng. *Carbon* 44(2) (2006) 216–224.

6 M. Toupin, T. Brousse, D. Belanger. *Chemistry of Materials* 16(16) (2004) 3184–3190.

7 O. Abe, Y. Taketa. *Journal of Physics D: Applied Physics* 24(7) (1991) 1163–1171.

8 J. M. Fletcher, W. E. Gardner, B. F. Greenfie, M. J. Holdoway, M. H. Rand. *Journal of the Chemical Society. A—Inorganic Physical Theoretical* (1968) 653–657.

9 P. K. Sharma, G. J. Moore, F. Zhang, P. Zavalij, M. S. Whittingham. *Electrochemical and Solid State Letters* 2(10) (1999) 494–496.

10 S. C. Pang, M. A. Anderson, T. W. Chapman. *Journal of the Electrochemical Society* 147(2) (2000) 444–450.

11 J. N. Broughton, M. J. Brett. *Electrochimica Acta* 49(25) (2004) 4439–4446.

12 X. P. Dong, W. H. Shen, J. L. Gu, L. M. Xiong, Y. F. Zhu, Z. Li, J. L. Shi. *Journal of Physical Chemistry. B* 110(12) (2006) 6015–6019.

13 A. E. Fischer, K. A. Pettigrew, D. R. Rolison, R. M. Stroud, J. W. Long. *Nano Letters* 7(2) (2007) 281–286.

14 M. Toupin, T. Brousse, D. Belanger. *Chemistry of Materials* 14(9) (2002) 3946–3952.

15 V. Subramanian, H. W. Zhu, B. Q. Wei. *Chemical Physics Letters* 453(4–6) (2008) 242–249.

16 P. Ragupathy, D. H. Park, G. Campet, H. N. Vasan, S. J. Hwang, J. H. Choy, N. Munichandraiah. *Journal of Physical Chemistry. C* 113(15) (2009) 6303–6309.

17 R. N. Reddy, R. G. Reddy. *Journal of Power Sources* 124(1) (2003) 330–337.

18 S. C. Pang, M. A. Anderson. *Journal of Materials Research* 15(10) (2000) 2096–2106.

19 T. Brousse, M. Toupin, R. Dugas, L. Athouel, O. Crosnier, D. Belanger. *Journal of the Electrochemical Society* 153(12) (2006) A2171–A2180.

20 S. Komaba, A. Ogata, D. Nakagawa. *ITE Letters* 8 (2007) 400–403.

21 S. Komaba, A. Ogata, T. Tsuchikawa. *Electrochemistry Communications* 10(10) (2008) 1435–1437.

22 K. W. Nam, K. B. Kim. *Journal of the Electrochemical Society* 153(1) (2006) A81–A88.

23 B. Djurfors, J. N. Broughton, M. J. Brett, D. G. Ivey. *Journal of the Electrochemical Society* 153(1) (2006) A64–A68.

24 B. Messaoudi, S. Joiret, M. Keddam, H. Takenouti. *Electrochimica Acta* 46(16) (2001) 2487–2498.

25 A. Ogata, S. Komaba, R. Baddour-Hadjean, J. P. Pereira-Ramos, N. Kumagai. *Electrochimica Acta* 53(7) (2008) 3084–3093.

26 V. Khomenko, E. Raymundo-Pinero, F. Beguin. *Journal of Power Sources* 153(1) (2006) 183–190.

27 T. Brousse, M. Toupin, D. Belanger. *Journal of the Electrochemical Society* 151(4) (2004) A614–A622.

28 M. S. Hong, S. H. Lee, S. W. Kim. *Electrochemical and Solid State Letters* 5(10) (2002) A227–A230.

30

SURFACE FUNCTIONALIZATION OF DIAMOND ELECTRODES

TAKESHI KONDO

30.1 INTRODUCTION

Conductive boron-doped diamond (BDD) has drawn attention as a novel functional electrode material due to its unique electrochemical properties, such as a wide potential window and low background current, in addition to extreme chemical and physical stability and biocompatibility. Therefore, diamond electrodes are expected to be used for highly sensitive electrochemical sensors [1]. Surface modification of diamond electrodes would effectively enable application in a wide variety of fields. For example, the reaction of some electroactive species, such as alcohols, sugars, and hydrogen peroxide, is not active at a diamond electrode and thus requires a large overpotential. If surface functional groups or molecules that act as active sites for electron transfer reactions are introduced to a diamond electrode, then highly sensitive electrochemical detection of a target analyte would be expected at the modified electrode. There have been many reports for surface modification of diamond with functional molecules as shown in this chapter. This chapter describes the surface modification of diamond electrodes with covalent molecular monolayers and electroanalytical applications of the modified diamond electrodes.

Electrical Phenomena at Interfaces and Biointerfaces: Fundamentals and Applications in Nano-, Bio-, and Environmental Sciences, First Edition. Edited by Hiroyuki Ohshima.

30.2 SURFACE MODIFICATION OF DIAMOND ELECTRODE WITH COVALENT MOLECULAR MONOLAYER

30.2.1 Silane Coupling Method

Surface modification of diamond using a silane coupling agent is a simple method that can be performed under mild conditions. As an example of the silane coupling method, surface modification of diamond with 3-aminopropyltriethoxysilane (APTES) is illustrated in Scheme 30.1. Typically, the diamond electrode surface is oxidized by anodic treatment (+2.4 V vs. Ag/AgCl in 0.1 M H_2SO_4 for 60 minutes) to generate surface OH groups. The oxidized sample is then immersed in 2 vol % APTES/ethanol solution for a few minutes and then thoroughly rinsed with clean ethanol, followed by heating to 110°C to completely remove ethanol [2]. Surface modification with APTES provides amino ($-NH_2$) groups on the diamond surface, which can be used for covalent linking with DNA, enzymes, and other functional molecules via amide bond formation [3]. The introduction of surface –SH [4] and perfluoroalkyl [5] groups to diamond via the silane coupling method has also been reported. However, compared with another method based on C–C bond formation described later, the modified surface may be poorly stable due to possible hydrolysis of the C–O–Si bond. Coverage of the silane coupling agent on the modified diamond surface is affected by the coverage of surface –OH groups on the diamond substrate prior to modification. The density of surface –OH groups generated by oxidative treatment may be dependent on the oxidation method and the type of crystal facet. X-ray photoelectron spectra of anodically treated single-crystal diamond surfaces indicated that both singly and doubly bonded oxygen groups (C–O and C=O) were generated on the (100) surface, whereas only singly bonded oxygen groups (C–O) were generated on the (111) surface [6]. Combined with results of the electrochemical investigation of APTES-modified single-crystal BDD electrodes, surface –OH groups were found to be more densely generated at the (111) facet than at the (100) facet. Therefore, coverage of the silane coupling agent on a polycrystal-

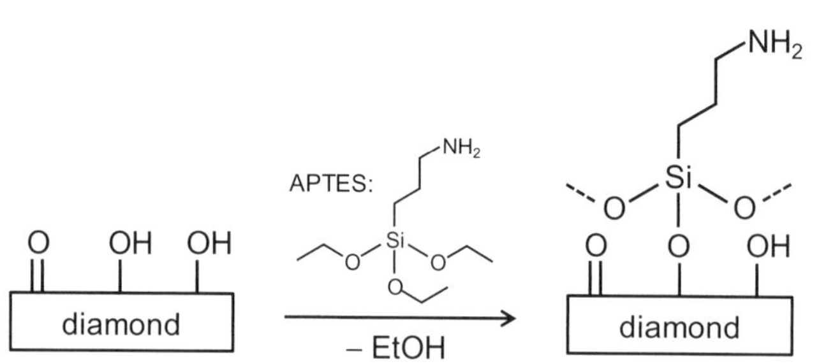

Scheme 30.1. Surface modification of oxidized diamond with APTES.

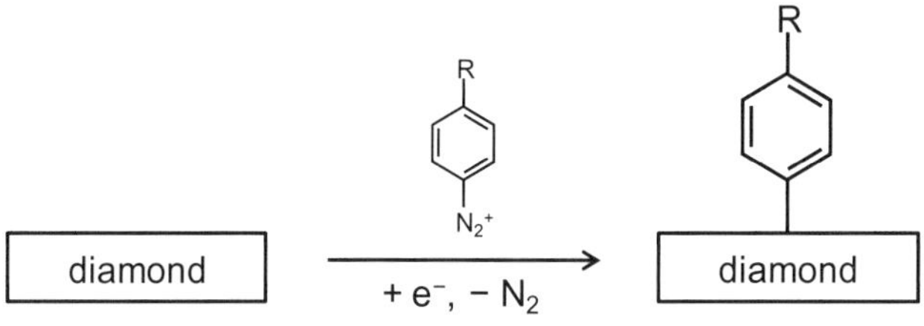

Scheme 30.2. Surface modification of diamond with aryldiazonium.

line diamond surface is considered to be inhomogeneous and dependent on the exposing crystal facet.

30.2.2 Surface Grafting with Aryldiazonium Salts

Surface grafting of diamond with aryldiazonium salts is one useful method to form a covalent molecular monolayer on a diamond surface. Surface grafting with diazonium salts is initiated with chemical or electrochemical reduction of the diazonium ion, as shown in Scheme 30.2. The resulting covalent C–C bond between the diamond surface and phenyl groups provides a basis for the introduction of various functional groups. For example, covalent modification of a conductive diamond surface with nitrophenyl monolayer [7, 8] can be achieved via electrochemical reduction of 4-nitrophenyl diazonium tetrafluoroborate in acetonitrile (AN) at a constant electrode potential of −0.2 V versus Ag/Ag^+ [9]. The nitro group can be reduced electrochemically to generate the amino group, which is useful for further functionalization with biorelated molecules such as DNA [9] and enzymes [10]. Quinone derivatives [11], polymer brushes [12], ruthenium complexes [13], and biphenyl derivatives via Suzuki coupling [14] are also introduced onto the diamond surface by surface grafting with aryldiazonium salts. Initial reduction of the diazonium salts can proceed spontaneously, for example, under ultrasonication; therefore, diazonium salts can also be used for the surface grafting of diamond nanoparticles [15–17].

30.2.3 Photochemical Modification Using Terminal Alkenes

Photochemical surface modification of diamond using terminal alkenes has been known as a useful method to fabricate stable molecular monolayers on the surface via C–C covalent bonds. In this method, the hydrogen-terminated diamond surface is typically covered with liquid alkene and illuminated with UV light (254 nm or 4.9 eV, which is smaller than the band gap energy of diamond [ca. 5.5 eV]) under an inert atmosphere (N_2, Ar). The presently accepted mechanism for this surface modification reaction is illustrated in Scheme 30.3 [18]. UV light illumination induces electron transfer from the

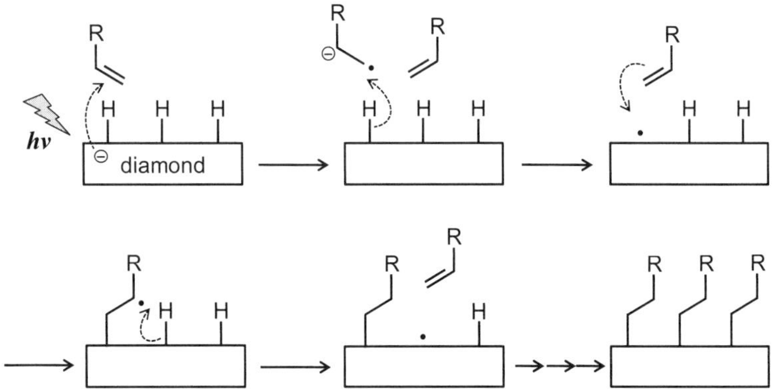

Scheme 30.3. Possible mechanism of photochemical modification of a diamond surface with terminal alkene. See text for details.

hydrogen-terminated diamond surface to alkene molecules, which leads to the generation of radical anions. A radical anion abstracts a hydrogen atom from the diamond surface to form a surface carbon dangling bond. The surface dangling bond reacts with another alkene molecule, which results in the formation of a C–C bond. The abstraction of surface hydrogen and C–C bond formation propagates in a chain reaction-like manner. Nebel et al. observed the growth of 10-amino-dec-1-ene islands protected with trifluoroacetic acid group (TFAAD) monolayers on a single-crystal diamond surface, which was most likely a result of the chain reaction [18].

Examples of liquid alkenes used for the photochemical modification of diamond surfaces are 1-dodecene [19–21], TFAAD [22–27], and 10-undecenoic acid (UA) [28, 29]. The TFAAD-modified diamond surface can be converted to an amino-terminated surface after deprotection. Surface modification with UA provides carboxylic acid groups on the diamond surface. These functionalities are useful for further surface modification with, for example, DNA [9, 20, 23, 30] and enzymes [24] via amide bond formation.

For the photochemical modification method using terminal alkenes, solution of alkene can be used instead of liquid alkenes. The use of alkene solution expands the choice of alkenes providing a wide variety of modified diamond surfaces. Such modified diamond surfaces were examined qualitatively and quantitatively by the use of vinylferrocene (VFC) as a modifier molecule [31]. Hydrogen-terminated polycrystalline or single-crystal BDD was immersed in a 5 mM VFC/mesitylene solution in a polytetrafluoroethylene (PTFE) reaction cell. After purging the atmosphere in the cell with argon gas, the sample surface was irradiated with UV light (254 nm) from a 40-W low-pressure mercury lamp through the quartz window of the cell.

Figure 30.1 shows typical cyclic voltammograms (CVs) in 0.1 M tetraethylammonium perchlorate (TEAP)/AN at a vinylferrocene-modified boron-

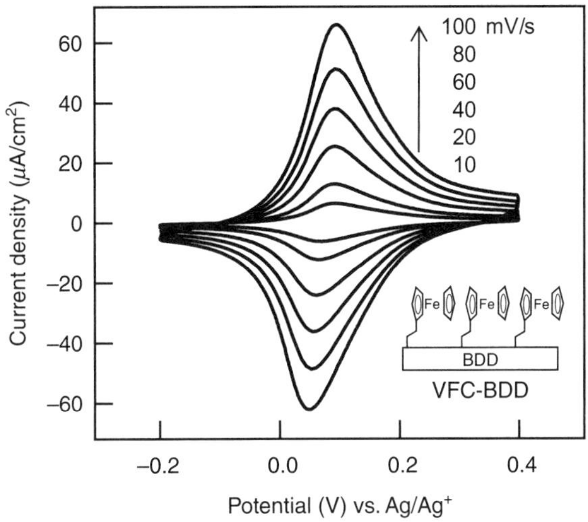

Figure 30.1. CVs in 0.1 M tetraethylammonium perchlorate (TEAP)/acetonitrile (AN) at a VFC-BDD electrode. Potential sweep rates were 10, 20, 40, 60, 80, and 100 mV/s.

doped diamond (VFC-BDD) electrode prepared with 3 hours of UV irradiation. A reversible peak pair at ca. +0.1 V versus Ag/Ag^+ showed the presence of ferrocene derivatives on the electrode surface. The anodic–cathodic peak separation (ΔE_p) was 25 mV at a potential sweep rate of 20 mV/s, much smaller than the theoretical value for a diffusion-controlled one-electron reversible redox system (57 mV). The anodic and cathodic peak current densities (i_{pa} and i_{pc}) were directly proportional to the potential sweep rate, and the i_{pa}/i_{pc} ratio was close to unity. These characteristics indicated that the redox process on these electrodes was controlled by charge transfer kinetics, confirming that the ferrocene derivatives were immobilized on the BDD surface. The ferrocene coverage (Γ, molecules per square centimeter) of the VFC-BDD electrode was estimated from the charge of the CV anodic peak assuming a one-electron reaction, from the following relationship:

$$\Gamma = (Q/F)N_A, \qquad (30.1)$$

where Q (coulomb per square centimeter) is the charge calculated from the CV anodic peak area, F is the Faraday constant, and N_A is the Avogadro constant. For a polycrystalline BDD sample, the ferrocene coverage increased with increasing irradiation time, saturating at ~7×10^{14} cm^{-2} upon irradiation for 3 hours or longer. From the two-dimensional closest packing of the ferrocenyl moiety on a flat surface, calculated assuming a hexagonal packing of the moiety with a diameter of 6.6 Å [32], the coverage was found to be 2.7×10^{14} cm^{-2}. The estimated ferrocene coverage of the VFC-BDD electrode

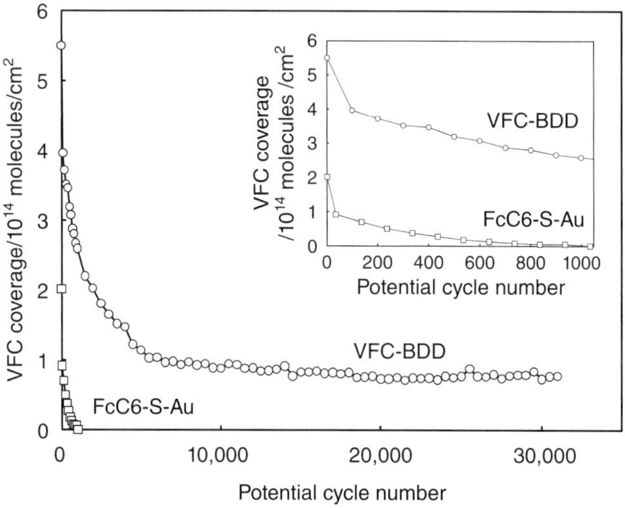

Figure 30.2. Surface VFC coverage of a (○) VFC-BDD and a (□) FcC6SH SAM-modified Au surface as a function of potential sweep cycle number. Potential sweeps were carried out in 0.1 M TEAP/AN over the potential range –0.2 to +0.4 V versus Ag/Ag^+ at 100 mV/s. The inset illustrates data up to 1000 cycles.

was larger than this value. In the case of the single-crystal (100) BDD sample, however, the saturation coverage was $3–4 \times 10^{14}$ cm^{-2}, which corresponded reasonably well to the monolayer coverage of VFC on this substrate. The factor of 2–3 difference between the estimated ferrocene coverage of polycrystalline and single-crystal samples was attributed to surface roughness.

The electrochemical stability of VFC-BDD was examined by successive potential cycling in 0.1 M TEAP/AN from –0.2 to +0.4 V versus Ag/Ag^+ with a potential sweep rate of 100 mV/s. Figure 30.2 shows the surface ferrocene coverage of a VFC-BDD electrode estimated from the CV peak area as a function of potential cycle number. Initially, the coverage was found to decrease as the cycle number increased. However, the coverage became almost constant at 8×10^{13} cm^{-2} after ~5000 cycles, up to 30,000 cycles. The stability to potential cycling of the VFC-BDD electrode was compared to that of a ferrocenylhexanethiol (FcC6SH) self-assembled monolayer (SAM)-covered Au electrode. The CV peak current decreased with increasing cycle number, and the peaks disappeared after 500 cycles. Because of the good solubility of ferrocenylalkanethiol in AN, it was expected that these ferrocene derivatives would be removed from the electrode surface during potential cycling even though they were immobilized on gold by covalent Au-S bonds. However, although VFC is also quite soluble in AN, removal of VFC from the diamond surface was possibly inhibited by the more stable C–C bond between VFC and diamond.

30.3 ELECTROANALYTICAL APPLICATIONS OF SURFACE-MODIFIED DIAMOND ELECTRODES

Conductive polycrystalline BDD films have been recognized as a promising candidate electrode for sensitive electroanalysis based on their wide potential window and low background current. In this section, electroanalytical applications of surface functionalized BDD electrodes prepared with the photochemical method using terminal alkene are described.

30.3.1 Oxalic Acid Detection at Quaternary Ammonium-Terminated Diamond Electrode

Sensitive detection of oxalate dianion ($C_2O_4^{2-}$) in biological materials, such as urine, is desirable because it can form insoluble salts with assimilated calcium and contributes to the formation of renal calculi. In order to detect the oxalate anion electrochemically with high sensitivity and to improve the stability of the electrochemical signals, surface modification of the electrode with positively charged organic monolayers is expected to be effective. Thus, a quaternary ammonium group was introduced onto the diamond surface by the use of a photochemical surface modification method with allyltriethylammonium bromide (ATAB) [33].

The anodic CV peak for oxalate oxidation was observed at ca. +1.5 V versus Ag/AgCl with the unmodified (H-terminated) BDD surface, but no discernable peak was obtained when a surface treated by anodic oxidation was used (Fig. 30.3). The peak potential of oxalate oxidation at an ATAB-BDD was

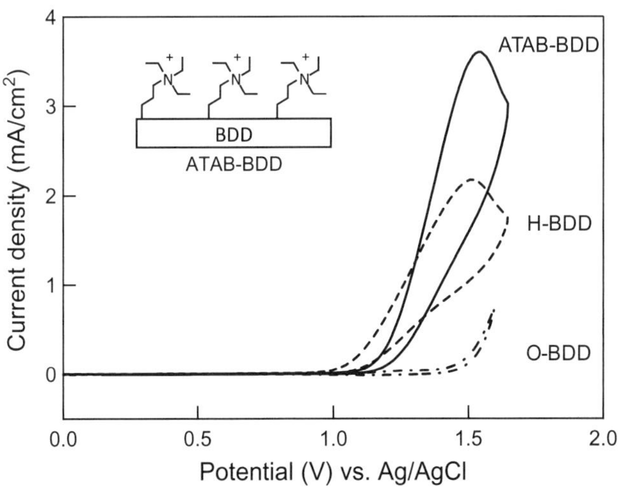

Figure 30.3. CVs for 7 mM oxalate in 0.1 M phosphate buffer solution (PBS, pH 7) at hydrogen-terminated, oxidized, and ATAB-modified BDD electrodes. The potential sweep rate was 100 mV/s.

observed at ca. +1.5 V versus Ag/AgCl, similar to that at the unmodified surface. However, the anodic peak current at the ATAB-BDD electrode was ca. 1.7 times larger than that for the peak obtained at the unmodified BDD electrode. The terminal group of the ATAB-BDD surface is considered to be positively charged ($-NEt_3^+$), whereas oxalate is in the dianion form ($C_2O_4^{2-}$) in neutral solution (pK_{a2} = 4.19) [34]. Thus, an electrostatic interaction between the anion and the electrode surface may be assumed to be attractive, which should lead to facile electron transfer compared to that at an electrically neutral hydrogen-terminated BDD surface, even though there is a positive dipole.

In order to estimate the applicability of ATAB-BDD electrodes to the electrochemical detection of oxalate, flow injection analysis-amperometric measurements (FIA-EC) were investigated at a fixed electrode potential (+1.15 V vs. Ag/AgCl). Figure 30.4A shows amperograms for 0.1 M phosphate buffer solution (PBS) containing oxalate in the concentration range of 0.8–100.0 μM. The background-subtracted signal current showed good linearity versus oxalate concentration in the range of 5.0–80.0 and 0.8–80.0 μM at the unmodified and ATAB-modified BDD electrodes, respectively (Fig. 30.4B). The slope of the calibration curve at the ATAB-BDD electrode was nearly a factor of 2 higher than that at the unmodified electrode. The lowest detection limit (signal-to-noise ratio of 3, nominal noise level of 0.2 nA) at the ATAB-BDD electrode based on the calibration curve was evaluated to be 32 nM, while that at the unmodified BDD was 125 nM. The difference in sensitivity of the signal current to oxalate concentration between the ATAB-BDD and the unmodified BDD electrodes was consistent with the results of CV. The ratio of the signal current of the nth to the initial injection ($I_{p,n}/I_{p,1}$) for repeated sample injections of 10 mM oxalate was investigated at a fixed potential of +1.150 V versus Ag/AgCl. After 40 consecutive injections, the signal current decreased to almost half of the initial value at the unmodified BDD electrode. Since the fixed potential was relatively positive, gradual surface oxidation of the electrode may have occurred during the measurement. In comparison, at the ATAB-BDD electrode, the fractional current reduction was only 0.18. Thus, the covalent attachment of appropriate molecular monolayers to BDD was found to be an effective means of improving the electrochemical stability of such electrode surfaces, as well as the sensitivity, in the electrochemical detection.

30.3.2 Dopamine (DA) Detection in the Presence of Ascorbic Acid (AA) at Carboxylic Acid-Terminated Diamond Electrodes

DA is an important neurotransmitter in mammalian central nervous systems. Electrochemical techniques are attractive for the determination of DA concentrations because of their high sensitivity and speed, as well as their applicability to real-time *in vivo* monitoring of DA in brain tissues. A major problem in the electrochemical determination of DA is interference from AA, which is ubiquitous in biological fluids. The problem encountered with the simultane-

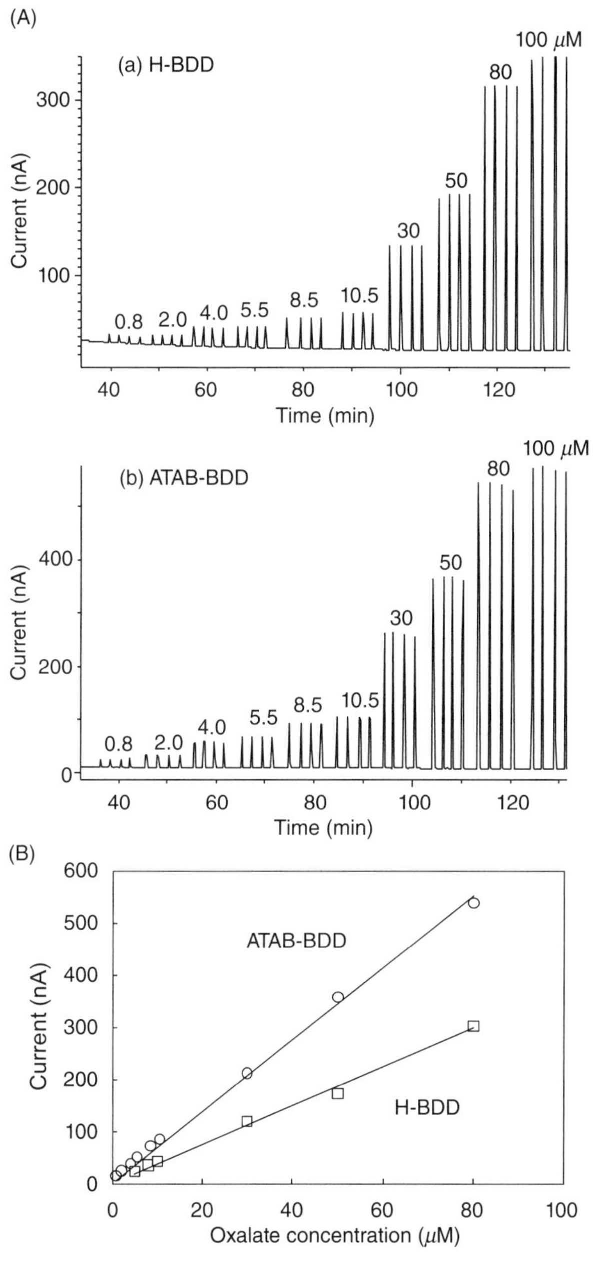

Figure 30.4. (A) FIA amperograms for 20 mL injections of 0.1 M PBS containing various concentrations of oxalate at (a) H-BDD and (b) ATAB-BDD electrodes. Four successive injections were recorded for each given sample concentration at 2-minute intervals. The electrode potential was fixed at +1.15 V versus Ag/AgCl. The flow rate was 1 mL/min. (B) Calibration curves created from the data sets of Figure 30.4A.

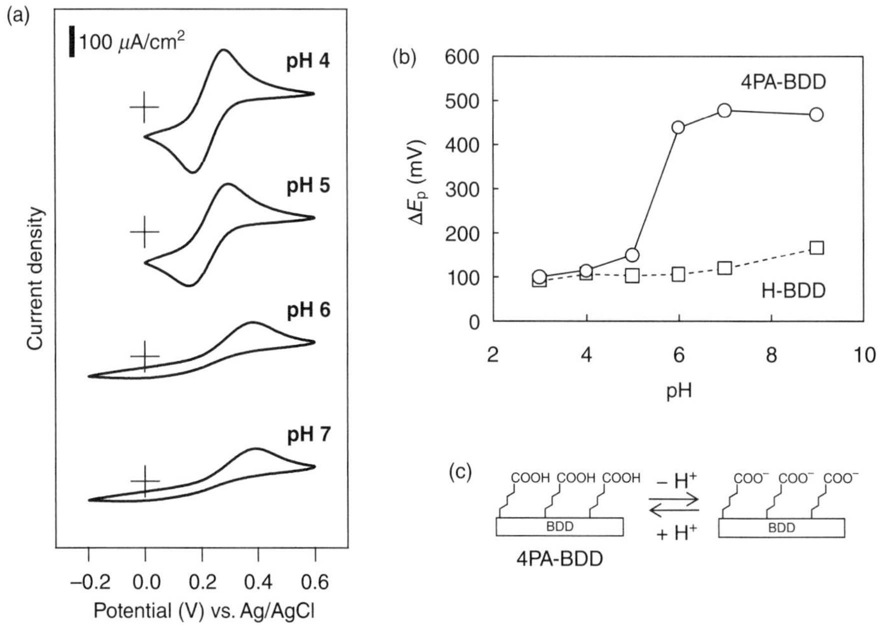

Figure 30.5. (a) CV for 1 mM $K_3Fe(CN)_6$ in 0.1 M Na_2SO_4 with various pH values at H-BDD and 4PA-BDD electrodes. The potential sweep rate was 100 mV/s. (b) pH dependence of the anodic–cathodic CV peak separation (ΔE_p). (c) Schematic illustration of 4PA-BDD surface.

ous presence of DA and AA is well-known: Both can be oxidized at very similar potentials at most solid electrode materials, causing an overlapping voltammetric response. In order to electrostatically attract the cationic DA while repelling the anionic AA, the BDD surface was photochemically modified with 4-pentenoic acid (4PA) to introduce negatively charged carboxyl groups on the surface [35].

The surface modification of BDD with 4PA were also characterized by the examination of the CV behavior of the $Fe(CN)_6^{3-/4-}$ redox probe. In Figure 30.5, the CV anodic–cathodic peak separation (ΔE_p) is represented as a function of electrolyte solution pH. The ΔE_p values were found to change greatly between pH 5 and 6 at the 4PA-BDD electrode, while they showed little change at the H-BDD electrode. The electron transfer rate of this redox pair has been shown to be sensitive to the electrostatic characteristics of the electrode surface. Thus, the great change in ΔE_p at pH 5–6 should be based on the fact that the terminal carboxyl groups of 4PA-BDD are protonated in the pH 3–5 region and are deprotonated in the pH 6–9 region. As such, a significant coverage of carboxyl groups is considered to exist on the 4PA-BDD surface.

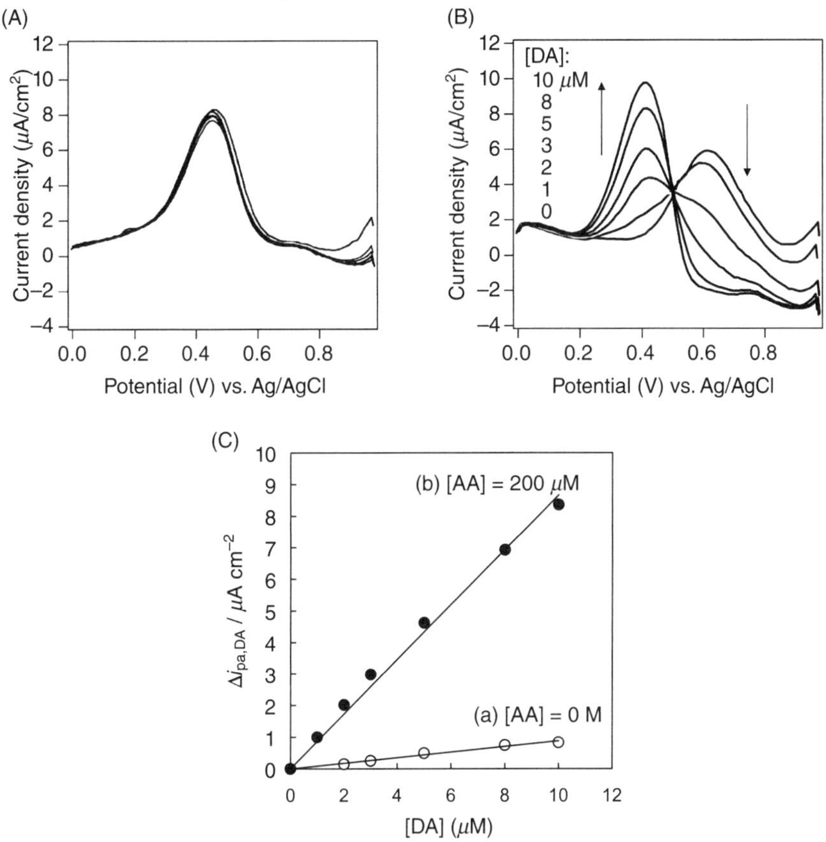

Figure 30.6. DPV curves for 200 μM AA in 0.1 M PBS (pH 7) containing various concentrations of DA at (A) H-BDD and (B) 4PA-BDD electrodes ([DA] = 0, 1, 2, 3, 5, 8, and 10 μM). (C) Calibration curves for DA determination (a) in the absence and (b) in the presence of 200 μM AA at a 4PA-BDD electrode. The increase in DPV peak current density ($\Delta i_{\mathrm{pa,\,DA}}$) as a function of [DA] is indicated.

Figure 30.6 shows differential pulse voltammetric (DPV) curves for various concentrations of DA in 0.1 M PBS containing 200 μM AA. At the H-BDD electrode, closely overlapping peaks for DA and AA were observed at 0.45 V versus Ag/AgCl (Fig. 30.6A). However, at 4PA-BDD, the DPV peaks of DA and AA were observed separately around +0.36 V versus Ag/AgCl (Fig. 30.6B). This separation is based on the respective oxidation potentials of DA (0.30 V) and AA (0.43 V) when they were measured separately at the 4PA-BDD electrode. More interestingly, the AA peak current density ($i_{\mathrm{pa,\,AA}}$) was found to decrease with increasing concentration of DA ([DA]), while the (AA) in the electrolyte was held constant. The increase of the DPV peak current density of DA ($\Delta i_{pa,\,DA}$) was found to be directly proportional to the [DA] in the range of 1–10 μM, and the slope of the calibration curve was found to increase by the addition of AA (Fig. 30.6C). In other words, the sensitivity of DA detection with DPV at the 4PA-BDD electrode was amplified by the addition of AA. Conventionally, AA has been recognized as an electroactive interferent for DA detection. However, the results described here indicate that AA can indeed play a role as an electrochemical sensitizer for DA detection with DPV at 4PA-BDD electrodes. The electrochemical amplification of DA detection by AA is considered to arise from the catalytic EC′ mechanism [36]:

$$\mathrm{DA} \leftrightarrows \mathrm{DA\ semiquinone} + 2\mathrm{H}^+ + \mathrm{e}^- \tag{30.2}$$

and

$$2\mathrm{DA\ semiquinone} + \mathrm{AA} \rightarrow 2\mathrm{DA} + \mathrm{DHA}, \tag{30.3}$$

where dehydroascorbate (DHA) is the principal oxidation product of AA. The DA semiquinone is one of the oxidation products of DA. Therefore, the purposeful use of the AA-induced catalytic electrochemical reaction at carboxyl-terminated BDD electrodes should allow determination of low concentrations of DA in samples containing AA at high concentrations, which is the practical case of usual biological fluid samples. The current amplification was also observed in chronoamperometry in the timescale within 50 ms. This indicates that the amplification phenomenon can be applied to flow analytical systems such as high-performance liquid chromatography (HPLC).

30.3.3 Hydrogen Peroxide Detection at Cobalt Phthalocyanin (CoPc)-Modified Diamond Electrodes

Hydrogen peroxide (H_2O_2) is one of the most important analytes in the field of electroanalysis because it is cytotoxic, an indicator of oxidative stress, and a product of several enzyme-catalyzed reactions. Thus, detection of H_2O_2 with high sensitivity at low concentrations is desirable in the food industry, environmental waste management, and medical diagnostics. In order to enable high-sensitivity detection of H_2O_2 at low concentrations, CoPc was immobilized on a BDD surface. Metal phthalocyanines and their derivatives are relatively stable compared to biological molecules such as enzymes. For the fabrication of CoPc-modified boron-doped diamond (CoPc-BDD), the BDD surface was covalently modified with 4-vinylpyridine (4VP) to introduce surface terminal pyridine moieties by the photochemical method. The pyridine-terminated BDD surface was then employed for the fabrication of the CoPc-BDD electrode (Scheme 30.4) [37].

Figure 30.7a shows the CV results of H-BDD, 4VP-BDD, and CoPc-BDD electrodes in a 0.01 M PBS with and without 97 μM H_2O_2. No redox peak to indicate the electrochemical reaction of H_2O_2 at the H-BDD and 4VP-BDD

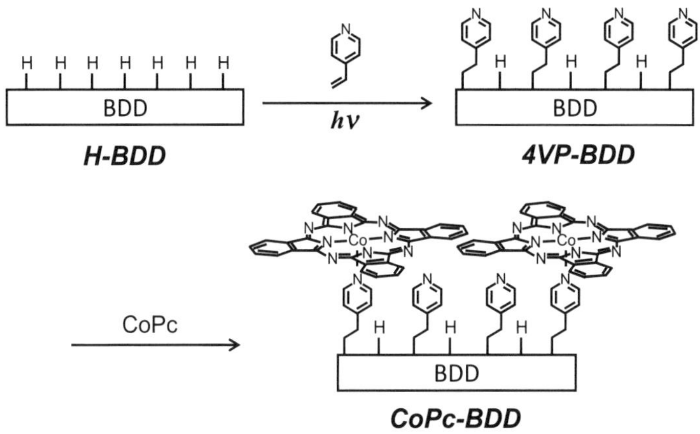

Scheme 30.4. Fabrication of CoPc-BDD electrode.

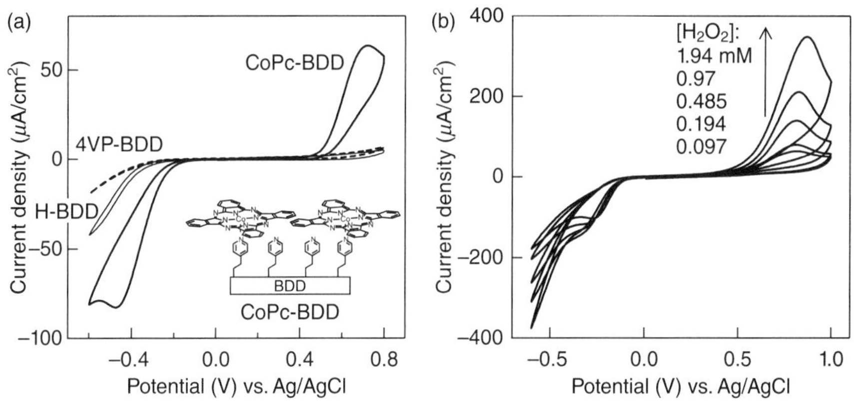

Figure 30.7. (a) CVs for 97 μM in 0.01 M PBS at H-BDD (thin solid line), 4VP-BDD (dashed line), and CoPc-BDD electrodes (thick solid line). (b) CVs in 0.01 M PBS containing various concentrations of H_2O_2 at the CoPc-BDD electrode. The potential sweep rate was 50 mV/s.

electrodes was observed. However, at the CoPc-BDD electrode, anodic and cathodic peak currents at +0.75 and –0.25 to –0.3 V versus Ag/AgCl, respectively, significantly increased in the presence of H_2O_2. CoPc is a catalyst for the electrochemical oxidation and reduction of H_2O_2 by the following mechanisms: [38]

(oxidation)

$$\begin{gathered} \mathrm{Co^{II}Pc \rightarrow Co^{III}Pc + e^-} \\ \mathrm{Co^{III}Pc + 1/2H_2O_2 \rightarrow Co^{II}Pc + 1/2O_2 + H^+} \end{gathered} \tag{30.4}$$

and

(reduction)

$$\begin{aligned} \mathrm{Co^{II}Pc + e^- \rightarrow Co^{I}Pc} \\ \mathrm{Co^{I}Pc + 1/2H_2O_2 + H^+ \rightarrow Co^{II}Pc + H_2O.} \end{aligned} \tag{30.5}$$

The anodic and cathodic peak current densities increased linearly as the H_2O_2 concentration in the electrolyte solution increased (Fig. 30.7b), which indicates that the CoPc-BDD electrode can be used for the detection of H_2O_2. The sensitivity of the CoPc-BDD electrode for the electrochemical detection of H_2O_2 was investigated using an FIA-EC system. Figure 30.8 shows the FIA amperograms at various concentrations of the injected H_2O_2 samples. At the H-BDD electrode, only a small signal current was observed after the injection of the 9.7 μM H_2O_2 sample (approximately 0.45 nA). However, at the CoPc-BDD electrode, a large signal current (31.8 nA) was obtained from the same sample injection, which was >70 times larger than that observed at the H-BDD electrode. The signal current of a lower concentration (9.7 nM) of the H_2O_2 sample was slightly larger than that of a blank injection (0.01 M PBS). The

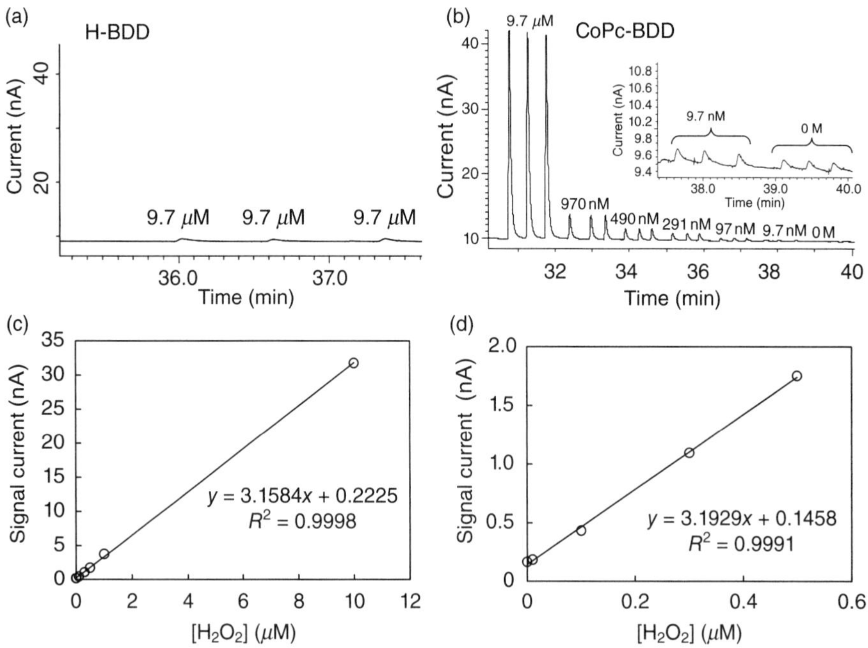

Figure 30.8. FIA amperograms for H_2O_2 detection at the (a) H-BDD and (b) CoPc-BDD electrodes. The electrode potential was +600 mV versus Ag/AgCl. Calibration curves for H_2O_2 detection using the FIA-EC system with the CoPc-BDD electrode in the ranges of (c) 0–9.7 and (d) 0–0.485 μM.

limit of detection (LOD) estimated from 3σ of the blank peaks was 4.9 nM. In contrast to the enhanced signal current, the background current was almost the same for the H-BDD and the CoPc-BDD electrodes. This indicates the usefulness of the surface functionalization of the BDD electrode, by which catalytic activity can be imparted to a BDD electrode without increasing the background current. A calibration curve obtained from the FIA-EC measurement displayed excellent linearity over the wide concentration range of 9.7 nM to 9.7 μM, with a sensitivity of 3.16 nA/μM (Fig. 30.8c,d). This also demonstrates the suitability of the CoPc-BDD electrode for H_2O_2 detection. The results of this study indicate that molecularly functionalized BDDs are promising electrode materials for sensitive electrochemical detection, especially at low concentrations.

30.4 CONCLUSION

This chapter described the covalent surface modification of diamond, especially via photochemical modification using terminal alkenes, and the application of the modified diamond electrodes to electroanalysis. Covalent surface modification was found to be one of the powerful methods to fabricate a functional electrolyte/diamond interface, which enables, for example, sensitive and selective electrochemical detection of biorelated analytes. In the future, as the use of nanostructured diamond including nanoparticles increases, the diamond surface/interface control should become highly important to fabricate novel diamond-based functional materials as well as electrode materials.

REFERENCES

1 A. Fujishima, Y. Einaga, T.N. Rao, D. A. Tryk (eds.), *Diamond electrochemistry*. Elsevier-BKC, Tokyo, 2005.

2 H. Notsu, T. Fukazawa, T. Tatsuma, D. A. Tryk, A. Fujishima. *Electrochem. Solid-State Lett.* 4 (2001) H1.

3 H. Notsu, T. Tatsuma, A. Fujishima. *J. Electroanal. Chem.* 523 (2002) 86.

4 T. Kondo, S. Aoshima, K. Honda, Y. Einaga, A. Fujishima, T. Kawai. *J. Phys. Chem. C* 111 (2007) 12650.

5 T. Tsubota, S. Tanii, T. Ishida, M. Nagata, Y. Matsumoto. *Diamond Relat. Mater.* 14 (2005) 608.

6 T. Kondo, K. Honda, D. A. Tryk, A. Fujishima. *J. Electrochem. Soc.* 152 (2005) E18.

7 T. C. Kuo, R. L. McCreery, G. M. Swain. *Electrochem. Solid-State Lett.* 2 (1999) 288.

8 H. Uetsuka, D. Shin, N. Tokuda, K. Saeki, C. E. Nebel. *Langmuir* 23 (2007) 3466.

9 B. Rezek, D. Shin, C. E. Nebel. *Langmuir* 23 (2007) 7626.

10 J. Wang, J. A. Carlisle. *Diamond Relat. Mater.* 15 (2006) 279.

11 A. Sarapuu, K. Helstein, D. J. Schiffrin, K. Tammeveski. *Electrochem. Solid-State Lett.* 8 (2005) E30.

12 T. Matrab, M. M. Chehimi, J. P. Boudou, F. Benedic, J. Wang, N. N. Naguib, J. A. Carlisle. *Diamond Relat. Mater.* 15 (2006) 639.

13 C. Agnes, J.-C. Arnault, F. Omnes, B. Jousselme, M. Billon, G. Bidan, P. Mailley. *Phys. Chem. Chem. Phys.* 11 (2009) 11647.

14 Y. L. Zhong, K. P. Loh, A. Midya, Z.-K. Chen. *Chem. Mater.* 20 (2008) 3137.

15 C. Mangeney, Z. Qin, S. A. Dahoumane, A. Adenier, F. Herbst, J.-P. Boudou, J. Pinson, M. M. Chehimi. *Diamond Relat. Mater.* 17 (2008) 1881.

16 S. A. Dahoumane, M. N. Nguyen, A. Thorel, J.-P. Boudou, M. M. Chehimi, C. Mangeney. *Langmuir* 25 (2009) 9633.

17 W. S. Yeap, S. Chen, K. P. Loh. *Langmuir* 25 (2009) 185.

18 C. E. Nebel, N. J. Yang, H. Uetsuka. *Surf. Sci.* 603 (2009) L5.

19 K. Y. Tse, B. M. Nichols, W. S. Yang, J. E. Butler, J. N. Russell, R. J. Hamers. *J. Phys. Chem. B* 109 (2005) 8523.

20 B. Sun, S. E. Baker, J. E. Butler, H. Kim, J. N. Russell, L. Shang, K. Y. Tse, W. S. Yang, R. J. Hamers. *Diamond Relat. Mater.* 16 (2007) 1608.

21 X. Wang, P. E. Colavita, K. M. Metz, J. E. Butler, R. J. Hamers. *Langmuir* 23 (2007) 11623.

22 T. Strother, T. Knickerbocker, J. N. Russell, J. E. Butler, L. M. Smith, R. J. Hamers. *Langmuir* 18 (2002) 968.

23 W. S. Yang, O. Auciello, J. E. Butler, W. Cai, J. A. Carlisle, J. Gerbi, D. M. Gruen, T. Knickerbocker, T. L. Lasseter, J. N. Russell, L. M. Smith, R. J. Hamers. *Nat. Mater.* 1 (2002) 253.

24 A. Härtl, E. Schmich, J. A. Garrido, J. Hernando, S. C. R. Catharino, S. Walter, P. Feulner, A. Kromka, D. Steinmüller, M. Stutzmann. *Nat. Mater.* 3 (2004) 736.

25 B. M. Nichols, K. M. Metz, K. Y. Tse, J. E. Butler, J. N. Russell, R. J. Hamers. *J. Phys. Chem. B* 110 (2006) 16535.

26 D. Shin, B. Rezek, N. Tokuda, D. Takeuchi, H. Watanabe, T. Nakamura, T. Yamamoto, C. E. Nebel. *Phys. Stat. Sol. A* 203 (2006) 3245.

27 N. Yang, H. Uetsuka, H. Watanabe, T. Nakamura, C. E. Nebel. *Diamond Relat. Mater.* 17 (2008) 892.

28 V. Vermeeren, S. Wenmackers, M. Daenen, K. Haenen, O. A. Williams, M. Ameloot, M. vande Ven, P. Wagner, L. Michiels. *Langmuir* 24 (2008) 9125.

29 Z. Remes, A. Kromka, H. Kozak, M. Vanecek, K. Haenen, S. Wenmackers. *Diamond Relat. Mater.* 18 (2009) 772.

30 C. E. Nebel, D. Shin, D. Takeuchi, T. Yamamoto, H. Watanabe, T. Nakamura. *Diamond Relat. Mater.* 15 (2006) 1107.

31 T. Kondo, H. Hoshi, K. Honda, Y. Einaga, A. Fujishima, T. Kawai. *J. Phys. Chem. C* 112 (2008) 11887.

32 C. E. D. Chidsey, C. R. Bertozzi, T. M. Putvinski, A. M. Mujsce. *J. Am. Chem. Soc.* 112 (1990) 4301.

33 T. Kondo, Y. Niwano, A. Tamura, T. A. Ivandini, Y. Einaga, D. A. Tryk, A. Fujishima, T. Kawai. *Electroanalysis* 20 (2008) 1556.

34 A. Berná, A. Rodes, J. M. Feliu. *J. Electroanal. Chem.* 563 (2004) 49.

35 T. Kondo, Y. Niwano, A. Tamura, J. Imai, K. Honda, Y. Einaga, D. A. Tryk, A. Fujishima, T. Kawai. *Electrochim. Acta* 54 (2009) 2312.

36 S.-Y. Yi, H.-Y. Chang, H.-H. Cho, Y. C. Park, S. H. Lee, Z.-U. Bae. *J. Electroanal. Chem.* 602 (2007) 217.

37 T. Kondo, A. Tamura, T. Kawai. *J. Electrochem. Soc.* 156 (2009) F145.

38 P. N. Mashazi, K. I. Ozoemena, T. Nyokong. *Electrochim. Acta* 52 (2006) 177.

31

QUANTUM ELECTROCHEMICAL STUDY OF BENZENE DERIVATIVES

1. ELECTRONIC STRUCTURE AND EVALUATION OF THE ANTIOXIDANT ACTIVITY OF ASPIRIN AND PARACETAMOL

Kazunaka Endo

Quantum electrochemistry is the application of quantum mechanical tools such as density functional theory (DFT) to the study of electrochemical processes, including electron transfer at electrodes. We present a study of the electronic structure, chemical bonding, and antioxidant activity of phenolic antioxidants (aspirin and paracetamol). X-ray photoelectron spectra of the antioxidants have been simulated by deMon DFT calculations for the molecules. The chemical bonding of aspirin is characterized by the formation of oxygen "lone pair" orbitals, which can neutralize free radicals and thus can be related to antioxidant properties of the drug. In the case of paracetamol, the additional nitrogen lone pair is formed, which can explain the toxicity of the drug. We propose an evaluation method of antioxidant activity based on the relationship between experimental half-wave oxidation potential ($E_{p/2}$) and calculated ionization potentials (I_p) by DFT calculations, and can conclude that paracetamol has a higher antioxidant activity than aspirin.

Electrical Phenomena at Interfaces and Biointerfaces: Fundamentals and Applications in Nano-, Bio-, and Environmental Sciences, First Edition. Edited by Hiroyuki Ohshima.

In Chapter 32, in order to clarify the electronic state of polyaniline synthesized electrochemically, we analyzed valence X-ray photoelectron spectra (XPS) and core–electron binding energies of N, N′-diphenyl-1,4-phenylenediamine (DPPD) and polyaniline by deMon DFT calculations using the molecule and the model molecule, respectively. Polyaniline has been synthesized electrochemically from two different thin films on platinum plates in oxidation and reduction states, respectively. Valence XPS of the conductive polymers were analyzed by deMon DFT calculations using model tetramers of the four chemical structural formulae (1S, 2S, 1A, and 2A), which MacDiarmid and coworkers proposed. (In this part, the author referred from the manuscript entitled "Analysis of X-ray Photoelectron Spectra of Electrochemically Prepared Polyaniline by DFT Calculations Using Model Molecules," written by K. Takaoka, T. Otsuka, K. Naka, A. Niwa, T. Suzuki, C. Bureau, S. Maeda, K. Hyodo, K. Endo, and D.P. Chong.)

We present a study of the electronic structure, chemical bonding, and antioxidant activity of phenolic antioxidants (aspirin and paracetamol). X-ray photoelectron spectra of the antioxidants have been simulated by deMon DFT calculations for the molecules. The chemical bonding of aspirin is characterized by the formation of oxygen "lone pair" π-orbitals, which can neutralize free radicals and thus can be related to antioxidant properties of the drug. In the case of paracetamol, the additional nitrogen lone pair is formed, which can explain the toxicity of the drug. We propose an evaluation method of antioxidant activity based on the relationship between experimental half-wave oxidation potential ($E_{p/2}$) and calculated ionization potentials (I_p) by DFT calculations, and can conclude that paracetamol has the higher antioxidant activity than aspirin.

31.1 INTRODUCTION

Antioxidants are widely believed to be an important suppressant of oxidative damage, which plays an important role in a range of diseases including cancer, cardiovascular disease, arthritis, and aging [1]. Biological antioxidants are generally divided into three groups: enzymes, inhibitors of radical formation, and free radical quenching agents (such as aspirin [2]; paracetamol [3]; vitamins A, C, and E [4]; and beta-carotene). Antioxidants have the function to react and therefore neutralize with free radicals, which can attack a variety of targets including lipids, fats, and proteins, at a rate faster than the substrate. We know that the function of antioxidants relates to redox potentials, which is an important parameter for monitoring.

Some studies have been performed on the antioxidant activity that involves proton and electron transfer mechanisms. In the former case, the effectiveness of the antioxidant in ArOH depends on the bond dissociation energy (BDE) since the weaker OH bond is directly linked to the activity of the antioxidant [5–7]. For the latter electron transfer mechanism, the activity relies upon the ionization potential (I_p) [7, 8] and half-wave oxidation potential ($E_{p/2}$) [9]. Therefore, we thought that the redox potentials play an important role for the antioxidant activity in biochemistry and medicine.

In this chapter, we perform a valence XPS experiment for phenolic antioxidants (aspirin and paracetamol) in order to examine the electronic structure and chemical bonding in antioxidants. We compare the experimental valence X-ray photoelectron spectra of oxidants with simulated ones by deMon DFT calculations using molecules. Furthermore, we propose a method to evaluate antioxidant activity based on the relationship between the experimental half-wave oxidation potential ($E_{p/2}$) and the ionization potential (I_p) by DFT calculations.

31.2 EXPERIMENTAL

The XPS measurements were carried out with a PHI 5600ci multitechnique spectrometer using monocromatized Al Kα radiation (E_{exc} = 1486.6 eV). The estimated energy resolution is 0.35 eV and the base pressure in the vacuum chamber during measurements was about 5×10^{-9} torr.

X-ray photoelectron measurement of aspirin and paracetamol were carried out on single crystals prepared by the method described in Reference 10.

We used commercially available aspirin (019-12362, Wako Chemical Co. Inc.) and paracetamol (015-13942, Wako Chemical Co., Osaka, Japan), respectively. They were pressed into the pellets and were kept in the nitrogen gas atmosphere before XPS and X-ray emission spectroscopy (XES) measurements.

31.3 THEORETICAL BACKGROUND

In order to calculate the ionization potential I_p, we used the generalized transition-state (GTS) method [11] proposed by Williams and coworkers corresponding to Slater's transition-state (TS) method [12]. We consider an electron removal process for a single molecule and the ionization potential is approximated by the following equation:

$$I_j = E(1) - E(0) \approx \{F(0) + 3F(2/3)\}/4, \tag{31.1}$$

where $F(\lambda) = \partial E(\lambda)/\partial\lambda$, and $\lambda(0 < \lambda < 1)$ is assumed to be a continuous variable and represents thc fraction number of electrons removed from the Kohn–Sham molecular orbital (MO) ψ_j, with $E(0)$ and $E(1)$ denoting the total energies of the neutral molecule and the cation molecule, respectively. According to Janak's theorem [13], $F(\lambda)$ becomes the negative KS orbital energy, $\varepsilon(\lambda)$. For the calculation of I_p, the "unrestricted" generalized transition-state (uGTS) method has been applied, and for the uGTS method, two-thirds of one α electron is removed from the highest occupied KSMO, ε_{HOMO}.

31.3.1 Valence Photoelectron Spectra

In order to simulate the XPS spectra, we used the restricted generalized diffuse ionization (rGDI) method [14, 15] and did the extension of the GTS method for the vertical ionization potentials (VIPs). In the rGDI method, a two-third electron is removed evenly from all α and β valence molecular orbitals (MOs), and the VIP (I_k), which is the one electron ionization energy from the kth orbital, was approximated and obtained by Equation 31.1. The intensity of valence XPS is estimated from the relative photoionization cross section (PICS) for Al Kα radiation using the Gelius intensity model [16, 17]:

$$I_j^{XPS} = N\sum_{A,\mu}|C_{A\mu j}|^2 \sigma_{A\mu}^{AO}, \tag{31.2}$$

where μ refers to the atomic subshell on the central atom A, $|C_{A\mu j}|^2$, and $\sigma_{A\mu}^{AO}$ represents the electron density populations and the PICS of the atomic orbitals (AOs), $\varphi_{A\mu}$, respectively. For the relative atomic PICS, we use the theoretical values from Yeh [18]. In previous studies [14, 15], we construct from a superposition of peaks centered on each VIP, I_k, and each peak is represented by Gaussian line shape functions of an approximate line width (0.10 I_k).

31.3.2 Relationship between the Experimental Half-Wave Oxidation Potential and the Calculated Ionization Energy

As discussed in some studies [5–8], we know that molecular properties, BDE, and ionization potential (I_p) are most affected by the activity of phenolic antioxidants. From the viewpoint of the proton transfer pathway, the energy of dissociation reaction (ArOH $\rightarrow$ ArO$^\bullet$ + H$^\bullet$) is determined by a following equation:

$$\mathrm{BDE}_{\mathrm{ArOH}} = (E_{\mathrm{ArO}^\bullet} + E_{\mathrm{H}^\bullet}) - E_{\mathrm{ArOH}}, \tag{31.3}$$

where $E_{\mathrm{ArO}\bullet}$, $E_{\mathrm{H}\bullet}$, and E_{ArOH} are total energies of radicals (ArO$^\bullet$, H$^\bullet$) and ArOH, respectively.

For the pathway of the electron transfer, we can consider the ionization reaction (ArOH $\rightarrow$ ArOH$^+$ + e$^-$) and use the half-wave oxidation potential ($E_{p/2}$), which corresponds to the ionization potential (I_p) [9]. The energy change of the ionization reaction is the enthalpy one (ΔH) and is expressed by the following equation:

$$\Delta H = -I_p. \tag{31.4}$$

From the standpoint of change of free energy ($\Delta G°$), we can employ the free energy expression relative to the standard hydrogen electrode and the separation of the solvent effect. This results in the following equation:

$$\Delta G° = (G^{\circ}_{\mathrm{ArOH}})_{\mathrm{gas}} - (G^{\circ}_{\mathrm{ArOH^+}})_{\mathrm{gas}} + (G^{\circ}_{\mathrm{electron}})_{\mathrm{H}_g} + \Delta\Delta G^{\circ}_{\mathrm{solv}}, \tag{31.5}$$

where $\Delta\Delta G^{\circ}_{\mathrm{solv}}$ is the difference of the free energy between solvent effects of ArOH and $\mathrm{ArOH^+}$. In Equation 31.5, if the entropy term of the free energy is disregarded, Equation 31.5 can be rewritten as

$$\Delta G° = I_{\mathrm{p}} + (G^{\circ}_{\mathrm{electron}})_{\mathrm{H}_g} + \Delta\Delta G^{\circ}_{\mathrm{solv}}. \tag{31.6}$$

On the other hand, the half-wave oxidation potential $E_{\mathrm{p/2}}$ is given as

$$E_{\mathrm{p/2}} = \frac{\Delta G°}{F} - \frac{RT}{F}\ln\frac{D_{\mathrm{ArOH}}}{D_{\mathrm{ArOH^+}}}, \tag{31.7}$$

where D_{ArOH} and $D_{\mathrm{ArOH+}}$ are diffusion constants of ArOH and $\mathrm{ArOH^+}$, respectively, and F is the Faraday constant. In molecules of the hydrocarbon series, diffusion constants D_{ArOH} and $D_{\mathrm{ArOH+}}$ can be expected to have almost equal values. Thus, Equation 31.7 is expressed by

$$E_{\mathrm{p/2}} \cong \frac{1}{F}I_{\mathrm{p}} + C \quad \left(C = \frac{(G^{\circ}_{\mathrm{electron}})_{\mathrm{Hg}} + \Delta\Delta G^{\circ}_{\mathrm{solv}}}{F} \right), \tag{31.8}$$

where C will be a constant when both terms, $G^{\circ}_{\mathrm{electron}}$ and $\Delta\Delta G^{\circ}_{\mathrm{solv}}$, are constant.

Therefore, Equation 31.8 denotes a linear relationship between $E_{\mathrm{p/2}}$ and I_{p}.

31.4 COMPUTATIONAL DETAILS

Aspirin and paracetamol molecules (Fig. 31.1) and the radicals were geometry optimized by a semiempirical AM1 method [19, 20]. The deMon-KS program calculations [21, 22] were performed with the exchange-correlation potential labeled as B88/P86, made from Becke's exchange functional [23] and Perdew's correlation functional [24]. In the deMon-KS program, we used an "extrafine" and "nonrandom" grid and various basis sets to calculate BDE and I_{p} of the model molecules with auxiliary fitting functions labeled (3,1; 3,1) for H, (4, 4; 4, 4) for C, N, O, F, and (5, 4; 5, 4) for S and Cl.

We use various basis sets, STO-3G, the double-ζ valence polarization (DZVP), DZVP2, the triple-ζ valence polarization (TZVP), and correlation-consistent polarized valence triple-ζ (cc-pVTZ). We performed

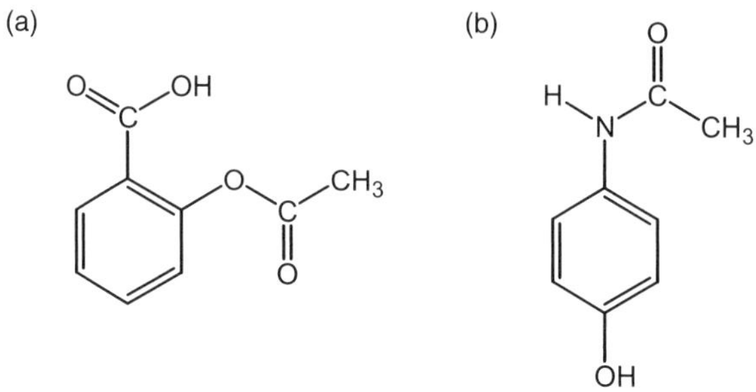

Figure 31.1. Chemical structure of (a) aspirin; (b) paracetamol.

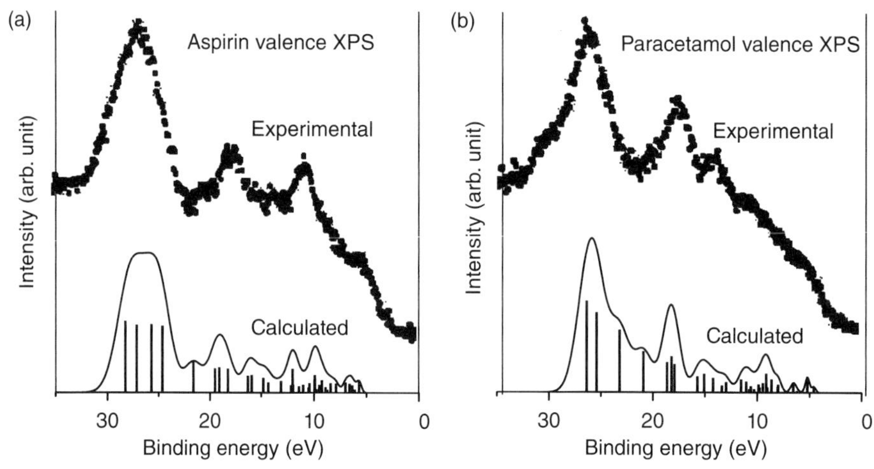

Figure 31.2. Simulated valence XPS and experimental spectra of (a) aspirin; (b) paracetamol.

test calculations with various bases to determine the dependency of the basis sets for BDE and I_p.

All calculations of the BDE and ionization potential I_p were performed by the deMon-KS DFT programs.

31.5 RESULTS AND DISCUSSION

31.5.1 Electronic State of Aspirin and Paracetamol

Figure 31.2 shows a comparison of simulated valence photoelectron spectra of aspirin and paracetamol obtained by deMon DFT calculations with the experimental ones. For aspirin, the details of the observed and calculated peaks were summarized with the main contributions of atomic PICS, orbital nature, and the functional groups in Table 31.1. The peak around 27.0 eV in Figure 31.2a is due to the sσ(O2s–C2s) bonding of (–O–C) functional groups, and the peak at around 18.0 eV results from sσ(C2s–C2s) bonding of benzene rings. The lower broadening peak in the range of 12–16 eV is divided into two components, and the peaks at around 15.0 and 13.5 eV are due to the pσ(C2s–C2p) and pσ(C2p–C2p) bonding of the benzene ring, respectively. The peak at around 10.5 eV is mainly due to the pπ(C2p–C2p) bonding of the benzene ring and pπ(C2p–O2p) of (–C=O) double bonding.

In the case of paracetamol in Figure 31.2b, the peaks at around 26.0 and 17.5 eV are due to sσ(O2s-, N2s-C2s) bonds and sσ(C2s–C2s) bonds in the benzene rings, respectively. The peak at around 6.0 eV is due not only to lone pairs of oxygen but also to lone pairs of nitrogen atoms. The details of the valence photoelectron spectra of paracetamol are shown in Table 31.2.

31.5.2 Evaluation of Antioxidant Activity

We examine the dependency of basis sets in the DFT calculations for BDE and ionization potential (I_p) of small molecules in order to obtain reasonable values of the BDE and I_p [25–30]. Table 31.3 shows the calculated BDE values with the experimental ones. In the table, we obtained 2.1 kcal/mol as the average absolute deviation (AAD) value of BDE obtained from Equation 31.3 by the DFT calculations with DZVP basis, although the value of BDE calculations with the cc-pVTZ basis set is 5.9 kcal/mol. In Table 31.4, we give the calculated I_p for aromatic compounds and the experimental values [31–41]. The I_p values are evaluated from using the uGTS method with DZVP basis. The best result was also 0.16 eV for the ADD with the DZVP basis. We determined the BDE and I_p of aspirin and paracetamol by the DFT calculations with DZVP basis.

As described by Equation 31.8 and as mentioned earlier, the relationship between the half-wave oxidation potential ($E_{p/2}$) and I_p is linear. In Figure 31.3, we show the experimental half-wave oxidation potential ($E_{p/2}$) [9] and our calculated ionized potentials for many aromatic compounds. In the figure, a linear plot of $E_{p/2}$ versus I_p was, thus, given as

$$E_{p/2} = 0.320 I_p - 1.962. \tag{31.9}$$

From Equation 31.9, we are able to predict two components ($E_{p/2}$ values and antioxidant activity) of the phenolic compounds. Note that the aromatic compounds with high antioxidant activity have low $E_{p/2}$ values [9]. Table 31.5 summarizes the calculated values of BDE, I_p, and $E_{p/2}$ for aspirin and paracetamol. In the table, the $E_{p/2}$ values are obtained to be 0.91 and 0.29 V from the I_p values of 8.98 and 7.04 eV for aspirin and paracetamol, respectively. From our

TABLE 31.1. Observed Peaks, VIPs, Main AO Photoionization Cross Section (PICS), Orbital Nature, and the Functional Group for Valence XPS of Aspirin

Peaks (eV)	VIPs (eV)	Main AO PICS	Orbital Nature[b]	Functional Group
27.0	{31.3; 30.2;	O2s	sσ(O2s–C2s)–B	–O–C
(22.0–30.0)[a]	28.7; 27.7}			
18.0	{21.2–24.7}	C2s	sσ(C2s–C2s)–B	–C=C (benzene ring)
(16.0–22.0)[a]				–C–C
15.0	{17.4–19.3}	O2p, C2s	pσ(O2p–C2s)–B	–O–C
(14.0–16.0)[a]		C2s, C2p	pσ(C2s–C2p)–B	–C=C (benzene ring)
13.5	{14.1–16.2}	C2p	pσ(C2p–C2p)–B	–C=C (benzene ring)
(12.0–14.0)[a]		C2s, C2p	pσ(C2p–C2s, C2p)–B	–C–C
10.5	{11.8–13.5}	O2p, C2p	pσ(O2p–C2p)–B	–C=O
(8.0–12.0)[a]		C2p	pπ(C2p–C2p)–B	–C=C (benzene ring)
		O2p, C2p	pπ(O2p–C2p)–B	–C=O
5.5	{8.7–11.4}	C2p	pπ(C2p–C2p)–B	–C=C (benzene ring)
(4.0–7.0)[a]		O2p	pπ(O2p)–NB	–O–, –C=O (lone pair)

[a] Shows the peak range (the shift between the observed and calculated VIP is 3.0 eV).
[b] B and NB mean bonding and nonbonding, respectively.

TABLE 31.2. Observed Peaks, VIPs, Main AO Photoionization Cross Section (PICS), Orbital Nature, and the Functional Group for Valence XPS of Paracetamol

Peaks (eV)	VIPs (eV)	Main AO PICS	Orbital Nature[b]	Functional Group
26.0	{29.4; 28.5;	O2s, C2s	sσ(O2s–C2s)–B	–O–C
(23.0–30.0)[a]	26.2}	N2s, C2s	sσ(N2s–C2s)–B	–N–C
17.5	{23.9; 21.7;	C2s	sσ(C2s–C2s)–B	–C=C (benzene ring)
(16.0–19.0)[a]	21.2; 21.0}			–C–C
14.0	{16.0–18.8}	O2p, C2s, C2p	pσ(O2p–C2s, C2p)–B	–O–C
(12.0–15.5)[a]		C2s, C2p	pσ(C2s–C2p)–B	–C=C (benzene ring)
11.0		N2p, C2s, C2p	pσ(N2p–C2s, C2p)–B	–N–C
(8.0–12.0)[a]	{12.9–14.6}	C2s, C2p	pσ(C2s–C2p)–B	–C–C
6.0		C2p	pσ(C2p–C2p)–B	–C=C (benzene ring)
(4.0–8.0)[a]	{11.0–12.8}	C2p	pπ(C2p–C2p)–B	–C=C (benzene ring)
		O2p, C2p	pπ(O2p–C2p)–B	–C=O
	{9.6; 9.2}	O2p	pπ(O2p)–NB	–O– (lone pair)
	{8.3; 7.6}	N2p	pπ(N2p)–NB	–N– (lone pair)

[a] Shows the peak range (the shift between the observed and calculated VIP is 3.0 eV).
[b] B and NB mean bonding and nonbonding, respectively.

TABLE 31.3. Calculated BDE (in kcal/mol) with Experimental Ones for Various Basis Sets of Small Molecules

Molecule	Exp. (Reference)			Calc.							
				DZVP		DZVP2		TZVP		cc-pVTZ	
HS–H	91.2	±0.7	[28]	92.6	(1.4)	97.3	(6.1)	97.2	(6.0)	97.6	(6.4)
NH_2–H	108.2	±0.3	[27]	111.3	(3.1)	118.3	(10.1)	118.3	(10.1)	117.1	(8.9)
CH_3O–H	104.0	±1.0	[29]	103.1	(0.9)	113.2	(9.2)	110.4	(6.4)	110.1	(6.1)
C_2H_5O–H	104.6	±0.8	[30]	101.5	(3.1)	109.2	(4.6)	108.9	(4.3)	108.7	(4.1)
C_2H_5S–H	87.3	±0.6	[28]	87.7	(0.4)	92.7	(5.4)	92.5	(5.2)	92.8	(5.5)
CH_3NH–H	100.0	±2.5	[28]	95.5	(4.5)	103.3	(3.3)	103.7	(3.7)	102.0	(2.0)
HOO–H	88.2	±1.0	[28]	89.2	(1.0)	97.8	(9.6)	96.5	(8.3)	94.5	(6.3)
CH_3OO–H	84.0	±2.0	[29]	86.5	(2.5)	94.1	(10.1)	93.0	(9.0)	91.0	(7.0)
C_6H_5O–H	87.3	±1.0	[31]	85.1	(2.2)	92.6	(5.3)	92.6	(5.3)	92.5	(5.2)
C_6H_5S–H	79.1	±2.0	[30]	80.4	(1.3)	85.2	(6.1)	85.4	(6.3)	85.7	(6.6)
C_6H_5NH–H	92.3	±2.0	[32]	89.8	(2.5)	96.7	(4.4)	97.4	(5.1)	96.7	(4.4)
$C_6H_5CH_2$–H	88.5	±1.5	[27]	92.6	(4.1)	95.9	(7.4)	95.2	(6.7)	94.5	(6.0)
AAD				2.1		6.9		6.5		5.9	

TABLE 31.4. Calculated I_p (in Electronvolt) with Experimental Ones for Various Basis Sets of Small Molecules

			Calc.							
Molecule	Exp. (Reference)		DZVP		DZVP2		TZVP		cc-pVTZ	
C_6H_6	9.24	[34]	9.30	(0.06)	19.20	(0.04)	9.25	(0.00)	9.22	(0.03)
C_6H_5OH	8.51	[35]	8.46	(0.05)	8.39	(0.12)	8.41	(0.10)	8.35	(0.16)
C_6H_5SH	8.30	[36]	8.08	(0.22)	8.04	(0.26)	8.05	(0.25)	8.01	(0.29)
$C_6H_5NH_2$	7.72	[37]	7.59	(0.13)	7.53	(0.19)	7.56	(0.16)	7.50	(0.22)
$C_6H_5CH_3$	8.83	[38]	8.83	(0.00)	8.75	(0.08)	8.78	(0.05)	8.74	(0.09)
$C_6H_5NO_2$	9.93	[39]	9.68	(0.25)	9.62	(0.31)	9.57	(0.36)	9.52	(0.41)
C_6H_5CN	9.73	[40]	9.63	(0.10)	9.58	(0.15)	9.59	(0.14)	9.55	(0.18)
$C_6H_5CF_3$	9.69	[41]	9.90	(0.21)	9.86	(0.17)	9.86	(0.17)	9.77	(0.08)
C_6H_5CHO	9.49	[42]	9.05	(0.44)	8.97	(0.52)	8.83	(0.66)	9.09	(0.40)
C_6H_5COOH	9.47	[42]	9.21	(0.26)	9.13	(0.34)	8.46	(1.01)	9.13	(0.34)
C_6H_5Cl	9.06	[42]	8.93	(0.13)	8.86	(0.20)	8.87	(0.19)	8.83	(0.23)
$C_6H_5OCH_3$	8.21	[42]	8.24	(0.03)	8.17	(0.04)	8.19	(0.02)	8.14	(0.07)
AAD			0.16		0.20		0.26		0.20	

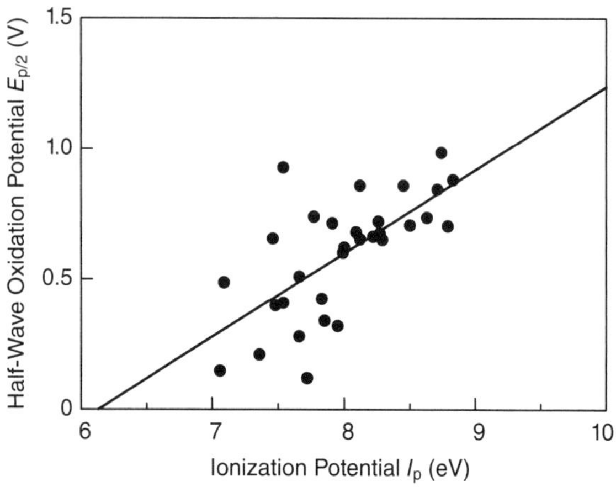

Figure 31.3. Correlation between experimental halfwave oxidation potential ($E_{p/2}$) and calculated ionization potential (I_p) for phenolic compounds.

TABLE 31.5. Calculated Bond Dissociation Energy (BDE) and Ionization Potential (I_p) of Aspirin and Paracetamol, and the Half-Wave Oxidation Potential Obtained by Equation 31.9

Molecules	BDE (kcal/mol)	I_p (eV)	$E_{p/2}$ (V)
Aspirin	97.79	8.98	0.91
Paracetamol	79.11	7.04	0.29

calculations, paracetamol has higher antioxidant activity than aspirin. This result corresponds well with experimental study [42].

31.6 CONCLUSION

We performed a study on the electronic state of phenolic antioxidants (aspirin and paracetamol) from a combination analysis for observed X-ray photoelectron spectra and simulated spectra by deMon DFT calculations using molecules. Furthermore, the antioxidant activity for the phenolic compounds was discussed in comparison with experimental BDE and ionization potential (I_p) with calculated values by DFT calculations using simple molecules:

1. Simulated X-ray valence photoelectron spectra and emission spectra of aspirin and paracetamol obtained by deMon DFT calculations showed

good agreement with our observed spectra. The chemical bonding of aspirin is governed by the formation of oxygen lone pair π-orbitals, which can neutralize free radicals and thus are important for the antioxidant property of the drug. In the case of paracetamol, an additional nitrogen lone pair is formed, which can explain the toxicity of the drug.

2. We proposed an evaluation method of antioxidant activity due to the relationship between the experimental half-wave oxidation potential ($E_{p/2}$) and the calculated ionization potential (I_p) by DFT calculations and showed that paracetamol has higher antioxidant activity than aspirin.

REFERENCES

1 J. P. Kehrer. *Crit. Rev. Toxicol.* 23 (1993) 21.

2 X. Shi, M. Ding, Z. Dong, F. Chen, J. Ye, S. Wang, S. S. Leonard, V. Castranova, V. Vallyathan. *Mol. Cell. Biochem.* 199 (1999) 93.

3 G. F. Merrill, E. Goldberg. *Basic Res. Cardiol.* 96 (2001) 423.

4 M. Marquez, C.E. Yepez, R. Sutil-Naranjo, M. Rincon. *Invest. Clin.* 43 (2002) 191.

5 M. Lucarini, G. F. Pedulli, M. Cipollone. *J. Org. Chem.* 59 (1994) 5063.

6 G. A. DiLabio, D. A. Pratt, A. D. LoFaro, J. S. Wright. *J. Phys. Chem. A* 103 (1999) 1653.

7 J. S. Wright, E. R. Johnson, G. A. Dilabio. *J. Am. Chem. Soc.* 123 (2001) 1173.

8 G. A. DiLabio, D. A. Pratt, J. S. Wright. *Chem. Phys. Lett.* 311 (1999) 215.

9 I. Vedernikova, J. P. Tollenaere, A. Haemers. *J. Phys. Org. Chem.* 12 (1999) 144.

10 R.A. Granberg. *J. Cryst. Growth* 198–199 (1999) 1287.

11 A. R. Williams, R. A. deGoot, C. B. Sommers. *J. Chem. Phys.* 63 (1975) 628.

12 J. C. Slater. *Advan. Quantum Chem.* 6 (1972) 1.

13 J. F. Janak. *Phys. Rev. A* 18 (1978) 7165.

14 K. Endo, Y. Kaneda, H. Okada, D. P. Chong, P. Duffy. *J. Phys. Chem.* 100 (1996) 19455.

15 T. Otsuka, K. Endo, M. Suhara, D. P. Chong. *J. Mol. Struct.* 522 (2000) 47.

16 U. Gelius, K. Siegbahn. *Faraday Discuss. Chem. Soc.* 54 (1972) 257.

17 U. Gelius. *J. Electron Spectrosc. Relat. Phenom.* 5 (1974) 985.

18 J.-J Yeh. *Atomic calculation photoionization cross section and asymmetry parameters*. Gordon and Breach, Newark, NJ, 1993.

19 M. J. S. Dewar, E. G. Zoebisch, E. F. Healy, J. J. P. Stewart. *J. Am. Chem. Soc.* 107 (1985) 180.

20 M. J. S. Dewar, E. G. Zoebisch. *Theochem* 1 (1988) 180.

21 A. St Amant, D. R. Salahub. *Chem. Phys. Lett.* 169 (1990) 387.

22 A. St-Amant. PhD thesis, University of Montreal, 1991.

23 A. D. Becke. *Phys. Rev. A* 38 (1988) 3098.

24 J. P. Perdew. *Phys. Rev. B* 33 (1986) 8822.

25 J. Berkowitz, G. B. Ellison, D. Gutman. *J. Phys. Chem.* 94 (1994) 2744.

26 D. R. Lide (ed.), *CRC handbook of chemistry and physics*, 76th ed. CRC Press, Boca Raton, FL, 1995.

27 D. F. McMillen, D. M. Golden. *Ann. Rev. Phys. Chem.* 33 (1982) 493.

28 F. G. Bordwell, X.-M. Zhang, A. V. Satish, J. -P. Cheng. *J. Am. Chem. Soc.* 116 (1994) 6605.

29 D. D. M. Wayner, E. Lusztyk, D. Page, K. U. Ingold, P. Mulder, L. J. J. Laarhoven, H. S. Aldrich. *J. Am. Chem. Soc.* 117 (1995) 8737.

30 F. G. Bordwell, X. -M. Zhang, J. -P. Cheng. *J. Org. Chem.* 58 (1993) 6410.

31 C. R. Brundle, M. B. Robin, N. A. Kuebler, H. J. Basch. *J. Am. Chem. Soc.* 94 (1972) 1451.

32 G. I. Nemeth, H. L. Selzle, E. W. Schlag. *Chem. Phys. Lett.* 215 (1993) 151.

33 R. J. Lipert, S. D. Colson. *J. Chem. Phys.* 92 (1990) 3240.

34 J. D. Faulk, R. C. Dunbar, C. Lifshitz. *J. Am. Chem. Soc.* 112 (1990) 7893.

35 J. Hager, M. Smith, S. Wallace. *J. Chem. Phys.* 83 (1985) 4820.

36 K. -T Lu, G. C. Eiden, J. C. Weisshaar. *J. Phys. Chem.* 96 (1992) 9742.

37 J. W. Rabalais. *J. Chem. Phys.* 57 (1972) 960.

38 M. Araki, S. I. Sato. *J. Phys. Chem.* 100 (1996) 10542.

39 D. W. Berman, D. S. Bomse, J. L. Beauchamp. *Int. J. Mass Spectrom. Ion Phys.* 39 (1981) 263.

40 R. G. McLoughlin, J. C. Traeger. *Org. Mass Spectrom.* 14 (1979) 434.

41 S. G. Lias, J. E. Bartmess, J. F. Liebman, J. L. Holmes, R. D. Levin, W. G. Mallard. *J. Phys. Chem. Ref. Data* 17(Suppl. 1) (1988).

42 D. S. Maharaj, K. S. Saravanan, H. Maharaj, K. P. Mohanakumar, S. Deya. *Neurochem. Int.* 44 (2004) 355.

32

QUANTUM ELECTROCHEMICAL STUDY OF BENZENE DERIVATIVES

2. ANALYSIS OF X-RAY PHOTOELECTRON SPECTRA OF ELECTROCHEMICALLY PREPARED POLYANILINE BY DFT CALCULATIONS USING MODEL MOLECULES

Kazunaka Endo

In order to clarify the electronic state of polyaniline synthesized electrochemically, we analyzed the valence X-ray photoelectron spectra (XPS) and the core–electron binding energies (CEBEs) of N, N′-diphenyl-1,4-phenylenediamine (DPPD) and polyaniline by deMon density functional theory (DFT) calculations using the molecule and the model molecule, respectively. Polyaniline has been synthesized electrochemically from two different thin films on platinum plates in oxidation and reduction states, respectively. The valence XPS of the conductive polymers were analyzed by deMon DFT calculations using model tetramers of the four chemical structural formulae (1S, 2S, 1A, and 2A), which MacDiarmid and coworkers proposed. We can conclude that the oxidized polyaniline corresponds to the 2A type, while the reduction form approximated the 1S type.

Electrical Phenomena at Interfaces and Biointerfaces: Fundamentals and Applications in Nano-, Bio-, and Environmental Sciences, First Edition. Edited by Hiroyuki Ohshima.

32.1 INTRODUCTION

Polyaniline became an important material of various conductive polymers, which attracted considerable attention due to its high electrical conductivity, good electrical and chemical stability at ambient conditions, and electrochemical redox behavior. The polymer shows great potential for use in a number of device applications including electrodes for rechargeable batteries [1], electrochromic displays [2], photovoltaics [3], electrochemical switches [4], and chemical sensors [5].

Polyaniline can be made electrochemically in aqueous or organic solutions [6–8], as well as chemically [9]. The chemical properties of polyaniline include both redox and acid–base reactivity, with the scheme "1S" ⇔ "2S" ⇔ "2A" ⇔ "1A" ⇔ "1S" proposed by MacDiarmid and coworkers [10]. The intrinsic oxidation state and protonation characteristics of polyaniline were shown to be readily elucidated using X-ray photoelectron spectroscopy [11–17]. Few of these works provided detailed analyses of the XPS, although the technique was also used in some studies of polyaniline derivatives. This study, thus, aims to clarify the electronic state of polyaniline films in oxidation and reduction types from the analysis of the experimental XPS by DFT calculations using model molecules.

In previous works [18–24] using the deMon DFT program [25], which uses the idea of transition state [26], we analyzed the XPS of 70 polymers by the DFT calculations using model oligomers. The studies showed excellent assignments of XPS of polymers using deMon DFT calculations because we used the energy shift ***WD*** (work function [***W***] and other energies [delta, ***D***]) to account for solid-state effects. The simulation of the valence spectra was performed on the model molecules using standard convolution with Gaussian line shape and using the Gelius intensity model [27] for molecular photoionization cross sections. The line width of each peak of ionization energy I_k was taken to be 0.10 I_k (proportional to the ionization energy) as stated in previous works [18, 28].

In this study, we synthesized polyaniline electrochemically from two different thin films on platinum plates in oxidation and reduction types. The valence XPS of the conductive polymer films were analyzed by deMon DFT calculations using the model tetramers of the structural formulae, which MacDiarmid and coworkers proposed. We will determine the reasonable formula for the oxidation and reduction of polyaniline films from the four chemical formulae. For DPPD as the polyaniline oligomer, we will calculate the valence XPS and CEBEs of C1s and N1s by the deMon DFT program in order to clarify the electronic state of polyaniline synthesized electrochemically. Furthermore, the CEBEs of the reasonable model molecules for oxidation and reduction of polyaniline films are calculated by the deMon program with polarized valence double-ς (DZVP) basis. With the reasonable ***WD*** values of DPPD, oxidation and reduction of polyaniline films will be estimated from the difference between the calculated and experimental CEBEs of the compounds.

32.2 EXPERIMENTAL

All the chemicals used were commercially available reagent grade and were used without further purification. Polyaniline films were deposited on platinum foils as a working electrode by sweeping the potential of the electrode in 100 mL of 0.5 M H_2SO_4 (1 M = 1 mol/dm^3) aqueous solution containing 1 mL of aniline. The size of platinum foils was 1 cm^2. The potential was cycled between –200 and 750 mV (vs. saturated calomel electrode) at the scanning rate of 50 mV/s for the synthesis of polyaniline. After the synthesis at 750 mV for 4 minutes (less than 10 mA of the current level), the films were kept at 750 mV in 0.5 M H_2SO_4 aqueous solution for 2 minutes for polyaniline to produce the oxidized types. For the reduction types, the film was held at –200 mV for polyaniline in the same solution for 1 minute. The samples in both states were immersed in an acetonitrile solution to rinse the surface. After that, we brought each sample to the spectrometer for a few minutes to measure its XPS.

The XPS of the films were obtained on a PHI 5400MC ESCA spectrometer using monochromatized Al Kα radiation. The spectrometer was operated at 600 W, 15 kV, and 40 mA. Photon energy was 1486.6 eV. A pass energy of 37.75 eV was used for high resolution scans in a valence band analysis (50-eV range). The angle between the X-ray source and analyzer was fixed at 45°. The spot size was 3 × 1 mm. Dispersion compensation yielded an instrumental resolution of 0.5 eV from full width at half-maximum for the Ag3d line of silver. Multiscan averaging on a multichannel analyzer was used for the valence band region, although a very low photoelectron emission cross section was observed in this range. The thickness of the films was estimated as about 0.5 μm from scanning electron microscope (SEM) observation.

32.3 THEORETICAL BACKGROUND

For the comparison between calculations for a single molecule of the model and experiments on a solid, we must shift each computed vertical ionization potential (VIP), I'k, by a quantity ***WD*** as I_k(EF) = I'k – ***WD***, to convert to ionization energy I_k(EF) relative to the Fermi level. This quantity ***WD*** denotes the sum of the work function of the sample and other energy effects, as stated in our previous studies [18, 28].

32.3.1 Core–Electron Binding and VIPs for XPS

In order to obtain accurate CEBEs, we used the generalized transition-state (GTS) method. In the GTS method, Williams and coworkers [29] proposed the extension of Slater's transition-state method [26] and approximated the endothermicity $\Delta E = E(1) - E(0)$ by

$$\Delta E = [F(0) + 3F(2/3)]/4, \tag{32.1}$$

where $F(x) = \partial E(x)/\partial x$, and $x(0 < x \leq 1)$ is assumed to be a continuous variable, with $E(0)$ and $E(1)$ denoting the energies of the initial and final states, respectively. For example, for the ionization of an electron from molecular orbital (MO) ϕ_k of interest, x represents the fraction of electron removed and, according to the Janak theorem [30], $F(x)$ is the negative orbital energy $\varepsilon_k(x)$. This procedure is applied in the following way. In the unrestricted generalized transition-state (uGTS) method, we removed two-third α electron from MO ϕ_k of interest.

For the VIPs of the valence regions, we use the so-called restricted diffuse ionization (rDI) model, which Asbrink et al. [31] proposed in the HAM/3 method. In the rDI model, half of an electron is removed evenly from the valence MOs and the negative of the resulting orbital energies corresponds to calculated VIPs. This allows us to obtain all the valence VIPs in a single calculation.

32.3.2 Intensity of Valence XPS

The intensity of valence XPS was estimated from the relative photoionization cross section for Al Kα radiation using the Gelius intensity model [27]. For the relative atomic photoionization cross section, we used the theoretical values from Yeh [32].

32.4 CALCULATIONS

It is generally accepted that geometry optimized by the AM1 method [33] corresponds well to the experiment. In the case of solving the nonlinear equations involving a Hartree–Fock-like DFT one, we think it is the best way to select a better initial condition. As indicated in previous works [18–24], we have performed the better assignments of XPS for 70 polymers by DFT calculations using the model molecules after the geometric optimization with a semiempirical AM 1 method [33].

The model tetramers of four types (1S, 1A, 2S, and 2A) for polyaniline were calculated by deMon DFT programs. For the geometry of the molecules in Figure 32.1, we used the optimized Cartesian coordinates from a semiempirical AM1 (version 6) method [33]. DeMon calculations were performed with the exchange-correlation potential labeled as B88/P86, made from Becke's exchange functional [34] and Perdew's correlation functional [35]. In running the deMon program, we used a nonrandom grid and DZVP basis of (621/41/1*) for C and N, (41) for H with auxiliary fitting functions labeled (4, 4; 4, 4) for C and O, and (3, 1; 3, 1) for H.

In order to compare the theoretical valence XPS for DPPD with the experimental one, we also used the DZVP basis. We calculated the CEBEs of aniline, DPPD, and reasonable model molecules for the oxidation and reduction of polyaniline films using the uGTS model. For aniline and DPPD, we tested the

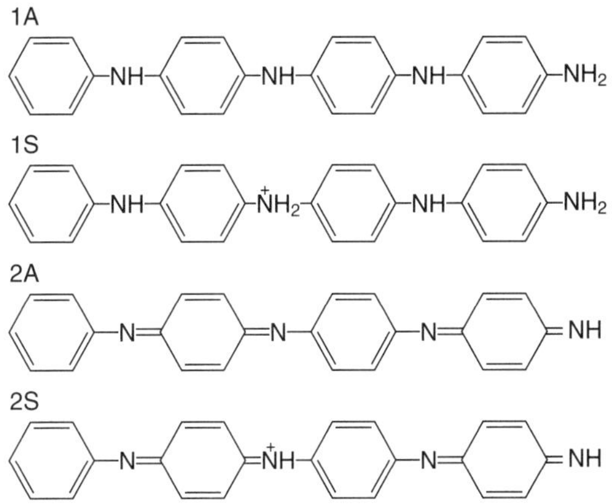

Figure 32.1. Chemical structure of model tetramer molecules in four types (1S, 1A, 2S, 2A) for polyaniline.

basis set dependence with a much better basis set, the scaled polarized valence triple-zeta (scaled pVTZ) set [36]. For a 1S-type molecule in the reduced film, we calculated the CEBEs by adding the chloride anion near the nitrogen cation to reproduce the electric neutral charge in the experimental reduced polyaniline film. In the case of the DPPD and aniline, we cited the experimental CEBEs by Beamson and Briggs [37] and Jolly and coworkers [38], respectively.

32.5 RESULT AND DISCUSSION

A feature of this study is the use of the rDI model using the deMon program for the simulation of the valence spectra of polyaniline and DPPD since we were already able to calculate the valence XPS of 70 polymers. The main purpose of our work is to clarify the electronic state of polyaniline films in oxidation and reduction types from the analysis of the experimental XPS by deMon DFT calculations using the model molecules. On the other hand, we calculate accurate CEBEs of aniline, DPPD, and reasonable model molecules of polyaniline in oxidized and reduced types by the deMon DFT program using scaled pVTZ (scaled pVTZ and DZVP) and DZVP bases, respectively, in order to obtain the reliable ***WD*** of DPPD and polyaniline in oxidized and reduced types from the difference between the calculated CEBEs and experimental values.

32.5.1 DPPD and Aniline

We showed the theoretical valence spectrum of DPPD by deMon DFT calculations with the experimental one in Figure 32.2. Our simulation for the valence

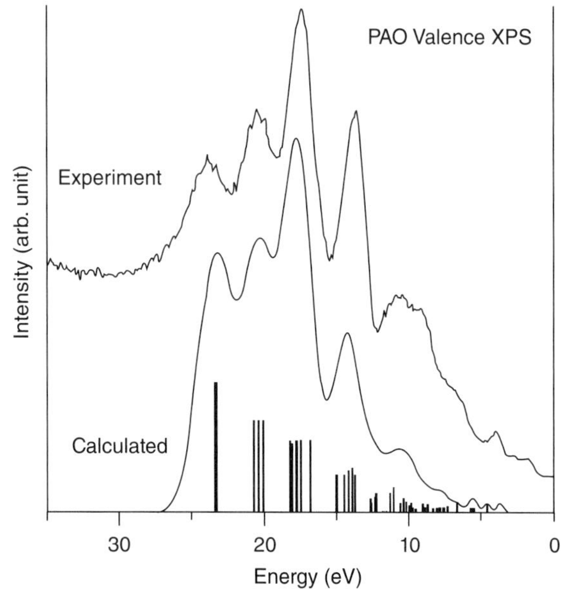

Figure 32.2. Theoretical valence X-ray photoelectron spectrum of DPPD calculated by deMon DFT program with experimental valence spectrum obtained by Beamson and Briggs..

spectra of DPPD using deMon calculation is in good accordance with the experimental spectra [37]. Table 32.1 summarized the experimental peaks, calculated VIPs, main atomic orbital (AO) photoionization cross sections, the orbital nature, and the functional group.

These detailed simulations of the valence spectra for DPPD is very useful in clarifying the electronic state of polyaniline synthesized electrochemically from two different thin films on platinum plates in oxidation and reduction types.

In recent studies using the deMon DFT program, Chong and coworkers [36, 39–41] offered the method of calculating accurate CEBEs of isolated gas-phase molecules. Then, we calculated the CEBEs of aniline and DPPD using the uGTS model with DZVP and pVTZ bases in Table 32.2. The calculated CEBEs of aniline using DZVP and pVTZ are in a good accordance with the experimental values [37]. The latter calculated values also correspond very well to the values calculated by Takahata and Chong [42]. For DPPD, we calculated the CEBEs of the N1s and six nonequivalent (C1s)s in a gas molecule and cited the experimental CEBEs of the N1s and two kinds of (C1s)s in the solid state. In order to obtain the reasonable ***WD*** value, we showed the averaged calculation values of CEBEs in the last column in Table 32.2. Thus, the ***WD*** of DPPD is estimated as 5.1 eV from the difference between the calculated CEBEs of (N1s and C1s) and the experimental CEBEs.

TABLE 32.1. Calculated Peaks, VIP, Main AO Photoionization Cross Section, Orbital Nature, and the Functional Group for Valence XPS of DPPD

Observed Peak (eV)	Calculated Peak (VIP) (eV)	Main AO Photoionization Cross Section	Orbital Nature	Functional Group
24.0 (22–26)	27.8, 26.9	$C_{2S}(0.5)$, $N_{2S}(0.5)$	sσ(C_{2s}-C,N_{2s})-B	C–C, C–N
20.0	24.4	C_{2s}	sσ(C_{2s}-C_{2s})-B	C–C
(19–22)	23.7	$C_{2s}(0.9)$, $N_{2s}(0.1)$	sσ(C_{2s}-C,N_{2s})-B	C–C, C–N
	23.1	C_{2s}	sσ(C_{2s}-C_{2s})-B	C–C
18.0	20.9, 20.8, 20.4	C_{2s}	sσ(C_{2s}-C_{2s})-B	C–C
(15–19)	20.0, 19.4	C_{2s}, N_{2s}	sσ(C_{2s}-C,N_{2s})-B	C–C, C–N
	18.5	C_{2s}, C_{2p}	pσ(C_{2s}-C_{2p})-B	C–C
13.0	16.5, 16.3	C_{2s}, C_{2p}	pσ(C_{2s}-C_{2p})-B	C–N
(12–15)	16.2, 15.9, 15.7	C_{2s}, C_{2p}, N_{2p}	pσ(C_{2s}-C, N_{2p})-B	C–C–N
	14.3, 14.1, 13.8	C_{2p}, N_{2p}	pσ, π(C_{2p}-C,N_{2p})-B	C=N, C=C
10.0	13.6, 13.5	C_{2s}, C_{2p}	pσ(C_{2s}-C_{2p})-B	C–C
(6.0–12)	12.8–12.1	C_{2p}, N_{2p}	pσ, π(C_{2p}-C,N_{2p})-B	C=C, C=N
	11.4, 11.3	C_{2p}	pπ(C_{2p}-C_{2p})-B	C=C
4.0 (4–6)	11.0–10.2 9.9–9.6	C_{2p}, N_{2p}	pπ(C_{2p}-C, N_{2p})-B	C=C,C=N
2.5 (2–4)	8.4, 8.3	N_{2p}	pπ(lone pair)-NB	–N–

WD (difference between observed and calculated peaks) was assumed as 3.5 eV.

32.5.2 Polyaniline

In Figure 32.3a,b, the valence spectrum of reduced polyaniline films seems to correspond well to that spectrum of DPPD, although the spectrum of polyaniline is much broader than that of DPPD. In the figure, we can see the characteristic differences between the valence spectra of model tetramers in 1A and 1S types as introduced by McDiarmid and coworkers. The spectrum of the model in the 1S type is much closer to the observed one in the reduction type than that of 1A: The spectrum in the 1S type has the most intense peak at around 18 eV and is broader owing to the positive charge effect than that of the 1A type.

For oxidized polyaniline film, we observed a broader valence spectrum due to the positive charge effects in Figure 32.4a,b. By analyzing the valence spectrum of DPPD, in the oxidized polyaniline film, the most intensive peak at around 24 eV is due to photoionization cross sections from {sσ(C2s-C2s), sσ(C2s-N2s)} bonding orbitals, which results from C-C and C-N functional groups, and the next intense peak at around 18 eV is related mainly with the sσ(C2s-C2s) bonding orbital and partially with {sσ(C2s-N2s) and pσ(C2s-C2p)} bonding orbitals of C-C and (C-N, C-C) groups, respectively. In Figure 32.4a,b, we can also see the characteristic differences between the valence spectra of model tetramers in 2A and 2S types. The spectrum of the model

TABLE 32.2. Observed and Calculated CEBEs of Aniline and DPPD

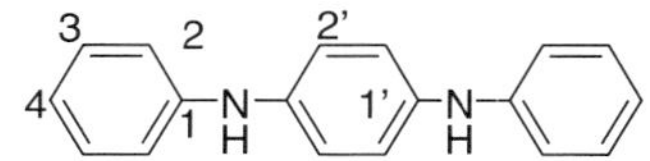

	Aniline					DPPD						
		Calc.					Calc.		(Calc.)$_{av}$			
	Obsd (eV)	DZVP	Delta$_1$[b]	pVTZ	Delta$_1$[b]	Obsd	DZVP	PVTZ	DZVP	Delta$_2$[c]	PVTZ	Delta$_2$[c]
CEBE(N1s)												
[–N–C]	405.3	405.18	0.1	405.38	0.1	399.9	404.96	405.21	405.0	0.0	405.2	0.2
CEBE(C1s)												
C1	291.3	291.56	0.3	291.35	0.1	285.9	291.23	291.11	291.1[a]	0.1	291.0[a]	0.0
C1'							290.95	290.83				
C2	289.9	290.10	0.2	289.92	0.0	284.7	289.79	289.65	289.9[a]	0.1	289.7[a]	–0.1
C2'							289.88	289.74				
C3	290.2	290.26	0.1	290.11	–0.1		290.10	289.94				
C4	289.7	289.86	0.2	289.74	0.0		289.59	289.48				
Average CEBE(C')												
	290.0	290.1	0.1	290.0	0.0							

[a] Value was given as the averaged CEBEs.
[b] Delta$_1$ is the deviation between observed (obsd) and calculated (calc.) CEBEs.
[c] Delta$_2$ is the deviation between observed and modified calculated CEBEs ((calc.)$_{av}$ – 5.1 eV).

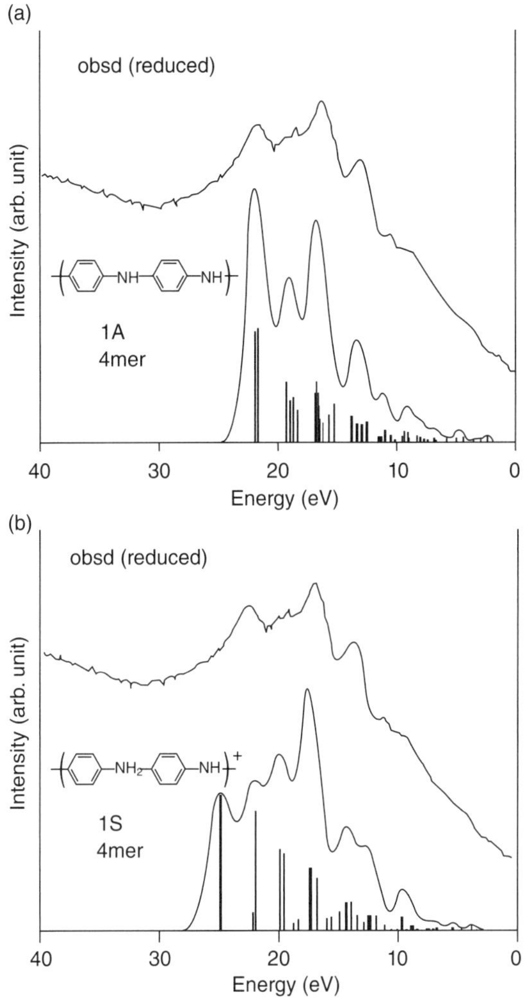

Figure 32.3. (a) Theoretical valence X-ray photoelectron spectrum of the model tetramer in 1A type of polyaniline calculated by deMon DFT program with experimental valence spectrum of polyaniline in reduced type. (b) Theoretical valence X-ray photoelectron spectrum of the model tetramer in 1S type of polyaniline calculated by deMon DFT program with experimental valence spectrum of polyaniline in reduced type.

in the 2A type is much closer to the observed one in the oxidized type than that of 2S.

We observed the CEBEs of C1s and N1s for polyaniline films in both reduction and oxidized types in Table 32.3. The observed CEBEs for polyaniline film correspond well to the values of DPPD, as obtained by Beamson and Briggs [37]. We then estimated the ***WD*** of polyainiline in both types as 5.1 eV since the differences between the calculated CEBEs of DPPD and the

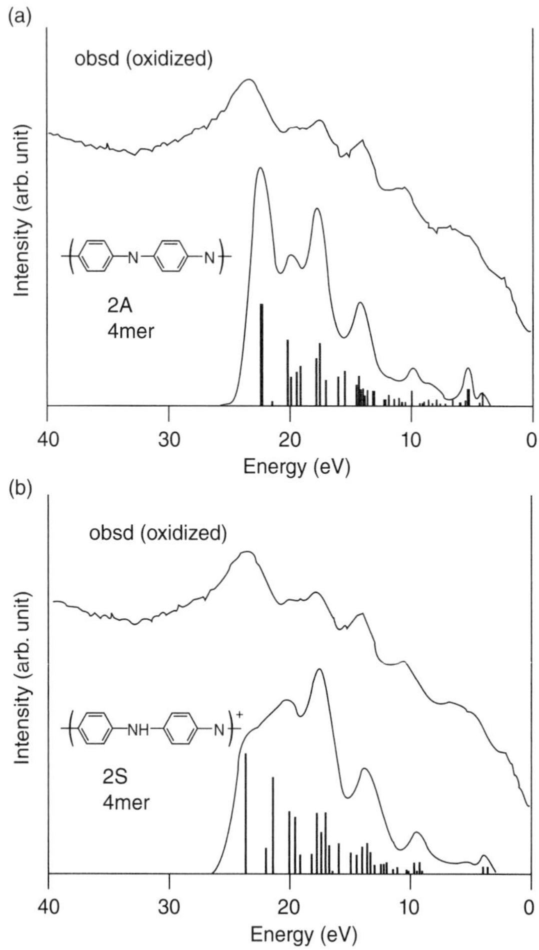

Figure 32.4. (a) Theoretical valence X-ray photoelectron spectrum of the model tetramer in 2A type of polyaniline calculated by deMon DFT program with experimental valence spectrum of polyaniline in oxidized type. (b) Theoretical valence X-ray photoelectron spectrum of the model tetramer in 2S type of polyaniline calculated by deMon DFT program with experimental valence spectrum of polyaniline in oxidized type.

experimental CEBEs of the polyaniline are considered to reflect the value of the reasonable ***WD***. Thus, in order to compare the reasonable ***WD*** for polyaniline in both types, we calculated the CEBEs of model tetramers for 2A and 1S types in the oxidation and reduction of polyaniline films using the uGTS model with the DZVP basis. For the 1S type model tetramer in the reduced film, we calculated the CEBEs by adding the chloride anion near by nitrogen cation to reproduce the electric neutral charge in the experimental reduced polyaniline film. In Table 32.3, we could estimate the deviations of C1s from

TABLE 32.3. Observed and Calculated CEBEs of DPPD and Polyaniline in Oxidized and Reduced Types

	DPPD			Polyaniline					
				Oxidized			Reduced		
	(***WD*** = 5.1 eV)			(***WD*** = 5.1 eV)			(***WD*** = 5.1 eV)		
	Obsd. (eV)	Calc.[a]	$Delta_3$[c]	Obsd.	Calc.[a]	$Delta_3$[c]	Obsd.	Calc.[a]	$Delta_3$[c]
					2A			1S	
CEBE(N1s)									
[–N–C]	399.9	399.9	0.0	399.7	398.4[a]	–1.3	399.8	399.2[a]	–0.6
CEBE(C1s)									
[–C–N]	285.9	286.0[b]	0.1	285.9	285.9[b]	0.0	285.9	286.1[b]	0.2
[–C=C–C–]	284.7	284.8[b]	0.1	284.8	285.1[b]	0.3	284.9	284.6[b]	–0.3

[a] Calculated CEBEs are modified by the reasonable ***WD*** (5.1 eV).

[b] Value was given as the averaged CEBEs.

[c] $Delta_3$ is the deviation between observed and modified calculated CEBEs (calc. – ***WD***).

the reasonable ***WD*** as ±0.3 eV, although the values for the N1s were obtained within −1.3 eV in a much larger deviation than those of C1s.

As indicated in previous works [18–24], our theoretical results of XPS by DFT calculations using model molecules for XPS of DPPD and polyaniline film are in reasonable agreement with experimental results. This is an important result for understanding the electronic states of conductive polymer films on platinum plates.

32.6 CONCLUSIONS

In order to clarify the electronic state of polyaniline synthesized electrochemically, we analyzed the valence X-ray photoelectron spectrum and CEBEs of DPPD by deMon DFT calculations using a molecule. Polyaniline has been synthesized electrochemically from two different thin films on platinum plates in oxidation and reduction states, respectively. Valence XPS of the conductive polymers were analyzed by deMon DFT calculations using model tetramers of the four chemical structural formulae (1S, 2S, 1A, and 2A), which MacDiarmid and coworkers proposed. We have determined the bond nature and the electron structures of DPPD and polyaniline from the analysis of valence XPS by DFT calculations. It was concluded from our analysis that the oxidized polyaniline corresponds well to the 2A type, while the reduction form is very close to the 1S type.

From calculations of the CEBEs of isolated gas-phase molecules using the deMon DFT program, we obtained accurate CEBEs of aniline and DPPD using the uGTS model with DZVP and pVTZ bases. With the reasonable ***WD*** values of DPPD, oxidation and reduction of polyaniline films are estimated as 5.1 eV from the difference between the calculated CEBEs of DPPD and the experimental ones of the compounds.

REFERENCES

1 R. J. Waltman, A. F. Diaz, J. Bargon. *J. Electrochem. Soc.* 131 (1984) 1452.

2 F. Garnier, G. Tourillon, M. Gazard, J. C. Dubois. *J. Electroanal. Chem.* 148 (1983) 299.

3 R. Noufi, A. J. Frank, A. J. Nozik. *J. Am. Chem. Soc.* 103 (1981) 1849.

4 H. S. White, G. P. Kittlesen, M. S. Wrighton. *J. Am. Chem. Soc.* 106 (1984) 5375.

5 C. Nylander, M. Armrgarth, I. Lundstrom. *Anal. Chem. Symp. Ser. (Chem. Sens.)* 17 (1983) 159.

6 M. Doriomedoff, F. Hautiere-cristofini, R. Surville, M. Josefowicz, L. T. Yu, R. Buvet. *J. Chim. Phys.* 68 (1971) 1055.

7 A. F. Diaz, J. A. Logan. *J. Electroanal. Chem.* 111 (1980) 111.

8 T. Ohsaka, Y. Onuki, N. Oyama, G. Katagiri, K. Kamisako. *J. Electroanal. Chem.* 161 (1984) 399.

9 A. Volkov, G. Tourillon, P. Lacaze, J. E. Dubois. *J. Electroanal. Chem.* 115 (1980) 279.

10 A. G. MacDiarmid, S. L. Wu, N. L. D. Somasiri, W. Wu, S. Y. Yaniger. *Mol. Cryst. Liq. Cryst.* 121 (1985) 173.

11 P. Snauwaert, R. Lazzaroni, J. Riga, J. J. Verbist. *Synth. Met.* 16 (1986) 245.

12 T. Nakajima, M. Harada, R. Osawa, T. Kawagoe, Y. Furukawa, I. Harada. *Macromolecules* 22 (1989) 2644.

13 E. T. Kang, K. G. Neoh, S. H. Khor, K. L. Tan, B. T. G. Tan. *J. Chem. Soc. Chem. Commun.* (1989) 695.

14 K. L. Tan, B. T. G. Tan, E. T. Kang, K. G. Noeh. *Phys. Rev. B* 39 (1989) 8070.

15 P. Snauwaert, R. Lazzaroni, J. Riga, J. J. Verbist, D. Gonbeau. *J. Chem. Phys.* 92 (1990) 2187.

16 S. N. Kumar, F. Gaillard, G. Bouyssoux, A. Sarte. *Synth. Met.* 36 (1990) 11.

17 K. G. Noeh, E. T. Kang, K. L. Tan. *J. Phys. Chem.* 96 (1992) 6777.

18 K. Endo, Y. Kaneda, H. Okada, D. P. Chong, P. Duffy. *J. Phys. Chem.* 100 (1996) 19455.

19 K. Endo, S. Maeda, M. Aida. *Polym. J.* 29 (1997) 171.

20 K. Endo, S. Maeda, Y. Kaneda. *Polym. J.* 29 (1997) 255.

21 K. Endo, D. P. Chong. *J. Surf. Anal.* 3 (1997) 618.

22 S. Kuroki, K. Endo, S. Maeda, D. P. Chong, P. Duffy. *Polym. J.* 30 (1998) 142.

23 K. Endo, D. P. Chong. *J. Surf. Anal.* 4 (1998) 50.

24 T. Otsuka, K. Endo, M. Suhara, D. P. Chong. *J. Mol. Struct.* 522 (2000) 47.

25 (a) A. St-Amant, D. R. Salahub. *Chem. Phys. Lett.* 169 (1990) 387; (b) A. St-Amant. Ph.D. Thesis, University of Montreal, 1991.

26 J. C. Slater. *Advan. Quantum Chem.* 6 (1972) 1.

27 (a) U. Gelius, K. Siegbahn. *Faraday Discuss. Chem. Soc.* 54 (1972) 257; (b) U. Gelius. *J. Electron. Spectrosc. Relat. Phenom.* 5 (1974) 985.

28 K. Endo, Y. Kaneda, M. Aida, D. P. Chong. *J. Phys. Chem. Solids* 56 (1995) 1131.

29 A.R. Williams, R.A. deGroot, C.B. Sommers. *J. Chem. Phys.* 63 (1975) 628.

30 J. F. Janak. *Phys. Rev.* A18 (1978) 7165.

31 L. Asbrink, C. Fridh, E. Lindholm. *Chem. Phys. Lett.* 52 (1997) 69.

32 J.-J. Yeh. *Atomic calculation of photoionization cross section and asymmetry parameters*. Gordon and Breach Science Publishers, Langhorne, PA, 1993.

33 (a) M. J. S. Dewar, E. G. Zoebisch. *Theochem* 180 (1988) 1; (b) M. J. S. Dewar, E. G. Zoebisch, E. F. Healy, J. J. P. Stewart. *Am. Chem. Soc.* 107 (1985) 3902.

34 A. D. Becke. *Phys. Rev.* A38 (1988) 3098.

35 J. P. Perdew. *Phys. Rev.* B33 (1986) 8822.

36 D. P. Chong, C. H. Hu, P. Duffy. *Chem. Phys. Lett.* 249 (1996) 491.

37 G. Beamson, D. Briggs. *High Resolution XPS of Organic Polymers. The Scienta ESCA 3000 Database*. Wiley, Chichester, 1992.

38 W. L. Jolly, K. D. Bomben, C. J. Eyermann. Atomic data and nuclear data table. *Atomic Data Nuclear Data* 31 (1984) 434.

39 D. P. Chong. *Chem. Phys. Lett.* 232 (1995) 486.

40 D. P. Chong. *J. Chem. Phys.* 103 (1995) 1842.

41 C. Bureau, D. P. Chong. *Chem. Phys. Lett.* 264 (1997) 186.

42 Y. Takahata, D.P. Chong. *Bull. Chem. Soc. Jpn.* 73 (2000) 2453.

33

SYNTHESIS AND SOLUTION PROPERTIES OF FLUOROCARBON–HYDROCARBON HYBRID SURFACTANTS

YUKISHIGE KONDO

33.1 INTRODUCTION

In mixtures of a fluorocarbon surfactant (FCS) and a hydrocarbon surfactant (HCS), FCS- and HCS-rich micelles are generally formed because of immiscibility between the fluorocarbon and hydrocarbon chains [1, 2]. To solve this problem, new surfactants with fluorocarbon and hydrocarbon chains in the molecule, that is, *hybrid surfactants*, were first synthesized in 1992 by Guo et al. [3]. These surfactants [$F(CF_2)_mCH(OSO_3Na)(CH_2)_nH$] form aqueous micelles, and their critical micelle concentrations (CMCs) follow the Klevens equation. However, they are so susceptible to hydrolysis that their aqueous solutions must be handled within 20 hours after preparation. This is because an electron-attractive fluorocarbon chain and a hydrophilic group are bound to the same carbon atom in the molecule.

In 1995, we developed sulfonate-type fluorocarbon hybrid surfactants [$F(CF_2)_m\Phi COCH(SO_3Na)(CH_2)_nH$; $m = 4, 6, 8$, $n = 4, 6, 8$; $\Phi = p$-phenylene] that are highly resistant to hydrolysis [4]. They show an emulsification ability for the ternary mixture of *n*-octane (hydrocarbon oil)/perfluoropolyether (fluorocarbon oil)/water, which should allow the preparation of water-based fluoropaints [4]. We also reported various anomalous properties. For example, the 10 wt % aqueous solution of the surfactant with $m = 6$ and $n = 4$ has a very

Electrical Phenomena at Interfaces and Biointerfaces: Fundamentals and Applications in Nano-, Bio-, and Environmental Sciences, First Edition. Edited by Hiroyuki Ohshima.

high viscosity at the human body temperature (37°C) [5–7], and the surfactants show intramicellar phase separation between the fluorocarbon and hydrocarbon chains [8, 9]. Neither homogeneously double-chained analogs nor general surfactants provide the above-mentioned properties [10], which suggests that the hybridity of the hydrophobic part of the hybrid surfactant molecule plays an important role in its unique properties.

In this chapter, we describe recent research on hydrocarbon/fluorocarbon hybrid surfactants, mainly focusing on our own work.

33.2 SYNTHESIS AND BASIC SOLUTION PROPERTIES OF NEW HYBRID SURFACTANTS

Over the past 5 years, three types of anionic hybrid surfactants [11–13] and one cationic hybrid surfactant [14] have been prepared; quite recently, we synthesized nonionic hybrid surfactants [15]. The synthetic pathways are shown in Schemes 33.1–33.4.

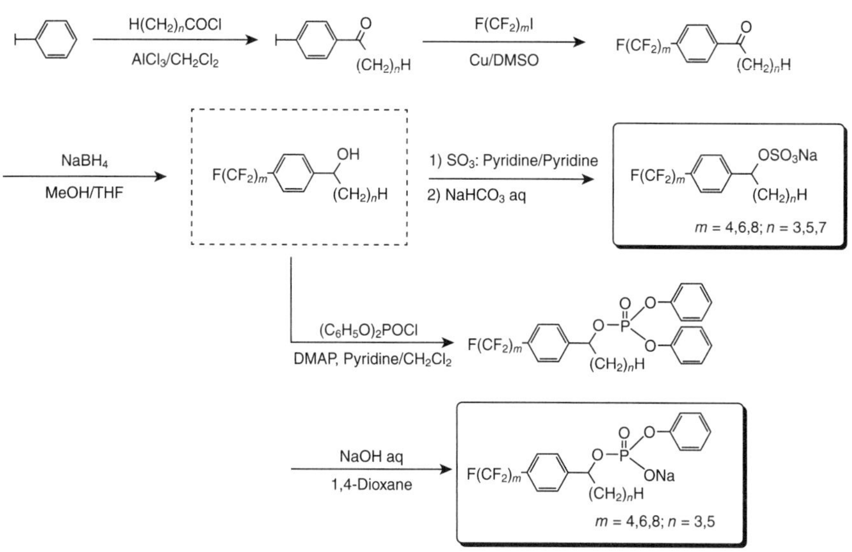

Scheme 33.1. Synthesis of anionic hybrid surfactants with a benzene ring.

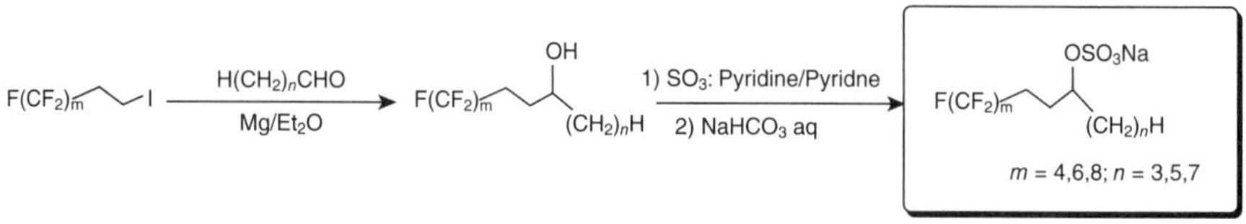

Scheme 33.2. Synthesis of anionic hybrid surfactants with an ethylene spacer.

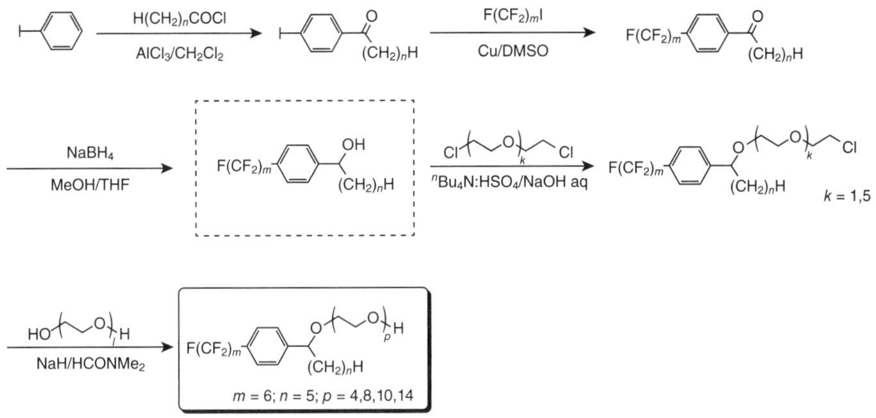

Scheme 33.3. Synthesis of nonionic hybrid surfactants with a benzene ring. DMSO, dimethyl sulfoxide; THF, tetrahydrofuran.

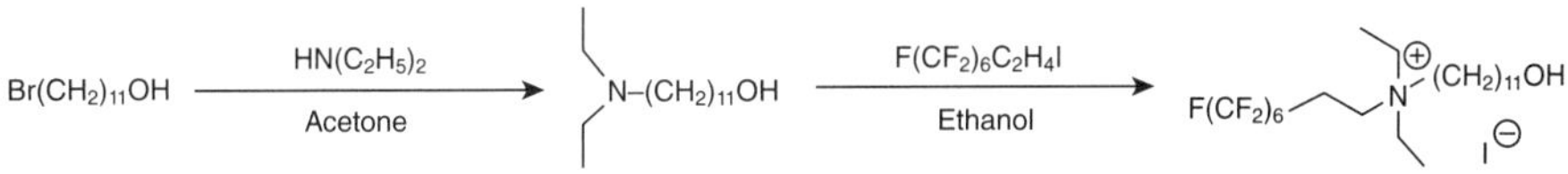

Scheme 33.4. Synthesis of cationic hybrid surfactants with a hydroxyl group at the molecular end.

All of the anionic and nonionic surfactants with a phenylene group are synthesized via hybrid fluorocarbon/hydrocarbon hybrid alcohols, that is, $F(CF_2)_m\Phi CHOH(CH_2)_nH$ (m = 4, 6, 8; n = 3, 5, 7), using iodobenzene as a starting material (Schemes 33.1 and 33.3). Unfortunately, the total yields are low. The yield for the anionic surfactants $F(CF_2)_m\Phi CH(OSO_3Na)(CH_2)_nH$ is only about 30%, and that for the nonionic surfactants $F(CF_2)_m\Phi CH(OSO_3Na)(CH_2)_nH$ is less than 10%. Fluorocarbon reactants are generally more expensive than hydrocarbon analogs, and more efficient synthetic methods with high yields can thus be expected. In contrast, the synthesis of anionic surfactants with an ethylene spacer, that is, $F(CF_2)_mC_2H_4CH(OSO_3Na)(CH_2)_nH$, consists of only two steps (Scheme 33.2) and gives a total yield of ca. 70%. Based on the synthetic yield, $F(CF_2)_mC_2H_4CH(OSO_3Na)(CH_2)_nH$ can be commercialized.

The surfactants shown in Schemes 33.1–33.3 are all resistant to hydrolysis in pure water. The stability of $F(CF_2)_6C_2H_4CH(OSO_3Na)(CH_2)_7H$ in acidic or basic conditions has been investigated in 1 M HCl, CH_3COOH, or NaOH [13]. The surfactant is barely susceptible to hydrolysis in 1 M CH_3COOH and 1 M NaOH, while it is hydrolyzed in a strong acidic medium to produce the alcohol derivative $F(CF_2)_6C_2H_4CHOH(CH_2)_7H$.

On the other hand, Aydogan et al. prepared an unsymmetrical bolaform-type cationic hybrid surfactant consisting of a diethylammonium group and a hydroxyl group as hydrophilic groups (Scheme 33.4). However, they did not mention the yield and stability for the hydrolysis of the surfactant [14].

33.3 BASIC SOLUTION PROPERTIES OF HYBRID SURFACTANTS

The solution properties of hybrid surfactants are summarized in Table 33.1 in terms of CMC, surface tension at CMC (γ_{CMC}), Krafft points, and cloud points.

At the same fluorocarbon (m) and hydrocarbon chain lengths (n), the Krafft points of anionic hybrid surfactants decrease in the following order: $F(CF_2)_m\Phi CH[OP(ONa)(OC_6H_5)](CH_2)_nH$ > $F(CF_2)_m\Phi CH(OSO_3Na)(CH_2)_nH$ > $F(CF_2)_mC_2H_4CH(OSO_3Na)(CH_2)_nH$. In contrast, CMC decreases in the following order: $F(CF_2)_mC_2H_4CH(OSO_3Na)(CH_2)_nH$ > $F(CF_2)_m\Phi CH(OSO_3Na)(CH_2)_nH$ > $F(CF_2)_m\Phi CH[OP(ONa)(OC_6H_5)](CH_2)_nH$. The CMC of anionic hybrid surfactants follows the Klevens law; thus, the following equation relates the CMC to the fluorocarbon (m)/hydrocarbon (n) chain length:

$$\log[\mathrm{CMC}] = A - Bm\ (\text{or } n). \tag{33.1}$$

A and B are constants, and the value of B shows how much one CF_2 or CH_2 group in the hydrophobic group contributes to the reduction of the CMC. For anionic hybrid surfactants, the contribution of the CF_2 group is larger than that of the CH_2 group because the fluorocarbon chain is more hydrophobic than the hydrocarbon chain.

The values of γ_{CMC} for the anionic and nonionic surfactants are in the range of 18–24 mN/m, which is similar to those for general FCSs.

The CMC of the cationic hybrid surfactant with a hydroxyl group at the molecular end, that is, $[F(CF_2)_6C_2H_4N(C_2H_5)_2(CH_2)_{11}OH]I$, is insensitive to salt concentrations in aqueous solutions, which is a strange property for ionic surfactants. This surfactant also gives a much lower γ_{CMC} than its analog, $[HO(CH_2)_{11}N(CH_3)_3]Br$, which is an unsymmetrical bolaform surfactant without a fluorocarbon chain [γ_{CMC} = 47 mN/m in 0.1 M LiBr (pH = 2) at 40°C].

There is interest as to whether the fluorocarbon and hydrocarbon chains in the hybrid molecule are miscible in the micelle state. The diameters of the micelles of the anionic hybrid surfactants are shown in Table 33.2. The micellar diameter increases with an increase in the fluorocarbon or hydrocarbon chain length except for $F(CF_2)_6\Phi CH(OSO_3Na)(CH_2)_5H$, which forms small micelles with diameters of 1.2 nm. Inoue et al. investigated in detail the micelle formation of the phosphate-type surfactant $F(CF_2)_m\Phi CH[OP(ONa)(OC_6H_5)](CH_2)_nH$ as a function of temperature; they reported that the hydrophobicity of the CH_2 group in the hydrocarbon chain in the surfactant is considerably reduced compared with that in general HCSs [16]. They believe that this may

TABLE 33.1. Basic Solution Properties of Hybrid Surfactants

Surfactant		Temperature (°C)	CMC (mM)	γ_{CMC} (mN/m)	Krafft Point or Cloud Point (°C)	Reference
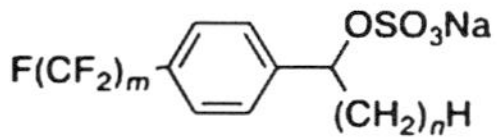	$m = 4$ $n = 3$	27	7.0	19	<0	11
	$m = 6$ $n = 3$		0.90	18	<0	
	$m = 8$ $n = 3$		0.08	20	16	
	$m = 4$ $n = 5$		3.0	19	<0	
	$m = 6$ $n = 5$		0.34	20	14	
	$m = 8$ $n = 5$		n/a	n/a	32	
	$m = 4$ $n = 7$		1.3	20	8	
	$m = 6$ $n = 7$		n/a	n/a	43	
	$m = 8$ $n = 7$		n/a	n/a	76	
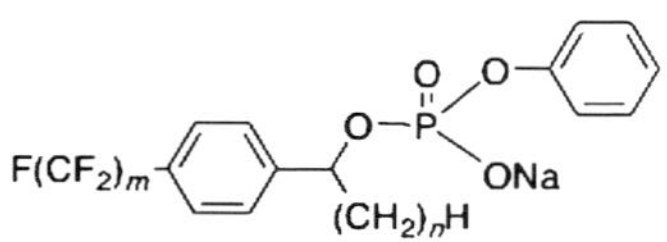	$m = 4$ $n = 3$	25	1.4	24	<0	12
	$m = 6$ $n = = 3$		0.35	23	13	
	$m = 8$ $n = 3$		0.07	22	22	
	$m = 4$ $n = 5$		0.97	24	<0	
	$m = 6$ $n = 5$	30	0.21	23	27	
	$m = 8$ $n = 5$		n/a	n/a	37	
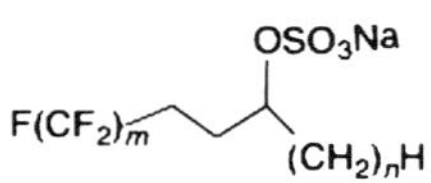	$m = 4$ $n = 3$	25	25	23	<0	13
	$m = 6$ $n = 3$		4.0	19	<0	
	$m = 8$ $n = 3$		0.47	20	<0	
	$m = 4$ $n = 5$		9.1	22	<0	
	$m = 6$ $n = 5$		1.2	21	<0	
	$m = 8$ $n = 5$		0.16	22	<0	
	$m = 4$ $n = 7$		4.1	18	<0	
	$m = 6$ $n = 7$		0.50	18	<0	
	$m = 8$ $n = 7$		0.07	19	<0	

(*Continued*)

TABLE 33.1 (*Continued*)

Surfactant		Temperature (°C)	CMC (mM)	γ_{CMC} (mN/m)	Krafft Point or Cloud Point (°C)	Reference
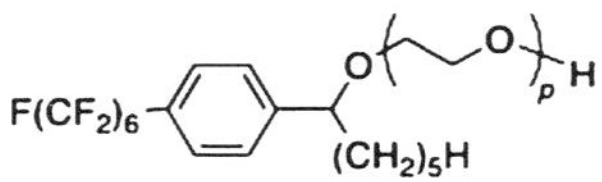	$p = 4$	25	n/a	n/a	<0	15
	$p = 8$		n/a	n/a	8	
	$p = 10$		0.63	20	28	
	$p = 14$		5.4	22	54	
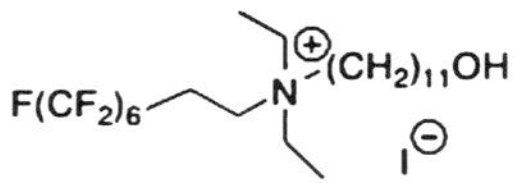		40 (0.1 M LiBr, pH = 2)	0.450	25	n/a	14
		40 (1 mM LiBr, pH = 2)	0.4	29	n/a	

TABLE 33.2. Micelle Diameters of Anionic Hybrid Surfactants

Surfactant		Temperature (°C)	Diameter (nm)	Reference
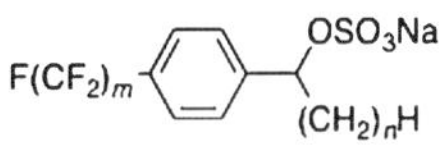	$m = 4$ $n = 3$	30	2.8	11
	$m = 6$ $n = 3$		3.6	
	$m = 4$ $n = 5$		4.4	
	$m = 6$ $n = 5$		1.2	
	$m = 4$ $n = 7$		6.1	
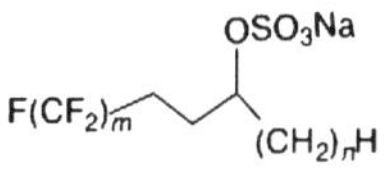	$m = 4$ $n = 3$	30	1.2	13
	$m = 6$ $n = 3$		2.4	
	$m = 8$ $n = 3$		2.8	
	$m = 4$ $n = 5$		1.8	
	$m = 6$ $n = 5$		3.0	
	$m = 4$ $n = 7$		2.8	
	$m = 6$ $n = 7$		3.2	

be caused by the intramolecular fluorocarbon–hydrocarbon interaction being characteristic to hybrid surfactants. Thus, the small hydrophobicity of the CH_2 group seems to be due to the contact of a hydrocarbon chain with water in the micelle state. Micelles of hybrid surfactants may have an intramicellar phase-separated structure in which the micelle interior is packed by fluorocarbon chains, while hydrocarbon chains cover the micelles. This micelle model does not contradict the results for the diameters of the hybrid surfactant micelles shown earlier.

33.4 APPLICATIONS OF HYBRID SURFACTANTS

33.4.1 Novel Emulsifiers for a Water-in-Supercritical CO_2 ($scCO_2$) Microemulsion System

One application of hybrid surfactants is as an emulsifier for $scCO_2$. $scCO_2$ has attracted much attention as a nontoxic, inflammable, environmentally friendly, and low-cost solvent over the past decade. However, $scCO_2$ is not suitable for dissolving polar substances; therefore, Sagisaka et al. [17] and other research groups [18, 19] have made efforts to prepare a water-in-$scCO_2$ microemulsion system. So far, FCSs have been found to be great emulsifiers for its preparation [20, 21]. On the other hand, Eastoe et al. noted that "one of the main hurdles for CO_2 technology is the cost impact of the surfactants" [22]. From this point of view, hybrid surfactants may be good candidates because they not only have less of the fluorinated hydrophobic part than general FCSs but they also show interfacial properties similar to FCSs. Sagisaka et al. and Eastoe et al.

succeeded in preparing water-in-scCO_2 microemulsions using hybrid surfactants synthesized in the 1990s: $F(CF_2)_6\Phi COCH(SO_3Na)(CH_2)_nH$ ($n = 2, 4, 6, 8$) and $F(CF_2)_mCH(OSO_3Na)(CH_2)_nH$ ($m = 7, 8$; $n = 4, 7, 8$) [17, 22]. Unfortunately, the former has a problem with low synthetic yields, and the latter is easily hydrolyzed [3]. Novel hybrid surfactants with an ethylene spacer, that is, $F(CF_2)_6C_2H_4CH(OSO_3Na)(CH_2)_nH$, may overcome these two problems since they have high synthetic yields and are resistant to hydrolysis [13].

33.4.2 Tooth Surface Modifiers

Teranaka et al. modified the surfaces of hydroxyapatite (HAp) pellets, which are modeling materials for teeth, by using the phosphate-type surfactants $F(CF_2)_8\Phi CH[OP(ONa)(OC_6H_5)](CH_2)_nH$ ($n = 3, 5$); they reported that the hybrid surfactants give high hydro- and lipophobicity onto the HAp surfaces for more than 90 days [23]. We previously modified bovine teeth using fluorocarbon silanes and demonstrated that highly hydro- and lipophobic teeth surfaces restrain dental plaque formation [24]. This result suggests that phosphate-type hybrid surfactants are also useful for the prevention of cavities.

33.4.3 Other Possible Applications

Nonionic hybrid surfactants such as $F(CF_2)_m\Phi CH[O(C_2H_4O)_kH](CH_2)_nH$ may be useful for the treatment of idiopathic respiratory distress syndrome (IRDS) of newborns. In order to deal with this disease, the liquid ventilation method, which compensates pulmonary respiration by filling the lungs with fluorinated solvents such as 1-perfluorooctyl bromide and 1-perfluorooctylethane, has been investigated. However, these solvents are hydrophobic as well as lypophobic; thus, they cannot dissolve many pharmacologically active substances. Krafft et al. succeeded in preparing water-in-fluorocarbon emulsions that are stable for long periods of time by using a nonionic hybrid surfactant with morpholine groups, that is, $C_8F_{17}C_{10}H_{20}OPO[N(C_2H_4)_2O]_2$; they demonstrated that the emulsion can dissolve water- and oil-soluble drugs [25]. Nonionic surfactants with ethylene oxide chains such as Tweens are generally harmless to the human body. Therefore, $F(CF_2)_m\Phi CH[O(C_2H_4O)_kH](CH_2)_nH$ are surfactants nontoxic to the human body that can be used to prepare water-in-fluorocarbon emulsions for dissolving water- and oil-soluble drugs in IRDS treatment.

33.5 UNUSUAL PROPERTIES OF HYBRID SURFACTANTS

33.5.1 Unusual Viscosity Behavior

As shown in Figure 33.1, the 10 wt % aqueous solution of $F(CF_2)_6\Phi COCH(SO_3Na)(CH_2)_4H$ shows very high viscosity at human body temperature [6].

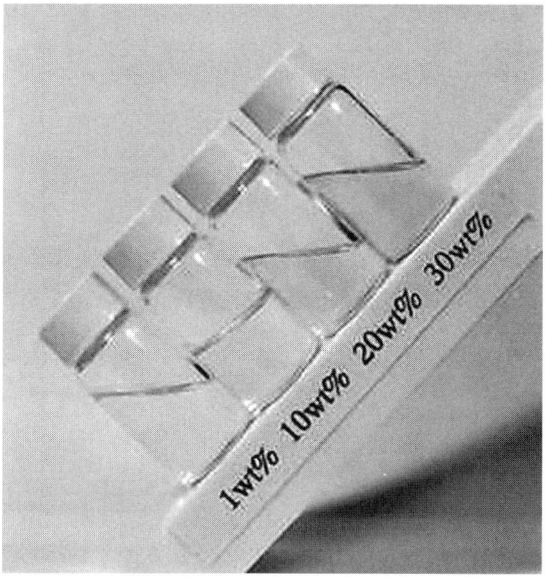

Figure 33.1. Aqueous solutions of $F(CF_2)_6\Phi COCH(OSO_3Na)(CH_2)_4H$ with different concentrations. The 10 wt % solution became highly viscous.

Figure 33.2 shows the relationship between the viscosity of $F(CF_2)_6\Phi COCH(SO_3Na)(CH_2)_4H$ aqueous solutions at different concentrations of 10, 20, and 30 wt % and temperature. The viscosity of the 10 wt % solution reaches its maximum (100 Pa·s) at 37°C, whereas that of the 20 and 30 wt % solutions monotonically decreases with increasing temperature. Furthermore, the novel hybrid surfactant $F(CF_2)_8C_2H_4CH(OSO_3Na)(CH_2)_3H$ also forms a highly viscous gel-like liquid at 10 wt % and 31°C [13]. However, the mechanism for such thermoresponsive viscosity behavior still remains to be revealed. Recently, Talmon et al. investigated the 10 wt % solution of $F(CF_2)_6\Phi COCH(SO_3Na)(CH_2)_4H$ by means of cryogenic transmission electron microscope (TEM) and self-diffusion nuclear magnetic resonance (NMR) [26]. They reported that significantly large aggregates are formed around 40°C compared with aggregates formed at lower and higher temperatures.

We synthesized the HCS $H(CH_2)_6\Phi COCH(SO_3Na)(CH_2)_4H$, which is structurally analogous to the hybrid surfactant $F(CF_2)_6\Phi COCH(SO_3Na)(CH_2)_4H$; we then measured the temperature dependence of the viscosity for the 10 wt % aqueous solution of $H(CH_2)_6\Phi COCH(SO_3Na)(CH_2)_4H$. The viscosity decreased monotonically with increasing temperature, and the viscosity at 37°C was 0.002 Pa·s, which is much smaller than that of the hybrid surfactant $F(CF_2)_6\Phi COCH(SO_3Na)(CH_2)_4H$ at 37°C. These results indicate that the unusual viscosity behavior for hybrid surfactants comes from the hybridity in

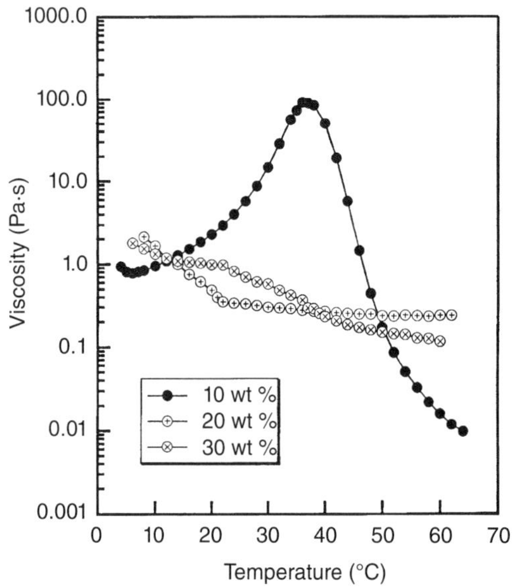

Figure 33.2. Relationship between the viscosity of aqueous solutions of $F(CF_2)_6\Phi COCH(OSO_3Na)(CH_2)_4H$ with different concentrations and temperature. Only the viscosity of the 10 wt % solution exhibited a maximum value.

the hydrophobic part of the hybrid surfactant, that is, the coexistence of a fluorocarbon chain and a hydrocarbon chain in a surfactant molecule.

33.5.2 Formation of Micelles with Long Lifetimes

The exchange rate of surfactants in monomer and micelle states is usually fast, and the micelle lifetime is on the order of 10^5–10^6 s. We found that the anionic surfactant $F(CF_2)_6\Phi CH(OSO_3Na)(CH_2)_5H$ shows a slow exchange of the surfactant between the two states on an NMR timescale [27]. This finding is the first example of an unusually slow exchange of anionic surfactants. The micelle has a lifetime about 10^2–10^3 times longer than that of general surfactant micelles. The pulsed-gradient spin echo (PGSE) experiment showed that the hybrid surfactant forms very small micelles with a hydrodynamic radius of 0.6 nm; this value is only about half the length from the ω-CF_3 group to the head group. The interdigitation between the surfactant molecules in the micelle contributes to the unusually long lifetime, that is, slow surfactant exchange on the NMR timescale.

A hybrid surfactant micelle with a long lifetime can be utilized as an aqueous asymmetric organic synthesis site. There have been many studies on asymmetric synthesis in aqueous micelles formed from chiral surfactants [28]. However, no high enantioselectivity of organic products has been observed.

This result is because micelles are dynamic aggregates. The lifetime of a micelle is so short that reaction intermediates cannot have stereochemical orientation inside the micelles. The hybrid surfactant $F(CF_2)_6\Phi CH(OSO_3Na)(CH_2)_5H$ is a racemate. Provided that enantiomerically pure (*R*)- or (*S*)-$F(CF_2)_6\Phi CH(OSO_3Na)(CH_2)_5H$ is prepared, its micelles may be an aqueous asymmetric organic synthesis site.

33.6 CONCLUSION

Hybrid fluorocarbon/HCSs show some anomalous solution properties that are not observed for general fluorocarbon or HCSs, but they have almost the same surface tension as FCSs. Hybrid surfactants can be used as substitutes for convenient FCSs, and their unique properties may be applied in industrial, medical, and dental fields in the near future. In order to promote the utilization of hybrid surfactants, it is important for more efficient synthetic pathways to be developed. In this respect, anionic hybrid surfactants with ethylene spacers and prepared in high yields have great potential.

REFERENCES

1 P. Mukerjee, K. J. Mysels, in K. L. Mittal (ed.), *Colloidal dispersions and micellar behavior*. American Chemical Society, Washington, DC, 1975, p. 239.
2 N. Funasaki, in K. Ogino, M. Abe (eds.), *Mixed surfactant systems*. Marcel Dekker, New York, 1992, p. 145.
3 W. Guo, Z. Li, B. M. Fung, E. A. O'Rear, J. H. Harwell. *J. Phys. Chem.* 96 (1992) 6738.
4 N. Yoshino, K. Hamano, Y. Omiya, Y. Kondo, A. Ito, M. Abe. *Langmuir* 11 (1995) 466.
5 A. Ito, K. Kamogawa, H. Sakai, Y. Kondo, N. Yoshino, M. Abe. *J. Jpn. Oil Chem. Soc.* 45 (1996) 479.
6 M. Abe, K. Tobita, H. Sakai, Y. Kondo, N. Yoshino, Y. Kasahara, H. Matsuzawa, M. Iwahashi, N. Momozawa, K. Nishiyama. *Langmuir* 13 (1997) 2932.
7 K. Tobita, H. Sakai, Y. Kondo, N. Yoshino, M. Iwahashi, N. Momozawa, M. Abe. *Langmuir* 13 (1997) 5054.
8 A. Ito, H. Sakai, Y. Kondo, N. Yoshino, M. Abe. *Langmuir* 12 (1996) 5768.
9 A. Ito, K. Kamogawa, H. Sakai, K. Hamano, Y. Kondo, N. Yoshino, M. Abe. *Langmuir* 13 (1997) 2935.
10 Y. Kondo, M. Hamasaki, K. Tobita, H. Sakai, M. Abe, N. Yoshino. *J. Jpn. Oil Chem. Soc.* 48 (1999) 707.
11 H. Miyazawa, K. Igawa, Y. Kondo, N. Yoshino. *J. Fluor. Chem.* 124 (2003) 189.
12 H. Miyazawa, H. Yokokura, Y. Ohkubo, Y. Kondo, N. Yoshino. *J. Fluor. Chem.* 125 (2004) 1485.
13 H. Miyazawa, Y. Kondo, N. Yoshino. *J. Oleo Sci.* 54 (2005) 167.

14 N. Aydogan, N. Aldis, O. Guvenir. *Langmuir* 19 (2003) 10726.
15 H. Miyazawa, Y. Wakatsuki, Y. Kondo, N. Yoshino. *J. Oleo Sci.* 54 (2005) 361.
16 T. Inoue, N. Nakashima, G. Sugihara, N. Yoshino. *J. Oleo Sci.* 54 (2005) 95.
17 M. Sagisaka, T. Fujii, Y. Ozaki, S. Yoda, Y. Takebayashi, Y. Kondo, N. Yoshino, H. Sakai, M. Abe, K. Otake. *Langmuir* 20 (2004) 2560.
18 B. H. Bunton, J. M. Perera, F. Grieser, G. W. Stevens. *Colloids Surf. A Physicochem. Eng. Asp.* 189 (2001) 177.
19 J. Eastoe, A. Paul, S. Nave, D. C. Steytler, B. H. Robinson, R. Rumsey, M. Thorpe, R. K. Heenan. *J. Am. Chem. Soc.* 123 (2001) 988.
20 C. T Lee Jr., P. Bhargava, K. P. Johnson. *J. Phys. Chem. B* 104 (2000) 4448.
21 D. Lee, J. C. Hutchison, J. M. DeSimone, R. W. Murray. *J. Am. Chem. Soc.* 123 (2001) 8406.
22 A. Dupont, J. Eastoe, L. Martin, D. C. Stytler, R. K. Heenan, F. Cuittard, E. T. de Givenchy. *Langmuir* 20 (2004) 9960.
23 N. Omoto, T. Nihei, S. Kurata, Y. Kondo, K. Umemoto, N. Yoshino, T. Teranaka. *Jpn J. Conserv. Dent.* 48 (2005) 128.
24 N. Yoshino, Y. Yamamoto, T. Teranaka. *Chem. Lett.* (1993) 821.
25 V. M. Sadtler, M. P. Krafft, J. G. Riess. *Angew. Chem. Int. Ed. Engl.* 35 (1996) 1976.
26 D. Danino, D. Weihs, R. Zana, G. Orädd, G. Lindblom, M. Abe, Y. Talmon. *J. Colloid Interface Sci.* 259 (2003) 383.
27 Y. Kondo, H. Miyazawa, H. Sakai, M. Abe, N. Yoshino. *J. Am. Chem. Soc.* 124 (2002) 6516.
28 S. I. Goldberg, N. Baba, R. L. Green, R. Pandian, J. Stowers, R. B. Dunlap. *J. Am. Chem. Soc.* 100 (1978) 6768.

34

ELECTROCHEMICAL DYNAMIC CONTROL OF SELF-ASSEMBLIES FORMED BY REDOX-ACTIVE SURFACTANTS

KOJI TSUCHIYA, HIROFUMI YAJIMA, HIDEKI SAKAI, AND MASAHIKO ABE

34.1 INTRODUCTION

Amphiphilic molecules exhibit various interfacial properties and self-assemble into various aggregates, such as spherical micelles, rodlike micelles, vesicles, and lyotropic liquid crystals (cubic, hexagonal, and lamellar phases). Recently, many researchers have focused on active and reversible control of interfacial properties and self-assemblies by a variety of external stimuli, for example, photochemical reactions [1–4], pH changes [5–7], temperature changes [8], and redox reactions [9,10]. To obtain active and reversible control of self-assemblies for surfactant systems, numerous surfactants whose hydrophilic–lipophilic balance or structural conformation can be changed by external stimuli have been synthesized. In this chapter, we introduce recent approaches toward active control of surfactant systems based on redox-active surfactants.

Redox-active surfactants contain functional groups such as ferrocene, viologen, or *N*-alkylated nicotinamides [11]. Because ferrocene is chemically stable

Electrical Phenomena at Interfaces and Biointerfaces: Fundamentals and Applications in Nano-, Bio-, and Environmental Sciences, First Edition. Edited by Hiroyuki Ohshima.

over a wide range of solution conditions, and because the redox potential of ferrocene does not lead to electrolysis of water and is appropriate for controlling the oxidation state of ferrocene in water [12], the majority of redox-active surfactant systems have used ferrocene as the redox-active functional group. The ferrocene group is electrically neutral and has no solubilizing ability in water. However, when oxidized, it forms the hydrophilic ferrocenium cation. The redox reactions using the ferrocenium cation thus permit the control of aggregation states such as micelles [12–23] and vesicles [24–29], interfacial properties [9, 30–32], and orientations of liquid crystals with a ferrocene-decorated electrode [33].

Saji and coworkers [12–18] developed a method to prepare organic thin films on electrodes by the electrolysis of ferrocenyl surfactant micellar solutions, in which hydrophobic dyes or pigments are solubilized. They named this method the "micellar electrolysis method." Solubilized hydrophobic compounds are released as a result of disruption of ferrocenyl surfactant micelles and are deposited on the electrode to form uniform thin organic films. Gokel et al. [24–27] studied the control of vesicle formation using newly synthesized ferrocenyl surfactants. For several ferrocenyl surfactants, electrochemical oxidation induces a remarkable change in surface tension (>20 mN/m) [10, 32]. Abbott and coworkers [9, 30–32] described a looped configuration of ferrocenyl surfactants on an air–liquid interface, which means both ends of the molecule are directed to the aqueous phase, and thus control the interface. They also reported that double-tailed surfactants containing ferrocenyl groups can be exploited to control the formation of complexes with DNA (lipoplexes) [34–36]. These redox-active complexes were useful as DNA delivery systems to achieve active, spatial, and temporal control by the oxidation of the ferrocenyl surfactant. Abe et al. [21–23] investigated the electrochemical reversible control of micellar solubilization of oily substances using various ferrocenyl surfactants.

This chapter deals with two recent topics regarding dynamic control of self-assemblies using redox reactions of the ferrocenyl surfactant: (i) vesicles spontaneously forming when a cationic redox-active ferrocenyl surfactant and an anionic surfactant are mixed in water [37], and (ii) wormlike micelles consisting of a cationic ferrocenyl surfactant and an organic counterion (sodium salicylate) [38].

34.2 REVERSIBLE CONTROL OF VESICLE FORMATION USING A REDOX REACTION

Vesicles are closed bilayer shells that encapsulate an aqueous interior in which water-soluble substances can be solubilized. They are widely used as functional drug carriers in drug delivery systems (DDSs), model membranes, and microreactors for specialized chemistry. If vesicle formation and disruption can be controlled by redox potential, it would be useful for the controlled release of water-soluble drugs, dyes, and perfumes and for the preparation of inorganic thin films based on the micellar electrolysis method reported by Saji et al. [12–18]. Gokel et al. [24] reported that a redox-active amphiphilic molecule was synthesized by introducing a ferrocene into dihydrocholesterol (cholestanol) and that the vesicles consisting of the oxidized compound produced by ultrasonication were disrupted by electrolytic reduction. However, they failed in the reversible reconstitution of the vesicles by only oxidation because external physical force (ultrasonication) was necessary at the time of vesicle formation.

Although vesicles are often formed spontaneously *in vivo*, they have been observed only rarely to be formed *in vitro* without applying a considerable amount of external mechanical energy (such as sonication or high-pressure filtration through filters of small pore sizes) or elaborate chemical treatments (detergent dialysis or reverse-phase evaporation). Vesicles thus formed (as per Gokel's preparation) are metastable and eventually revert to planar multilamellar structures. Kaler and coworkers [39–41] recently reported spontaneous vesicle formation without external force in aqueous mixed solutions of cationic and anionic surfactants. The vesicles thus formed are thermodynamically stable for a long time (up to several years). If a redox-active surfactant is used as one of the species for the mixed system, vesicle formation and disruption are therefore expected to be reversibly controlled by the redox reaction.

The present section describes our studies to reversibly control the formation and disruption of vesicles in the mixed system of (11-ferrocenylundecyl) trimethylammonium bromide (FTMA, Scheme 34.1), a redox-active cationic

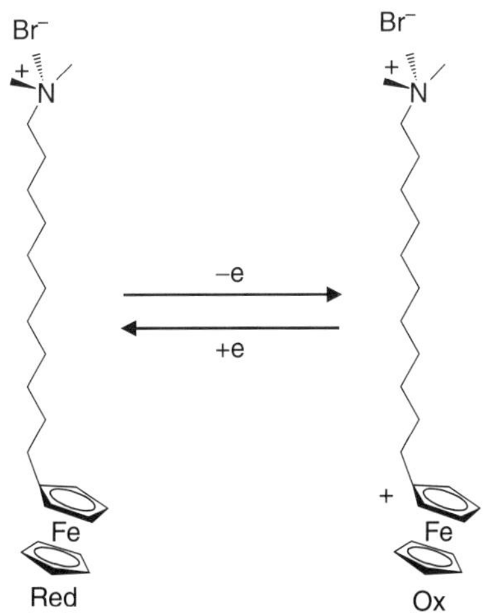

Scheme 34.1. Redox reaction of (11-ferrocenylundecyl)trimethylammonium bromide (FTMA).

surfactant, and sodium dodecylbenzenesulfonate (SDBS), an anionic surfactant, by potentiostatic electrolysis. FTMA has a redox-active ferrocenyl group positioned at the end of its alkyl chain [37]. The ferrocenyl group (reduced form) in the molecule is hydrophobic, whereas the ferrocenium cation (oxidized form) acts as a hydrophilic group. This causes a remarkable change in the hydrophilic–lipophilic balance of FTMA, such that the redox reaction of FTMA can be expected to reversibly control the formation and disruption of vesicles formed in aqueous FTMA/SDBS mixtures.

34.2.1 Electrochemical Control of Vesicles in Aqueous Mixtures of FTMA and SDBS

A solution of FTMA (reduced form) : SDBS mixture (7:3 in volume ratio) at 1 wt % surfactant concentration was oxidized at +0.6 V versus saturated calomel electrode (SCE), and the solution of the oxidized form thus obtained was reduced at 0.0 V versus SCE. Figure 34.1 shows the results of optical microscopic observation under differential interference on the reduced, oxidized, and re-reduced solutions. Vesicles were formed in the reduced mixture as evidenced by a doughnutlike structure, characteristic of vesicles, observed in Figure 34.1a. No self-assembly was observed in the oxidized solution (Fig. 34.1b) by optical microscopy, whereas vesicles were again observed in the re-reduced solution (Fig. 34.1c).

The results obtained in this study demonstrate that the redox reaction of the FTMA molecules constituting vesicles by potentiostatic electrolysis causes the vesicles to disrupt and reform reversibly.

Figure 34.2 shows the hydrodynamic diameters as measured by dynamic light scattering (DLS) on molecular assemblies constituted by the reduced, oxidized, and re-reduced forms of FTMA/SDBS mixtures in solution. A peak distribution around 35 nm probably due to FTMA/SDBS mixed micelles

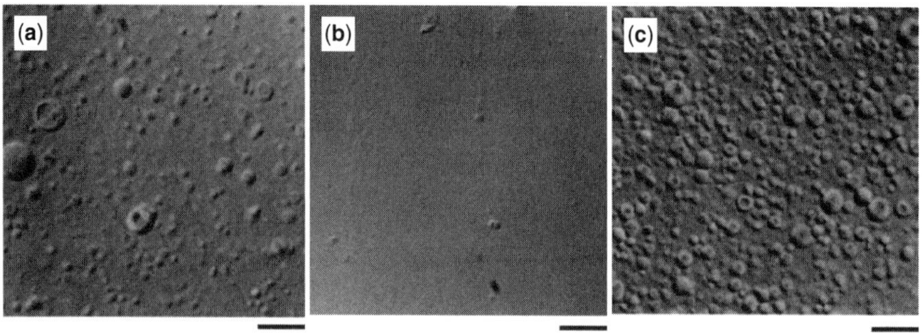

Figure 34.1. Optical micrographs of mixed solutions of FTMA and SDBS (1 wt % total surfactant concentration, FTMA : SDBS = 7:3) containing (a) reduced, (b) oxidized, and (c) re-reduced forms of FTMA prepared by electrolysis. The scale bar corresponds to 17 μm.

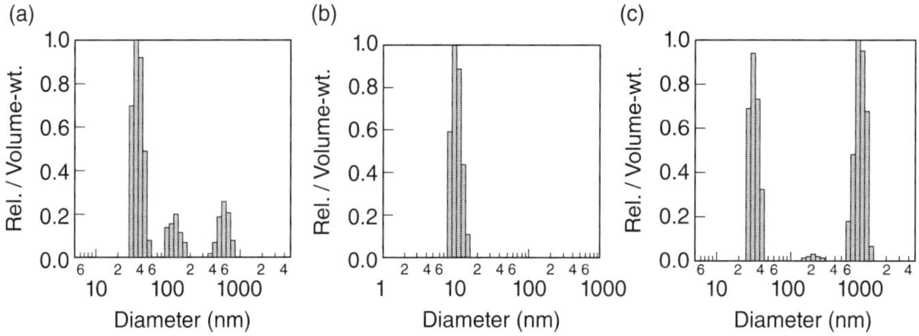

Figure 34.2. Particle size distribution measured by dynamic light scattering (DLS) for aqueous FTMA/SDBS mixtures (1.0 wt % total surfactant concentration, FTMA : SDBS = 7:3) containing (a) reduced, (b) oxidized, and (c) re-reduced forms of FTMA prepared by electrolysis.

(swollen micelles) and peaks around 110 and 570 nm due to vesicles were obtained for the reduced form (Fig. 34.2a), whereas a peak around 10 nm alone was observed for the oxidized form (Fig. 34.2b). Larger vesicles observed in the optical microscopic image of the reduced form (Fig. 34.1a) could not be measured because dispersed particles larger than about 5 μm cannot be detected with light scattering techniques. This change in aggregation size indicates that the oxidation of FTMA induces the disruption of vesicles and the formation of smaller assemblies (micelles). The re-reduced form gave a size distribution concentrated around 30 nm due to swollen micelles and around 190 and 830 nm due to vesicles (Fig. 34.2c). This result suggests the reformation of vesicles caused by the reduction of FTMA.

Consideration of the mechanism of vesicle disruption induced by oxidation suggests that vesicles are disrupted by the dissolution of the FTMA/SDBS complex constituting their bimolecular membranes into the bulk phase as a result of its largely increased hydrophilicity (Scheme 34.1). The FTMA molecule becomes a bolaform amphiphile (a molecule in which two hydrophilic groups are located at both ends of the hydrophobic chain) with ferrocenium ion by oxidation, whereas the ferrocenyl group at the terminal of the hydrophobic group is electrically neutral in the reduced form. Thus, dramatical changes in the hydrophilicity and hydrophobicity of FTMA caused by the redox reaction bring about the disruption and reformation of vesicles (Fig. 34.3).

34.3 REVERSIBLE CONTROL OF VISCOELASTICITY USING A REDOX REACTION

Significant attention has recently been paid to the active control of viscosity or viscoelasticity of fluids using external stimuli. The approach of controlling

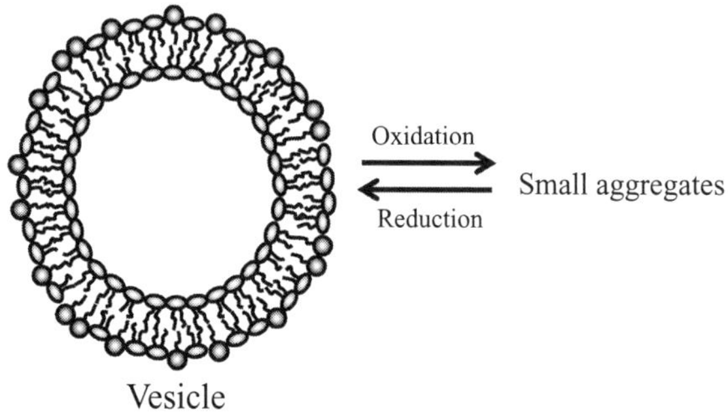

Figure 34.3. Formation and disruption model of vesicle induced by the redox reaction of FTMA.

fluid viscosity can be applied to clutches for transmissions [42], valves [43], precise control of mechanical positions [44], and controlled release of substances such as dyes, perfumes, and drugs. Electrorheological fluids (ERFs) [45, 46] are colloidal dispersions, the viscosity of which can be controlled by applied electric potential. Conventional ERFs are roughly classified into two types of dispersion systems: fine particle dispersions [46, 47] and liquid crystal dispersions [48, 49]. Application of an alternating electric field to low-viscosity ER dispersions produces a viscosity rise mainly in the direction perpendicular to the field because the dispersed particles align to form strings in the direction parallel to the field due to the attraction of induced dipoles [45]. The electrically induced properties of fine particle and liquid crystal dispersions generally correspond to Bingham and Newtonian fluids, respectively. These conventional ERFs, however, have some disadvantages in terms of their industrial applications because their stability is usually poor, and a high voltage (usually above 1 kV/mm) and a high dispersoid concentration (usually above 10 vol %) are required to gain a useful viscosity change.

The present section deals with our recent studies [38] on a novel ERF whose viscoelasticity can be controlled using a "redox reaction" and *low* (~0.5 V) electrochemical potentials. Our system is based on a cationic "redox-switchable" ferrocenyl surfactant that self-assembles into wormlike micelles. Wormlike micelles cause the system to exhibit a remarkable viscoelastic behavior as a result of the formation of a three-dimensional network consisting of entangled micelles. Wormlike micelles are thermodynamically *stable* in water. The viscoelastic behavior of wormlike micellar solutions is similar to that of concentrated polymer solutions, except that the micelles can pass through themselves by opening and reclosing (phantom network) [50, 51]. Some cationic surfactants, such as cetyltrimethylammonium bromide (CTAB) or cetyl pyridinium bromide, are known to form wormlike micelles in water even at *low* concentrations when organic additives such as sodium salicylate

(NaSal) [50, 51], tosylate [52, 53], butyl benzene sulfonate [54–56], and hydroxynaphthalenecarboxylate [57–59] are added to the solution.

In this study, FTMA, a cationic redox-switchable ferrocenyl surfactant, was mixed with NaSal to obtain viscoelastic wormlike micellar solutions. It is possible to reduce the viscoelasticity of aqueous mixtures of FTMA (reduced form) and NaSal by controlling the degree of entanglement of wormlike micelles through the oxidation of FTMA.

34.3.1 Phase and Rheological Behavior of Aqueous Mixtures of FTMA (Reduced Form) and NaSal

Figure 34.4 shows a photograph of the appearance of an aqueous solutions containing mixture of FTMA/NaSal with molar ratios of [NaSal]/[FTMA] = 0, 0.2, and 0.4. Tilting of the solutions by 90° caused the solutions of FTMA alone (Fig. 34.4a) and FTMA/NaSal mixture at the molar ratio of [NaSal]/[FTMA] = 0.2 (Fig. 34.4b) to flow immediately, whereas it produced no flow of the solution of the mixture at the molar ratio of [NaSal]/[FTMA] = 0.4 (Fig. 34.4c). This suggests the formation of wormlike micelles in the solution at the higher molar ratio. This solution was also found to be spinnable, indicating that it was elastic as well as viscous.

We then conducted dynamic viscoelasticity measurements at 25°C on the aqueous mixture at the molar ratio ([NaSal]/[FTMA]) of 0.4. The mixture was found to be a non-Newtonian fluid. Figure 34.5 shows the storage modulus G′

(c)

(b)

(a)

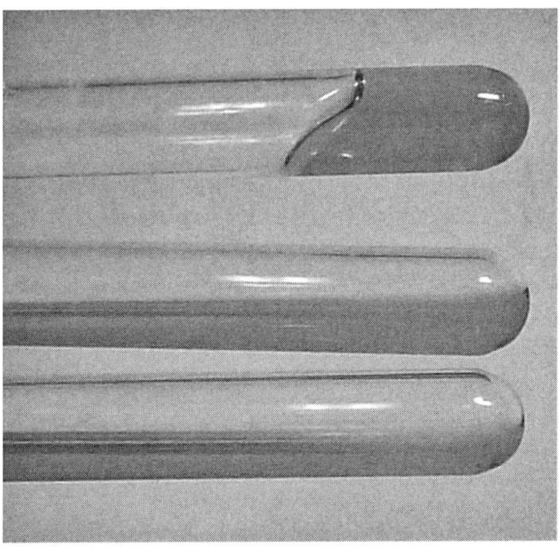

Figure 34.4. Photograph showing the appearance of an aqueous FTMA (reduced form)/NaSal mixture at a fixed FTMA concentration of 50 mM ([NaSal]/[FTMA] = (a) 0, (b) 0.2, and (c) 0.4) tilted by an angle of 90°.

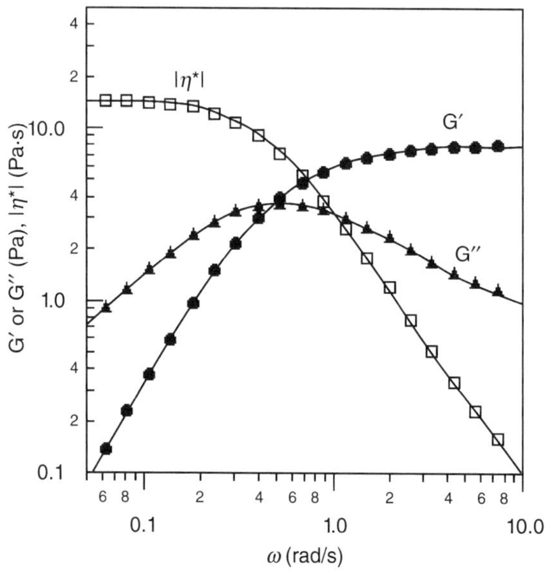

Figure 34.5. Storage modulus G′ (solid circles), loss modulus G″ (solid triangles), and the real part of the complex viscosity |η*| (open squares) as a function of angular frequency ω for aqueous FTMA (reduced form)/NaSal mixture (50 mM FTMA, [NaSal]/[FTMA] = 0.4).

(Pa) and the loss modulus G″ (Pa) as a function of angular frequency ω (rad/s). The G′ curve exhibited a definite plateau at high frequencies and G″ was higher than G′ at low frequencies. The viscoelastic behavior of the mixture shown in Figure 34.5 is similar to that of concentrated colloidal polymer systems with sufficient entanglement. Figure 34.6 shows that the relationship between G″ and G′, a Cole–Cole plot, for the mixture is in accordance with a semi-circular-shaped curve, indicating a behavior consistent with a Maxwell model with a single relaxation [51, 53, 58, 60, 61]. The viscoelastic behavior in Figure 34.6 is similar to that for aqueous solutions of CTAB, a typical cationic surfactant, with added NaSal [51]. We estimated the zero shear viscosity (η_0) of the mixture. The real part of the complex viscosity $|\eta^*|$ is also shown in Figure 34.5. The zero shear viscosity is given by $\lim_{\omega \to 0} |\eta^*|$. The value of η_0 obtained for the mixture was about 15 Pa·s, which is four orders of magnitude higher than the viscosity of water (about 1×10^{-3} Pa·s). The viscoelastic behavior of the aqueous FTMA/NaSal mixture contrasts with conventional ERFs, which show Bingham or Newtonian properties.

The aggregation state of the mixture was investigated by freeze–fracture transmission electron microscopy (FF-TEM). Extremely elongated aggregates that were flexible and entangled with each other were observed on the FF-TEM micrograph (Fig. 34.7). These aggregates had diameters of 8–10 nm

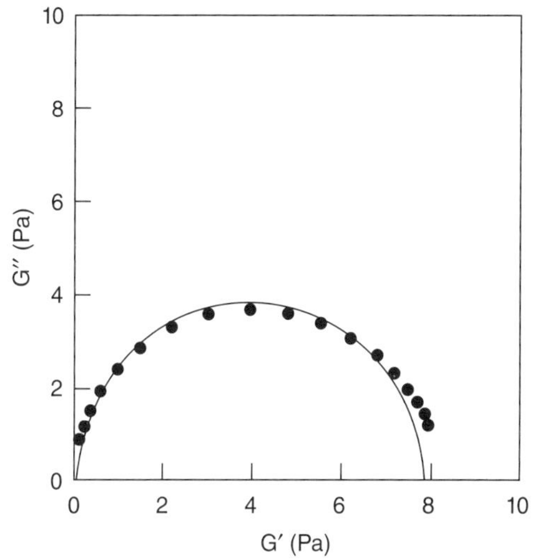

Figure 34.6. Cole–Cole plots (G″ vs. G′) for aqueous FTMA (reduced form)/NaSal mixture (50 mM FTMA, [NaSal]/[FTMA] = 0.4). The semicircle traces the behavior of a Maxwell-type liquid.

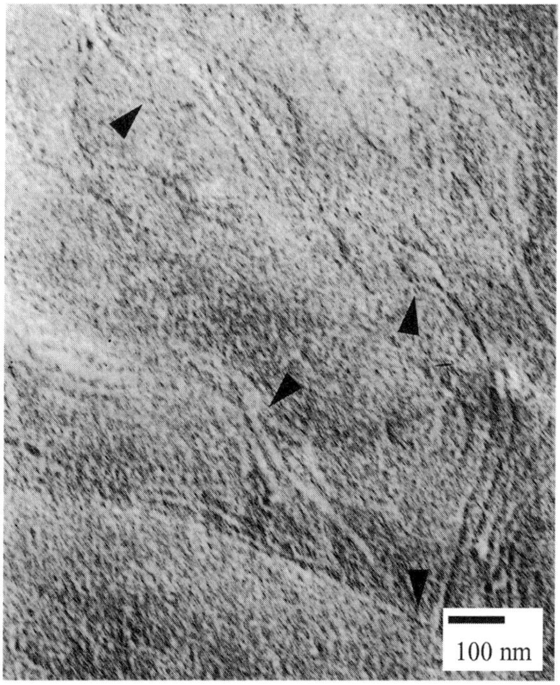

Figure 34.7. Freeze–fracture transmission electron microscopy micrograph of aqueous FTMA (reduced form)/NaSal mixture (50 mM FTMA, [NaSal]/[FTMA] = 0.4).

and lengths in the micrometer range. The micrograph clearly shows that the aggregates are wormlike micelles even though their diameter is slightly larger than that (5 nm) of wormlike micelles formed in CTAB/NaSal aqueous mixtures [61]. The high viscoelasticity observed for the mixed solution is therefore responsible for an entangled network of wormlike micelles of FTMA.

34.3.2 Electrochemical Control of Viscoelasticity

The highly viscoelastic aqueous FTMA/NaSal mixture (50 mM FTMA, [NaSal]/[FTMA] = 0.4, total volume = 50 mL) was electrolyzed at +0.5 V versus SCE, which is much more positive than the equilibrium potential (+0.15 V vs. SCE) of pure FTMA.

After electrolytic oxidation, a peak assigned to the ferrocenium cation (oxidized form) [13] appeared at 625 nm and a peak assigned to the ferrocenyl group (reduced form) [13] at 440 nm decreased considerably in the UV–vis spectrum. This means that FTMA molecules in the sample were oxidized by the electrolysis. Figure 34.8 is a photograph of the mixture taken before (Fig. 34.8a) and after (Fig. 34.8b) potentiostatic oxidation. Upon oxidation of FTMA, the solution was observed to undergo a pronounced reduction in viscosity. When the sample was tilted by 90°, the oxidized form of the mixture (dark blue) easily flowed (Fig. 34.8b), whereas the reduced solution (yellow) hardly flowed (Fig. 34.8a). We then examined the viscoelastic behavior of the oxidized mixture with a double-concentric cylinder-type rheometer. The oxidized mixture behaved similarly to water. The shear stress plots against the shear rate were linear with a zero intercept, and no hysteresis was observed

(b)

(a)

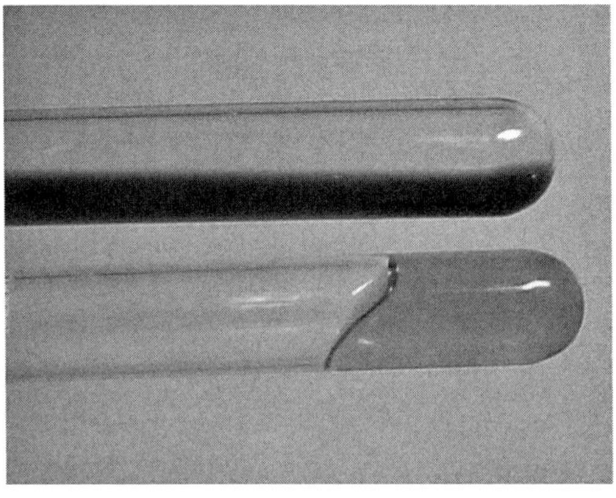

Figure 34.8. Photograph showing the appearance of aqueous FTMA/NaSal mixture (50 mM FTMA, [NaSal]/[FTMA] = 0.4) (a) before and (b) after electrolytic oxidation. See color figure on companion ftp site.

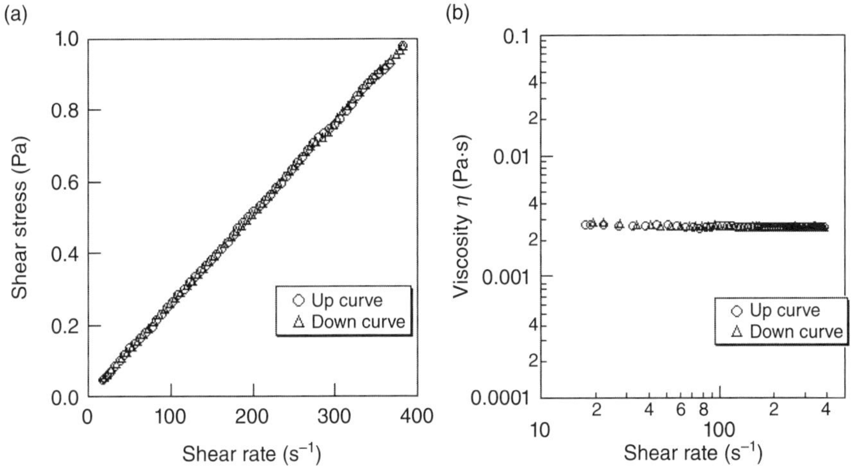

Figure 34.9. (a) Flow curve and (b) viscosity curve for FTMA/NaSal mixture (50 mM FTMA, [NaSal]/[FTMA] = 0.4) after electrolytic oxidation.

between the flow curves obtained by increasing and then decreasing the shear rate (Fig. 34.9a). The oxidized sample is therefore a Newtonian fluid, in which the viscosity is independent of shear stress. The viscosity is 2.5×10^{-3} Pa·s (Fig. 34.9b), which is about 1/6000 of the zero shear viscosity (15 Pa·s) of the reduced sample (Fig. 34.5). This remarkable viscoelasticity decrease is caused by a significant change in the aggregation state of the FTMA/NaSal mixture, that is, from wormlike micelles to monomers and/or other smaller aggregates (probably short rodlike micelles). Abbott et al. [10] reported the limiting surface areas (A_{min}) occupied by the reduced and oxidized forms of FTMA at the surface of aqueous solutions. The reported value of 75 ± 5 Å^2 for the oxidized form is very close to the value for the reduced form of 85 ± 5 Å^2, despite an appreciable difference in the hydrophilicity between the two forms of the surfactant. This is because both reduced and oxidized forms of FTMA behave as bolaform surfactants (both forms direct both ends of the molecule to the aqueous phase to form a looped configuration), thereby adopting similar conformations at the surface [10]. This means that the critical packing parameter (CPP), a parameter that relates the properties of a surfactant to its preferred aggregation form, of the oxidized form of FTMA is almost identical with that of the reduced form. Abbott et al. also reported the absence of micellization at concentrations of 30 mM for the oxidized form of FTMA compared with a critical micelle concentration (CMC) of 0.07 mM for reduced FTMA [10]. This suggests the dissolution into monomers of a portion of the molecules forming wormlike micelles upon oxidation. The number of aggregates formed in the mixture decreases and elliptical aggregates whose length is much shorter than the wormlike micelles are likely to be formed after oxidation of the surfactant.

Because the entanglement of elliptical aggregates to form a network is unlikely to occur, a large decrease in viscoelasticity results from oxidation. The results mentioned earlier demonstrate that control of the dynamic viscoelasticity of the system is possible via electrochemical control of a redox reaction. The present method of viscoelasticity control has several advantages over conventional ERFs. We can significantly decrease the viscoelasticity (down to 1/6000) of aqueous *dilute* FTMA/NaSal solutions ([FTMA] = 2.3 wt %) by applying an extremely *low* DC voltage (+0.5 V vs. SCE). In addition, aqueous FTMA(reduced form)/NaSal mixture is viscoelastic in all directions, whereas conventional ERFs exhibit their viscosity only in the direction perpendicular to the applied electric field. Under our electrolytic conditions, the time needed for the viscosity of the system to decrease is rather long compared with that for conventional ERFs. The time can be reduced, however, if the total sample volume is decreased and the surface area of the electrode is increased. FTMA molecules located near the electrode will be quickly oxidized, which in turn will cause a sudden decrease of viscoelasticity in the vicinity of the electrode. Hence, thin-layer cells typically employed for conventional ERF systems are far more useful for our system to control the viscoelasticity.

The achievements described in this section are the key to a method of efficiently controlling the rheological behavior of the system. This method would be applicable to the controlled release of substances such as dyes and perfumes entrapped in an entangled network of wormlike micelles, inkjet printers, flow rate controlling systems using electrical signals, and clutches for transmissions and valves.

34.4 CONCLUSIONS

This chapter described several approaches leading to dynamic control of colloidal self-assemblies that contain redox-active surfactants. Ferrocenyl surfactants allow changes in self-assemblies by potentiostatic electrolysis because the ferrocenyl group (reduced form) is hydrophobic, while the ferrocenium cation (oxidized form) behaves as a hydrophilic group.

When the reduced form of the ferrocenyl surfactant was mixed with an anionic surfactant in water, vesicles were spontaneously formed. Electrolytic oxidation induced the disruption of the vesicles, and subsequent reduction of the oxidized solution caused reformation of the vesicles. This reversible formation/disruption of vesicles could be applied to the controlled release of water-soluble substances (drugs, perfumes, dyes, etc.).

With respect to wormlike micellar systems, the oxidation of the ferrocenyl surfactant caused a pronounced change in the solution viscoelasticity because the three-dimensional entanglement of wormlike micelles formed in the reduced solution was collapsed into monomers and/or smaller aggregates due to the enhancement of the hydrophilicity of the ferrocenyl surfactant. This system is a promising candidate for a new ERF that could be applied to ink for inkjet printers and novel controlled release systems.

Further targeted designs and usage of redox-active surfactants should produce progress in new and interesting applications.

REFERENCES

1 S. Shinkai, K. Shigematsu, Y. Kusano, O. Manabe. Photoresponsive crown ethers. Part 3. Photocontrol of ion extraction and ion transport by several photofunctional bis(crown ethers). *J. Chem. Soc. Perkin Trans. 1* (1981) 3279–3283.

2 S. Shinkai, T. Ogawa, Y. Kusano, O. Manabe, K. Kikukawa, T. Goto, T. Masuda. Photoresponsive crown ethers. 4. Influence of alkali metal cations on photoisomerization and thermal isomerization of azobis(benzocrown ethers). *J. Am. Chem. Soc.* 104(7) (1982) 1960–1967.

3 H. Sakai. Photocontrol of solution properties. *Mater. Technol.* 19(4) (2001) 157–167.

4 L. Yang, N. Takisawa, T. Hayashita, K. Shirahama. Colloid chemical characterization of the photosurfactant 4-ethylazobenzene 4′-(oxyethyl)trimethylammonium bromide. *J. Phys. Chem.* 99(21) (1995) 8799–8803.

5 H. Tsukube. Artificial transport of amino-acid, oligopeptide, and related anions by macrocyclic polyamine-transition metal complex carriers. *J. Chem. Soc. Perkin Trans. 1* (1983) 29–34.

6 R. M. Izatt, J. D. Ramb, R. T. Hawkins, P. R. Brown, S. R. Izatt, J. J. Christensen. Selective M^+-H^+ coupled transport of cations through a liquid membrane by macrocyclic calixarene ligands. *J. Am. Chem. Soc.* 105(7) (1983) 1782–1785.

7 S. Yusa, Y. Shimada, T. Yamamoto. pH-responsive micellization of amphiphilic diblock copolymers synthesized via reversible addition–fragmentation chain transfer polymerization. *Macromolecules* 36(11) (2003) 4208–4215.

8 S. Shinkai, S. Nakamura, S. Tachiki, O. Minabe, T. Kajiyama. Thermocontrol of ion permeation through ternary composite membranes composed of polymer/liquid crystal/amphiphilic crown ethers. *J. Am. Chem. Soc.* 107(11) (1985) 3363–3365.

9 N. Aydogan, N. L. Abbott. Effect of electrolyte concentration on interfacial and bulk solution properties of ferrocenyl surfactants with anionic headgroups. *Langmuir* 18(21) (2002) 7826–7830.

10 B. S. Gallardo, K. L. Metcalfe, N. L. Abbott. Ferrocenyl surfactants at the surface of water: principles for active control of interfacial properties. *Langmuir* 12(17) (1996) 4116–4124.

11 P. Anton, J. Heinze, A. Laschewsky. Redox-active monomeric and polymeric surfactants. *Langmuir* 9(1) (1993) 77–85.

12 K. Hoshino, K. Suga, T. Saji. Electrolyte effects on the electrochemical activity of micelle-solubilized substance. *Chem. Lett.* 15(6) (1986) 979–982.

13 T. Saji, K. Hoshino, Y. Ishii, M. Goto. Formation of organic thin films by electrolysis of surfactants with the ferrocenyl moiety. *J. Am. Chem. Soc.* 113(2) (1991) 450–456.

14 T. Saji, K. Hoshino, S. Aoyagui. Reversible formation and disruption of micelles by control of the redox state of the head group. *J. Am. Chem. Soc.* 107(24) (1985) 6865–6868.

15 K. Hoshino, T. Saji. Electrochemical formation of an organic thin film by disruption of micelles. *J. Am. Chem. Soc.* 109(19) (1987) 5881–5883.

16 T. Saji, K. Hoshino, S. Aoyagui. Reversible formation and disruption of micelles by control of the redox state of the surfactant tail group. *J. Chem. Soc. Chem. Commun.* (1985) 865–866.

17 K. Hoshino, T. Saji. Electrochemical formation of thin film of viologen by disruption of micelles. *Chem. Lett.* 16(7) (1987) 1439–1442.

18 T. Saji. Electrochemical formation of a phthalocyanine thin film by disruption of micellar aggregates. *Chem. Lett.* 17(4) (1988) 693–696.

19 Y. Takeoka, T. Aoki, K. Sanui, N. Ogata, M. Yokoyama, T. Okano, Y. Sakurai, M. Watanabe. Electrochemical control of drug release from redox-active micelles. *J. Control. Release* 33(1) (1995) 79–87.

20 Y. Takeoka, T. Aoki, K. Sanui, N. Ogata, M. Watanabe. Electrochemical studies of a redox-active surfactant. correlation between electrochemical responses and dissolved states. *Langmuir* 12(2) (1996) 487–493.

21 Y. Kakizawa, H. Sakai, M. Abe, Y. Kondo, N. Yoshino. Redox control of solubilization with a single-tailed cationic surfactant bearing ferrocenyl moiety. *Mater. Technol.* 19(6) (2001) 259–266.

22 Y. Kakizawa, H. Sakai, A. Yamaguchi, Y. Kondo, N. Yoshino, M. Abe. Electrochemical control of vesicle formation with a double-tailed cationic surfactant bearing ferrocenyl moieties. *Langmuir* 17(26) (2001) 8044–8048.

23 T. Takei, H. Sakai, Y. Kondo, N. Yoshino, M. Abe. Electrochemical control of solubilization using a ferrocene-modified nonionic surfactant. *Colloids Surf. A* 183–185 (2001) 757–765.

24 J. C. Medina, I. Gay, Z. Chen, L. Echegoyen, G. W. Gokel. Redox-switched molecular aggregates: the first example of vesicle formation from hydrophobic ferrocene derivatives. *J. Am. Chem. Soc.* 113(1) (1991) 365–366.

25 S. Muñoz, G. W. Gokel. Redox-switched vesicle formation from two novel, structurally distinct metalloamphiphiles. *J. Am. Chem. Soc.* 115(11) (1993) 4899–4900.

26 K. Wang, S. Muñoz, L. Zhang, R. Castro, A. E. Kaifer, G. W. Gokel. Organometallic amphiphiles: oxidized ferrocene as headgroup for redox-switched bilayer and monolayer membranes. *J. Am. Chem. Soc.* 118(28) (1996) 6707–6715.

27 K. Wang, G. W. Gokel. Redox-switched amphiphiles: oxidized ferrocene derivatives form stable vesicles when either one or two alkyl tails are present. *J. Phys. Org. Chem.* 10(5) (1997) 323–334.

28 Y. Kakizawa, H. Sakai, K. Nishiyama, M. Abe, H. Shoji, Y. Kondo, N. Yoshino. Solution properties of double-tailed cationic surfactants having ferrocenyl groups in their hydrophobic moieties. *Langmuir* 12(4) (1996) 921–924.

29 H. Sakai, H. Imamura, Y. Kakizawa, M. Abe, Y. Kondo, N. Yoshino, J. H. Harwell. Active control of vesicle formation using a redox-active surfactant. *Denki Kagaku oyobi Kogyo Butsuri Kagaku* 65(8) (1997) 669–672.

30 N. Aydogan, N. L. Abbott. Comparison of the surface activity and bulk aggregation of ferrocenyl surfactants with cationic and anionic headgroups. *Langmuir* 17(19) (2001) 5703–5706.

31 N. Aydogan, B. S. Gallardo, N. L. Abbott. A molecular-thermodynamic model for Gibbs monolayers formed from redox-active surfactants at the surfaces of aqueous solutions: Redox-induced changes in surface tension. *Langmuir* 12(3) (1999) 722–730.

32 B. S. Gallardo, N. L. Abbott. Active control of interfacial properties: a comparison of dimeric and monomeric ferrocenyl surfactants at the surface of aqueous solutions. *Langmuir* 13(2) (1997) 203–208.

33 Y. -Y. Luk, N. L. Abbott. Surface-driven switching of liquid crystals using redox-active groups on electrodes. *Science* 301(5633) (2003) 623–626.

34 N. L. Abbott, C. M. Jewell, M. E. Hays, Y. Kondo, D. M. Lynn. Ferrocene-containing cationic lipids: Influence of redox state on cell transfection. *J. Am. Chem. Soc.* 127(33) (2005) 11576–11577.

35 C. L. Pizzey, C. M. Jewell, M. E. Hays, D. M. Lynn, N. L. Abbott, Y. Kondo, S. Golan, Y. Talmon. Characterization of the nanostructure of complexes formed by a redox-active cationic lipid and DNA. *J. Phys. Chem. B* 112(18) (2008) 5849–5857.

36 M. E. Hays, C. M. Jewell, Y. Kondo, D. M. Lynn, N. L. Abbott. Lipoplexes formed by DNA and ferrocenyl lipids: Effect of lipid oxidation state on size, internal dynamics, and ζ-potential. *Biophys. J.* 93(12) (2007) 4414–4424.

37 H. Sakai, H. Imamura, Y. Kondo, N. Yoshino, M. Abe. Reversible control of vesicle formation using electrochemical reaction. *Colloids Surf. A* 232(2–3) (2004) 221–228.

38 K. Tsuchiya, Y. Orihara, Y. Kondo, N. Yoshino, T. Ohkubo, H. Sakai, M. Abe. Control of viscoelasticity using redox reaction. *J. Am. Chem. Soc.* 126(39) (2004) 12282–12283.

39 E. W. Kaler, A. K. Murthy, B. E. Rodriguez, J. A. N. Zasadzinski. Spontaneous vesicle formation in aqueous mixtures of single-tailed surfactants. *Science* 245(4924) (1989) 1371–1374.

40 M. T. Yatcilla, K. L. Herrington, L. L. Brasher, E. W. Kaler, S. Chiruvolu, J. A. Zasadzinski. Phase behavior of aqueous mixtures of cetyltrimethylammonium bromide (CTAB) and sodium octyl sulfate (SOS). *J. Phys. Chem.* 100(14) (1996) 5874–5879.

41 E. W. Kaler, K. L. Herrington, A. K. Murthy, J. A. N. Zasadzinski. Phase behavior and structures of mixtures of anionic and cationic surfactants. *J. Phys. Chem.* 96(16) (1992) 6698–6707.

42 D. L. Hartsock, R. F. Novak, G. J. Chaundy. ER fluid requirements for automotive devices. *J. Rheol.* 35(7) (1991) 1305–1326.

43 H. Block, J. P. Kelly. Electro-rheology. *J. Phys. D Appl. Phys.* 21(12) (1988) 1661–1677.

44 P. J. Burchill. Electrorheological fluids and their applications. *Mater. Forum* 15(3) (1991) 197–204.

45 W. M. Winslow. Induced fibration of suspensions. *J. Appl. Phys.* 20(12) (1949) 1137–1140.

46 T. C. Halsey. Electrorheological fluids. *Science* 258(5083) (1992) 761–766.

47 A. Yethiraj, A. van Blaaderen. A colloidal model system with an interaction tunable from hard sphere to soft and dipolar. *Nature* 421(6922) (2003) 513–517.

48 M. Miesowics. The three coefficients of viscosity of anisotropic liquids. *Nature* 158(4001) (1946) 27.

49 I. -K. Yang, A. D. Shine. Electrorheology of a nematic poly(n-hexyl isocyanate) solution. *J. Rheol.* 36(6) (1992) 1079–1104.

50 H. Shikata, H. Hirata, T. Kotaka. Micelle formation of detergent molecules in aqueous media: Viscoelastic properties of aqueous cetyltrimethylammonium bromide solutions. *Langmuir* 3(6) (1987) 1081–1086.

51 T. Shikata, S. J. Dahman, D. S. Pearson. Rheo-optical behavior of wormlike micelles. *Langmuir* 10(10) (1994) 3470–3476.

52 Z. Lin, J. J. Cai, L. E. Scriven, H. T. Davis. Spherical-to-wormlike micelle transition in CTAB Solutions. *J. Phys. Chem.* 98(23) (1994) 5984–5993.

53 J. F. A. Soltero, J. E. Puig, O. Manero. Rheology of the cetyltrimethylammonium tosilate–water system. 2. Linear viscoelastic regime. *Langmuir* 12(11) (1996) 2654–2662.

54 S. Imai, T. Shikata. Viscoelastic behavior of surfactant threadlike micellar solutions: Effect of additives 3. *J. Colloid Interface Sci.* 244(2) (2001) 399–404.

55 M. Bhat, V. G. Gaikar. Characterization of interaction between butyl benzene sulfonates and cetyl trimethylammonium bromide in mixed aggregate systems. *Langmuir* 15(14) (1999) 4740–4751.

56 M. Bhat, V. G. Gaikar. Characterization of interaction between butylbenzene sulfonates and cetyl pyridinium chloride in a mixed aggregate system. *Langmuir* 16(4) (2000) 1580–1592.

57 P. A. Hassan, B. S. Valaulikar, C. Monohar, F. Kern, L. Bourdieu, S. J. Candau. Vesicle to micelle transition: rheological investigations. *Langmuir* 12(18) (1996) 4350–4357.

58 R. Oda, J. Narayanan, P. A. Hassan, C. Monohar, R. A. Salkar, F. Kern, S. J. Candau. Effect of the lipophilicity of the counterion on the viscoelasticity of micellar solutions of cationic surfactants. *Langmuir* 14(16) (1998) 4364–4372.

59 E. Mendes, J. Narayanan, R. Oda, F. Kern, S. J. Caudau, C. Manohar. Shear-induced vesicle to wormlike micelle transition. *J. Phys. Chem. B* 101(13) (1997) 2256–2258.

60 T. Shikata, M. Shiokawa, S. Imai. Viscoelastic behavior of surfactant threadlike micellar solutions: effects of additives, 4. *J. Colloid Interface Sci.* 259(2) (2003) 367–373.

61 T. Shikata. Rheological studies of threadlike micellar solutions and spherical particle suspensions. *Nihon Reoroji Gakkaishi* 25(5) (1997) 255–265.

35

PHOTOINDUCED MANIPULATION OF SELF-ORGANIZED NANOSTRUCTURE OF BLOCK COPOLYMERS

Kunihiko Okano and Takashi Yamashita

35.1 INTRODUCTION

Self-assembly of molecules to form ordered structures is an attractive issue in biology and materials science. For example, in proteins, intramolecular and intermolecular self-assembly plays a key role in enzyme or carrier systems. In materials science, the organization of molecules is quite important for mechanical, optical, and electrical properties. Block copolymers are fascinating condensed soft materials that are made possible after the discovery of living anionic polymerization in the middle of the last century. The molecular architecture of a block copolymer, containing two or more chemically distinct polymer segments connected to each other, leads to microphase separation of dissimilar polymer domains into periodic structures. Typical sizes of these domains range from 5 to 50 nm, which are useful periodic values in the semiconductor industry for future integrated circuits.

Block copolymers prepared by controlled polymerization are considered in this chapter. So far, several types of block copolymers have been investigated (Fig. 35.1). Shape and the number of distinct homopolymer sections determine the molecular architecture of a block copolymer; diblock, triblock, multiblock, star–shaped, and cyclic copolymers are possible. Their chemical and physical

Electrical Phenomena at Interfaces and Biointerfaces: Fundamentals and Applications in Nano-, Bio-, and Environmental Sciences, First Edition. Edited by Hiroyuki Ohshima.

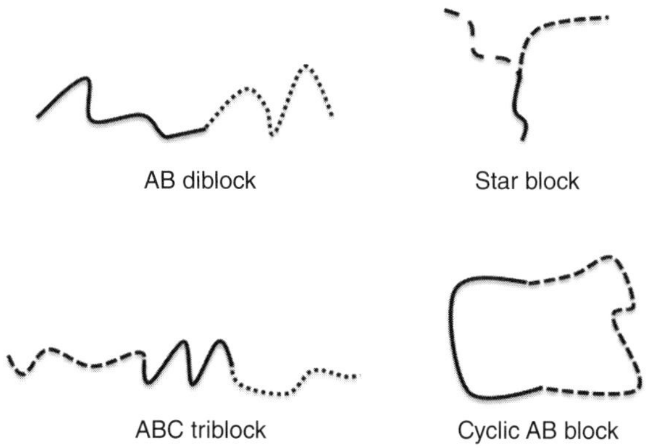

Figure 35.1. Block copolymer architectures.

properties make it possible to precisely control the size and shape of microdomains afforded by simply changing their molecular weights and compositions.

Broad potential uses of block copolymers for versatile nanotechnologies have been proposed based on their unique nanopatterns. However, the main subject of using block copolymers lies with control of microstructure. A variety of mechanical, electrical, and magnetic forces and surface interactions have been suggested as means to manipulate the nanostructures of block copolymers.

35.2 SYNTHESIS

Presently, block copolymers are obtained from anionic polymerization and controlled radical polymerization. A primary example of a living polymerization is an anionic system. Anionic polymerization is still used widely in the industrial manufacture of block copolymers, including poly(styrene-butadiene-styrene) (SBS) thermoplastic elastomers. Recently, exotic molecular architectures such as mixed arm stars [1–3] and cyclic block copolymers [4] have been reported.

The main advance in the synthesis of block copolymers in the last decade has been the development of living radical polymerization. In the process of living radical polymerization, atom transfer radical polymerization (ATRP) has been used most broadly to prepare block copolymers. ATRP has been used to prepare AB diblock and ABC triblock copolymers [5]. The technique has been used to create block copolymers based on polystyrene and various polyacrylates.

35.3 PHASE BEHAVIOR

To discuss the phase behavior of block copolymers, diblock copolymers are useful examples because these have the simplest molecular architecture and have been studied extensively from experimental and theoretical perspectives [6–11]. In diblock copolymers, the phase separation occurs in binary mixtures of the copolymers and results in nanoscale structural organization of each block. The structure of phase separation is determined by the strength of the repulsive interaction as characterized by the product χN, where χ is the Flory–Huggins interaction parameter [12, 13] and N is the number of monomers in the diblock copolymer. When this value exceeds a critical value, phase separation can occur. Figure 35.2 shows a typical microphase separation in diblock copolymers. These structures are determined by a delicate energy balance between the stretching energy of the polymer chains and the interaction energies at the interface of each domain. The morphology of the microdomains ranges from spheres to lamellae depending on the volume fraction of one polymer segment. The sizes of the microdomains depend on the molecular weight of polymer segments. In symmetric diblock copolymers exhibiting

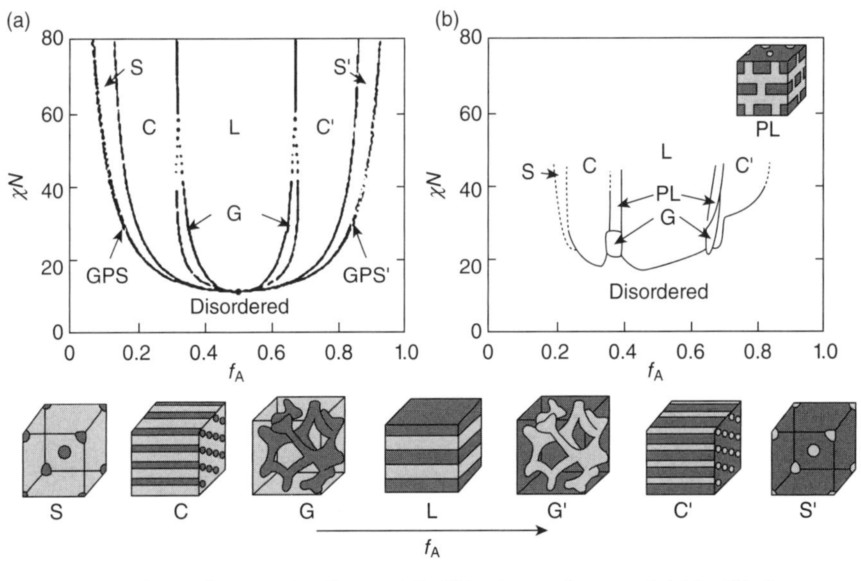

Figure 35.2. Phase diagram for linear AB diblock copolymers. (a) Equilibrium morphologies predicted by self-consistent mean-field theory: spherical (S), cylindrical (C), gyroid (G), and lamellar (L). (b) Experimental phase diagram of poly(styrene-b-isoprene) diblock copolymers: perforated layers (PL). Bottom: a representation of the equilibrium microdomain structures as f_A is increased for fixed χN [74].

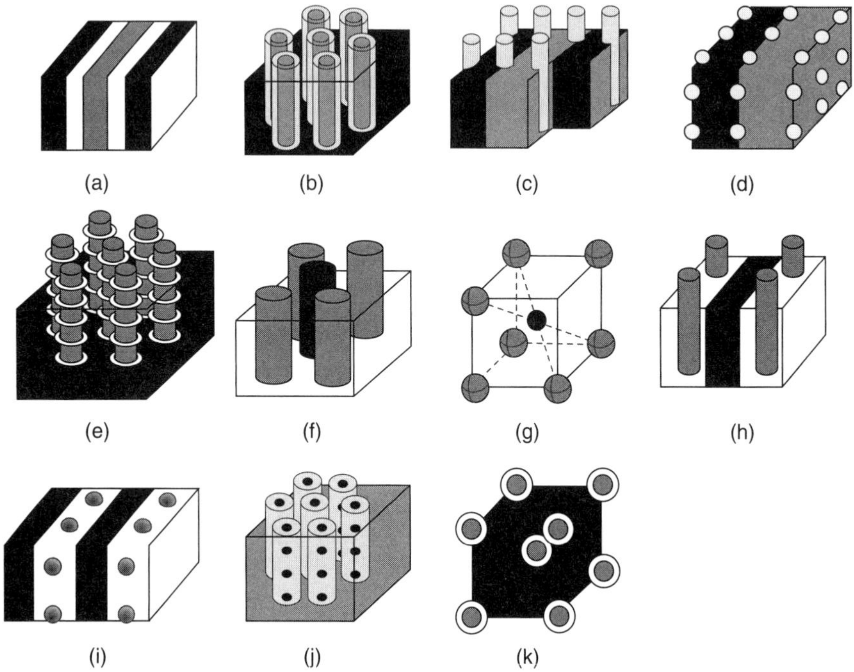

Figure 35.3. Schematic representation of all the phases considered. Each shade indicates domains of phase separations. (a) Lamellar phase, (b) coaxial cylinder phase, (c) lamella–cylinder phase, (d) lamella–sphere phase, (e) cylinder–ring phase, (f) cylindrical domains in a square lattice structure, (g) spherical domains in the CsCl-type structure, (h) lamella–cylinder, (i) lamella–sphere, (j) cylinder–sphere, (k) cocentric spherical domain in the bicontinuous cubic (bcc) structure [14].

a lamellar morphology, the scale of lamellar microdomains depends on the degree of segregation.

The microphase separation of triblock copolymers is more complicated than that of diblock copolymers. In triblock copolymers, there are three interaction parameters, between AB, BC, and CA blocks. Thus, the morphology depends not only on the temperature, molecular weight, and fraction species of each block but also on the sequence of polymer segments. These parameters yield a complex morphology of microdomains in triblock copolymers (Fig. 35.3). A theoretical calculation of the morphological phase diagram for ABC triblock copolymers in the strong segregation limit was performed by Zheng and Wang [14].

The phase structures of block copolymers with more complex architectures, $(AB)_n$ multiblock copolymers [15], have been studied. These molecular structures extend beyond the AB diblock and ABC triblock systems. For example, Matsushita and coworkers have reported systematic studies of the hierarchical microdomain structure in a multiblock copolymer [16–19].

35.4 PHASE BEHAVIOR OF COPOLYMER/ HOMOPOLYMER BLENDS

Self-organization of block copolymers has yielded morphologies that in some cases differ significantly from those observed in comparable linear copolymer systems [20, 21]. A morphological transformation in conventional block copolymer/homopolymer blends provides significant versatility in terms of tailored polymer nanostructures. Whereas a homopolymer will tend to localize within its host microphase, an added copolymer will tend to be more spatially confined. To precisely control the morphology, the use of two block copolymers enlarges the parameter space that can be feasibly explored. The segment species (e.g., A, B, or C blocks), composition, and molecular weight of each copolymer in a given copolymer/copolymer blend affect its morphology and yield complex nanostructures. There are two types of blends composed of block copolymers: two diblock copolymers possessing a common chemical species or an ordered diblock copolymer mixed with an ordered triblock copolymer.

35.5 THIN-FILM MORPHOLOGY

The morphology of a microdomain is crtically influenced both by film thickness and by the energetics of interaction at the air/polymer and polymer/ substrate interfaces [22–26]. Generally, interfacial energies differ for blocks of different chemical compositions. For example, lamellar microdomains of symmetric (styrene-methylmethacrylate) block copolymer (PS-*b*-PMMA) deposited on the native oxide surface of a silicon wafer are oriented parallel to the substrate, with polymethylmethacrylate (PMMA) wetting the substrate surface, while PS lamellae assemble at the air interface. This ordering behavior of PS-*b*-PMMA films on a selective substrate has been investigated systematically by Anastasiadis et al. using secondary ion mass spectroscopy, interference microscopy, and neutron reflectivity [26].

35.6 MANIPULATION OF MORPHOLOGIES BY EXTERNAL STIMULI

In this section, we review various approaches to manipulate the microdomains of block copolymer films that are important to realize their potential for nanotechnologies. Controlling microdomain orientation and defects plays a crucial role for optimization of the resultant physical properties of nanostructures. As mentioned earlier, microdomain control of the molecular structure can be achieved through the choice of molecular weight, composition, and chain architecture. Coupling of external stimuli via a mechanical field, an electric field, a temperature gradient, or photoirradiation can create the preferred orientation out of the set of energy-equivalent arrangements. Each example of external bias for a change in microdomain will be discussed in the following sections.

35.6.1 Mechanical Fields

Keller et al. reported the first example of orientation of a cylinder morphology in block copolymers by a shearing procedure [27, 28]. After this, other types of procedures and materials were utilized to develop different textured microdomains, for example, coupling of liquid crystalline mesogenes to one of the blocks to induce a global orientation of microdomains [29, 30]. These works propose that the hierarchical structure formed by the liquid crystalline phase and the microdomains of block copolymers could provide opportunities for manipulating the microstructures by external stimuli.

Another microstructure orientation technique called roll casting has been developed by Thomas et al. [31–34]. In this procedure, the polymer solution is applied to counter-rotating rolls, and this permits the processing of high-molecular-weight block copolymers below their degradation temperature. Roll casting is also used on side-chain liquid crystalline block copolymers to study the relative alignment of liquid crystalline mesogenes with respect to the microdomains in well-ordered sample films.

35.6.2 Direct Crystallization

Polymer microdomains are grown directionally according to temperature gradients as well as single-crystal silicon ingots. A typical method is addition of seed crystal to a melting sample. By applying this methodology, bulk lamellar systems form well-oriented structures without shear [35, 36]. Further application of this technique to thin films provides an additional procedure to control macroscopic orientation [37].

Alternatively, Thomas et al. have shown that benzoic acid, which exhibits crystallization in suitable polymers, forms aligned microdomains perpendicular to the substrates [38]. However, it not clear that suitable solvents can be found for a wide variety of copolymer systems. Moreover, addressability of the cylinders is rendered difficult because the grain size of the cylinder is small.

35.6.3 Electric Fields

It has been reported that an electric field can be applied to align polymeric microdomains. There have been two main techniques for electric alignment in films: in-plane and out-of-plane electrodes. For example, Morkved and Amundson demonstrated the possibility of microfabricated in-plane electrodes, which can generate strong electric fields to align cylindrical microdomains [39–41]. Furthermore, in this system, impressive ordered domain was shown by annealing above the glass transition temperature under the electric field. Out-of-plane electrodes can force cylindrical polystyrene microdomains to orient perpendicular to the substrate. While the technique is successful in aligning cylinders perpendicular to the substrate, the field has little or no effect on the packing of the cylinders. Morkved et al. investigated a morphological change from lamellar structures paralleling the substrate to perpendicular domains as the film thickness varies within the range [42].

35.6.4 Combination of Top-Down and Bottom-Up Approaches

Block copolymer nanofabrication provides large-area periodic functional patterns or objects with preferable sizes on the order of tens of nanometers. To precisely construct microdomains, the convergence of the lithographic top-down approach with block copolymer self-assembly within the same system has received attention.

There are generally two strategies for achieving spatial control of nanostructures: First, the substrates are patterned with lithographic techniques to generate top-down patterns, on top of which the block copolymer bottom-up nanopatterns can be directed; second, the combination of top-down and bottom-up patterning occurs.

35.7 PHOTOCONTROL OF MICROPHASE SEPARATIONS

At a molecular level, chromophore molecules are well-known for photoinduced alignment change, and such ordering can be enhanced by cooperative motions of the molecular alignment in a liquid crystalline phase. On the other hand, supramolecular ordered nanostructures that are self-assembled in block copolymer films have been widely used as templates to prepare nanomaterials [43–47]. In the microphase-separated processes, the supramolecular function between incompatible phases plays an important role in the formation of diverse nanostructures such as spheres, cylinders, double gyroids, and lamellae [43]. The orderings of one phase domain might endow the other microphase-separated domain with regularity by the supramolecular cooperative copolymers and might transfer the molecular ordering. Therefore, well-ordered nanostructures in the chromophore-containing block motion might be obtained by the photocontrol method. In the next section, we discuss the photoresponsive behavior of azobenzene chromophores in the polymer. We focus especially on the photochemical phase transition and the photoinduced alignment change during the liquid crystalline phase.

35.7.1 Photochemical Phase Transition of Azobenzene Liquid Crystalline Polymer

Azobenzene derivatives cause *trans–cis* photoisomerization and *cis–trans* thermal isomerization. The photochromic reactions in azobenzene are very fast, occurring on a timescale of picoseconds. Therefore, a fast laser pulse with

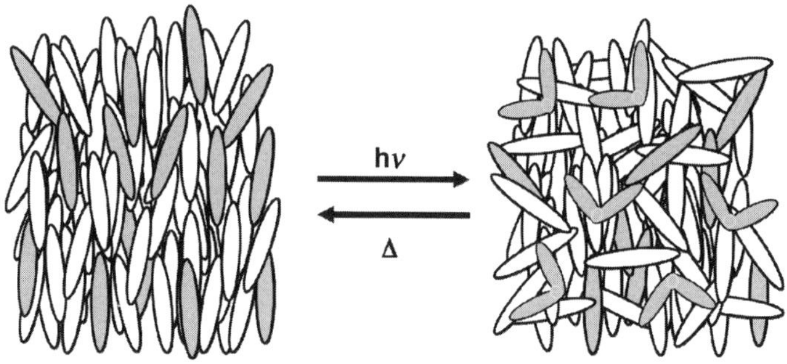

Figure 35.4. Photochemical phase transition of azobenzene/liquid crystal systems.

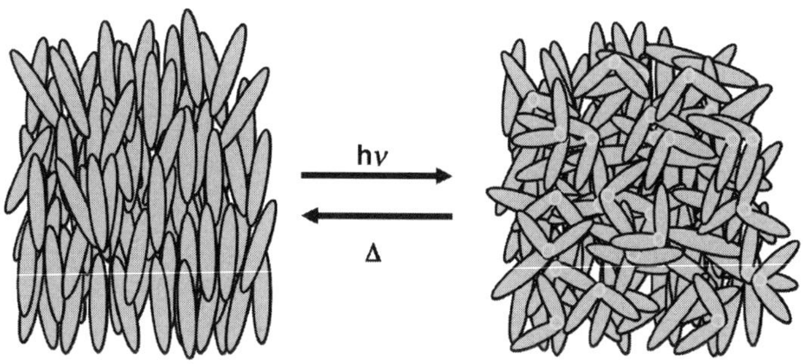

Figure 35.5. Photochemical phase transition of azobenzene liquid crystals.

a width of picoseconds is used as an excitation source for the photochemical phase transition of guest–host systems (Fig. 35.4). However, photoinduced alignment changes in azobenzene/liquid crystalline host systems depend strongly on an orientational relaxation of mesogenes. Liquid crystalline polymer doped with azobenzenes especially possesses high viscosity, so the orientational relaxation requires a relatively long time.

To overcome this problem, a new system has been developed in which every mesogene in liquid crystals or liquid crystalline polymers is provided with a chromophore [48–54]. It has been found that some azobenzene compounds exhibit liquid crystalline behavior. In these azobenzene liquid crystals, the azobenzene moiety could act both as a mesogene and as a photoresponsive moiety. On the basis of this concept, the photoresponses of azobenzene liquid crystals were examined with short laser pulses, and fast responses of azobenzene liquid crystals in nomadic–isotropic phase transitions have been demonstrated. It was found that in polymer azobenzene liquid crystals, the nematic–isotropic phase transition is induced in 200 ns under optimized conditions [48] (Fig. 35.5). This fast response of azobenzene liquid crystals is an encouraging result from the viewpoint of photonic devices.

35.7.2 Photoinduced Alignment Change of LCs by Linearly Polarized Light

trans-Azobenzenes exhibit π–π^* transition moments parallel to the direction of the transition moment. Therefore, *trans*-azobenzenes induce angular-dependent absorption of linearly polarized light; that is, *trans*-azobenzene molecules with their transition moments parallel to the polarization direction of linearly polarized light are activated very effectively to their excited states, followed by *trans–cis* isomerization, and molecules with their transition moments perpendicular to the polarization direction of actinic light are inactive toward isomerization. After repetition of isomerization cycles, once *trans*-azobenzene molecules have aligned perpendicular to the polarization direction of the actinic light, they become inactive toward the incident radiation. This means that at the end of the multicycles, there will be a net population of azobenzene molecules aligned perpendicular to the light polarization (Fig. 35.6). This is called the "Weigert effect" [55]. Since the first report of the phenomenon, there has been much research published on this photoinduced anisotropy in the alignment of dye molecules, For example, the work by Todorov et al. on a dye-doped optical system shows the possibility of using a polymer containing azobenzene molecules as an optical recording medium [56].

Important factors for the reorientation process found through intensive studies include azobenzene contents [57], morphology of samples before irradiation [58, 59], intensity of actinic linearly polarized light [60], and spacer length of side chains [61]. In addition, the structural effects of azobenzene moieties on photoinduced alignment behavior have been explored systematically in a series of side-chain liquid crystalline polymers possessing different azobenzene moieties (Fig. 35.7) [62–64].

As mentioned earlier, one can control the alignment direction of azobenzene-containing polymers in a two-dimensional manner (in-plane) by changing the polarization direction of the linearly polarized light. In the case of unpolarized

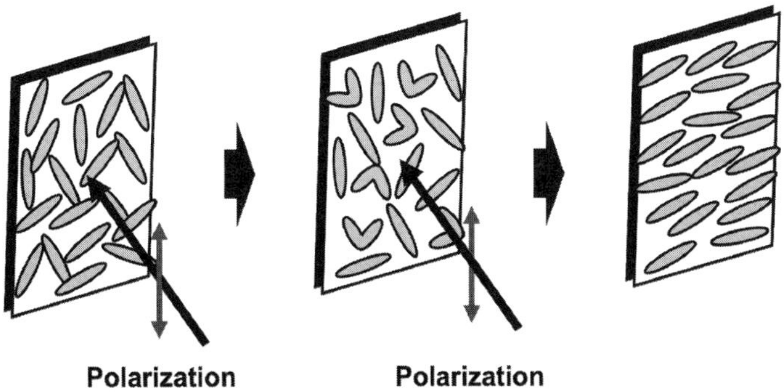

Figure 35.6. Weigert effect in the alignment change of azobenzene moieties.

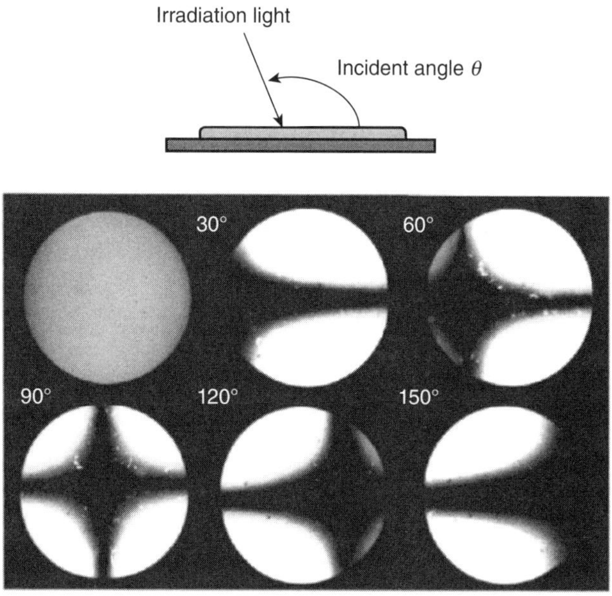

Figure 35.7. Conoscopic observation of the alignment of the azobenzene mesogenes as a function of the incident angle of the unpolarized actinic light.

light, only the propagation direction is perpendicular to the electric vector of the light. Thus, it is expected that the azobenzene moieties become aligned only in the propagation direction of the actinic light when unpolarized light is employed. In fact, several studies have been published concerning out-of-plane alignment behavior when using unpolarized light [65–67]. From the viewpoint of three-dimensional manipulation of molecules by light, promising results have been obtained for liquid crystalline polymers. For example, one can control the alignment of azobenzene and mesogenic moieties in a three-dimensional fashion, precisely along the propagation direction of the actinic light by changing the incident direction of the actinic light. This result is expected to form the basis for a new procedure to manipulate molecules in a three-dimensional manner with light.

35.7.3 Photoinduced Alignment Change in Azobenzene Liquid Crystalline Block Copolymers

As mentioned earlier, in the past two decades, electric or magnetic field, temperature gradient, crystallization, and shearing have been explored as means to control the nanostructure of block copolymers. Recently, several studies on the photoinduced alignment change in microphase separations through photoisomerization of azobenzene moieties have been reported. Owing to its physical properties of intensity, polarization, phase, and wavelength, light can

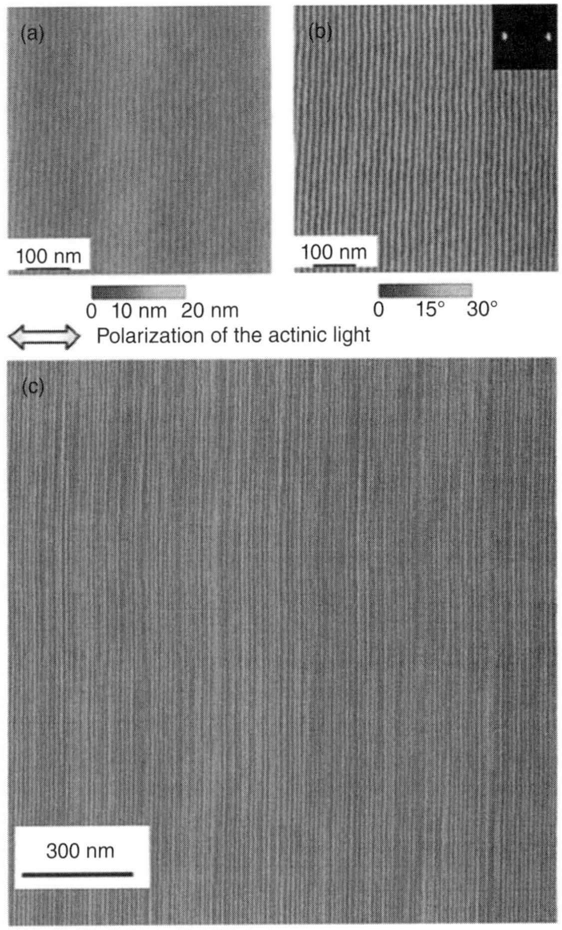

Figure 35.8. AFM images of the annealed block copolymer films in an irradiated area: (a) AFM topological image and (b) phase image. Inset of panel (b) shows the fast Fourier transform (FFT) image. (c) The atomic force microscopy (AFM) phase image in an area of 1.5 × 1.5 μm.

be used in the manipulation of microphase separations in order to manipulate nanostructures precisely.

As shown in Figure 35.8, Ikeda et al. proposed an optical method to control a parallel patterning of polyethylene oxide nanocylinders in an amphiphilic diblock liquid crystalline copolymer [68]. The polyethylene oxide nanocylinders are aligned perpendicular to the polarization direction of the actinic light by the supramolecular cooperative motions between the ordered azobenzene liquid crystals and the microphase separation. A macroscopic parallel array of nanocylinders can be easily obtained by the simple method of photocontrol, and the noncontacted proved a novel procedure to control the nanostructures.

Seki et al. succeeded in three-dimensional alignment control of the microphase-separated structure of a polystyrene-based block copolymer [69]. Moreover, memory effects on the orientation of azobenzene-containing liquid crystalline phase were observed. The same research group indicated the active and reversible photocontrol of two-dimensional microphase separation in a monolayer formed by a triblock copolymer with an azobenzene in the side chain [70].

35.7.4 Photoresponsive Liquid Crystalline Polymer with a Block Mesogenic Side Group

As discussed earlier, liquid crystalline block copolymers containing an azobenzene moiety were developed and examined to control structures of nanophase separation by light (Fig. 35.9a). Recently, Okano et al. proposed a novel molecular architecture of an azobenzene liquid crystalline polymer with a block mesogenic side group (Fig. 35.9b,c) [71]. The block mesogene is an amphiphilic molecule containing a hydrophobic mesogenic core and hydrophilic oxyethylene moieties. This liquid crystalline polymer exhibits a nanophase separation because the monomer has amphiphilic segments and a mesogenic core in the same unit (Fig. 35.10a). In fact, some amphiphiles consisting of a mesogenic core and attached flexible chains that are incompatible with each other have been shown to form complex liquid crystalline phases [72, 73]. The surface morphology can be changed into one of alignment upon irradiation of linearly polarized light due to the Weigert effect (Fig. 35.10b). Furthermore, a photo-

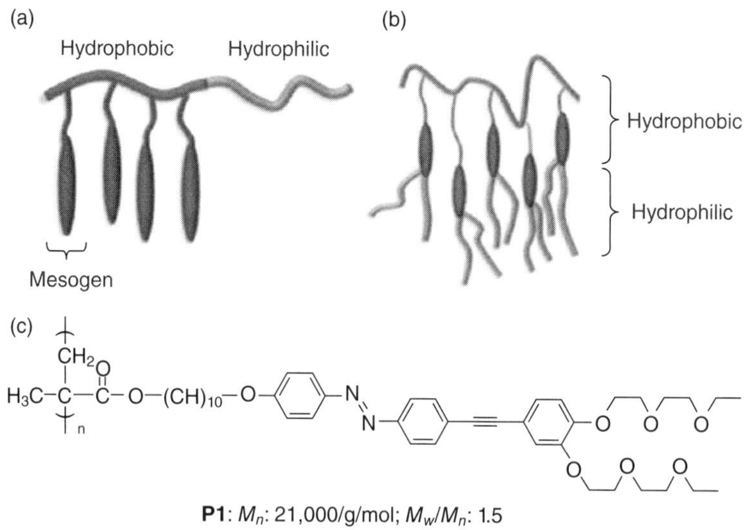

Figure 35.9. Illustrations of polymers that exhibit nanophase separation: (a) general structure of liquid-crystalline block copolymer (LCBCP) (previous studies); (b) liquid crystalline polymer (LCP)-containing block mesogene core in the side chain (this study); and (c) chemical structure and properties of the polymer.

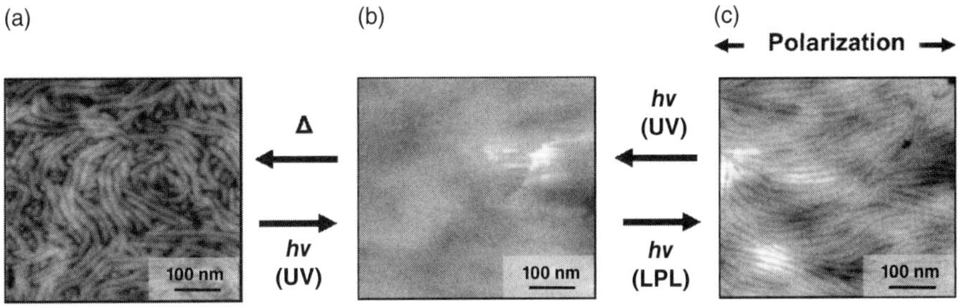

Figure 35.10. AFM images of the polymer film in an area of 500 nm × 500 nm: (a) after annealing in an LC3e phase for 40 hours; (b) photoinduced alignment change upon irradiation of linearly polarized light; the inset shows the molecular alignment of azotolanes in the phase-separated structure; (c) photochemical phase transition with nonpolarized UV light; the inset shows the scheme of the change in molecular alignment by light. LPL, linearly polarized light.

chemical phase transition induced by unpolarized UV irradiation erases the surface morphology of the film. By combining their photochemical and thermal properties, the surface morphology of the liquid crystalline polymer film becomes rewritable. This new polymer system could create highly complex segregated structures through an increased number of molecular segments of a block mesogene moiety in the side chain; the morphology can be manipulated by means of photoirradiation.

REFERENCES

1 M. Pitsikalis, S. Pispas, J. W. Mays, N. Hadjichristidis. *Adv. Polym. Sci.* 135 (1998) 1.
2 N. Hadjichristidis. *J. Polym. Sci. A Polym. Chem.* 37 (1999) 857.
3 N. Hadjichristidis, N. Pispas, M. Pitsikalis, H. Iatrou, C. Vlahos. *Adv. Polym. Sci.* 142 (1999) 71.
4 Z. -G. Yan, Z. Yang, C. Price, C. Booth. *Makromol. Chem. Rapid Commun.* 14 (1993) 725.
5 K. A. Davis, K. Matyjaszewski. *Macromolecules* 34 (2001) 2101.
6 L. Leibler. *Macromolecules* 13 (1980) 1602.
7 A. N. Smenov. *Macromolecules* 25 (1986) 4967.
8 T. Ohta, K. Kawasaki. *Macromolecules* 19 (1986) 2621.
9 E. Helfand. *Macromolecules* 8 (1975) 552.
10 T. Hashimoto, M. Shibayama, H. Kawai. *Macromolecules* 13 (1980) 1237.
11 F. S. Bates, G. H. Fredrickson, H. Kawai. *Macromolecules* 41 (1990) 525.
12 J. Fried. *Polymer science and technology*, 2nd ed. Prentice Hall, Upper Saddle River, NJ, 2003.
13 P. J. Flory. *Principles of polymer chemistry*. Cornell University Press, Ithaca, NY, 1953.

14 W. Zheng, Z. -G. Wang. *Macromolecules* 28 (1995) 7215.
15 L. F. Wu, T. P. Lodge, F. S. Bates. *Macromolecules* 37 (2004) 8184.
16 Y. Nagata, J. Masuda, A. Noro, D. Y. Cho, A. Takano, Y. Matsushita. *Macromolecules* 38 (2005) 10220.
17 A. Noro, D. Cho, A. Takano, Y. Matsushita. *Macromolecules*. 38 (2005) 4371.
18 J. Masuda, A. Takano, Y. Nagata, A. Noro, K. Hayashida, Y. Matsushita. *Macromolecules* 40 (2007) 4023027.
19 J. Matsuda, A. Takano, Y. Nagata, A. Noro, Y. Matsushita. *Phys. Rev. Lett.* 97 (2006) 098301.
20 D. Phochan, S. P. Gido, S. Pispas, J. W. Mays, A. J. Ryan, J. P. A. Fairclough, I. W. Hamley, N. J. Terrill. *Macromolecules* 29 (1996) 5091.
21 D. J. Pochan, S. P. Gido, S. Pispas, J. W. Mays. *Macromolecules* 29 (1996) 5099.
22 T. P. Russel, G. Coulon, V. R. Deline, D. C. Miller. *Macromolecules* 22 (1989) 4600.
23 M. J. Folkes, A. Keller, F. P. Scalis. *Colloid Polym. Sci.* 251 (1973) 1.
24 A. Keller, E. Pedemonte, F. M. Willmouth. *Nature* 225 (1970) 538.
25 Z. -H. Cai, K. Huang, P. A. Montano, T. P. Russell, J. M. Bai, G. W. Zajac. *J. Chem. Phys.* 98 (1993) 2367.
26 S. H. Anastasiadis, T. P. Russel, S. K. Satija, C. F. Majkrzak. *J. Chem. Phys.* 92 (1990) 5677.
27 G. Hadziioannou, A. Mathis, A. Skoulios. *Colloid Polym. Sci.* 257 (1979) 136.
28 S. Okamoto, K. Saijo, T. Hashimoto. *Macromolecules* 27 (1994) 5547.
29 C. Osuji, Y. Zhang, G. Mao, C. Ober, E. Thomas. *Polymer* 41 (2000) 8897.
30 C. Osuji, J. Chen, G. Mao, C. Ober, E. Thomas. *Macromolecules* 32 (1999) 7703.
31 R. Albalak, E. Thomas. *J. Polym. Sci. Part B Polym. Phys.* 32 (1993) 37.
32 R. Albalak, E. Thomas, M. Capel. *Polymer* 38 (1997) 3819.
33 C. Honeker, E. Thomas. *Chem. Mater.* 8 (1996) 1702.
34 Y. Cohen, R. Albak, B. Dair, M. Capel, E. Thomas. *Macromolecules* 41 (2000) 6502.
35 T. Hashimoto, J. Bodycomb, Y. Funaki, K. Kishimura. *Macromolecules* 32 (1999) 952.
36 J. Bodycomb, Y. Funaki, K. Kishimura, T. Hashimoto. *Macromolecules* 33 (2000) 5760.
37 J. C. Meredith, A. Karim, E. J. Amis. *Macromolecules* 33 (2000) 5760.
38 C. Park, C. De Rosa, E. L. Thomas. *Macromolecules* 34 (2001) 2602.
39 K. Amundson, E. Helfand, X. Quan, S. Hudson, S. D. Smith. *Macromolecules* 27 (1994) 6559.
40 K. Amundson, E. Helfand, X. Quan, S. D. Smith. *Macromolecules* 26 (1993) 2698.
41 K. Amundson, E. Helfand. *Macromolecules* 26 (1993) 1324.
42 T. L. Morkved, M. Lu, A. M. Urbas, E. E. Ehrichs, H. M. Jaeger, P. Mansky, T. P. Russel. *Science* 273 (1996) 931.
43 S. O. Kim, H. H. Solak, M. P. Stoykovich, N. J. Ferrier, J. J. Pablo, P. F. Nealey. *Nature* 424 (2003) 411.
44 W. A. Lopes, H. M. Jaeger. *Nature* 414 (2001) 735.

45 C. A. Breen, T. Deng, T. Breiner, E. L. Thomas, T. M. Swager. *J. Am. Chem. Soc.* 125 (2003) 9942.

46 A. Haryono, W. H. Binder. *Small* 2 (2006) 600.

47 M. R. Bockstaller, R. A. Mickiewicz, E. L. Thomas. *Adv. Mater.* 17 (2005) 1331.

48 T. Ikeda, O. Tsutsumi. *Science* 268 (1995) 1873.

49 O. Tsutsumi, T. Shiono, T. Ikeda, G. Galli. *J. Phys. Chem. B* 101 (1997) 1332.

50 T. Ikeda. *Mol. Cryst. Liq. Cryst.* 364 (2001) 187.

51 O. Tsutsumi, T. Kitsunai, A. Kanazawa, T. Shiono, T. Ikeda. *Macromolecules* 31 (1998) 355.

52 A. Shishido, O. Tsutsumi, A. Kanazawa, T. Shiono, T. Ikeda, N. Tamai. *J. Phys. Chem. B* 101 (1997) 2806.

53 A. Shishido, O. Tsutsumi, A. Kanazawa, T. Shiono, T. Ikeda, N. Tamai. *J. Am. Chem. Soc.* 119 (1997) 7791.

54 A. Shishido, A. Kanazawa, T. Shiono, T. Ikeda, N. Tamai. *J. Mater. Chem.* 9 (1999) 2211.

55 T. Weigert. *Verh. Phys. Ges. Berlin* 21 (1919) 485.

56 T. Todorov, N. Tomova, L. Nikolova. *Opt. Commun.* 47 (1983) 123.

57 J. Stumpe, L. Lasker, Th. Fischer, M. Rutloh, S. Kotromin, R. Ruhmann. *Thin Solid Film* 284 (1996) 252.

58 L. Lasker, J. Stumpe, Th. Fischer, M. Rutloh, S. Kostromin, R. Ruhmann. *Mol. Cryst. Liq. Cryst.* 261 (1995) 371.

59 L. Lasker, Th. Fisher, J. Stumpe, S. Kostromin, S. Ivanov, V. P. Shibaev, R. Ruhmann. *Mol. Cryst. Liq. Cryst.* 253 (1994) 1.

60 Y. Wu, Y. Demachi, O. Tsutsumi, A. Kanazawa, T. Shiono, T. Ikeda. *Macromolecules* 31 (1998) 4457.

61 Y. Wu, Q. Zhang, A. Kanazawa, T. Shiono, T. Ikeda. *Polymer* 40 (1999) 4787.

62 Y. Wu, Q. Zhang, A. Kanazawa, T. Shiono, T. Ikeda, Y. Nagase. *Macromolecules* 32 (1999) 3951.

63 H. J. Haitjema, G. L. von Morgen, Y. Y. Tan, G. Challa. *Macromolecules* 27 (1994) 6201.

64 M. Pfaadt, C. Boeffel, H. W. Spiess. *Acta. Polym.* 47 (1996) 35.

65 K. Ichimura, S. Morino, H. Akiyama. *Appl. Phys. Lett.* 73 (1998) 921.

66 Y. Wu, T. Ikeda, Q. Zhang. *Adv. Mater.* 11 (1999) 300.

67 Y. Wu, J. Mamiya, A. Kanazawa, T. Shiono, T. Ikeda, Q. Zhang. *Macromolecules* 32 (1999) 8829.

68 H. Yu, T. Iyoda, T. Ikeda. *J. Am. Chem. Soc.* 128 (2006) 11010.

69 Y. Morikawa, S. Nagano, K. Watanabe, K. Kamata, T. Iyoda, T. Seki. *Adv. Mater.* 18 (2006) 883.

70 Y. Kadota, K. Aoki, S. Nagano, T. Seki. *J. Am. Chem. Soc.* 127 (2005) 8266.

71 K. Okano, Y. Mikami, T. Yamashita. *Adv. Funct. Mater.* 19 (2009) 3804.

72 M. Kolbel, T. Beyersdorff, X. H. Cheng, C. Tschierske, J. Kain, S. Diele. *J. Am. Chem. Soc.* 123 (2001) 6809.

73 B. Glettner, F. Liu, X. Zeng, M. Prehm, U. Baumeister, G. Ungar, C. Tschierske. *Angew. Chem. Int. Ed.* 120 (2008) 6169.

74 S. F. Bates, G. H. Fredrickson. *Phys. Today* 52 (1999) 32.

36

APPLICATIONS OF ELECTRICAL PHENOMENA IN MEMBRANES AND MEMBRANE SEPARATION PROCESSES

KAZUHO NAKAMURA

36.1 INTRODUCTION

The membrane filtration process is widely used in many industries, including wastewater treatment, food processing, and medical and pharmaceutical processes. The use of the membrane filtration process for water treatment has increased especially in the last decade because of the demand for drinking water under conditions of a global shortage of water and water pollution and because of the high performance of the membrane filtration process for water quality [1, 2]. In this chapter, membranes and the membrane filtration process will be discussed with respect to electrokinetic phenomena.

The separation performance of a membrane mainly depends on pore size. The main separation mechanism is the difference between the size of species in the feed stream and the pore size of the membrane. The larger species compared with the membrane pore size will be retained by the pore structure, and the smaller species and liquid will pass through the membrane. Microfiltration (MF) is used for the separation of species having a size from 0.1 to 10.0 μm (yeast and bacteria). Ultrafiltration (UF) is used for the separation of species from 2 to 100 nm (protein and viruses), and nanofiltration (NF) and reverse osmosis (RO) are used for the separation of species from 0.2 to 2.0 nm (sugars, salts).

Electrical Phenomena at Interfaces and Biointerfaces: Fundamentals and Applications in Nano-, Bio-, and Environmental Sciences, First Edition. Edited by Hiroyuki Ohshima.

A membrane filtration process consists of a membrane, a membrane module, and related equipment. Various organic (cellulose acetate, polyamide, polyvinylidene fluoride, etc.) or inorganic (alumina, silica, etc.) materials are used as membrane materials. Membrane modules, which are the physical units that house the membranes, are designed in consideration of the mass transfer characteristics of the device. Pressure gradient across the membrane is used as the driving force in the membrane filtration process. In the process design, the membrane is usually selected in terms of separation performance and chemical resistance, and the membrane module is designed to satisfy mechanical, hydrodynamic, and economical requirements.

Membrane fouling is considered one of the major factors limiting the use of the membrane filtration process in many applications and is the irreversible alteration in the membrane caused by specific physical and/or chemical interactions between the membrane and various components present in the process stream [3]. In actual processes, fouling is usually observed as the increase in pressure drop and an alteration in separation during the process operation, and physical or chemical cleanings or replacement of the membrane are eventually required. Figure 36.1 shows simple models of attached foulants on membrane pore structure. The fouling configuration can be classified into three simple mechanisms: pore narrowing by adsorption, pore blocking/plugging, and cake/gel layer formation. Although the actual fouling phenomena are very complex because these mechanisms will occur simultaneously and sequentially, this simple classification is useful for understanding fouling phenomena.

The physical or chemical interactions between the solutes or particles in the feed stream and the membrane will cause membrane fouling [3–5]. But in many actual processes, the species causing the dominant fouling effects is not always known because of the complexity of the feed stream. Identifying and

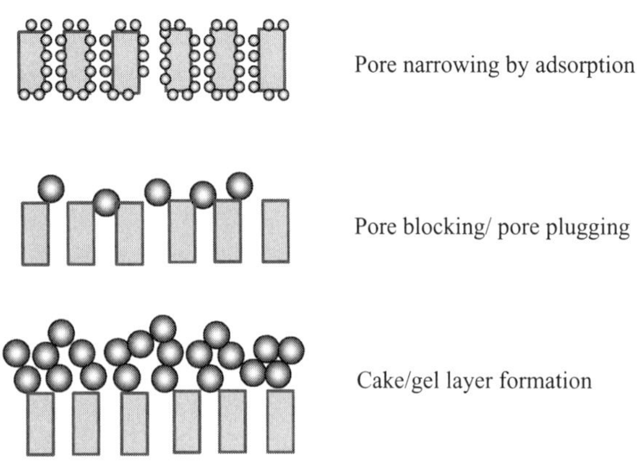

Figure 36.1. Membrane fouling mechanism.

characterizing the actual foulant are desired to understand fouling phenomena and to adopt countermeasures for the fouling.

Electrostatic and hydrophobic interactions are recognized as the major interactions causing fouling [4]. Surface modification of the membrane such as hydrophilication is usually employed for avoiding fouling by hydrophobic interaction. The zeta potential of the foulant or the membrane, that is, the electrical potential at the surface of shear between the solid and the liquid, can be used for the prediction of the electrostatic interaction. The zeta potential can be determined by electrophoresis for particulate materials and by streaming potential (SP) for porous membranes [3, 4].

In this chapter, the membrane and the membrane filtration process are discussed in terms of electrokinetic phenomena. A general description of membrane filtration and SP is presented first. Application of the zeta potential for the characterization of fouling phenomena and the application of SP for membrane characterization are then discussed.

36.2 PRESSURE DROP AND SP IN THE MEMBRANE FILTRATION PROCESS

36.2.1 Water Permeation and Pressure Drop

In the membrane filtration process, hydraulic flux is an important indicator of process performance. The flux, J_V, is the volume flow of liquid through the membrane per unit filtration area and is described by Darcy's law:

$$J_V = k_d \frac{P_{\mathrm{TM}}}{\mu L}, \tag{36.1}$$

where k_d is a constant, hydraulic permeability; P_{TM} is the pressure difference across the membrane; L is the thickness of the membrane; and μ is the viscosity of the filtrate. The pore structure of a membrane is usually modeled as a bundle of cylindrical capillaries of uniform radius, r, and uniform length, L, and J_V is given by the Hagen–Poiseuille equation:

$$J_V = \frac{N\pi r^4}{8} \frac{P_{\mathrm{TM}}}{\mu L}, \tag{36.2}$$

where N is the number of capillaries per unit area. A membrane or cake layer can be also modeled as a packed bed of uniform spheres, and J_V is given by the Kozeny–Carman relation:

$$J_V = \frac{1}{k_c S_0^2} \frac{\varepsilon^3}{(1-\varepsilon)^2} \frac{P_{\mathrm{TM}}}{\mu L}, \tag{36.3}$$

where k_c is the Kozeny constant, S_0 is the specific surface area, and ε is the porosity. The hydraulic resistance R is usually used as an index of the permeability of the membrane and the fouling layer:

$$J_V = \frac{1}{R}\frac{P_{\mathrm{TM}}}{\mu}. \tag{36.4}$$

36.2.2 SP

In the membrane filtration process, electric potential difference can be measured in two different directions. One is the transmembrane direction, which is the filtration direction, and the other is the tangential direction [4], which is the cross-flow direction. In this chapter, only the transmembrane direction is discussed.

When the electrolyte solution is forced to pass through the membrane at a transmembrane pressure, P_{TM}, it generates an electric potential difference of E_{TM} across the membrane. The SP is defined as the potential difference per unit transmembrane pressure, that is, $\Delta E_{\mathrm{TM}}/\Delta P_{\mathrm{TM}}$.

The apparent zeta potential, ζ_{obs}, can be determined using the Helmholtz–Smoluchowski (HS) equation from the measured SP, assuming the ideal conditions that the pore has a cylindrical capillary with laminar flow and that the surface conductivity and the electrical double-layer overlapping are negligible [3, 4, 6]:

$$\zeta_{\mathrm{obs}} = \frac{\mu\lambda}{\varepsilon_0\varepsilon_{\mathrm{r}}}\frac{\Delta E_{\mathrm{TM}}}{\Delta P_{\mathrm{TM}}}, \tag{36.5}$$

where λ is the conductivity of the solution, ε_0 is the permittivity of free space, and ε_{r} is the relative dielectric constant of the solution.

The space charge (SC) model, which is a model of electrokinetic phenomena for a charged capillary, is also used for analyzing SP in a membrane filtration system [7–11]. An SC model relates the SP to the pore radius and surface charge density. Although a numerical calculation is required to solve the SC model accurately, an analytical solution is useful for this application. The SP is given by the following equations [12]:

$$\mathrm{SP} = \frac{\Delta E_{\mathrm{TM}}}{\Delta P_{\mathrm{TM}}} = \frac{q_{\mathrm{p}}}{\lambda_{\mathrm{eff}}\kappa\mu}\frac{I_2(\kappa r_{\mathrm{p}})}{I_1(\kappa r_{\mathrm{p}})}, \tag{36.6}$$

$$\kappa^{-1} = \left(\frac{\varepsilon_0\varepsilon_{\mathrm{r}}R_{\mathrm{g}}T}{F^2\sum_i z_i c_i}\right)^{1/2}, \tag{36.7}$$

$$\lambda_{\mathrm{eff}} = \lambda_{\mathrm{avg}} + \frac{q_{\mathrm{p}}^2}{\mu}\left(1 - \frac{I_0(\kappa r_{\mathrm{p}})I_2(\kappa r_{\mathrm{p}})}{I_1^2(\kappa r_{\mathrm{p}})}\right), \tag{36.8}$$

and

$$\lambda_{\text{avg}} = \lambda_0 \left(1 + \frac{q_{\text{p}} \kappa^{-1} F}{\varepsilon_0 \varepsilon_{\text{r}} RT} \left(\frac{2}{\kappa r_{\text{p}}} - \frac{1}{I_1(\kappa r_{\text{p}})} \right) \frac{\sum_i z_i^3 u_i c_i}{\sum_i z_i^2 u_i c_i} \right), \tag{36.9}$$

where q_p is the surface charge density of the membrane pore; λ_{eff} is the effective conductivity of the electrolyte solution in the pore; I_1 and I_2 are the modified Bessel functions of the first kind of orders 1 and 2, respectively; κ^{-1} is the Debye length; R_g is the ideal gas constant; T is absolute temperature; F is the Faraday constant; z_i is the valence of species i; and c_i is the concentration of species i on the center line of the capillary. Details of the description are discussed by Newman [12].

36.3 MEASUREMENT OF SP

Figure 36.2 shows the experimental apparatus used to measure SP on a membrane. Generally, E_{TM} is measured with a digital multimeter, and P_{TM} is

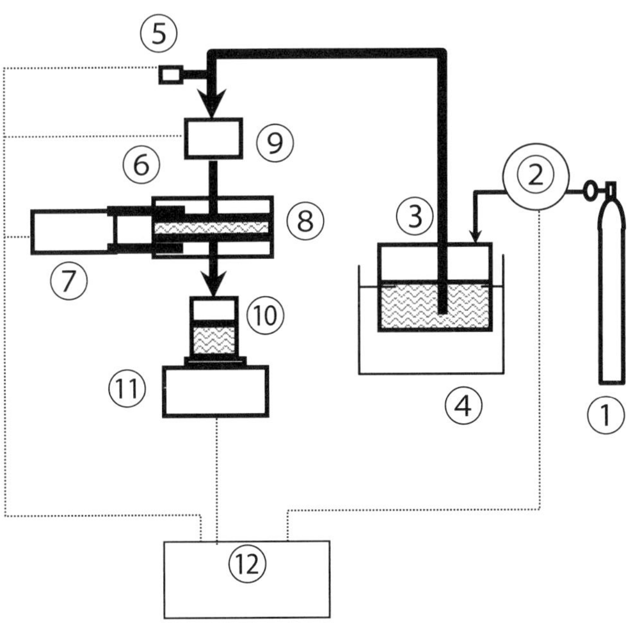

Figure 36.2. Experimental apparatus for streaming potential measurement. 1, N_2 gas cylinder; 2, pressure controller; 3, pressure holding tank; 4, temperature control bath; 5, pressure transducer; 6, Pt black electrodes; 7, digital multimeter; 8, membrane holder; 9, flow cell for conductivity measurement; 10, filtrate reservoir; 11, electric balance; and 12, personal computer [14].

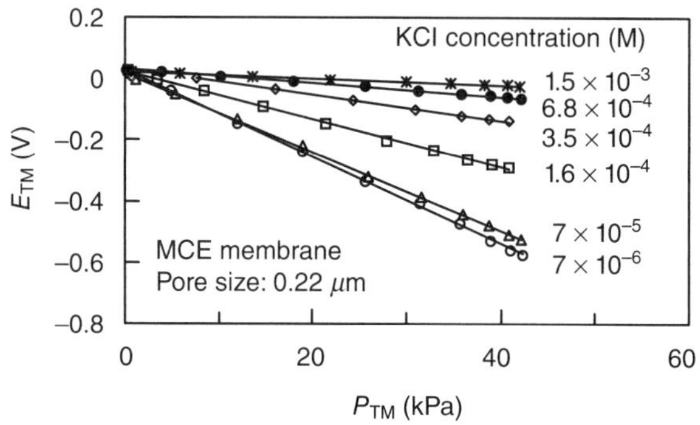

Figure 36.3. Relationship between the E_{TM} and P_{TM} [14]. MCE, mixed cellulose ester.

measured with a pressure transducer. Ag/AgCl electrodes are commonly used because of their low asymmetry potential and high chemical stability [13]. In Liu et al. [14], we discussed the effects of the pressurizing method and the electrode material on E_{TM} measurement, and it was shown that stable measurement of E_{TM} was achieved by continuously increasing the pressure, regardless of the electrode material, and Pt black was the most useful electrode material because of its high physical stability in addition to its low asymmetry potential and high chemical stability.

Figure 36.3 shows the relationship between the E_{TM} and P_{TM} for the membrane having a pore size of 0.22 μm at various KCl concentrations [15]. The validity of measurement can be checked by the linearity of P_{TM} versus E_{TM} plots. The SP, $\Delta E_{TM}/\Delta P_{TM}$, is determined as the slope of the linear regression line for the plots.

36.4 APPLICATION OF THE ZETA POTENTIAL TO THE CHARACTERIZATION OF MEMBRANE FOULING

During membrane filtration, the surface charge condition and the part dominating the total filtration resistance will be changed by fouling. In this section, how the fouling phenomena affect the apparent zeta potential of the membrane is discussed.

36.4.1 Protein Adsorption on MF Membranes

Protein adsorption on the surface or within pores of an MF membrane is a major cause of membrane fouling in the MF of protein-containing solutions. Sun et al. completed a wide review of the studies on protein adsorption phenomena and control methods [16]. The protein adsorption of an MF or UF membrane is used as an indicator of the membrane performance. Many researchers have focused on the elucidation of protein adsorption mechanisms

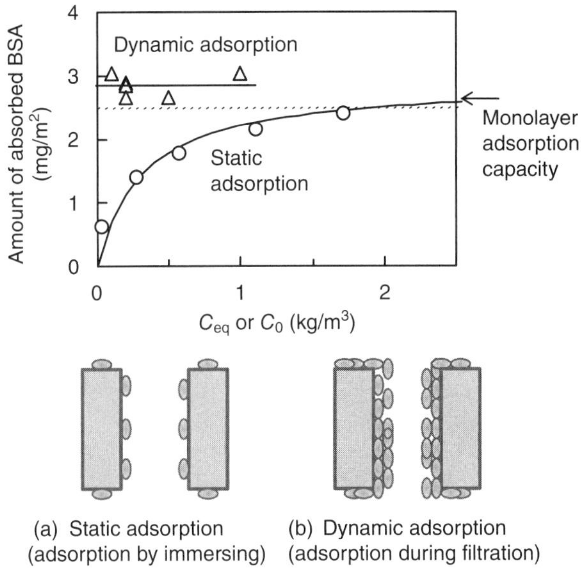

Figure 36.4. Adsorption isotherms for the static and dynamic adsorption method [18].

or control methods of the adsorption phenomena [17–20]. But the protein adsorption on these membranes is a phenomenon that is still not well understood because it depends on many factors such as operating conditions [21, 22], solution chemistry [23–31], surface physicochemical conditions [31, 32], or pore morphology [33, 34].

Adsorption phenomena can be characterized by the adsorption isotherm. Figure 36.4 shows the adsorption isotherms of the static and dynamic adsorptions of bovine serum albumin (BSA) on a porous glass (PG) membrane (pore size: 0.1 μm) [18]. Static adsorption means adsorption by immersing the membrane into the protein solution, and dynamic adsorption means adsorption during filtration. In this case, the size of the protein is sufficiently smaller than the pore size of the membrane, so the protein molecules will adsorb onto the pore surface of the membrane. The only difference between these adsorption methods is the mass transfer condition, which is the diffusion for static adsorption and the volumetric flow in the pores for dynamic adsorption. The adsorption isotherm of the static adsorption was a Langmuir type, and the maximum adsorption capacity was lower than the monolayer adsorption capacity. On the other hand, the adsorption isotherm of the dynamic adsorption was a multilayer type. It was shown that the volumetric flow in the pores will affect the adsorption equilibrium in addition to the mass transfer condition in the pores [18].

Figure 36.5 shows the change in the zeta potential before and after the BSA filtration of the PG membranes, which are hydrophilic (unmodified) and hydrophobic (octadecylsilane [ODS]/tetramethylsilane [TMS] modified) membranes [5]. Before the BSA filtration, both membranes had almost

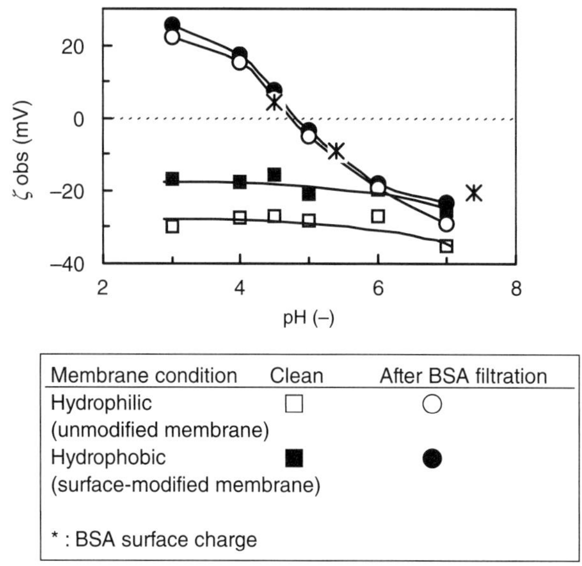

Figure 36.5. Zeta potential of a membrane before and after BSA filtration [5].

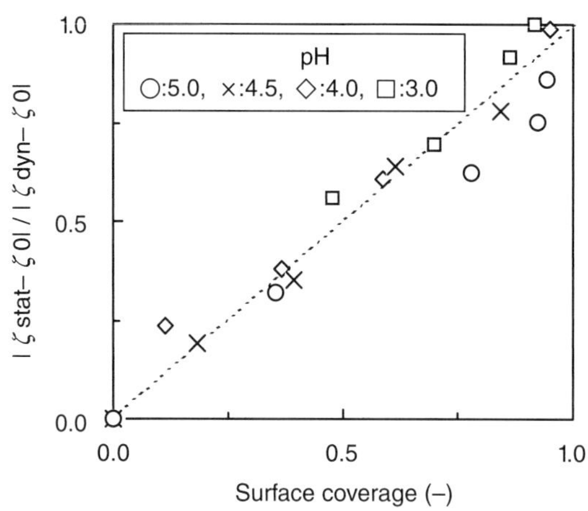

Figure 36.6. The relationship between the change ratio of the zeta potential and the surface coverage of BSA [35].

constant negative values at all pH values at about –30 and –20 mV, respectively. After the BSA filtration, where the pore surface might be completely covered with BSA molecules, the pH dependence of the zeta potential changed to almost the same value as that of the BSA molecules regardless of the hydrophobicity of the membrane [5].

Figure 36.6 shows the effect of surface coverage of BSA on the change ratio of the zeta potential, $|\zeta_{stat} - \zeta_0|/|\zeta_{dyn} - \zeta_0|$ [35]. ζ_0 is the zeta potential of the

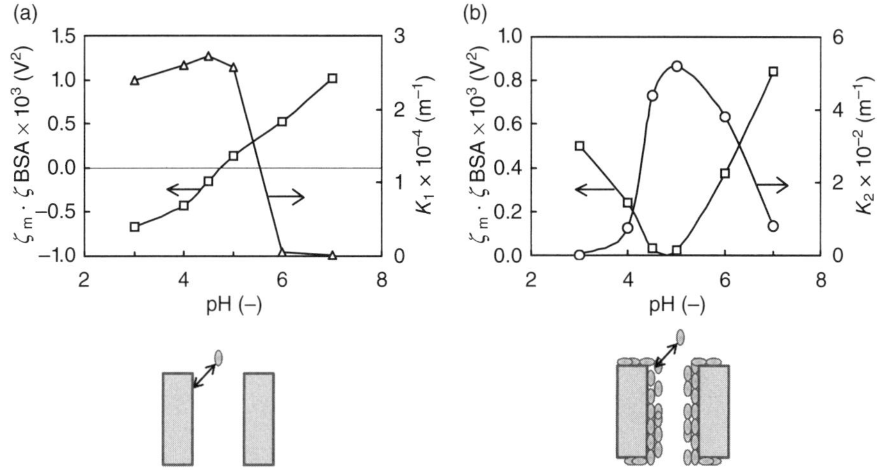

(a) Interaction between clean pore surface and BSA (monolayer adsorption)

(b) Interaction between adsorbed membrane and BSA (multiayer adsorption)

Figure 36.7. The interaction between BSA and (a) clean pore surface, (b) adsorbed BSA [5].

clean membrane; ζ_{stat} is the zeta potential of the membrane after static adsorption; and ζ_{dyn} is the zeta potential of the membrane after dynamic adsorption, which is almost equal to the zeta potential of the BSA. The change ratio of the zeta potential is directly proportional to the surface coverage regardless of pH. This result shows that the change in the zeta potential by BSA adsorption will be determined by the proportion of the pore surface area occupied by BSA compared to the unoccupied pore surface area regardless of the sign or level of charge of the BSA molecules [35]. These observations show that the change in the zeta potential is useful to characterize protein adsorption at the pore surfaces of a membrane.

The electrostatic interaction between the pore surface and the BSA molecules can be evaluated by the product of both the zeta potential of the membrane, ζ_m, and BSA molecules ζ_{BSA}, $\zeta_m \cdot \zeta_{BSA}$ [5, 20, 29, 36]. Figure 36.7 shows the electrostatic interaction for monolayer adsorption (primary adsorption) and multilayer adsorption (secondary adsorption) [5]. The electrostatic interaction is compared with adsorption coefficient K, which is obtained from the breakthrough curve and is an index of the magnitude of the overall adsorption interaction between protein molecules and their adsorption site [5, 35, 37].

The electrostatic interaction of monolayer adsorption (Fig. 36.7a) was attractive at pH 4.5 and below and was repulsive at pH 5 and above. The electrostatic interaction, $\zeta_m \cdot \zeta_{BSA}$, changed linearly with an increase in pH. Within the pH range from 3 to 5, where the electrostatic force changes from attractive to slightly repulsive, K_1 showed an almost constant value regardless of the change in the electrostatic force, and at pH 6 and 7, where the electrostatic repulsive force is expected, K_1 decreased drastically compared to values in the lower pH range. The observation suggests that the electrostatic repulsive force can reduce K_1, whereas the electrostatic attractive force has no influence on K_1. A possible explanation for this observation is as follows. K_1 could reflect the combination of the rates of the true adsorption step and the mass transfer step for the radial direction in the pores, which will be dominated by the Brownian motion. These steps might be much faster regardless of the existence of an electrostatic attractive force, and the electrostatic repulsive force could behave as an energetic barrier for these steps [5].

The electrostatic interaction of multilayer adsorption (Fig. 36.7b) was repulsive all over the pH range and showed a minimum at pH 5, which is the isoelectric point (IEP) of BSA. The pH dependence of K_2 showed a clear correlation with that of the electrostatic repulsive force. At pH values away from the IEP of BSA, K_2 would decrease due to the energetic barrier provided by the electrostatic repulsive force for the interaction between BSA, and BSA adsorbed on the pore surface, and at a pH around the IEP, the adsorption interaction of the multilayer adsorption was hardly disturbed by the electrostatic interaction. Although the promoting force for the multilayer adsorption is not clear, the hydraulic shear stress in membrane pores can be considered a cause of the promoting force [5, 35].

From these observations, it is clear that the electrostatic repulsive force should provide an energetic barrier for the interaction between BSA molecules and their adsorption site, and the electrostatic attractive force might have no significant influence on the overall adsorption interaction.

36.4.2 The Location Dominating the Zeta Potential in a Fouled MF Membrane

Membrane fouling is observed as an increase in pressure drop across the membrane, and the filtration resistance is used as quantitative analysis. Figure 36.8 shows the typical status of particles captured by a porous membrane. The total filtration resistance R_{total} can be expressed by the sum of each local filtration resistances:

$$R_{total} = R_m + R_b + R_c, \tag{36.10}$$

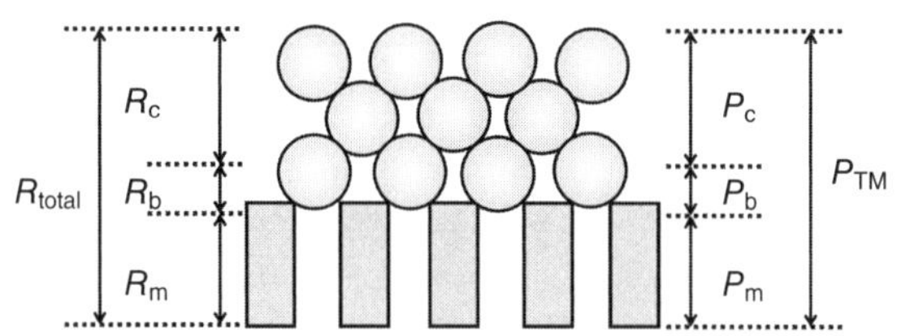

Figure 36.8. The filtration resistance and pressure drop in particle filtration by microfiltration membrane.

where R_m is the hydraulic resistance of the clean membrane, R_b is the hydraulic resistance after pore plugging, and R_c is the hydraulic resistance of the cake layer. The location dominating the pressure drop can be analyzed with this resistance analysis. The local pressure drop can be written as follows:

$$P_m = \frac{R_m}{R_{total}} P_{TM}, \quad P_b = \frac{R_b}{R_{total}} P_{TM}, \quad P_c = \frac{R_c}{R_{total}} P_{TM}, \tag{36.11}$$

where P_m, P_b, and P_c are the pressure drop at the membrane, the pore plugging, and the cake layer, respectively. The zeta potential would be expressed as the following equation because the partial pressure drop will generate electrical difference at each layer:

$$\zeta_{obs} = \frac{R_m}{R_{total}} \zeta_m + \frac{R_b}{R_{total}} \zeta_b + \frac{R_c}{R_{total}} \zeta_c, \tag{36.12}$$

where ζ_m, ζ_b, and ζ_c, are the zeta potentials of the membrane, the pore plugging part, and the cake layer, respectively.

Figure 36.9 shows the change in filtration resistance and ζ_{obs} during the cake filtration and the pore plugging filtration as typical results [38]. The ζ_m and ζ_b

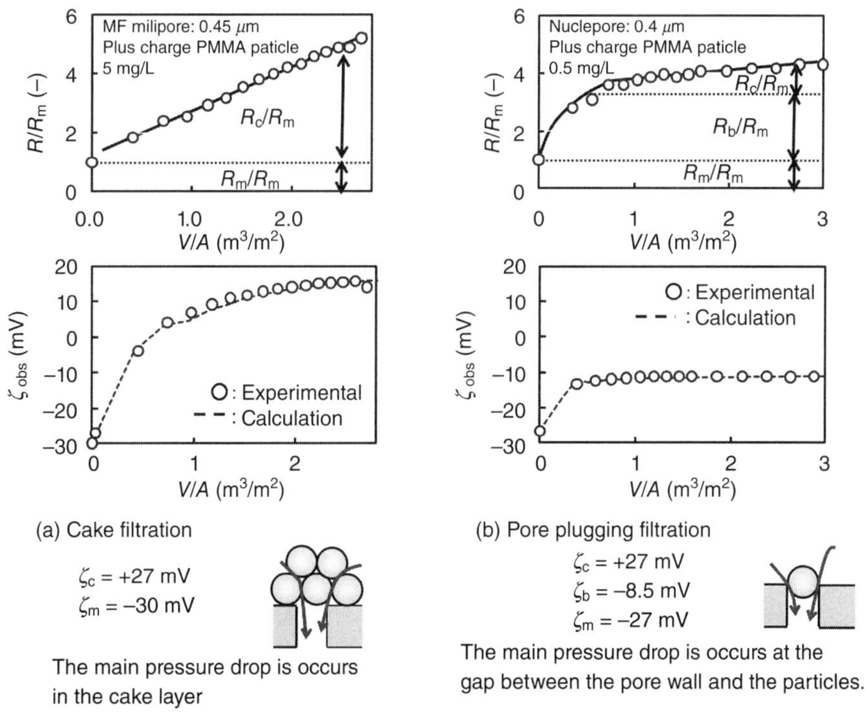

Figure 36.9. The changes in filtration resistance and ζ_{obs} during (a) cake filtration and (b) pore plugging filtration [38]. PMMA, polymethyl methacrylate.

were determined by preliminary experiments. In the cake filtration, where R_b is negligible, the filtration resistance increased linearly by the accumulation of the cake layer, and ζ_{obs} changed from ζ_m to ζ_c according to Equation 36.12. In the pore plugging filtration, the filtration resistance increased rapidly at the early stage of the filtration at pore plugging and then increased gradually at the accumulation of the cake layer. During the filtration, R_b generated by the volumetric flow in the gap between pore walls and captured particles is a large part of total filtration resistance. ζ_{obs} increased rapidly at the early stage and then became almost constant. This change of ζ_{obs} was well fitted by Equation 36.12, and ζ_b was determined to be –8.5 mV as a fitting parameter. The value of ζ_b reflects both the zeta potentials of the pore wall and the particle surface captured at the surface of the membrane.

These observations demonstrate that ζ_{obs} reflects the surface charge condition at the part dominating the total filtration resistance.

36.5 APPLICATION OF SP TO THE CHARACTERIZATION OF PORE SIZE AND SURFACE CHARGE DENSITY OF MF/UF MEMBRANES

According to Equation 36.6, the SP depends on pore radius, surface charge density, and electric conductivity. Figure 36.10 shows the SP calculated using Equation 36.6 with constant surface charge density and constant pore size. At the constant surface charge density ($q_p = 1.0 \times 10^{-2}$ C/m^2), the SP showed a unique curve depending on the pore radius. The absolute value of the SP decreased with a decrease in the pore radius. At the constant pore size

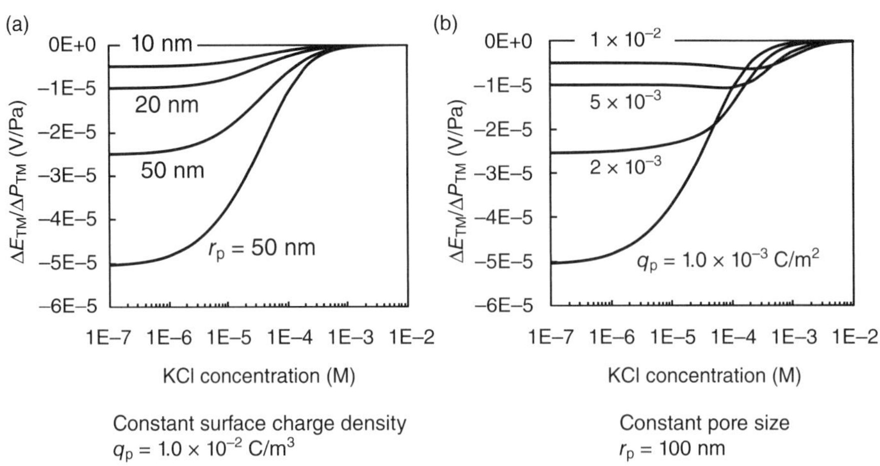

Figure 36.10. Effects of surface charge density and pore size on streaming potential.

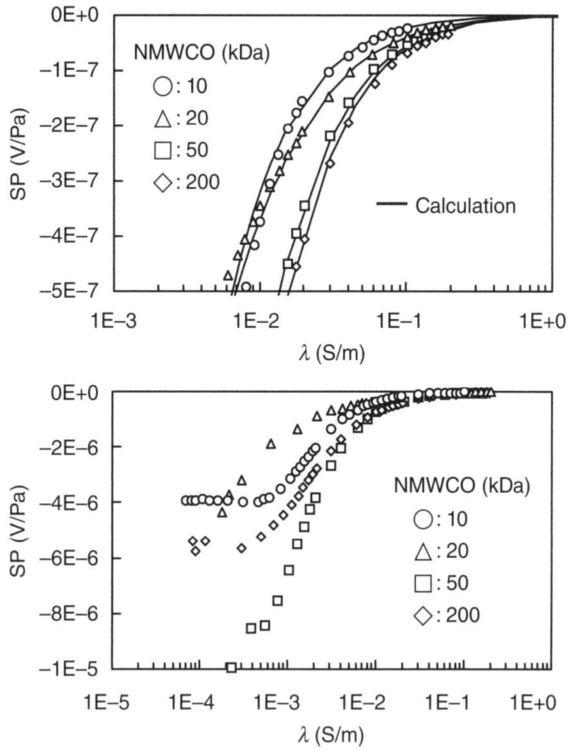

Figure 36.11. Effect of KCl concentration on the streaming potential of UF membranes [39].

(r_p = 100 nm), the SP shows a complex behavior against the KCl concentration, where the higher surface charge density does not always provide higher SP in the same KCl concentration. These observations also show that the pore size or surface charge density cannot be determined from one SP value.

Figures 36.11 and 36.12 show the effect of the KCl concentration on the SP of UF membranes having nominal molecular weight cutoff (NMWCO) from 10 to 200 kDa [39] and MF membranes having pore sizes from 0.025 to 0.8 μm [15]. The experimental plots showed the unique curve depending on the NMWCO or the pore size. The calculation results of Equation 36.6 were well fitted at higher KCl concentrations (the upper panels in Figs. 36.11 and 36.12) and were not fitted for the lower KCl concentration region, which means Equation 36.6 is valid only for higher KCl concentrations. At lower KCl concentrations, the surface charge density is not constant because chloride ion adsorption is not saturated. The r_p and q_p could be determined by the curve fitting as fitting parameters and are listed in Table 36.1. In MF membranes, $2r_p$, which is pore size determined from SP measurement, increased

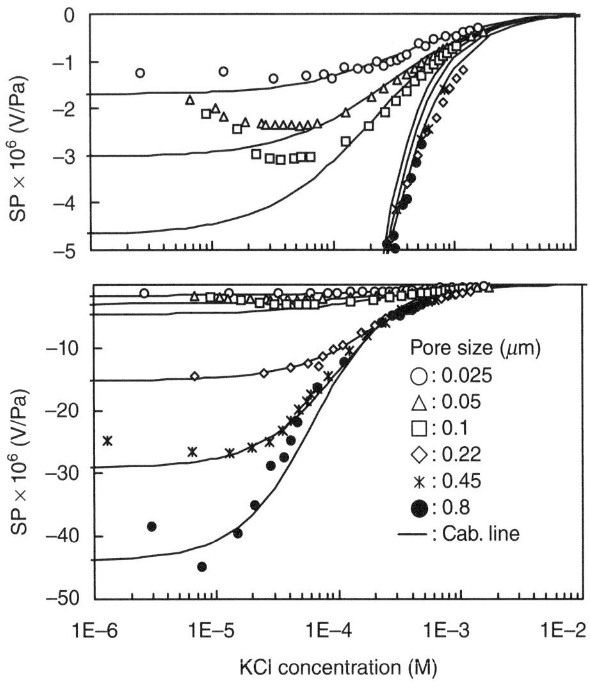

Figure 36.12. Effect of KCl concentration on the streaming potential of MF membranes [15].

TABLE 36.1. r_p and q_p Obtained by Curve Fitting Using Equation 36.6 for UF/MF Membranes

NMWCO (kDa)	Material	$2r_p$ (nm)	q_p (mC/m^2)	MWCO (kDa)
10	Polysulfone	23.2	–1.43	3.9
20	Aromatic Polyamide	31.6	–1.95	6.8
50	Polysulfone	46.8	–1.45	10
200	Polysulfone	69.2	–1.43	41
Nominal Pore Size (nm)	**Material**	**$2r_p$ (nm)**	**q_p (mC/m^2)**	
25	Mixed	21	–3.0	
50	cellulose	33	–2.6	
100	ester (MF	44	–2.2	
220	millipore)	140	–2.1	
450		220	–1.5	
800		300	–1.2	

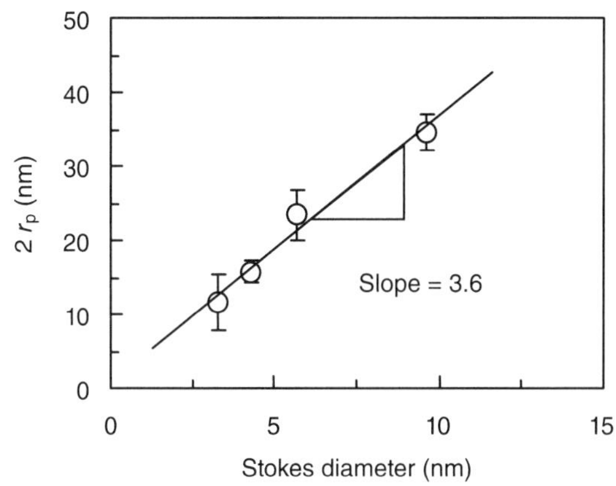

Figure 36.13. The relationship between $2r_p$ and actual separation performance of UF membranes [39].

with an increase in the nominal pore size, although $2r_p$ was always smaller than the nominal pore size.

From these observations, it is demonstrated that the pore size and the surface charge density of MF or UF membranes can be determined from the KCl concentration dependency of the SP using Equation 36.6.

In order to confirm the relationship between $2r_p$ and the actual separation performance of UF membranes, the Stokes diameter, which means the cutoff diameter of the UF membrane, was estimated from the molecular weight cutoffs (MWCOs) of UF membranes measured with polyethylene glycol (PEG). Figure 36.13 shows the relationship between $2r_p$ and the Stokes diameter. The relation showed good linearity, while the absolute value was different. These observations demonstrate that the r_p obtained from the KCl concentration dependency of SP using Equation 36.6 is useful for the prediction of the separation performance of UF membranes.

36.6 CONCLUSIONS

The phenomena occurring during membrane filtration are very difficult to understand clearly, but various monitoring techniques have been developed. The SP can provide fundamental information about the filtration process because the pressure difference across the membrane is the driving force of both SP and separation by filtration. In this chapter, it is demonstrated that the change in the apparent zeta potential will reflect fouling phenomena, and the pore size and surface charge density can be determined by an analysis of the conductivity dependence of the SP. The application of these values will be of interest for membrane process monitoring and control technologies.

NOMENCLATURE

c_i	Concentration of species *i* on the center line of the pore (mol/m^3)
E_{TM}	Electric potential difference across the membrane (V)
F	Faraday constant (C/mol)
J_V	Flux (m/s)
k_c	Kozeny constant (–)
k_d	Hydraulic permeability (m)
K_1	Adsorption coefficient of the primary adsorption (m^{-1})
K_2	Adsorption coefficient of the secondary adsorption (m^{-1})
L	Membrane thickness (m)
N	Number of capillaries per unit area (m^{-2})
P_{TM}	Pressure difference across the membrane (Pa)
P_{m}	Pressure drop at the membrane (Pa)
P_{b}	Pressure drop at pore plugging (Pa)
P_{c}	Pressure drop at the cake layer (Pa)
q_{p}	Surface charge density (C/m^2)
r	Radius (m)
r_{p}	Pore radius (m)
R	Filtration resistance (m^{-1})
R_{b}	Filtration resistance at pore plugging (m^{-1})
R_{c}	Filtration resistance of the cake layer (m^{-1})
R_{g}	Ideal gas constant (J/mol/K)
R_{m}	Filtration resistance of the membrane (m^{-1})
R_{total}	Total filtration resistance (m^{-1})
S_0	Specific surface area (m^{-1})
SP	Streaming potential (V/Pa)
T	Absolute temperature (K)
u_i	Mobility of species *i* (m^2 mol/J/s)
z_i	Valence of species *i* (–)
ε	Porosity (–)
ε_0	Permittivity of free space (F/m)
ε_{r}	Relative dielectric constant of the solution (–)
ζ_0	Zeta potential of the clean membrane (V)
ζ_{b}	Zeta potential at pore plugging (V)
ζ_{BSA}	Zeta potential of BSA (V)
ζ_{c}	Zeta potential at the cake layer (V)
ζ_{dyn}	Zeta potential of the membrane after dynamic adsorption (V)
ζ_{m}	Zeta potential of the membrane (V)
ζ_{obs}	Apparent zeta potential (V)
ζ_{stat}	Zeta potential of the membrane after static adsorption (V)
κ^{-1}	Debye length (m)
λ	Electric conductivity of the bulk solution (S/m)
λ_{avg}	Average electric conductivity of the solution in the pore (S/m)
λ_{eff}	Effective electric conductivity of the solution in the pore (S/m)

λ_0 Electric conductivity of the solution on the center line of the pore (S/m)

μ Viscosity of the solution (Pa·S)

REFERENCES

1 P. Le-Clech, V. Chen, T. A. G. Fane. *J. Memb. Sci.* 284 (2006) 17–53.
2 A. W. Zularisam, A. F. Ismail. *Razman Salim* 194 (2006) 211–231.
3 L. J. Zeman, A. L. Zydney. *Microfiltration and ultrafiltration principles and applications*. Marcel Dekker, New York, 1996.
4 A. V. Delgado. *Interfacial electrokinetics and electrophresis*. Marcel Dekker, New York, 2002, Section 20.
5 K. Nakamura, K. Matsumoto. *J. Memb. Sci.* 280 (2006) 363–374.
6 M. Nyström, A. Pihlajamäki, N. Ehsani. *J. Memb. Sci.* 87 (1994) 245–256.
7 H. J. M. Hijinen, J. A. M. Smit. *J. Memb. Sci.* 99 (1995) 285–302.
8 X. -L. Wang, T. Tsuru, S. Nakao, S. Kimura. *J. Memb. Sci.* 103 (1995) 117–133.
9 S. Basu, M. M. Sharma. *J. Memb. Sci.* 124 (1997) 77–91.
10 A. Shukla, A. Kumar. *J. Memb. Sci.* 237 (2004) 119–130.
11 A. Szymczyk, P. Fievet, B. Aoubiza, C. Simon, J. Pagetti. *J. Memb. Sci.* 161 (1999) 275–285.
12 J. S. Newman. *Electrochemical systems*. Prentice-Hall, Englewood Cliffs, NJ, 1991.
13 C. Guell, M. Ferrando, F. Lopez. *Monitoring and visualizing membrane-based process*. Wiley-VCH, Weinheim, 2009.
14 H. Liu, K. Nakamura, K. Matsumoto. *Membrane* 28(6) (2003) 278–287.
15 K. Nakamura, H. Liu, K. Matsumoto. *Membrane* 29(3)(2004) 180–187.
16 S. Sun, Y. Yue, X. Huang, D. Meng. *J. Memb. Sci.* 222 (2003) 3.
17 C. Velasco, M. Ouammou, J. I. Calvo, A. Hernandez. *J. Colloid Interface Sci.* 266 (2003) 148–152.
18 K. Nakamura, K. Matsumoto. *J. Memb. Sci.* 145 (1998) 119–128.
19 C. Herrero, P. Pradanos, J. I. Calvo, F. Tejerina, A. Hernandez. *J. Colloid Interface Sci.* 187 (1997) 344–351.
20 A. Martin, F. Martinez, J. I. Calvo, P. Pradanos, L. Palacio, A. Hernandez. *J. Memb. Sci.* 207 (2002) 199–207.
21 T. J. Su, J. R. Lu, Z. F. Cui, R. K. Thomas. *J. Memb. Sci.* 173 (2000) 167–178.
22 K. L. Jones, C. R. O'Melia. *J. Memb. Sci.* 165 (2000) 31–46.
23 L. Ricq, S. Narcon, J. -C. Reggiani, J. Pagetti. *J. Memb. Sci.* 156 (1999) 81–96.
24 K. J. Kim, A. G. Fane, M. Nystrom, A. Pihlajamaki. *J. Memb. Sci.* 134 (1997) 199–208.
25 L. Ricq, A. Pierre, J. -C. Reggiani, S. Zaragoza-Piqueras, J. Pagetti, G. Daufin. *J. Memb. Sci.* 114 (1996) 27–38.
26 C. Causserand, M. Nyström, P. Aimar. *J. Memb. Sci.* 88 (1994) 211–222.
27 W. M. Clark, A. Bansal, M. Sontakke, Y. H. Ma. *J. Memb. Sci.* 55 (1991) 21–38.

28 W. R. Bowen, Q. Gan. *J. Colloid Interface Sci.* 144 (1991) 254–262.

29 F. Martinez, A. Martin, P. Pradanos, J. I. Calvo, L. Palacio, A. Hernandez. *J. Colloid Interface Sci.* 221 (2000) 254–261.

30 S. Saksena, A. L. Zydney. *Biotechnol. Bioeng.* 43 (1994) 960–968.

31 S. H. Ye, J. Watanabe, Y. Iwasaki, K. Ishihara. *J. Memb. Sci.* 249 (2005) 133–141.

32 M. Taniguchi, G. Belfort. *J. Memb. Sci.* 231 (2004) 147–157.

33 Z. -P. Zhao, Z. Wang, S. -C. Wang. *J. Memb. Sci.* 217 (2003) 151–158.

34 J. Hosch, E. Staude. *J. Memb. Sci.* 121 (1996) 71–82.

35 K. Nakamura, K. Matsumoto. *J. Memb. Sci.* 285 (2006) 126–136.

36 I. H. Huisman, P. Pradanos, A. Hernandez. *J. Memb. Sci.* 179 (2000) 79–90.

37 K. Nakamura, K. Matsumoto. *J. Chem. Eng. Jpn.* 31 (1998) 536–544.

38 K. Nakamura, T. Orime, K. Matsumoto. *Proceedings of The Fifth Conference of Aseanian Membrane Society*, P8-8 (2009).

39 K. Nakamura, W. Hirayama, T. Nittami, K. Matsumoto. *Membrane (Maku)* 31(6) (2004) 180–187.

PART III

APPLICATIONS IN BIOSCIENCES

37

DIELECTRIC DISPERSION IN COLLOIDAL SYSTEMS: APPLICATIONS IN THE BIOLOGICAL SCIENCES

MARÍA LUISA JIMÉNEZ, SILVIA AHUALLI, RAÚL A. RICA, AND ÁNGEL V. DELGADO

37.1 INTRODUCTION

From cells to bacteria, DNA helica, fungi and microorganisms, a wide variety of samples of biological systems are composed of nano- or micrometric heterogeneities immersed in water solutions. In this size regime, the dynamics of bioparticles and their interaction with the surrounding medium in processes such as microbial adhesion, binding of (heavy) metals to microorganisms, biocorrosion, bacterial infection, biofilm formation, drug adsorption and encapsulation, cell division, or DNA conformation are determined by the properties and interactions at the interface. On the other hand, the manipulation of these systems at the nanoscale level requires the knowledge of the structure, composition, and dynamics of the interface. These are just a few examples; many more can be mentioned where the measurement and modeling of interfacial properties of bioparticles are of utmost importance.

In particular, interaction forces between the bioparticles are mainly governed by the electrical properties of the bioparticle surface. The first attempt to characterize such properties dates back to one century ago, when Höber [1] compared the low- and high-frequency conductivity of erythrocytes and

Electrical Phenomena at Interfaces and Biointerfaces: Fundamentals and Applications in Nano-, Bio-, and Environmental Sciences, First Edition. Edited by Hiroyuki Ohshima.

demonstrated the existence of the cellular membrane. The works of Maxwell [2], Wagner [3], Fricke [4], and Pauly and Schwan [5] set down the basis for the research in this field. Since then, much has been done [6–15], not only in the fundamental research area but also keeping in mind many technological and biomedical applications of biological process characterization due to the noninvasive character of conductivity measurements and their sensitivity to the structure, shape, size, and composition of the different phases of the system. For example, changes have been detected in the dielectric properties of cells affected by cancer [16], malaria [17], or in different states of the cell life cycle [18].

When particles or nano- and microorganisms are placed in contact with water or aqueous solutions, they usually acquire charge at their surfaces, mostly (but not only) due to the dissociation of surface groups. In the case of cells, the particle charge has mainly two origins. First, inside the cell, there is a net charge that sustains the transmembrane potential. This is due to selective ion channels and active ion pumps embedded in the membrane, which maintain different ionic concentrations on the intracellular and extracellular sides of the membrane. Second, the hydrophilic head groups of the lipid molecules forming the membrane are partially dissociated. Hence, the cell is modeled as a homogeneous or heterogeneous conductive core (the cytoplasm, subscript "c") surrounded by an isolated charged membrane (subscript "m").

Whatever the charging mechanism, the concentration of ions of the same sign (coions) and opposite sign (counterions) is perturbed around the particle. In the equilibrium, the coion and counterion concentration profile is controlled by the balance between the electric force and the thermal diffusion, so there is a layer around the particle (the electric double layer [EDL]) enriched in counterions and depleted in coions [19]. The thickness of this region (the Debye length, κ^{-1}) is governed by the ionic strength, while the concentration at any distance from the surface depends also on the particle surface charge density. If the particle charge is homogeneously distributed, the EDL will have the same geometry as the particle. This means that the centres of the particle charge and the counterion charge in the EDL coincide. If this center symmetry is disrupted by an external action, for example, by applying an electric field, an electric multipole is induced [20–21]. In fact, for spherical particles, the application of an electric field induces a dipole, while in the other cases, far from the particle, higher order multipoles are negligible as compared to the dipole [22–23]. Hence, the local electric potential Ψ will have two components, one coming from the external electric field **E** and the other one coming from the induced dipole **d** [20]:

$$\Psi(r,\theta) = -E_0 r\cos\theta + \frac{d}{4\pi\varepsilon_0\varepsilon_s r^2}\cos\theta, \tag{37.1}$$

where r and θ are spherical coordinates, ε_0 is the vacuum permittivity, and ε_s is the relative electric permittivity of the solution. The induced dipole is proportional to the applied electric field. Thus, the magnitude of interest is the dipole coefficient

$$C = \frac{d}{E_0 4\pi\varepsilon_0\varepsilon_s a^3}$$

(a is the particle radius) [24]. If we analyze the relation between the applied electric field and the electric displacement, we will obtain an enhanced electric permittivity, ε. This enhancement is related to the dipole coefficient of the individual particles through the equation [20, 24]

$$\varepsilon = \varepsilon_s[1+3\phi C], \tag{37.2}$$

where ϕ is the particle volume fraction. In this way, the measurement of the electric permittivity can give us information on the electrical state of the nanoscale interface between the particle and the solution. This is exploited in different ways in many applications, a number of them in biology, despite its applicability being straitened by some experimental difficulties, mainly electrode polarization (EP) [25–28], which we will deal with in Section 37.3.

If there are dielectric losses, then the dipole will have an out-of-phase component. Also, the ionic conduction contributes to this out-of-phase component since it produces a redistribution of charges in the medium and, hence, also a modification of the local electric field. Accordingly, it is usual to define a complex relative electric permittivity [20, 29]:

$$\varepsilon^* = \varepsilon' - i\varepsilon'' - i\frac{K}{\omega\varepsilon_0}, \tag{37.3}$$

where ε' and $\varepsilon''$are the real and imaginary components of the relative electric permittivity of the system, respectively; K is the conductivity; and ω is the angular frequency. In Equation 37.3, we note that the imaginary component includes the dielectric losses (represented by ε'') and ohmic contributions (represented by K).

If we place the suspension between two electrodes, the system behaves as a capacitor with a complex capacity [24]. If we can measure the capacity, we will obtain the complex electric permittivity, although we cannot separate the ohmic and dielectric loss contribution to this capacity unless a model is used.

When an electric field is applied to a system composed by a solution and some bioparticles, various mechanisms of polarization of the particles will activate at different timescales [29–31]. If the electric field oscillates with amplitude E_0 and frequency ω, the active mechanisms will depend on the frequency of the applied field and the electric permittivity will have an out-of-phase component. The faster the electric field oscillation, the fewer the mechanisms that have time to develop between two field reversals, and the electric permittivity will decrease, showing successive relaxation phenomena.

Starting from the faster polarization mechanisms, in Section 37.2, we will summarize the main relaxation phenomena and how they relate to some interfacial properties. Some remarks about the experimental techniques will be given in Section 37.3. Finally, in Section 37.4, we will show some examples of how this technique can be exploited with biological systems. These are in turn of particular interest in biology for many reasons, mainly because they simulate many living organisms and because their surface can retain substances around the colloidal particle and thus can serve as, for instance, a drug delivery system [32–34].

In virtue of the Kramers–Krönig relations [35], the real and imaginary parts of the electric permittivity provide the same information and can be obtained from one another. Hence, except in some cases maintained intentionally, we will show mainly features of the real part.

37.2 BASIS OF THE DIELECTRIC PHENOMENA IN COLLOIDAL SYSTEMS

In order to get an overview of the polarization mechanisms responsible for the frequency dependence of the permittivity, or dielectric dispersion, we have to study the dynamics of the system at the nanoscale level. We can describe these phenomena by analyzing the permittivity behavior under the action of oscillating electric fields of different frequencies. Although both descriptions are equivalent, it is conceptually easier to work in the time domain to begin our study. Thus, the fastest polarization mechanism is associated with the induction of molecular dipoles as the electronic clouds deform [29]. Then, associated permanent dipoles like those of water molecules orient and the electric permittivity reaches its characteristic $\varepsilon_s \approx 80\varepsilon_0$ ($T = 25°C$) value. The characteristic time for such orientations is ~1 ns, so that in the gigahertz frequency range, the relaxation (so-called γ-relaxation) of water molecules occurs.

In this range, there is no polarization associated with the interface between particles and solution; hence, it has no interest for our purposes. Far below these frequencies, the molecular polarization is well developed and the dielectric losses associated with it can be neglected, so that both the permittivity and conductivity of the aqueous medium are frequency independent. Hereafter, we will consider that this is the case for the different media that form the system since, as we will see, the relaxation phenomena associated with the dispersed particles are in the kilohertz and megahertz ranges. Also, in this section, we will focus on the most simple cases that still display all the dielectric phenomena, that is, colloidal homogeneous particles and cells composed by a homogeneous conductive core surrounded by an insulating membrane of thickness $h << a$. Hence, the relative electric permittivities of cytoplasm ε_c^* and insulating membrane ε_m^* of the cells, the solution ε_s^*, and the homogeneous particles ε_p^* are

$$\varepsilon_{c}^{*} = \varepsilon_{c} - i\frac{K_{c}}{\omega\varepsilon_{0}}, \tag{37.4}$$

$$\varepsilon_{m}^{*} = \varepsilon_{m}, \tag{37.5}$$

$$\varepsilon_{s}^{*} = \varepsilon_{s} - i\frac{K_{s}}{\omega\varepsilon_{0}}, \tag{37.6}$$

and

$$\varepsilon_{p}^{*} = \varepsilon_{p} - i\frac{K_{p}}{\omega\varepsilon_{0}}. \tag{37.7}$$

We will start with the fastest polarization phenomenon associated with the presence of the particle/solution interface and related to the mismatch between particle and solution conductivities: the Maxwell–Wagner (MW) relaxation [2–3]. Then, there is a phenomenon associated with the presence of the membrane: the β-relaxation [29]. Finally, we will analyze the slowest process: the α-relaxation [20, 29, 36–42].

37.2.1 The MW Relaxation

For frequencies below the relaxation of the fluid permittivity, the dipole coefficient is solely determined by the relative permittivity values of the particle and of the electrolyte solution. They determine the degree of polarization. A bound charge is formed at the interface of the particle and the solution due to the molecular orientation at both sides of the interface, as shown in Figure 37.1. As a consequence, a dipole characterized by a coefficient C_{MW} is produced [2–3, 20, 29]:

$$C_{MW}(\infty) = \frac{\varepsilon_{i} - \varepsilon_{s}}{\varepsilon_{i} + 2\varepsilon_{s}}, \tag{37.8}$$

where ε_i is the particle (ε_p) or cell core (ε_c) permittivity.

When the frequency is decreased, the ionic current produces an accumulation and depletion of counterions and coions at the particle interface. This is due to the different conductivities of the particle and the solution. A schematic view is represented in Figure 37.2.

In the situation represented in the figure, the contribution to the ionic currents of ions in the particle is larger than in the solution, and thus the redistribution of ions at the interface is produced as represented in Figure 37.2. These ionic profiles generate a dipole in the same direction of the external field, that is, a positive dipole coefficient that increases the local electric field, and, as a consequence, also the ionic fluxes near the particle. This process ceases when the local electric field is such that all ions reaching the surface from one side leave the surface from the other side. For cells, in the inner

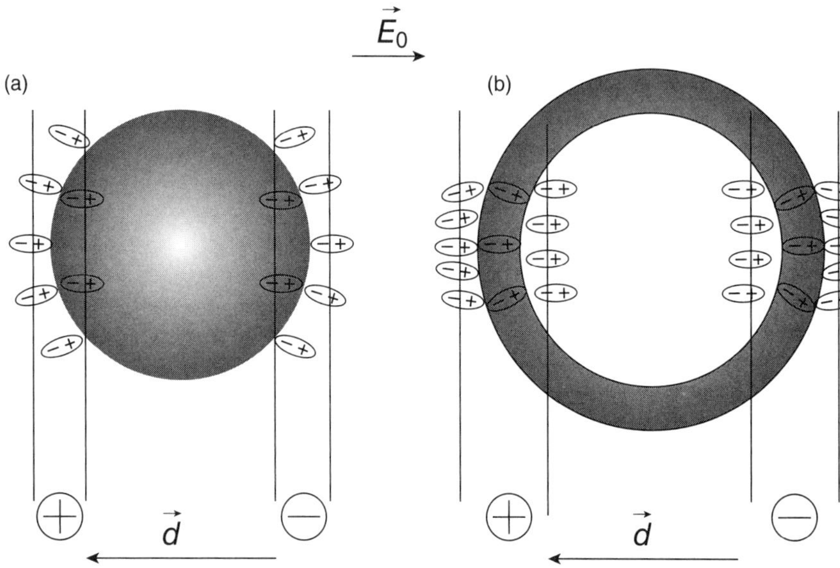

Figure 37.1. Scheme of the induced dipole due to the permittivity contrast. (a) Homogeneous particle ($\varepsilon_p < \varepsilon_s$). (b) Cell ($\varepsilon_m < \varepsilon_c < \varepsilon_s$). The electric permittivity inside the particle and membrane is smaller (thus, the electric field is stronger) than in the other parts of the system.

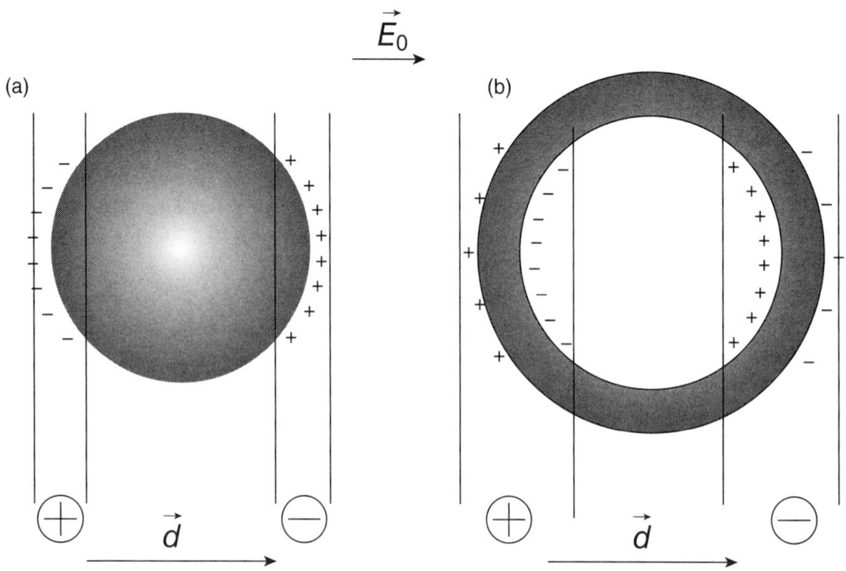

Figure 37.2. Scheme of the induced dipole originated in the conductivity contrast. (a) Homogeneous particle. (b) Cell.

interface, the charge building does not stop until the electric field inside the core vanishes. This leads to another relaxation that is absent in the case of homogeneous particles and that will be described in the next section.

For colloids, at sufficiently low frequencies, the dipole coefficient only depends on the conductivity values of the particle and the electrolyte solution [2–3, 20, 29]:

$$C_{\mathrm{MW}}(0) = \frac{K_p - K_s}{K_p + 2K_s}. \tag{37.9}$$

The characteristic time in which the dipole coefficient reaches the value given by Equation 37.9 is

$$\tau_{\mathrm{MW}} = \varepsilon_0 \frac{\varepsilon_p + 2\varepsilon_s}{K_p + 2K_s}. \tag{37.10}$$

At frequencies around $1/\tau_{\mathrm{MW}}$, ions start to accumulate around the particle, the charge density is out of phase with respect to the local field, and this leads to the appearance of an out-of-phase (imaginary) part of the dipole coefficient. At frequencies below this so-called MW dispersion, the period of the applied field becomes so large that all the free charge density shifts in-phase with the applied field. Finally, the frequency dependence of the dipole coefficient is expressed by

$$C^* = \frac{\varepsilon_p^* - \varepsilon_s^*}{\varepsilon_p^* + 2\varepsilon_s^*}. \tag{37.11}$$

With Equations 37.2 and 37.11, the MW relaxation of the electric permittivity is obtained:

$$\varepsilon^*(\omega) = \varepsilon_s^* \frac{2(1-\phi)\varepsilon_s^* + (1+2\phi)\varepsilon_p^*}{(2+\phi)\varepsilon_s^* + (1-\phi)\varepsilon_p^*}, \tag{37.12}$$

which is a Debye-type relaxation,

$$\delta\varepsilon^*(\omega) = \varepsilon^*(\omega) - \varepsilon(\infty) = \frac{\delta\varepsilon_{MW}(0)}{1 + i\omega\tau_{MW}} \tag{37.13}$$

of characteristic time in Equation 37.10 and amplitude $\delta\varepsilon_{\mathrm{MW}}$:

$$\delta\varepsilon_{\mathrm{MW}}(0) = \varepsilon(0) - \varepsilon(\infty) = 9\phi\varepsilon_s \frac{\left[\dfrac{K_p}{K_s} - \dfrac{\varepsilon_p}{\varepsilon_s}\right]^2}{\left(\dfrac{\varepsilon_p}{\varepsilon_s} + 2\right)\left[\dfrac{K_p}{K_s} + 2\right]^2}, \tag{37.14}$$

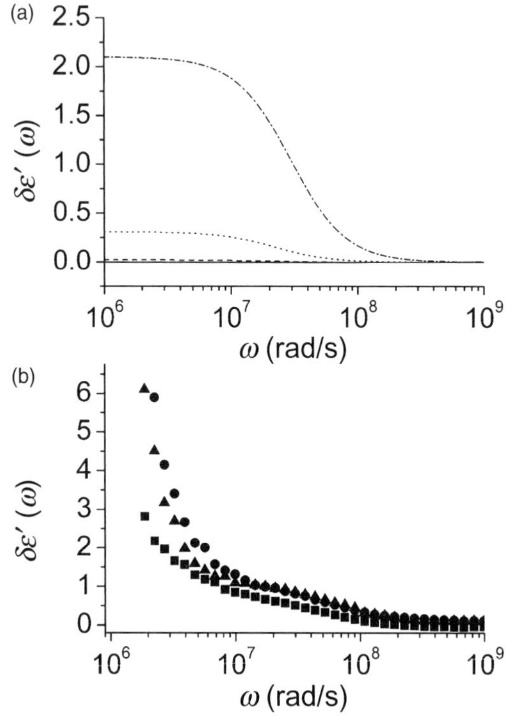

Figure 37.3. Real part of the electric permittivity as a function of the frequency for the Maxwell–Wagner relaxation. (a) From Equations 37.9, 37.11, and 37.12. $\alpha = 1$ μm, 0.14 mM NaCl, and $K_p/K_m = 0.03$ (solid line), 0.15 (dashed line), 0.7 (dotted line), and 3.9 (dash-dotted line). (b) Experimental data for a 1% suspension of spherical latex particles of diameter 530 nm immersed in a 1 mM NaCl solution at 25°C (squares), 35°C (circles), and 40°C (triangles). Reproduced with permission from Jiménez et al. [84].

where quantities without an asterisk correspond to the real parts of their complex counterparts. In Figure 37.3a, an example of the MW relaxation is given. The most noticeable fact is the small value of the amplitude as compared to the gamma relaxation. This is due to the fact that the MW relaxation is due to ions, which are much less than the water molecules involved in the gamma relaxation. This has also been experimentally observed (Fig. 37.3b), although in such case, the kilohertz range relaxation partially masks the MW dispersion.

The MW theory was generalized for the case of spheroidal particles by Fricke [4] and Sillars [43]. Since the dipole moment is linear with the external electric field, the dipole moment of a randomly oriented particle is the superposition of the ones of parallel and perpendicular orientations. Thus, the relative electric permittivity is the sum of the permittivities in the three possible orientations:

$$\varepsilon^* = \varepsilon_s^* + \frac{\phi}{3}\varepsilon_s^* \sum_{i=a,b,c} \left(\varepsilon_{p,i}^* - \varepsilon_s^*\right) \frac{1}{\varepsilon_s^* + A_i\left(\varepsilon_{p,i}^* - \varepsilon_s^*\right)}, \tag{37.15}$$

where a, b, and c are the three possible orientations of the ellipsoid; $\varepsilon^*_{\mathrm{p},i}$ is the relative electric permittivity of the ellipsoid with the axis i (a, b, or c) oriented parallel to the electric field, and A_i is their depolarization factor. Accordingly, for spheroidal particles, MW splits into two Debye-type relaxation phenomena associated with the parallel and perpendicular orientations.

Generally, colloidal particles or cell membranes are charged and insulating, thus preventing ionic flows inside. However, beyond the surface, the excess of counterions produce an excess of ionic fluxes. If the ionic strength is sufficiently high (as it is in biological systems), the EDL is very thin as compared to the particle size, and the excess of ionic currents inside the EDL is usually modeled as an excess of surface conductivity. For example, a usual value of the plasmatic NaCl concentration is 1 mM. This means that the EDL thickness is 10 nm. For cells 1 μm in size, this is 100 times smaller than the cell size. The relation between this surface conductivity and the electric properties of the particle is given by the Bikerman equation, which is, for symmetrical electrolytes [44],

$$K^{\sigma} = \frac{2e^2z^2n}{k_{\mathrm{B}}T\kappa}\left[D^{+}\left(\mathrm{e}^{-\frac{ze\zeta}{2k_{\mathrm{B}}T}}-1\right)\left(1+\frac{3m^{+}}{z^2}\right)+D^{-}\left(\mathrm{e}^{\frac{ze\zeta}{2k_{\mathrm{B}}T}}-1\right)\left(1+\frac{3m^{-}}{z^2}\right)\right], \tag{37.16}$$

where e is the electron charge, z is the valency of the electrolyte, n is the ionic bulk number concentration, k_{B} is the Boltzman constant, T is the absolute temperature, $D^{\pm}$ is the ionic diffusion coefficient, ζ is the zeta potential (the electric potential at the slipping plane, i.e., the boundary between the fluid that is stagnant on the particle surface and the mobile part of the EDL), and $m^{\pm}$ is the dimensionless quantity indicating the contribution of electroosmosis to surface conductivity. Exceptions to this equation are concentrated suspensions, particles with ionic conduction inside the stagnant layer, as usually occurs with porous particles, and thick double layers. The first case is usually found in biological suspensions, while the second and third ones have the important example of DNA. In fact, the observed dielectric spectrum in the latter case is related to the fluctuation of condensed counterions. These are the counterions that are very close to the chain, and electric attraction is so strong that thermal agitation is not enough to make ions move away from the particle. As a consequence, the applied field provokes only fluctuations along distances of the order of the characteristic lengths of the particle [45]. Also, the diameter of the helix is comparable to the EDL thickness. This particular case has been extensively studied in the literature [42, 46–51] and will be treated in more detail in Section 37.3.

It is possible to take into account the existence of surface conductivity as an effective particle conductivity in the MW model. O'Konski [52] showed that in the simplest case, that is, a spherical particle with a homogeneous surface charge, the particle generates the same perturbation in the local electric field as a neutral conducting particle with a bulk conductivity, $K_p = 2K^\sigma/a$. This generalization of the MW description is then called the Maxwell–Wagner–O'Konski (MWO) approach. If the particles are cells, the particle conductivity is given by $K_p = 2K^\sigma/a + K_c$ [29]. For geometries other than spheres, the dipolar fields generated by the particle depend on its orientation. Hence, the relation between K_p and K^σ becomes more complex and orientation dependent [21–23, 40–41, 53]:

$$K_{p\|,\perp} = K_p + K^\sigma f_{\|,\perp} \tag{37.17}$$

being $f_{\|,\perp}$ geometrical factors. O'Konski gave also approximate relations for $f_{\|,\perp}$ in some limiting cases, but the most general functions were obtained in Reference 21. Equation 37.17 in combination with Equation 37.15 results in two Debye relaxations for the sometimes called extended MW relaxation [23]. However, since the amplitude is very small and the characteristic frequencies are very close, only one (broad maybe) relaxation can be detected. An example is shown in Figure 37.4, where the data for elongated hematite particles (aspect ratio 6.1) exhibit only one relaxation in addition to the asymptotic tail of the low-frequency relaxation (explained later).

All these formulae are valid for dilute suspensions. To remove this limitation, Bruggeman, and later Hanai [54], calculated a mixture formula, which is the result of the reiterative application of Equation 37.12 starting from a solution to which an infinitesimal amount of particles is added; Equation 37.12 is used to calculate the permittivity of the suspension. In the next step, again an infinitesimal amount of particles is added to a homogeneous solution with the

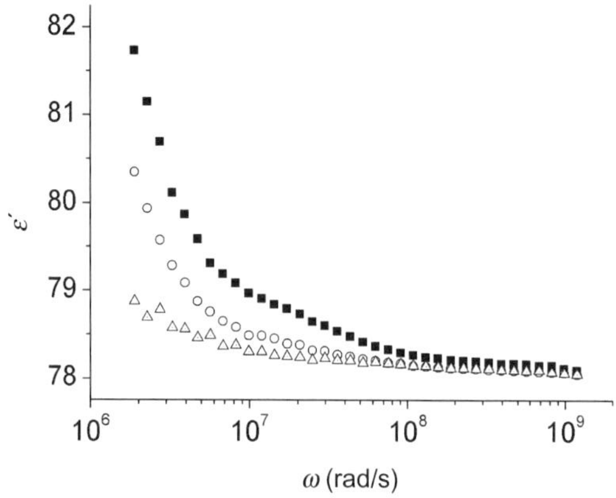

Figure 37.4. Real part of the experimental electric permittivity of a 0.8% suspension of hematite particles (length 551 nm, aspect ratio 6.1) in a 0.5 mM NaCl solution. Reproduced with permission from Jiménez et al. [31].

new permittivity. Finally, the approximate Bruggeman–Hanai equation is obtained:

$$\frac{\varepsilon^* - \varepsilon_p^*}{\varepsilon_s^* - \varepsilon_p^*}\left(\frac{\varepsilon_s^*}{\varepsilon^*}\right)^{1/3} = 1 - \phi. \tag{37.18}$$

Equation 37.18 has been extensively applied to obtain the electric permittivity of individual cells from data of their suspensions. In this respect, a more accurate way to reach this relation is provided by the cell models [39, 55]. The suspension is substituted by a single particle surrounded by a finite volume such that the volume fraction is preserved. In this model, the interactions between particles are accounted for by appropriate boundary conditions at the external border of the cell.

37.2.2 The β-Relaxation

As mentioned, the insulating nature of the membrane prevents the existence of ionic current from the cell core to the external solution. Hence, the situation described in Figure 37.2b is not stationary. On the contrary, inside the core, the ionic redistribution stops only when the electric field vanishes inside. This situation is analogous to the load process of a capacitor (the membrane) between two resistors (the core and the external solution) in series. For frequencies high enough, the external ionic distribution is stationary and the dipole coefficient depends on the conductivities of both media:

$$C_\beta(\infty) = \frac{K_c + \dfrac{2K^\sigma}{a} - K_s}{K_c + \dfrac{2K^\sigma}{a} + 2K_s}. \tag{37.19}$$

At lower frequencies, when ions cease to accumulate in the inner interface, the dipolar fields do not depend anymore on the inner conductivity and the dipole coefficient strongly decreases to

$$C_\beta(0) = \frac{\dfrac{2K^\sigma}{a} - K_s}{\dfrac{2K^\sigma}{a} + 2K_s} \tag{37.20}$$

because there is no electromigration inside the core. The β-relaxation consists of a Debye-type process with a characteristic time given by

$$\tau_\beta = \frac{\varepsilon_0 \varepsilon_m a}{h}\left(\frac{1}{K_c} + \frac{1}{\dfrac{2K^\sigma}{a} + 2K_s}\right), \tag{37.21}$$

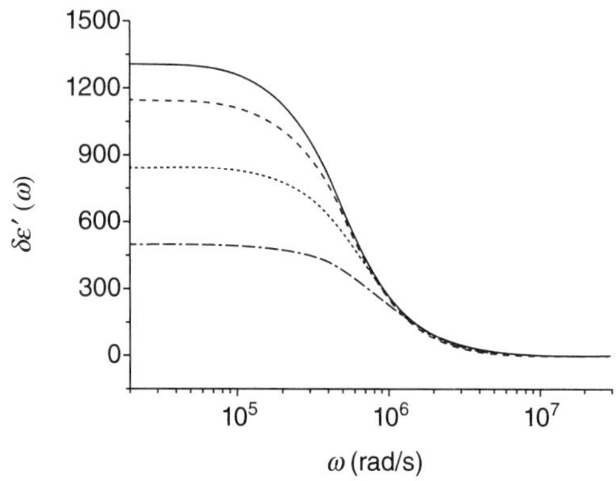

Figure 37.5. Theoretical values of the permittivity increment of the β-relaxation for different values of the zeta potential. Parameters: a = 1000 nm; h = 10 nm; NaCl concentration 0.14 mM; K_c = 0.01 S/m; ζ = –50 mV (solid line), –100 mV (dashed line), –150 mV (dotted line), and –200 mV (dash-dotted line).

with h as the membrane thickness. The amplitude of the relaxation in the permittivity spectrum is given by

$$\delta\varepsilon_\beta(0) = \frac{9\phi}{4}\frac{\varepsilon_m \dfrac{a}{h}}{\left(1+\dfrac{\dfrac{K^\sigma}{a}}{K_s}\right)}. \tag{37.22}$$

This permittivity increment is very large as compared to that of the MWO relaxation amplitude due to the high capacity of the cell membrane (Fig. 37.5). This increment is diminished by the surface conductivity that partially shields the cell. For frequencies lower than the β-relaxation, the cell behaves as it is a charged, insulating, and homogeneous particle.

37.2.3 The α-Relaxation

If the particle has no charge, the ionic flux distribution around the particle becomes stable. But if particles are charged, another phenomenon occurs in the kilohertz range. Due to the different transport numbers inside and outside the EDL for both coions and counterions, the phenomenon of concentration polarization takes places. Beyond the EDL, the neutral electrolyte concentration becomes perturbed as schematically shown in Figure 37.6, and this in turn provokes large diffusion fluxes that oppose this neutral electrolyte perturbation [20, 37].

These electrolyte clouds extend approximately the size of the particle plus its EDL. For thin double layers, the latter can be neglected and the character-

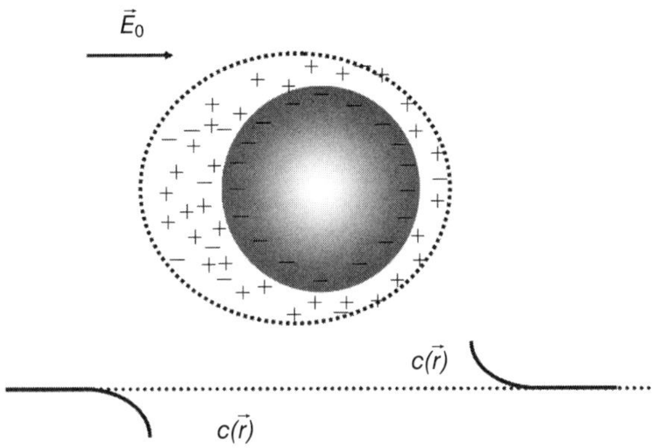

Figure 37.6. Schematic representation of the neutral electrolyte gradients and ionic fluxes around a colloidal particle. The dashed line is the limit of the EDL, which is deformed due to the neutral electrolyte gradient (the EDL becomes thinner in the region where the concentration is elongated).

istic time for the formation of the concentration polarization is the diffusion time:

$$\tau_\alpha = \frac{a^2}{2D_{\text{eff}}}, \tag{37.23}$$

where

$$D_{\text{eff}} = \frac{D^+ D^-}{D^+ + D^-}$$

for a 1-1 electrolyte. Accordingly, the diffusion fluxes and the concentration polarization formation itself are largely delayed with respect to the external field.

Concentration polarization produces changes in the dipole coefficient because of two reasons: (i) The asymmetric electrolyte concentration at both sides of the particle produces an asymmetric EDL thickness and, hence, the centers of the particle surface charge and the EDL charge are separated; and (ii) the diffusion fluxes oppose the electromigratory fluxes in the EDL and diminish the dipole moment originated in the conductivity mismatch. This provokes a large out-of-phase component of the dipolar field, which is responsible for a huge increase of the electric permittivity below the so called α-relaxation. In effect, if we consider in Equation 37.1 that ε_s and C are complex quantities, we have

$$\varepsilon' = \varepsilon_s + 3\phi\varepsilon_s\left(C' - \frac{K_s}{\omega\varepsilon_0}C''\right). \tag{37.24}$$

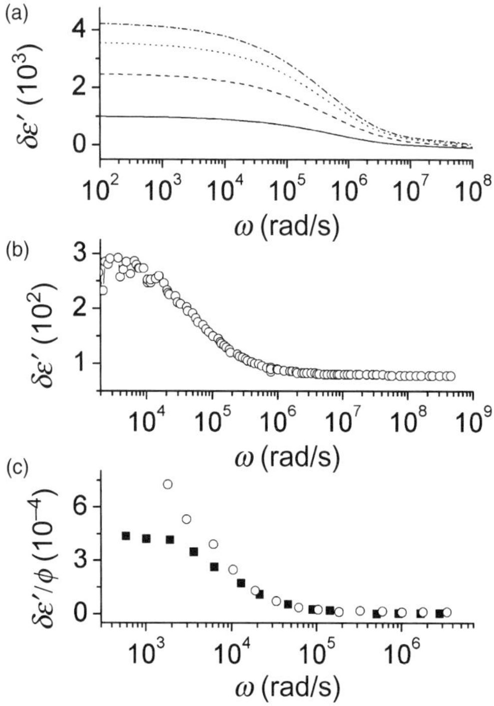

Figure 37.7. (a) Increment of the relative electric permittivity of a suspension of 55-nm spheres in a 1 mM NaCl solution and zeta potentials –25 mV (solid line), –50 mV (dashed line), –75 mV (dotted line), and –100 mV (dashed-dotted line) calculated with the numerical model in Reference 36. (b) Experimental relative electric permittivity data obtained for phospholipid vesicles. Reproduced with permission from Tirado et al. [30]. (c) Same for bacterial cell suspensions. Full squares: *Corynebacterium* sp. strain DSM 44016; open circles: *Corynebacterium* sp. strain DSM 6688. Reproduced with permission from van der Wal et al. [70].

We can see that for high frequencies, the dominant part is the in-phase component of the dipolar coefficient, while for low frequencies, the second addend in Equation 37.24 becomes larger than the first one.

This phenomenon has been extensively studied in the literature [56]. Analytical approximations are available for simplified situations like spherical particles with thin double layers immersed in symmetrical electrolyte solutions [20, 29, 36]. Also, numerical solutions are available for general cases when particles are spherical [37–39, 42]. Some results are shown in Figure 37.7. A large relaxation amplitude is observed both theoretically and experimentally. In the case of cells (see Fig. 37.7b,c), the large size of the particles produces a very low relaxation frequency.

For nonspherical particles, Fixman [42] provided a general numerical solution, while there is an analytical approximation only for the amplitude and

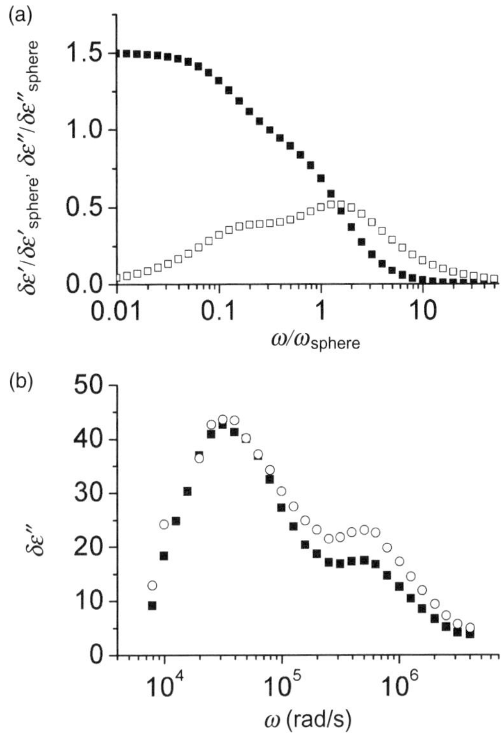

Figure 37.8. (a) Real (full squares) and imaginary (open squares) parts of the relative electric permittivity of a suspension of elongated particles with aspect ratio 13 divided by that of spheres with the same volume with $K^{\sigma}/K_sR = 1$. The frequency is normalized by the inverse of the characteristic time of the α-relaxation of the spheres. Reproduced with permission from Grosse et al. [41]. (b) Imaginary part of the relative electric permittivity as a function of the frequency for a 1% spheroidal hematite (length 551 nm and aspect ratio 6.1) in a 0.5 mM NaCl solution and pH 8 (squares) and 8.8 (circles). Reproduced with permission from Jiménez et al. [31].

characteristic frequency of the relaxation and not for the whole frequency dependence [40, 41]. As it occurs in the case of the MWO relaxation, the α process splits in the case of spheroidal particles due to the two principal size parameters of the particle. However, contrary to the case of MWO, these two relaxations can be separated in the spectrum in cases of low particle charge and large aspect ratio. As shown in Figure 37.8, this split has been observed both theoretically and experimentally [31, 40, 41].

Both in Figures 37.7 and 37.8 and in other examples in the literature [37, 40, 57, 58], it has been observed that the α-dispersion is very sensitive to small changes in the particle and EDL structure and morphology. However, the technique has not been widely exploited in the literature for bioparticles since, as it will be shown in Section 37.3, their small characteristic frequency limits the practical applicability of dielectric dispersion measurements due to the electrode polarization (EP), which will be described below.

In the case of cells, if the cell wall is insulating, then the scene is the same as in colloids, and the electric properties of the inner part do not affect this frequency range spectrum. However, in Reference 59, it is shown that if the ion exchange properties of the cell wall is taken into account, then the dielectric spectra differ from those of a homogeneous particle. The reason is that while the electric field is the same around the cell and the homogeneous particle, the cell membrane is charged and hence stores more energy.

37.3 THE DIELECTRIC DISPERSION MEASUREMENT

Broadband dielectric spectra show a rich phenomenology that provides a lot of information about the particle/solution interface. Among all the dispersions studied in Section 37.2, the richest one is by no doubt the α-relaxation. For this reason, a lot of effort has been put in raising the accuracy of its determination.

For conducting samples, the principal limitation is the EP, that is, the accumulation of ions around the electrodes when an electric field is applied. These ions screen the applied field, and hence, the potential drop in the sample is greatly reduced in an uncontrolled way. Despite this phenomenon being already recognized in the 1960s [25], it has no definite solution yet, although it may provoke the complete masking of the sample contribution to the real part of the electric permittivity. Since the dielectric losses (which in turn could give us some information of the sample) are also hidden by the contribution of the electrolyte conductivity, the applicability of this technique has not been as widespread as it would have been desirable.

In any case, although the accessible frequency range is limited, there is still a broad frequency interval in which the electric permittivity can be corrected from the EP by some techniques [26, 27, 60–63]. In the majority of cases, the correction is a time-consuming and tedious task. Hence, when monitorization of an *in vivo* process (e.g., cells in division) is the aim, we do not have this time. In this sense, a new method was proposed in Reference 28 and was successfully applied in many occasions [31, 55, 64–66]. It is based on the similarity between the imaginary part of the electric permittivity and the logarithmic derivative (LD) of the real part (ε''_D). This makes the LD the most suitable quantity to characterize the electric permittivity spectra of the suspension since (i) contrary to the imaginary part of the electric permittivity, the LD is free of the conductive contribution, and (ii) if electrodes are detached enough, the EP contribution can be better separated from the real spectra of the suspension than in the case of the real part [28]. In Figure 37.9 an example of ε''_D evaluation in an experimental low-frequency spectrum is shown.

In the last few years, the ability in microfabrication has led to a number of applications in the so-called lab-on-chip technology. An example is the electric

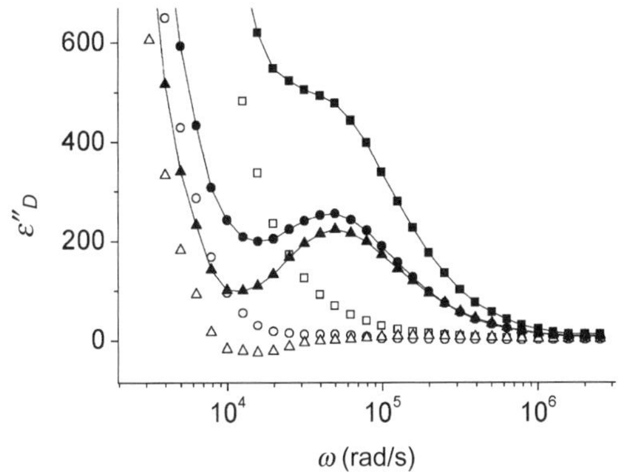

Figure 37.9. Logarithmic derivative of the relative electric permittivity of a 0.5 mM NaCl solution (open symbols) and of a suspension of latex spheres 530 nm in diameter, in a 0.5 mM NaCl solution (full symbols), for different electrode spacings. Reproduced with permission from Jiménez et al. [28].

impedance spectroscopy on single cells, that is, the direct study of the electrical properties of a single cell instead of calculating them from those of a population [67]. The method involves the measurement of the capacitance of single cells while passing through a microfluidic device. Successful application of this technique to obtain the DNA content of single eukaryotic cells in a noninvasive way has been reported [67].

37.4 APPLICATION TO BIOLOGICAL SYSTEMS

37.4.1 Liposomes and Cells

As mentioned, cells differ from colloids in their internal structure and in the presence of the membrane with particular electric and mechanical properties. The transmembrane potential that distinguishes living cells from nonviable ones appears as a consequence of the ionic selectivity of the membrane, and this can be detected by dielectric spectroscopy [68]. Also, the cell membrane is not rigid and it can deform under certain conditions. Such geometrical changes can also be detected by dielectric spectroscopy. This is the case of cells in division: One argument for the applicability of the dielectric spectroscopy is the noninvasive character of this procedure. Under strong electric fields, liposomes and cells loose their spherical shape and elongate along the field direction. In the extreme case, they form pearl chains and, eventually, the so-called electroporation occurs (the formation of pores in the membrane) [69]. In dielectric spectroscopy, we are restricted to low electric fields, and deformation, pearling, and electroporation are prevented. Hence, any change in the

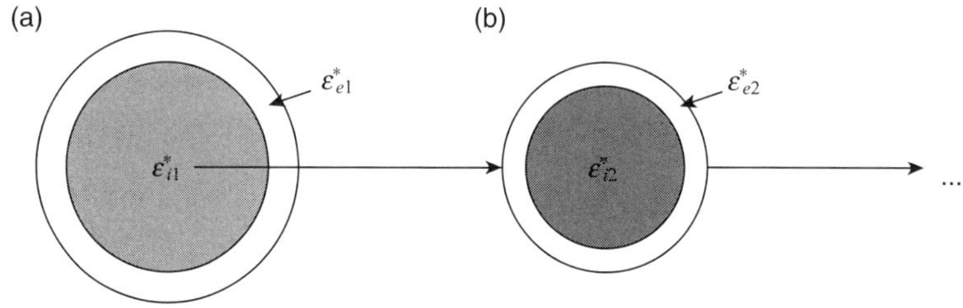

$$\varepsilon^* = \varepsilon^*_{e1} \frac{2(1-\phi_1)\varepsilon^*_{e1} + (1+2\phi_1)\varepsilon^*_{i1}}{(2+\phi_1)\varepsilon^*_{e1} + (1-\phi_1)\varepsilon^*_{i1}} \qquad \varepsilon_{i1}{}^* = \varepsilon^*_{el} \frac{2(1-\phi_2)\varepsilon^*_{e2} + (1+2\phi_2)\varepsilon^*_{i2}}{(2+\phi_2)\varepsilon^*_{e2} + (1-\phi_2)\varepsilon^*_{i2}}$$

Figure 37.10. Scheme of the multishell model (see text).

dielectric spectrum is due to a change in the structure of both the particle and the interface, and dielectric spectroscopy has been successfully applied to cells.

Particularly, experiments have focused on the 0.1-MHz to 1-GHz frequency range and the permittivity spectra of the cells, ε^*_p, is calculated by using the MW or Hanai equations. These spectra are subsequently analyzed in the light of the β-dispersion-type equations. Only a few works have dealt with the kilohertz range spectra in which the α-dispersion takes place, mainly because of the experimental uncertainties that the EP provokes [26, 70].

The permittivity of the single cell is the manifestation of the polarization due to the mismatch between the conductivities and permittivities of the different parts of the system. Using a multishell model [12], the electrical properties of the different parts of the cell are calculated. In this model, the MW formula (Eq. 37.12) is used in various steps. In the first step, the particle is formed by a homogeneous subparticle of electric permittivity ε^*_{i1} surrounded by a shell of permittivity, ε^*_{e1} (see Fig. 37.10a). The permittivity of the subparticle is calculated from that of the particle by using the MW formula in Figure 37.10a. In the next step, it is considered that the subparticle is not homogeneous but formed by another subparticle of electric permittivity ε^*_{i2} surrounded by a shell of permittivity, ε^*_{e2} (see Fig. 37.10b). The electric permittivity ε^*_{i2} can be calculated with the MW formula of Figure 37.10b. This procedure can be repeated until the complex cell is reproduced. In the simplest case, that is, a cell with a nucleus, these two steps are enough to obtain the electric properties of the nucleus.

This is done, for example, in Reference 71, where the experimental electric dispersion of lymphocytes and erythrocytes is analyzed. These cells are similar in size but differ in the core composition. While erythrocytes do not have any organelle, lymphocytes have a nucleus that occupies 40–70% of the total cell volume. The electric permittivity spectra exhibit only one relaxation phenomenon, which fits well in the single-shell model, but in the case of lymphocytes, the superposition of two relaxation phenomena is observed, a behavior well reproduced by a double-shell model. The same authors generalized the model

[12, 72] for spheroidal particles by using the extended MW equation together with Equation 37.17.

There are other situations in which two relaxation phenomena are observed. In Reference 12 the cell division of yeast produces a modification of the shape of the electric dispersion from one to two relaxation processes. Both relaxations are better separated when the liposome is more elongated, and this is in agreement with Equations 37.15 and 37.17. The same authors [73] proposed a model with two relaxations associated with the long and short axes of the particle. This model, with strong analogies to the extended MWO approach, predicts a change in the dielectric relaxation amplitude and the characteristic frequency of the long axis while the axial ratio increases. Later, a more general model was proposed, in which any shape can be implemented [74]. The Laplace equation is numerically solved in a complex cell with an irregular shape and applied to yeast cells in division, showing that by using the electric permittivity, information about the cell shape can be obtained in real time.

Finally, many examples can be found in the literature for the experimental analysis of the dielectric properties of biological cultures. To mention a few, the usefulness of this method to monitor the interaction between virus and cell has been suggested in Reference 75; fermentation with *Escherichia coli* (the change of carbon source of *E. coli* from glycerol to proteins) can clearly be identified by analyzing the measured capacitance and conductance during the fermentation [76]; the membrane potential is calculated from dielectric spectroscopy (DS) data in Reference 77.

37.4.2 DNA Characterization

The abundant literature existing on the dielectric spectra of DNA stems from the richness of those spectra, the numerous processes in which DNA is involved, and the necessity to monitor its electric polarizability in view of various applications, such as dielectrophoretic deposition [78, 79].

DNA can be viewed as a rodlike charged polyelectrolyte when its contour length is smaller than the persistence length. The dependence of its dielectric spectrum on the geometry of the particle has been widely studied in the literature [48]. Although the dielectric spectra can be analyzed on the basis of α and MW processes, it is usually interpreted in the light of diffusion processes over the typical lengths of the chain. This means that one can study the interaction between DNA and other species by analyzing the dielectric relaxations. In Mandel's model [45], the DNA coil is modeled as a partially flexible chain made of rigid subunits of length b, which is a measure of the interaction between the DNA subunits. Condensed counterions are allowed to fluctuate along the particle and generate two dielectric relaxations associated with the diffusion along the two main distances of the chain: the gyration radius and b.

For example, Zuzzi et al. [50] studied the liposome-induced compaction of DNA, by studying its DS upon addition of lipid to DNA-rich solutions. Such addition provokes the change between elongated coils to globular shapes.

According to Mandel's model, they could calculate the length b from the amplitude and characteristic frequency of the relaxation. The same procedure was used by Bonincontro et al. [51], who studied the interaction of cetyltrimethylammonium bromide (CTAB) with DNA. From the dielectric spectra and with the Mandel model, they calculated the changes in the length b and in the concentration of condensed counterions, which is provoked by the interaction of the DNA helix with other species. Tomic et al. [47, 80, 81] studied the dielectric spectra of DNA in Na^+ solutions. They found that the spectra of the imaginary part exhibit two peaks, which they associated with Na^+ diffusion in two characteristic lengths, namely, the correlation and persistence lengths.

37.4.3 Soft Particles

By soft particles, it is meant particles having a deformable coating permeable to fluid and ions, and made of flexible chains in a more or less random configuration. This is an idealization of many microorganisms that bear a diffuse surface layer where the physical and chemical properties gradually change from the medium to the inner part. The properties and behavior of this layer are essential to understand many biological processes in which the interaction between microbes and their medium plays a role. On the other hand, polyelectrolyte layers are often used for stabilizing nanoparticles in high ionic strength environments or to provide them with a biodegradable or biocompatible coating. This nanostructure can be designed for loading and releasing drugs or other therapeutic agents. Understanding the structure and behavior of polyelectrolyte chains in these environments is essential in designing the final devices, with biological or other applications in mind. The dielectric spectroscopy technique seems to be perfect to examine this situation.

In this kind of particle, the structure of the surface implies some fundamental differences from rigid colloids; basically, (i) the charge is not at a surface but distributed in a 3-D region, and (ii) the liquid and ionic motion inside the soft layer is modified: The shear plane looses its physical meaning. The electric permittivity response can be modeled by using the same characterization of the soft layer described in other chapters of this volume.

This modification of the structure of the interface does not introduce any new phenomenon different from those already described. However, the electric permittivity spectra greatly differ from those of rigid particles. In Reference 82, it is demonstrated that no difference is expected in the relaxation frequency. However, a rather large permittivity decrease is observed experimentally [83]. In the available theoretical models, the effect of the soft layer is an enhancement of the α-relaxation amplitude as shown in Figure 37.11. In this figure, theoretical data were obtained with the model in Reference 82. The reason for this enhancement is the large electroosmotic fluxes (liquid flow driven by ionic motion) inside the soft layer, which produces concentration polarization gradients of larger amplitudes than in rigid particles with only surface conductivity. This explains the strong dependence of the electric per-

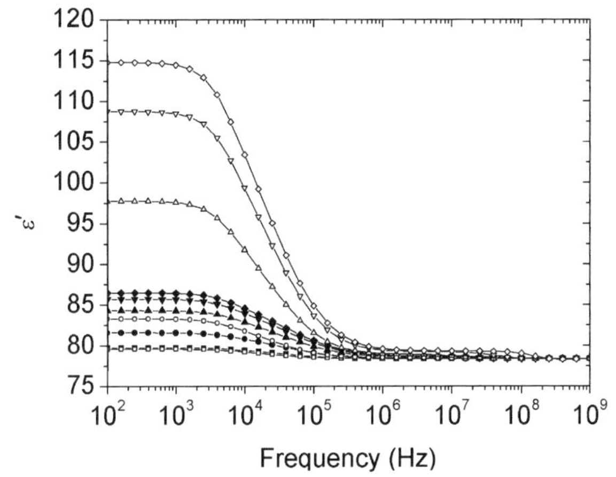

Figure 37.11. Relative electric permittivity as a function of frequency for a 0.1% suspension of hard (full symbols) and soft (open symbols) particles with a core radius of 100 nm and a soft shell of 150-nm thickness. Particle charges: 10^{-15} C (squares), 2×10^{-15} C (circles), 5×10^{-15} C (up triangles), 10×10^{-15} C (down triangles), and 20×10^{-15} C (diamonds). For hard particles, the charge is at the particle surface, and for the soft particles, the charge is in the polymer layer. Other parameters: penetration length 50 nm, KCl concentration 1 mM.

mittivity dispersion amplitude on the soft layer charge. Furthermore, since it also depends on the core surface charge, this technique turns out to be more suitable than any other technique to obtain information on the inner properties of soft particles.

37.5 CONCLUDING REMARKS

There are few colloidal systems where dielectric spectroscopy techniques find deeper application than in suspensions of biomolecules, cells, microbes, or, very generally, soft particles. This is because dielectric spectroscopy is extremely sensitive to practically every feature of the system: charge distribution on the particles, ionic composition of the medium (also the internal medium, if any), shape, heterogeneity, characteristic dimensions, and so on. These possibilities are, in addition, based on well-established models, which date back to Maxwell. Once a dielectric spectrum is obtained (in general for the suspension, but even for an individual cell), the use of suitable theories can provide the information sought, regarding mainly the overall geometry and the charge distribution of the cell or the bioparticle. The main limitation for a wider use of this technique comes from the precision of measurements at low frequency (in the kilohertz region). But even this can be overcome for two main reasons. One is the growing accuracy of methods for correcting the perturbing effects of EP. The other reason is that the characteristic distance for many ionic transport processes in cells and microbes is comparable to the membrane thickness, and this leads to relaxations well in the megahertz region. In summary, dielectric spectroscopy is a promising technique that, perhaps complemented by microscopic observations, can yield very valuable information on the structure and dynamics of biomolecules or cell interfaces.

ACKNOWLEDGMENT

Financial support from Junta de Andalucía, Spain, under project PE2008-FQM-3993 is gratefully acknowledged.

REFERENCES

1 (a) R. Höber. *Arch. Ges. Physiol* 133 (1910) 237; (b) R. Höber. *Arch. Ges. Physiol* 148, (1912) 189; (c) R. Höber. *Arch. Ges. Physiol* 150 (1913) 15.

2 J. C. Maxwell. *Electricity and magnetism*, vol. 1. Clarendon Press, Oxford, 1892.

3 K.W. Wagner. *Arch. Elektrotech.* 2 (1914) 371.

4 (a) H. Fricke. *Phys. Rev.* 24 (1924) 575; (b) H. Fricke. *Phys. Rev.* 26 (1925) 678; (c) H. Fricke. *Phys. Rev.* 26 (1925) 682.

5 H. Pauly, H. P. Schwan. *Z. Naturforschung* 14b (1959) 125.

6 R. Pethig, D. B. Kell. *Phys. Med. Biol.* 32 (1987) 933.

7 S. Takashima. *Electrical properties of biopolymers and membranes*. Adam Hilger, Bristol, 1989.

8 K. R. Foster, H. P. Schwan, in C. Polk, E. Postow (eds.), *Handbook of biological effects of electromagnetic fields*, 2nd ed. CRC Press, Boca Raton, FL, 1996.

9 C. M. Mihai, M. Mehedintu, E. Gheorghiu. *Bioelectrochem. Bioenerg.* 40 (1996) 187.

10 G. H. Markx, C. L. Davey. *Enzyme Microb. Technol.* 25 (1999) 161.

11 J. Gimsa, in A. V. Delgado (ed.), *Interfacial electrokinetics and electrophoresis*. Marcel Dekker, New York, 2002.

12 K. Asami. *Prog. Polym. Sci.* 27 (2002) 1617 and references therein.

13 Y. Feldman, I. Ermolina, Y. Hayashi. *IEEE Trans. Dielectr. Electr. Insul.* 10 (2003) 728.

14 V. Raicu, T. Saibara, A. Irimajiri. *Bioelectrochem. Bioenerg.* 47 (1998) 325.

15 R. D. Stoy, K. R. Foster, H. P. Schwan. *Phys. Med. Biol.* 27 (1982) 501.

16 J. Yang, Y. Huang, X. B. Wang, F. F. Becker, P. R. C. Gascoyne. *Biophys. J.* 72 (1997) MP450.

17 P. Gascoyne, C. Mahidolb, M. Ruchirawatb, J. Satayavivadb, P. Watcharasitb, F. F. Becker. *Lab Chip* 2 (2002) 70.

18 K. Asami, K. Takahashi, K. Shirahige. *Yeast* 16 (2000) 1359.

19 R. J. Hunter. *Foundations of colloid science*. Oxford University Press, Oxford, 2001.

20 S. S. Dukhin, V. N. Shilov. *Dielectric phenomena and the double layer in disperse systems and polyelectrolytes*. Wiley, New York, 1974.

21 S. S. Dukhin. *Adv. Colloid Interface Sci.* 44 (1993) 1.

22 T. Bellini, F. Mantegazza, V. Degiorgio, R. Avallone, D. A. Saville. *Phys. Rev. Lett.* 82 (1999) 5160.

23 T. Bellini, F. Mantegazza. *Interfacial electrokinetics and electrophoresis*. Marcel Dekker, New York, 2002.

24 A. V. Delgado, F. J. Arroyo, in A. V. Delgado (ed.), *Interfacial electrokinetics and electrophoresis*. Marcel Dekker, New York, 2002.

25 H. P. Schwan. In W. L. Nastuk (ed.), *Physical techniques in biological research*, vol. 6. Academic Press, New York, 1963.

26 H. Fricke, H. J. Curtis. *J. Phys. Chem.* 41 (1937) 729.

27 C. Grosse, M. Tirado. *IEEE Trans. Inst. Meas.* 50 (2001) 1329.

28 M. L. Jiménez, F. J. Arroyo, J. van Turnhout, A. V. Delgado. *J. Colloid Interface Sci.* 249 (2002) 327.

29 C. Grosse, in A. V. Delgado (ed.), *Interfacial electrokinetics and electrophoresis*. Marcel Dekker, New York, 2002.

30 M. Tirado, C. Grosse, W. Schrader, U. Kaatze. *J. Non-Cryst. Solids* 305 (2002) 373.

31 M. L. Jiménez, F. J. Arroyo, F. Carrique, U. Kaatze. *J. Phys. Chem. B* 107 (2003) 12192.

32 H. Ohshima. *Theory of colloid and interfacial electric phenomena*. Elsevier, Amsterdam, 2006.

33 J. F. L. Duval, F. Gaboriaud. *Curr. Opin. Colloid Interface Sci.* 15 (2010) 184.

34 J. L. Arias, M. A. Ruiz, V. Gallardo, A. V. Delgado. *J. Control. Rel.* 125 (2008) 50.

35 C. J. F. Böttcher, P. Bordewijk. *Theory of electric polarization*, vol. II. Elsevier Scientific, Amsterdam, 1978.

36 M. Fixman, *J. Chem. Phys.* 78 (1983) 1483 .

37 E. H. B. DeLacey, L. R. White. *J. Chem. Soc. Faraday Trans. 2* 77 (2007) 1981.

38 (a) C. S. Mangelsdorf, L. R. White. *J. Chem. Soc. Faraday Trans.* 86 (1990) 2859; (b) C. S. Mangelsdorf, L. R. White. *J. Chem. Soc. Faraday Trans.* 94 (1998) 2441; (c) C. S. Mangelsdorf, L. R. White. *J. Chem. Soc. Faraday Trans.* 94 (1998) 2583.

39 F. Carrique, F. J. Arroyo, M. L. Jiménez, A. V. Delgado. *J. Chem. Phys.* 118 (2003) 1945.

40 C. Grosse, V. N. Shilov. *J. Colloid Interface Sci.* 193 (1997) 178.

41 C. Grosse, S. Pedrosa, V. N. Shilov. *J. Colloid Interface Sci.* 220 (1999) 31.

42 M. Fixman. *J. Chem. Phys.* 124 (2006) 214506.

43 R. W. J. Sillars. *J. Inst. Elec. Engrs.* 80 (1937) 378.

44 J. Lyklema. *Fundamentals of interface and colloid science*, vol. 2. Academic Press, New York, 1995.

45 (a) M. Mandel. *Mol. Phys.* 4 (1961) 489; (b) M. Mandel, T. Odijk. *Ann. Rev. Phys. Chem.* 35 (1961) 75.

46 S. Omori, Y. Katsumoto, A. Yasuda, K. Asami. *Phys. Rev. E* 73 (2006) 050901.

47 S. Tomić, S. D. Babić, T. Ivek, T. Vuletić, S. Krča, F. Livolent, R. Podgornik. *Europhys. Lett.* 81 (2008) 68003.

48 Y. Katsumoto, S. Omori, D. Yamamoto, A. Yasudo, K. Asami. *Phys. Rev. E* 75 (2007) 011911.

49 D. J. Bakewell, I. Ermolina, H. Morgan, J. Milner, Y. Feldman. *Biochim. Biophys. Acta* 1493 (2000) 151.

50 S. Zuzzi, C. Cametti, G. Onori, S. Sennato. *Phys. Rev. E* 76 (2007) 011925.

51 A. Bonincontro, S. Marchetti, G. Onori, A. Rosati. *Chem. Phys.* 312 (2005) 55.

52 C. T. O'Konski. *J. Phys. Chem.* 64 (1960) 605.

53 V.N. Shilov, in S. P. Stoylov, M. V. Stoimenova (eds.), *Molecular and colloidal electro-optics*. Taylor and Francis, Boca Raton, FL, 2007.

54 (a) T. Hanai, in P. Sherman (ed.), *Emulsion science*. Academic Press, London, 1968; T. Hanai. *Kolloid-Z* 171 (1960) 23.

55 S. Ahualli, M. L. Jiménez, A. V. Delgado, F. J. Arroyo, F. Carrique. *IEEE Trans. Dielectr. Electr. Insul.* 13 (2006) 657.

56 C. Grosse, A. V. Delgado. *Curr. Opin. Colloid Interface Sci.* 15 (2010) 145 and references therein.

57 A. V. Delgado, F. Gonzalez-Caballero, F. J. Arroyo, F. Carrique, S. S. Dukhin, I. A. Razilov. *Colloid Surf. A Physicochem. Eng. Asp.* 131 (1998) 95.

58 F. J. Arroyo, F. Carrique, M. L. Jiménez-Olivares, A. V. Delgado. *J. Colloid Interface Sci.* 229 (2000) 118.

59 V. Zimmerman, C. Grosse. *J. Phys. Chem. B* 108 (2004) 12617.

60 L. A. Rosen, D. A. Saville. *Langmuir* 7 (1991) 36.

61 J. Kijlstra, R. A. J. Wegh, H. P. van Leeuwen. *J. Electroanal. Chem.* 366 (1994) 37.

62 H. Wakamatsu. *Hewlett-Packard J.* 48 (1997) 37.

63 M. M. Springer, A. Korteweg, J. Lyklema. *J. Electroanal. Chem.* 153 (1983) 55.

64 B. H. Bradshaw-Hajek, S. J. Miklavcic, L. R. White. *Langmuir* 25 (2009) 961.

65 A. D. Hollingsworth, D. A. Saville. *J. Colloid Interface Sci.* 280 (2004) 279.

66 M. J. Han, K. S. Zhao. *J. Phys. Chem. C* 112 (2008) 19412.

67 L. L. Sohn, O. A. Saleh, G. R Facer, A. J Beavis, R. S Allan, D. A Notterman. *Proc. Natl. Acad. Sci. USA* 97 (2000) 10687.

68 C. Prodan, E. Prodan. *J. Phys. D Appl. Phys.* 32 (1999) 335.

69 Z. A. Schelly. *Colloids Surf. A Physicochem. Eng. Asp.* 209 (2002) 305.

70 A. Van der Wal, M. Minor, W. Norde, A. J. B. Zehnder, J. Lyklema. *J. Colloid Interface Sci.* 186 (1997) 71.

71 K. Asami, Y. Takahashi, S. Takashima. *Biochim. Biophys. Acta* 1010 (1989) 49.

72 K. Asami, T. Yonezawa, H. Wakamatsu, N. Koyanagi. *Bioelectrochem. Bioenerg.* 40 (1996) 141.

73 K. Asami, T. Yonezawa. *Biochim. Biophys. Acta* 1245 (1995) 317.

74 A. Di Biasio, C. Cametti. *Bioelectrochem* 71 (2007) 149.

75 O. Fedchuk, A. Fdchuk, I. Bartsykovska, A. Fedchuk. *Antiviral Res.* 78 (2008) A38.

76 M. Hofmann, M. Funke, J. Buchs, W. Mokwa, U. Schnakenberg. *Sens. Actuators B Chem.* 147 (2010) 93.

77 C. T. Bot, C. Prodan. *Biophys. Chem.* 146 (2010) 133.

78 S. Tuukkanen, J. J. Toppari, V. P. Hytonen, A. Kuzyk, M. S. Kulomaa, P. Torma. *Int. J. Nanotechnol.* 2 (2005) 280.

79 H. Dalir, Y. Yanagida, T. Hatsuzawa. *Sens Actuators B Chem.* 136 (2009) 472.

80 S. Tomić, T. Vuletić, S. D. Babić, S. Krča, D. Ivanković, L. Griparić, R. Podgornik. *Phys. Rev. Lett.* 97 (2006) 098303.

81 S. Tomić, S. D. Babić, T. Vuletić, S. Krča, D. Ivanković, L. Griparić, R. Podgornik. *Phys. Rev. E* 75 (2007) 021905.

82 S. Ahualli, M. L. Jiménez, F. Carrique, A. V. Delgado. *Langmuir* 25 (2009) 1986.

83 S. Ahualli. PhD Dissertation, University of Granada, 2008.

84 M. L. Jiménez, F. J. Arroyo, F. Carrique, U. Kaatze, A. V. Delgado. *J. Colloid Interface Sci.* 281 (2005) 503.

38

ELECTROKINETIC METHODS IN BIOLOGICAL INTERFACES: POSSIBILITIES AND LIMITATIONS

Ángel V. Delgado, Julián López-Viota, and Fernando González-Caballero

38.1 INTRODUCTION

Ever since the discovery of electrophoresis and electroosmosis in the early 19th century by Reuss [1, 2], the appreciation of the importance of electrokinetic phenomena and their associated methods in the characterization of solid/liquid and liquid/liquid interfaces has grown almost exponentially. Using comparatively accessible techniques, it has become possible to gain a great deal of information about the electrical state (and, more indirectly, about the physicochemical structure) of the interface.

In the classical Gouy–Chapman description [3–6], the interface is envisaged as a rigid, ideal surface separating two media, most often solid and liquid, and endowed with a homogeneous charge distribution, located on the surface, and characterized by a constant surface charge density, σ^0. The electroneutrality of the system is guaranteed by a volume charge distribution $\rho(r)$ produced by an unequal distribution of counterions and coions (both considered as point charges), subjected to electrostatic attraction and repulsion from the surface, respectively, and to thermal motion by collisions with the solvent molecules (Fig. 38.1).

The distribution of electric potential $\Psi(r)$ and of ionic concentrations (number per volume) $n_i(r)$ ($i = 1, 2, \ldots, N$; N is the total number of ionic

Electrical Phenomena at Interfaces and Biointerfaces: Fundamentals and Applications in Nano-, Bio-, and Environmental Sciences, First Edition. Edited by Hiroyuki Ohshima.

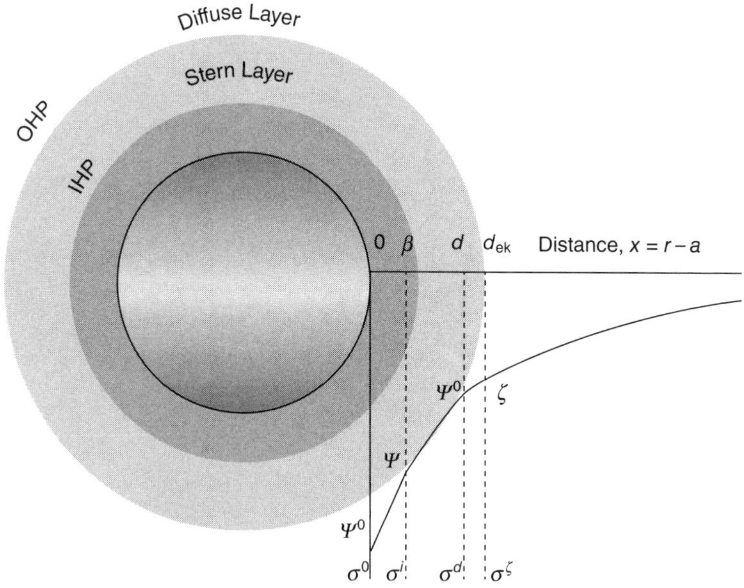

Figure 38.1. Classical description of the electrical double layer around a spherical interface.

species in solution, and it equals 2 in the simplest case) is given by the Poisson and Boltzmann equations, respectively:

$$\nabla^2\Psi(r) = -\frac{\rho(r)}{\varepsilon_m} = -\frac{1}{\varepsilon_m}\sum_{i=1}^{N} z_i e n_i(r) \tag{38.1}$$

and

$$n_i(r) = n_i^0 \exp[-z_i e\Psi(r)/k_B T], \tag{38.2}$$

where ε_m is the electric permittivity of the dispersion medium; z_ie and n_i^0 are the charge and bulk concentration, respectively, of the type i ions; and k_BT is the thermal energy.

The electroneutrality condition is often expressed in terms of a virtual surface charge density σ^d (diffuse charge density) as follows:

$$\begin{aligned}&\sigma^0 + \sigma^d = 0\\&\sigma^d \equiv \frac{1}{a^2}\int_a^\infty \rho(r) r^2 dr,\end{aligned} \tag{38.3}$$

a being the particle radius. The special charge distribution just described is known as the electric double layer (EDL), while the electrolyte region surrounding the particle, where $\rho(r) \neq 0$, is the diffuse layer or ionic atmosphere.

For completeness, let us mention that the origin of electric potentials is taken in the bulk solution, far from the interface, where the ionic concentrations are essentially unaffected by the presence of the particle. With that reference, the potential at the surface itself is known as surface potential Ψ^0, and its value at the beginning of the diffuse layer is the Stern potential Ψ^d. It will be clear that $\Psi^0 = \Psi^d$ in the Gouy–Chapman approach.

The partial differential Equation 38.1 does not have a general analytical solution. This can be found in a number of cases, some of them of considerable interest, fortunately. For example [7, 8], in the case of particles with low (as compared to the thermal potential k_BT/e) surface potential, a simple expression can be given:

$$\Psi(r) = \Psi^0 \frac{a}{r} \exp[-\kappa(r-a)], \tag{38.4}$$

where κ is the Debye–Hückel parameter, the reciprocal of which is the Debye length or EDL thickness:

$$\kappa = \left(\frac{\sum_{i=1}^{N} z_i^2 e^2 n_i^0}{\varepsilon k_B T} \right)^{1/2}. \tag{38.5}$$

This description was improved by Stern and Grahame considering that at least a layer of ions, partially or totally dehydrated, must reside very close to the solid surface, interacting with it so strongly that the effects of collisions with the (stagnant) solvent can be neglected. The geometric locus of the centers of these ions (mainly counterions) is called inner Helmholtz plane (IHP) ($r - a = \delta$), whereas the beginning of the truly diffuse region of the ionic distribution is the outer Helmholtz plane (OHP) (Fig. 38.1). The equivalent surface charge densities of these planes are σ^i and σ^d, and the electric potentials (again relative to the bulk) are denoted as Ψ^i and Ψ^d, respectively. Electroneutrality is now ensured by the equality $\sigma^0 + \sigma^i + \sigma^d = 0$.

Most of these quantities are far from being experimentally accessible (see, however, Reference 3 for details about titration procedures for estimating σ^0). The gap can be filled by electrokinetics: If the particles in Figure 38.1 are pinned in the laboratory frame of reference and an electric field is applied, the charged fluid in the EDL will be set in motion, and far from the particle, a constant fluid velocity parallel (if the particle charge is negative) or antiparallel (positively charged particle) to the field can be measured. It can be shown [2, 4, 9] that for not too high fields (i.e., fields provoking perturbations in the local concentration profiles much smaller than the equilibrium concentrations), the fluid velocity (electroosmotic velocity, v_{eo}) is proportional to the field strength. Except in microchannels [10, 11], direct determination of this velocity is difficult, and it is preferred to carry out measurements of the associated electroosmotic flow rate. The process described (electroosmosis) is one of the so-called electrokinetic phenomena [2–4]. These are relative motions of the two phases separated by the charged interface, forced by the action of an external field, very often an electric potential gradient, but also a pressure gradient, a gravitational or centrifugal field, and so on.

In many instances, it is more practical to determine the particle velocity under the external field, with respect to the liquid at rest. This is by far the most widely used electrokinetic technique, and it is named electrophoresis. The particle velocity will be equal and opposite to the electroosmotic velocity described earlier, and hence the two phenomena are reciprocal in some sense. In any case, either velocity will be controlled by the value of the potential at the (ideal, again) surface of separation between the mobile and immobile portions of fluid. Such a surface is not necessarily (in fact it is not in almost any case) located exactly at the interface between the two media, for instance, on the particle surface in the case of electrophoresis. More often than not, it will be at some (unknown) distance (d_{ek}) from it, perhaps in the vicinity of the OHP [12]. The surface (slip surface or slip plane) will correspond to some value of the equilibrium potential, named electrokinetic or zeta potential (ζ), and the region extending from the solid surface to the slip plane is known as the stagnant layer (Fig. 38.1).

This description is made very brief in order to avoid repetition of material already presented in this and in other texts [3, 4, 6, 8]. We will on purpose not consider at this moment deep reformulations, dealing particularly with the fact that ions have finite size and that, in addition to interacting with the charged interface and colliding with solvent molecules, they experience interactions (electrostatic plus hard sphere) among themselves. This manifests in ionic correlations and alters the concentration and potential profiles (with respect to the point-charge approach), in addition to solving the paradox of unrealistically high point ionic concentrations in the vicinity of highly charged surfaces. Numerical simulations of the equilibrium [13, 14] and nonequilibrium [15, 16] EDL have been recently reported.

In this chapter, we rather intend to consider what is different in interfaces of biological interest, including, for instance, those of living cells in organisms and of artificial particles engineered to get into contact with them. Then, we will discuss how and to what extent electrokinetic techniques can be used to gain information from such interfaces. The chapter will be organized as follows. In Section 38.2, the special features of biological interfaces and, specifically, the existence of a complex membrane will be considered. Section 38.3 will give details on how such structures affect the electrokinetics of the particle or cell, and some applications will be revised in Section 38.4. Also, Section 38.5 considers the case of artificial particles coated by molecules chosen for improving the interfaces of biological systems. Section 38.6 will include our conclusions.

38.2 SPECIAL FEATURES OF BIOLOGICAL INTERFACES

We will consider next the features of the double layers (and their response to external fields) in the case of cells in contact with the extracellular space. We must leave aside the case where the cell is physiologically active [17–19]: Think of protein adsorption and folding, nerve impulse generation and transmission, membrane channel opening or closing in response to changes in some ion concentrations, and so on. We will just consider the existence of a membrane with a specific physicochemical structure (Fig. 38.2), based on ionizable molecules responsible for the charge generation on it, and hence giving rise to an electrical double layer, with some similarities and some differences with respect to those shown in Figure 38.1.

As observed in Figure 38.2, the phospholipid head groups of the membrane will bear a certain amount of surface charge, depending on its molecular constitution and on the pH of the medium in which the cell is immersed. The main groups responsible for the membrane charge[1] are carboxylic, phosphate, and amino ones; the latter are typically less numerous, and thus the net charge of the membrane is negative at neutral pH. These groups are part of the phospholipids and proteins forming the membrane; this is organized in such a way that the hydrophobic regions of these molecules are oriented toward the membrane interior, whereas the polar, hydrophilic parts are exposed to the

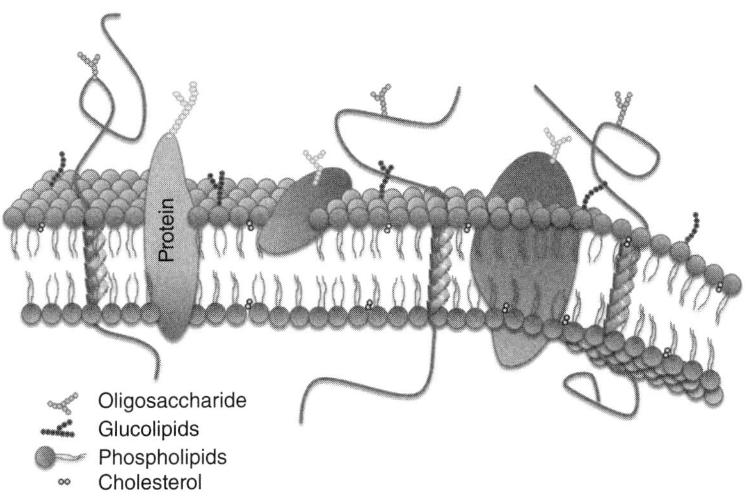

Figure 38.2. Schematic representation of a cellular membrane.

[1] As noted by Kleijn and Leeuwen [17], the membrane cannot be considered as an ideal surface but rather as a phase with finite thickness (tens of nanometers), and hence they prefer to speak of "interphase" rather than "interface" for describing the membrane region. In this chapter, we will keep the traditional denomination of interface, though keeping in mind the structure depicted in Figure 38.2.

solutions inside the cell (the cytoplasm) and outside it (the extracellular medium). In a classical article, Singer and Nicolson [20] first proposed a membrane structure consisting of a lipidic bilayer in which the proteins and saccharides are implanted.

As in the EDLs described in the previous section, the presence of the charged groups on the membrane produces a charged ionic atmosphere, mainly constituted by counterions, but now they can partly penetrate the membrane, whereas coions are mostly excluded from it. The ideal solid/liquid interface is now changed into a complex region containing, in addition to the solvent (water), ionizable groups and their countercharges. These are not only in the membrane but are also partly in the diffuse layer, formed in the solution in contact with the cell wall.

The charge distribution described resembles the one characteristic of so-called soft colloids, that is, those characterized by the existence of a polyelectrolyte layer grafted on a solid nucleus, or even just a polyelectrolyte brush, with no rigid nucleus in it [7, 8, 21–23]. The basic issue of this model is the consideration of an EDL description extended to consider a Donnan layer (Fig. 38.3). This is reached by considering that the charged membrane layer

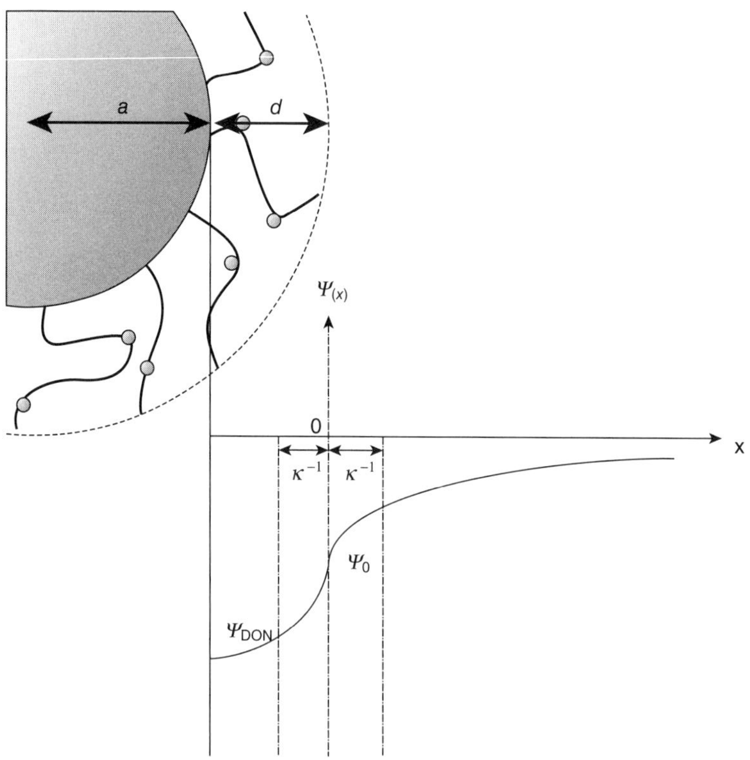

Figure 38.3. Schematic representation of the potential and charge distributions in a soft particle.

can be modeled as a homogeneous density (see, however, References 24 and 25 for the case of other profiles of ionizable groups) of fully ionized groups, giving a fixed volume charge density ρ_{fix} to the layer. If, for simplicity, we consider a flat interface, and z is the coordinate normal to it. Poisson equation for a charged layer extending from $z = -d$ to $z = 0$ must be splitted as follows:

$$\begin{aligned} \frac{d^2\psi(z)}{dz^2} &= -\frac{\rho(z)}{\varepsilon}, \quad z>0 \\ \rho(z) &= \sum_{i=1}^{N} n_i(z) z_i e = \sum_{i=1}^{N} n_i^0 z_i e \exp[-z_i e \Psi(z)/k_B T] \\ \frac{d^2\psi(z)}{dz^2} &= -\frac{\rho(z)+\rho_{\text{fix}}}{\varepsilon}, \quad -d<z<0. \end{aligned} \tag{38.6}$$

The versions corresponding to spheres or other geometries can be found in References 23–27. The effect of particle concentration, of interest in suspensions with moderate- to high-volume fractions of solids, has been reported, for example, by Ahualli et al. [28]. Equations 38.6 can be solved for given bulk concentrations of ions in solution and given charge densities of dissociable groups, subject to the following boundary conditions (see References 21–23 for details):

$$\begin{aligned} \left.\frac{d\Psi}{dz}\right|_{z=-d^+} &= -\frac{\sigma}{\varepsilon} \\ \Psi(0^-) &= \Psi(0^+) \\ \left.\frac{d\Psi}{dz}\right|_{z=0^-} &= \left.\frac{d\Psi}{dz}\right|_{z=0^+} \\ \Psi(z\to\infty) &= 0 \\ \left.\frac{d\Psi}{dz}\right|_{z=\to\infty} &= 0. \end{aligned} \tag{38.7}$$

Equation 38.6 can be simplified in the case of a symmetrical electrolyte with valency v and ion bulk concentrations $n_1^0 = n_2^0 = n^0$, to read

$$\begin{aligned} \frac{d^2\psi(z)}{dz^2} &= \frac{2ven^0}{\varepsilon}\sinh\left(\frac{ve\Psi}{k_B T}\right), \quad z>0 \\ \frac{d^2\psi(z)}{dz^2} &= \frac{2ven^0}{\varepsilon}\sinh\left(\frac{ve\Psi}{k_B T}\right) - \frac{\rho_{\text{fix}}}{\varepsilon}, \quad -d<z<0, \end{aligned} \tag{38.8}$$

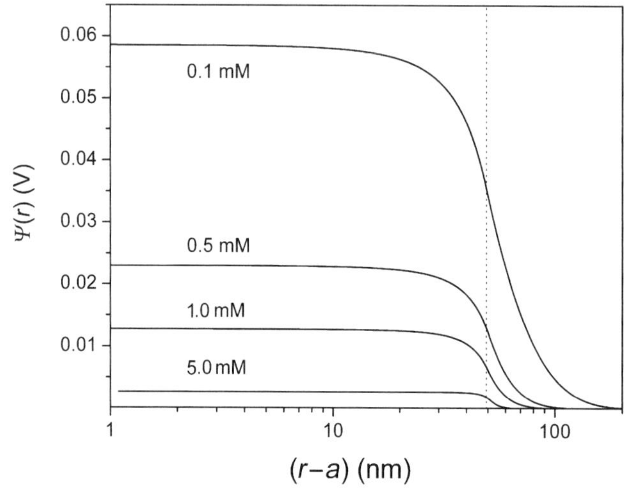

Figure 38.4. Effect of ionic strength (KCl concentration) on the electrical potential profiles around a spherical soft particle (uncharged core) with the following parameters: core radius $a = 100$ nm; charged layer thickness $d = 50$ nm; total charge in the layer: 10^{-15} C; dimensionless friction parameter $\lambda a = 100$.

or in dimensionless form,

$$\begin{aligned} \frac{d^2y}{dz^2} &= \kappa^2 \sinh y, \quad z > 0 \\ \frac{d^2y}{dz^2} &= \kappa^2 \left(\sinh y - \frac{\rho_{\text{fix}}}{2evn^0} \right), \quad -d < z < 0 \end{aligned} \tag{38.9}$$

with $y \equiv ve\Psi/k_BT$. Although semianalytical procedures do exist for solving Equation 38.9, and analytical expressions have been found for low potentials [23], the most general approach is the numerical one, illustrated in Figure 38.3, for the case of dilute suspensions of spheres, and different concentrations of bulk electrolyte and given layer charge. Note how, if the EDL thickness is much smaller than the soft layer dimension ($\kappa^{-1} << d$), the counterions penetrating the layer very efficiently screen the fixed charge, and the potential is constant (independent of the position). This is the Donnan potential, Ψ_{DON} (Fig. 38.3); its value in the case of a flat interface can be found by setting equal to zero the derivative in Equation 38.8:

$$\Psi_{\text{DON}} = \frac{k_BT}{ve} \sinh^{-1}\left(\frac{\rho_{\text{fix}}}{2evn^0} \right) \tag{38.10}$$

or, in the case of low fixed charge density,

$$\Psi_{\mathrm{DON}}\big|_{\rho_{\mathrm{fix}} \ll en^0} \cong \frac{\rho_{\mathrm{fix}}}{\varepsilon\kappa^2}. \tag{38.11}$$

Calling Ψ^0 the potential at the interface between the charged membrane and the external solution, it is possible to obtain the electric potential distribution in terms of the two mentioned potentials [23]:

$$\begin{aligned}\Psi^0 &= \Psi_{\mathrm{DON}} - \frac{k_B T}{ve}\tanh\left(\frac{ve\Psi_{\mathrm{DON}}}{2k_{\mathrm{B}}T}\right)\\ \Psi(z) &= \Psi_{\mathrm{DON}} + \left(\Psi^0 - \Psi_{\mathrm{DON}}\right)\exp[-\kappa_{\mathrm{m}}|z|],\end{aligned} \tag{38.12}$$

where κ_{m} has been termed the effective Debye–Hückel parameter, and it includes the contribution of the fixed charges to the specification of the ionic atmosphere inside and outside the membrane. Its value is

$$\kappa_{\mathrm{m}} = \kappa\left[1+\left(\frac{\rho_{\mathrm{fix}}}{2evn^0}\right)^2\right]^{1/4}. \tag{38.13}$$

38.3 ELECTROKINETICS OF PARTICLES WITH SOFT INTERFACES

38.3.1 Basic Equations

Having described the equilibrium structure of the double layers in the cell/solution interface (or interphase), we will consider the changes required in general electrokinetic theory in order to account for the response of cell suspensions to external fields. Essential contributions in this respect were published by Ohshima [8, 22, 24, 29–31], Hill et al. [32, 33], Hill and Saville [34], and Lopez-Garcia et al. [27]. We will for the moment assume that the particle is subjected to a constant electric field **E**, weak enough for the particle to undergo electrophoretic motion with a velocity (the electrophoretic velocity $\mathbf{v}_e$) proportional to the field, the electrophoretic mobility u_e being the constant of proportionality ($\mathbf{v}_e = u_e\mathbf{E}$). This is the experimentally accessible quantity, and the problem now is to theoretically relate its value to the parameters of the EDL described earlier, in particular, to the charge density of the core and the membrane, the thickness of the latter, and the ionic composition of the medium. This is not the place to give the details of the derivation, already reported in the cited references, but rather to show the basic equations and some results. A spherical particle of radius a with a membrane of thickness d (total size of the colloidal unit: $b = a + d$) will be assumed. From the solution to this problem, it is easy to find that pertaining to the flat geometry. The reference system is considered fixed on the particle center.

The starting equations are usual in the electrokinetics of particles:

1. Poisson equation for the potential distribution:

$$\nabla^2\Psi(\mathbf{r}) = \begin{cases} -\sum_{i=1}^{N} \frac{ez_i n_i(\mathbf{r})}{\varepsilon} & \text{if } d < r < b \\ -\sum_{i=1}^{N} \frac{ez_i n_i(\mathbf{r})}{\varepsilon} - \frac{\rho_{\text{fix}}}{\varepsilon} & \text{if } a < r < d \end{cases} \quad (38.14)$$

2. Continuity equations for each ionic species:

$$\nabla \cdot \left(-D_i \left(\nabla n_i(\mathbf{r}) + \frac{n_i(\mathbf{r}) e z_i \nabla \Psi(\mathbf{r})}{k_B T} \right) + n_i(\mathbf{r})\mathbf{v}(\mathbf{r}) \right) = 0, \quad (38.15)$$

   where D_i is the diffusion coefficient of ionic species i, and $\mathbf{v}(\mathbf{r})$ is the fluid velocity distribution.

3. Navier–Stokes equations for a viscous, incompressible fluid (viscosity η and density ρ):

$$\rho \frac{\partial \mathbf{v}(\mathbf{r})}{\partial t} = \begin{cases} -\nabla P(\mathbf{r}) + \eta \nabla^2 \mathbf{v}(\mathbf{r}) - \sum_{i=1}^{N} e z_i n_i(\mathbf{r}) \nabla \Psi(\mathbf{r}) & \text{if } d < r < b \\ -\nabla P(\mathbf{r}) + \eta \nabla^2 \mathbf{v}(\mathbf{r}) - \sum_{i=1}^{N} e z_i n_i(\mathbf{r}) \nabla \Psi(\mathbf{r}) - \gamma \mathbf{v}(\mathbf{r}) & \text{if } a < r < d \end{cases} \quad (38.16)$$

$$\nabla \cdot \mathbf{v} = 0,$$

   where $P(\mathbf{r})$ is the pressure field, and it is assumed that the membrane exerts an extra drag on the liquid, represented by the friction parameter γ. It has become usual to combine the two friction parameters (γ, η) into a single one $\lambda = (\gamma/\eta)^{1/2}$ [23].

All the quantities $\Xi(\mathbf{r})$ involved in the description of the EDL dynamics in the presence of an electric field (pressure, potential profile, fluid velocity, and ionic concentrations) are first of all expressed as the sum of an equilibrium quantity and a perturbation term, linear with the field ($\Xi(\mathbf{r}) = \Xi^0(r) + \delta\Xi(r, \theta)$). In addition, the angular and radial variables (a system of spherical coordinates centered in the particle is used) are separated, taking into account the spherical symmetry of the problem (see References 22–24, 27, and 28 for the details). We will limit ourselves next to the specification of the boundary conditions, enclosing the physics of the model, and of the main results.

38.3.2 Boundary Conditions

An essential ingredient of the model is the specification of the boundary conditions for the equilibrium quantities and their perturbations. These are as follows:

1. For the electric potential:
   - In equilibrium, the application of the Gauss law leads to

$$\sigma^0 = -\varepsilon \left. \frac{d\Psi^0(r)}{dr} \right|_{r=a}$$
$$\left. \frac{d\Psi^0(r)}{dr} \right|_{r\to\infty} = 0. \tag{38.17}$$

   - The continuity of the normal component of the displacement vector and of the electric potential can be written as

$$\left. \frac{\partial[\delta\Psi(r,\theta)]}{\partial r} \right|_{r=a} - \frac{\varepsilon_{\mathrm{p}}}{\varepsilon a}\delta\Psi(a,\theta) = 0, \tag{38.18}$$

   ε_p being the permittivity of the particle's core.
   - The potential and the normal component of the displacement vector must also be continuous at the cell/solution interface:

$$\delta\Psi(b^-,\theta) = \delta\Psi(b^+,\theta)$$
$$\left. \frac{\partial[\delta\Psi(r,\theta)]}{\partial r} \right|_{r=b^-} = \left. \frac{\partial[\delta\Psi(r,\theta)]}{\partial r} \right|_{r=b^+}. \tag{38.19}$$

   - Far from the particle, the gradient of the potential perturbation is minus the applied field:

$$\nabla\delta\Psi(\mathbf{r}) = -\mathbf{E}. \tag{38.20}$$

2. For the fluid velocity $\mathbf{v}$ and pressure $P(\mathbf{r})$:
   - The velocity must be zero on the core surface and it should tend to minus the electrophoretic velocity $\mathbf{v}_e$ far from the cell:

$$\mathbf{v}|_{r=a} = \mathbf{0}$$
$$\mathbf{v}|_{r\to\infty} = -\mathbf{v}_e. \tag{38.21}$$

   - It must be continuous at the surface layer/solution limit, $r = b$:

$$\mathbf{v}(b^-) = \mathbf{v}(b^+). \tag{38.22}$$

- The continuity of the hydrodynamic stress tensor at the membrane limit leads to the continuity of both the vorticity and the pressure:

$$\begin{aligned} \nabla \times \mathbf{v}\big|_{r=b^-} &= \nabla \times \mathbf{v}\big|_{r=b^+} \\ P(\mathbf{r})\big|_{r=b^-} &= P(\mathbf{r})\big|_{r=b^+}. \end{aligned} \tag{38.23}$$

3. For the ionic concentrations and velocities:

- Ions cannot penetrate the core:

$$\mathbf{v}_j(\mathbf{r})\cdot\hat{\mathbf{r}} = 0, \quad j = 1, 2, \ldots, N. \tag{38.24}$$

- The membrane/solution interface is assumed to be permeable to ions. Their concentrations and velocities are continuous at $r = b$:

$$\begin{aligned} n_j(\mathbf{r})\big|_{r=b^-} &= n_j(\mathbf{r})\big|_{r=b^+} \\ \mathbf{v}_j(\mathbf{r})\big|_{r=b^-} &= \mathbf{v}_j(\mathbf{r})\big|_{r=b^+} \\ j &= 1, 2, \ldots, N \end{aligned} \tag{38.25}$$

- The concentrations perturbations are negligible far from the interface:

$$\delta n_j(\mathbf{r})\big|_{r\to\infty} = 0 \quad j = 1, 2, \ldots, N. \tag{38.26}$$

The solution of the linearized system of equations subject to the boundary conditions can be carried out analytically or semianalytically in some cases [22], but the general solution requires the use of numerical methods. MATLAB® routines have been used in our group with success in a wide variety of situations, including in particular the case of concentrated suspensions and alternating (AC) fields [28]. Some results on the DC (direct current, i.e., constant applied electric field) electrophoretic mobility of soft particles will be briefly discussed next.

38.3.3 Electrophoretic Mobility Data

Figure 38.5 shows the velocity distribution (in the tangential direction, for $\theta = \pi/2$ and a unit electric field) around a spherical particle 100 nm in a radius coated with a 100-nm-thick membrane. The total charge on the core is zero, and in the membrane, it is specified to be equal to 10^{-15} C, and the KCl concentration is 1 mM. With this, the value of the Debye length is 9.6 nm, and $\kappa a = 10.4$ and $\kappa d = 5.2$. In this figure, we check for the effect of changing the value of the parameter λ (an increase in this parameter means more friction in the membrane, i.e., a more compact layer, and in fact, $\lambda \to \infty$ is equivalent to a rigid particle of radius b). Note how the mobility attains a constant value at a distance from the surface close to λ^{-1}, which means that this parameter is

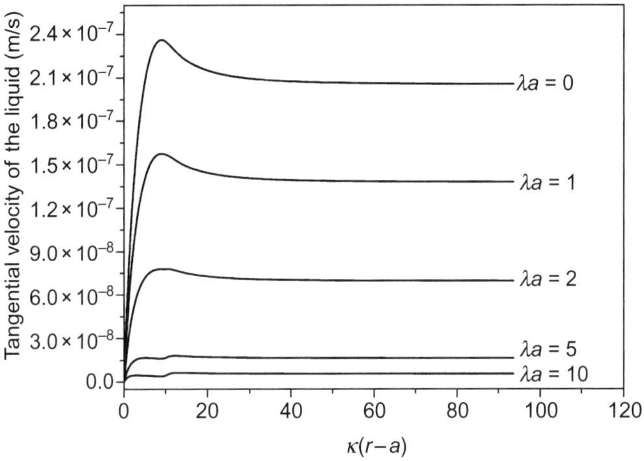

Figure 38.5. Effect of the friction parameter λ on the fluid velocity profile (tangential velocity at an angle $\theta = 90°$ around a spherical cell with a core 100-nm radius coated by a 100-nm-thick membrane, in the case that the total charge on the particle is zero, and that in the membrane is 10^{-15} C. The ionic concentration is 1 mM KCl.

the shielding length of the polyelectrolyte layer: The fluid velocity with respect to the particle is negligible for distances lower than λ^{-1} to the surface. In the figure, it can also be observed that the fluid velocity far from the particle differs from that in the membrane layer, and that the long-distance value is attained at a distance comparable to the EDL thickness in the external limit of the membrane. This is schematically shown in Figure 38.6.

We continue this brief analysis by considering the effect of varying the cell charge (Fig. 38.7). Note how the mobility of the cell is strongly controlled by the amount and sign of the membrane charge. It can be shown [28] that, although the charge on the particle also has a significant effect, even small opposite charges of the membrane suffice to provoke mobility inversion.

Concerning the effect of ionic strength, Figure 38.8 illustrates the variation of the mobility with KCl concentration. The main result is the tendency of the mobility to saturate at a value (u_e^∞ in Fig. 38.6) that is essentially controlled by the membrane charge. This is a distinct feature of the electrokinetics of particles with a penetrable membrane: The ionic strength alone is unable to completely stop the electrophoretic motion, no matter the concentration of electrolyte used. In fact, Ohshima has reported values of the limiting mobility for different conditions [22] in which this is explicitly demonstrated:

$$u_e^\infty = \frac{\rho_{\text{fix}} A}{\eta \lambda^2 B}\left(1 - \frac{a^3}{b^3}\right), \quad A = \frac{1}{2}\exp[\lambda(b-a)]\left(1 - \frac{a^3}{b^3}\right), \quad B = \frac{1}{2}\exp[\lambda(b-a)]$$

$$u_e^\infty = \frac{\rho_{\text{fix}}}{\eta \lambda^2}: \quad \kappa a \gg 1, \lambda a \gg 1, \kappa d \gg 1, \lambda d \gg 1. \tag{38.27}$$

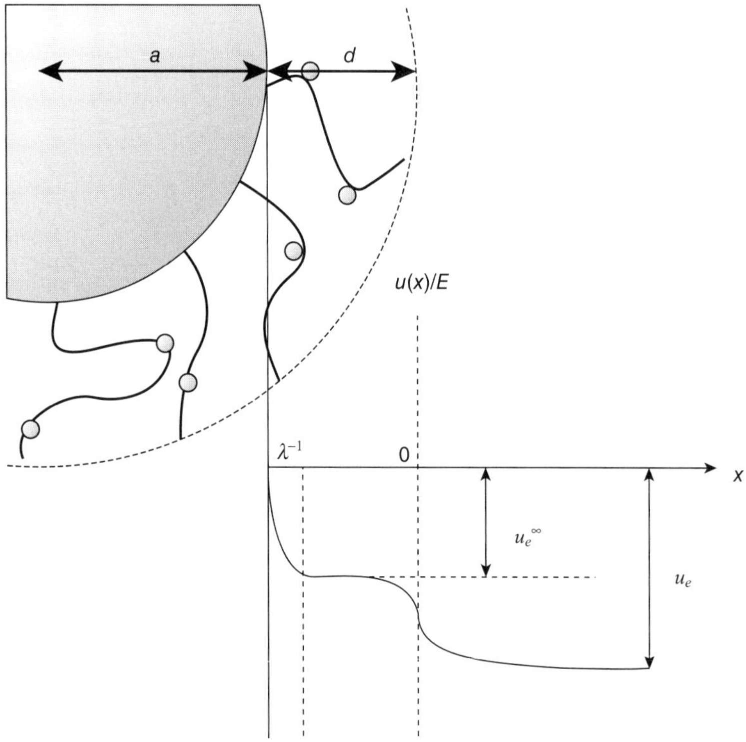

Figure 38.6. Schematic variation of the fluid velocity per unit field around the soft particle.

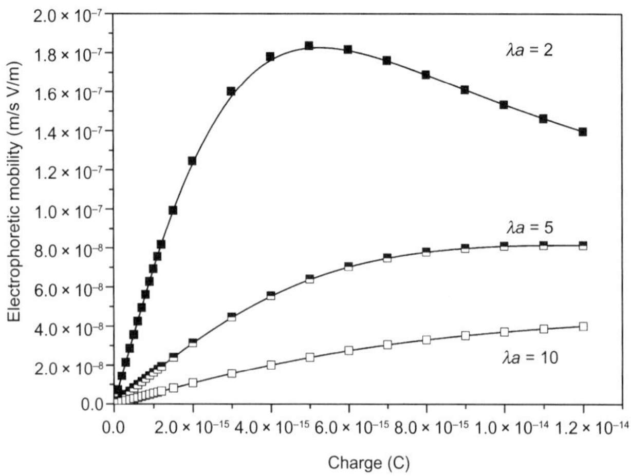

Figure 38.7. Effect of the membrane total charge (for uncharged core) on the electrophoretic mobility of spherical cells, for the indicated values of λa. Other parameters are the same as in Figure 38.5.

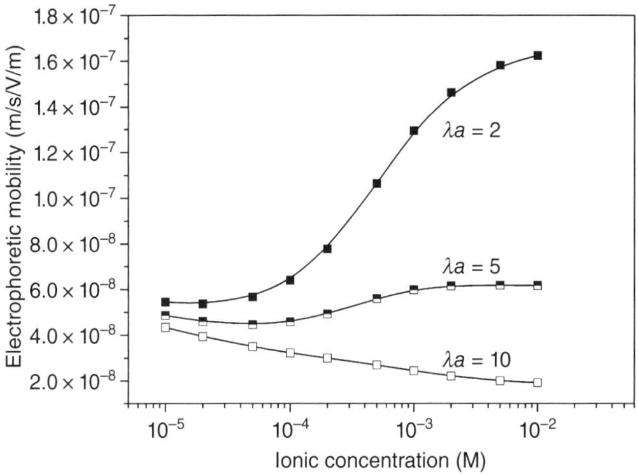

Figure 38.8. Effect of ionic concentration (KCl) on the electrophoretic mobility of spherical cells. Conditions: $a = 100$ nm, zero core charge; polymer charge 1.7×10^{-15} C; $d = 50$ nm, $\kappa a = 10$; λa values as indicated.

38.3.4 Other Electrokinetic Phenomena in Cell Suspensions: Dielectrophoresis and Electrorotation

Although electrophoresis is the most widely employed method in the electrokinetic characterization of cell suspensions, other methods can be cited. In other contributions of this book, the use of methods based on the application of AC fields is dealt with. An interesting review was publised by Gimsa rather recently [35]. It is a fact that these methods have progressively substituted DC-field-based techniques in biotechnology, in particular, in such fields as membrane conductance or cytoplasmic electrical properties, for example.

A number of techniques can be cited where the action of an AC field on the polarization of the cell and the interaction of the cell with the field are registered. In some cases, the information is obtained indirectly, observing the collective response of the suspension of cells (electric light scattering, electro-optics, electric birefringence, etc.), but here we will focus on those situations where direct observation of the cells is employed.

38.3.4.1 Dielectrophoresis. This phenomenon is the motion of cells or particles in the presence of a nonhomogeneous electric field. The motion of the particles can occur toward or away from the high-field region, depending on the direction of the induced dipole, which can be either parallel or antiparallel to the field, depending on the frequency, the cell charge, or the ionic strength. This can be understood by considering the expression for the dielectrophoretic force on the particles [36, 37]:

$$\mathbf{F}_{DP} = 2\pi\varepsilon R^3 \,\mathrm{Re}(C_0^*)\nabla^2 \mathbf{E}_{\mathrm{rms}}, \tag{38.28}$$

where R is the cell size; $C_0^*(\omega)$ is the induced dipole coefficient for a field of frequency ω; and $\mathbf{E}_{rms}$ is the root mean square amplitude of the field. It is important to consider [38] that the field also exerts force on the surrounding liquid, and this force is transmitted to the particle. As mentioned, experimental determinations of the dielectrophoretic displacement are often based on "collective" measurements, namely, the evaluation of the impedance of the suspension [39] or the changes in the light scattered by the suspension upon application of the field. Such changes are in fact related to the dielectrophoretic aggregation of the particles. Interpretation of these results requires of the knowledge of the individual cell geometry and of the conditions for the formation of aggregates of a certain configuration (chainlike, disk-like, etc.) [40].

The dielectrophoretic motion has found special application in the manipulation and sorting of cells [41]. Figure 38.9 is an example of the kind of electrode arrangement used by Kadaksham et al. [42] and its application to cell clustering. Sorting or separation [43–47] are also possible based on the different positions and velocities in the nonhomogeneous field of cells with different polarizabilities, sizes, geometries, and so on.

A practical realization of dielectrophoresis is the so-called traveling-wave dielectrophoresis (TWD) [48, 49]. In this technique, the particle undergoing dielectrophoresis is made to move by a traveling electric field, which is applied by a comb of interdigitated electrodes where the voltages are applied with specified phase differences. The related phenomenon of AC electroosmosis [50, 51] has received considerable attention recently because of its applications in microfluidics. Cen et al. [52] have designed a microchip where dielectrophoresis, TWD, and electrorotation are used for the characterization of malignant tumor cells. As an example, Figure 38.10 shows the TWD velocity of viable and nonviable NCI-H929 myeloma cells. A similar study was performed by Kua et al. [53] and was applied to the separation of viable and nonviable *Saccharomyces cerevisiae* yeast cells. As a final, very important example, due the malignancy of this cancer, the work of Menachery et al. [54] showing the possibilities of dielectrophoresis in recognizing the variants of the small-cell lung cancer cells is worth considering as an additional proof of the interest of electrokinetics in biomedical fields.

Kang and Li [55] have recently reviewed the developments of electrokinetic techniques in microchannels for the manipulation of cells, including electrophoresis, dielectrophoresis, and induced-charge electrokinetics. These authors also considered the possibility of producing cell lysis when applying high electric fields in electrokinetic sorting and manipulation. The process is due to the extra fields produced in the membrane, which may alter the flow in membrane ion channels and hence the normal physiological activity of the cell.

38.3.4.2 Electrorotation. In this technique, a nonhomogeneous field is again applied to the cell [35, 56–58]. A circularly polarized, rotating electric field is applied to the cell suspension, and microscopic inspection (or light

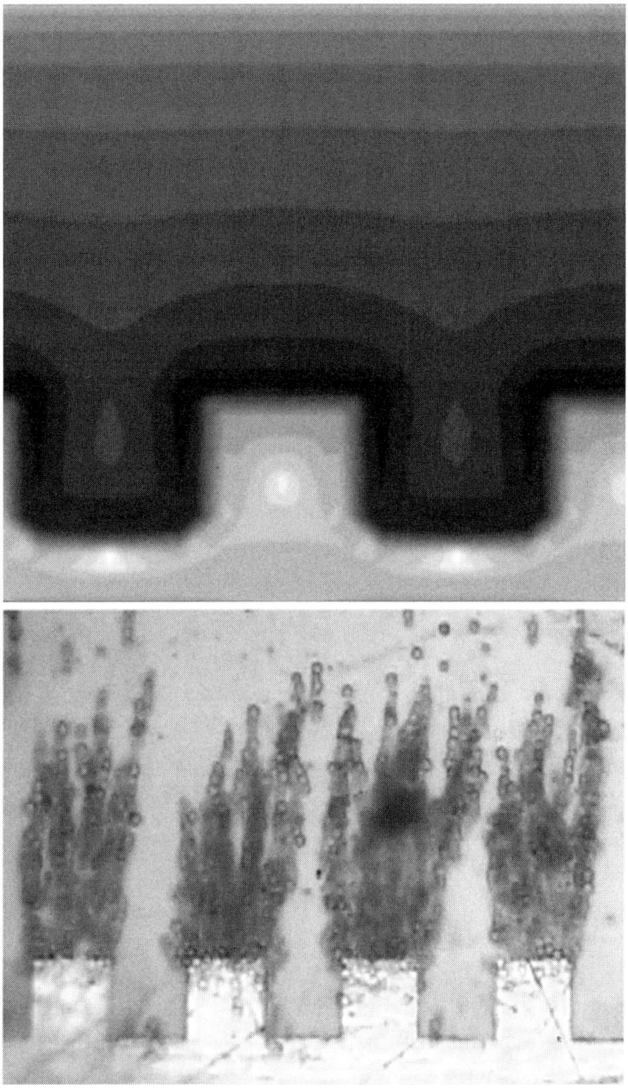

Figure 38.9. Electric field contour plot around a periodic microelectrode system (separation 50 μm) structure (top). Microscopic visualization of the chaining and migration of viable yeast cells after application of an electric field of 1 kHz during 90 seconds (bottom). Reproduced with permission from Kadaksham et al. [42]

scattering techniques; see Reference 59) allows to determine the velocity and direction of the induced cell rotation. The phenomenon can be explained at first glance by considering that the field induces a dipole in the cell and that this dipole will eventually get out of phase with respect to the field because of the finite time required by polarization processes in the cell. As a consequence, a torque will be exterted on the cell, producing its rotation with

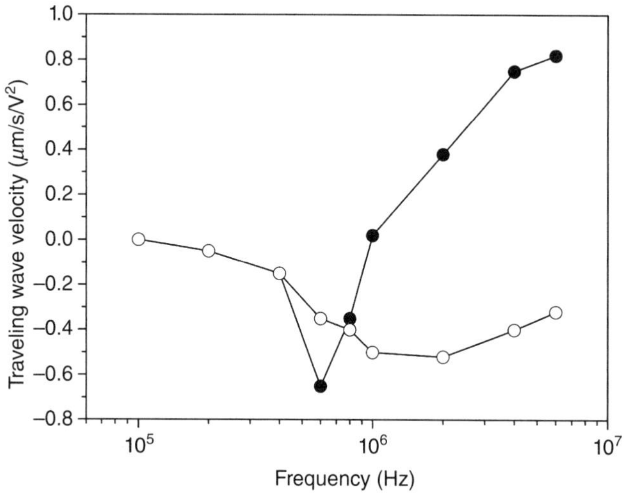

Figure 38.10. Traveling wave dielectrophoretic velocity of viable (full symbols) and nonviable (open symbols) NCI-H929 cells as a function of the frequency of the applied field. Dispersion medium: 5% dextrose solution. Reproduced with permission from Cen et al. [52]

angular velocity ω, either in favor or against the direction of the field rotation:

$$\Omega = -\frac{\varepsilon E_0^2}{2\eta}\mathrm{Im}(C_0^*). \tag{38.29}$$

Again it is very important to consider corrections to this simple model [37, 60]: The field also exerts its action on the charged liquid of the EDL, just like in ordinary electrophoresis. Nevertheless, this correction has only been applied to rigid particles, and no theory exists, to the authors' knowledge, dealing with a full description of the electrorotation of soft particles or cells. In spite of these limitations, electrorotation has been widely applied in biological investigations, a fact that can be explained by considering that typical cell sizes are in the micrometer range, very suited for microscopic observation, and that it is very informative on the cell polarization mechanisms and its frequency dependence. In addition, it has been demonstrated that the electrorotation behavior allows to distinguish, for instance, between alive and nonviable cells [61]. Figure 38.11 is an example of the electrorotation of chicken red blood cells in a dispersion medium containing phosphate buffer and sucrose [62].

38.4 APPLICATIONS

In addition to the examples cited earlier, the use of electrokinetics for the characterization of the cell/environment interface (or interphase) has proved

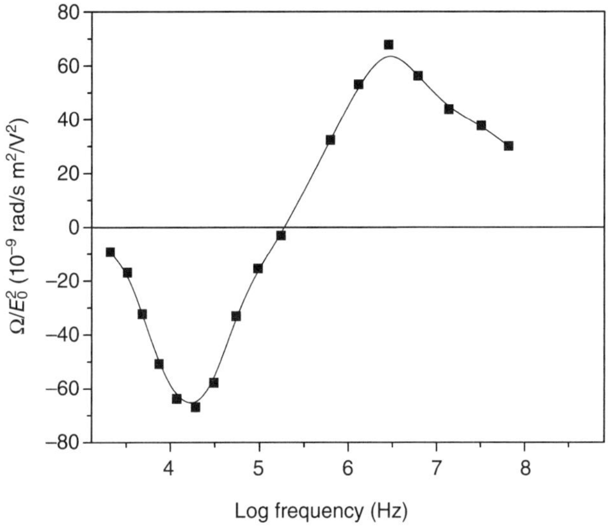

Figure 38.11. Electrorotation speed (divided by E_0^2) of chicken red blood cells dispersed in a phosphate buffer/sucrose solution with 0.015 S/m conductivity. Reproduced with permission from Lippert and Gimsa [62]

useful when working with many microbial organisms. This is because of the tendency of microbes to undergo surface colonization and adhesion, the case of dental biofilms of bacteria being among the best studied [25, 63]. Although in some instances the standard electrokinetic theory (and hence the zeta potential as determinant of it) has been applied for the evaluation of the electrostatic component of the total interaction energy (Derjaguin–Landau–Verwey–Overbeek [DLVO] forces), it has become clear that this is not correct, because of the soft surface structures of many cells, including gram-positive bacteria [25, 64]. This was explicitly proved by Tsuneda et al. [65], who evaluated the adhesion efficiencies of several heterotrophic bacteria to glass beads and compared such results with electokinetic data. The agreement with the theory of soft particles was qualitatively demonstrated.

A more quantitative approach was reported by Hayashi et al. [66]: these authors used the soft particle theory for the analysis of electrophoretic mobility obtained on the bacteria: *Escherichia coli*, *Pseudomonas putida*, *Alcaligenes faecalis*, and *Alcaligenes* sp. The results (Fig. 38.12 is a summary) indicate that the parameters ρ_{fix} and $1/\lambda$ differ for bacterial cells at different stages of their growth and that electrokinetics can also be used for evaluating the cell growth phase.

The application of the models described earlier for the electrokinetics of soft particles to cells with complex membrane structures is still incomplete for one main reason: It is an oversimplified picture considering that the charged coating cannot be described as a homogeneous one, even if well characterized

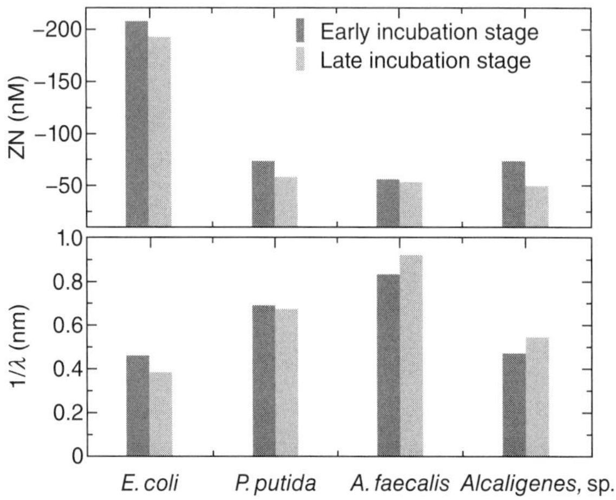

Figure 38.12. Characteristics of the soft membrane of the bacteria indicated at short and long incubation times. *ZN* is equivalent to ρ_{fix} (the former can be converted into the latter by multiplication by the Faraday constant). Reproduced with permission from Hayashi et al. [66]

from the electrohydrodynamic point of view. As mentioned by Duval and Gaboriaud [25], using such a picture is equivalent to the assumption of perfectly flat interfaces in the case of rigid particles. Like roughness and even excess conductance on the rigid particle/liquid interface cannot be ignored, possible heterogeneities of the soft coating around the cell core should be recognized in a more complete model that has been named diffuse soft particle electrokinetic (DSPE) formalism by Duval and coworkers [24, 25]. This approach has proved useful for the evaluation of the electrokinetics of many kinds of microbes.

A practical way to take these features into account was suggested in Reference 24 by proposing that the screening or friction parameter should change with the radial coordinate *r* as follows:

$$\lambda(r) = \lambda_0\sqrt{f(r)}$$
$$f(r) = \frac{1}{2}\omega\left[1 - \tanh\left(\frac{r-b}{\alpha}\right)\right], \tag{38.30}$$

where α is the measure of the separation between a homogeneous, steplike profile of coating ($\alpha = 0$, $\lambda = \lambda_0$) to a highly diffuse soft layer ($\alpha \to 1$). With this, complex cells with multilayered interfaces as *Streptococcus salivarius* HB or some viruses [25, 67]. Interestingly, it was found that the structure of the innermost layers could not be probed by electrophoretic mobility determinations if $\lambda_0 d > 1$, and that the sensitivity of the approach increases as the ionic strength is reduced. It can be suggested that AC electrokinetic techniques would be applicable in this case as the presence of conducting layers in the

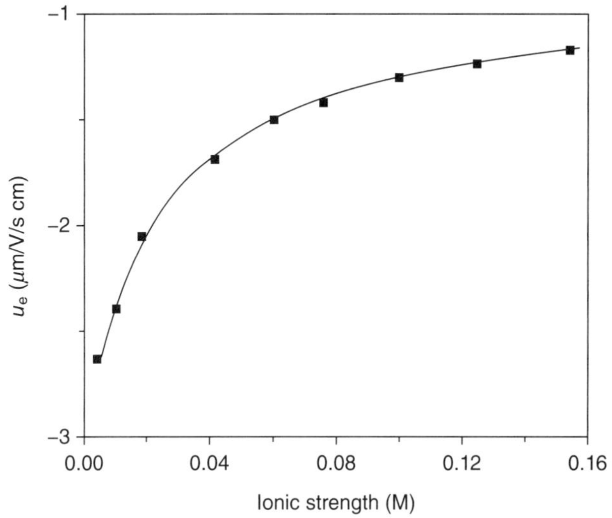

Figure 38.13. Electrophoretic mobility of human red blood cells as a function of ionic strength, at pH 7.4 and 37°C. The solid line is the best fit to a modified soft-particle theory with two layers in the membrane. Parameters of the fitting: inner layer, $\rho_{fix} = -8 \times 10^7$ C/m^3; outer layer, $\rho_{fix} = +1.9 \times 10^7$ C/m^3; $d = 1$ nm; $1/\lambda = 0.841$ nm. Reproduced with permission from Makino [69]

EDL can be detected by, for instance, electric permittivity or dynamic mobility experiments. Gaboriaud et al. [68] have also shown an interest in completing electrokinetic analysis with a direct interphasial observation through atomic force microscopy (AFM).

Investigations are not limited to microbial cells. Makino [69] investigated the membrane structure of red blood cells obtained from sheep blood by measuring their electrophoretic mobility as a function of ionic strength, controlled by dilution of Dulbecco's buffer solution with water. A modified Ohshima theory considering a charged layer consisting of two oppositely charged sublayers was used and an excellent agreement was found between theory and experiments, as shown in Figure 38.13. These studies were further used for explaining the adsorption of the red cells on different polymeric microcapsules, reaching the interesting conclusion that microcapsules with soft surfaces did not interact with the red cells in spite of being oppositely charged, whereas adsorption is favored onto more rigid surfaces. Electrokinetics was indirectly used in these investigations for proving that electrostatic interactions are surpassed by surface softness in determining the adsorption of red blood cells on surfaces.

A completely different application of soft particle electrokinetics to molecules of biological interest was explored by Ducel et al. [70] in dealing with the interactions involved in the formation of complex coacervates between plant proteins (pea globulin and gliadin) and polysaccharides (Arabic gum and alginate) in solution. These authors applied the theory to the determination of the polymer charge density and the depth of the layer accessible by counterions at different pH conditions. The obtained results in combination with light scattering data showed that when a protein bears a higher charge, a spherical shape favors the interaction with rodlike polysaccharide (i.e., pea globulin). On the contrary, a poorly charged protein needs a softer structure to facilitate the interaction with polysaccharide (i.e., alpha gliadin). In addition, it seems also that the charge and shape of polysaccharides used to coacervate with proteins are key factors to avoid precipitation. It is concluded that with plant proteins, a relatively poorly charged polysaccharide having a rodlike structure is more convenient than a strongly charged and spherical polysaccharide.

38.5 THE CASE OF ARTIFICIAL PARTICLES IN BIOLOGICAL ENVIRONMENTS

The surface functionalization of colloidal particles is essential for biomedical applications, as that procedure allows targeting against diseased areas or interacting selectively with cells or biomolecules. These comprise organic compounds that vary significantly both in chemical nature and molecular weight: from amino acids, short linear and cyclic peptides to proteins; from nucleic acid bases, nucleosides, and mononucleotides to ribonucleic acid (RNA) and deoxyribonucleic acid (DNA); from simple sugars to high-molecular-weight, poorly soluble polysaccharides, vitamins, and so on. These are just a few examples.

In many instances, synthetic colloidal particles are coated with some of the cited molecules in order to improve their biocompatibility, surface charge, or perhaps to tune their specific interactions with certain tissues or cells. Depending on the particular molecular structure and environmental conditions like pH or ionic strength, adsorption properties of the functionalizing molecules can vary drastically, even within the same class of compounds. Here again, electrokinetics can be of help. The adsorption of biomolecules on the surface of colloidal particles can be examined by using electrokinetic measurements. In particular, electrophoresis is a very useful tool for determining (at least qualitatively) the coating efficiency. A clear example can be seen comparing the electrophoretic mobility data of the bare and the coated particles.

Viota et al. [71] studied the functionalization of gold nanoparticles with two biomolecules: creatine (amino acid derivative) and human serum albumin (protein). From an electrokinetic point of view, the differences between the trends of u_e with either pH or ionic strength in the case of gold, gold–creatine and gold–creatine–albumin can be associated with the adsorption of the amino acid and protein, respectively. The results in Figures 38.14 and 38.15 show that the coated particles behave as soft colloids, displaying the characteristic mark of these systems, that is, the tendency of the mobility toward a nonzero value

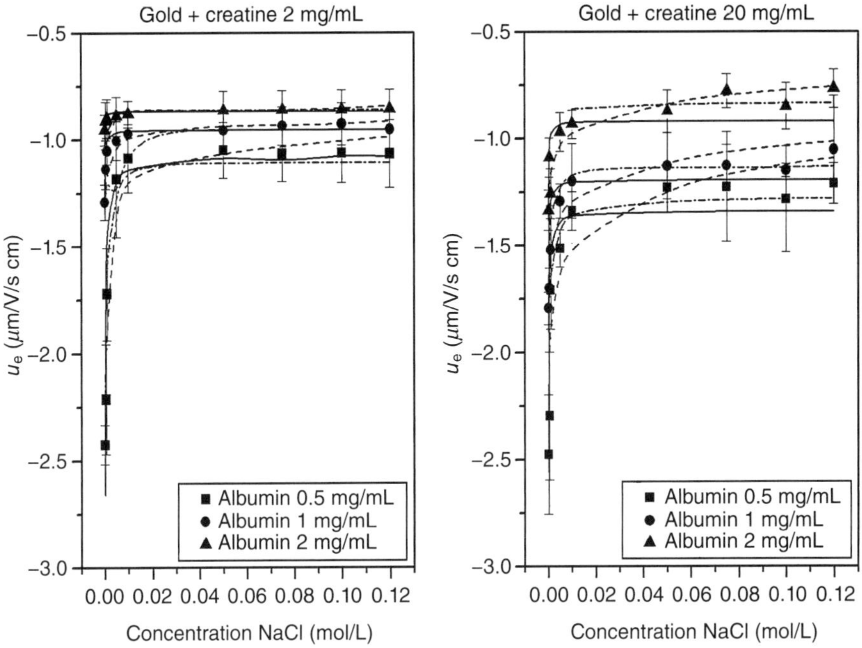

Figure 38.14. Electrophoretic mobility of albumin-coated gold–creatine particles as a function of NaCl concentration, for 2 and 20 mg/mL creatine, at the indicated concentrations of albumin (0.5–1.0 and 2.0 mg/mL). The lines correspond to fittings to different soft particle electrokinetic models (see text). Reproduced with permission from Viota et al. [71]

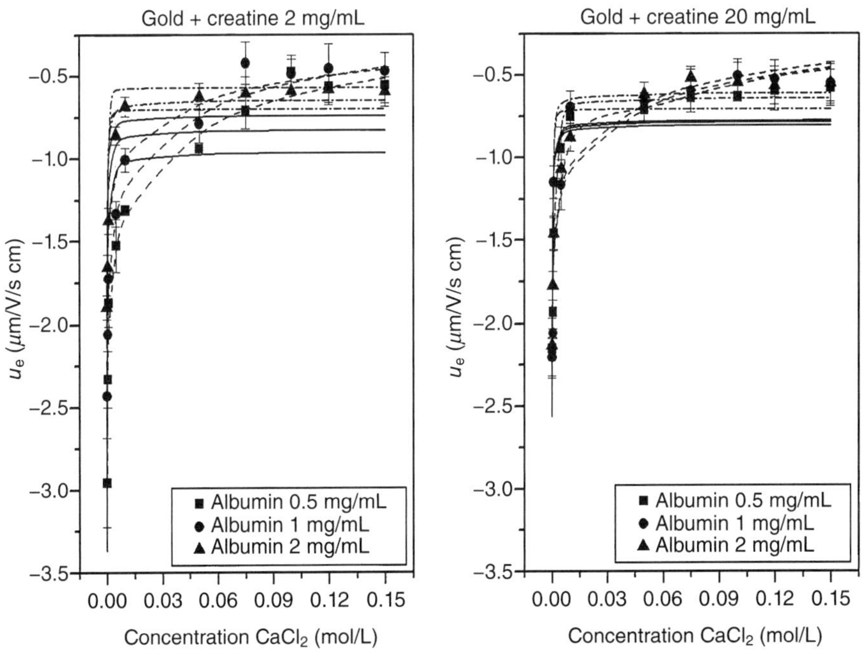

Figure 38.15. Electrophoretic mobility of albumin-coated gold–creatine particles as a function of $CaCl_2$ concentration, for 2 and 20 mg/mL creatine, at the indicated concentrations of albumin (0.5–1.0 and 2.0 mg/mL). The lines correspond to fittings to different soft particle electrokinetic models (see text). Reproduced with permission from Viota et al. [71]

when the ionic strength is high enough. In these studies, use was made of the approximate Ohshima model (solid lines) [21, 22], as well as a more complete model (dash-dotted lines) [72], and the Duval–Ohshima complete theory (dashed lines) [24]. It must be remarked that the agreement between the model and the experimental data is reasonable in all cases, although it appears more satisfactory when the DSP formalism is used. The best-fit parameters take reasonable values, with screening lengths κ^{-1} around 11 nm in the case of NaCl and 9.5 nm in the presence of $CaCl_2$. In addition, it is also worthy to mention that the values of the Donnan (Ψ_{DON}) and surface potentials (Ψ_0) are always on the order of a few millivolts, far from the values of tens of millivolts typical of zeta potentials. For instance, in the cases depicted in Figure 38.14, they range from −7.7 to −1.1 mV (Ψ_{DON}), and from −2.8 to −1.8 mV (Ψ_0). The values are slightly lower in the presence of $CaCl_2$, as expected [71]. The gold particles coated as described by creatine and albumin appeared as excellent candidates to become drug vehicles capable of passing through the blood–brain barrier, a challenge in the design of drug carriers suitable for the treatment of tumors and other brain diseases, like Alzheimer's.

It must be pointed out that the electrokinetic behavior of colloidal particles coated with biomolecules does not always render an electrokinetic behavior similar to that of soft particles. The size, liquid permeability, and stiffness of the coating molecules will be determinant for this. Viota et al. [73] investigated how superparamagnetic iron oxide nanoparticles can be coated by different amino acids (L-arginine, glycine, L-lysine, L-glutamine). The adsorption of the amino acids onto the surface of magnetite cores was checked by comparing the electrophoretic mobility of the amino acid-coated magnetite particles, with the ones obtained for the unmodified magnetite cores. A shift of the isoelectric point of the magnetite particles toward more basic pH values was found, indicating the adsorption of positive species. For a 3% concentration of magnetite particles, it is necessary to work with concentrations of amino acids below 0.1 mg/mL in order to observe a clear effect of the concentration of amino acid on the electrophoretic mobility of the functionalized magnetite particles. For such low concentrations, the four amino acids provoke charge inversion and extensive overcharging of the magnetite surface. Finally, because of the relatively small size of the amino acids, even though the surface of the magnetic cores is covered by amino acid, the particles do not display a soft electrokinetic behavior; in contrast, their electrophoretic mobility goes effectively to zero as the ionic strength is raised to around 0.1 mol/L, as illustrated in Figure 38.16.

38.6 SUMMARY AND CONCLUSIONS

In this chapter, we have given a brief account of the electrokinetic techniques that can be used in the electric characterization of dispersed systems of interest in the biological sciences. We have started by stressing the special character-

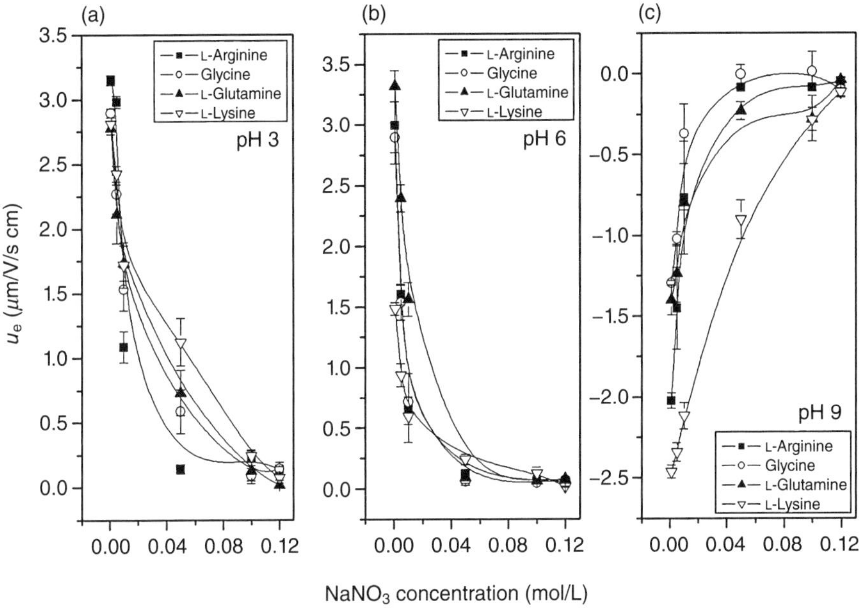

Figure 38.16. Electrophoretic mobility of magnetite particles coated with the amino acids indicated at pH 3 (a), 6 (b), and 9 (c), as a function of $NaNO_3$ molar concentration. Amino acid concentration used for coating: 0.5 mg/mL. The pH of the dispersion medium was fixed at the value used for coating. Reproduced with permission from Viota et al. [71]

istics of the kind of systems that can be encountered in this field. The main one in the case of cells, microbes, or artificial particles designed for interacting with them is the identification of an external coating (the cell membrane, the biomolecules functionalizing the particles, etc.), which is permeable to the dispersion solvent and ions. The denomination of soft particles reserved for such systems is justified, in contrast with that of rigid particles where an impermeable and undeformable interface is their main characteristic. The existence of a number of electrokinetic theoretical models of various degrees of rigor and sophistication is crucial for obtaining information about the mechanical and electrohydrodynamic properties of the external coating. This has proved true in spite of the necessity of eventually completing electrokinetic (mostly electrophoretic mobility) determinations with other techniques, like light scattering or AFM, for instance.

Nevertheless, electrokinetics can hardly provide a full characterization of the interfacial (or rather interphasial) region [17]. These limitations can be understood by considering that the presence of charged groups that are not homogeneously distributed in the membrane may give rise to tangential fields, which can alter the potential distribution (assumed spherically symmetric in the models described). In addition, the membranes are able to adjust their permeability to certain kinds of ions as well as their transmembrane potential, in the function of the local concentrations of certain ions. In general, the lipidic core of the membrane does not allow the passage of ions unless active transport molecules are present. No electrokinetic model comprising all these features has been elaborated, but electrokinetic techniques are, however, extremely useful in the characterization of such complex systems.

ACKNOWLEDGMENTS

Financial support from Junta de Andalucía, Spain, under project PE2008-FQM-3993 is gratefully acknowledged. The Juan de la Cierva grant (Spanish Ministry of Science and Innovation) awarded to J.L.V. is also gratefully acknowledged.

REFERENCES

1 S. Wall. *Curr. Opin. Colloid Interface Sci.* 15 (2010) 119.
2 A. V. Delgado, F. González-Caballero, R. J. Hunter, L. K. Koopal, J. Lyklema. *J. Colloid Interface Sci.* 309 (2007) 194.
3 J. Lyklema. *Fundamentals of interface and colloid science*, vol. 2. Academic Press, New York, 1995, ch. 4.
4 R. J. Hunter. *Zeta potential in colloid science. Principles and applications*. Academic Press, London, 1981.
5 R. J. Hunter. *Foundations of colloid science*. Oxford University Press, Oxford, 2000.
6 A. V. Delgado, F. J. Arroyo, in A. V. Delgado (ed.), *Interfacial electrokinetics and electrophoresis*. Marcel Dekker, New York, 2002, Chapter 1.
7 H. Ohshima, K. Furusawa. *Electrical phenomena at interfaces, fundamentals, measurements and applications*. Marcel Dekker, New York, 1998.
8 H. Ohshima. *Theory of colloid and interfacial electric phenomena*. Elsevier, Amsterdam, 2006.
9 S. S. Dukhin, B. V. Derjaguin, in E. Matijević (ed.), *Surface and colloid science*, vol. 7, Wiley, New York, 1974, ch. 3.
10 M. J. Kim, A. Beskok, K. D. Kihm. *Exp. Fluids* 33 (2002) 170.
11 S. J. Lee, S. Kim. *Microfluid. Nanofluidics* 6 (2009) 577.
12 S. S. Dukhin, R. Zimmermann, C. Werner. *Colloids Surf. A* 195 (2001) 103.
13 E. Wernersson, R. Kjellander, J. Lyklema. *J. Phys. Chem. C* 114 (2010) 1849.
14 M. M. Hatlo, L. Lue. *Soft Matter* 5 (2009) 125.
15 R. Kjellander. *J. Phys. Condens. Matt.* 21 (2009) 424101.
16 A. Martin-Molina, C. Rodriguez-Beas, R. Hidalgo-Alvarez, M. Quesada-Perez. *J. Phys. Chem. B* 113 (2009) 6834.
17 J. N Kleijn, H. P. van Leeuwen, in A. Baszkin, W. Norde (eds.), *Physical chemistry of biological interfaces*. Marcel Dekker, New York, 2000, ch. 2.

18 T. M Devlin. *Biochemistry with clinical correlations*. Wiley, New York, 2006.

19 A. P. Demchenko, S. O. Yesilevskyy. *Chem. Phys. Lipids* 160 (2009) 83.

20 S. I. Singer, G. L. Nicolson. *Science* 175 (1972) 720.

21 H. Ohshima, T. Kondo. *Biophys. Chem.* 39 (1991) 191.

22 H. Ohshima. *Adv. Colloid Interface Sci.* 62 (1995) 189.

23 H. Ohshima. *Sci. Technol. Adv. Mater.* 10 (2009) 063001.

24 J. F. L. Duval, H. Ohshima. *Langmuir* 22 (2006) 3533.

25 J. F. L. Duval, F. Gaboriaud. *Curr. Opin. Colloid Interface Sci.* 15 (2010) 184.

26 H. Ohshima. *J. Colloid Interface Sci.* 323 (2008) 313.

27 J. J. Lopez-Garcia, C. Grosse, J. Horno. *J. Colloid Interface Sci.* 301 (2006) 651.

28 S. Ahualli, M. L. Jiménez, F. Carrique, A. V. Delgado. *Langmuir* 25 (2009) 1986.

29 H. Ohshima. *J. Colloid Interface Sci.* 228 (2000) 190.

30 H. Ohshima. *Electrophoresis* 285 (2006) 1411.

31 H. Ohshima. *Colloid Polym. Sci.* 285 (2007) 1411.

32 R. J. Hill, D. A. Saville, W. B. Russel. *J. Colloid Interface Sci.* 258 (2003) 56.

33 R. J. Hill, D. A. Saville, W. B. Russel. *J. Colloid Interface Sci.* 263 (2003) 478.

34 R. J. Hill, D. A. Saville. *Colloids Surf. A* 267 (2005) 31.

35 J. Gimsa, in A. V. Delgado (ed.), *Interfacial electrokinetics and electrophoresis*. Marcel Dekker, New York, 2002, ch. 13.

36 B. Prüger, P. Eppmann, E. Donath, J. Gimsa. *Biophys J.* 72 (1997) 1414.

37 V. Zimmermann, C. Grosse, V. N. Shilov. *Colloids Surf. A* 159 (1999) 299.

38 O. Shramko, V. N. Shilov, T. Simonova. *Colloids Surf. A* 140 (1998) 385.

39 D. W. E. Allsopp, K. R. Milner, A. P. Brown, W. B. Betts. *J. Phys. D Appl. Phys.* 32 (1999) 1066.

40 B. Khusid, A. Acrivos. *Phys. Rev. E* 54 (1996) 5428.

41 H. Morgan, T. Sun, D. Holmes, S. Gawad, N. G. Green. *J. Phys. D Appl. Phys.* 40 (2007) 61.

42 J. Kadaksham, P. Singh, N. Aubry. *Electrophoresis* 26 (2005) 3738.

43 J. Yang, Y. Huang, X. Wuang, F. F. Becker, P. R. C. Gascoyne. *Anal. Chem.* 71 (1999) 911.

44 J. Suehiro, R. Pethig. *J. Phys. D Appl. Phys.* 31 (1998) 3298.

45 W. H. Arnold. *J. Micro Nanolithogr. Mems Moems* 8 (2009) 011001.

46 H. Morgan, M. P. Hughes, N. G. Green. *Biophys. J.* 77 (1999) 516.

47 R. Pethig, A. Menachery, S. Pells, P. De Sousa. *J. Biomed. Biotech.* 2010 (2010) 182581.

48 S. N. Higginbotham, D. R. Sweatman. *J. Phys. D Appl. Phys.* 41 (2008) 175503.

49 P. R. C. Gascoyne, J. V. Vykoukal, J. A. Schwartz, T. J. Anderson, D. M. Vykoukal, K. W. Current, C. McConaghy, F. F. Becker, C. Andrews. *Lab Chip* 4 (2004) 299.

50 H. K. Yang, H. Y. Jiang, A. Ramos, P. Garcia-Sanchez. *Microfluid. Nanofluidics* 7 (2009) 767.

51 A. Gonzalez, A. Ramos, P. Garcia-Sanchez, A. Castellanos. *Phys. Rev. E* 81 (2010) 016320.

52 E. G. Cen, C. Dalton, Y. Li, S. Adamia, L. M. Pilarski, K. V. I. S. Kaler. *J. Microbiol. Methods* 58 (2004) 387.

53 C. H. Kua, Y. C. Lam, I. Rodriguez, C. Yang, K. Youcef-Toumi. *Anal. Chem.* 80 (2008) 5454.

54 A. Menachery, S. Chappell, R. J. Errington, D. Morris, P. J. Smith, M. Wiltshire, E. Furon, J. Burt, in M. Laudon, B. Romanowicz (eds.), *NSTI Nanotech 2008*, vol. 3. CRC Press, Boca Raton, 2008.

55 Y. Kang, D. Li. *Microfluid. Nanofluidics* 6 (2009) 431.

56 J. Gimsa. *Colloids Surf. A* 149 (1999) 451.

57 P. Eppmann, B. Prüger, J. Gimsa. *Colloids Surf. A* 149 (1999) 443.

58 B. Prüger, P. Eppmann, J. Gimsa. *Colloids Surf. A* 136 (1998) 199.

59 J. W. Choi, A. Pu, D. Psaltis. *Opt. Express* 14 (2006) 9780.

60 C. Grosse, V. N. Shilov. *Colloids Surf. A* 140 (1998) 199.

61 A. D. Goater, J. P. H. Burt, R. Pethig. *J. Phys. D Appl. Phys.* 30 (1997) L65.

62 S. Lippert, J. Gimsa, in P. Kostarakis (ed.), *Proceedings of the 2nd International Workshop on the Biological Effects of EMFs*. World Health Organization, Rhodes, 2002, p. 830.

63 H. J. Busscher, M. Rinastiti, W. Siswomihardjo, H. C. van der Mei. *J. Dent. Res.* 89 (2010) 657.

64 R. Sonohara, N. Muramatsu, H. Ohshima, T. Kondo. *Biophys. Chem.* 55 (1995) 273.

65 S. Tsuneda, J. Jung, H. Hayashi, H. Aikawa, A. Hirata, H. Sasaki. *Colloids Surf. B* 29 (2003) 181.

66 H. Hayashi, H. Seiki, S. Tsuneda, A. Hirata, H. Sasaki. *J. Colloid Interface Sci.* 264 (2003) 565.

67 J. F. L. Duval, H. J. Busscher, B. van de Belt-Gritter, H. C. van der Mei, W. Norde. *Langmuir* 21 (2005) 11268.

68 F. Gaboriaud, E. Dague, S. Bailet, F. Joran, J. F. L. Duval, F. Thomas. *Colloids Surf. B* 52 (2006) 108.

69 K. Makino, in A. V. Delgado (ed.), *Interfacial electrokinetics and electrophoresis*. Marcel Dekker, New York, 2002.

70 V. Ducel, P. Saulnier, J. Richard, F. Boury. *Colloids Surf. B* 41 (2005) 95.

71 J. L. Viota, S. Mandal, A. V. Delgado, J. L. Toca-Herrera, M. Möller, F. Zanuttin, M. Balestrino, S. Krol. *J. Colloid Interface Sci.* 332 (2009) 215.

72 H. Ohshima. *Electrophoresis* 27 (2006) 526.

73 J. L. Viota, F. J. Arroyo, A. V. Delgado, J. Horno. *J. Colloid Interface Sci.* 344 (2010) 144.

39

MOLECULAR MECHANISMS OF MEMBRANE FUSION

Shinpei Ohki and Klaus Arnold

39.1 INTRODUCTION

Membrane fusion is an essential molecular event involved in many cellular processes, such as exocytosis, endocytosis, intracellular vesicle trafficking, fertilization, and viral infection of target cells. In spite of extensive studies of membrane fusion, however, the basic molecular mechanisms in biological systems are not well understood, probably because of the complex nature of biological membranes and the variety of molecular pathways for membrane fusion. Since biological membranes are made of lipid bilayers and proteins, membrane fusion studies have explored not only membrane systems composed of lipid alone but also those composed of lipid/protein combined. Even in a simple lipid membrane system, there are different methods of producing membrane fusion, such as ion-induced, macromolecular-induced, protein-induced, and externally applied, electric field-induced lipid membrane fusion. However, all types of membrane fusion seem to consist of the following successive processes: (i) close approach and adhesion of two membranes, (ii) deformation in the lipid bilayers of the interacting membranes, (iii) destabilization and intermixing of lipid components through the deformed region of the membranes, and (iv) restabilization of the membranes. Even though the initial induction processes of membrane fusion may be different, the final stage seems to involve a common feature: deformation and destabilization of

Electrical Phenomena at Interfaces and Biointerfaces: Fundamentals and Applications in Nano-, Bio-, and Environmental Sciences, First Edition. Edited by Hiroyuki Ohshima.

TABLE 39.1. Membrane Fusion

Lipid membrane systems
Ion-induced fusion
Macromolecule-induced fusion
Protein-induced fusion
Some biological membrane systems:
Virus membrane fusion
Synaptic vesicle fusion in exocytotic systems

lipid membranes and successive intermixing and reformation of new lipid membranes.

In this chapter, we attempt to describe mainly the molecular mechanisms of ion-induced lipid membrane fusion followed by brief descriptions of possible molecular pathways for the different membrane fusion processes given in Table 39.1.

39.2 LIPID MEMBRANE FUSION

39.2.1 Ion-Induced Lipid Vesicle Membrane Fusion

It has been found that acidic phospholipid vesicles promote aggregation and fusion when fusogenic cations are increased to a certain concentration in the vesicle suspension medium. Also, the changes in some membrane properties (e.g., increase in membrane interfacial tension, decrease in membrane surface dielectric constant, dehydration of surface structured water, and alteration of membrane surface hydrophobicity) induced by fusogenic ions are highly correlated with the fusion threshold level. These experimental results are listed in Tables 39.2 and 39.3. We would like to examine how these quantities are involved in the mechanism of membrane fusion in these membrane systems.

The forces acting between two lipid vesicles generally consist of the van der Waals attractive force due to the interaction between the hydrophobic membrane cores, the electrostatic repulsive forces when the membranes possess electrical charges, and the hydration interaction forces. The sign of the latter forces depends upon the nature of the membrane surfaces. For most lipid vesicle systems, the hydration interaction force is exerted as a major repulsive force between two interacting membranes, thus preventing close approach of lipid vesicles. However, it has been proposed that such repulsive interaction forces can be reversed and become attractive by changing the nature of membrane surfaces. For example, when divalent or multivalent cations bind to acidic lipid membranes, they alter the membrane surface to a more hydrophobic nature. This alteration would reduce the hydration repulsive interaction force or even change its repulsive force to an attractive one. Consequently, at a certain hydrophobicity of the membrane surface, the total interaction forces could bring two interacting membranes closely together in a manner that could

TABLE 39.2. Threshold Concentrations of Various Ions to Induce Fusion for Phospholipid Vesicles Prepared in 0.1 M NaCl at pH 7.4 and Associated Membrane Properties (Data from Ohki and Arnold [2])

Addend Ions	Phospholipid Vesicles	Fusion Threshold Concentration of Ions	Increased Increase in Interfacial Tension (dyn/cm)[a,c]	Surface Dielectric Constant[b,c]	Hydrated Water per Phospholipid[d]
PS vesicles					
Na^+	SUV	None (up to 1 M)	0	32	9
	LUV	None (up to 1 M)	0	32	
K^+	SUV	None (up to 1 M)	0	32	
	LUV	None (up to 1 M)	0	32	
H^+	SUV	2 mM	7	12	
Mg^{2+}	SUV	7 mM	6.5	14	
	LUV	None (up to 20 mM)			
Ca^{2+}	SUV	1 mM	7.5	12	1–2
	LUV	2.5 mM	8.5		
Mn^{2+}	SUV	0.6 mM	7.5	12	
	LUV	1.5 mM			
La^{3+}	SUV	3–5 μM	7.5	11–14	1–2
Tb^{3+}	SUV	3–5 μM	7.5	11–13	
PA vesicles					
Na^+	SUV	None	0	32	
	LUV	None	0		
K^+	SUV	None	0	32	
	LUV	None	0		
H^+	SUV	2 mM	7	12	
Mg^{2+}	SUV	0.2 mM	7	12	
Ca^{2+}	SUV	0.2 mM	7	12	
	LUV	1 mM	7		
PI vesicles[e]					
Na^+	SUV	None		35	
	LUV	None			
Mg^{2+}	SUV	None (up to 20 mM)			
	LUV	None (up to 50 mM)			
Ca^{2+}	SUV	10 mM		12.5	
	LUV	None up to 50 mM			

[a] Ohki [119].
[b] Ohki and Arnold [26].
[c] Ohki and Zschörnig [120].
[d] K. Arnold and S. Ohki (unpublished data).
[e] M. Müller et al. (unpublished data).
PA, phosphatidic acid; PI, phosphatidylinositol.

TABLE 39.3. Various Physicochemical Properties for Phosphatidylserine Vesicle Membranes (Adapted from Ohki and Arnold [2]

Solutions	Area/ Molecule (Å^2/ Molecule)	Surface Potential, $\psi 0$ (mV)	Interfacial Tension (dyn/cm)	Hydrophobic Index (p)	W_{min}^{not} (erg/cm^2)	Interfacial Tension (dyn/cm) at the Rim Area	Hydrophobic Index at the Rim Area
0.1 M NaCl	80 (SUV)	–77	6.0	0.12	–0.2	–	–
	68 (LUV)	–84	2.5	0.05	None	–	–
0.1 M NaCl/ 1.0 mM $CaCl_2$	80 (SUV)	–455.	13.5	0.27	–4.8	19	0.38
	68 (LUV)	–48	9.5	0.19	–2.6	10	0.2
0.1 M NaCl/ 2.5 mM $CaCl_2$	80(SUV)	–42	16.0	0.32	–6.3	21	0.42
	68 (LUV)	–47	12.5.	0.25	–4.5	18	0.36

deform the shape of the two interacting membranes. Then, the deformed regions of the membranes assume an energetically higher state than the rest of the membrane and thus become more unstable. Such regions of the membrane—the rim of the closely adhered area of the membranes—would thus become a possible initiation site for membrane fusion. This scheme of membrane fusion has been theoretically developed [1–2] and is supported by a body of experimental data. We explore the theory first.

39.2.1.1 Theory of Membrane Adhesion. Let us consider two small lipid vesicles of spherical shape that are interacting with each other at a separation distance, R (Fig. 39.1a). The total interaction energy, W^{tot}, may be expressed

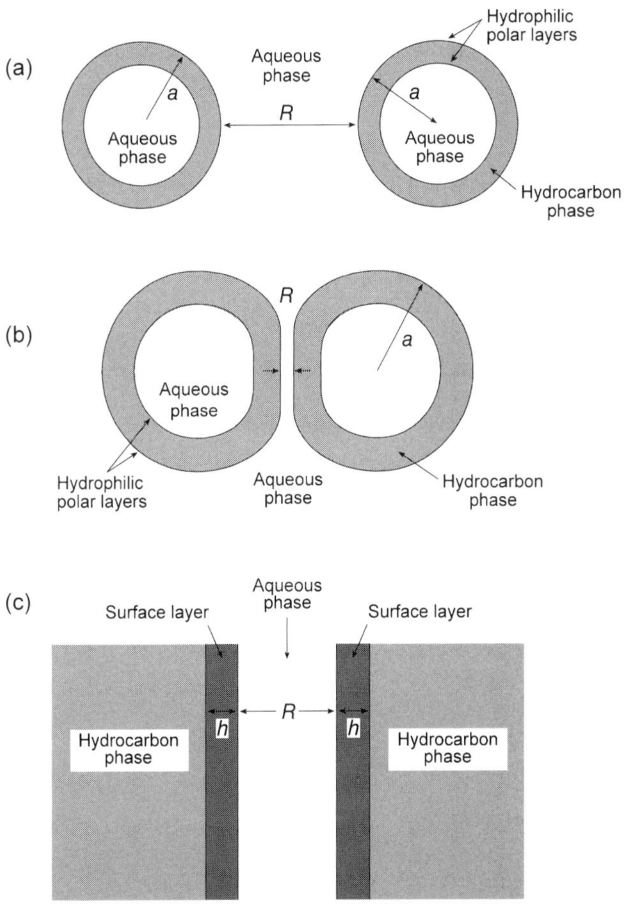

Figure 39.1. (a) A schematic diagram for two lipid vesicles of radius a interacting with each other at separation distance R in an aqueous solution. (b) A schematic diagram of closely adhered lipid vesicles separated by an aqueous phase. (c) A schematic diagram of closely adhered lipid membranes separated by an aqueous phase.

as a sum of the van der Waals attractive interaction energy, U_A, the electrical repulsive interaction energy, V^{ele}, and the hydration interaction energy, V^{hd}:

$$W^{tot} = U_A + V^{ele} + V^{hd}. \tag{39.1}$$

When the separation distance between two interacting membranes is small enough compared to the size of the lipid vesicle (e.g., $R \le 15$ Å with a ≥200 Å, respectively), the interaction energies of the interacting membranes may be replaced with that of two interacting plate membranes of infinite extent having surface polar layers of thickness, h, usually of the order of 3 Å (see Fig. 39.1c).

39.2.1.1.1 The van der Waals Interaction Energy. The van ver Waals interaction energy for such interacting membranes per unit area is expressed by

$$\begin{aligned} U_A(R) = -1/12\pi[(\sqrt{A_P} - \sqrt{A_W})^2/R^2 + 2(\sqrt{A_p} - \sqrt{A_W})(\sqrt{A_H} - \sqrt{A_p})/(R+\mathrm{h})^2 \\ + (\sqrt{A_H} - \sqrt{A_p})^2/(R+2h)^2], \end{aligned} \tag{39.2}$$

where R is the separation distance of the two interacting membranes, h is the thickness of the surface layer, A is the Hamaker constant [3] associated with the different phases of the interacting system. For water, the Hamaker constant $A_w = 3.0 \times 10^{-13}$ erg, while A_H for the hydrocarbon phase is 6.0×10^{-13} erg [4, 5]. For the surface layer of varying hydrophobicity, $A_p \equiv pA_H + (1 - p)A_w$, in which p is the degree of hydrophobicity (hydrophobic index) of the surface layer ($0 \le p \le 1$): The case of $p = 0$ corresponds to the hydrophobicity of the waterlike phase and $p = 1$ corresponds to that of the hydrocarbon-like phase. The values of p for various lipid membranes are shown in a latter section. By using Hamaker constants and different hydrophobic indexes, p, we can calculate the van ver Waals interaction energy for two interacting membranes as a function of interseparation distance R.

39.2.1.1.2 Electrostatic Interaction Energy. Electrostatic interaction energy of the same membrane as above having surface charges is expressed [6] by

$$\begin{aligned} V^{ele}(R) &= (64nkT/\kappa)\gamma_0^2 \exp(-\kappa R) \\ \gamma_0 &= (\exp(ze\psi_0/2kT) - 1)/(\exp(ze\psi_0/2kT) + 1), \end{aligned} \tag{39.3}$$

where n is the number density of counterions in an electrolyte solution far from the surface, ψ_0 is the surface potential of the membrane surface, z is the counterion valency, and κ is the Debye constant. In this equation, n, z, and κ are known for a given experimental solution and the surface potential ψ_0 is a quantity obtainable by various experimental methods [7–11]. The surface potential can also be calculated for a given electrolyte solution by knowing the binding constants of ions with charge sites of the membrane surface and

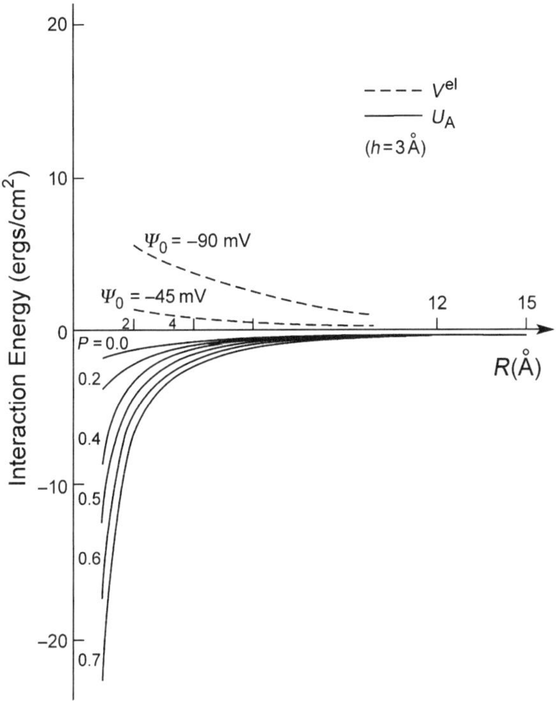

Figure 39.2. The van der Waals interaction energy (solid lines) calculated for various hydrophobic indices (p = 0, 0.2, 0.4, 0.5, 0.6, and 0.7) and electrostatic interaction energies (dashed lines) as a function of intermembrane distance R. The surface potential (φ_0 = −84 and −45 mV) corresponding to those of a phosphatidylserine membrane in 0.1 M NaCl, pH 7.4, and 0.1 M NaCl/1 mM $CaCl_2$, pH 7.4, respectively.

the surface charge density of the membrane according to the method described in published articles [7–9]. Although this formula is a linearized approximation, when the surface potential is small compared with kT/e (as is the case at the fusion threshold point), the formula can be used as a good approximation. For membranes having different hydrophobic indexes and electrical surface potentials, ψ_0, the van der Waals interaction energy and electric interaction energies are shown in Figure 39.2. For separation distance less than a few angstroms, the calculated values are not exact.

39.2.1.1.3 Hydration Interaction Energy. We can consider the hydration interaction energy in the following manner. The hydration repulsive interaction force between two interacting hydrophilic lipid bilayers has been studied well [12–15]. It is found that the interacting force between two hydrated parallel membranes decreases exponentially with a short decay constant, l (l = 1.7 − 2.5 Å) with the separation distance:

$$P = P_0 \exp(-R/l), \tag{39.4}$$

where P is the force exerting on the interacting bilayer surfaces per unit area. The energy bringing the two interacting surfaces from the infinite separation distance to a distance R will be

$$V_R^{\rm hd}(R) = -\int_{\infty}^{R} P dR = P_0 l \exp(-R/l). \tag{39.5}$$

According to published work [13], P_0 is estimated to be 7×10^9 dyn/cm^2 for phosphilipid membranes (e.g., phosphatidylcholine [PC]) in a 0.1 M NaCl solution. Therefore, the energy for removing the interlayer water (e.g., to bring the two membranes to molecular contact) per unit area would be

$$V_R^{\rm hd}(0) = -\int_{\infty}^{0} P dR = P_0 l. \tag{39.6}$$

If the value of l is taken to be 2.0 Å, $P_0 l = 7 \times 10^9$ dyn/cm$^2 \times 2 \times 10^{-8}$ cm = 140 dyn/cm.

However, if the surfaces are hydrophobic in nature, the interaction force exerted between two such membranes would be attractive [16–19].

According to the data in the literature [5], in such cases, the force decay constant l' is about 10 Å. Thus, such a force is expressed by

$$P' = P_0' \exp(-R/l'), \tag{39.7}$$

where P_0' is a negative value. Thus, in such a case, the energy to bring such membrane surfaces from the infinite distance to distance R would be

$$V_A^{\rm hd}(R) = P_0' l' \exp(-R/l') = V_A^{\rm hd}(0) \exp(-R/l'). \tag{39.8}$$

Now, if we assume that a membrane surface consists of a mixture of hydrophobic and hydrophilic surfaces at a fractional ratio, p, then the total hydration interaction energy per unit area will be expressed as the weighted sum of the hydrophilic and hydrophobic surface interaction energies:

$$\begin{aligned} V^{\rm hd}(R) &= (1-p) V_R^{\rm hd}(R) + p V_A^{\rm hd}(R) \\ &= (1-p) P_0 l \exp(-R/l) + p P_0' l' \exp(-R/l'), \end{aligned} \tag{39.9}$$

and

$$V^{\rm hd}(0) = (1-p) P_0 l + p P_0' l', \tag{39.10}$$

where $1 - p$ is the fractional portion of the membrane surface, which is hydrophilic, and p is that which is hydrophobic.

According to a previous work [20], the energy to remove water from the interlayer space is expressed in terms of interfacial tension:

$$E_{\rm hd} = 2(\gamma_{\rm m} - \gamma_{\rm m/w}), \tag{39.11}$$

where γ_m and $\gamma_{m/w}$ are the surface tension of the membrane and the interfacial tension of the membrane against the water phase, respectively. Thus, the energy should be equal to $V^{hd}(0)$:

$$E_{hd} = V^{hd}(0). \tag{39.12}$$

The values of γ_m and $\gamma_{m/w}$ are not easily obtainable quantities, although they can be measured to a limited degree. Thus, for simplicity, we assume that the surface of the membrane consists of a mixture of waterlike phase and hydrocarbon-like phase at a certain ratio, which varies depending on the membrane surface:

$$\gamma_m = p\gamma_h + (1-p)\gamma_w, \tag{39.13}$$

where p is the surface hydrophobic index ($0 \le p \le 1$) representing the degree of hydrophobicity of the membrane surface, and γ_h and γ_w are the surface tensions of hydrocarbon and waterlike phases, respectively, which are known quantities. The hydrophobic index of lipid membranes can be defined as follows: When the interfacial tension of the membrane against the aqueous phase is 0 dyn/cm, the hydrophobic index p of the membrane surface is zero, and when it is 50 dyn/cm, which is the interfacial tension of the hydrocarbon/water interface, the hydrophobic index p is 1.0. We can denote $\gamma_{m/w}$ by $p\gamma_{h/w}$.

Therefore, the dehydrated energy E_{hd} (Eq. 39.11) is rewritten in terms of γ_m and γ_w using Equation 39.13:

$$\begin{aligned} E_{hd} &= 2\{p\gamma_h + (1-p)\gamma_w - p\gamma_{h/w}\} \\ &= 2p(\gamma_{h/a} - \gamma_{h/w}) + 2(1-p)\gamma_{w/a} \\ &= pV_A^{hd}(0) + (1-p)V_R^{hd}(0), \end{aligned} \tag{39.14}$$

where $2p(\gamma_{h/a} - \gamma_{h/w})$ is the hydrophobic portion of energy and $2(1-p)\gamma_{w/a}$ is the hydrophilic portion of energy, respectively. Here, $\gamma_{h/a}$ is ~20 dyn/cm, $\gamma_{h/w}$ is ~50 dyn/cm, and $\gamma_{w/a}$ is ~72 dyn/cm [21]. Incidentally, when $p = 0$, E_{hd} is to be calculated to be about 144 dyn/cm, which is approximately equal to $V_R^{hd}(0) = 140$ dyn/cm obtained from Equation 39.6. In the case of $p = 1.0$, $E_{hd}(0)$ will be about −60 dyn/cm and $V_R^{hd}(0) = -60$ dyn/cm. Therefore, with Equations 39.5, 39.8, 39.12, and 39.14, the hydration interaction energy $V^{hd}(R)$ can be calculated as a function of the interlayer separation distance R for a given hydrophobic index, p, of the membrane surfaces:

$$\begin{aligned} V^{hd}(R) &= (1-p)V_R^{hd}(0)\exp(-R/l) + pV_A^{hd}(0)\exp(-R/l') \\ &= 2(1-p)\gamma_{w/a}\exp(-R/l) + 2p\{\gamma_{h/a} - \gamma_{h/w})\exp(-R/l'). \end{aligned} \tag{39.15}$$

Thus, Figure 39.3 shows the hydration interaction energy for two interacting membrane surfaces having various surface hydrophobic indices, p, as a function of the interlayer distance R using Equation 39.15. From Figures 39.2 and

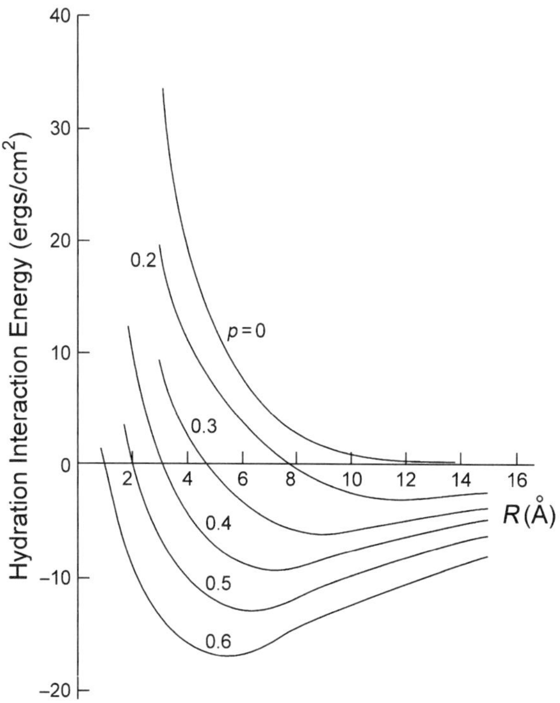

Figure 39.3. Hydration interaction energy for interacting parallel membranes with various hydrophobic indices for membrane surfaces as a function of separation distance *R*.

39.3, the hydration interaction energy is a major portion of these interaction energies for lipid membranes.

39.2.1.1.4 The Total Interaction Energy. The total interaction energy per unit area, $W^{tot}(R) = U_A(R) + V^{ele}(R) + V^{hd}(R)$, can be obtained using Equations 39.2, 39.3, and 39.15 with specific parameters for specfic membranes (e.g., hydrophobic index p and electrical surface potential ψ_0).

For example, the total interaction energies (per unit area) of various phosphatidylserine (PS) bilayer membranes are shown in Figure 39.4 using the specific parameters for the membranes as shown in Table 39.3. For case (a) in Figure 39.4, the situation approximately corresponds to that of large PS vesicles (large unilamellar vesicle [LUV]) in 0.1 M NaCl. For small PS vesicle (small unilamellar vesicle [SUV]) membranes, the membrane surface hydrophobicity is approximately $p = 0.12$ in 0.1 M NaCl, from the observed interfacial tension (6.0 dyn/cm) of the PS monolayer formed at the oil/water interface at an area per molecule of 80 Å^2. The vesicle membrane is in a more expanded state than the 68 Å^2 for the LUVs or flat membranes. In this case, the hydrated repulsion is still quite large (Fig. 39.4b) and the minimum interaction energy

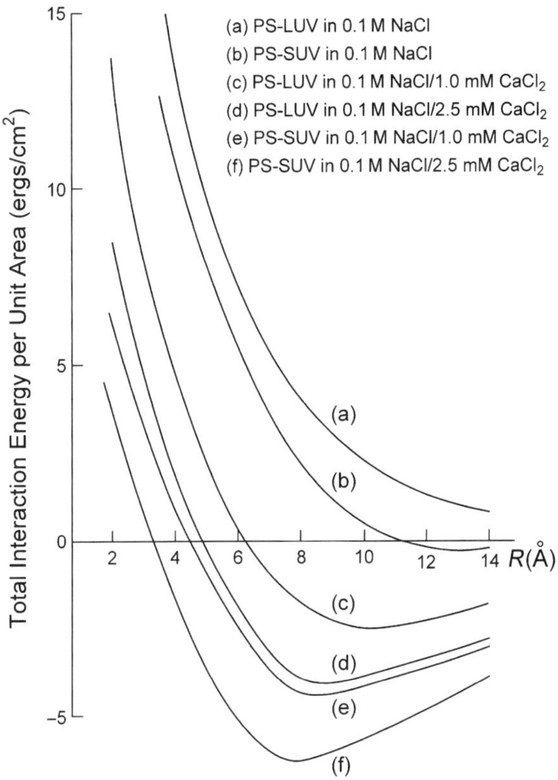

Figure 39.4. The total interaction energies for closely adhered phosphatidylserine membranes at various physical states of membranes and various ionic (NaCl and $CaCl_2$) concentrations at pH 7.4: (a) PS-LUV in 0.1 M NaCl, (b) PS-SUV in 0.1 M NaCl, (c) PS-LUV in 0.1 M NaCl/1.0 mM $CaCl_2$, (d) PS-LUV in 0.1 M NaCl/2.5 mM $CaCl_2$, (e) PS-SUV in 0.1 M NaCl/1.0 mM $CaCl_2$, and (f) PS-SUV in 0.1 M NaCl/2.5 mM $CaCl_2$.

is too small (~0.2 erg/cm^2) to have a close approach. When the small PS vesicles are suspended in 0.1 M NaCl/1 mM $CaCl_2$, the interfacial tension of the membranes is about 13.5 dyn/cm (6 + 7.5 = 13.5 dyn/cm) from Table 39.2, which corresponds to a surface hydrophobic index of about 0.27 (see Table 39.3). The interlayer interaction energy is shown in Figure 39.4e. The minimum attractive energy is about −4.8 erg/cm^2 at the interlayer distance $R = 8$ Å. Therefore, the two membranes would adhere to each other. However, as the interlayer distance decreases, the interaction energy increases and creates a net repulsive force. Therefore, even at the fusion threshold condition, the parallel-adhered membranes are not energetically ready to mix with each other. However, if the hydrophobic index becomes more than about 0.4, the energy barrier between the two interacting membranes is greatly reduced even at molecular distances. This condition may induce lipid membrane fusion.

39.2.1.2 Possible Fusion Site for Lipid Vesicles. As indicated earlier, the interaction energy calculated for the interacting parallel membranes at the fusion threshold condition for SUVs still shows a significantly high energy barrier at a molecular distance (Fig. 39.4e), which should prevent molecular mixing between the two apposed membranes. However, two membranes could be adherent to each other in the parallel area with a boundary region between the adherent and nonadherent portion of the membranes. This boundary region would involve more highly curved membrane areas (rim regions) as indicated in Figure 39.5. These areas are more deformed and physically more stressed, and have a higher hydrophobic surface than the other parts of the membranes. According to the previous studies of these curved regions [22], a line tension would be exerted upon such curved regions in addition to the intrinsic surface tension. The surface tension exerted on these curve regions at the adhering plane, x, is expressed as

$$x = \sigma_0 + \tau / r, \tag{39.16}$$

where σ_0 is the intrinsic interfacial tension of the vesicle membranes, τ is the line tension, and r is the radius of the adherent circle at the rim regions of the

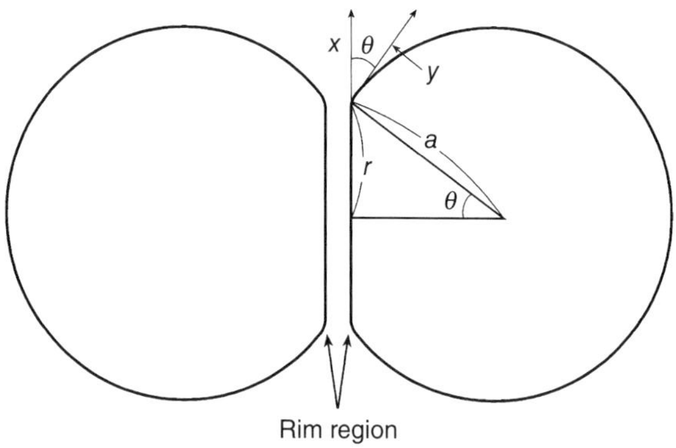

$$2y \cos\theta = 2y - W_{min}^{tot}$$

$$y \cos\theta = \sigma_0 + \tau/r$$

$$W_{min}^{tot} \cos\theta/(2(1-\cos\theta)) = \sigma_0 + \tau/(a \sin\theta)$$

Figure 39.5. A schematic diagram of the tension exerted at the boundary (rim) regions between adhered two membranes, where y is the interfacial surface tension of a lipid vesicle membrane at the rim and x is the component of the tension of y, which is in the adhered plane at the membrane boundary. The radii of the vesicle and the curvature of the membrane boundary in the adhesion plane are denoted by a and r, respectively, and $\theta = \sin^{-1}(r/a)$.

adhesion plane of the membranes. The line tension is a quantity obtainable from an experimental measurement of the curved surface [23]. Because strong attractive forces exerted on the two membranes bring them close together against the interfacial tension and cause deformation of the membrane at the boundary region mentioned earlier, the line tension develops to counterbalance the membrane deforming forces. At the same time, the interfacial tension forces exerted at this boundary should be in balance with the cohesive energy. With reference to Figure 39.5, the following relationship (Young's equation) can be defined:

$$2y\cos\theta = 2y - W_{\min}^{\mathrm{tot}}, \tag{39.17}$$

where y is the interfacial tension of the membrane at the membrane boundary and $W_{\min}^{\mathrm{tot}}$ is the minimum value of the total interaction energy W^{tot}. The interseparation distance of the two interacting membranes is the distance corresponding to the minimum point of the total interaction energy. Since two such membrane surfaces are involved in membrane contact, the factor 2 appears in Equation 39.17. From Equation 39.17, we have

$$y = W_{\min}^{\mathrm{tot}}/2(1-\cos\theta). \tag{39.18}$$

On the other hand, the component of y parallel to the adhesion plane is equal to x:

$$y\cos\theta = x = \sigma_0 + \tau/r. \tag{39.19}$$

When Equations 39.18 and 39.19 are combined, we have

$$W_{\min}^{\mathrm{tot}}\cos\theta/(2(1-\cos\theta)) = \sigma_0 + \tau/r = \sigma_0 + \tau/(a\sin\theta), \tag{39.20}$$

where a is the radius of the lipid vesicle. For a given set of values for $W_{\min}^{\mathrm{tot}}$, σ_0, τ, and a, a mechanical equilibrium angle, θ, for adhesion of two interacting spherical membranes will be determined from Equation 39.20. Once the mechanical equilibrium position θ is obtained, the interfacial tension at the boundary region, y, can be calculated from Equation 39.18 (shown in Table 39.3).

We can use this type of analysis for lipid vesicle fusion events. We analyzed the PS vesicles at their fusion threshold conditions. As we described earlier, at the fusion threshold condition, the total repulsive interaction energy for even small PS vesicles (PS-SUV) is still rather high between two parallel membranes at a separation distance less than a few angstroms, and consequently, molecular mixing and membrane fusion would not be induced. However, the rim regions at the two membrane boundaries have higher interfacial tension than the rest of the vesicle membrane, which is a function of vesicle radius. It was shown that the difference of interfacial tension for PS vesicles increases with smaller radius a. For example, for a decrease in radius from 800 to 200 Å,

y increases about 5~6 dyn/cm, and the hydrophobic index p increases by about 0.10–0.12. As a result, the rim regions' hydrophobic index could be about 0.4 (0.27 + 0.12 = 0.39). The hydrophobic index at the rim region is sufficiently high that molecular mixing occurs in that region and consequently fusion will be initiated from such rim regions (Table 39.3).

In order to support the lipid membrane fusion theory described here, fusion of lipid membranes with different radii was measured as a function of the concentration of lysolipids incorporated in these membranes [24]. When a certain amount (e.g., 10%) of lysophosphatidylcholine was incorporated in the PS membranes of vesicles with small radii, calcium ion-induced membrane fusion was well suppressed. When the same amount of lysolipids was incorporated onto the vesicle membrane with a larger radius, however, membrane fusion was not so well suppressed as in the case of smaller radius vesicles. Since lysolipids are energetically stabler in positively curved membrane surfaces, they enter into the curved surfaces, which are the fusion sites of usual phospholipid membranes, and by decreasing the surface hydrophobicity, they suppress membrane fusion.

39.2.2 Other Lipid Membrane Fusion Theories

In addition to this theoretical model, several other ideas for lipid membrane fusion have been considered, such as molecular deformity on the membranes [25–29]. As we will show later, small macromolecules such as polypeptides and amphiphilic proteins [30, 31] and annexins [32] can induce lipid membrane fusion. These suggest that local perturbations of the hydrophilic–hydrophobic boundary or of the phospholipid packing surface act to induce membrane fusion [33] or that the increase in membrane curvatures [25, 26, 28, 29, 34] is a possible site for membrane fusion. However, these ideas have not been well developed as a unique theory. Among them, the most well-adopted theory for lipid membrane fusion so far is the so-called stalk hypothesis [36, 37]. In this theoretical model, the fusion process occurs by an ordered sequence of steps, where the first step involves a merging of the proximal monolayers of two interacting bilayer membranes, with subsequent stalk formation (see Fig. 39.6a). The stalk formation between the two membranes may occur due to undulation of lipid molecules [38] or local binding of lipid bilayers, which results in the formation of a local region of outer monolayer fusion (Fig. 39.6a,b). Then, as a second step, the stalk induces hemifusion intermediates followed by a third step, in which a fusion pore (Fig. 39.6a) opens through the hemifused area, allowing intermixing between the two membrane contents. The stalk hypothesis can be described by macroscopic models treating bilayers and monolayers as homogeneous elastic surfaces. However, the structure of the nonbilayer transition states during fusion is unclear. In the elastic models, the high curvature of the nonbilayer intermediates creates void spaces, resulting in an unrealistically high activation energy [39] (Fig. 39.6b). This high activation energy was reduced by allowing the lipid to be tilted substantially

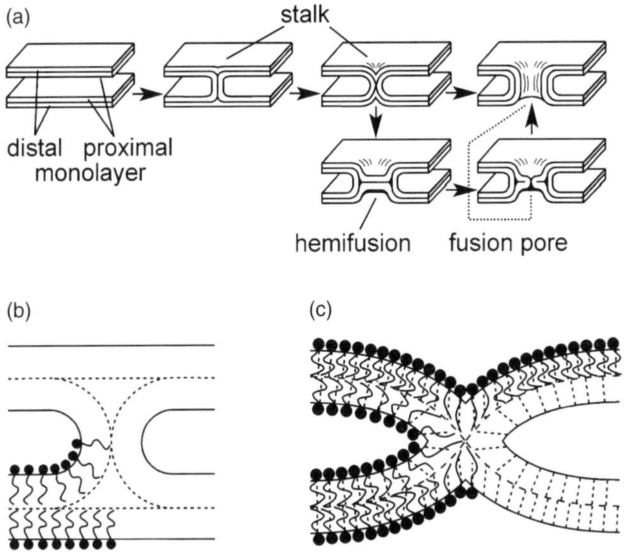

Figure 39.6. Transition states in membrane fusion according to the Stalk hypothesis. (a) First step: The proximal monolayers fuse to make a membrane (stalk). Second step: The stalk expands and forms a hemifused membrane. Third step: The hemifused membrane forms a fusion pore (adapted from Jahn et al. [43]. (b) The initial model for the stalk formation (adapted from Kozlovsky and Kozlov [41]). (c) The later model for the stalk formation (adapted from Kozlovsky and Kozlov [41]).

in the later modified stalk model (see Fig. 39.6c) [40–42], but the models still involve unphysical discontinuities [43]. Many authors dealing with virus fusion and some concerned with synaptic vesicle fusion with presynaptic membranes consider the lipidic stalk to be a standard way to deal with lipid membrane fusion. The lipid stalk is considered as one of the possible intermediates of membrane fusion. However, we think the lipid stalk formation is of rather low probability to be formed against the high repulsive interaction energy barrier for ordinal lipid membranes in biological systems, which possess high hydrophilic surfaces. Also, the current structural techniques do not easily measure stalk formation as a membrane fusion site [44].

39.3 LIPID MEMBRANE FUSION INDUCED OR MODULATED BY MACROMOLECULES

Here, we would like to limit macromolecules to be only water-soluble polymers and proteins.

The effects of water-soluble polymers on membranes are strongly dependent on the mode of interaction of the polymer with the membrane surfaces. Therefore, we will describe the processes of adsorption or exclusion of the

TABLE 39.4. Polymer and Protein-Induced Fusion (Adapted from Arnold et al. [46])

	Lipid Vesicles	Type
Uncharged polymers		
Polyethylene glycol	PC, PS	SUV, LUV
Dextran	PC, PS	SUV, LUV
Polyanions		
Dextran sulfate	PC, PC/PE (+Ca^{2+})	SUV
	PC (+Ca^{2+})	SUV, LUV
	PC/stearylamine	SUV
	PC/cholesterol	REV
Heparine	PC (+Ca^{2+})	SUV, LUV
DNA	PC/cholesterol	REV
PASP	PC/Chol/DEBDA	
Polycations		
Polylysine	PC/PS	SUV
	PC/P/CL	SUV
	PC/PE/PS	LUV
	PE/PA, PE/CL (+mannitol)	LUV
Polyhistidine	PC/PA, PE/CL, PE/PG	SUV
	PS, PG, CL, PE/PS	
Ca^{2+}-dependent potein-induced membrane fusion		
Synexin (Annexin VII)	Acidic lipids	SUV, LUV
Annexin (I–VII)	Acidic lipids	
Amphiphilic proteins		
Polypeptides		

CL, cardiolipin.

polymers on membrane surfaces. The molecular interrelations of the aggregation processes of the cell by the polymers are introduced, followed by a discussion of the membrane fusion processes by these polymers. Such studies should contribute to the understanding of the mechanisms of protein-induced membrane fusion [45]. It has been reported that several water-soluble polymers are able to induce the aggregation and fusion of cells and phospholipid membrane vesicles (Table 39.4) [46].

39.3.1 Vesicle Aggregation by Polar Polymers

Steric stabilization is a term used to explain the phenomenon of stabilization of hydrophobic particles in water by means of adsorption or covalent attachment of polar polymers. Napper [47] has provided an extensive discussion of the theoretical concepts of this field. Macromolecular structures at cell surfaces have a great impact on cell–cell interactions. During the approach of polymers to another cell surface, complex rearrangements and interactions of

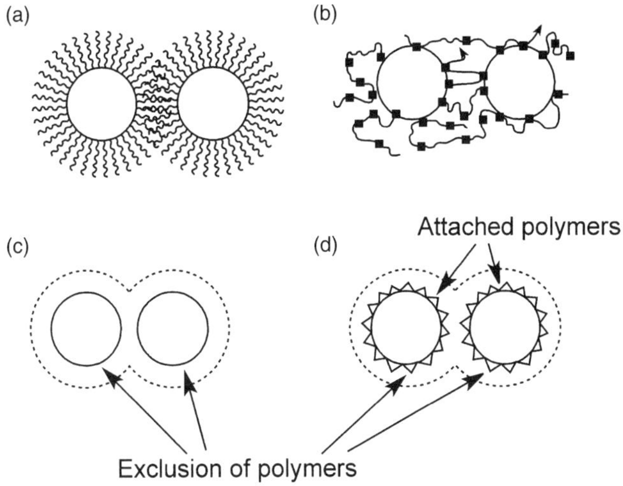

Figure 39.7. Polymer-induced (a) steric stabilization, (b) bridging formation, (c) depletion attachment, and (d) modulated-depletion attachment of lipid vesicles (adapted from Arnold et al. [46]).

the cell surface take place resulting in repulsion as well as attraction depending on the properties of the surface layer and solution. Different possibilities of polymer surface interaction are shown in Figure 39.7. They are steric repulsion and attraction (Fig. 39.7a), bridging attraction (Fig. 39.7b), and depletion attraction (Fig. 39.7c).

1. *Steric repulsion.* The adsorption or covalent attachment of polar polymers on membranes where the polymer chains extend into the aqueous phase leads to a repulsion of membrane surfaces due to steric hindrance, changes in the conformational entropy, and excluded volume effects [47]. For charged polymers, an additional contribution results from the electrostatic repulsion. More recently, the importance of polar (hydration) repulsion between the polymer layers was realized as a mechanism of steric repulsion [48–50]. The repulsive effect of attached polymers can occur only if the polymers have a high solubility in the medium of the cell or vesicle suspensions. In poor solvents, segments of the polymer can attract each other due to Van der Waals interaction. If two polymer-coated surfaces approach under these conditions, the attraction between the outermost segments of layers results in a surface attraction. On further approach, repulsion occurs due to the steric overlapping of polymer segments (see Fig. 39.7a). Aggregates formed by the segment–segment attraction cannot lead to the close surface contact necessary for membrane fusion.
2. *Bridging attraction.* If the interaction between polymer segments and the membrane surfaces is attractive, cross-binding of surfaces occurs readily

under appropriate conditions (Fig. 39.7b). As a rule, this bridging mechanism can occur only at low polymer concentrations when the surfaces are covered to a lesser extent. At high polymer concentrations, strong electrostatic, steric, or polar repulsive forces due to the coverage of the surface prevent the mutual approach of the surfaces. Charged polymers are very potent candidates for the aggregation of phospholipid vesicles by the formation of polymer bridges. Electrostatic attraction between the charged groups of the polymer and the membrane surfaces can overcome repulsive components. For example, positively charged polypeptides (e.g., polylysine) aggregate negatively charged vesicles, and another type of electrostatic interaction between sites of equal sign of negatively charged surfaces and polymers is the formation of cation bridging between the polymer and the surface.

The important point in the bridging mechanism is that even for a relatively small adsorption energy per each binding site of the polymer, the cumulative action of the binding sites along a polymer chain will result in a substantial energy of surface interaction, especially for polymers of high molecular weight.

3. *Depletion attraction.* If mutual repulsion between the free polymer and the membrane surface occurs, polymer molecules are excluded from the surface. A depletion layer, characterized by a depletion profile, is formed in the vicinity of the membrane surface (Fig. 39.7c). As demonstrated experimentally, polar polymers such as polyethylene glycol (PEG), dextran, and other polysaccharides are repelled by phospholipid bilayers [51–53] and biological membranes [50, 54–56]. Once two depleted surfaces are closer to each other than the thickness of the depletion layer, surfaces are pushed together due to the higher osmotic pressure of polymers dissolved in the bulk relative to the osmotic pressure in the center of the gap (*depletion flocculation*). This is opposite to the behavior predicted in the case of polymer adsorption, as discussed before. The depletion attraction is effective if the repulsion between the polymer molecules, as well as between the polymer molecules and the particle surfaces, is higher than the particle–particle repulsion. This effect usually occurs at a certain critical concentration of such a polymer.

The influence of PEG on the Ca^{2+}-induced aggregation of small PS vesicles is shown as follows: The aggregation of PS vesicles occurs for 1 mM Ca^{2+} concentration in the absence of PEG. The addition of PEG reduces the threshold of Ca^{2+} for aggregation. Without Ca^{2+}, the vesicle aggregation starts at about 15 wt % PEG 6000. Compared to that required for PC vesicle aggregation, the PEG concentration for the aggregation of PS vesicles is higher due to the electrostatic repulsion of PS vesicles.

Depletion attraction can also occur if parts of the polymer molecules attach themselves to the particle surface (Fig. 39.7c). A depletion layer resulting from the repulsion between free and attached polymers occurs between the polymer surface layer and the bulk polymer. Such phenomena were observed for the effect of dextran on the aggregation of erythrocytes [50, 54]. In some cases, depletion attraction was followed by depletion repulsion at higher concentrations when the surface concentration of the polymer is higher than its bulk concentration [50].

39.3.2 Lipid Vesicle Fusion Induced or Modulated by Polymers

39.3.2.1 Uncharged Polymer-Induced Lipid Vesicle Fusion. Unlike other uncharged polymers, PEG is able to induce fusion of SUVS of charged as well as uncharged phospholipids. The threshold concentration of PEG needed for lipid vesicle fusion, depending on the molecular weight and vesicle composition and amounts, is about 20–30 wt % of PEG 6000 [25, 26, 28, 29, 35, 57–61]. PEG is capable of inducing fusion of PC vesicles, which are strongly hydrated and usually difficult to fuse. Fusion of acidic vesicles, when incorporating PC, is usually inhibited. Dextran can induce fusion of large vesicles, but at a much slower rate than PEG. PEG is considered to force vesicles together due to the *depletion attraction* process. As mentioned earlier, these depletion attractions of lipid vesicles occur in the presence of cations and these polymers. At the threshold concentrations of the polymers, vesicles are flattened due to the withdrawal of free water from the vesicles by PEG, and fusion occurs at the highly curved boundaries as noted by some authors [25, 26, 35]. Two lipid vesicles might fuse due to the disruption of lipid packing exposing hydrophobic regions. The detergent-like properties of PEG are another possible cause of bilayer destabilization [62]. Lentz et al. [28] have found that high bilayer curvature encourages fusion of closely apposed membrane bilayers. However, the most highly curved vesicles did not mix their internal contents but ruptured and resealed, leading to the formation of large vesicles.

39.3.2.2 Polyanion-Induced Vesicle Fusion. Glycosaminoglycan comprises negatively charged polymers such as heparin and chondroitin sulfate, which are important components of the extracellular matrix. Dextran sulfate has a structure very similar to glycosaminoglycans and has been used instead of glycosaminoglycans by many investigators [63–66]. The interest in this polymer was increased because it was used as a potential drug against HIV infection [67]. The glycosaminoglycans bind to neutral phospholipid vesicles (PC, PC/phosphatidylethanolamine [PE]) only in the presence of Ca^{2+}. If positively charged stearylamine or other detergent-like chemicals are incorporated in the vesicle, binding occurs without Ca^{2+} [68]. Negatively charged phospholipids do not bind these polymers, even in the presence of some Ca^{2+}. As discussed by Kim and Nishida [69], divalent cations can form bridging between the phosphate group of the lipids and the negatively charged sites of these polyanions. If positively charged amphiphiles are incorporated in the bilayer, binding also occurs in the absence of Ca^{2+}. Positively charged membranes are the prerequisite for the binding of these polyanions. The vesicle aggregation induced by polyanions is assumed to be mediated by a bridging mechanism (*bridging attachment*). At high concentrations of these polymers, vesicle disaggregation occurs. The polymer concentration necessary for maximum aggregation is slightly lower than the polyanion concentration for the saturation of polymer adsorption. As expected, the bridge mechanism can only occur at low polymer concentrations where the surface of the vesicles is covered to a lesser extent.

So far, vesicle fusion induced by polyanions was only measured by the use of lipid mixing assays. Internal content mixing fusion assays have failed to show the internal content mixing due to the strong leakage of internal contents, indicating the occurrence of defects in the bilayer. Therefore, a more probable interpretation of the lipid mixing is that lipid vesicle ruptured and resealed, leading to the formation of larger vesicles.

The detailed mechanisms of polyion-induced vesicle fusion are unclear. After the aggregation of vesicle membranes, it may be that cross-linking and membrane destabilization are the crucial points in this process. In all cases, polyion-induced membrane fusion is accompanied by a leakage of the lipid vesicle contents, which indicates the occurrence of defects in the membranes. Such defects may arise at the boundaries between domains formed by the adsorption. A phase separation or domain formation within the membrane was also observed for all polyanion-induced membrane fusions.

39.3.2.3 Cation-Induced Membrane Fusion Modulated by Proteins. It is generally accepted that proteins are involved in the different stages of biological membrane fusion.

Protein-induced aggregation and fusion can involve both electrostatic and hydrophobic interactions between the protein and the participating membranes. Charged polypeptides promote aggregation of membranes mainly by electrostatic interaction. Polylysine-induced fusion has been reported [70–73]. The extent of vesicle aggregation and fusion depends on the percentage of negatively charged lipids in the vesicles and the relative amount of polylysine. This protein induces cross-linking of negatively charged vesicles, similar to the effect of charged polymers on vesicles (Fig. 39.7). The aggregating activity of polylysine on lipid vesicles shows a maximum effect at a certain concentration of polylysine, above which the vesicle aggregation declines at higher concentrations of the polypeptides. Once the vesicles are aggregated, the excess polylysine inhibits close approach and fusion due to *steric* and electrostatic effects [73]. Optimal aggregation occurs when charge neutralization is possible without complete coverage of the vesicles with polypeptides.

The fusion activity of charged polypeptides is usually pH dependent. For example, the fusion of PC/PS vesicles induced by polylysine increases strongly below pH 4.0, suggesting a correlation with the protonation of the PS-carboxyl group. In another case, the fusion of vesicles induced by polyhistidine, changes of the protein properties are responsible for the pH dependence [74]. It can be concluded that for fusion to occur, each component of the lipid–polypeptide–lipid complex has to be optimized for close aggregation. As a result, the major function of the charged polypeptides in vesicle fusion is the promotion of aggregation through charge neutralization of the aggregated system.

39.3.2.4 Ca^{2+}-Dependent Membrane Fusion Induced by Proteins. The aggregation and fusion of acidic vesicles occurs usually at millimolar Ca^{2+} concentration. However, Ca^{2+}-dependent natural membrane fusion processes require only micromolar Ca^{2+} concentrations. A simple explanation for this discrepancy in the Ca^{2+} threshold is that other factors are involved for increasing the fusion activity of Ca^{2+}. Some water-soluble proteins were identified, which are able to associate with vesicles in the presence of Ca^{2+}, thereby lowering the threshold of Ca^{2+} concentration for aggregation and fusion.

Synexin was the first cytosolic protein whose activity in Ca^{2+}-induced vesicle fusion was studied [75–78]. Without Ca^{2+}, synexin is not able to induce aggregation and fusion of lipid vesicles. The activities of synexin are triggered by Ca^{2+}, but not by Mg^{2+}. Negatively charged phospholipid vesicles of different compositions can be aggregated and fused by synexin/Ca^{2+}. In general, synexin enhances the initial rate of fusion and lowers the Ca^{2+} threshold. The largest effect by synexin is observed in the case of PA/PE vesicles, in which the initial rate of fusion is increased by two orders of magnitude, and the threshold for Ca^{2+} is lowered from about 1 mM to 10 μM.

Ca^{2+} appears to bind to synexin at two different sites with different affinities [79, 80]. The site of highest affinity regulates the binding of the protein to the membrane surface. Contrary to the Ca^{2+}-dependent polymers and cationic polypeptides, which directly link to a vesicle surface (Fig. 39.7), synexin shows a maximum aggregating effect even at high concentrations. This indicates a different mechanism of bridging. The common property of these systems (annexin family, including synexin) is the existence of multiple binding sites of the polymers and proteins for Ca^{2+}. Annexins with multiple binding sites aggregate vesicles in either their monomeric or polymeric form, involving annexin–annexin interactions. After the first contact of the vesicles, a closer approach of these vesicles can result from subsequent binding of annexin monomers or polymers. However, in order to enable intervesicle phospholipid–phospholipid contact, annexins must be displaced from the region of close contact. This model made some assumption about the mode of binding of annexins to membrane surfaces; a high flexibility of the attachment of annexins to surfaces occurs and the interbilayer distances relevant to the formation of the Ca^{2+} trans-complex can be realized. It has been suggested that synexin can undergo a conformational change, which allows it to insert completely into the lipid bilayer [81].

Amphiphilic proteins consisting of a polar and hydrophobic part can facilitate aggregation and fusion of vesicles by the penetration of the hydrophobic part in one vesicle and the binding of a polycationic or Ca^{2+}-binding site to the adjacent vesicle.

A cluster of arginine and lysine residues with interspersed electrically neutral amino acids can bind a significant fraction of cytoplasmic protein to the acidic vesicle membranes if the cluster contains more than five basic amino acid residues [82]. Many proteins were found to act in this way on vesicle aggregation and fusion (insulin, colicin, cardiotoxin, etc.). The relatively rigid helix structure of melitin may constrain the basic residues to point away from the surface. Such proteins will adsorb strongly to bilayer membranes by means of hydrophobic interaction. The fusogenic abilities of basic amphipathic α-helical peptides depend on the content of α-helical structures of the peptides in the presence of vesicles [30]. Low pH and Ca^{2+} are often known to trigger the required conformational change of the proteins. The conformational change is followed by a penetration of the hydrophobic parts of the proteins into the hydrophobic core of the lipid bilayer. The insertion of protein segments may provide the perturbation of bilayer stability necessary for the formation of transient intermediates for bilayer merging and fusion. The pH-induced fusion of acidic vesicles by lysozyme exhibits properties of such a mechanism [83].

39.4 BIOLOGICAL MEMBRANE FUSION

There are many membrane fusion processes occurring in biological cells. They are observed to occur between two plasma membranes, intracellular membranes, or intracellular and plasma membranes. Among them, many experiments have been done with the fusion of a virus with cell membranes and the fusion of synaptic vesicles with plasma membranes. We would like to limit our discussion to only these two subjects, virus membrane fusion and synaptic exocytotic membrane fusion. We briefly summarize how these membrane fusion processes progress based upon information obtained so far. Virus membrane fusion occurs without involvement of Ca^{2+}, while synaptic exocytosis needs calcium ions for membrane fusion.

39.4.1 Virus Membrane Fusion

Infection by viruses having lipid bilayer envelopes occurs through the fusion of the viral membrane with the target cell membrane. When they meet in a proper solution and environment, a viral binding protein interacts with a receptor on the target cell membrane, which causes adsorption of the virus to the target membrane. Some of the virus-binding proteins and the receptors of the target membranes are listed in Table 39.5a. Viral binding proteins are normally located very close to or are part of the viral fusion proteins (linked with disulfide bonds). After the initial adsorption between the virus and the target membranes at some distance (about 100 Å), viral fusion proteins facilitate the interaction process. The fusion proteins have a variety of molecular architectures, but all seem to have a common mechanism of action in which a ligand-triggered, large conformational change in the fusion protein is coupled with the receptors on the target membrane, merging the two lipid bilayers together. Thus, the membrane-enveloped viruses introduce their internal con-

TABLE 39.5a. Various Viruses: Class I and Class II Virus Fusion Proteins (Adapted from White [92] and Hoekstra [121])

Virus Family	Examples	Binding Proteins	Fusion Proteins	Molecular Weight	Oligomer	Fusion Peptides	Active pH	Receptors
Class I								
Orthomyxoviridae	Influenza virus	HA1	HA1 + HA2	44 + 30	Homotrimer	Yes	<5.5	Sialic acid lipids & proteins
Paramyxoviridae	Sendai virus	HA	HA, F1 + F2	66, 52 + 11	Homotetramer	Yes	Neutral	Sialic acid, lipids, and proteins
Human immunodeficiency viruses	HIV	gp120	gp120 + gp41	120 + 41	Homotrimer	Yes	Neutral	CD4
Class II								
Flaviviridae	Flavi virus	G	G	61	Homotrimer	yes	<6.2	PS and ganglioside
Togaviridae	Semilki Forest	E1	E1, E2 + E3	51, 52 + 11	Dimer/trimer	yes	≤6.2	MHC

tents (their DNA or RNA genomes) into the host cell cytoplasm. A few viruses are listed in Table 39.5a. Most viral fusion proteins have short chains of fusion peptides in their N-terminal regions. For influenza viruses, the fusion protein (hemagglutinin [HA]), which is a homotrimer, binds to a receptor (sialic acid residues on lipids and proteins) on the target cell membranes. When the pH of their medium becomes lower than pH 5.5, a conformational change of viral fusion protein (HA2) occurs; the binding proteins (HA1) are detrimerized and the α-helix of the fusion protein (HA2) unwinds to form an extended three-chain coiled coil in a trimer with fusion peptides attached to the target membrane (Fig. 39.8b). Then, collapse of the extended intermediate complex occurs

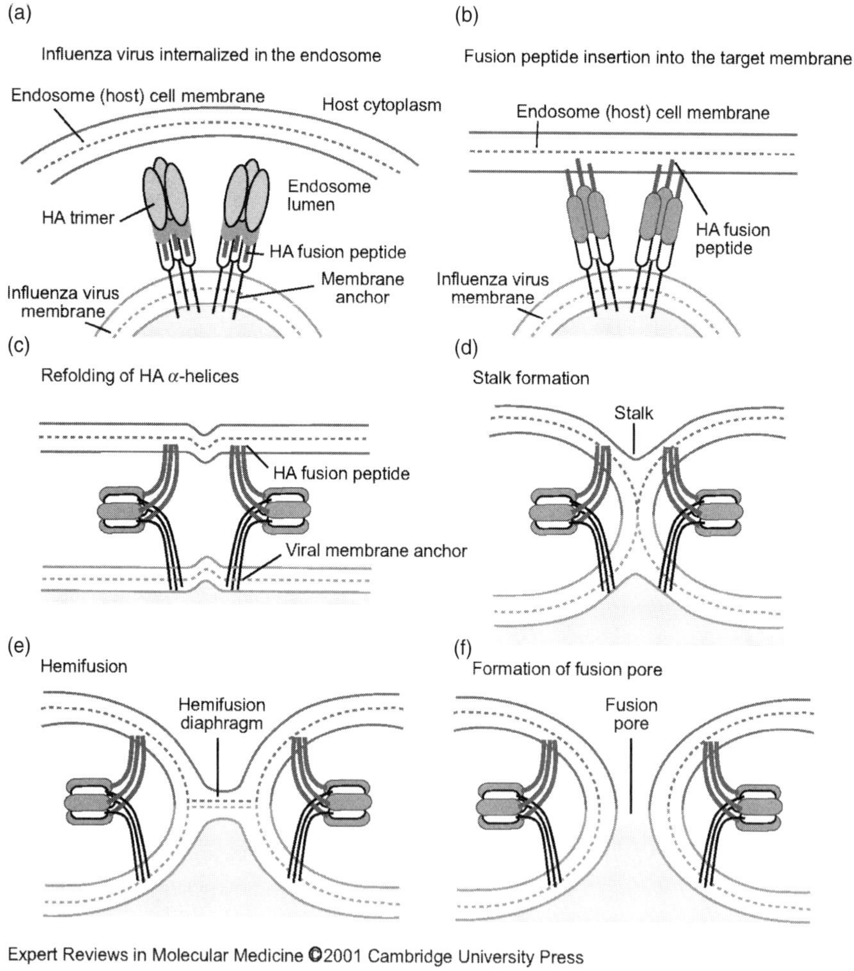

Figure 39.8. Stalk–pore hypothesis for virus fusion model by influenza hemagglutinin (HA) (this figure was copied from Cross et al. [96], with permission from the publisher).

to generate the postfusion conformation (the trimer of hairpin complex), which is shown as a "horizontal" orientation in Figure 39.8c. As a result, the two lipid membranes are drawn close to each other. The outer lipid bilayer leaflets of the two membranes are pulled together to form a stalk formation (Fig. 39.8d), which proceeds to a hemifusion intermediate (Fig. 39.8e) followed by a fusion pore formation (Fig. 39.8f). The viral internal contents (transcript complex, genomes and nucleocapsid proteins) are then transferred into the cytoplasm of the target cells. The influenza virus usually does not induce membrane fusion with cell membranes. The viruses are first internalized into the target cell by a receptor-mediated endocytotic process, and they fuse with the endosomal membrane by acidification and release the internal content into the cell cytoplasmic space. Many other viruses, such as paramyxoviruses and retroviruses, after binding with their receptors on the target cell surfaces, fuse with the plasma membranes and introduce their internal contents (genomes and nucleocapsid proteins) into the cell cytosol directly. During their fusion, similar structural changes in fusion proteins occur as with influenza viruses described earlier. Most viruses have their specific receptors for binding (shown in Table 39.5a). However, the type of conformational change for fusion proteins occurs similarly in almost all virus fusion processes [84], apart from some detailed differences. For example, the role of the α-helix in the influenza virus is taken by β-sheet complexes in Semliki Forest virus (SFV) (shown in Table 39.5b). Most authors prefer to use the lipid stalk formation as an intermediate process for virus membrane fusion process [85–87]. Recent research does not explore how many fusion proteins are needed for each virus fusion process, although many such questions about fusion pores were raised during 1989–2000 [88–93]. Some have suggested that the fusion pore for the influenza virus consists of at least three, or most likely four, HA trimers [94], and others [95] have mentioned about six HA trimers from the analysis of fusion kinetics data. Other consider that the actual numbers of these fusion proteins may be different from cell type to cell type [84]. Almost all authors [84, 96–98] in recent review articles support the stalk intermediate form [42, 99] for viral membrane fusion. We think that the number of fusion proteins (e.g., HA) for a single viral fusion process or one pore formation is important and there may be other fusion schemes, where local bending bilayers produced by the viral fusion protein pores may be possible fusion sites without forming the stable intermediate structure.

39.4.2 Synaptic Vesicle Fusion with the Presynaptic Membrane

Synaptic vesicles contain chemical transmitter substances that act to transfer nerve signals to other nerves. Through fusion with the presynaptic membranes, the internal contents (chemical transmitters) of synaptic vesicles are released into the external environment, where the chemical transmitters interact with receptors of the postsynaptic membranes of other nerve cells. Synaptic vesicles are about 500 Å in diameter and vesicle membranes are highly enriched in

TABLE 39.5b. Comparison of Class I and Class II Virus Membrane Fusion Proteins (Adapted from Kielian and Rey [98])

	Conformational Change During Fusion	Predominant Secondary Structure	Postfusion Structure	Maturation to Perfusion State through	Fusion Peptide in Metastable Structure
Class I Influenza virus (HA2)	Metastable fusion protein trimer to stable fusion protein trimer	α-helix peptide	Trimer of hairpins with central α-helical coiled coil	Proteolytic processing of fusion protein	N-terminal buried in trimer interface
Class II Semilki Forest (E1)	Metastable dimmer to stable protein trimer	β-sheet	Trimer of hairpins composed of β-structure	Proteolytic processing companion protein	Internal loop at fusion protein tip, capped by dimmer interaction

TABLE 39.6. SNARE Proteins

v-SNAREs: synaptobremins (15 KDa), one α-helix, synaptotagmin
t- SNAREs: syntaxin-1 (40 KDa), one α-helix and SNAP-25 (30 KDa), one α-helix
Synaptotagmines (130 KDa): family of membrane-trafficking proteins

proteins that mediate vesicle fusion at the active zone of presynaptic membranes. A synaptic vesicle membrane contains more than 10 different proteins and a small vesicle membrane has about 200 protein molecules. Several other proteins are near the active zone of the presynaptic membranes. Many studies have been directed toward establishing which proteins are responsible for vesicle fusion. Among many proteins, v-SNARE and t-SNARE groups have been found to be essential proteins for vesicle fusion [100–101]. On the synaptic vesicles, v-SNAREs consist of synaptobrevins (VAMPs) and synaptotagmins, and on the presynaptic membranes, t-SNAREs consist of syntaxin-1 and SNAP-25. The molecular weights of these proteins are somewhere between 10 and 45 KDa except for synaptotagmins (see Table 39.6), which are small, abundant, mostly membrane-bound proteins. Although they vary considerably in structure and size, all share a segment in their cytosolic domain, which consists of 60–70 amio acids, which are capable of reversible assembly into tight, four α-helix bundles called "*trans*"-SNARE complexes (in Fig. 39.9). When they form the *trans*-SNARE complexes, they bring the two (vesicle and plasma) membranes close together, and during this process, the hemifusion structure may be formed between vesicle and plasma membranes, which will proceed to form a fusion pore according to the stalk hypothesis. Upon fusion, the chemical transmitters are released from the synaptic vesicles to the external environment. The chemical transmitters diffuse and bind with their own receptors on the postsynaptic membrane and induce the postsynaptic potential change.

By doing so, one nerve excitation signal is transmitted to another nerve chemically.

During the synaptic vesicle fusion, Ca^{2+} is needed. Whether the Ca^{2+} acts to promote transformation of the SNARE complexes or Ca^{2+} acts through a lipid complex to modulate lipid membrane interaction is not clearly known. Following synaptic vesicle fusion, the *trans*-SNARE complex is altered to the *cis*-SNARE complex. This change initiates recycling of these SNARE proteins with the help of d-SNAP and NSF. These transactions are shown in Figure 39.9b.

At this moment, it is well established that the minimum number of the proteins needed for synaptic vesicle fusion is three: synaptobrevin, syntaxin, and SNAP-25 [99–101]. During vesicle fusion, the four-α-helix *trans*-SNARE complex (one synaptobrevin and one syntaxin and two SNAP-25's) is required. Some authors consider that after the four α-helix trans-complex formation, fusion starts with stalk formation and then proceeds through the hemifusion

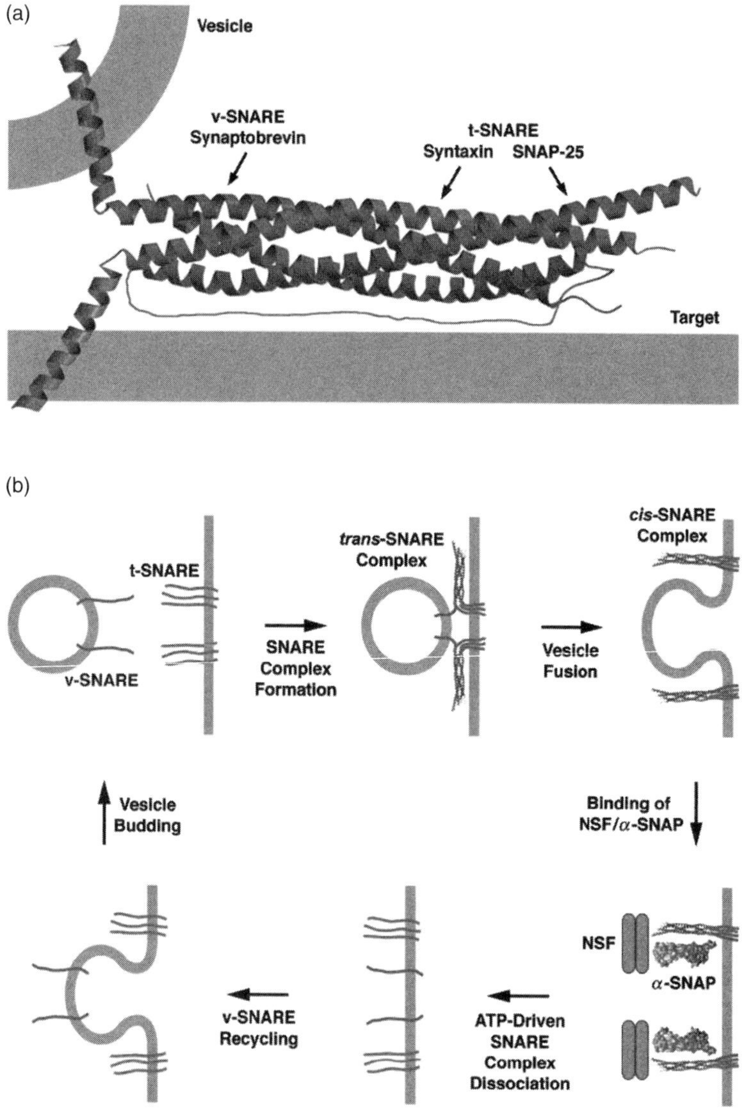

Figure 39.9. Schematic diagrams of structure and function of SNAREs at a synaptic vesicle and a presynaptic membrane (this figure was copied from Bonifacino and Glick [104], with permission from the publisher).

intermediate, followed by final fusion pore formation [105–107]. Other authors have considered the possibility of a fusion pore lined by fusion proteins. This idea was proposed earlier by Fernandez et al. [108] and others [89, 109–111]. Recently, Jackson and Chapman [112] have proposed that the pores formed with proteins from the vesicular and target membranes involve a gap junction-like channel that spans the lipid bilayers of both the plasma and vesicle membranes. They observed the transmembrane segments of syntaxin forming a lining of the fusion pore of Ca^{2+}-triggered exocytosis [113]. Since Ca^{2+} has a special interaction with various proteins and lipid bilayers, fusion may occur through Ca^{2+}–protein interactions (as discussed in the earlier section as well as here) including synaptotagmins as well as the Ca^{2+}–lipid membrane interaction. As an alternative view, the experimental results reported by a group of workers [114], which stress the importance of the Ca^{2+}–lipid complex induced by the SNARE complex for synaptic vesicle fusion, suggesting membrane fusion mechanisms other than the stalk hypothesis. Quite recently, it was reported that one SNARE complex is sufficient for this membrane fusion event [115].

It seems that synaptic vesicle fusions with presynaptic membranes may be of complex nature due to not only lipid bilayer but also protein pore involvement. Further studies on the pore formation of fusion proteins are needed.

39.5 CONCLUDING REMARKS

As we know that Ca^{2+} has a specific interaction with phosphate groups of molecules, we would like to add some information to our membrane fusion theory because we used our fusion experiments with Ca^{2+} and specific lipids, acidic phospholipid membranes. There must be some effects of Ca^{2+}–phosphate complex in our membrane fusion systems. It has been studied that the Ca^{2+} binds to acidic phospholipid membranes, and eventually, it forms the Ca^{2+} trans-complexes, which have been measured by X-ray diffraction methods [116]. For the experimental procedures, the prepared lipid vesicles are exposed to Ca^{2+} and then, the external (cellular) water of vesicle suspension is dehydrated and after that process, the X-ray diffraction measurements of lipid bilayers have been done in a few hours. During that time, the Ca^{2+}–lipid trans-complex would be formed. Our fusion measurements were done usually within a few minutes after the addition of Ca^{2+} to the lipid vesicle suspension and so as other experimental results (interfacial surface tension, surface dielectric constants) were obtained within a few minutes' duration, except for the determination of bound water by nuclear magnetic resonance (NMR). Our fusion theory is not for the Ca^{2+}–lipid trans-complexed samples. The experimental samples to be used for our theory are hydrated samples and are not kept for a long time (e.g., for hours). However, without dehydrating the samples, if the sample is kept for a long time in a suspension solution, the lipid membranes would form the Ca^{2+} complex in time. Indeed, we measured the increased leakage of Na^+ or K^+ across such lipid vesicles with time [117]. The ion leakage did not start immediately but increased significantly with time, in 0.5 or 1.0 hour at the fusion threshold concentration of Ca^{2+}. It depends on the Ca^{2+} concentrations. We thought that this leakage was due to some Ca^{2+}–lipid complex formation in time. The lipid membrane samples for our membrane

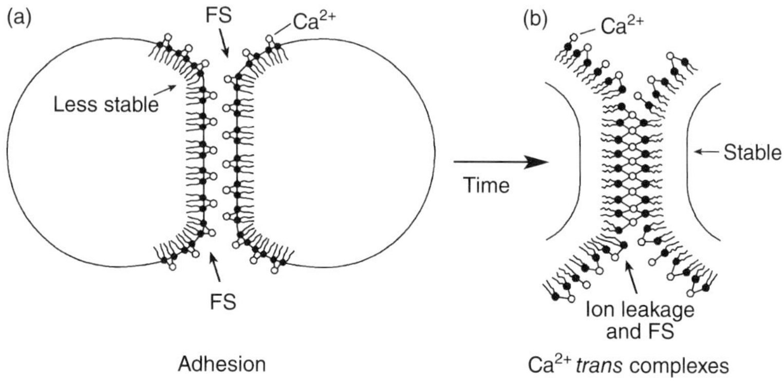

Figure 39.10. Formation of (a) Adhered phospholipid (acidic lipid) vesicles with Ca^{2+} to (b) Ca^{2+}–lipid trans-compex (adapted from Ohki and Arnold [26, 35]; Ohki [118]). FSs, fusion sites.

fusion theory are related to Figure 39.10a and are not the samples with the Ca^{2+}–lipid complex.

As far as the stalk hypnosis for lipid membrane fusions is concerned, it may be possible for some lipid membrane systems. However, there are two questions: Usually, lipid membranes in biological systems possess hydrophilic surfaces. It seems since hydrophilic surfaces have a strong hydration repulsive interaction forces [12, 13, 15], flat membrane surfaces do not easily adhere together to form the stalk formation. Second, when the membrane fusion occurs, we should be able to measure the membrane shape, such as the stalk intermediate, easily. Some experimentalists have mentioned that the current membrane structural techniques do not easily measure the stalk formation as the membrane fusion sites [44].

In order to find what types of lipid membrane fusion are involved for viral membrane fusion and synaptic vesicle membrane fusion, we need to find how many fusion proteins are involved in a single fusion process and what size, shape, and surface nature (hydrophobicity or hydrophilicity) of protein pores are formed, and what the surface nature of interacting fusion proteins is for each membrane fusion. These are important matters to know as a means of either modification or inhibition of these fusion events.

As a final comment, we would like to stress that the important properties for lipid membrane fusion involve (i) surface hydrophobic properties of the membranes, (ii) nessesssity of curvature at the membrane fusion sites, and (iii) close approach between the two interacting membranes.

ACKNOWLEDGMENT

The authors express their many thanks to Drs. R.A. Spangler and C. Richard Zobel, Department of Physiology and Biophysics, SUNY at Buffalo, for reading the manuscript and for valuable comments on this work.

REFERENCES

1 S. Ohki. *Membrane* 19 (1994) 48–60.

2 S. Ohki, K. Arnold. *Colloid Surf. B Biointerfaces* 18 (2000) 83–97.

3 H. C. Hamaker. *Physica* 23 (1937) 1058–1072.

4 D. J. Shaw. *Introduction to collod and surface chemistry*. Butterworths, London, 1980.

5 J. N. Israelachvili. *Intermolecular and surface forces*, 2nd ed. Academic Press, New York, 1992.

6 E. J. W. Verway, L. Th. G. Overbeek. *Theory of the stability of lyophobic colloids*. Elsevier, Amsterdam, 1948.

7 S. Ohki, R. Sauve. *Biochim. Biophys. Acta* 511 (1978) 377–387.

8 S. Ohki, R. Kurland. *Biochim. Biophys. Acta* 645 (1981) 170–176.

9 S. Ohki, H. Ohshima. *Biochim. Biophys. Acta* 812 (1985) 147–154.

10 S. Nir, C. Newton, D. Papahadjopoulod. *Bioelectrochem. Bioenerg.* 5 (1978) 110–133.

11 M. Eisenberg, T. Gresalfi, T. Ricco, S. McLaughlin. *Biochemistry* 18 (1979) 5213–5223.

12 D. M. LeNeveu, R. P. Rand, V. A. Parsegian. *Nature (London)* 259 (1975) 601–603.

13 R. P. Rand. *Ann. Rev. Biophys. Bioeng.* 10 (1981) 277–314.

14 T. J. McIntosh, A. D. Magid, S. A. Simon, in S. Ohki (ed.), *Biological and model membrane interactions*. Plenum Press, New York, 1991, p. 249.

15 R. P. Rand, V. A. Parsegian, in P. Yeagle (ed.), *The structure of biological membranes*. CRC Press, Boca Raton, FL, 1992, p. 251.

16 J. Israelachvili, R. Pashley. *Nature* 300 (1982) 341–342.

17 R. M. Pashley, P. M. McGuiggan, B. W. Ninham, D. F. Evans. *Science* 229 (1985) 1088–1089.

18 Y. I. Rabinovich, B. V. Derjaguin. *Colloids Surf.* 30 (1988) 243–251.

19 H. K. Christenson, J. Fang, B. W. Ninham, J. L. Parker. *J. Phys. Chem.* 94 (1990) 8004–8006.

20 S. Ohki. *Stud. Biophys.* 110 (1985) 95–104.

21 J. T. Davies, E. K. Rideal. *Interfacial phenomena*. Academic Press, New York, 1961.

22 I. B. Ivanov, P. A. Kralchevsky, in I. B Ivanov (ed.), *Thin liquid films*. Marcel Dekker, New York, 1988, p. 49.

23 J. K. Angarska, J. D. Tachev, I. B. Ivanov, P. A. Kralchevsky, E. F. Leonard, in S. Ohki (ed.), *Cell and model membrane interactions*. Plenum Press, New York, 1990, pp. 199–213.

24 S. Ohki, K. Arnold. *Colloid Surf. B Biointerfaces* 63 (2008) 276–281.

25 R. I. MacDonald. *Biochemistry* 24 (1985) 4058–4966.

26 S. Ohki, K. Arnold. *J. Memb. Biol.* 114 (1990) 195–203.

27 D. S. Dimitrov, R. Blumenthal, in S. Ohki (ed.), *Cell and model membrane interactions*. Plenum Press, New York, 1991, p. 229.

28 B. R. Lentz, G. F. McIntyre, D. J. Parks, J. C. Yates, D. Massenburg. *Biochemistry* 31 (1992) 2643–2653.

29 B. R. Lentz, J. Prevratil, D. Parks, J. R. Wu. *Biophys. J.* 61 (1992) A497.

30 S. Lee, R. Aoki, O. Oishi, H. Aoyagi, N. Yamasaki. *Biochim. Biophys. Acta* 1103 (1992) 157–162.

31 M. Otter-Nilloson, R. Hendriks, E. L. Pecheur-Huet, D. Hoekstra, T. Nilsson. *EMSO J.* 18 (1999) 2074–2083.

32 S. C. Hung, M. Srivastava, H. B. Pollard. *Proc. Natl. Acad. Sci. U. S. A.* 93 (1996) 10797–10802.

33 G. Cevc, H. Richardsen. *Adv. Drug Deliv. Rev.* 38 (1999) 207–232.

34 R. M. Epand, R. F. Epand. *Biopolymers* 55 (2000) 358–363.

35 S. Ohki, K. Arnold, in R. Glaser, D. Gingell (eds.), *Biophysics of the cell surface*. Springer-Verlag, Berlin, 1990, p. 193.

36 M. M. Kozlov, V. S. Markin. *Biofizika* 28 (1983) 242–247.

37 L. V. Chernomordik, G. B. Melikan, Y. A. Chizmadzhev. *Biochim. Biophys. Acta* 906 (1987) 309–352.

38 W. Helfrich. *Z. Naturforsch.* 33A (1978) 305–315.

39 D. P. Siegel. *Biophys. J.* 65 (1993) 2124–2140.

40 P. I. Kuzmin, J. Zimmerberg, Y. A. Chizmadzhev, F. S. Cohen. *Proc. Natl. Acad. Sci. U. S. A.* 98 (2001) 7235–7240.

41 Y. Kozlovsky, M. M. Kozlov. *Biophys. J.* 82 (2002) 882–895.

42 Y. Kozlovsky, L. V. Chernomordik, M. M. Kozlov. *Biophys. J.* 83 (2002) 2634–2651.

43 R. Jahn, T. Lang, T. C. Südhof. *Cell* 112 (2003) 519–533.

44 A. Mittal, J. Bentz, in P. L. Yeagle (ed.), *The structure of biological membranes*, 2nd ed. CRC Press, Boca Raton, FL, 2005, p. 511.

45 K. Arnold, in R. Lipowsky, E. Sackmann (eds.), *Handbook of biological physics*. Elsevier, Amsterdam, 1995, p. 903.

46 K. Arnold, M. Krumbiegel, O. Zschörnig, D. Barthel, S. Ohki, in S. Ohki (ed.), *Cell and model membrane interactions*. Plenum Press, New York, 1991, p. 63.

47 D. H. Napper. *Polymeric stabilization of colloidal dispersions*. Academic Press, New York, 1983.

48 C. J. Van Oss, in R. Glaser, D. Gingell (eds.), *Biophysics of the cell surface*. Springer-Verlag, Berlin, 1990, p. 131.

49 C. J. Van Oss, K. Arnold, R. J. Good, K. Grawrisch, S. Ohki. *J. Macromol. Sci. Chem.* A27 (1990) 563–580.

50 C. J. Van Oss, K. Arnold, W. T. Coakley. *Cell Biophys.* 17 (1990) 1–10.

51 K. Arnold, O. Zschörnig, W. Herold, D. Barthel. *Mol. Cryst. Liq. Cryst.* 152 (1987) 357–362.

52 K. Arnold, O. Zschörnig, L. Pratsch, E. Donath, D. Barthel, A. Herrmann. *Stud. Biophys.* 127 (1988) 113–120.

53 K. Arnold, O. Zschörnig, D. Barthel, W. Herold. *Biochim. Biophys. Acta* 47 (1990) 949–954.

54 H. Bäumler, E. Donath. *Stud. Biophys.* 120 (1987) 113–122.

55 L. Pratsch, E. Donath. *Studia Biophys.* 123 (1988) 101–108.
56 K. Arnold, O. Zschörnig. *Biomed. Biochim. Acta* 47 (1988) 949–964.
57 C. P. S. Tilcock, D. Fisher. *Biochim. Biophys. Acta* 688 (1982) 645–652.
58 L. T. Boni, J. S. Han, S. W. Hui, P. Mukherjee, J. T. Ho, C. Y. Jung. *Biochim. Biophys. Acta* 775 (1984) 409–418.
59 S. W. Burgess, T. J. McIntosh, B. R. Lentz. *Biochemistry* 31 (1992) 2653–2661.
60 A. M. J. Blow, G. M. Botham, D. Fisher, A. H. Goodall, C. P. S. Tilcook, J. A. Lucy. *FEBS Lett.* 94 (1978) 305–310.
61 R. A. Parente, B. R. Lentz. *Biochemistry* 25 (1986) 6678–6688.
62 K. Arnold, A. Herrmann, L. Pratsch, K. Gawrisch. *Biochim. Biophys. Acta* 815 (1985) 515–518.
63 K. Arnold, T. D. Flanagan, S. Ohki. *Biomed. Biochim Acta* 7 (1990) 633–635.
64 S. Ohki, K. Arnold, N. Srinivasakumar, T. Flanagan. *Antiviral Res.* 18 (1992) 163–177.
65 A. Herrmann, T. Korte, K. Arnold, B. Hillebrechi. *Antiviral Res.* 19 (1992) 295–311.
66 M. Krumbiegel, K. Arnold. *Chem. Phys. Lipids* 54 (1990) 1–7.
67 H. Mitsuya, D. J. Looney, S. Kuno, R. Ueno, F. Wong-Staal, S. Broder. *Science* 240 (1988) 646–649.
68 O. Zschörnig, K. Arnold, W. Richter, S. Ohki. *Chem. Phys. Lipids* 63 (1992) 15–22.
69 Y. C. Kim, T. Nishida. *J. Biol. Chem.* 252 (1977) 1243–1249.
70 J. G. Stollery, W. J. Vail. *Biochim. Biophys. Acta* 471 (1977) 372–390.
71 A. E. Gad, B. C. Silver, G. D. Eytan. *Biochim. Biophys. Acta* 690 (1982) 124–132.
72 J. Bondeson, R. Sundler. *Biochim. Biophys. Acta* 1026 (1990) 186–194.
73 A. Walter, C. J. Steer, R. Blumenthal. *Biochim. Biophys. Acta* 861 (1986) 319–330.
74 P. S. Uster, D. W. Deamer. *Biochemistry* 24 (1985) 1–8.
75 C. E. Creutz, W. J. Zaks, H. C. Hamman, W. H. Martin, in A. Sowers (ed.), *Cell fusion*. Plenum Press, New York, 1978, p. 45.
76 J. Wilschut, D. Papahadjopoulos. *Nature* 281 (1979) 690–692.
77 K. Hong, P. R. Meers, N. Düzgunes, D. Papahadjopoulos, in J. Wilschut, D. Hoekstra (eds.), *Membrane fusion*. Marcel Dekker, New York, 1991.
78 K. Hong, N. Düzgunes, D. Papahadjopoulos. *Biophys. J.* 37 (1982) 297–305.
79 C. E. Creutz, D. C. Sterner. *Biochem. Biophys. Res. Commun.* 114 (1983) 355–364.
80 P. Meers, J. Bentz, D. Alford, S. Nir, D. Papahadjopoulos, K. Hong. *Biochemistry* 27 (1988) 4430–4439.
81 P. Meers, T. Mealy, N. Pavlotsky, A. I. Tauber. *Biochemistry* 31 (1992) 6372–6382.
82 M. Mosior, S. McLaughlin. *Biochemistry* 31 (1992) 1767–1773.
83 K. Arnold, D. Hoekstra, S. Ohki. *Biochim. Biophys. Acta* 1124 (1992) 88–94.
84 S. C. Harrison. *Nature Struct. Mol. Biol.* 15 (2008) 690–698.

85 H.-S. Yin, X. Wen, R. G. Paterson, R. A. Lamb, T. S. Jardentzky. *Nature* 439 (2006) 38–44.
86 W. Weissenhorn, A. Hinz, Y. Gaudin. *FEBS Lett.* 581 (2007) 2150–2155.
87 K. Stiany, S. Brandler, C. Kossl, F. X. Heinz. *J. Virol.* 81 (2007) 11526–11531.
88 T. Stegmann, J. M. Whit, A. Helenius. *EMBO J.* 9 (1990) 4231–4241.
89 A. E. Spruce, A. Iwata, W. Almers. *Proc. Natl. Acad. Sci. U. S. A.* 88 (1991) 2623–2627.
90 H. Ellens, J. Bentz, D. Mason, F. Zhang, J. M. White. *Biochemistry* 29 (1990) 9697–9707.
91 M. J. Clague, C. Schoch, R. Blumenthal. *J. Virol.* 65 (1991) 2402–2407.
92 J. M. White. *Science* 258 (1992) 917–924.
93 T. Kanaseki, K. Kawasaki, M. Murata, Y. Ikeuchi, S.-I. Ohnishi. *J. Cell Biol.* 137 (1997) 1041–1056.
94 T. Danieli, S. L. Pelletier, Y. I. Henis, J. M. White. *J. Cell Biol.* 133 (1996) 559–569.
95 R. Blumenthal, D. P. Sarkar, S. Durrell, D. E. Howard, J. Morris. *J. Cell Biol.* 35 (1996) 63–71.
96 K. J. Cross, L. M. Burleigh, D. Steinhauer. *Expert Rev. Mol. Med.* Vol 3 (2001) 1–18.
97 D. M Eckert, P. S. Kim. *Annu. Rev. Biochem.* 70 (2001) 777–810.
98 M. Kielian, F. A. Rey. *Nature Rev. Microbiol.* 4 (2006) 67–76.
99 L. V. Chernomordik, M. M. Kozlov. *Nature Struct. Mol. Biol.* 15 (2008) 675–683.
100 T. Söllner, M. K. Bennet, S. W. Whiteheart, R. H. Scheller, J. E. Rottman. *Cell* 75 (1993) 409–418.
101 T. Söllner, S. W. Whiteheart, M. Brunner, H. Erdjument-Bromage, S. Geromanos, P. Tempst, J. E. Rothman. *Nature* 362 (1993) 318–324.
102 R. B. Sutton, D. Fasshauer, R. Jahn, A. Brunger. *Nature* 395 (1998) 347–353.
103 T. Weber, B. V. Zemelman, J. A. McNew, B. Westermann, M. Gmachl, F. Parlati, T. H. Söllner, J. E. Rothman. *Cell* 92 (1998) 759–772.
104 J. S. Bonifacino, B. S. Glick. *Cell* 116 (2004) 153–166.
105 Y. A. Chen, R. H. Scheller. *Nat. Rev. Mol. Cell Biol.* 2 (2001) 98–106.
106 W. Wickner, R. Schekman. *Nat. Struct. Mol. Biol.* 15 (2008) 658–664.
107 X. Lu, F. Zhang, J. A. McNew, Y.-K. Shin. *J. Biol. Chem.* 280 (2005) 30538–30541.
108 J. M. Fernandez, E. Neher, B. D. Gomperts. *Nature* 312 (1984) 453–455.
109 W. Almers, F. W. Tse. *Neuron* 4 (1990) 813–818.
110 J. R. Monck, J. M. Fernandez. *J. Cell Biol.* 119 (1992) 1395–1404.
111 A. E. Spruce, L. J. Breckenridge, A. K. Lee, W. Almers. *Neuron* 4 (1990) 643–654.
112 M. B. Jackson, E. R. Chapman. *Nat. Struct. Mol. Biol.* 15 (2008) 684–689.
113 X. Han, C. T. Wang, J. Bai, E. R. Chapman, M. R. Jackson. *Science* 304 (2004) 289–292.
114 B. P. Jena. *Exp. Biol. Med.* 230 (2005) 307–319.

115 G. van den Bogaart, M. G. Holt, G. Bunt, D. Riedel, F. S. Wouters, R. Jahn. *Nat. Struct. Mol. Biol.* 17 (2010) 358–364.

116 A. Portis, C. Newton, D. Papahadjopoulos. *Biochemistry* 18 (1979) 780–790.

117 D. Papahadjoloupos, S. Ohki, in J. F. Johnson, R. S. Porter (eds.), *Liquid crystals and ordered fluids*. Plenum Press, New York, 1970, pp. 13–32.

118 S. Ohki, in S. Ohki (ed.), *Cell and model membrane interactions*. Plenum Press, New York, 1991, p. 267.

119 S. Ohki. *Biochim. Biophys. Acta* 689 (1982) 1–11.

120 S. Ohki, O. Zschörnig. *Chem. Phys. Lipids* 65 (1993) 193–204.

121 D. Hoekstra. *J. Bioenerg. Biomemb.* 27 (1990) 121–155.

40

DRUG DELIVERY SYSTEM

KIMIKO MAKINO

40.1 INTRODUCTION

The minimum needed amount of medicine should be released at the therapeutic site of action only when it is required. However, when the medicine is administered by a usual method, only a small amount of the administered dose of medicine reaches the site, while the major amount of medicine is distributed to the rest of the body, which increases the administered dose and causes side effects. Therefore, controlled release and targeting of medicine are required for the efficient treatment of diseases. To reach the goal of targeting, colloidal drug delivery systems have been studied and developed. Some devices that have micro- and nanosized drug delivery systems have been studied since the 1980s. This chapter includes drug delivery systems using microspheres and nanospheres.

40.1.1 Controlled Release of Medicine Using Microspheres

LUPRON DEPOT® is a product of Takeda Pharmaceutical Company Limited and is a medicine-releasing leuprolide acetate, luteinizing hormone-releasing hormone (LH-RH) analog, for 1 or 3 months with a constant ratio and keeping the plasma concentration of the medicine constant for the term after subcutaneous injection, as shown in Figures 40.1 and 40.2. This medicine has been used to treat prostate cancer, breast cancer, and certain benign gynecological disorders. Poly(lactide-co-glycolide) (PLGA) is a biocompatible and biodegradable polymer, which is hydrolytically degraded into oligomers

Electrical Phenomena at Interfaces and Biointerfaces: Fundamentals and Applications in Nano-, Bio-, and Environmental Sciences, First Edition. Edited by Hiroyuki Ohshima.

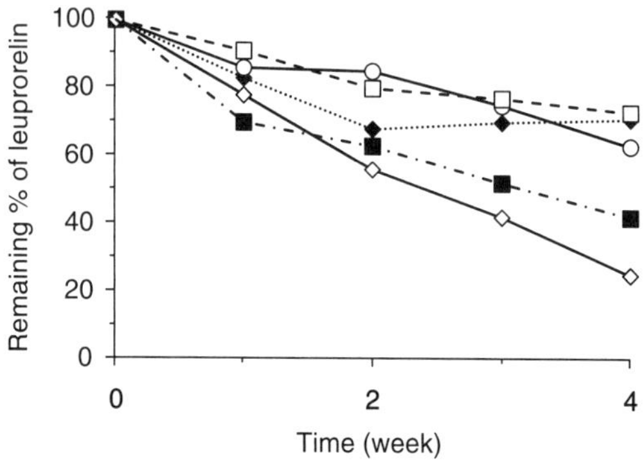

Figure 40.1. *In vitro* release of leuprorelin from polylactide (PLA) and PLGA microspheres:◆,PLA-22000;○,PLA-12000;□,PLGA(90/10)-21000;■,PLA-6000;◇,PLGA(75/25)-14500.

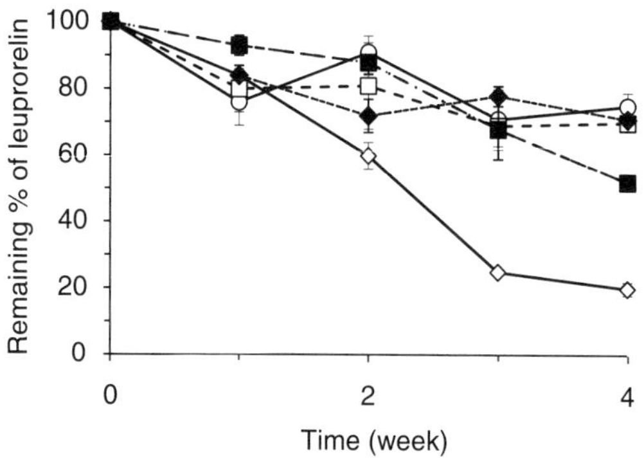

Figure 40.2. *In vivo* release of leuprorelin from PLA and PLGA microspheres. Symbols are the same as those shown in Figure 40.1.

and monomers in water, as shown in Figure 40.3. LUPRON DEPOT is a product mainly composed of PLGA microspheres with diameters between 100 and 500 μm containing an LH-RH analog. During the degradation of PLGA, the LH-RH analog is released in a constant ratio [1–3].

40.1.2 Nanomedicine

Nanomedicine is medical treatment at the "nano" scale of about 100 nm or less. In the 1980s, progress in developing nanosized hybrid therapeutics and drug delivery systems has been remarkable and products have been approved for clinical use. Most are anticancer therapies, polymer-coated liposomes (Doxil®/Caelyx®), antibodies (Herceptin®, Avastin®), and a nanoparticle con-

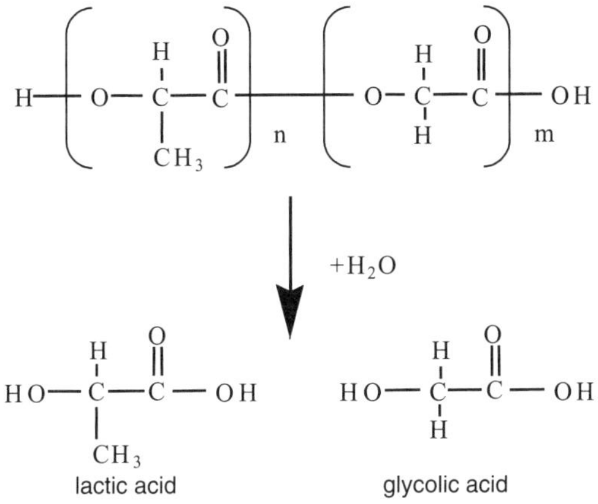

Figure 40.3. Hydrolytic degradation of PLGA.

taining paclitaxel (Abraxane™). The concepts of antibody conjugates, liposomes, nanoparticles, and polymer micelles stemmed from the 1970s. Each had advantages and disadvantages. Antibodies have potential for selective targeting but can be immunogenic. Liposomes are biocompatible drug carriers but easily release drugs quickly or do not release drugs and are sometimes captured by the reticuloendothelial system (RES), even when the liposome surfaces are coated by hydrophilic polymer layers. Particles with diameters larger than 200 nm are easily recognized by RES and are digested by macrophages after being intravenously administered. To escape from the recognition by RES, many studies have been reported. For this purpose, synthetic biocompatible polymers have been developed. Nanosized particles have high surface-to-volume ratio, could be especially dangerous, although they are less effectively taken up by macrophages, and can reach the brain passing through the blood–brain barrier (BBB). Any toxicity of nanoparticles depends on the route and frequency of administration and on the polymer used to prepare the particles.

40.1.2.1 Doxorubicin-Loaded PEGylated Liposomes. PEGylated liposomal formulation of drugs extends circulation time to increase the probability of accumulating at sites compared with free drug molecules [4–9]. This type of formulation containing doxorubicin was applied for the treatment of AIDS-related Kaposi's sarcoma. Kaposi's sarcoma is a vascular tumor of the skin and viscera and is the common malignancy seen in patients with human immunodeficiency virus infection (AIDS). Many therapies have been demonstrated to be effective in the treatment of AIDS-related Kaposi's sarcoma, including alitretinoin gel, interferon alpha, and various forms of cytotoxic chemotherapy. Pegylated liposomal doxorubicin has been established as the treatment of choice for patients with AIDS-associated Kaposi's sarcoma in Western countries. Preclinical and clinical evidence of this formulation (Doxil/Caelyx) has

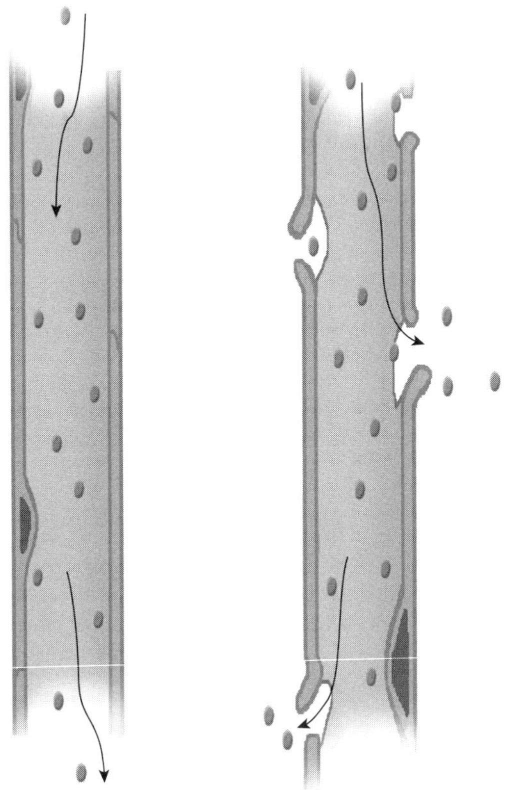

Figure 40.4. Pegylated liposomal doxorubicin in normal (left) and tumor vessels (right).

demonstrated that the nanoparticle, especially PEGylated liposome, delivery system leads to greater localization of doxorubicin to the tumor site and consequently improved efficacy, as well as reduced toxicity. For vascularized tumors, the selective accumulation and retention of liposomes is a result of the combination of "leaky" tumor neovasculature and malfunctioning lymphatics, integrated in the enhanced permeability and retention (EPR) effect, as shown in Figure 40.4. The modification of liposome surfaces with polyethylene glycol (PEGylation) is useful for the long circulation of liposomes in the blood flow, escaping from RES. The surfaces of conventional bear liposomes are easily recognized by the immune system, such as macrophages, and are taken up by phagocytosis, then the drugs are easily metabolized. In general, long-circulating liposomes show dose independent, nonsaturable, long linear kinetics and increased bioavailability. These characteristics of pegylated liposomal doxorubicin (Fig. 40.5) have been highly recognized through the clinical success of Doxil/Caelyx in the treatment of skin, cervical, prostate, breast, ovarian, and some other cancers. The encapsulation of doxorubicin in

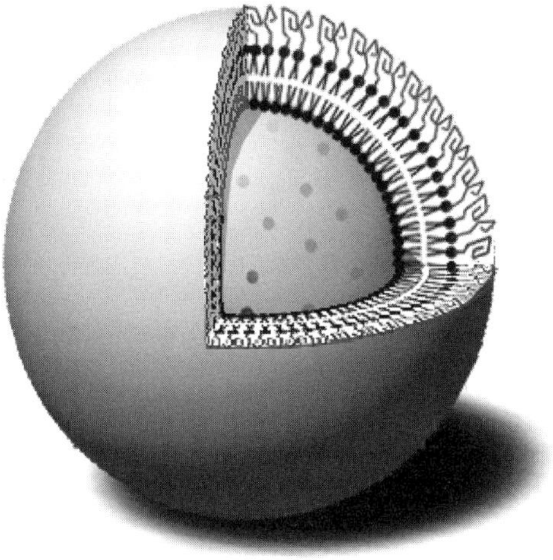

Figure 40.5. PEGylated liposomal doxorubicin.

PEGylated liposomes has increased the concentration of doxorubicin in the tumor site via EPR effect by passive targeting and has decreased its volume of distribution and, thus, decreased the therapy-associated cardiotoxicity.

40.1.2.2 Human Albumin-Stabilized Nanoparticle Paclitaxel Formulation. Albumin-stabilized paclitaxel nanoparticles (Abraxane) have shown significantly less toxicity than usual Cremophor-containing paclitaxel [10–16]. Paclitaxel is an anticancer agent effective for the treatment of breast, ovarian, lung, head, and neck cancers. Paclitaxel has been formed with the micelle-forming vehicle Cremophor EL (CrEL) to enhance drug solubility since paclitaxel is highly hydrophobic and is not soluble in water. Polyoxyethylated castor oil derivative CrEL is the most viable option of paclitaxel. Currently, paclitaxel is commercially marketed in a formulation that contains a solvent system of CrEL and dehydrated ethanol. CrEL has been widely used as a vehicle for the solubilization of a number of hydrophobic drugs including anesthetics, vitamins, photosensitizers, and immune suppressants. The amount of CrEL in Taxol® per administration was relatively high, and therefore its toxicological and pharmacological behavior in the context of chemotherapeutic treatment with paclitaxel is of major importance. The addition of CrEL results in hypersensitivity reactions, neurotoxicity, and altered pharmacokinetics of paclitaxel. To circumvent these unfavorable effects caused by CrEL, a new delivery system, Cremophor-free, human albumin-stabilized nanoparticle paclitaxel formulation has been developed, Abraxane. The human albumin-stabilized paclitaxel nanoparticles were prepared by nanomization of paclitaxel in human albumin solution using a high-pressure nanomizer, and the average size of the formulation is 130 nm. The comparative properties of Abraxane and the

TABLE 40.1. Comparative Properties of Nanoparticle Albumin-Bound Paclitaxel and Cremophor EL-Containing Paclitaxel

	Nanoparticle Albumin-Bound Paclitaxel	Cremophor EL-Based Paclitaxel (e.g., Taxol)
Hypersensitivity reactions	No	Yes
Leaches plasticizer from intravenous tubing	No	Yes
Forms micelles in plasma	No	Yes
Transported via gp60 receptor on vascular endothelial cells	Yes	No
Nonlinear pharmacokinetics	No	Yes
Drug–drug interactions	Not known	Yes

conventional paclitaxel CrEL-containing formulation are summarized in Table 40.1.

40.1.3 Pulmonary Drug Delivery Systems

The lung (adjectival form: pulmonary) is the essential respiration organ, and two lungs are located in the chest on either side of the heart. Their principal function is exchange of oxygen and carbon dioxide, transporting oxygen from the atmosphere into the bloodstream and releasing carbon dioxide from the bloodstream to the atmosphere by the passage of air through the mouth to the alveoli. The air progresses through the mouth or nose, travels through the oropharynx, nasopharynx, the larynx, the trachea, the primary bronchiole, the secondary bronchiole, the terminal bronchiole, the respiratory bronchiole, and finally reaches the alveolar duct where the gas exchange of CO_2 and O_2 takes place. Recently, there have been many attempts to improve the systemic delivery of peptide and protein drugs by routes of administration other than injection [17–19]. The drug delivery in these studies have included nasal, rectal, buccal, and respiratory routes of administration [20–26]. Because of its unique physiological characteristics, the lung is an attractive port of entry to the systemic circulation for the administration of drugs; that is, the alveoli present a large surface area for adsorption of about 100 m^2. A very thin diffusion path separates the airspace form the blood stream; that is, the alveolar epithelium, the vascular endothelium, and their respective basal membranes are less than 0.5 μm thick. Also, the high blood flow of about 5 L/min of the pulmonary circulation rapidly distributes molecules throughout the body without first-pass hepatic metabolism, and the metabolic activity locally in the lungs is relatively low. Together with the success of the design of new inhalers, pulmonary delivery of small drugs and proteins has reached clinical trials of drugs such as insulin, calcitonin, interferon, and hormone.

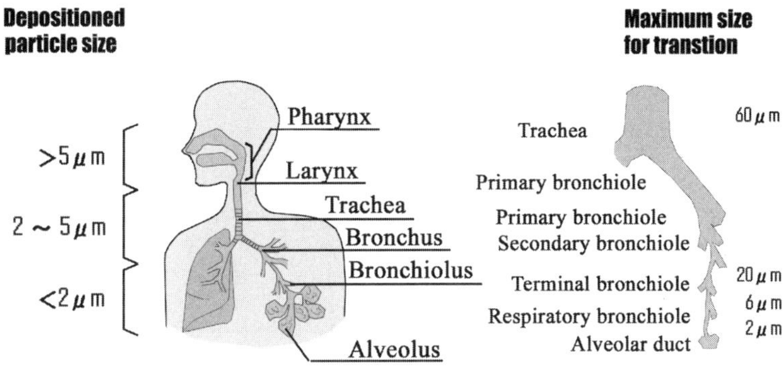

Figure 40.6. Depositioned particle size in the respiratory tract.

The environment in the lungs is very moist, and the humidity in the respiratory tract is almost 100%. To reach the alveolar through the respiratory tract, the medicine should have the proper size and density, shown as an aerodynamic diameter. As shown in Figure 40.6, particles with aerodynamic diameters between 2 and 5 μm can efficiently reach the alveoli. Particles smaller than 1 μm are easily inhaled by respiration but are exhausted from the lungs without deposition in the alveoli, like tobacco smoke. The aerodynamic diameter of the particle, d_{aer}, is defined as Equation (40.1), which is simply derived from the Stokes equation:

$$d_{ae} = d_p\sqrt{\frac{\rho_p}{\rho_0}}, \tag{40.1}$$

where d_p is the diameter of the particle, which is usually measured using laser diffraction; ρ_p is the density of the particle; and ρ_0 is the density of water at the same temperature [27–29].

As mentioned before, the environment in the lungs is very moist, which makes it hospital for bacteria and causes infectious diseases in the lungs. For the treatment of these infectious diseases, direct delivery of antimicrobe agents to the lungs through the respiratory tract has been considered to be effective. This includes local injection of medicine to the lungs. Also, this concept has been applied to the treatment of lung carcinoma.

40.1.3.1 Pulmonary Delivery of Antitubercle Agents to Treat Lung Tuberculosis (TB). TB is recognized as a global public health emergency by the declaration of the World Health Organization on April 23, 1993. The pathogen of TB is *Mycobacterium tuberculosis* and it enters the body through the respiratory tract by inhalation. Once the pathogens reach the alveoli, they are phagocytosed by alveolar macrophages but are not digested and remain in macrophages and grow inside. Live *M. tuberculosis* inhibits the fusion of phagosomes and lysosomes after being internalized in alveolar macrophages, although the mechanism is not completely understood. The current treatment of TB is directly observed treatment, short course (DOTS) of oral administration of several anti-TB medicines such as isoniazid (INH), ethambutol (EB), and rifampicin (RFP). More than 60 years after the first treatment appeared, little has changed for patients with drug-resistant TB. Pills are still difficult to swallow, have an unpleasant taste, and must be taken daily for 6 months. The long-term treatment of these medicines results in low compliancy and develops several multidrug-resistant strains of *M. tuberculosis*.

To kill *M. tuberculosis* remaining inside macrophages, the development of a dosage form/formulation of drug delivery system making pathogens more easily to be phagocytosed by macrophages is needed. Also, the activation of the phagocytotic capacity of pathogen-infected macrophages is important. It is known that the phagocytosis of *M. tuberculosis* increases the phagocytotic activity of alveolar macrophages. Then, alveolar macrophages infected with *M. tuberculosis* take up more effectively *M. tuberculosis* than uninfected macrophages, which means the infected macrophages more effectively trap foreign substances than normal ones. Using this property, anti-TB medicine can be effectively delivered to infected macrophages than to normal ones. The application of microsphere technology to the treatment of patients infected with *M. tuberculosis* has been studied so that multiple drug-resistant strains do not develop [30–45]. Slow release of RFP from microspheres localized in alveolar macrophages is ideal to treat the disease since tubercle bacillus is a facultative intracellular parasite in alveolar macrophages, and microspheres can penetrate into the infected macrophages by phagocytosis. Also, the slow release of RFP from microspheres locally injected into the lung is ideal to maintain both high drug level in alveolar macrophages and low drug level in blood. PLGA is a biocompatible and biodegradable polymer, which is hydrolytically degraded into nontoxic oligomers or monomers, lactic acid, and glycolic acid, and has been widely used as an implantable drug carrier. Depending on the ratio of lactic acid to glycolic acid and on the molecular weight, the release ratio of drugs loaded in PLGA microspheres changes. From RFP-loaded PLGA microspheres containing RFP and PLGA with a molecular weight of 10,000 and a lactic acid-to-glycolic acid ratio of 75:25, almost 90% of RFP molecules loaded in PLGA microspheres were slowly released for 3 days, as shown in Figure 40.7 at pH 7.4. On the other hand, less than 5% of RFP molecules loaded in PLGA microspheres were released for 3 days at pH 4.0. This means that RFP is released in cytosol but is not released in phagolysosome. RFP-loaded PLGA microspheres with average diameters of 3 μm were most effectively taken up by alveolar macrophages, as shown in Figure 40.8. Also, RFP-loaded PLGA microspheres accelerate the phagocytotic function of alveolar macrophages, as shown in Figure 40.9. Also, the amount of RFP released from RFP-loaded PLGA microspheres inside alveolar macrophages were compared to free RFP powder. As shown in Figure 40.10, an almost 20 times larger amount of RFP was delivered inside alveolar macrophages using RFP-loaded PLGA microspheres compared to using free RFP powder, when

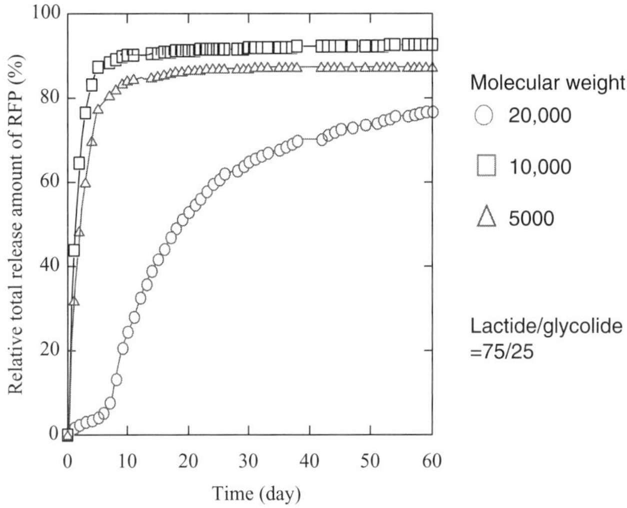

Figure 40.7. Effects of molecular weight of PLGA on RFP release from RFP-loaded PLGA microspheres at pH 7.4.

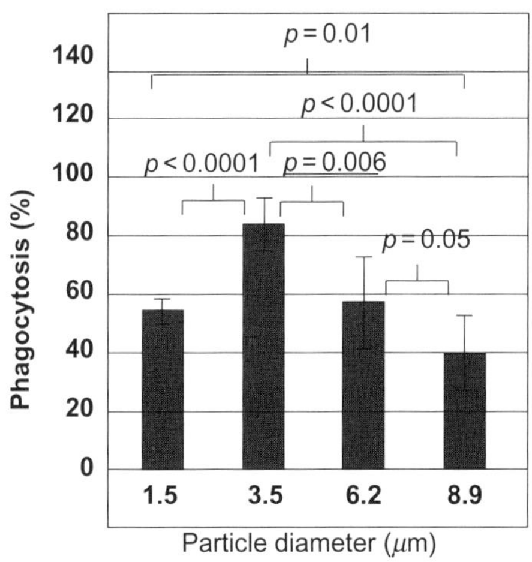

Figure 40.8. Relationship between particle size and phagocytotic activity.

40 μg of RFP was added to 5×10^5 cells/well/mL. Therefore, if RFP-loaded PLGA microspheres reach the alveoli, the microphseres are effectively uptaken by alveolar macrophages infected with *M. tuberculosis* and kill the *M. tuberculosis* inside macrophages. Dry powder inhaler (DPI) works to deliver RFP-loaded PLGA powder deep in the lungs. RFP-loaded PLGA

RFP/PLGA MS (μg/mL)	Phagocytotic ratio of FITC-PSt latex beads (%)
0	31
0.012	45
0.12	47
1.2	52

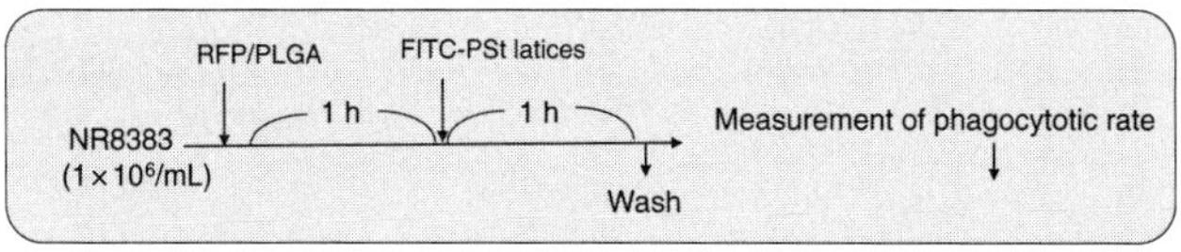

Figure 40.9. Accelerated phagocytotic function by phagocytosis of RFP-loaded PLGA microspheres.

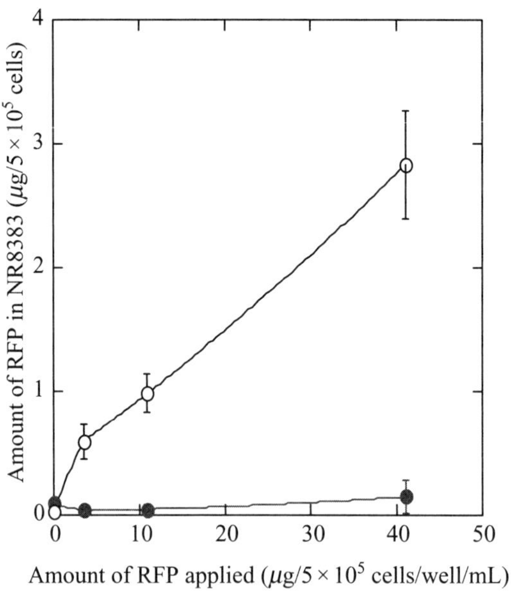

Figure 40.10. Effective intercorporation of RFP into alveolar macrophages using RFP-loaded PLGA microspheres. RFP-loaded PLGA microspheres, ○, and RFP molecules, •, were added to the medium.

microspheres were prepared using a spray dryer. In Figure 40.11, a scanning electron microscope (SEM) image of 10% RFP-loaded PLGA microspheres is shown. The average diameter was determined as almost 3 μm from the size distribution, as shown in Figure 40.12. The pulmonary deposition of the dry powders was assessed in an Andersen cascade impactor equipped with an

Figure 40.11. SEM image of 10% RFP-loaded PLGA microspheres.

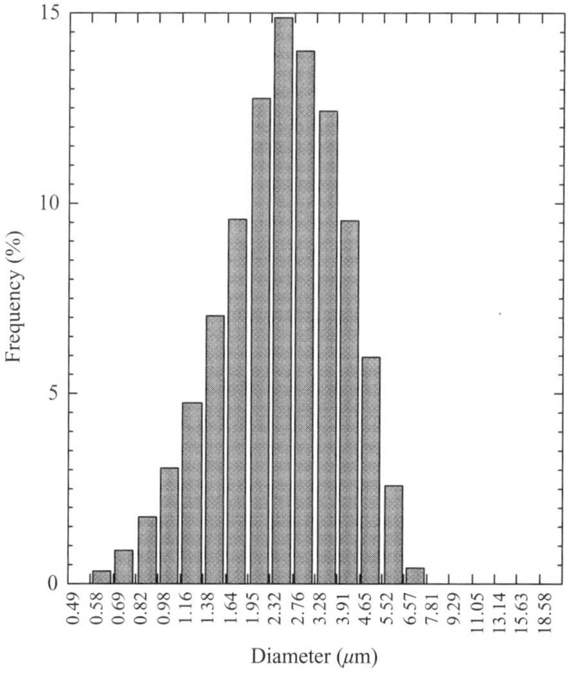

Figure 40.12. Particle size distribution of RFP-loaded PLGA microspheres in air.

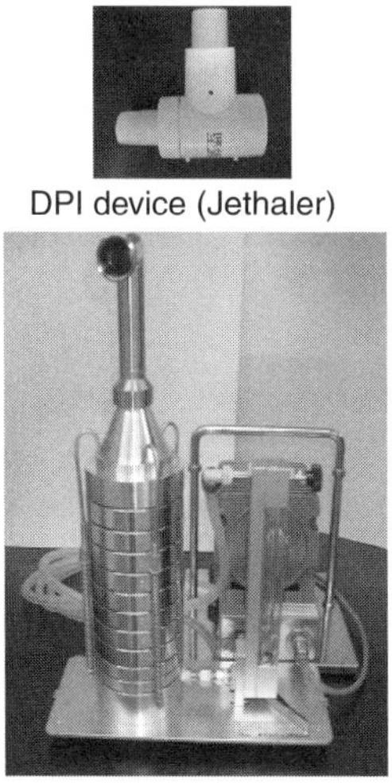

Figure 40.13. Andersen nonviable impactor.

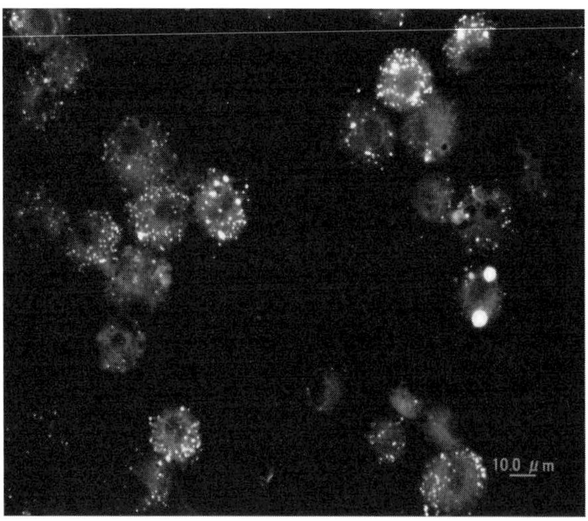

Figure 40.14. Alveolar macrophages taken recovered by BALF from rat lungs 6 hours after administration of coumarin-loaded PLGA microspheres.

induction port (eight-stage nonviable cascade impactor) using a DPI device, Jethaler (Fig. 40.13) at a flow rate of 28.3 L/min for 5 seconds, following the United States Pharmacopeia (USP). The powder deposited on the different levels was recovered and the amount was measured. The value of fine particle fraction (FPF) was the ratio of the powder deposited between levels 3 and 7, the aerodynamic diameter between 0.65 and 4.7 μm, to the total amount of powder applied for the measurements. The FPF value of the RFP-loaded PLGA microspheres was about 20%. Then, the RFP-loaded PLGA microspheres were administered through the respiratory tract using Dry Powder Insufflators™ (Penn-Century, Wynomoor, PA) to rats. Coumarin 6-loaded PLGA microspheres were administered to the rat lungs and the alveolar macrophages were recovered from bronchoalveolar lavage fluid (BALF), obtained from the lungs 6 hours after administration. The results are shown in Figure 40.14 and the incorporation of the microspheres inside alveolar macrophages was observed. However, the distribution of the RFP-loaded PLGA powders is not uniform in the lungs. Improvements of devices and powders are needed.

ACKNOWLEDGMENT

This work is partly supported by MEXT (2004-2008) and MEXT (2010-2014).

REFERENCES

1 H. Okada, M. Yamamoto, T. Heya, Y. Inoue, S. Kamei, Y. Ogawa, H. Toguchi. *Journal of Controlled Release* 28 (1994) 121.

2 H. Okada, Y. Doken, Y. Ogawa, H. Toguchi. *Pharmaceutical Research* 11 (1994) 1143.

3 H. Okada, Y. Doken, Y. Ogawa, H. Toguchi. *Pharmaceutical Research* 11 (1994) 1199.

4 A. Green, P. G. Rose. *International Journal of Nanomedicine* 1 (2006) 229.

5 A. Udhrain, K. M. Skubitz, D. W. Northfelt. *International Journal of Nanomedicine* 2 (2007) 345.

6 T. A. ElBayoumi, V. P. Torchillin. *Clinical Cancer Research* 15 (2009) 1973.

7 A. B. Madhankumar, B. S. Webb, X. Wang, Q. X. Yang, D. A. Antonetti, P. A. Miller, J. M. Seehan, J. R. Connor. *Molecular Cancer Therapeutics* 8 (2009) 648.

8 E. Andeopoulou, D. Gailotti, E. Kim, A. Downey, D. Mirchandani, A. Hamilton, A. Jacobs, J. Coutin, F. Muggia. *Annals of Oncology* 18 (2007) 716.

9 T. A. Elbayoumi, V. P. Torchillin. *International Journal of Pharmaceutics* 357 (2008) 272.

10 N. K. Ibrahim, N. Desai, S. Legha, P. Soon-Shiong, R. L. Theriault, E. Rivera, B. Esmaeli, S. E. Ring, A. Bedikian, G. N. Hortobagyi, J. A. Ellerhorst. *Clinical Cancer Research* 8 (2002) 1038.

11 A. Sparreboom, C. D. Scripture, V. Trieu, P. J. Williams, T. De, A. Yang, B. Beals, W. D. Figg, M. Hawkins, N. Desai. *Clinical Cancer Research* 11 (2005) 4136.

12 N. Desai, V. Trieu, Z. Yao, L. Louie, S. Ci, A. Yang, C. Tao, T. De, B. Beals, D. Dykes, P. Noker, R. Yao, E. Labeo, M. Hawkins, P. Soon-Shiong. *Clinical Cancer Research* 12 (2006) 1317.

13 M. R. Green, G. M. Manikhas, S. Orlov, B. Afanasyev, A. M. Makhson, P. Bhar, M. J. Hawkins. *Annals of Oncology* 17 (2006) 1263.

14 N. Wiedenmann, D. Valdecanas, N. Hunter, S. Hyde, T. A. Buchholz, L. Milas, K. Mason. *Clinical Cancer Research* 13 (2007) 1868.

15 C. D. Scripture, W. D. Figg, A. Sparreboom. *Therapeutics and Clinical Risk Management* 1 (2005) 107.

16 E. Miele, G. P. Spinelli, E. Miele, F. Tamao, S. Tamao. *International Journal of Nanomedicine* 4 (2009) 99.

17 M. Amidi, K. M. Krudys, C. J. Snel, D. J. A Crommelin, O. E. D. Pasque, W. Hennink, W. Jiskoot. *Journal of Controlled Release* 127 (2008) 257.

18 A. Aadjei, P. Gupta. *Journal of Controlled Release* 29 (1994) 361.

19 H. Yamamoto, Y. Kuno, S. Sugimoto, H. Takeuchi, Y. Kawashima. *Journal of Controlled Release* 102 (2005) 373.

20 C. Lombry, C. Bosquillon, V. Preat, R. Vanbever. *Journal of Controlled Release* 83 (2002) 331.

21 J. Fiegel, J. Fu, J. Hanes. *Journal of Controlled Release* 96 (2004) 411.

22 S. Chono, T. Tanino, T. Seki, K. Morimoto. *Journal of Controlled Release* 127 (2008) 50.

23 R. F. Donnelly, P. A. McCarron, C. M. Cassidy, J. S. Elborn, M. M. Tunney. *Journal of Controlled Release* 117 (2007) 217.

24 M. E. Falagas, A. Michalopoulos, E. I. Metaxas. *International Journal of Antimicrobial Agents* 35 (2010) 101.

25 L. Garcia-Contreras, A. J. Hickey. *Advanced Drug Delivery Reviews* 54 (2002) 1491.

26 C. Amrianecci, D. Paolino, C. Celia, M. Fresta, M. Carafa, F. Alhaique. *Journal of Controlled Release* 147 (2010) 127.

27 A. K. Oskouie, K. E. Noll, H. C. Wang. *Aerosol Science* 34 (2003) 501.

28 P. Tang, H. K. Chan, J. A. Raper. *Powder Technology* 147 (2004) 64.

29 P. C. Seville, T. P. Learoyd, H. Y. Li, I. J. Williamson, J. C. Birchall. *Powder Technology* 178 (2007) 40.

30 P. Muttil, J. Kaur, K. Kumar, A. B. Yadav, R. Sharma, A. Misra. *European Journal of Pharmaceutical Sciences* 34 (2008) 351.

31 J. Kaur, P. Muttil, R. K. Verma, K. Kumar, A. B. Yadav, R. Sharma, A. Misra. *European Journal of Pharmaceutical Sciences* 34 (2008) 56.

32 P. Muttil, J. Kaur, K. Kumar, A. B. Yadav, R. Sharma, A. Misra. *European Journal of Pharmaceutical Sciences* 32 (2007) 140.

33 R. Sharma, D. Saxena, A. K. Dwivedi, A. Misra. *Pharmaceutical Research* 18 (2001) 1405.

34 C. Wang, A. J. Hickey. *International Journal of Pharmaceutics* 396 (2010) 99.

35 N. Sivadas, D. O'Rourke, A. Tobin, V. Buckley, Z. Ramtoola, J. G. Kelly, A. J. Hickey, S. A. Cryan. *International Journal of Pharmaceutics* 358 (2008) 159.

36 M. J. Telko, J. Kujanpää, A. J. Hickey. *International Journal of Pharmaceutics* 336 (2007) 352.

37 K. Makino, T. Nakajima, M. Shikamura, F. Ito, S. Ando, C. Kohchi, H. Inagawa, G. I. Soma, H. Terada. *Colloids Surfaces B: Biointerfaces* 36 (2004) 35.

38 K. Makino, T. Nakajima, M. Shikamura, F. Ito, S. Ando, C. Kochi, H. Inagawa, G.-I. Soma, H. Terada. *Colloids and Surfaces. B, Biointerfaces* 36 (2004) 35–42.

39 M. Hino, M. Oda, A. Yoshida, K. Nakata, C. Kohchi, T. Nishizawa, H. Inagawa, H. Hori, K. Makino, H. Terada, G.-I Soma. *In Vivo* 19 (2005) 821–830.

40 K. Tomoda, S. Kojima, M. Kajimoto, D. Watanabe, T. Nakajima, K. Makino. *Colloids and Surfaces. B, Biointerfaces* 45 (2005) 1–6.

41 T. Hasegawa, K. Hirota, K. Tomoda, F. Ito, H. Inagawa, C. Kochi, G.-I Soma, K. Makino, H. Terada. *Colloids and Surfaces. B, Biointerfaces* 60 (2007) 221–228.

42 K. Hirota, T. Hasegawa, H. Hinata, F. Ito, H. Inagawa, C. Kochi, G.-I. Soma, K. Makino, H. Terada. *Journal of Controlled Release* 119 (2007) 69–76.

43 T. Hasegawa, K. Iijima, K. Hirota, T. Nakajima, K. Makino, H. Terada. *Colloids and Surfaces. B, Biointerfaces* 63 (2008) 209–216.

44 K. Hirota, T. Hasegawa, T. Nakajima, H. Inagawa, C. Kohchi, G. I. Soma, K. Makino, H. Terada. *Journal of Controlled Release* 142 (2010) 339.

45 T. Onoshita, Y. Shimizu, N. Yamaya, M. Miyazaki, M. Yokoyama, N. Fujiwara, T. Nakajima, K. Makino, H. Terada, M. Haga. *Colloids and Surfaces. B, Biointerfaces* 76 (2010) 151–157.

41

ON-CHIP CELL ELECTROPHORESIS AND EVALUATING CELLULAR FUNCTIONS

Takanori Akagi and Takanori Ichiki

41.1 INTRODUCTION

Cell electrophoretic mobility (EPM), which can be measured in a noninvasive manner by a cell electrophoresis experiment, is a characteristic physical property that reflects the electrical and mechanical properties of the cell surface [1, 2]. Expression of cell surface molecules such as glycoproteins and glycolipids is regulated by signals from inside the cell. Hence, the biological condition of the cells can be determined using cell EPM as an index. Since 1960, a few studies concerning the correlation between cell EPM and biological condition have been carried out. Yamada and Yamada experimentally showed the relationship between cell EPM variation and the increase in the negative surface charge density of malignantly transformed cells [3]. Later, with the advance in the understanding of colloid physics, Makino et al. showed that the EPM variation of cancer cells is also attributable to changes in the mechanical properties of the surface glycocalyx as well as in the surface charge density [4]. However, the importance of cell EPM has been underestimated for a long time mostly because of the technical difficulties associated with its reliable measurement. When cell electrophoresis experiments were performed using traditional instruments, electroosmosis generated the circulating flow inside a closed chamber called an electrophoretic tank. Hence, considerable skill and time

Electrical Phenomena at Interfaces and Biointerfaces: Fundamentals and Applications in Nano-, Bio-, and Environmental Sciences, First Edition. Edited by Hiroyuki Ohshima.

were required to identify the stationary plane in the tank and to measure the speed of the cells that move in the stationary plane where the effect of electroosmosis was absent [1]. Moreover, the optical measurement cannot exclude the artifactual data caused by agglomerated cells or dust particles. In 2000, a cell electrophoresis system using a microcapillary electrophoresis (μCE) chip was developed [5]. Such system facilitates the precise measurement of cell EPM since the electroosmosis flow (EOF) in a microcapillary with open ends is a simple plug flow. Using such system, the authors studied cell EPM changes associated with some important biological phenomena and revealed distinct correlations between cell EPM and cell condition [6–11]. In this chapter, a brief overview of the experimental studies of cell electrophoresis is introduced and our recent research achievements on the μCE chip-based cell electrophoresis are discussed.

41.2 A CHIP-BASED CELL ELECTROPHORESIS SYSTEM

A chip-based cell electrophoresis system developed by the authors and its operation procedure are described. As shown in Figure 41.1, the system consists of a μCE chip equipped with a microfluidic channel with small reservoir tanks on both ends, platinum electrodes, a microscope, a high-speed charge-coupled device (CCD) camera, and a personal computer with custom-developed particle tracking velocimetry (PTV) software. Poly(dimethylsiloxane) (PDMS) was used as the material for μCE chips. Rectangular channels (length: 1 cm, width: 100 μm, height: 50 μm) in PDMS μCE chips were fabricated by

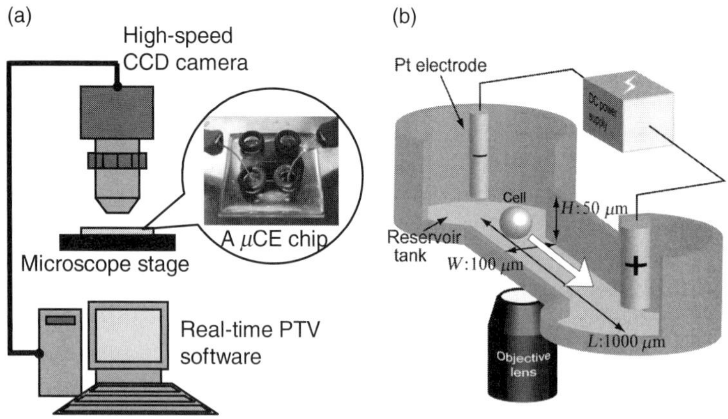

Figure 41.1. A chip-based cell electrophoresis system. (a) An on-chip cell electrophoresis system consists of a microcapillary electrophoresis chip, a microscope, a high-speed charge-coupled device (CCD) camera, and a personal computer. Movies of migrating cells were automatically analyzed using particle tracking velocimetry (PTV) software. (b) Schematic of cell electrophoresis on the chip.

photolithography and molding methods [12]. Subsequently, the inner surface of the microchannels was coated with a biomimetic polymer such as 2-methacryloyloxyethyl-phosphorylcholine (MPC) polymer to reduce the nonspecific adsorption [13]. After forming surface silanol by photochemical oxidation upon vacuum ultraviolet (VUV) irradiation, the phospholipid copolymer containing MPC and 3-methacryloxyethyl triethoxysilane (METESi) was coated inside microchannels via chemical bond with the surface silanol [14].

Ca- and Mg-free phosphate-buffered saline (PBS) (pH 7.4) (ionic strength 0.16) (1.06 mM KH2PO4, 155.17 mM NaCl, 2.97 mM Na2HPO4•7H2O) was used as the migration buffer. Since pH 7.4 is almost the isoelectric point of the MPC polymer, EOF generation in the channel is also suppressed. Before each electrophoresis experiment, the difference in the water level between reservoirs was carefully adjusted to prevent cell migration due to hydropressure. The electrodes were dipped into the sample reservoirs and cell electrophoresis was performed on a microscope stage to enable observation. The electric field of 50 V/cm was applied for 30 seconds to perform the electrophoresis experiments. Movies of migrating biological cells were recorded using a high-speed CCD camera and a personal computer, and the velocity of the migrating cells was automatically analyzed using the PTV software. The measured velocity of the migrating cells (U_m) includes the EOF velocity (U_{eo}) attributable to the microchannels (see Fig. 41.2). The true electrophoretic velocity of the cells (U_{ep}) is calculated as

$$U_{ep} = U_m - U_{eo}. \tag{41.1}$$

The EOF velocity was measured using charge-free beads of 500-nm diameter, which are latex particles coated with hydroxypropyl cellulose, in the control experiment. Hence, the true EPM of the cells (μ_{ep}) is calculated using

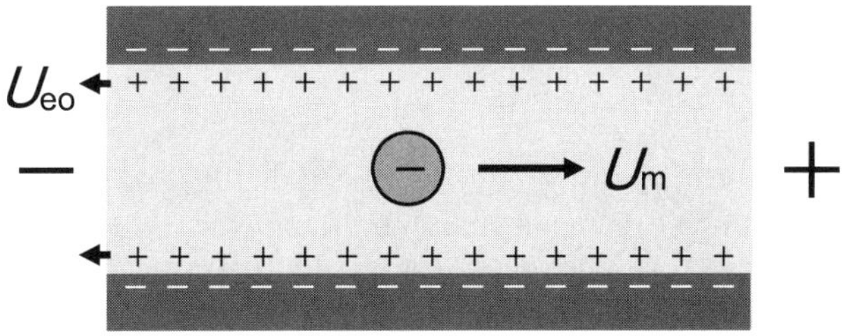

Figure 41.2. Cell electrophoresis in a microcapillary. Cells move at the speed corresponding to their surface charge density when an electric field is applied to the solution. The measured velocity of migrating cells (U_m) includes EOF (U_{eo}) velocity attributable in the microchannels. The true electrophoretic velocity of the cells (U_{ep}) is calculated by subtracting the U_{eo} from U_m. The true EPM of the cells (μ_{ep}) is calculated by dividing the U_{ep} by the electric field strength (E).

$$\mu_{ep} = \frac{U_{ep}}{E} = \frac{U_m - U_{eo}}{E}, \tag{41.2}$$

where E is the electric field strength (volt per centimeter).

41.3 CELL EPM AND CELLULAR FUNCTIONS

41.3.1 Cell Cycle

The cell cycle is one of the basic functions of all living things, which is necessary to replicate chromosomes and to transfer them to daughter cells (see Fig. 41.3a). Gene expression and response to drugs are known to be different depending on the cell cycle phase. Since an accurate evaluation of the cell cycle is a key issue in cell-based screening and therapy, cell EPM changes throughout the cell cycle were studied, and consequently, a strong correlation between EPM and the cell cycle phase was found. The EPM of human leukemia (HL60) cells treated with various synchronizing drugs was measured and compared to that of untreated cells [6]. The EPMs of nonsynchronized cells were distributed between 0 and −2.75 ($\times 10^{-4}$ cm^2/V/s). In contrast, the EPMs of the cells synchronized at the G_1 phase of the cell cycle using leptomycin B were distributed between −0.25 and −1.5 ($\times 10^{-4}$ cm^2/V/s) (see Fig. 41.3b). When similar experiments were carried out using other synchronizing drugs that induced S or G_2/M cell cycle arrest, their EPMs were distributed within a range narrower than that of nonsynchronized cells. It was revealed that the peak value of the EPM distribution differs depending in the synchronizing drug used. An additional study was carried out to investigate the cell EPM variation with the progress of the cell cycle, and a positive correlation between the cell EPM variation and the cell cycle phase was found (see Fig. 41.3c). These results suggest that the cell cycle phase can be determined using the cell EPM as an index.

41.3.2 Apoptosis

Apoptosis, a genetically controlled form of cell death, is one of the fundamental processes by which organisms maintain homeostasis (see Fig. 41.4a). Since cancer is caused by the failure of apoptosis control, a molecular-level understanding of apoptosis can lead to the discovery of anticancer drugs. Molecular targeted drugs that act on apoptosis-related genes have been developed one after another and numerous candidate substances are now under test. High-throughput screening in the preclinical studies of candidate substances is important to expedite anticancer drug development. On the basis of these backgrounds, the relationship between cell EPM and apoptosis was studied.

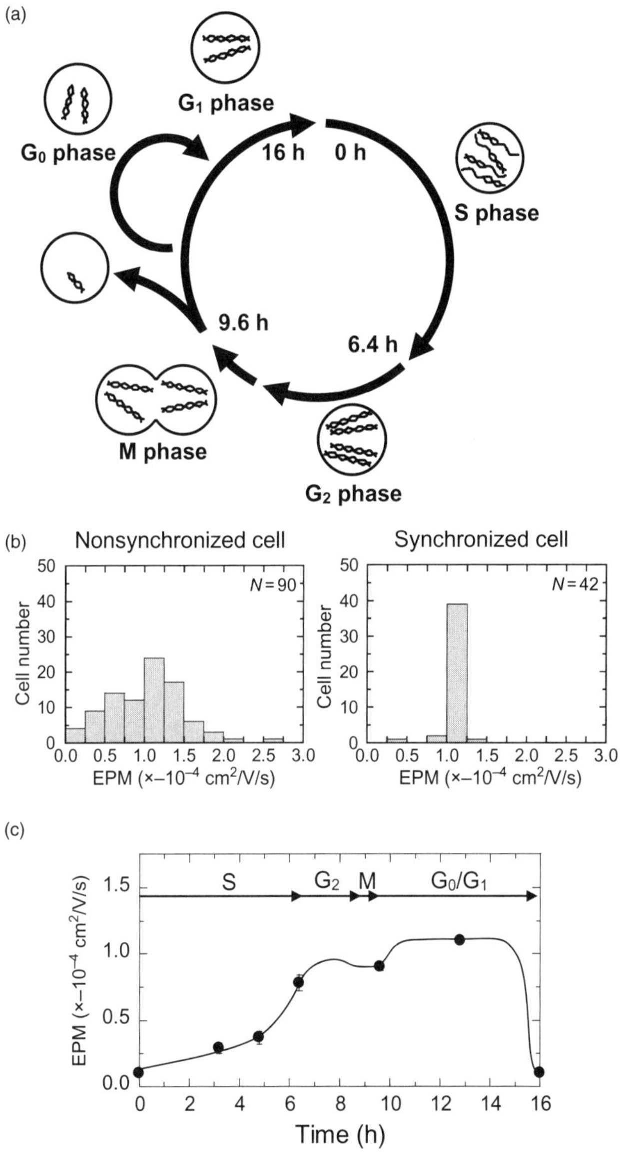

Figure 41.3. Effect of cell cycle on EPM of HL-60 cells. (a) Schematic of the cell cycle. (b) EPM histograms of cells after treatment with PBS (left) and leptomycin B (right) for 24 hours. (c) Cell EPM variation through the cell cycle. The line is drawn for visual convenience.

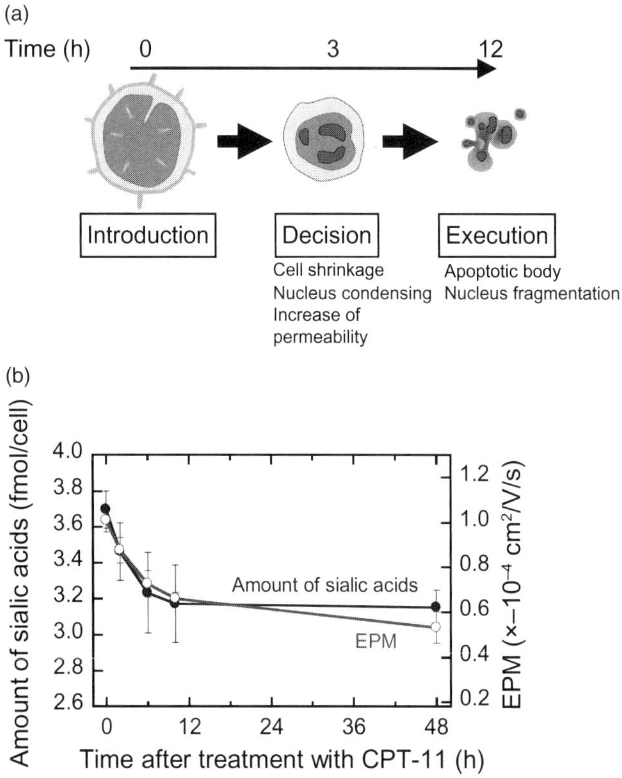

Figure 41.4. Effect of CPT-11-induced apoptosis on the EPM of HEK cells. (a) Schematic of the apoptosis of HEK cells. (b) Time course of average EPM and amount of sialic acids of HEK cells after treatment with CPT-11. Each value represents the mean of 10–69 experiments (average EPM) and 4 experiments (amount of sialic acids).

The EPM of human embryonic kidney (HEK) cells was measured after apoptosis induction by treatment with irinotecan hydrochloride (CPT-11), an anticancer drug that has a DNA synthesis inhibitory effect [9]. Significant changes in cell shape were not observed upon optical observation even 10 hours after treatment with CPT-11. In contrast, the cell EPM was found to begin decreasing upon drug application. In order to investigate the mechanism of cell EPM reduction, the amount of sialic acids in a cell was measured by reverse-phase high-performance liquid chromatography (HPLC) after labeling them with 1,2-diamino-4,5-methyleneoxybenzene. As shown in Figure 41.4b, the amount of sialic acids in cells decreased with decreasing cell EPM. The decrease in the negative charge of the cell surface was confirmed. Hence, cell EPM can be used as an indicator of changes in the cell surface induced by apoptosis.

41.3.3 Enzyme Reaction

Cell dissociation agents such as proteases are widely used to detach cells from a culture dish during the cultivation of adherent cells. However, proteases may harm the cell surface, and it is not easy to estimate the degree of damage. Therefore, a quantitative study was performed to evaluate the effects of a protease on the cell surface. Following treatment with 0.5% trypsin, the effects on human epithelioid carcinoma (HeLa) cells (1×10^5 cells) were evaluated using a phase contrast microscope by observing proliferation potency. Soon after trypsin addition into a culture dish, dramatic changes in the pseudopodia of the cells were observed over time, and in 5 minutes, the cells detached from the culture plate. The proliferation activity of the cells (see Fig. 41.5a) decreased

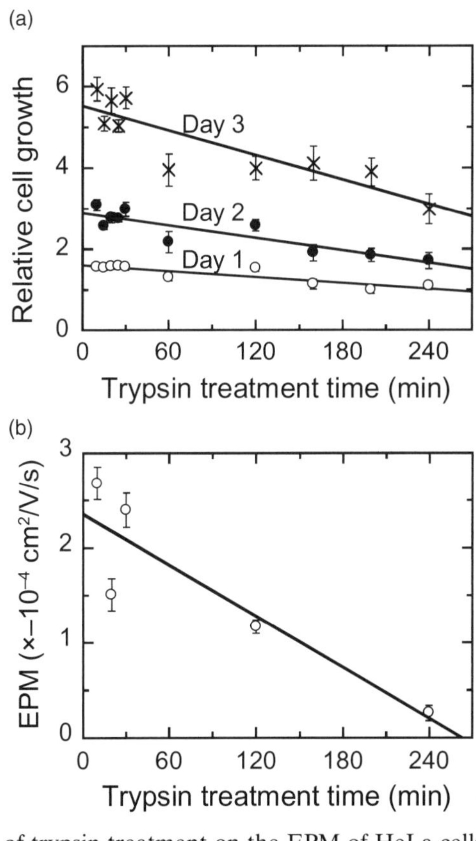

Figure 41.5. Effect of trypsin treatment on the EPM of HeLa cells. (a) Cell proliferation after trypsin treatment. The cell number was determined 0, 1, 2, and 3 days after trypsin treatment and normalized with the cell number on day 0. Each point represents the mean of 20 experiments. (b) EPM of cells after trypsin treatment. Each value represents the mean with an SE of 30–77 experiments.

gradually with the duration of trypsin treatment. These changes in the shape and proliferation activity of the cells after trypsin treatment indicate that the cell surface was damaged. On the other hand, the EPM of the cells decreased with increasing trypsin treatment duration as shown in Figure 41.5b [7]. Note that the cell EPM, as well as the proliferation activity of the cells, monotonically varied with increasing trypsin treatment duration. Thus, these results suggest the feasibility of cell EPM measurement to evaluate enzyme-induced changes on the cell surface, not only qualitatively but also quantitatively.

41.3.4 Leukocyte Differentiation

Leukocytes are immune system cells that eliminate a foreign substance. Different types of leukocytes such as lymphocytes, granulocytes, and monocytes circulate in the blood. Leukemia cell lines are often used for investigating interactions between biomaterials and leukocytes. However, leukemia cell lines and primary leukocytes differ. Hence, it is necessary to measure the EPM of various types of leukocytes to accumulate fundamental knowledge concerning their surface charge. The EPMs of four types of mouse leukemia cell line and three types of mouse primary leukocyte were measured by on-chip cell electrophoresis in order to investigate their surface charges [10]. Figure 41.6

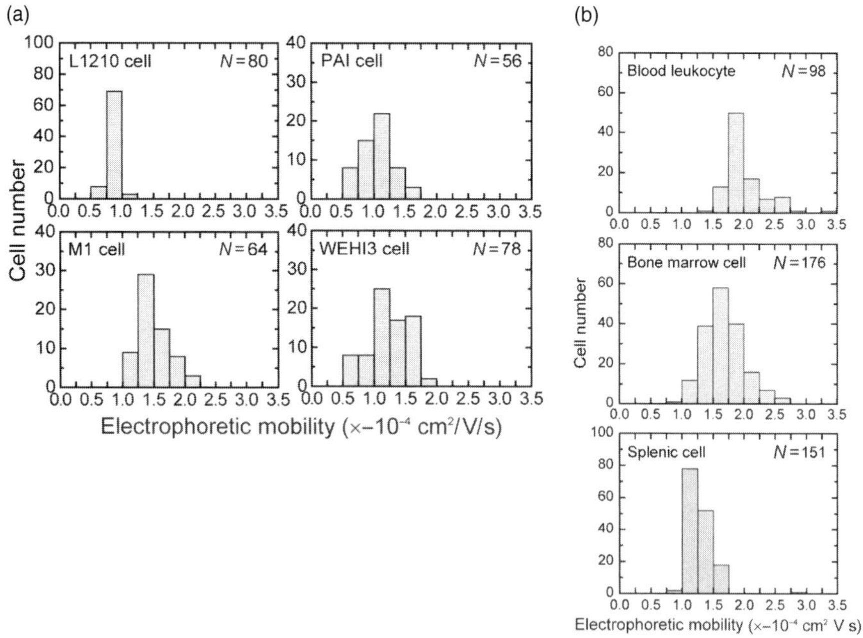

Figure 41.6. (a) EPM histograms of leukemia cell lines: L1210, PAI, M1, and WEHI3 cells. (b) EPM histograms of mouse primary leukocytes: blood leukocyte, bone marrow cells, and splenic cells.

TABLE 41.1. Average EPMs Derived from the Histogram Data Presented in Figure 41.6

Cells	Disease	Mice	EPM ($cm^2/V/s$) (Mean with SD)
L1210 cell	Acute lymphatic leukemia	DBA	$-0.83 \times 10^{-4} \pm 0.085 \times 10^{-4}$
PAI cell	Myeloma	BALB/c	$-1.0 \times 10^{-4} \pm 0.24 \times 10^{-4}$
M1 cell	Myelocytic leukemia	SL	$-1.5 \times 10^{-4} \pm 0.25 \times 10^{-4}$
WEHI3 cell	Myelomonocytic leukemia	BALB/c	$-1.2 \times 10^{-4} \pm 0.32 \times 10^{-4}$
Blood leukocyte	Normal	C57BL	$-2.0 \times 10^{-4} \pm 0.32 \times 10^{-4}$
Bone marrow cell	Normal	C57BL	$-1.7 \times 10^{-4} \pm 0.32 \times 10^{-4}$
Splenic cell	Normal	C57BL	$-1.3 \times 10^{-4} \pm 0.20 \times 10^{-4}$

shows the EPM distributions of the cells. Table 41.1 lists the statistical parameters derived from the histogram data presented in Figure 41.6. The average EPMs of L1210, PAI, M1, and WEHI3 cells are -0.83×10^{-4}, -1.0×10^{-4}, -1.5×10^{-4}, and -1.2×10^{-4} $cm^2/V/s$, respectively. The average EPMs of primary blood leukocytes, bone marrow cells, and splenic cells are -2.0×10^{-4}, -1.6×10^{-4}, and -1.3×10^{-4} $cm^2/V/s$, respectively. The EPM was different for each type of leukocyte. These quantitative data concerning cell surface charge are useful fundamental knowledge for understanding the interactions between cells and biomaterials, and for designing biomaterials.

Further experiment was performed in order to investigate whether such system was applicable to the monitoring of the progress of cellular differentiation. The EPM of human promyelocytic leukemia cell lines (HL60) during neutrophil differentiation induced by all-trans retinoic acid (ATRA) and dimethyl sulfoxide (DMSO) was measured. In order to confirm the action of the two differentiation inducers, the cells were observed using a light microscope at 2, 4, and 6 days after the initiation of the neutrophil differentiation. In each observation, 100 sample cells were stained by Wright–Giemsa stain and were classified morphologically to examine the ratio of differentiated cell subpopulations. Ninety-four percent of the undifferentiated HL60 cells were revealed to be myelocyte. In contrast, the composition of cells cultured with ATRA or DMSO showed temporal changes. Subsequently, the EPM of HL60 cells was measured using an on-chip cell electrophoresis system before and after neutrophil differentiation. The EPM changed gradually with the progress of the neutrophil differentiation. In order to investigate the relationship between the differentiational state and EPM, the experimental data were analyzed by principal component analysis (PCA). As shown in Figure 41.7, it was revealed that there is a strong correlation between morphological classification and EPM during the neutrophil differentiation. The result suggests

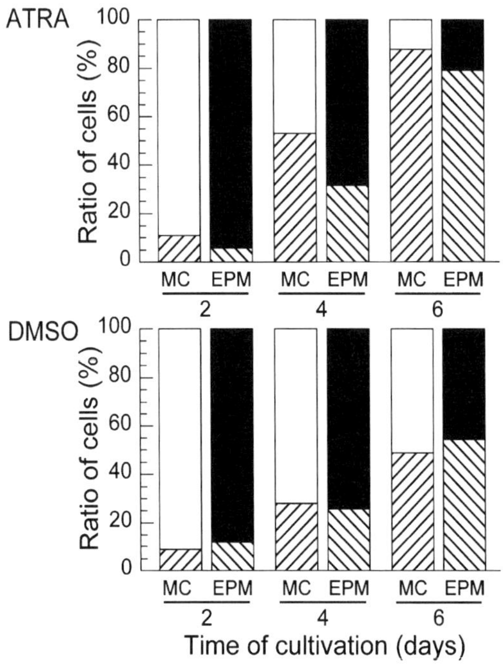

Figure 41.7. Correlation between morphology and cell EPM during neutrophilic differentiation. □, myelocytes and promyelocytes; ▨, metamyelocytes, band, and segmented neutrophils; ▧, cells whose EPM were between 0 and -1.1×10^{-4} cm^2/V/s; ■, cells whose EPMs were between -1.1×10^{-4} and -3.0×10^{-4} cm^2/V/s. MC, morphological classification.

that on-chip EPM measurement systems can be used as a monitoring tool for neutrophil differentiation.

41.3.5 Mesenchymal Stem Cell (MSC) Differentiation

MSCs are used or are under study for clinical applications since they can self-renew and can differentiate into various cell types *in vivo* and *in vitro*, such as regenerate a skeleton. Considering the safety of medical treatment, it is considered that the early establishment of a noninvasive inspection method of all cells before use becomes necessary in order to check the contamination, infection, and unfavorable transformation during *in vitro* cultivation. Hence, the measurement of cell EPM changes with the progress of differentiation from human mesenchymal stem cells (hMSCs) to osteogenic reparative cells was studied [11].

To induce the osteogenic differentiation of human umbilical cord blood-derived MSCs with human telomerase reverse transcriptase (hUCBTERT-21 cells) (JCRB Cell Bank, Osaka, Japan), the cells were cultured for 21 days

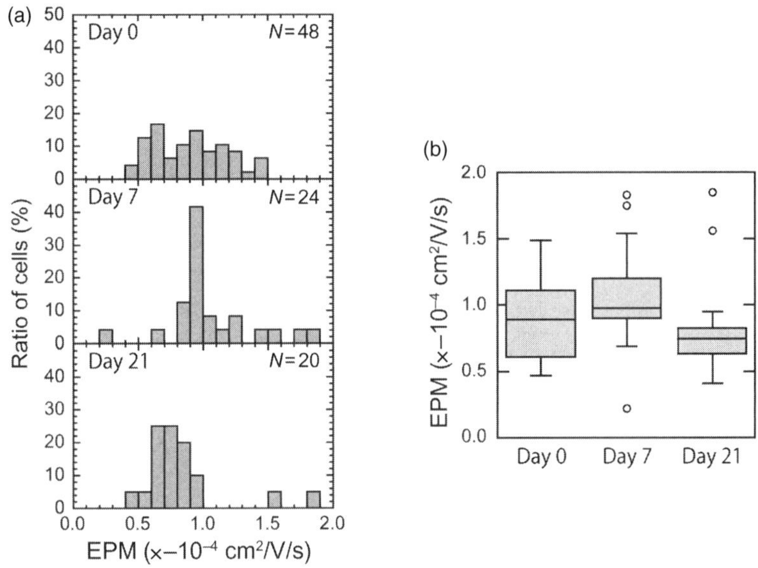

Figure 41.8. (a) EPM histograms of hUCBTERT-21 cells at 0, 7, and 21 days after the initiation of osteogenic differentiation. The EPMs of the cells were significantly different among the groups ($p < 0.01$, Kruskal–Wallis test). (b) EPM data represented by a box plot. Each box represents the interquartile range (IQR). The bottom and top of the box are the lower quartile (25%) and the upper quartile (75%), respectively. The band in the box is the median (50%). Whiskers show the sample minimum and sample maximum. Circles indicate the outliers that are lower and higher data without 1.5 IQR of the lower and upper quartiles, respectively.

following the protocol described in Reference 15. After confirmation of the osteogenic differentiation of hUCBTERT-21 cells by the appearance of calcium phosphate precipitation in the cultivation dish, the effect of osteogenic differentiation on the EPM of the cells was examined. As shown in Figure 41.8a, the EPM of undifferentiated cells (day 0) followed a trapezoidal-like distribution. In contrast, the EPM changed gradually into a bell-shaped distribution with the progress of differentiation. When the EPM distributions were compared statistically using Kruskal–Wallis tests, they were significantly different among the groups ($p < 0.01$), but the biological meaning is unclear.

For a comprehensive analysis, the data in Figure 41.8a are replotted in a box plot, which is a graphical representation of data distribution developed by J.W. Tukey (Fig. 41.8b). Here, each box represents the interquartile range (IQR), which is the difference between the 25% and 75% of the distribution. The band in the box is the median (50%). Whiskers show the sample minimum and sample maximum, and circles indicate the outliers that are lower and higher data without 1.5 IQR of the lower and upper quartiles, respectively. It is noteworthy that the range of the distribution decreased gradually with the progress of differentiation. Recently, Chang et al. have studied how blood stem cells commit to become white or red blood cell progenitors and reported significant findings that show that stem cell populations maintain a built-in variability [16]. The time-dependent decrease in EPM heterogeneity observed in this study seems to reflect the process by which stem cells lose their built-in variability with the progress of differentiation. Although some studies have demonstrated that expression of surface proteins, for example, osteocalcin, osteonection, collagen I, osteopontin, and CD106, of hMSC change after osteogenic differentiation, it was unclear about cell surface charge density. It was found for the first time that the EPM distribution of hMSCs significantly changed with the progress of osteogenic differentiation. The present result suggests that the EPM distribution can be potentially used as a good index of the phenotypic variability of stem cell populations.

REFERENCES

1 J. N. Mehrishi, J. Bauer. *Electrophoresis* 23 (2002) 1984–1994.
2 G. G. Slivinsky, W. C. Hymer, J. Bauer, D. R. Morrison. *Electrophoresis* 18 (1997) 1109–1119.
3 T. Yamada, M. Yamada. *Nature* 244 (1973) 297–299.
4 K. Makino, T. Taki, M. Ogura, S. Handa, M. Nakajima, T. Kondo, H. Ohshima. *Biophys. Chem.* 47 (1993) 261–265.
5 T. Ujiie, T. Kikuchi, T. Ichiki, Y. Horiike. *Jpn. J. Appl. Phys.* 39 (2000) 3677–3682.
6 T. Akagi, M. Suzuki, T. Ichiki. *Jpn. J. Appl. Phys.* 45 (2006) L1106–L1109.
7 T. Akagi, M. Suzuki, T. Sato, T. Ichiki. *Jpn. J. Appl. Phys.* 46 (2007) 6404–6409.
8 T. Akagi, A. Minamino, T. Ichiki. *Electron. Comm. Jpn.* 91 (2008) 40–45.
9 T. Akagi, T. Ichiki. *Anal. Bioanal. Chem.* 391 (2008) 2433–2441.
10 T. Akagi, Y. Hamada, T. Ichiki. *Trans. Mater. Res. Soc. Jpn.* 35 (2010) 259–262.
11 R. Matsuhashi, T. Akagi, T. Ichiki. *Trans. Mater. Res. Soc. Jpn.* 35 (2010) 263–266.
12 G. M. Whitesides. *Nature* 442 (2006) 368–373.
13 K. Ishihara, H. Oshida, Y. Endo, T. Ueda, A. Watanabe, N. Nakabayashi. *J. Biomed. Mater. Res.* 26 (1992) 1543–1552.
14 Y. Hamada, T. Ono, T. Akagi, K. Ishihara, T. Ichiki. *J. Photopolymer Sci. Technol.* 20 (2007) 245–249.
15 M. Terai, T. Uyama, T. Sugiki, X. K. Li, A. Umezawa, T. Kiyono. *Mol. Biol. Cell* 16 (2005) 1491–1499.
16 H. H. Chang, M. Hemberg, M. Barahona, D. E. Ingber, S. Huang. *Nature* 453 (2008) 544–547.

42

SURFACE CHARACTERISTICS AND ATTACHMENT BEHAVIORS OF BACTERIAL CELLS

Hisao Morisaki

In natural environments, most bacteria are attached to various surfaces. Attachment leads to the formation of biofilm, which can influence the surrounding environment either beneficially or harmfully [1, 2]. For example, the biofilm formed on a solid surface can purify environmental water [3–5], whereas that on medical instruments can lead to nosocomial infections [6, 7]. Elucidating the mechanism underlying bacterial attachment is crucial for promoting or preventing biofilm formation, as well as for understanding bacterial ecology.

Under typical conditions (ca. pH 7), both the bacterial cells and the substratum surface are usually negatively charged. Therefore, under these conditions, two opposite kinds of interaction exist: electrostatic repulsive interaction and van der Waals attractive interaction. For the cell attachment, the surface properties of bacterial cells relating to these interactions have important influences.

In this chapter, first, the bacterial cell surface properties, in terms of surface charge and hydrophobicity, are described by relating with cell attachment. Second, the mechanism of cell attachment, including recent advances, is discussed. Third, several aspects noteworthy for future study are mentioned.

Electrical Phenomena at Interfaces and Biointerfaces: Fundamentals and Applications in Nano-, Bio-, and Environmental Sciences, First Edition. Edited by Hiroyuki Ohshima.

42.1 SURFACE PROPERTIES OF BACTERIAL CELLS AND CELL ATTACHMENT

42.1.1 Cell Surface Charge and Hydrophobicity

42.1.1.1 Cell Surface Charge. Bacterial cell surface charge originates from dissociation or protonation of carboxyl, phosphate, and amino groups and consequently depends on pH [8, 9]. At physiological pH values (between pH 5 and 7), bacteria in aqueous suspension are almost always negatively charged due to an excess of carboxyl and phosphate groups than amino groups [10] with a few exceptions in which bacterial cells exhibit a positive charge at their physiological conditions [11–14].

The bacterial surface charges vary depending on the surface chemistry of these functional groups. The cell surface chemistry can be analyzed with methods such as chemical modification [15], Fourier transform infrared (FT-IR) [16], and X-ray photoelectron spectroscopy (XPS) analysis [17].

42.1.1.2 Cell Surface Hydrophobicity. Bacterial hydrophobicity is the tendency of a bacterial cell to interact with cells of similar hydrophobicity as opposed to water [18], and in the case of bacterial cells, this refers to the hydrophobic properties of the surface of the cell. The hydrophobicity of a bacterial cell is largely influenced by the residues and structures on the surface of the cell, which can be either hydrophilic or hydrophobic [18]. The hydrophobicity or hydrophilicity of bacterial cells affects greatly cell attachment as well as surface charge [19].

42.1.2 Surface Properties and Cell Attachment

Under low ionic strength conditions, electrostatic repulsion generally dominates [20] to hinder bacterial attachment because mostly both bacterial cells and substrate surfaces are negatively charged in normal conditions, and these charges are less shielded in lower ionic strengths.

In numerous cases, increased hydrophobicity of bacterial surfaces or substrate surfaces tended to result in an increased number of attached cells [10, 21–23]. A hydrophobic surface with less adsorbed water molecules seems to need less energy for two surfaces contact together.

The role of electrostatic interaction and hydrophobicity on bacterial cell attachment can be elucidated by the studies, as shown next, investigating cell attachment under various conditions different in pH, ionic strength and materials used.

The attachment of a marine *Pseudomonas* sp. to a variety of surfaces was studied, and the number of attached bacterial cells was related to the surface charge and degree of hydrophobicity of the substratum [24]. Large numbers of bacteria attached to hydrophobic plastics with little or no surface charge (Teflon, polyethylene, polystyrene, poly[ethylene terephthalate]); moderate numbers attached to hydrophilic metals with a positive (platinum) or neutral (germanium) surface charge; and very few attached to hydrophilic, negatively charged substrata (glass, mica, oxidized plastics). The results suggest that both electrostatic and hydrophobic interactions are involved in bacterial attachment.

Satou et al. [25] reported that oral streptococci adhered most to the surface of the least negative potential. Van der Mei et al. [26] studied the adhesion of six coagulase-negative staphylococci strains to a positively and negatively charged acrylate surfaces with similar wettability. They found a decrease in attachment with increasing electrostatic repulsion between bacteria and negatively charged surfaces. Piette and Idziak [27] reported that *Pseudomonas fluorescens* attached to meat surface (tendon slice) nonspecifically in a larger number at higher ionic strength. At high ionic strength (0.15 M phosphate-buffered saline), bacterial (*Pseudomonas fragi* and *Brochothrix thermosphacta*) attachment to lean beef tissue was reduced, but not at 0.015 M [28].

Takashima and Morisaki [29] studied the attachment of *Pseudomonas syringae* cells to a glass plate as a function of pH. The surface charge of the cells and the glass plate was measured as the electrophoretic mobility (EPM) of the cells and electroosmotic liquid velocity at the surface of the glass plate. It was revealed that more cells attached at a lower electrostatic repulsion between the cells and the glass plate.

The swimming cells of *Escherichia coli* near a glass surface attached in larger numbers irreversibly at higher ionic strengths than at lower ionic strengths [30]. The cells of *Listeria monocytogenes*, *Salmonella typhimurium*, *Staphylococcus aureus*, and *E. coli* attached much more to hydrophobic or positively charged surfaces than to hydrophilic noncharged surfaces [31].

42.1.3 Cell Surface Properties and Attachment for a Large Number of Strains

Although many studies have been done concerning bacterial cell surface properties (surface charge and hydrophobicity) to correlate with cell attachment, majority of them comprise only two or fewer strains [32]. Thus, the studies as shown next including relatively many strains may be useful to deduce the relationship between the cell surface properties and cell attachment.

The adhesion of 10 strains of rhizobacteria to rice root and polystyrene was examined [33]. The adhesion of gram-negative rhizobacteria was virtually absent for strains with zeta potentials of more than –25 mV and increased for the strains with less negative potential. However, the adhesion of gram-positive rhizobacteria was low regardless of their zeta potentials.

Among 16 bacterial strains, more hydrophobic cells adhered to sulfated polystyrene to a greater extent than hydrophilic cells [34]. Among 50 strains of *Lactococcus lactis*, hydrophobic or low negatively charged strains exhibited much more adhesion to polystyrene plates at a low ionic strength of 1.5 mM [35]. It is also noteworthy that the increase in ionic strength from 1.5 to

150.0 mM led to a significant increase in adhesion for the strains belonging to the electronegative hydrophilic group. This is presumably due to the attenuation of electrostatic repulsion between the negatively charged bacterial cells and the negatively charged polystyrene surface.

Rhodococcus spp. cells of 52 strains were investigated with the relation between colony morphotype and extracellular polysaccharide (EPS, a hydrophilin) production [36]. Strains of rough colony morphotype were less in EPS production, more hydrophobic, adhered well to various materials, and formed cell aggregates.

The diversity and a tendency of the surface charge of 101 bacterial strains isolated from a grassland soil were reported [37]. As shown in Figure 42.1a, the fast-growing bacteria showed greater negative charge (measured as EPM) than slow-growing bacteria. The pH dependence of the EPM of the fast-growing bacteria showed greater change than that of slow-growing ones. These are the case for the 109 bacterial strains isolated from paddy field soil [38].

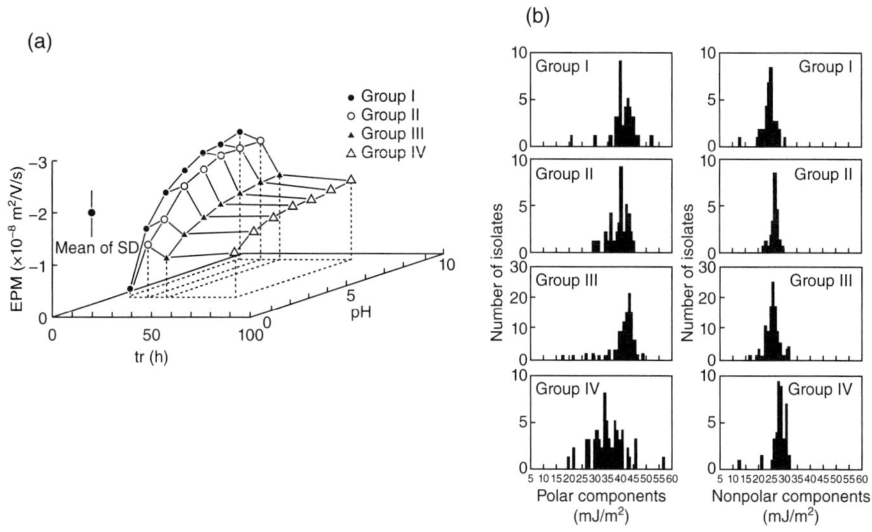

Figure 42.1. (a) Changes in the electrophoretic mobility of grassland bacteria as a function of pH and tr (colony-forming time; the larger, the slower growing). The mean of tr values were 9.3, 17.7, 28.2, and 63.2 hours for groups I, II, III, and IV, respectively. •, mean of 22 isolates of group I; ○, 23 isolates of group II; ▲, 22 isolates of group III; Δ, 34 isolates of group IV, ⸸, mean of standard deviation (SD) of the EPM of four groups at seven different pH values (0.46×10^{-8} m^2/V/s, $n = 28$). Reprinted with permission from the Journal of General and Applied Microbiology, The cell surface charge of fast- and slow-growing bacteria isolated from grassland soil. Vol. 39 (1993) 65–74. (b) Distribution of the polar and nonpolar components of surface tension for the isolates divided into four groups according to their colony-forming time on diluted nutrient broth agar plate. The mean values of tr were 9.7, 21.2, 35.5, and 80.0 hours for groups I, II, III, and IV, respectively. Reprinted with permission from the Journal of General and Applied Microbiology, Hydrophobicity of the cells of fast- and slow-growing bacteria isolated from a grassland soil. Vol. 39 (1993) 381–388.

Two hundred fifty isolated strains from grassland soil were examined for their cell hydrophobicity, and it was found that slow growers are more hydrophobic at their cell surfaces than fast growers as shown in Figure 42.1b [39].

Bacterial strains (ca. 90 strains) isolated from sand particles showed interesting relationships between cell surface properties and attachment strength [40]. The fraction of cells obtained by a strong detachment force, sonication, contained slow growers in a higher proportion. The slow growers were proved to be less negative and more hydrophobic, as well as for bacterial strains isolated from grassland soil and paddy soil [37–39]. The slow growers attached much more to a glass and plastic surface compared with fast growers. The slow growers may have such cell surface properties as to attach firmly to survive the struggle against the fast growers, which weakly attach and easily move to a new habitat and compete with other organisms by growing fast.

42.1.4 Attachment Force

The measurement of interaction force between bacterial cells and substratum surface is an important subject for developing the methods to promote or prevent bacterial attachment as well as for advancing theory by comparing the theoretical prediction with the force actually measured.

If the charge properties of the substratum surface can be changed without affecting other substratum surface properties, we can know the role of electrostatic interaction on cell attachment quantitatively. However, the charge properties of substratum surface are difficult to vary independently [20]. Morisaki et al. [41] circumvented this difficulty by using an electrode as a substratum surface for cell attachment. The cell attachment of *P. syringae* to transparent electroconductive glass plates in a parallel-plate flow chamber in an external electric field applied perpendicular to the substratum surface was studied. A microbial cell suspension was flowed through the chamber. The time course of the cell attachment in a 10 mM phosphate buffer solution (pH 7.0) to the glass plate was investigated at various electric field strengths ranging from –4.2 to +4.1 V/cm (Fig. 42.2). The attachment rate and the maximum number of attached cells increased linearly with the increase in the strength of the positive electric field. In contrast, the rate and the number of cells decreased linearly with the decrease in the strength of the negative electric field. By extrapolating the attachment data, the negative electric field strength where the cell attachment was fully impeded was determined. From this value and the bacterial zeta potential, the electrostatic repulsion fully preventing bacterial attachment, that is, the microbial attachment force, was calculated to be 5.0×10^{-11} N per cell of average size (0.56×13.4 μm). In a slightly different experimental setup, detachment of the bacterial cells from the quartz plate was studied [42]. The force necessary to remove the bacterial cells of *Bacillus* sp. attached to a quartz plate was measured by using a well-defined liquid flow caused by electroosmosis. The force ranged about 10^{-14} to 10^{-12} N per cell and revealed to be greater at lower pH values due to a decrease in the negative

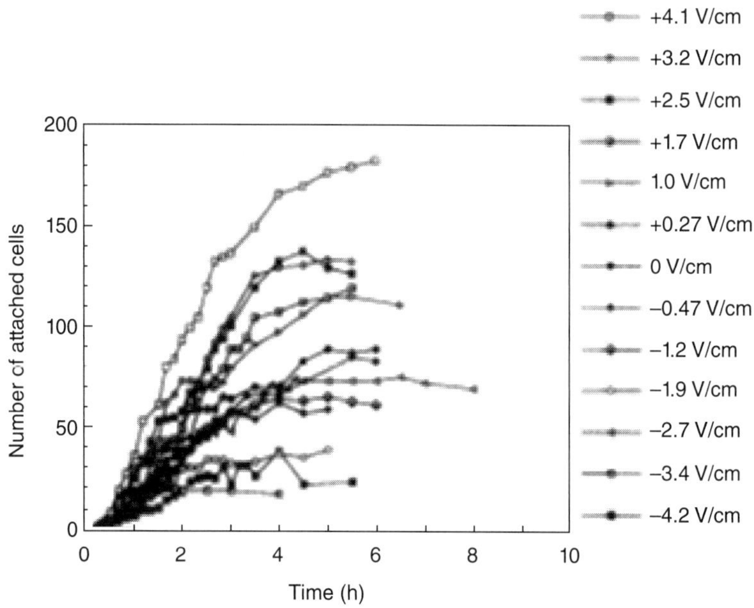

Figure 42.2. The time course of cell attachment to the electroconductive glass surface under various electric field strengths. For field strengths of +1.0, 0, and −1.2 V/cm, the mean values of two experiments are plotted. Reprinted with permission from Elsevier from Hisao Morisaki, Kunihiko Nakagawa, Haruki Shiraishi, Measurement of attachment force of microbial cells. Colloids and Surfaces B: Biointerfaces, 6 (1996) 347–352.

charge of both the surface of bacterial cells and the quartz plate. An exponential relation was observed between the removal force and the number of removed cells in media of pH 7–9 (ionic strength: 10 mM); more cells were removed by a smaller force.

Using a parallel-plate flow chamber, the hydrodynamic shear force to detach adhered bacterial cells from hydrophilic glass and hydrophobic dimethyldichlorosilane-coated glass was measured in a 10 mM potassium phosphate buffer [43]. The force varied from 0.31 to over 19.64×10^{-12} N per cell of various species, that is, *Staphylococcus epidermidis*, *Raoultella terrigena*, *Streptococcus thermophilus*, and *Pseudomonas aeruginosa*, generally greater for the hydrophobic plate. With the other studies included, it was summarized as follows [43]: (i) The shear force required to stimulate detachment of adhered cells is generally higher (ranging from 0.1 to 42.4×10^{-12} N per cell of various bacterial cells and substratum surfaces [10, 44–46]) than the force to prevent the adhesion [46–52], suggesting strengthening of attachment after initial attachment; (ii) perpendicularly oriented interaction forces measured with optical tweezers remain in piconewton (10^{-12} N) per cell range [53–55] as well as the forces obtained by other methods, while atomic force microscopy yields stronger forces, generally in nanonewton (10^{-9} N) range [56–59].

Attachment force measured with various methods scattered over a wide range, 0.1–49,500 × 10^{-12} N per cell according to Boks et al. [43]. The study under well-defined experimental conditions such as bacterial species and strains, culture conditions, suspending medium conditions (pH, ionic strength, and temperature), flowing velocity of the medium, and substratum solid for attachment, will give the values that can be utilized for comparison with theory and that are useful to deepen our understanding about the attachment mechanism.

42.2 ATTACHMENT MECHANISM

42.2.1 Theory of Bacterial Attachment

Bacterial attachment phenomena in connection with cell surface properties have been important subjects in various fields such as food processing and medical biotechnology and in environmental engineering of bioremediation and wastewater treatment. Here, the theories of the bacterial attachment are outlined, and a recent progress treating the bacterial cells as soft living colloidal particles and the consequent reconsideration about the bacterial cell attachment mechanism are introduced and discussed: For details of the development of the cell attachment mechanism, refer to several reviews [9, 10. 20, 60–62].

42.2.1.1 The Thermodynamic Approach. When a bacterial cell (b) suspending in a medium (m) contacts with the surface (s), a new interface is formed between (b) and (s) and two interfaces are lost between (b) and (m) and between (s) and (m). So, the free energy change caused by the cell attachment (ΔG_{att}) is expressed as follows:

$$\Delta G_{att} = \gamma_{bs} - \gamma_{bm} - \gamma_{sm},$$

where γ_{bs}, γ_{bm}, and γ_{sm} are the bacterial cell–substratum, bacterial cell–medium, and substratum–medium interfacial free energy, respectively. The thermodynamic approach helps to explain a common observation: hydrophobic (hydrophilic) bacterial cells prefer a hydrophobic (hydrophilic) substratum surface for attachment [63]. However, obviously, the equation gives no information for the cell approaching the surface.

42.2.1.2 The Classical Derjaguin–Landau–Verwey–Overbeek (DLVO) Theory. When a bacterial cell in a medium approaches a substratum surface, various interactions will occur between the cell and the surface depending on the distance between them. The classical DLVO theory describes the interaction energy between the interacting surfaces based on Lifshitz–van der Waals and electrostatic interactions and their decay with separation distance [64, 65]. Many researchers have found a link between decreasing bacterial attachment and decreasing ionic strength, which is consistent with the classical DLVO theory [30, 57, 66–72].

42.2.1.3 Extended DLVO Theory. Van Oss et al. [73] introduced a so-called extended DLVO theory by including short-range Lewis acid–base interactions in the classical DLVO theory. By the inclusion of acid–base interactions, attractive hydrophobic interactions and repulsive hydration effects can be accounted for. In some cases, the extended DLVO theory seems to qualitatively predict experimental adhesion results better than the classical DLVO theory [20, 60, 74]. However, the distance dependence of acid–base interactions is also relatively short ranged; a distance less than 5 nm is required for acid–base interactions can become operative [20].

42.2.2 Bacterial Cells as Soft Living Colloidal Particles

42.2.2.1 Cell Motility and Attachment Rate. Kogure et al. [75] reported that the attachment rate of *Vibrio alginolyticus* cells to a glass surface is dependent on the swimming speed of the cells. This was also the case for other bacterial strains [76]. Although a long-standing question, if motility facilitates attachment, has been answered, another unsolved problem remained: how the bacteria can overcome the energy barrier, estimated to be about 10^2 times greater than the thermal kinetic energy of the bacterial cell or about 10^6 times greater than the cell swimming energy according to the DLVO theory [77] due to electrostatic repulsion between the cells and glass. To solve this problem, a characteristic electrokinetic property of bacterial cells must be carefully considered as discussed next.

42.2.2.2 Soft Colloidal Particles, Bacterial Cells. For colloidal particles, including bacterial cells, their zeta potential (ζ) has been calculated from their EPM (μ) by the Smoluchowski equation: $\mu = \varepsilon_r\varepsilon_0\zeta/\eta$, where ε_r is the relative permittivity of the medium in which the cells are suspended, ε_0 is the permittivity of a vacuum, and η is the viscosity of the medium. The zeta potential has been assumed to be the surface potential of the particle. The surface potential (Ψ_0) is related with the surface charge density (σ) by the following equation: $\Psi_0 = \sigma/\varepsilon_r\varepsilon_0\kappa$, where κ is the Debye–Hückel parameter, depending mainly on the square root of electrolyte concentrations. The combination of these equations gives $\mu = \sigma/\eta\kappa$. This equation predicts that the EPM will approach to zero value by decreasing proportionally to the square root of the increasing ionic strength for particles like bacteria whose surface charge is due to the ionization of functional groups. However, the observations for various kinds of biological cells [29, 78–81] hardly obey this prediction. The cell EPM of a bacterial strain gradually approached a nonzero value with increasing ionic strength.

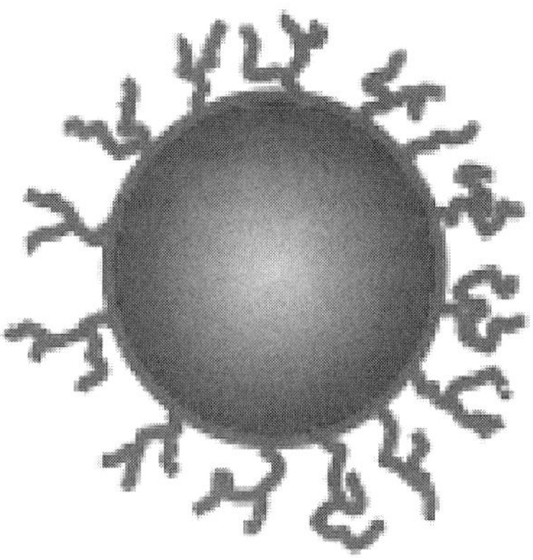

Figure 42.3. A schematic illustration of a soft colloidal particle covered with a polymer layer.

42.2.2.3 Ohshima's Soft Particle Theory. Ohshima [82, 83] has pointed out that the Smoluchowski equation cannot be applied to "soft" particles having polymers at their surfaces as schematically illustrated in Figure 42.3: The equation seems to be limited to "rigid" particles having no polymer. He solved the problem concerning soft particles whose EPM approaches a nonzero value with increasing ionic strength by presenting a theory formulated as follows:

$$\mu = (\varepsilon_r \varepsilon_o/\eta)[(\Psi_o/\kappa_m + \Psi_{DON}/\lambda)/(1/\kappa_m + 1/\lambda)] + (zeN/\eta\lambda^2),$$

where Ψ_{DON} is the Donnan potential of the polymer layer, κ_m can be interpreted as the Debye–Hückel parameter of the polymer layer, λ is a parameter whose reciprocal has the dimension of length corresponding to a softness of the polymer layer, z is the valence of the charged group in the polymers, e is the electric unit charge, and N is the density of the charged groups in the polymer layer.

A number of electrophoresis measurements have been carried out on various types of soft particles, which include hydrogel-coated latex particles and biological cells, showing that the electrophoretic behaviors of these particles can be explained well by the electrophoresis theory of soft particles [84].

42.2.2.4 Disappearance of the Energy Barrier for Soft Colloidal Particles, Bacterial Cells. *V. alginolyticus* cells attached to a glass surface depending on their swimming speed as if there were no energy barrier between the cell and the glass surface [75, 76]. The surface potential of *V. alginolyticus* obtained from Ohshima's theory was surprisingly small compared to that calculated by the Smoluchowski equation [76] (the surface potentials of three *V. alginolyticus* strains were –0.5 to –0.7 mV by Ohshima's theory and –21 to –23 mV by the Smoluchowski equation at an ionic strength 0.05 M); similar calculations

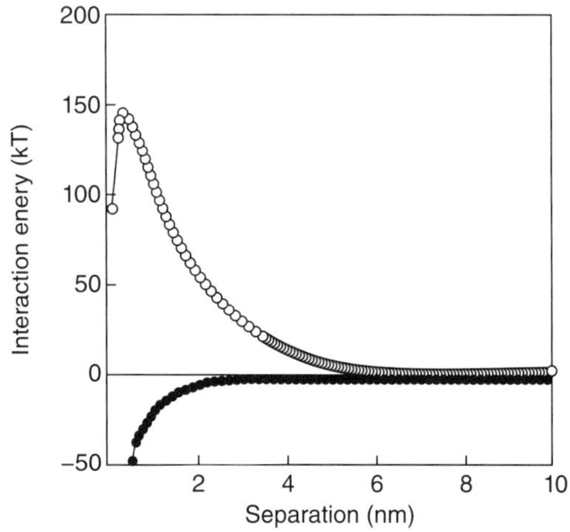

Figure 42.4. Interaction energy between microbial cells and a glass surface as a function of separation between them at an ionic strength 0.05 M. The values in two cases, a greater negative level of surface potential (−20 mV, ○) and a smaller negative potential (−1 mV, •), were plotted. The interaction energy was divided by the thermal kinetic energy (kT) at 25 °C, where k is the Boltzmann constant and T is the absolute temperature. Reprinted with permission from the Society for General Microbiology, from Hisao Morisaki, Shinichi Nagai, Hiroyuki Ohshima, Eiko Ikemoto, Kazuhiro Kogure, The effect of motility and cell-surface polymers on bacterial attachment. Microbiology 145 (1999) 2797–2802.

have been reported in other studies [85–88]. This leads to a conclusion that we have overestimated the height of the energy barrier between bacterial cells and surfaces. Figure 42.4 exemplifies the difference in the energy profile between two cases: a larger negative level of cell surface potential obtained from the Smoluchowski equation and a smaller surface potential from Ohshima's theory. In the case wherein the energy barrier becomes smaller than the thermal kinetic energy of the cells, the attachment rate of the cells to a substratum surface depends on the frequency of cell collision to the surface, hence on the swimming speed of the cells, as reported [75, 76].

42.2.2.5 Cell Attachment over a Wide Range of Ionic Strength. The height of the energy barrier between colloidal particles and substratum surfaces varies depending on the ionic strength of the medium. When the ionic strength is quite low, a small energy barrier may remain even for a soft colloidal particle, while the barrier disappears at higher ionic strength.

Suppose bacterial cells (concentration c) are suspended and a glass plate is immersed in a medium. When the cells travel a distance, L, on average in a given time, t, along a direction toward a glass plate, the frequency of cell collision to the glass surface (area S) in the given time is $cSLF_1$ (F_1 is the fraction of the cells moving toward the glass), in case of no energy barrier between the cells and the glass plate. If an energy barrier $\Delta G^{\neq}$ remains, a part of the cells, because of Boltzmann distribution, can surmount the barrier. In this case, the number of cells able to surmount the barrier and hit the glass surface is given by $ce^{-\Delta G^{\neq}/RT}SLF_1$, where R is the gas constant and T is the absolute temperature. If a part of the collisions, F_2, results in cell attachment to the surface, the number of cells, $A(t)$, attached to the surface area S in the given time is expressed as follows:

$$A(t) = ce^{-\Delta G^{\neq}/RT}SLF_1F_2.$$

When the cells can only move by Brownian motion, L in the given time t will be $(2Dt)^{1/2}$ according to Einstein's diffusion theory, where D is the diffusion constant of the cells in the medium. For these cells, the number of attached cells is $A(t) = ce^{-\Delta G^{\neq}/RT}S(2Dt)^{1/2}F_1F_2$. This equation predicts the following: (i) The number of attached cells will increase linearly along with the square root of time; (ii) the number of attached cells will increase along with increasing ionic strength due to the decrement in the energy barrier; (iii) the number of attached cells will reach a plateau over a specific ionic strength where the energy barrier disappears. These predictions match well with the experimental results obtained for bacterial cells whose movement is limited to the Brownian motion [89].

In the case of motile cells, L will be determined by both the Brownian motion and cell motility; thus, $A(t)$ will be

$$A(t) = ce^{-\Delta G^{\neq}/RT}S((2Dt)^{1/2} + V)F_1F_2,$$

where V is the distance over which the cells travel during time t by their motility toward a substratum surface on average. This equation, although waiting for experimental verification, can be simplified as $A(t) = cS((2Dt)^{1/2} + V)F_1F_2$ under the conditions of no energy barrier and well explains the attachment behaviors of *V. alginolyticus* cells [76] as discussed earlier.

42.3 FUTURE STUDY

For future study in cell attachment, several aspects as follows are necessary to be considered and carefully controlled:

1. *Physiological State of Cells.* It seems to be difficult to find generalities in the subject of cell attachment if the organisms are used without defining their physiological state; for example, cell surface properties may change along with growth phase [90, 91].
2. *Attachment via Cell Appendages.* Many kinds of bacteria have filamentous cell appendages, such as flagella and pili. These appendages may influence the electrokinetic property of bacterial cells [92, 93] and cell attachment by bridging the cell and substratum surface via appendages [8]. Unique cell surface structures found in *Caulobacter crescentus* [94],

Archaea cells [95], and *Acinetobacter* sp. [96] causing a strong attachment of these cells are interesting subjects for further study.

In the case when specific interactions exist at the attaching site of the appendage to the surface, both the long-range nonspecific interactions and the short-range specific interactions must be taken into consideration for the cell approaching then attaching to the surface [20].

3. *Substratum Surface for Cell Attachment.* When a solid is immersed in a medium, ions and molecules dissolved in the medium adsorb to the solid surface to form a so-called conditioning film. The properties of the conditioning film must be clarified to elucidate its influence on the subsequent cell attachment. In the case when soft polymer layers develop on the surface, Ohshima's soft particle theory seems to be applicable to characterize the surface and to analyze the interactions between the cells and the surfaces. Bacteria can attach not only to a flat surface but also to small size particles [97]; thus, these particles must be considered as cell attachment sites.
4. *Methods to Study Bacterial Attachment.* Bacterial cells attaching weakly to surfaces can easily be removed by rinsing the surface. Such experimental conditions are still frequently neglected in the design of methods to study microbial attachment [20]. These experimental conditions must be carefully designed to enable us to interpret and enhance comparison of results from different researchers. The methods to study microbial adhesive interactions were summarized and critically evaluated by Bos et al. [20].

REFERENCES

1 J. W. Costerton, H. M. Lappin-Scott. Microbial biofilms, in H. M. Lappin-Scott, J. W. Costerton (eds.), *Introduction to microbial biofilms*. Cambridge University Press, Cambridge, UK, 1995, pp. 1–11.

2 T. R. Garrett, M. Bhakoo, Z. Zhang. Understanding adhesion of bacteria and biofilms on surfaces. *Prog. Nat. Sci.* 18 (2008) 1049–1056.

3 R. Singh, D. Paul, R. K. Jain. Biofilms: implications in bioremediation. *Trends Microbiol.* 14 (2006) 389–397.

4 Y. Tsuchiya, M. Ikenaga, A. Kurniawan, A. Hiraki, T. Arakawa, R. Kusakabe, H. Morisaki. *Microbes Environ.* 24 (2009) 43–51.

5 A. Hiraki, Y. Tsuchiya, Y. Fukuda, T. Yamamoto, A. Kurniawan, H. Morisaki. Analysis of how a biofilm forms on the surface of the aquatic macrophyte *Phragmites australis*. *Microbes Environ.* 24 (2009) 265–272.

6 T. -F. C. Mah, G. A. O'Toole. Mechanisms of biofilm resistance to antimicrobial agents. *Trends Microbiol.* 9 (2001) 34–39.

7 D. Lindsay, A. von Holy. Bacterial biofilms within the clinical setting: what healthcare professionals should know. *J. Hosp. Infect.* 64 (2006) 313–325.

8 K. C. Marshall. *Interfaces in microbial ecology*. Harvard University Press, Cambridge, MA, 1976.

9 A. T. Poortinga, R. Bos, W. Norde, H. Busscher. Electric double layer interactions in bacterial adhesion to surfaces. *Surf. Sci. Rep.* 47 (2002) 1–32.

10 M. Katsikogianni, Y. F. Missirlis. Concise review of mechanisms of bacterial adhesion to biomaterials and of the techniques used in estimating bacteria-material interactions. *Eur. Cells Mater.* 8 (2004) 37–57.

11 H. J. Busscher, M. -N. Bellon-Fontaine, N. Mozes, H. C. van der Mei, J. Sjollema, A. J. Léonard, P. G. Rouxhet, O. Cerf. An interlaboratory comparison of physico-chemical methods for studying the surface properties of microorganisms—Application to *Streptococcus thermophilus* and *Leuconostoc mesenteroides*. *J. Microbiol. Methods* 12 (1990) 101–115.

12 B. A. Jucker, H. Harms, A. J. B. Zehnder. Adhesion of the positively charged bacterium *Stenotrophomonas* (*Xanthomonas*) *maltophilia* 70401 to glass and Teflon. *J. Bacteriol.* 178 (1996) 5472–5479.

13 R. E. Martinez, O. S. Porovsky, J. Schott, E. H. Oelkers. Surface charge and zeta-potential of metabolically active and dead cyanobacteria. *J. Colloid Interface Sci.* 323 (2008) 317–325.

14 R. E. Martinez, E. Gardes, O. S. Porovsky, J. Schott, E. H. Oelkers. Do photosynthetic bacteria have a protective mechanism against carbonate precipitation at their surfaces? *Geochim. Cosmochim. Acta* 74 (2010) 1329–1337.

15 J. E. Heckels, B. Blackett, J. S. Everson, M. E. Ward. The influence of surface charge on the attachment of *Neisseria gonorrhoeae* to human cells. *J. Gen. Microbiol.* 96 (1976) 359–364.

16 W. Jiang, A. Saxena, B. Song, B. B. Ward, T. J. Beveridge, S. C. B. Myneni. Elucidation of functional groups on gram-positive and gram-negative bacterial surfaces using infrared spectroscopy. *Langmuir* 20 (2004) 11433–11442.

17 Y. E. Dufrene, A. van der Wal, W. Norde, P. G. Rouxhet. X-ray photoelectron spectroscopy analysis of whole cells and isolated cell walls of gram-positive bacteria: comparison with biochemical analysis. *J. Bacteriol.* 179 (1997) 1023–1028.

18 H. C. van der Mei, M. Roseberg, H. J. Busscher. Assessment of microbial cell surface hydrophobicity, in N. Mozes, P. S. Handley, H. J. Busscher, P. G. Rouxhet (eds.), *Microbial cell surface analysis: structural and physicochemical methods*. VCH Publishers, New York, 1991, pp. 265–289.

19 R. J. Doyle, M. Rosenberg (eds.), *Microbial cell surface hydrophobicity*. American Society for Microbiology, Washington, DC, 1990.

20 R. Bos, H. C. van der Mei, H. J. Busscher. Physico-chemistry of initial microbial adhesive interactions—Its mechanisms and methods for study. *FEMS Microbiol. Rev.* 23 (1999) 179–230.

21 A. H. Hogt, J. Dankert, J. Feijen, J. A. De Wries. Cell surface hydrophobicity of *Staphylococcus* species and adhesion onto biomaterials. *Antonie van Leeuwenhoek* 48 (1982) 496–498.

22 A. Yousefi Rad, H. Ayhan, E. Pi _UNDEFINEDkin. Adhesion of different bacterial strains to low-temperature plasma-treated sutures. *J. Biomed. Mater. Res.* 41 (1998) 349–358.

23 A. Roosjen, H. J. Busscher, W. Norde, H. C. van der Mei. Bacterial factors influencing adhesion of *Pseudomonas aeruginosa* strains to a poly(ethylene oxide) brush. *Microbiology* 152 (2006) 2673–2682.

24 M. Fletcher, G. I. Loeb. Influence of substratum characteristics on the attachment of a marine pseudomonad to solid surfaces. *Appl. Environ. Microbiol.* 37 (1979) 67–72.

25 J. Satou, A. Fukunaga, N. Satou, H. Shintani, K. Okuda. Streptococcal adherence on various restorative materials. *J. Dent. Res.* 67 (1988) 588–591.

26 H. C. van der Mei, P. Brokke, J. Dankert, J. Feijen, H. J. Busscher. Influence of electrostatic interactions on the deposition efficiencies of coagulase-negative staphylococci to collector surfaces in a parallel plate flow chamber. *J. Dispers. Sci. Technol.* 13 (1992) 447–458.

27 J.-P. G. Piette, E. S. Idziak. A model study of factors involved in adhesion of *Pseudomonas fluorescens* to meat. *Appl. Environ. Microbiol.* 58 (1992) 2783–2791.

28 S. Bouttier, K. G. Han, C. Ntsama, M. N. Bellon-Fontaine, J. Fourmat. Role of electrostatic interactions in the adhesion of *Pseudomonas fragi* and *Brochothrix thermosphacta* to meat. *Colloids Surf. B Biointerfaces* 2 (1994) 57–65.

29 S. Takashima, H. Morisaki. Surface characteristics of the microbial cell of *Pseudomonas syringae* and its relevance to cell attachment. *Colloids Surf. B Biointerfaces* 9 (1997) 205–212.

30 M. A. S. Vigeant, R. M. Ford. Interactions between motile *Escherichia coli* and glass in media with various ionic strengths, as observed with a three-dimensional-tracking microscope. *Appl. Environ. Microbiol.* 63 (1997) 3474–3479.

31 D. Cunliffe, C. A. Smart, C. Alexander, E. N. Vulfson. Bacterial adhesion at synthetic surfaces. *Appl. Environ. Microbiol.* 11 (1999) 4995–5002.

32 D. P. Bakker, B. R. Postmus, H. J. Busscher, H. C. van der Mei. Bacterial strains isolated from different niches can exhibit different patterns of adhesion to substrata. *Appl. Environ. Microbiol.* 70 (2004) 3758–3760.

33 W. Achouak, F. Thomas, T. Heulin. Physico-chemical surface properties of rhizobacteria and their adhesion to rice roots. *Colloids Surf. B Biointerfaces* 3 (1994) 131–137.

34 M. C. M. Van Loosdrecht, J. Lyklema, W. Norde, G. Schraa, A. J. B. Zehnder. The role of bacterial cell wall hydrophobicity in adhesion. *Appl. Environ. Microbiol.* 53 (1987) 1893–1897.

35 E. Giaouris, M.-P. Chapot-Chartier, R. Briandet. Surface physicochemical analysis of natural *Lactolactis* strains reveals the existence of hydrophobic and low charged strains with altered adhesive properties. *Int. J. Food Microbiol.* 131 (2009) 2–9.

36 N. Iwabuch, M. Sunairi, H. Anzai, H. Morisaki, M. Nakajima. Relationships among colony morphotypes, cell-surface properties and bacterial adhesion to substrata in *Rhodococcus*. *Colloids Surf. B Biointerfaces* 30 (2003) 51–60.

37 H. Morisaki, Y. Kasahara, T. Hattori. The cell surface charge of fast- and slow-growing bacteria isolated from grassland soil. *J. Gen. Appl. Microbiol.* 39 (1993) 65–74.

38 R. Shingkai, K. Gorlach, T. Hattori, K. Samukawa, H. Morisaki. The cell surface charge of fast- and slow-growing bacteria isolated from a paddy soil. *J. Gen. Appl. Microbiol.* 40 (1994) 469–475.

39 Y. Kasahara, H. Morisaki, T. Hattori. Hydrophobicity of the cells of fast- and slow-growing bacteria isolated from a grassland soil. *J. Gen. Appl. Microbiol.* 39 (1993) 381–388.

40 H. Morisaki, H. Toda. Surface characteristics of bacterial cells isolated from river sand grains and their relevance to attachment. *Microbes Environ.* 13 (1998) 9–16.

41 H. Morisaki, K. Nakagawa, H. Shiraishi. Measurement of attachment force of microbial cells. *Colloids Surf. B Biointerfaces* 6 (1996) 347–352.

42 H. Morisaki. Measurement of the force necessary for removal of bacterial cells from a quartz plate. *J. Gen. Microbiol.* 137 (1991) 2649–2655.

43 N. P. Boks, W. Norde, H. C. van der Mei, H. J. Busscher. Forces involved in bacterial adhesion to hydrophilic and hydrophobic surfaces. *Microbiology* 154 (2008) 3122–3133.

44 N. Mohamed, M. A. Teeters, J. M. Patti, M. Hook, J. M. Ross. Inhibition of *Staphylococcus aureus* adherence collagen under dynamic conditions. *Infect. Immun.* 67 (1999) 589–594.

45 N. F. Owens, D. Gingell, P. R. Rutter. Inhibition of cell-adhesion by a synthetic-polymer adsorbed to glass shown under defined hydrodynimic atress. *J. Cell Sci.* 87 (1987) 667–675.

46 P. R. Rutter, B. Vincent. Attachment mechanism in the surface growth of microorganisms, in M. J. Bazin, J. J. Prosser (eds.), *Physiological models in microbiology*. CRC Press, Boca Raton, FL, 1988, pp. 87–107.

47 W. E. Thomas, L. M. Nilsson, M. Forero, E. V. Sokurenko, V. Vogel. Shear-dependent "stick-and roll" adhesion of type 1 fimbriated *Escherichia coli*. *Mol. Microbiol.* 53 (2004) 1545–1557.

48 J. M. Higashi, I. W. Wang, D. M. Shlaes, J. M. Anderson, R. E. Marchant. Adhesion of *Staphylococcus epidermidis* and transposon mutant strains to hydrophobic polyethylene. *J. Biomed. Mater. Res.* 39 (1998) 341–350.

49 M. S. Shive, S. M. Hasan, J. M. Anderson. Shear stress effects on bacterial adhesion, leukocyte adhesion, and leukocyte oxidative capacity on a polyetherurethane. *J. Biomed. Mater. Res.* 46 (1999) 511–519.

50 I. W. Wang, J. M. Anderson, M. R. Jacobs, R. E. Marchant. Adhesion of *Staphylococcus epidermidis* to biomedical polymers—Contributions of surface thermodynamics and hemodynamic shear conditions. *J Biomed. Mater. Res.* 29 (1995) 485–493.

51 N. Mohamed, T. R. Rainier, J. M. Ross. Novel experimental study of receptor-mediated bacterial adhesion under the influence of fluid shear. *Biotechnol. Bioeng.* 68 (2000) 628–636.

52 J. E. Duddridge, C. A. Kent, J. F. Laws. Effect of surface shear-stress on the attachment of *Pseudomonas fluorescens* to stainless steel under defined flow conditions. *Biotechnol. Bioeng.* 24 (1982) 153–164.

53 E. Fallman, S. Schedin, J. Jass, M. Andersson, B. E. Uhlin, O. Axner. Optical tweezers based force measurement system for quantitating binding interactions: system

design and application for the study of bacterial adhesion. *Biosens. Bioelectron.* 19 (2004) 1429–1437.

54 K. H. Simpson, M. G. Bowden, M. Hook, B. Anvari. Measurement of adhesive forces between *S. epidermidis* and fibronectin-coated surfaces using optical tweezers. *Lasers Surg. Med.* 31 (2002) 45–52.

55 K. H. Simpson, A. G. Bowden, S. J. Peacock, M. Arya, M. Hook, B. Anvari. Adherence of *Staphylococcus aureus* fibronectin binding protein a mutants: an investigation using optical tweezers. *Biomol. Eng.* 21 (2004) 105–111.

56 T. Cao, H. Y. Tang, X. M. Liang, A. F. Wang, G. W. Auner, S. O. Salley, K. Y. S. Ng. Nanoscale investigation on adhesion of *E. coli* surface modified silicone using atomic force microscopy. *Biotechnol. Bioeng.* 94 (2006) 167–176.

57 N. I. Abu-Lail, T. A. Camesano. Role of ionic strength on the relationship of biopolymer conformation, DLVO contributions, and steric interactions to bioadhesion of *Pseudomonas putida* KT2442. *Biomacromolecules* 4 (2003) 1000–1012.

58 A. Mendez-Vilas, A. M. Gallardo-Moreno, M. L. Gonzalez-Martin. Nanomechanical exploration of the surface and sub-surface of hydrated cells of *Staphylococcus epidermidis*. *Antonie Van Leeuwenhoek* 89 (2006) 373–386.

59 W. R. Bowen, A. S. Fenton, R. W. Lovitt, C. J. Wright. The measurement of *Bacillus mycoides* spore adhesion using atomic force microscopy, simple counting methods, and a spinning disk technique. *Biotechnol. Bioeng.* 79 (2002) 170–179.

60 M. Hermansson. The DLVO theory in microbial adhesion. *Colloids Surf. B Biointerfaces* 14 (1999) 105–119.

61 R. M. Goulter, I. R. Gentle, G. A. Dykes. Issues in determining factors influencing bacterial attachment: a review using the attachment of *Escherichia coli* to abiotic surfaces as an example. *Lett. Appl. Microbiol.* 49 (2009) 1–7.

62 K. Hori, S. Matsumoto. Bacterial adhesion: from mechanism to control. *Biochem. Eng. J.* 48 (2010) 424–434.

63 Y. H. An, R. J. Friedman. Concise review of mechanisms of bacterial adhesion to biomaterial surfaces. *J. Biomed. Mater. Res.* 43 (1998) 338–348.

64 P. R. Rutter, B. Vincent. The adhesion of microorganisms to surfaces, physicochemical aspects, in R. C. W. Berkeley, J. M. Lynch, J. Melling, P. R. Rutter, B. Vincent (eds.), *Microbial adhesion to surfaces*. Ellis Horwood, London, 1980, pp. 79–91.

65 Th. F. Tadros. Particle-surface adhesion, in R. C. W. Berkeley, J. M. Lynch, J. Melling, P. R. Rutter, B. Vincent (eds.), *Microbial adhesion to surfaces*. Ellis Horwood, London, 1980, pp. 93–113.

66 M. C. M. van Loosdrecht, J. Lyklema, W. Norde, A. J. B. Zehnder. Bacterial adhesion: a physicochemical approach. *Microb. Ecol.* 17 (1989) 1–15.

67 C. H. Bolster, A. L. Mills, G. M. Hornberger, J. S. Herman. Effect of surface coatings, grain size, and ionic strength on the maximum attainable coverage of bacteria on sand surfaces. *J. Contam. Hydrol.* 50 (2001) 287–305.

68 C. R. Bunt, D. S. Jones, I. G. Tucker. The effects of pH, ionic strength and polyvalent ions on the cell surface hydrophobicity of *Escherichia coli* evaluated by the BATH and HIC methods. *Int. J. Pharm.* 113 (1995) 257–261.

69 D. G. Jewett, T. A. Hilbert, B. E. Logan, R. G. Arnold, R. C. Bales. Bacterial transport in laboratory columns and filters: influence of ionic strength and pH on collision efficiency. *Water Res.* 29 (1995) 1673–1680.

70 Q. Li, B. E. Logan. Enhancing bacterial transport for bioaugmentation of aquifers using low ionic strength solutions and surfactants. *Water Res.* 33 (1999) 1090–1100.

71 K. Otto, H. Elwing, M. Hermansson. Effect of ionic strength on initial interactions of *Escherichia coli* with surfaces, studied online by a novel quartz crystal microbalance technique. *J. Bacteriol.* 181 (1999) 5210–5218.

72 A. Zita, M. Hermansson. Effects of ionic strength on bacterial adhesion and stability of flocs in a wastewater activated sludge system. *Appl. Environ. Microbiol.* 60 (1994) 3041–3048.

73 C. J. Van Oss, R. J. Good, M. K. Chaudhury. The role of van der Waals forces and hydrogen bonds in "hydrophobic interactions" between biopolymers and low energy surfaces. *J. Colloid Interface Sci.* 111 (1986) 378–390.

74 J. M. Meinders, H. C. Van der Mei, H. J. Busscher. Deposition efficiency and reversibility of bacterial adhesion under flow. *J. Colloid Interface Sci.* 176 (1995) 329–341.

75 K. Kogure, E. Ikemoto, H. Morisaki. Attachment of *Vibrio alginolyticus* to glass surfaces is dependent on swimming speed. *J. Bacteriol.* 180 (1998) 932–937.

76 H. Morisaki, S. Nagai, H. Ohshima, E. Ikemoto, K. Kogure. The effect of motility and cell-surface polymers on bacterial attachment. *Microbiology* 145 (1999) 2797–2802.

77 K. C. Marshall, R. Stout, R. Mitchell. Mechanism of the initial events in the sorption of marine bacteria to surfaces. *J. Gen. Microbiol.* 68 (1971) 337–348.

78 M. C. M. van Loosdrecht, J. Lyklema, W. Norde, G. Schraa, A. J. B. Zehnder. Electrophoretic mobility and hydrophobicity as a measure to predict the initial steps of bacterial adhesion. *Appl. Environ. Microbiol.* 53 (1987) 1898–1901.

79 S. Kawahata, H. Ohshima, N. Muramatsu, T. Kondo. Charge distribution in the surface region of human erythrocytes as estimated from electrophoretic mobility data. *J. Colloid Interface Sci.* 138 (1990) 182–186.

80 H. H. M. Rijnaarts, W. Norde, E. J. Bouwer, J. Lyklema, A. J. B. Zehnder. Reversibility and mechanism of bacterial adhesion. *Colloids Surf. B Biointerfaces* 4 (1995) 5–22.

81 R. Sonohara, N. Muramatsu, H. Ohshima, T. Kondo. Differences in surface properties between *Escherichia coli* and *Staphylococcus aureus* as revealed by electrophoretic mobility measurements. *Biophys. Chem.* 55 (1995) 273–277.

82 H. Ohshima. Electrophoretic mobility of soft particles. *J. Colloid Interface Sci.* 163 (1994) 474–483.

83 H. Ohshima. Electrophoresis of soft particles. *Adv. Colloid Interface Sci.* 62 (1995) 189–235.

84 H. Ohshima. Electrophoretic mobility of a charged spherical colloidal particle covered with an uncharged polymer layer. *Electrophoresis* 23 (2002) 1993–2000.

85 N. I. Abu-Lail, T. A. Camesano. Role of lipopolysaccharides in the adhesion, retention, and transport of *Escherichia coli* JM109. *Environ. Sci. Technol.* 37 (2003) 2173–2183.

86 S. Tsuneda, J. Jung, H. Hayashi, H. Aikawa, A. Hirata, H. Sasaki. Influence of extracellular polymers on electrokinetic properties of heterotrophic bacterial cells examined by soft particle electrophoresis theory. *Colloids Surf. B Biointerfaces* 29 (2003) 181–188.

87 S. Tsuneda, H. Aikawa, H. Hayashi, A. Hirata. Significance of cell electrokinetic properties determined by soft-particle analysis in bacterial adhesion onto a solid surface. *J. Colloid Interface Sci.* 279 (2004) 410–417.

88 A. T. Poortinga, R. Boss, H. J. Busscher. Electrostatic interactions in the adhesion of an ion-penetrable and ion-impenetrable bacterial strain to glass. *Colloids Surf. B Biointerfaces* 20 (2001) 105–117.

89 H. Morisaki, H. Tabuchi. Bacterial attachment over a wide range of ionic strengths. *Colloids Surf. B Biointerfaces* 74 (2009) 51–55.

90 R. J. Neufeld, J. E. Zajic, D. F. Gerson. Cell surface measurements in hydrocarbon and carbohydrate fermentation. *Appl. Environ. Microbiol.* 39 (1980) 511–517.

91 T. Ozawa, S. Tokuda, Y. Komai. Correlation between competetive nodulation ability and cell surface hydrophobicity of *Bradyrihizobium japonicum*. *Bull. Jpn. Soc. Microb. Ecol.* 6 (1991) 19–23.

92 H. C. van der Mei, H. J. Busscher, R. Bos, J. de Vries, C. J. P. Boonaert, Y. F. Dufrêne. Direct probing by atomic force microscopy of the cell surface softness of a fibrillated and nonfibrillated oral streptococcal strain. *Biophys. J.* 78 (2000) 2668–2674.

93 S. Okuda, R. Igarashi, Y. Kusui, Y. Kasahara, H. Morisaki. Electrophoretic mobility of bacillus subtilis knockout mutants with and without flagella. *J. Bacteriol.* 185 (2003) 3711–3717.

94 W. A. Corpe. 1972). Microfouling: the role of primary film forming marine bacteria. International Congress on Marine Corrosion and Fouling, National Bureau of Standards, Gaithersburg, pp. 598–609.

95 C. Moissl, R. Rachel, A. Briegel, H. Engelhardt, R. Huber. The unique structure of archaeal "hami," highly complex cell appendages with nano-grappling hooks. *Mol. Microbiol.* 56 (2005) 361–370.

96 K. Hori, H. Watanebe, S. Ishii, Y. Tanji, H. Unno. Monolayer adsorption of a "bald" mutant of the highly adhesive and hydrophobic bacterium *Acinetobacter* sp. strain Tol 5 to a hydrocarbon surface. *Appl. Environ. Microbiol.* 74 (2008) 2511–2517.

97 H. W. Paerl. Microbial attachment to particles in marine and freshwater ecosystems. *Microb. Ecol.* 2 (1975) 73–83.

43

DESIGN AND FABRICATION OF STERICALLY STABILIZED LIPOSOMES DISPERSED IN AQUEOUS SOLUTIONS BY UTILIZING ELECTROSTATIC INTERACTIONS FOR USE IN BIOMEDICAL APPLICATIONS

Katsumi Uchida and Hirofumi Yajima

43.1 INTRODUCTION

Liposomes form spherical lipid vesicles with a bilayered membrane structure consisting of amphiphilic lipid molecules and are composed of nanocontainers with diameters of 10–1000 nm. They are good biodegradable and biocompatible cell membrane models and play roles in protecting encapsulated active ingredients from hostile external environments and in sustained drug release [1, 2]. They have been extensively employed as drug carriers encapsulating bioactive molecules such as proteins, enzymes, DNA, and various drug molecules. However, both *in vivo* and *in vitro*, liposomes are apt to fuse and/or aggregate with one another and thus their encapsulated contents are easily leaked. This feature has limited their use as practical drug carriers in biomedical applications. Hence, many studies have been performed for the improvement and control of vesicle stability in aqueous solutions with regard to their

Electrical Phenomena at Interfaces and Biointerfaces: Fundamentals and Applications in Nano-, Bio-, and Environmental Sciences, First Edition. Edited by Hiroyuki Ohshima.

functionality as practical drug carriers. In this chapter, recent approaches, including our own, are presented for the improvement and control of steric stabilization of charged liposomes by their complexation with oppositely charged materials through electrostatic interactions.

43.2 COMPLEXATION OF LIPOSOMES WITH CHARGED POLYMERS FOR CONSTRUCTING STERICALLY STABILIZED VESICLE DISPERSION SYSTEMS

As one strategy for stabilization, cholesterol has been incorporated into the bilayer in order to enhance the rigidity of the bilayer vesicles, reducing the mobility of the phospholipid molecules [3]. Charged or hydrophilic molecules have also been incorporated into the bilayer or adsorbed on the vesicle outer phase by means of electrostatic stabilization or repulsion. In this section, we will summarize research on stabilized vesicle dispersion systems by utilizing polymers absorbed on the liposome surface through electrostatic interactions.

Bronich et al. [4] demonstrated that a cationic graft copolymer (PEO-g-PEI) consisting of polyethylene oxide (PEO) and polyethylenimine (PEI) was adsorbed on the surface of negatively charged liposomes composed of dipalmitoylphosphatidylcholine and diphosphatidylglycerol through electrostatic interactions. The PEO chains attached to the liposome surface through PEI mediation and then formed a nonionic hydrophilic layer that was involved in sterical stabilization of the liposome. These electroneutral complexes of cationic PEO-g-PEI with anionic liposomes remained stable in the dispersed state, compared to homopolymeric PEI/liposome complexes, which tended to aggregate. Binding of the cationic copolymer did not significantly alter the structure of the lipid bilayer.

Sideratou et al. [5] reported that dihexadecylphosphate (DHP) liposomes interacted with guanidinium cation and its derivatives through electrostatic and/or hydrogen-bonding forces. Moreover, they [6] demonstrated that the interaction between phosphatidylcholine (PC)-cholesterol liposomes incorporating DHP as a molecular recognizable lipid with complementary guanidinylated diaminobutane poly(propylene imine) dendrimers of the fourth and fifth generation afforded liposome/dendrimer composites that formed aggregates. These aggregates could be redispersed by the addition of excess phosphate buffer. Interestingly, the steric stability of the liposomal membrane was retained in the aggregation and redispersion processes.

Volodkin et al. [7] reported the complexation of negatively charged liposomes with cationic polypeptide poly(L-lysine) (PLL). The resulting complexes underwent a structural change from isolated polyion-coated liposomes to vesicle aggregates. Fabrication of stable single PLL-covered liposomes depended on a variety of parameters, mainly the PLL/lipid ratio and the PLL molecular weight. The stability of the lipid membranes dramatically decreased above the main transition temperature. The PLL coating did not induce any changes in stabilization by adsorption on the liposomal surface, but the electrostatic interaction had a strong effect on the membrane integrity of the fluid liposomes due to lipid mobility.

Kawakami et al. [8] investigated the effect of various hydrophilic polymers on the dispersity of liposomes composed of hydrogenated lecithin. The interaction between the hydrophilic polymers and liposome membranes was responsible to some extent for the stabilization of liposome dispersions because the lack of such interaction led to depletion flocculation. Sodium alginate possesses a favorable ability to stabilize the dispersion of liposomes over a wide concentration range due to its charged moieties, which interact electrostatically with the membranes. When there is no interaction or very weak interactions between the polymers and the membranes, the concentration of the polymers must be carefully adjusted so as not to destabilize the dispersion. These observations suggested strategies to select polymers to stabilize liposome dispersions.

Yaroslavov et al. [9] reported the adsorption of cationic poly-(*N*-ethyl-4-vinylpyridine) (PEVP) on the surface of negatively charged liquid and solid PC-cardiolipin liposomes. The cationic PEVP polymer adsorbed on the surface of liquid liposomes, resulting in a decrease of the surface charge, loss of stability of the liposomes, and aggregation. The addition of polyacrylate anions resulted in a complete release of the polymer cations from the surface into the solution through interpolyelectrolyte complexation with polycations, accompanied by a redispersion of the liposomes. On the other hand, a polycation possessing side-chain cetyl fragments formed a stable complex with the liposome due to the incorporation of cetyl fragments into the liquid bilayer. In this case, the polyanion did not release the polycation from the surface, and a complex of liposome, polycation, and polyanion was formed, which was stabilized by the negative charge of the polyanion. Thus, the modification of a polycation with side-chain alkyl fragments enhanced complex stability against a strong negatively charged competitor because of the incorporation of alkyl fragments into the liquid bilayer. Furthermore, Yaroslavov et al. [10] investigated the interaction between zwitterionic polymers composed of quaternizing polyvinylpyridine and bromoacids ($Br(CH_2)_n$–COOH, where $n = 1$, 2, 3, and 5) with negatively charged liposomes. For $n = 1$, no interaction between the polymer and liposomes was detected. For $n = 2$, electrostatic attraction induced polymer–liposome association. Aggregation of the liposomes accompanied polymer adsorption, but the addition of NaCl collapsed the aggregates. For $n = 3$, electrostatic binding occurred along with aggregation. However, the addition of NaCl was unable to induce polymer–liposome dissociation because of the strong electrostatic binding. For $n = 5$, there was both polymer binding and NaCl-promoted dissociation but no substantial aggregation. These results revealed that the interaction between zwitterionic polymers and anionic liposomes was dependent on the length of the spacer separating the cationic and anionic charged groups.

Zhang and Granick [11] proposed a simple strategy to produce stabilized liposome particles without aggregation by mixing phospholipid liposomes (diameter of 200 nm) with carboxyl-modified negatively charged polystyrene latex (diameter of 20 nm) under sonication conditions. The driving force of the interaction was believed to be the charge–dipole attraction between the charged nanoparticles and the P–N (phosphorus–nitrogen) dipoles of the phospholipid head groups, which preferentially orient nearly parallel to the local bilayer membrane surface [12].

We have reported the polyethylene glycol (PEG)-mediated steric stabilization of complexes of negatively charged liposomes with folate-conjugated poly(amidoamine) (FA-PAMAM) dendrimers [13]. In view of the topological effects of polyions [14], we have focused on dendrimers with three-dimensional architectures and terminal functional groups. Poly(amidoamine) (PAMAM) dendrimers with large cationic charge densities are promising functional nanomaterials for use as drug carriers and as cell transfection agents [15, 16]. Drugs are retained inside the PAMAM dendrimer, and the terminal amino groups of the dendrimer are involved in the introduction of target ligand molecules such as folic acid [17, 18]. Folate receptors are well-known to be overexpressed in cancer cells. Thus, it is expected that FA-PAMAM would recognize only specific cancer cells and be incorporated into these cells through endocytosis. In order to develop an active drug carrier on the basis of these findings, we fabricated supramolecular complexes composed of anionic liposomes with cationic FA-PAMAM. To prevent the complexes from aggregating through intermolecular cross-linked FA-PAMAM and to elevate the dispersion stability of the complexes, we introduced PEG mono-oleyl ether composed of hydrophilic PEG and a hydrophobic alkyl chain as a component of the anionic liposome. Figure 43.1 shows that the complexes of an anionic liposome with a cationic dendrimer were stably dispersed in aqueous solution. Figure 43.2 shows the effect of the concentration of FA-PAMAM on the cumulant average diameter and ζ-potential of liposomes with PEG. The diameter of the liposomes increased slightly with FA-PAMAM concentration up to $R \approx 1.2$ (where R is the ratio of FA-PAMAM to PC; R = [FA-PAMAM]/[PC] [w/w]) and then leveled off. The size increased in increments of about 7–8 nm, and the polydispersity index was less than 0.1 despite the addition of FA-PAMAM. With increasing cationic FA-PAMAM concentration up to $R \approx 4.0$, the ζ-potential of the anionic liposomes was shifted from a negative (–9 mV) to a positive value (+10 mV), with a zero value occurring at $R \approx 1.2$. Interestingly, even if the ζ-potential of the complex with PEG was zero, the neutralized complex was stably dispersed, which revealed that the steric repulsion between PEG chains prevented the liposome from aggregating. This plays an important role in the stability of the liposome/dendrimer dispersion in water. For comparison, as a control, the ζ-potential of the cationic liposome with a PEG chain was measured as a function of the weight concentration ratio of FA-PAMAM

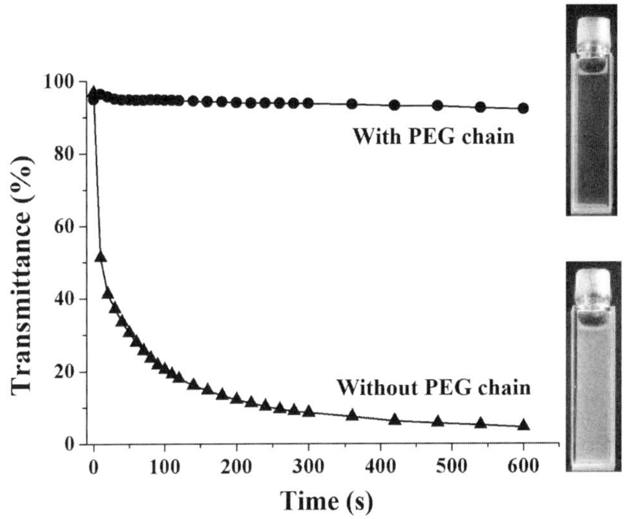

Figure 43.1. Changes in the optical transmittance of the anionic liposomes without PEG chain (triangles) and with PEG chain (circles) in water as a function of incubation time after the addition of cationic dendrimer (FA-PAMAM). The inset photographs (upper photo: the liposome with PEG chain; lower photo: the liposome without PEG chain) show the liposome suspensions after incubation of FA-PAMAM for 3 hours (Ohori et al. [13]).

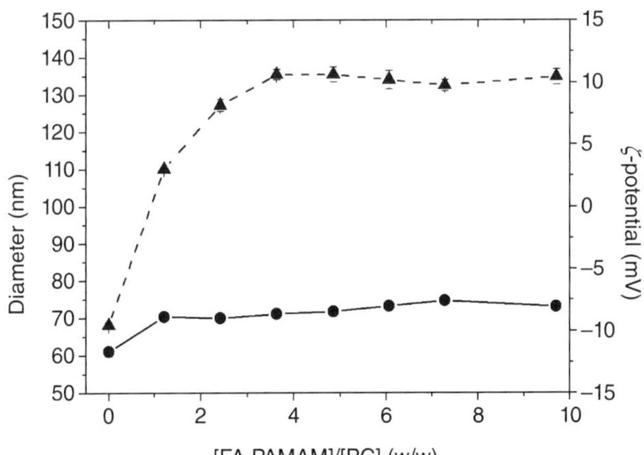

Figure 43.2. Cumulant diameter (circles) and ζ-potential (triangles) of the anionic liposome with PEG chain as a function of the weight concentration ratio of FA-PAMAM to PC ($n = 3$) (Ohori et al. [13]).

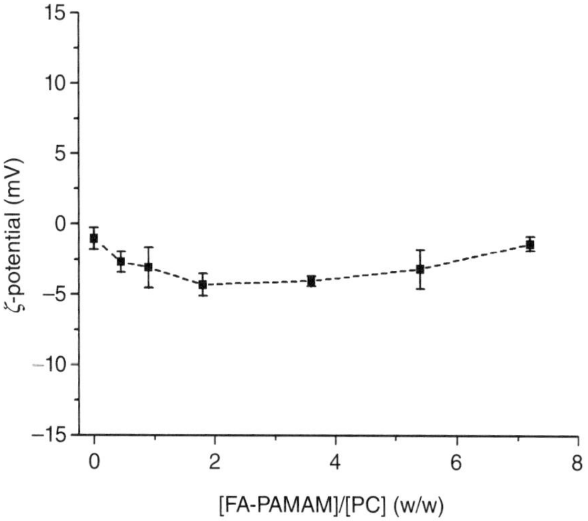

Figure 43.3. Change in the ζ-potential of the cationic liposome with PEG chain as a function of the weight concentration ratio of FA-PAMAM to PC.

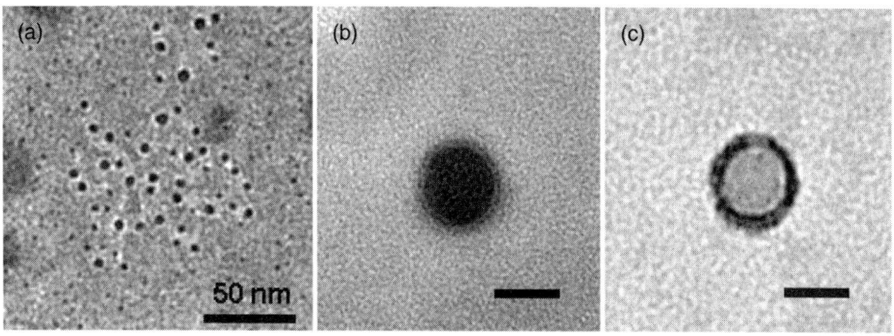

Figure 43.4. TEM images of (a) FA-PAMAM, (b) the anionic liposome with PEG chain before the addition of FA-PAMAM, and (c) the anionic liposome with PEG chain after the addition of FA-PAMAM. The scale bar in the TEM images is 50 nm.

and PC (Fig. 43.3). The ζ-potential scarcely changed upon the addition of FA-PAMAM. These results confirm the formation of complexes of anionic liposomes with cationic FA-PAMAM. Figure 43.4 shows transmission electron microscope (TEM) images in which the outer surface of the liposome is coated with a monolayer of dendrimers. The vesicle structure of these liposomes was confirmed to be stable during the adsorption processes for FA-

PAMAM by studying the leakage of sulforhodamine B (SRB) entrapped in the liposomes.

43.3 CONTROL OF THE FUSOGENIC ACTIVITY OF LIPOSOMES

Because fusogenic liposomes can introduce their contents into the cytoplasm by fusing with cellular membranes such as plasma and endosomal membranes, these liposomes are very important as drug delivery systems of membrane-impermeable molecules, such as proteins, genes, and oligonucleotides. In this section, some studies with regard to control of the fusogenic activity of liposomes will be discussed.

One approach to preparing fusogenic liposomes is to conjugate fusogenic molecules to liposome membranes. Kono et al. [19] designed liposomes modified with poly(glycidol) comprising cationic β-alanine residues, that is, a PEG derivative with positively charged groups. PEG is a well-known fusogenic polymer. Modification of the liposomal membrane with a cationic PEG derivative imparted fusogenic ability to the stable liposomes [20] and provided strong interactions with the cell surface. The fusogenic ability of the modified liposomes increased with the amount of polymer fixed on the liposome. Furthermore, Kono et al. [21] prepared thermosensitive poly(*N*-acryloylpyrrolidine) (PAPr)-modified cationic liposomes whose affinity for negatively charged membranes was controlled by environmental temperature. PAPr has a low critical solution temperature (LCST) at around 52°C. The ζ-potential of the unmodified cationic liposomes did not change regardless of temperature. The polymer-modified cationic liposomes exhibited much lower ζ-potential values below the LCST of the polymer compared with the unmodified cationic liposomes. Above the LCST, their ζ-potential significantly increased. The unmodified cationic liposomes formed aggregates and fused intensively with anionic liposomes due to electrostatic interactions. In contrast, the aggregation and fusion of the polymer-modified cationic liposomes with anionic liposomes were strongly suppressed below the LCST, whereas they were promoted above the LCST.

Pornpattananangkul et al. [22] controlled the fusogenic activity of liposomes by adsorbing carboxyl-modified gold nanoparticles on the outer surface of cationic liposomes. At neutral pH, small gold nanoparticles with deprotonated carboxyl groups were able to bind to cationic liposomes through electrostatic attractions, thereby stabilizing the liposomes. When the pH was dropped below 5, the carboxylic group was protonated. The gold nanoparticles with the resulting neutral carboxyl groups were detached from the liposome surface due to the lack of binding forces, thereby freeing the liposomes (Fig. 43.5). The relative fusion efficiency of gold nanoparticle-stabilized cationic liposomes for anionic liposomes was ~25% at pH 7 compared to ~80% at pH 4. This result indicates that carboxyl-modified gold nanoparticles can

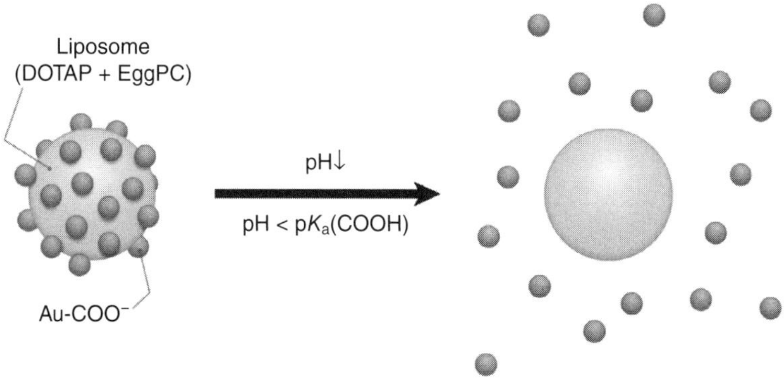

Figure 43.5. Schematic illustration of a carboxyl-modified gold nanoparticle (AuC)-stabilized liposome (1,2-di-(9*Z*-octadecynoyl)-3-trimethylammonium propane (DOTAP) and EggPC) and its destabilization at acidic pH. The liposome is stabilized by deprotonated AuC ($Au–COO^-$) at neutral pH. When the pH drops below the pK_a value of the carboxylic group ($pK_a \sim 5$), $Au–COO^-$ nanoparticles are protonated to form Au–COOH, which subsequently detach from the liposome, resulting in the formation of a bare liposome with resumption of fusion activity (Pornpattananangkul et al. [22]).

control the fusogenic activity of liposomes by the adjustment of the environmental pH.

43.4 FUNCTIONALIZED DRUG CARRIERS BASED ON LIPOSOMES

Finally, in this section, some information about the fabrication of functionalized drug carriers based on liposomes will be presented.

Fujimoto et al. [23] prepared hollow particles using liposomes as a template for the layer-by-layer deposition of biopolymers. They carried out the alternating deposition of PLL and poly-L-aspartic acid (P_{ASP}) onto a negatively charged liposome composed of dilauroyl phosphatidic acid (DLPA) and dimyristoylphosphatidylcholine (DMPC) (Fig. 43.6). The deposition was governed by electrostatic attractions, although the liposome was a fluid and extremely brittle template. The ζ-potentials alternated between positive and negative charges at each deposition. The amount of the deposited polymer was dependent on the concentration of the polymer aqueous solution and the ratio of DLPA in the liposome. The generated polymer layer allowed the storage and release of encapsulated substances. The release rate of encapsulated substances from the liposome by adding a surfactant was tunable by coverage of the first PLL layer.

Yuasa et al. [24] proposed a design for an anticancer drug delivery system based on electrostatic binding of negatively charged liposomes and cationic metalloporphyrins under physiological conditions. Cationic porphyrins have

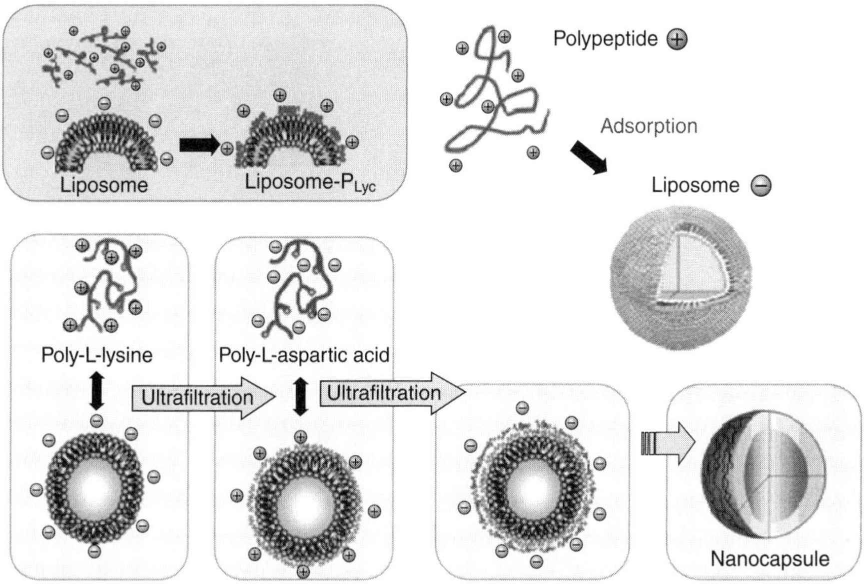

Figure 43.6. Schematic illustrations of nanocapsules by the layer-by-layer deposition of polypeptides onto a liposome template (Fujimoto et al. [23]).

attracted a great deal of attention because of the promise they have for use in photodynamic, antiviral, and anticancer therapies [25,26]. However, cationic porphyrins are water soluble and therefore readily pass through glomeruli. Thus, to overcome this difficulty, liposomes were used as retentive nanocarriers of cationic porphyrins. Iron(III) porphyrin-loaded liposomes lacked cytotoxicity but efficiently generated a toxic hydroxyl radical from a superoxide anion radical through iron(III)-catalyzed dismutation and Fenton-like reaction. This caused a targeted necrosis of tumor cells. The concentration of the superoxide anion radical was locally increased as a result of the reduced activity of superoxide dismutase (SOD) and catalase in these cells.

Among many recently developed drug carriers, inorganic nanomaterials, mesoporous silica nanoparticles in particular, have been very attractive because of their biocompatibility combined with high surface area, pore volume, uniform and tunable pore diameter, and surface chemistries [27]. Through simple electrostatic interactions, drugs can be loaded into silica nanoparticles at very high concentrations. However, metabolites and ions in body fluid can displace the loaded drugs, resulting in premature drug release. To avoid this, molecular-gating strategies based on nanoparticles have been designed, wherein drugs are released only upon gate opening or removal [28]. Liu et al. [29] fused liposomes on silica cores followed by successive steps of electrostatically mediated lipid exchanges between silica-supported bilayers and oppositely charged free liposomes, to reduce bilayer

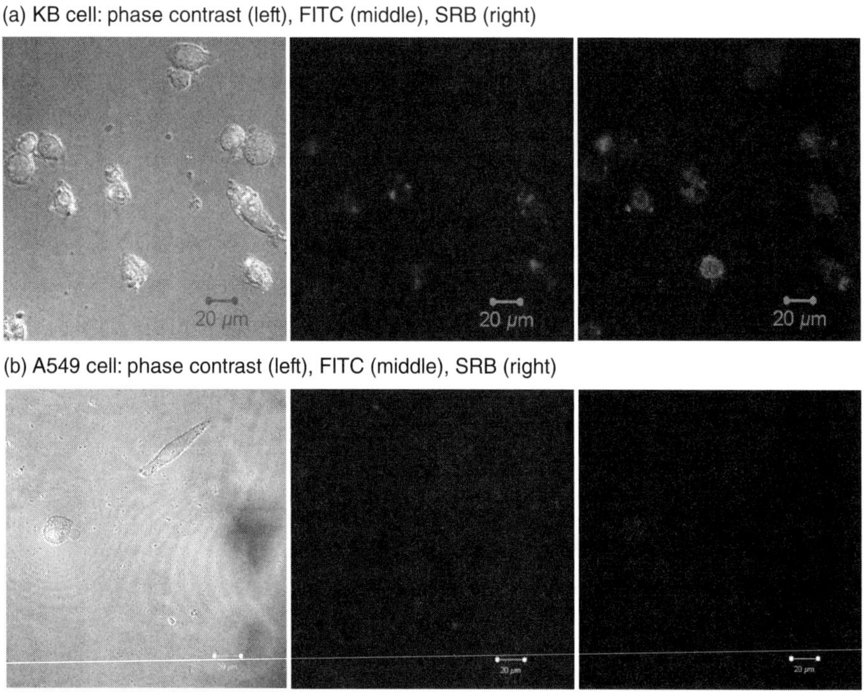

Figure 43.7. Phase contrast and fluorescence images by confocal laser microscopy of (a) KB cells and (b) A549 cells after the addition of the complexes. SRB is entrapped in the liposome and FITC is bound to the PAMAM dendrimer. The scale bar in the images is 20 μm.

defects and control surface charge, allowing cargo retention, delivery, and release inside cells.

As mentioned in Section 43.2, we have prepared complexes of an anionic liposome composed of PEG mono-oleyl ether with folate-conjugated cationic poly(amidoamine) dendrimers (FA-PAMAM) [13] in order to develop a new biomaterial for use as an active drug carrier. The uptakes of fluorescence-labeled complexes into KB cells and A549 cells were evaluated by confocal laser microscopy (Fig. 43.7). KB cells express excessive folate receptors, whereas A549 cells possess few folate receptors. SRB, a fluorescent reagent, was entrapped in the liposomes, and fluorescein isothiocyanate (FITC), another fluorescent reagent, was bound to PAMAM dendrimer. Figure 43.7 reveals that the complexes were drawn only into KB cells. Consequently, the complexes are expected to be effective for active drug targeting. The effect of the complexes on accelerating drug uptake into tumor tissues is currently being investigated.

REFERENCES

1 K. Kostarelos, Th. F. Tadros, P. F. Luckham. *Langmuir* 15 (1999) 369–376.

2 D. D. Lasic, F. J. Martin. *Stealth liposomes*. CRC Press, Boca Raton, FL, 1995.

3 S. C. Semple, A. Chonn, P. R. Cullis. *Biochemistry* 35 (1996) 2521–2525.

4 T. K. Bronich, S. V. Solomatin, A. A. Yaroslavov, A. Eisenberg, V. A. Kabanov, A. V. Kabaov. *Langmuir* 16 (2000) 4877–4881.

5 Z. Sideratou, D. Tsiourvas, C. M. Paleos. *Langmuir* 16 (2000) 9186–9191.

6 Z. Sideratou, J. Foundis, D. Tsiourvas, I. P. Nezis, G. Papadimas, C. M. Paleos. *Langmuir* 18 (2002) 5036–5039.

7 D. Volodkin, H. Mohwald, J. C. Voegel, V. Ball. *J. Control. Release* 117 (2007) 111–120.

8 K. Kawakami, Y. Nishihara, K. Hirano. *J. Phys. Chem. B* 105 (2001) 2374–2385.

9 A. A. Yaroslavov, V. Y. Kul'kov, A. A. Efimova, M. O. Ignatiev. *Thin Solid Films* 265 (1995) 66–70.

10 A. A. Yaroslavov, T. A. Sitnikova, A. A. Rakhnyanskaya, Y. A. Ermakov, T. V. Burova, V. Y. Grinberg, F. M. Menger. *Langmuir* 23 (2007) 7539–7544.

11 L. Zhang, S. Granick. *Nano Lett.* 6 (2006) 694–698.

12 D. A. Pink, M. Belaya, V. Levadny, B. Quinn. *Langmuir* 13 (1997) 1701–1711.

13 R. Ohori, K. Uchida, A. Saito, H. Yajima. *Chem. Lett.* 37 (2008) 324–325.

14 B. P. Devlin, D. A. Tirrell. *Macromolecules* 19 (1986) 2465–2466.

15 C. Kojima, K. Kono, K. Maruyama, T. Takagishi. *Bioconjugate Chem.* 11 (2000) 910–917.

16 J. Haensler, F. C. Szoka, Jr. *Bioconjugate Chem.* 4 (1993) 372–379.

17 S. Wang, R. J. Lee, C. J. Mathias, M. A. Green, P. S. Low. *Bioconjugate Chem.* 7 (1996) 56–62.

18 K. Kono, M. Liu, J. M. J. Fréchet. *Bioconjugate Chem.* 10 (1999) 1115–1121.

19 K. Kono, M. Iwamoto, R. Nishikawa, H. Yanagie, T. Takagishi. *J. Control. Relase* 68 (2000) 225–235.

20 Q. F. Ahkong, D. Fisher, W. Tampion, J. A. Lucy. *Nature* 253 (1975) 194–195.

21 K. Kono, A. Henmi, T. Takagishi. *Biochim. Biophys. Acta* 1421 (1999) 183–197.

22 D. Pornpattananangkul, S. Olson, S. Aryal, M. Sartor, C.-M. Huang, K. Vecchio, L. Zhang. *ACS Nano* 4 (2010) 1935–1942.

23 K. Fujimoto, T. Toyoda, Y. Fukui. *Macromolecules* 40 (2007) 5122–5128.

24 M. Yuasa, K. Oyaizu, A. Horiuchi, A. Ogata, T. Hatsugai, A. Yamaguchi, H. Kawakami. *Mol. Pharm.* 1 (2004) 387–389.

25 F. X. G. Han, R. T. Wheelhouse, L. H. Hurley. *J. Am. Chem. Soc.* 121 (1999) 3561–3570.

26 T. R. Cech. *Angew. Chem. Int. Ed.* 39 (2000) 34–43.

27 V. Sokolova, M. Epple. *Angew. Chem. Int. Ed.* 47 (2008) 1382–1395.

28 S. Giri, B. G. Trewyn, M. P. Stellmaker, V. S. Y. Lin. *Angew. Chem. Int. Ed.* 44 (2005) 5038–5044.

29 J. Liu, X. Jiang, C. Ashley, C. J. Brinker. *J. Am. Chem. Soc.* 131 (2009) 7567–7569.

44

CELL REGULATION THROUGH MEMBRANE RAFTS/CAVEOLAE

YOHEI SAITO, TOSHIYUKI OWAKI, AND FUMIO FUKAI

44.1 INTRODUCTION

All living creatures are made of cells. Cells are surrounded by a membrane called "plasma membrane," which is composed mainly of phospholipids and proteins embedded there. Cells can define their extent and maintain the essential differences between the contents and environment by compartmentalizing with the plasma membrane. The plasma membrane serves not only as a passive barrier but also as a highly active apparatus that transduces external stimuli into the inside. Thus, the plasma membrane is considered to be a biointerface that plays an essential role in maintaining cell vitality.

The fluid mosaic model proposed by Singer and Nicolson [1] has long been believed to best explain properties of the plasma membrane, which is a simple continuous bilayer of phospholipid molecules. However, the "membrane microdomains" was proposed by Stier and Sackmann [2]. Thereafter, Simons and Meer developed the concept of membrane microdomains, enriched with cholesterol, glycolipids, and sphingolipids, and designated these microdomains as "lipid rafts" [3]. Lipid rafts become stabilized by the presence of a family of specific structural proteins, caveolin (caveolin-1, -2, and -3) and forms small caves or flask-shaped invaginations (50–100 nm in size) of the plasma membrane, so-called caveolae. Caveolins are localized at the cytoplasmic surface of the caveolae but do not penetrate the caveolar membrane. Caveolin-1 plays

Electrical Phenomena at Interfaces and Biointerfaces: Fundamentals and Applications in Nano-, Bio-, and Environmental Sciences, First Edition. Edited by Hiroyuki Ohshima.

a key role in both maintaining the caveolae structure and interacting with a variety of signaling molecules, such as Src, Grb, Raf, MEK/ERK, and protein kinase C [4]. It has been postulated that membrane rafts/caveolae serve as platforms for cell signaling [5]. Rafts/caveolae have received increasing attention as biological devices for cell regulation.

The extracellular matrix (ECM) is not only necessary for tissue maintenance but also as a storehouse of signals for controlling various cellular processes, such as proliferation, survival, differentiation, and gene expression. Additionally, ECM proteins including fibronectin (FN) have biologically active cryptic sites, termed matricryptic sites [6]. These matricryptic sites could be often exposed by proteolytic degradation [7, 8]. Davis et al. [6] have coined the term "matricryptin" to express biologically active proteolytic fragments of ECM proteins. Matricryptic sites and matricryptins of ECM proteins have been implicated in a variety of cellular processes [6], particularly in pathological processes such as inflammation. We previously found that FN has a matricryptic site opposing cell adhesion [9]: A synthetic peptide derived from the 14th FN-III repeat, termed peptide FNIII14, suppresses the $\beta 1$-integrin-mediated cell adhesion to FN [10]. Peptide FNIII14 thus expresses the antiadhesive activity by inactivating the $\beta 1$-integrins. The C-terminal sequence YTIYVIAL of FNIII14, which is an active site, is cryptic in plasma FN molecule but exposed by either FN processing with MMP-2 or FN interaction with heparin [11]. By negatively modulating the cell–ECM interaction, peptide FNIII14 is capable of influencing physiological cellular processes such as survival [12], differentiation [13, 14], and gene expression [15], as well as pathological events including tumor metastasis [16].

There are several sequences analogous to the YTIYVIAL sequence of FN in other ECM proteins including TN-C [17]. An analogous sequence, YTITIRGV, is present in the FN-III repeat A2 of the alternative splicing region of human tenascin (TN)-C molecule. This gets us interested in the function of TN-C since TN-C is characterized by its adhesion modulatory function: It acts as an adhesive substrate for some cell types but is antiadhesive for many other cell types [18–20]. The antiadhesive effect of TN-C may be attributed to this sequence. In this chapter, we show that the YTITIRGV-containing TN-C peptide, termed peptide TNIIIA2, is capable of promoting/inducing cell adhesion to FN, in sharp contrast to the effect of peptide FNIII14 [21]. This TNIIIA2 activity is due to induction of the conformational change in $\beta 1$-integrin from resting to active. Syndecan-4, a cell surface heparan sulfate proteoglycan (HSPG), is shown to mediate the effect of TNIIIA2 [21]. Moreover, the $\beta 1$-integrin activation by TNIIIA2 essentially requires the integration of $\beta 1$-integrin into lipid rafts/caveolae of plasma membrane. This provides a new mechanism for the $\beta 1$-integrin activation through a lateral interaction between membrane molecules in rafts/caveolae. We also show that a variety of leukemic cell lines undergo apoptotic death during a prolonged adhesion to FN forced by peptide TNIIIA2 [22]. This apoptosis is also dependent on membrane rafts/caveolae. Thus, data presented in this chapter prompt us to speculate that modulation of membrane rafts/caveolae may enable us to control fundamental cellular processes, including cell survival and growth.

44.2 TNIIIA2, A TN-C-DERIVED PEPTIDE, STIMULATES CELL ADHESION TO FIBRONECTIN

We first examined the effect of peptide TNIIIA2 on cell adhesion to FN in the presence or absence of peptide TNIIIA2 using normal mouse fibroblasts NIH3T3. As shown in Figure 44.1a ("control"), in the absence of peptide, many cells attached to the FN but remained to spread 60 minutes after seeding. In the presence of peptide TNIIIA2, most cells already spread on FN at that time ("+TNIIIA2" of Fig. 44.1a,b). A control peptide TNIIIA2scr, in which the YTITIRGV sequence is shuffled (RITYITGV), showed no effect on cell adhesion. Peptide TNIIIA2 stimulated cell adhesion in a concentration-dependent manner, and which was reversed by function-blocking mAbs directed to integrin subunits $\alpha5$ and $\beta1$ (Fig. 44.1b) but weakly by anti- $\alpha4$ Ab.

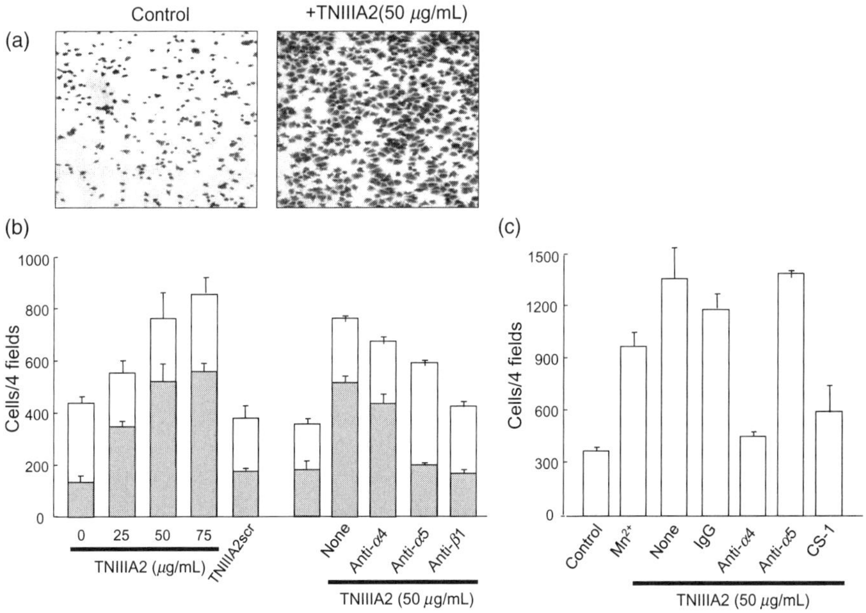

Figure 44.1. Effects of peptide TNIIIA2 on cell adhesion to FN. NIH3T3 (a,b) and Ramos (c) cells suspended in the serum-free medium in the presence of peptide TNIIIA2 or TNIIIA2scr (50 μ/mL) were seeded with or without Mn^{2+} (1 mM) or anti-integrin mAb (20 μg/mL) on a 96-well plate coated with FN. Cells were allowed to adhere for 1 hour, and the number of cells either attached and spread (b, gray bars) or attached (b,c; open bars) were counted, as described earlier [9, 10]. Data represent the mean ± SD of three determinations.

Peptide TNIIIA2 also stimulated adhesion of nonadherent hematopoietic progenitor cell line Ramos, which mainly expresses $\alpha 4\beta 1$ as $\beta 1$ class integrin, whose integrins are in an inactive state. Ramos cells showed specific adhesion to FN only in the presence of Mn^{2+}, a well-known integrin activator (Fig. 44.1c). Peptide TNIIIA2 induced Ramos cell adhesion to FN even without Mn^{2+}, and which was reversed by anti-$\alpha 4$ mAb but not by anti-$\alpha 5$ mAb (Fig. 44.1c). Thus, peptide TNIIIA2 stimulated adhesion of nonadherent as well as adherent cell types to FN via FN receptors ($\alpha 4\beta 1$ and $\alpha 5\beta 1$), suggesting that this peptide might induce the FN receptors in functional activation.

44.3 CELL ADHESION INDUCED BY TNIIIA2 IS ATTRIBUTED TO FUNCTIONAL ACTIVATION OF β1-INTEGRINS

The activation of the FN receptor can be judged by its binding with anti-$\beta 1$-integrin mAbs, such as AG89 [23] and HUTS-4 [24] because these mAbs selectively bind to the active conformation-specific epitope of $\beta 1$-integrin subunit regardless of the α subunit. We therefore evaluated the effect of peptide TNIIIA2 on the conformational status of $\beta 1$-integrins, as judged by flow cytometric analysis using these mAbs.

Ramos cells without any stimulation showed a low accessiblity with AG89, while stimulation with Mn^{2+} increased expression of the AG89 epitope (Fig. 44.2). Peptide TNIIIA2 induced a marked increase in the expression of the AG89 epitope even without Mn^{2+} (Fig. 44.2). No significant change in expression of integrin $\alpha 4\beta 1$ was observed with peptide TNIIIA2, as confirmed by immunoblot analysis (data not shown). Expression of the active $\beta 1$-specific epitope induced by stimulation with peptide TNIIIA2 was observed similarly by using another active conformation-specific mAb HUTS-4 (data not shown).

Peptide TNIIIA2 thus induced the functional activation of $\beta 1$-integrins by changing its conformation from an inactive to an active one.

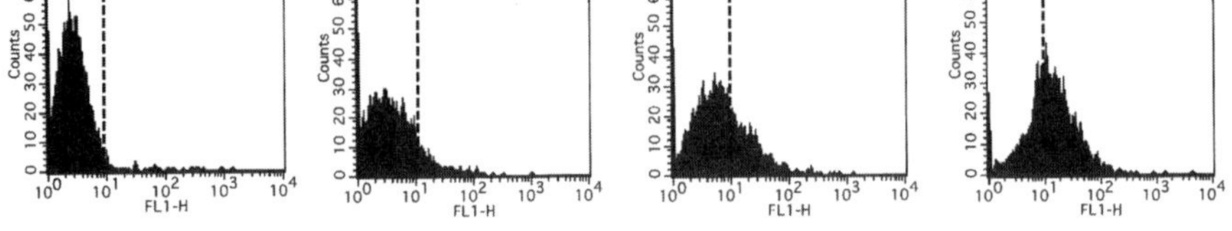

Figure 44.2. Peptide TNIIIA2 induces the conformational change of the $\beta 1$-integrins. Ramos cells were treated with (a) nonimmune mouse IgG, (b) Mn^{2+}, and (c) 25 or(d) 50 μg/mL of TNIIIA2 (50 μg/mL), respectively. Cells were incubated with fluorescein isothiocyanate (FITC) AG89 and then were subjected to flow cytometry.

44.4 CATIONIC PROPERTY OF PEPTIDE TNIIIA2 IS CRUCIAL FOR THE ACTIVATION OF β1-INTEGRINS

The alanine scanning mutagenesis was performed to identify amino acid residues of the active sequence YTITIRGV essential for the β1-integrin activation. Table 44.1 represents the mutant peptides examined here and summarizes the results of their activities evaluated by flow cytometric analysis using AG89 (data not shown). The β1-integrin activation by TNIIIA2 was highly sensitive to the single alanine replacement of two Ile (#16 and 18) and Val (#21) residues. Note that the replacement of Arg#19 to Ala resulted only in a partial reduction of the TNIIIA2 activity, while the replacement of Arg#19 to Glu, instead of Ala, elicited a complete loss of TNIIIA2 activity. The TNIIIA2 activity was also abrogated by replacing the basic amino acid Lys#9 to Glu. The mutant peptide K9R19K with substitutions of Lys#9 to Arg and Arg#19 to Lys, in which the cationic property of peptide TNIIIA2 remained almost unchanged, caused no reduction of the TNIIIA2 activity.

Thus, the scanning mutagenesis indicated not only the sequential specificity but the essential role of cationic property of peptide TNIIIA2 in β1-integrin activation. These results suggested that a certain acidic cell surface molecule(s) such as proteoglycan (PG) might be involved in the expression of TNIIIA2 activity.

44.5 TNIIIA2 INDUCES β1-INTEGRIN ACTIVATION THROUGH BINDING WITH SYNDECAN-4 IN A MEMBRANE RAFTS/ CAVEOLAE-DEPENDENT MANNER

To answer the question raised earlier, we examined whether free glycosaminoglycan (GAG) chains can influence the β1-integrin activation induced by TNIIIA2. The TNIIIA2 activity was abrogated completely by addition of heparin, partially by chondroitin sulfate, hardly by hyaluronic acid [21]. The data suggest the involvement of the sulfated GAG chain in the expression of the TNIIIA2 activity. In fact, treatment of NIH3T3 cells with heparitinase I abrogated the β1-integrin activation by peptide TNIIIA2 [21].

To examine the possibility that TNIIIA2 could induce β1-integrin activation through binding with a membrane type heparan sulfate (HS) PG, an attempt to probe directly cell surface molecule biding to peptide TNIIIA2 was carried out. A TNIIIA2-derivative containing a photoreactive amino acid L-p-benzoylphenylalanine (Bpa) and biotin, biotinylated TNIIIA2-Bpa, was developed. The biotinylated TNIIIA2-Bpa retained the TNIIIA2 activity (data not shown). By labeling human fibroblastic cells WI38VA13 using this photoreactive probe, a broad band of high molecular mass (>200 kDa) and two clear bands of 110 and 70 kDa were detected with avidin-conjugated peroxidase (Fig. 44.3a). When the photolabeling was performed in the presence of heparin, the high-molecular-mass broad band disappeared (Fig. 44.3a, arrowhead). The

TABLE 44.1. Scanning Mutagenesis of Peptide TNIIIA2[a]

Peptides	Sequence Positions																					TNIIIA2 Activity[b]
	1	2	3	4	5	6	7	8	9	10	11	12	13	14	15	16	17	18	19	20	21	
TNIIIA2 (N)	R	S	T	D	L	P	G	L	K	A	A	T	H	Y	T	I	T	I	R	G	V (C)	++
I16A	R	S	T	D	L	P	G	L	K	A	A	T	H	Y	T	A	T	I	R	G	V	–
I18A	R	S	T	D	L	P	G	L	K	A	A	T	H	Y	T	I	T	A	R	G	V	–
R19A	R	S	T	D	L	P	G	L	K	A	A	T	H	Y	T	I	T	I	A	G	V	+
R19E	R	S	T	D	L	P	G	L	K	A	A	T	H	Y	T	I	T	I	E	G	V	–
K9E	R	S	T	D	L	P	G	L	E	A	A	T	H	Y	T	I	T	I	R	G	V	–
K9R19K	R	S	T	D	L	P	G	L	R	A	A	T	H	Y	T	I	T	I	K	G	V	++

Sequence positions mutated are underlined.
[a] Data taken from a previous study [21].
[b] TNIIIA2 activity evaluated by flow cytometric analysis using mAb AG89 recognizing the active β1-integrin-specific epitope: –, no reactivity; +, moderate (about a half) reactivity; ++, high (over about 70%) reactivity.

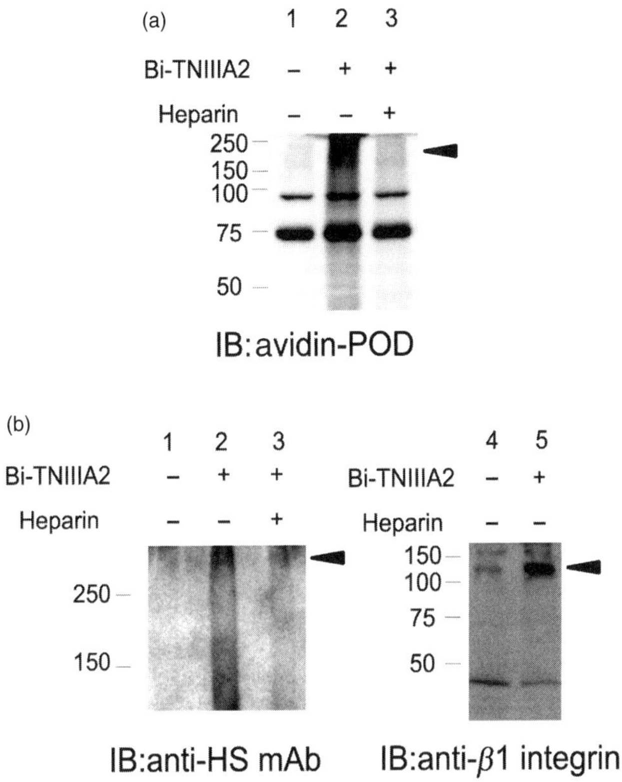

Figure 44.3. Photoaffinity labeling of cell surface molecule mediating the TNIIIA2 activity. (a) Photoaffinity labeling of WI38VA13 cells without (lane 1) or with biotinylated TNIIIA2-Bpa was performed in the absence (lane 2) or presence (lane 3) of heparin as describe previously [21]. Biotinylated TNIIIA2-Bpa-proved molecules were detected by immunoblot (IB) analysis using avidin–peroxidase (POD). (b) WI38VA13 cells photoproved as prepared above (lanes 1–3 of panel a) were dissolved, precipitated with avidin beads, and then subjected to immunoblot analysis using anti-HS antibody (lane 1–3) or anti-β1 integrin mAb (DE9) (lanes 4 and 5).

TNIIIA2-binding molecule(s) labeled with biotinylated TNIIIA2-Bpa was precipitated, taking advantage of the specific interaction between biotin and avidin. When the precipitate was subjected to immunoblot analysis, the high-molecular-mass broad band was visualized also by using anti-HS mAb (Fig. 44.3b, lane 5). Importantly, the β1-subunit was detected in the precipitates as a sharp band migrating at a position quite different from the broad band detected with anti-HS mAb (Fig. 44.3B). This suggested that peptide TNIIIA2 bound specifically to HSPG, not to β1-integrin, and might promote the binding of β1-integrin with HSPG.

To verify further the involvement of syndecan-4 in the expression of TNIIIA2 activity, we examined the effect of reduced expression of syndecan-4

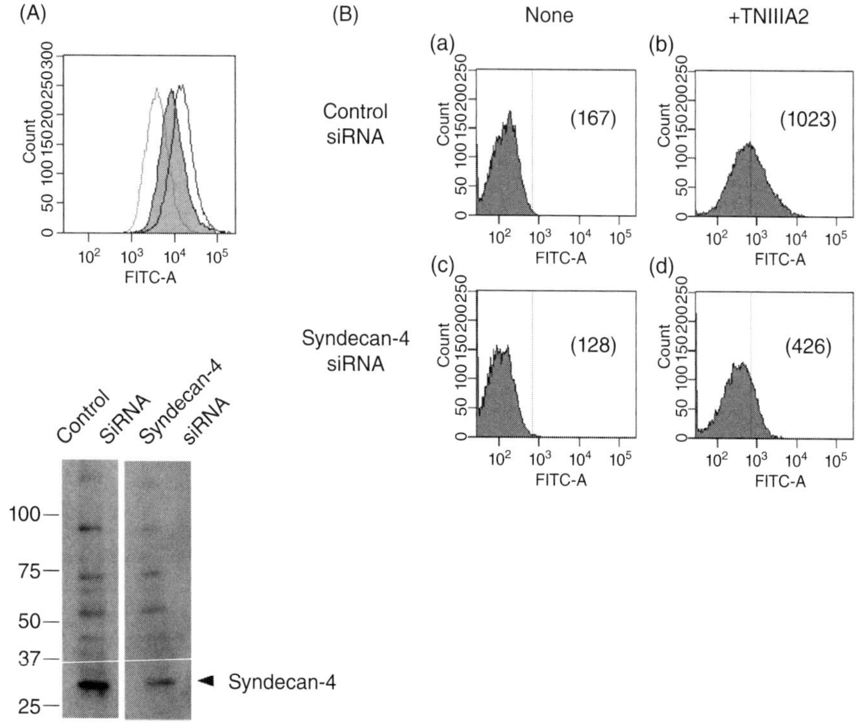

Figure 44.4. Syndecan-4 mediates the TNIIIA2-induced β1-integrin activation. Expression of syndecan-4 in WI38VA13 cells was evaluated by flow cytometry using anti-syndecan-4 antibody (A): thin line—without Ab, solid line—control siRNA, filled—syndecan-4 siRNA. WI38VA13 cells transfected with control siRNA (B-a, b) or syndecan-4 siRNA (B-c,d) were treated with (B-b,d) or without peptide TNIIIA2 (25 μg/mL) and then were subjected to flow cytometric analysis using AG89 (B).

by RNA interference. Flow cytometric analysis showed that the syndecan-4 expression on WI38VA13 cells was reduced about 30% by the introduction of siRNA targeted against syndecan-4 core protein (Fig. 44.4A). In untreated WI38VA13 cells, peptide TNIIIA2 remarkably promoted expression of the AG89 epitope, while reduced expression of syndecan-4 resulted in a reduction of the TNIIIA2-induced promotion of the AG89 epitope expression (Fig. 44.4B). Transfection with control siRNA did not show any significant change in the expression of the AG89 epitope (Fig. 44.4B). We next observed a forced expression of the syndecan-4 core protein with complete deletion of the cytoplasmic domain that is indispensable for the syndecan-4-based signaling [25]. However, the β1-integrin activation in response to peptide TNIIIA2 remained unchanged [21]. This suggested that the TNIIIA2 activity might depend on the extracellular domain of a syndecan-4 molecule but not on its cytplasmic domains.

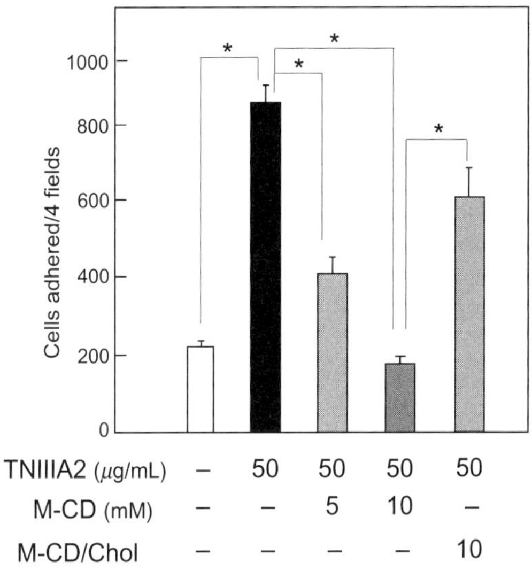

Figure 44.5. TNIIIA2-induced stimulation of cell adhesion requires integration of membrane rafts/caveolae. Ramos cells suspended in the serum-free medium in the presence of peptide TNIIIA2 (50 μg/mL) were seeded with or without β-methyl-cyclodextrin (M-CD) (5–10 mM) or M-CD (10 mM)-cholesterol inclusion complex on a 96-well plate coated with FN. Cells were allowed to adhere for 1 hour, and the number of cells either attached was counted, as described earlier. Data represent the mean ± SD of three determinations. * $P < 0.001$ compared with control.

Recent studies have demonstrated that cholesterol-enriched membrane rafts/caveolae act as a platform for signal transduction. To further explore the mechanism of β1-integrin activation following syndecan-4 ligation with peptide TNIIIA2, we examined the effect of disruption of membrane rafts/caveolae by depletion of cholesterol with β-methyl-cyclodextrin (M-CD). As shown in Figure 44.5, cholesterol depletion by the addition of M-CD prevented the TNIIIA2-stimulated adhesion of Ramos cells to the FN. This prevention was reversed by resupplementation of cholesterol with the M-CD–cholesterol inclusion complex (Fig. 44.5). It was thus suggested that TNIIIA2 induces β1-integrin activation through biding with syndecan-4 in a membrane rafts/caveolae-dependent manner.

44.6 FORCED CELL ADHESION BY TNIIIA2 TO FIBRONECTIN SUBSTRATE LEADS LEUKEMIC CELLS TO APOPTOTIC DEATH IN A MEMBRANE RAFTS/CAVEOLAE-DEPENDENT MANNER

As mentioned earlier (see Fig. 44.1), Ramos cells can adhere to fibronectin substrate in the presence of TNIIIA2. Ramos cell adhesion to FN induced by

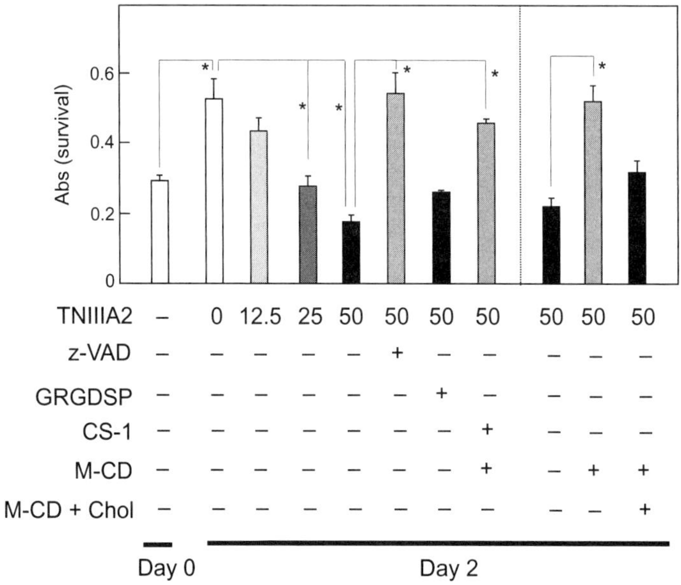

Figure 44.6. Apoptosis induction of Ramos cells through the rafts/caveolae-dependent activation of the FN receptors. Ramos cell suspension (3×10^4 cells) with or without TNIIIA2 (12.5–50 μg/mL) was cultured in the presence or absence of a general caspase inhibitor, z-VAD fmk (50 μM), or CS-1 or GRGDSP peptide (200 μg/mL), or β-methyl-cyclodextrin (M-CD) (10 mM), or M-CD–cholesterol inclusion complex. After 2 days in culture, viable cells were detected by Cell Counting Kit (Dojindo, Tokyo, Japan). Each point represents the mean ± SE of triplicate determinations. One of three individual experiments is shown. * $P < 0.001$ compared with control.

TNIIIA2 seemed to be maintained at least 2 days. During this forced adhesion, proliferation of Ramos cells was suppressed depending on added concentration of TNIIIA2 (Fig. 44.6). TS2/16 or Mg^{2+}, which is known to act as a β1-integrin activator, also had the ability to suppress Ramos cell growth (data not shown) [22]. TNIIIA2-induced suppression of Ramos cell growth was blocked almost completely in the presence of z-VAD fmk, a general caspase inhibitor, z-VAD fmk (Fig. 44.6), suggesting that forced adhesion to fibronectin leads Ramos cells to caspase-dependent apoptosis. Supporting this observation, Ramos cells showed a change in nuclear morphology characteristic of apoptotic cells, so-called nuclear condensation and its fragmentation [22]. Apoptosis was further confirmed by the appearance of phosphatidylserine on cell surfaces, detected by binding with annexin V [22]. This TNIIIA2-induced apoptosis was reversed by pretreating cells with CS-1 peptide, an antagonist for $\alpha 4\beta 1$, but not with GRGDSP peptide, an antagonist for $\alpha 5\beta 1$, suggesting that TNIIIA2 induced apoptosis in Ramos cells through adhesion via $\alpha 4\beta 1$ to the FN. Importantly, this adhesion-dependent apoptosis of Ramos cells was prevented through disruption of membrane rafts/caveolae by depletion of cholesterol with M-CD (Fig. 44.6, right panel). This prevention was reversed by the resupplementation of cholesterol (Fig. 44.6, right panel).

These results suggest that prolonged adhesion of leukemic cells to a fibronectin substrate forced by β1-integrin activation leads Ramos cells to apoptosis in a membrane rafts/caveolae-dependent manner.

44.7 ADHESION-DEPENDENT APOPTOSIS IN OTHER HEMATOPOIETIC TUMOR CELLS

To ascertain whether adhesion-dependent apoptosis is inducible also in other types of hematopoietic tumor cells, a variety of leukemic cell lines, which express different levels of fibronectin-receptors ($\alpha4\beta1$ and $\alpha5\beta1$) and syndecan-4 as a receptor of TNIIIA2, were tested. Similar to Ramos cells, TNIIIA2-induced, adhesion-dependent apoptosis in other leukemic cell lines was shown to depend on integration of membrane rafts/caveolae (data not shown). Results were summarized in Table 44.2. Integrin activators, such as TS2/16 and Mg^{2+}, were capable of inducing adhesion to fibronectin, resulting in apoptosis after prolonged adhesion. These integrin activators induced apoptosis in monocytes and lymphocytes from peripheral blood. In contrast to these integrin activators, TNIIIA2 requires syndecan-4 expression as a membrane receptor for β1-integrin activation; normal peripheral blood cells such as neutrophils, monocytes, and lymphocytes, which poorly expressed syndecan-4, are almost insusceptible to TNIIIA2-induced apoptosis. The data suggest that leukemic cells expressing $\alpha4\beta1$ undergo apoptosis after prolonged adhesion to fibronectin in a membrane rafts/caveolae-dependent manner, while TNIIIA2 additionally requires syndecan-4 for the apoptosis induction.

44.8 CONCLUSION

Here, we showed that peptide TNIIIA2 derived from the FN-III repeat A2 of TN-C molecule is capable of stimulating cell adhesion to FN through functional activation of β1-integrins [21]. This TNIIIA2 activity was shown to be mediated by syndecan-4. TNIIIA2 promoted the physical association of syndecan-4 with β1-integrins. Moreover, the extracellular domain of syndecan-4 was sufficient for the TNIIIA2-induced β1-integrin activation [21]. These results suggest that the TNIIIA2-induced β1-integrin activation is attributed to the lateral interaction of syndecan-4 with β1-integrins. The activation of β1-integrins through its lateral interaction with syndecan-4 may be more effectively achieved when these membrane molecules are accumulated into a certain area of plasma membrane. Supporting the assumption, the TNIIIA2-stimulated Ramos cell adhesion to the FN was abrogated through disruption of membrane rafts/caveolae by cholesterol depletion.

TABLE 44.2. Expression of $\alpha4\beta1$, $\alpha5\beta1$, and Syndecan-4, and Effects on Adhesion-Dependent Apoptosis of Hematopoietic Progenitor Cell Lines, Fresh Leukemic Cells from Acute Myelogenous Leukemia (AML) Patients, and Peripheral Blood Cells from Normal Adults (Taken from the Data Reported by Saito et al. [22])

	Expression[a]			Adhesion Induced by[b]		Apoptosis Induced by[c]	
Cells	$\alpha4\beta1$	$\alpha5\beta1$	Syndecan-4	Mg^{2+}	TNIIIA2	Mg^{2+}	TNIIIA2
Cell lines							
B cell							
Ramos	96.5	3.2	92.1	++	+++	++	+++
Rajii	94.3	65.3	2.3	++	–	+	–
Erythroid							
K562	9.3	97.2	66.7	++	+++	–	–
Myeloid							
U937	98.1	98.2	87.5	++	+++	+++	++
Fresh AML cells							
Patient A	98.2	88.8	48.8	++	++	++	+
Peripheral blood cells							
Neutrophil	6.7	ND	3.2	–	–	–	–
Monocyte	48.6	ND	2.0	++	–	++	–
Lymphocyte	40.5	ND	4.4	++	–	–	–

Data show the fraction of cells adhered or died of total cells: +++, > about 70%; ++, around a half; +, < about 30%; –, almost negative or none.

[a] Expression of VLA-4, VLA-5, and syndecan-4 was analyzed by flow cytometry using corresponding Abs.

[b,c] Effects of Mg^{2+} (10 mM) or TNIIIA2 (50 μg/mL) on cell adhesion to the FN ([b]) and consequent apoptosis ([c]) were evaluated by the cell adhesion assay and the DNA degrdadation, respectively.

ND, not determined.

Accumulation of syndecan-4 and $\beta1$-integrins into membrane rafts/caveolae might enable a lateral interaction of $\beta1$-integrins with syndecan-4. This might induce the $\beta1$-integrin activation, resulting in stimulation/induction of cell adhesion.

It has been postulated that $\beta1$-integrins in their inactive conformation are involved in the disordered proliferation of hematopoietic tumor cells [26]. In general, leukemic cells, including Ramos Burkitt's lymphoma cells, can proliferate in suspension without anchoring on a solid substrate such as FN. We found that Ramos cells undergo apoptosis when kept adhering to FN by stimulating with TNIIIA2. Since apoptosis was also induced by using other integrin activators, Mg^{2+} and TS2/16 (an integrin-activating antibody) [22], it was presumed that potentiated and sustained adhesion to FN causally induced apoptosis. Results using other hematopoietic tumor cell lines expressing different levels of fibronectin receptors ($\alpha4\beta1$ and $\alpha5\beta1$) showed that adhesion via $\alpha4\beta1$-integrin was attributed to the signaling responsible for apoptosis [22]. TNIIIA2 needed syndecan-4 to activate $\beta1$-integrins. This might be the reason why TNIIIA2 induced apoptosis in hematopoietic tumor cells that expressed both VLA-4 and syndecan-4, but hardly in normal peripheral blood cells, such as neutrophils, monocytes, and lymphocytes, which were poor in syndecan-4 expression. The TNIIIA2-related matricryptic site of TN-C might contribute, once exposed, to preventing the prolonged survival of hematopoietic malignant progenitors through potentiated and sustained activation of $\alpha4\beta1$. Although the precise molecular mechanism responsible for the modulation of $\beta1$-integrin activation by peptide TNIIIA2 is not fully understood, this peptide factor would be a useful tool for therapeutic intervention.

REFERENCES

1 S. J. Singer, G. L. Nicolson. The fluid mosaic model of the structure of cell membranes. *Science* 175 (1972) 720–731.

2 A. Stier, E. Sackmann. Spin labels as enzyme substrates. Heterogeneous lipid distribution in liver microsomal membranes. *Biochim. Biophys. Acta* 311 (1973) 400–408.

3 K. Simons, G. Meer. Lipid sorting in epithelial cells. *Biochemistry* 27 (1982) 6197–6202.

4 R. G. Anderson. The caveolae membrane system. *Ann. Rev. Biochem.* 67 (1998) 199–225.

5 K. Simons, D. Toomre. Lipid I rafts and signal transduction. *Nature Rev. Mol. Cell Biol.* 1 (2000) 31–39.

6 G. E. Davis, K. J. Bayless, M. J. Davis, G. A. Meininger. Regulation of tissue injury response by the exposure of matricryptic sites within extracellular matrix molecules. *Am. J. Pathol.* 156 (2000) 1489–1498.

7 W. G. Stetler-Stevenson. Dynamics of matrix turnover during pathologic remodeling of the extracellular matrix. *Am. J. Pathol.* 148 (1996) 1345–1350.

8 E. H. Sage. Bioactive fragments of extracellular proteins as regulators of angiogenesis. *Trends Cell Biol.* 7 (1997) 182–186.

9 F. Fukai, H. Takahashi, Y. Habu, N. Kubushiro-Kamiya, T. Katayama. Fibronectin harobors anti-celladhesive activity. *Biochem. Biophys. Res. Commun.* 220 (1996) 394–398.

10 F. Fukai, S. Hasebe, M. Ueki, M. Mutoh, C. Ohgi, H. Takahashi, K. Takeda, T. Katayama. Identification of the anti-adhersive site within the heparin-binding domain of fibronectin. *J. Biochem. (Tokyo)* 121 (1997) 189–192.

11 K. Watanabe, H. Takahashi, Y. Habu, N. Kamiya-Kubushiro, S. Kamiya, H. Nakamura, H. Yajima, T. Ishii, T. Katayama, K. Miyazaki, F. Fukai. Interaction with heparin and matrix metalloproteinase 2 cleavage expose a cryptic anti-adhesive site of fibronectin. *Biochemistry* 39 (2000) 7138–7144.

12 F. Fukai, M. Mashimo, K. Akiyama, T. Goto, S. Tanuma, T. Katayama. Modulation of apoptotic cell death by extracellular matrix proteins and a fibronectin-derived antiadhesive peptide. *Exp. Cell Res.* 242 (1998) 92–99.

13 S. Kamiya, R. Kato, M. Wakabayashi, T. Tohyama, I. Eanami, M. Ueki, H. Yajima, T. Ishii, H. Nakamura, T. Katayama, J. Takagi, F. Fukai. Fibronectin peptides derived from two distinct regions stimulates adipocyte differentiation by preventing fibronectin matrix assembly. *Biochemistry* 41 (2002) 3270–3277.

14 R. Tanaka, T. Owaki, S. Kamiya, T. Matsunaga, K. Shimoda, H. Kodama, R. Hayashi, T. Abe, Y. P. Harada, M. Shimonaka, H. Yajima, H. Terada, F. Fukai. VLA-5-mediated adhesion to fibronectin accelerates hemin-stimulated erythroid differentiation of K562 cells through induction of VLA-4 expression. *J. Biol. Chem.* 284(30)(2009) 19817–19825.

15 S. Kamiya, T. Kawaguchi, S. Hasebe, S. Kamiya, Y. Saito, S. Miura, S. Wada, H. Yajima, T. Katayama, F. Fukai. A fibronectin fragment induces tumor necrosis factor production of rat basophilic leukemia cells. *Biochim. Biophys. Acta.* 1675 (2004) 87–94.

16 R. Kato, T. Ishikawa, S. Kamiya, F. Oguma, M. Ueki, S. Goto, H. Nakamura, T. Katayama, F. Fukai. A new type of antimetastatic peptide derived from fibronectin. *Clin. Cancer Res.* 8 (2002) 2455–2462.

17 S. Miura, S. Kamiya, Y. Saito, S. Wada, R. Hayashi, J. Taira, H. Kodama, H. Yajima, M. Ueki, F. Fukai. Antiadhesive sites present in the fibronectin type III-like repeats of human plasma fibronectin. *Biol Pharm Bull.* 30(5)(2007) 891–897.

18 M. A. Bourdon, E. Ruoslahti. Tenascin mediates cell attachment throiugh an RGD-dependent receptor. *J. Cell Biol.* 108 (1989) 1149–1155.

19 R. Chiquet-Ehrismann, P. Kalla, C. A. Pearson, K. Beck, M. Chiquet. Tenascin interferes with fibronectin action. *Cell* 53 (1988) 383–390.

20 P. Joshi, C. Y. Chung, I. Aukhil, H. P. Erickson. Endothelial cells adhere to the RGD domain and the fibrinogen-like terminal knob of tenascin. *J. Cell Sci.* 106 (1993) 389–400.

21 Y. Saito, H. Imazeki, S. Miura, T. Yoshimura, H. Okutsu, Y. Harada, T. Ohwaki, O. Nagao, S. Kamiya, R. Hayashi, H. Kodama, H. Handa, T. Yoshida, F. Fukai. A peptide derived from tenascin-C induces beta1 integrin activation through syndecan-4. *J. Biol. Chem.* 282(48)(2007) 34929–34937.

22 Y. Saito, T. Owaki, T. Matsunaga, M. Saze, S. Miura, M. Maeda, M. Eguchi, R. Tanaka, J. Taira, H. Kodama, S. Goto, Y. Niitsu, H. Terada, F. Fukai. Apoptotic death of hematopoietic tumor cells through potentiated and sustained adhesion to fibronectin via VLA-4. *J. Biol. Chem.* 285(10)(2010) 7006–7015.

23 J. Takagi, T. Isobe, Y. Takada, Y. Saito. Structural interlock between ligand-binding site and stalk-like region of beta1 integrin revealed by a monoclonal antibody recognizing conformation-dependent epitope. *J. Biochem. (Tokyo)* 121 (1997) 914–921.

24 A. Luque, M. Gomez, W. Puzon, Y. Takada, F. Sanchez-Madrid, C. Cabanas. Activated conformations of very late activation integrins detected by a group of antibodies (HUTS) specific for a novel regulatory region (355-425) of the common beta1 chain. *J. Biol. Chem.* 271 (1996) 11067–11075.

25 E. S. Oh, A. Woods, S. T. Lim, A. W. Theibert, J. R. Couchman. Syndecan-4 proteoglycan cytoplasmic domain and phosphatidylinositol 4,5-bisphosphate coordinately regulate protein kinase C activity. *J. Biol. Chem.* 273 (1998) 10624–10629.

26 R. W. Hurley, J. B. McCarthy, C. M. Verfaillie. Direct adhesion to bone marrow stroma via fibronectin receptors inhibits hematopoietic progenitor proliferation. *J. Clin. Invest.* 96 (1995) 511–519.

45

OXIDOREDUCTASES: ASYMMETRIC REDUCTION USING PHOTOSYNTHETIC ORGANISMS

Tetsuo Takemura and Kaoru Nakamura

45.1 INTRODUCTION

Synthesis of optically active compounds has been one of the important targets in organic synthesis especially for drug synthesis since more than half of drug candidate molecules have more than one chiral centers [1]. To synthesize them in environmentally friendly methods, catalysts play an important role, and both chemical and biological catalysts need to be developed because they are complementary to each other. Now, about 10% of the total drug synthesis depends on biocatalysts. Biocatalytic processes used in the synthesis of chiral intermediates for pharmaceuticals have been reviewed [2].

Reduction is an important biocatalytic reaction since in asymmetric reduction, optically active compounds can be obtained directly from achiral ketones in one step. With recent remarkable progress, the number of commercially available and easy-to-handle oxidoreductases is increasing. For example, 38 of various types of reducing enzymes (ChiralScreen™ OH) are available from Daicel [3]. Codexis also prepares a large number of various types of screening kit for the reduction of ketones, diketones, and keto esters as well as for the reduction of olefin [4]. With the increased number of available enzymes, suitable biocatalysts having high enantioselectivities for a variety of substrates are easier to find. Coenzymes such as nicotinamide or flavin play

Electrical Phenomena at Interfaces and Biointerfaces: Fundamentals and Applications in Nano-, Bio-, and Environmental Sciences, First Edition. Edited by Hiroyuki Ohshima.

an important role in biotransformation since almost all oxidoreductases use them and coenzymes stoichiometrically reduce or oxidize substrates through electron transfer with the aid of enzymes. In this chapter, a brief outline of the reaction mechanism of oxidoreductase and the recycling of coenzymes are described, and the use of photosynthetic organisms for asymmetric reduction of ketones is reviewed. Previous outlooks on biocatalytic reduction are reviewed in recent reports [5].

45.2 REACTION MECHANISM

45.2.1 Mechanism for Nicotinamide Coenzymes

A dehydrogenase usually requires a coenzyme such as nicotinamide adenine dinucleotide (NADH or NAD^+) or nicotinamide adenine dinucleotide phosphate (NADPH or $NADP^+$). Enzymes and coenzymes work together to catalyze the reduction or oxidation reaction of substrates. For example, the reduction of a ketone with the reduced form of the coenzyme (NAD(P)H) proceeds as follows:

1. NAD(P)H, a coenzyme, and a ketone bind to a dehydrogenase.
2. The ketone is reduced, while the coenzyme is oxidized. Thus, a hydride transfers from NAD(P)H (the reduced form of the coenzyme) to the ketone with the aid of a proton, giving $NAD(P)^+$ (the oxidized form of the coenzyme) and the corresponding alcohol (Fig. 45.1).
3. The coenzyme and product dissociate from the enzyme.

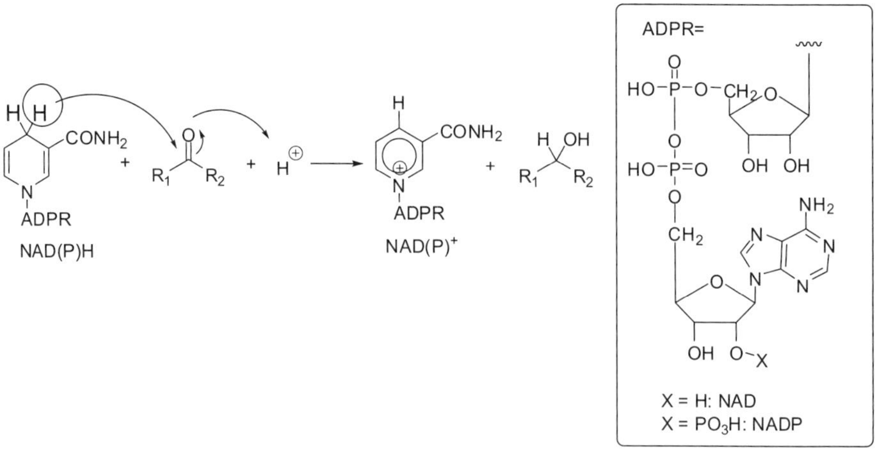

Figure 45.1. Reduction of a ketone with NAD(P)H in a dehydrogenase. ADPR, adenosine diphosphate ribose.

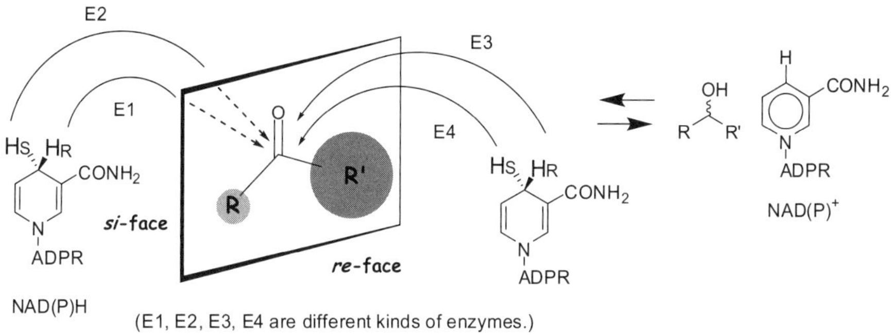

Figure 45.2. Stereochemistry for an NAD(P)H-dependent dehydrogenase.

45.2.2 Stereochemistry for NAD(P)H-Dependent Dehydrogenases

The reaction mechanism of an NAD(P)H-dependent dehydrogenase has been studied in detail. There are four stereochemical patterns for the transfer of a hydride from a coenzyme, NAD(P)H, to a substrate as shown in Figure 45.2. The hydride attacks either *si*-face or *re*-face of the carbonyl group depending on the orientation of the binding of the substrate to the enzyme, which results in the formation of (*R*) and (*S*)-alcohols, respectively. On the other hand, the enzyme transfers either the pro-(*R*)-hydride or the pro-(*S*)-hydride of the coenzyme depending on the kind of enzyme.

The crystal structure of an active site of an NADH-dependent enzyme, (*R*)-2-hydroxyglutarate dehydrogenase (HGDH), is determined at the resolution of 1.98 Å, as shown in Figure 45.3 [6]. HGDH catalyzes the reduction of 2-oxoglutarate to (*R*)-2-hydroxyglutarate. Arg235 interacts with the substrate's α-carboxylate and carbonyl groups, having a dual role in both substrate binding and activation, and the γ-carboxylate group can dock at an arginine cluster composed of Arg76, Arg52, and Arg9. The proton relay system built up by Glu264 and His297 permits His297 to act as an acid–base catalyst, and the pro-(*R*)-hydrogen from NADH is transferred as hydride to the carbonyl group *si*-face, leading to the formation of (*R*)-2-hydroxyglutarate (Fig. 45.3).

45.2.3 Mechanism for Flavin-Dependent Enzymes

Reaction mechanisms of the Baeyer–Villiger oxidation of ketones by phenylacetone monooxygenase from *Thermobifida fusca* is described here as an example for the flavin-dependent enzyme. The enzyme has flavin as the cofactor and also requires NADPH as the second cofactor [7]. As shown in Figure 45.4, the first step is the reduction of flavin fixed to the enzyme with NADPH. Then, the addition of oxygen to the reduced form of flavin occurs, giving hydroperoxide anion, which reacts with the ketone substrate to form a Criegee intermediate. At last, the ester product is formed, and H_2O, product, and $NADP^+$ dissociate from the enzyme sequentially. Overall, molecular oxygen

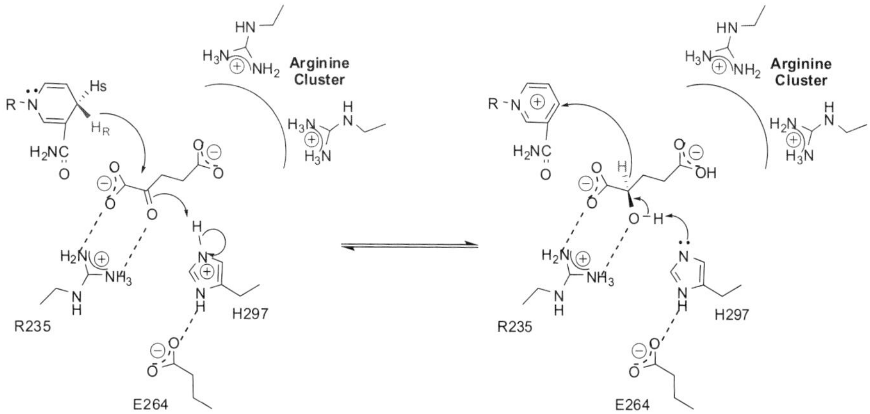

Figure 45.3. Active site of an NADH-dependent dehydrogenase transferring a hydride between the substrate and the coenzyme [6].

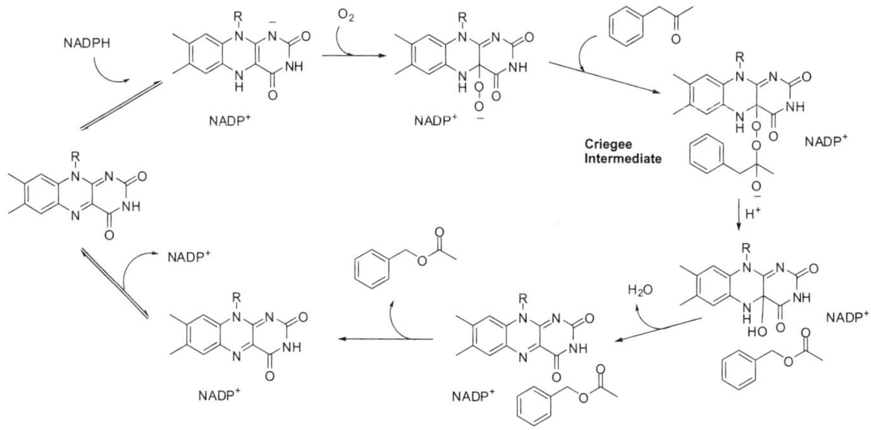

Figure 45.4. Reaction mechanism of the Baeyer–Villiger oxidation of ketones by monooxygenase with flavin and NADPH [7].

and NADPH are necessary. Only one oxygen atom of an oxygen molecule is used for the oxidation of the substrate, and another is emitted as a water molecule.

45.3 HYDROGEN SOURCE FOR THE REGENERATION OF THE REDUCED FORM OF THE COENZYME

Enzymes that catalyze the reduction of carbonyl groups require a coenzyme from which a hydride is transferred to the carbonyl carbon. Since a coenzyme is too expensive as a throwaway reagent, the oxidized form of the coenzymes has to be transformed to the reduced form for the next cycle of the reaction. Hydrogen sources are necessary to perform this reduction reaction. Alcohols such as ethanol and 2-propanol, sugars such as glucose, glucose-6-phosphate (G6P), and glucose-6-sulfate, formic acid, and amino acids such as glutamic acid and dihydrogen can be used for this purpose. Power sources such as electric and light power are also possible to use. Examples are shown in this section.

45.3.1 Alcohols

Alcohols such as ethanol, 2-propanol, and so on, have been widely used to recycle a coenzyme for reduction catalyzed by alcohol dehydrogenase since the enzyme catalyzes both reduction and oxidation. Usually, an excess amount of the hydrogen source is used to push the equilibrium to the formation of product alcohols. For example, 2-propanol was used to reduce hexanone to (*S*)-2-hexanole in 72% yield with >99% enantiomeric excess (ee) by alcohol dehydrogenase from *Geotrichum candidum* with NAD^+ as shown in Figure 45.5 [8].

45.3.2 Sugars

Glucose and G6P has been widely used as a reducing energy. Recent examples show that a recombinant *Escherichia coli*, which coexpresses a carbonyl reductase from yeast, *Saccharomyces cerevisiae* (SCR), and a glucose dehydrogenase from *Bacillus megaterium*, was used for recycling NADPH for the reduction of acetoxyacetophenone to the corresponding (*S*)-alcohol in 98% ee (Fig. 45.6) [9].

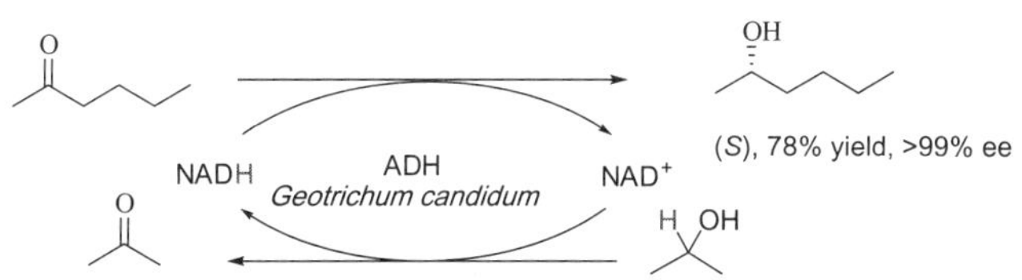

Figure 45.5. Use of 2-propanol as a hydrogen source for recycling NADH [8].

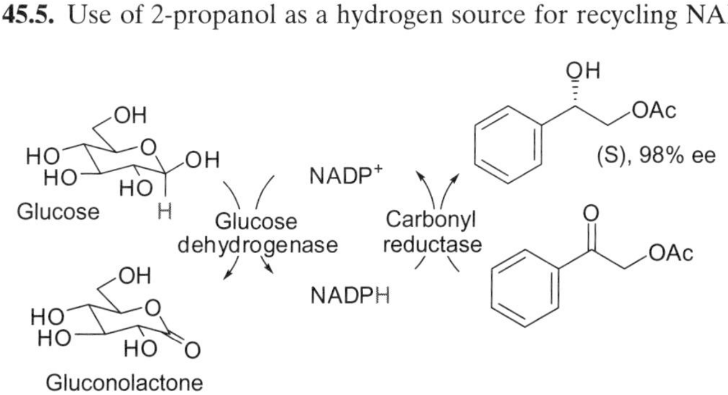

Figure 45.6. Use of glucose as a hydride source for recycling NADPH [9].

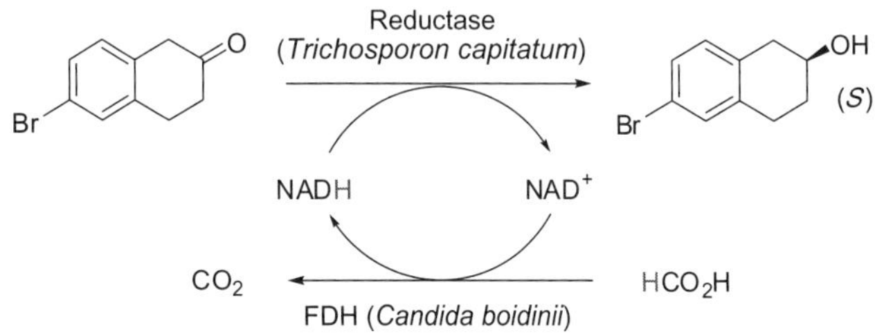

Figure 45.7. Use of formate as a hydrogen source for recycling NADH [10].

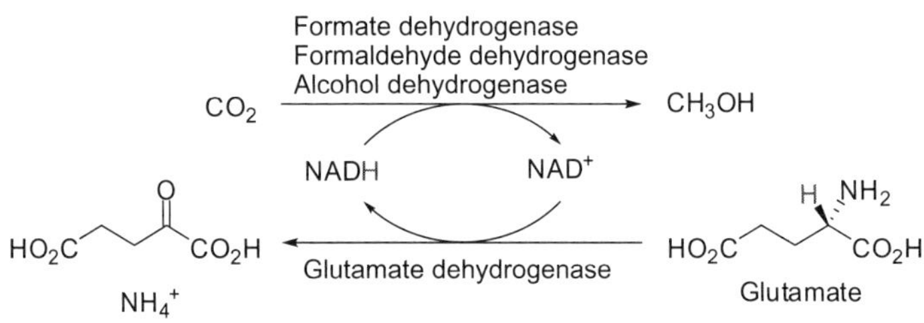

Figure 45.8. Use of glutamate as a hydrogen source for recycling NADPH [11].

45.3.3 Formate

Formate is one of the most representative hydrogen sources for biocatalytic reduction because CO_2 formed by the oxidation of formate is released easily from the reaction system. For example, a formate–formate dehydrogenase (FDH) system was applied in the reduction of 6-bromotetralone to (*S*)-6-bromotetralol, a potential pharmaceutical precursor, with the NADH-dependent ketone reductase from *Trichosporon capitatum* (Fig. 45.7) [10].

45.3.4 Amino Acid

Glutamate dehydrogenase has been frequently used for the recycling of NADH. For example, in the reduction of carbon dioxide to methanol through formic acid and formaldehyde by a multienzymatic system, the oxidized form of the cofactor (NAD^+) was recycled to the reduced form (NADH) by glutamate dehydrogenase and glutamic acid as shown in Figure 45.8 [11].

45.3.5 Molecular Hydrogen

Molecular hydrogen has been used for the recycling of coenzymes [12]. For example, the soluble hydrogenase I (H_2: $NADP^+$ oxidoreductase, EC 1.18.99.1) from the marine hyperthermophilic strain of the archaeon *Pyrococcus furiosus* (PF H2ase I) has been used as a biocatalyst in the enzymatic production and regeneration of NADPH utilizing molecular hydrogen. Utilizing the thermophilic NADPH-dependent alcohol dehydrogenase from *Thermoanaerobium*

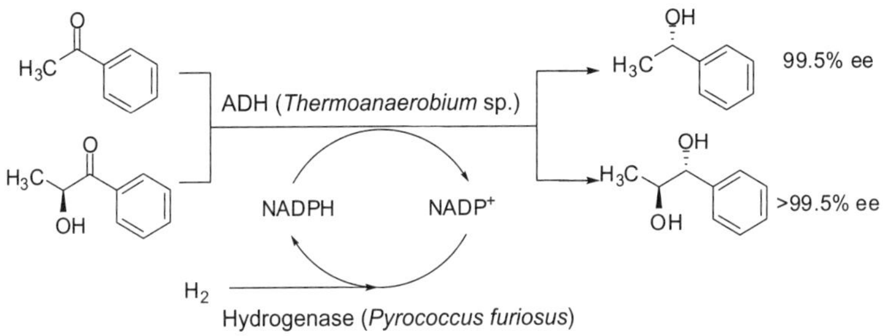

Figure 45.9. Use of molecular hydrogen as a hydrogen source for recycling NADPH [12].

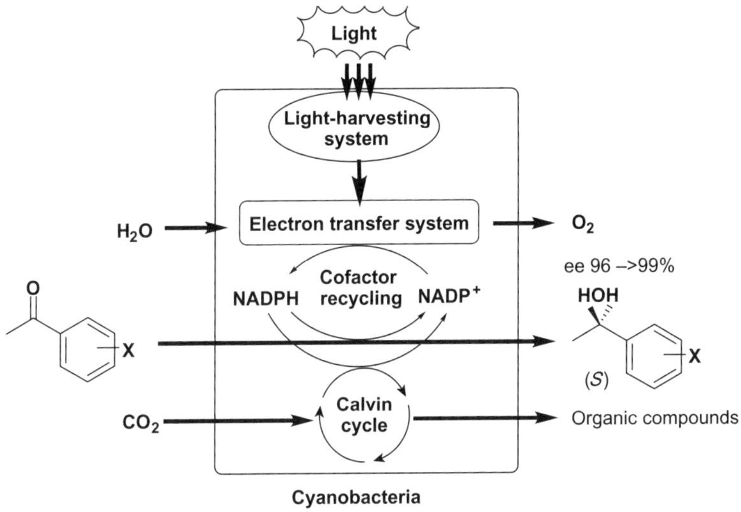

Figure 45.10. Use of light energy as a hydrogen source for recycling NADPH [13].

sp. (ADH M) coupled to the PF H2ase I *in situ* NADPH-regenerating system, (2*S*)-hydroxy-1-phenyl-propanone was quantitatively reduced to the corresponding (1*R*,2*S*)-diol in >98% de (diastereomeric excess), with total turnover numbers (ttn: mole product per mole consumed cofactor $NADP^+$) of 160 could be reached (Fig. 45.9) [12].

45.3.6 Light Energy

Photochemical methods [13] have been developed to provide an environmentally friendly system that employs light energy to regenerate NAD(P)H, for example, by the use of a cyanobacterium, a photosynthetic biocatalyst. Using the biocatalysts, the reduction of acetophenone derivatives occurred more effectively under illumination than in the dark (Fig. 45.10). [13b,c] The light energy harvested by the cyanobacterium is converted into chemical energy in

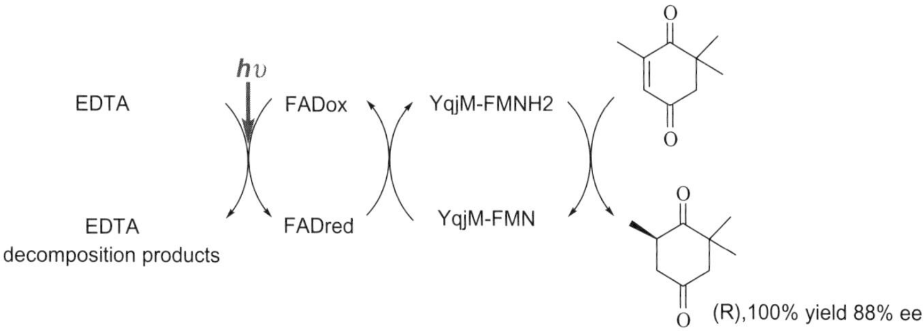

Figure 45.11. Use of light energy as a hydrogen source for recycling flavin [14].

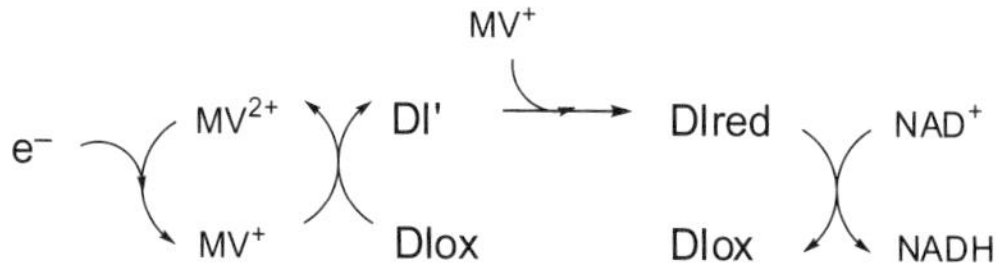

Figure 45.12. Use of electric power and diaphorase as a hydrogen source for recycling NADH [16a].

the form of NADPH through an electron transfer system, and, consequently, the chemical energy (NADPH) is used to reduce the substrate to chiral alcohol (96–>99% ee). The light energy, which is usually utilized to reduce CO_2 to synthesize organic compounds in the natural environment, was used to reduce the artificial substrate in this case. Details are presented in Section 45.4.

A reduced form of flavin can also be regenerated using light energy. For example, light-driven reduction of ketoisophoron was reported (Fig. 45.11) [14]. Using old yellow enzyme homologue, YqiM, from *Bacillus subtilis* to reduce substrate and using FAD and EDTA to recycle enzyme-bound flavin with light energy, ketoisophoron was reduced to the corresponding (*R*)-product in quantitative yield with 88% ee (Fig. 45.11).

45.3.7 Electric Power as a Hydrogen Source of Reduction

Electrochemical regeneration of NAD(P)H has long been recognized as a potentially powerful technology in the viewpoint of green chemistry as it would not require a second enzyme and cosubstrate [15]. However, the method is not effective because of the necessity of high overpotentials with direct reduction of the coenzyme, electrode fouling, dimerization of the coenzyme, and the fact that only the enzyme in the immediate vicinity of the electrode will be productive. Viologen-diaphorase (lipoamide dehydrogenase) was used for the reduction of $NAD(P)^+$, where viologens were used as a mediator of an electron from electrode to diaphorase (Fig. 45.12) [16a].

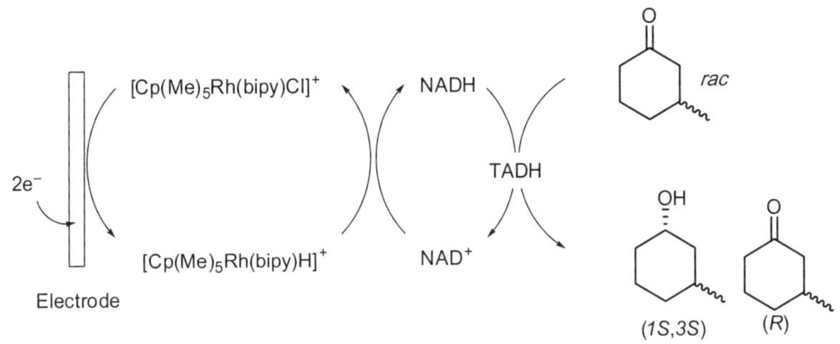

Figure 45.13. Use of electric power and rhodium complex as a hydrogen source for recycling NADH [16b].

Organometallics such as rhodium complex was also used for the electrochemical regeneration of NAD(P)H from electrode [16b], as shown in Figure 45.13. The oxidized form of the coenzyme was reduced by the reduced form of rhodium complex and the resulting oxidized form of the rhodium complex was reduced electrochemically at the electrode. *Thermus* sp. alcohol dehydrogenase (TADH)-catalyzed reduction of racemic 3-methylcyclohexanone afforded (1*S*,3*S*)-3-methyl cyclohexanol and (*R*)-3-methylcyclohexanone.

45.4 PHOTOSYNTHETIC ORGANISM-MEDIATED ASYMMETRIC REDUCTION OF KETONES

45.4.1 Light-Mediated Cofactor Recycling System in the Biocatalytic Asymmetric Reduction of Ketone

Biocatalytic reduction is a useful tool for obtaining optically active alcohols, and many scientists have studied reactions using isolated enzymes, microbes, and plant cell cultures as biocatalysts [5]. Reduction of substrates usually requires a large input of energy, and in microbial reductions, carbohydrates such as sugars have been used to recycle the coenzyme. These carbohydrates are generated through photosynthesis with sunlight energy. In other words, we have been *indirectly* using light energy for asymmetric reduction.

On the contrary, the *direct* use of light energy for such reactions is plausible if we use a biocatalyst that falls into a new category, the phototroph.

For the majority of redox enzymes, nicotinamide adenine dinucleotide, [NAD(H)], and its respective phosphate, [NADP(H)], are required. These cofactors are prohibitively expensive if used in stoichiometric amounts. Since it is only the oxidation state of the cofactor that changes during the reaction, it may be regenerated *in situ* by using a second redox reaction to allow it to reenter the reaction cycle. As described before, usually, simple alcohols such as ethanol [17] and 2-propanol [8], glucose [9], and formate [10] are used to generate the oxidized form of the coenzyme to the reduced form. These reductants originally stem from bioproducts of CO_2 by the aid of sunlight with phototrophs. Phototrophs such as algae and plants capture light energy to generate NADPH from $NADP^+$ through photosynthetic electron transfer reactions. Subsequently, CO_2 is converted into sugar, generally using NADPH [18]. It is expected that the reducing power of NADPH generated through photosynthesis can also be used in the reduction of exogenous substrates such as unnatural ketones to yield useful optically active alcohols. Thus, cofactor recycling is no problem when photosynthetic living cells are used as biocatalysts for reduction. Accordingly, we can use solar energy directly for bioconversion of artificial substrates. Cyanobacteria (a kind of microalgae) are likely to be desired biocatalysts for this purpose since they belong to both phototroph and microbe categories; in other words, they are plantlike photosynthetic bacteria. Therefore, the growth rate of cyanobacteria is as high as that of typical microbes. The problem of using cultured plant cells is that they usually grow very slowly.

Now, a new system for cofactor recycling is proposed, in which cyanobacteria converts light energy to reduce power. The advantage of this reaction is that light energy, which is a cheap resource, can be used since the microalgae possess a system that is required for the production of reduced coenzymes, NADPH. It is already reported that aryl methyl ketones are reduced to the corresponding *S*-alcohols by *Synechococcus elongatus* PCC 7942, which is one of the cyanobacteria, that is, a photosynthetic prokaryote, with high enantioselectivity (>99% ee) [13a]. When pentafluoroacetophenone is used as a substrate, the reaction proceeds in an excellent chemical yield (>90%) with high enantioselectivity (>99% ee). An advantage of the proposed biotransformation system is its high substrate/biocatalyst (s/b) ratio (=2.0), which is comparable to a low s/b ratio of the other biocatalysts (baker's yeast: 0.003–0.02 and plant cell cultures of *Marchantia polymorpha*: 0.001); therefore, a cyanobacterium-catalyzed reaction is very effective. This supports the idea that the reaction proceeds by a cofactor recycling system that uses light energy. To clarify the proposal, the effect of light and the cofactor dependency on the reduction of ketones is investigated. Figure 45.15 illustrates the time course of the yields of the alcohol. The reaction rate under illumination (the line with the squares) is higher than that under darkness (the line with the diamonds). Actually, the initial rate under illumination is about four times higher than that in the dark. Furthermore, the line with the triangles shows the yields in the reduction under illumination after an initial 2 days under darkness. Apparently, the reaction increased rapidly under illumination. This graph supports the assumption of a direct effect on the reaction using light (Fig. 45.14) [13b].

45.4.2 Light-Mediated Regulation of Asymmetric Reduction

Microbes and plant cell cultures are more useful than isolated enzymes due to the existence of many enzymes that catalyze various reactions. Furthermore,

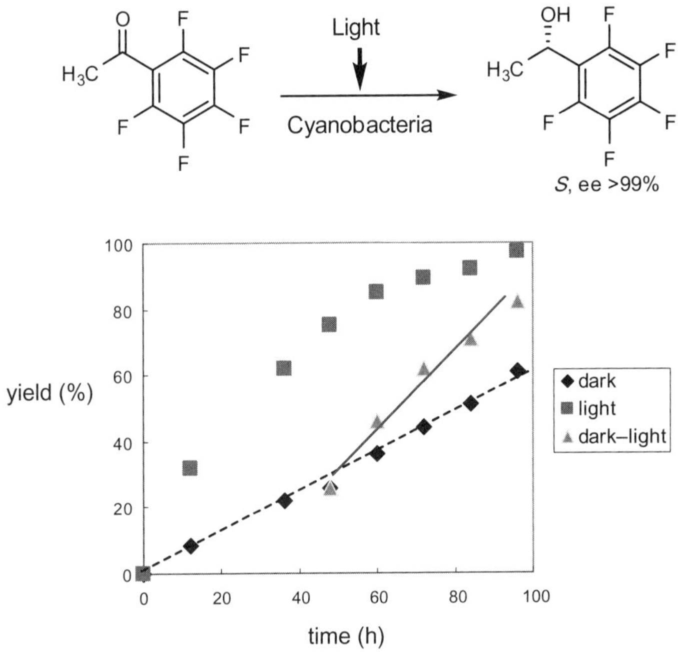

Figure 45.14. Effect of light on reduction of pentafluoroacetophenone with *Synechococcus elongatus* PCC 7942 [13b].

in the asymmetric reduction of substrates, whole living cell systems do not require cofactors and cofactor regeneration systems since they already have these requirements. On the other hand, generally, enantiomeric purities for isolated enzymatic systems are high compared to whole-cell reactions because in the latter systems, several enzymes in the cell participate in the reaction and the stereochemical preferences are not the same for all of the different enzymes: One enzyme may afford one stereoisomer, while another enzyme may produce its antipode, and as the result, the reduction system afforded unsatisfactory stereoselectivities. To date, several methods have been developed for improving the selectivity of whole-cell-catalyzed reactions by modifying the reaction conditions: addition of inhibitors of a specific enzyme [19], reaction in organic solvents [20], use of hydrophobic polymers such as XAD-resin [21], and the treatment of microbial cells with acetone [22]. Although these methods are effective for increasing the stereoselectivity of microbial reductions and several methods can be applied for a practical synthesis, many reactions are still unsuited to these types of stereochemical control and new strategies to increase the stereoselectivities of whole cell-mediated reactions are therefore desirable. Recently, a novel effective method for stereochemical control was proposed: light-mediated stereochemical control of asymmetric reduction.

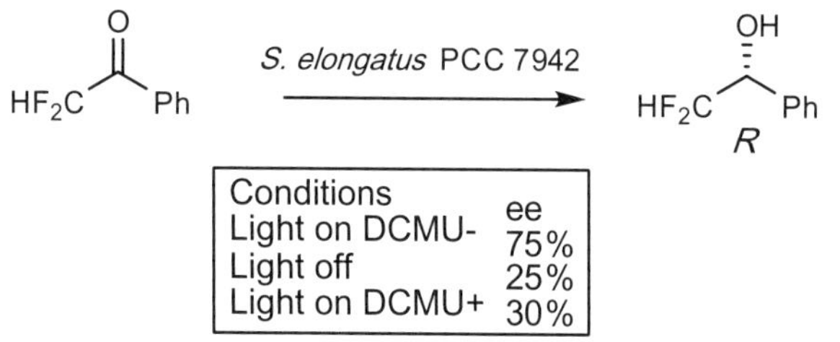

Figure 45.15. Light-mediated stereochemical control of alga-mediated asymmetric reduction [13c].

When α,α-difluoroacetophenone was added to a suspension of *S. elongatus* PCC 7942 and shaken (140 rpm) in the dark, under standard conditions for microbial reduction, the corresponding *R*-alcohol was obtained with poor enantioselectivities (20–30% ee). However, as shown in Figure 45.15, when the cells were incubated and reacted under illumination (13.4 μmol photons/m^2/s), (*R*)-alcohol was obtained with 75% ee [13c].

To investigate the effect of light precisely, 3-(3,4-dichlorophenyl)-1,1-dimethylurea (DCMU), an inhibitor of photosynthesis, was used. As the result, even if the reaction was conducted at light conditions, the addition of DCMU surely decreased the ee down to 30%, almost the same selectivity at dark conditions. Furthermore, oxidation of the reduced form of coenzymes by cell free extract of the microbe in the presence of the ketone reveals that NADPH is effective but NADH is less effective. The result indicates that the real factor for increasing the selectivity is the photosynthetic process conducted by light. The mechanism of stereochemical control is proposed as follows: Under illumination, chlorophyll in cyanobacteria captures light energy and generates the reducing agent NADPH. The *R*-enzymes (the enzymes that selectively produce *R*-alcohol) used thus generated NADPH selectively, while the *S*-enzymes (the enzymes that produce *S*-alcohol) would not use the same coenzyme. Light also changes the physiology of cyanobacteria through the photosynthetic process. These physiological changes induce expression and/or activation of enzymes. Thus, enzymes that produce the *R*-alcohol become active. We are not able to determine which mechanism is plausible at the present time. Under darkness, the reducing ability may stem from the oxidation of alcohols such as saccharides; the *S*-enzymes preferentially use the thus generated reduced form of the coenzyme. The *R*-enzymes can also participate in the reduction under dark, and consequently, both *R*-and *S*-alcohols are produced. Under illumination, only the *R*-enzymes are activated and, as stated earlier, these produce the *R*-alcohol selectively.

45.4.3 Microalgae-Mediated Asymmetric Reduction

As described before, microalgae are useful biocatalysts for asymmetric reduction. Examples are listed in Figure 45.16.

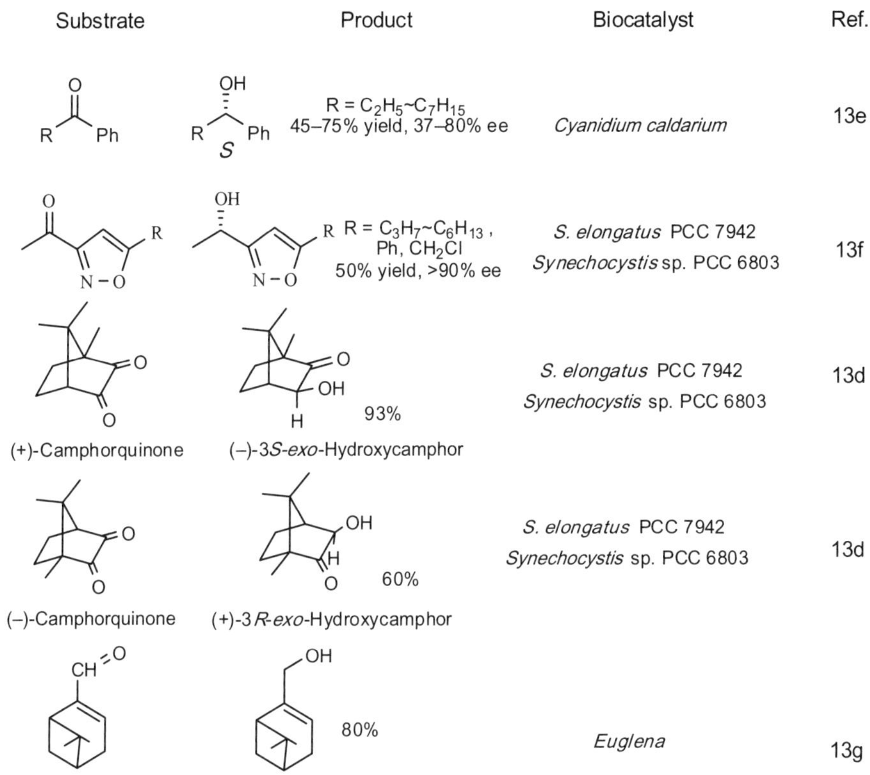

Figure 45.16. Microalgae-mediated asymmetric reduction.

45.4.4 Use of Deleted Mutant of a Cyanobacterium for Stereochemical Control

Gene technology is useful for stereochemical control of asymmetric reduction. If a low enantioselectivity in asymmetric reduction is due to the result of participating multiple active dehydrogenases (i.e., *R*- and *S*-enzymes), then the deletion of a gene of the microbe will change the stereoselectivity of the reduction. Hopefully, reduction by the mutant microbe will improve the enantioselectivity of the product alcohol. *Synechocystis* sp. PCC 6803 was chosen as the microbe because all gene sequences of the microbe can be obtained from "cyanobase" [23]. Several dehydrogenase gene-deleted mutants were prepared and the effect of deletion on the asymmetric reduction of ketones was investigated.

The reduction of t-butyl acetoacetate with the wild-type microbe gave the alcohol in 32% yield with 39% ee (*R*). On the contrary, the *slr0942* (gene coding alcohol dehydrogenase, aldo-keto reductase) knockout mutant afforded the corresponding alcohol in 30% yield with 53% ee (*R*). Thus, the reduction

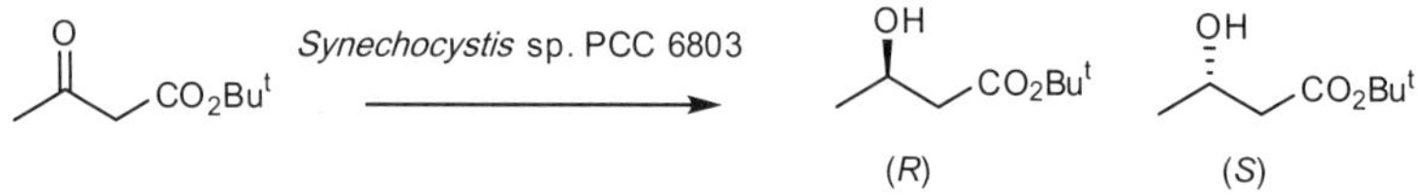

Mutant	yield,%	ee, % (Config.)
Wild type	32	39(*R*)
slr0942-deleted	30	53(*R*)
slr0315-deleted	6	91(*S*)

slr0942: Alcohol dehydrogenase, aldo-keto reductase
slr0315: 3-Oxyacyl-(acyl-carrier protein) reductase

Figure 45.17. Reduction of ketones with gene-deleted mutant of cyanobacteria [24].

with the knockout mutant increased enantioselectivity; that is, *slr0942* gene expression gave the enzyme that contributed to the reduction of t-butyl acetoacetate to give the corresponding (*S*)-alcohol. On the contrary, the *slr0315* knockout mutant gave the antipode (*S*)-alcohol in 6% yield with 91% ee, although the reduction scarcely proceeded. The enzyme expressed from the *slr0315* gene may be the dehydrogenase that reduced t-butyl acetoacetate to the corresponding (*R*)-alcohol. Thus, control of stereoselectivities was achieved using deletion of a dehydrogenase gene as shown in Figure 45.17 [24].

45.4.5 Effect of Carbon Dioxide Concentrations on the Asymmetric Reduction of Ketones with Plant-Cultured Cells

Generally, heterotrophic plant-cultured cells were used as biocatalysts and the corresponding phototrophic cells were not applied because of the difficulty of preparing these types of cells. However, the use of phototrophic cells can open up a new method for stereochemical control of asymmetric reduction.

Thus, the use of high atmospheric CO_2 concentrations in the reaction system using phototrophic *Nicotiana tabacum* cells effectively increases the enantioselectivity of asymmetric reduction of ketones.

In the case of air conditions, The reduction of t-butyl acetoacetate gave a low ee for the corresponding (*S*)-alcohol (46% chemical yield and 13% ee) under air and dark conditions; however, both the chemical yield and the ee were enhanced under the illumination of fluorescent light (83% chemical yield with 75% ee). The selectivity was increased by light conditions and the selectivity was further increased using the same cells cultured and reacted at high CO_2 concentrations. The ee of the (*S*)-alcohol was increased according to the atmospheric CO_2 concentrations from 0.03% (global atmospheric CO_2 concentration, 75% ee) to 98% ee under 1.54% CO_2 concentration (Fig. 45.18). On the contrary, the reduction under dark conditions in the presence of glucose gave the antipode (*R*)-alcohol in 83% chemical yield with 82% ee. Thus, both enantiomers of the alcohol were obtained by changing the condi-

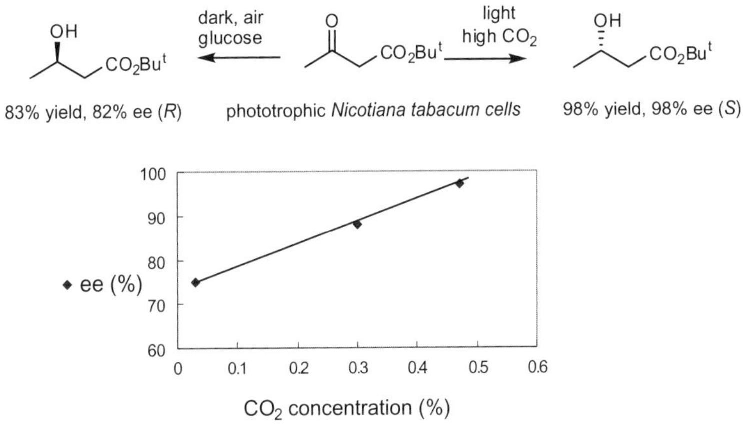

Figure 45.18. Effect of CO_2 concentrations to enantioselectivities of reduction [25].

tions: Light-high CO_2 conditions gave the (*S*)-alcohol and dark-low CO_2 concentrations in the presence of glucose afforded the (*R*)-alcohol [25].

45.4.6 A Germinated Plant as a Novel Biocatalyst

Until now, many kinds of biocatalysts have been applied for useful biotransformations, and for asymmetric reductions, isolated enzymes, microbes such as yeasts and fungi, and plant cell cultures have been used [5]. Microbes and plant cell cultures are more useful than isolated enzymes due to the existence of many enzymes that catalyze various reactions. Furthermore, in the asymmetric reduction of substrates, whole living cell systems do not require cofactors and cofactor regeneration systems since they already have these requirements. However, it is not easy to obtain these cell cultures for organic chemists who are not familiar with cell cultivation. On the other hand, another category of biocatalysts, baker's yeast and vegetables, has been applied to organic synthesis because these biocatalysts are easily obtainable from markets and are easily manipulated by organic chemists. However, an essential problem, the reproducibility of experiments with raw baker's yeast and vegetables, has remained since the conditions of these biocatalysts are easily changeable depending on the places they come from. Moreover, even from the same area, it is difficult to obtain the biocatalyst with the same condition constantly throughout the year. To solve these problems, recently, a novel biocatalyst that does not depend on the place and/or time of isolation and that is still universally available has been developed. This is a sprout prepared from a vegetable seed. The merits of using vegetable seeds as prebiocatalysts are as follows: (i) Vegetable seeds are obtainable all over the world; (ii) seeds are well preserved for a long time; and (iii) seeds have an ability to germinate at any time of the year if suitable conditions for the germination are provided.

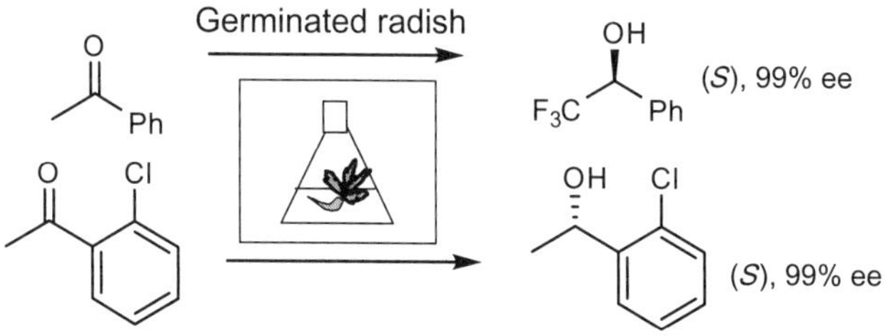

Figure 45.19. Reduction of ketones with a germinated plant [26].

Radish seeds were obtained commercially, and the surface of the seed was sterilized by rinsing with water and 1% sodium hypochlorite and was soaked in water for a while. Then, the seeds were germinated to obtain a uniform radish sprout. The radish sprout was used as a biocatalyst to reduce trifluoroacetophenone under illumination, which resulted in affording the corresponding (*S*)-alcohol in 30–40% yield with 90–100% ee. With the same method, o-chloroacetophenone was reduced and afforded the corresponding (*S*)-alcohol in >99% ee. Interestingly, although (*S*)-alcohols were obtained in both reactions, the absolute configurations are different according to definition. The difference in the stereoselectivities of the reduction is due to the existence of plural enzymes that participate in the reduction (Fig. 45.19) [26].

45.5 CONCLUSION

Dehydrogenases have been widely used as biocatalysts for asymmetric reduction. Dehydrogenases generally require coenzymes for enzyme action, and recycling of the oxidized coenzymes, which are concomitantly brought about by the reduction of substrates, is necessary. Several methods for the regeneration of coenzymes are described. Recent progress on the asymmetric reduction of ketones by photosynthetic organisms such as microalgae, plant cell cultures, and germinated plants is also described.

REFERENCES

1 J. S. Carey, D. Laffan, C. Thomsonc, M. T. Williamsd. *Org. Biomol. Chem.* 4 (2006) 2337–2347.

2 R. Patel, N. Coord. *Chem. Rev.* 252 (2008) 659–701.

3 Daicel Co. http://www.daicelchiral.com/en/contents/chiralscreen/oh/index.html, 2011.

4 (a) Codexis Co. http://www.codexis.com/pdf/Product_List.pdf, 2001; (b) D. Kalaitzakis, J. D. Rozzell, S. Kambourakis, I. Smonou. *Org. Lett.* 7 (2005) 4799–4801.

5 (a) T. Matsuda, R. Yamanaka, K. Nakamura. *Tetrahedron Asymmetry* 20 (2009) 513–557; (b) K. Nakamura, R. Yamanaka, T. Matsuda, T. Harada. *Tetrahedron Asymmetry* 14 (2003) 2659–2681; (c) K. Nakamura, T. Matsuda. *Curr. Org. Chem.* 10(11)(2006) 1217–1246; (d) K. Nakamura, T. Matsuda, in V. Gotor, I. Alfonsa, E. Garcia-Uradiales (eds.), *Asymmetric organic synthesis with enzymes*. Wiley-VCH Verlag GmbH, Weinheim, 2008, pp. 193–228; (e) K. Nakamura, T. Matsuda, in L. Ulf (ed.), *Organic reactions in water—principles, strategies and methods*. Blackwell Publishing, 2007, pp. 301–349.

6 B. M. Martins, S. Macedo-Ribeiro, J. Bresser, W. Bucke, A. Messerschm. *FEBS J.* 272 (2005) 269–281.

7 E. Malito, A. Alfieri, M. W. Fraaije, A. Mattevi. *Proc. Natl. Acad. Sci. U. S. A.* 101 (2004) 13157–13162.

8 T. Matsuda, Y. Yamagishi, S. Koguchi, N. Iwai, T. Kitazume. *Tetrahedron Lett.* 47 (2006) 4619–4622.

9 T. Ema, H. Yagazaki, N. Okita, M. Takeda, T. Sakai. *Tetrahedron* 62 (2006) 6143–6149.

10 V. J. Shorrock, M. Chartrain, J. M. Woodley. *Tetrahedron* 60 (2004) 781–788.

11 B. Ei-Zahab, D. Donnelly, P. Wang. *Biotech. Bioeng.* 99 (2008) 508–514.

12 R. Mertens, L. Greiner, E. C. D. van den Ban, H. B. C. M. Haaker, A. Liese. *J. Mol. Catal. B Enzym.* 24–25 (2003) 39–52.

13 (a) K. Nakamura, R. Yamanaka, K. Tohi, H. Hamada. *Tetrahedron Lett.* 41 (2000) 6799–6802; (b) K. Nakamura, R. Yamanaka. *Chem. Commun.* (2002) 1782–1783; (c) K. Nakamura, R. Yamanaka. *Tetrahedron Asymmetry* 13 (2002) 2529–2533; (d) T. Utsukihara, W. Chai, N. Kato, K. Nakamura, C. A. Horiuchi. *J. Mol. Catal. B Enzym.* 31 (2004) 19–24; (e) T. Utsukihara, O. Misumi, N. Kato, T. Kuroiwa, C. A. Horiuchi. *Tetrahedron Asymmetry* 17 (2006) 1179–1185; (f) K. Itoh, H. Sakamaki, K. Nakamura, C. A. Horiuchi. *Tetrahedron Asymmetry* 16 (2005) 1403–1408; (g) Y. Noma, H. Takahashi, Y. Asakawa. *Phytochemistry* 30 (1991) 1147–1151.

14 (a) A. Taglieber, F. Schulz, F. Hollmann, M. Rusek, M. T. Reetz. *Chembiochem* 9 (2008) 565–572; (b) T. B. Fitzpatrick, N. Amrhein, P. Macheroux. *J. Biol. Chem.* 278 (2003) 19891–19897.

15 W. A. van der Donk, H. Zhao. *Cur. Opin. Biotechnol.* 14 (2003) 421–426.

16 (a) S. Kim, S.-E. Yun, C. Kang. *J. Electroanal. Chem.* 465 (1999) 153–159; (b) V. Hollrigl, K. Otto, A. Schmid. *Adv. Synth. Catal.* 349 (2007) 1337–1340.

17 L. K. P. Lam, I. A. Gair, J. B. Jones. *J. Org. Chem.* 53 (1988) 1611–1615.

18 F. B. Salisbury, C. W. Ross. *Plant Physiology*, 4th ed. Wadsworth publishing, Belmont, 1992, Chapter 10.

19 K. Nakamura, Y. Kawai, N. Nakajima, A. Ohno. *J. Org. Chem.* 56 (1991) 4778–4783.

20 (a) K. Nakamura, Y. Inoue, T. Matsuda, I. Misawa. *J. Chem. Soc. Perkin 1* (1999) 2397–2402; (b) O. Rotthaus, D. Krüger, M. Demuth, K. Schaffner. *Tetrahedron* 53 (1997) 935–938; (c) K. Nakamura, S. Kondo, N. Nakajima, A. Ohno. *Tetrahedron* 51 (1995) 687–694; (d) K. Nakamura, Y. Inoue, A. Ohno. *Tetrahedron Lett.* 36 (1995) 265–266.

21 (a) K. Nakamura, M. Fujii, Y. Ida. *J. Chem. Soc. Perkin 1* (2000) 3205–3211; (b) P. D'Arrigo, C. Fuganti, G. Pedrocchi, P. Fantoni, S. Servi. *Tetrahedron* 54

(1998) 15017–15026; (c) P. D'Arrigo, M. Lattanzio, G. Pedrocchi, P. Fantoni, S. Servi. *Tetrahedron Asymmetry* 9 (1998) 4021–4026.

22 (a) T. Matsuda, Y. Nakajima, T. Harada, K. Nakamura. *Tetrahedron Asymmetry* 13 (2002) 971–974; (b) T. Matsuda, T. Harada, N. Nakajima, K. Nakamura. *Tetrahedron Lett.* 41 (2000) 4135–4138; (c) K. Nakamura, T. Matsuda. *J. Org. Chem.* 63 (1998) 8957–8964.

23 Y. Nakamura, T. Kaneko, M. Hirosawa, N. Miyajima, S. Tabata. *Nucleic Acids Res.* 26 (1998) 63–67.

24 T. Takemura, K. Akiyama, K. Umeno, Y. Tamai, H. Ohta, K. Nakamura. *J. Mol. Catal. B Enzym.* 60 (2009) 93–95.

25 H. Kojima, A. Okada, S. Takeda, K. Nakamura. *Tetrahedron Lett.* 50 (2009) 7079–7081.

26 K. Matsuo, S. Kawabe, Y. Tokuda, T. Eguchi, R. Yamanaka, K. Nakamura. *Tetrahedron Asymmetry* 19 (2008) 157–159.

46

SURFACE ORGANIZATION OF POLY (ETHYLENE GLYCOL) (PEG)-BASED BLOCK COPOLYMERS FOR BIOMEDICAL APPLICATIONS

HIDENORI OTSUKA

46.1 INTRODUCTION

Block copolymers in a selective solvent have a tendency to self-assemble at surfaces and into micelles [1–4]. At an aqueous interface, the amphiphilic property of block copolymers composed of hydrophilic and hydrophobic segments can provide the distal end of the hydrophilic chain to extend into the bulk aqueous solution, anchoring the hydrophilic block to the substrate surface through hydrophobic segments [1, 2]. In an aqueous solution, micelles with core–shell structures are formed through the segregation of insoluble blocks into the core, which is surrounded by a hydrophilic shell composed of hydrophilic blocks [3, 4]. This interfacial activity of amphiphilic block copolymers provides high utility in the biomedical field as colloidal dispersants, surface modifiers, and drug carriers, prompting many studies of block copolymer adsorption on solid surfaces [5–8], force measurements between tethered layers [9–11], and the characterization of micelle properties [12–14]. This review describes the recent progress in the field of block copolymer assembly on the surface and in the solution, focusing on the biological and biomedical application of polyethylene glycol (PEG)-based block copolymers. PEG chains tethered on a surface or forming the corona of nanometer-scaled micelles

Electrical Phenomena at Interfaces and Biointerfaces: Fundamentals and Applications in Nano-, Bio-, and Environmental Sciences, First Edition. Edited by Hiroyuki Ohshima.

exhibit the ability to sterically exclude other macromolecules and particles (steric stabilization) related to high flexibility and the large exclusion volume of PEG strands in water, and this is particularly useful for preventing the adsorption of proteins and the adhesion of cells. In this regard, these supramolecular structures involving PEG-based block copolymers should be of substantial importance for the development of blood-contacting biomaterials, which are expected to play a key role in such fields as cell and tissue engineering, biosensing, and drug delivery systems (DDSs).

46.2 CONSTRUCTION OF PEG-BRUSHED LAYER USING BLOCK COPOLYMERS

PEG coating has been used to minimize nonspecific fouling of material surfaces with biocomponents, particularly plasma proteins. For example, a PEGylated surface, which means the surface is covered with tethered chains of PEG using the functionality of PEG end groups, extremely reduces protein adsorption [15, 16], resulting in a high blood compatibility [17, 18]. PEG coating can be performed using various methods such as covalent grafting of PEG having reactive chain ends to the surface [19, 20], graft copolymerization of a PEG macromonomer onto the surface [21, 22], and direct adsorption of PEG onto surfaces in the form of a surfactant or a block copolymer in which one of the blocks is a PEG [23]. Adsorption of amphiphilic PEG-containing block copolymers on the surface allows temporary passivation of the surfaces toward cell and protein adsorption, but such coatings are not stable and the density of PEG achievable on the surface is typically lower than that required for complete passivation. Further, most of the PEG-coated surfaces possess no reactive group on the PEG chain end. To provide additional functionality on the PEG-coated surface, we designed PEG/polylactide (PLA) block copolymers (PEG–PLA) having an end-functionalized PEG (α-acetal-PEG) segment.

Recently, a facile and quantitative synthetic method has been developed for heterobifunctional PEG [24–28], which denotes PEG having different functional groups at each of both chain ends. When one of the functional end groups in the heterobifunctional PEG selectivity initiates the polymerization of a hydrophobic monomer, a new heterobifunctional AB block copolymer can be prepared, keeping the other functional group at the PEG chain end available [29, 30]. Particularly, lactide was chosen as the hydrophobic segment because PLAs are biodegradable and nontoxic polymers, which are widely utilized as implant materials or tissue engineering scaffolds. Moreover, both PEG and PLA were approved for clinical use by the Food and Drug Administration (FDA).

Reactive block copolymers of α-acetal-PEG/PLA can be utilized as surface modifiers of biodegradable PLA to provide the reactive sites on a PEGylated surface (Fig. 46.1) [31–33]. One of the objectives in this study is to

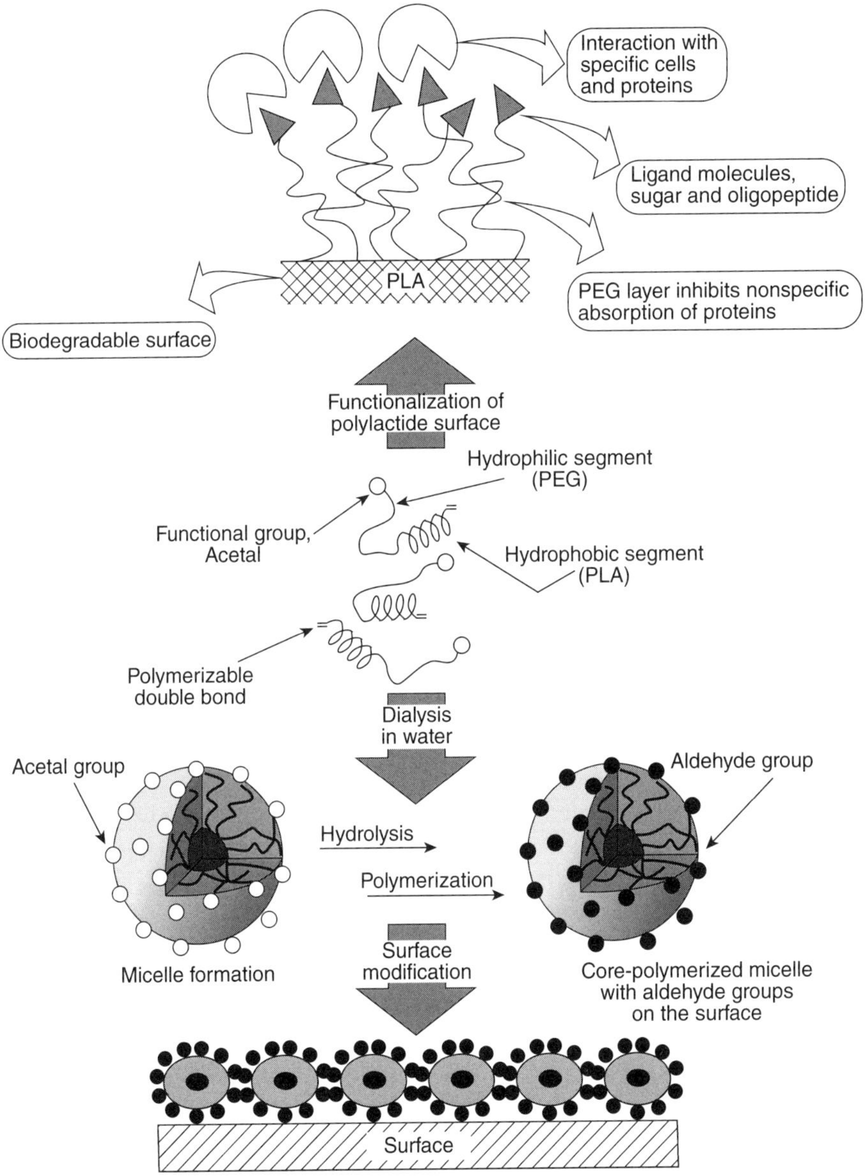

Figure 46.1. Schematic representation of the application of heterobifunctional PEG/PLA block copolymer for the construction of functional PEG layer at the materials interface.

investigate the effect of the variation in PEG chain length on surface properties. This includes assays on protein adsorption and cellular attachment to get a biochemical insight into the behavior of tethered PEG under biological conditions. For this purpose, numerous acetal-PEG/PLAs with different lengths of both PEG and PLA were synthesized. Molecular weights (MWs) of PEG/PLA segments were abbreviated as follows: PEG/PLA(0.65/11.0, 1.8/7.0, 3.3/5.4, 5.0/4.6, 8.7/6.9), where the numbers in parentheses denote the MW of the PEG segments and PLA segments in kilogram per mole, respectively. To prepare the PEGylated surface, the glass substrates, which were cleaned with 1% (w/v) solution of sodium hydroxide/water and then a boiling mixture of sulfuric acid and hydrogen peroxide, were placed in 2% (v/v) solution of 3-(trimethoxysilyl)propyl methacrylate/ethanol for 2 hours. The glass substrates were dried at 160°C for 24 hours under vacuum. The PEG-brushed layer was constructed on this silanized glass surface by the spin coating of 4% (w/v) solution of PLA solution/toluene, followed by the 2% (w/v) solution of acetal-PEG/PLA/toluene. In this way, α-acetal-PEG/PLA can be incorporated into the surface layer of PLA using a common solvent for PEG/PLA and the substrate PLA, allowing to form a dense and stable PEG layer on the PLA matrix to inhibit cell adhesion even in the presence of serum. The stability of incorporated block copolymers is enhanced compared to that of adsorbed molecules because the copolymer molecules become physically entangled with substrate PLA strands and the cohesive force between the PLA segments contributes to prevent the copolymer molecules from leaching out into the aqueous medium.

The ζ-potential of silanized glass substrates covered with various acetal-PEG/PLA block copolymers having different PEG chain length was measured in phosphate buffer solution as a function of pH (5.8–8.0), the ionic strength of the medium being equal to 0.01 (Fig. 46.2). The ζ-potential of a cleaned glass surface showed clearly a negative charge (–50 to –70 mV). The silane coupling of the glass increased the ζ-potential, but the still negative charge indicated incomplete silanation. When a PLA homopolymer (MW 20,000) was spin coated on a silanized glass surface, the ζ-potential was decreased again. This is due to the negative charge of the PLA polymer itself as reported by Dunn et al. [34]. Coating of PEG/PLA block copolymers onto PLA increased the ζ-potential, and further, a progressive increase is obvious with increasing PEG MW, indicating a screening of the surface charge. This means that the formation of the PEG-brushed layer shifts the position of the slipping plane into the solution side. Hydrophilic PEG segments are considered to expand away from the surface due to their strong hydration power. A similar trend was also reported by Suzawa and Shirahama [35]. Figure 46.3 shows the pH dependence of ζ-potentials for polystyrene (PS) and poly (styrene/2-hydroxyethyl methacrylate) (P(St/HEMA)) lattices. The ζ-potential of P(St/HEMA) particles was considerably smaller than that of PS particles over the whole pH range measured and decreased with an increase in the amount of copolymerized HEMA. This discrepancy could probably be attributed to the structures

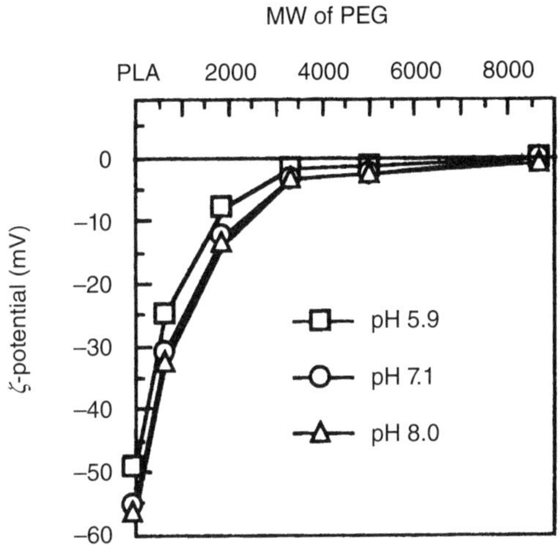

Figure 46.2. ζ-Potential variation of glass substrates as a function of MW of PEG in phosphate buffer solution.

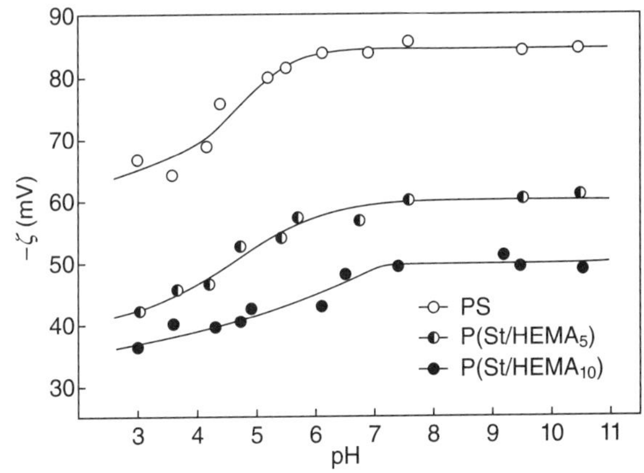

Figure 46.3. ζ-Potentials of PS and P(St/HEMA) latex particles as a function of pH (25°C, ionic strength 0.01).

of the electrical double layer for these lattices, namely, the surface charge density (σ) represents the amount of all ionized groups existing on the latex surface, while the ζ-potential is the potential at the slipping plane of the electrical double layer. To clarify this, they attempted to estimate the position (t) of the slipping plane by using the following Eversole–Boardman equation [36]:

$$\text{In}\tanh(ze\,\zeta/4kT) = \text{In}\tanh(ze\,\phi_\delta/4kT) - \kappa t, \tag{46.1}$$

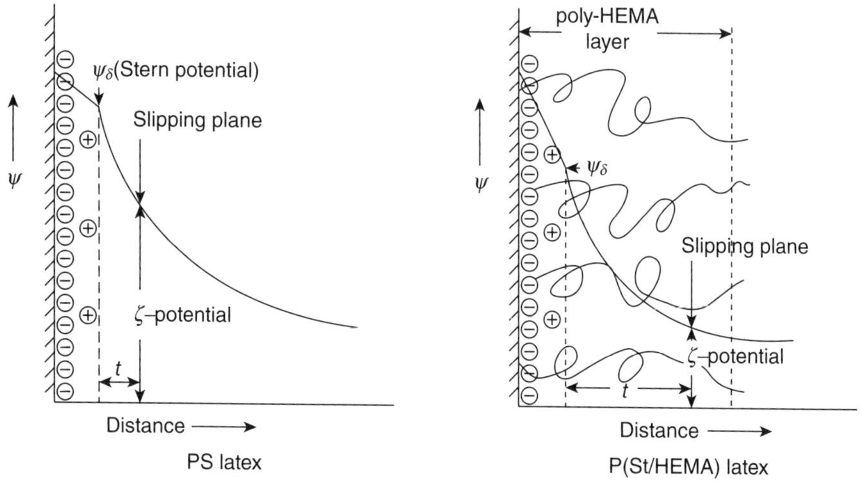

Figure 46.4. Schematic representation of the electrical double layers for PS and P(St/HEMA) latex surfaces.

where z is the valence of ions, e is the charge of an electron, k is the Boltzmann constant, T is the absolute temperature, ϕ_δ is the Stern potential, and κ is the Debye–Hückel parameter. The values of t for P(St/HEMA) lattices are about two times greater than that for PS latex, and t increases with an increase in the content of copolymerized HEMA. From the above-mentioned results, a schematic representation of the electrical double layer for PS and P(St/HEMA) lattices can be given (see Fig. 46.4); that is to say, a hydrated poly-HEMA layer (water-soluble polymer layer) exists on the latex surface, and this layer shifts the position of the slipping plane away from the latex surface. Thus, in spite of having much the same surface charge density, the potential at the slipping plane (i.e., ζ-potential) for the P(St/HEMA) latex could be lower compared with that for the PS latex. Also, the smaller Stern potential (ϕ_δ) for the P(St/HEMA) latex may contribute to the decrease in the ζ-potential of this latex.

The wettability of the surface covered with PEG/PLA block copolymers was estimated both in air and in water by contact angle measurement (Table 46.1). In water-in-air measurements, coating of PEG/PLA block copolymer onto a PLA surface increased its wettability with increasing PEG MW, as indicated by a decrease in static contact angle. A similar trend was observed in air-in-water measurements. The contact angle of a water droplet in air decreased remarkably in the range between PEG/PLA(0.65/1.10) and PEG/PLA(5.0/4.6). The decrease became moderate at the region with higher PEG MWs. Since the top few angstroms can be sensed by a contact angle measurement, the relatively high contact angles on the surfaces containing the lower MW PEG is most likely to be attributed to an incomplete coverage of the

TABLE 46.1. Static/Dynamic Contact Angle (CA) Variations on PLA and PEG/PLA-Coated Surfaces

	Static CA		Dynamic CA		
Polymer Code	Water in Air	Air in Water	Adv.	Rec.	Hysteresis
PLA	80.2	123.4	82.1	61.9	20.2
PEG/PLA(0.65/1.1)	78.9	128.6	84.5	59.4	25.1
PEG/PLA(1.8/7.0)	75.6	131.3	73.2	46.3	26.9
PEG/PLA(3.3/5.4)	59.7	139.8	67.8	27.5	40.3
PEG/PLA(5.0/4.6)	47.2	142.3	62.2	25.3	36.9
PEG/PLA(8.7/6.9)	47.9	147.4	49.1	26.2	22.9

The contact angle of each sample was measured on at least 10 spots, and these values were averaged.

uppermost surface by PEG chains. The dynamic contact angle was then measured to estimate the dynamics of the uppermost surface. The coating of PEG/PLA block copolymers reduces both the advancing and the receding angles of the substrates, although the change depends on the PEG MW, which is consistent with the result of the static contact angle. The maximum hysteresis was observed for the substrate with medium PEG chain length such as PEG/PLA(3.3/5.4). Hysteresis in the dynamic contact angle may be caused by the hydration of PEG segments. In the dry state, the PEG chain should assume a conformation flat to the surface experienced by the advancing contact line. Upon hydration, however, the PEG chain should extend from the surface due to the hydration of PEG chains. As a result, the receding contact line experiences a more hydrophilic surface than the advancing contact line. It is likely that this is the origin of the hysteresis observed on these surfaces.

The most serious problem encountered with surfaces designed for the purpose of biospecific recognition such as the antigen–antibody and the sugar–lectin interactions is the nonspecific adsorption of other proteins. In this regard, the nonfouling property provided by PEG coating is certainly a great advantage. As demonstrated in Figure 46.5, inhibition of protein (bovine serum albumin [BSA]) adsorption was achieved on PLA surfaces modified with α-acetal-PEG/PLA copolymers, depending on the PEG MW [33]. This is mainly due to the steric stabilization by PEG chains. Moreover, minimum adsorption was obtained at a medium PEG chain length, that is, PEG/PLA(3.3/5.4); note that this surface revealed the maximum hysteresis in dynamic contact angle measurement. Protein adsorption may be related to the hysteresis observed in the dynamic contact angle, which is likely to depend on particular surface

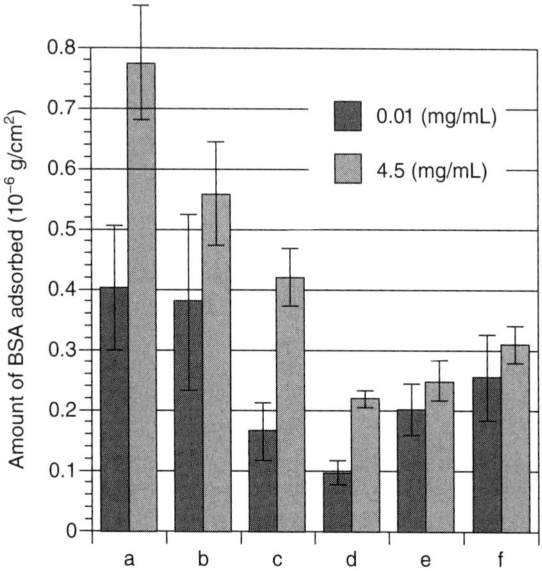

Figure 46.5. BSA adsorption from Dulbecco phosphate-buffered saline (PBS) (–) solution on PLA and various α-acetal-PEG/PLA surfaces at room temperature after incubation for 90 minutes: PLA homopolymer (a), and α-acetal-PEG/PLA block copolymers of different PEG/PLA segments (b–f); (b) PEG/PLA(0.65/11.5), (c) PEG/PLA(1.8/7.0), (d) PEG/PLA(3.3/5.4), (e) PEG/PLA(5.0/4.6), and (f) PEG/PLA(8.7/6.9), where the numbers in parentheses denote the molecular weight of the PEG segments and PLA segments in kilogram per mole, respectively. On a PLA surface, BSA was significantly adsorbed, while on PEG-coated surfaces, BSA adsorption clearly decreased. The less-adsorptive character against BSA was more pronounced at the region with higher PEG molecular weight, especially at the PEG/PLA(3.3–5.4) surface, showing minimum adsorption of BSA.

properties such as the density and mobility of tethered PEG chains on the surface.

Furthermore, ligands including proteins, peptides, and sugars can be immobilized to the distal end of these PEG chains utilizing aldehyde functionality, which converted from α-acetal groups, constructing substrates that recognize a specific molecule with a least nonspecific adsorption of other components. After the construction of the PEG/PLA surface, the acetal groups at the PEG chain end were successfully transformed into aldehyde end groups. An aldehyde group reacts smoothly with amino groups forming a Schiff base, a chemical path that can be employed for conjugation of proteins and peptides. The conversion of the acetal end groups into aldehyde end groups was conducted directly on the surface by immersing the surface into aqueous media adjusted to pH 2 using hydrochloric acid for 6 hours (Fig. 46.6). Actually, the presence as well as the reactivity of the aldehyde group on the surface was experimentally confirmed by employing a model reaction of an aldehyde group with an

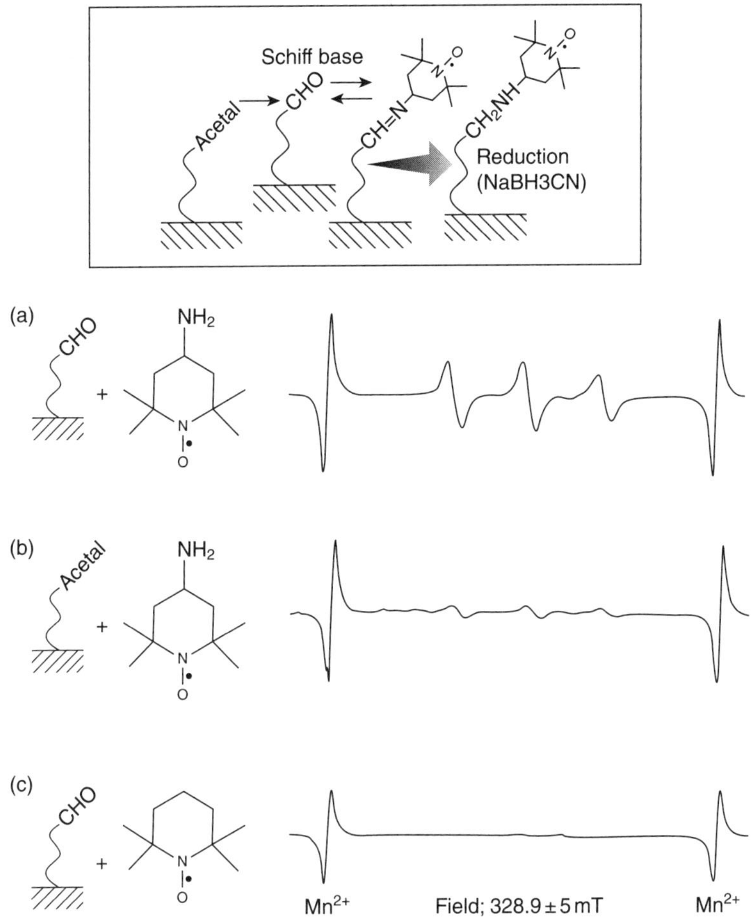

Figure 46.6. ESR spectra after the reaction between the PEG/PLA surface and TEMPO derivatives. Three typical signals were clearly observed when 4-amino-TEMPO was used as the surface modification reagent of the aldehyde surface, indicating the effective covalent-conjugation of 4-amino-TEMPO with the aldehyde group at the end of PEG on the surface (Fig. 46.6a). When the acetal surface was treated with 4-amino-TEMPO as a control experiment, only a slight signal was observed probably due to the physical adsorption of 4-amino-TEMPO on the surface (Fig. 46.6b). When the aldehyde surface was treated with TEMPO having no functional (amino) group, no ESR signal was observed (Fig. 46.6c).

electron spin resonance (ESR) probe, 2, 2, 6, 6-tetramethyl-1-piperidinyloxy (TEMPO) derivative [33]. We further showed that α-acetal-PEG/PLA modified surfaces, which are not adhesive for hepatocytes, can be converted into the selective substratum for hepatocytes by covalent linkage of a carbohydrate ligand specific for the hepatocyte asialoglycoprotein receptor to the distal end of PEG chains [37]. This approach may be generally useful for developing regionally selective microarchitectured scaffolds fabricated from biodegradable polymers for spatial organization of cocultured cells with different functions. The similar surface engineering was performed by Langer et al., who carried out the synthesis of a PEG/PLA block copolymer with a biotinylated PEG end group, which can be spin coated on the substrate [38]. This polymer is biodegradable and resistant to nonspecific protein adsorption, and the biotin moiety at the distal end of the PEG segment allows the surface to be biospecific by binding avidin-conjugated ligands. Cannizzaro et al. also reported the development of a novel biodegradable polymer matrix designed to present bioactive motifs on the surface using a block copolymer of biotinylated PEG with PLA [39]. Surface engineering is achieved using avidin as a bridge between the biotinylated polymer matrix and biotinylated ligand molecules. In this way, biotinylated peptides (GRGDS) were immobilized on the biotinylated PEGylated surface through the avidin bridge. These GRGDS-immobilized surfaces undergo an appreciable interaction with cells expressing integrin receptors and provoke desired cellular responses including spreading of endothelial cells [39].

Han and Hubbell investigated the new lactide-based PEG polymer networks (glycerol center [GL]–PEGs) prepared by UV photopolymerization of two nontoxic macromers, triacrylated lactic acid oligomer emanating from a GL and monoacrylated PEG [40, 41]. By derivatizing the terminal hydroxyl function of the incorporated PEG with a bioactive peptide, these degradable networks become useful as polymer scaffolds for tissue engineering.

The assembly of the polymer film onto the surface based on the electrostatic interaction of the positively charged polymer backbone and the negatively charged metal oxide surfaces was reported using poly(L-lysine)-g-PEG, consisting of a poly(L-lysine) (PLL) backbone with PEG side chains [42, 43]. Since this immobilization is based on electrostatic interactions, pH is clearly an important parameter for the tolerability of such systems. The resultant surfaces were found to exhibit drastic reduction of protein adsorption.

Whitesides et al. have used self-assembled functionalized monolayers (SAMs) of ethylene glycol oligomers (–EG_nOH, $n = 2$–6, and –EG_6OCH_3) with an alkanethiolate tail on gold substrate for the biospecific adsorption of a particular protein with a minimal nonspecific adsorption of other proteins [44, 45]. It is not clear whether the mechanism for the protein-repellent property of SAMs having short oligo(ethylene glycol) chains is similar to that for high-MW PEG. The extensive solvation as well as large exclusion volume of PEG in water seems to play a critical role in inhibiting protein adsorption for the high-MW PEG, yet SAMs having densely packed oligo(ethylene glycol) groups probably have only an insufficient volume to accomodate extensive solvation. They have used SAMs having both tri(ethylene glycol) groups and those with benzenesulfonamide groups as model substrates to study the biospecific adsorption of carbonic anhydrase (CA), which is a well-characterized monomeric protein (MW = 30,000) that binds parasubstituted benzenesulfonamide ligands with equilibrium dissociation constants (K_d) of approximately

10^{-6} to 10^{-9} M [46]. When a complex mixture containing nine different protein species with no specific binding property to the benzenesulfonamide group (2 mg/mL total concentration) was introduced into the flow cell equipped with a surface plasmon resonance (SPR) detector, there was almost no change in the resonance, indicating essentially no protein adsorption; however, when CA was present in this protein mixture, SPR detected the binding of CA with no disturbance due to the other proteins. This system thus may provide a convenient method for biophysical studies of biointerfacial recognition.

46.3 CONCLUSIONS

Recently, tremendous progress has been attained in the characterization and application of nanostructured materials using block copolymers. Nanostructure fabrication from block copolymers involves polymer design, synthesis, self-assembly, and derivatization. Block copolymers self-assembled into micelles or adsorbed on the surface both in brush and micelle form afford a powerful means of manipulating the characteristics of surfaces and interfaces, and therefore are expected to have novel applications. Particularly, biomedical applications have been explored in a variety of research areas from the physicochemical as well as biological points of view. In this review, we highlighted how amphiphilic block copolymers having self-assembling properties form brushes on the surface. Novel approaches for the construction of functionalized PEG layers on surfaces were discussed to achieve the biospecific adsorption of a target protein through an appropriate ligand tethered on PEG layers without non-specific adsorption of other proteins. These surfaces functionalized in brush or in micelle form can be used in diverse fields of medicine and biology to construct high-performance medical devices including scaffolds for tissue engineering and matrices for DDSs.

REFERENCES

1 V. V. Tsukruk. Assembly of supramolecular polymers in ultrathin films. *Prog. Polym. Sci.* 22 (1997) 247–311.

2 B. Zhao, W. J. Brittain. Polymer brushes: surface-immobilized macromolecules. *Prog. Polym. Sci.* 25 (2000) 677–710.

3 Z. Tuzar, P. Kratochvil. Micelles of block and graft copolymers in solutions. *Surf. Colloid Sci.* 15 (1993) 1–83.

4 C. Allen, D. Maysinger, A. Eisenberg. Nano-engineering block copolymer aggregates for drug delivery. *Colloids Surf.* 16 (1999) 3–27.

5 E. Parsonage, M. Tirrell, H. Watanabe, R. G. Nuzzo. Adsorption of poly(2-vinylpyridine)-poly(styrene) block copolymers from toluene solutions. *Macromolecules* 24 (1991) 1987–1995.

6 T. Vladkova, N. Krasteva, A. Kostadinova, G. Altankov. Preparation of PEG-coated surfaces and a study for their interaction with living cells. *J. Biomater. Sci. Polym. Ed.* 10 (1999) 609–620.

7 I. I. Potemkin, E. Yu Kramarenko, A. R. Khokhlov, R. G. Winkler, P. Reineker, P. Eibeck, J. P. Spatz, M. Moller. Nanopattern of diblock copolymers selectively adsorbed on a plane surface. *Langmuir* 15 (1999) 7290–7298.

8 E. Yu Kramarenko, I. I. Potemkin, A. R. Khokhlov, R. G. Winkler, P. Reineker. Surface micellar nanopattern formation of adsorbed diblock copolymer system. *Macromolecules* 32 (1999) 3495–3501.

9 J. Klein, P. F. Luckham. Long-range attractive forces between two mica surfaces in an aqueous polymer solution. *Nature* 308 (1984) 836–837.

10 P. G. De Gennes. Polymers at an interface: a simplified view. *Adv. Colloid Interface Sci.* 27 (1987) 189–209.

11 G. J. C. Braithwaite, A. Howe, P. F. Luckham. Interactions between poly(ethylene oxide) layers adsorbed to glass surfaces probed by using a modified atomic force microscope. *Langmuir* 12 (1996) 4224–4237.

12 Y. Y. Won, H. T. Davis, F. S. Bates, M. Agamalian, G. D. Wignall. Segment distribution of the micellar brushes of poly(ethylene oxide) via small-angle neutron scattering. *J. Phys. Chem. B* 104 (2000) 7134–7143.

13 S. Cammas-Marion, T. Okano, K. Kataoka. Functional and site-specific macromolecular micelles as high potential drug carriers. *Colloids Surf. B* 16 (1999) 207–215.

14 D. E. Bergbreiter. Self-assembled, sub-micrometer diameter semipermeable capsules. *Angew. Chem. Int. Ed.* 38 (1999) 2870–2872.

15 K. Holmberg, K. Bergstrom, C. Brink, E. Osterberg, F. Tiberg, J. M. Harris. Effects on protein adsorption, bacterial adhesion and contact angle of grafting PEG chains to polystyrene. *J. Adhes. Sci. Technol.* 7 (1993) 503–517.

16 L. K. Ista, H. Fan, O. Baca, G. P. Lopez. Attachment of bacteria to model surfaces: oligo(ethylene glycol) surfaces inhibit bacterial attachment. *FEMS Microbiol. Lett.* 142 (1996) 59–63.

17 J. H. Lee, H. B. Lee, J. D. Andrade. Blood compatibility of polyethylene oxide surfaces. *Prog. Polym. Sci.* 20 (1995) 1043–1079.

18 S. Jo, K. Park. Surface modification using silanated poly(ethylene glycol)s. *Biomaterials* 21 (2000) 605–616.

19 S. Saneinejad, M. S. Shoichet. Patterned glass surfaces direct cell adhesion and process outgrowth of primary neurons of the central nervous system. *J. Biomed. Mater. Res.* 32 (1998) 13–19.

20 C. R. Deible, E. J. Beckman, A. J. Russell, W. R. Wagner. Creating molecular barriers to acute platelet deposition on damaged arteries with reactive polyethylene glycol. *J. Biomed. Mater. Res.* 41 (1998) 251–256.

21 J. P. Lens, P. F. H. Harmsen, E. M. Ter Schegget, J. G. A. Terlingen, G. H. M. Engbers, J. Feijen. Immobilization of functionalized alkyl-poly(ethylene oxide) surfactants on poly(ethylene) surfaces by means of an argon plasma treatment. *J. Biomater. Sci. Polym. Ed.* 8 (1997) 963–982.

22 Y. X. Qie, D. Klee, W. Pluster, B. Severich, H. Hocker. Surface modification of polyurethane by plasma-induced graft polymerization of poly(ethylene glycol) methacrylate. *J. Appl. Polym. Sci.* 61 (1996) 2373–2382.

23 J. H. Lee, J. Kopecek, J. D. Andrade. Protein-resistant surfaces prepared by PEO-containing block copolymer surfactant. *J. Biomed. Mater. Res.* 23 (1989) 351–368.

24 S. Cammas, Y. Nagasaki, K. Kataoka. Heterobifunctional poly(ethylene glycol): synthesis of α-methoxy-ω-amino and α-hydroxy-ω-amino PEOs with the same molecular weights. *Bioconjug. Chem.* 6 (1995) 226–230.

25 Y. Nagasaki, T. Kutsuna, M. Iijima, M. Kato, K. Kataoka. Formyl-ended heterobifunctional poly(ethylene oxide):synthesis of poly(ethylene oxide) with a formyl group at one end and a hydroxyl group at the other end. *Bioconjug. Chem.* 6 (1995) 231–233.

26 Y. Nagasaki, M. Iijima, M. Kato, K. Kataoka. Primary amino-terminal heterobifunctional poly(ethylene oxide). Facile synthesis of poly(ethylene oxide) with a primary amino group at one end and a hydroxyl group at the other end. *Bioconjug. Chem.* 6 (1995) 702–704.

27 Y. Nagasaki, R. Ogawa, S. Yamamoto, M. Kato, K. Kataoka. Synthesis of heterotelechelic poly(ethylene glycol) macromonomers. Preparation of poly(ethylene glycol) possessing a methacryloyl group at one end and a formyl group at the other end. *Macromolecules* 30 (1997) 6489–6493.

28 Y. Akiyama, H. Otsuka, Y. Nagasaki, M. Kato, K. Kataoka. Selective synthesis of heterobifunctional poly(ethylene glycol) derivatives containing both mercapto and acetal terminals. *Bioconjug. Chem.* 11 (2000) 947–950.

29 Y. Nagasaki, T. Okada, C. Scholz, M. Iijima, M. Kato, K. Kataoka. The reactive polymeric micelle based on an aldehyde-ended poly(ethylene glycol)/poly(lactide) block copolymer. *Macromolecules* 31 (1998) 1473–1479.

30 M. Iijima, Y. Nagasaki, T. Okada, M. Kato, K. Kataoka. Core-polymerized reactive micelles from heterotelechelic amphiphilic block copolymers. *Macromolecules* 32 (1999) 1140–1146.

31 H. Otsuka, Y. Nagasaki, K. Kataoka. Novel approaches for the construction of functionalized poly(ethylene glycol) (PEG) layer on surfaces using heterobifunctional PEG/polylactide(PLA) block copolymers and their micelles, in R. Gross, C. Scholz (eds.), *Polymers from renewable resources: biopolyesters and biocatalysis.* ACS Symposium Series 764. American Chemical Society, Washington, DC, 2000, pp. 311–327.

32 H. Otsuka, Y. Nagasaki, K. Kataoka. Dynamic wettability study on the functionalized PEGylated layer on polylactide surface constructed by the coating of aldehyde-ended poly(ethylene glycol) (PEG)/polylactide (PLA) block copolymer. *Sci. Technol. Adv. Mater.* 1 (2000) 21–29.

33 H. Otsuka, Y. Nagasaki, K. Kataoka. Surface characterization of functionalized polylactide through the coating with heterobifunctional poly(ethylene glycol)/polylactide block copolymers. *Biomacromolecules* 1 (2000) 39–48.

34 S. E. Dunn, A. G. A. Coombes, M. C. Garnett, S. S. Davis, M. C. Davies, L. Illum. In vitro cell interaction and in vivo biodistribution of poly(lactide-co-glycolide) nanospheres surface modified by poloxamer and poloxamine copolymers. *J. Control. Release* 44 (1997) 65–76.

35 T. Suzawa, H. Shirahama. Adsorption of plasma proteins onto polymer lattices. *Adv. Colloid Interface Sci.* 35 (1991) 139–172.

36 W. G. Eversole, W. W. Boardman. The effect of electrostatic forces on electrokinetic potentials. *J. Chem. Phys.* 9 (1941) 798–801.

37 H. Otsuka, Y. Nagasaki, K. Okano, K. Kataoka. Functionalization of polylactide (PLA) surface using reactive block copolymer of poly(ethylene glycol) (PEG/PLA) for tissue engineering. *Proceedings of the Sixth World Biomaterials Congress, Kamueala, Hawaii*, 2000, Society of Biomaterials, Minneapolis, MN, p.702.

38 F. E. Black, M. Hartshorne, M. C. Davies, C. J. Roberts, S. J. B. Tendler, P. M. Williams, K. M. Shakessheff, S. M. Cannizzaro, I. Kim, R. Langer. Surface engineering and surface analysis of a biodegradable polymer with biotinylated end groups. *Langmuir* 15 (1999) 3157–3161.

39 S. M. Cannizzaro, R. F. Padera, R. Langer, R. A. Rogers, F. E. Black, M. C. Davies, J. B. Tendler, K. M. Shakessheff. A novel biotinylated degradable polymer for cell-interactive applications. *Biotech. Bioeng.* 58 (1998) 529–534.

40 D. K. Han, J. A. Hubbell. Lactide-based poly(ethylene glycol) polymer networks for scaffolds in tissue engineering. *Macromolecules* 29 (1996) 5233–5235.

41 D. K. Han, J. A. Hubbell. Synthesis of polymer network scaffolds from L-lactide and poly(ethylene glycol) and their interaction with cells. *Macromolecules* 30 (1997) 6077–6083.

42 G. L. Kenausis, J. Voros, D. L. Elbert, N. Huang, R. Hofer, L. Ruiz-Taylor, M. Textor, J. A. Hubbell, N. D. Spencer. Poly(L-lysine)-g-poly(ethylene glycol) layers on metal oxide surfaces: attachment mechanism and effects of polymer achitecture on resistance to protein adsorption. *J. Phys. Chem. B* 104 (2000) 3298–3309.

43 D. L. Elbert, J. A. Hubbell. Self-assembly and steric stabilization at heterogeneous, biological surfaces using adsorbing copolymers. *Chem. Biol.* 5 (1998) 177–183.

44 E. Ostuni, L. Yan, G. M. Whitesides. The interaction of proteins and cells with self-assembled monolayers of alkanethiolates on gold and silver. *Colloids Surf. B* 15 (1999) 3–30.

45 C. Roberts, C. S. Chen, M. Mrksich, V. Martichonok, D. E. Ingber, G. M. Whitesides. Using mixed self-assembled monolayers presenting RGD and (EG)3OH groups to characterize long-term attachment of bovine capillary endotherial cells to surfaces. *J. Am. Chem. Soc.* 120 (1998) 6548–6555.

46 M. Mrksich, J. R. Grunwell, G. M. Whitesides. Biospecific adsorption of carbonic anhydrase to self-assembled monolayers of alkanethiolates that present benzene-sulfoamide groups on gold. *J. Am. Chem. Soc.* 117 (1995) 12009–12010.

47

PEGYLATED NANOPARTICLES FOR BIOLOGICAL AND PHARMACEUTICAL APPLICATIONS

HIDENORI OTSUKA

47.1 INTRODUCTION

Block copolymers with amphiphilic character, having a large solubility difference between hydrophilic and hydrophobic segments, have a tendency to self-assemble into micelles in a selective solvent [1–4]. In an aqueous solution, micelles with core–shell structure are formed through the segregation of insoluble hydrophobic blocks into the core, which is surrounded by a shell composed of hydrophilic blocks. This self-assembling property of amphiphilic block copolymers provides their high utility in the biomedical field as drug carriers, surface modifiers, and colloidal dispersants [5–8]. This review describes the recent progress in the field of block copolymer assembly in the solution and on the surface, focusing on the biological and biomedical application of polyethylene glycol (PEG)-based block copolymers. PEG chains as hydrophilic polymers with a flexible nature can be selected as shell-forming segments, which assemble into dense palisades of tethered chains to achieve unique properties. The biocompatibility was guaranteed by the dense PEG shell, which endows the micelle with a stealth character in the blood compartment, achieving a long circulation [9]. PEG chains attached to a surface or

Electrical Phenomena at Interfaces and Biointerfaces: Fundamentals and Applications in Nano-, Bio-, and Environmental Sciences, First Edition. Edited by Hiroyuki Ohshima.

forming the corona of a nanosphere exhibit rapid chain motion in an aqueous medium and have a large excluded volume. The steric repulsion resulting from a loss of conformational entropy of the bound PEG chains upon the approach of a foreign substance and the low interfacial free energy of PEG in water contributes to the extraordinary physiological properties of nanospheres covered with PEG [10–16]. PEG grafted to surfaces of biomedical devices also proved to increase their biocompatibility and to reduce thrombogenicity [17–24]. Furthermore, surface organization of reactive micelles with cross-linking cores was described, allowing the surface to have an extremely high nonfouling character and working as a reservoir for hydrophobic agents, as summarized in Figure 47.1. Core segregation from aqueous milieu is the direct driving force for micellization and proceeds through a combination of intermolecular forces, including hydrophobic interaction [12, 15, 25–30], electrostatic interaction [31–34], metal complexation [35, 36], and hydrogen bonding [37] of constituent block copolymers. A variety of drugs including genes and proteins, metals, and semiconductors with diverse characteristics can be incorporated into the core-forming segment of the block copolymer so that one can expect a sufficiently strong interaction with core-incorporated molecules. In order to prepare the drug delivery system for site specificity, the outer shell of the polymeric micelle was built in such a way that it was covered with functional groups, which react readily with potential pilot molecules or target-specific antibodies. These strategies to construct functionalized PEG layers was further applied to metal and semiconductor nanoparticles, which have recently attracted much interest in biological assay systems due to their unique photochemical and photophysical properties [38–42]. These photonic properties depend on the particle size and composition, which can be varied with the method of preparation, including the use of Langmuir–Blodgett films [43, 44], reverse micelles [45, 46], vesicles [47], and various polymer networks [48–50], yet the resulting nanoparticles were not effective in preventing nonspecific aggregation in an aqueous medium unless their surface was modified with hydrophilic coatings including PEGylation. Accordingly, the surface organization of PEG on these nanoparticles may open new opportunities for their use in the biological fields.

47.2 POLYMERIC MICELLES FOR DRUG DELIVERY

Drug targeting for efficient accumulation in the body is often hampered by the rapid recognition of a carrier system by the reticuloendothelial system (RES) and by the subsequent kidney and/or hepatic elimination. Moreover, for modulated drug delivery to solid tumors, which locate outside the blood compartment, the carrier is required to exhibit not just a sufficient half-life in the blood compartment but also the capability of extravasation at the tumor site. Recent developments led to the design of drug carriers with prolonged circulation in the vascular system [9]. Cancer chemotherapy may cause severe side effects, leaving patients under extreme distress. To overcome this problem,

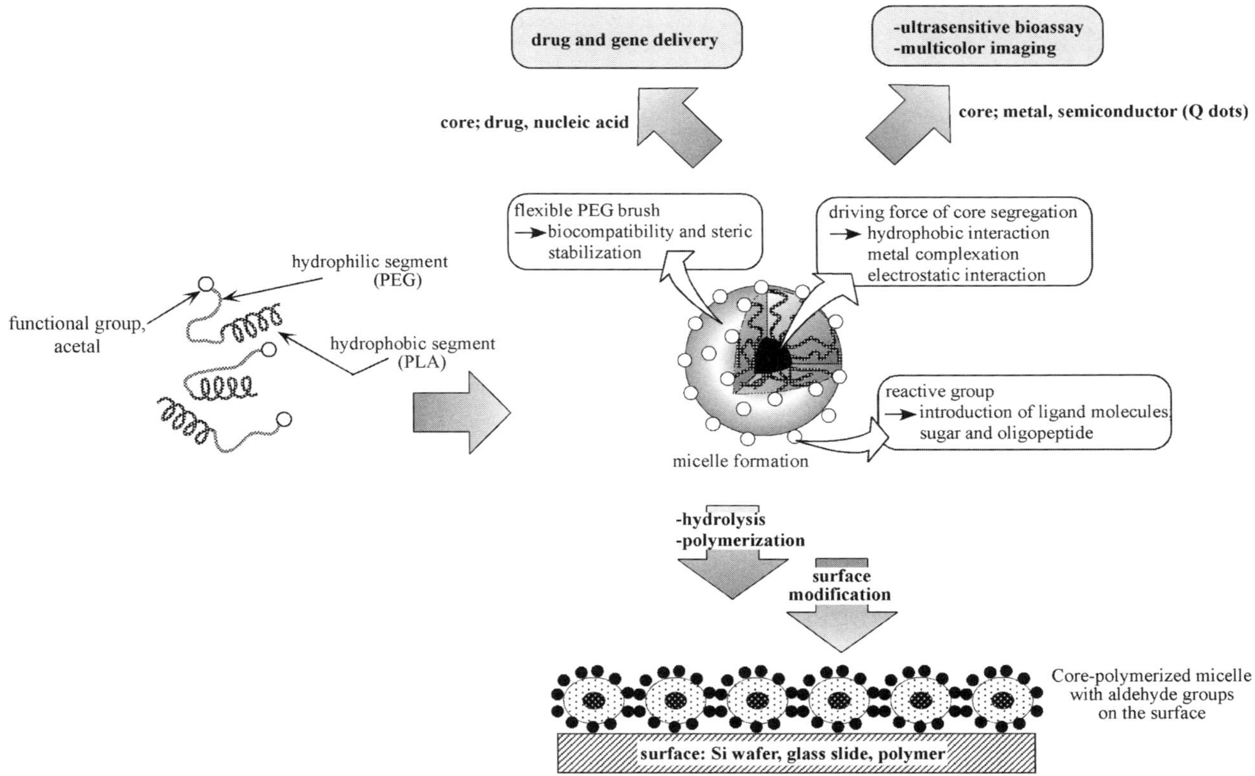

Figure 47.1. Schematic representation of the application for multimolecular assembly using heterobifunctional PEG and their block copolymer.

an interest has been raised in the application of block copolymer micelles as a novel carrier system for anticancer agents because of the high drug-loading capacity of the inner core as well as of the unique disposition characteristics in the body [9–12]. Compared to surfactant micelles, polymeric micelles are generally more stable, with a remarkably lowered critical micelle concentration (CMC), and have a slower rate of dissociation, allowing retention of loaded drugs for a longer period of time and, eventually, achieving higher accumulation of a drug at the target site [9]. Polymeric micelles have a size of 30–50 nm in diameter, ranging closely to that of viruses, and apparently, this size range is favorable for extravasation to achieve so-called enhanced permeation retention (EPR) effect [51].

Special focus is focused here to polymeric micelles formed from heterobifunctional block copolymers. A challenge in the development of novel micellar carrier systems is to design targetable polymeric micelles in which pilot molecules are installed on their surface to achieve a specific-binding property to target cells. Of particular importance in this regard is the establishment of a novel and effective synthetic route for end-functionalized amphiphilic block copolymers with appreciable biocompatibility and biodegradability, allowing conjugation of the pilot molecules at the tethered end of the hydrophilic segment.

As reported previously, we have developed a facile and quantitative synthetic method for heterobifunctional PEG, which denotes PEG having different functional groups at the α- and ω-ends [52–56]. This polymerization procedure is further applicable to the preparation of an end-functionalized block copolymer by extending a second polymer segment from the ω-end of the thus-prepared hetero-PEG, keeping the other functional group at the PEG chain end (α-end) to be available for further reaction. An introduction of an aldehyde group at the PEG end was achieved by the use of potassium 3,3-diethoxypropanolate (PDP) as an initiator for the polymerization of ethylene oxide (EO). The acetal group at the α-end of resultant PEG can be easily converted into an aldehyde group by acid treatment. In this way, α-acetal-poly(ethylene glycol)-*block*-poly(D,L-lactide) (α-acetal-PEG/PLA) was synthesized by a one-pot anionic ring-opening polymerization of EO followed by D,L-lactide (LA) initiated with PDP as an initiator at room temperature under argon (Scheme 47.1) [21–24, 57]. Because the PEG/polylactide (PLA) block copolymer has an amphiphilic character, it forms polymeric micelles in an aqueous milieu. A dialysis method was employed to prepare the polymeric micelle; that is, after the block copolymer was dissolved in a good solvent for both segments, such as dimethylacetamide (DMAc), the solution was dialyzed against water [57, 58]. The size and shape of the polymeric micelles were estimated by dynamic light scattering (DLS)/static light scattering (SLS), revealing the micelle to be ~30 nm with a very low polydispersity. The conversion of the acetal end group into the aldehyde group was conducted after the micelle formation by adjusting the pH of the medium to pH 2 with an addition of a

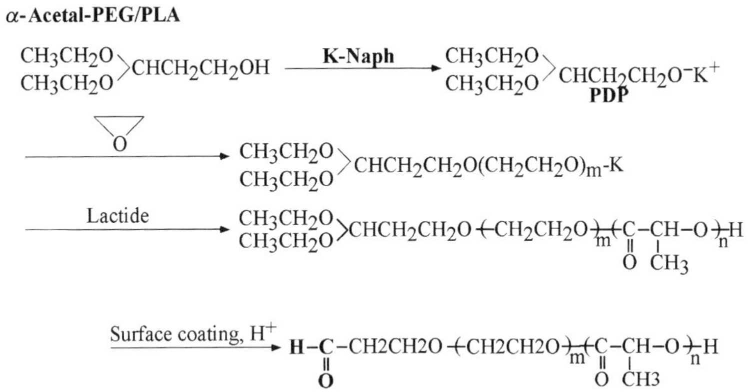

Scheme 47.1. Synthetic procedure of end-functionalized block copolymer [α-acetal-poly(ethylene glycol)-poly(D,L-lactide) block copolymer].

small quantity of hydrochloric acid. From the ^{1}H-NMR analysis, the conversion of acetal to aldehyde was estimated as up to 90%.

The availability of the aldehyde groups positioned at the micelle's periphery for further modification was then confirmed through the reaction of the aldehyde groups with ligand molecules such as tyrosyl-glutamic acid (Tyr-Glu) as a model peptidyl ligand as well as saccharide moieties including glucose, galactose, and lactose [58, 59].

The concept of polymeric micelle stabilization through the formation of a hydrophilic palisade surrounding a water-incompatible core can be extended to include the case of macromolecular association through electrostatic interaction. It was shown that a pair of block copolymers with an oppositely charged polyelectrolyte segment, poly(ethylene glycol)-*block*-poly(L-lysine) (PEG-*b*-PLL) and poly(ethylene glycol)-*block*-poly(α,β-aspartic acid) (PEG-*b*-P(Asp)), spontaneously associates to form micelles with a core composed of a polyion complex (PIC) of poly(L-lysine) and poly(α,β-aspartic acid) segments. The PIC micelles opened the way to incorporate charged macromolecules of synthetic and biological origins including proteins and nucleic acids into the micelles and have been developed as nonviral DNA delivery systems [32, 60–62]. The strategy to coat the PIC core of a polycation and DNA with a hydrophilic segment of the cationic block copolymer was applied to prepare a variety of nanoassociates. These include PEG-*b*-PLL [31,61,62], poly(ethylene glycol)-*block*-poly(ethyleneimine) (PEG-*b*-PEI) [63], and poly(ethylene glycol)-*block*-poly(dimethylaminoethyl methacrylate) (PEG-*b*-PAMA) [64], which were polymerized from corresponding monomers using PEG with the terminal functional group as a macroinitiator. Synthesis of these cationic block copolymers as well as the physicochemical and biological properties of their micelles with DNA are reviewed in detail elsewhere [5, 65].

47.3 SURFACE MODIFICATION WITH POLYMERIC MICELLES FOR THE DESIGN OF A FUNCTIONAL BIOINTERFACE

Providing polymeric micelles are able to be immobilized on the surface by maintaining their core–shell architecture, the aforementioned advantages of micelles in the solution, particularly nonfouling property, should be also expected on the surface. However, physical coagulation forces may not be stable enough to maintain the micelle structure in the process of surface fixation. A disruption of the micelle upon attachment to the surface is reported both experimentally and theoretically [66–68]. Johner and Joanny [68] used scaling arguments to show the disruption of the micelles during their adsorption process. The disruption results in the formation of loops and trains on the surface, leading to the formation of a loosely packed layer structure. This discrepancy may be solved by the preparation of reactive and structurally stabilized micelles. Heterobifunctional block copolymers of PEG/PLA that possess a reactive group at the PEG end and a polymerizable group at the PLA end were successfully prepared. The polymerization of PLA end after the micellization resulted in the formation of core-polymerized micelles with high stability in harsh environments [69].

Then, the core-stabilized reactive micelles were covalently linked to the surface, forming a single layer of micelle as well as a multilayered, highly organized micelle hydrogel [70–72]. Although the star polymers of PEGs with micelle-like structures have been utilized for the surface modification to resist protein adsorption and to apply for the scaffold in tissue engineering by Merrill et al. and Griffith-Cima et al. [73, 74], the polymeric micelle is expected to present a more extended application for its easy preparation and unique structure. Since PEG density in the shell of the micelle is appreciably high, the surfaces treated with PEG/PLA micelles are eventually to have a dense layer of tethered PEG chain, which may provide an effective nonfouling property. Further, by incorporating hydrophobic drugs into the core of surface-immobilized micelles, one can create surfaces releasing drugs in a controlled manner.

47.3.1 Adsorption of Protein on a Micelle-Coated Surface

The aminated glass slides and polypropyrene (PP) films were coated with monolayer and multilayers of micelles mediated by polyallylamine (PAlAm). The substrate was immersed in the micelle solution in 2-(4-(2-hydroxyethyl)-1-piperazinyl) ethanesulfonic acid (HEPES) (pH 6.7) containing $NaCNBH_3$ for 2 hours. After the light rinsing with Milli-Q (Millipore) water, the micelle-coated substrate was immersed into 0.6% (w/v) PAlAm in HEPES (pH 6.7) containing $NaCNBH_3$ for 2 hours. The above-mentioned procedure was repeated until the desired number of coatings was obtained.

The surface covered with core-polymerized reactive micelle was evaluated by the protein adsorption measurements. The quantitative analysis of adsorbed

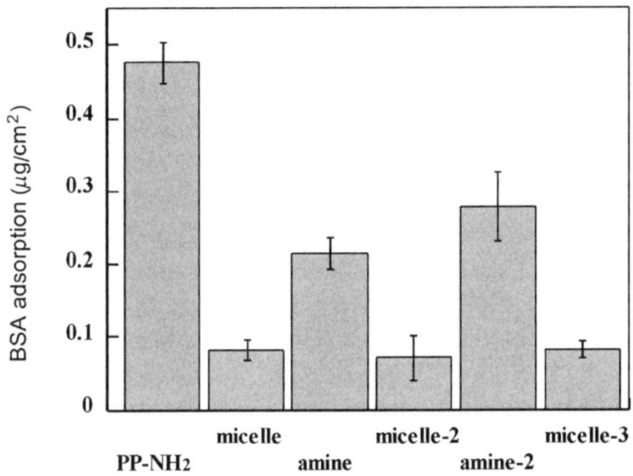

Figure 47.2. Adsorption of BSA to aminated polypropylene surfaces modified with grafted PEG-aldehyde and reactive micelles. The numbers denote the numbers of layers of micelle and polyallylamine (PAlAm).

proteins on the surface was carried out by the micro bicinchoninic acid (BCA) method that determines small amounts of protein in an aqueous solution by the absorption of Cu^+ chelated with BCA [23, 75]. Figure 47.2 shows the equilibrated amount of bovine serum albumin (BSA) on the PP films after exposure to 45 mg/mL of BSA solution for 60 minutes. The micelle coating exhibited a significantly reduced amount of BSA compared with bare and plasma-treated PP plates (the surfaces aminated by plasma treatment necessarily for micelle coating through the reaction between the amine and aldehyde groups on the micelle). The nonfouling property against BSA was more significant than the surface grafted with PEG-aldehyde under the same condition as the micelle coating. It should be noted that the protein adsorption was reduced by increasing the number of micelle coating.

47.3.2 Release of Pyrene from Micelle-Coated Surfaces

The micellar gel has another unique property that is not typical of conventional hydrogels. The gel consists of micelles that have hydrophobic cores of ~10 nm in size. The micelles on a surface are expected to hold drugs in the same manner as the ones in the solution and to release them in a controlled manner. As a model drug, pyrene was incorporated into the micelle by mixing pyrene with the micellar solution, following the procedure by Kwon et al. [76] The pyrene-incorporated micelles were coated on an aminated glass slide in the aforementioned manner. The sample was then exposed to an excess amount of Milli-Q water. By measuring the fluorescence at $\lambda_{ex} = 336$ nm, the release of pyrene from the micelle-coated surface was monitored. Note that the initial amount of pyrene, as well as the rate of release, was dependent on the number of micelle coatings; as the number of coatings increased, the initial fluorescence was higher and the release rate was slower. Repetitive loading and release of pyrene were also successfully accomplished with the micelle-gel layer. After the hexapoid coatings of pyrene-free micelles, the samples were exposed to the micelle solution containing pyrene for 12 hours to transfer pyrene from the solution to the gel phase. The loading and release of pyrene can be repeated reproducibly. The third cycle of the pyrene release carried out at 4°C showed a slower decline of fluorescence presumably due to the change in the mobility of the PLA segment with temperature, suggesting a variation in the release rate obtained by the use of block copolymers with a hydrophobic segment that is different from PLA. These results indicate that the loading capacity and the release rate of a drug can be controlled by the number of coatings.

47.4 METAL AND SEMICONDUCTOR NANOPARTICLES AS BIOLOGICAL LABELS

Metal and semiconductor nanoparticles are of considerable current interest because of their unique size-dependent properties [77–80]. By controlling the structure precisely at nanoscale dimensions, one can control and tailor the properties of nanoparticles. In addition, one can modify the nanostructures to better suit biological systems, for example, modifying their surface layer for enhanced aqueous solubility, biocompatibility, and biorecognition [81]. These abilities to systematically modify the properties of nanoparticles by controlling their structure and their surface properties at a nanoscale level makes them extremely attractive candidates for use in biotechnological systems, from fundamental scientific studies to commercially viable technologies [82]. The use of exquisite recognition properties of biomolecules in organizing nonbiological inorganic objects into functional materials becomes an important frontier in materials science. Recent research by several groups has linked colloidal nanoparticles to biomolecules such as sugars [83–86], peptides [87], proteins [88–90], and DNA [91–93]. These nanoparticle bioconjugates are being used for assembling new materials [94, 95], for developing homo- and heterogeneous bioassays [96–98], particularly multicolor fluorescent labels for ultrasensitive and high-throughput detection and imaging [88–92]. Many advantages are expected by replacing conventional molecular tags, such as fluorescent chromophores, with nanoparticles. These include higher quantum efficiencies, greater scattering or absorbance cross sections, optical activity over more biocompatible wavelengths, and significantly increased chemical or photochemical stability. The availability of these new nanoparticles will greatly facilitate *in situ* probes and sensor methods.

47.4.1 Metal Nanoparticles

The use of gold colloid in biological applications began in 1971, when Faulf and Taylor invented the immunogold staining procedure. Since that time, the labeling of targeting molecules, especially proteins, with gold nanoparticles has revolutionized the visualization of cellular or tissue components by electron microscopy [99]. The optical and electron beam contrast qualities of gold colloid have provided excellent detection qualities for imaging technologies in electron microscopy. Although metal and inorganic nanoparticles can be prepared from various materials by several methods [79, 100–103], the coupling and functionalization with biological components have only been carried out with a limited number of chemical methods.

To apply gold colloids in newly developed biomedical assay systems, a simple and facile means of anchoring different ligand biomolecules onto particle surfaces is strongly required, and the stability in the physiological condition should be improved. Particularly, color changes induced by the association of nanometer-sized gold particles provide a basis of a simple yet highly selective method for detecting specific biological reactions between anchored ligand molecules and receptor molecules in the milieu. Mirkin and coworkers have shown that gold colloidal particles modified with oligonucleotides form large assemblies through the hybridization with complementary oligonucleotide strands, providing a new method for colorimetric detection of targeted DNA sequences [91–94, 96, 98]. With decreasing gold colloidal particle size, however, colloidal stability decreases significantly due to increased particle surface energy. Such gold nanoparticles aggregate in a high-ionic strength milieu as well as adsorb biomolecules such as proteins and DNA nonspecifically, resulting in reduced sensitivity and selectivity when used as colloidal sensor systems in biological fluids.

47.4.1.1 Metal Nanoparticles Synthesized in Amphiphilic Block Copolymer Micelles. Several methods have attempted to synthesize and stabilize nanoscale gold particles in an aqueous milieu. Most of them have utilized surfactant and/or polymer stabilizers yet were not effective in preventing aggregation of nanoparticles particularly under physiological conditions (concentrated salt medium) [104].

Block copolymer micelles could be successfully used as nanoreactors for noble metal colloid formation; in such micelles, chemical and physical reactions can be confined to the fluid micellar cores, the sizes of which are confirmed as of a nanometer scale. For the amphiphilic block copolymer in the nonpolar selective solvent, the unpolar blocks form the corona, which provides solubilization and stabilization, while the polar or hydrophilic and functionalized blocks form the core, which is able to dissolve metal compounds due to coordination, followed by the nucleation and growth of metal particles upon reduction. The important step involves the solubilization of inorganic compounds into the micellar core. As a guideline for optimum precursor materials and micellar core blocks, one can use Pearson's hard/soft acid/base (HSAB) concept [105], which has been generalized to include metals and semiconductors [106]. The general strategy is to start from weakly coordinated metals, for example, $Pd(OAc)_2$ or $Pd(ClO_4)_2$, which are complexes of a soft acid (the transition metal ions) and a hard base (acetates, perchlorates, etc.). The formation of a more stable complex of a soft acid with a softer base, for example, polyvinylpyridine, to assemble the micellar core is the driving force for solubilization. The polymer complex should not be too stable since overstabilization could prevent the formation of the desired colloid in the subsequent chemical reaction. Chemical reactions typically involve a reduction process to obtain noble metal colloids or the preparation of sulfides or oxides. Using polystyrene-poly-4-vinylpyridine (PS-b-P4VP) as the constituting amphiphilic block copolymer, Sidorov et al. prepared mono- and bimetallic colloids with sizes controlled by varying such parameters as species of metal salt, type of reducing agent, and block copolymer composition. However, the most study using amphiphilic block copolymers was successfully employed only in nonpolar organic solvents because appropriate polymers to show both the ability to form micelles in water and to bind metals in the core are not available. Use of water as a reaction medium becomes possible when a double-hydrophilic block copolymer is adopted, where both blocks are soluble in water, but only one block is able to coordinate with metal ions. The example of such an application was reported for the interaction of PEG-*block*-PEI with $AuCl_3$, $PdCl_2$, H_2PtCl_6, Na_2PtCl_6, K_2PtCl_6, and Na_2PdCl_4 [107]. Addition of the gold salt to a PEG-*b*-PEI solution resulted in the formation of polydisperse micelles, and PEG-*b*-PEI induced the reduction of the gold salt to form gold nanoparticles. Neither PEG nor PEI itself showed this behavior of autoreduction. Analytical ultracentrifugation confirmed that 75–80% of the gold was formed inside the micelle, suggesting particle formation around the PEI chains [107, 108]. Micelle formation upon salt addition to a PEG-*b*-PEI solution was also observed for $PdCl_2$ and K_2PtCl_4 with polydisperse, large, and unstable properties, but no self-reduction occurred in these cases. It was further found for PEG-*b*-PEI that branched copolymers with multiple PEG blocks attached to PEI are better stabilizers for metal nanoparticles as compared with the diblock system. Light scattering and transmission electron microscope (TEM) revealed the existence of large micellar aggregates for the diblock system. If stable micelles were formed with the metal salts (H_2PtCl_6, Na_2PtCl_6, and Na_2PdCl_4), efficient control of the nanoparticle growth and stabilization was possible, although the equilibration of the micelle architecture could be a slow process taking up to weeks [109]. Besides the identified parameter, such as the polymer/metal ratio and the type of reducing agent on the metal nanoparticle size and shape, complex ion geometry and also the pH of the solution were found to be of importance.

47.4.1.2 Gold Nanoparticles Synthesized by Heterobifunctional PEG. Recently, Wuelfing et al. reported that surface PEGylation of gold particles by CH_3O-PEG-SH significantly improved their dispersion stability in aqueous milieu due to the steric repulsion effects of tethered PEG strands [110]. Yet, the CH_3O-PEGylated nanoparticles possess no reactive groups to further immobilize ligand molecules. Thus, a system possessing both sufficient colloi-

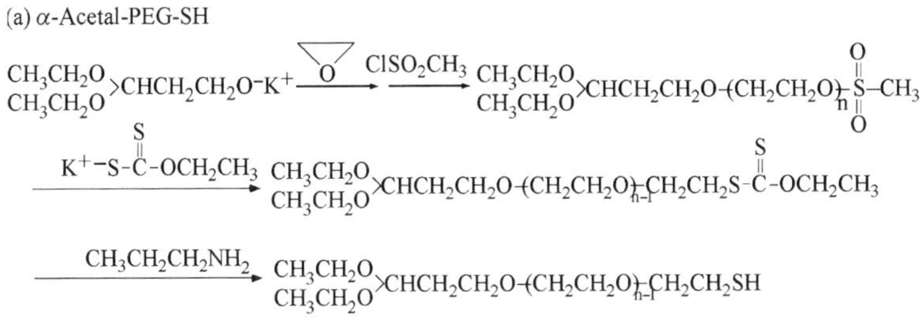

Scheme 47.2. Synthetic procedure of α-acetal-ω-mercapto-poly(ethylene glycol).

dal stability and biofunctionality is strongly desired [111, 112] to construct quantitative bioassays using gold nanoparticles. Recently, we have reported a facile route to functionalize PEGylated gold nanoparticles having a potential utility for colloidal sensor systems [113] by the use of newly designed heterobifunctional PEG derivatives.

The strategy to synthesize various types of heterobifunctional PEGs is based on the ring-opening polymerization of EO using a metal alkoxide initiator with a protected functional group [52–56]. The synthesis of a heterobifunctional PEG containing both mercapto- and acetal terminal groups (α-acetal-ω-mercapto-PEG, acetal-PEG-SH) was previously reported as shown in Scheme 47.2a. The acetal moiety can readily be transformed into a reactive aldehyde group by simple treatment with dilute acid. An aldehyde-functionalized PEGylation of gold nanoparticles using this acetal-PEG-SH (Mn = 3090) was successfully achieved, obtaining gold nanoparticles with high dispersion stability, particularly in a physiological milieu, and appreciable aldehyde reactivity to immobilize ligand molecules on the PEG coronas, which additionally impart sensitivity and selectivity due to their nonfouling properties. As a model ligand, lactose was successfully introduced in the distal end of the PEG chain to induce a lectin-mediated quantitative and reversible association of gold nanoparticles under physiologically relevant conditions, producing a concomitant color change (red $\rightarrow$ purple $\rightarrow$ red) as described in Scheme 47.3.

47.4.1.2.1 Dispersion Stability in Physiological Conditions. Gold nanoparticles were prepared by the reduction of metal salt ($HAuCl_4$) with $NaBH_4$ in the presence of acetal-PEG-SH. The formation of gold nanoparticles, with an average diameter of 8.9 nm as measured by TEM, was confirmed by the UV–vis spectrum of an absorption band near $\lambda = 520$ nm, assigned to a gold nanoparticle plasmon band (Fig. 47.3).

Acetal-PEG-SH coated gold nanoparticles were appreciably stable in buffers of elevated salt concentrations (•, ▴) and also in a serum-containing medium (○), while gold nanoparticles physically stabilized through the adsorption of acetal-PEG-OH (Mn = 3000) (▪, ▾) aggregated immediately even in

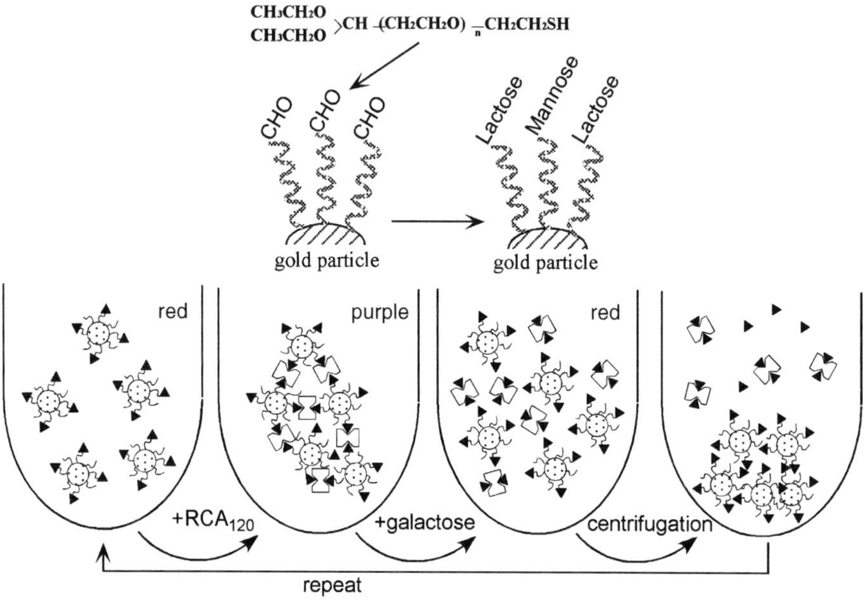

Scheme 47.3. Schematic representation of the reversible aggregation–dispersion behavior of Lac-PEGylated gold nanoparticles by the sequential addition of RCA_{120} lectin and galactose with an actual concomitant change in color from pinkish red → purple → pinkish red. See color figure on companion ftp site.

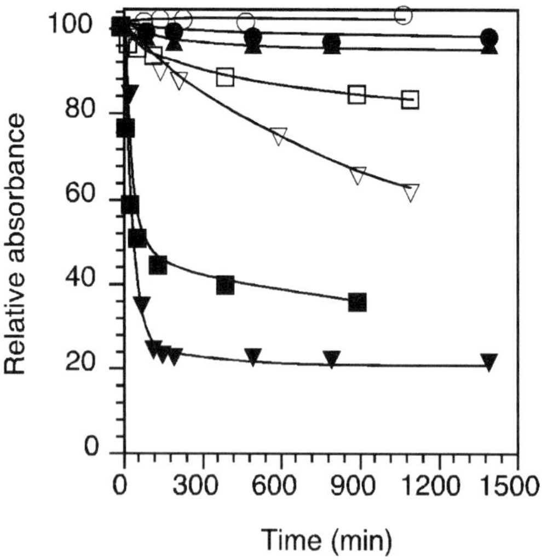

Figure 47.3. Dispersion stability of gold nanoparticles with time in various environments. Acetal-PEG-SH protected gold nanoparticles in 0.15 M PBS (•), 0.3 M PBS (▴), and 0.15 M PBS with 2%-serum (○); acetal-PEG-OH adsorbed gold nanoparticles in 0.03 M PBS (▪), 0.15 M PBS (▾), 0.03 M PBS with 2% serum (□), and 0.15 M PBS with 2% serum (▿).

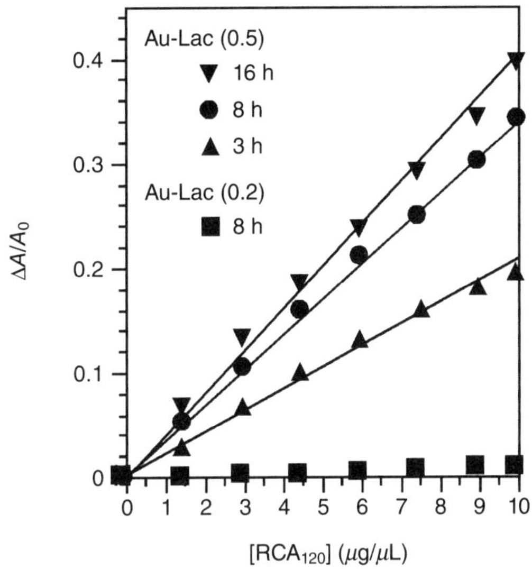

Figure 47.4. Change in the surface plasmon band intensity ($\Delta A(=A_0 - A)/A_0$) with RCA_{120} concentration ($[RCA_{120}]$) for nanoparticle Au-Lac(0.5) (▲, •, ▼) and Au-Lac(0.2) (■) at different times after the onset of the aggregation reaction. A_0 is the surface plasmon band intensity in the absence of RCA_{120}.

0.03 M phosphate-buffered saline (PBS) (Fig. 47.4). Highly enhanced stabilization as well as reduced nonspecific interactions with biological components including cells and proteins of gold nanoparticles bearing acetal-PEG-SH allows this system to be developed for diagnostics in blood serum-containing samples.

47.4.1.2.2 Quantitative and Reversible Lectin-Induced Association of Gold Nanoparticles. The PEG acetal terminal group was converted to aldehyde by gentle acid treatment, followed by the reaction with sugar derivatives having p-aminophenyl moieties at the C-1 position (p-aminophenyl-β-D-lactopyranoside: Lac) in the presence of $(CH_3)_2NHBH_3$. The reaction of these Lac-derivatized gold nanoparticles with bivalent galactose-binding lectin (*Ricinus communis* agglutinin, RCA_{120}) [114, 115] was followed as a function of time through optical changes in the surface plasmon band in the UV–vis spectrum. These solutions were initially a pinkish-red color due to the well-dispersed nature of the particles (Scheme 47.3). After the introduction of the RCA_{120} lectin in 0.15 M PBS, the color gradually changed from red to purple (Scheme 47.3). In line with this directly observable change in appearance, significant differences in the optical spectra over time were observed, specifically a broadening and red shift in the particle surface plasmon resonance (SPR) from 523 nm to a longer wavelength [116]. This is attributed to distance-dependent changes in the optical properties of three-dimensionally aggregated gold nanoparticles cross-linked by the RCA_{120} lectin, which recognizes lactose residues on the PEGylated gold surface. The lectin–particle system was well ordered and three-dimensionally linked, as deduced from TEM and small angle neutron scattering (SAXS) analysis. Images of two-dimensional, single-layer aggregates revealed close-packed assemblies of the colloids with uniform particle separations of about 20 nm, which correspond to the length of the combination of lectin and PEG linker. The aggregation by the RCA_{120} lectin was reversible, recovering the original dispersed phase and color by the addition of excess galactose. Notably, the degree of aggregation was proportional to the lectin concentration, allowing the system to be utilized to quantitate lectin concentration with nearly the same sensitivity as ELISA ($\approx 1\,\mu$g/mL [1 ppm]). This simple, yet highly effective derivatization of gold nanoparticles with heterobifunctional PEG provides a convenient method to construct practical colloidal sensor systems currently applied in bioassays and biorecognition.

47.4.1.3 Sensitivity Enhancement for SPR. Gold nanoparticles can also be applied to enhance detection limits in an SPR-based real-time biospecific interaction analysis [117, 118]. The dynamic enhancement of SPR biosensing with colloidal gold was initially observed in a sandwich immunoassay in which gold nanoparticles were coupled to a secondary antibody. This system was used to detect a primary antibody bound through specific immunosorption to the antigen immobilized on the gold sensor surface. Our recent study demonstrated that an appreciable enhancement in sensitivity was obtained to detect the avidin immobilized on the SPR sensor surface through biotin–avidin interaction using α-biotinylated PEG-functionalized gold nanoparticles. The immunosorptive binding of the colloidal gold to the sensor surface led to a large shift in plasmon angle, a broadened plasmon resonance, and an increase in minimum reflectance, thereby allowing picomolar detection of antigens [117]. Similarly, an improvement in sensitivity by about 1000-fold was obtained in a nucleic acid hybridization analysis, when a colloidal gold/oligonucleotide conjugate was used as a probe [118]. These results suggest that the detection limit of SPR [119] now begins to approach that of a traditional fluorescence-based DNA hybridization detection system.

47.4.2 Luminescent Quantum Dots (QDs) for Biological Detection

A unique property of semiconductor nanoparticles, known as QDs and metal nanoshells, is that they absorb and scatter light of the near infrared, a spectrum region where tissue is essentially transparent. These nanoparticles are often composed of atoms from II to VI or from III to V elements in the periodic table. QDs are highly light absorbable and luminescent with an absorbance onset and emission maximum shift to higher energy with decreasing particle size due to quantum confinement effects [88]. These nanoparticles are in the size range of 2–8 nm in diameter. Unlike molecular fluorophores, which typically have very narrow excitation spectra, semiconductor QDs absorb light over a very broad spectral range. This makes it possible to optically excite a broad spectrum of QD colors using a single excitation laser wavelength, which enables one to simultaneously probe several markers. Thus, bioconjugated QDs have been used in DNA hybridization [91–94], immunoassay [89], receptor-mediated endocytosis [89], and time-gated fluorescence imaging of tissue sections [120]. QDs are also emerging as a new class of fluorescent labels for *in vivo* cellular imaging. An important advantage is that the extremely high photostability of QDs allows real-time monitoring or tracking of intracellular processes over long periods of time. Another advantage is the ability to use multicolor nanoparticles to simultaneously image multiple targets inside living cells or on the cell surface.

A typical QD in these applications, CdS, has been synthesized via arrested precipitation from simple inorganic ions using polyphosphate and low-molecular-weight thiols as stabilizers [121, 122], from dimethylcadmium in trioctylphosphine using trioctylphosphine oxide (TOPO) as a stabilizer, and from cadmium 2-ethylhexanoate in dimethyl sulfoxide (DMSO) using ethylhexanoate as a stabilizer. However, these QDs are practically synthesized at high temperature (~300°C), and the following bioconjugation was applied using several methods: use of a bifunctional ligand such as mercaptoacetic acid for linking QDs to biomolecules [89], TOPO-capped QDs bound to a modified acrylic acid polymer by hydrophobic forces, and QD solubilization and bioconjugation using a mercaptosilane compound [88]. All of these require high temperature and many processes for both synthesis and bioconjugation, further resulting in QDs that seem to lack stability and selectivity in their physiological condition. In contrast, as well as in the case for the synthesis of gold nanoparticles, the PEG-*b*-PAMA solution containing Cd(II) salt and Na_2S, respectively, was simultaneously introduced to form CdS nanoparticles stabilized by PEG-*b*-PAMA at room temperature in an aqueous solution. These nanoparticles showed not only the stability in concentrated salt solution but also biorecognition properties, demonstrating the high utility in the biomedical applications, particularly high-throughput detection and imaging.

Interestingly, the CdS/dendrimer nanocomposite was synthesized by the arrested precipitation of CdS in the presence of Starburst (poly(aminoamine) [PAMAM]) dendrimers as stabilizers [123]. Recently, amino-derivatized polysaccharides (amino-dextran or Amdex) were employed as a stabilizer for the preparation of stable aqueous dispersions consisting of CdS nanoparticles, and the resultant Amdex–CdS nanoparticle complexes could be activated and conjugated with an antibody [124]. Due to the fact that the CdS/polymer nanocomposite underwent further aggregation in solution to produce micrometer-scale composite flocs, individual CdS nanoparticles cannot be isolated in these CdS/polymer nanocomposites, resulting in the loss of photochemical activity.

47.5 CONCLUSION

Recently, tremendous progress has been attained in the characterization and application of nanostructured materials using block copolymers. Nanostructure fabrication from block copolymers involves polymer design, synthesis, self-assembly, and derivatization. Block copolymers self-assembled into micelles afford a powerful means of manipulating the characteristics of surfaces and interfaces and, therefore, are expected to have novel applications. In this review, nanoparticle fabrication using heterobifunctional PEGs and their block copolymer is explored to construct functionalized PEG layers on surfaces, achieving the biospecific adsorption of a target protein through an appropriate ligand tethered on PEG layers without nonspecific adsorption of other proteins. These properties of polymeric micelles formed through the multimolecular assembly of block copolymers are highly useful as novel core–shell-type colloidal carriers for drug and gene targeting.

Surface organization of block copolymer micelles with a cross-linking core was also highlighted here. Since the thin hydrogel layer made by alternate coating of micelles and PAlAm exhibited nonfouling properties and worked as the reservoir for hydrophobic reagents, these surfaces can be used in diverse fields of medicine and biology to construct high-performance medical devices and drug delivery systems.

Furthermore, by controlling metal and semiconductor structure precisely through the concept to construct functionalized PEG layers, one can modify the nanostructures to better suit their integration with biological systems, for example, modifying their surface layer for enhanced aqueous solubility, biocompatibility, and, more importantly, biorecognition. The use of exquisite recognition properties of biomolecules in organizing nonbiological inorganic objects into functional materials has led to new applications including ultrasensitive bioassays and multicolor fluorescent labels for high-throughput detection and imaging.

REFERENCES

1 M. Moffitt, K. Khougaz, A. Eisenberg. Micellization of ionic block copolymers. *Acc. Chem. Res.* 29 (1996) 95–102.

2 P. Munk, K. Prochazka, Z. Tuzar, S. E. Webber. Exploiting polymer micelle technology. *Chemtech* 28 (1998) 20–28.

3 Z. Tuzar, P. Kratochvil. Micelles of block and graft copolymers in solutions. *Surf. Colloid Sci.* 15 (1993) 1–83.

4 C. Allen, D. Maysinger, A. Eisenberg. Nano-engineering block copolymer aggregates for drug delivery. *Colloids Surf.* 16 (1999) 3–27.

5 K. Kataoka, A. Harada, Y. Nagasaki. Block copolymer micelles for drug delivery: design, characterization and biological significance. *Adv. Drug Deliv. Rev.* 47 (2001) 113–131.

6 S. Cammas-Marion, T. Okano, K. Kataoka. Functional and site-specific macromolecular micelles as high potential drug carriers. *Colloids Surf. B* 16 (1999) 207–215.

7 E. Yu Kramarenko, I. I. Potemkin, A. R. Khokhlov, R. G. Winkler, P. Reineker. Surface micellar nanopattern formation of adsorbed diblock copolymer system. *Macromolecules* 32 (1999) 3495–3501.

8 M. Antonietti, C. Göltner. Superstructures of functional colloids: chemistry on the nanometer scale. *Angew. Chem. Int. Ed. Engl.* 36 (1997) 910–928.

9 K. Kataoka, G. S. Kwon, M. Yokoyama, T. Okano, Y. Sakurai. Block copolymer micelles as vehicles for drug delivery. *J. Control. Release* 24 (1993) 119–132.

10 K. Kataoka. Design of nanoscopic vehicles for drug targeting based on micellization of amphiphilic block copolymers. *J. Macromol. Sci. Pure Appl. Chem. A* 31 (1994) 1759.

11 M. Yokoyama, T. Okano, Y. Sakurai, H. Ekimoto, C. Shibazaki, K. Kataoka. Toxicity and antitumor activity against solid tumors of micelle-forming polymeric anticancer drug and its extremely long circulation in blood. *Cancer Res.* 51 (1991) 3229–3236.

12 M. Yokoyama, M. Mitauchi, N. Yamada, T. Okano, Y. Sakurai, K. Kataoka, S. Lnoue. Polymer micelles as novel drug carrier: adriamycin-conjugated poly(ethylene glycol)-poy(aspartic acid) block copolymer. *J. Control. Release* 11 (1990) 269–278.

13 M. T. Peracchia, C. Vauthier, F. Puisieux, P. Couvreur. Development of sterically stabilized poly(isobutyl 2-cyanoacrylate) nanoparticles by chemical coupling of poly(ethylene glycol). *J. Biomed. Mater. Res.* 34 (1997) 317–326.

14 K. Bergström, E. Österberg, K. Holmberg, A. S. Hoffman, T. P. Schuman, A. Kozlowski, J. H. Harris. Effects of branching and molecular weight of surface-bound poly(ethylene glycol) on protein rejection. *J. Biomater. Sci. Polym. Ed.* 6 (1994) 123–132.

15 R. Gref, Y. Minamitake, M. T. Peracchia, V. Torchilin, V. Trubetskoy, R. Langer. Biodegradable long-circulating polymeric nanosphere. *Science* 28 (1994) 1600–1603.

16 M. T. Peracchia, C. Vauthier, D. Desmaele, A. Gulik, J. C. Dedieu, M. Demoy, J. Angelo, P. Couvreur. Pegylated nanoparticles from a novel methoxypolyethylene glycol cyanoacrylate-hexadecyl cyanoacrylate amphiphilic copolymer. *Pharm. Res.* 15 (1998) 550–556.

17 K. Holmberg, K. Bergstrom, C. Brink, E. Österberg, F. Tiberg, J. M. Harris. Effects on protein adsorption, bacterial adhesion and contact angle of grafting PEG chains to polystyrene. *J. Adhes. Sci. Technol.* 7 (1993) 503–517.

18 L. K. Ista, H. Fan, O. Baca, G. P. Lopez. Attachment of bacteria to model surfaces: oligo(ethylene glycol) surfaces inhibit bacterial attachment. *FEMS Microbiol. Lett.* 142 (1996) 59–63.

19 C. R. Deible, E. J. Beckman, A. J. Russell, W. R. Wagner. Creating molecular barriers to acute platelet deposition on damaged arteries with reactive polyethylene glycol. *J. Biomed. Mater. Res.* 41 (1998) 251–256.

20 S. Jo, K. Park. Surface modification using silanated poly(ethylene glycol)s. *Biomaterials* 21 (2000) 605–616.

21 H. Otsuka, Y. Nagasaki, K. Kataoka. Novel approaches for the construction of functionalized poly(ethylene glycol) (PEG) layer on surfaces using heterobifunctional PEG/polylactide(PLA) block copolymers and their micelles, in R. Gross, C. Scholz (eds.), *Polymers from renewable resources: biopolyesters and biocatalysis*. ACS Symposium Series 764. American Chemical Society, Washington, DC, 2000, pp. 311–327.

22 H. Otsuka, Y. Nagasaki, K. Kataoka. Dynamic wettability study on the functionalized PEGylated layer on polylactide surface constructed by the coating of aldehyde-ended poly(ethylene glycol) (PEG)/polylactide (PLA) block copolymer. *Sci. Technol. Adv. Mater.* 1 (2000) 21–29.

23 H. Otsuka, Y. Nagasaki, K. Kataoka. Surface characterization of functionalized polylactide through the coating with heterobifunctional poly(ethylene glycol)/polylactide block copolymers. *Biomacromolecules* 1 (2000) 39–48.

24 H. Otsuka, Y. Nagasaki, K. Kataoka. Self-assembly of poly(ethylene glycol)-based block copolymers for biomedical applications. *Curr. Opin. Colloid Interface Sci.* 6 (2001) 3–10.

25 A. V. Kabanov, V. P. Chekhonin, V. Yu, V. Alakhov, E. V. Batrakova, A. S. Lebedev, N. S. Melik-Nubarov, S. A. Arzhakov, G. V. Morzov, E. S. Severn, V. A. Kabanov. The neuroleptic activity of haloperidol increases after its solubilization in surfactant micelles. *FEBS Lett.* 258 (1989) 343–345.

26 G. S. Kwon, M. Naito, M. Yokoyama, T. Okano, Y. Sakurai, K. Kataoka. Micelles based on AB block copolymers of poly(ethylene oxide) and poly(β-benzyl L-aspartate). *Langmuir* 9 (1993) 945–949.

27 D. Bazile, C. Prud'homme, M.-T. Bassoullet, M. Marland, G. Spenlehauer, M. Veillard, Me. Stealth. PEG-PLA nanoparticles avoid uptake by the mononuclear phagocytes system. *J. Pharm. Sci.* 84 (1995) 493–498.

28 S. A. Hagan, A. G. A. Coombes, M. C. Garnett, S. E. Dunn, M. C. Davis, L. Illum, S. S. Davis, S. E. Harding, S. Purkiss, P. R. Gellert. Polylactide-poly(ethylene glycol) copolymers as drug delivery systems. 1. Characterization of water dispersible micelle-forming systems. *Langmuir* 12 (1996) 2153–2161.

29 X. Zhang, J. K. Jackson, H. M. Burt. Development of amphiphilic block copolymers as micellar carriers of taxol. *Int. J. Pharm.* 132 (1996) 195–206.

30 K. Yasugi, Y. Nagasaki, M. Kato, K. Kataoka. Preparation and characterization of polymer micelle from poly(ethylene glycol)-poly(d,l-lactide) block copolymer as potential drug carrier. *J. Control. Release* 62 (1999) 89–100.

31 A. Hadada, K. Kataoka. Formation of polyion complex micelles in aqueous milieu from a pair of oppositely-charged block copolymers with poly(ethylene glycol) segments. *Macromolecules* 28 (1995) 5294–5299.

32 K. Kataoka, H. Togawa, A. Harada, K. Yasugi, T. Matsumoto, S. Katayose. Spontaneous formation of polyion complex micelles with nallow distribution from antisense oligonucleotide and cationic block copolymer in physiological saline. *Macromolecules* 29 (1996) 8556–8557.

33 T. K. Bronich, A. V. Kabanov, V. A. Kabanov, K. Yu, A. Eisenberg. Soluble complex from poly(ethylene oxide)-block-polymethecrylate anions and N-alkylpyridinium cations. *Macromolecules* 30 (1997) 3519–3525.

34 E. A. Lysenko, T. K. Bronich, A. Eisenberg, V. A. Kabanov, A. V. Kabanov. Block ionomer complexes from polystyrene-block-polyacrylate anions and N-cetylpyridinium catios. *Macromolecules* 31 (1998) 4511–4515.

35 M. Yokoyama, T. Okano, Y. Sakurai, S. Suwa, K. Kataoka. Introduction of cisplatin into polymeric micelle. *J. Control. Release* 39 (1996) 351–356.

36 N. Nishiyama, M. Yokoyama, T. Aoyagi, T. Okano, Y. Sakurai, K. Katakoka. Preparation and characterization of self-assembled polymer-metal complex micelle from cisdichlorodiamine platinum (II) and poly(ethylene glycol)-poly(a,b-aspartic acid) block coplymer in an aqueous medium. *Langmuir* 15 (1999) 377–383.

37 K. Kataoka, A. Ishihara, A. Harada, H. Miyazaki. Effect of secondary structure of poly(L-lysine) segments on the micellization of poly(ethylene glycol)-poly(L-lysine) block copolymer partially substituted with hydrocynnamoyl-group at the N-position in aqueous milieu. *Macromolecules* 31 (1998) 6071–6076.

38 H. Weller. Colloidal semiconductor Q-particle—Chemistry in the transition region between solid-state and molecules. *Angew. Chem. Int. Ed. Engl.* 32 (1993) 41–53.

39 R. F. Khairutdinov. Physical chemistry of nanocrystalline semiconductors. *Colloid J.* 59 (1997) 535–548.

40 H. H. Huang, F. Q. Yan, J. M. Kek, C. H. Chew, G. Q. Xu, W. Ji, P. S. Oh, S. H. Tang. Synthesis, characterization, and nonlinear optical properties of copper nanoparticles. *Langmuir* 13 (1997) 172–175.

41 A. L. Rogach, A. Eychmuller, A. Kornowski, H. Weller. Thiol-stabilized CdSe and CdTe nanocrystals in the size quantization regime: Synthesis, optical and structural properties. *Macromol. Symp.* 136 (1998) 87–89.

42 J. H. Fendler, Y. Tian, in J. H. Fendler (ed.), *Nanoparticles and nanostructured film*. Wiley, Weinheim, 1998.

43 E. S. Smotkin, L. Chongmok, A. J. Bard, A. Campion, M. A. Fox, T. E. Mallouk, S. E. Webber. Size quantization effects in cadmium-sulfide layers formed by a Langmuir–Blodgett technique. *Chem. Phys. Lett.* 152 (1988) 265–268.

44 L. S. Li, L. H. Qu, R. Lu, X. G. Peng, Y. Y. Zhao, T. J. Li. Preparation and structure of quantum-sized cadmium sulfide grown in amphiphilic oligomer Langmuir–Blodgett films. *Thin Solid Films* 329 (1998) 408–411.

45 P. Lianos, J. K. Thomas. Cadmium-sulfide of small dimensions produced in inverted micelles. *Chem. Phys. Lett.* 125 (1986) 299–302.

46 M. P. Pileni, L. Motte, F. Billoudet, C. Petit. Synthesized "in situ" in reverse micelles of silver-sulfide semiconductors. *Surf. Rev. Lett.* 3 (1996) 1215–1218.

47 H. -C. Youn, S. Baral, J. H. Fendler. Dihexadecyl phosphate vesicle-stabilized and insitu generated mixed CdS and ZnS semiconductor particles-preparation and utilization for photosensitized charge separation and hydrogen generation. *J. Phys. Chem.* 92 (1988) 6320–6327.

48 M. Moffitt, A. Eisenberg. Size control of nanoparticles in semiconductor-polymer composites. 1. Control via multiplet aggregation numbers in styrene-based random ionomers. *Chem. Mater.* 7 (1995) 1178–1184.

49 M. Moffitt, L. McMahon, V. Pessel, A. Eisenberg. Size control of nanoparticles in semiconductor-polymer composites. 2. Control via sizes of spherical ionic microdomains in styrene-based diblock ionomers. *Chem. Mater.* 7 (1995) 1185–1192.

50 M. Antonietti, E. Wenz, L. Bronstein, M. Seregina. Synthesis and characterization of noble metal colloids in block copolymer micelles. *Adv. Mater.* 7 (1995) 1000–1005.

51 Y. Matsumura, H. Maeda. A new concept for macromolecular therapeutics in cancer chemotherapy: mechanism of tumoritropic accumulation of proteins and the antitumor agent SMANCS. *Cancer Res.* 46 (1986) 6387–6392.

52 Y. Nagasaki, T. Kutsuna, M. Iijima, M. Kato, K. Kataoka. Formyl-ended heterobifunctional poly(ethylene oxide): synthesis of poly(ethylene oxide) with a formyl group at one end and a hydroxyl group at the other end. *Bioconjug. Chem.* 6 (1995) 231–233.

53 Y. Nagasaki, M. Iijima, M. Kato, K. Kataoka. Primary amino-terminal heterobifunctional poly(ethylene oxide). Facile synthesis of poly(ethylene oxide) with a primary amino group at one end and a hydroxyl group at the other end. *Bioconjug. Chem.* 6 (1995) 702–704.

54 Y. Nagasaki, R. Ogawa, S. Yamamoto, M. Kato, K. Kataoka. Synthesis of heterotelechelic poly(ethylene glycol) macromonomers. Preparation of poly(ethylene glycol) possessing a methacryloyl group at one end and a formyl group at the other end. *Macromolecules* 30 (1997) 6489–6493.

55 T. Nakamura, Y. Nagasaki, K. Kataoka. Synthesis of heterobifunctional poly(ethylene glycol) with a reducing monosaccharide residue at one end. *Bioconjug. Chem.* 9 (1998) 300–303.

56 Y. Akiyama, H. Otsuka, Y. Nagasaki, K. Kataoka. Selective synthesis of heterobifunctional poly(ethylene glycol) derivatives containing both mercapto and acetal terminals. *Bioconjug. Chem.* 11 (2000) 947–950.

57 Y. Nagasaki, T. Okada, C. Scholz, M. Iijima, M. Kato, K. Kataoka. The reactive polymeric micelle based on an aldehyde-ended poly(ethylene glycol)/poly(lactide) block copolymer. *Macromolecules* 31 (1998) 1473–1479.

58 K. Yasugi, T. Nakamura, Y. Nagasaki, M. Kato, K. Kataoka. Sugar-installed polymer micelles: synthesis and micellization of poly(ethylene glycol)-poly(D,L-lactide) block copolymers having sugar groups at PEG chain end. *Macromolecules* 32 (1999) 8024–8032.

59 Y. Yamamoto, Y. Nagasaki, M. Kato, K. Kataoka. Surface charge modulation of poly(ethylene glycol)-poly(D,L-lactide) block copolymer micelles: conjugation of charged peptides. *Colloids Surf. B Biointerfaces* 16 (1999) 135–146.

60 M. A. Woflert, E. H. Schacht, V. Toncheva, K. Ulbrich, O. Nazarova, L. W. Seymour. Characterization of vector for gene therapy formed by self-assembly of DNA with synthetic block co-polymers. *Hum. Gene. Ther.* 7 (1996) 2123–2133.

61 S. Katayose, K. Kataoka. Water soluble polyion complex associates of DNA and poly(ethylene glycol)-poly(L-lysine) block copolymer. *Bioconjug. Chem.* 8 (1997) 702–707.

62 S. Katayose, K. Kataoka. Remarkable increase in nuclease resistance of plasmid DNA through supramolecular assembly with poly(ethylene glycol)-poly(L-lysine) block copolymer. *J. Pharm. Sci.* 87 (1998) 160–163.

63 Y. Akiyama, A. Harada, Y. Nagasaki, K. Kataoka. Synthesis of poly(ethylene glycol)-block-poly(ethyleneimine) possessing an acetal group at the PEG end. *Macromolecules* 33 (2000) 5841–5845.

64 K. Kataoka, A. Harada, D. Wakebayashi, Y. Nagasaki. Polyion complex micelles with reactive aldehyde groups on their surface from plasmid DNA and end-functionalized charged block copolymers. *Macromolecules* 32 (1999) 6892–6894.

65 Y. Kakizawa, K. Kataoka. Block copolymer micelles for delivery of gene and related compounds, Adv. *Drug Deliv.* 54 (2002) 203–222.

66 J. P. S. Farinha, J. M. R. d'Oliveira, J. M. Martinoho, R. Xu, M. A. Winnik. Structure in tethered chains: polymeric micelles and chains anchored on polystyrene latex spheres. *Langmuir* 14 (1998) 2291–2296.

67 H. D. Bijsterbosch, M. A. Cohen Stuart, G. J. Fleer. Adsorption kinetics of diblock copolymers from a micellar solution on silica and titania. *Macromolecules* 31 (1998) 9281–9294.

68 A. Jonner, J. F. Joanny. Block copolymer adsorption in a selective solvent: a kinetic study. *Macromolecules* 26 (1990) 5299–5311.

69 M. Iijima, Y. Nagasaki, T. Okada, M. Kato, K. Kataoka. Core-polymerized reactive micelles from heterotelechelic amphiphilic block copolymers. *Macromolecules* 32 (1999) 1140–1146.

70 K. Emoto, Y. Nagasaki, K. Kataoka. Coating of surfaces with stabilized reactive micelles from poly(ethylene glycol)-poly(D,L-lactic acid) block copolymer. *Langmuir* 15 (1999) 5212–5218.

71 K. Emoto, Y. Nagasaki, K. Kataoka. A core-shell structured hydrogel thin layer on surfaces by lamination of a poly(ethylene glycol)-b-(D,L-lactide) micelle and polyallylamine. *Langmuir* 16 (2000) 5738–5742.

72 K. Emoto, M. Iijima, Y. Nagasaki, K. Kataoka. Functionality of polymeric micelle hydrogels with organized three-dimensional architecture on surfaces. *J. Am. Chem. Soc.* 122 (2000) 2653–2654.

73 S. J. Sofia, V. Premnath, E. W. Merrill. Poly(ethylene glycol) grafted to silicon surfaces: grafting density and protein adsorption. *Macromolecules* 31 (1998) 5059–5070.

74 S. T. Lopina, G. Wu, E. W. Merrill, L. Griffith-Cima. Hepatocyte culture on carbohydrate-modified star polyethylene oxide hydrogels. *Biomaterials* 17 (1996) 559–569.

75 P. K. Smith, R. I. Krohn, G. T. Hermanson, A. K. Mallia, F. H. Gartner, M. D. Provenzano, F. K. Fujimoto, N. M. Goeke, B. J. Olson, D. C. Klenk. Measurement of protein using bicinchoninic acid. *Anal. Biochem.* 150 (1985) 76–85.

76 G. S. Kwon, S. Suwa, M. Yokoyama, T. Okano, Y. Sakurai, K. Kataoka. Enhanced tumor accumulation and prolonged circulation times of micelle-forming poly(ethylene oxide-aspartate) block copolymer-adriamycin conjugates. *J. Control. Release* 29 (1994) 17–23.

77 A. Henglein. Small-particle research: physicochemical properties of extremely small colloidal metal nand semiconductor particles. *Chem. Rev.* 89 (1989) 1861–1873.

78 G. Schmid. Large clusters and colloids: metals in the embryonic state. *Chem. Rev.* 92 (1992) 1709–1727.

79 A. P. Alvisatos. Semiconductor clusters, nanocrystals, and quantum dots. *Science* 271 (1996) 933–937.

80 M. Nirmal, L. E. Brus. Luminescence photophysics in semiconductor nanocrystals. *Acc. Chem. Res.* 32 (1999) 404–414.

81 C. A. Mirkin, T. A. Taton. Semiconductors meet biology. *Nature* 405 (2000) 626–627.

82 C. M. Niemeyer. Nanoparticles, proteins, and nucleic acids: biotechnology meets materials science. *Angew. Chem. Int. Ed. Engl.* 40 (2001) 4128–4158.

83 H. Otsuka, Y. Akiyama, Y. Nagasaki, K. Kataoka. Quantitative and reversible lectin-induced association of gold nanoparticles modified with a-lactosyl-w-mercapto-poly(ethylene glycol). *J. Am. Chem. Soc.* 123 (2001) 8226–8230.

84 J. M. Fuente, A. G. Barrientos, T. C. Rojas, J. Rojo, J. Cañada, A. Fernández, S. Penadés. Gold glyconanoparticles as water-soluble polyvalent models to study carbohydrate interactions. *Angew. Chem. Int. Ed. Engl.* 40 (2001) 2258–2261.

85 T. C. Rojas, J. M. Fuente, A. G. Barrientos, S. Penadés, L. Ponsonnet, A. Fernández. Gold glyconanoparticles as building blocks for nanomaterials design. *Adv. Mater.* 14 (2002) 585–588.

86 C. -C. Lin, Y. -C. Yen, C. -Y. Yang, C. -L. Chen, G. -F. Chen, C. -C. Chen, Y. -C. Wu. Selective binding of mannose-encapsulated gold nanoparticles to type 1 pill in *Escherichia coli. J. Am. Chem. Soc.* 124 (2002) 3508–3509.

87 S. R. Whaley, D. S. English, E. L. Hu, P. F. Barbara, A. M. Belcher. Selection of peptides with semiconductor binding specificity for directed nanocrystal assembly. *Nature* 405 (2000) 665–668.

88 M. Jr. Bruchez, M. Moronne, P. Gin, S. Weiss, A. P. Alvisatos. Semiconductor nanocrystals as fluorescent biological labels. *Science* 281 (1998) 2013–2015.

89 W. C. W. Chan, S. M. Nie. Quantum dot bioconjugates for ultrasensitive nonisotopic detection. *Science* 281 (1998) 2016–2018.

90 H. Mattoussi, J. M. Mauro, E. R. Goldman, G. P. Anderson, V. C. Sunder, F. V. Mikulec, M. G. Bawendi. Self-assembly of CdSe-ZnS quantum dots bioconjugates using an engineered recombinant protein. *J. Am. Chem. Soc.* 122 (2000) 12142–12150.

91 G. P. Mitchell, C. A. Mirkin, R. L. Letsinger. Programmed assembly of DNA functionalized quantum dots. *J. Am. Chem. Soc* 121 (1999) 8122–8123.

92 Y. W. Cao, R. Jin, C. A. Mirkin. DNA-modified core-shell Ag/Au nanoparticles. *J. Am. Chem. Soc.* 123 (2001) 7961–7962.

93 S. -J. Park, A. A. Lazarides, C. A. Mirkin, R. L. Letsinger. Directed assembly of periodic materials from protein and oligonucleotide-modified nanoparticles building blocks. *Angew. Chem. Int. Ed. Engl.* 40 (2001) 2909–2912.

94 C. A. Mirkin, R. L. Letsinger, R. C. Mucic, J. J. Storhoff. A DNA-based method for rationally assembling nanoparticles into macroscopic materials. *Nature* 382 (1996) 607–609.

95 A. P. Alvisatos, K. P. Johnsson, X. Peng, T. E. Wilson, C. J. Loweth, M. P. Jr. Bruchez, P. G. Schultz. Organization of nanocrystal molecules using DNA. *Nature* 382 (1996) 609–611.

96 R. Elghanian, J. J. Storhoff, R. C. Mucic, R. L. Letsinger, C. A. Mirkin. Selective colorimetric detection of polynucleotides based on the distance-dependent optical properties of gold nanoparticles. *Science* 277 (1997) 1078–1081.

97 B. Dubertret, M. Calame, A. J. Libchaber. Single-mismatch detection using gold-quenched fluorescent oligonucleotides. *Nat. Biotechnol.* 19 (2001) 365–370.

98 R. A. Reynolds, C. A. Mirkin, R. L. Letsinger. Homogeneous, nanoparticle-based quantitative colorimetric detection of oligonucleotides. *J. Am. Chem. Soc.* 122 (2000) 3795–3796.

99 M. Hayat (ed.), *Colloidal gold: principles, methods and applications*. Academic Press, San Diego, CA, 1989.

100 G. Schmid. *Clusters and colloids*. VCH, Weinheim, 1994.

101 U. Kreibig. *Optical properties of metal clusters*. Springer, New York, 1995.

102 D. L. Feldheim, C. D. Keating. Self-assembly of single electron transistors and related devices. *Chem. Soc. Rev.* 27 (1998) 1–12.

103 A. N. Shipway, E. Katz, I. Willner. Nanoparticle arrays on surfaces for electronic, optical, and sensor applications. *ChemPhysChem* 1 (2000) 18–52.

104 T. Sato, R. Ruch. *Stabilization of colloidal dispersions by polymer adsorption*. Surfactant Science Series, No. 9. Marcel Dekker, New York, 1980.

105 R. G. Pearson. Hard and soft acids and bases. *J. Am. Chem. Soc.* 85 (1963) 3533–3539.

106 L.-H. Lee (ed.), *Fundamentals of adhesion*. Plenum Press, New York, 1991.

107 L. Bronstein, M. Sedlak, J. Hartmann, M. Breulmann, H. Cölfen, M. Antonietti. Interaction of metal compounds with "double-hydrophilic" block copolymers in aqueous medium and metal colloid formation. *Polym. Prepr. (Am. Chem. Soc., Div. Polym.)* 76 (1997) 54.

108 L. M. Bronstein, S. N. Sidorov, P. M. Valetsky, J. Hartmann, H. Cölfen, M. Antonietti. Induced micellization by interaction of poly(2-vinylpyridine)-block-poly(ethylene oxide) with metal compounds. Micelle characteristics and metal nanoparticle formation. *Langmuir* 15 (1999) 6256–6262.

109 S. N. Sidorov, L. M. Bronstein, P. M. Valetsky, J. Hartmann, H. Cölfen, H. Schnablegger, M. Antonietti. Stabilization of metal nanoparticles in aqueous medium by polyethyleneoxide-polyethyleneimine block copolymers. *J. Colloid Interface Sci.* 212 (1999) 197–211.

110 W. P. Wuelfing, S. M. Gross, D. T. Miles, R. W. Murray. Nanometer gold clusters protected by surface-bound monolayers of thiolated poly(ethylene glycol) polymer electrolyte. *J. Am. Chem. Soc.* 120 (1998) 12696–12697.

111 L. Quaroni, G. Chumanov. Preparation of polymer-coated functionalized silver nanoparticles. *J. Am. Chem. Soc.* 121 (1999) 10642–10643.

112 P. A. Buining, B. M. Humbel, A. P. Philipse, A. J. Verkleij. Preparation of functional silane-stabilized gold colloids in the (sub)nanometer size range. *Langmuir* 13 (1997) 3921–3926.

113 H. Otsuka, Y. Akiyama, Y. Nagasaki, K. Kataoka. Quantitative and reversible lectin-induced association of gold nanoparticles modified with a-lactosl-w-mercapto-poly(ethylene glycol). *J. Am. Chem. Soc* 123 (2001) 8226–8230.

114 I. J. Goldstein, R. D. Poretz. 2. Isolation, physicochemical characterization, and carbohyderate-binding specificity of lectins, in I. E. Liener, N. Sharon, I. J. Goldstein (eds.), *The lectins. properties, functions, and applications in biology and medicine*. Academic Press, Orlando, FL, 1986, pp. 35–244.

115 E. C. Sweeney, A. G. Tonevitsky, D. E. Temiakov, I. I. Agapov, S. Saward, R. A. Palmer. Preliminary crystallographic characterization of ricin agglutinin. *Proteins* 28 (1997) 586–589.

116 G. Mie. Contribution to the optics of turbid media, particularly of colloidal metal solutions. *Ann. Phys.* 25 (1908) 377.

117 L. A. Lyon, M. D. Musick, M. J. Natan. Colloidal Au-enhanced surface plasmon resonance immunosencing. *Anal. Chem.* 70 (1998) 5177–5183.

118 L. He, M. D. Musick, S. R. Nicewarner, F. G. Salinas, S. J. Benkovic, M. J. Natan, C. D. Keating. Colloidal Au-enhanced surface plasmon resonance for ultrasensitive detection of DNA hybridization. *J. Am. Chem. Soc.* 122 (2000) 9071–9077.

119 (a) For reviews on surface plasmon resonance measurement, see J. M. Brockman, B. P. Nelson, R. M. Corn. Surface plasmon resonance imaging measurement of ultrathin organic films. *Annu. Rev. Phys. Chem.* 51 (2000) 41–63; (b) C. Williams. Biotechnology matchmaking: screening orphan ligands and receptors. *Curr. Opin. Biotechnol.* 11 (2000) 42–46; (c) R. L. Rich, D. G. Myszka. Advances in surface plasmon resonance biosencor analysis. *Curr. Opin. Biotechnol.* 11 (2000) 54–61.

120 M. Dahan, T. Laurence, F. Pinaud, D. S. Chemla, A. P. Alivisatos, M. Sauer, S. Weiss. Time-gated biological imaging by use of colloidal quantum dots. *Opt. Lett.* 26 (2001) 825–827.

121 X. G. Peng, J. Wickham, A. P. Alivisatos. Kinetics of II-VI and III-V colloidal semiconductor nanocrystal growth: “focusing” of size distribution. *J. Am. Chem. Soc.* 120 (1998) 5343–5344.

122 J. K. Lorenz, A. B. Ellis. Surfactant-semiconductor interfaces: perturbation of the photoluminescence of bulk cadmium selenide by adsorption of tri-n-octylphosphine oxide as a probe of solution aggregation with relevance to nanocrystal stabilization. *J. Am. Chem. Soc.* 120 (1998) 10970–10975.

123 J. Huang, K. Sooklal, C. J. Murphy. Polyamine-quantum dot nanocomposites: linear versus starburst stabilizer architecture. *Chem. Mater.* 11 (1999) 3595–3601.

124 I. Sondi, O. Siiman, S. Koester, E. Matijevic. Preparation of amidodextran-CdS nanoparticle complexes and biologically active antibody-amidodextran-CdS nanoparticle conjugates. *Langmuir* 16 (2000) 3107–3118.

INDEX

Electrical Phenomena at Interfaces and Biointerfaces: Fundamentals and Applications in Nano-, Bio-, and Environmental Sciences, First Edition. Edited by Hiroyuki Ohshima.